9th Conference on Interdisciplinary Problems in Environmental Protection and Engineering (EKO-DOK 2017)

E3S Web of Conferences Volume 17 (2017)

Boguszow-Gorce, Poland
23 - 25 April 2017

Editors:

Bartosz Kazmierczak
Malgorzata Kutylowska
Katarzyna Piekarska
Agnieszka Trusz-Zdybek

ISBN: 978-1-5108-4120-8

Printed from e-media with permission by:

Curran Associates, Inc.
57 Morehouse Lane
Red Hook, NY 12571

Some format issues inherent in the e-media version may also appear in this print version.

Printed by Curran Associates, Inc. (2017)

For additional information, please contact EDP Sciences – Web of Conferences
at the address below.

EDP Sciences – Web of Conferences
17, Avenue du Hoggar
Parc d'Activité de Courtabœuf
BP 112
F-91944 Les Ulis Cedex A
France

Phone: +33 (0) 1 69 18 75 75
Fax: +33 (0) 1 69 28 84 91

contact-edps@webofconferences.org

Additional copies of this publication are available from:

Curran Associates, Inc.
57 Morehouse Lane
Red Hook, NY 12571 USA
Phone: 845-758-0400
Fax: 845-758-2633
Email: curran@proceedings.com
Web: www.proceedings.com

TABLE OF CONTENTS

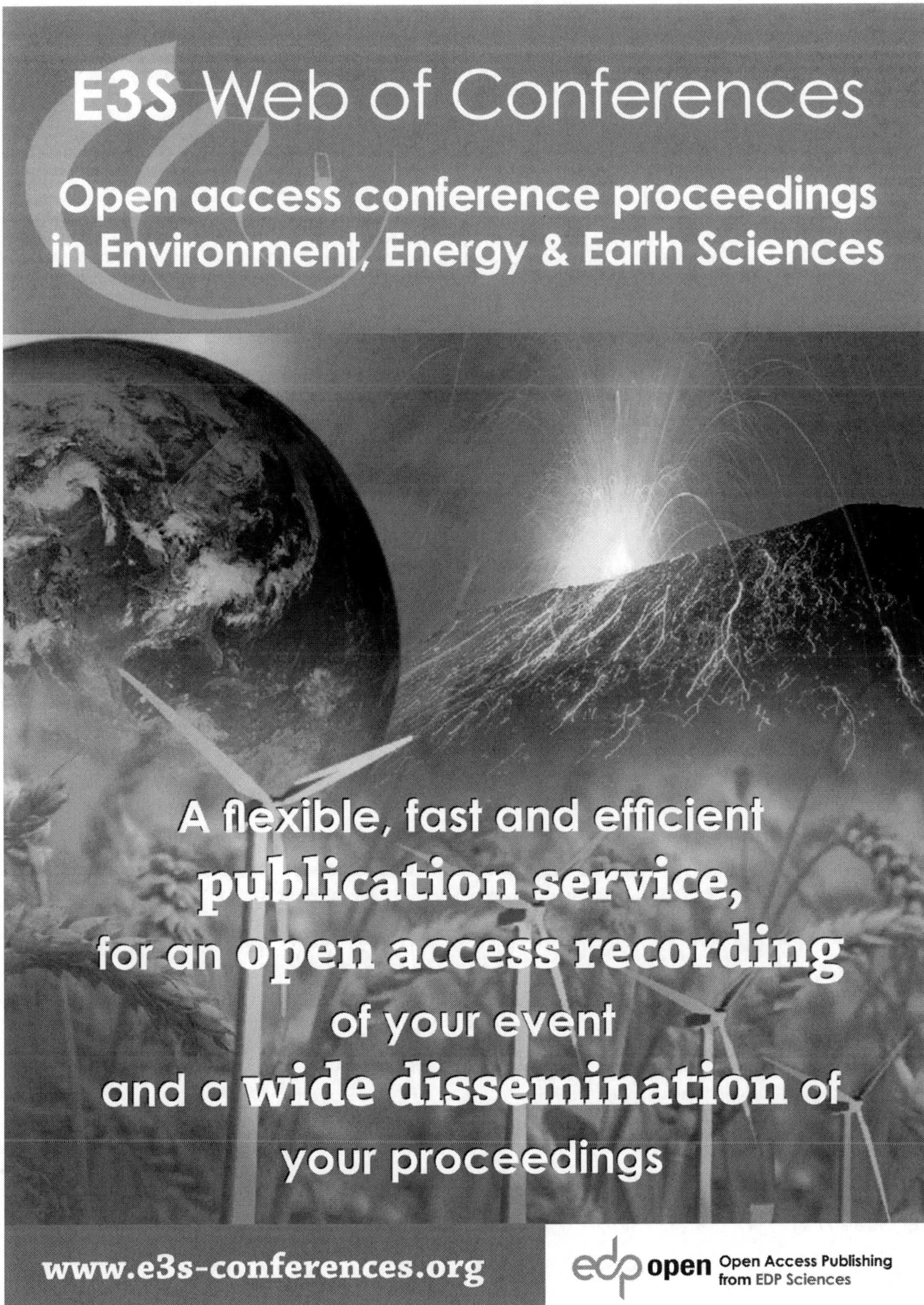
E3S Web of Conferences
Open access conference proceedings
in Environment, Energy & Earth Sciences
A flexible, fast and efficient
publication service,
for an open access recording
of your event
and a wide dissemination of
your proceedings
www.e3s-conferences.org
edp open Open Access Publishing from EDP Sciences

ISSN (electronic edition): 2267-1242

sciences

Statement of Peer review

In submitting conference proceedings to E3S Web of Conferences, I certify to the Publisher that I adhere to the **Policy on Publishing Integrity** of the journal in order to safeguard good scientific practice in publishing.

1. All articles have been subjected to peer review administered by the proceedings editors.
2. Reviews have been conducted by expert referees, who have been requested to provide unbiased and constructive comments aimed, whenever possible, at improving the work.
3. Proceedings editors have taken all reasonable steps to ensure the quality of the materials they publish and their decision to accept or reject a paper for publication has been based only on the merits of the work and the relevance to the journal.

Title, date and place of the conference

9th Conference on Interdisciplinary Problems in Environmental Protection and Engineering EKO-DOK 2017
23-25 April 2017
Boguszów-Gorce, Poland

Proceedings editor(s):

Bartosz Kaźmierczak, Wrocław University of Science and Technology,
Małgorzata Kutyłowska, Wrocław University of Science and Technology,
Katarzyna Piekarska, Wrocław University of Science and Technology,
Agnieszka Trusz-Zdybek, Wrocław University of Science and Technology

Date and editor's signature

08.05.2017
08.05.2017

08.05.2017
08.05.2017

The determination of petroleum hydrocarbons in sewer sediment

Ewa Badowska [1,*], and *Anna* Bąk[1]

[1]Lodz University of Technology, Institute of Environmental Engineering and Building Installations, 90-924 Lodz, Al. Politechniki 6, Poland

Abstract. Based on the available standards it is possible to analyze the concentration of petroleum hydrocarbons in sewer sediment. Guidelines included in the standard PN-EN 14039:2008 – Characterization of waste – Determination of hydrocarbon content in the range of C10 to C40 by gas chromatography and in standard PN-EN ISO 16703:2011 – Soil quality – Determination of content of hydrocarbon in the range C10 to C40 by gas chromatography are useful here. The results of analysis concerning elaboration of the methodology of petroleum hydrocarbons determination in sewer sediment based on available literature are presented in the article. According to carried out literature review and the possibilities of available data use, five temperature programs were initially selected. All temperature programs were tested on a standard sample (oil mixture). In the next stage one of prepared method to determine the amount of petroleum hydrocarbons in sewer sediment was applied for sewer sediment come from the combined sewer system. The verification of the sample preparation procedures according to available standards on real sediment revealed the difficulties. Finally, preliminary value of petroleum hydrocarbons concentrations in sewer sediment from the combined sewage system was determined.

1 Introduction

Petroleum hydrocarbons are one of indicators of rain sewage quality. Unfortunately, despite their negative impact on the environment [1–3] very often the rain sewage is characterized by other indicators. The reason for this may be both procedure and costs of carrying out this type of analyses. The range of determining petroleum hydrocarbons is various, but it is usually in the range between C_{10} and C_{40}, which results from the composition of products of crude oil processing [4].

Sewer sediment, which is washed out from sewer system during increased sewer flows, may effect on the quality of rain sewage, and particularly combined sewer. The sudden release of accumulated pollution in sediment increases the pollution load directed to the receiver, and thus increases the negative impact on environment. The amount of sewer sediment in sewer system depends on many factors, for example on hydraulic conditions in the section of the system, the character of the rain phenomenon, or the period of dry

* Corresponding author: ewa.badowska@p.lodz.pl

weather conditions. Localization of system sections exposed to the accumulation of sediment and sewer sediment qualitative characteristics may contribute to a better, more environmentally-friendly exploitation of sewage system.

2 Methodology for determination of petroleum hydrocarbons in sewer sediment

In the performed analysis the methodology for determining the amount of petroleum hydrocarbons in sewer sediment was based on the available standards for determining the amount of petroleum hydrocarbons in waste (Characterization of waste – Determination of hydrocarbon content in the range of C10 to C40 by gas chromatography) [5] and soils (Soil quality – Determination of content of hydrocarbon in the range C10 to C40 by gas chromatography) [6]. Quoted standards [5, 6], because of the lack of standard directly dedicated to sewer sediment, provided the basis for the process of determining the amount of petroleum hydrocarbons in sewer sediment. Two standards were applied because the range of the analysed compounds, the basic methodology for the preparation of samples and the formula for calculating the final amount of petroleum hydrocarbons in both standards are similar.

Separate stage in the preparation of searching the methodology was determination of chromatographic analysis conditions. In the table below the summary literature review concerning the temperature conditions applied in the determination of petroleum hydrocarbons in sediment, soils and waste are presented.

Table 1. The temperature conditions in the process of determination of petroleum hydrocarbons, by means of gas chromatography, according to literature review.

Parameter/ Reference	1[7]	2[8]	3[9]	4[10]	5[11]	6[12]	7[6]	8[5]
Detector temperatue [°C]		300	300	315	310	340	360	360
Initial temperature [°C]	60	40	35	60	50	40	80	80
Linear temperature increase [°C /min]	3	4	14	1	2	0.5	1	1
Final temperature [°C]	320	300	320	290	310	340	360	300/ 360
Time [min]	20	10	10	7	18.5	0.5	15	10/15
Injection temperature [°C]		250	250	285	300	350		
Application	sewer sedi-ment	bottom sedi-ment	C_{12}-C_{40} water and soil	C_9-C_{40} water, soil, sedi-ment	soil and sedi-ment	water, soil, sedi-ment	soil	waste

After a preliminary analysis of the literature review five temperature programs [7–12], possible to analyse, were extracted. All analyses using the Bruker 450 gas chromatograph equipped with a flame ionization detector (FID) and column Varian Capillary Column Select TM Mineral Oil LVI were conducted. The capillary column 15 m in length and 0.32 mm in diameter was used. There were used: nitrogen as carrier gas, air and hydrogen as auxiliary gases.

Each of the selected temperature program was modified several times to obtain a chromatogram, similar in shape to presented in standard [5, 6]. While changes in temperature programs were tested, the samples prepared using the standard containing a mixture of oils (RIVM-NMi-001), in which the amount of the concentration of petroleum hydrocarbons was in the range from 0.5 to 12.0 mg/ml, were applied. Achieving chromatogram for standard samples the in shape similar to that presented in standard (Fig. 1) confirms the appropriate conditions of chromatographic analysis and allows using analysed method to determine the amount of petroleum hydrocarbons in real samples.

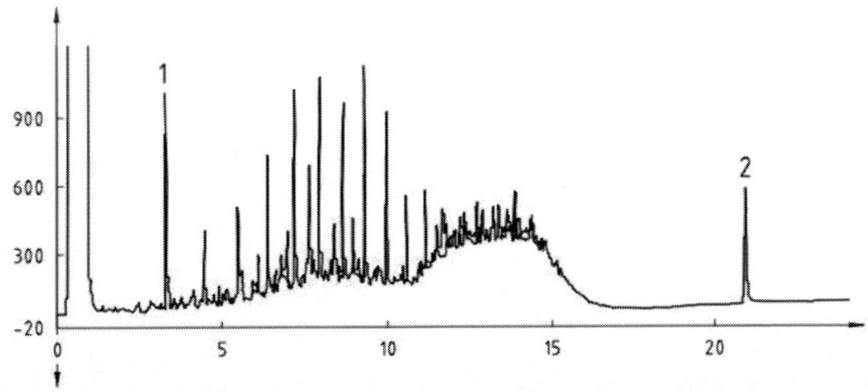

Fig. 1. Gas chromatogram of the calibration mixture consisting of equal parts of diesel fuel and lubricanting oil [5, 6].

3 Results

The results obtained during the research, after every change in temperature program, were analyzed and on this base further modifications in individual temperature program were determined. In case it is ascertained that changes carried out, to certain time (change of temperature, time) do not give the desired result, for example, there is inadequate separation of peaks, resigned from further changes were not made. Finally, temperature program, which was used to analyse the quality of sewer sediment from the combined sewer system, was achieved. In this way the procedure of the preparation and analysis on real samples was verified.

3.1 The selected temperature program

Temperature program that best fulfilled set conditions was program 2 [8] with the initial settings (40°C for 4 min / 20°C / min to 300°C / 300°C for 10 min). On the figure below (Fig. 2) the chromatogram of the sample at a concentration of 3.0, 5.0, and 8.0 mg/ml for the initial setup of method 2 [8] is presented.

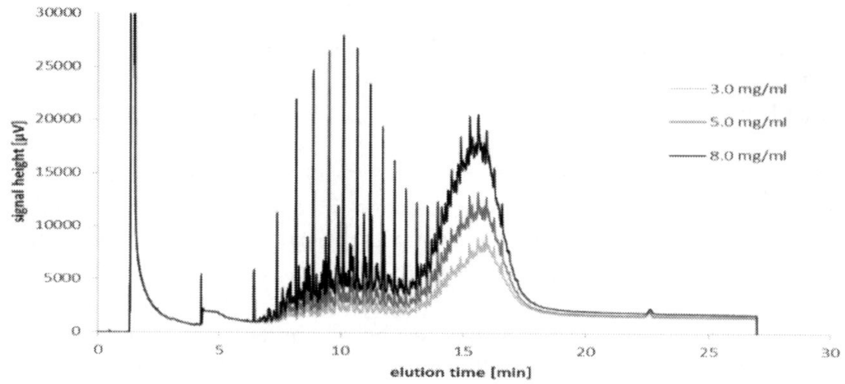

Fig. 2. Gas chromatograms of standard (mixture of mineral oil) of 3.0, 5.0 and 8.0 mg/ml concentration analyzed using method 2.

In the chromatograms presented above the peaks limiting the range of analysis (n-decane and n-tetracontane) are clearly visible. In addition, the left side of the graph is characterized by well-separated single peaks, and the right has a characteristic hump. Method 2 [8] worked well for both low and high concentrations of standard sample. However, attention was paid to retention time of n-decane, which amounted more than 6 minutes. Too long retention time unnecessarily lengthens the entire analysis, which affects the quantity of used gas. In order to optimize the process the initial temperature of 40°C was changed to 60°C and 80°C. Increasing the initial temperature achieved the expected effect of shortening the retention time of n-decane. It is also noted that at increased initial temperature the baseline of individual peaks was decreased. Total analysis time could be reduced due to the shift of peaks of n-tetracontane. Better results were obtained for an initial temperature of 80°C. Figure below shows the chromatogram for a saple of 1 mg/ml concentration obtained after changing the initial temperature to 80°C.

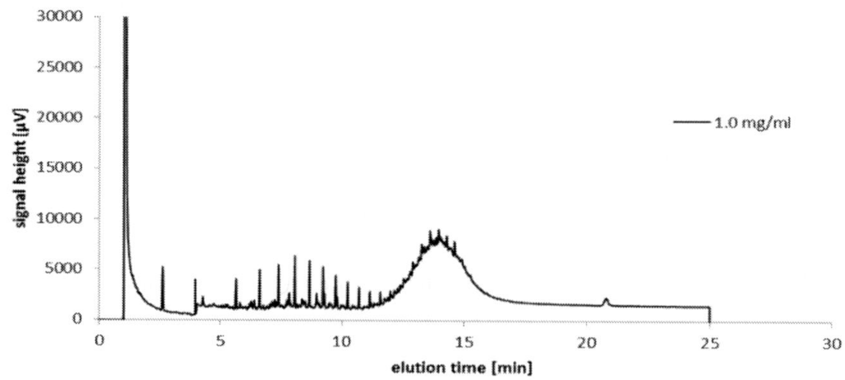

Fig. 3. Gas chromatogram of standard sample of 1 mg/ml concentration obtained after changing initial temperature to 80°C.

In the next stage, it was decided to make changes that would result in a better separation of the individual peaks. For this purpose, increase linear of temperature from 20°C/min to 10°C/min was reduced. Unfortunately, this change was associated with prolongation of the total analysis time. In the figure below, the effects of rate changes of temperature increase are shown.

 DOI: 10.1051/e3sconf/20171700001

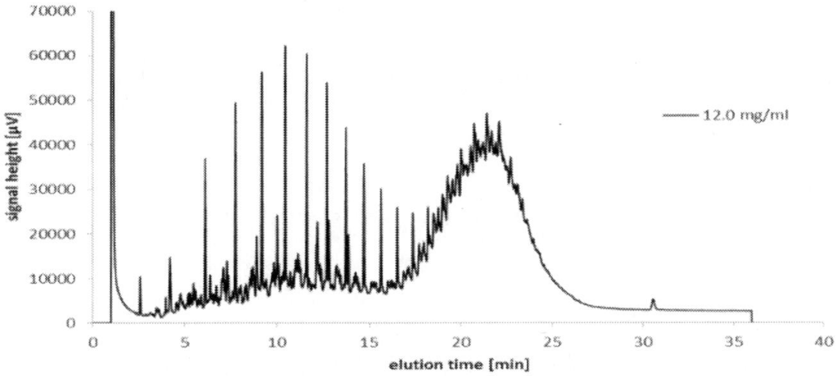

Fig. 4. The gas chromatogram of sample standard solution of 12mg/ml concentration obtained after changing temperature increase to 10°C/min.

It was decided, on the basis of figure shape (Fig. 4) that the gas chromatogram meets the basic conditions classifying the temperature program as possible to use in determining the concentration of petroleum hydrocarbons. Clearly visible peaks n-decane and n-tetracontane (compounds limiting the scope of the analysis), occuring well-separated individual peaks and occuring hump point it out.

In the next stage, the reproducibility of results through 5 times solution injection of 4 mg/ml concentrations was verified. On the basis of figures, the average area of the peaks of the analyzed solution and the corresponding its standard deviation, which amounted to 2.7%, was calculated (PN-EN 14039: 2008 and PN-EN ISO 16703:2011 standards allow standard deviation max 5%).

Determined and verified the temperature program was used to plot calibration curves using a hydrocarbon solvent, n-hexane and n-heptane. It was decided to use the two hydrocarbon solvents, in order to verify later their usefulness in determining the amount of petroleum hydrocarbons in sewer sediment in real samples. The calibration curves using the following standard solution of 0.2, 0.4, 0.6, 0.8, 1.0, 4.0 and 8.0 mg/ml concentrations were made. Respectively calibration curve equation for n-hexane (1) and n-heptane (2) for analyzed concentrations were obtained. The determination coefficient in both cases was approximately R^2=0.99.

$$y = 26.714x + 1322.5 \quad (1)$$

$$y = 29.772x + 628.17 \quad (2)$$

3.1 Sample preparation

Preparation of real samples for chromatographic analysis is an important step which has great impact on the final result. Correct sampling is the first step in the determination of petroleum hydrocarbons in sewer sediment. The samples of sewer sediment from the selected location (combined sewer) were as far as possible (the amount of sediment in the channel) taken from the upper layer. In some cases, because of the precipitation frequency, the amount of sewer sediment in the channel was so small that the collection of the top layer was not possible. A small amount of sewer sediment in the channel also influenced in the next stage the procedure for sample preparation (homogenization). In cases, where there was little amount of sewer sediment before homogenization process, it was necessary to remove large solids (stone, glass) from a sample. Then, the sample after addition of the appropriate reagents (according to standard) and shaking was transferred to a separatory funnel and washed to remove the acetone. Shaking process provides sedimentation of solids and allows the collection of supernatant which is used in the later stages of the analysis. Finally, prepared in this way samples are treated on Florisil columns (diatomaceous substance) and ready for chromatographic analysis. The methodology of sample preparation for the analysis of the amount of petroleum hydrocarbons in sewer sediment turned out to be a difficult process. The quality of sewer sediment sample (its consistency, quantity) has the biggest impact.

3.3 The amount of petroleum hydrocarbons in sewer sediment from combined sewer system

The conducted analysis concerning establishing the conditions for determining the amount of petroleum hydrocarbons in sewer sediment were tested on seven sample series coming from combined sewer system of Lodz. The collection point was located just before the combined sewer overflow J-1, which is located in the southern part of the city. The catchment assigned to combined sewer overflow J-1 is the urbanized area where there is a multi-family housing, public buildings and numerous points of service.

The conducted research can be divided into two parts. The first part concerns the analysis carried out in July 2016 year (5 collections), while the second contains two series of samples that have been made at the turn of 2016 and 2017. In the first part of the research, except for one case (the first series), from which 3 samples with hexane and 1 with heptane were prepared, two samples using n-hexane and n-heptane as hydrocarbon solvent were prepared. In the second part of the analysis three samples using n-hexane and n-heptane were made. The table below shows the values of amount of petroleum hydrocarbons in the analyzed sewer sediment.

Table 2. The summary results of the contents of petroleum hydrocarbons in sewer sediment.

Collection date	Solvent	Average value mg/kg drs	Efficiency %
14.07.2016	n-hexane	98.64	27.37
	n-heptane	43.70	27.79
18.07.2016	n-hexane	87.18	21.94
	n-heptane	74.18	22.25
21.07.2016	n-hexane	283.84	16.17
	n-heptane	64.85	62.27
25.07.2016	n-hexane	245.83	5.76
	n-heptane	198.80	9.21
28.07.2016	n-hexane	91.79	19.08
	n-heptane	Rejected sample	
21.12.2016	n-hexane	630.25	2.39
	n-heptane	612.00	2.73
03.01.2017	n-hexane	395.34	5.66
	n-heptane	426.52	5.37

From the summary table it clearly results that in the initial analysis the procedure of sample preparation was inaccurate. This is evidenced by the values shown in the last column of Table 2. These values relating to the blank values in accordance with the standard [5, 6] should not exceed 10%. The first samples which were prepared correctly (a value of 10% was not exceeded) were sewer sediment samples collected on 25.07.2016.

This results from the improvement of the process of sample preparation for chromatographic analysis. The samples from next collection (28.07.2016) prepared with n-heptane were rejected, whereas in samples prepared with n-hexane the value 10% (19.08) were exceeded. The main issue, in this series of samples, determined sediment which for sample with n-heptane not sediment completely, which caused difficulties in later part of analysis leading finally to the rejection of samples. On July 28, 2016, the amount of sewer sediment in the channel was great, but it was not a sewer sediment which accumulated in the channel successively (a small amount of sewer sediment and the wet weather conditions on 25.07.2016). In the other two samples series carried out in the second part of the research, after conducted another corrections in the samples preparation, the values fulfil the requirement of 10% [5, 6]. It can therefore be concluded that efficiency of sample preparation for determination of petroleum hydrocarbons in sewer sediment using gas chromatography was achieved. The results obtained, meeting the guidelines (10%), fall into the range of determination specified in the standards [5, 6] amounting between 100–10000 mg/kg dry matter, therefore, the selected standards could be used in the present case. In the second part of the analysis the sewer sediment quality was significantly better, which greatly simplified the samples preparation (mainly the homogenisation process). Also a higher value of amount of petroleum hydrocarbons in the analyzed sewer sediment up to 630 mg/kg dms (n-hexane) and 612 mg/kg dms (n-heptane) was received. This amount of petroleum hydrocarbons in the sewer sediment cause disqualification of sewer sediment as a safe and possible for the storing on landfill, where the maximum limit value amounts 500 mg/kg dms [14].The figure below shows the gas chromatogram of sewer sediment sample collected on 21.12.2016 prepared using n-hexane as the hydrocarbon solvent.

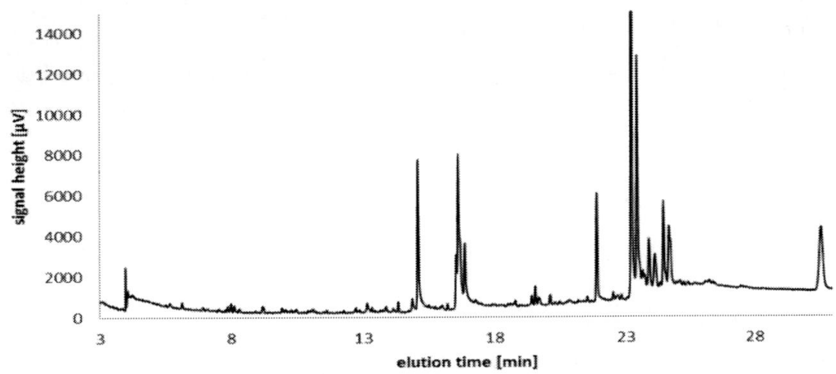

Fig. 5. Gas chromatogram of sewer sediment sample collected on 21.12.2016.

In carried out research also the impact of dry weather conditions period on the value of petroleum hydrocarbons in the sewer sediment was taken into consideration. Unfortunately due to the very short periods of dry weather conditions no connection between accumulation of petroleum hydrocarbons in sewer sediment during dry weather conditions was found. It can only be noted that the sewer sediment characterized by good consistency, which enables conducting the analysis, is formed after longer periods of dry weather conditions.

4 Conclusions

1. Using the available standards [5, 6] allows the determination of petroleum hydrocarbons in sewer sediment.

2. The temperature program (80°C for 4 min / 10°C/min to 300°C / 300°C for 10 min.) allowing the analysis of the study compounds using the available equipment was determined.
3. Both n-hexane and n-heptane can be used as hydrocarbon solvents during the determination process of petroleum hydrocarbons in sewer sediment.
4. The quality of the sewer sediment (the consistency) has a significant impact on the preparation of samples for further analysis.
5. Improving the procedure for sample preparation ensured obtaining 10% efficiency of the determination process.
6. Some parts of the analyzed sewer sediment samples contain more than 500mg/kg dms, which prevents their storage [13].
7. The determination of the impact of dry weather conditions on the amount of petroleum hydrocarbons in sewer sediment requires many sample series.

References

1. S.J. Varjani, Bioresour. Technol., **223**, 277 (2017).
2. R.N. Montagnolli, P.R. Matos Lopes, E.D. Bidoia, Arch. Environ. Contam. Toxicol., **68**, 342 (2015).
3. ATSDR, *Toxicological profile for total petroleum hydrocarbons (TPH)*, (1999).
4. J. Surygała, *Ropa naftowa a środowisko przyrodnicze*, (Oficyna Wydawnicza Politechniki Wrocławskiej, 2001).
5. PN-EN 14039:2008, *Characterization of waste – Determination of hydrocarbon content in the range of C10 to C40 by gas chromatography*, (2008).
6. PN-EN ISO 16703:2011, *Soil quality – Determination of content of hydrocarbon in the range C10 to C40 by gas chromatography*, (2011).
7. E. Hartmann, W.A. Bischoff, M. Kaupenjohann, *Untersuchung von Klärschlamm auf ausgewählte Schadstoffe und ihr Verhalten bei der landwirtschaftlichen Klärschlammverwertung* (Berlin, 2004).
8. M. Gąsiorek, *Zeszyty Naukowe. Inżynieria Środowiska* (Oficyna Wydawnicza Uniwersytetu Zielonogórskiego, 2009).
9. Tennessee State Government, *Method for determination of extractable petroleum hydrocarbons by GC/FID.*
10. G. Toffoli, *Extractable petroleum hydrocarbons methodology*, (2010).
11. P. Kuráň, The Scientific World Journal, **2014**, 1 (2014).
12. F. David, K. Jacq, B. Tienpont, *High Throughput Mineral Oil Analysis. (Hydrocarbon Oil Index) by GC-FID using the Agilent Low Thermal Mass*, (2011).
13. Regulation of the Minister of the Economy on acceptance of waste to store on landfill, Jaw of Laws 2015, pos. 1277, (2015).

Hydrochemical transformations of river waters during the flow in the reception basin on the basis of Olechówka River in Łódź

Tomasz Bagrowicz[1], *Anna* Fortuniak[2], *Michał* Górecki[3,*], *Milena* Lewandowska[1], and *Maciej* Ziułkiewicz[2]

[1]Third year of full-time studies in the field of Geomonitoring, the Faculty of Geographical Sciences, University of Łódź, 90-129 Łódź, ul. Narutowicza 88, Poland
[2]Laboratory of Geology, University of Łódź, 90-129 Łódź, ul. Narutowicza 88, Poland
[3]Second year of doctoral studies, the Faculty of Geographical Sciences, University of Łódź, 90-129 Łódź, ul. Narutowicza 88, Poland

Abstract. This work presents the results of the research experiment of conducting hydrochemical observation of the Olechówka River in Łódź and its flow from the river source until its estuary. The main aim of the research was to set tendencies of changes in waters flowing down from municipal reception basin with developed rain drain system and in waters flowing into bathing areas at the same time. Along with the flow of the river, 12 measuring-research points were established. In each of those points, measurements of the discharge rate and mean flow velocity of water in the channel were performed, which enabled to set average time of the flow of water between individual measuring-research points. The time of storing water in reservoirs was taken into consideration. The total time of the flow of water in the Olechówka River amounted to 856 h and 15 min. The analyses included in situ measurements (T, pH, SEC), analytical determinations Cl^-, NH_4^+, NO_3^-, TN, PO_4^{3-}, TP and Oxidability. The interchangeability of values of tested indicators and discharge point out to their decrease along with the increase of water flowing in the riverbed. There were self-cleaning processes identified in the Olechówka River: dilution and denitrification, along with the accumulation of total phosphorus in the river flow.

1 Introduction

Changes in the water regime caused by urbanization make the functioning of rivers gain additional local economic importance. Three basic sections can be distinguished on rivers flowing through the city: above the city – flow approximate to natural; within water intakes for the city – flow reduced by consumption; below the city – flow increased by sewage discharge [1]. Small rivers have the additional function of feeding water reservoirs established on their courses, including those with recreational function. In the area of the

* Corresponding author: michal.gorecki@unilodz.eu

Łódź urban agglomeration, the Ner River serves as an example of a river with a complex economic function. In its upper course, by infiltration, it supplies urban underground water intakes in the vicinity of Rzgów. Further along its course, it feeds the recreational reservoirs in Ruda Pabianicka – so called Stefański ponds. Finally, from Chocianowice, it becomes the receiver of first stormwater, and then sanitary sewage. The Olechówka River belongs to the fluvial system of the Ner River. It is connected to it through the Jasień River. Its entire course is within the boundaries of the city of Łódź, running through its southern districts. The two primary functions of the river are recreational and as a receiver of sewage. The intensity of management of the Olechówka River catchment makes it impossible to designate its sections with specific functions – both of the aforementioned functions overlap.

The far-reaching transformation of water relations in the urban catchment and in the urban river channel itself provided the basis for the distinguishing [2] of a new type of river regime, namely anthropogenic-nival-rain regime. In the case of such a regime, outflow culminations can occur in any month of the hydrological year. Research conducted in the Jasień River catchment [3] showed that in the conditions of 2–5 times lower infiltration, 0.5–2 times lower groundwater retention, 2–3 times higher seasonal variability of precipitation amount, and 2–3 times lower density of the fluvial network, the rate of surface discharge decreased 50-100 times with a simultaneous significant increase in its amplitude. In such circumstances, the urban river starts resembling a mountain stream. It becomes a very dynamic object, not only in terms of hydrological, but also hydrochemical, and qualitative characteristics.

Considering the above, in 2015 and 2016, students of major Geomonitoring at the Faculty of Geographical Sciences of the University of Łódź, under the scientific supervision of the Laboratory of Geology of the University of Łódź, conducted research experiments involving the application of two recognised methods of research on the quality of river waters. The first of the experiments, implemented in 2015, involved the preparation of the hydrochemical image, i.e. performance of simultaneous measurements of physico-chemical parameters, and water sampling for analyses along the entire course of the river. The results were presented in a scientific conference "Hydrology of urbanised areas", Warsaw University of Life Sciences, Warsaw, September 2016 [4]. The second experiment, conducted in 2016, involved hydrochemical research with consideration of the time of water flow in the channel of the Olechówka River from its "sources" to the place where it flows into the Jasień River. Both of the experiments were conducted in selected measurement and control sites (mcs). The objective of the works was to verify which of the research methods is more relevant for small urban rivers[1]. It seems that the rules of monitoring of surface waters flowing with imprecise flow reflect the specificity of development of the quality of waters of such hydrographic objects.

2 Study object

The Olechówka River is an order V watercourse belonging to the Odra River catchment. It is connected to the Odra River through Warta, Ner, and Jasień. The river has a length of 12.5 km, and catchment with an area of 35.2 km^2. The drainage area in its upper course covers areas of multi-family residential building development (residential districts of Widzew Wschód and Olechów), industrial development, as well as post-industrial and recreational areas (Dąbrowa). In its lower course, it covers single-family and service

[1]Understood as a watercourse that begins and ends in the city and whose catchment substantially coincides with the urban area.

DOI: 10.1051/e3sconf/20171700002

building development (Ruda Pabianicka). Due to this, the sewage system of the catchment covers 32.6 km^2 (approximately 93%), and surface seal is reaching 13.4 km^2 (38%). Between strongly urbanised areas, in its middle course, the river flows through areas of agricultural fallow land in Stare Chojny. At the end of this section, it flows into Stawy Jana – the largest retention reservoir on its course with a volume of 44.5 thousand m^3. Smaller reservoirs include: Górna (Upper) Olechówka (7.7 thousand m^3), Tomaszowska (17.5 thousand m^3), and Młynek (31 thousand m^3). The two largest ones have the recreational function, and are dried in the autumn-winter period. The remaining ones regulate discharges, and water is retained in them all year round. Apart from the establishment of the reservoirs, the valley and channel itself were subject to strong transformations aimed at the adjustment of the river to fulfilling the role of precipitation sewage collector. The course of the river was corrected. Sedimentation tanks were constructed at the outlets of larger stormwater collectors. The banks and bottom of the channel were reinforced with concrete. The natural spring niches were subject to building development, and the course of the river currently has its beginning at the outlet from the stormwater collector from the residential district Łódź-Olechów. The Olechówka River has become an integral part of the storm sewage system of the city, and the main collector of stormwaters from 70 sewage outlets [4] – Fig. 1.

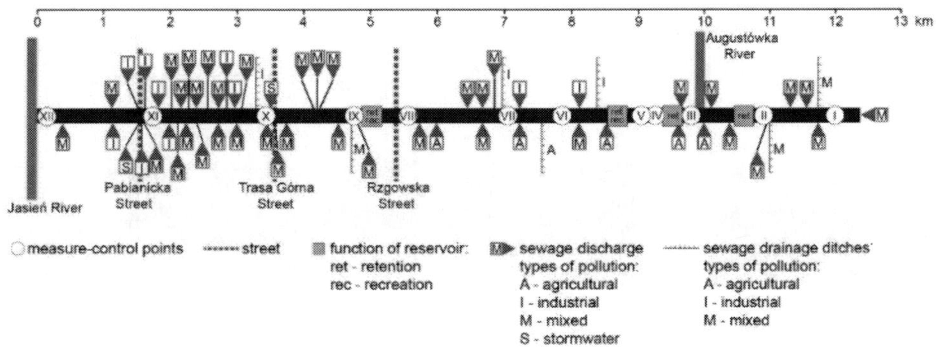

Fig. 1. Scheme of the Olechówka River.

3 Material and methods

The implementation of the adopted research objective required the performance of hydrometric measurements for the determination of the discharge rate (Q) and mean flow velocity of water in the channel ($\bar{\upsilon}$) at selected study sites. This involves the application of a method proposed by Mańczak, which is based on a statistical analysis of the relationship Q with water quality indicators [5]. The location of measurement-control sites is presented in Fig. 1. Their location was determined based on the criteria binding for the monitoring of flowing surface waters [6].

The time of water flow over the distance between subsequent mcs was calculated taking into account the distance determined by the mileage of the river [7] and $\bar{\upsilon}$. The calculations also considered time of water retention in the reservoirs, determined based on their volume, and Q at sites locates directly below the reservoirs.

Due to the specificity of urban pollutants, the hydrochemical research on the Olechówka River was conducted in the scope of salinity indices – standardized electrolytic conductivity (SEC) and chloride ion (Cl^-); eutrophication indices – mineral forms of nitrogen (NH_4^+, NO_2^- and NO_3^-), phosphates (PO_4^{3-}), and total nitrogen and phosphorus (TN and

TP). Total suspension (TS) was considered as an indicator of mechanical pollution and Oxidability (COD-Mn) was also considered as an indicator of organic pollution. Samples collected for the determinations of TN, TP, and TS were subject to analyses at the laboratory of the Collective Sewage Treatment Plant in Łódź. The remaining determinations were performed at the laboratory of Laboratory of Geology. Measurements of water temperature and SEC were performed by means of Elmetron CP315 and CX742 devices with an Eg-60 electrode. Chemical analyses were performed in accordance with the guidelines of [8], and according to the method PN-EW 872:2007+Ap.1:2007 in the case of TS.

The statistical processing of results, like a data clustering with Ward method of agglomeration, involved the application of STATISTICA 7.1. software.

4 Results and discussion

4.1 Hydrometeorological conditions

Calculations of the time of water flow in the channel of the Olechówka River from the "sources" to its mouth show that in total it amounts 856 hours (13.04.2016, 7.00 a.m. – 19.05.2016, 11.23 p.m.) partial results of the time of water flow are presented in Fig. 2.

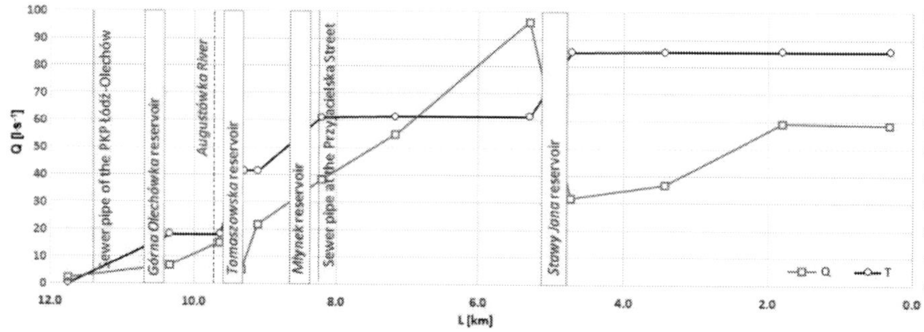

Fig. 2. Accumulated time of water flow in the channel of the Olechówka River and changes in the discharge rate in the Olechówka River in the period of the hydrochemical research.

Measurements of Q performed during sampling document a gradual increase in the amount of water flowing in the Olechówka River (Fig. 2). A reduction of discharges in reservoir Tomaszowska and Stawy Jana was recorded. According to materials obtained from the Department of Meteorology and Climatology of the University of Łódź, in the first week of the research period, low precipitation was recorded (<10 mm), resulting in a slight wetting of the Olechówka catchment area. During the conducted research, small and relatively short rainfalls occurred, in the majority of cases very short or in the form of a drizzle. They were observed more frequently and in higher amounts in the middle and lower course of the river than along its upper course. This suggests that the recorded increase in discharges in the Olechówka River was primarily caused by drainage of groundwaters in the catchment, either directly by the channel, or by the stormwater system.

4.2 Hydrochemical conditions

The hypothesis adopted in the methodological assumption, regarding the implemented scope of research, stating the possibility of occurrence of characteristic types of pollutants in the Olechówka River, can be subject to preliminary verification through the calculation of values of coefficients of correlation between the determined hydrochemical elements. This was performed for standardised data, with use the mean value and standard deviation of each data set. The results of the calculations are presented in Table 1.

Table 1. Matrix of correlation of hydrochemical elements and discharge rate and time of water flow in the channel of the Olechówka River; value of Pearson correlation coefficient (r), significant cases (p<0,05) were bolded.

	Time	Temp.	SEC	Cl^-	PO_4^{3-}	NO_3^-	TS
Q	**0.662/ $p = 0.019$**	**0.696/ $p = 0.012$**	−0.528/ $p = 0.078$	−0.453/ $p = 0.14$	−0.451/ $p = 0.141$	**−0.586/ $p = 0.045$**	−0.255/ $p = 0.425$
Time	–	0.43/ $p = 0.163$	**−0.61/ $p = 0.035$**	**−0.592/ $p = 0.043$**	**−0.677/ $p = 0.016$**	**−0.857/ $p = 0.000$**	−0.082/ $p = 0.8$
SEC			–	**0.978/ $p = 0.000$**	**0.828/ $p = 0.001$**	**0.79/ $p = 0.002$**	−0.095/ $p = 0.769$
Cl^-				–	**0.827/ $p = 0.001$**	**0.761/ $p = 0.004$**	−0.132/ $p = 0.682$
TP					−0.232/ $p = 0.468$	−0.233/ $p = 0.466$	**0.657/ $p = 0.02$**
PO_4^{3-}					–	**0.89/ $p = 0.000$**	−0.068/ $p = 0.834$
COD-Mn							**0.918/ $p = 0.000$**

The discharge rate shows a significant positive correlation with water temperature, and negative with the concentration of nitrates. The time of water flow is positively correlated with Q and negatively with SEC and Cl^- as well as PO_4^{3-} and NO_3^-. Based on statistically significant values of coefficients of correlation between hydrochemical elements, their two groups with the strongest correlations can be distinguished: SEC, Cl^-, PO_4^{3-}, and NO_3^-, as well as COD-Mn, TS and TP. The first group represents mineral substances in ion forms, and the second one – suspended substances and total phosphorus. Only the elements of the first group show a significant correlation with the time of water flow in the channel of the Olechówka River – Table 1. Changes in the values of representatives of both of the groups are presented in Fig. 3. The association of the analyzed elements into such groups corresponds with the results of observations of Olechówka from 2015. Based on the observations it was documented that pollutants supplied to the catchment show a certain sequentiality. The obtained hydrochemical image of a precipitation flood showed runoff from the catchment of first dissolved substances, then suspension, and derivatives of organic substances [4].

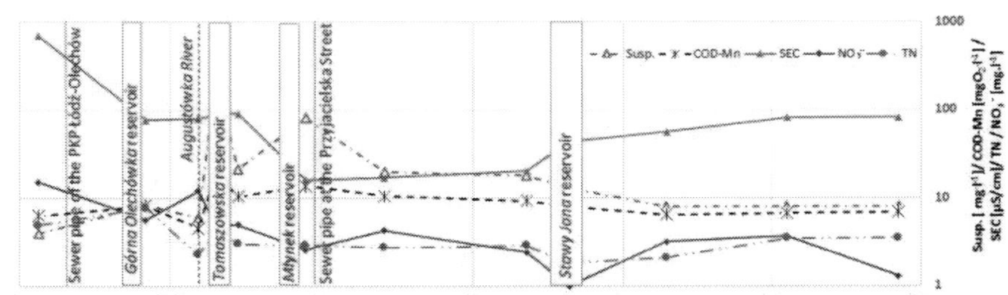

Fig. 3. Changes of selected hydrochemical elements.

 DOI: 10.1051/e3sconf/20171700002

The hydrochemical element showing an evident correlation with hydrological characteristics (Q and T) in the Olechówka River is nitrate ion. Negative values of the coefficient of correlation suggest that concentrations of NO_3^- decrease with an increase in the volume of flowing water and increase in the duration of such movement – Fig. 4.

a/ b/

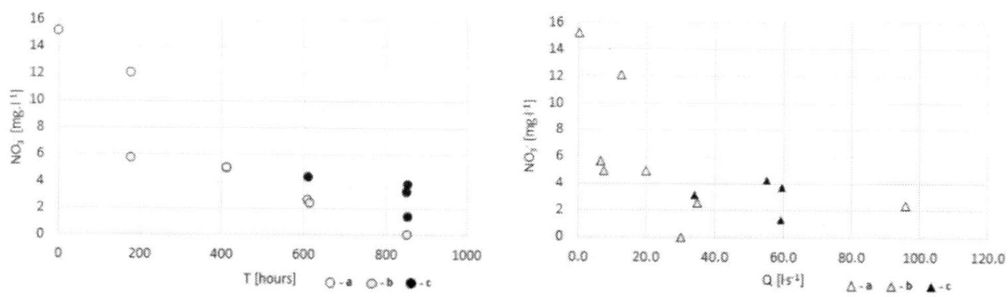

Fig. 4. Changes of concentrations of nitrates in relation to the time of water flow in the channel of the Olechówka River (a.) and discharge rate value at particular mcs (b.); explanation of the symbols a – mcs above the retention reservoirs, b – mcs directly below the reservoirs, c – mcs on the river.

The consideration in the reduction of nitrate concentrations of the presence of retention reservoir on the course of the Olechówka River and the resulting changes in discharge conditions suggests that the basic role of elimination of this form of mineral nitrogen results from water retention and therefore assimilation by phytoplankton or denitrification, depending on the conditions occurring in the reservoir [9]. The graphic image of co-variability of NO_3^- and Q suggests that the Olechówka River is a strongly polluted river according to the classification of rivers by Mańczak [5]. At low discharges, concentrations of pollutants are high, and decrease as a result of dissolution with waters successively supplied to the river channel along its course.

It is interesting that such dissolution does not result in a response of indices of content of mineral salts. The correlation of SEC and Cl^- ion with hydrochemical elements only concerns time of water flow, and is weaker than biogenic ions – Table 1. Changes in SEC (Fig. 3) suggest the impact of strong sources of release of salts in the upper part of the catchment, and weaker in its lower part. The middle section of the Olechówka River seems to be devoid of such an impact, and the retention reservoirs contribute to changes in SEC both in plus and in minus.

COD-Mn is a representative of the group showing no correlations with the hydrological characteristics of the Olechówka River. The hydrochemical profile (Fig. 3) shows the impact of the largest sewage collector, located at the outlet of the Olechówka River and reservoir. Both of the objects contribute to a significant increase in values of COD-Mn and TS. This evidences a direct effect of the stormwater system on the river, and indirect effect of the retention reservoir. The biological assimilation of NO_3^- mentioned before results in the emission of an increased load of organic matter and total nitrogen (Fig. 3) and (organic?) suspension. An increase in the concentration of TN which occurred in the Olechówka River after its flowing through the reservoir suggests that also a release of certain amounts of nitrogen occurs from bottom sediments. Analogical changes occur in the case of reservoir Górna Olechówka. In reservoirs where waters are seasonally drained out, this type of changes are not recorded. Notice that Stawy Jana contributes to the reduction of both of the forms of nitrogen and suspension.

The complex image of changes in the waters of the Olechówka River in the scope of hydrological and hydrochemical characteristics of particular mcs is provided by the similarity analysis, conducted in the Euclidean space by means of the Ward method. In a river representing the continuum of hydrological and hydrochemical changes, functioning in a space subject to anthropopressure to a weaker degree than an urban catchment, subsequent mcs should show a considerable degree of similarity, and be ordered in accordance with the actual distribution of mcs in the course of the river, reflecting among others the course of the self-cleaning process. The result of the similarity analysis of waters flowing in the channel of the Olechówka River was presented in Fig. 5.

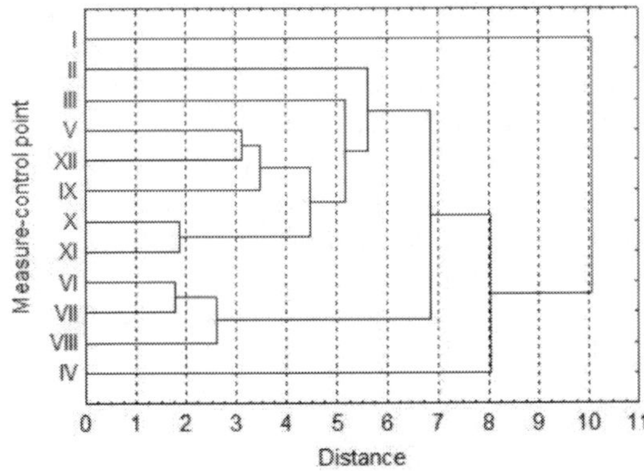

Fig. 5. Dendrogram of similarity of particular mcs in hydrochemical terms.

The greatest similarity occurs in the case of mcs VI and VIII. Together with mcs No. VIII, they constitute the only group of sites ordered in the diagram (Fig. 5) in the same way as in reality (Fig. 1). The sites are distinguished by average values of hydrochemical elements at high discharge progression. On the section between mcs VI and VIII, Olechówka flows through agricultural fallow land, in an unsealed channel, where possible channel drainage is possible, and the number of outlets of the sewage system is lower than on other sections (Fig. 1). Mcs I is distinguished among all the others. It is locates on the initial section of the Olechówka River. High index values are observed here at simultaneous very low discharge. Mcs IV is also characterized by high degree of individualism. It is located below reservoir Tomaszowska. This evidences the importance of the reservoir for the maintenance of the continuity of processes occurring in the channel. The reservoir can be considered as a specific hydrochemical barrier. The remaining reservoirs also contribute to the disturbance of the hydrochemical continuum of the river, as manifested in sifting sites VI and IX in the sequence, far from sites located above the reservoirs (V and VIII).

Due to the function of water supply to recreational reservoirs by the Olechówka River, its waters should represent a specific state of quality. The verification was performed in relation to normatives of waters of class II [10][2]. The waters of the Olechówka River are not suitable for bathing over the entire length, as determined by the values of Oxidability and levels of ammonia nitrogen. Moreover, in the upper section, the water is disqualified by excessively high concentrations of nitrate nitrogen and total phosphorus.

5 Summary

In the Olechówka River, during water flow from the "sources" to the river's mouth to the Jasień River, the presence of two types of pollutants was determined. The first one includes mineral salts, including forms of nitrogen and phosphorus, developing its

[2]Existing provisions for water intended for bathing organization do not include hydrochemical elements that were studied. Therefore they resorted to the provisions of the surface water classifications in force until 2004.

concentrations in relation to the hydrological parameters of the studied river, i.e. discharge rate, and particularly time of water flow. Retention reservoirs play an important role in the recorded hydrochemical transformations. The scale of changes determined by reservoirs allows for recognising them as specific hydrochemical barriers, whereas it is of importance whether water in the reservoirs is maintained all year round, or only seasonally.

The second group of hydrochemical elements is total suspension, Oxidability, and TP. The indicators do not refer to hydrological changes in the channel of the Olechówka River, but respond to the presence of retention reservoirs, particularly reservoir Tomaszowska, and water supply from the stormwater system. Both of the hydrotechnical objects should be treated as important sources of pollution of the studied river.

The study shows that the hydrochemical transformations of waters during their flow in the river channel over a length of 12 km for 36 days are subject to changes basically caused by the presence of retention reservoirs. Due to the limited amount of atmospheric precipitation, the influence of the stormwater system was limited, and only manifested itself in the case of the largest sewage collector.

References

1. M.J. Łoś, Wiadomości Melioracyjne i Łąkarskie, **38** (1995)
2. S. Czaja, *Zmiany stosunków wodnych w warunkach silnej antropopresji (na przykładzie konurbacji katowickiej)* (Wydawnictwo Uniwersytetu Śląskiego, Katowice, 1999)
3. R. Dębski, *Komunikat o ukończeniu tematu naukowo-badawczego pt. zakłócenia cykli hydrologicznych w miastach na przykładzie Łodzi jako wynik ich gospodarczego rozwoju i przeciwdziałania ujemnym zjawiskom. Materiały na konferencję na temat: Wpływ aglomeracji przemysłowych na środowisko człowieka na przykładzie makroregionu łódzkiego* (1987)
4. M. Ziułkiewicz, A. Fortuniak, A. Waack-Zając, M. Górecki, S. Grzędzińska, B. Małecka, Monografie Komitetu Gospodarki Wodnej Polskiej Akademii Nauk, **39** (2016)
5. H. Mańczak, *Techniczne podstawy ochrony wód przed zanieczyszczeniem* (Politechnika Wrocławska, Wrocław, 1972)
6. K. Gołębiowska, A. Niespodziewany, T. Reczek, *Wskazówki metodyczne do projektowania regionalnego monitoringu wód powierzchniowych płynących* (Biblioteka Monitoringu Środowiska, Warszawa, 1994)
7. UM Łódź, *Projekt generalny rzeki Olechówki* (Wydział Gospodarki Komunalnej Urzędu Miasta Łodzi, Łódź, 2000, unpublished source)
8. S. Witczak, J. Kania, E. Kmiecik, *Katalog wybranych fizycznych i chemicznych wskaźników zanieczyszczeń wód podziemnych i metod ich oznaczania* (Biblioteka Monitoringu Środowiska, Warszawa, 2013)
9. R. Gołdyn, *Zmiany biologicznych i fizyczno-chemicznych cech jakości wody rzecznej pod wpływem jej piętrzenia we wstępnych, nizinnych zbiornikach zaporowych* (Wydawnictwo Naukowe UAM, Poznań, 2000)
10. Rozporządzenie Ministra Ochrony Środowiska, Zasobów Naturalnych i Leśnictwa z dnia 5 listopada 1991 r. w sprawie klasyfikacji wód oraz warunków, jakim powinny odpowiadać ścieki wprowadzane do wód lub do ziemi (Dz.U. 1991 nr 116 poz. 503)

Sludge settling characterization for the mathematical modelling of sidestream treatment processes

Piotr Balbierz[1,*], and *Katarzyna* Rucka[1]

[1]Wroclaw University of Science and Technology, Faculty of Environmental Engineering, Wybrzeze Wyspianskiego 27, 50-370 Wroclaw, Poland

Abstract. More stringent effluent limits, especially with regard to total nitrogen, force wastewater treatment plants to consider sidestream nitrogen removal from digested sludge dewatering liquor which can contain up to 20% of incoming nitrogen load. Deammonification, the most promising biological technology for sidestream nitrogen removal, is based on partial short-cut nitrification and the Anammox process. Both steps of this process are conducted by slowly growing autotrophic bacteria, namely nitrifies and Anammox, which however may form different types of sludge: flocculent and granular, respectively. Nevertheless, ensuring effective sludge separation from treated effluent is crucial for the ability to retain both kinds of slowly growing biomass in the system. This paper presents the principles of sludge settling modelling, the results of a series of column tests conducted on sludge from pilot-scale two-stage deammonification treatment facility, as well as an attempt to estimate settling parameters in order to fit a sedimentation model.

1 Introduction

Sedimentation is a gravitational solids-liquid separation, based on the differences in density between solid particles and the liquid. In conventional suspended-growth activated sludge systems, sedimentation is one of the most crucial physical unit operations, as it directly affects the quality of the treated effluent [1].

Secondary clarifiers are designed to achieve two goals which are critical to operation and performance of activated sludge system: clarification of the final effluent to comply with the discharge limits and thickening of the sludge to minimize the stream recirculated to bioreactors. Both settler functions are dependent on sludge settling properties, which tend to be highly variable and difficult to predict. Therefore, poor biomass separation is not uncommon, which may lead to severe biological process disturbances [2].

Depending on the concentration and properties of solid particles four types of settling behaviour can be distinguished at different layers along the height of primary and secondary settling tanks, which are discrete particle settling, flocculent particle settling, hindered settling and compression settling [1].

* Corresponding author: piotr.balbierz@pwr.edu.pl

1.1 Sedimentation in sidestream treatment processes

More stringent effluent limits, especially with regard to total nitrogen, force wastewater treatment plants to consider sidestream nitrogen removal from digested sludge dewatering liquor which can contain up to 20% of incoming nitrogen load.

Deammonification, the most innovative and promising biological technology for sidestream nitrogen removal, is based on partial short-cut nitrification and the anammox process (short for ANaerobic AMMonium OXidation). In the partial nitrification step approximately half of the initial ammonium is oxidized to nitrite by strictly aerobic ammonium oxidizing bacteria. In the anammox step ammonium (electron donor) and nitrite (electron acceptor) are converted by strictly anaerobic anammox bacteria into dinitrogen gas and small amounts (~10%) of nitrate without external organic carbon source. Therefore, deammonification allows to decrease oxygen demand by 60%, and organic carbon demand and sludge production by 90%, compared to conventional nitrification/denitrification systems [3, 4].

Due to lower investment costs, deammonification is usually operated in one-stage systems, where both steps with different requirements take place in one reactor, which makes the process control more complicated. An alternative are two-stage systems, where partial nitritation and anammox are split in two reactors in series, which allows to optimize process condition for each of the steps separately. Moreover both attached-growth (biofilm, moving bed biofilm reactor and rotating biological contactor) and suspended growth (sequencing batch reactor and upflow anaerobic sludge blanket reactor) systems are in operation at full-scale [4].

Partial nitritation and anammox are conducted by slowly growing autotrophic bacteria with a long doubling time, which, however, in the suspended-growth systems may form different types of sludge: flocculent and granular, respectively. In favourable conditions (short settling time and adequate shear force) can grow forming dense, well settling granules, which can settle with velocities over 100 m/h, compared with 1–10 m/h for flocculent sludge [5, 6].

Nevertheless, ensuring effective sludge separation from treated effluent is crucial for the ability to retain both kinds of slowly growing biomass in the system, especially because solids retention, settling and separation is among the most common operating problems in full-scale deammonification installations [4].

1.2 Sedimentation modelling

Mathematical description of activated sludge sedimentation is usually based on the solids flux theory, which states that sludge movement in the thickening zone of the settler results from gravity settling and convection due to sludge recirculation underflow stream. The solids flux caused by gravity settling is a function of sludge concentration and sludge zone settling velocity (ZSV), which characterizes hindered settling. ZSV is dependent on the sludge concentration and expressed by the classic Vesilind equation [3]:

$$v_s = v_0 \, e^{-kX} \tag{1}$$

where v_s is sludge settling velocity, v_0 is maximum sludge settling velocity, X is sludge concentration and k is empirical settling parameter.

However, it should be noted that Vesilind equation is valid for the hindered settling only, as, when the sludge concentration goes to zero, the predicted settling velocity reach values, which are not realistic. Therefore, Takacs *et al.* [7] presented a modification of

Vesilind equation, which accounts for the fact that settling velocity decreases when sludge concentration approaches zero, and is expressed by the following set of equations:

$$v_s = v_0\left(e^{-r_h(X-X_{\min})} - e^{-r_f(X-X_{\min})}\right) \tag{2}$$

$$X_{\min} = f_{ns}X_0 \tag{3}$$

$$0 \le v_s \le v_{0,\max} \tag{4}$$

where r_h is hindered zone settling parameter, r_f is flocculent zone settling parameter, X_{min} is minimum attainable suspended solids concentration, X_0 is mixed liquor suspended solids entering the settler, f_{ns} is non-settleable fraction of mixed liquor suspended solids and $v_{0,\, max}$ is an upper limit of sludge settling velocity.

The traditional models describing activated sludge settling velocity are no longer valid for granular sludge, therefore some theoretical models based on the Stock or Allen equations have been proposed [8, 9]. However, they require the knowledge about such parameters as granule diameter, density, sphericity and mass shape factor that are usually not readily available.

On the other hand Liu *et al.* [10] proposed a Vesilind-based generalized model for settling velocity of aerobic and anaerobic granular sludge, which is a function of mean size of granules (d_p), sludge volume index (SVI), and sludge concentration (X):

$$v_s = \alpha \frac{d_p^2}{SVI} e^{-\beta X} \tag{5}$$

where α and β are constant coefficients. For small enough particles model reduces to the classic Vesilind equation.

The aim of this paper is to characterize and compare settling properties of activated sludge, from a full-scale biological nutrient removal plant and two-stage deammonification pilot-scale plant, based on the results of a series of column tests, as well as to attempt to estimate settling parameters in order to fit a sedimentation model.

2 Materials and methods

This study was conducted at the pilot plant used for assessment of different reject water treatment processes, equipped with four 150 dm^3 SBR-type reactors and fed with real reject water from full-scale sludge dewatering station at Wroclaw WWTP in south-west Poland.

2.1 Source of sludge

Three different kinds of activated sludge were used in the experiment.

First sludge sample (BNR) was taken from a full-scale biological nutrient removal Wrocław WWTP (approx. 1 000 000 p.e.) operated in winter conditions (10°C) at a long solids retention time (SRT) of 35 d and mixed liquor suspended solids (MLSS) of 5.5 kg TSS/m^3.

The second sludge sample (SNC) was taken from a pilot-scale short-cut nitrification SBR fed with real sludge dewatering reject water and operated at temperature of 25°C, an SRT of 3 d and MLSS of 1.5 kg TSS/m^3. At start-up this SBR was seeded with the BNR

DOI: 10.1051/e3sconf/20171700003

sludge, and during later stable 6 month operation a daily dose of BNR sludge was added to keep the desired sludge concentration and improve settling properties.

The last sludge sample (ANX) was taken from a pilot-scale Anammox SBR fed with partially nitrified real sludge dewatering reject water and operated at temperature of 23°C, an SRT of 50 d and MLSS of 2.0 kg TSS/m^3. At start-up this SBR was seeded with granular sludge from an UASB Anammox reactor, without further addition during stable 6 months operation.

2.2 Batch settling test procedure

In order to characterize settling properties of sludge samples and to determine zone settling velocity as a function of sludge concentration a series of batch settling tests was carried out. Batch settling test are considered to be a good source of data for estimation of the sedimentation model parameters as the obtained result depend only on the physical properties of the settling column (which are constant for a given set-up) and the settling properties of the sludge [11].

Each sludge sample was thickened for an hour to achieve initial concentration of approximately 9 kg TSS/m^3. From this thickened sample a series of samples with more or less equidistant sludge concentrations was prepared by diluting the sludge with an appropriate effluent. The range of concentrations was chosen to ensure that sufficient settling velocity and a distinct sludge/water interface could be observed.

Batch settling tests were performed at room temperature (approx. 22°C) in settling columns with a height of 50 cm and diameter of 8 cm without a stirrer (Fig. 1.), by pouring a well-mixed sample into a column and recording the height of sludge blanket for a period of 20–60 minutes depending on the concentration with the recording interval varied from 20 s to 5 min.

For each test the recorded interface height was plotted against time, and the settling velocity was determined from the slope of the linear part of curve.

The sludge concentration was determined after each test according to EN-872 – standard direct method for suspended solids on glass fibre filters with 1.2 m pore size.

The obtained relationship between settling velocity and sludge concentration was used to estimate parameters of the Takacs sedimentation model [7], as it does not require the knowledge of sludge morphology and is commonly used in commercially available simulators.

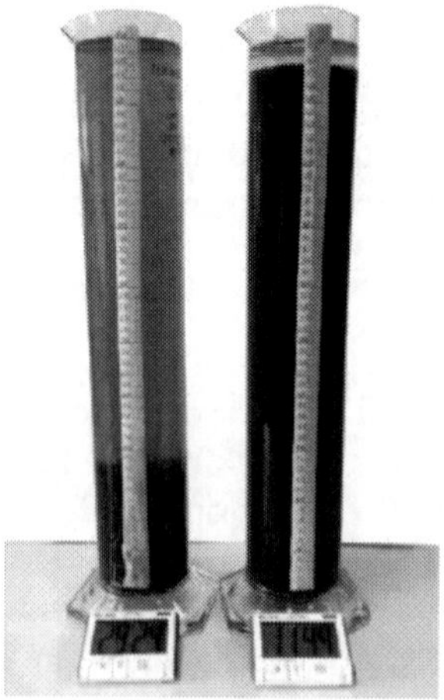

Fig. 1. Experimental set-up.

DOI: 10.1051/e3sconf/20171700003

3 Results and discussion

The results of the batch settling test series with different ANX sludge concentration presented as the change of sludge blanket height in time are show in Fig. 2. As the time in which the linear part of the curve is observed varies significantly with sludge concentration, it is important to carefully choose the test length and recording interval to obtain reliable data.

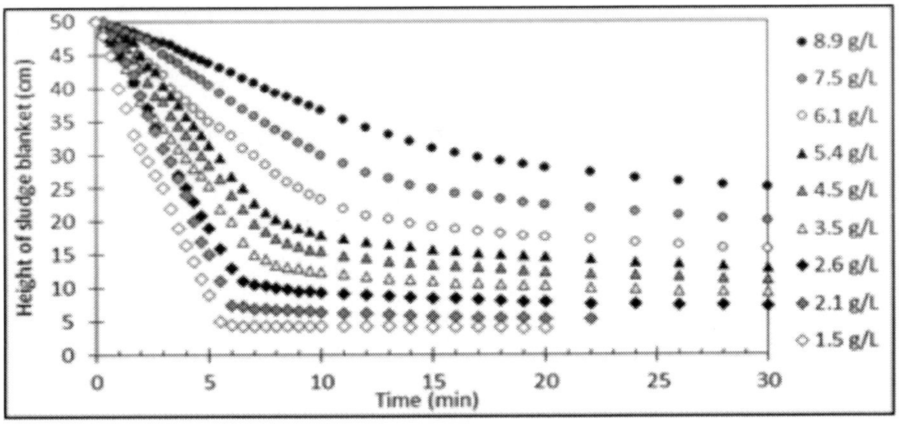

Fig. 2. Measured sludge blanket height in time from the test series with ANX sludge.

The comparison of determined zone settling velocities as a function of the BNR, SCN and ANX sludge concentration is presented in Figure 3. The settling velocities for the BNR sludge (1.6–6.3 kg/m^3) were in range of 0.04–5.6 m/h, for the SCN sludge (1.4–5.3 kg/m^3) 0.66–5.5 m/h and for the ANX sludge (1.5–8.9 kg/m^3) 0.89–4.85 m/h, and fall in the standard range for the flocculent sludge 1–10 m/h [5, 6]. The BNR and SCN sludge showed similar characteristics at low concentration, but in case of the BNR sludge a rapid decrease of settling velocity above 3.5 kg/m^3 was observed, which is a known phenomenon at lab-scale due to elevated wall friction effect and can be mitigated by introduction of slow stirring (1-2 rpm) [12]. The ANX sludge showed more flat characteristics, and generally significantly higher sedimentation velocities, especially at higher concentrations.

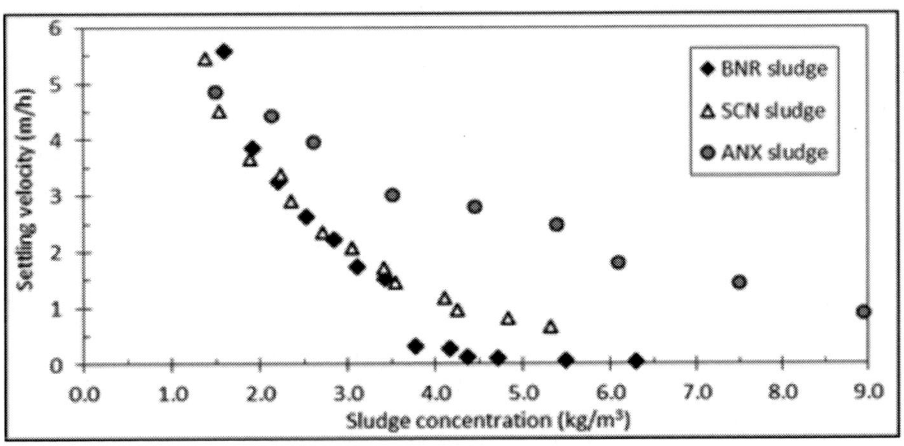

Fig. 3. Comparison of determined settling velocities for BNR, SCN and ANX sludge.

The best fit of the Takacs sedimentation model to the experimentally determined zone settling velocities for the BNR, SCN and ANX sludge is presented in Figure 4, and the estimated model parameters are listed in Table 1.

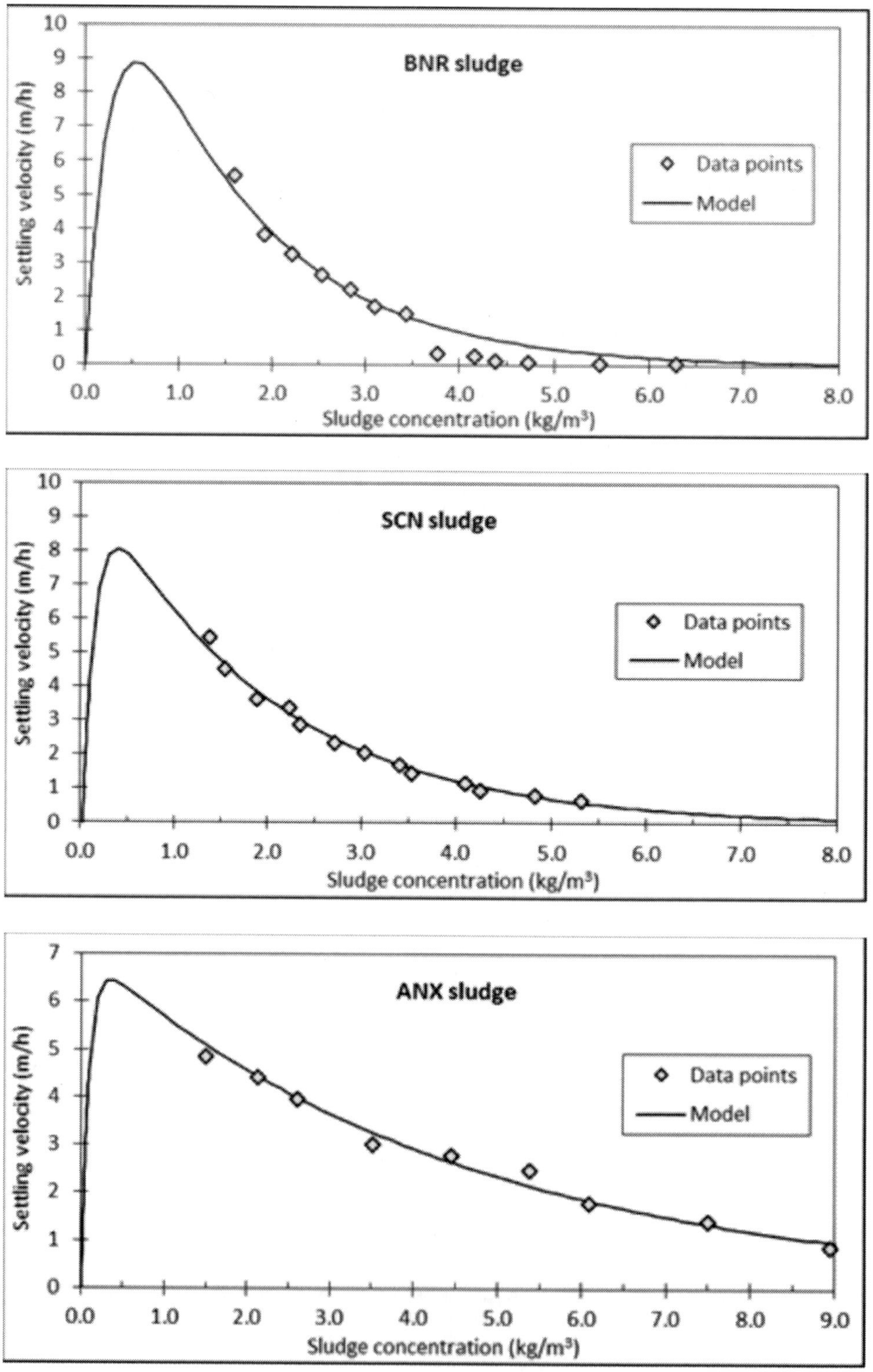

Fig. 4. Best fit of the Takacs model for BNR (top), SNC (middle) and ANX sludge (bottom).

Table 1. Estimated Takacs model parameters for the BNR, SCN and ANX sludge.

Parameter	Unit	BNR sludge	SCN sludge	ANX sludge
X	kg/m^3	1.6–6.3	1.4–5.3	1.5–8.9
v_s	m/d	0.04–5.6	0.66–5.5	0.89–4.85
v_0	m/d	15.5	10.5	7.08
r_h	m^3/kg	0.692	0.537	0.221
r_f	m^3/kg	4.0	8.0	12
f_{ns}	-	0.0005	0.02	0.005
X_0	kg/m^3	5.5	1.5	2.0

The values of the v_0, r_h and r_f parameters were estimated based on the performed series of batch settling test, whereas the values of the f_{ns} and X_0 parameters were assumed based on the typical values of effluent suspended solids and MLSS in the appropriate reactors (not shown). The values of the r_f parameter are initial estimates, as the model is relatively insensitive to this parameter in the sludge concentration range that can be tested in the batch settling tests. Therefore, the final values of the r_f parameter, which impacts the predicted settling velocities at low sludge concentrations, need to be established during clarifier model fine tuning based on the data from the real system.

In general, a good fit of the Takacs sedimentation model to the experimental data points of all the three kinds of tested activated sludge was achieved, except for the data points of the BNR sludge with concentration above the 3.5 kg/m^3, which were rejected from model fitting, due to the observed wall effect disturbances mentioned earlier.

The settling properties of the SCN sludge are similar to the BNR sludge, which was to be expected as the SCN reactor is bioaugmented with the BNR sludge to improve the settling characteristics of the otherwise poorly settling nitrifiers dominated activated sludge. However, slightly higher settling velocities were observed at higher concentration, which may be attributed to the high concentration of nitrite in the SNC reactor. Nitrite, especially in its unionized form the free nitrous acid, is a known strong biocidal agent [13], which may selectively inactivate filamentous bacteria which tend to extend from the sludge flocs, thus improving sludge settling.

Sludge from the Anammox reactor, despite the initial seed with the Anammox granular sludge, exhibited flocculent-like behaviour, which may be attributed to the fact the SBR was operated for an extended period with long sedimentation phase (60 min.), which removed a strong factor promoting granulation. Moreover, some amount of flocculent sludge was fed to the Anammox SBR with effluent from the partial nitritation SBR (the first stage of 2-stage deammonification). Presence of nitrifiers in the Anammox reactor may be beneficial, as nitrifiers will remove any dissolved oxygen, which inhibits Anammox bacteria at very low concentration. The visual inspection of the ANX sludge showed that the sludge consisted of small red granules covered with brown flocs, which may explain the flocculent-like behaviour. Nevertheless, the ANX sludge was characterized by a significantly higher settling velocities, especially at higher sludge concentration, compared to typical flocculent sludge (see Fig. 3.).

DOI: 10.1051/e3sconf/20171700003

4 Conclusions

For the purpose of mathematical modelling and simulation of sidestream treatment processes, batch settling tests allow for a proper characterization of sludge settling properties, and the Takacs sedimentation model provides a good prediction of sludge settling behaviour, if the sludge has a flocculent-like nature.

Acknowledgements – The authors gratefully acknowledge co-funding from the National Centre for Research and Development (grant no. PBS2/B9/25/2014) and the Municipal Water and Sewage Company in Wroclaw.

References

1. Metcalf & Eddy, *Wastewater Engineering: Treatment and Resource Recovery* (G. Tchobanoglous, F.L. Burton, H.D. Stensel, ed.), McGraw-Hill Education (2014)
2. *Clarifier Design: WEF Manual of Practice No. FD-8*, McGraw Hill Professional, 2005
3. J. Makinia, *Mathematical modelling and computer simulation of activated sludge systems*, London ; New York, IWA Publishing (2010)
4. S. Lackner, E.M. Gilbert, S.E. Vlaeminck, A. Joss, H. Horn, M.C.M. van Loosdrecht, Water Res. **55** (2014)
5. W.R. Abma, C.E. Schultz, J.W. Mulder, W.R.L. van der Star, M. Strous, T. Tokutomi, M.C.M. van Loosdrecht, Water Sci.Technol. **55**, 8 (2007)
6. A. Nor Anuar, Z. Ujang, M.C.M. van Loosdrecht and M.K. de Kreuk, Water Sci. Technol. **56**, 7 (2007)
7. I. Takacs, G. Patry, D. Nolasco, Water Res. **25**, 10 (1991)
8. Y.-H. Liu, Y.-L. He, S.-C. Yang, Y.-Z. Li, Biotechnol. Lett. **28** (2006)
9. H-F. Lu, P. Zheng, Q-X. Ji, H-T. Zhang, J-Y. Ji, L. Wang, S. Ding, T-T. Chen, J-Q. Zhang, C-J. Tang, J-W. Chen, Bioresource Technol. **123** (2012)
10. Y. Liu, Z-W. Wang, Y-Q. Liu, L. Qin, J-H. Tay, Biotechnol. Progr. **21** (2005)
11. A. Vanderhasselt, P.A. Vanrolleghem, Water Res. **34**, 2 (2000)
12. A.-E. Stricker, I. Takacs, A. Marquot, Water Sci. Technol. **56**, 12 (2007)
13. Y. Law, L. Ye, Q. Wang, S. Hu, M. Pijuan, Z. Yuan, Chem. Eng. J. **259** (2015)

E3S Web of Conferences **17**, 00004 (2017) DOI: 10.1051/e3sconf/20171700004
EKO-DOK 2017

Stationarity of maximum low-flow periods duration in the right-bank area of the Upper Vistula catchment – Mann-Kendall versus Spearman test

Katarzyna Baran-Gurgul[1,*]

[1]Cracow University of Technology, Institute of Water Engineering and Water Management, ul. Warszawska 24, 31-155 Kraków, Poland

Abstract. This study compares two methods of analyzing the stationarity of time series – the Spearman test and the Mann-Kendall test, with the formula adjusted by Hamed and Rao. Daily discharge series collected at 78 gauging stations were compared for a period of 30 years. The study area consisted of the right-bank area of the Upper Vistula River catchment. Low-flow periods were identified via the pit under threshold method (PUT). Threshold discharge values in the study were represented by $Q_{70\%}$, $Q_{80\%}$, and $Q_{90\%}$. Maximum annual durations $T_{maxR,i}$, (i = 1, 2, ..., 30 low-flow periods) were identified for assumed threshold discharge values based on low-flow period duration series for selected gauging stations. Research on the stationarity of T_{maxR} series for most of the studied water gauging stations has shown an absence of a basis for the rejection of the hypothesis of a lack of a trend for peak low-flow period duration relative to time, independently of assumed threshold discharge. Most of the detected trends are decreasing trends. The lower the threshold discharge value, the larger the number of T_{maxR} series being non-stationary. The Spearman test detected more non-stationary series than the Mann-Kendall test for the studied gauging stations, independently of assumed threshold discharge values.

1 Introduction

The evaluation of seasonal and multi-annual variability of river discharge, including variability of the occurrence and characteristics of low-flow periods, is one of the most important themes in modern hydrology. Multi-annual variability of discharge time series, including periods of low flow, has been analyzed in a large number of papers. Trend analyses included a number of different tests including non-parametric tests such as the Mann-Kendall test and the rank correlation test by Spearman [1–7]. Work by Yue et al. [8] has shown that these two tests are almost equally powerful in the identification of trends in time series.

* Corresponding author: Katarzyna.Baran-Gurgul@iigw.pk.edu.pl

 DOI: 10.1051/e3sconf/20171700004

A low-flow event represents a process of continuous water discharge at a given gauging station below an assumed threshold level [9–13]. For the pit under threshold method (PUT) [14], the onset t_p of a low-flow period occurs at the moment when discharge decreases below a threshold value Q_g and the end t_k of a low-flow period occurs when discharge once again exceeds the value Q_g. The low-flow duration t_n is measured in 24-hour periods in the form: $t_n = t_k - t_p + 1$. In order to eliminate short durations with discharge less than Q_g that are considered irrelevant from the perspective of water shortages, the following assumption is also employed: minimum low-flow duration t_{min}. In the research literature, this value ranges broadly from 5 days to 21 days [9, 15–19]. If two adjacent low-flow periods are separated by a few days when discharge is greater than the threshold value Q_g, then one may consider these periods to be dependent in order to combine them into a longer time period. In most cases, durations for each low-flow period are added together [10, 15–19].

In this paper, a maximum annual low-flow period is defined as a low-flow period with a longest duration over the course of the hydrologic year. If a low-flow period starts during one year and ends during another, then it is assigned to the year when its middle happens to occur.

The purpose of the paper is to compare two methods of analyzing the stationarity of time series for maximum low-flow periods measured at 78 water gauging stations in the catchment of the Upper Vistula River in southeastern Poland. The two methods are the rank correlation test by Spearman and the Mann-Kendall test adjusted by Hamed and Rao, which considers the issue of autocorrelation in time series [20].

2 Materials

The study is based on daily discharge series for the period from 1.11.1983 to 31.10.2013. This yields a total of 30 hydrologic years and 10,958 discharge values obtained at each of 78 water gauging stations located across the right-bank area of the Upper Vistula River. The data were obtained from the Institute of Meteorology and Water Management - National Research Institute (IMWM-NRI). The location of the sites is shown in Fig. 1.

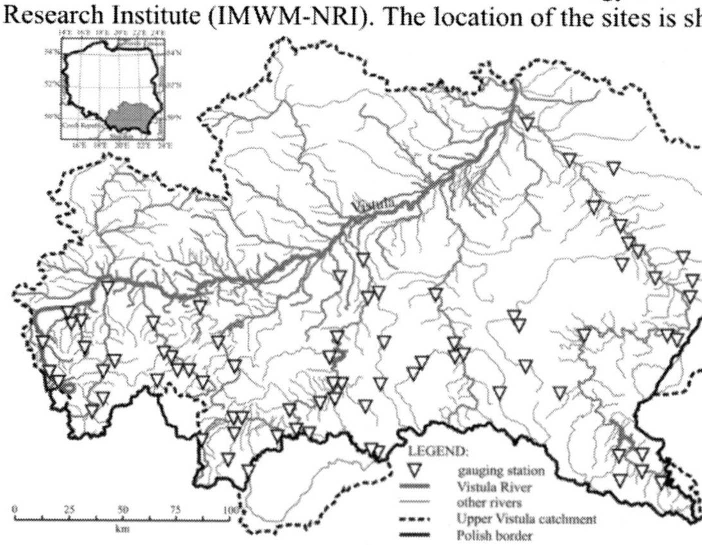

Fig. 1. Location of selected water gauging stations in the catchment of the Upper Vistula.

Most of the rivers in the study (68 of 78) featuring water gauging stations are second-order and third-order rivers. About half of the studied catchments (42 of 78) have surface areas ranging from 100 to 1,000 km^2. Most of the water gauges used in the study are located at sites found below 400 meters above sea level.

 DOI: 10.1051/e3sconf/20171700004

Low-flow periods were identified using the PUT method. Threshold discharge values for this study were assumed to be ones with an exceedance probability of 0.7, 0.8, and 0.9 ($Q_{70\%}$, $Q_{80\%}$, $Q_{90\%}$), as read from the flow duration curve. It was also assumed that a single low-flow period must last at least 7 days. Furthermore, if the time gap between 2 consecutive low-flow periods was shorter than 4 days, then the two periods were combined into one period.

Maximum durations time series $T_{maxR,i}$ (i = 1, 2, ..., 30 maximum annual low-flow periods) were identified for assumed threshold discharge values based on a series of low-flow period durations observed at selected water gauging stations in the study area.

The maximum mean duration $\overline{T_{maxR}}$ for low-flow periods based on data obtained at all the water gauging stations – and independently of assumed threshold discharge values – were noted for the years 1987, 1994, 2003, and 2012 (Fig. 2). The longest low-flow durations or those over 200 days occurred with threshold discharge at $Q_{70\%}$ (Fig. 3) in the years 1984 (gauge no. 10), 1993 (gauges no. 5, 6), and 2003 (gauges no. 7, 20, 47).

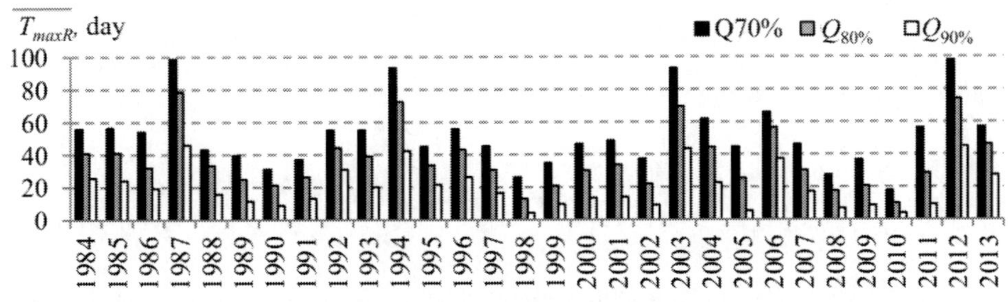

Fig. 2. Mean durations $\overline{T_{maxR}}$ for annual low-flow periods, in the study period, at 78 water gauging stations in the Upper Vistula catchment varying with assumed threshold discharge.

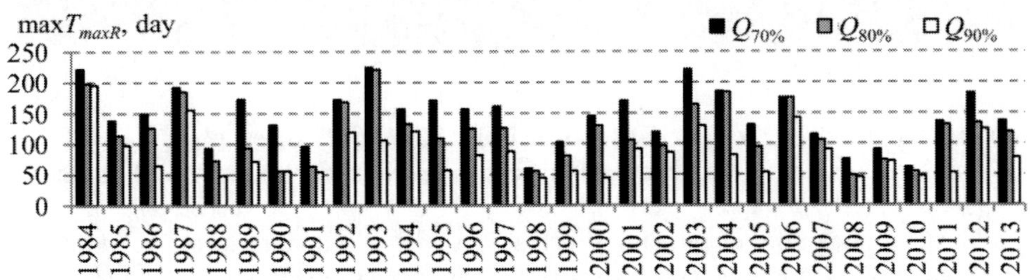

Fig. 3. Maximum durations maxT_{maxR} for annual low-flow periods, in the study period, at 78 water gauging stations in the Upper Vistula catchment varying with assumed threshold discharge.

3 Methods

Trend analysis was performed via both the Mann-Kendall and Spearman tests for every time series for maximum low-flow periods $T_{maxR,i}$, i = 1, 2, …, 30. Both tests are non-parametric, and are designed to verify hypothesis H_0 on the absence of a trend in a time series $T_{maxR,i}$, and more precisely that variables $T_{maxR,i}$ are independent and feature identical distributions, with an alternative hypothesis H_1 on the presence of a monotonic trend in time series $T_{maxR,i}$.

4.1 The Mann-Kendall trend test

What is given is a time series X_i, i = 1, 2, …, n. The test statistic is defined as [5, 8, 11, 20, 21]:

$$S = \sum_{i=1}^{n-1} \sum_{j=i+1}^{n} \operatorname{sgn}(x_j - x_i) \tag{1}$$

where:

$$\operatorname{sgn}(\theta) = \begin{cases} 1 & \text{if } \theta > 0 \\ 0 & \text{if } \theta = 0 \\ -1 & \text{if } \theta < 0 \end{cases} \tag{1}$$

When H_0 is true, the mean and variance of S are given by:

$$E(S) = 0 \tag{2}$$

$$\operatorname{var}(S) = \frac{n(n-1)(2n+5)}{18} \tag{3}$$

The above equality (4) occurs when there are no repeated elements of a time series (so-called ties). If there are more than a few repeating elements, then the formula (4) needs to be adjusted, which yields the following form [21]:

$$\operatorname{var}(S) = \frac{n(n-1)(2n+5)}{18} - \frac{\sum_{i=1}^{k} t_i \cdot i(i-1)(2i+5)}{18} \tag{4}$$

where t_i, $i = 1, 2, \ldots, k$, indicates that there exists a number of groups with k repeating elements.

Kendall [22] proved that when $n > 10$, then the test statistic:

$$U = \frac{S - \operatorname{sgn}(S)}{\sqrt{\operatorname{var}(S)}} \tag{5}$$

approximately follows the standard normal distribution N(0;1). If the p_v-value of the hypothesis test is equal to the chosen significance level ($\alpha = 5\%$), the null hypothesis must be rejected.

The basic assumption when using this test is a lack of autocorrelation in the data series. If this condition is not met and autocorrelation is positive, then the variance var(S) is underestimated. Bayley and Hammersley [20] introduced a variance adjustment:

$$\operatorname{var}^*(S) = \frac{n}{n_S^*} \operatorname{var}(S) \tag{6}$$

when using "an effective number of observations n_S^*," which Hamed and Rao [20] suggest to determine using the following formula:

$$\frac{n}{n_S^*} = 1 + \frac{2}{n(n-1)(n-2)} \sum_{i=1}^{n-1} (n-i)(n-i-1)(n-i-2)\rho_S(i) \tag{7}$$

where n is the number of observations and $\rho_S(i)$ is the autocorrelation function for observation ranks. The variance adjustment is only calculated for data whose autocorrelations are significant at 5% level.

The variance var(S) in (6) is replaced by var*(S) from (7) and the rest of the testing process is the same as in (6) (i.e. comparing the values of U with the quantiles of N(0;1) corresponding to the type I error of the test).

4.2 The Spearman trend test

Let X_i, $i = 1, 2, \ldots, n$, be a time series. The test statistic is defined as [8, 10, 11, 21, 23]:

$$R_S = 1 - \frac{6\sum_{i=1}^{n}[R(X_i) - i]^2}{n(n^2-1)} \tag{8}$$

where $R(X_i)$ is the rank of observation X_i. Repeating values (or ties) are assigned a mean rank.

If $n \geq 10$ [24], then the test statistic

$$t_S = R_S\sqrt{\frac{n-2}{1-R_S^2}} \tag{9}$$

follows Student's distribution with $n - 2$ degrees of freedom.

If the p_v-value of the hypothesis test is equal to the chosen significance level (α = 5%), the null hypothesis must be rejected.

4 Results

The Mann-Kendall and Spearman tests were used to analyze multi-annual variability trends for time series $T_{maxR,i}$ (i = 1, 2, …, 30) for low-flow periods, given three threshold values of discharge ($Q_{70\%}$, $Q_{80\%}$, $Q_{90\%}$) and data from 78 water gauging stations spread across the right bank of the Upper Vistula catchment. The p_v values obtained for the hypotheses of both examined tests ($p_{v,MK}$ – Mann-Kendall test, $p_{v,S}$ – Spearman test) resemble one another for all the considered cases (Fig. 4).

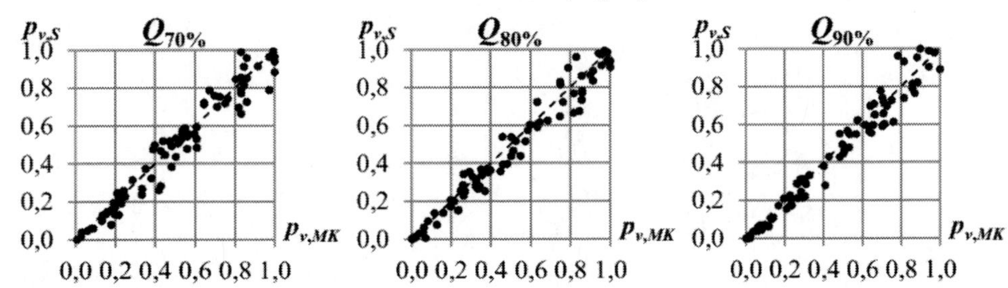

Fig. 4. Relationship between $p_{v,MK}$ values and $p_{v,S}$ obtained from the Mann-Kendall test and the Spearman test, respectively, for threshold discharge $Q_{70\%}$, $Q_{80\%}$ and $Q_{90\%}$ at 78 water gauging stations in the right-bank area of the Upper Vistula catchment.

Research on the stationarity of T_{maxR} series at most water gauging stations in the study area has shown the lack of a basis for the rejection of the hypothesis on the absence of a trend in the maximum duration T_{maxR} of low-flow periods relative to time, independently of the assumed threshold discharge value. Figure 5 shows only cases where at least one test showed non-stationarity. At the threshold discharge $Q_{70\%}$ both the $p_{v,MK}$ value of the Mann-Kendall test and the $p_{v,S}$ value of the Spearman test are smaller than the herein assumed level of significance (0.05) in three cases (marked as dots); a trend was indicated only once by the Spearman test (marked as asterisk). The lower the threshold discharge, the larger the number of cases where T_{maxR} series are non-stationary; at the threshold discharge $Q_{80\%}$ both tests showed a trend in five cases, while at $Q_{90\%}$ the number of cases was 11. The Spearman test yields a larger number of series characterized by non-stationarity than the Mann-Kendall test for the studied water gauging stations, independently of the assumed threshold discharge value ($Q_{70\%}$: one more case, $Q_{80\%}$ and $Q_{90\%}$: two more cases).

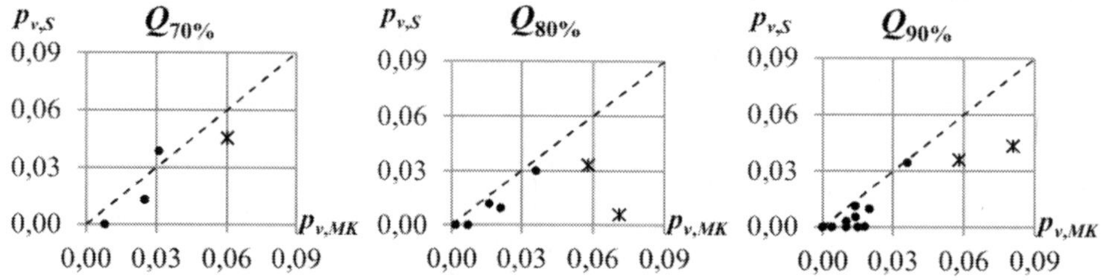

Fig. 5. Relationship between $p_{v,MK}$ values and $p_{v,S}$ obtained from the Mann-Kendall test and the in the Spearman test right-bank area of the Upper Vistula catchment in the special case when $p_{v,MK} < 5\%$ and $p_{v,S} < 5\%$ (dots), $p_{v,MK} \geq 5\%$ and $p_{v,S} < 5\%$ (asterisks).

Figure 6 shows the spatial distribution of water gauging stations and associated stationarity information for time series for the studied maximum T_{maxR} low-flow periods.

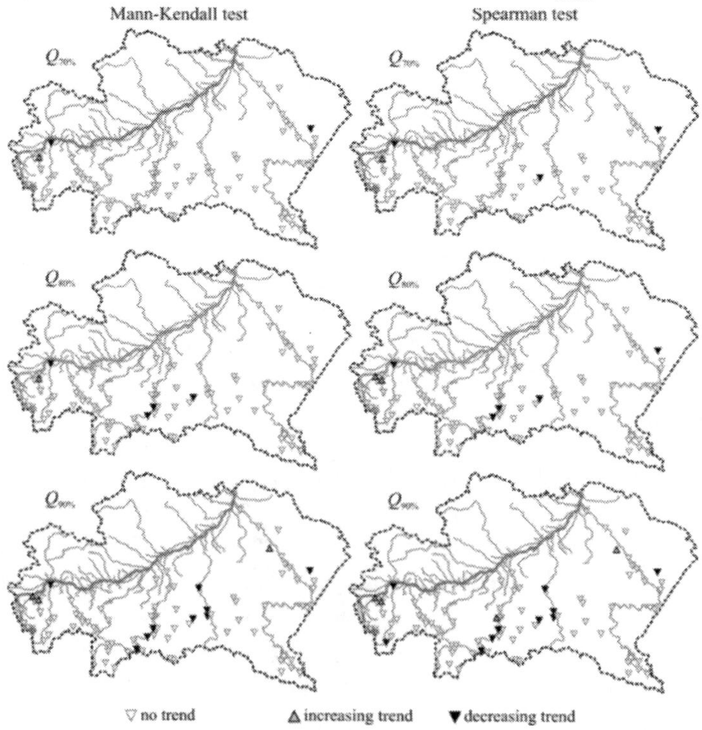

Fig. 6. Location of water gauging stations and information on series stationarity for maximum low-flow periods T_{maxR} with respect to the test employed and the threshold discharge value.

Figure 7 shows the multi-annual T_{maxR} distribution for all cases where at least one of the tests yields a trend for a T_{maxR} series, or more specifically when $p_{v,MK} < 5\%$ and $p_{v,S} < 5\%$ (frameless graphs) and $p_{v,MK} \geq 5\%$ and $p_{v,S} < 5\%$ (framed graphs). The observed trends are mostly decreasing trends. Both tests indicated the presence of a trend only in the case of the Czechowice-Bestwina station located on the Biała River (gauge no. 7), independently of the assumed threshold discharge value, with the p_v-value of the hypothesis test at less than 1%.

In the case of 14 of 78 gauging stations, at least one of the examined tests showed the presence of a trend for a T_{maxR} series for at least one threshold discharge value.

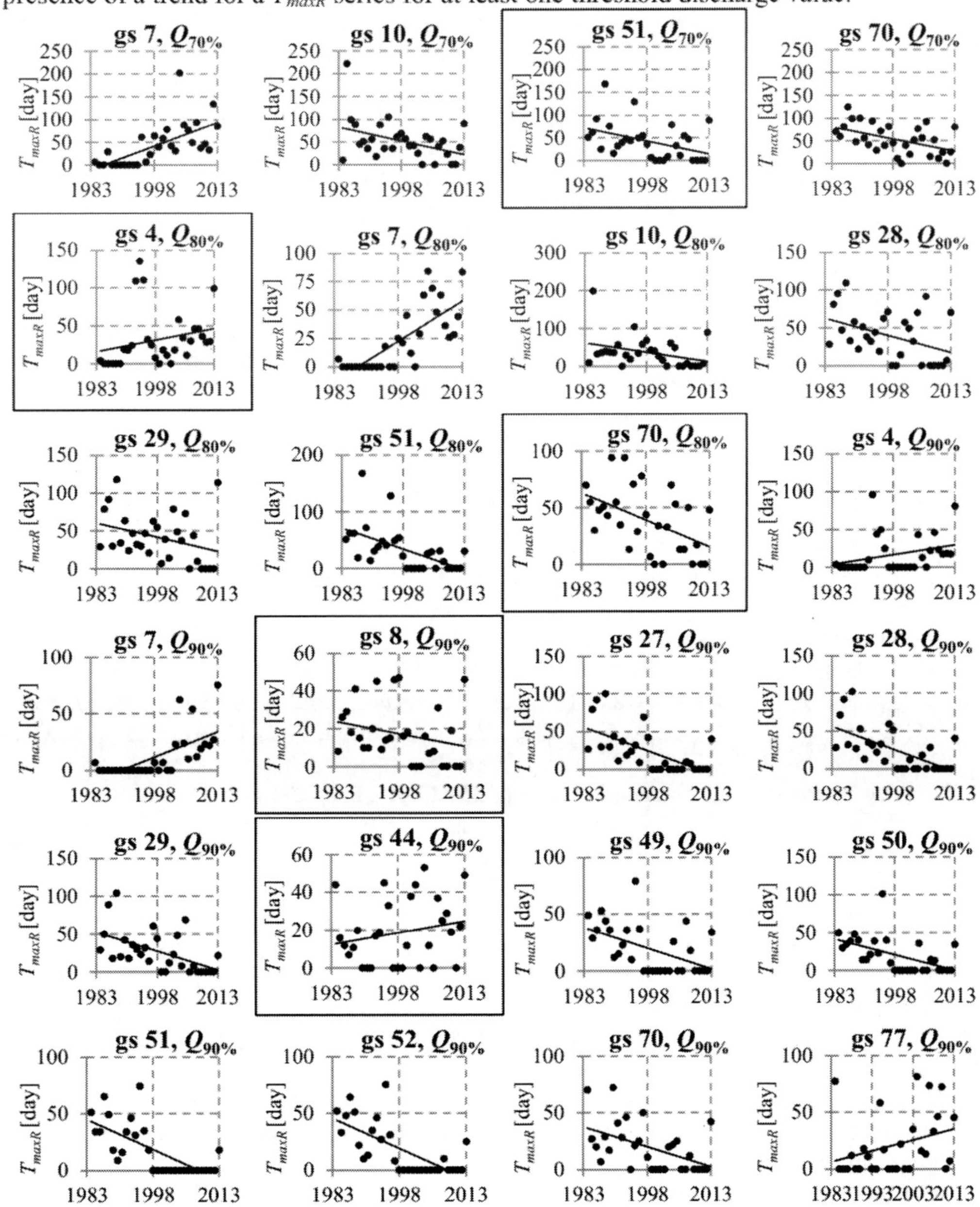

Fig. 7. T_{maxR} values over the long term for cases where at least one of two tests indicated the presence of a trend – frameless graphs: $p_{v,MK} < 5\%$ and $p_{v,S} < 5\%$; framed graphs: $p_{v,MK} \geq 5\%$ and $p_{v,S} < 5\%$; abbreviation gs 7 denotes river gauging station number 7. The lines are constructed using OLS method.

5 Final remarks

Research on the stationarity of T_{maxR} series – for most water gauging stations – has shown a lack of a basis for the rejection of the hypothesis on the absence of a trend for maximum duration T_{maxR} for low-flow periods with respect to time, independently of threshold discharge values assumed in the study. In the case of 14 of 78 water gauging stations, at least one of the two tests examined did indicate non-stationarity for at least one threshold discharge value. Most of the trends noted in this study are decreasing trends.

Comparative analysis has shown that the lower the threshold discharge value, the higher the number of cases where T_{maxR} series are non-stationary. The Spearman test yields a larger number of series characterized by non-stationarity than the Mann-Kendall test for the studied water gauging stations, independently of the assumed threshold discharge value. The inflation of the variance of the test statistic caused by positive autocorrelation may be the reason. Although the tests are very similar, only the MK test takes into account this autocorrelation. The largest probability of non-stationarity was noted for the Czechowice-Bestwina gauging station on the Biała River, with a very low (less than 1%) p_v-value of the null hypothesis.

This study was part of a larger research endeavor: „The influence of anthropogenic factors on qualitative and quantitative properties of hydrological processes in the catchment", task 3: Analysis of low flow properties in the Carpathian area of the Upper Vistula River basin. The data has been provided by the Institute of Meteorology and Water Management, National Research Institute.

References

1. E. M. Douglas, R. M. Vogel, C. N. Knoll, J. Hydrol., **240**, 90–105 (2000)
2. J. Hannaford, T. Marsh, Int. J. Climatol., **26**, 1237–1253 (2006)
3. E. Kahya, S. Kalaycı, J. Hydrol., **289,** 128–144 (2004)
4. A. Rutkowska, Acta Sci. Pol., Form. Cir., **12** (4), 85-94 (2013)
5. A. Rutkowska, M. Ptak, Stud. Geotech. Mech. **34** (1), 51–63 (2012)
6. C. Svensson, Z.W. Kundzewicz, T. Maurer, Hydrolog. Sci. J., **50** (5), 811–824 (2005)
7. M. Zeleňáková, P. Purcz, T. Soľáková, B. Demeterová, Acta U. Agr. Et Silv. Mend. Brun., **LX** (5), 265–274 (2012)
8. S. Yue, P. Pilon, G. Cavadias, J. Hydrol., **259** (1–4), 254–271 (2002)
9. H. Hisdal, L. M. Tallaksen, *Drought Event Definition* (Technical Report No. 6, University of Oslo, 2000)
10. M. Ozga-Zielińska, J. Brzeziński, *Hydrologia stosowana*. (PWN, Warszawa 1997)
11. J. Pociask-Karteczka (ed.), *Zlewnia. Właściwości i procesy*. (Wyd. UJ, Kraków, 2006)
12. V. U. Smakhtin, J. Hydrol., **240**, 147–186 (2001)
13. S. Węglarczyk, *Słownik terminów związanych z gospodarowaniem zasobami wodnymi*. (Wyd. PK, Kraków, 2006).
14. L. Gottschalk, Y. Kun-xia, E. Leblois, L. Xiong, J. Hydrol., **481**, 204-219 (2013)
15. B. Ratomska, *Ocena suszy w aspekcie hydrologicznym* (Ph.D. Thesis, PK, 1993)
16. E. Zelenhasić, A. Salvai, Water Resou. Res., **23** (1), 156–168 (1987)
17. W. Jakubowski, *Rozkłady prawdopodobieństwa w ocenie suszy hydrologicznej*. (monogr., UP, Wrocław, 2011)
18. T. Tokarczyk, Acta Geophys, **61** (2), 404–421 (2013)
19. A. K. Fleig, L. M. Tallaksen, H. Hisdal, S. Demuth, S., Hydrol. Earth Syst. Sci. **10** (4), 535–552 (2006)
20. K. H. Hamed, A. R. Rao, J. Hydrol., **204** (1–4), 182–196 (1998)
21. S. Węglarczyk, *Statystyka w inżynierii środowiska* (Wyd. PK, Kraków, 2010)
22. M. G. Kendall, Biometrika, **30**, 81–93 (1938)
23. D. R. Helsel, R. M. Hirsch, *Statistical Methods in Water Resources* (USGS—TWRI Book 4, Chapter A3, 2002)
24. R. H. McCuen, *Modeling Hydrologic Change: Statistical Methods* (Lewis Publishes, U. S. CRC Press, 2002)

Does Dredging Of Floodplain Lakes Affects The Structure Of The Macrophytes And Epiphytic Fauna Inhabiting *Stratiotes Aloides*?

Martyna Bąkowska[1,*], *Krystian* Obolewski[1] and *Ryszard* Wiśniewski[1]

[1]Kazimierz Wielki University, Department of Hydrobiology, 85-064 Bydgoszcz, Poland

Abstract. Hydrotechnical activities on the Słupia river above the town of Słupsk (northern Poland) were performed against floods in the valley. The results based on the data on fauna and flora gathered before (in 2003, *T+0*) and after (in *T+2*, *T+6* and *T+11*) the reclamation showed that the river canalization and oxbow lake dredging can significantly disturb the ecology of water bodies adjacent to the river. Observations of the oxbow lake (OLS5), showed the decrease in the number of macrophyte phytocenoses from 11 before the dredging to 6 very valuable *Potametea phytocenoses*, after wards. The development of common pleustonic phytocenoses of *Lemnetea minoris* and the massive development of green algae resulted from the bottom sediment removal. In the case of population, *Stratiotes aloides* was rebuilt after elimination during dredging already in *T+2*. Before hydrotechnical works, the oxbow lake was characterized by 12 invertebrate taxa inhabiting *S.aloides*, whereas only 9 were found first period after the dredging, 15 taxa in second period and 23 in a recent study. During both observation periods, dominated by Chironomidae larvae (*T+0*), Trichoptera larvae (*T+2* and *T+6*) and again Chironomidae larvae (*T+11*). The hydrotechnical activities significantly altered in the first period the structure of fauna and flora inhabiting the water body. Such radical interference in natural values of oxbow lakes is un-suitable for the preservation of ecosystems due to the significant decrease in their biodiversity. Achieved on a similar or higher level of macrophytes diversity and epiphytic invertebrates was noticeable ten years after the dredging of a water body.

1 Introduction

Oxbow lakes belong to those important aquatic ecosystems in floodplains which due to lentic character create suitable habitats for a significant number of macrophytes and invertebrates communities preferring stagnant water. Natural processes of succession in such water bodies lead to changes in the structure of fauna and flora. The intensive development of floodplain lakes can be restrained, to some extent, through reclamation activities, starting from the least invasive method as the oxbow lake re-connection with the river to the most radical method as the bottom sediment removal (dredging). The change of

* Corresponding author: bakowska@ukw.edu.pl

environmental conditions as a result of dredging is substantial for the hydrobionts existence [1]. One of the symptoms of ecological decease of the ecosystem is the withdrawal of floating macrophytes including *Stratiotes aloides* L. (water soldier), a perfect natural substrate for epiphytic fauna in eutrophic and hyper-eutrophic water bodies [2–5].
Immersed or floating macrophytes play a particularly important role in intensively eu-trophicated water body where anaerobic conditions at the bottom disable the existence of zoobenthos [6]. They become then "a second bottom" inhabited willingly by oxybionts [7, 8]. Another important role of floating aquatic plants is delivery of food for aquatic fauna [9].

The paper is aimed at effects of oxbow lake dredging on fauna and flora inhabiting *S.aloides* as well as assessment of biological diversity of oxbow lakes under the influence of the reclamation works conducted in the river valley.

2 Material and Methods

The oxbow lake chosen for the research (called OLS5) is situated in the district of the town of Słupsk city (upstream N54°C 26'54" E17°C 01'52", downstream N54°C 26'59" E17°C 01'57"). In 2003, the object was dredged and deepened as an element of the complex restoration of flood-protection systems in the valley. As a result of dredging, the average depth of the water body increased from 1 to 2.5 m, and the excavated material was placed along the banks of the floodplain lake. Similarly, to this water body, the other oxbow lakes along the river section were "re-freshed," as well.

Aquatic vegetation was studied twice, both before reclamation works in the period from April to July of 2003 (*T+0*) and after wards in the same months in 2005 (*T+2*), 2009 (*T+6*) and 2014 (*T+11*). Classification of the distinguished communities was con-ducted according to the commonly used methods [10]. Due to practical reasons, plant communities were distinguished with no syntaxonomical range. They are the base for the assessment of the dredging effects on the habitat conditions for hydrobionts.
Samples of floating *S.aloides* as well as epiphytic fauna inhabiting this species were collected in spring-summer periods 2003 and 2014. Each of eight samples collected from overgrown and then dredged oxbow lake consisted of 12 -14 representatives of *S.aloides.*

The samples were large enough to take off phytophilic fauna and the fauna from water solder. Calculations of the densities and biomass were done according to Obolewski et al. [8]. The Shannon index (H') was also determined to show effects of hydrotechnical activities on the diversity of the studied water body. The significant differences between structure of phytofauna amounts before and after the dredging were determined using two-way ANOVA tests at $p<0.05$.

3 Results

3.1 Change in a vegetation cover

The dredged oxbow lake was characterized by 11 plant communities in *T+0* (Tab. 1) and represented by 3 classes: *Lemnetea minoris*, *Potametea* and *Phragmitetea*. The highest share among them had phytocenoses of *Potametea* R.Tx. et Prsg. The dominating community in the oxbow lake was *Stratiotetum aloidis* (Now. 1930) Miljan 1933. Characteristic species for that class was Stratiotes aloides and represented the highest share in the phytocenoses. It created a widespread compact cover, mainly at places of shallow and intensively insolated water. Besides the characteristic species, the society was abundant

DOI: 10.1051/e3sconf/20171700005

in the frog bit *Hydrocharis morsus-rane*, the greater duckweed *Spirodela polyrhiza*, the pond-weed *Potamogeton natans* and the canadian waterweed *Elodea canadiensis*. The other phytocenoses of *Potametea* class took relatively smaller areas and were represented by commonly exisitingexisting the broad leaf pondweed *Potametum natantis* Soo 1927, as well as valuable floristic societies of water lilies *Nupharo-Nymphaeetum albae* Tomasz. 1977 and the frog lily *Hottonietum palustris* R.Tx. 1937.

Table 1. List of systematic macrophytes communities in the floodplain lake before and after the hydrotechnical activities. (Explanation: + found occurrence, - no occurrence).

	Syntaxon	Dredged			
		Before	After		
		T+0	*T+2*	*T+6*	*T+11*
	Cl. *Lemnetea minoris*				
	O. *Lemnetalia minoris*				
	All. *Lemnion gibbae*				
1	*Spirodeletum polyrhizae*	+	++	++	+
	All. Riccio fluitantis-Lemnion trisulcae				
2	*Lemnetum trisulcae*	+	++	++	++
	Cl. *Potametea*				
	O. *Potametalia*				
	All. *Potamion*				
3	*Ranunculetum circinati*	+	-	-	-
4	*Como. Elodeetum canadensis*	+	++	+	+
5	*Comm. Ceratophyllum demersum*	-	+	+	+
	All. *Nymphaeion*				
	Hydrocharitetum morsus-ranae	+	-	-/+	+
6	*Stratiotetum aloidis*	++	-/+	+	+
7	*Potametum natantis*	+	-	-	-/+
8	*Nupharo-Nymphaeetum albae*	+	-	+	+
	All. *Hottonion*				
9	*Hottonietum palustris*	+	-	-	-
	Cl. *Phragmitetea*				
	O. *Phragmitetalia*				
	All. *Phragmition*				
10	*Phragmitetum australis*	+	-	+	+
11	*Glycerietum maximae*	+	-	-	-
	Other communities				
	The communities of filamentous algae	-	++	-/+	-

The presence of 2 societies and 2 plant communities as well as intensive development of chlorophytes (tab. 1) was stated after the dredging. The changes in light and thermal conditions caused the intensive development of pleustonic phytocenoses of *Lemnetea minoris* R. Tx. 1955. After the hydrotechnical activities, in the society of *Spirodeletum polyrhizae* dominated the lesser duckweed *Lemna minor* and the ivy leaf duckweed *Lemna trisulca* with sporadic share of the greater duckweed *Spirodela polyrhiza*. It shows a double–layer structure of vegetation with the bottom-rooted common hornwort *Ceratophyllum demersum*, that composed at many places its own phytocenoses, not noted before dredging.

As the consequence of dredging, 6 plant communities of *Potametea* and 2 of *Phragmitetea* classes disappeared. Phytocenoses with *Stratiotetum aloidis* were saved from the dredging only at blind side-canals. Probably, the dredging was too deep to maintain valuable macrophytes, including water lilies and the frog lily. The communities are particularly distinguishing due to esthetic values and the presence of endangered and

DOI: 10.1051/e3sconf/20171700005

protected species. Moreover, lakeside reeds of *Phragmitetea* R.Tx. et. Prsg. 1942 class were completely removed during the reclamation works.

3.2 Change in the fauna colonizing *Stratiotes aloides*

The highest density of invertebrates inhabiting *S.aloides* was stated in *T+0*, when in July it amounted to 102 ind. kg^{-1}, and after dredging in July, *T+2* (114 ind. kg^{-1}). The maximal biomass was noted before dredging in May, when achieved 63.6 gWW kg^{-1} and, after hydrotechnical works in July *T+2*, reaching (53.5 gWW kg^{-1}). Gastropoda and Chironomidae larvae determined the density and created groups of subdominants. However, only snails determined the biomass. During consecutive months, the share of snails in epiphytic fauna got lower as their amount decreased (Fig.1). Before dredging, the snails were represented by: *Lymnaea stagnalis*, *Lymnaea tranculata*, *Lymnaea corvus*, *Acroloxus lacustris*, *Planorbarius corneus*, *Ancylus fluviatilis*, *Anisus leucostoma* and *Anisus vorliculus*, whereas after the hydrotechnical works (*T+2*) the following species disappeared, *A. fluviatilis*, *A.leucostoma* and *A.vorliculus*. In *T+6* also appeared *P. corneus, L. stagnalis, Bithynia tentaculata, Viviparus viviparous* and in *T+11* as well *Lymnaea auricularia, Segmentina nitida* and *V.contectus*.

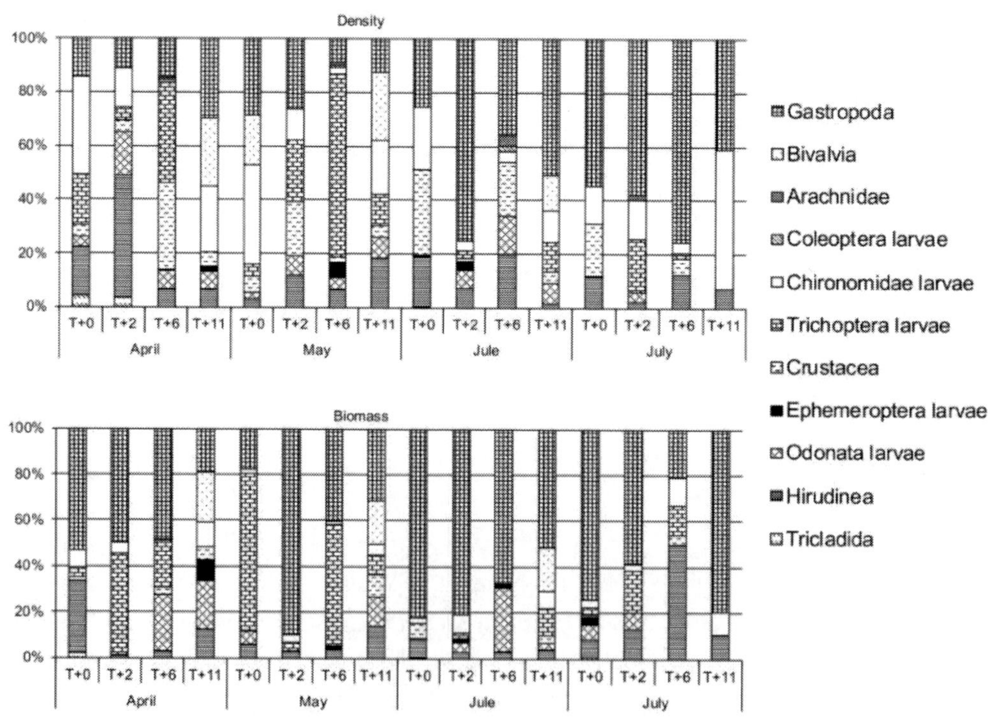

Fig. 1. The share of individual taxa of epiphytic fauna inhabiting *Stratiotes aloides* in the study floodplain lake (*T+0*) before and (*T+2, T+6, T+11*) after dredging.

In *T+0*, before the hydrotechnical works, the Hirudinea class was represented by *Glossiphonia complanata* L., *Erpobdella* sp., *Helobdella stagnalis* L. and *Alboglossiphonia* sp. The last species did not appeared after the reclamation. OLS5, when overgrown, created favourable conditions for the leech in April and the least convenient at the turn of April and May. After the dredging, the share of the leech in the epiphytic fauna was the lowest in June. The reclamation influenced the structure of leeches inhabiting *S. aloides* and caused the decrease in the leech density by 2 times and the decrease in its biomass by 3.5 times (Fig. 2).

Isopoda, represented by *Asellus aquaticus*, occurred only in the unchanged water body comprising both 13% of the total density (before 17%, after 8%) and the structure of epiphytic fauna. The dredging caused the complete withdrawal of *A. aquaticus* from *S. aloides*, particularly evident in *T+2*. The hydrotechnical works caused the decrease in the crustaceans density by 3 times and the decrease its biomass by 2 times.

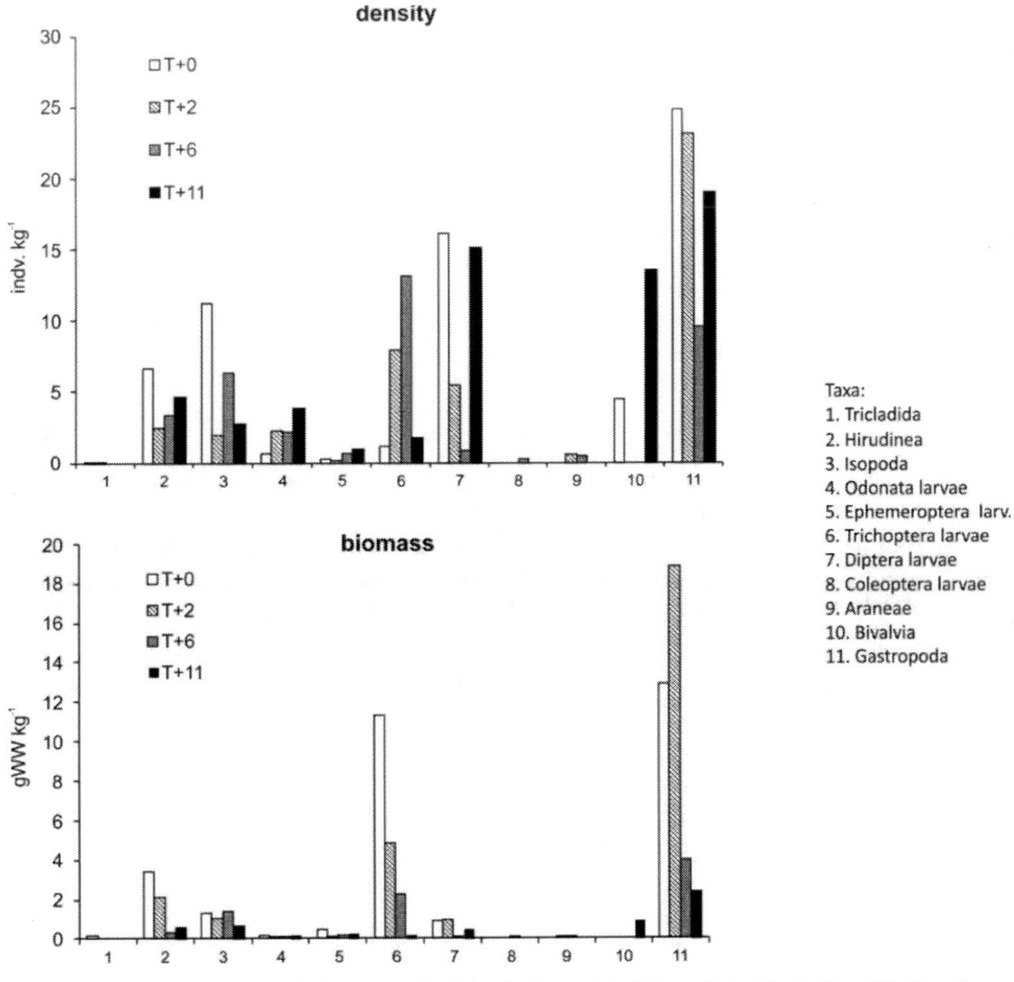

Fig.2. Changing the density and biomass of epiphytic fauna inhabiting *S.aloides* before (*T+0*) and after (*T+2, T+6, T+11*) dredging of floodplain lake.

The average share of dragonflies in the density and biomass of phytophilic fauna during both periods of observations was low (Fig.1). *S.aloides*, in the oxbow lake was permanently inhabited by dragonflies, which found the most convenient conditions for the development in April. Since that month, their share in epifauna systematically got lower, but increased in July. The density and biomass of dragonflies increased after the dredging (respectively 4 and 1.5 time, Fig.2). The reclamation of OLS5 seems to had negative influence on qualitive structure of Odonata represented by *Ischnura* sp., *Lestes* sp., *Libellula* sp. and *Aeschna grandis*.

Larvae of caddies-flies (Trichoptera) were developing the most intensively in May when they were an important element of fauna colonized *S.aloides* (Fig.1). Hydrotechnical works occurred not to threaten species of this order: after the dredging increased the density

(6.5-times) and decreased biomass (6-times) of Trichoptera larvae (Fig. 2). The order represented *Ecnomus tenellus*, *Limnephilus* sp. and *Lepidostoma* sp. both before and after the oxbow lake dredging.

A significant element of epifauna existing outside and inside of *S.aloides* were representatives of Diptera larvae (Fig. 1). In April *T+0* we found the highest biomass of these organisms, whereas in May their maximal abundance occurred. The share of Diptera larvae representatives decreased significantly after the dredging: more than 2.3 times in terms of abundance and almost 2 times in terms of biomass. In *T+11* there was a reconstruction of Diptera abundance (~15-16 indiv. kg^{-1}), but at low biomass (Fig.2). The representatives of Diptera, dominated in inhabiting *S.aloides* in both periods of observations. It was mainly larvae of *Bezzia* sp., *Eucricotopus* sp., *Psectocladius* sp. and *Chironomus* sp.

In the collected samples from OLS5 we found also a representative of the Arachnids - *Argyroneta aquatic*. The species were characterized by small biomass, taking only some per cent of total density and biomass of epifauna. The other systematic groups, including the ephemerid larvae (*Caenis* sp., *Clëon* sp., *Leptophlebia* sp.), Triclotida (*Dendrocoelum lacteum*), Coleoptera (Dytiscidae) and Bivalvia (*Acroloxus lacustris*, *Sphaerium* sp.), were sporadic on the biotic substrate. All of the taxa, excluding Triclotida, developed just after the biotic substrate was removed from the floodplain lake.

Radical interference in natural values of oxbow lakes is unsuitable for the preservation of ecosystems, due to the decrease diversity of fauna. The differences between number of taxa, densities and biomass of the fauna inhabiting the water body before and after the hydrotechnical works were not statistically significant in two-way ANOVA at $p<0.05$ (Table 2). But there were significant differences in diversity ($p=0.03$) before and after the hydrotechnical works, and years of research. Comparison of the number of taxa and their density for the periods before and after dredging, vs. the years of study, showed statistically significant differences ($p<0.05$). Despite visible differences in biomass of epiphytic fauna, they not shown to be statistically significant.

Table 2. Mean values (± SE in parenthesis) of epiphytic macrofauna of floodplain lake during before dredging (BD) and after dredging (AD) time and results of two-way ANOVA tests evaluating the effects of time (BD vs. AD), year (Y), and time vs. Years.

	Before dredging	After dredging	Source of variation (F; p)		
	(BD)	(AD)	Time (BD×AD) df=1	Year (Y) df=3	Time × Y df=17
No of taxa	10	23	3.91; 0.06	1.12; 0.38	2.41; **0.04**
Diversity (nat ind^{-1})	1.46±0.04	1.38±0.12	6.00; **0.03**	2.61; **0.04**	1.90; 0.09
Total density (inds m^{-2})	65.3±5.4	47.3±8.1	1.15; 0.29	0.70; 0.56	3.04; **0.01**
Total biomass (gWW m^{-2})	30.67±4.85	14.00±5.87	1.46; 0.24	0.92; 0.52	0.93; 0.56

Bold values indicate significant differences

4 Discussion

Oxbow lakes in the Słupia river valley are overgrown by macrophytes preferring lentic and fertile water, surrounded by a zone of rushes. In deeper parts, immerse Potamion phytocenoses developed. They also play a pioneer role in the natural processes of getting shallow and overgrowing. During a further process, the phytocenoses are replaced by societies that belong to Nymphaeion, represented by the yellow water lily and the water lily Nupharo–*Nymphaeetum albae*, mainly. In OLS5, we noticed that the phytocenoses of *Nupharo–N.albae* were replaced by *Stratiotetum aloidis*. Before dredging, *S.aloidis* created a widespread and dense cover, mainly over shallow water, at easily heated, wind and wave protected places. Shortly after dredging water soldier gave way to, but rebuild the population after *T+6* (Table 1).

Natural evolution of oxbow lakes is convenient for the development of the variety of fauna and flora species regardless of the stage of the floodplain lake overgrowing. The ecological role of macrophytes increases since they form not only a habitat for fauna but also their purifying role in the nutrient cycling in a lake [11]. Once, we noted that short-time but drastic hydrotechnical activities easily disturbed this natural succession of a water body. Thus, extremely important to the ecosystem functioning is not to break this natural habitat. Achieved on a similar or higher level of diversity, density and biomass of epiphytic invertebrates was noticeable after ten years after the dredging of the water body.

An incrasing eutrophication and intensive biogeochemical processes occurring at the bot-tom of floodplain lakes causes the qualitative and quantitative loss in benthos species in such a water body [11, 12]. Those unfavourable conditions make the macroinvertebrates species settle rushes and floating macrophytes [8, 13].

S.aloides, dominating among plants with floating leaves in OLS5, was inhabited by various taxa of aquatic fauna. Due to oxygen deficit in the bottom zone, the plant created the only alternative for the oxybionts existence [13–15]. Before dredging, *S.aloides* was inhabited by 8 – 10 taxa consisting of species characterized for macroivertebrates at amounts of 65.3 ind. kg^{-1} and biomass of 31 $gWWkg^{-1}$ of dry mass of the water soldier (tab. 2). As a result of dredging, bottom fauna species lost a significant substrate for the development, though the plant was replaced by green algae. As soon as bottom sediments were removed, oxygen conditions improved over the bottom. Hereafter, representatives of bottom fauna started settling their previous habitats [1]. Water soldier in the reclaimed OLS5 was inhabited by 15–23 taxa, which initially (*T+2* and *T+6*) density decreased by 1.5–1.8 times, but the double increase in the biomass was limited by the Mollusca development.

The comparison of results on epifauna inhabiting *S.aloides* in different oxbow lakes showed that the most convenient conditions for the fauna settlement represent those water bodies with a permanent exchange of water from the rivers. It is confirmed by presenting the densities and Shannon diversity indices for other floodplain lakes [5, 16]. After dredging Shannon's index value decreased from 2.49 (*T+0*) to 2.22 (*T+2*) and then grew into value 2.60 in *T+6* and 2.57 in *T+11*. All of the studied oxbow lakes are dominated by Gastropoda, intensively inhabiting *S.aloides*, except first period for an oxbow lake re-connected to the river called OLS5 (Trichoptera larvae) adjacent to the Słupia river. Malacofauna dominated in terms of biomass in each water body. Analysis of two-way ANOVA indicated no significant differences in epifauna densities and biomass in OLS5 before and after the dredging and only in case of diversity. Other comparisons showed no significant differences in means at $p<0.05$, even between dredged OLS5 and overgrown phase this oxbow lake. It indices that the hydraulic connectivity as a key factor influencing epifauna in-habiting *S.aloides*. The re-connection and permanent exchange of water between an oxbow lake and the river are safe and life-supporting also for many macrophyte phytocenoses. Any hydrotechnical activities within natural oxbow lakes should be preceded by a detailed analysis of biological and hydrological conditions in each water body to avoid losses of valuable elements of the ecosystems in river valleys. Changes in habitat quality affect biotic components of aquatic ecosystems [8, 17] and are reflected in the biodiversity of plant and animal assemblages [18, 19].

5 Conclusions

1. River regulation, mainly dredging, conducted in the Słupia river valley destructively influenced macrophytes and the density of epifauna inhabiting aquatic plants, that caused the decrease in biological diversity of the oxbow lakes.
2. Due to phytocenotic and faunistic values, the use of naturally overgrowing oxbow lakes against flooding or retention purposes should be limited.
3. The hydrotechnical activities significantly altered in the first period (*T+2* and *T+6*) the structure of fauna and flora inhabiting the water bodies. Such radical interference in natural values of oxbow lakes is unsuitable for the preservation of ecosystems due to the significant decrease in their biodiversity. Achieved on a similar or higher level of diversity of epiphytic invertebrates was noticeable after ten years after the dredging of the water body.
4. Any hydrotechnical activities within natural oxbow lakes should be preceded by a detailed recognition of biological and hydrological conditions in each water body to avoid losses of valuable elements of the ecosystems.

References

1. K. Obolewski, Oceanol. Hydrobiol. St., **40**(4), 52–63 (2011)
2. J. Kornatowski, Pol. Arch. Hydrobiol. **23**, 365–376 (1976)
3. E.S. Brammer, Acta Univ. Ups. **540**, 1–22 (1979)
4. J. Linhart, Acta Univ. Palacki. Olomuc., Fac. Rer. Nat., Biol. **37**, 67–76 (1999)
5. K. Obolewski, K. Glińska-Lewczuk, N. Jarząb, P. Burandt, Sz. Kobus, R. Kujawa, T. Okruszko, M. Grabowska, S. Lew, A. Goździejewska, A. Skrzypczak, Pol. J. Environ. Stud., **23**, 1679–1689 (2014)
6. S. Lalonde, J.A. Downing, Can. J. Fish. Aquat. Sci. **49**, 992–1000 (1992)
7. L.W.G. Higler, Verh. Internat. Verein. Limnol. **19**, 2773–2777 (1975)
8. K. Obolewski, A. Strzelczak, K. Glińska-Lewczuk, Ecol. Eng., **66**, 72–81 (2014)
9. M. Tarkowska-Kukuryk, Pol. J. Ecol. **54**, 441–451 (2006)
10. W. Matuszkiewicz, *Guide to marking the plants assemblages of the Poland* (PWN. Warszawa, 2001)
11. J. V. Ward, K. Tockner, D. B. Arscott, C. Claret, Freshwat. Biol. **47**, 517–539 (2002)
12. K. Glińska-Lewczuk, A. Bieniek, P. Sowiński, K. Obolewski, P. Burandt, C.M. Timofte, J. Elem., **19**, 361–376 (2014)
13. E. Suutari, J. Salmela, L. Paasivirta, M.J. Rantala, K. Tynkkynen, M. Luojumäki, J. Suhonen, J. Insect Conserv. **13**, 413–419 (2009)
14. M. Strzałek, P. Koperski, Aquat. Bot. **90**, 1–6 (2009)
15. M.J. Rantala, J. Ilmonen, J. Koskimäki, J. Suhonen, K. Tynkkynen, Aquat. Ecol. **38**, 77–82 (2004)
16. K. Obolewski, K. Glińska-Lewczuk, Sz. Kobus, J. Elementol., **14**, 119–135 (2009)
17. E. Wilk-Woźniak, S. Ligęza, E. Shubert, Clean – Soil, Air, Water **42**, 421–427 (2014)
18. M.L. Pedersen, N. Friberg, J. Skriver, A. Baattrup-Pedersen, S.E. Larsen, Ecol. Eng. **30**, 145–156 (2007)
19. L. Sartori, S. Canobbio, R. Cabrini, R. Fornaroli, V. Mezzanotte, J. Limnol. **74**, 335–345 (2015)

Cytotoxicity properties of PM2.5 collected in Wrocław agglomeration

Maciej Krzysztof Bełcik[1,*], and *Katarzyna* Piekarska[1]

[1]Faculty of Environmental Engineering, Wrocław University of Science and Technology, Wybrzeże S. Wyspiańskiego 27, 50-370 Wrocław, Poland

Abstract. WHO report published in May 2016 presents the cities with the highest concentrations of particulate matter fraction 10 μm and 2.5 μm on the earth. Report presents that many of Polish cities are on the top of polluted regions in European Union. One of the cities mentioned in the report is Wrocław. Wrocław is the urban and industrial agglomeration, where air pollutions comes from three main sources: low-emissions, industrial emissions and urban traffic. In recent years, reduction in annual average concentrations of particulate matter has been observed. However Wrocław is still one of the cities in Poland, which have problems with smog situations during winter season. This article presents results of researches conducted on the particulate matter fraction 2.5 μm collected in Wrocław agglomeration. Pollutions have been tested using PAN-I assay kit, which contains 4 single toxicity tests allows to check different parameters defining toxic effects on human cells. Cytotoxicity tests which are conducted are: LDH test to evaluate the integrity of cell membrane, XTT test to assess mitochondrial activity, NR test for the evaluation of the lysosomal activity and SRB test to evaluate the total protein content. Tests were performed on adenocarcinomic human alveolar basal epithelial cell line (A549), which allows to simulate conditions as close as real.

1 Introduction

There is more than 2,000 chemical substances in the ambient air. Some of them are in a gas form, the other are absorbed on the surface of the particulate matter. Among this ones which are absorbed on the surface of particulate matter there are inter alia polycyclic aromatic hydrocarbons, phenols, organic compounds and many more [1, 2].

Particulate matter with all absorbed substances can penetrate the human respiratory system may cause health problems among which it is worth mentioning asthma, pneumonia, decreased lung function and cardiovascular system. The smaller particulate matter can penetrate deeper parts of airways and reach directly into alveoli. All these factors cause health problems induced by dust pollutions are observed not only in groups exposed to environmental risk (children and elderly people) but also in health part of population [3, 4].

* Corresponding author: maciej.belcik@pwr.edu.pl

Recently in May 2016 the World Health Organization present the report with the average concentrations of particulate matter fraction 10 μm and 2.5 μm obtained in cities among the world. The data presented by WHO comes from local air quality monitoring centres. In a report appears many of Polish cites which are on the top the polluted regions in European Union. One of the mentioned cities is Wrocław – the urban and industrial agglomeration, where air pollutions comes from three main sources: low-emissions, industrial emissions and urban traffic. Although the air quality in our city was improved in recent years and it has been observed reduction in annual average concentrations of particulate matter, there are still happened smog situations especially in winter season. The biggest problem in particulate pollution in Wrocław agglomeration is low emission caused by low quality coal combusted for heating houses not connected to the central heating network [5].

2 Cytotoxicity

Cytotoxicity of chemical substances is called properties that can cause disturbances in the functioning of the cells of living organisms, organs or whole organism, which could lead to their death [6].

There are described many cytotoxicity tests in the literature, but in the studies of particulate matter extracts so far mainly researches conducted by directly contact method with monolayer cell cultures were reported. The most commonly used to this type of tests were cell lines A549 obtained from human lungs and L929 obtained from rat fat tissue [7, 8].

With the development of environmental toxicity tests it was created new methods of assessment, which are characterized by considerably reliability, speed and simplicity in conducting and evaluating test results. One of new test is PAN-I assay, which allows to examine four different parameters of cytotoxicity using the same cell culture. Using PAN-I assay allows to evaluate the integrity of cell membrane, assess mitochondrial activity, evaluate lysosomal activity and the total protein content. All this single tests can be conducted one by one on the same cells [9].

3 Materials and methods

3.1 Particulate matter collection and extraction

Material for research was particulate matter air pollutions fraction 2.5 μm collected in Wrocław agglomeration (Poland). In Wrocław lives about 630 thousand residents with average density of population 2.1 thousand people per square kilometre. Samples were collected using high-flow air aspirators on glass filters from two places located in Wrocław over all year long. This allows to gather samples from four different seasons: spring, summer, autumn and winter.

The first location (1) was area of water treatment plant at the Na Grobli Street which is located at a certain distance from busy traffic routes and houses. However, this place are located in the area of impact of low-emission. Second location (2) was within a city centre, directly at the Wiśniowa Street. In the neighbourhood of intake there are placed single- and multi-family residential buildings the majority of which are not connected to the city heating network. In the Figure 1 was marked location of samples collection on the Wrocław map.

Before and immediately after the collection filters were weighed and differences in their weight was the mass of gathered particulate matter. Extraction of pollutants absorbed on the surface of dust was conducted in Soxhlet apparatus without absence of light and using dichloromethane as a dissolvent. After extraction the solvent were vaporized to dryness in vacuum evaporator [10].

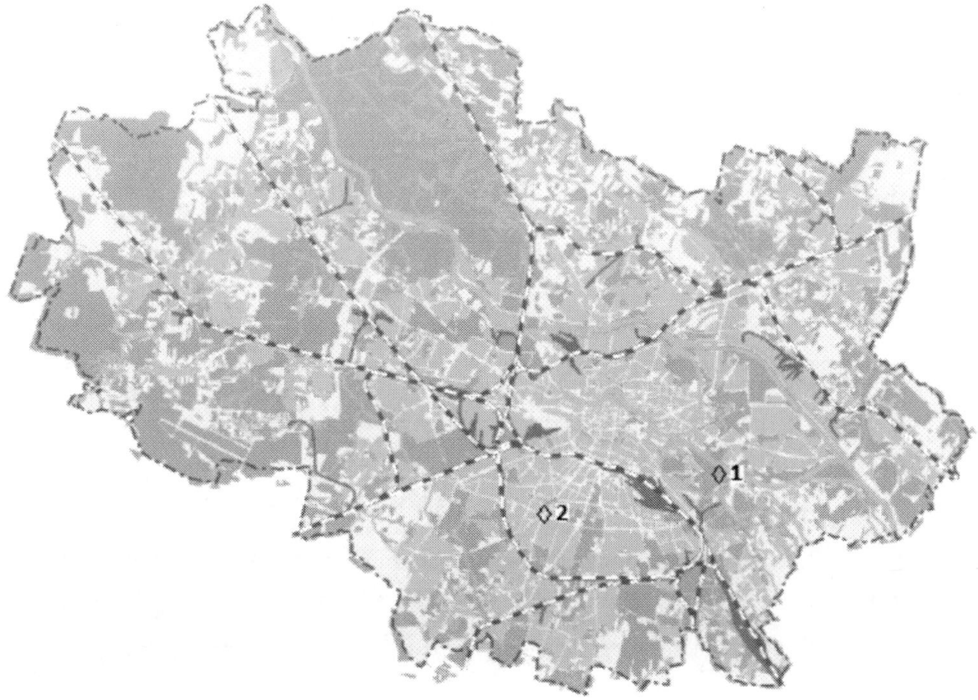

Fig. 1. Location of sample points on the Wrocław map.

3.2 A549 cell line

Cytotoxicity tests were performed on the A549 cell line which are adenocarcinomic human alveolar basal epithelial cells (American Type Culture Collection, Cell Culture Line 185). The choice of adherent A549 cells allows to stimulate conditions as close as in real ingress finest fractions of particulate matter pollutions to the alveoli. A549 cell line is widely used as an in vitro model for various studies, including cytotoxicity studies of air pollutions [7, 11].

Breeding of A549 cell line was carried out in culture medium Dulbeco containing 10% addiction of calf serum, 2mM L-glutamine, 100 units/cm^3 penicillin and 100 µg/cm^3 streptomycin. Cell culture was incubated at 37°C and 5% of carbon dioxide atmosphere.

3.3 PAN-I cytotoxicity test

Cytotoxicity tests were performed using PAN-I assay containing four single tests to evaluate different parameters defining toxic effects on cells. Parameters measured by PAN-I assay was: integrity of cell membrane (LDH test), mitochondrial activity (XTT test), lysosomal activity (NR test) and total protein content (SRB test). All tests were performed on the same breeding of A549 cells in the 96-well plates. On the monolayer cell culture it was applied particulate matter fraction 2.5 µm extracts with appropriate dilutions. The exposure time was 48 hours. After that it was performed all toxicity tests starting from LDH according to the instruction of use. As a positive controls was used solvent (DMSO) and as a negative control Triton X-100.

Only LDH test was performed in the supernatant of the culture medium with using kinetic results reading by the time 1 hour at 22°C and 2 min interval. Other tests were used on the original cells washed after each one with PBS solution. Results of the PAN-I assay was reading by the spectrophotometer μQuant with a specific wavelengths. The used wavelength was as follow: 340 nm for LDH test, 480 nm for XTT test and 540 nm for NR and SRB tests. In XTT, NR and SRB test it was also used reference wavelength of 690 nm.

4 Results

Results of cytotoxicity tests are shown as EC_{50} parameter. EC_{50} means a concentration which causing 50% inhibition of observed effect. Table 1 presents concentration of particulate matter fraction 2.5 μm causing 50% inhibition at an appropriate wavelength light absorbance. The lowest concentration presented in table means more toxic effects of particulate matter.

Table 1. EC_{50} concentrations of particulate matter extracts.

Location point	Season	LDH	XTT	NR	SRB
Na Grobli Street	Spring	14.000	6.258	3.938	3.622
	Summer	36.000	9.048	4.793	4.566
	Autumn	>50.000	5.129	2.283	2.483
	Winter	6.000	3.413	1.541	1.207
Wiśniowa Street	Spring	23.000	11.682	5.687	4.674
	Summer	>50.000	30.757	18.773	15.570
	Autumn	26.000	5.120	2.769	2.810
	Winter	43.000	17.300	6.045	5.634

5 Discussion

Comparing the EC_{50} results obtained in different cytotoxicity tests it is clearly shown higher values in case of LDH test then in other ones. The fact of much higher result values in LDH assay is connected with method of results reading. Disorders in spectrophotometry results is related with colour of the particulate matter extract sample. A LDH test is only one which reading is conducted in supernatant. This fact is the reason that the colour of the samples can disturb results of spectrophotometer reading at wavelength 340 nm. Described problem is not presented in publications conducted by other authors on A549 cell line, but the testing procedure in this cases is different from PAN-I instruction [12–14].

Values of EC_{50} concentrations obtained in other three tests are more objective and similar to each other. It should be also noted, that the results in XTT assay are slightly higher than those obtained in NR and SRB test. Presented results confirmed cytotoxicity properties of particulate matter fraction 2.5 μm in relation to the adenocarcinomic human alveolar basal epithelial cells A549. Impact of the particulate matter and organic extracts were also investigated by other authors. Among many results of researches it was measured apoptosis an viability of cells and cytotoxic effects using different parameters tests [11, 14–16].

Obtained results shows differences in toxicity of samples collected in different seasons of the year. Particulate matter collected on Na Grobli Street during autumn and winter season shows cytotoxicity properties in lower concentration then collected in spring and summer season. Similar situations shows results obtained on Wiśniowa Street, but in this case spring and winter concentrations are close to each other. This situation is connected with the more intensity of low-emission of particle pollutions connected to the combustion of the poor quality coal and waste burning.

Presented results shows also differences in toxicity depending on consumption place. Lower cytotoxicity properties showed samples collected near Wiśniowa Street, than near Grobla Street. This may be due to collection point near Wiśniowa gather particulate point coming typically from local sources of formation. The overwhelming amount of dust in this point is traffic pollutions and particulate matter coming from the combustion of nearly objects. Collecting point located near Na Grobli Street, which is at some distance from the sources of contamination, gather mixture of particulate matter comes from sources all over the city. This differences types of substances absorbed on the particulate matter fraction 2.5 μm can create complicated mixtures with unknown properties and functions.

6 Conclusions

The presented test results allow to present following conclusions:

1. It has been observed differentiation between an obtained results depending on the place and the season of collection.
2. It was proved higher cytotoxicity properties of particulate matter fraction 2.5 μm collected during autumn-winter season than in spring-summer season.
3. It was found that particulate matter collected on the Wiśniowa Street showed higher values of EC_{50} than samples collected on Na Grobli Street.
4. It has been shown inaccuracy of LDH test in conducted formula. It is proposed to change the method of carrying out, to not read the results in supernatant.
5. It has been shown similarity the results of EC_{50} obtained in the NR test and SRB test.
6. It has been observed lower sensitivity of the XTT test with respect to the NR test and SRB test.
7. It has been confirmed effectiveness of using A549 cell line in researches of particulate matter fraction 2.5 μm.

The investigations were co-financed within the framework of the order No. 0401/0068/16 with the specific subsidy granted for the Faculty of Environmental Engineering Wroclaw University of Science and Technology (W-7) by the Minister of Science and Higher Education.

References

1. D.K.M Alves, F. Kummrow, A.A. Cardoso, D.A. Morales, G.A. Umbuzeiro, Environ. Mol. Mutagen., **57**, 41 (2015)
2. L.D. Claxton, P.P. Matthews, S.H. Warren, Mutat. Res, **567**, 347 (2004)
3. J. Burns, H. Boogaard, R. Turley, L.M. Pfadenhauer, /a./m. van Erp, A.C. Rohwer, E. Rehfuess, Cochrane Database of Systematic Reviews, *Interventions to reduce particulate matter air pollution and their effect on health*, Cochrane Library (2014)

4. S.S. Lim et al., Lancet, **380**, 2224 (2014)
5. Global Urban Ambient Air Pollution, WHO Report, May 2016.
6. R. Walker, Toxicologia prospective y seguridad quimica, 147 (1992)
7. K. Piekarska, M. Zaciera, A. Czarny, E. Zaczyńska, Environ. Prot. Eng., **35**, 37 (2009)
8. K. Piekarska, M. Zaciera, A. Czarny, E. Zaczyńska, Environ. Prot. Eng, **37**, 85 (2011)
9. M. Kamber, European Environmental Mutagen Society September 9th – 13th, 2007 Pharmazentrum University Basel, *Comparison of 4 cytotoxicity assays performed sequentially on the same cell sample*, Basel, Switzerland (2007)
10. B. Binková, M. Černá, A. Pastorková, R. Jelínek, I. Beneš, J. Novák, R.J. Šrám, Mutat. Res.-Fund. Mol. M., **525**, 43 (2003)
11. M.E. Salcido-Neyoy, Y. Sánchez-Pérez, A.R. Osornio-Vargas, M.E. Gonsebatt, J. Meléndez-Zajgla, R. Morales-Bárcenas, P. Petrosyan, E.D. Molina-Servin, E. Vega, N. Anzano-León, C.M. Garcia-Cuellar, Environ. Pollut., **203**, 175 (2015)
12. H. Líbalowá, S. Krčková, K. Uhlířová, A. Milcová, J. Schmuczerová, M. Ciganek, J. Kléma, M. Machala, R.J. Šrám, J. Topianka, Toxicology letters, **225**, 350 (2014)
13. D. Breznan, S. Karthikeyan, M. Phaneuf, P. Kumarathasan, S. Cakmak, M.S. Denison, J.R. Brook, R. Vincent, Particle and Fibre Toxicology, **13**, 41 (2016)
14. M. Dergham, C. Lepers, A. Verdin, F. Cazier, S. Billet, D. Courcot, P. Shirali, G. Garçon, Environ. Res., **137**, 256 (2015)
15. U.S. Akhtar, N. Rastogi, R.D. McWhinney, B. Urch, C.W. Chow, G.J. Evans, J.A. Scott, Toxicology Reports, **1**, 145, (2014)
16. G. Mazzarella, A. Lucariello, A. Bianco, C. Calabrese, T. Thanassoulas, L. Savarese, A. Fiumarella, V. Esposito, A. de Luca, In Vivo, **28**, 557 (2014)

Effect of the presence of *Actinomycetes* in the activated sludge on the quality of the treated wastewater

Elżbieta Bezak-Mazur[1], *Renata* Stoińska[1,*], *Jolanta* Rajca[1], and *Beata* Łatasiewicz[1]

[1]Kielce University of Technology, Faculty of Environmental, Geomatic and Energy Engineering, Department of Water Supply and Sewage Technology, al. Tysiąclecia Państwa Polskiego 7, 25-314 Kielce.

Abstract. The aim of the study was to determine the effect of the *Actinomycetes* proliferation in the activated sludge on the quality of the treated wastewater and the sewage receiver. The river which is the sewage receiver flows near the wastewater treatment plant. The study was performed on the wastewater (raw and treated) and on the river water samples (collected before and after wastewater discharge). The analysis of the research results, such as the content of total organic carbon (TOC), total phosphorus and oxygen consumption in the examined samples, permit the conclusion that the presence of relatively large population of the *Actinomycetes* in sewage sludge has a negative impact on the quality of the sewage receiver. Determining the effect of the *Actinomycetes* in the activated sludge on the wastewater treatment process involved the application of specific analyses, such as Sludge Biotic Index (SBI) and the identification of filamentous bacteria in the activated sludge. The analysis of studies indicates that the presence of the *Actinomycetes* in the activated sludge adversely affects the efficiency of the wastewater treatment process.

1 Introduction

One of the most common problems faced by researchers studying sewage treatment plants is poor sludge sedimentation. It results from, i.a., the excessive proliferation of filamentous microorganisms. This phenomenon is defined as "filamentous bulking". Approximately 20 types of filamentous microorganisms are identified in activated sludge [1]. These include microorganisms belonging to the *Actinomycetes* group, which comprises a number of different bacteria with similar morphology (*Nocardia amarae, Nocardia pinensis, Rhodococcus, Gordona, Skermania*). These organisms are characterised by true branching and the formation of branched colonies primarily inside or around the sludge flocs. These microorganisms are also distinguished by hydrophobic cell walls, which are responsible for the adverse effect on the process of mechanical wastewater treatment, since they adhere to air bubbles and float together with sludge flocs on the surface of the settling tank thus preventing the sludge sedimentation. Owing to the difficulty in eliminating this

* Corresponding author: rstoinska@tu.kielce.pl

type of filamentous microorganisms in the activated sludge, it is crucial to recognize the threats to the environment that they may pose.

The aim of the study was to determine the effect of the *Actinomycetes* proliferation in the activated sludge on the quality of the treated wastewater and the sewage receiver.

2 Methodology of research

Determining the effect of the *Actinomycetes* proliferation in the activated sludge on the quality of the treatment process involved the application of the following analyses:

- The identification of filamentous microorganisms of the *Actinomycetes* genus in the activated sludge and in the floated sludge flocs in the secondary clarifier. The identification was performed by the method proposed by Eikelboom [2] .
- Microscopic analysis of sludge flocs floating on the wastewater surface in the secondary clarifier. Microscopic analysis of sludge flocs floating on the surface of the wastewater in the secondary settling tank was performed based on the sludge biotic index (SBI) [1]. The microscopic observations were performed using a phase-contrast microscope Delta Optical Evolution 100 coupled to the camera HDCE-X5.
- Phosphorus speciation analysis of the suspended solids contained in the treated wastewater. Phosphorus speciation analysis of the suspended solids was performed by the Golterman method [3]. This method is a sequential extraction scheme where each extraction stage is followed by filtering the sample and its further treatment with a reagent of increasing extraction force [4]. In the first stage the suspended solids sample which was passed through a cellulose filter was subjected to extraction with a solution of $CaNa_2$-EDTA for 4 h. In the second stage the residue formed after the first stage was subjected to extraction with a solution of Na_2-EDTA for 18h. The next stage involved the extraction of the sample with a solution of H_2SO_4 for 2 h. The fourth, last step of the sequential extraction lasted for 2 hours and was performed with the application of NaOH solution. The total phosphorus concentration was determined in the resulting filtrates. The determination of phosphorus was performed by spectrophotometric method using a *PERKIN ELMER spectrophotometer UV-VIS Lambda 25.*
- The content of total phosphorus in raw and treated wastewater and in the receiver before and after the inflow of the treated wastewater. The river which is the sewage receiver flows near a wastewater treatment plant. The average flow of the river is $0.06 m^3/s$ and the nominal capacity of waste water treatment plant is $900\ m^3/d$. The concentrations of orthophosphates and total phosphorus were determined by spectrophotometry [5] using a *PERKIN ELMER spectrophotometer UV-VIS Lambda 25.*
- The determination of the chemical oxygen demand (COD) in the sewage and in the wastewater receiver was performed by the permanganate method in accordance with the PN-85/C-04578.02 standard.
- The determination of total organic carbon in the wastewater and in the river water was conducted using the *AutoLab analyzer.*

3 Results Analysis

The analysis started with the macroscopic observation of the activated sludge. For this purpose, the sludge was poured into a graduated cylinder and examined after 30 minutes with regard to its settling properties (Fig. 1).

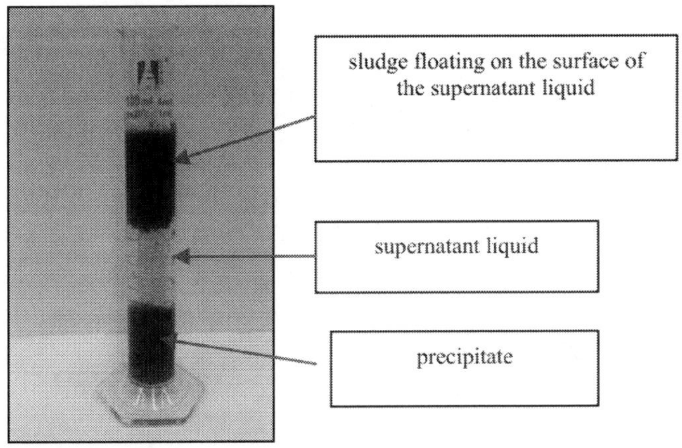

Fig.1. Activated sludge after a 30-minute sedimentation.

It was observed that even after 30 minutes the sludge did not completely settle. The precipitate layer at the bottom of the cylinder amounted to 25 ml and the sludge layer floating on the surface of the supernatant liquid amounted to as much as approximately 30 ml. Sludge stratification is a very detrimental phenomenon in the process of mechanical wastewater treatment. The sludge floating on the surface of the supernatant liquid can reach the receiver together with the treated wastewater.

Filamentous bulking caused by an excessive proliferation of filamentous organisms in the sediment is the most common cause of poor sludge sedimentation, therefore, the next step involved the identification of filamentous microorganisms in the activated sludge and in the sludge flocs floating on the surface of the wastewater in the secondary clarifier. The dominant filamentous microorganisms identified in the examined samples belong to the *Actinomycetes* group. The group of these bacteria stain dark blue when Gram's method is applied and do not change colour in Neisser staining [1]. Figure 2 shows a microscopic picture of the *Actinomycetes* microorganisms identified in the sludge floating on the surface of the treated wastewater in the secondary settling tank.

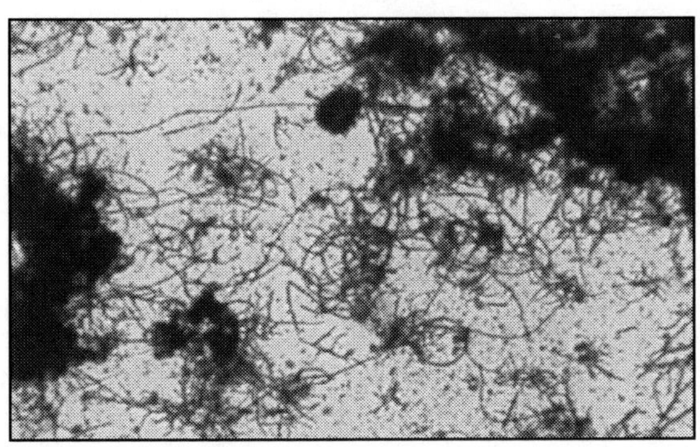

Fig. 2. *Actinomycetes* microorganisms after the application of Gram staining (400-fold magnification).

Bacteria belonging to the group of *Actinomycetes* have hydrophobic cell walls, resulting in their adherence to air bubbles and elevation to the surface with sludge flocs. These microorganisms also produce hydrophobic substances. Figure 3 shows a microscopic image of the sludge collected from the surface of the supernatant liquid (Figure 3). The microscopic image shows non-wetting liquid around the sludge floc on the microscope slide. These are probably hydrophobic substances synthesized by the identified filamentous microorganisms.

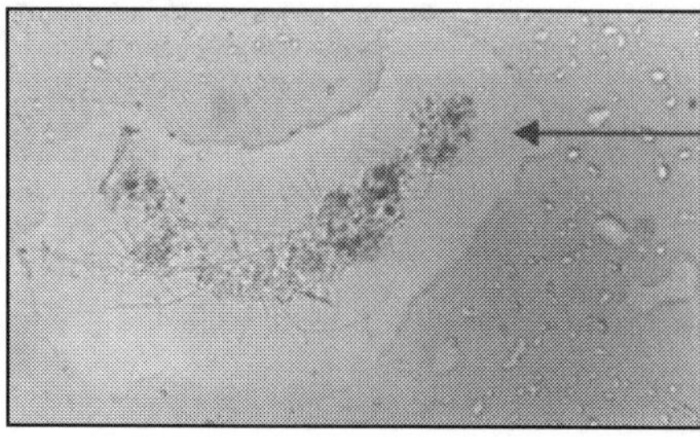

Fig. 3. Microscopic image of a sludge floc floating on the surface of the supernatant liquid.

Therefore the microscopic analysis of the activated sludge and the sludge flocs floating on the surface of the wastewater in the secondary settling tank was performed based on the Sludge Biotic Index [1]. This microscopic analysis allows to define a biological sludge quality on a scale of 0 to 10 and classify the sludge into one of four classes describing the quality of the treatment process. The sludge index value is affected by the density and the kind of the dominant group as well as the biodiversity of the sludge (the number of the protozoa and invertebrates taxa). The microscopic analysis of the sludge showed that the anchored ciliates and testate amoebae constitute the dominant group of the protozoa. They contribute to improving the sludge quality, but their concentration in the sludge does not exceed 1×10^6 microorganisms/dm^3. Moreover, the examined activated sludge does not reveal significant biodiversity. In total there are four identified taxa: testate amoebae, rotifers, anchored ciliates and small swimming flagellates (Fig. 4).

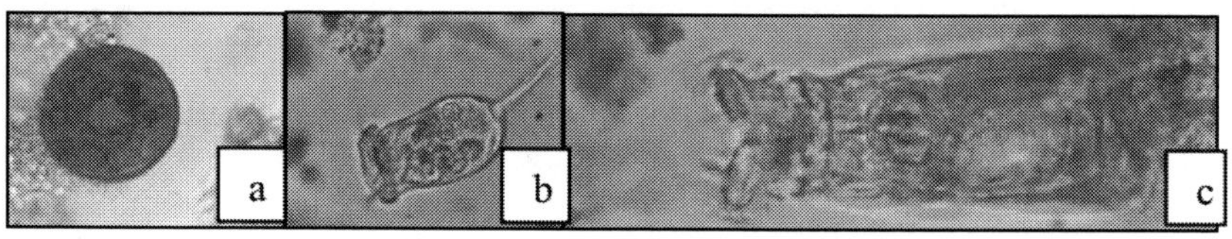

Fig. 4. Taxa of the microorganisms identified in the activated sludge: (a) testate amoeba, (b) anchored ciliate, (c) rotifer.

The value of SBI was estimated as 4 with the use of the table designed for determining the Sludge Biotic Index. It is a low value on the 0-10 scale and the sludge quality is defined as class III. Class III comprises the sludge in which the biological treatment in the aeration tank is insufficient and the sludge is poorly colonized. The analysis of the Biotic Sludge Index was also performed on the sludge composed of flocs floating on the wastewater surface in the secondary settling tank. It appeared to contain even fewer microorganisms, but the analysis of the sludge floating on the wastewater surface that enters the river shows that the lowest possible number of microorganisms would be advisable. On the other hand, protozoa and rotifers provide good filtration of the organic matter and therefore can positively affect portions of the sludge carried away with filamentous bacteria. The transported sludge flocs were dominated by small flagellates. It also contained testate amoeba and rotifers. The Sludge Biotic Index in the examined sample is 1 and thus the sludge is classified as having the lowest, class IV quality, characterized by a very low efficiency of biological wastewater treatment. The microscopic observation of the carried out sludge revealed also the presence of parasite eggs (Figure 5). It was estimated that their concentration is about 12×10^3 eggs /dm^3.

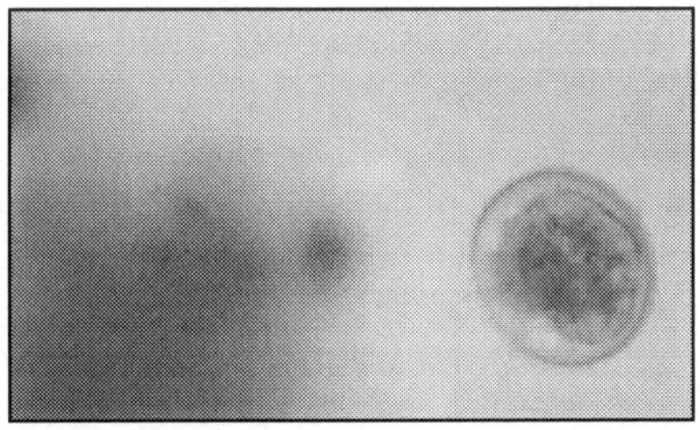

Fig. 5. Fertilized nematode egg.

The picture of nematode egg presented in Fig. 5 was not taken in the course of parasitic egg identification conducted in accordance with standard methodology but during the analysis of the Sludge Biotic Index. Therefore, the egg is not clearly visible in the preparation, and it is difficult to accurately identify its taxonomic affiliation. Nevertheless, the mere presence of potentially parasitic nematode eggs in the sludge discharged into the river is a considerable sanitary threat.

In the secondary clarifier the rising sludge contains not only filamentous micro-organisms but also portions of organic matter with a considerable amount of nutrients. The penetration of a high concentration of biogenic elements to the sewage receiver poses a threat to the aquatic environment. The increased concentration of nutrients contributes to water eutrophication. Therefore, the conducted investigation examined the concentration of phosphates, total phosphorus, total organic carbon and chemical oxygen demand in the raw and treated wastewater and in the river before and after the inflow of the treated sewage (Table 1).

Table 1. Measured parameters of sewage and water river.

	Raw sewage	Treated sewage	River water before the sewage inflow	River water after the sewage inflow
Total organic carbon (TOC) [mg C/dm^3]	85.63	41.79	21.23	36.38
COD-Mn [mg O_2/dm^3]	114.1	24.2	11.0	18.8
Phosphates [mg PO_4/dm^3]	11.72	1.47	0.15	0.40
Phosphates [mg P/dm^3]	3.82	0.48	0.05	0.13
Total phosphorus [mg P/dm^3]	8.32	2.54	0.17	0.43

The conducted analysis of phosphorus compounds in the river shows that their amount in the receiver was over twice as high after the wastewater discharge as before. The Regulation of the Minister of Environment dated 23 December 2002 on the criteria for designation of water bodies sensitive to pollution by nitrogen compounds from agricultural sources [6] determined the limit values of basic nutrients in water bodies, the exceedance of which indicates the occurrence of eutrophication. According to the provided values, eutrophication in the flowing surface waters occurs when i.a. the total phosphorus content exceeds 0.25 mg P/dm^3. This indicates that when the treated sewage enters the river water the total phosphorus content in the river exceeds the limit value of the eutrophication indicator. Moreover, the analysis of the remaining indicators presented in Table 1 permits the conclusion that the quality of the river water has deteriorated due to the discharge of the treated wastewater. Table 2 presents selected limit values of water quality indicators for surface water quality classes presented in Annex 1 to the Regulation of the Minister of

Environment dated 11 February 2004 establishing the classification for the presentation of surface waters and groundwater condition, their monitoring and interpretation of results, and the presentation of the state of these waters [7].

Table 2. Selected limit values of water quality indicators for surface water quality classes presented in Annex 1 to the Regulation of the Minister of Environment dated 11 February 2004 [7].

Water quality indicator	Limit values in classes I-IV				
	I	II	III	IV	V
COD-Mn [mgO_2/dm^3]	3	6	12	24	>24
Total Organic Carbon [mg C/dm^3]	5	10	15	20	>20
Phosphates [mg PO_4/dm^3]	0.2	0.4	0.7	1.0	>1.0
Total phosphorus [mg P/dm^3]	0.2	0.4	0.7	1.0	>1.0

Comparing the results of river water examination (Table 1) and the limit values of the water quality in surface water quality classes (Table 2), it can be said that the values of the total organic carbon indicate a decrease from class IV to V class of water quality, whereas the values of phosphates and total phosphorus indicate a decrease from class I to class II of water quality. Merely by analyzing the selected parameters one can notice a deterioration in the surface water quality due to the discharge of the treated wastewater.

The research study of the potential occurrence of eutrophication in the wastewater receiver involves the analysis of the total phosphorus and orthophosphates values. The orthophosphates are determined in the sewage passed through the cellulose filter, whereas total phosphorus is determined in the mineralized and non-filtered wastewater. Subtracting the values of orthophosphates from the total phosphorus value allows the estimation of the phosphorus compounds content in the total suspension. In the case of significant amounts of sludge being elevated in the secondary clarifier the total suspended solids in the treated wastewater will primarily include the particles of the elevated sludge. It was calculated that in the investigated wastewater the value of total suspended solids amounted to 1900 mg/dm^3. The value of this parameter is very high considering that in the treated wastewater it usually does not exceed 1000 mg $/dm^3$ [8].

Therefore, in order to determine the amount of phosphorus that can be assimilated by the microorganisms in the river water a phosphorus speciation analysis was performed in the examined suspended solids contained in the treated wastewater. The results of speciation analysis conducted by the Golterman method are shown in Table 3. The phosphorus content in the individual fractions are expressed in mg P/g.d.w. In the examined treated wastewater the total suspended solids constituted 1.9 g.d.w/dm^3, and therefore the content of particular phosphorus fractions was also expressed in mg P/dm^3. According to Golterman the bioavailable phosphorus forms are represented by the fractions Ca-EDTA and Na-EDTA bound to sludge particles [4].

Table 3. Results of the speciation analysis of the suspended solids conducted by the Golterman method in the treated wastewater.

Fraction	Phosphorus content [mg P/g d.w.]	Content of particular phosphorus fractions in 1 dm^3 of wastewater [mg P/dm^3]
Ca-EDTA	0.436	0.828
Na-EDTA	0.101	0.192
H_2SO_4	0.174	0.331
NaOH	0.320	0.608
Total	1.031	1.959

The amount of phosphorus, which is potentially assimilated by the river phytoplankton can be estimated by adding together the content of bioavailable phosphorus forms (fractions Ca-EDTA and Na-EDTA) contained in the sludge (Table 3) and the orthophosphate concentration expressed as elemental phosphorus (Table 1) contained in the treated wastewater:

The concentration of bioavailable phosphorus in treated wastewater =
$$=C_{PO4(P)}+C_{Ca\text{-}EDTA}+C_{Na\text{-}EDTA}$$

Calculated concentration of bioavailable phosphorus in treated wastewater =
$$=0.48 \text{ mg P/dm}^3+0.828 \text{ mg P/dm}^3+0.192 \text{ mg P/dm}^3=1.50 \text{ mg P/dm}^3$$

In the case of a large amount of organic matter infiltrating the surface water with the treated sewage, the execution of phosphorus speciation analysis in the separated suspension allows more accurate estimation of the concentration of phosphorus that can be assimilated by the microorganisms and cause eutrophication. The content of total phosphorus in the treated wastewater was 2.54 mg P/dm^3 with the amount of phosphorus that can directly cause eutrophication amounting to 1.50 mg P/dm^3. The amount of bioavailable phosphorus which can directly cause eutrophication makes as much as 64% of the amount of phosphorus in the treated wastewater (Fig. 6).

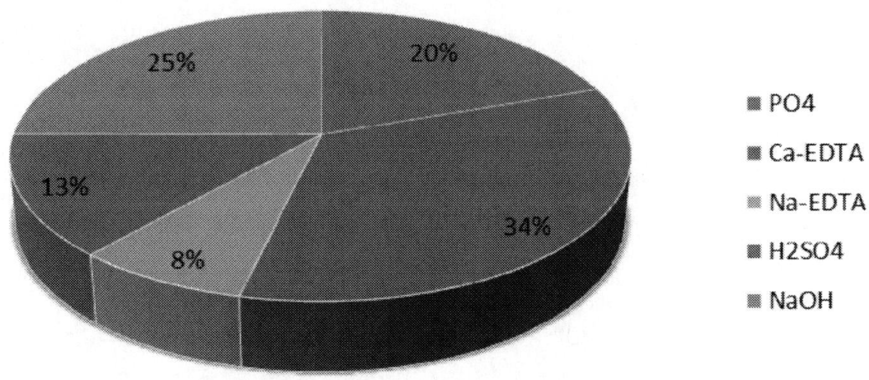

Total phosphorus = orthophosphates + phosphorus contained in the suspended solids (with the specified speciation forms: Ca-EDTA, Na-EDTA, H_2SO_4, NaOH)

Fig. 6. Percentage shares of phosphorus forms determined in the treated sewage.

4 Conclusions

Based on the conducted research study it can be concluded that the multiplication of the *Actinomycetes* microorganisms in the activated sludge adversely affects wastewater treatment processes and significantly deteriorates the quality of the treated wastewater, thus endangering the aquatic environment. The results of the speciation analysis of the suspended solids contained in the treated wastewater indicate that the forms which can be assimilated by living organisms causing water eutrophication represent more than half of the phosphorus. Thus it is advisable to perform speciation analysis in the treated sewage containing significant quantities of suspended solids since it allows more accurate assessment of the threat of excessive growth of phytoplankton in the wastewater receiver.

References

1. E. Fiałkowska, J. Fyda, A. Pajdak-Stós, K. Wiąckowski, *Osad czynny. Biologia i analiza mikroskopowa* (Wyd. Seidel-Przywecki, 2009)
2. D.H. Eikelboom, J.J. Buijsenh, *Handbuch für die mikroskopische Schlammuntersuchung,* (F. Hirthammer Verlag, München, 1983)
3. H L. Golterman, Hydrobiologia, **335** (1996)
4. E. Bezak-Mazur, R. Stoińska, Technologia Wody, **44**, 6 (2015)
5. The determination of phosphorus. Spectrophotometric method with ammonium molybdate, BS EN ISO 6878:2004 (2004)
6. Polish Journal of Laws of 2002, item 2093- (*The Regulation of the Minister of Environment dated 23 December 2002 on the criteria for designation of water bodies sensitive to pollution by nitrogen compounds from agricultural sources*, 2002)
7. Polish Journal of Laws of 2004, item 284-(*The Regulation of the Minister of Environment dated 11 February 2004 establishing the classification for the presentation of surface waters and groundwater condition, their monitoring and interpretation of results, and the presentation of the state of these waters*, 2004)
8. G. Kaczor, Infrastructure and Ecology of Rural Areas, **3** (2008)

Mineral status of plants and soil in silver fir stands of the Silesian and Żywiec Beskidy (Western Carpathians)

Jacek Borgulat[1,*], *Tomasz* Staszewski[1], *Włodzimierz* Łukasik[1], and *Łukasz* Jałowiecki[1]

[1]Institute for Ecology of Industrial Areas, Kossutha 6, 40-844 Katowice, Poland

Abstract. For last decades the Silesian and Żywiec Beskidy have been affected by different types of anthropopression. Up to the nineties of the 20th century, forests of the area were affected by a high level of air pollutants transported mainly from the Upper Silesian and Ostrava industrial regions. From 2004 intensive felling and skidding works in declining forest stands combined with their reconstruction have been carried out. Among different species silver fir has been introduced on a large scale. In the paper the results of investigation of plants and soil carried out 7 sites located in Beskidy Mountains (49°41'6"N, 19°1'48"E) are presented. Contents of K, Mg, Ca and N in needles of silver fir (*Abies alba* Mill.) and leaves of bilberry (*Vaccinium myrtillus* L.) were determined which was accompanied by a chemical characteristics of forest topsoil from the same sites. Moreover, in both environment compartments, the concentration of S, Pb, Cd, and Zn was also determined to assess the level of anthropopression in individual sites. In general, the obtained results show relatively good mineral status of plants. Some disturbances in the ratios of elements in plants were found in sites located near the Polish and Czech border. In plants and soil in these sites an elevated level of heavy metals was found which testifies the transboundary transport of air pollutants. The bilberry seems to be a better bioindicator of heavy metal pollution than silver fir.

1 Introduction

Development of Silesian industry in the 19th century became the reason for clear-cutting thousand hectares of forests. Monocultures of quickly growing spruce were planted on habitats of beech, beech-fir and beech-fir-spruce forests within the area of the Western Carpathians. Spruce seeds of alien origin were often used for establishment of artificial forests promoted by economy. However, they were not adapted to climatic-habitat conditions and the quality of these forests became getting worse because of the local air pollution. Several successive years of droughts followed by the borer gradation resulted in the spruce forest dieback. At the beginning of the 21st century, a large-scale felling of artificial spruce (*Picea abies*) stands has begun. This situation needs the stand reconstruction which is assessed to comprise 5500 ha. It is assumed that such approach will

* Corresponding author: j.borgulat@ietu.katowice.pl

DOI: 10.1051/e3sconf/20171700008

result in adequacy of species composition with habitat. The intensive introduction of beech and fir in the Silesian and Żywiec Beskid area [1] is one of the most important measures.

The aim of this study was to characterize the soil which new seedlings of silver fir were planted in and mineral status of silver fir (*Abies alba* Mill.) needles and leaves of bilberry (*Vaccinium myrtillus* L.) which dominates plant life forms in the herbaceous layer of the investigated area. An attempt was also made to compare the ability of both plant species to accumulate elements of anthropogenic origin (Pb, Cd, Zn, S). The accumulation of elements in a foliage of bilberry has been widely used in environmental monitoring [2–5] but information on the use of fir needles as a bioindicator of anthropogenic emissions is very scarce [6, 7].

2 Materials and methods

2.1 The study area and field sampling

2.1.1 Study area

Plants and soil were collected from 2 Forest Districts – Wisła and Ujsoły. Both sites belong to the Beskid Śląski and Żywiec Landscape Parks (Fig. 1).

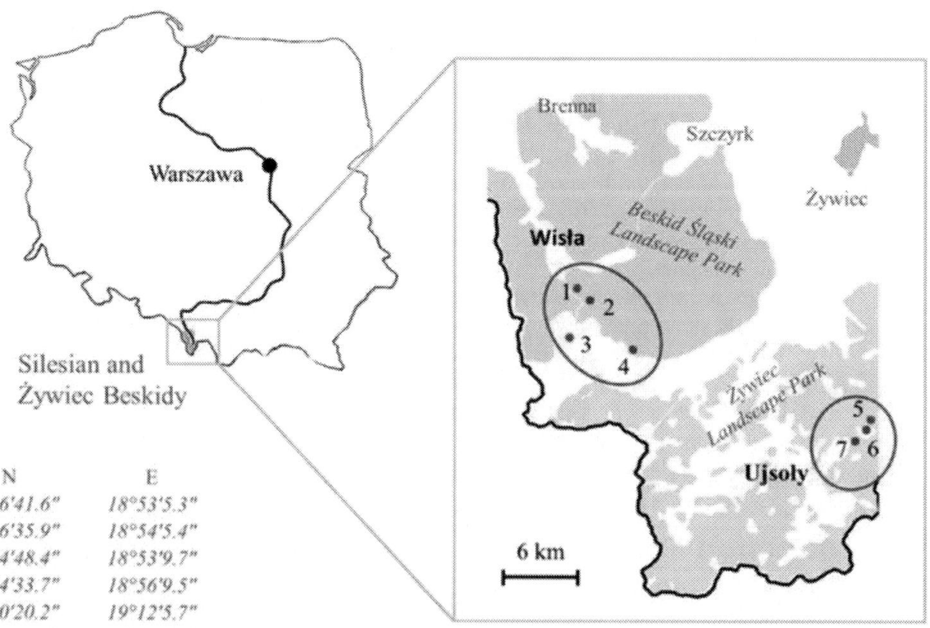

	N	E
1	*49°36'41.6"*	*18°53'5.3"*
2	*49°36'35.9"*	*18°54'5.4"*
3	*49°34'48.4"*	*18°53'9.7"*
4	*49°34'33.7"*	*18°56'9.5"*
5	*49°30'20.2"*	*19°12'5.7"*
6	*49°30'53.1"*	*19°12'5.8"*
7	*49°31'10.6"*	*19°12'5.9"*

Fig. 1. Location of study area.

In Wisła Forest District samples were taken from four sites where intensive reconstruction of stands towards silver fir (*A. alba* Mill.) has been carried out. The investigated area is affected by emissions from domestic heating, tourist traffic and long distance emissions from Ostrava (Czech Republic) and Upper Silesian Industrial Region.

In Ujsoły Forest District soil and plants were collected in 3 areas where silver fir seedlings were planted. The area is located close to the Polish and Slovak border and it is considered as a "clean region" where air pollutants origin only from local domestic heating.

2.1.2 Sample collection

In each site, after litter and humus removal, 20 cm layers of soil (A-horizon) were collected. Six soil sub-samples were combined in a composite sample. It was accompanied by the sampling of plant material in the neighbourhood of the sampled soil. Current year needles were sampled from six approximately 15-year-old silver fir (*Abies alba* Mill.) trees. An average sample was taken for the analysis. The fully matured, undamaged leaves were detached from 20 shoots of bilberry (*Vaccinium myrtillus* L.) and pooled into one sample per site. All samples were collected in September.

2.2 Physicochemical Analyses

2.2.1 Soil

The soil material taken from A horizons was air-dried and sieved through a sieve with 1 mm mesh. Each mixed sample was examined for the following physico-chemical parameters: the pH values (pH in H_2O and pH in 1M KCl) were measured by a pH-meter according to PNISO 10390:1997, determination exchangeable cations (Na, K, Mg, Ca) according to PN-ISO 23470, Digestion of the samples in order to obtain total content of heavy metals (Zn, Cd, Pb) was performed using *aqua regia*, and the metals were determined by ICP method according to PB-05/1.

2.2.1 Plant

Plant material was ground and mineralized with mixture of perchloric acid and nitric acid (1:4) in the MDS 2000 microwave system. The contents of Na and K were determined by the AAS method (SpectraAA 300 Varian). The contents of Cd, Zn, Pb, Mg and Ca were determined by the ICP method with an ultrasonic nebulizer, total N was measured using the Kjeldahl method according to PN-ISO 11261:2002, S was measured using an IR detector (according to PN-G-04558:2001 and PN-EN 15407:2011, respectively).

The analyses of plant and soil were carried out in a certified laboratory of the Institute for Ecology of Industrial Areas, Katowice (Certificate No AB 325). Statistical analysis was made by STATISTICA (StatSoft, Inc. 2011, ver.12).

3 Results and Discussion

3.1 The content of -micro and -macro elements in leaves and soil

Nutritional status of trees is commonly determined by chemical analysis of leaves. In contrast with such species like spruce and beech, the most frequently planted species in the Beskids forests, there is very scarce information on the optimal or deficit concentrations of micro- and macroelements in silver fir needles.

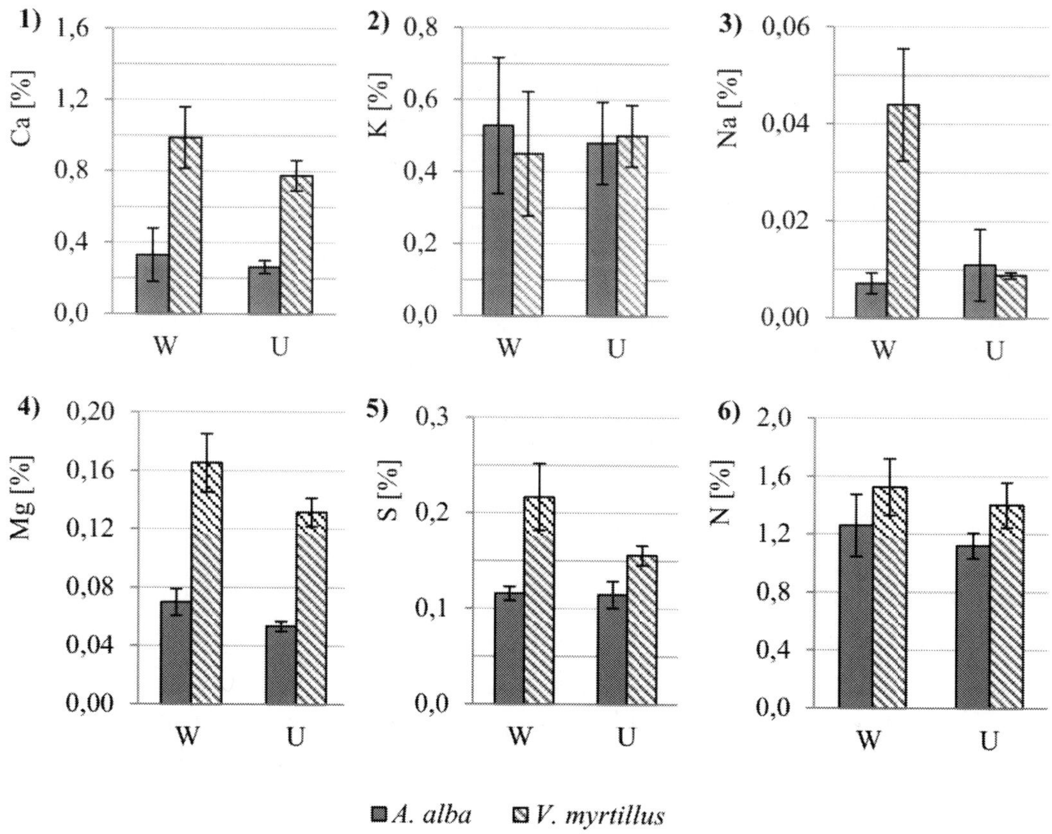

Fig. 1–6. Contents of Ca, K, Na, Mg, S and N in needles of silver fir (*Abies alba* Mill.) and leaves of bilberry (*Vaccinium myrtillus* L.) (Mean ±SD; W–Wisła Forest District; U–Ujsoły Forest District).

Contents of macroelement in bilberry and silver fir leaves is presented in Figures 1–6. N is an essential nutrient for plant growth, development and reproduction. Its deficiency most often results in stunted growth, slow growth, and chlorosis.

Similar N concentrations in silver fir needles was found for both sites. They amounted to 1.22% and 1.12% in Wisła and Ujsoły Forest Inspectorate, respectively (Fig. 3). These values are in the range of nitrogen concentration found by other authours. In Šumava where silver fir has been also reintroduced, N concentration ranged from 0.95% to 2.15% (mean – 1.39%) [6]. The concentration in the range of 1.1% to 1.25% were found in Germany and Slovakia [8, 9].

The regulation of water regime is one of the most important functions of K and this is also important in the resistance of individual tree species to different stresses (e.g. resistance to frost) [6]. It was found that needles taken from weakened firs contain less K than the vital ones [10]. K content in silver fir needles from Bohemian Forest ranged from 0.3% to 1.1% [6]. In other investigations the mean K content in needles varied from 0.6% (Slovakia) to 0.8% (Poland) [9] and in Germany, the mean value in the fir needles was found to be 0.77% [8, 11].

Hüttl [12] reported a deficiency limit value of 0.4–0.45% for potassium in silver fir. The obtained results for Wisła and Ujsoły - 0.48 and 0.52%, respectively show that the concentration of K is close to a deficiency limit (Fig. 2.), which can result in weak plant development of this species.The concentration of K in the investigated soils (4.3–4.9 mg·100g^{-1}) does not show its shortage for plants. It was addressed by Zawada [10] that despite the availibility of K in the soil, the weakned silver fir trees do not uptake this element in a sufficient quantity. It should be expected that other factor can be responsible for the deficit of this element in silver fir needles.

Tab. 1. Soil physicochemical analysys (SD-standard deviation, W–Wisła Forest District; U–Ujsoły Forest District).

Site		Ca	K	Na	Mg	Zn	Cd	Pb	pH	pH
		[mg·100g^{-1}]				[mg·kg^{-1}]			H_2O	KCl
W	mean	7.88	4.90	0.82	1.78	35.3	0.54	69.1	4.2	3.3
	±SD	3.3	2.4	0.4	1.1	16	0.2	13	0.2	0.2
U	mean	11.3	4.30	0.63	2.85	67.6	0.52	39.3	5.0	4.0
	±SD	6.9	1.7	0.1	2.1	11	0.1	21	0.2	0.2

Mg is one of the most important nutrients involved in many enzyme activities and the structural stabilization of tissues. Its deficiency results in shorter roots, smaller shoots and necrotic spots on leaves and in impaired carbon metabolism and decline of chlorophyll and carbon fixation.

The Mg concentration in silver fir needles was slightly higher in Wisła Forest District (0.07%) when compared to Ujsoły Forest District (0.05%) (Fig. 4). The higher Mg concentration in fir needles (0.15% to 0.19%) was found in the mountains of Poland and Slovakia [6, 9]. The problem of Mg in forest ecosystems was studied e.g. by Hüttl [12] and Hüttl and Schaaf [13]. Their determined threshold for Mg deficiency was 0.07–0.08%. On the basis of silver fir investigation in about 100 sites within the natural range of the species in Poland, it was found that the contents of Mg in needles of silver fir trees with a low increment coefficient was lower than 0.12% [10]. The concentration of Mg found in needles of silver fir growing in Wisła and Ujsoły Forest District can show the symptoms of Mg deficit.

Ca is an essential plant nutrient. It is required for structural roles in cell wall and membranes. The Ca concentration in silver fir needles amounting to 0.34% and 0.26% for Wisła and Ujsoły Forest Districts, respectively can be assessed as a relatively low when compared to another authors who reported the values in the range of 0.76-1.09% [6, 10]. The low level of Ca in the investigated sites can negatively affect the silver fir growth. However, the concentration of Ca in the soil of both Forest Districts is in the range of 6.53–12.5 mg·100g^{-1} which was found in the soil of fir stands with normal vitality located in the Carpathian Mountains [10].

The Na and S concentrations in the silver fir needles were similar in both sites and they were close to the values obtained in other mountain sites in Poland and the Czech Republic [6, 10, 14].

In case of bilberry, the concentrations of Ca, Na and S in leaves correspond with values from other unpolluted areas [15, 16]. The concentrations of K and Mg in bilberry leaves, such as silver fir needles, were lower than the values found by the above mentioned authors.

3.2 The content of heavy metals in leaves and soil

Contents of heavy metals in soil are presented in Table 4. In the case of Pb in Wisła Forest District higher average contents of this element was found when compared to Ujsoły Forest District. Concentrations of Pb in Wisła are close to the permissible value for the protected natural areas [17], which can be the effect of car fumes (due to tourist attendance) or industrial emissions presumably in the 90s of the 20th century. Poland was among the most polluted countries in Europe with a deposition of Cd and Pb amounting to 27 t·year^{-1} and 1100 t·year^{-1}, respectively [18]. Despite this, concentrations of Cd in both sites are close to the average level of this metal in Polish soils (0.5 mg·kg^{-1}) [19]. More than twice Zn contents was found for Wisła Forest District than Ujsoły Forest District but in both cases they do not exceed permissible value for the protected natural areas [17].

Heavy metals measured in aerial parts of trees (leaves, needles) have two potential origins: transfer from soil and atmospheric deposition [20, 21]. Plant-to-soil concentration ratio can be used as a rough indicator of the magnitude of element transfer from soil to plant [22, 23].

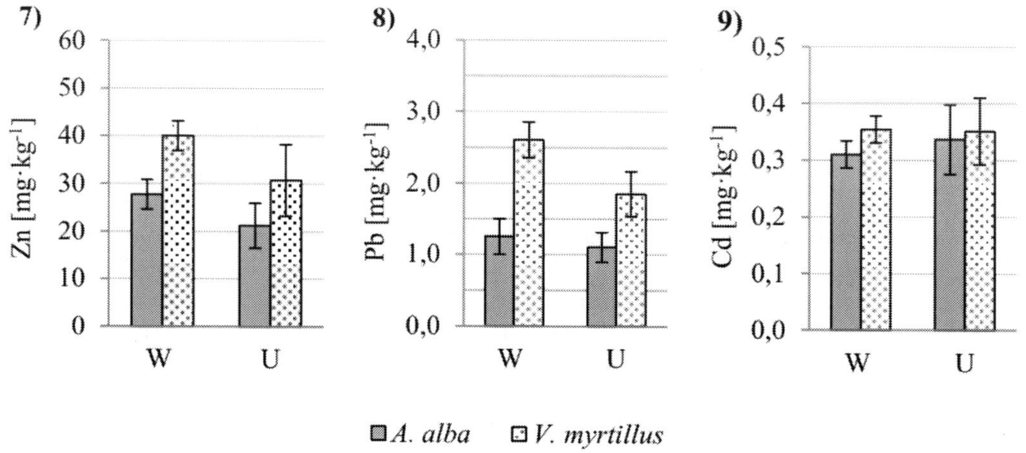

Fig. 7–9. Contents of Zn, Pb and Cd in needles of silver fir (*Abies alba* Mill.) and leaves of bilberry (*Vaccinium myrtillus* L.) (Mean ±SD; W–Wisła Forest District; U–Ujsoły Forest District).

Contents of heavy metals in soil are presented in figure 7–9.

Zn plays an important role in the formation of chlorophyll and it takes part in many metabolic processes. Plants uptake actively Zn proportionally to its concentration in the soil [24]. The critical deficiency contents of Zn amounts to >15-20 mg·kg^{-1} and the toxic level is 100-300 mg·kg^{-1}. Other authors have found at the concentration of above 100 mg·kg^{-1} the majority of plants show metabolism disorders [25]. The contents of Zn in leaves shows differences depending on plant species. In this study the concentrations of Zn in leaves of both species did not exceed physiological needs.

Zn concentration in silver fir needles from Jodłowice and Gola Wielka was 35 and 3mg·kg^{-1}, respectively [14]. In other investigations the mean Zn contents in needles was more than 40 mg·kg^{-1} [7]. Zn contents in the analyzed samples from Bohemian Forest ranges from 12 to 73 mg·kg^{-1} with the mean value 41 mg·kg^{-1} [6]. In all the said works, concentration of Zn in needles was higher than the one obtained in this paper. In *V. myrtillus* the quantity of Zn in leaves was higher than in *A. alba* and it was 40 mg·kg^{-1} in Wisła Forestry and 30 mg·kg^{-1} in Ujsoły Forestry. Smaller Zn values in leaves of *V. myrtillus* were obtained by Mróz and Demczuk [16] in the polluted areas. Similar values were obtained by Kozancka – 20 mg·kg^{-1} [15].

The critical limit of Pb toxicity for plants is 30-300 mg·kg^{-1} and a wide range of values can be contributed to high tolerance of some species (mainly growing in contaminated areas) to Pb. For example, native species of *Betula pendula* and *Pinus sylvestris* have a fairly good condition and have the ability of spontaneous spreading by seeds under high contamination of Pb and Zn [26]. In hereby examination, contents of Pb in silver fir does not exceed 2.6 mg·kg^{-1}. In Szymura's research [14] in current year needles of *A. alba* 9.8 and 10 mgPb·kg^{-1} was noticed in Jodłowice and Gola Wielka. Higher values of Pb were noticed in *V. myrtillus*. When comparing coefficients of Pb bioaccumulation (Pb contents in leaves/Pb contents in soil) in both species (*V. myrtillus* 0.047 > *A. alba* 0.026), higher ability to bioaccumulation of Pb at *V. myrtillus* can be noticed, which indicates that it is a better bioindicator for this element. Additionally, in Wisła Forestry higher values of Pb in soil were noticed than in Ujsoły, which is reflected in the results.

Symptoms of the harmful effect of Cd in plants occur at the concentration of 5–30 $mg{\cdot}kg^{-1}$ [22], in sensitive plants the value of 5 $mg{\cdot}kg^{-1}$ seems to be toxic [27].The 0.2 $mg{\cdot}kg^{-1}$ level in plants is considered to be a highly elevated concentration according to the scale proposed by Arndt [28]. The mean Cd concentration in spruce needles collected from all 23 Polish national parks was 0.18 $mg{\cdot}kg^{-1}$ [29]. In hereby paper, concentration of Cd in leaves of both species was similar (about 0.3$mg{\cdot}kg^{-1}$). The obtained results show the area of research was or still is submitted to anthropopression. Smaller values for bilberry from the ones obtained in this paper were noticed by Mróz and Demczuk [16] in the unpolluted areas 0.02–0.06 $mg{\cdot}kg^{-1}$. Szymura [14] noticed in current year needles of *A. alba* contents at the level of 1.02 $mg{\cdot}kg^{-1}$ in Jodłowice and 2.17 $mg{\cdot}kg^{-1}$ in Gola Wielka.

3.3 Conclusions

- Concentration of macronutrients found in leaves of bilberry and silver fir needles is relatively low (near a deficit value) when compared to literature data.
- A slight exceedance of Pb concentration level recommended for the protected areas was found in the soil of the Wisła Forest District.
- Higher level of Pb accumulation in bilberry leaves when compared to silver fir needles was found which suggests the use of bilberry leaves as a bioindicator of this element in Silesian and Żywiec Beskidy.

The study was carried out under the project "Determination of habitat conditions and vitality of silver fir (*Abies alba* Mill.) with recommendation of areas suitable for its cultivation in Silesian and Żywiec Beskidy" ordered by the Regional Directorate of State Forests in Katowice and supported by the Voivodeship Fund for Environmental Protection and Water Management, Katowice.

References

1. T. Staszewski, P. Kubiesa, Beskydy **1**, 77–84 (2008)
2. C. Reimanna, F. Kollerb, G. Kashulinac, H. Niskavaarad, P. Englmaiere, Environ Pollut **115**, 239–252 (2001)
3. M. Salemaa, J. Derome, H.S. Helmisaari, T. Nieminen, I. Vanha-Majamaa, Sci Total Environ **324**, 141–160 (2004)
4. A. Brekken, E. Steinnes, Sci Total Environ **326**,181–195 (2004)
5. D. Białońska, A. Zobel, M. Kuraś, T. Tykarska, K. Sawicka-Kapusta, Water Air Soil Poll **181**, 123–133 (2007)
6. R. Novotný, D. Černý, V. Šrámek, J Trop For Sci **56**, 381-388 (2010)
7. L. Gandois, A. Probst, Chemosphere **87**, 204-210 (2012)
8. R. Bäumler, T. Goerttler, W. Zech, Forstwiesenschaftliches Centralblatt **114**, 30–39 (1995)
9. B. Maňkovská, B. Godzik, O. Badea, Y. Shparyk, P. Moravčík, Env Poll **130**, 41–54 (2004)
10. J. Zawada, Prace Instytutu Badawczego Leśnictwa A, **849** (1998)

11. M. Musio, N. Augustin, H.P. Kahle, A. Krall, E. Kublin, R. Unseld, K. Wilpert, Ecol Model **179**, 307–316 (2004)
12. R.F. Hüttl, Kali-Briefe **17**, 1–7 (1986)
13. R.F. Hüttl, W. Schaaf,*N Deficiency in Forest Ecosystems* (London, Kluwer Academic Publishers: 362, 1997)
14. T. Szymura, Trees **23**, 211–217 (2009)
15. T. Kozanecka, J. Chojnicki, W. Kwasowski, Pol J Environ Stud **11**, 395–399 (2002)
16. L. Mróz, M. Demczuk, Pol J Ecol **58**, 475–486 (2010)
17. Official Journal of Law (2002). No. 165, Clause 1359
18. K.D. Van den Hout, D.J. Bakker, J.J.M. Berdowski, J.A. Van Jaarsweld, G.J. Reinds, J. Bril, A. Breeuwsma, J.E. Groenenberg, W. De Vries, J.A. Van Pagee, M. Villars, C.J. Sliggers, Water Air Soil Poll **109**, 357–396 (1999)
19. J. Lis, A. Pasieczna, *Atlas geochemiczny Polski 1:2 500 000* (PIG, Warszawa 1995).
20. A. Kabata-Pendias, Geoderma **122**, 143–149 (2004)
21. J. Klaminder, R. Bindler, O. Emteryd, I. Renberg, Geochim Cosmochim Ac **69**, 2485–2496 (2005)
22. A. Kabata-Pendias, H. Pendias, *Trace elements in soils* (3rd Ed. Boca Raton, London, New York, CRC Press, 413, 2001)
23. S. Ehlken, G. Kirchner, J Environ Radioact **58** (2002)
24. B. Hafeez, Y.M. Khanif, M. Salem, American Journal of Experimental Agriculture **3**, 374-391 (2013)
25. M.H. Zenk, Gene **179**, 21–30.(1996)
26. K. Skubała, Arch Environ Prot **37**, 57-74 (2011)
27. I. Ahmad, M.J. Akhtar, Z.A. Zahir, A. Jamil, Pak J Bot **44**, 1569–1574 (2012)
28. U. Arndt, W. Nobel, B. Schweizer, *Bioindikatoren. Möglichkeiten, Grenzen und neue Erkenntnisse* (Ulmer Verlag, Stuttgart 1987)
29. T. Staszewski, W. Łukasik, P. Kubiesa, Environ Monit Assess **184**, 4597–4608 (2012)

The content of heavy metals in soils and plants around the waste landfill in Siechnice (Poland)

Maciej Borowczak[1], and *Anna* Hołtra[1,*]

[1]Wroclaw University of Science and Technology, Faculty of Environmental Engineering, Wybrzeze Wyspianskiego 27, 50-370 Wroclaw, Poland

Abstract. Metal concentrations (Cu, Zn, Pb, Ni, Cr) in soils and plants around the metallurgical waste landfill of Siechnice, Poland, have been determined. The soil samples were collected from the surface layer of 0-25 cm. The soils and plants were digested in HNO_3, 60%, using the Microwave Digestion System (Milestone Start D). The analysis of metals has been carried out through the Flame Atomic Absorption Spectroscopy (FAAS) method. The pH and the conductivity of the soil solutions were measured. The pollution index (Wn) and the bioaccumulation coefficient (WB) have been determined. The permissible content of Cu, Zn, Ni and Cr in soils were exceeded compared to the geochemical background in uncontaminated soils of Poland. Chromium concentrations have also been exceeded relative to the standard for soils, according to the Polish Ministry of Environment Regulation from September 1st, 2016. The assessment of the bioaccumulation coefficients of metals in plants (*Plantagolanceolata* L. and *Zea mays* L.) has shown significant bioaccumulation of copper and zinc compared to the permissible concentrations of undesirable substances in feed in agreement with the Polish Ministry of Agriculture and Rural Development Regulation from January 23rd, 2007.

1 Introduction

The main sources of heavy metals in the environment are the anthropogenic activities. Branches of significant impact are those from the industry: mining, metallurgy, energy, and also transport. In the beginning of operation of Siechnice ironworks produced carbide and welding electrodes. Since 1932, it had been producing ferrochromium [1]. Table 1 shows the emission of pollutants from the organized and unorganized sources released by the Siechnice ironworks during the last years of its operation. The waste landfill is a remnant of the closure of the plant in 1995. The heap of slags with admixture of metallic phase is located in the south-east of Wroclaw, in the Siechnice municipality, by the provincial road no. 95 from Wroclaw to Olawa. In the north-east direction, at a distance of approx. 300 meters from the landfills, are located the water-bearing areas of Wroclaw. Moreover, in the vicinity there is mainly agricultural wasteland. The only exception is the farmland in the south side of the landfill, by the provincial road no. 95. Within the area there is a municipal sewage treatment plant, concrete plant and shipping companies.

* Corresponding author: anna.holtra@pwr.edu.pl

Table 1. Emission from the ironworks of Siechnice [1].

Metal	Emission [Mg/a]		Precipitation [$Mg/km^2 \cdot a$]
	Organized	Unorganized	
Cr	95	126	0.02 – 4.5
Fe	9.5	12.6	0.01 – 2.3
As, Mn, Pb, Ni, V, Cu, Cd	43	3	0.005 – 0.12

The landfill is located on sandy soils and loam soils, similarly as the Wroclaw water-bearing areas. They are quaternary formations, and the aquifer level consists of gravel and coarse, medium and fine sands. They are characterized by a high rate of infiltration. Groundwater is isolated by clays and poorly drained loams. This is a barrier against effluents that could penetrate deep into the soil together with the rainwater.

On an area of approx. 12 hectares lies approx. 2 million Mg of waste in the form of slag and dust caused by the production of ferrochromium (Tab. 2). In order to secure the landfill site against the elution of heavy metals into the soil, the waste has been stabilized with boron and calcium compounds - the works have been carried out since 1981 [1]. Since 2012 the exploitation of waste located on a heap is being carried out to recover the chromium.

Table 2. The composition of waste [1].

Component	Percentage of composition [%]
CaO	35 – 56
SiO_2	16.6 – 41.3
Al_2O_3	0.5 – 9.5
Cr_2O_3	0.5 – 10.5
Fe_2O_3	0.25 – 3.8
MgO	5.4 – 14.5
metallic ferrochromium	no data

The metallic pollutants from soils are accumulated in the trophic chain. Plants, depending on the species, show a typical threshold of heavy metal tolerance. The excess of metal results the inhibition of growth, disturbances of the photosynthesis process and deformation of the root system in plants [2, 3]. The English plantain (*Plantagolanceolata* L.) and maize (*Zea mays* L.) are able to accumulate metals in amounts exceeding their physiological needs, without harm to the organism. They were used in phytoextraction of metals from floatation tailings from the copper industry. *Plantagolanceolata L.* collects the elements in the stem and leaves. Study showed the presence of 53 mg Cu/kg of dry matter and 100 mg Zn/kg of dry matter in English plantain. In the process of phytoextraction of zinc and nickel *Zea mays* L. was used. Maize accumulates impurities mainly in the root, to a lesser extent in the stem and leaves. The content of zinc, which deforms the root system and inhibits the growth of corn, is 250 mg Zn/kg of dry matter. Toxic concentration of zinc in the soil for maize is 750 mg/kg of dry matter [4–6].

2 Materials and methods

In the summer season of 2016 14 environmental samples were collected from the area around the heap in Siechnice. In one location, three soil samples were collected each time within the distance of approx. 100-200 cm from each other. The soils were collected from the surface layer of 0–25 cm. The soil and plant samples no. 1 were collected at the border of Wroclaw water-bearing areas (towards the north-east of the landfill of metallurgic waste). Samples of no. 2 came from the area in the east of the Siechnice heap, 4 meters from the water reservoir, between the eastern ring road of Wroclaw and the Stawowa Street. The soil samples no. 3a, 3b and 3c were picked up from cropland close to the landfill (the southern direction). Samples no. 3c were collected from the location closest to the heap. Soil and plant samples no. 3b were picked up from a place located in a greater distance from the landfill. Samples no. 3a were collected by the provincial road no. 95. Also, samples no. 4 came from a cultivated field located on the opposite side of the provincial road no. 95. Samples no. 5 came from the roadside of West Street, which is the access road for the landfill in Siechnice (west side of the landfill). Samples of maize (*Zea mays* L.) were collected from the farmland (3a, 3b, 3c), while English plantain samples (*Plantagolanceolata* L.) were collected from the remaining locations.

Each time three of the air-dried soil samples were sifted and averaged. Measurements of pH and electrical conductivity in the soil solutions with a stoichiometry ratio of 1:2.5 (m:v) were made according to PN-ISO 10390:1997. Prepared in accordance with PN-ISO 11465:1999, the soil and plant samples were digested (ca. 0.2 g of dry matter) in 8 ml of HNO_3, 60%. The microwave mineralization process was conducted with use of the Start D device (Milestone). Metal concentrations (Zn, Cu, Ni, Cr, Pb) have been determined through the FAAS method with use of the iCE 3500 Solaar Thermo device (Thermo Scientific) [PN-ISO 11047:2001]. Arithmetic mean and standard deviation were calculated with three replicates of the sample.

Chemical analyses have been carried out in the certified Laboratory of Toxycology and Environmental Research in the Faculty of Environmental Engineering at the Wroclaw Uniwersity of Science and Technology.

3 Results and discussion

3.1 The physicochemical parameters

Soil solutions no. 1, 2, 4 and 5 are alcaline (pH above 7.50 in 1 M KCl and 7.90 in H_2O). Soil samples no. 5 have the most alkaline character, pH of 8.52 in 1 M KCl, pH of 8.92 in H_2O. That may be due to the influence of metallurgical waste currently being transported from the heap in Siechnice. Other samples may also be contaminated with dust from the heap, because of the destruction of carbonate layer which protected the metallurgic waste against the migration of metals into the ground depth together with rainwater.

Soil samples from the cultivated field which is located in the immediate vicinity of the heap (3a, 3b, 3c) have slightly acidic character. In the 1 M KCl solution , the pH values ranged from 6.00 to 6.38, and were also below 7.00 in H_2O. The acidification of soils may be caused by the ongoing agrotechnical treatments and the fertilizing of soils.

The measurements of conductivity indicate slight and medium salinity of soils around the waste landfill in Siechnice, ranging from 13.23 to 40.26 mS/m (lower limit 25 mS/m [7]). The only exception is sample no. 3a with recorded EC value of 102.32 mS/m. Such high salinity of the soil can be a result of the decline of the land, causing the gravitational runoff of the rainwater with mineral compounds and their accumulation in the trough. In addition, the factor fostering the salinity is also sprinkling the provincial road no. 95 with salt in winter. The salinity of the soil sample no. 5 (40.26 mS/m) may result from the contamination of land during the exploitation of the heap and transport of the metallurgical waste from the heap in Siechnice. In this location, the pH of the soil was the highest. In the soil sample no. 4 from the cropland (31.60 mS/m) salinity may be the effect of the use of mineral fertilizers. The lowest value of the EC (14.20 mS/m) was observed in sample no. 2, collected from the greatest distance from the heap and roads.

3.2 Spectroscopy analyses

The concentration of chromium was exceeded at one measurment point with respect to an acceptable level of metal in soils resulting from the Polish Ministry of Environment Regulation from 1. September, 2016 [8] (Tab. 3, Fig. 1). By the landfill's exit road (sample no. 5) the concentration of chromium amounts to 276.49 mg/kg of dry matter. High content of chromium could be the result of the ongoing exploitation and disposal of excavated material from the heap, as well as the long-term works of the Siechnice ironworks. Also may be due to the proximity of roads, including a motorway bypass of Wroclaw, used for transporting the waste from the heap in Siechnice. Exceeding of chromium in the soil may be a consequence of sliding of the excavated material from trucks on the road. In the immediate vicinity of the water-bearing area of Wroclaw the concentration of chromiun was 119.66 mg Cr/kg of dry mtter (soil sample no. 4). Movement of the dust determined by the direction of winds blowing in this area from the south-west to the north-east direction.

Table 3. The content of heavy metals in soils.

No.	Cu [mg/kg]		Zn [mg/kg]		Pb [mg/kg]		Ni [mg/kg]		Cr [mg/kg]	
	x	s	x	s	x	s	x	s	x	s
1	23.05	0.58	77.25	1.31	17.5	0.92	5.35	0.20	119.66	1.54
2	3.15	0.77	19.35	1.16	b.o.	b.o.	3.11	0.57	47.84	0.68
3a	10.14	0.80	47.65	1.85	6.71	0.78	3.53	0.19	85.29	0.79
3b	7.59	0.25	47.36	1.04	9.45	0.57	2.75	0.77	104.44	2.97
3c	7.05	0.34	45.98	0.74	12.73	0.62	3.14	0.74	62.65	1.45
4	16.40	0.84	38.03	0.66	4.17	0.78	2.27	0.42	8.16	0.31
5	16.03	0.63	55.23	0.72	no	no	13.71	0.59	276.49	1.02

x - arithmetic mean; s – standard deviation for 3 measurements, no - below the limit of Pb quantification with FAAS method.

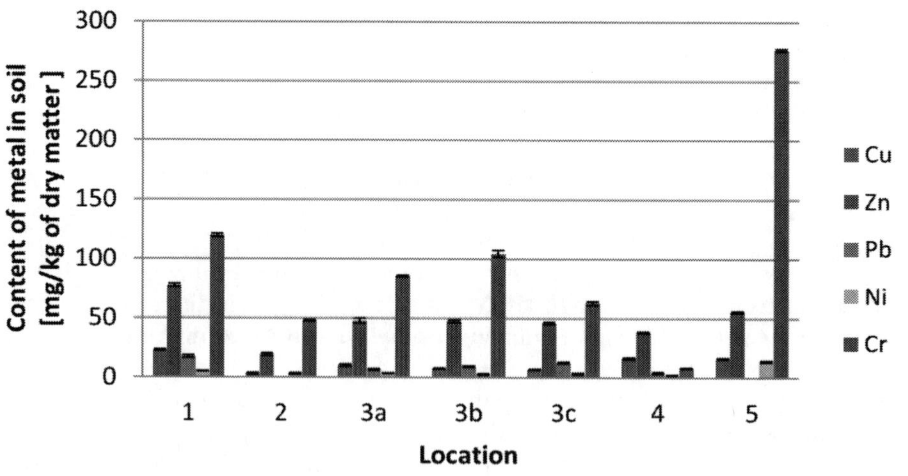

Fig. 1. The content of heavy metals in soils.

The concentration of other metals in soils are below the limit values set by the standard [8], but there are higher than the geochemical background in Poland. The average content of metals in the surface level of soils for the geochemical background of the whole Poland are, respectively: Cu – 6.5 mg/kg; Zn – 33 mg/kg; Pb – 18 mg/kg; Ni – 11 mg/kg; Cr – 40 mg/kg [9]. The increased concentrations of heavy metals in soil samples may be the result of contamination of the area during operating period of the Siechnice ironworks and transferring of dust in the wind according to the isolines of the dust fall [1].

Copper in soil samples is within the range of 3.15 to 23.05 mg/kg of dry matter (Tab. 3). The lowest concentration was recorded at the measurement point no. 2, and the maximum content of zinc was observed in the sample no.1. Zinc is in the range of 19.35 to 77.25 mg/kg of dry matter. Similarly to the copper, the zinc content is lowest at the location no. 2, and the highest concentration is in the soil sample no.1. The lead concentration does not exceed the geochemical background in soil samples. The minimum content of lead is in the soil sample no. 4 (4.17 mg/kg of dry matter) and the highest concentrations were reported in location no. 1 (17.5 mg/kg of dry matter). Higher concentration of nickel compared to the geochemical background was observed only in sample no. 5 (13.71 mg/kg of dry matter). The minimum concentration of Ni is in sample no. 4 (2.27 mg/kg of dry matter). Low concentrations of heavy metals in location no. 4 are due to a remote location of sampling in the south-west of the landfill, on the opposite side of the road no. 95. The influence of dust moving with the wind in this direction is the smallest.

During the comparison of the results (Tab. 4) of acceptable concentration of heavy metals in forage plants [10], exceedings of the content of copper were observed, except for the seeds of maize (*Zea mays* L). The English plantain (*Plantagolanceolata* L.) accumulates from 15.33 to 20.17 mg Cu/kg of dry matter in the tissues (limit for the forage plants: 10 mg Cu/kg of dry matter [10]). In the case of maize (*Zea mays* L.) the highest concentration of metal was observed in the root (from 19.64 to 21.4 mg Cu/kg of dry matter). The leaves of *Zea mays* L. accumulated from 16.63 to 20.09 mg Cu/kg of dry matter. The lowest concentration of the element was observed in the seed of maize (from 6.51 to 7.65 mg Cu/kg of dry matter), therefore it is possible to use the seeds in the animal feed.

Table 4. The content of metals in plants.

Plantagolanceolata L.	Cu [mg/kg]		Zn [mg/kg]	
	x	s	x	s
1	16.73	0.24	63.32	1.08
2	15.33	0.29	54.55	0.20
4	16.28	0.09	52.67	0.33
5	20.17	0.13	37.72	0.54
Zea mays L. - leaf				
3a	17.68	0.41	46.75	0.73
3b	19.64	0.30	32.96	0.22
3c	16.63	0.17	39.08	0.16
Zea mays L. - root				
3a	21.40	0.13	42.47	0.41
3b	20.09	0.61	100.56	1.37
3c	21.96	0.33	43.15	0.22
Zea mays L. - seed				
3a	6.51	0.02	5.50	0.13
3b	7.28	0.18	23.31	0.73
3c	7.65	0.23	13.35	0.17

x - arithmetic mean; s – standard deviation for 3 measurements.

The content of zinc in plants generally does not exceed the acceptable concentration of 100 mg Zn/kg of dry matter in the feed plants [10]. The only exception is the sample no. 3b, where the root of maize (*Zea mays* L.) reported a slightly over-normative value – 100.56 mg Zn/kg of dry matter. Similarly to the case of copper, the lowest concentration of zinc was observed in the seeds of maize (from 5.50 to 23.31 mg Zn/kg of dry matter). The results show that the test plants can be used in the phytoremediation process of areas contaminated with zinc and copper. In *Plantagolanceolata* L. the accumulation of zinc is 2–3 times higher than copper. In maize a similar correlation was observed (2-fold higher accumulation of Zn). In the roots of *Zea mays* L. 4-fold higher concentration of zinc was observed.

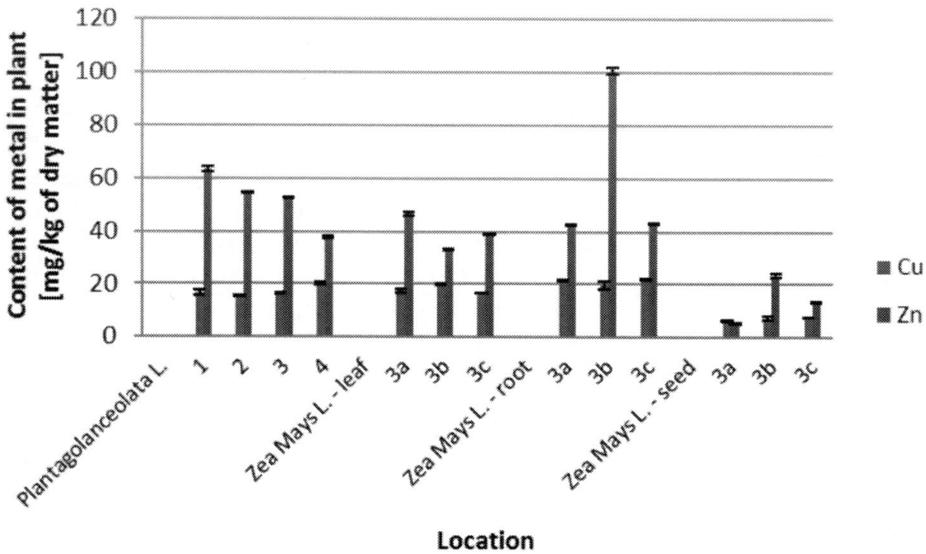

Fig. 2. The content of copper and zinc in plants.

3.3 Pollution index

The pollution index (Wn) is expressed as the ratio of metal in the soil to the geochemical background in uncontaminated soils of Poland. The calculated coefficients of the accumulation of metals in the soil are shown graphically in Figure 3.

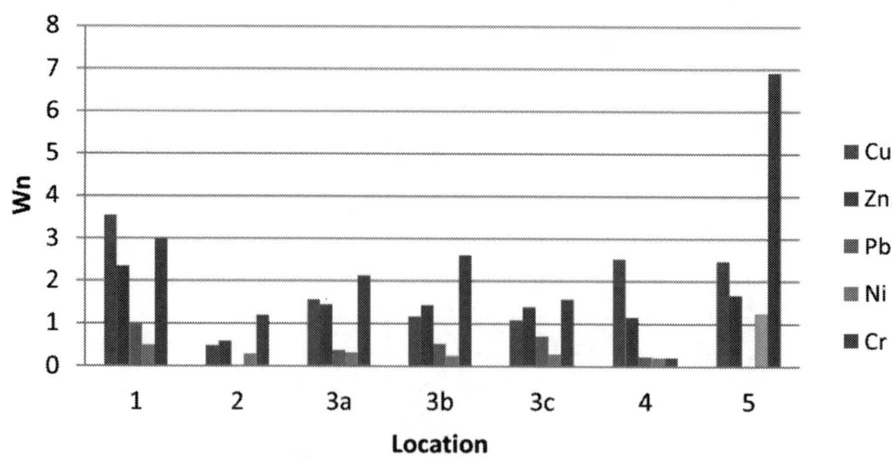

Fig. 3. The pollution indexes (Wn) of metals in soils.

The highest, almost 7-fold enrichment of the soil with chromium, relative to the average chromium content in the geochemical background in Polish [9], was observed in sample no. 5. Such high value of Wn 6.91 may indicate the influence of the waste from the landfill and the contamination of the soil. Other soil samples are also enriched with chromium (Wn 1.20 – 2.99). Only sample no. 4 is not enriched with this metal (Wn 0.2). Enrichment of the soil with copper and zinc was observed for the same soil samples. The maximum values of Wn for copper and zinc were in sample no. 1 and amounted of 3.55 and 2.34 for Cu and Zn respectively. The remaining samples were also enriched with copper and zinc. Wn indexes for copper were in the range of 1.08 – 2.52 and for zinc from 1.15 to 1.67. Only soil sample no. 2 was not enriched with copper (Wn 0.48) or zinc (Wn 0.59). Soil sample no. 5 was enriched with nickel and the pollution index is 1.25. None of the soils contained concentration of lead above the geochemical background.

3.3 Bioaccumulation coefficient

The bioaccumulation coefficient (WB) represents the ratio of the content of the metal in plant (its organs) in relation to the content of the metal in the soil. The parameter determines the ability of the plant to collect the components from the soil and also indicates the size and speed of metal movement from the soil solution to the aerial parts of plants. The calculated bioaccumulation coefficients of copper and zinc (WB) in maize (*Zea mays* L.) and English plantain (*Plantagolanceolata* L.) are shown graphically in Figure 4.

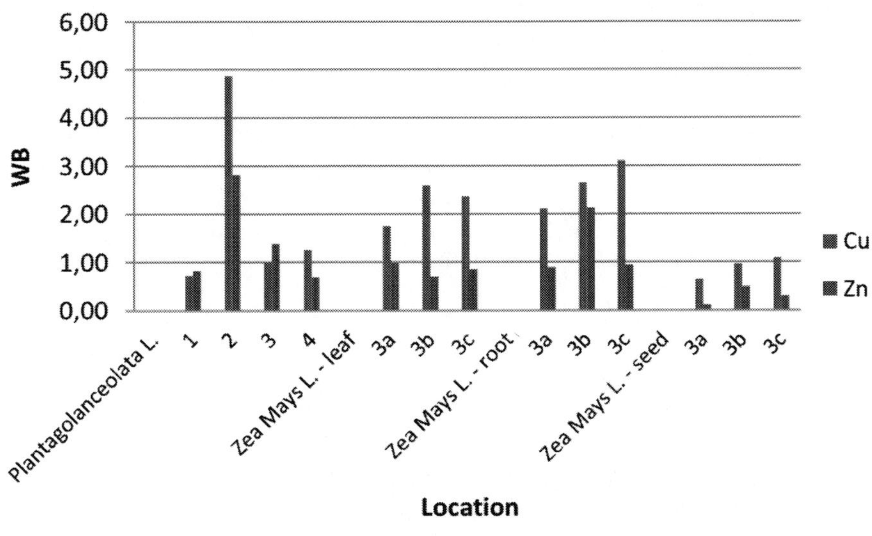

Fig. 4. The bioaccumulation coefficients of metals in *Plantagolanceolata* L. and *Zea mays* L.

The bioaccumulation coefficient values show that plants accumulate higher amounts of copper (WB 0.64–4.87) than zinc (WB 0.12–2.82). Some plants have accumulated metals in the tissues in quantities which could endanger the trophic chain. In sample no. 2 of the English plantain (*Plantagolanceolata* L.) the maximum value of WB for both copper and zinc, respectively 4.87 and 2.82 (Fig. 4), was recorded. High bioaccumulation coefficient values of copper were also observed in sample no. 5 (WB 1.26) and sample no. 4 (WB 0.99). A higher concentration of zinc in English plantain than in the soil was also recorded in sample no. 4 (WB 1.38).

During the analysis of the values of the bioaccumulation coefficients of copper and zinc in maize (*Zea mays* L.) it was observed that copper is essentially cumulated in biomass. The majority of the accumulation of copper takes place in the root of maize (WB 2.11–3.11) and in its leaves (WB 1.74–2.59). In the seeds of maize the copper concentration (WB 1.09) is comparable to copper content in the soil. In the case of zinc, only in one sample no. 3b higher accumulation was noted in the roots of maize than in the soil (WB 2.12).

4 Conclusions

- Soils from the area around the landfill of metallurgical waste in Siechnice have alkaline character. Such pH limits the ability of the activation and migration of heavy metals contained in the soils deeper into the soil profile. The only exception is sample no. 3 from the cropland near the heap of metallurgical waste. The agrotechnical treatments and the fertilizing of the soil are the reasons for the slightly acidic character of soil no. 3. This creates a potential risk of migration of heavy metals into the soil profile.
- Essentially, the tested soils are of slight and medium salinity. The exception is sample no. 3a, of which the EC value was of 102.32 mS/m. Such high salinity may be due to the accumulation of mineral compounds, coming from the rainwater flowing of the neighborhood areas and also the provincial road no. 95, in the decline of land.
- Chromium has been exceeded in sample of soil no. 5 (276.49 mg Cr/kg of dry matter) compared with the Polish Ministry of the Environment Regulation from 2016, which may be a result of the ongoing works on the heap of metallurgic waste in Siechnice.
- In the six soil samples chromium, copper and zinc were in excess as compared to the geochemical background (Wn). In soil no. 5 exceeding of the concentration of chromium is nearly 7-fold and nickel was exceeded about 1.25-times. Exceeding of copper and zinc in soil no. 1 is approx. 3.5-fold and 2.3-fold, respectively. There were no exceedances of lead in the soil samples.
- In plants excess of the content of copper and zinc in the biomass was observed, in comparison to the limit concentrations of these metals in forage plants [10]. In sample no. 2 of the English plantain (*Plantagolanceolata* L) almost 5-fold higher concentration of copper and almost 3-times higher content of zinc were reported. The values of the bioaccumulation coefficient (WB) of copper in maize (*Zea mays* L.) are lower. Exceeding of metal in roots and leaves of the maize was approx. 3-fold. The content of zinc was exceeded above 2-times only in the roots of *Zea mays* L. in sample no. 2.

References

1. A. Biłyk, A. L. Kowal, Environmental Pollution Control, **1-2** (1993)
2. A. Hołtra, D. Zamorska-Wojdyła, Environ Prot Eng **41**, 3 (2015)
3. A. Hołtra, D. Zamorska-Wojdyła, Environ Prot Eng **40**, 4 (2014)
4. A. Kabata-Pendias, H. Pendias, *Trace elements in soil and plants* (CRC Press, Inc., USA, 2011)
5. J. Antonkiewicz, Cz. Jasiewicz, Acta Sci. Pol. Technol., **61** (2003)
6. A. Baran, Proceedings of ECOpole **5**, 1 (2011)
7. J. Minczewski, Z. Marczenko, *Analytical chemistry* (PWN, Warszawa, 2011)
8. The Polish Ministry of Environment Regulation from September 1st, 2016, Journal of Laws 2016 pos. 1395
9. A. Kabata-Pendias, H. Pendias, *Biogeochemistry of trace elements* (PWN, Warsaw, 1999)
10. The Polish Ministry of Agriculture and Rural Development Regulation from January 23rd, 2007, Journal of Laws 2007 no. 20 pos. 119

Forecasting water use in selected water supply system

Krzysztof Boryczko[1,*]

[1]Rzeszow University of Technology, al. Powstancow Warszawy 6, 35-959 Rzeszow, Poland

Abstract. The study analyzed data on water consumption in selected municipal water supply system. The forecast was made based on daily water consumption from the period of five years (2012 ÷ 2016). The results are a valuable information source for municipal water supply system operator. Water consumption forecast allows to schedule the amount of water that should be produced e.g., taking into account day of the week, month, and the least onerous terms for water consumers to make repair or modernization of water supply system elements.

1 Introduction

By definition, forecasting, other word: prediction, is the rational prediction of events in a scientific manner. Otherwise, it is a possibility to conclude about so far unknown events based on known events that occurred in the recent past [1]. The primary function of the forecast is to provide the most objective solutions, which relate to the anticipated event occurring in the future [2-4]. Basic functions of the forecasts are [6, 7]:

- dissecting, which creates prerequisites to take rational decisions,
- activating - stimulating to take action that favours the execution of forecast,
- information.

The forecast can be presented in view of the so-called forecast horizon, i.e. the period for which it was made. In this classification we distinguish:

- immediate, direct forecast which does not exceed 1 month,
- short-range forecast, covering 1 - 3 months,
- medium-range forecast which does not exceed 2 years,
- long-range forecast, made for more than 2 years.

Classification of forecasts can also be shown in view of the nature (e.g. simple or complex, single or repeatable, quantitative or qualitative), in view of the degree of details (general or detailed) or in view of the purpose of conducting (research, normative or active or passive) [7].

The issue raised in the work, associated with forecasting, requires an appropriate reference to Polish and foreign literature [7-25]. In recent years, the methods of forecasting rapidly develop and are used in practice more and more often. The essential problem of the subject - forecasting water use, turned out to be sporadically reported in the literature. In the

* Corresponding author: kb@prz.edu.pl

study we analysed data on water consumption in the municipality in south-east of Poland. Details of daily water consumption, on the basis of which the simulation was made, are for 5 recent years (2012 – 2016). To forecast water consumption the method of exponential smoothing with seasonal additive was used.

2 Description of the water supply system

Water supply system in the municipality includes equipment for taking, treatment and distribution of water for the household, economic and industrial purposes. Municipal water supply system has the radial structure, branched, typical for water supply in rural areas – areas characterized by dispersed housing. Water supply system takes water from three underground sources and supplies water to about 16 thousand residents.

Water supply system in the municipality consists of:

- 9 tanks with total capacity 2050 m^3.
- main - the length of 2 km, diameters 315 mm, 225 mm, 160 mm,
- distribution - 158.1 km length, diameter 110 mm, 90 mm.

Whole network is made of PVC.

3 The method of exponential smoothing with seasonal additive

The equation describing the exponential smoothing model with additive seasonality:

$$\begin{cases} F_t = \alpha \cdot (y_t - C_{t-r}) + (1-\alpha) \cdot F_{t-1} \\ C_t = \gamma \cdot (y_t - F_t) + (1-\gamma) \cdot C_{t-1} \end{cases} \quad (1)$$

where:

y_t - the forecast variable at a moment or period of time t,
F_t - smoothed evaluation of the level (mean value) at a moment or period of time t,
C_t - evaluation of seasonality index at a moment or period of time t,
α - smoothing parameter of forecast variable level with values from the range (0, 1],
γ - evaluation parameter of seasonal index with values from the range (0, 1],
r - the length of a seasonal cycle (number of phases of each cycle).

The purpose of equation (1) is to determine a smoothed value of the entire time series and to determine the index of seasonality.

The parameters α and γ can not be equal to 0 and the sum of the weights should be equal to one.

The use of this model on the basis of the time series consisting of n-elements relies on calculating the following equation:

$$y_t^* = F_n + C_{T-r}, \; for \; T = n-1,...,n+1 \quad (2)$$

wherein the coefficients F_n and C_{T-r} are calculated from the following formula:

$$F_n = \alpha \cdot (y_n - C_{n-r}) + (1-\alpha) \cdot F_{n-1} \quad (3)$$

$$C_{T-r} = \gamma \cdot (y_{T-r} - F_{T-r}) + (1-\gamma) \cdot C_{T-2r} \quad (4)$$

To perform the forecast, the necessary are values F_n and C_n, which are dependent on the F_{n-1} and C_{n-r}. When calculating the forecast y_t* the expired forecasts must be determined:

DOI: 10.1051/e3sconf/20171700010

$$y_t^* = F_{t-1} + C_{t-r},\ for\ t > r \tag{5}$$

where:

$$F_{t-1} = \alpha \cdot (y_{t-1} - C_{t-1-r}) + (1-\alpha) \cdot F_{t-2} \tag{6}$$

$$C_t = \gamma \cdot (y_t - F_t) + (1-\gamma) \cdot C_{t-r},\ for\ 2r \geq t > r \tag{7}$$

$$C_{t-r} = \gamma \cdot (y_{t-r} - F_{t-r}) + (1-\gamma) \cdot C_{t-2r},\ for\ t > 2r \tag{8}$$

Before the calculation, the initial values $C_1, \ldots, C_r$ and F_r should be taken. The values should be adopted for the first cycle in accordance with the equation:

$$C_i = y_i - \frac{y_1 + \ldots + y_r}{r},\ for\ i = 1, \ldots, r \tag{9}$$

The total sum of values of $C_1, C_2, \ldots, C_r$ should be equal to 0:

$$\sum_{i=1}^{r} C_i = 0 \tag{10}$$

Value F_r is calculated according to the expression:

$$F_r = \frac{y_1 + \ldots + y_r}{r} \tag{11}$$

Table 1. shows the mechanism of calculations using EXCEL.

$$F_7 = \frac{y_1 + y_2 + y_3 + y_4 + y_5 + y_6 + y_7}{r}$$

$$C_1 = y_1 - \frac{y_1 + y_2 + y_3 + y_4 + y_5 + y_6 + y_7}{r}$$

Table 1. The mechanism of calculation

no.	day	month	water consumption, y	F_t	C_t	y_t*
1	1	January 2012	936	-	-161	-
2	2		381	-	-191	-
3	3		1228	-	53	-
4	4		980	-	-167	-
5	5		1552	-	405	-
6	6		961	-	-106	-
7	7		1042	1097	-13	-
8	8		1458	1202	48	936
9	9		1005	1200	-193	1011
10	10		1032	1156	-36	1253
11	11		981	1155	-170	989
12	12		987	1040	176	1559
13	13		883	1030	-126	935
14	14		810	989	-96	1016
15	15		1396	1060	192	1036

$$F_8 = \alpha \cdot (y_8 - C_1) + \\ +(1-\alpha) \cdot F_7$$

$$y_8^* = F_8 + C_1$$

1462	1	January 2015	745	1024	-32	1361
1463	2		1049	1079	-142	770
…						
1822	27	December 2016	1020	1216	-143	1153
1823	28		1307	1233	39	1220
1824	29		1112	1214	-63	1210
1825	30		1126	1171	40	1340
1826	31		1301	1153	184	1391
Forecast						
1827	1	January 2017	-	-	-	991
1828	2		-	-	-	882
1829	3		-	-	-	1189
1830	4		-	-	-	912
1831	5		-	-	-	902
1832	6		-	-	-	963
1833	7		-	-	-	1099

$$y^*_{1827} = F_{1826} + C_{1462}$$

$$y^*_{1828} = F_{1826} + C_{1463}$$

4 Forecasting water consumption in selected municipality

Data on water consumption for the period 2012 ÷ 2016 and the forecast for 2017 are characterized by fluctuations in water consumption with clearly shaping seasonality in the summer months.

Table 2 shows the average value of the data (on the basis of which the forecast for 2012 ÷ 2016 was carried out) for each day of the week.

Table 2. Average value of the water consumption, variations, mean deviation for 2011 – 2015.

day of the week	water consumption, m^3	σ^2	σ
Mon	1044	94288	307
Tue	1196	106206	326
Wed	1200	100183	317
Thu	1210	105111	324
Fri	1200	112284	335
Sat	1187	138509	372
Sun	1557	118600	344

On the basis of Table 2, it can be observed that the increase in water consumption took place at the weekend and the highest value for the week was reached on Sunday. This is due to lifestyle of the recipients – the longer time spent at home, compared to weekdays, and house chores. Then, on Monday, water consumption was reduced from 1557 m^3 to 1044 m^3, and successively during a week it was about 1,200 m^3.

Four years data was used for building the model and forecast for 2016 was made to compare it the forecasted data with historical data from 2016 (Fig. 1).

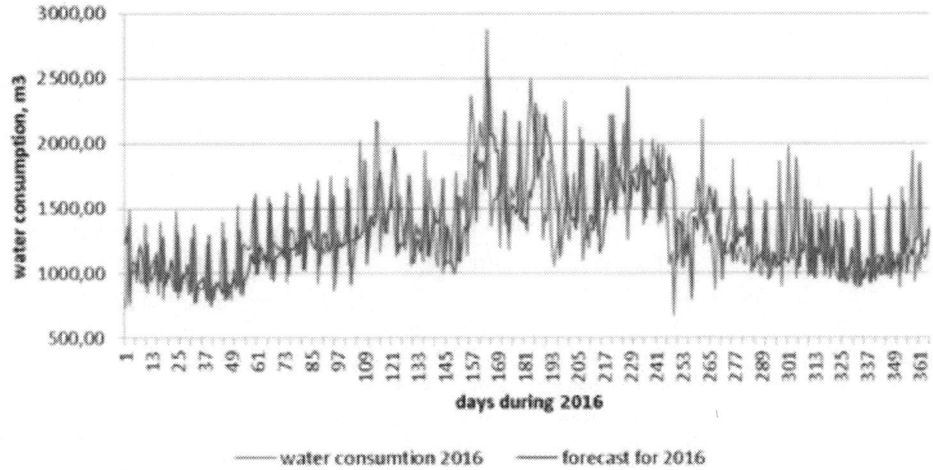

Fig. 1. Average water consumption during forecast period (2017).

Prepared model correctly predicts future water consumption. Figure 2 presents water consumption forecast for 2017.

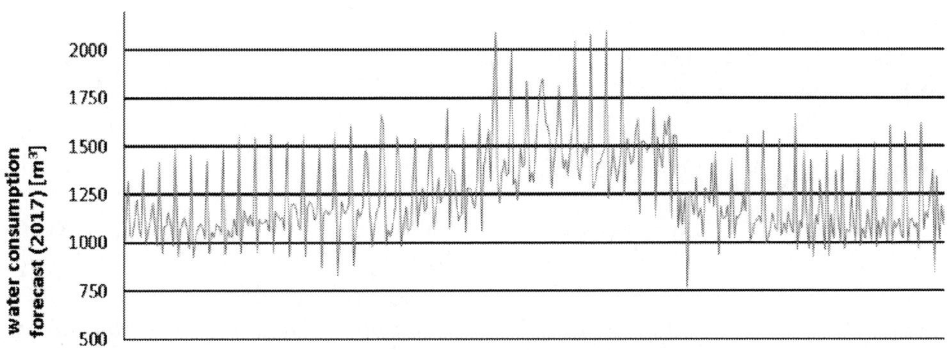

Fig. 2. Average water consumption during forecast period (2017).

For comparison, the same data as Table 2. was created for the forecast of water consumption for 2017 on Fig. 3.

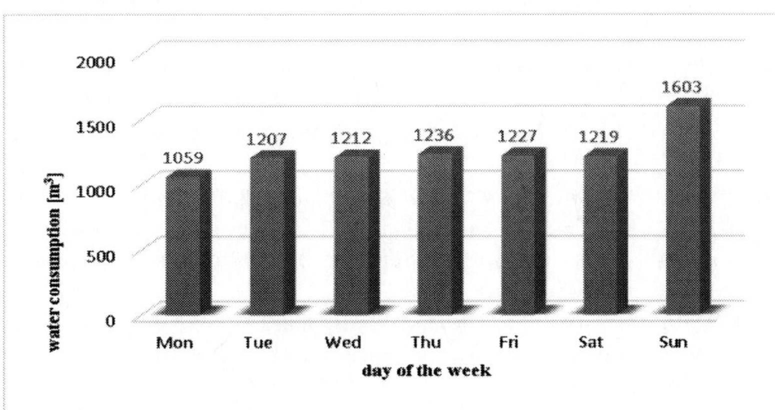

Fig 3. Average water consumption during forecast period (2017).

5 Conclusions

Data on water consumption for the period 2012–2016 and the forecast for 2017 are characterized. After analysing the results and comparing them to the previous water consumption, it was stated that the forecasting model was correct. The results of the method of exponential smoothing with seasonal additive as well as water consumption in the past, oscillate around the trend line and the apparent increase in water consumption also takes place during the summer (May–August).

The method meets the principle of the easy availability of output data that is convenient from a practical point of view. These data do not include external factors (for example environmental) – variables, but they are based on chronological data of the past.

In summary, this paper presents a model of the time series, which is a simple and flexible tool for forecasting. The basic argument, which supports the use of the above models is mainly good quality of water consumption forecast and hence a better quality of water distribution services. With the appropriate series of past data forecasting demand for water is possible. Choice of the correct methods and properly conducted forecast allow to draw the appropriate conclusions, thanks to which it is possible to improve the operation of water supply systems.

The paper presents the simple forecasting model that is a universal and fast tool to predict the demand for water in water supply system on the basis of the collected data. Additionally described model can be used during the operation of the water companies to plan, among others:

- the purchase of reagents for treatment
- repairs related to the shutdown of water supply
- costs of pumping water.

References

1. B. Pawełek, A. Zeliaś, S. Wanat, *Prognozowanie ekonomiczne, Teoria przykłady zadania* (Wydawnictwo Naukowe PWN, Warszawa, 2003)
2. K. Pietrucha-Urbanik, B. Tchórzewska–Cieślak, M. Urbanik, *Analysis of the gas network failure and failure prediction using the Monte Carlo simulation method*, Eksploat Niezawodn **18**, 254-259 (2016)
3. K. Pietrucha-Urbanik, *Failure analysis and assessment on the exemplary water supply network*, Eng Fail Anal **57**, 137-142 (2015)
4. J.P. Leitao, A. Martins, C. Amado, *Comparative Study of Three Stochastic Models for Prediction of Pipe Failures in Water Supply Systems*, J Infrastruct Syst **19**, 442-450 (2013)
5. X. Jing, *An integrated prediction model for water supply-demand ability*, 4th International Conference on Advanced Materials and Information Technology Processing (AMITP), ACSR-Advances in Comptuer Science Research, Guilin, 2016, 60, 528-531.
6. J. Han, *The Analysis and Prediction of Water Supply and Demand for Beijing in future*, 3rd International Conference on Education, Management and Computing Technology (ICEMCT), Advances in Social Science Education and Humanities Research, Hangzhou, 2016, 59, 1050-1053
7. J. Łomotowski, Z. Siwoń, W. Cieżak, P. Licznar, J. Cieżak, *Analizy i prognozowanie rozbiorów wody w systemach wodociągowych* (Komitet Inżynierii Lądowej i Wodnej PAN, Warszawa, 2008)
8. J. Ciezak, W. Ciezak, *Routine forecasting of the daily profiles of hourly water distribution in cities. An effectiveness analysis*, Environ Prot Eng **41**, 179-186 (2015)

9. Z. Siwoń, W. Cieżak, J. Cieżak, *Water demand modeling for housing estates*, Och Śr **30**, 23-28 (2008).
10. J. Cieżak, Z. Siwoń, W. Cieżak, *Praktyczne aspekty badań strat wody w sieciach wodociągowych*, Och Śr **4**, 25-30 (2004)
11. A. Studzinski, K. Pietrucha-Urbanik, Analysis of water pipe breakage in Krosno, Poland, *Environmental Engineering IV*, Dudzińska M. R. Pawłowski L., Pawłowski A., Editor, London: Taylor & Francis Group, 59-62 (2013)
12. B. Tchórzewska-Cieślak, D. Szpak, *A Proposal of a Method for Water Supply Safety Analysis and Assessment*, Och Śr **37**, 43-47 (2015)
13. B. Tchórzewska-Cieślak, Urban Water Safety Management, *Chemical Engineering Transactions*, V; DeRademaeker Cozzani, E; Pierucci, S; Klemes, JJ, Editor, Milano, ITALY: AIDIC SERVIZI SRL, 201-206 (2012)
14. B. Tchórzewska-Cieślak, *Estimating the acceptance of bearing the cost of the risks associated with the management of water supply system*, Och Sr **29**, 69-72 (2007)
15. J. Rak, B. Kucharski, *Slude management in water treatment plants*, Environ Prot Eng **35**, 15-21 (2009)
16. K. Boryczko, B. Tchórzewska-Cieślak, I. Piegdoń, Possibilistic risk analysis of failure in water supply network, *Safety and Reliability: Methodology and Applications*, Nowakowski et al. (Eds), Editor, London: Taylor & Francis Group, 1473-1480 (2015)
17. K. Boryczko, *Water age in the water supply network as health risk factor associated with collective water supply*, Ecol Chem Eng A **23**, 33-43 (2016)
18. A. Kotowski, B. Kaźmierczak, M. Wdowikowski, Trend Analysis of Annual and Seasonal Precipitation Amounts in the Upper Odra Catchment, Och Sr **36**, 49-54 (2014)
19. M. Kutylowska, Neural network approach for failure rate prediction, Eng Fail Anal **47**, 41-48 (2015)
20. A. Studzinski, K. Pietrucha-Urbanik, *Selected issues of costs and failure of pipes in an exemplary water supply system*, Rocz Ochr Sr **18**, 616-627 (2016)
21. L. Liu, B. Wang, G. H. Huang, et al., *Forecast-based analysis for regional water supply and demand relationship by hybrid Markov chain models: a case study of Urumqi, China*, J Hydroinform **18**, 905-918 (2016)
22. K. Boryczko, A. Pasierb, *Method for forecasting the failure rate index of water pipelines Environmental Engineering V*, M. Pawlowska, L. Pawłowski Editor, London: Taylor & Francis Group, 15-24 (2017)
23. J. Kaleta, D. Papciak, A. Puszkarewicz, *Removal of Ammonia Nitrogen from Groundwater on Chalcedony Deposits in Two-stage Biofiltration Process*, Rocz Ochr Sr **15**, 1352-1366 (2013)
24. B. Tchórzewska-Cieślak, K. Boryczko, *Analysis of risk of failure in water main pipe network and of developing poor quality water*, Environ Prot Eng **40**, 77-92 (2014)
25. J. Bajer, Economical and reliability criterion for the optimization of the water supply pumping stations designs, *Environmental Engineering IV*, Dudzińska M. R. Pawłowski L., Pawłowski A., Editor, London: Taylor & Francis Group,jhio9yu6j 21-28 (2013)
26. E. Szymura, I. Zimoch, *Operator reliability in risk assessment of industrial systems function,* Przem Chem **93**, 111-116 (2014)

The studies on waste biodegradation by *Tenebrio molitor*

Magdalena Bożek[1], *Beata* Hanus-Lorenz[1], and *Justyna* Rybak[1,*]

[1]Wrocław University of Science and Technology, Department of Environmental Engineering, Wybrzeże Wyspiańskiego 27, 50-370 Wrocław,

Abstract. As cities are growing in size with a rise in the population, the amount of plastic waste generated is increasing and becoming unmanageable. The treatment and disposal of plastic waste is an urgent need of our present and future. It has been proved recently that mealworms, the larvae of *Tenebrio molitor* Linnaeus, are able eat styrofoam, a common polystyrene product. Polystyrene is one of the most widely used plastics, the scale of its production being several million tons per year. *Tenebrio molitor* is one of the largest pests found in stored-grain products. The insect is indigenous to Europe, but is currently cosmopolitan in distribution. The styrofoam is efficiently degraded in the larval gut by microorganisms. We have used the larvae of *T. molitor* to biodegrade three types of food packaging plastics: polystyrene (PS), polyvinyl chloride (PVC) and polylactide (PLA). PVC is a thermoplastic made of 57% chlorine (derived from industrial grade salt) and 43% carbon (derived predominantly from oil /gas via ethylene). It is the world's third-most widely produced synthetic plastic polymer, which is not biodegradable easily. On the other hand, PLA is an easily biodegradable and bioactive thermoplastic aliphatic polyester derived from corn and tapioca starch or sugarcane. Three groups of larvae were fed selected types of polymers as an only food, while a control population was fed on oatmeal. The mass loss, dry matter content and biochemical composition of mealworms were assessed in the performed laboratory experiments. The protein concentration in homogenates of the larvae was determined by the Bradford method. To determine the level of hydrolized carbohydrates we used anthrone method. The classical sulfo-phospho-vanillin assay (SPVA) was used to quantitate total lipids in mealworms. The results allowed to compare the decomposition efficiency of selected polymer materials by mealworms and to recognize the mechanism of decomposition contributing to the future use of these animals for the treatment and disposal of plastic waste.

1 Introduction

One of the most persistent pollutants which are very resistant to biodegradation in the environment is plastics. Their accumulation in terrestrial habitats can be dangerous as

* Corresponding author: justyna.rybak@pwr.edu.pl

plastics pose the serious threat to the environment health. Plastics can be degraded by chemical, thermal, photo or biological degradation [1].

Polystyrene (PS) is one of the most difficult material for biodegradation. There are only few examples of its biodegradation, but at the same time it is commonly used and widespread. The production of PS reaches several million tons per year [2]. In the recent studies scientists have been proved that styrofoam can be degraded effectively by bacteria which were discovered in the gut of larvae of *Tenebrio molitor* Linnaeus [3]. This is the unique evidence that mealworms can be employed in the biodegradation of plastics.

Polyvinylchloride (PVC) is a very strong material, resistant to abrasion PVC consisting of 57% chlorine (from industrial grade salt) and 43% carbon (from oil /gas via ethylene). It is commonly used in pipes, hoses and floor coverings. PVC can be decomposed by chemical and phytodegradation but degradation with the use of bacteria have been observed extremely rare [1]. The experiments concerning the biodegradation of PVC with bacteria which are present in the gut of *T. molitor* have never been performed.

Thus, the aim of our studies was to compare the efficiency of biodegradation of two plastics which belong to the most resistant and almost non-biodegradable materials. The mealworms were employed for the waste biodegradation. As a comparison the third material, polylactide (PLA), was used. PLA is a natural, biodegradable plastic, made from renewable resources (such as starch, corn or sugarcane) which can be naturally produced or synthesized from renewable resources [4]. This is a sustainable alternative to non-renewable petroleum- based synthetic biodegradable plastics. The lactides present in this material can be easily produced by the microorganisms in the fermentation process of agricultural by-products, mainly the carbohydrate rich substances [4]. The mealworms were fed these three materials (PS, PVC, PLA). Here, we report evidence that biodegradation of PS and PVC in comparison to PLA and observation made for control insects does occur. This observation was based on the mass loss of studied waste. What is more, feeding on these food sources can theoretically bring energy to mealworms, thus the effect on lipid, total sugar and carbohydrates contents of larvae was also examined under laboratory conditions. Additonally, we also studied the starved mealworms as a comparison to larvae fed biodegradable and non-biodegradable plastics.

2 Materials and methods

2.1 Materials

Tenebrio molitor is a species of darkling beetles which belongs to Tenebrionidae family. Like all holometabolic insects they undergo four life stages: egg, larva, pupa and imago. Larvae measure about 2.5 cm, whereas adults have 1.25–1.8 cm length. Adults individuals lay eggs in the soft ground and then die. The optimum temperature for breeding is within the range of 25–35°C. In natural conditions, beetles are found in decaying and rotting bark of deciduous trees. The larvae and adult beetles feed on dead organisms and live in warehouses cereal products, pantries, poultry farms and dovecotes. In this experiment larvae of *T. molitor* came from the inbreeding farm in Lublin (Poland).

2.2 Methods

Experiment was carried out in the parallel 5 groups (20 specimens each, the parameter for selecting 20 specimens in each group was their similar mass). Larvae were weighted weekly. The variants used in tests were as follows:

- larvae eating oatmeal *ad libitum* (control);

- larvae eating polystyrene (PS) in the form of 1mm thick plates manufactured in the laboratory (raw material);
- larvae eating polylactide (PLA) in the form of 1mm thick plates manufactured in the laboratory (raw material);
- larvae eating polyvinyl chloride (PVC) for medical purposes, as commercially available tubing for oxygen supply cut in 10mm pieces;
- starved larvae.

After 3 weeks larvae were homogenized using cold mortar and pestle with ice-cold 0.1M phosphate buffer pH 7.4, 1mM EDTA (1:10 w/v). For carbohydrates assay aliquots of homogenates were frozen directly for further analysis. Required aliquots for lipid assay were treated with chloroform: methanol (1:2 v/v). The rest of homogenate were centrifuged at 11 000 x g, 5 min and frozen in aliquots at -20°C in 2ml Eppendorf vials for protein assay.

Firstly, we determined the mass loss of studied larvae and materials used. Secondly, we studied biochemical composition of mealworms. Protein content was assessed with the Bradford assay [5]. To determine the level of hydrolized carbohydrates we used anthrone reagent [6] and measure the absorbance of samples at 625 nm. The classical sulfo-phospho-vanillin assay (SPVA) was used to quantitate the total lipids in mealworms [7]. Absorbance was measured spectrophotometrically at 525 nm. In all assays multipoint standard curves with high linear regression (R2: protein curve = 0.91; carbohydrate curve = 0.99; lipids curve = 0.99) were used. All statistical analyses were performed in Statistica ® software package. Differences in mass change of mealworms (for five food types) and mass loss for three types of materials with the duration of exposure were tested for each factor using analysis of variance (ANOVA).

3 Results and discussion

3.1 Mass loss measurements

We determined the percentage of mass changes of larvae and mass loss of used materials after 7, 14 and 21 days from the beginning of the experiment. The results are presented in appropriate Tables (Tab. 1, 2) and Figures (Fig. 1).

Table 1. Mass changes of mealworms caused by waste eating and starvation with the comparison of control larvae (g) "–" means decrease of the mass.

	Mass changes of *T. molitor* after 21 days (%)
Control	45
PS	-18
PLA	-15
PVC	-19
Starved larvae	-16

Table 2. Mass loss of materials used in studies (g).

	Mass (g) 1 day	Mass (g) 21 day	Mass loss after 21 days (%)
PS	1,239	1,132	9
PLA	0,537	0,474	12
PVC	2,052	1,995	3

The results show the biggest mass change (19%) for studied larvae fed PVC (Tab. 1, Fig. 1). We also observed the lowest mass loss concerning the mass of PVC (Tab. 2). This is probably due to unavailability of this material to mealworms. PVC for medical purpose was characterised by the smooth surface which make difficult to chew and eat this material by mealworms. They started eating on cutting edges as tubing was served in the form of 10mm pieces. Although, we recorded small but noticeable mass loss of PVC (3%) after 21 days of exposure to mealworms eating, which is interesting and promising in the context of further studies. The masses loss of studied materials were not significantly different ($P > 0.05$). The significant increasing trend in mass change (45%) was observed in the control sample (Tab. 1) which is not surprising. As it was expected, we observed the lowest mass change of larvae for PLA consumption (Table 1) and the highest mass loss of PLA alone (Tab. 2) (Fig. 1). The significant differences were observed only between control mealworms eating oatmeal and larvae fed PVC ($P < 0.05$). What is suprising, the mass of larvae is greater in starved condition compared to mass of larvae fed on PVC. This is probably due to unavailability and difficulty in chewing of PVC, the larvae lost more energy for make the PVC available for consumption and as a result they lost more mass than starved ones which were waiting for the food without unnecessary energy loss.

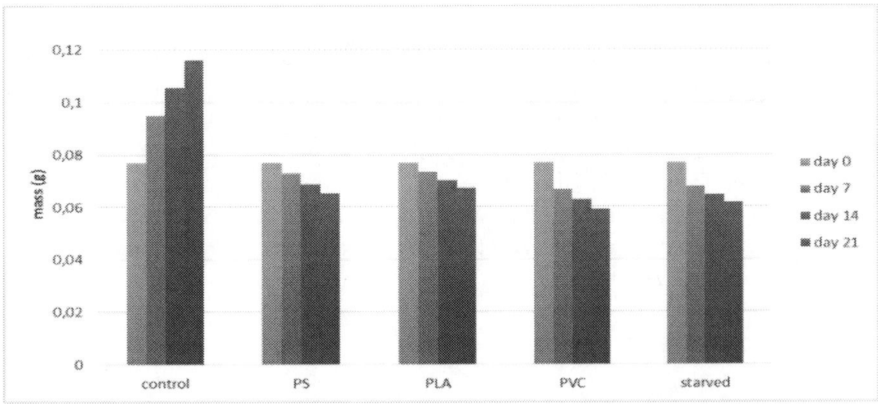

Fig. 1. The average mass of larvae after 7, 14 and 21 days of the experiment (g).

3.2 Biochemical characteristic of mealworms

3.2.1 Protein content

The protein content in the tissues of mealworms not varied significantly depending on the food type. The results are shown in Fig. 2. As the proteins are the main building material for the body of insect they can be used in the hydrolyzed form as fuel for metabolism or converted to glucose through gluconeogenesis. As our results show protein levels protein content varied between 16.01–23.38 ng/mg and were not significantly affected by the food type. Some authors noticed that protein synthesis decreases during food deprivation [8, 9], leading to a decrease in protein levels as observed here in starved larvae and mealworms fed PS, PVC and PLA. Surprisingly, the highest amount of protein (excluding control) was recorded in larvae fed PLA which may be a result of sparing the energy as a form of adaptation to the difficult conditions and food unavailability. As it was suggested by Renault at al. [8] an insect's recovery from starvation may depend upon its mobility and readiness for searching food, thus a compromise between the demand for carbohydrates and the preservation of muscle proteins is needed. On the other hand, such materials like PVC and PS are not a good source of energy at all which lead to the depletion of proteins in mealworms fed both types of waste.

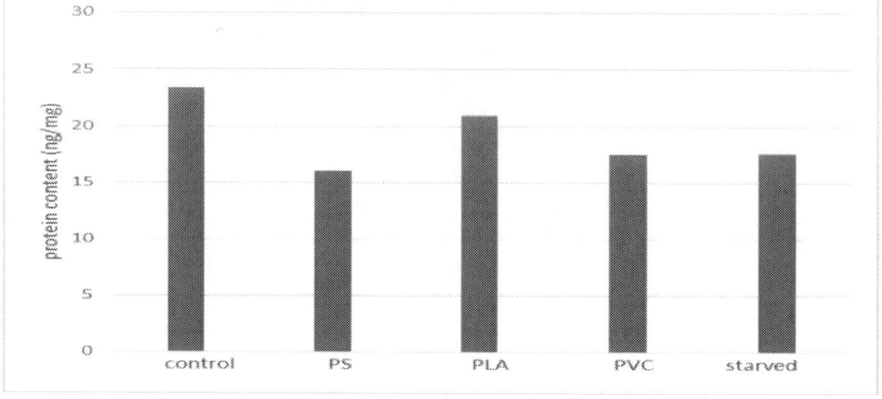

Fig. 2. The protein content in mealworms fed different types of food and starved (ng/mg).

3.2.2 Carbohydrates content

Carbohydrates provide most of the energy during insect life. Some insects metabolize glycogen during the initial stages of starvation, then switch to lipid and protein metabolism when carbohydrates are gone. Other species, however, depend more heavily on lipid metabolism during exposition to stress factors [10]. The highest amount of carbohydrates was registered in mealworms fed PLA where the lowest mass change of larvae was observed simultaneously. The lowest carbohydrates content was registered for mealworms fed PVC an PS as these materials are not reliable food source for development and life cycle completion (Fig. 3). As it was shown in the experiment the carbohydrates content was not significantely depleted comparing to the control mealworms fed oatmeal.

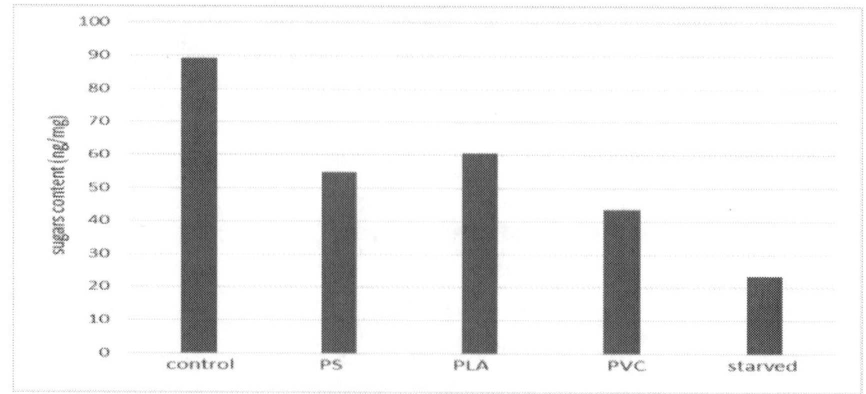

Fig. 3. The carbohydrates content in mealworms fed different types of food and starved (ng/mg).

3.2.3. Lipid content

The most significant changes during stress and starvation are usually observed in the levels of triglycerides, as the level of these compounds decreases significantly. The most common source of energy in insects during starvation period is the oxidation of fatty acids stored in the form of triglyceride, large amounts of lipids are accumulated in insect fat body, with fat content exceeding 50% of the total dry mass in selected studied insects [10]. Levels of stored fat significantly help to survive during starvation and environmental stress, the level of triglyceride is always connected with the duration of starvation period [10]. In our studies the amount of lipids is the highest in the control mealworms fed oatmeal which is in the line with the observations of other authors suggesting that starvation is connected with the rapid fat reserves burning [10] (Fig. 4). Larvae fed PLA have the highest amount of lipids comparing to mealworms fed PVC and PS which suggest that although they decompose these materials they do not ingest it and use the energy effectively for life activities.

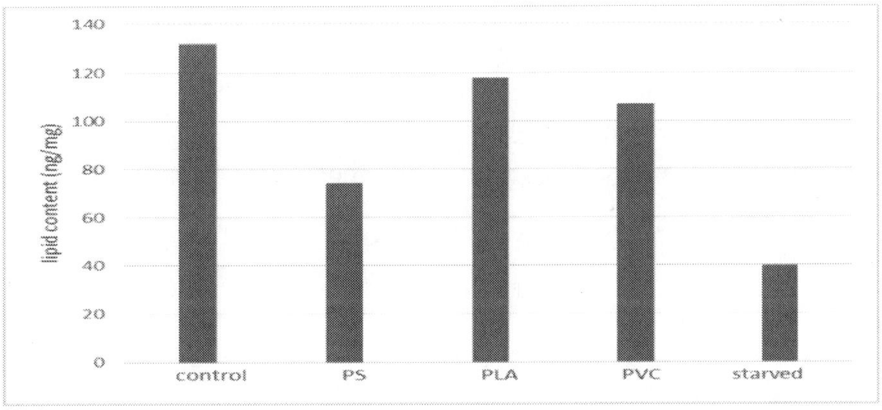

Fig. 4. The lipid content in mealworms fed different types of food and starved (ng/mg).

4 Conclusions

In our tests on processing commonly used plastics with *T. molitor* larvae we recorded that all specimens survived three weeks of experiment period, although they required much more effort to maintain their vital functions feeding PS, PVC, PLA or starving. The conducted biochemical assays confirmed that stored energy in the form of lipids and carbohydrates is used for keeping the basic metabolic rates, chewing and eating polymers and depredating them by mealworms. We are aware that the ability of plastics bioprocessing by beetle larvae is strictly connected with the adaptive predisposition of gut microbiome of mealworms. The mass loss of all studied materials (PS, PVC, PLA) suggest that they are eaten by larvae and decomposed efficiently by microorganisms presented in the gut of mealworms. On the other hand, the mass changes of studied specimens indicate that all of these materials are not very efficient source of energy for mealworms, they only allow them to survive but not to deliver a sufficient energy as mealworms lost weight constantly. As it was expected, the lowest mass loss was recorded in the mealworms fed PLA, which is a biodegradable polyester. Although the PVC was not eaten and degraded greatly by mealworms, the obtained results suggest that this material is also biodegradable by bacteria which inhabit the gut of mealworms. Such studies have never been performed yet.

The interest of scientists in *T. molitor* larvae is mainly due to their nutritional potential. Additionaly, their ability to process biodegradable materials as well as hardly biodegradable and even non-biodegradable products can be a key feature in further studies and in the future application of these animals for the treatment and disposal of plastic waste which is extremely beneficial for environment health.

References

1. M. Aruna, V. Shanthi, Int. J. Pharm. Sci. Rev. Res. **36**, 5 (2015)
2. J. Maul, B.G. Frushour, J.R. Kontoff, H. Eichenauer, K-H. Ott, C. Schade, *Ullmann's Encyclopedia of Industrial Chemistry* (Wiley-VCH, Weinheim, 2007)
3. Y. Yang, J. Yang, W-M. Wu, J. Zhao, Y. Song, L. Gao, R. Yang, L. Jiang, Environ. Sci. Technol. **49**, 6 (2015)
4. K. M. Nampoothiri, N. R. Nair, R. P. John, Bioresource Technol. **101**, 22 (2010)
5. M.M. Bradford, Anal. Biochem. **72**, 6 (1976)
6. E. Van Handel, JAMCA **1**: 1 (1985)
7. E. Van Handel, JAMCA **1**: 2 (1985)
8. D. Renault, F. Hervant, P. Vernon., Physiolog. Entomol. **27**, 10 (2002)
9. G. Bosquet, Comp. Biochem. Physiol. **58**, 5 (1977)
10. M.T. Marron, T.A. Markow, K.J. Kain, A.G. Gibbs, J. Insect Physiol. **49**, 9 (2003)

Higher efficiency of the cooling unit

Wojciech Cepiński[1,*]

[1]Wrocław University of Science and Technology, Faculty of Environmental Engineering, Department of Air Conditioning, Heating, Gas Engineering and Air Protection, 4/6 Norwida st., 50-373 Wrocław, Poland

Abstract. The paper describes the issue of the combined work of the cooling unit with the air direct-contact, gravel, ground heat exchanger (Polish acronym BGWCiM), patented for the Wroclaw University of Science and Technology. The options of the combined work have been discussed; the document also presents schemes and proposes the algorithms of work. The cost-efficiency of the energy harvesting from the ground by the means of the proposed solution have been analyzed and compared for two work modes, with and without the BGWCiM. Moreover, the research results and experimental measurements were presented and discussed.

1 Introduction

In a world where there is a lack of sources for energy, along with the major usage of traditional fossil fuels, it is imperative to find better solutions to use more natural energy, while replenishing the resources that are used.

When planning, designing, and innovating projects, it is important to remember the following: legal requirements, the growing knowledge and trend of saving and renewing energy, along with the positive or negative impact it will cause, regardless of how significant.

It is worth noting that out of all the energy produced in Mid-European climate conditions, 40% is used for heating and cooling buildings.

1.1 Air direct-contact, gravel, ground heat exchanger

For more than 30 years, the Wroclaw University of Science and Technology has studied the effectiveness of extraction of cheap, natural, renewable energy from a small depth of the soil, in the air direct-contact, gravel, ground heat exchanger (Polish acronym BGWCiM) for the purposes of ventilation and air-conditioning [1]

* Corresponding author: wojciech.cepinski@pwr.edu.pl

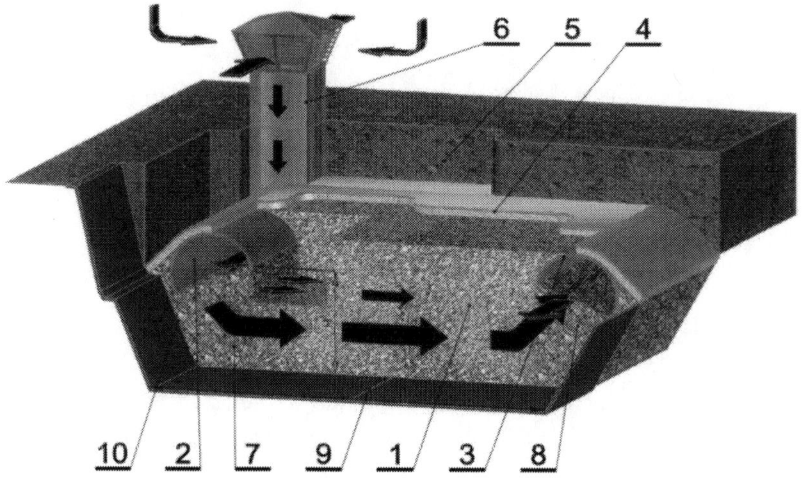

Fig.1. The air direct-contact, gravel, ground heat exchanger (Polish acronym BGWCiM)
1 – accumulating bed, 2 – conducting channel, 3 – collecting channel, 4 – heat- moisture insulation, 5 – covering blanket, 6 – draving air inlet, 7 – air distribution bed, 8 – air collecting bed, 9 – ground, 10 – geotextile [1,4].

In the Mid-European climate, the ground temperature at 4–5m deep is almost static throughout the year and is close to the average outdoor temperature (around+10 (± 1,5) °C). A device to heat in the winter and cool in the summer can be constructed by placing the heat exchanging gravel bed of appropriate volume at a smaller depth or even at a ground surface level. This is how the BGWCiM works – shown in Figure 1. It was built and patented in the 1980's by prof. Gerard Besler at the Wroclaw University of Science and Technology, and has been used in Poland and overseas. The construction is very simple and inexpensive to build.

The exchanger can be installed in practically any place of the property (Fig. 2).

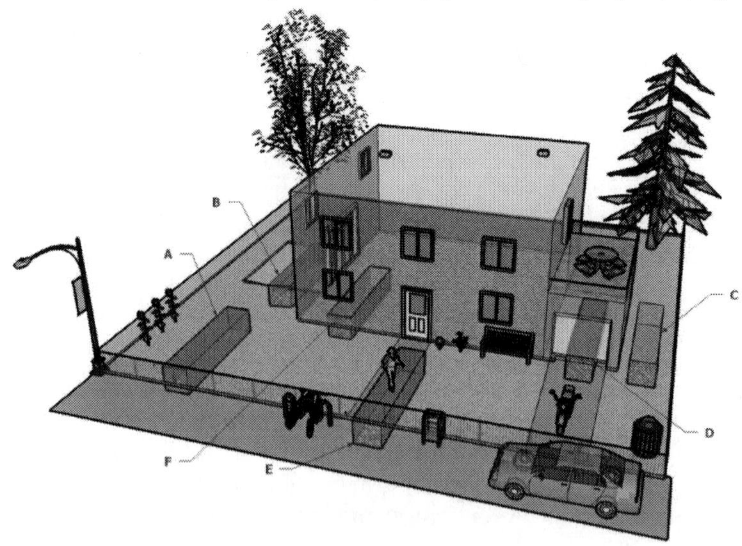

Fig. 2. Possible locations of the exchanger: A – in the lawn, B – below the terrace, C – along fences, D – below the garage, E – below the sidewalk, F – under the building [3].

In the exchanger, the outside air is supplied horizontally through a gravel bed length of 3–5 m (fig. 1) When the air flows between the gravel, the air temperature reaches the temperature of the gravel. Because of this, the temperature increases from -18°C to around 0°C in the winter, and decreases from +30°C to around +20°C in the summer. The air pressure drop is relatively small, only 100 to 250 Pa.

Figure 3 presents the average air temperature after passing through the BGWCiM, as well as the average outdoor temperatures in Wroclaw. The numbers are derived from many years of statistical measurements.

This figure can be used in the designing process, and serve as a reference to analyze the device's performance for the entire year.

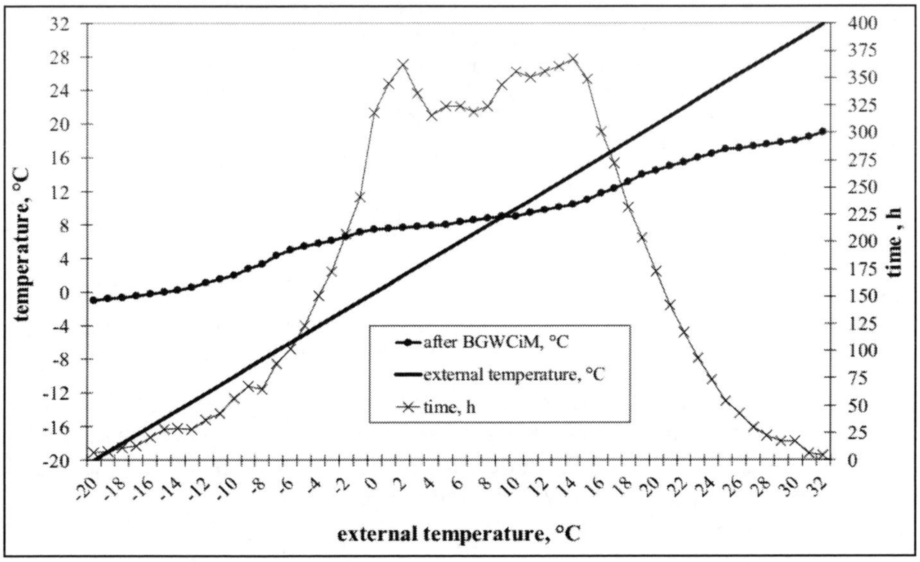

Fig. 3. Mean values of air temperature after the ground heat exchanger and occurrence time of given outdoor temperatures during the year [2, 4].

The exchanger's efficiency decreases during cold and warm periods. This is due to the changing temperature of the ground through outdoor temperature function.

This should be taken into account during the calculation, design, and installation of the BGWCiM.

1.2 Cooling units

Currently, the most popular air conditioning and cooling device that is used in Poland and overseas is the compressor unit with a condenser that is directly or indirectly cooled with atmospheric air. The condenser may be cooled directly by outside air or by water prepared in a wet cooling tower or in a dry cooler.

2 Solution idea

The original idea was to conjoin the compressor unit with the air direct-contact, gravel, and ground heat exchanger to increase the effectiveness of cooling energy production. Diagram of such a solution is shown in Figure 4.

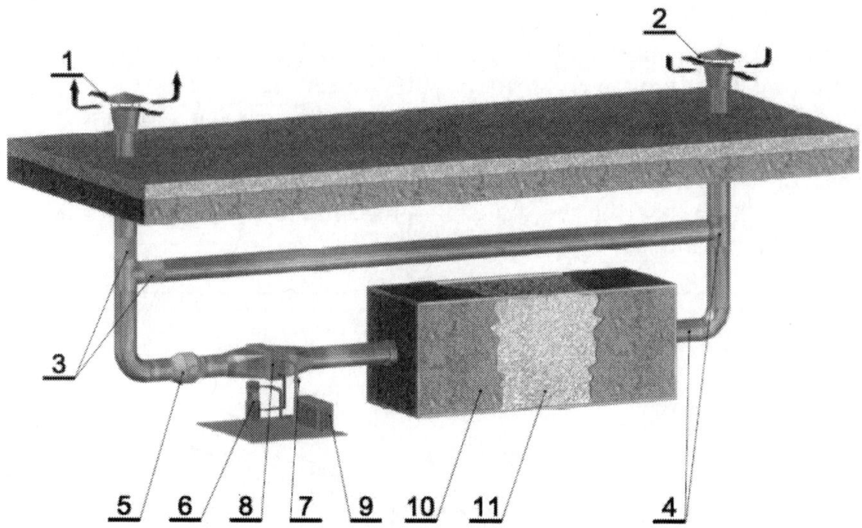

Fig. 4. Schematic diagram of the cooling unit with BGWCiM working in combination with each other, 1 – exhaust air, 2 – supply air, 3 – air damper, 4 – air ducts, 5 – fan, 6 – compressor, 7 – expansion valve, 8 – condenser, 9 – evaporator, 10 – heat exchanger, 11 – accumulating bed [3].

As a result of this, fluctuations of air temperature entering the condenser, both in the daily and seasonal cycle are considerably smaller, and improve the operating conditions and efficiency of the device. Due to the reduction of extreme parameters of the operation, it significantly increases the service life and reliability of the unit.

3 Analysis

The purpose of showing the collaboration between the cooling unit and BGWCiM is to show the fundamental benefits and risks.

The analysis was made for a single-family building located in Wroclaw. The cooling load of the building is 8,5kW.

During the warm period, the average external temperature was used to study and compare how the cooling unit performed with the BGWCiM, and how the same cooling unit worked with only external air.

The sum of internal heat gains, heat conduction, ventilation and maximum solar radiation for each outdoor temperature caused a demand for cooling energy.

3.1 Power of the cooling unit

Figure 5 presents the cooling capacity as the function of external temperature for operation with and without BGWCiM. Also, it shows the cooling demand of the building as a function of external temperature. When the cooling unit is linked with the BGWCiM, the cooling capacity increases substantially. It is noticeable during the entire warm period and even grows up to 20% during the hottest period. The cooperation of the cooling unit with BGWCiM enables to choose a smaller and cheaper unit.

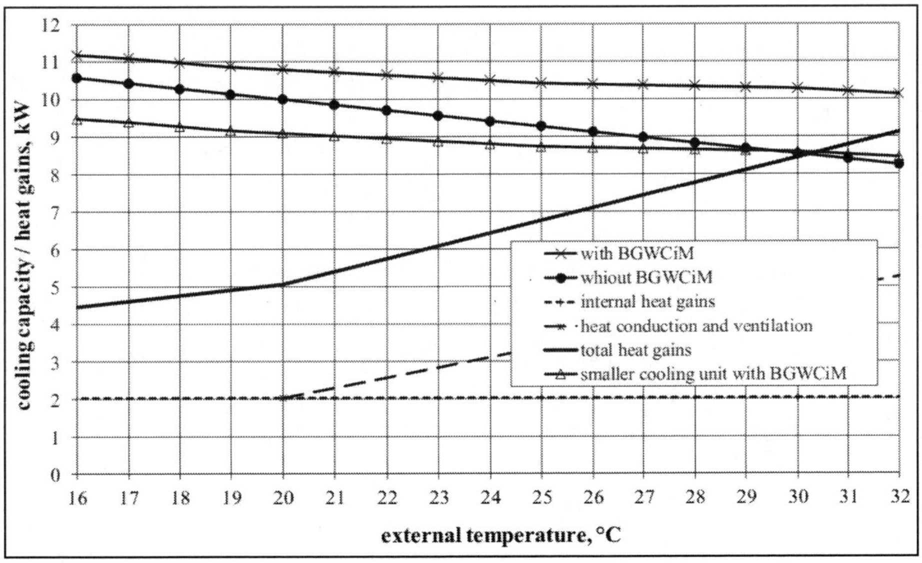

Fig. 5. Cooling capacity as the function of outdoor temperature.

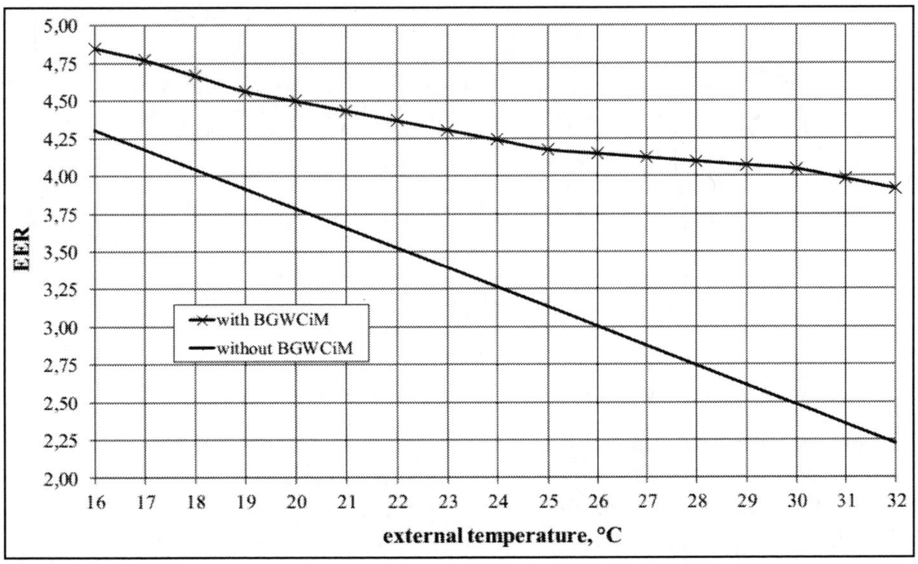

Fig. 6. EER coefficient as the function of outdoor temperature.

3.2 Energy Efficiency Ratio EER

The cooling unit's efficiency is defined by the energy efficiency ratio (EER). The EER is the ratio of the cooling capacity to the power input. The value of this coefficient depends mainly on the required temperature of the installation and the temperature of the source of energy, more specifically from the difference of these temperatures [5].

Increasing the energy efficiency ratio of the cooling unit in cooperation with BGWCiM in relation to the operation using the external air is significant for all operating temperatures. In the warmest moments EER is almost doubled and then cooling needs are highest. Figure 6 presents the EER coefficient as the function of external temperature.

4 Experimental data

The measurements were performed in the laboraory located in Wrocław. The graphs below present some examples of the experimental results obtained within the 2-year period of exploitation.

Figure 7 and 8 display exemplary results of the measurements from two different temperature phases during a summer period - the first and second half of the summer. These periods were characterized by typical summer temperature variabilities of atmospheric air over several days.

The graphs show that temperature variation is the result of cyclical operation.

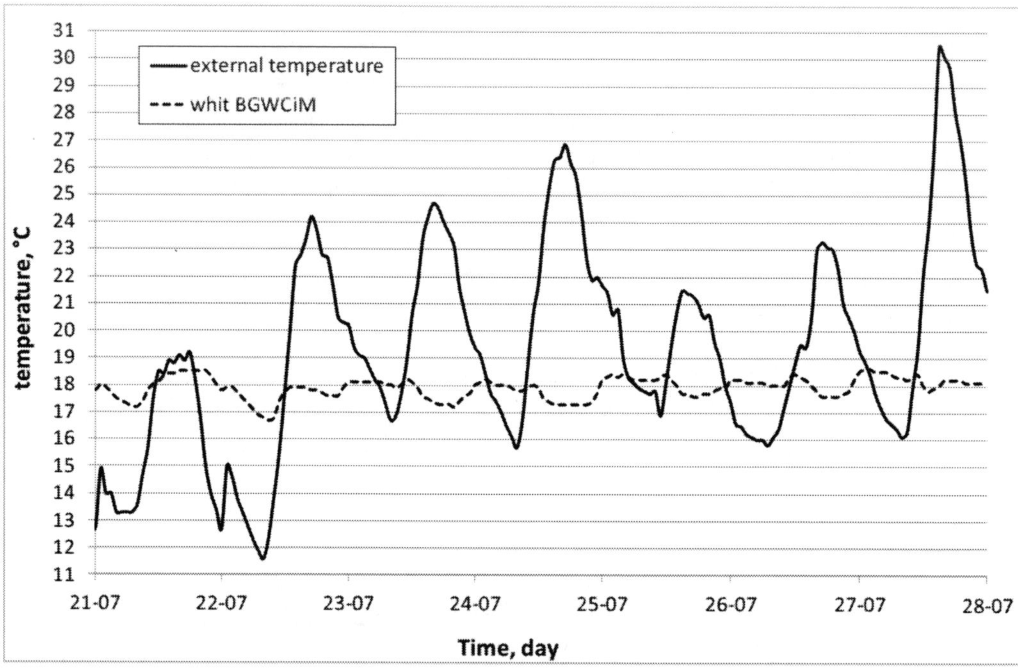

Fig. 7. Cooperation cooling unit with BGWCiM - selected week in July [4].

Research shows that the temperature of the air leaving the heat exchanger when working with a cooling unit, change in the manner established at the stage of theoretical considerations. It is characterized by small amplitude changes over a single day and multi-day period. The lower temperature of the air leaving the heat exchanger in the first half of the study (16°C - 17°C) relative to the temperature of the second half (18°C - 20°C) is the result of accumulated cooling, which is a consequence of the cooling soil in the winter period and a slow discharge with the progress of the time.

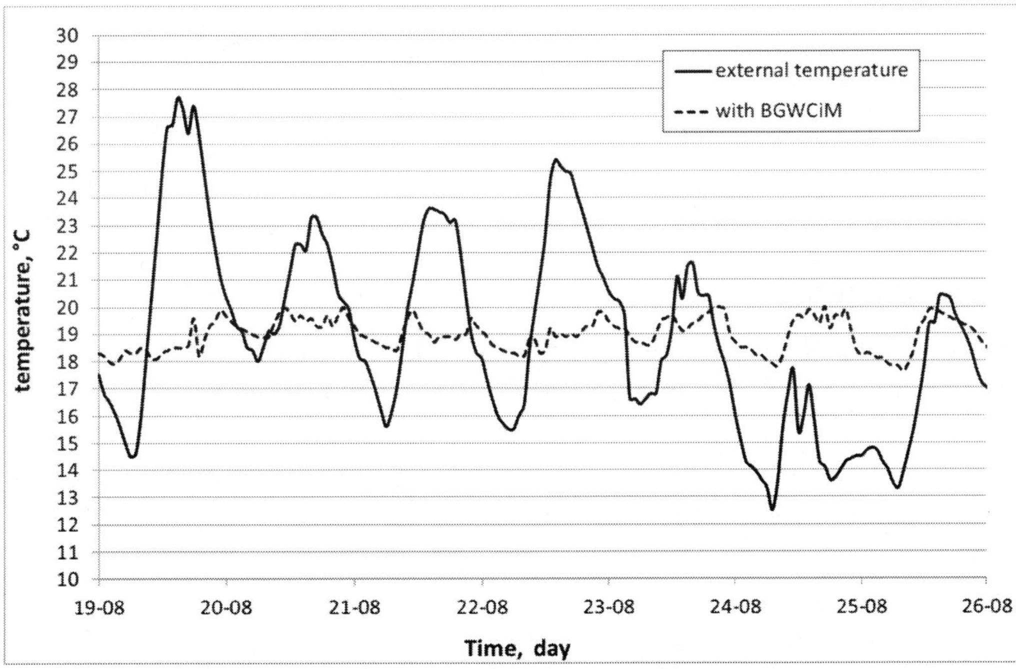

Fig. 8 Cooperation cooling unit with BGWCiM - selected week in August [4].

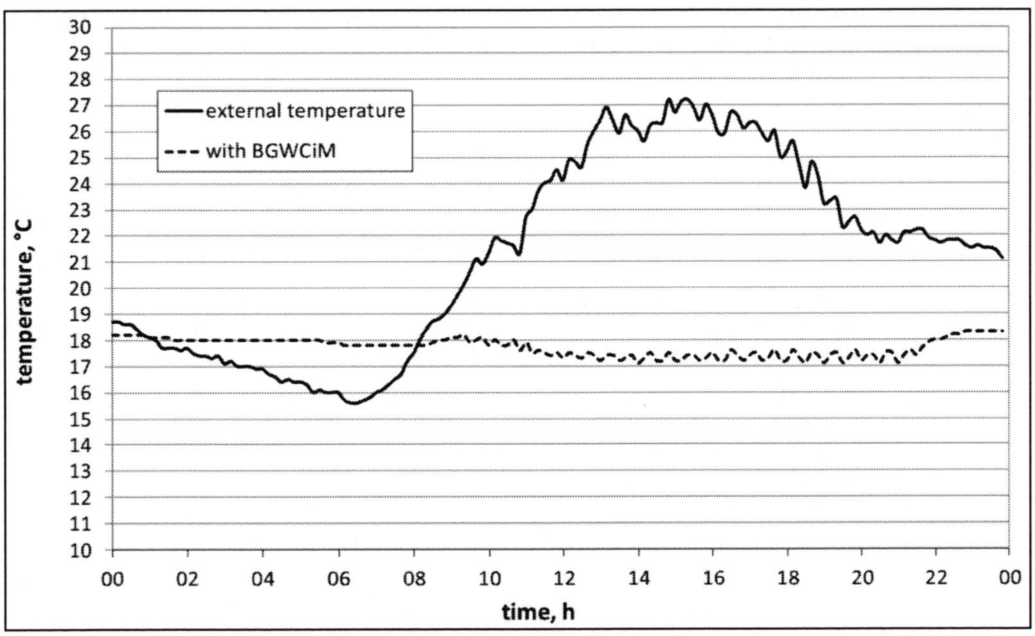

Fig. 9. Operation of the system in a typical summer day [4].

Figure 9 shows the operation of the system on a typical summer day. After a night's stopover, due to a lack of cooling demand during this period, and nightly bed regeneration at 9:30 due to the increase of temperature in the room which is the result of the growth heat gains, the cooling unit starts to work. As we can see, a system using the cooling unit ensures the necessary parameters of the microclimate in the room and lowers the the duration of time it needs to be used in a day.

Over the analyzed summer period, during the use of the cooling unit operation:

- Air temperature after passing through the BGWCiM, flowing to the condenser over a single day has practically not changed, and in the perspective of the several days was associated with fluctuations in external conditions and ground load,
- Air temperature after passing through the BGWCiM changed in the range from 15°C to 20°C, did not rise above 20°C, and was lower than the outside air temperature 80% of the time.

5 Conclusions

Performed analysis and received results shows that using the BGWCiM together with the cooling unit can reduced the use of energy by 10 % during the warm period.

When the cooling unit is conjoined with the BGWCiM it:

- causes much more favorable working condition devices, which results in smaller loads of components due to the reduction of temperature fluctuations.
- cuts off energy peaks which creates a lower failure rate, longer service life, durability and reliability
- increases the cooling capacity and give the possibility of buying a smaller cooling unit
- increases the average energy efficiency ratio
- shortens work time and reduces consumption of electric energy while maintaining the same conditions of the microclimate

BGWCiM conjoined with the cooling unit has many benefits. It produces renewable energy, is environment-friendly, and uses less electric energy.

References

1. G.J. Besler i inni, Bezprzeponowy gruntowy wymiennik ciepła i masy, Patent No. 128261 PRL Wroclaw University of Science and Technology Wrocław (1980)
2. M. Besler, Badania efektywności wykorzystania energii gruntu w inżynierii środowiska, PhD thesis, Wrocław (1997)
3. W. Cepiński, Efektywność pozyskiwania energii naturalnej. PhD thesis, Wrocław (2010)
4. S. Przydrożny, Wentylacja, Wydawnictwo Politechniki Wrocławskiej, Wrocław (1991)
5. M. Rubik, Pompy ciepła Poradnik, Ośrodek Informacji, Technika instalacyjna w budownictwie, Warszawa (2006)

Analysis of concentrations trends and origins of PM_{10} in selected European cities

Anna Chlebowska-Styś[1,2,*], *Izabela* Sówka[1], *Dominik* Kobus[3], and *Łukasz* Pachurka[1]

[1]Wroclaw University of Science and Technology Faculty of Environmental Engineering, Unit of Ecologistics and Environmental Risk Management, Wyb. Wyspianskiego 27, 50-370 Wroclaw, Poland
[2]Voivodship Inspectorate of Environmental Protection in Poznan, 4 Czarna Rola Str., 61-625 Poznan, Poland
[3]Institute of Environmental Protection – National Research Institute, Air Quality Monitoring Division, Krucza 5/11d, 00-548 Warsaw, Poland

Abstract. Particulate matter is generated in many natural and anthropogenic processes and, therefore, taking part in chemical reactions and physical processes in the atmosphere could affect human health, climate, and natural environment. The dust, after emission to the atmosphere, remains there in a floating state to form an atmospheric aerosol with different properties – depending on the morphology, the fraction, shape, surface, and chemical composition of the particles. The chemical composition of particulate matter is heterogeneous and depends on the type of emission sources and the time of residence in the atmosphere. The contribution of individual types of sources in the total dust emission is different in European countries. The aim of the study was to analyze the concentrations of PM_{10} at urban background and traffic stations in the chosen 17 European cities. For this purpose, the data from the database of the European Environment Agency and Chief Inspectorate of Environmental Protection was used. The annual average concentrations of PM_{10} and daily average concentrations of PM_{10} in 2014 will be presented to confirm that in the European cities there is a problem about high PM_{10} concentrations, especially in Poland and other eastern countries. In the case of daily average PM_{10} concentration, the highest concentrations of pollutants in the winter season suggests that for the low air quality in these cities the phenomenon of so- called low-stack-emissions from the municipal sector is responsible.

1. Introduction

Nowadays the air pollution is a huge problem, especially in the cities. Both, politicians and social organizations try to fight with the air pollution in many different ways (for example legal norms, an environmental education etc.). Air pollution is any substance which changes normal composition of the air. The air pollution can have a different origin (natural or anthropogenic), a different physical state (gas, solid or aerosols), a different chemical compositions, a various residence time in the air and even a different ability to

* Corresponding author: anna.chlebowska@pwr.edu.pl

long-term transport. Also, some of air pollutants can react with other substances in the air to a new form, sometimes to more dangerous pollution [7]. Particulate matter, also known as dust or aerosols affect human health, environment and climate causing large economic losses [2, 5–7, 11, 13, 14, 16, 19, 20].

Some particulates occur naturally (primary PM), originating from volcanic eruptions, deserts, dust storms, trees, living vegetation, grassland or forest natural fires, and sea spray. Also particulate matter can be produced in the atmosphere through chemical reactions of gaseous precursors (secondary PM). There is a lot of sources of dust emission. The main anthropogenic sources, which are predominant in urban areas, are production processes and fossil fuels combustions (in vehicles, power plants and industrial processes). Metallurgical, energy, chemistry and mining industry are one of the biggest sources of their emission. Also, transport is the important to source of emission (tire and brake and road wear). Domestic heating for households plays the important role in air pollution in Poland [1, 7, 10, 12, 18].

Europe has a serious problem with the air pollution, especially a particulate matter pollution. Generally, the worst situation is in big cities. According to the data of the European air-quality database in 2014 PM_{10} concentrations continued to be above the UE Limit Value in the large part of Europe [4, 20]. Monitoring stations with measured concentrations of PM_{10} above daily limit value were situated in 21 EU Member States in 2014. About 94% of the cases were observed in urban or suburban areas [7]. Concentration above the PM_{10} annual limit value was monitored in 2014 in 4% of all the reporting to EEA stations. 93% of these stations were located in urban areas. According to the WHO guidelines, annual mean PM_{10} was exceeded at 55% of the total amount of the stations and in 31 countries. The decrease in concentrations of PM_{10} (about 0.6-0.9 $\mu g/m^3$ per year) were observed in 75% stations in Europe in 2000-2014 [7].

Trends of changes in particulate matter PM_{10} concentrations in selected European cities in 2010-2014 is presented in the paper. The aim of the study is to analyze the concentration and origin of PM_{10} measured at the urban traffic and background monitoring stations, in order to identify the main sources of particulate matter emissions in European cities.

2. Materials and methods

Data from the databases of the European Environment Agency and Chief Inspectorate of Environmental Protection was used for analysis. Poznan, Wroclaw, Warsaw, Cracow, Gdansk, Wien, Brussel, Praha, Berlin, Copenhagen, Barcelona, Helsinki, Paris, London, Budapest, Roma, and Stockholm are the chosen cities. The analysis of the concentration of particulate matter in these places should help to answer, what are the main sources of PM emission – traffic or so-called low-stack emissions (emissions from sources with a height lower than 40 m.

The analysis of concentraction in relation to the limit values is based on the measurement at fixed sampling points. Sampling points chosen to analysis are located at two types of sites: urban background and traffic related locations. The AQD [4] sets the data quality objective as a minimum data capture of 90% (valid data out of all possible in a year) for official assesment purposes. For presented analysis the more relaxed coverage criterion of 50% has been used. It allows taking into account more stations without loss of representativeness. It should be noted that the vast majority of data series taken into consideration have completeness in excess of 75%. The analysis covers the period of 2010–2014.

In accordance with the provisions of the AQD and implementing legislation Member States shall report to the EEA provisional data continuously. This includes primarily the results of automatic measurements (so-called up-to-date UTD data), which are thus available to the public on the European Air Quality Portal. The annual series of verified and validated measurement data are delivered to the end of September in the year following the year of measurements performing. They are reported together with information about the assessment system and the results of annual air quality assessment: declarations of compliance with the air quality standards and descriptions of exceedance cases.

European standards for reference measurements of PM_{10} and $PM_{2.5}$ were concluded in PN-EN 12341: 2014 Ambient air - Standard gravimetric measurement method to determine the concentration of mass fractions PM_{10} or $PM_{2.5}$ particulate matter [15]. Basing on weight results the concentration of PM is specified [17].

3. Measurements results analysis and observations on PM_{10} concentrations trends through selected European cities

Annual indexes calculated for each selected cities are used in the presented analysis. These indexes have been calculated separately for urban background and traffic stations, as arithmetic averages of the annual mean values from measuring stations of the specific type in subsequent years. Referring the calculated index to the standards set by the European Commission for the annual PM_{10} concentration it can be indicated that the limit values has been exceeded nineteen times in 3 of 18 selected cities. All cases of exceedances were observed in Polish cities: 10 times in Cracow (5 times at the traffic station and 5 background stations), 3 times in Warsaw for a traffic station and 3 times in Wroclaw (once at the background stations and 2 times for the traffic station) (Tab. 1). PM_{10} concentrations approaching the specified standards were observed in Paris and Budapest at traffic stations and in other analyzed Polish cities. The highest and the lowest index value was observed in 2010 at the traffic station in Cracow and in 2012 at background station in Helsinki, respectively. These cities differ in terms of climate and culture that influence the quality of the air. In all of the analyzed cities, the higher concentrations of PM_{10} were observed for traffic stations, which are generally located in a close vicinity of the streets with high traffic level. It can indicate the dominant influence of transport emissions on air quality in the central parts of the cities, where the traffic stations are often situated. Referring to the average daily value recommended by the WHO (20 $\mu g/m^3$) it should be noted, that the population of most European cities are exposed to air pollution by PM_{10} levels dangerous to health. In 2010-2014, the value of 20 $\mu g/m^3$ was not exceeded only 17 times in 6 cities (Helsinki, Stockholm, Gdansk, London, Copenhagen, Paris).

The Institute of Environmental Protection – National Research Institute (IEP-NRI) study shows the general decrease in PM_{10} averaged concentrations in Poland since 2010, both of average annual and daily PM_{10} concentrations (Fig. 1) [9]. Since 2007, the calculated index of average annual concentration of PM_{10} for Polish does not exceed 40 $\mu g/m^3$. The situation in Polish cities is much worse than the PM_{10} pollution expressed as a population weighted index calculated for the whole country using the methodology described in [8]. Concentrations in Cracow and Warsaw is about 20 $\mu g/m^3$ higher than the average concentration calculated for the whole country. Although the annual average concentrations in the area of Poland remain several years of below standard values, every year the number of permitted days per year with a value greater than 50 $\mu g/m^3$ is exceeded. In Poland, air pollution with PM_{10} is generally a seasonal problem – exceedances occur mainly in autumn-winter period, hence the annual average value does not exceed the standard. On the other hand, a large number of days with high concentrations of PM_{10} in winter badly affects the quality of life. According to the IEP-NRI study in 2012 standard value of the 24-hour concentration was exceeded on at least one station in 21 of the 28 European Union countries. In 2012 the exceedance of the normative value determined on the basis of percentile values of S90,4 from the daily PM_{10} concentrations (above 50 $\mu g/m^3$)

was observed at 21.4% of measuring stations in EU countries (including Poland). In Poland, the exceedance was concerned for 71.6% of the stations. In a case of the limit value established for the annual average concentrations of PM_{10}, the exceedances occurred at 5.1% in the EU and 24.2% in Polish stations, respectively [3].

Table 1. Annual concentration of PM_{10} in selected European cities in 2010–2014.

Country	City name	Type of measurement stations	Annual concentration of PM_{10} [µg/m^3]				
			2010	2011	2012	2013	2014
Austria	Wien	Background	27	27	23	23	21
Austria	Wien	Traffic	31	32	26	26	23
Belgium	Brussel	Background	26	27	25	24	23
Belgium	Brussel	Traffic	27	27	24	24	23
Czech Republic	Prague	Background	25	26	24	25	26
Czech Republic	Prague	Traffic	32	30	27	27	28
Germany	Berlin	Background	25	23	21	21	23
Germany	Berlin	Traffic	33	30	27	27	31
Denmark	Copenhagen	Background	12	nd	17	nd	19
Denmark	Copenhagen	Traffic	26	32	28	27	28
Spain	Barcelona	Background	27	28	27	21	22
Spain	Barcelona	Traffic	33	36	35	26	28
Finland	Helsinki	Background	14	13	11	12	15
Finland	Helsinki	Traffic	23	22	18	21	21
France	Paris	Background	26	24	27	26	18
France	Paris	Traffic	39	39	37	36	29
United Kingdom	London	Background	17	21	18	18	19
United Kingdom	London	Traffic	26	29	25	26	25
Hungary	Budapest	Background	32	33	29	28	29
Hungary	Budapest	Traffic	36	38	28	31	36
Italy	Roma	Background	28	31	28	26	27
Italy	Roma	Traffic	34	35	34	31	30
Poland	Gdansk	Background	27	25	22	20	26
Poland	Cracow	Background	48	54	53	44	46
Poland	Cracow	Traffic	79	76	66	60	64
Poland	Poznan	Background	38	39	33	27	32
Poland	Warsaw	Background	35	35	37	32	32
Poland	Warsaw	Traffic	52	49	39	40	42
Poland	Wroclaw	Background	48	39	39	36	35
Poland	Wroclaw	Traffic	62	53	nd	nd	nd
Sweden	Stockholm	Background	13	15	13	15	14
Sweden	Stockholm	Traffic	25	29	24	28	22

Institute of Meteorology and Water Management – National Research Institute (IMGW) data indicates that since 2010 in Poland as well in the summer seasons the higher temperatures, as in the autumn-winter season are registered (Fig. 2). Referring PM_{10} concentrations reported in Poland in the years 2010-2014 there is observed a clear converse relationship between the air temperature and the concentration of PM_{10} in the air (e.g. anomaly cold and very cold periods). This is another proof that the particulate matter in the air in Poland is affected by low-stack emission from the municipal sector. Low temperatures force Polish cities inhabitants to heat homes and fuel burning of poor quality coal (it is associated with their economic situation and the low environmental awareness) affects the deterioration of air quality in the urban areas. Moreover, in winter season often a low wind speed, a high atmospheric pressure and temperature inversion situations can favor the occurrence of smog situation.

For selected European cities average daily concentrations of PM_{10} in 2014 were analyzed in the way to demonstrate the dependence of PM_{10} concentrations on the season. The results are presented in the form of a mosaic in Figure 3. Each grid in the figure shows the daily average PM_{10} concentration. The value of the annual average PM_{10} concentrations and the number of days with a value of more than 50 $\mu g/m^3$ during the year was added to the graph. Presented data show that there is a strong seasonal concentrations change in the Polish cities – in autumn-winter season daily average concentrations and number of days with exceedance of 50 $\mu g/m^3$ reach the highest values. A city located in the north of Europe (Copenhagen, Helsinki, Stockholm, London) have concentrations of PM_{10} at a similar level throughout the year. Additionally, there are low PM_{10} concentrations levels - below the European standards. It shows that an influence of the climate of these cities and also the heating system of buildings is quite different than in cities of Central and Eastern Europe. In other analyzed cities (Vienna, Brussels, Prague, Paris, Berlin, Rome, Barcelona) variation in concentrations within a year was also not observed. It can be assumed that the main source of emissions of PM_{10} in those parts of Europe is transport. Similar conclusions can be drawn based on the analysis of hourly PM_{10} concentrations averaged for every the same hour within the calendar year (Fig. 4). The analysis of data from the selected background stations shows that in the Polish cities the highest daily PM_{10} concentrations are observed in the evening and late evening (18.00–23.00) and then again in the morning (6.00–9.00). It is correlated with the rhythm of home heating – residents (especially owners of old central heating) start heating homes after work (in the afternoon) and then repeat this step in the morning. Also, some influence of worse pollution dispersion situations which sometimes occur during the night hours should be taken into consideration. In other selected cities (Paris, Prague, Berlin, Vienna and Brussels) a slight increase in the concentrations of PM_{10} in the morning at the time when people are traveling to work was observed.

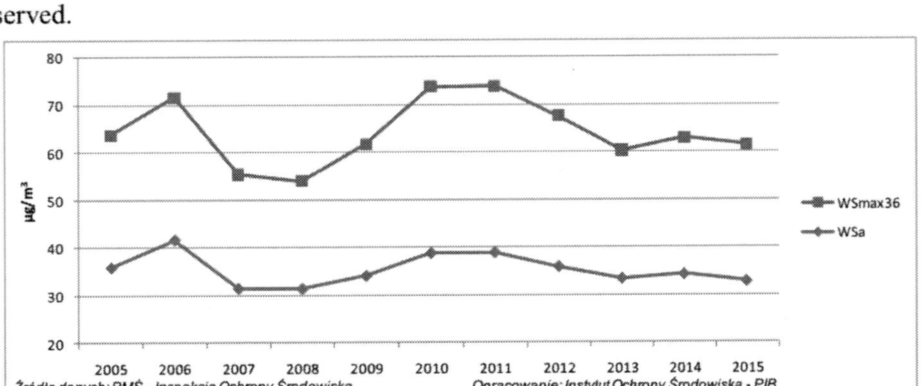

Fig. 1. Variability in annual and daily average concentrations of PM10 at urban and suburban background monitoring stations in Poland in the years 2005-2015 [9].

 DOI: 10.1051/e3sconf/20171700013

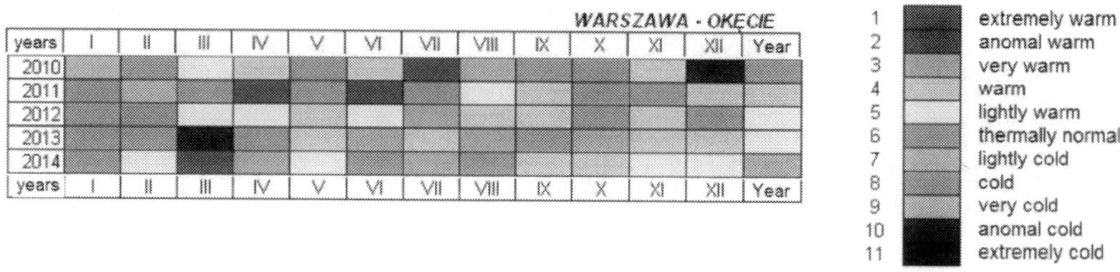

Fig. 2. Thermal classification of months and years (2010-2014) *[source: own compilation based on IMGW-PIB]*.

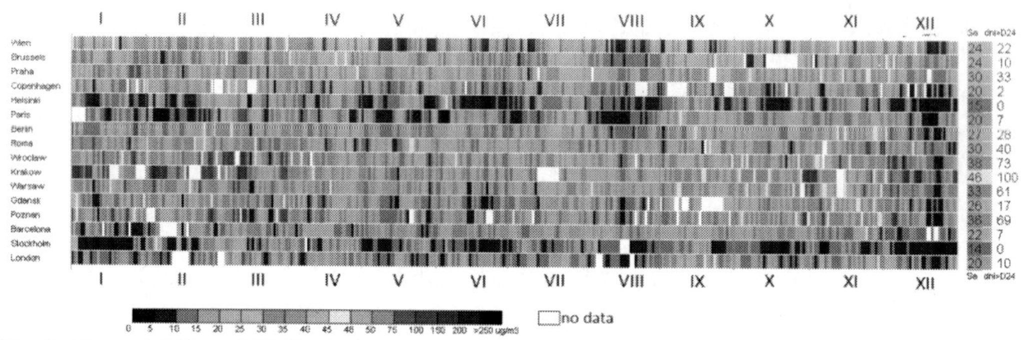

Fig. 3. The variability of PM10 daily mean concentrations measured at selected urban background stations in 2014 *[source: own elaboration based on data published by the EEA]*.

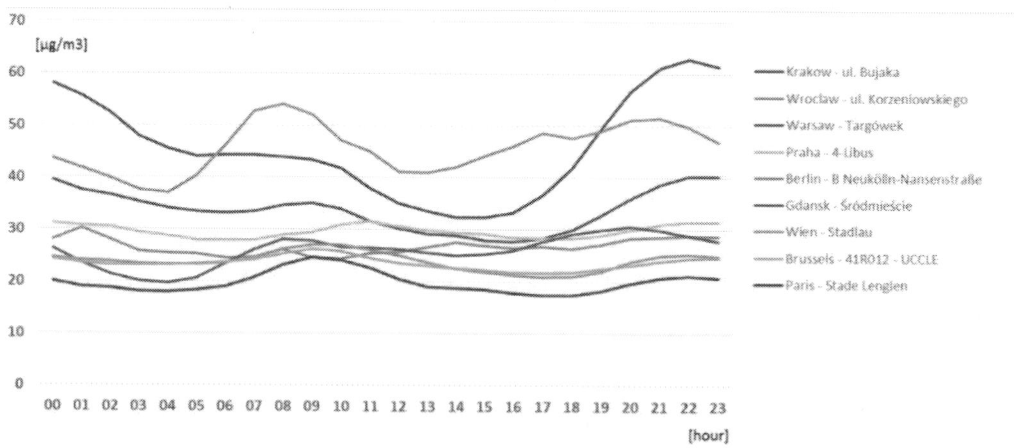

Fig. 4. Averaged diurnal variability of 1-hour PM10 concentrations at selected urban background monitoring stations in 2014 *[source: own elaboration based on data published by the EEA]*.

In 2014 the maximum daily PM_{10} concentration in Cracow and Wroclaw was above 200 $\mu g/m^3$ (Fig. 5). In Poland, the appropriate services have the obligation to inform the public about the occurrence of health hazards due to high concentrations of PM_{10} in a situation of daily mean value above 200 $\mu g/m^3$. The high PM_{10} concentrations episodes (above 100 $\mu g/m^3$) were also observed in Paris, Poznan, Brussels, Rome, Berlin, and Poznan. The intensity of smog situations (height of concentrations) in Poland is often much greater than in other analyzed countries. Even short-term exposure to such high concentrations of PM_{10} could pose a serious threat to human health, especially vulnerable groups (children, the elderly, pregnant women and people suffering from lung diseases) [2].

DOI: 10.1051/e3sconf/20171700013

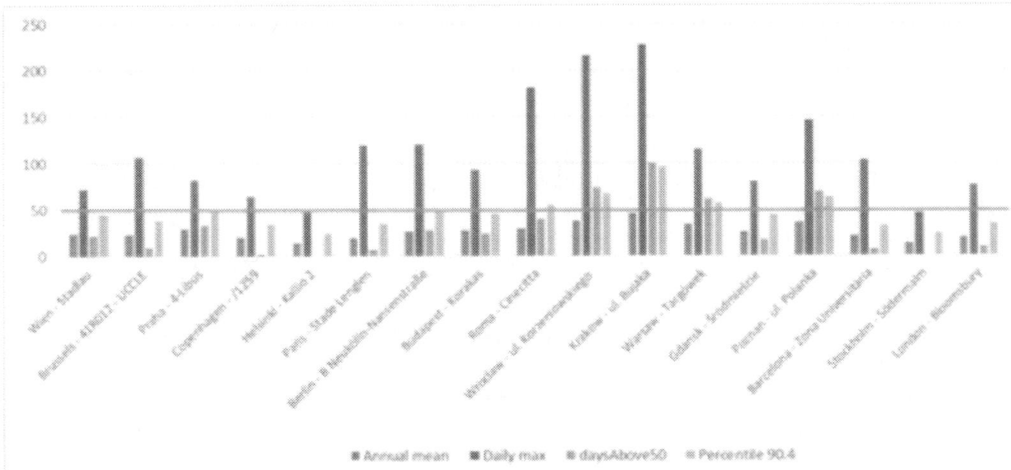

Fig. 5. The values of the selected statistical parameters for the PM10 concentration data series at urban background stations in 2014 *[source: own elaboration based on data published by the EEA]*.

5. Summary

The analysis of the data shows that European cities face the problem of air pollution by particulate matter (PM). Every year, millions of inhabitants are exposed to oversize PM concentrations, in the case of PM_{10} annually states exceedance of the maximum PM concentrations levels established for the protection of human health. Practically, in each of the analyzed cities the average concentration recommended in the WHO Guidelines (20 $\mu g/m^3$) was exceeded, and there were only a few cities with background pollution below that level. Definitely, the worse situation is observed in central and east-central Europe, where dominates the culture of apartments heated using the coal, often of poor quality. It can also be seen that location, climate, and terrain of the cities have a huge impact on their air quality. EEA data shows that the highest concentration of PM10 in Europe in Poland, Italy, Bulgaria and Slovakia is registered [3]. While the location of the city in the mountain sites affects the deterioration of air quality (Cracow), cities located in well-ventilated terrain, on the shores of seas and boasting a large area of green spaces in their borders have good air quality (Helsinki, Gdansk).

The investigations were co-financed within the framework of the order No. 0402/0088/16 with the specific subsidy granted for the Faculty of Environmental Engineering Wroclaw University of Science and Technology (W-7) by the Minister of Science and Higher Education to conduct research and development work and related tasks contributing to the development of young scientists and doctoral students in 2016/2017.

References

1. A. Degórska, J. Bartnicki, *Udział Polski w atmosferycznym transporcie zanieczyszczeń powietrza na obszarze Europy*, Institute of Environmental Protection – National Research Institute (2011)
2. C.A. Pope, D. W. Dockery, J. Air Waste Manage Asso, **56**, 709–742 (2006)
3. D. Kobus, G. Mitosek, and J. Iwanek, *Pyły Drobne w Atmosferze. Kompendium Wiedzy o Zanieczyszczeniu Powietrza Pyłem Zawieszonym w Polsce*, Główny Inspektorat Ochrony Środowiska, 230–239 (2016)

4. *Directive 2008/50/EC of the European Parliament and of the Council of 21 May 2008 on ambient air quality and cleaner air for Europe*, Official Journal of the European Union, 11.6.2008, L152/1
5. EEA, *Air quality in Europe – 2014 report. EEA Report No 5/2014*, European Environment Agency EEA (2014)
6. EEA/JRC, *Environment and human health, EEA Report No 5/2013*, European Environment Agency and the European Commission's Joint Research Centre (2013)
7. EEA, *Air quality in Europe – 2016 report,* European Environment Agency (2016)
8. J. Iwanek and D. Kobus, Przemysł Chemiczny **95.3**, 406–411 (2016)
9. J. Iwanek, D. Kobus., G. Mitosek, R. Parvi, *Jakość powietrza w Polsce w roku 2015 w świetle wyników pomiarów prowadzonych w ramach Państwowego Monitoringu Środowiska*; Institute of Environmental Protection – National Research Institute (2016)
10. M. Kanakidou et al., Atmos Chem Phys, **5**, 1053–1123 (2005)
11. M. Lippmann, *Environmental toxicants: human exposure and their health effects*, John Wiley & Sons (2008)
12. M. Reizer, *Pyły Drobne w Atmosferze. Kompendium Wiedzy o Zanieczyszczeniu Powietrza Pyłem Zawieszonym w Polsce*, Główny Inspektorat Ochrony Środowiska, 11-21 (2016)
13. N. Englert, Air Qual Atmos Health Toxicol Lett, **13(3)**, 149:235 (2004)
14. P. Paasonen et all., Nat Geosci, **6**, 438–442 (2013)
15. PN-EN 12341:2014-07 Ambient air – Standard gravimetric measurement method to determine the concentration of mass fractions PM_{10} or $PM_{2.5}$ particulate matter
16. R. W. Atkinson, I. C. Mills, H. A. Walton, H. R. Anderson, J. Expo Sci Environ Epidemiol, **25**, 208–214 (2015)
17. T. Frączkowski, *Pyły Drobne w Atmosferze. Kompendium Wiedzy o Zanieczyszczeniu Powietrza Pyłem Zawieszonym w Polsce*, Główny Inspektorat Ochrony Środowiska, 106-110 (2016)
18. US EPA, *Integrated Science Assessment for Particulate Matter. Raport EPA/600/R-08/139F*, United States Environmental Protection Agency (2009)
19. WHO, *Review of evidence on health aspects of air pollution – REVIHAAP project technical report*, World Health Organization (2013)
20. WHO, *Air quality guidelines for Europe, 2nd end*, World Health Organization (2000)
21. WHO, Health *risks of particulate matter from long-range transboundary air pollution*, World Health Organization (2006)

Domestic hot water consumption in multi-apartment buildings

Agnieszka Chmielewska[1,*], *Małgorzata* Szulgowska-Zgrzywa[1], and *Jan* Danielewicz[1]

[1]Wroclaw University of Science and Technology, Faculty of Environmental Engineering, Wybrzeże Wyspiańskiego 27, 50-370 Wroclaw, Poland

Abstract. The article presents the analysis of the hot water consumption in apartment buildings. This study was conducted basing on the data gathered from 626 apartments during two following years. By statistical analysis of the collected data, the correlation between the DHW consumption and both the size of the apartments and the number of occupants in the dwellings was evaluated. The dependencies between the hot water consumption and these two parameters were calculated. The results were compared to the data available in the literature. The article lists the areas of the greatest inaccuracies in the calculations basing on the available indicators and proposes an alternative method that corrects them. The results were used to develop a calculation model for daily hot water usage in multi-apartment buildings.

1 Introduction

Climate changes, rising energy costs and concerns about the energy security require careful analysis of the energy consumption. Residential building sector has consumed 11–35% of total energy at the national levels in most of the countries [1], which is also responsible for higher emission of greenhouse gas. A large part of the final energy supply in residential building is used for the DHW purposes. In 2012 in Poland, the energy consumption in this sector constituted at 18% [2]. However, in new buildings the energy for domestic hot water preparation is an important component, so important is also a description of the hot water consumption. The size of the hot water consumption has a significant impact on the energy consumption.

Consumption of hot water in residential buildings depends on many factors, sociological and technical, that are often impossible to determine. A typical variable used to describe water consumption is the average consumption of water per capita. It is debatable, because it is difficult to determine the number of people living in a building. The main parameters, which are known and can be used as a basis for estimating the consumption of hot water is the number of flats in a multi-apartment building, size of the housing and the sanitary equipment of dwellings. Researchers use different approaches to this problem, as shown in the next section of the article.

* Corresponding author: agnieszka.chmielewska@pwr.edu.pl

2 Modeling the consumption of hot water

The number of occupants as a parameter of the hot water utilization is the most often analyzed factor in many countries. Parker [3] noticed that the occupants' count is the key determinant of the DHW consumption and used the occupancy pattern in modeling of the national energy consumption. Evarts and Swan [4] found higher DHW consumption in households of three people or less. The authors of the publication also noted higher unit consumption of hot water in homes with fewer residents, which is confirmed by the analysis Ahmed, Pylsy and Kurnitski [5], Merrigan [6], Georg, Pearre and Swan [7] and Hendron & Burch [8]. The authors found a linear relationship between the water consumption and the number of occupants. In the first publication, the authors presented that the consumption increases by 45 l/day for each additional person where the occupant number was above 2, and in the second publication by 39 l/day for each additional person. According to the Canadian Mortgage and Housing Corporation [9], Canadian seniors' buildings consume less than 44% of the DHW comparing to the standard family apartment buildings. In another study the author noted that the presence of children caused a higher consumption rate of DHW [5]. Foekema and Engelsma [10] found that there is also a relationship between the frequency of use and duration of a shower and the age and size of households in Dutch residential building. Unfortunately, despite of many studies, the calculation method used in Poland [11] or Germany [12] is very simplified. Polish regulations [11] concerning calculation of the average daily consumption of hot water uses the total space of a building as a parameter. In case of the German regulation, we have a reference to the number of occupants. However, the authors also added the factor that allows the conversion of the total building area to the estimated number of inhabitants. A very interesting approach can be found in SAP 2012 [13], relating to the dwelling with an area of $450m^2$. The authors proposed a function that allows calculation of the number of users using the daily consumption of hot water, and then applied the correction factors to reduce the expected level of water consumption during the summer and to increase it during the winter. When creating simulations of energy use, it is important to utilize the profiles based on data from the actual country and the current economic situation. For this reason, in the further part of the study the authors analyzed the statistical data from the area of Dolny Śląsk and on this basis proposed a different computational model of the hot water consumption in residential buildings.

3 Analysis of the measured data

3.1. The measured data

The article presents the analyzes of the DHW consumption. The analyzes were performed basing on the measurement data from 16 multi-apartment buildings (626 apartments) located in Wroclaw, Poland. Each of the units (the apartments) has been equipped with a residential water meter for measuring the cold and hot water consumption. The data were being gathered every six months. In addition, the analyzes were supported by the data obtained from the estate manager and on the data from the survey regarding the number of people staying in the apartments. Comprehensive data about the apartments' floor area and the annual consumption of hot water came from 626 apartments. The data about the number of installation users was achieved from 78 apartments. The study covered the period from January 2012 to December 2013.

 DOI: 10.1051/e3sconf/20171700014

3.2 Analysis method

Basing on the measurement data, the analysis of the correlation between the number of users and the floor area of the apartments and the daily consumption of hot water was performed. For each of the apartments the annual consumption of domestic hot water was calculated (V_{DHW}, m^3/year), as a difference between the measured values at the beginning and at end of the year. Then, on the basis of the equation 1, the daily water consumption ($V_{d,DHW}$) was calculated for each apartment.

$$V_{d,DHW} = V_{DHW} / 365 \text{ [m}^3\text{/day]} \quad (1)$$

The main part of the analysis was focusing not only on the apartments themselves, but on the whole buildings and their hot water consumption model. The annual and daily hot water consumption was determined for each of the 16 buildings. Because all the 16 buildings came from the same residential complex, a new database was created in order to simulate buildings of unusual distributions of housing. The database was created using the measurements of water consumption of 626 apartments. The database consists of randomly simulated buildings, each consisting of randomly selected flats in different configurations. To produce the additional sample data, the bootstrap method was used. As a result, the authors obtained the sample buildings consisting of 4 to 60 flats (in the range of the buildings area from 150 to 3000 m^2).

3.3 Results discussion

In the paper, the ability to describe the daily consumption of hot water in an apartment building by the number of users or by the housing area was analyzed. Basing on the data provided in Table 1 it can be concluded that the average daily consumption of hot water in apartments increases with the number of users, where the number of people does not exceed 3. For the flats with four users, the increase of the consumption of hot water was not observed. However, due to the very small sample size, the apartments with four users cannot be taken as authoritative in this analysis.

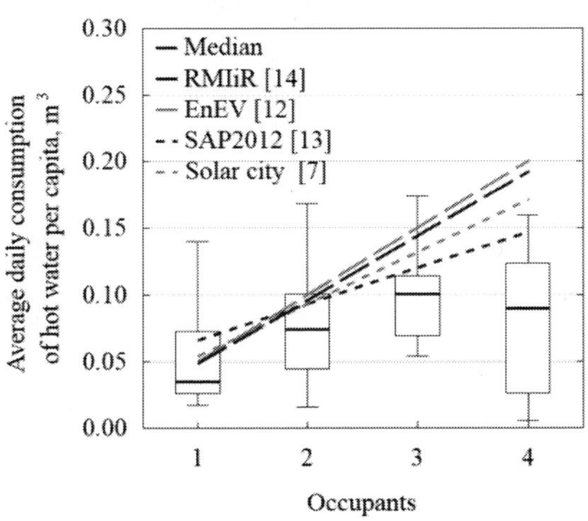

Fig. 1. Occupancy and average daily DHW consumption.

Table 1. Occupancy and average daily DHW consumption.

Occupants	Number of homes	Median	Average
1	15	0.035	0.052
2	41	0.076	0.083
3	17	0.101	0.102
4	5	0.089	0.081
Total	78	0.035	0.081

In the literature, we can find many attempts to describe the DHW demand as depending on the number of users. Figure 1 shows the selected models applied to the measured data. The RMIiR model [14] and the [12] model describe well only the apartments with one user.

This function assumes that each additional person in the household uses exactly the same amount of hot tap water. It does not take into account the fact that some part of the water is consumed on common goals, i.e. housekeeping or cooking. The hot water consumption in flats is well described by the relationship shown in SAP2012 [13], which was developed in order to analyze the energy being consumed to prepare hot tap water in homes in the UK. Similar results were also obtained using the model developed with the data from the Canadian housing in the Solar City [7].

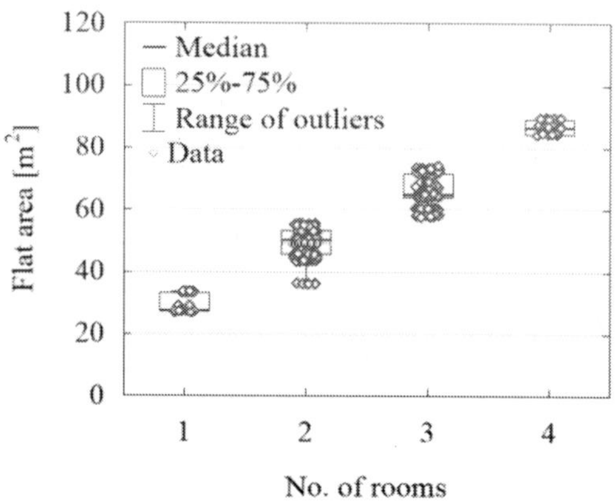

Fig. 2. Correlation between the flat area and the number of rooms.

Table 2. Correlation between the flat area and the number of rooms.

Symbol	No. of rooms	Area flat
M1	1	≤ 35 m^2
M2	2	35.1 – 56.5 m^2
M3	3	56.5 – 79 m^2
M4	4	≥ 79.1 m^2

Additionally, the ability to describe the daily consumption of hot water by the size of the apartments was analyzed. For individual residential units, it is not a good analytical approach. Most of the models assume a linear relationship between the consumption of hot water and the floor area of the flat. For more detailed information about the relationship between the consumption of hot water and the size of a dwelling (which presumably could be correlated with population growth) the relationship between the floor area of flats and the number of rooms in 259 apartments was examined. The data shown in Figure 2 illustrate a strong correlation between the area of a flat and the number of rooms. For further analysis, the classification of housing presented in Table 2 was adopted.

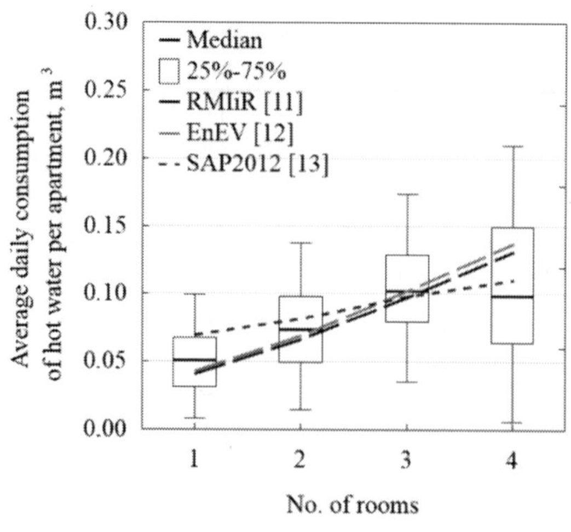

Fig. 3. Number of rooms and average daily DHW consumption.

Table 3. Number of rooms and average daily DHW consumption.

No. of rooms	Number of homes	Median	Average
1	57	0.051	0.051
2	327	0.074	0.074
3	200	0.102	0.104
4	42	0.099	0.105
Total	626	0.082	0.084

Then, the correlation between the number of rooms and the daily consumption of hot water was examined. The results are shown in Table 3 and in Figure 3. The analysis of the average daily hot water consumption shows that with the increase of the number of rooms, and thus with the increase of the floor area of the housing, the consumption of hot water also increases. However, in three- and four-room apartments, this value is stable. This phenomenon may be related to the constant average number of users in the apartments of this size. In addition, figure 3 shows the relationship that is currently being used to calculate the DHW consumption for the purposes of estimating the demand for energy in buildings in Poland, Germany and the UK. The results are shown for an average housing area in each group. The first two models: [11] and [12] do not take into account the fact that in 4-room apartments the water consumption does not differ from 3-room apartments. The method presented in SAP2012 [13] describes the actual average daily consumption of hot water significantly better.

The conducted analysis shows that both variables similarly describe the change in the daily consumption of hot water. Unfortunately, the number of users, which is the most common variable in all sorts of analyzes, is difficult to determine. Therefore, the later studies were focusing mostly on the area and the number of rooms in apartments.

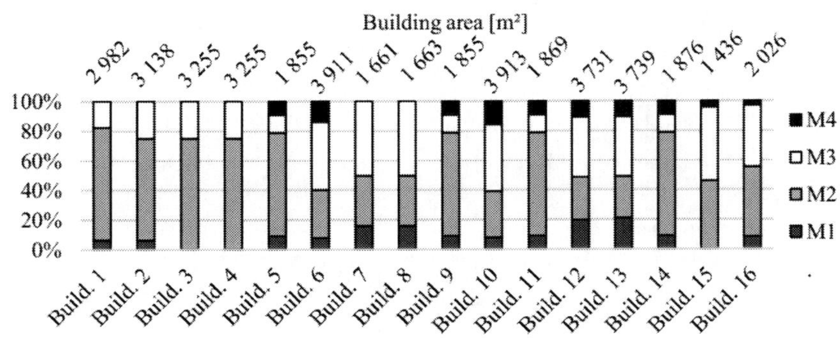

Fig. 4. Specification of the areas and the structures of flats in the analyzed buildings.

The analysis of hot water consumption was performed for 16 buildings located in Wroclaw, Poland. In Figure 4 the buildings' mix of the flat types and their area were presented. Additionally, in Figure 5 the dependency between the daily hot water consumption and the building area was presented. The results were related to the dependency function from the actual Polish procedure of certification. As one can see on the graph, the dependency function from the procedure fits the trend of the data, but significant differences are also visible: from 2% to even 37%. This discrepancy may be a result of the structure of flats in buildings. This is particularly visible in the buildings 1, 2, 3, 7 and 8. In these buildings there are no 4-room apartments. In such cases, the daily hot water consumption is often higher than in other buildings.

Figure 6 shows the hot water demand in the buildings with an unusual distribution of flats. To generate such data the bootstrap method was used, utilizing the measured data of the appropriate apartments. The first chart shows only the buildings with 1-room apartments, the second chart – the buildings with 2-room apartments, third – 3-room apartments and the last – 4-room apartments. For the buildings with 3-room flats, the function proposed in the Regulations [11, 12] coped with the estimation of the hot water consumption quite well. However, in case of the buildings with 1, 2- and 4-room flats, the slope of the function is not as expected. These comparisons suggest that it would be useful to develop a method of estimating the consumption of hot water, which takes into account the structure of a building. Such a method will be proposed later in the study.

DOI: 10.1051/e3sconf/20171700014

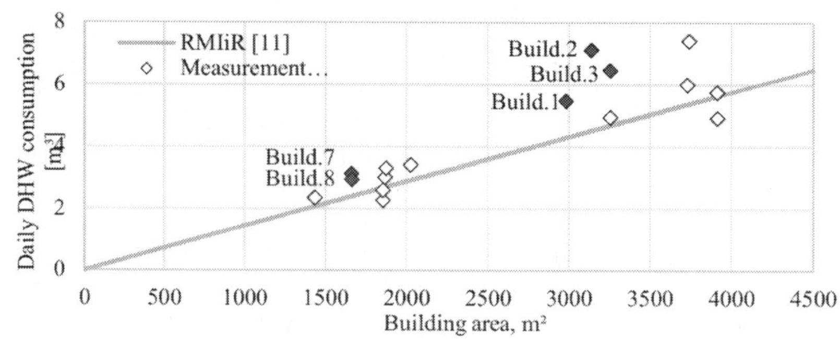

Fig. 5. Building area and average daily DHW consumption.

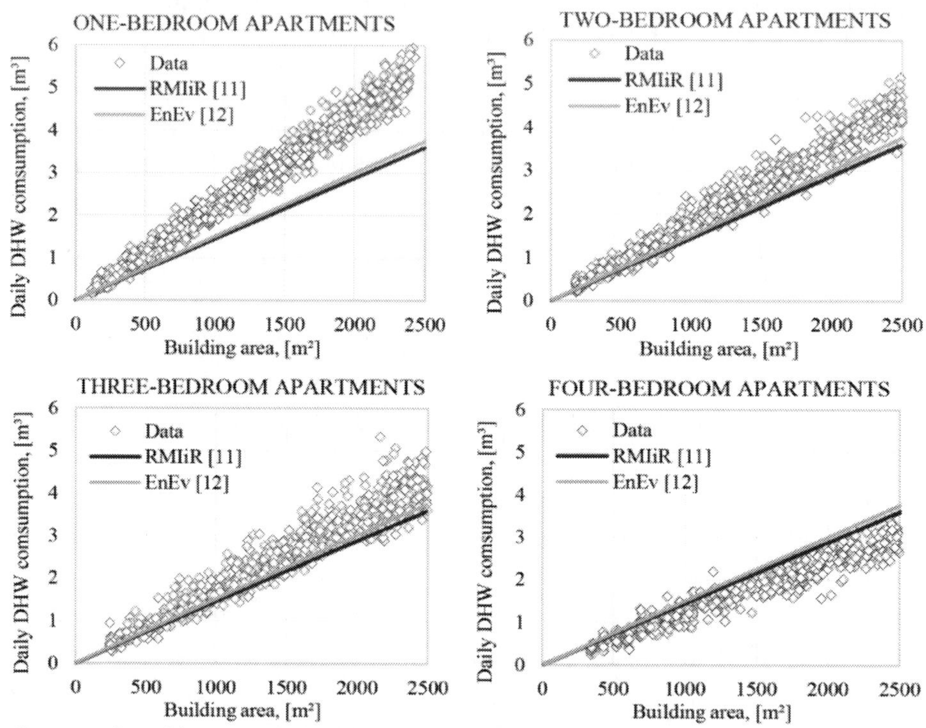

Fig. 6. Consumption of hot water in buildings of unusual size apartments.

4. Average daily model of hot water consumption

After the examination of the measurement data, a refinement of the model of hot water consumption in apartment buildings was proposed. The model is indented to determine the daily average consumption of hot water in the buildings with the desired structure. The model assumes that the number of apartments in a building is more than 4. Smaller buildings and individual flats must be considered separately.

To develop the model, 1600 cases (400 buildings with one-bedroom apartments, two-room apartments, three-room apartments and four-room apartments) obtained using the bootstrap technique have been used. The draws were repeated to check the influence of the randomly selected sample on the outcome of the model. No association was found. In all the consecutive draws the model parameters remained unchanged. For each designed case, the following calculations ware made: the total area of 1-room apartments, 2-room apartments, 3-room apartments and 4-room apartments (A_{M1}, A_{M2}, A_{M3}, A_{M4}) and the daily hot water consumption ($V_{d,DHW}$). The multiple regression analysis performed using

 DOI: 10.1051/e3sconf/20171700014

Statistica package estimated the model's parameters. The final model is described by the equation 2. The coefficient of determination of the model (R^2) is 0.95.

$$V_{d,DHW} = 0.0022 \cdot A_{M1} + 0.0018 \cdot A_{M2} + 0.0017 \cdot A_{M3} + 0.0012 \cdot A_{M4} + 0.00122\ [m^3/day] \quad (2)$$

where: A_{m1} – area one-room apartments [m^2], A_{m2} – area two-room apartments [m^2], A_{m3} – area tree-room apartments [m^2], A_{m4} – area four-room apartments [m^2], $V_{d,DHW}$ [m^3/day] – daily hot water consumption.

The results of the model taking into account the building structure were shown in fig. 7. The relative error of the estimates of the proposed model equals to 10%, while the corresponding error of the model proposed in the Regulation [11] was at the level of 17%, and the model from the Regulation [12]: 14%. The verification was conducted using a sample from an independent draw. Also, when applied to the actual data (that are presented in Figure 5), the proposed model was showing better results, as its relative error was at the level of 12% comparing to 15% and 16% in case of the currently available models.

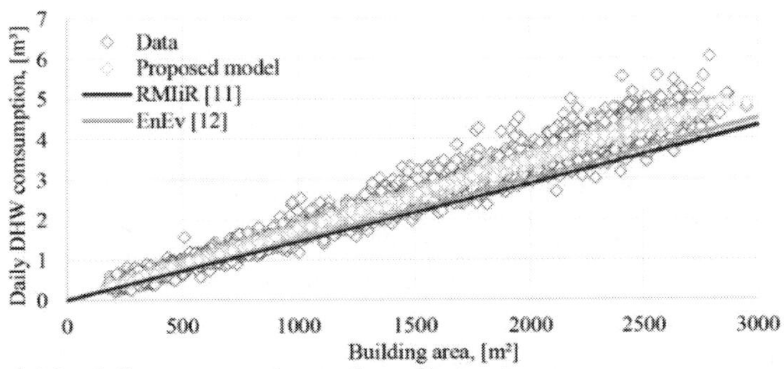

Fig. 7. Proposed model for daily average consumption of hot water.

Figure 8 presents the percentage change of the consumption of hot water in the buildings with different proportion of the floor area of four-room apartments comparing to the two-room apartments. The reference point is a building with 50% share of two-room apartments and 50% share of four-room apartments. Depending on the building's share, its estimated hot water consumption can vary by 20% despite of the same total building area. In case of one-room apartments, the analysis difference may be even greater.

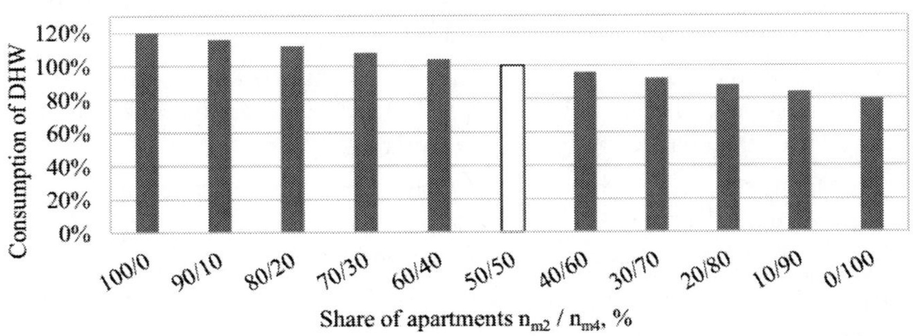

Fig. 8. The percentage change in the consumption of hot water depending on the share of two-bedroom apartments with respect to four-room apartments.

5. Conclusions

The article presents the results of measurements of the hot water consumption of 626 apartments of different sizes, number of rooms and number of users. The researchers have demonstrated that it is possible to describe the daily consumption of hot water with the area of a housing and the number of users. However, the number of users of an apartment is often difficult to determine. Adopting the area of a building as the determining variable without regard to the number of users, e.g. as in the model described in [11], results in large inaccuracies in estimating the consumption of hot water. In this publication, the authors demonstrated the ability to describe the hot water consumption of a flat basing on the number of rooms in the flat and its area, considering that such information are much easier to obtain. As the final result of the data analysis, the authors proposed the calculation model of the hot water consumption of a building, where the variables are the total areas of 1-, 2-, 3- and 4-room flats. The model showed a better fit to the analyzed sample of data than the models proposed in the current Regulations in Poland [11] and Germany [12]. The relative error of the simulation was 10% compared to 17% error of the model from the Regulation [11].

References

1. Energy Efficiency Trends and Policies in the Household and Tertiary Sectors, An Analysis Based on the ODYSSEE and MURE Databases, June 2015
2. Efektywność wykorzystania energii w latach 2002-2012, Główny Urząd Statystyczny, Warszawa, 2014.
3. D. Parker, Energy Building, p. 863–876, 35 (9) (2003)
4. J.C. Evarts, L.G. Swan, Energy Build. 58, 58–65 (2013)
5. K. Ahmed, P. Pylsy, J. Kurnitski, Energy and Buildings, 97, 77-85 (2015)
6. T. Merrigan, Residential hot water use in Florida and North Carolina, ASHRAE, Transactions 94 (1), 1099–1109 (1988)
7. D. George, N. S. Pearre, L. G. Swan, Energy and Buildings, p. 304–315 (2015)
8. R. Hendron, J. Burch, Development of standardized domestic hot water event schedules for residential buildings, in: Proceedings of the ASME 2007 Energy Sustainability Conference, Long Beach, CA, pp. 531–539 (2007)
9. Canadian Mortgage and Housing Corporation (CMHC), Analysis of the Annual Energy and Water Consumption of Apartment Buildings in the CMHC HiS-TAR Database , Canada, Research Highlights (1999)
10. H. Foekema, O. Engelsma, Eenanderconsumptiepatroon het waterverbruikthuis 2001, VEWIN, Amsterdam, Netherlands (2001)
11. Rozporządzenie Ministra Infrastruktury i Rozwoju z dnia 27 lutego 2015 r. w sprawie metodologii wyznaczania charakterystyki energetycznej budynku lub części budynku oraz świadectw charakterystyki energetycznej
12. Verordnung über energiesparenden Wärmeschutz und energiesparende Anlagentechnik bei Gebäuden (Energieeinsparverordnung - EnEV) Anlage 3: Methode und Berechnung der Gesamtenergieeffizienz von Gebäuden (2012)
13. BREDEM 2012, A technical description of the bred omestic energy model, version 1.1, January 2015, www.bre.co.uk/sap2012
14. Rozporządzenie Ministra Infrastruktury z dnia 6 listopada 2008 r. w sprawie metodologii obliczania charakterystyki energetycznej budynku stanowiącej samodzielną całość techniczno-użytkowa oraz sposobu sporządzania i wzorów świadectw ich charakterystyki energetycznej.

E3S Web of Conferences 17, 00015 (2017) DOI: 10.1051/e3sconf/20171700015
EKO-DOK 2017

TTC- Based Test as an Efficient Method to Determine Antibiofilm Activity of Silver Nanoparticles

Joanna Chojniak[1,*], *Izabela* Biedroń[1], and *Grażyna* Płaza[1]

[1]Institute for Ecology of Industrial Areas, Microbiology Department, Kossutha 6, 40-844 Katowice, Poland

Abstract. Among metal nanoparticles, silver nanoparticles are a widely used in various life sectors such as in biomedical applications, air and water purification, food production, cosmetics, garments and in various household products. There are several methods for production of silver nanoparticles. Generally, silver nanoparticles can be prepared by chemical methods such as chemical reduction and electrochemical techniques, physical methods, and biological methods such as the use of microorganisms. The biological route of synthesis provides a great diversity in choice for its raw materials such as bacteria, algae, fungi and plants. The aim of the study was to evaluate the tetrazolium/formazan test as a method to determine antibiofilm activity of biological synthetized silver nanoparticles. In this study *Bacillus subtilis* grown on brewery effluent and produced biosurfactant was used for silver nanoparticles (Ag-NPs) synthesis. The culture supernatants were used in synthesis of Ag-NPs. The formation of nanoparticles accompanied by colour changes of the used reaction system was confirmed by UV-vis spectroscopy. The bacteria isolated from the biofilm of water supply system were used in the evaluation of the antibiofilm activity of biologically synthetized Ag-NPs. To compare the results the commonly used crystal violet assay (CV) for biofilm analysis was applied.

1 Introduction

1.1 Biosynthesis and characterization of silver nanoparticles

Recently subject of many investigations have become metal nanoparticles. Most frequently used methods of synthesis of metal nanoparticles are chemical and physical methods. In this methods microwaves, ultrasound, irradiation, mechanical grinding are used. Both methods have a negative features, for example, using toxic solvents to the synthesis, during the process of hazardous intermediates are produced, and the processes are often high energy [1, 2]. The alternative way to the processes is biological synthesis of metal nanoparticles (biosynthesis). In this method microbial activities of bacteria, fungi,

* Corresponding author: chojniak@ietu.katowice.pl

yeast are used, and plant extract are applied. The synthesis of silver nanoparticles used microbiological method is the important part of nanobiotechnology, connecting the range of biotechnology research with nanotechnology. However, despite some reports, still many aspects of nanoparticles biosynthesis should be studied in detail It was observed that biosurfactants produced by microorganisms play very important role in aggregation and stabilization process [3–13]. They adsorbed onto Ag-NPs surface stabilizing the nanoparticles and prevent their aggregation. Biosurfactants of biological origin have a lot of positive properties: most of them are rapidly degraded, they are not toxic, their activity is in a wide range of temperature, salinity and acidity [14, 15, 17–21].

1.2 Biofilm in water drinking distribution

The drinking water distribution system is the environment where biofilm is formed, regardless of used disinfectants. Control of the development of this structure is important for both aesthetic , health and financial reasons. Formation of biofilm in water distribution system consists of several stages (surface conditioning, irreversible attachment, colonization, detachment) and it is regulated by physical, chemical and biological processes. The first phase is a non-specific and reversible process of binding the bacteria cell to the surface. In the next phase the bacteria secrete the insoluble extracellular polymeric substances (EPS) and form the hydrogel matrix in which cells are embedded. With the increasing amount of EPS a mature biofilm is developing [21, 23]. Bacteria which make up biofilm on the water-pipes' internal surfaces produce great amounts of extracellular polymer substances (EPS). It is a complex mixture of proteins, polysaccharides, lipid acids and nucleic acids. Composition and structure of EPS influences on biofilm integrity, both in terms of physiology and structure as well as have direct impact on physical and chemical properties as well as biological properties of biofilm. Polymer substances essentially improve adhesion as well as make up a barrier which protects bacteria cells from the impact of adverse factors such as disinfecting agents [24]. Microorganisms resistant to antibiotics in water distribution system also exhibit resistance to chlorination. Environmental stress caused by the disinfection process may consequently increase the resistance to antibiotics [25]. Biofilm forming microorganisms are far more resistant to environmental stress, thus the first selection emerged microorganism whose high ability for adhesion and simultaneous resistance to antibiotics, which confirms the direct descent of biofilm. In earlier studies were selected 11 strains accepted as a model group for biofilm formation in water distribution system. Selection of microorganisms based on the identification of high ability to adhere and presenting resistance against antibiotics and chemotherapeutics. Ability to adhesion is a source of information on the actual response of the microorganism to conditions in the environment.

The aim of the study was to evaluate the impact of biological synthetized Ag-NPs on the biofilm formation by the TTC method. Also, the CV assay was used to compare the results.

2 Materials and Methods

2.1 Biosynthesis and characterization of silver nanoparticles

Bacillus subtilis isolated form the petroleum contaminated soil was used in the synthesis of Ag-NPs [26]. The strain was identified and characterized in earlier studies [27]. The strain was grew on the brewery wastewater at 30°C for 96 hours, under shaking. After incubation the culture was centrifuged by 10 min, 5 000 rpm. The free-cell supernatant was used for biosynthesis of Ag-NPs. $AgNO_3$ was added to the supernatant in final concentration 1 mM. The solution was incubated at 30°C for 48 hours. The presence of nanoparticles was monitored by UV-Vis spectrophotometer at a wavelength of 300–700 nm. The concentration of Ag-NPs measured by atomic absorption spectroscopy was around 165 mg/L.

2.2 Bacterial strains

Biofilm samples were collected from drinking water distribution system in Wroclaw. Morphological characterization (colony colour, shape and size) was recorded on selective media and biochemical characterization was tested according to API Zym test [28]. The identification of selected strains was carried out by the sequencing and comparative analysis of 16S rDNA,. The isolates were detected representatives of the genus: *Advanella Microbacterium, Kocuria, Moraxella, Enhydrobacter, Micrococcus, Pseudomonas, Bacillus, Psychrobacter*. Selected strains were grown on LB agar plates at 22°C .The single isolated colony was inoculated in sterile LB broth and incubated at 22°C in rotary shaker (150 rpm) for 24–48 hours. After incubation, culture was centrifuged at 5000 rpm for 10 minutes.

2.3 Evaluation of Ag-NPs impact on the biofilm formation by selected bacteria

2.3.1 Crystal violet assay

The bacterial suspension was adjusted to 0.5 McFarland's standard. A volume of 100 µl of bacterial suspension was added to each well on microtiter plate. The silver nanoparticles were added to the suspensions and incubated 48h at 22°C. After incubation the medium was discarded from the each well and thoroughly washed with PBS (Phosphate-buffered saline). A volume of 100 µl of 0.1% crystal violet was added to each well and left for 30 min at room temperature. The stain was discarded and plate was thoroughly washed by PBS. The remaining in biofilm structure stain was solubilized with 200 µl of 95% ethanol. The concentration of CV was determined by measuring the optical density of solution at 620 nm [15].

2.3.2 TTC assay

The bacterial suspension was adjusted to 0.5 McFarland's standard. A volume of 100 µl of bacterial suspension was added to each well on microtiter plate. A volume of 10 µl of silver nanoparticles solution was added to the suspensions and incubated 24 hours and 72 hours at 22°C. After incubation, the medium was discarded from every well and thoroughly washed twice with PBS to remove unbound cells. A volume of 100 µl of 0.01% solution of TTC was added to each well and incubated for 24 hours at 22°C. After incubation, TTC solution was removed and microplate was air dried. Bound red formazan was dissolved using 20% acetone and 80% ethanol. The concentration of reduced TTC was determined by measuring the optical density of solution at 500 nm [19]. Each experiment was done in 3 replicates. Cell-free supernatant contained only biosurfactant and AgNPs + biosurfactant solution were used as the controls. The impact AgNPs was detected in indirect way, eg.

$$OD_{AgNPs} = OD_{AgNPs+\ biosurfactant\ solution} - OD_{cell\text{-}free\ supernatant}$$

Also, the bacterial cultures were used to evaluate the biofilm formation.

3 Results and Discussion

During the synthesis of silver nanoparticles the colour of the solution was changed. The presence of silver nanoparticles was monitored by UV-Vis spectrophotometer in the range of 300-700 nm (Fig. 1). The typical peak between 400–410 nm for Ag-NPs was observed.

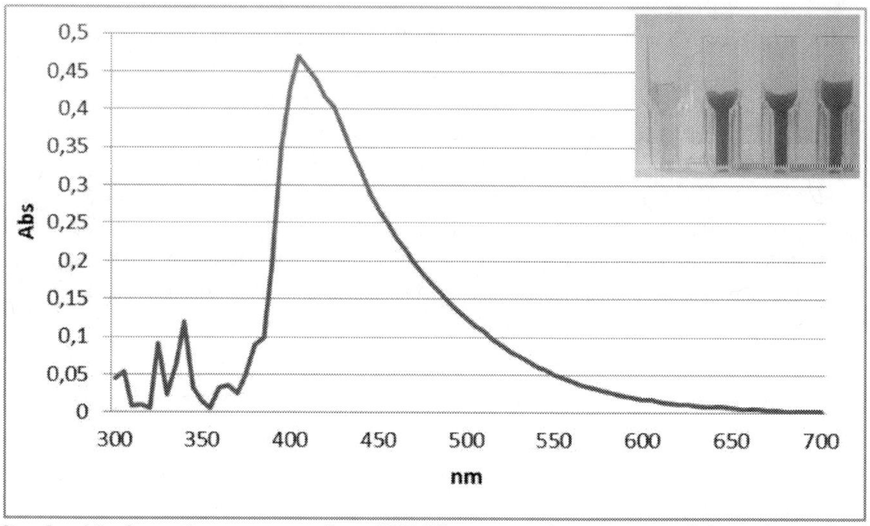

Fig. 1. Graph with absorption spectrum from UV-Vis spectrophotometer and the colour of solution.

Crystal violet (CV) staining was used as colorimetric method to measure biofilm formation in the presence of silver nanoparticles [16]. Based on the results obtained a greater inhibition of the biofilm formation was observed in the solution containing silver nanoparticles and biosurfactant. Nine strains showed a reduced level of adhesion after 24 hours of incubation as a result of silver nanoparticles solution (Fig. 2). In the case of two strains: *Bacillus thuringiensis* (5) and *Moraxella osloensis* or *Enhydrobacter aerosaccus* (2), where there was no observed the reduction in biofilm formation. The strains are Gram positive and produce significant amount of EPS composed mainly from proteins [28]. Also strain belong to the genus *Microccocus* (10) showed no significant decrease of biofilm formation. The largest adhesion inhibition was observed in two strains: *Advenella incenata* (6) – 65,31%and strain belong to the genus *Microccocus* (9)- 80,92%. After 72 hours of incubation, 10 of 11 strains showed 40% or more reduction in adhesion as a result of silver nanoparticles. The highest inhibition of biofilm formation was observed, for *Advenella incenata* (1), 86.85% (Fig. 3). It was a strain which showed very high antibiotic resistance [28]. No impact of silver nanoparticles was observed for strain 3- *Kocuria rhizophila* or *Psychrobacter sp.*

DOI: 10.1051/e3sconf/20171700015

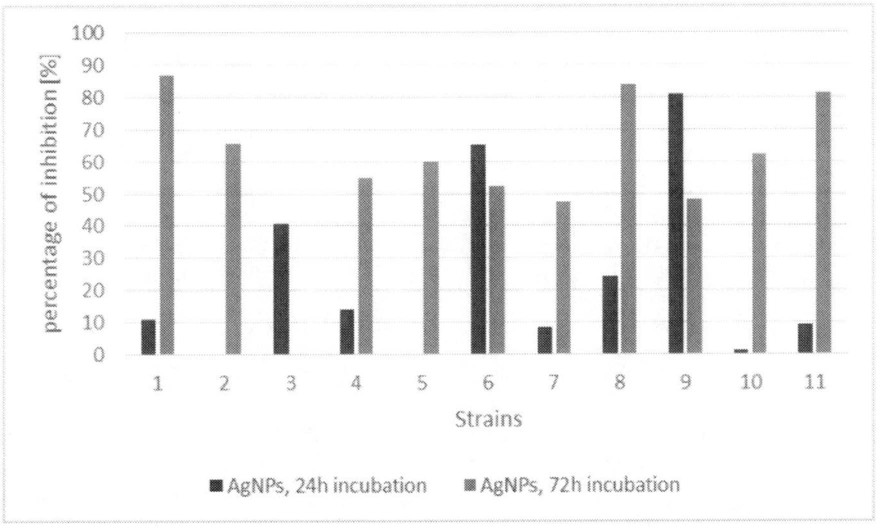

Fig. 2. Percentage of inhibition formation biofilm in Crystal Violet (CV) method. Strains: 1. *Advenella incenata,* 2. *Moraxella osloensis* or *Enhydrobacter aerosaccus,* 3. *Kocuria rhizophila or Psychrobacter sp.,* 4. *Microbacterium oxydans,* 5. *Bacillus thuringiensis,* 6. *Advenella incenata,* 7. genus *Advenella,* 8. *Micrococcus luteus,* 9. genus *Micrococcus,* 10. genus *Micrococcus,* 11. genus *Pseudomonas.*

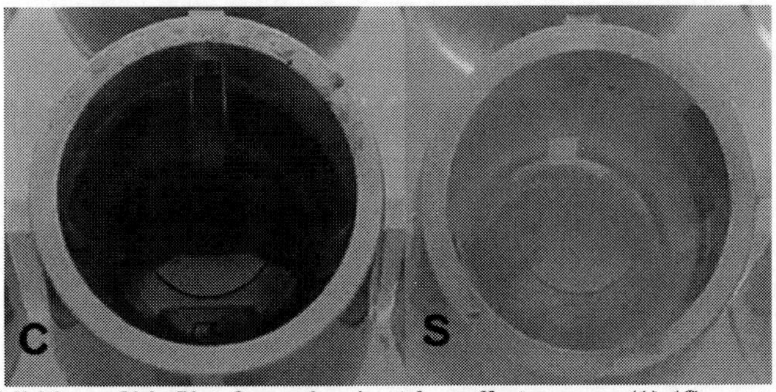

Fig. 3. The highest inhibition of biofilm formation in *Advenella incenata* (1) (C- control, S- sample).

The TTC assay is quantification techniques based on the metabolic activity of biofilm cells. In this method tetrazolium salt is converted by metabolically active cells to a colored formazan. The TTC assay results indicate minimum inhibition of dehydrogenase activity of biofilm. Significantly inhibition was observed in *Microbacterium oxydans* (4) after 72h of incubation. There was no impact of nanoparticles for dehydrogenase activity in strain (1) *Advenella incenata,* wherein the highest inhibition of biofilm formation was observed in CV assay.

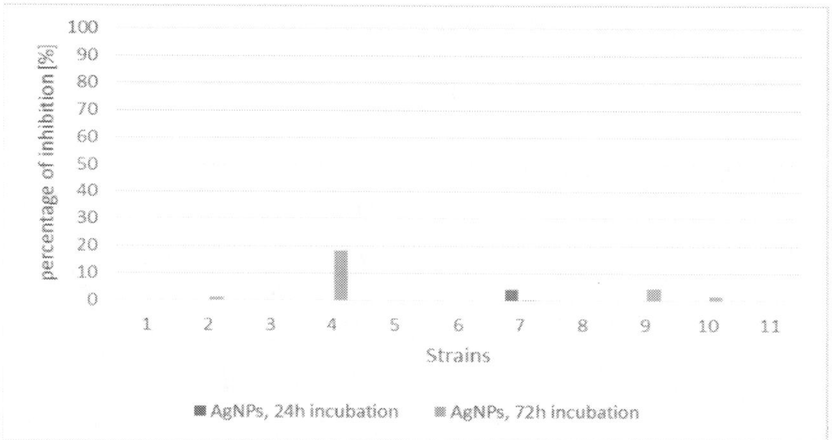

Fig. 4. Percentage of inhibition cellular respiration in TTC method. Strains: 1. *Advenella incenata,* 2. *Moraxella osloensis* or *Enhydrobacter aerosaccus,* 3. *Kocuria rhizophila or Psychrobacter sp.,* 4. *Microbacterium oxydans,* 5. *Bacillus thuringiensis,* 6. *Advenella incenata,* 7. genus *Advenella,* 8. *Micrococcus luteus,* 9. genus *Micrococcus,* 10. genus *Micrococcus,* 11. genus *Pseudomonas*.

From last few years studies of AgNPs as antibacterialn and antibiofilm agents have increased. Nano-sized silver particles can be prevalent for much longer time than silver ions and act on sites within bacterial cells that are critical to physiological function (e.g. cell wall, DNA/RNA, protein syntheses and electron transport). It is said that the effect of AgNPs on biofilm removal was size-dependent, smaller nanoparticles had the higher impact on biofilm structure [29]. In our study we tested nanoparticles on environmental strains, in contrast to research on the model strains. Perez-Diaz et al. and Martinez – Gutierrez observed antibiofilm and antibacterial properties of AgNPs [30, 31]. We confirmed activity of AgNPs as anti-biofilm agent, but we weren't observed decrease in respiratory activity by TTC assay, suggesting biocidal properties. Due to the non-specific efficient of AgNPs we think that nanoparticles have especially influence to metabolic reactions that controls the production of EPS compounds [31]. We have obtained results indicating that nanoparticles inhibited biofilm production by blocking the formation of exopolysacharides, which allows us to claim that to assessment antibiofilm properties, CV assay is more appropriate.

4. Conclusions

1. Probably, the silver nanoparticles inhibited biofilm production by blocking the formation of EPS. Therefore the crystal violet method is more reliable in determining of the biofilm inhibition than TTC assay.
2. Each strain has its own resistance against the silver nanoparticles.
3. The results of silver nanoparticles activity were observed after prolonger incubation.

The study was supported by the National Science Centre (Poland) (Project No. 2013/09/B/NZ9/01759 (decision no. 2013/09/B/NZ9/01759)

References

1. K. Patel, S. Kapoor, D.P. Dave, T. Ukherjee, J. Chem. Sci. **117**, 311 (2007)
2. P. Sileikaite, I. Prosycevas, J. Puiso, A. Juraitis, A. Guobiene, Mater Sci. **12**, 287 (2006)

3. A.K. Mittal, Y. Chisti, U.C. Banerjee, Biotechnol. Adv. **31,** 346 (2013)
4. A.S. Reddy, C.Y. Chen, C.C. Chen, J.S. Jean, Fan C.W., H.R. Chen, J.C. Wang, V.R. Nimje, J.Nanosc.Nanotech. **9,** 6693 (2009b)
5. A.S. Reddy, C.Y. Chen, S.C. Baker, C.C. Chen, J.C. Jean, C.W. Fan, H.R. Chen, J.C. Wang, Mat. Letters **63**, 1227 (2009a).
6. B.R. Singh, S. Dwivedi, A.A. Al-Khedhairy, J. Musarrat, Coll.Surf.B: Bioint. **85**, 207 (2011)
7. C. Tamuly, M. Hazarika, S.Ch. Borah, M.R. Das, M.P. Boruah, Coll. Sur.B: Biointerfaces **102**, 627 (2013)
8. G.G. Kumar, S.K. Mamidyala, Coll.Sur.B: Bioint. **84**, 462 (2011)
9. G.S. Kiran, A. Sabu, J. Selvin, J. Biotechnol. **148**, 221 (2010)
10. G.V. Kumar, D.S. Gokavarapu, A. Rajeswari, S.T. Dhas, V. Karthick, Z. Kapadia, T. Shrestho, A.I. Barathy, A. Roy, S. Sinna, Colloids Surfaces B. Biointerface. **87**, 159 (2011)
11. I.M. Banat, A. Franzetti, I. Gandolfi, M.G. Martinotti, L. Fracchia, T.J. Smyth, R. Marchant, Microbial Biosurfactants Production, Applications And Future Potential, Appl. Microbiol. Biotechnol., **87**, 427 (2010)
12. X. Zhang, S. Yan, R.D. Tyagi, R.Y. Surampalli, Chemosphere **82**, 489 (2011)
13. Y.J. Somg, K-H Jang, S.B. Kim, Progress Biochemistry. **44**, 1133 (2009)
14. D. Bhattacharya, R.K. Gupta, Crit. Rev. Biotechnol. **25**, 199 (2005)
15. D.K. Bozanic, S. Dimitrijevic- Brankovic, N. Bibic, A.S. Luyt , V. Djokovic, Morphology, Optical And Antimicrobial Properties. Carbohyd. Polymers, **83**, 883 (2011).
16. D. Dusane, V. Pawar, Y. Nancharaiah, V. Venugopalan, A. Kumar, S. Zinjarde, Biofouling, vol. **27**, 645 (2011)
17. K.B. Narayanan, N. Sakthivel, Add. Coll. Int. Sc. **156**, 1 (2010)
18. L. Kvitek, A. Panacek, J. Soukupova, M. Kolar, R. Vecerova, R. Prucek, M. Holecova, R. Zboril, J. Phys. Chem. C **112**, 5825 (2008)
19. M. Gericke, A. Pinches, Hydrometallurgy, **83**, 132 (2006)
20. P.S. Chandran, M. Chaudhary, R. Pasricha, A. Ahmad, M. Sastry, Biotechnol. Prog. **22**, 277 (2006)
21. T. Leung, Ch.K. Wong, Y. Xie, Mat. Chem.Physics, **121**, 402 (2010)
22. L. F. Melo, F. J. Mergul, vol. **2011**, 1414 (2014)
23. M. J. Lehtola, I. T. Miettinen, M. M. Keinänen, T. K. Kekki, O. Laine, A. Hirvonen, T. Vartiainen, and P. J. Martikainen, Water Res., vol. 38, no. **17**, 3769 (2004)
24. Z. Xue, V. R. Sendamangalam, C. L. Gruden, and Y. Seo, Environ. Sci. Technol., vol. **46**, 13212 (2012)
25. C. Xi, Y. Zhang, C. F. Marrs, W. Ye, B. Foxman, J. Nriagu, C. Xi, Y. Zhang, C. F. Marrs, W. Ye, C. Simon, B. Foxman, and J. Nriagu, Appl. Environ. Microbiol. Vol. **75** nr. 17, 5714 (2009)
26. C.J. Berry, S. Story, D.J. Altman, R. Upchurch, W. Whitman, D. Singleton, G. Płaza, R.L. Brigmon, *Biological Treatment of Petroleum and Radiological Contaminated Soil* (Oxford University Press., 2006)

27. G.A. Płaza, M. Pacwa - Płóciniczak, Z. Piotrowska - Seget, R. Brigmon, E. Król, *Characterization of Bacillus strains producing biosurfactants* (Springer Science Business Media, 2015)
28. I. Biedroń, T. Traczewska, G. Płaza, Rocz. Ochr. Śr.,**18(2)**, 86 (2016)
29. C.Y. Loo, P.M. Young, R. Cavaliere, C.B. Whitchurch, W.H. Lee, R. Rohanizadeh, Drug Dev Ind Pharm. **40(6)**, 719 (2014)
30. F. Martinez-Gutierrez, L. Boegli, A. Agostinho, E.M. Sánchez, H. Bach, F. Ruiz, G. James, Biofouling. **29(June)**, 651 (2013)
31. M.A. Pérez-Díaz, L. Boegli, G. James, C. Velasquillo, R. Sánchez-Sánchez, R.E. Martínez-Martínez, G.A. Martínez-Castañón, Martinez-F. Gutierrez, Mater Sci Eng C., **55**, 360 (2015)

Decision support for optimal location of local heat source for small district heating system on the example of biogas plant

Bartłomiej Ciapała[1,*], *Jakub* Jurasz[2], and *Mirosław* Janowski[1]

[1]AGH University of Science and Technology, Faculty of Geology, Geophysics and Environmental Protection, Department of Fossil Fuels, 30 Mickiewicza Av., 30-059 Kraków, Poland
[2]AGH University of Science and Technology, Faculty of Management, Department of Engineering Management, 30 Mickiewicza Av., 30-059 Kraków, Poland

Abstract. Developing a new district heating system requires making decisions affecting entire range of following activities and wellness of the system. The article presents methodics of choosing optimal location and crucial customers with three approaches to optimisation process, discuss obtained results and obstacles. Further improvements and potential applications are named.

1 Introduction

Establishing a new, local district heating (DH) network is a complex and long-lasting project, including cooperation and communication with both local community and individual potential heat receivers. Presented article describes an algorithm useful in the first stages of designing a DH network. The algorithm may support decision making and developing conceptual network's framework. Although the tool was primary developed to help owners of the biogas plants to increase waste heat utilisation, it is also fully applicable for any other DH heat source.

2 Algorithm description

2.1 Input data

As the algorithm is performed with Solver in Microsoft Excel, the input data must have been "translated" into a matrix of digits so that the problem might be given, "understood" by this extension and solved. First steps were performed out of Excel software, based on available terrain maps. Each building was marked and superimposed with regular, square network, what allowed to assign proper number of buildings for each square. Here, for a simplicity it has been assumed that all consumers exhibit the same level of power/heat

* Corresponding author: bciapala@agh.edu.pl

DOI: 10.1051/e3sconf/20171700016

demand, meaning that a digit "1" in given square corresponds to a heat demand equal to 43 GJ. Square with greater numbers have accordingly greater heat demand (see Fig. 2.).

If known was an exact annual energy demand, value of each square might be a sum of thermal energy amount required for households within certain cell. In this process useful is GIS software, which, beside fast and precise computing, allows for creating informative maps with customised layout (see Fig.1.) as well as tables significantly expediting creating the input matrix for the algorithm.

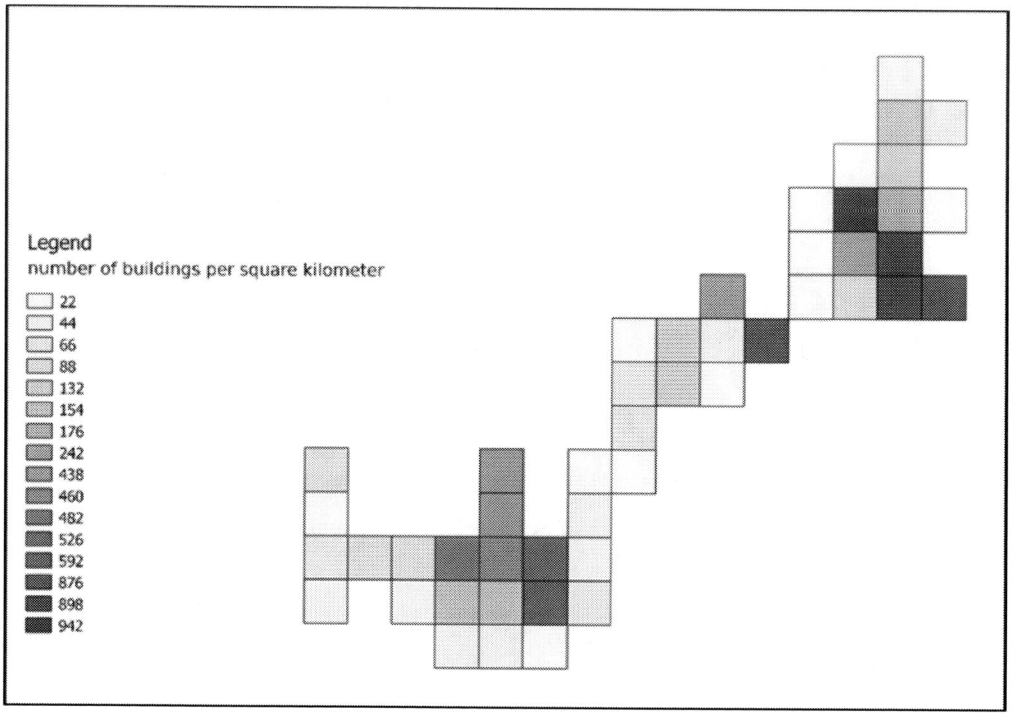

Fig. 1. A simplified map visualizing the heat demand/number of households over considered area.

2.2 Biogas plant location

In order to find the optimal location for the biogas power plant and to select heat receivers which should be connected a mixed integer nonlinear mathematical model has been introduces. The problem of finding the optimal location for a biogas plant is definitely a multi objective problem. Here we introduce three simple but meaningful objective functions. Literature commonly make of two units: GJ*km and GJ/km (called linear heat density) [1–3] alternative way of expressing energy rated to a relevant parameter is GJ/km^2 [3], nevertheless this approach is not useful in present consideration.

2.2.1 Minimizing the total length of the hot water pipes:

The following equation (1) describes objective function for this criterion, id est check for each consumer j and choose m receivers, which may be connected with minimal length of pipeline.

$$\min Z^{I} = \sum_{i=1}^{m} x_i d_{i,j} \quad \forall j \tag{1}$$

where: x_i – is a binary variable (0,1) which value is estimated based on Eq. 2, $d_{i,j}$ – is a distance between heat source i and the heat consumer j.
The x_i parameter is determined as 0, when in location i has no heat source within or 1, when there is a heat source in the location i (equation (2).

$$x_i = \begin{cases} 1\ if\ heat\ source\ has\ been\ located\ in\ location\ i \\ 0\ otherwise \end{cases} \tag{2}$$

2.2.2 Minimizing the heat transferred over given distance [GJkm] pipes:

The equation (3) objective is to pick heat receivers, which are supposed to be connected to the pipeline. Each combination of set of connected receivers and heat source's location should be checked, what is extremally long-lasting.

$$\min Z^{II} = \sum_{i=1}^{m} \sum_{j=1}^{n} y_j c_j x_i d_{i,j} \tag{3}$$

where: c_j – is a heat demand from consumer j, y_j – is a binary variable calculated based on equation (4).

The following equation determines, that $y_j$variable is 1 when the heat receiver is connected to the network and equals 0, when building is not considered as prospective in the combination.

$$y_j = \begin{cases} 1\ if\ consumer\ j\ has\ been\ connected\ to\ the\ biogas\ plant \\ 0\ otherwise \end{cases} \tag{4}$$

2.2.3 Maximizing the heat transferred per unit of distance [GJ/km] pipes:

$$\max Z^{III} = \frac{\sum_{j=1}^{n} y_i c_j}{\sum_{i=1}^{m} x_i d_{i,j}} \quad \forall j \tag{5}$$

2.2.4 Constraints

Naturally, objective functions are subject to the following constraints:

$$\sum_{i=1}^{m} x_i = 1 \tag{6}$$

which constitutes that only one location for heat source is allowed, and

$$\sum_{j=1}^{n} y_j c_j = P \tag{7}$$

providing that total required energy amount does not exceed capability of heat source, where: P – is the maximal volume of heat delivered by given biogas plant [GJ].

Constrains given above have been implemented in our example. Also additional constrains might be applied, for example:

$$\sum_{i=1}^{m} \sum_{j=1}^{n} d_{i,j} \leq D \tag{8}$$

where: D – is the given maximal length of hot water pipes.

It is important to bear in mind that such "facility location" problem is very specific. Meaning that connecting additional heat receiver will always lead to an increase in overall objective function. Therefore it was necessary to introduce an additional constraint which will ensure that the algorithm will always connect some number of buildings. This constraint is given in Eq. 9, where N is the minimal number of connected heat receivers.

$$\sum_{j=1}^{n} y_j > N \tag{9}$$

2.2.5 Algorithm operation

For selecting the optimal location of a local heat source, the *brute-force* method may be applied – in case of this research we have used an add-in to the MS Office Excel in form of Solver with its inbuilt Evolutionary and GRG methods for solving the nonlinear problems. Therefore not always the optimal configuration has been found for given objective function (as shown in Tab. 1). The step by step procedure of finding the optimal location of the heat source as well as selecting the subset of heat receivers which will be connected to the hot water heat pipe is as follows:

1. Determine the potential locations where the heat source can be located – a potential location in case of investigated model is a site free from heat receivers. Please bear in mind that an additional constraint may be introduced that ensures that the heat source is not closer to the next heat receiver than a given distance.
2. Select one of the location and create a distance matrix between heat source and each of the heat consumers.
3. Select a subset of heat consumers in case of which all constraints are satisfied and objective function reaches its minimum. Memorize the configuration and the results.
4. Go back to the second step and select another location. Repeat procedure from point 3. If all potential locations have been investigated choose that one for which the objective function is optimal.
5. Display resulting configuration and resulting objective function.

3. Results

The applicability of the proposed models has been tested for an exemplary commune (see Fig. 2. for its layout) which consist of 146 buildings. The colourful squares represent areas where buildings (heat consumers) are located, whereas blank squares are free zones where a biogas plant might be located. The red square represents area, where heat source should be optimally placed. As it was mentioned above, presented is simplified model, with buildings of the same heat demand, meaning that a digit in given square corresponds to a heat demand equal to 43 GJ multiplied by square's value. The maximal volume of heat which can be delivered by the considered biogas plant amounts to 4300 GJ. Here it is important to underline that we refer only to the heat – meaning the energy, not the biogas heating plant momentary power output. It is an important issue which should be addressed in future works, because heat demand is not constant and exhibit significant variations even during the heating (autumn – spring) period. Apart from the above, the minimal number of connected households should be not smaller than N = 50. Additionally it is important to add that each square represents an area equal to 10 000 m^2.

Hereunder, on Fig. 3–5 are presenting optimal configuration for three investigated objective functions Eq. 1, 3 and 5. In further analysis we will refer to them as cases A, B and C, where in A the objective function was minimize the total length of the hot water pipes, in B the objective was to minimize the value of "GJkm" parameter and in C the objective was the find the configuration which will result in the maximal value of "GJ/km" parameter. Results are summarized in Table 1.

 DOI: 10.1051/e3sconf/20171700016

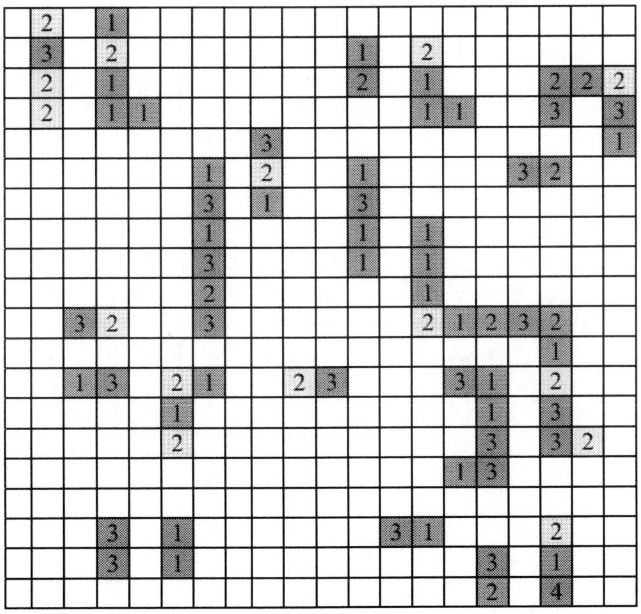

Fig. 2. Village layout.

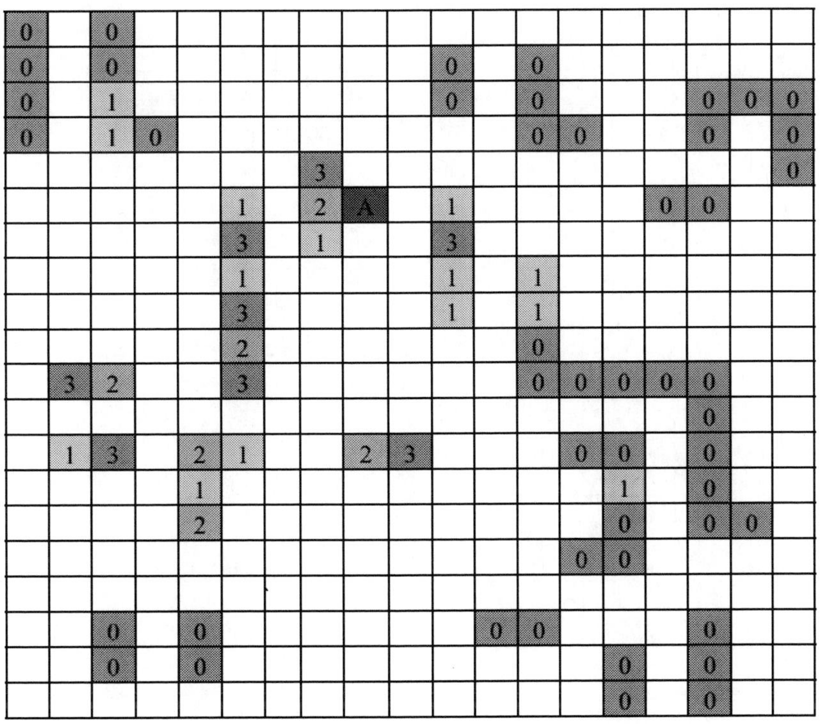

Fig. 3. Heat source location (A) and number of selected heat consumers – for the minimal length of the hot water pipes objective function.

 DOI: 10.1051/e3sconf/20171700016

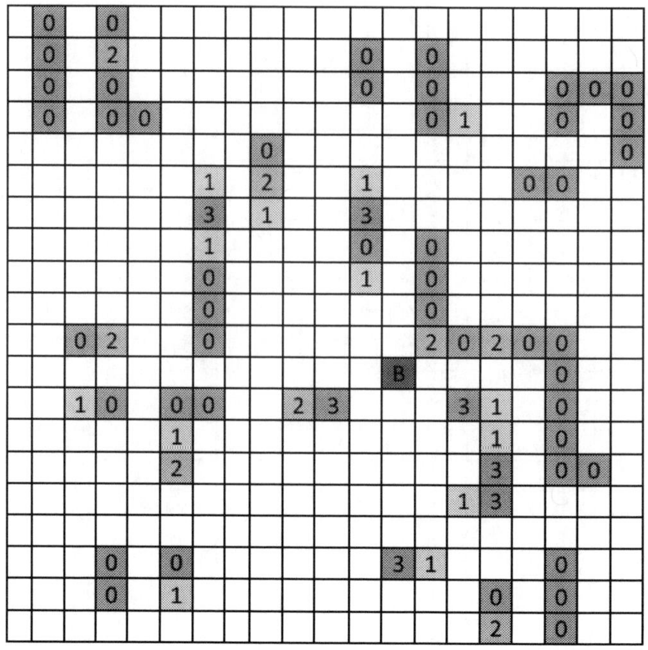

Fig. 4. Heat source location (B) and number of selected heat consumers for the minimal value of "GJkm" objective function.

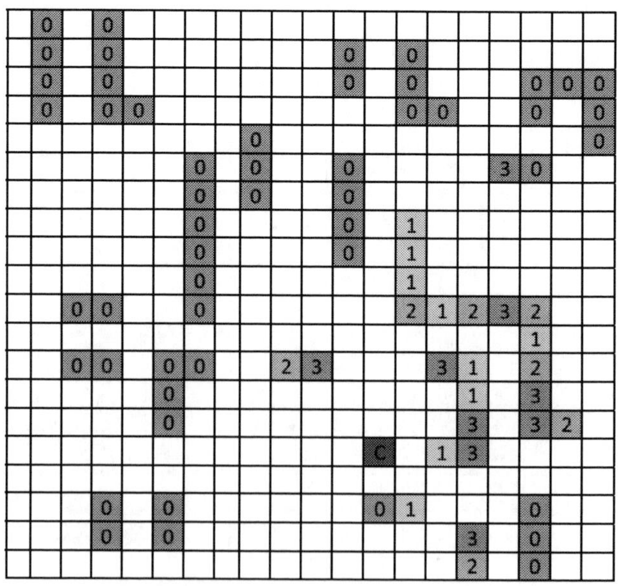

Fig. 5. Heat source location (C) and number of selected heat consumers – for the maximal value of the "GJ/km" objective function.

Table 1. Results for the conducted optimization
(value in shadowed cell is the subject of the optimisation).

	Objective function		
Case	Minimal length of the hot water pipes [km]	Minimal value of "GJkm"	Maximal value of "GJ/km"
A	2.9	6235	741.37
B	3.64	7826	590.65
C	**2.01**	**4321.5**	**1069.65**

4. Discussion

One of the most obvious aspects of proposed method is why length of the pipeline is counted as a sum of distance between each heat receiver and thermogenic plant meanwhile buildings might be grouped in a single pipeline. Main reason of this approach is, that chosen receivers will certainly be the optimal set, as simulation conducts worst-case scenario. What is more, even grouped receiver require the pipeline of increased diameter, thus investment costs are increased per meter of such pipeline. It is worth to emphasize, that the algorithm does not provide proposition of the pipeline tracing, neither takes into consideration a shape of the parcels or existing buildings/pipelines.

The A case may seem to be surprising, as expected would be rather circular shape of network range, nevertheless the second constrain - required number of 50 connected dwellings - interferences typical behaviour of the algorithm. Actually, it tends to connect required number of heat receivers in as numerous groups as possible. Such an approach may be useful, if target was to connect maximal number of dwellings within strongly limited budget.

This and further results clearly show, that proposed solutions of this highly complex and multi-optional problem are not optimal, due to limited time of computation and, thus, incomplete collections of possible solutions, which number amounts to:

$$\frac{L!}{B!(L-B)!} \times N \tag{10}$$

where L is number of all buildings, B is number of buildings to be connected and N is number of available heat source positions.

In future considerations, authors will pursue to perform full brute force analysis. Of course, differences in obtained results (A, B, C) come from different applied objective functions. A case described in 2.2.2 subsection (B) tends to indicate as prospective groups of buildings, even if they are placed farther, whilst method from subsection 2.2.3 (C) tends to point location close to selected buildings. The C objective function provides high linear energy density, id est comparatively favourable investment return time, whilst the B objective function is supposed to minimise the share of energy that is lost over the pipeline, although it is disputable.

Although the algorithm was primarily developed to point the best position of the biogas plant, it is fully useful also in considerations of other heat sources, with properly stated constrains, even geothermal energy sources (with directional drilling wells).

References

1. S. Nielsen, B. Moller, Energy, **69**, 458–468 (2013)
2. H. C. Gils, J. Cofala, F. Wagner, W. Schopp, Energy, **58**, 318–329 (2013)
3. M. Kimming, C. Sundberg, A. Nordberg, P.-A. Hansson, Energ Policy, **78**, 51–61 (2015)
4. S. Pusat, H. H. Erdem, Energ Buildings, **72**, 177–185 (2014)

Stability analysis of a dike constructed of anthropogenic materials

Andrzej Danilewicz[1,*]

[1]Gdansk University of Technology, Faculty of Civil and Environmental Engineering, Poland

Abstract. Numerical analyses of stability are presented for a hypothetical dikes constructed of dredged materials and bottom ash, in varying proportions. The paper is related to the international research project DredgDikes, which focused on the use of anthropogenic materials in dikes construction and flood protection. The aim of this project was to investigate the possibility of using anthropogenic materials from rivers and sea dredging and coal combustion products in dikes construction. The influence of different proportions of ash and dredged materials on the stability of dike is discussed. Mechanical and hydraulic parameters of various dredged material-ash mixtures are based on the results of laboratory and field tests performed in DredgDikes project. Numerical calculations are carried out using geotechnical software Plaxis. Cementation phenomenon and its effect on the performance of ash and dredged material mixtures is discussed.

1 Introduction

A dike is a construction made to prevent flooding, regulate river course or protect land against the sea. The design of a dike cross-section, cubage and high depends on the local variation in water levels. The flood protection in Poland is regulated under the Water Law [1] and technical requirements of dikes and other flood protection structures are defined in a ministerial directive [2]. Generally local materials are used to construct dike, but anthropogenic materials like sediments removed from rivers and sea during dredging works and also coal combustion products can be used under previous geotechnical supervision. Dredged materials are natural and integral components of the ecosystem. Coal combustion products (CCPs) are by-products of coal-fired power plants. The scope of the international project called DredgDikes was to propose how to plan and build dikes with the use of ripened dredged materials, coal combustion products and geosynthetics [3]. Dredged materials (and CCPs as well) are considered as a waste according to European Waste Catalogue [4]. Well grained dredged materials and CCPs have been used before in landscaping as well as other recultivation projects with good geotechnical experience. CCPs are also widely used in road constructions (especially in Poland).

The use of CCPs and dredged materials was a new idea in DredgDikes project. Stability mechanisms differ from dike types [7]. The stability of a dike slopes is one of the main

* Corresponding author: adanilew@pg.gda.pl

issue to be taken under account during engineering design. During this analysis the dike and subsoil have to be investigated according to EC7 [5] and national recommendations [6].

2 Test dike description

During the DredgDikes Project two dikes were constructed - one in Rostock in Germany and the other one in Trzcińsko, near Gdańsk in Poland (see Fig. 1.). One of the main scope of polish part of the project was to investigate the possibility of using mixed dredge materials with coal combustion products for river dikes constructing. For this part of the project, the dredge material from the bottom of Vistula river was adopted. As a CCPs the fly ash from nearest Gdańsk power plant was used.

Fig. 1. Test dike in Trzcińsko - near Gdańsk [8].

The construction process of dike in Trzcińsko was ended in May 2012. The dike is 3m high with inclination 1:2. The dike body is built from dredged material and compacted bottom ash mixture. As a result of preliminary studies, the construction material for the dike body consists of 30% of dredged material and 70% of ash. The dike is built on the compacted clay for environmental isolation. The inner side of the dike is covered with Tefra material – commercial mix of fly-ash and sand. The water level can be maintained about 2.5 meters above the ground level. The phreatic line for steady state flow was measured at elevation 1.62m. The dimensions of the test dike are presented in Fig. 2.

2.1 Laboratory tests

The parameters of soil layers used for the test dike construction was investigated in laboratory during the project and published in [9] (see Tab. 1). The series of laboratory tests have determinated mechanical parameters and physical properties of the clay and dredged material-ash bottom mixture. To obtain internal friction angle and cohesion direct shear tests were performed. Permeability coefficients were obtained for the different mixtures as well.

 DOI: 10.1051/e3sconf/20171700017

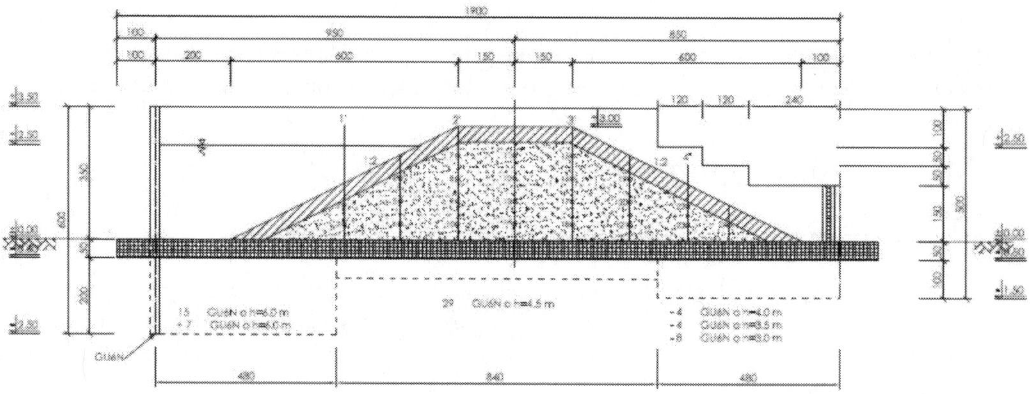

Fig. 2. Cross section of the polish test dike [8].

Table 1. Geotechnical parameters for clay and Tefra layer [9].

		Clay	Tefra
unit weight	[kN/m^3]	18	16
saturated unit weight	[kN/m^3]	21	20
permeability coefficient	[m/s]	1E-7	1E-7
undrained shear	[kPa]	50	30
effective cohesion	[kPa]	30	50
effective angle of friction	[°]	25	20
Young's module	[MPa]	15	30
Poisson's ratio	[-]	0,35	0,25

2.1 Mixture parameters

One of the goals in this tests was to choose the optimum dredged material–CCPs mixture ratio to be applied in field test. The mixture with 70% content of dredged sand from Vistula river mouth and 30% of bottom ash from EC Gdańsk power plant was chosen. Obtained values with their interpolation are presented in Fig. 3. according to proposition given in [10].

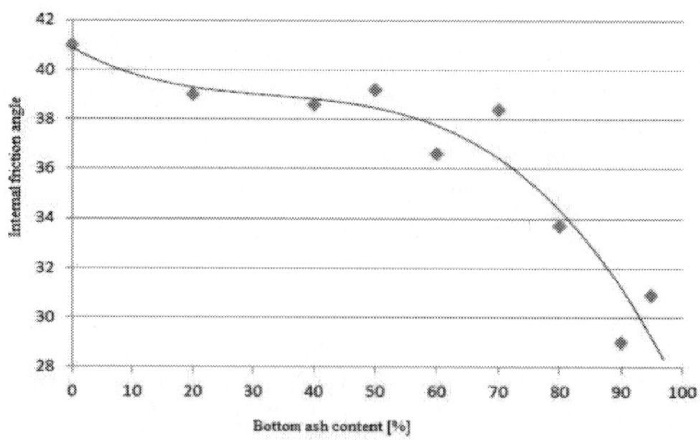

Fig. 3. Internal friction angle for various mixtures [10].

According to recommendation from DredgDikes project only mixtures with unit weight higher than 12 kN/m^3 should be used for the dike construction, so mixtures 90/10 and 100/0 are skipped in further considerations. Some geotechnical properties for different mixtures are presented in Table 2.

Table 2. Geotechnical parameters for various mixtures [10].

		0/100	10/90	20/80	30/70	40/60	50/50	60/40	70/30	80/20
Unit weight	[kN/m^3]	16.87	16.68	16.48	15.60	14.72	13.83	13.05	12.36	12.07
Saturated unit weight	[kN/m^3]	20.43	20.11	19.79	19.05	18.33	17.62	16.39	16.39	16.01
Permeability coefficient	[m/24h]	8.64	7.52	6.41	5.30	4.19	3.08	1.97	0.86	0.86
Effective cohesion	[kPa]	0	0	0	0	0	0	2.6	5.5	8.8
Effective internal friction angle	[°]	41	40	39.30	38.80	38.60	39.50	38.50	38	34.50
Young's module	[MPa]	36.50	34.50	32.50	30.50	28.50	26.50	24.50	22.50	20.50
Poisson's ratio	[-]	0.25	0.25	0.25	0.25	0.25	0.25	0.25	0.25	0.25
Cementation	[kPa]	0	2.2	4.4	6.6	8.8	11	13.3	15.5	17.7

3 Numerical analysis

As a first step numerical calculation of steady state flow through polish test dike and stability analysis were performed using Plaxis program. This sort of calculation was performed before and published in [7],[11]. Numerical model of the polish test dike is shown in Fig. 4.

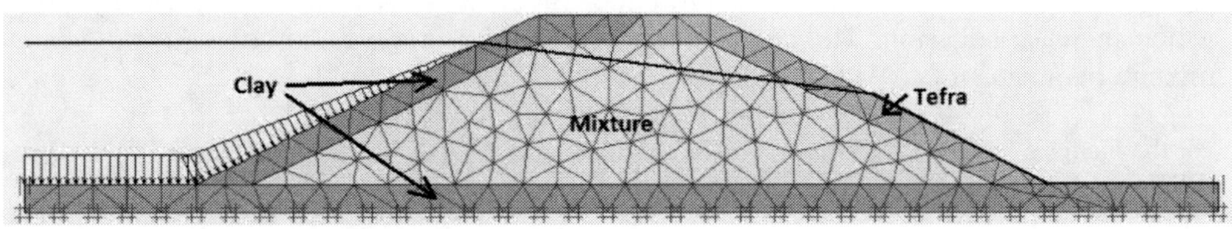

Fig. 4. Numerical model with FEM discretization of the polish test dike.

3.1 *Phi-c reduction* calculation

Phi-c reduction calculation process is an option in Plaxis software to compute safety factors. In this kind of calculation strength parameters of soil i.e. tangent of internal friction angle and cohesion are reduced step by step until failure occurs.

As a first step stability of the polish test dike was repeated with outcome of safety factor equal 2.667. This value is slightly different from values obtained in [7, 11] and the difference (about 10%) is due to calculation method in other programs (SLIDE 5.0).

After preliminary calculation, the dike stability analysis for different mixture dredged material and ash bottom proportion were performed. The failure surface has similar form in all cases. The list of obtained safety factors are presented in Tab. 3 and Tab. 4.

Table 3. Safety factors Msf for different mixtures.

	0/100	10/90	20/80	30/70	40/60	50/50	60/40	70/30	80/20
Safety factor Msf [-]	2.02	1.98	1.96	1.95	1.94	1.97	2.33	2.67	3.20

Table 4. Safety factors Msf for different mixtures after cementation process.

	0/100	10/90	20/80	30/70	40/60	50/50	60/40	70/30	80/20
Safety factor Msf [-]	2.02	2.02	2.57	2.88	3.21	3.62	3.97	4.17	4.56

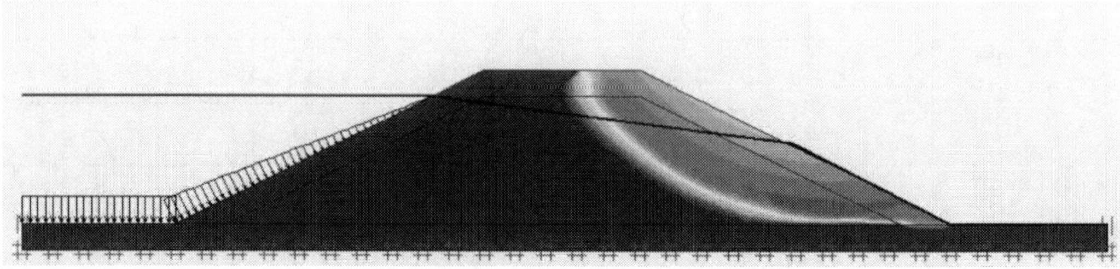

Fig. 5. Failure mechanism for the polish test dike for steady state flow.

4 Conclusions

All analysis proved dike slope stability according to minimum value of Msf≥1.3 (and recommended value of Msf≥1.5). The Eurocode 7 requirements have also been met. It has to be mentioned that all calculations were performed for characteristic values. Cementation process highly increase slope dike stability. One of the most important characteristic of dredged material and coal combustion products mixture is its ability to cementing process which increase cohesion. The contribution of cohesion is more important from the other mixture parameters.

All the analysis were realized thanks to DredgDikes project (2011-2013) part-financed by EU funds (ERDF) in the frame of South Baltic Programme.

References

1. Polish Water Law 2005, Journal of Laws 05.239.2019
2. Polish Directive of the Ministry of Environment of 20 April 2007 on the technical requirements to be met by hydrotechnical structures and their location, Journal of Laws 07.86.579
3. F. Saathoff, S. Cantré, Z. Sikora, *Application of Dredged Materials, Coal Combustion Products and Geosynthetics in Dike Construction* (2015)
4. EC 2000. European Waste Catalogue (EWC) 2000/532/EC
5. EN 1997-1. Eurocode 7 - Geotechnical design - Part 1: General rules (1997)
6. PN-EN 1997-1:2008. Eurokod 7 Projektowanie Geotechniczne, Polish national implementation of the Eurocode 7, 152 pp
7. L. Bałachowski, Stability of the Inner Slope of the Polish DredgDikes Research Dike at Stationary Flow, *Proceedings of the South Baltic Conference on Dredged Materials in Dike Construction*, Rostock: 105-110 (2014)

8. R. Duszński, M. Cudny, *Projekt stanowiska badawczego wału przeciwpowodziowego wykonanego z mieszanki urobku czerpalnego i popiołów*. Politechnika Gdańska (2012)
9. L. Bałachowski, Z. Sikora, *Mechanical properties of bottom ash – dredged material mixturs in laboratory tests*. Studia Geotechnica et Mechanica, p. 3-11, Vol. XXXV, No. 3, 2013, Oficyna Wydawnicza Politechniki Wrocławskiej (2013)
10. S. Sorek, *Analiza porównawcza stateczności wału wykonanego z różnych proporcji popiołu i refulatu*, Master thesis, Gdańsk University of Technology, Gdańsk, (2015)
11. I. Pessato, *Steady state and transient analysis of the flow through the test dike*. Master thesis, University of Padova, Padova (2014)
12. Plaxis 8.2 manuals: www.plaxis.nl

Full-scale biodrying process of municipal solid waste

Marlena Dębicka[1*], and *Maria* Żygadło[1]

[1]Kielce University of Technology, Faculty of Environmental, Geomatic and Energy Engineering, Division of Waste Management, 25-314 Kielce, al. Tysiąclecia PP 7, Poland

Abstract. The paper presents the results obtained in the full-scale waste biodrying reactor. The studied facility includes in the biological stage a rectangular-shaped, galvanized steel reactor equipped with a module for active aeration connected with a stove and a bio-filter for removing odours. The undersize fraction (Ø <80 mm) of the municipal solid waste (MSW) that undergoes mechanical pretreatment is treated by 14th days in the 150 m^3 capasity reactor. Initial moisture content of the untreated waste was 54.65%. Moisture content was declined gradually during biodrying process. Temperature changes during 14th days of biodrying process were monitored with the maximum temperature 70°C. To assess the degree of stabilization of the biodried waste, the determination of the O_2 uptake was measured. The oxygen demand of untreated waste was 53.16 mg O_2/g d.m. and after 14th days for biodried waste oxygen consumption was 19.78 mg O_2/g d.m. The results obtained in studies by respirometric dynamic method of oxygen uptake (expressed as O_2/96h parameter) had been compared to the results performed using the static method, where AT_4 is the applied indicator.

1 Introduction

Common and widely studied mechanical-biological treatment (MBT) of aerobic waste systems are composting and stabilization [1]. Biodrying is a new variation of MBT processes that removes convective moisture allowing to produce highly calorific output [2, 3]. Wastes after biodrying process with low moisture content and enhanced calorific value, are not completely biologically stable, but can be used as a source of refuse derived fuel (RDF). These RDFs should be used immediately, without storage [4–7].

Drying technology has been developed within the scope of food, agricultural, pharmaceutical, paper and other industries [8]. The basic parameters that affect municipal solid waste (MSW) biodrying are process temperature, moisture content, aeration efficiency and reactor design. So far biodrying research in laboratory- and full- scale are limited. Investigations on 480 dm^3 lab-scale reactor resulted in increasing the sorting efficiency of post product by ~ 40%, higher heavy metals concentration in dry mass by 60% and lower heating value by ~160% [9, 10]. Experience of a 240 dm^3 lab-scale reactor included

* Corresponding author: mdebicka@tu.kielce.pl

searching for design of a drying tunnel allowed to produce high calorific biofuel [11]. In the literature it has been reported studies on the influence of temperature and air-flow rate on the biostabilization–biodrying of municipal solid waste in the 148 dm^3 lab-scale reactor [12, 13]. It was estimated, that temperature optimal for biodrying process is 45°C, while in the composting and the stabilization the temperature can rise up to 70°C.

In order to understand the science and engineering of biodrying processes adequately, it is necessary to make investigations in full-scale facilities. The article presents the results of temperature changes, microbial activity and moisture content variation observed in full-scale biodrying facility. Results obtained in our study have been compared to results reported for other full-scale MSW biodrying experiments.

2. Materials and methods

The layout of the technology is schematically presented in Figure 1. The biodrying stage has the following specifications:

— feedstock characteristics: 0-80 mm fraction of municipal solid waste, containing ~40% of biodegradable waste, ~40% of mineral contaminates (glass, ceramic, metals, other minerals), ~12% of plastic, foil, textiles and ~8% of fraction <10 mm.
— reactor capacity and retention time: 50–60 Mg treated by 14th days,
— major equipment: the rectangular-shaped reactor made from galvanized steel,
— remaining equipment: the bio-filter that removes odours, the stove used optionally in specific winter conditions, rotating drum,
— aeration system: the supply ventilator (air flow 4500 m^3 h^{-1} using 7.5 kW power) and the exhaust ventilator (air flow 3000 m^3 h^{-1} using 4 kW power),
— final products: 20-80 mm fraction biodried waste can be used as a refuse derived fuel RDF and finally combusted, 0-20 mm fraction directed to stabilization process and would be finally landfilled.

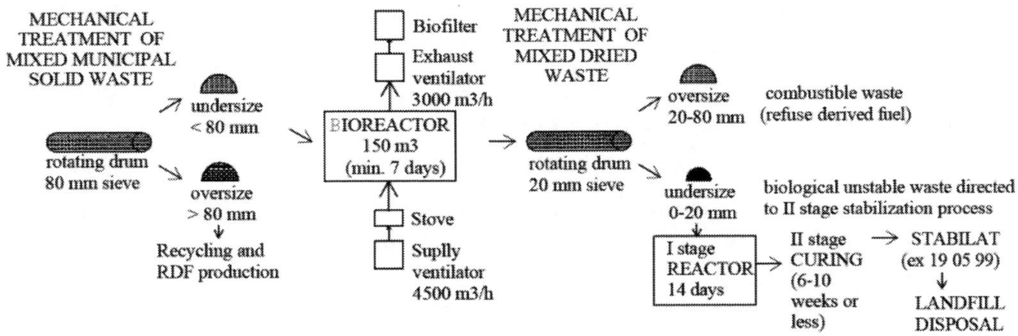

Fig.1. Technology concept in the studied MBT plant.

Samples were taken according to regulatory requirements [14]. The water content was calculated based on the weight loss in the 1000 g sample at 105°C [15]. The measurement of the loss on ignition (LOI) was determined in a muffle furnace at 550°C by 6 hours using 5 g samples according to the Polish Standard [16]. The temperature inside the biomass was measured 5 times per day in the different reactor parts using 2-metre lance. The biological stability was determined by dynamic aerobic test using a respirometer Oxymax ER-10, which allows to measure the CO_2 production and the O_2 uptake during the time. The respiration activity should be measured in an oxidic atmosphere, for the 30 g minimum mass sample with a humidity of 40–50% [17]. By means of a respirometer, a number of measurements were taken in time intervals every 5 minutes, recording the net increase or decrease in the concentration of the monitored gases. The microbial activity AT_4 was investigated in accredited laboratory in Pszczyna according to internal procedure.

3 Results and discussion

3.1 Temperature and moisture variation

Initial moisture content of the untreated waste was 54.65%. Moisture content (MC) were declined gradually during biodrying process. The main mechanism of water removal was the evaporation of water as a consequence of microbial heat generation. In our study the biodrying processes were completed with 35.04% of moisture content. Thus the drop of moisture at 20% content was obtained during studied biodrying process. Similar results had been reported by Zawadzka et al. and Hurka & Malinowski [11, 18]. According to [19] the continuous decrease in the feedstock humidity could be additional indicator of the organic matter decomposition, because in result of biodrying process about 28% of initial waste mass was evaporated as a CO_2 and H_2O. Similar results had been reported in publications [20, 21].

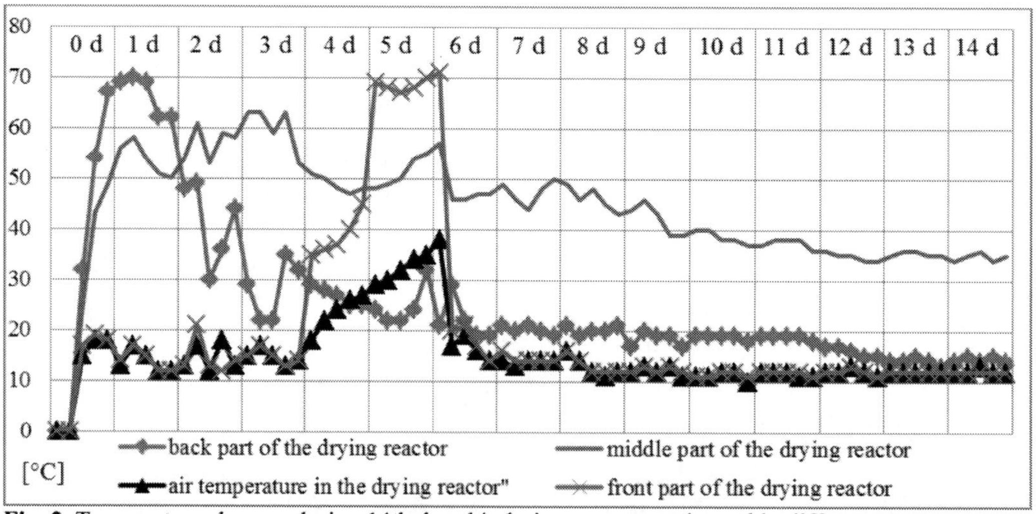

Fig. 2. Temperature changes during 14th days biodrying process registered in different reactor areas.

Temperature changes registered in the different reactor parts are presented in Fig. 2. The feedstock temperature was monitored 5 times per day in the front, middle and back part of the drying reactor. Temperature is also the factor that allows to assess the microbial activity. The highest temperature about 70°C was recorded in the middle part of the reactor while the average level was 43.45°C, similar to Dziedzic et al. [22]. The lowest temperatures were registered in the front part of the reactor. Temperatures in the middle part in the reactor was optimal for the biodrying process as was estimated by Tambone et al. [23] (about 45°C). Rapid increase in the temperature values was recorded during 5th day, what needed to switch off the aeration system. After 6th day temperature was contentiously decreased and the aeration system was switched on again.

3.2 Organic matter degradation

The reactivity/stability of the MBT products can be determined by aerobic and anaerobic methods. Respiration techniques based on the measurement of the loss of O_2 are the most accepted because of the precise information about the real biological activity of the samples [24]. The parameter O_2/96 h obtained during respiration technics presents the microbiological oxygen consumption over 4 days and determines the ability of the waste for further biological decomposition. Table 1 summarizes the data respiration activity of the raw (untreated MSW) 0-80 mm fraction and the dried 0-80 mm fraction. In Poland (till 2016) waste after MBT process should fulfill the requirement of respiration activity level $AT_4 < 10$ mg O_2/g d.m. to be called ''stabilat'' and than it is allowed to be landfilled [25].

Table 1. The microbial activity measured as changes examined for 0-80 mm MSW fraction before and after biodrying process in industrial reactor.

Fraction / Unit	Accumulative consumption O_2 [mg 96 h^{-1}]	Accumulative production CO_2 [mg 96 h^{-1}]	Consumption O_2/96 h [mg O_2 g d.m.$^{-1}$]	AT_4 [mg O_2 g d.m.$^{-1}$]
Untreated waste	3117.47	3578.19	53.16	not measured
Biodried waste	1782.15	1939.48	19.78	14.30

The results obtained in the studies by respirometric dynamic method of oxygen uptake (expressed as O_2/96h parameter) had been compared to the results performed using the static method, where AT_4 is the applied indicator. Indicator AT_4 is not satisfied, because its value exceeds 10 mg O_2/g d.m.

According to the data presented in Table 1 the starting microbial activity measured as oxygen demand for untreated waste, expressed as O_2/96h parameter, was 53.16 mg O_2/g d.m. That suggests high degree of microbial activity and capability to biodegradation of organic matter. Biodried waste was characterized by lower oxygen consumption and expressed as O_2/96h parameter, was 19.78 mg O_2/g d.m. Decreasing of oxygen demand parameter allows to make the conclusion that the biodried waste undergone stabilization. It was accomplished in 14 days at a certain level. It is called ''drying effect'' that occurs when the amount of readily biodegradable matter is completed or when low moisture conditions inhibit microbial activity [26].

Our results had been compared to reports regarding the microbial activity of biodrying waste in the 240 dm^3 lab-scale reactor, where O_2/96h parameter was decreased from 8.01 to 0.31 mg O_2/g d.m.. The difference in our and that study can be explained by reactor design and intensity of aeration [27]. Results from 19 full-scale MBT plants (2 with biodrying reactors) showed even average 90% decrease in the respiration activity index [28].

The results obtained in our studies by respirometric dynamic method of oxygen uptake (expressed as O_2/96h parameter) had been compared to the results performed using the static method, where AT_4 is the applied indicator. Indicator AT_4 is not satisfied, because its value exceeds 10 mg O_2 /g d.m.

According to Figure 3 the amount of organic matter (measured as LOI) was decreased during biodrying process. Every point in Fig. 3 is the average of three repetitions. Higher organic matter losses were observed in the middle part of the reactor. The biodegradation in the middle reactor area was faster than in other areas. It is connected with the highest temperature profile and intensified microbial activity. The degradation process slows down due to decreasing microbial activity, that is caused by the lowering water content [22].

LOI results from our study were compared with results obtained in a two-stage MBT process in the full-scale 36 m^3 reactor using 4500 m^3/h aeration [22]. In the first stage waste was stabilized for 14th days, than in the second stage 7-days biodrying was applied. Starting LOI value was 77.1% d.m., while after stabilization and biodrying amounted to 51.2% d.m., 48.3% d.m. respectively. LOI was reduced by 30%, similar to our study but it was still above the value limited by law requirements where LOI should be lower than 35% d.m. [25].

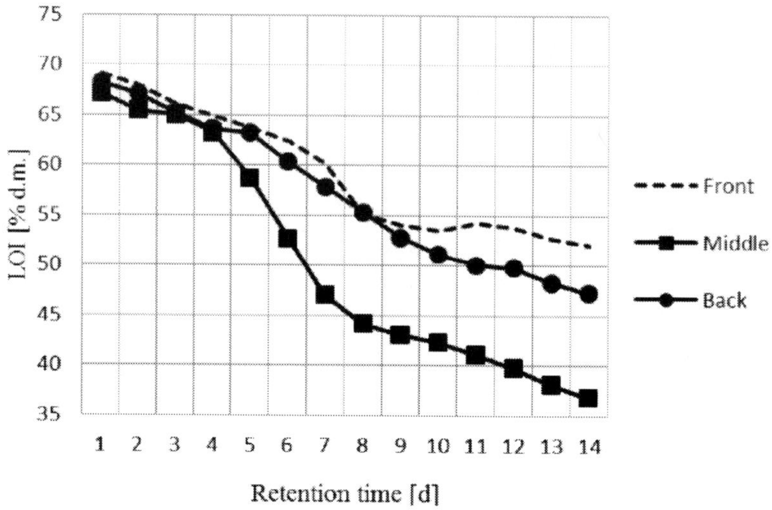

Fig. 3. The loss of ignition measured in the front, middle and back parts of the biodrying reactor during 14th days retention time.

4 Conclusion

The MSW undergone the drying process for 14th days in the full-scale reactor. Moisture content was reduced by 20%, while organic matter determined as loss of ignition was reduced by 35%. The highest difference between starting and final values of LOI was observed in the middle part of the reactor where temperatures were about 70°C.

The respiration activity after biodrying was higher than the limited level of 10 mg O_2/g d.m, expressed as AT_4 . However, the amount of O_2/96 h parameter was decreased by 63%.

Summing up the results we must conclude that it is advisable to extend the biodrying process to get AT_4 at the required level. For this purpose, it is reasonable to attempt to intensify the level of aeration of reactor after a period of 7 days.

Investigations were led using: OXYMAX respirometer ER-10 (Columbus Instruments). Equipment was founded by MOLAB Project ''The development of laboratories of public universities in Świętokrzyskie Region''. Number of Project in Kielce University of Technology: POIG 02.02.00-26-023/08-00.

References

1. B. Bilitewski, *Mechanical Treatment: Unit Processes. Solid Waste Technology & Management*, **1** (A. John Wiley and Sons, UK, 2011)
2. M. Żygadło, M. Dębicka, Structure & Environment, **4**, 37-72 (2014)
3. E.C. Rada, M. Ragazzi, A. Badea, U.P.B. Sci. Bull, **74**, 209-216 (2012)
4. F. Adani, F. Tambone, A. Gotti, Waste Manage, **24**, 775-783 (2004)

5. R. M. Negoi, M. Regazzi, T. Apostol, E.C. Rada, C. Marculescu, U.P.B. Sci. Bull, **71** , 193-204 (2009)
6. A. Elnaas, A. Belherazem, W. Müller, A. Nassour, M. Nelles, at: http://uest.ntua.gr/tinos2015/proceedings/pdfs/elnaas_et_al.pdf
7. M. Malinowski, J. Sikora, Proceedings Ecopole, **1**, 223-230 (2014)
8. C.A. Velis, P.J. Longhurst, G.H. Drew and R. Smith, S.J.T. Pollard, Bioresource Technol, **11**, 2747-2761 (2009)
9. D. Zhang, H. Zhang, C. Wu, L. Shao, P. He, Waste Manage, **31**, 2816-2823 (2011)
10. L. Shao, Z. Ma, H. Zhang , D. Zhang, P. Ge, Waste Manage, **30**, 1165-1170 (2010)
11. A. Zawadzka, L. Krzystek, S. Ledakowicz, Environ Prot Eng, **35**, 123-133 (2009)
12. F. Adani, D. Baido, E. Calcaterra, P. Genevini, Bioresource Technol, **83**, 173-179 (2002)
13. M. Sugni, Calcaterra E., F. Adani, Bioresource Technol, **26**, 1331-1337 (2005)
14. *Solid recovered fuels. Methods for sampling.* PN-EN 15442:2011
15. *Municipal solid waste. Examination of combustible property. Determination of the total moisture.* PN – 93/Z15008
16. *Characterization of waste – Determination of loss on ignition in waste, sludge and sediments.* PN – EN 15169:2007
17. R. Barrena, G. d'Imporzano, S. Pons´a, T. Gea, A. Artola, F. V´azquezc, A. S´ancheza, F. Adani, Hazard Mater, **162**, 1065-1072 (2009)
18. M. Hurka, M. Malinowski, Infrastructure And Ecology Of Rural, **4**, 1127-1136 (2013)
19. A. Domińczyk, L. Krzystek, S. Ledakowicz, Inz. Ap. Chem, **51**, 115-116 (2012)
20. M. Dębicka, M. Żygadło, J. Latosińska, Ecol Chem Eng A, **12,** 1461-1470 (2013)
21. A. Tom, A. Haridas, R. Pawels, Procedia Technology, **25**, 130-137 (2016)
22. K. Dziedzic, B. Łapczyńska-Kordon, M. Malinowski, M. Niemiec, J. Sikora, Chem Process Eng, **36**, 381-384 (2015)
23. F. Tambone, B. Scagalia, S. Scotti, F. Adani, Bioresource Technol, **102**, 7443-7450 (2011)
24. R.B. Gomez, F.V. Lima, A.S. Ferrer, Waste Manage Res, **37**, 37-47 (2006)
25. *The Decree of the Minister of the Environment on the municipal solid waste treatment in mechanical biological plants* [Journal of Laws of 2012, No. 0, item 1052]
26. D.M. Griffin, *Water and microbial stress. Advances in microbial ecology*, (Springer, US, 1981)
27. Ł. Abramczyk, A. Domińczyk, R. Ślęzak, R. A. Cichowicz, Acta Innovations, **13**, 5-12 (2014)
28. J. Tinter, E. Smidt, K. Böhm, E. Binner, Waste Manage, **30**, 1903-1907 (2010)

Effects of aluminium oxide nanoparticles on bacterial growth

Nina Doskocz[1,*], *Katarzyna* Affek[1], and *Monika* Załęska-Radziwiłł[1]

[1]Warsaw University of Technology, Faculty of Building Services, Hydro and Environmental Engineering, Department of Biology, Nowowiejska 20, 00-653 Warsaw, Poland

Abstract. Production and wide application of nanomaterials have led to nanotechnology development but their release to environment and the induction of toxic reactions, affects the natural microbial communities. Therefore, studies on the impact of nanoparticles on microorganisms and environment are required and needed. The aim of this study was to assess the impact of aluminium oxide nanoparticles on the growth of *Pseudomonas putida*. To compare the harmfulness of different forms of aluminium oxide, the ecotoxicity of its macro-forms was also evaluated in the study. Research showed that the exposure to nanoparticles can negatively influence microorganisms. The EC_{50}-16h determined in this study was 0.5 mg/l, and NOEC equaled 0.19 mg/l. Nano-Al_2O_3 proved to be more toxic to *P. putida* than aluminium oxide. This indicates that the nano-form of a given substance demonstrates different properties and may constitute a far greater danger for the environment than the same substance in the large form. According to EU and US EPA criteria, nano-Al_2O_3 proved to be very toxic and highly toxic, respectively. Changes in bacterial communities caused by nanoparticles may affect the normal biological, chemical and nutrient cycle in the ecosystem and the effect triggered by nanomaterials in relation to other organisms is unpredictable.

1 Introduction

Nanoparticles (NPs) are molecules of dimensions below 100 nm. They are of colloidal particle size and are often smaller than bacterial and eukaryotic cells. Due to their unique physicochemical properties and morphology, such as: high surface to volume ratio, high chemical reactivity, ability to form aggregates, diffusivity, and mechanical strength, NPs have become an attractive material for commercial and technological application. NPs are currently used in many products. Silver (Ag) NPs are incorporated into textiles, clothing, food packaging and other materials for elimination of bacteria, cerium oxide NPs are used in electronics, biomedical supplies, energy, and fuel additives, nano-TiO_2 may be found in sunscreens, cosmetics, coatings and paints. Carbon nanotubes have a wide range of uses, ranging from electronic components to composites used to produce of vehicles. Many nanomaterials are used as membrane filters to remove pollution from water [1, 2, 3, 4, 5].

* Corresponding author: nina.doskocz@pw.edu.pl

Production and use of NPs results in their dissemination in the environment through air, wastewater and surface waters. NPs used in cosmetics in Sweden and Denmark are detected in wastewater at about 50 g per capita and day. Considering city with one million inhabitants the content of these compounds in wastewater would probably reach about 50 t/d [3]. If not captured during wastewater treatment they can occur in water intended for human consumption. Deposition of NPs in bottom sediments of water reservoirs and their migration through environmental compartments may also have dangerous consequences. Literature data indicates that their increasingly common application and release to environment may trigger toxic reactions, among others, in natural microbial communities.

NPs can change the properties of bacterial cell membrane, influence the permeability and respiration of the bacterial cells, cause damage to DNA and release the toxic ions, but the exact mechanism by which NPs promote this antibacterial ability, is so far the subject of debate [6]. Silver NPs based on the graphene were tested in relation to *Pseudomonas aeruginosa*. The minimum bactericidal concentration (MBC) was 20 mg/ml, and after 1 h disintegration of the cell wall was observed [6]. Tamayo et al. observed that silver and copper NPs may penetrate the wall and cell membrane of bacteria *Listeria monocytogenes*. This led to the separation of the cytoplasmic membrane of the cell wall [8]. Literature data also indicate that NPs proved to reveal the antibacterial properties against *Staphylococcus aureus*, *Escherichia coli*, *Bacillus subtilis* (nano-Cu, nano-ZnO, and nano-Ag) [9, 10, 11], *Klebsiella pneumoniae* (nano-ZnO) [10], *P. aeruginosa* (nano-Ag) [12], *Enterobacter cloacae* (Cu–SiO_2 nanocomposite) [13], *Salmonella typhimurium* (Cu NPs/chitosan) [14], *L. monocytogenes* (Cu NPs) [15], *Enterococcus faecalis* (nano-TiO_2) [11], and *Aeoromonas hydrophila* (nano-Al_2O_3, nano-ZrO_2) [16]. The impact of NPs on yeast *Candida albicans*, and mold *Aspergillus niger* was also observed [11].

Furthermore, NPs show negative impact on microorganisms involved in wastewater treatment. Zheng et al. proved that zinc oxide NPs had adverse influence on biological processes of nitrogen removal as well as on denitrifying bacteria and microorganisms responsible for phosphorus elimination called PAO (*polyphosphate accumulating organisms*). Nitrogen removal efficiency decreased from 81.5% to 75.6% and 70.8%, when concentrations of nano-ZnO equaled 10 mg/l and 50 mg/l, respectively [17]. These results indicate that zinc oxide NPs lead to the disruption of the activated sludge treatment process. In turn, a study published by Choi showed that during wastewater treatment, 1 mg/l of nanosilver inhibited the growth of autotrophic nitrifying bacteria by about 80% [18].

In order to determine their potential environmental fate and improve the understanding of the toxicity effects of the aluminum oxide NPs (nano-Al_2O_3, < 50 nm), the present study examines the antibacterial activity of these NPs against *P. putida* bacteria. The study compares the toxicity of NPs with their bulk counterparts (compounds of the macro-form – Al_2O_3). Only scarce information is available on fate, transport and effects in the environment of metal-based particles such as nano-Al_2O_3. Aluminum oxide NPs are currently used by the military and commercial industries as components of coatings and propellants and their widespread use may result in the release into the environment [19, 20, 21].

2 Materials and methods

2.1 Chemicals

Aluminum oxide NPs (nano-Al_2O_3), nanopowder < 50 nm of specific surface of > 40 m^2/g and aluminum oxide (Al_2O_3) of purity of > 98% were obtained from Sigma-Aldrich. Initial solutions of the tested compounds were prepared in deionised water. Because of the tendency of aggregate formation by the compounds, the obtained solutions of the compounds were sonicated for 30 min using an ultrasonic disintegrator of MDM-10 type (0.4 kW at a frequency of 20 kHz). The solutions were diluted (using a medium according to ISO 107122-1994 test procedure for *P. putida* [22]) with the ratio of geometric progression of dilutions (q) of 2 obtaining solutions ranging from 0.17 mg/l to 200 mg/l.

2.2 Bacterial strains

Heterotrophic gram-negative rods of *P. putida* were isolated from the laboratory activated sludge (Department of Biology, Faculty of Building Services, Hydro and Environmental Engineering, Warsaw University of Technology). Aseptic technique was used throughout the testing process. Constant bacterial cultures were maintained throughout the experimentation and incubations for all tests were conducted for 16 h at 26°C.

Bacterial cultures were then subjected to biochemical identification by API procedure. A Gram stain was used to determine whether the bacteria were positive or negative. Oxidase test was performed by adding bacterial smear to filter paper containing oxidase reagent (a mixture of dimethyl-4-phenylenediamine hydrochloride and α-naphthol). Colour development was observed within 1 min. Catalase test was performed by adding one drop of 30% hydrogen peroxide (Aflofarm, Pabianice, PL) to a slide that contained bacterial smear. Bubbling reaction was observed within 1 min. API 20 NE bacterial identification was performed according to manufacturer's instruction (bioMérieux, Durham, NC) [23].

2.3 Growth inhibition test with *Pseudomonas putida*

Growth test with *P. putida* was performed in accordance with ISO 107122-1994 [22]. Growth inhibition (I) was assessed on the basis of optical density values obtained for the samples with $\lambda = 610$ nm at the beginning and at the end of the 16 h test and was calculated according to Eq. 1.

$$I = \frac{B_c - B_n}{B_c - B_0} \cdot 100\,[\%] \qquad (1)$$

I – Growth inhibition [%],
B_c – Optical density of suspension in control sample after 16 h,
B_n – Optical density of suspension in the sample examined after 16 h,
B_0 – Optical density of suspension in control sample in time 0.

2.4 Calculation of EC_{50} and NOEC

Effective concentrations (EC_{50}) were calculated using probit analysis with 95% confidence intervals [25]. No observed effect concentrations (NOEC) were determined using ANOVA and Tukey's test [26].

2.5 Toxicity assessment of compounds

The assessment of toxicity of the test samples in relation to bioindicators was performed on the basis of the EU criteria—Directive 93/67/EEC (Table 1) and the U.S. Environmental Protection Agency (U.S. EPA) criteria (Table 2) [25, 26].

Table 1. Assessment of toxicity of chemicals in relation to the criteria of their harmfulness to aquatic biocenoses according to EU.

$LC(EC)_{50}$ [mg/l]	Toxicity assessment
< 0.1	Extremely toxic
0.1 - 1	Very toxic
>1 - 10	Toxic
>10 - 100	Harmful
>100	Non-toxic

Table 2. Assessment of toxicity of chemicals in relation to the criteria of their harmfulness to aquatic biocenoses according to US EPA.

$LC(EC)_{50}$ [mg/l]	Assessment of toxicity of chemical
≤ 1	Highly toxic
> 1 -100	Moderately toxic
>100	Slightly toxic

3 Results and discussion

The results of growth inhibition test with *P. putida* are shown on Figure 1. The assessments of ecotoxicity of NPs according to the European Union Directive (93/67/EEC) [25] and the US Environmental Protection Agency (US EPA) criteria [26] are presented in Table 3.

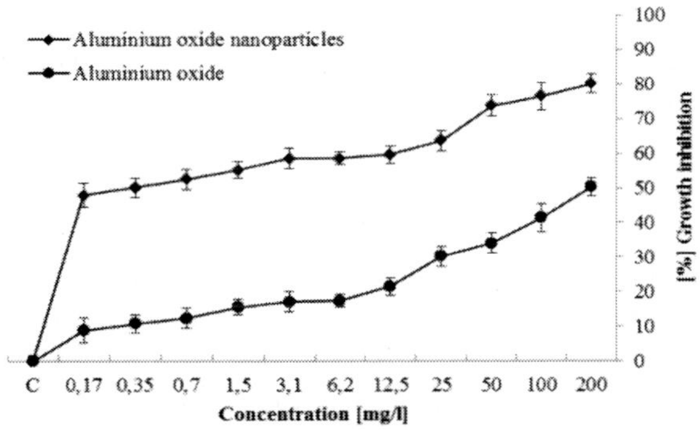

Fig. 1. Effect of nano-Al_2O_3 and Al_2O_3 on the growth of *P. putida*. C – means the control.

Table 3. Effect of nano-Al_2O_3 and Al_2O_3 on *P. putida*. 95% confidence intervals are shown.

Compound	Toxicity assessment criteria	Time [h]	EC_{50} (95% confidence intervals) [mg/l]	NOEC [mg/l]	Ecotoxicity assessment	
					UE	US EPA
Nano-Al_2O_3	Growth	16	0.5 (0.22-1.03)	0.19	Very toxic	Highly toxic
Al_2O_3			295.7 (143.21-241.22)	1.9	Non-toxic	Slightly toxic

Growth inhibition of *P. putida* was observed during the study. Aluminium oxide NPs inhibited the process in above 50% at all concentrations tested, except for the lowest (0.17 mg/l). In the highest tested concentration, inhibition of growth was 80.13% (Figure 1). EC_{50}-16h was 0.5 mg/l and NOEC – 0.19 mg/l (Table 3). The obtained values of EC_{50}-16h and NOEC-16h showed that *P. putida* was sensitive to the tested NPs and indicate that aluminium oxide in macro-form has a less impact on bacteria than the compound in the nano-form.

It is worth mentioning that very limited research concerns nano-Al_2O_3 antimicrobial evaluation as scientists have mostly explored the antimicrobial properties of Ag NPs [27, 28, 29]. EC_{50} in studies of Hachicho et al. [30] causing 50% growth inhibition for Ag NPs was about 250 mg/l. In turn, the experiments conducted by Matzke et al. showed that *P. putida* reacted very sensitively when exposed to nano-silver. EC_{50} value was between 0.25 μg/l and 3.46 μg/l for the different Ag NPs [27]. An experiment conducted by Gajjar et al. also showed some inhibition of *P. putida* in the presence of NPs. Toxic doses of the Ag NPs were 0.2 mg/l, for Cu NPs - 10 mg/l, and for Zn NPs - 7-10 mg/l [29].

Ecotoxicity assessment on the basis of EC_{50} showed that nano-Al_2O_3 are very toxic to bacteria. According to EU criteria, nano-Al_2O_3 was very toxic to *P. putida*. However, according to US EPA criteria aluminium oxide NPs were highly toxic (Table 3).

The obtained results as well as data from literature confirm that the effects of nano-Al_2O_3 on *P. putida* growth were stronger than those observed for their bulk counterparts (Table 3). The effect of Al_2O_3 on *P. putida* was only slight. EC_{50}-16h was 295.7 mg/l and NOEC – 1.9 mg/l. Similar conclusions were obtained by Jiang [30] and Jośko and Oleszczuk [31]. Tested Nanoparticles: Al_2O_3, SiO_2, TiO_2, ZnO and Ni demonstrated higher toxicity than their bulk counterparts.

Nanocompounds, as a result of their exceptional characteristics, can penetrate cell walls and afterwards the ions can exert toxic effects on cell organelles. The nano size of NPs determines their greater specific surface area than that of their bulk counterparts, due to which they can be more effective in their interactions with the components of bacterial cells and can provide higher risk to the environment than the same substances in their macro-forms [21, 30, 31, 32].

4 Conclusion

The aim of this study was to investigate the effect of aluminium oxide NPs on the growth of *P. putida* and the obtained data allowed to formulate the conclusion that nano-Al_2O_3 may inhibit bacterial growth. In accordance with the EU criteria, tested NPs were very toxic to the bacteria. However, according to US EPA, they were highly toxic.

The obtained results clearly indicate the necessity for ecotoxicity studies of NPs in relation to microorganisms. The study also shows that the currently available ecotoxicity data about compounds in bulk counterparts cannot be used to assess the harmfulness of their nano-form counterparts.

Increasing production of NPs lead to their accumulation in the environment, especially in landfills and landfill leachates and may result in non-target effects on the populations of microbes that play beneficial role in the environment. Negative consequences may concern disturbances in element cycling (carbon, sulfur, nitrogen, etc.), pollutants' degradation and plant growth promotion [29]. There is a constant need for further studies, including not only conventional but also multispecies and molecular tests, in order to explain the mechanisms of NPs' effects on microorganisms and to ensure safety of beneficial microorganisms in the environment.

References

1. https://www.epa.gov/chemical-research/research-nanomaterials
2. M. Łebkowska, M. Załęska-Radziwiłł, Ochr. Sr. **33**, 23-26 (2011)
3. S.K. Brar, V. Mausam, R.D. Tyagi, R.Y. Surampalli, Waste Manage., **30**, 504-520 (2008)
4. G. Bystrzejewska-Piotrowska, J. Golimowski, P.L. Urban, Waste Manage. **29**, 2587-2595 (2009)
5. P. Mosdorf, E. Karwowska E, A.R. Kunicki, G. Karolczak, Rocz. Ochr. Sr., **13**, 1635-1651 (2011)
6. J. R. Morones, J. L. Elechiguerra, A. Camacho, K. Holt, J. B. Kouri, J. T. Ramírez, M. J. Yacaman, Nanotechnology, **16**, 2346-2353 (2005)
7. T. He, H. Liu, Y. Zhou, J. Yang, X. Cheng, H. Shi, Biometals., **27**, 673-678 (2014)
8. L. A. Tamayo, P. A. Zapata, N. D. Vejar, M. I. Azócar, M. A. Gulppi, X. Zhou, G. E. Thompson, F. M. Rabagliati, M. A. Páez, Mat. Sci. Eng. C-Mater., **40**, 24-31 (2014)
9. J. P. Ruparelia, A. K. Chatterjee, S. P. Duttagupta, S. Mukherji, Acta Biomater., **4**, 707-716 (2008)
10. A. S. H. Hameed, Ch. Karthikeyan, A. P. Ahamed, N. Thajuddin, N.S. Alharbi, S.A. Alharbi, G.Ravi, Sci. Rep., **6**, 1-11 (2016)
11. D. Mitoraj, A. Jańczyk, M. Strus, H. Kisch, G. Stochel, P. B. Heczko, W. Macyk, Photochem. Photobiol. Sci., **6**, 642-648 (2007)
12. N. Perkas, G. Amirian, S. Dubinsky, S. Gazit A Gedanken, J. Appl. Polym. Sci., **104**, 1423-1430 (2007)
13. Y.H. Kim, D.K. Lee, H.G. Cha, C.W. Kim, Y.S. Kang, J. Phys. Chem. C, **111**, 3629-3635 (2007)
14. G. Cárdenas, J. Diaz, M.F. Meléndrez, C. Cruzat, A. García, Polym. Bull., **62**, 511-524 (2009)
15. N. Cioffi, L. Torsi, N. Ditaranto, G. Tantillo, L. Ghibelli, L. Sabbatini, T. Bleve-Zacheo, M. D'Alessio, P.G. Zambonin, E. Traversa, Chem. Mater., **17**, 5255-5262, (2005)
16. N. Chrzanowska, M. Załęska-Radziwiłł, *Interdyscyplinarne zagadnienia w inżynierii i ochronie środowiska* (Oficyna Wydawnicza Politechniki Wrocławskiej, 2012)
17. X. Zheng, R. Wu, Y. Chen, Environ. Sci. Technol., **45**, 2826-2832 (2011)

18. O. Choi, K. K. Deng, Jr. Kim, L. Ross, R. Y. Surampalli. Z. Hu, Water Res., **42**, 3066-3074 (2008)
19. N. Chrzanowska, M. Załeska-Radziwiłł, Desalination Water Treat., **52**, 3680-3689 (2014)
20. J. G. Coleman, D. R. Johnson, J. K. Stanley, A. J. Bednar, C. A. Weiss Jr., R. E. Boyd, J. A. Steevens, Environ. Toxicol. Chem., **7**, 1575-1580 (2010)
21. M. Załęska Radziwiłł, N. Doskocz, Desalination Water Treat., **57**, 1573-1581, (2016)
22. ISO 107122-1994 Water quality – *Pseudomonas putida* growth inhibition test (*Pseudomonas* cell multiplication test).
23. BioMe´rieux, Durham, NC. Instructions for use, API® 20E – 18-24 hour identification of Enterobacteriacae and other non-fastidious gram negative bacteria
24. P. M. Berthouex, L. C. Brown, *Statistic for environmental engineers* (Lewis Publishers, C.R.C. Press Inc, 1994)
25. Commission Directive 93/67/EEC of 20 July 1993 laying down the principles for assessment of risks to man and the environment of substances notified in accordance with Council Directive 67/548/EEC
26. US EPA (1998). Guidelines for Ecological Risk Assessment, EPA/630/R-95/002F, Risk Assessment Forum, Washington
27. N. Hachicho, P.Hoffmann, K. Ahlert, H. J. Heipieper, Res. Lett., **335**, 71-77 (2014)
28. M. Matzke, K. Jurkschat, T. Backhaus, Ecotoxicology, **23**, 818-829 (2014)
29. P. Gajjar, B. Pettee, D. W. Britt, W. Huang, W. P. Johnson, A. J. Anderson, J. Biol. Eng., **3** (2009)
30. W. Jiang, *Bacterial Toxicity of Oxide Nanoparticles and Their effects on bacterial surface biomolecules* (University of Massachusetts - Dissertation, 2011)
31. I. Jośko, P. Oleszczuk, Environ. Sci. Process. Impacts, **15**, 296-306 (2013)
32. M. Załęska-Radziwiłł, N. Doskocz, Desalination Water Treat., **57**, 1443-1450 (2016)

Arsenic removal from industrial wastewater

Jacek Dziubek[1,*]

[1]Gelsenwasser Polska GmbH, Dresden, Germany

Abstract. The article presents current research on the removal of arsenic in one of copper smelter in Legnica-Głogów Copper District. Arsenic removal technology applied in the results of previous author's researches has been described. Also, the content of arsenic in wastewater discharged into the receiver in 2012–2016 was examined. Additionally, the concentration of heavy metals in these wastewater was checked.

1 Introduction

The problem of excessive arsenic content in the air in the last three years in the cities belonging to Legnica-Głogów Copper District has been often the subject of intense debate in the local press or at meetings of municipal councils of each village of the region[1,2]. The problem is not trivial, because the research station of the Regional Environmental Protection Inspectorate located in Głogów has shown exceeded arsenic concentration in almost every month over the past few years, and they reach 250% of the permissible norm. Information about the risk of exceedances of permitted levels of substances in the air for 2017 shows the towns Głogów and Legnica as at risk of exceeding the annual average target level of arsenic[3]. The public opinion indicates KGHM Polska Miedź SA, a large producer of non-ferrous metals, as a potential perpetrator of exceedances. The problem of air pollution is therefore subject not only lively discussion, but also analysis and research. However, the issue of arsenic in wastewater from KGHM and the way of its treatment should be also analysed. This work is a continuation of the author's research in the years 2005–2011 and is an attempt at such analysis.

2. Ecotoxicology of arsenic

Arsenic belongs to the group of trace elements. This means that it is essential in the organism in some range in which.When exceeded it becomes to be toxic. The best-known and best described cases of mass poisoning by arsenic come from Asia (Bangladesh, India, Taiwan). In Bangladesh, 27% ofshallow wells contained

* Corresponding author: jacek.dziubek@gelsenwasser.pl

oversized, more than 50 mg / l, amount of arsenic (current WHO recommendation is 10 mg / l). Approximately 30-35 million people in this country and 6 million people in West Bengal (India) were exposed to high concentrations of arsenic in water (above 50 mg / l), while even57 million people were exposed to concentrations between 10 and 50 mg / l. The most common symptoms are eczema, rashes and tumours leading to skin cancer [4].Other studies indicate that the standardized mortality rate (which is the ratio of the observed death rate to the expected quantity, expressed as a percentage) for cancer of the respiratory system for persons exposed to arsenic with the contact time of more than 25 years is 833.3% [5].

Among many examples of the harmful effects of this element to the environment, national studies are worth quoting [6]. The research was conducted in the closed arsenic mine in Złoty Stok and in the immediate vicinity (the concentration of arsenic were dozens of times higher than accepted as normal for soils occurring there). A number of anomalies were showed. The most common were discoloration of plants, stunting and deformation of a whole or its parts.

A small number of indigenous species of trees and shrubs, low amount of specimen, and also the fact that the area of the mine was not inhabited by plants typical for this area after a few decades of closing the plant, is also characteristic for this area.At least 21 species of vascular plants avoid studied subsoils. Teratology changes were observed at 36.6% of vascular plants in this area.

3 Arsenic in the technological line of KGHM Polska Miedź

The deposits of copper sulphide LGOM contain a considerable amount of accompanying elements.Some of them are recovered in the technological processes used in KGHM Polska Miedź. Others, including arsenic, are a serious source of environmental and technological problems. Its content is complicated by, among others, processes of smelting, refining and disposal of SO_2 to sulphuric acid. Therefore, arsenic belongs to the group of associated metals, where content in metallurgical processes must be regulated. While being removed from the technology it goes to the flue dust, sewage, solid waste and is secreted in the form of a sponge Cu-As. Figure 1 illustrates a block diagram of copper production and the place of arsenic output.

Arsenic in metallurgical phases of concentrate processing accumulates in:
- by-products of galvanizing processes (dusts and sludges)
- primary production of galvanizing phase (copper matte, blister copper and finally copper anode)
- electrolyte.

Trace amounts of arsenic are present in the shaft slag, electric furnace slag and the anode sludge. Arsenic goes to the sewage system mainly from two directions: Sulfuric Acid Plants, where gases from the furnace processes are recycled and electrolyte treatment plant.

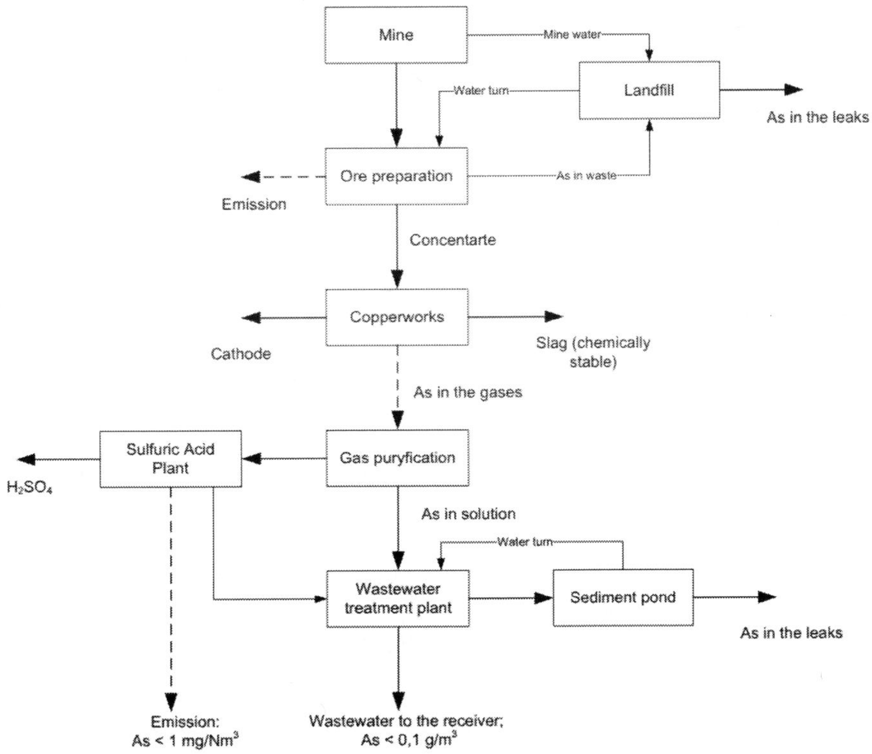

Fig. 1. Main cycles of arsenic in copper production process [based on 7].

4 Methodsof arsenic disposal tested in the sewage treatment plants of KGHM Polska Miedź

Due to differences in the metallurgical technology, forms and concentration of the occurring arsenic, in the configuration of the sewage system, as well as wastewater treatment technology, the removal of arsenic is carried out differently by each of the three wastewater treatment plants. Since the arsenic content in the sewage is high at all three ironworks, the concentration of arsenic is reduced in each of the wastewater treatment plant in several processes.

In order to bring all wastewater treatment plant to such state. a number of methods for removing arsenic were tested.

Due to shortage of wastewater treatment plant where such high concentrations of arsenic could be removed, methods of sewage treatment were searched for in hydrometallurgy, chemistry sulfuric acid and water purification.Relatively low cost and easy insertion into existing wastewater treatment plants wererequired.

Several methods of arsenic removal were tested:

- preparations sulphide precipitation,
- $Ca(OH)_2$ precipitation,
- iron salts precipitation,
- Mineral Like Precipitation (MLP),
- electro impulse method of wastewater treatment,
- use of waste materials with a high content of iron (II),
- salts of barium precipitation.

 DOI: 10.1051/e3sconf/20171700020

During application of some methods (MLP, $Ca(OH)_2$ precipitation and the precipitation of the iron salts) oxidation of As (III) to As (V) was tested.

The results of some experiments have been already shown[8,9,10,11]. Generally, the usefulness of these methods for removing high concentrations of arsenic in laboratory and in real scale for all of the three treatment KGHM Polska Miedź SA was demonstrated.

4.1 Methodology

4.1.1 Flowsheet

Figure 2 shows a flowsheet for the removal of arsenic in the wastewater treatment plant of Copperworks Głogów II (HMG II).

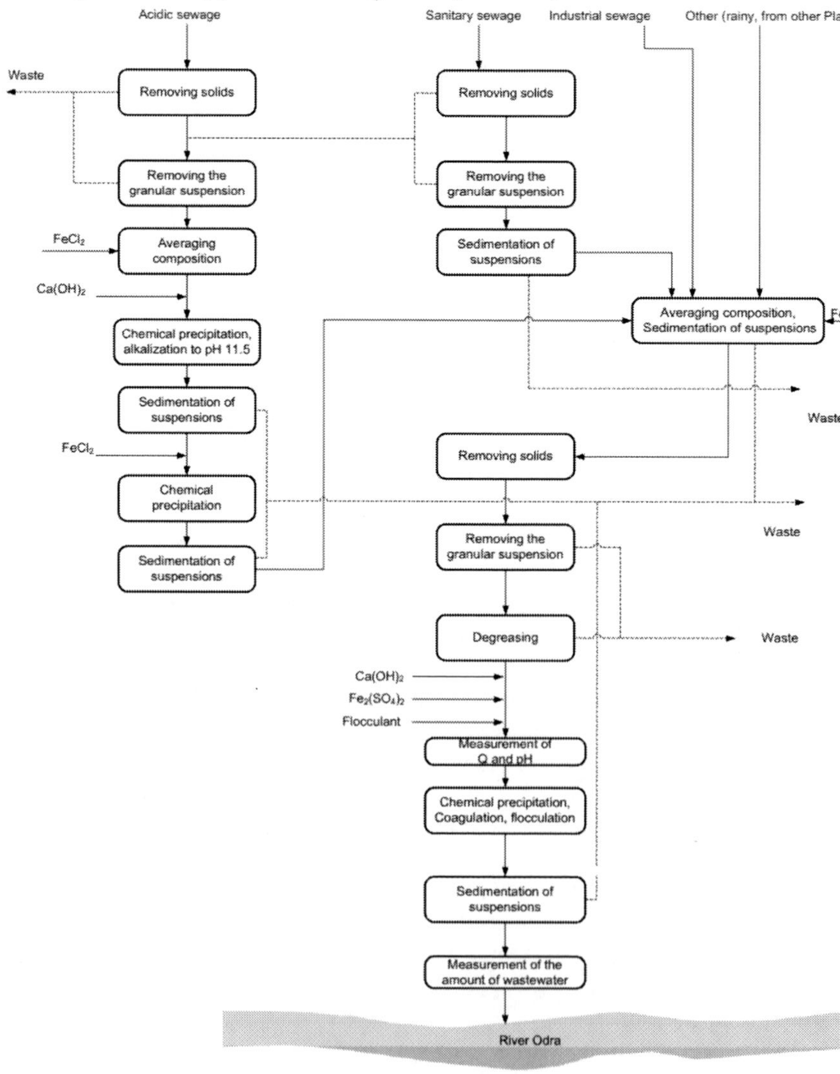

Fig. 2. Schematic diagram of sewage treatment plant HMG II.

4.1.2 Materials, reagents

Calcium hydroxide.To test the precipitation of arsenic by calcium and wastewater alkalization, the 7% milk of lime (a suspension of $Ca(OH)_2$ in water) was used. Milk of lime was prepared on site from hydrated lime. Supplier of hydrated lime was Lhoist Poland.

Iron salts.To test the precipitation of arsenic by iron salts, iron sulphate (PIX 113) and ferric chloride (PIX 109) (produced by Kemipol Ltd) was used.

- PIX 113 is an aqueous solution of ferrous sulphate - chemical formula $Fe_2(SO_4)_3$. The total iron content is 11.8% ± 0.4% and the iron (II) 0.4% ± 0.3%. The density of the solution varies in the range of 1.50 – 1.57 g / cm^3. The pH is less than 1.
- PIX 109 is an aqueous solution of ferric chloride, chemical formula FeC13. The total iron content is 10.5% ± 0.5% and the iron (II) 0.5% max. The density of the solution ranges from 1.27 – 1.37 g/cm^3. The pH is about 1.

The liquid waste with a high content of Fe (II).In studies for arsenic removal by means of liquid waste with a high content of Fe (II) used technological baths from galvanizing processes from nearby galvanizing steel were used. They can be characterized by a low pH value = 0.7, Fe (III) content of 1100 g and 150 000 g Fe (II) per 1 m^3. They are also contaminated with heavy metals (nickel, lead, copper).

4.1.3 The composition of waste water used in tests

The composition of raw sewage has been shown in Table 1. Other wastewater groups are not characterized by such high concentrations of arsenic (sporadically exceeding 1 g/m^3, max 4g/m^3) or other pollutants, therefore their compositions was neglected. Acidic wastewater is responsible for 95 % of the arsenic load flowing into the wastewater treatment plant.

Table 1. Composition of acid sewage from HMG II (2016)
[Own study based on the database from Energetyka Ltd.].

	Unit	**Average**	**Maximum**
pH	-	<2.0	<2.0
Cl	[g Cl/m^3]	4876	8453
SO_4	[g SO_4/m^3]	10086	16299
Susp.	[g/m^3]	66	240
COD	[g O_2/m^3]	153	346
Cu	[g Cu/m^3]	16	62
Fe	[g Fe/m^3]	19	160
Pb	[g Pb/m^3]	9.9	22
Zn	[g Zn/m^3]	4.5	28
Ni	[g Ni/m^3]	3.1	19
As	[g As/m^3]	310	532

4.1.4 Analytical methods

Table 2 summarizes the analytical methods used, together with standards according to which the analysis was performed.

Table 2. Analytic methods.

Type of examination	Research method	Standard
COD	Spectrophotometric method	PN-ISO 15705:2005
Chloride	Titrimetric method	PN-ISO 9297:1994
pH	Electrometric method	PN-90/C-04540/01
Sulfate	Ion chromatography (IC) method	PN-EN ISO 10304
Total Arsenic	Inductive atomic emission spectrometry (ICP-OES) after mineralization	PN-EN 11885:2001
Zinc		
Copper		
Nickel		
Lead		
Iron		

4.1.5 A method of operation.

Removal of pollutants in sewage treatment plants HMG II was carried out according to the scheme on Fig. 2, using the following doses of chemicals and process parameters:

- Sanitary sewage - the elimination of solids and suspensions in physical processes.
- Acidic sewage - dosage to reservoirs of waste $FeCl_2$ at a dose of 1000 g/m^3. Alkalinisation carried out in reactors (until pH> 11.5) and the neutralization products of sedimentation in settling tanks Dora. Then precipitation / coagulation by waste liquid with high concentration of $FeCl_2$ - a dose of 500 g/m^3 and sedimentation of reaction products.
- Averaging of composition in reservoirs where all wastewater goes: sanitary and acidic - after preliminary treatment and industrial and rainy-cooling water - directly from the drains. In advance liquid waste with high concentration of $FeCl_2$ (dose of 150 g/m^3) is dosed.
- In the second stage treatment, after calibration with lime milk to pH 8.5-9.0, process of coagulation by assisted iron (III) sulphate, assisted by an anionic polyelectrolyte was conducted (coagulant dose - 40 g/m^3, flocculant dose 0.25 – 0.4 g/m^3).

4.2 Results

In the period precedingthe initiation of effective arsenic removal trials, arsenic content exceeded the permissible content of arsenic in sewage discharged into the Odra river. The wastewater treatment technology used in this period did not guarantee the attainable concentration of arsenic in the effluent from the treatment plant. The permissible concentration of arsenic, i.e. 0.1 g/m^3, was practically continuous. In some samples the excess was 3100% (!).

Laboratory research and technical trials led to the development of a technology that effectively eliminates arsenic from sewage. These technologies have been, therefore, implemented. As part of this work effectiveness of arsenic (and heavy metals) removal after the implementation period has been verified. The data is summarised in tables 3 and 4.

Table 3. Arsenic concentration in effluent from sewage treatment plant HMG II [Own study based on the database from Energetyka Ltd.].

Year	Arsenic concentration [g/m^3]	
	Average	Maximum
2012	0.025	0.098
2013	0.011	0.077
2014	0.007	0.059
2015	0.006	0.040
2016	0.035	0.090

Table 4. Heavy metal concentration in effluent from sewage treatment plant HMG II [Own study based on the database from Energetyka Ltd.].

Heavy Metal	Unit	Average	Maximum
pH	-	8.6	8.9
Cl	[g Cl/m^3]	940	1388
SO_4	[g SO_4/m^3]	740	1165
COD	[g O_2/m^3]	23.7	38.2
Cu	[g Cu/m^3]	0.048	0.075
Fe	[g Fe/m^3]	0.649	1.130
Pb	[g Pb/m^3]	0.118	0.190
Zn	[g Zn/m^3]	0.076	0.138
Ni	[g Ni/m^3]	0.054	0.119
Ag	[g Ag/m^3]	0.005	0.028
Cr	[g Cr/m^3]	0.005	0.058
Cd	[g Cd/m^3]	Below the determination threshold	
Hg	[g Hg/m^3]		

5 Summary

Several years of work on the implementation of effective method of arsenic removal in wastewater treatment plant HMG II were successfully ended. Initially, the concentration of this element exceeded the limits in wastewater discharged into a receiver from several dozen to several hundred times. several methods on a laboratory and technical scale were tested. Some of them were tested in real conditions of operation of the plant. It has been shown that it is possible to effectively remove arsenic present in high concentrations in industrial wastewater to the level required by the law in daily work conditions.

After the initial tests, it was checked whether the proposed methods, despite changing the conditions of the infestations, still fulfilled their role. The results were positive. Within 5 years the water permit was not exceeded once. The finally proposed technologies proved to be very effective. Maximum concentrations of arsenic in wastewater discharged to the receiver do not exceed 0.1 g/m^3, and average values are 0.006–0.035 g As $/m^3$.

It is worth pointing out that both arsenic and heavy metals in 2012–2016 in wastewater discharged to the receiver never exceeded the limit values and their mean concentrations are significantly lower than the limit values. This demonstrates the correct choice of arsenic removal technology.

References

1. http://regionfan.pl/glogow-arsen-wisi-w-powietrzu/, access: 29.03.2017
2. http://www.wroclaw.pl/10-faktow-o-czystosci-powietrza-na-dolnym-slasku, access: 19.03.2017
3. *Informacja o przekroczeniach i ryzyku przekroczeń poziomów dopuszczalnych i docelowych substancji w powietrzu na terenie województwa dolnośląskiego w 2017r.*http://www.wroclaw.pios.gov.pl/pliki/komunikaty/1_Inf_ryzyko_2017_02_21.pdf: access: 20.03.2017
4. *Arsenic. Medical and Biologic Effects of Environmental Pollutants.* National Academy of Sciences, Washington (1977)
5. W. Seńczuk, *Toksykologia.* Wydawnictwo Lekarskie PZWL, Warszawa (2004)
6. W. Giża, *Zmiany teratologiczne u roślin z podłoży arsenowych w Złotym Stoku.* Acta Universitias Wratislaviensis No 2037, Prace Botaniczne **LXXV,** Wydawnictwo Uniwersytetu Wrocławskiego, Wrocław (1997)
7. KGHM, *Monografia KGHM Polska Miedź S.A.* Lubin (2007)
8. J. Dziubek, D. Kwaśny, *Poprawa skuteczności usuwania arsenu ze ścieków przemysłu miedziowego po zastosowaniu metody elektroimpulsowej oraz przy wykorzystaniu żelaza odpadowego z procesów trawienia na przykładzie oczyszczalni ZGW H II.* X Seminarium - Metalurgia Metali Towarzyszących. Współczesne technologie i wyzwania (2005)
9. J. Dziubek, *Oczyszczanie elektroimpulsowe a elektrokoagulacja: podobieństwa i różnice na przykładzie usuwania arsenu ze ścieków przemysłowych.* Conference Eko-Dok III. Szklarska Poręba (2010).
10. J. Dziubek, *Usuwanie arsenu ze ścieków przemysłowych.* Doctoraldissertation, Politechnika Wrocławska (2012)
11. K. Majewska-Nowak, J. Dziubek, *Chemiczne i elektrochemiczne strącanie arsenu występującego w wysokich stężeniach w ściekach z przemysłu metali kolorowych.* Conference Eko-Dok IV. Szklarska Poręba (2012)

An experimental study of the air humidification process using a membrane contactor

Sebastian Englart[1,*]

[1]Wrocław University of Science and Technology, Faculty of Environmental Engineering, Department of Air Conditioning, Heating, Gas Engineering and Air Protection, 4/6 Norwida st., 50-373 Wrocław, Poland

Abstract. The article presents the results of the experimental examination of the effectiveness of air humidification using a membrane module. The construction of the membrane module and the measuring stand is also discussed. In order to assess the effectiveness of air humidification using the membrane module, the measurements of temperature and humidity at the membrane module's inlet and outlet, air flow rate, water flow rate and water temperature were taken. Based on the measurements, the effectiveness coefficients, *E*, have been determined. The power demand for the solution under study has also been discussed.

1 Introduction

The humidification of air in air conditioning is accomplished using either steam or water. While steam humidification is conducted using steam humidifiers, in the case of using water, there are three basic air humidification methods:
– air flow through a spray chamber, where water is sprayed with nozzles; humidification takes place upon the direct contact of air with the sprayed water,
– air flow over the wetted surface – this method is employed in evaporative humidifiers (e.g. in chambers with sprinkled surfaces),
– spraying water to form aerosol-size droplets (of a diameter of approx. 5 μm) either directly in the room or in the channel supplying air to the room – the air humidification method, whereby the sprayed water is completely evaporated and the air is adiabatically cooled, is accomplished using pneumatic nozzles, high-speed rotary discs, or ultrasounds.

Central air humidification devices used include spray chambers and sprinkled surface chambers, which in turn include fixed and movable sprinkled packing chambers [1, 2].

In addition to commonly used solutions for conducting air drying, humidifying or evaporative cooling processes, also membranes can be used [3]. For this purpose, chiefly synthetic membranes are utilized, which make a porous semipermeable barrier between the liquid and the air.

The use of membranes for air treatment can offer the following advantages, compared to traditional air treatment techniques [4-8]:

- no direct contact between the liquid and the air,

[*] Corresponding author: sebastian.englart@pwr.edu.pl

- elimination of liquid droplets from the air stream,
- due to the small sizes of its pores, the membrane prevents microorganisms, bacteria or spores from passing from one phase to another,
- a membrane prevents the contamination of the liquid with airborne dusts,
- a membrane provides a large exchange surface relative to the occupied space,
- good efficiency,
- low friction losses, both on the air and liquid sides.

The article has reported the results of the experimental examination of the effectiveness of air humidification using a membrane module.

2 Evaluation criteria

There are few evaluation criteria that can be used for comparing air humidification devices and processes. This can be for example humidification effectiveness, temperature effectiveness, average moisture permeability or electric power demand index.

2.1 Process effectiveness

The effectiveness of a process can be defined as the ratio of the actual process to the theoretical process. With contact between air and water in an ideal exchanger with an infinitely large surface area, the theoretical quantity of sensible heat exchanged in the system can be written as:

$$Q_{S,t} = m_a c_{pa}(t_1 - t_{w1}) \tag{1}$$

In a real exchanger, the quantity of exchanged sensible heat will be:

$$Q_S = m_a c_{pa}(t_1 - t_2) \tag{2}$$

This results in the effectiveness, E_t, defined as a dimensionless index:

$$E_t = \frac{Q_S}{Q_{S,t}} = \frac{t_1 - t_2}{t_1 - t_{w1}} \tag{3}$$

This coefficient can be defined as the temperature effectiveness. Similarly, for the latent heat associated with the exchanged mass flow, it can be written as follows:
• theoretical:

$$Q_{L,t} = m_a r(x_{w1} - x_1) \tag{4}$$

where : x_{wl} is equal to the air moisture content at the water temperature.
• actual:

$$Q_L = m_a r(x_2 - x_1) \tag{5}$$

So, the effectiveness, E_x, will take on the following form:

$$E_x = \frac{Q_L}{Q_{L,t}} = \frac{x_2 - x_1}{x_{w1} - x_1} \tag{6}$$

This coefficient can be defined as the humidification effectiveness.
In the above equations:
- index *1* refers to the parameters of air at the inlet,
- index *2* refers to the parameters of air at the device's outlet,
- index *t* denotes ideal (theoretical) conditions,

t – air temperature, K ,
t_w – water temperature, K,
x – moisture content, kg/kgs.p.,
m_a – air mass flow rate, kg/s,
c_{pa} – specific heat of air, J/(kgK),
r – heat of vaporization, J/kg.

Studies [9, 10] define efficiencies in a similar manner. That being said, the temperature effectiveness is described as the sensible cooling effectiveness. Another effectiveness indicator can be the average moisture permeability, which is determined as [11]:

$$Pe = \frac{m_a(x_2 - x_1)}{A_{tot}} \tag{7}$$

Where: A_{tot} – total area of the fiber outer surface, m^2.
Effectiveness is sometimes called efficiency and can be expressed in per cents.

2.2 Power demand

When selecting an air humidification method, in addition to the criterion of effectiveness or efficiency that can be achieved by that humidification method or a specific device, the power demand or energy consumption of this specific solution must be taken into account. Comparison of different air humidification devices can be made, for example, based on the electric power demand index as related to the quantity of sprayed water or steam. Basic energy expenditures associated with the operation of a membrane humidifier result from the transfer of air and water. Hence, the demand for electric power needed for air transfer can be expressed by the formula below:

$$N_a = \frac{m_a \Delta p_m}{\rho_a \eta} \tag{8}$$

For the pump water transfer system:

$$N_w = \frac{m_w \Delta p_p}{\rho_w \eta} \tag{9}$$

m_w – water mass flow rate, kg/s,
Δp_m – pressure drop in the flow of air through the membrane module, Pa,
Δp_p – pump delivery head, Pa,
ρ – density, w-water, a-air, kg/m^3,
η – device efficiency.

The total electric power demand for the operation of air humidification with the membrane module will be:

$$N = N_a + N_w \tag{10}$$

By relating the above power demand for the humidification degree, W_x, we can determine the relative power demand:

$$n = \frac{N}{W_x} \tag{11}$$

where:

$$W_x = PeA_{tot} \tag{12}$$

The index n can be used for comparing different types of air humidifiers.

3 Experimental setup

Tests of the air humidification process using the membrane module were carried out on a test stand (Fig. 1), which enabled the measurement of:

- ambient air temperature and humidity,
- temperature and humidity of air at the membrane module's inlet and outlet,
- air flow rate in the channel,
- temperature of water upstream/downstream the module,
- water flow rate.

For making the membrane module, a polypropylene (PP) capillary membrane with the following parameters was used:

- membrane inner diameter, 1.67mm,
- membrane outer diameter, 2.71mm,
- average pore size, 0.43μm,
- maximum pore size, 0.55μm.

In the module, 744 membranes of a total surface area of 3.04 m^2 and a specific surface area of 667 m^2/m^3 were used. The membranes were situated perpendicularly to the air flow direction.

Measurements were taken for variable air and liquid flows. For air, the flow was varied within the range of 60 -120 m^3/h, while for water, within 100-400 l/h. Examples of air transformation in the test installation are shown in the *i-x* diagram (Fig. 2). Transformation 1-2 was accomplished in the heater, while transformation 2-3, in the membrane module.

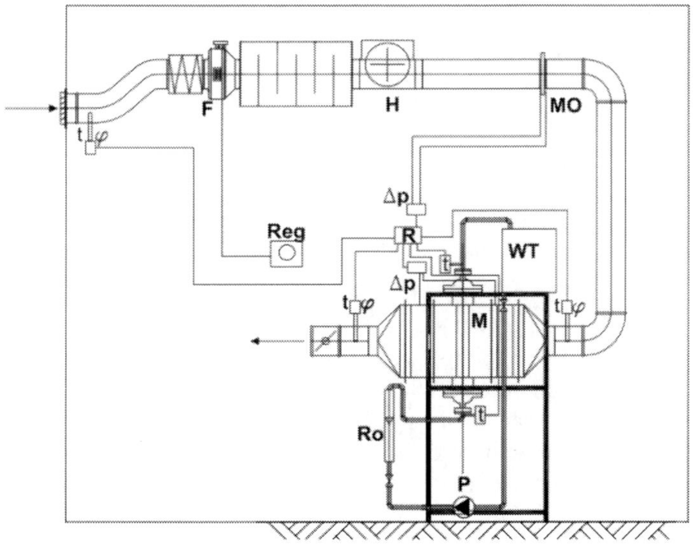

Fig. 1. Schematic diagram of the measuring stand. F-fan, H-heater, MO-measuring orifice, Ro-rotameter, R-recorder, WT-water reservoir, M-membrane module, P-pump, Reg-thyristor controller, sensors of: t-temperature, φ–humidity, Δp-differential pressure.

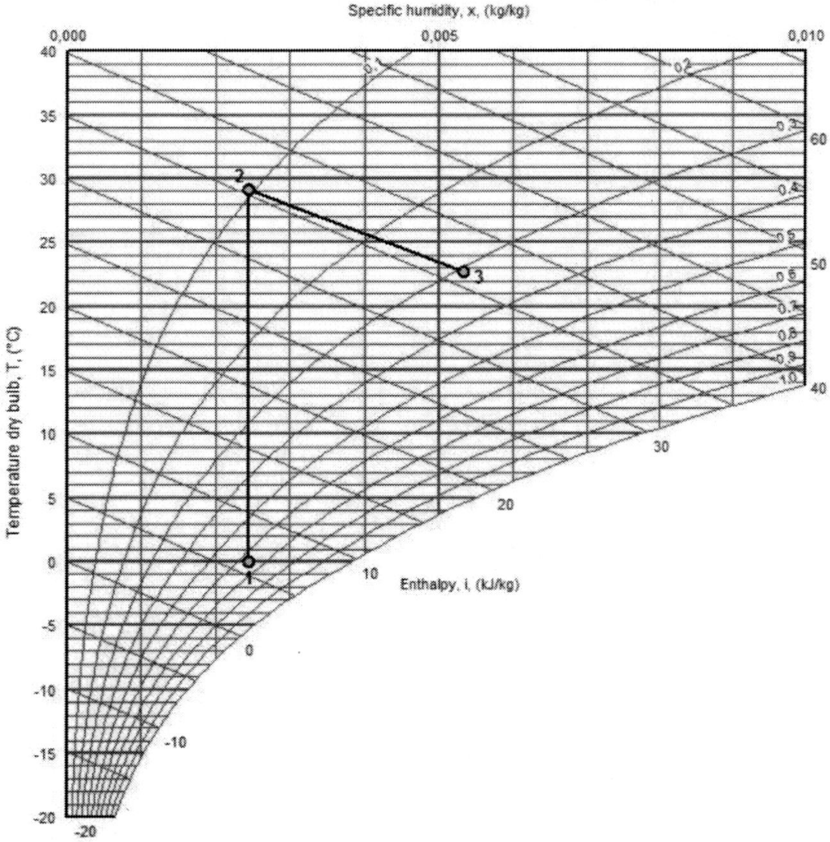

Fig. 2. Examples of air transformation in the experimental installation during air humidification.

4 Experimental results

In order to assess the effectiveness of air humidification using the membrane module, the measurements of temperature and humidity at the membrane module's inlet and outlet, air flow rate, water flow rate and water temperature were taken. Based on the measurements, the effectiveness coefficients, average moisture permeability and the relative power demand have been determined.

4.1 Effectiveness

The values of the effectiveness coefficients were determined, respectively, form Equations (3) and (6). The percentage effectiveness coefficient values as a function of air flow rate are represented in Fig. 3. The average effectiveness E_t within the examined air flow rate variability interval was 55.3%. Changes in the flow rate of water and air flowing through the module did not cause any significant changes in temperature effectiveness. The average effectiveness E_x amounted to 31%. In this case, a slight decrease in humidification effectiveness with the increase in air flow rate was observed. Practically, no effect of variations in the flow rate of water flowing through the module on the air humidification effectiveness was found.

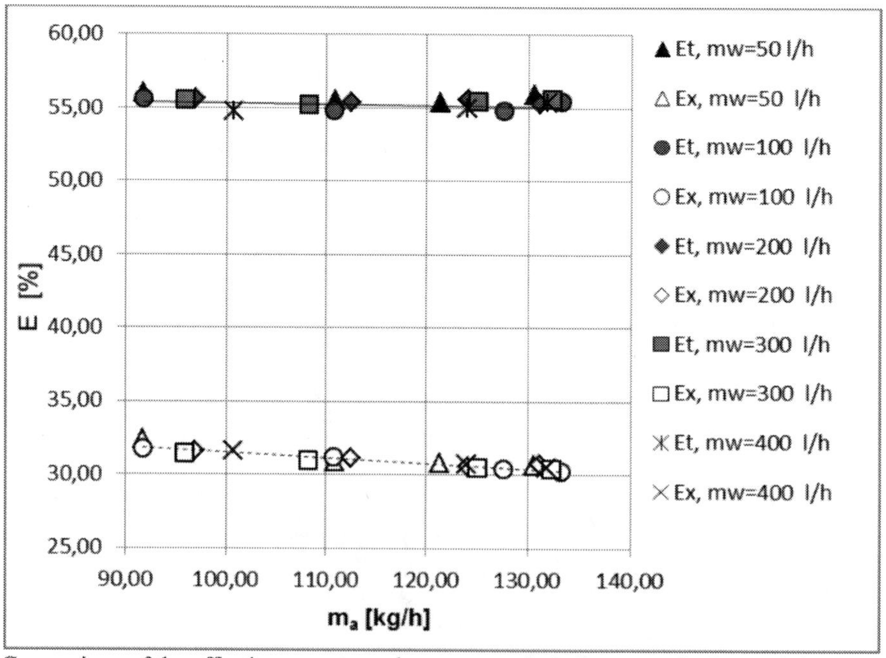

Fig. 3. Comparison of the effectiveness at varying water and air flow rates in the module.

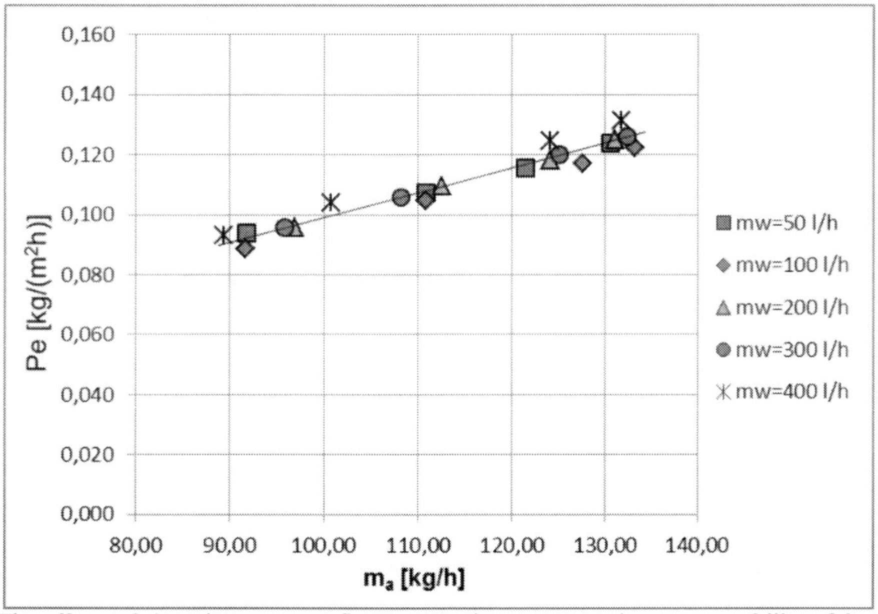

Fig. 4. The effects of air and water mass flow rate on the average moisture permeability of the module.

4.2 Average moisture permeability

The effect of variations in the flow rate of air and water flowing through the membrane module on the average moisture permeability value is illustrated in Fig. 4. With increasing air flow rate, the average moisture permeability increased.

4.3 Relative power demand

The relative power demands data for sample air humidifiers, provided in article [12], have been used to plot the diagram shown in Fig. 5. Four humidifiers were compared, including: a rotational humidifier, a pneumatic spray nozzle humidifier, and two steam humidifiers: with electric heaters and with immersed electrodes, respectively. The maximum relative power demand for the test module, as determined from Equation (11), amounted to 59 W/(kg/h). This index has been obtained with maximum air and water mass flow rate, minimal humidification degree, W_x=0,37 kg/h, and the total electric power demand N=22 W. Within these parameters the relative power demand was reduced by over 92% compared to steam humidifiers.

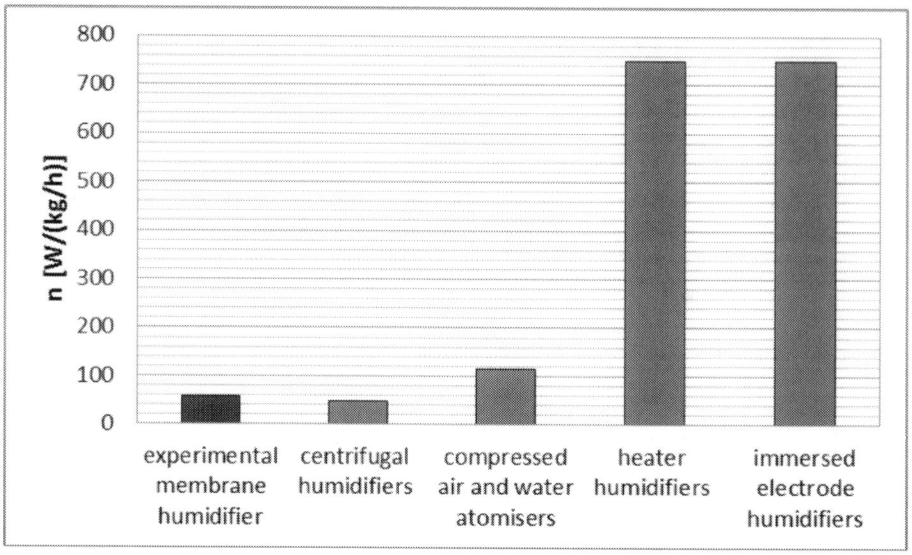

Fig. 5. Comparison of the relative power demand for different humidifiers.

5 Conclusions

Membranes are not commonly used for air treatment in air conditioning systems. However, by using them, the relative power demand could be reduced, compared to traditional solutions. The performed experimental tests have shown that the air humidification process can be conducted using a membrane module. Satisfactory effectiveness levels were achieved for the process carried out.

Membrane modules can find application in air conditioning technology as air humidification devices.

References

1. S. Englart, A. Kmieć, A. Ludwińska, Can. J. Chem. Eng. **87**, 185-192 (2009)
2. A. Kmieć, S. Englart, A. Ludwińska, Can. J. Chem. Eng. **87**, 163-168 (2009)
3. J. Woods, Renew. Sust. Energ. Rev. **33,** 290-304 (2014)
4. S. Bergero, A. Chiari, Appl. Therm. Eng. **21,** 1119-1135 (2001)
5. C. Isetti, E. Nannei, A. Magrini, Energ. Buildings, **25**, 185-193 (1997)
6. D. W. Johnson, C. Yavuzturk, J. Pruis, J. Membr. Sci., **227**, 159-171 (2003)
7. K. Kneifel K., S. Nowak, W. Albrecht, R. Hilke, R. Just, J. Membr. Sci., **276**, 241-251 (2006)
8. P. Scovazzo, J. Burgos, A. Hoehn, P. Todd, J. Membr. Sci., **149**, 69-81 (1998)
9. L.Z. Zhang, Z.X. Li, T.S. Zhong, L.X. Pei, J. Membr. Sci., **427**, 1-9 (2013)
10. L.Z. Zhang, Int. J. Heat. Mass. Tran., **55,** 5861-5869 (2012)
11. L.Z. Zhang, S.M. Huang, Int. J. Heat. Mass. Tran., **54,** 1055-1063 (2011)
12. A. Ludwińska, S. Englart, A. Kmieć, COW, **39**, 26-29 (2008)

E3S Web of Conferences 17, 00022 (2017) DOI: 10.1051/e3sconf/20171700022
EKO-DOK 2017

Regrowth of willow shoots in the second 4-year rotation at various fertilization with nitrogen

Diana Fijałkowska[1,*], and *Leszek* Styszko

[1]Koszalin University of Technology, Faculty of Civil Engineering, Environmental and Geodetic Sciences, Sniadeckich 2, 75-452 Koszalin, Poland

Abstract. The aim of the experiment was to assess the dynamics of regrowth of shoots of 10 genotypes of basket willow (*Salix viminalis* L.) in the second 4-year rotation, cultivated on sandy soil in the Middle Pomerania (16°24'N and 54°08'E) under various fertilization with nitrogen (0, 60, 120 and 180 kg N·ha^{-1}) and under rainfall-retention water system. The field experiment was established in 2007 using method of randomized sub-blocks in a subsidiary system with four replications, where on large plots four doses of nitrogen were drawn, and within doses – 10 genotypes of willow. In 2012-2015 the dynamics of regrowth of willow shoots in four separate dates for the first and second row of plot separately were evaluated. During first rotation (2008-2011) in the first row of plot shoots were mowed 2-times (after third and first year of regrowth), and the second row – once (after fourth year of regrowth). During the second 4-year rotation of the biggest variability in relation to the height and thickness of willow shoots and the number of live and dead shoots in snag was noted between years of shoots regrowth and for the amount of live and dead snags on the plot – between the frequencies of mowing.

1 Introduction

The biomass harvested for energy purposes is a significant source of renewable energy in Poland and in Europe [1, 2]. Its consumption in Polish power industry increased from 9641 t to 65,470 t in the years 2005-2011 [3]. Big quantities of biomass for heating are used by households [2]. On intensive field cultivations, during 3-year rotations, growth of wood is sometimes up to 10 times higher than in the forests [4]. Growths of willow biomass depend on adaptation of cultivation of willow genotypes to natural conditions (soil, water and thermal) where plantation is located, including a system of organic and mineral fertilization and irrigation [5]. In Sweden, about 50% of the willow cultivations is fertilized with municipal wastewater, and on the other objects mineral fertilization with dose of approx. 100 kg N·ha^{-1} is used. In Canada it is recommended to apply fertilization equivalent to 100-150 kg N·ha^{-1} taking into account also organic fertilization [6]. Most nitrogen is taken by willow in the period from April to October, and most intensively – from May to July [7]. Efficiency of use of nutrient elements is higher in longer production cycles than in one year cycle [8, 9]. It has been reported in the literature that during the

* Corresponding author: fijalkowska@wilsig.tu.koszalin.pl

second cycle of cultivation shoots biomass yields are higher than during the first [10, 11]. Willow clones vary in pace of achieving maximum production potential and its increase depends on the dose of fertilizer. Swedish and US experiments prove that the optimal willow biomass yield is obtained during cultivation in 3-6 year harvest cycle depending on the clone, habitat conditions, fertilization and meteorological conditions [12]. The most important factor limiting the growth of willows in Sweden and across temperate regions are, apart from water, nutrients, mainly nitrogen [13].

Research concerning dynamics of willow biomass growth during years of cultivation were carried out in Poland mostly in the first 3-4 year rotation [14, 15-19]. Such measurements may be used for prediction of willow biomass yield [16].

The aim of the experiment was to assess the dynamics of regrowth of shoots of 10 genotypes of basket willow (*Salix viminalis* L.) in the second 4-year rotation, cultivated on light soil in the Middle Pomerania under various fertilization with nitrogen (0, 60, 120 and 180 kg $N \cdot ha^{-1}$) and under rainfall-retention water system.

2 Materials and methods

Measurements of height and thickness of willow shoots, the number of living and dead stems in snag and quantities of live snags on the plot were conducted on a field experiment carried out in 2012-2015 on the experimental field of Technical University of Koszalin in Kościernica (16°24′N and 54°08′E). The soil under the experiment was light, of quality class RIVa-IVb, of good rye complex, appropriate podsolic – pseudopodsolic with composition of light loamy sand – to the depth of 100 cm, and of light loam – deeper. The content of humus in the layer 0-30 cm of soil was 1.41%. The experiment was established in 2007. The first 4-year cycle of shoots regrowth was in 2008-2011, and the second - in 2012-2015. As part of the experiment, within large plots, four doses of mineral nitrogen were drawn: **(a)** 0 kg $N \cdot ha^{-1}$, **(b)** 60 kg $N \cdot ha^{-1}$, **(c)** 120 kg $N \cdot ha^{-1}$ and **(d)** 180 kg $N \cdot ha^{-1}$, and within doses of nitrogen, 10 genotypes of basket willow (*Salix viminalis* L.): 1047, 1054, 1047D, Start, Sprint, Turbo, Ekotur, Olof, Jorr i Tordis. Nitrogen fertilization was applied according to experiment design, each year in April. In 2007, on the plot of 25.3 m^2 of area, in two rows 56 willow cuttings were planted per row, that is 22,134 pcs. per ha. In the first 4-years rotation the first row was mowed twice (after 3 years and after one year of regrowth), and the second row – once (after four years of regrowth). For examined factors standard analysis of variance was conducted and structure of variance components was determined, and significance of the effects was assessed with F test. Data on the weather pattern representative for Kościernica region were obtained from the meteorological station in Koszalin (Table 1). Annual rainfall in Koszalin was from 831.9 mm in 2012 to 588.9 mm in 2014, and during willow vegetation – from 536.6 mm in 2012 to 387.0 mm in 2014. Extreme hydrothermal conditions (extremely dry and very dry and very wet and extremely wet) indicated by Sielianinow coefficient (K) fall within the ranges < 0.7 and > 2.5. During vegetation period this coefficient ranged from 1.87 in 2012 to 1.34 in 2013 and 2014. Extremely dry and very dry conditions ($K < 0.7$) occurred in May 2012, July 2014 and August 2015, and very wet ($K > 2.5$) occurred in April 2014, June 2015 and September 2012.

Table 1. Rainfall [mm] and Sielianinow hydrothermal coefficient in Koszalin in the years 2012-2016 according to Institute of Meteorology and Water Management.

Month	Annual rainfall				Sielianinow coefficient (K) in the years			
	2012	2013	2014	2015	2012	2013	2014	2015
I-III	175.9	128.7	84.5	100.3	-	-	-	-
IV	39.1	42.2	72.1	18.4	1.92	2.07	2.50	0.83
V	15.8	51.1	52.5	37.3	0.37	1.19	1.37	1.06
VI	88.6	64.4	55.9	144.7	1.88	1.37	1.23	3.42
VII	119.4	54.5	37.9	96.3	2.14	0.98	0.59	1.79
VIII	99.8	52.3	83.6	13.3	1.78	0.93	1.56	0.21
IX	102.5	85.4	54.4	67.1	2.65	2.21	1.19	1.56
X	71.4	44.3	30.6	28.5	2.19	1.36	0.88	1.09
IV-X	536.6	394.2	387.0	405.6	1.87	1.34	1.34	1.42

3 Results and discussion

Factors used in the experiment (willow genotypes, nitrogen dose, dates of measurements and years of shoots regrowth) had significant impact on the variability of analyzed features (Table 2).

Table 2. Impact of examined factors on the structure of variance components in analyses of features.

Variance component[1]	Number of levels	Percentage structure of variance components in analyses				
		shoots		shoots in snag		live snags on the plot
		height	thickness	live	dead	
E	10	3.7***	5.5***	1.6***	1.1***	6.2***
D	4	0.7***	0.9***	1.0***	0.4***	9.0***
C	4	4.8***	3.7***	2.4***	2.3***	0.1***
B	2	4.5***	6.8***	15.4***	3.1***	21.0***
A	4	73.7***	68.0***	18.6***	55.2***	2.2***
Sum (A-E)		87.4	84.9	39.0	62.1	38.5
ExD		0.9***	1.4***	1.5***	1.4***	5.1***
ExC		0.2***	0.3***	0.2***	0.1*	0.0
DxC		0.0	0.3***	0.8**	1.4***	0.6***
ExB		0.6***	0.8***	3.0***	1.6***	15.4***
DxB		0.3***	0.1***	1.3***	1.0***	12.2***
CxB		0.1***	0.1***	0.3***	0.2***	0.0
ExA		1.2***	1.6***	2.8***	1.3***	0.9***
DxA		1.0***	1.2***	0.6***	4.0***	0.4***
CxA		1.7***	0.8***	7.5***	2.2***	0.1***
BxA		0.1***	0.1***	20.4***	2.3***	0.1***
Other interactions		6.5	8.4	22.6	22.4	26.7
Sum of interactions		12.6	15.1	61.0	37.9	61.5

[1] Designation of variance components and significance levels are given in Table 3

Variability caused by their actions ranged from 87.4% for the height of shoots to 38.5% for live snags on the plot. Particularly big effects were obtained for interactions: BxA

(20.4%) for live shoots in snag and ExB and DxB for the live and the dead snags on the plot.

Table 3 presents results of biometric measurements of will in the years 2012-2015, and Table 4, differences between the extreme values and their relative reference to an average from experiment within tested feature. The data given in Table 4 confirm that the parameters of willow canopy architecture characterized by height and thickness of shoots and number of live and dead shoots in snag the biggest differences were caused by years of shoots regrowth (respectively : 122.5%, 117.8%, 77.6% and 152.5%) and the lowest – dose of nitrogen (respectively: 8.9%. 12.8%, 17.4% and 14.2%).

Table 3. Impact of examined factors on results of willow biometric measusremnts.

Examined factors		Designation of analyses				
		shoots		shoots in snags		live snags on the plot [pcs.]
Factors	level	height [cm]	thickness [mm]	live [pcs.]	dead [pcs.]	
Years of shoots regrowth [A]	1	97.2	6.5	9.95	0.47	22.2
	2	240.6	14.8	6.08	1.44	21.2
	3	387.5	23.0	4.78	2.66	20.5
	4	460.2	27.7	4.95	3.58	18.8
	$NIR_{0.05}$	2.9***	0.2***	0.15***	0.06***	0.2***
Variant of shoots mowing [B]	I	268.3	15.9	4.89	1.81	17.5
	II	324.5	20.1	7.99	2.26	23.8
	$NIR_{0.05}$	2.0***	0.1***	0.11***	0.04***	0.1***
Dates of measurements [C]	31 V	248.3	15.4	7.36	1.81	20.9
	30 VI	276.4	17.1	6.97	1.81	20.8
	30 IX	323.4	19.3	5.91	2.12	20.6
	10 XI	337.4	20.3	5.51	2.40	20.4
	$NIR_{0.05}$	2.9***	0.2***	0.15***	0.06***	0.2***
Doses of nitrogen in kg $N{\cdot}ha^{-1}$ [D]	0	277.5	16.6	6.90	2.00	23.6
	60	313.5	18.7	6.89	2.20	22.1
	120	290.7	17.8	6.19	1.91	20.2
	180	303.9	18.9	5.78	2.03	16.8
	$NIR_{0.05}$	2.9***	0.2***	0.15***	0.06***	0.2***
Willow genotype [E]	1047	280.5	17.1	7.23	2.26	21.8
	1054	278.2	16.8	6.44	2.01	20.1
	1047D	294.0	17.9	7.30	2.04	19.2
	Start	244.0	14.4	5.12	1.81	15.8
	Sprint	275.4	16.2	6.23	2.17	18.8
	Turbo	277.6	16.5	6.45	1.74	21.5
	Ekotur	354.0	22.7	6.90	2.10	24.7
	Olof	316.9	19.4	5.43	1.82	20.4
	Jorr	289.8	17.2	6.89	2.08	21.3
	Tordis	356.4	22.2	6.40	2.31	23.1
	$NIR_{0.05}$	4.6***	0.3***	0.24***	0.10***	0.2***
Average		296.3	18.0	6.44	2.04	20.7

Significance at level: *α=0.05; **α=0.01; ***α=0.001

Table 4. Variability of examined factors and their relative reference to an average from experiment.

Main factors of the experiment		Designation of analyses				
		shoots		shoots in snags		live snags on the plot [pcs.]
Factors[1]	Difference between extreme values	height [cm]	thickness [mm]	live [pcs.]	dead [pcs.]	
A	difference	363	21.2	5.00	3.11	3.4
	percent	122.5	117.8	77.6	152.5	16.4
B	difference	56.2	4.2	3.1	0.45	6.3
	percent	19.0	23.3	48.1	22.1	30.4
C	difference	89.1	4.9	1.85	0.59	0.5
	percent	30.1	27.2	28.7	28.9	2.4
D	difference	26.4	2.3	1.12	0.29	6.8
	percent	8.9	12.8	17.4	14.2	32.9
E	difference	112.4	8.3	2.11	0.57	8.9
	percent	37.9	46.1	32.8	27.9	43.0
Average from experiment	difference	296.3	18.0	6.44	2.04	20.7
	percent	100.0	100.0	100.0	100.0	100.0

[1] Designation of experiment main components are given in Table 3

Impact of studied factors was different for live snags on the plot: the biggest differences caused willow genotypes (43.0%), slightly lower – dose of nitrogen (32.9%) and variants of shoots mowing (30.4%), small – years of shoots regrowth (16.4%), and the smallest – dates of measurements during vegetation (2.4%).

Table 5 presents data for interaction of variants of willow mowing with doses of nitrogen. The biggest differences between variant II and I of shoots mowing for all studied factors were at dose of nitrogen of 180 kg $N \cdot ha^{-1}$, and the smallest – for height and thickness of shoots at dose of 60 kg $N \cdot ha^{-1}$, and for other features – on control objects (0 kg $N \cdot ha^{-1}$).

Table 5. Impact of interaction between variants of willow mowing and doses of nitrogen on parameters of willow canopy architecture.

Nitrogen dose [kg $N \cdot ha^{-1}$]	Mowing variant[2]	Shoots height [cm]	Shoots thickness [mm]	Live shoots in snag [pcs.]	Dead shoots in snag [pcs.]	Live snags on the plot [pcs.]
0	I	250.8	14.7	5.93	1.87	22.9
	II	304.2	18.6	7.86	2.12	24.4
60	I	291.5	16.8	5.40	2.03	20.0
	II	335.5	20.6	8.38	2.38	24.1
120	I	263.9	15.8	4.50	1.71	16.6
	II	317.4	19.8	7.87	2.11	23.8
180	I	267.0	16.4	3.71	1.61	10.5
	II	340.8	21.5	7.85	2.45	23.2
$NIR_{0.05}$		4.1***	0.3***	0.21***	0.09***	0.2***

[2] Mowing variant: I – mowing after 3-year and 1-year regrowth, II – mowing after 4-year regrowth

Table 6 presents data concerning interaction between variants of willow mowing and willow genotypes. The biggest differences between variant II and I of shoots mowing for height and thickness of shoots were on plots of 1047 clone, and the lowest – for Ekotur variety. When taking into account live and dead shoots in snag the biggest differences were for Sprint variety, and the lowest, respectively for Olof and Ekotur varieties. At live willow snags on the plot, Start variety responded strongest to variants of mowing, and Ekotur and Olof varieties had no reaction to variants of mowing.

Table 6. Impact of interaction between variants of shoots mowing and willow genotypes on parameters of willow canopy architecture.

Willow genotype	Mowing variant[2]	Shoots height [cm]	Shoots thickness [mm]	Live shoots in snag [pcs.]	Dead shoots in snag [pcs.]	Live snags on the plot [pcs.]
1047	I	239.1	14.1	5.58	2.12	17.8
	II	322.0	20.1	8.89	2.40	25.8
1054	I	251.6	14.6	4.36	1.70	15.5
	II	304.8	19.0	8.53	2.32	24.8
1047D	I	260.3	15.7	5.25	1.64	15.4
	II	327.6	20.0	9.36	2.44	23.0
Start	I	209.2	12.0	2.96	1.52	8.7
	II	278.9	16.8	7.28	2.10	22.9
Sprint	I	239.6	13.7	3.87	1.64	11.9
	II	305.2	18.7	8.59	2.71	25.8
Turbo	I	240.1	13.8	4.89	1.53	19.1
	II	315.1	19.1	8.01	1.96	24.0
Ekotur	I	339.9	22.0	6.24	2.15	25.1
	II	368.1	23.5	7.55	2.05	24.3
Olof	I	300.0	17.7	5.18	1.77	20.9
	II	333.7	21.1	5.69	1.88	19.8
Jorr	I	257.0	14.7	5.28	1.90	18.2
	II	322.6	19.7	8.50	2.26	24.4
Tordis	I	346.0	21.1	5.26	2.11	22.5
	II	366.9	23.2	7.55	2.51	23.7
$NIR_{0.05}$		6.5***	0.4***	0.34***	0.14***	0.3***

[2]Mowing variant: I – mowing after 3-year and 1-year regrowth, II – mowing after 4-year regrowth

Own study showed that during the multi-year willow cultivation there are changes in the canopy architecture resulting from the impact of years of shoots regrowth, mowing pattern, fertilization with mineral nitrogen and varietal characteristics of willow. In Polish literature such effects are documented in reference to the first 2-, 3- or 4-year rotation, but still there are no such analyses for subsequent rotations [4, 14-18, 20-22]. The foreign literature reports that, in further rotations higher willow biomass yield is obtained than in the first [10, 11]. Styszko i Fijałkowska [21] have also proven that negative impact of follow-up fertilization of willow with compost an mineral nitrogen on shoots regrowth in height and thickness and on number of live shoots in snag and live snags on the plot is expected.

Measurements of dynamics of willow shoots increase during vegetation may be used for predicting biomass yield. Prognostic equations of an average yield of fresh biomass of 9 willow clones, acquired from light soil, fertilized with compost from municipal sewage sludge and various nitrogen doses (0, 90 and 180 kg $N \cdot ha^{-1}$) have been developed by the authors in previous work [16]. Equations include data from biometric measurements: length and thickness of shoots and number of shoots in snag from four dates of measurements and interaction of length and thickness of shoots. Those analyzes showed that incorporation of additional data relating to the weather pattern allows prediction of yield of fresh biomass of shoots with high precision (about 4.7%) during each year of cultivation. There are no similar analyses in the literature for the varieties and clones of willow fertilized with mineral nitrogen.

4 Conclusions

1. Number of years of shoots regrowth, variants of shoots mowing, dose of nitrogen fertilization and willow genotype affect canopy architecture in the second 4-year rotation assessed with length and thickness of shoots, number of live and dead shoots in snag and number of live snags on the plot.
2. Willow genotypes vary in response to years of cultivation, fertilization with mineral nitrogen and variants of mowing in the dynamics of shoots increase in length and thickness, the number of live and dead shoots in snag and number of live snags on the plot.
3. Increased fertilization with mineral nitrogen enhances growth of willow shoots in length and thickness, but reduces the number of live shoots in snag and the number of live snags on the plot.
4. Impact of willow mowing variant on growth of willow shoots in length and thickness, number of live shoots in snag and number of live snags on the plot is more visible at high than at low fertilization with mineral nitrogen.

The authors would like to thank the Lillohus AB company; 291 61 Kristianstad in Sweden for gratuitous transfer to Olof, Jorre and Tordis varieties and Mr. Przemysław Dobrzaniecki from Agrobränsle AB in Poznan for intermediation in this donation.

References

1. EurObserv'ER, 15th EurObserv'ER Report, 103 (2015)
2. GUS, *Energia ze źródeł odnawialnych w Polsce w 2015 r.* (GUS Warszawa, 2016)
3. Z. Grudziński, ROCZ OCHR SR **15**, 2249-2266 (2013)
4. M. Stolarski, S. Szczukowski, J. Tworkowski, Fragm. Agron. **19**(2), 39-51 (2002)
5. M. Stolarski, S. Szczukowski, J. Tworkowski, [In:] *Biomasa dla elektroenergetyki i ciepłownictwa – Szanse i problemy.* Wyd. Wieś Jutra Warszawa, 93-99 (2007)
6. W. Guidi, F.E. Pitre, M. Labresque, *Biomass Now – Suistainnable Growth and Use.* INTECH: 421-448. http://dx.doc.org./10.5772/51111 (2013)
7. M. Labrecque, T. Teodorescu, FOREST ECOL MANAG **150**, 223-239, (2001)
8. H. Adegbidi, T.A. Volk, E.H. White, L.P. Abrahamson, R.D. Briggs, D.H. Bickelhaupt, BIOMASS BIOENERG **20**, 399-411 (2001)
9. A. Trojanowska, [In:] A. Grzybek (ed.) *Modelowanie energetycznego wykorzystania biomasy.* Falenty-Warszawa. Wydawnictwo ITP, 76-82 (2010)
10. M. Labrecque, T. Teodorescu, BIOMASS BIOENERG **25**, 135-146 (2003)
11. W. Nissim, F. Pitre, T. Teodorescu, M. Labrecque, BIOMASS BIOENERG **56,** 361-369 (2013)
12. R. Kopp, L.P. Abramhamson, E.H. White, K.F. Burns, C.A. Nowak, BIOMASS BIOENERG **12**(5), 313-319 (1997)
13. M. Weih, N.-E. Nordh, BIOMASS BIOENERG **23**, 397-413 (2002)
14. W. Nowak, J. Sowiński, A. A. Jama, Fragmenta Agronomica **28**(2), 55-62 (2011)
15. M. Stolarski, *Agrotechniczne i ekonomiczne aspekty produkcji biomasy wierzby krzewiastej (Salix spp.) jako surowca energetycznego* (Wyd. UW-M, Olsztyn, 2009)
16. L. Styszko, D. Fijałkowska, M. Sztyma-Horwat, M. Ignatowicz, PAK, **5** 512-515 (2010a)
17. L. Styszko, D. Fijałkowska, M. Sztyma-Horwat, ROCZ OCHR SR **12,** 339-350 (2010b).
18. L. Styszko L., M. Ignatowicz, A. Borzymowska, Zeszyty Problemowe Postępów Nauk Rolniczych **Z. 564,** 237-245 (2011)

19. J. Tworkowski, S. Szczukowski, M. Stolarski, Fragmenta Agronomica **27**(4), 135-146 (2010)
20. M. Ignatowicz, L. Styszko, [In:] T.M. Traczewska (ed.) *Interdyscyplitarne zagadnienia w inżynierii i ochronie środowiska*, Oficyna Wyd. PW Wrocław, 203-212, (2013)
21. L. Styszko, D. Fijałkowska, Zeszyty Problemowe Postępów Nauk Rolniczych **Z. 582**, 73-80 (2015)
22. S. Szczukowski, J. Tworkowski, M. Stolarski, W. Fortuna, Fragmenta Agronomica **26**(3), 146-1 (2009)

Effectiveness of pollutants removal in hybrid constructed wetlands – different configurations case study

Magdalena Gajewska[1], *Krzysztof* Jóźwiakowski[2], and *Katarzyna* Skrzypiec[1,*]

[1]Gdańsk University of Technology, Faculty of Civil and Environmental Engineering, Narutowicza St. 11/12, 80-233, Gdańsk, Poland (mgaj@pg.gda.pl, katskrzy1@student.pg.gda.pl)
[2]University of Life Sciences in Lublin, Faculty of Production Engineering, Akademicka St. 13, 20-950, Lublin, Poland (krzysztof.jozwiakowski@up.lublin.pl)

Abstract. In recent years, an increase in interest in hybrid constructed wetland systems (HCWs) has been observed. The aim of the paper is to compare different HCW configurations in terms of mass removal rates and efficiency of pollutants removal. Analysed data have been collected at multistage constructed wetlands in Poland, which are composed by at least two beds: horizontal subsurface flow (SSHF) and vertical subsurface flow (SSVF). The evaluation was focused on hybrid constructed wetlands performance with HF+VF vs. VF+HF configuration, where influent wastewater of the same composition was treated. In analysed HCWs, the effective removal of organic matter from 75.2 to 91.6% COD was confirmed. Efficiency of total nitrogen removal varied from 47.3 to 91.7%. The most effective removal of TN (8.3 g m^{-2} d^{-1}) occurred in the system with VF+VF+HF configuration.

1 Introduction

Constructed wetlands (CWs) for wastewater treatment have been met with great interest during the past decades because of their performances, low investment and operational costs and positive effect on the environment. Subsurface CW systems are composed of filters with different modes of sewage flow: SS VF - vertical or SS HF - horizontal. Hybrid constructed wetlands combine the benefits of both types of beds, thus proving to achieve higher efficiency of pollutants removal, especially nitrogen compounds. These systems consist of two or more VF and HF filters, both types of beds have different working conditions and ensure different character of the processes. In vertical flow CW the dominating processes are organic matter mineralisation and ammonia nitrification due to better oxygen conditions in comparison with the horizontal flow CW, with claims of better final effluent quality (lower organic matter concentration, complete nitrification and partial denitrification) [1–6].

The pathways for nitrogen transformation in subsurface flow wetlands are complex and diverse, and include ammonification, sorption, plant and microbial uptake, ammonia

* Corresponding author: katskrzy1@student.pg.gda.pl

volatilization, nitrification, and denitrification. The prevailing processes differ in magnitude depending on the design and mode of operation. In general, SSVF wetlands are unsaturated units usually fed intermittently, resulting in high oxygen transfer capacity, thus favouring the nitrification process within the wetland bed. On the other hand, SSHF wetlands operate mostly under anoxic/anaerobic conditions due to the permanent saturation of the bed, which makes it a likely environment for denitrification. SSHF beds are also responsible for effective suspended solids and organic matter removal [4, 5, 7].

The objective of the paper is to evaluate performance of hybrid constructed wetlands with different configurations of the beds in respect of pollutants removal efficiency and MRR (mass removal rate) after each stage of treatment. Very unique results from the HCW built to treat domestic wastewater were obtained, in which the discharged wastewater was split into two streams treated in two configurations, enabling for reliable estimation.

2 Materials and methods

The investigations were carried out at five constructed wetlands in Poland: two for single family (SF TWs) in Dąbrowica I and Dąbrowica II, southeast part of the country, two for the local community (local TWs) in Darżlubie and Wiklino, located in Pomerania region, and one for reject water treatment from digested sludge dewatering in centrifuge at an existing activated sludge plant (Pilot Wschód RWC).

The analysed MTWs differ from one another in the order and number of subsequent stages. The characteristic of the systems with configuration of the beds is given in Table 1.

Table 1. The operation conditions of the MTWs.

Plant, location and type	Flow [m^3 day^{-1}] (pe)	Configuration	Area [m^2]	Hydraulic load [mm day^{-1}]
Darżlubie, Poland, local TW	56.6;	SSHF I	1200	47.3
	(650)	Cascade bed	400	141.2
		SSHF II	500	113.4
		SSVF	250	113.4
		SSHF III	1000	56.7
			Σ3350	
Wiklino, Poland, local TW	18.6;	SSHF I	1050	19.5
	(220)	SSVF	624	65.7
		SSHF II	540	38.0
			Σ2214	
Pilot Wschód, Poland, RWC	0.24;	SSVF I	7.5	3.2
	(5)	SSVF II	5.0	4.8
		SSHF	3.9	28.5
			Σ16.4	
Dąbrowica I, Poland, SF TW	0.3;	SSHF	24	12.0
	(3)	SSVF	24	
			Σ48	
Dąbrowica II, Poland, SF TW	0.3;	SSVF	24	12.0
	(3)	SSHF	24	
			Σ48	

At local TWs, configuration with first stage HF bed was applied. In case of Darżlubie facility, second stage filter was divided into two parallel beds working continuously while at Wiklino VF beds worked alternately with two weeks resting periods. The configuration performance with SSVF beds working as a first stage followed by SSHF was based on results achieved in Pilot Wschód RWC. The results for this configuration were compared with those achieved in Waizenfeld hybrid constructed wetland near Nuernberg of the same configuration (described in [3]). In case of SF TWs, opposite bed configurations used in these two systems (HF+VF vs. VF+HF) were very valuable for comparison of treatment efficiency due to the identical dimensions of the beds and the same composition of influent wastewater [2, 3].

Averaged samples of influent, effluent and after subsequent stages of treatment in all systems were analysed. In order to evaluate the removal efficiencies, the following parameters were measured: organic matter (COD), total nitrogen (TN), ammonium nitrogen ($N\text{-}NH_4^+$), nitrate nitrogen ($N\text{-}NO_3^-$), according to the methods of Polish Standards in accordance with the standard methods (APHA 2005) [8, 9]. Removal efficiency was calculated as a quotient of contaminants concentration difference in influent and effluent after subsequent stages of constructed wetland and concentration in influent. Mass removal rate (MRR) was calculated on the basis of the following equation:

$$MRR = (C_{in} \cdot Q_{in} - C_{out} \cdot Q_{out})/A \text{ [g m}^{-2}\text{ d}^{-1}] \quad (1)$$

A is the area of MTW [m^2], Q_{in} and Q_{out} are the average influent and effluent flow rates, respectively [$m^3\ d^{-1}$], C_{in} and C_{out} are average influent and effluent pollutant concentrations, respectively [$mg\ L^{-1}$].

In case of Darżlubie and Wiklino, combined spot samples were taken monthly at each stage of the HCWs from 1998 to 2006. Data from Pilot Wschód RWC was collected in 2010 and from HCW systems in Dąbrowica I and Dąbrowica II in 2007–2010 [2, 10].

3 Results and discussion

Treatment efficiency depends on many factors, among other such as susceptibility of wastewater to biodegradation, hydraulic load and organic load. In the analysed MTWs, the quality of discharged wastewater was unstable in time and varied significantly for both nitrogen and organic compounds (Table 2). An assessment of wastewater biodegradability is most often based on COD/BOD_5 ratio, and when effective nitrogen removal is required the BOD_5/total nitrogen (TN) ratio is considered [4, 10, 11].

Table 2. The quality of influent wastewater and biodegradability of wastewater in analysed MTWs.

Plant	TN [mg N L^{-1}] mean±σ min÷max	COD [mg O_2 L^{-1}] mean±σ min÷max	BOD_5 [mg O_2 L^{-1}] mean±σ min÷max	COD/BOD_5 [–] mean	BOD_5/TN [–] mean
Darżlubie	120.8±4.3 114.3÷128.9	843.8±40.7 791.4÷901.5	368.7±16.0 340.2÷390.5	2.4	3.1
Wiklino	130.4±9.2 118.4÷148.0	714.6±110.7 508.8÷932.5	382.1±72.0 280.6÷500.7	1.9	3.4
Pilot Wschód RWC	788.1±170.9 710.4÷1789.2	1022.7±93.5 880.0÷1260.0	378.9±87.0 270.8÷569.0	2.7	0.5
Dąbrowica I	134.0±25.4 109.0÷201.0	408.0±94.2 68.0÷300.0	169.0±65.9 99.0÷338.0	2.4	1.3
Dąbrowica II	134.0±25.4 109.0÷201.0	408.0±94.2 68.0÷300.0	169.0±65.9 99.0÷338.0	2.4	1.3

Only in one HCW, wastewater is characterised by COD/BOD5 below 2 and can be assumed as easily degradable and good source of electron donors for nitrogen removal in denitrification process. In case of reject water, the BOD5/TN ratios were far from recommended value, usually much smaller, showing a very high concentration of nitrogen in comparison to the available easily degradable organic matter.

Organic load applied in Darżlubie facility was the highest and amounted to 39.0 g COD m^{-2} day^{-1} while in Wiklino HCW it was three times lower. For HCW treating reject water this parameter was 22.8 g COD m^{-2} day^{-1}. The lowest organic matter loading rate (7.0 g COD m^{-2} day^{-1}) occurred in the smallest SF TWs.

Considerably high hydraulic load was used for local TWs, with the highest values in Darżlubie. In the Pilot Wschód facility the lowest hydraulic load was applied (3.2 mm day^{-1} on the first SSVF bed). Hydraulic load in all systems does not exceed the recommended value of 300 mm day^{-1} [4, 12].

Pollutants concentrations in wastewater in analysed hybrid constructed wetlands are higher than the most commonly occurring in other European countries. The quality of influent wastewater is comparable with raw wastewater without pretreatment (France) [2, 6, 7, 12, 13].

3.1 HF + VF configuration

Based on removed loads of pollutants, mass removal rates from each stage of treatment in analysed HCWs were calculated and presented in Table 3.

Table. 3. Mass removal rates (MRR, g m^{-2} d^{-1}) in local TWs.

Plant	Configuration	COD	TN	$N\text{-}NH_4^+$	N_{org}	$N\text{-}NO_3^-$
Darżlubie	SSHF I	20.6	3.4	3.6	+++	+++
	Cascade bed	18.0	2.9	2.2	1.2	+++
	SSHF II	7.0	0.8	0.4	0.1	0.3
	SSVF	10.8	0.7	1.4	0.3	+++
	SSHF III	5.7	0.5	0.2	0.1	0.2
	Total	13.1	1.9	1.8	0.1	0.0
Wiklino	SSHF I	6.9	1.1	1.1	0.0	0.9
	SSVF	5.0	0.6	0.5	0.5	+++
	SSHF II	2.5	0.8	0.5	0.3	0.0
	Total	5.3	0.9	0.8	0.2	0.0

+++ - an increase of pollutant parameter occurred

The highest COD and TN removal rates were noticed in first stage SSHF beds. In case of SSVF beds, MRR values were lower for organic matter (10.8 and 5.0 g COD m^{-2} d^{-1}, respectively) and total nitrogen (0.7 and 0.6 g TN m^{-2} d^{-1}, respectively). System in Darżlubie was characterized by higher average removal rates of almost all pollutants (13.1 g COD m^{-2} d^{-1}, 1.9 g TN m^{-2} d^{-1}) while in Wiklino facility such effects were not noted. An important issue appeared in relation to the removal of nitrate nitrogen as concentrations of $N\text{-}NO_3^-$ increased in local TW systems, especially after SSVF beds.

Pollutants removal efficiency in analysed HCWs is presented in Fig. 1.

Analysed HTWs were characterized by very effective organic matter removal, with 88.2% in Wiklino and 91.6% in Darżlubie. The highest TN removal was observed in Darżlubie (91.7%). Achieved results are similar to data described previously in [14, 6, 15, 16]. During measurement period facilities worked with various effectiveness of nitrogen removal. In HF beds TN was removed with higher effectiveness in comparison with VF beds. Removal efficiency of $N\text{-}NH_4^+$ in all the treatment stages were usually much higher compared to total nitrogen removal.

DOI: 10.1051/e3sconf/20171700023

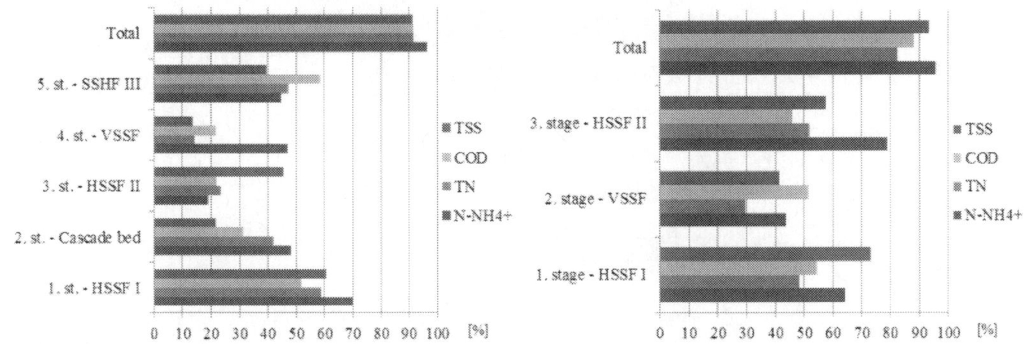

Fig. 1. Removal effectiveness of pollutants – Darżlubie (on the left) and Wiklino (on the right).

3.2 VF + HF configuration

Configuration with VF+HF was applied at pilot wetland system for reject water treatment. In Table 4 mass removal rates for subsequent beds of the analysed facility are given.

Table. 4. Mass removal rates (MRR, $m^{-2}\ d^{-1}$) in Pilot Wschód RWC.

	COD	TN	$N\text{-}NH_4^+$	N_{org}	$N\text{-}NO_3^-$
VSSF I	12.7	8.3	8.4	0.3	+++
VSSF II	6.8	5.5	5.5	0.7	+++
HSSF	12.2	6.1	5.2	0.4	0.1
Total	10.8	7.0	6.8	0.4	0.0

+++ - an increase of pollutant parameter occurred

Average mass removal rate of total nitrogen was significant high and confirmed effective removal (7.0 g $m^{-2}\ d^{-1}$). Organic matter removal rate in the CW system may be considered satisfactory (10.8 g $m^{-2}\ d^{-1}$).

Achieved results in analysed HCW were compared with those from HCW located in Waizenfeld, Germany (Table 5). In respect of ammonium nitrogen removal, more efficient nitrification process took place at polish facility and amounted to 8.4 g N m^{-2} day^{-1}, while in Waizenfeld the highest removal rate was 3.6 g N m^{-2} day^{-1}, both results were achieved in VSSF beds.

Table 5. The average contamination removal rates (g m^{-2} day^{-1}) in HCW in Germany [3].

Configuration	COD	TN	$N\text{-}NH_4^+$
	Waizenfeld		
VSSF	35.0	2.7	3.6
HSSF	0.54	0.68	0.21
Total	23.6	2.05	2.52

Configuration VF+HF created favourable conditions for nitrification process and effective removal of organic compounds, whilst in Waizenfeld total nitrogen was not so effectively removed as it could be expected. Based on data collected at Pilot Wschód RWC, effectiveness of performance was calculated and presented in Fig. 2.

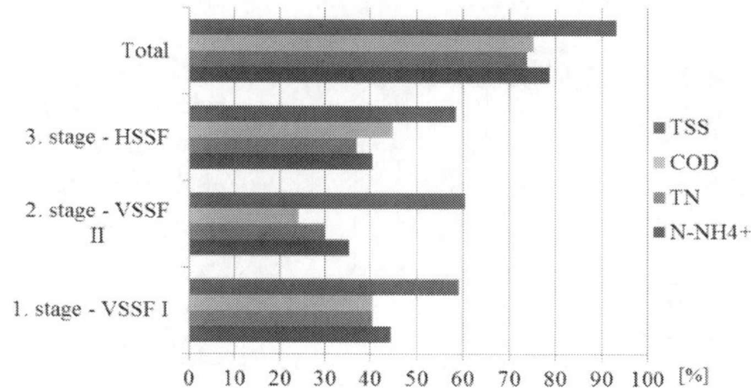

Fig. 2. Removal effectiveness of pollutants - Pilot Wschód RWC.

Total nitrogen removal was higher at Pilot Wschód in comparison with the German CW, where TN was removed with efficiency of 64.2%. On the other hand, the Waizenfeld hybrid constructed wetland was 96.7% efficient in ammonium removal while in polish HCW the efficiency of $N\text{-}NH_4^+$ removal was much lower (78.6%). In Waizenfeld the most efficient removal processes took place in VF bed situated at the first stage of treatment, and HF bed played less important role.

3.3 HF + VF vs. VF + HF configuration – comparison

SF TWs in Dąbrowica I and Dąbrowica II were fed with domestic wastewater of the same composition which allowed for comparison of the effectiveness of the HF+VF and VF+HF configurations. Calculated MRR are presented in Table 6.

Table. 6. Mass removal rates (MRR, g m^{-2} d^{-1}) in SF TWs.

Plant	Configuration	BOD_5	COD	TN
	HSSF	1.3	2.9	0.2
Dąbrowica I	VSSF	0.7	1.6	0.6
	Total	1.0	2.3	0.4
	VSSF	1.9	4.4	0.8
Dąbrowica II	HSSF	0.1	0.2	0.2
	Total	1.0	2.3	0.5

Analysed systems achieved satisfactory and equal total removal rates for organic matter expressed as BOD_5 and COD, although comparing subsequent stages in each system, first stage of treatment performed more effectively than the second one. TN removal in SSHF filters in both HCWs amounted to 0.2 g m^{-2} d^{-1}. Higher removal rates were achieved in SSVF beds (0.6 g m^{-2} d^{-1} in the filter working as second stage and 0.8 g m^{-2} d^{-1} in configuration as first stage).

In Fig. 3 pollutants removal efficiency of each stage in VF+HF and HF+VF configuration is shown.

Overall COD removal efficiency was very similar (around 90%). In VF beds, nitrogen compounds were removed with much higher efficacy compared to the HF beds. Performance of VF bed when applied as first stage is characterised by higher removal rates of all pollutants. Thus configuration with VF bed at the beginning of the system has significant impact on TN removal. Comparing horizontal flow beds working in different configurations, it was noted that second stage SSHF bed creates unfavourable conditions for nutrients removal, as effectiveness for TN did not exceed 25%. It can be associated with insufficient content of organic matter e.g. for denitrification process.

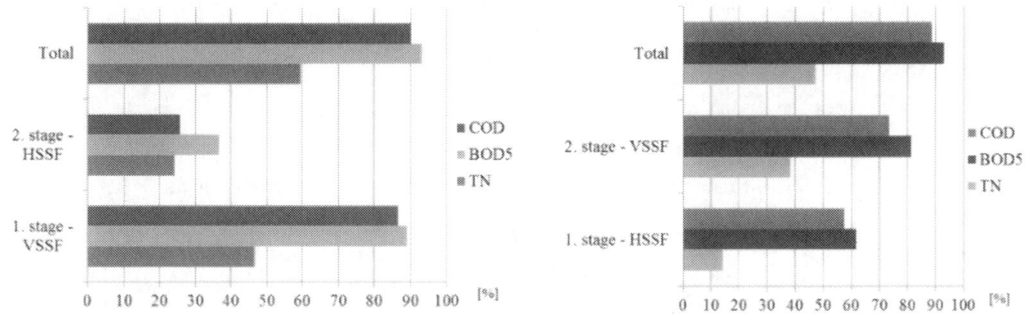

Fig. 3. Removal efficiency - Dąbrowica I (on the right) and Dąbrowica II (on the left).

4 Conclusions

Based on carried out analyses of efficiency removal and mass removal rate for each stage of treatment in hybrid constructed wetlands it could be concluded:

1. Hybrid constructed wetlands ensure stable in time removal of organics (up to 90%) and total nitrogen (up to 60%) irrespective of applied configuration.
2. The configurations with SSVF beds as a first stage of treatment were characterized by higher nitrogen removal rate in comparison to configuration with SSHF beds in front.
3. The case study of HCW working with the same composition of wastewater revealed that the configuration SS VF + SS HF ensured better removal of pollutants and in consequence better quality of final effluent.

The authors wish to thank Gdańsk Water Utilities participating in the project entitled "Interactive Water Management" with the acronym "IWAMA" co-funded by BSR INTERREG Programme 2014-2020, within which this paper is written.

References

1. C. Pelissari, C. Ávila, C. M. Trein, J. García, R. Dultra de Armas, P. H. Sezerino, Sci. Total Environ. **574** (2017)
2. M. Gajewska, H. Obarska-Pempkowiak, Environ. Prot. Eng. **37** (2011)
3. M. Gajewska, A. Tuszyńska, H. Obarska-Pempkowiak, Pol. J. Environ. Stud. **13**, 3 (2004)
4. R. H. Kadlec, S. Wallace, *Treatment wetlands, second edition* (CRC Press Taylor & Francis Group, Boca Raton, London, New York, 2009)
5. F. Masi, S. Caffaz, A. Ghrabi, Water Sci. Technol. **67**, 7 (2013)
6. P. Molle, S. Prost-Boucle, A. Lienard, Ecol. Eng. **34** (2008)
7. F. Masi, N. Martinuzzi, R. Bresciani, L. Giovanelli, G. Conte, Water Sci. Technol., **3** (2007)
8. American Public Health Association (APHA), *Standard methods for the examination of water and wastewater* (21st Ed. Washington, DC, 2005)
9. Polish standards according limits for discharged sewage and environmental protection from July, 24 2006 (No. 137, item 984) and January, 28 2009 (No. 27, item 169).
10. M. Gajewska, K. Jóźwiakowski, A. Ghrabi, F. Masi, Environ. Sci. Pollut. Res. **22** (2015)

11. K. Miksch, J. Sikora, *Biotechnologia ścieków (Biotechnology of wastewater)* (Wydawnictwo Naukowe PWN Warszawa, 2010)
12. J. Vymazal, Ecol. Eng. **25** (2005)
13. J. Zhai, Md. H. Rahaman, X. Chen, H. Xiao, K. Liao, X. Li, C. Duan, B. Zhang, G. Tao, Y. John, J. Vymazal, Ecol. Eng. **97** (2016)
14. Y. Millot, S. Troesch, D. Esser, P. Molle, A. Morvannou, R. Gourdon, D. P. L. Rousseau, Ecol. Eng. **97** (2016)
15. A. Morvannou, N. Forquet, S. Michel, S. Troesch, P. Molle, Water Sci. Technol. **71**, 9 (2015)
16. J. Paing, A. Guilbert, V. Gagnon, F. Chazarenc, Ecol. Eng. **80** (2015)

Monitoring the quality of water from isolated marshes resulting from opencast and underground mining of the Sieniawa lignite deposit

Grzegorz Galiniak[1,*], *Kazimierz* Różkowski[1]

[1]AGH University of Science and Technology, Al. Mickiewicza 30; 30-059 Krakow, Poland

Abstract. The „Sieniawa" lignite mine is one of the oldest Polish mines extracting lignite by means of both underground and opencast mining. The latter process is carried out until the present day. The transformation of the landscape is the most visible evidence of carrying out mining activities and therefore systematic approach to the reclamation of brownfield sites is a very important element of the mine's environmental policy. The authors present previous experiences and the course of reclamation of post-mining terrains created during the operation of the "Sieniawa" lignite mine. The greatest emphasis is placed on numerous small anthropogenic reservoirs formed as a result of spontaneous flooding due to underground exploitation (surface subsidence) and filling of open pits with water.

1 Introduction

Mining is one of those branches of industrial activities that bring significant economic benefits, but, on the other hand, inevitably lead to a conflict with the environment. The influence on the natural environment is manifested by a reduction in the amount of natural resources, degradation (sometimes devastation) of mining areas, hydrogeological disturbances, and emissions to the atmosphere. The current Polish law imposes an obligation to conduct reclamation in order to restore the utility values of the affected areas. It should be noted that the modern reclamation has more important goals than just bringing the utility value. Its purpose is to create complex and complete ecosystems in the post-mining areas. In this case, the open-pit mining (and lignite mining in particular) and modern reclamation may create new possibilities for the development of water, forest and agricultural ecosystems.

An example of proper approach to the issue of reclamation of post-mining areas and the so-called sustainable management of natural resources can be found in one of the longest operating mines in Poland, namely the "Sieniawa" lignite mine, which can be praised because of the effective and properly conducted reclamation works [1–3].

* Corresponding author: galiniak@agh.edu.pl

DOI: 10.1051/e3sconf/20171700024

2 The characteristic of the mine and the deposit

The "Sieniawa" lignite mine, as the last and at the same time one of the few mines carrying out both the common open-pit and underground mining (Fig. 1.), conducted underground exploitation until 1997. It is estimated that during the fifty years between 1950 and 2001 that preceded the liquidation of the mine and its acquisition by another entity, 6.1 million tonnes of lignite were extracted, from which around 5 million tonnes using underground method. Due to the complicated geological structure, a result of glaciotectonic deformation of young Tertiary sediments, the documented deposits are characterized by relatively small dimensions, strong folding, and significant geometric inconsistency of the folding structures - from symmetrical through overturned, recumbent to returned and scale folds [4, 5]. Specific geological and mining conditions were associated with the necessity to conduct exploitation within small, separated exploitation fields, containing individual fold structures. The average area of a mining excavation reached up to 5 ha, with depth of up to 40 m above the main water-bearing sublignite level [1, 2, 6, 7]. Usually, after few years of mining activity and depleting the resources, the exploited field was being abandoned and mining activities were being transferred to another field, usually located in close proximity. The abandoned open pits were subjected to reclamation activities characterized by varying degrees of involvement, generally increasing with experience and the growing ecological awareness. In the case of underground mining with roof caving, the empty spaces of the excavations were spontaneously tightened as a result of surface (hanging rocks) subsidence. The deformations propagating towards the surface (in the case of relatively shallow mining excavations) have resulted in the formation of surface troughs and sink-holes directly above the caving area.

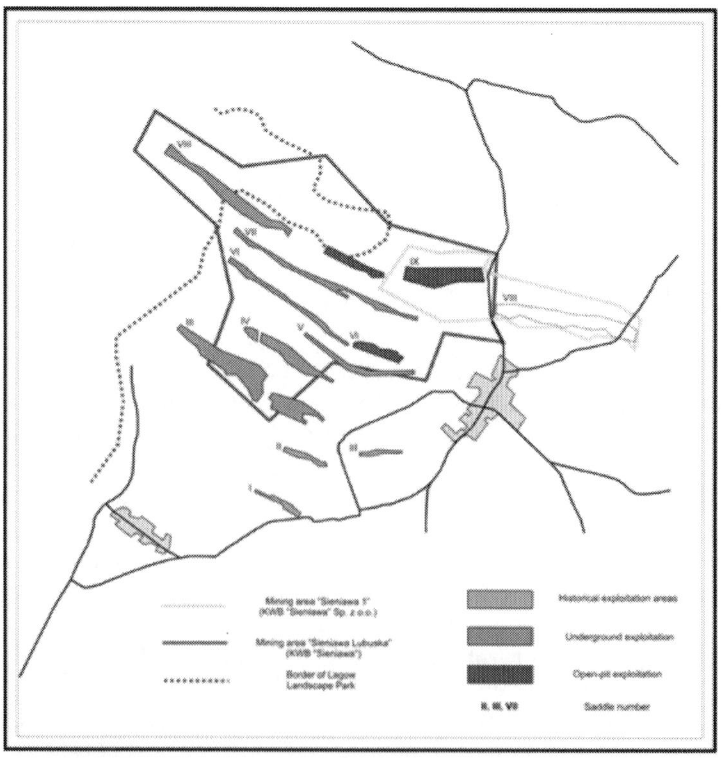

Fig. 1. Historical and present location of exploitation areas.

3 Mining and the transformation of the natural environment

The terrains acquired for both underground and open-pit exploitation were typical end moraine areas, characterized by a high variety of landscape forms. The majority of the area is covered by beech and mixed (beech-pine-birch) forests on brown and rusty soils. The remaining terrains were used as agriculture fields. Additionally, the area was characterized by a lack of developed hydrographic networks with the only exception of temporary (formed during heavy rainfall events) sink-holes filled with water occurring in the area of the No. III and VI saddles [1, 2, 6, 8]

The underground mining activity in the "Sieniawa" lignite mine was carried out in various periods of time in five saddles (No, III, IV, VI, VII, and VIII) while the open-pit method was used to produce lignite in the eastern part of the No. VI saddle and western part of the eastern wing of the No. VII and IX saddles. At present the exploitation is continued in the eastern part of the No. VIII saddle.

The areas of underground mining activities were previously deforested and excluded from forestry activity for around 15 to 20 years, depending on the time of mining works, resources available in the deposit, and character of the rock layers surrounding the seam. Taking into account the seam thickness, mining method (caving without backfilling) and the depth of the exploitation, the negative impact on the environment was manifested with numerous craters with diameter of up to several meters, ditches, and steps of up to a few meters deep and several meters long. As the result of deformations, the soil structure was damaged, especially in the areas of landslides, slopes, and steps. The fertile top layer of soil was scraped and the lower layers of sand and sandy clays were exposed. The silt and clay isolating layers above and below the lignite seam were the reason why the exploitation has not influenced the complex water conditions. In some of the sink-holes located over the silt basis, small water ponds (fitting very well into the existing landscape character) were formed.

Underground mining activities have resulted in anthropogenic transformation of an area reaching some 101.71 ha with almost 3245 meters of forest roads destroyed or damaged. This caused indirect but significant problems in transport in the area of ca. 500 ha located between the saddles and in the forest areas surrounding the mine.

The environmental impact of the open-pit mining activity has resulted in the occurrence of significant transformations of post-mining elements, namely pits and dumps. These mining objects are characterized are reaching up to a few hectares, while their depth (height) is up to several meters [6, 9, 10].

4 Description and assessment of the reclamation works

Taking into account that the majority of post-mining grounds were classified as class C and the fact that the areas acquired for mining activity were previously deforested it was decided that forest reclamation will be the main method of reclamation. Additionally, this decision was supported by the fact that in the 70s of 20th century the "Sieniawa" lignite mine passed over 20 ha of reclaimed lands (the No. III saddle) to State Forests National Forest Holding. During this time, the so-called artificial reforestation was carried out. The main activities included removing trees older than 35 years from the mining area and replacing them with young plants. The new forest more easily adapted itself to the surface deformations and new environmental conditions. Taking into account that some of the saddles were reclamated already during the mining process, it seemed reasonable to divide the reclamation process into two parts. The first one comprised of preparing the reclaimed areas for planting and the other included works and activities related to planting and nurturing the young trees.

The decision to choose forest reclamation as the leading method in the post-mining areas of the "Sieniawa" lignite mine was preceded by numerous consultations with the representatives of local authorities, State Forests National Forest Holding, and the managers of the mine. The local authorities demanded the reclamation of both the lands acquired for mining activity and those neighboring the mining area. However, no specified demands were given concerning the method of reforestation. Such requirements were, however, given by the representatives of State Forests; a special attention was paid to the fertile layer of soil and the introduction of productive species. The presented attitude is typical for State Forests National Forest Holding which in most cases is administering the reclaimed post-mining terrains.

In general, between 1957 and 1998, Świebodzin, Sieniawa and Łagów Forest Inspectorates passed over 136 ha of forest to the "Sieniawa" lignite mine, while between 1992 and 2002 the mine passed almost 127 ha of reclaimed grounds to the Świebodzin Forest Inspectorate, including 120 ha of areas ready for planting. The remaining part consisted of roads, ponds, and forest dividing-lines [1, 2]

In the more long-term perspective of 9 to 19 years after the reclamation works in particular saddles, it can be stated that the forest reclamation was carried out without significant complications and the results are satisfying. The mix of tree species was properly chosen. The reclaimed lands are not different than the surrounding areas. Simultaneously, they provide a habitat for animals, contributing to the increase in their population.

Additionally, small water reservoirs with defined morphology were created in the final lignite pits and in the specially formed troughs without run-off located in the reclaimed areas thanks to the inflow of groundwater and precipitation of water, leading to an increase in the habitat's attractiveness (Fig. 2). Simultaneously, in the areas of troughs above the underground cavings, isolated marshes reaching area of 0.78 ha (Jezioro Ciche) were formed (Fig. 3) as a result of ground subsidence below the shallow groundwater level.

Fig. 2. The anthropogenic reservoir created in a pit of a former open cast mine – saddle No. VI; photo by G. Galiniak.

DOI: 10.1051/e3sconf/20171700024

Fig. 3. The anthropogenic reservoir ("The Silent Lake") in a sink-hole created as a result of underground exploitation - saddle No. VII; photo by G. Galiniak.

When analyzing the effects of previously conducted reclamation works, a preliminary analysis, based on a limited range of physical and chemical parameters, of the chemical composition of water from the reservoirs created in the post-mining terrains, was performed. In total, water from eight different reservoirs, representing the previously-mentioned types, was analyzed (Fig. 4). The samples were equally split between the classes. In the case of two biggest ones, with depths, determined by probing, exceeding 5.0 and 7.5 meters, multiple sampling was used as three samples were collected: from the bottom, middle, and near-surface zones. For the sake of comparison, an additional sample was collected from a well drilled for mine supply, filtered within the Quaternary aquifer (Fig. 4).

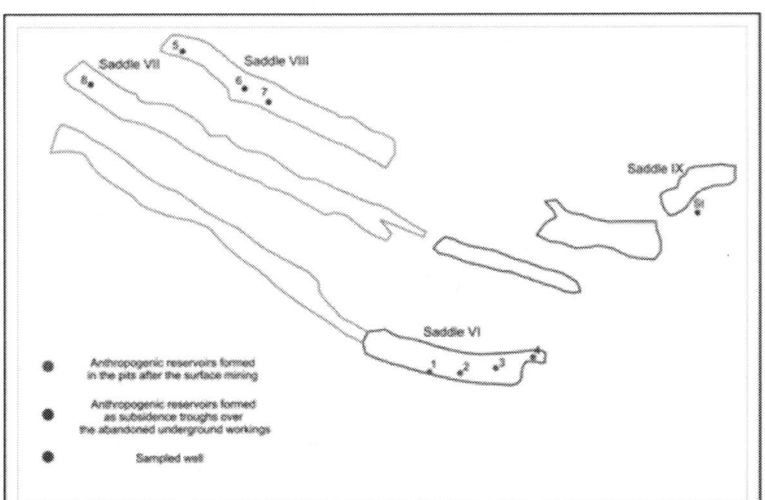

Fig. 4. The location of water samples collection.

Genetic dissimilarity between the individual types of the formed water reservoirs has found its reflection in the chemical variability of water filling particular troughs [11, 12]. The initial point was the different mineralization, although only fresh water was present within the area studied. Higher values in the range 161–431 mg/dm^3 were associated with the artificially formed reservoirs in the former open pits. Selective dumping of overburden material, which was used to form the reservoir structure, allowed avoiding a long-term contact between water and lignite residues. An analysis of ion composition of water indicated relatively low concentration of sulphate ions in comparison to the reservoirs filling the lignite pits known from the literature, marked within the range of 49.8–112.6 mg/dm^3. They constitute from 31.4 to 47.1% mval of anions, remaining in minority in comparison to bicarbonate ions commonly existing in surface and shallow groundwater. With water reaction oscillating around neutral, from weakly acid to weakly alkaline, the total iron concentration did not exceed 0.71 mg/dm^3. With the exception of iron and manganese, which concentrations an order of magnitude higher, and slightly elevated concentration of chlorides, the analyzed water samples were similar to those collected from the Quaternary well with a mineralization of 247.8 mg/dm^3. Similar characteristic was presented by A. Mizera who described the Quaternary water level in "The documentation of reclamation of post-mining areas in the No.VI and VIII saddle of the "Sieniawa" lignite mine (in liquidation) located in Sieniawa Lubuska".

The troughs formed as a result of subsidence of ground above tightened spaces created lowerings without run-off which, after some time, were filled with water. Because of a compact and loose character of the roof layers, partially with isolating character, accompanied with the lack of overburden layers of the disturbed structure and increased specific area, the filling waters are characterized with even lower mineralization than the samples collected from the terrains affected by the open-pit mining. Within the collected samples the total dissolved substance did not exceed 182 mg/dm^3, with pH value ranging from weakly acidic to weakly alkaline. The lowering of terrain, often significantly below the surrounding area, and deforestation of the border area, resulted in limiting the wind operation and decreased the water-mixing capability. With the limited mixing capability, the oxygen concentration in the water decreases quickly, leading to the formation of reducing environment. In the deepest reservoir, The Silent Lake, with bottom located significantly lower than the surrounding terrain, a typical stratification was created, which was emphasized by the drop of temperature from 18.9°C in the near-surface zone (epilimnion) to 8.3°C in the bottom zone (hypolimnion). The drop in total mineralization of the water is most likely related to the ongoing process of water demineralization, observed and described by A. Jędrczak and A. Gontaszewska in the area of Muskau Bend [9, 12, 13, 14, 15]. The progressive oxidation of sulphates has led to a drop in their concentration, well below the detection limit of the measuring equipment, with maximal detected concentration in the shallow waters reaching 1.29 mg/dm^3. The waters filling the subsidence troughs were characterized with simultaneous high concentrations of total iron, up to 2.85 mg/dm^3, in near-surface waters, and at the depth of over seven meters in the monimolimnion layer up to 15.55 mg/dm^3).

5 Summary

The „Sieniawa" lignite mine is a perfect example of harmonious cooperation with the surrounding environment over a 60-year period of the deposit exploitation. The reclamation carried out in the post-mining areas and the care for the natural environment, both in the neighboring terrains and in the mining area, have contributed to the establishment of the Łagów Landscape Park in 1985. At the moment, the mining is conducted on relatively low scale, without damage to the natural environment. The currently conducted reclamation considerable reduces the negative impact of mining. When compared to other Polish lignite mines, the production in Sieniawa is significantly lower, but the reclamation activities are as relevant as in the larger mines. Reclamation is conducted professionally and in line with the guidelines, which was confirmed by the National Fund for Environmental Protection and Water Management and the Ministries of Economy, State Treasury, and Environment. The aforementioned opinions are shared by foresters from all over the country visiting the mine and surrounding areas.

The work has been conducted on the AGH's Faculty of Mining and Geoengineering within the frame of Research Project No. 11.11.100.597.

References

1. G. Galiniak, J. Jarosz, R. Tomaszewski, Górnictwo i Geoinżynieria, **34 (4)** 167–178 (2010)
2. G. Galiniak, J. Jarosz, Zeszyty Naukowe Uniwersytetu Zielonogórskiego – Inżynieria Środowiska, **17 (137)** 180–188 (2010)
3. K. Różkowski, K. Polak, M. Cała, Górnictwo i Geoinżynieria, **34 (4)**, 517–525 (2010)
4. A. Gontaszewska, Hereditas Minariorum **2**, 51–65 (2015)
5. M. Piwocki, Węgiel Brunatny, **2**, 11–15 (2003)
6. L. Kozacki, Seria Geografia **21** (1980).
7. J. Żaba, Geologia, **1** (1977)
8. L. Kozacki, Badania geologiczno-inżynierskie dla potrzeb budownictwa na obszarach zaburzonych glacitektonicznie Ziemi Lubuskiej, (1974)
9. Gontaszewska, A. Kraiński., Prace Naukowe GIG. Górnictwo i Środowisko, **3**, 221–234 (2007)
10. A. Gontaszewska, A. Kraiński, Wybrane problemy badań geologicznych i hydrogeologicznych dla górnictwa i energetyki, 108–119 (2012)
11. D. Bedla, A. Misztal, Annual Set The Environment Protection, **16**, 421–439 (2014)
12. E. Jaruchiewicz, Environmental & Socio-economic Studies, **2 (1)**, 16–26 (2015)
13. A. Jędrczak, Wydawnictwo Wyższej Szkoły Inżynierskiej w Zielonej Górze (1992)
14. K. Labus, S. Skoczyńska, Geol Q, **57 (3),** 561–566 (2013)
15. S. Lutyńska, K. Labus, Arch Environ Prot, **41 (3),** 60–69 (2015)

Possible Co-fermentation of Water and Sewage Sludge

Justyna Górka[1,*], and *Małgorzata* Cimochowicz-Rybicka[1]

[1]University of Technology, Department of Environmental Engineering, 24 Warszawska, Cracow 31–155, Poland

Abstract. The article describes problems related to intensification of energy production at a wastewater treatment plant. The authors analyzed anaerobic co-digestion of sludge from both water treatment plant and sewage treatment plant. Water sludge is produced during coagulation, ozonation and backwashing of rapid anthracite filters. Its characteristic and properties depend on a raw water quality, treatment methods as well as types of chemicals used and their doses. According to the Polish Act of 4 December 2012 (Journal of Laws from 2013, item 21) the water sludge should be treated as hazardous waste. An alternative way to dispose water sludge and reduce its volume may be sludge reuse. The authors suggested a research methodology and analyzed the preliminary results, which showed that co-digestion of sewage and water sludge enhanced biogas production. The authors assume that the results of the study will provide a basis for development of methodology for sludge control and disposal.

1 Introduction

During water treatment (surface water in particular) a large amount of sludge and wastewater are produced. According to the waste classification (the Act of 14 December 2012 on waste [1]) sludge generated during water treatment processes should be treated as hazardous or other then hazardous waste. In Poland, management of this type of sludge has been significantly influenced by the legal requirements involving the country's membership in the European Union. This resulted in a ban on the storage of sludge, beginning from January 1, 2016. Therefore, we are forced to limit production of this type of waste or to seek other alternative ways of its elimination, with the possibility of its reuse as a raw material [2, 3].

The literature offers many publications on water sludge, in which its composition is described [4–7] as well as the methods of its cultivation and utilization [8–12]. Water sludge is mostly used as coagulant and phosphorus recovery from wastewater, as a coagulant in wastewater treatment, for heavy metal immobilization, production of ceramics or cement, land reclamation, sludge dewatering and also as a fertilizer in farming and forestry [8–10].

* Corresponding author: justynagrka@gmail.com

In contrast, there is little information on co-fermentation of water sludge and sewage sludge. Płonka [13] in her study used water sludge in co-fermentation with sewage sludge. The results showed that a higher amount of water sludge in samples resulted in a lower fermentation gas production. The presence of water sludge hindered the digestion process. However, it can be noted that the presence of water sludge improved wastewater sludge dewaterability. On the other hand, the authors' study [14] showed that co-fermentation of these two types of sludge increased a biogas production from 20 to 140% (depending on the sewage sludge/water sludge ratio in the samples), if compared to the digested sludge samples. Though, water sludge demonstrated a negative effect on filtration properties of sewage sludge.

The article tried to investigate a possibility of co-fermentation of water and sewage sludge as well as evaluate filtration properties of the sludge mixture. The study consisted of three phases:

- respirometric comparative analysis of biogas production,
- respirometric study of methanogenic activity,
- dewaterability - capillary suction time and specific resistance to filtration.

2 Methods

2.1 Respirometric test

Respirometric tests have been used for energy research to analyze both quantity and quality of a fermentation gas. Therefore, a respirometer and biogas production measurements were conducted to describe and evaluate biodegradability or activity in anaerobic processes. The respirometers used in anaerobic processes measure gas production as a gas volume determined either at a constant pressure or at a constant sample volume. Methane was selected as an indicator of the anaerobic process efficiency mostly due to its energetic properties, but also to its physical and chemical characteristics [15]. The test stand consisted of:

- Respirometer for aerobic-anaerobic tests (AER-208, manufactured by CHALLENGE SYSTEM),
- Water bath,
- Heating/cooling unit,
- Computer for online data processing (gas volumes measured every 2h).

Measurements were carried out over 20 days at mesophilic (35°C) conditions in three separate runs. Compositions of samples were selected based on the dry organic solids content and included water and digested sludge with a volume of samples: 1 dm^3. Co-substrate concentrations ranged from 3.0 to 6.0 g $VSS \cdot dm^{-3}$. To ensure the appropriate process conditions pH of samples was adjusted to 7.0 with NaOH. Before measurements, the sample were purged for three minutes with technical nitrogen. The laboratory analysis included: dry solids (VS), volatile suspended solids (VSS), chemical oxygen demand (COD), alkalinity, pH, total phosphorus (TP), total nitrogen (TN) and ammonia nitrogen (N_{NH4}).

2.2 Methanogenic activity test

Methanogenic activity was calculated based on the amount of methane produced from the sludge during the respirometric tests, and expressed in g COD_{CH4} g VSS $^{-1}$ d^{-1} [15–18]. The maximum methane production and the R factor (as $ml_{CH4} \cdot h^{-1}$) were determined based on the graphs showing the volume of methane produced per time unit. Fig. 1 shows the way R was determined for the digested sludge.

Table 1. Chemicals used during respirometric tests to determine the methanogenic activity.

Substrate concentration		Enrichment of samples – neutralized mixture of samples (pH 7.0): acidic and butter acids; COD of sample: 3.0 – 5.5 $gO_2 \cdot dm^{-3}$
Minerals	Solution 1	NH_4Cl (170 $g \cdot dm^{-3}$), $MgSO_4 \cdot 4H_2O$ (9 $g \cdot dm^{-3}$), KH_2PO_4 (37 $g \cdot dm^{-3}$), $CaCl_2 \cdot 2H_2O$ (8 $g \cdot dm^{-3}$)
	Solution 2	$FeCl_3 \cdot 4H_2O$ (2000 $mg \cdot dm^{-3}$), $(NH_4)_6Mo_7O_2 \cdot 4H_2O$ (90 $mg \cdot dm^{-3}$), $CoCl_2 \cdot 6H_2O$ (2000 $mg \cdot dm^{-3}$), $Na_2SeO_3 \cdot 5H_2O$ (100 $mg \cdot dm^{-3}$), $MnCl_2 \cdot 4H_2O$ (500 $mg \cdot dm^{-3}$), $NiCl_2 \cdot 6H_2O$ (50 $mg \cdot dm^{-3}$), $CuCl_2 \cdot 2H_2O$ (30 $mg \cdot dm^{-3}$), EDTA (1000 $mg \cdot dm^{-3}$), $ZnCl_2$ (50 $mg \cdot dm^{-3}$), HCl 36% ($1 cm^3$), H_3BO_3 (50 $mg \cdot dm^{-3}$), resazurin (500 $mg \cdot dm^{-3}$)
Yeast extract		0.2 ($g \cdot dm^{-3}$ of sludge sample)

The methanogenic activity of biomass was calculated according to the equation:

$$AKT = (R \cdot 24)/(W \cdot V \cdot VSS) \quad (1)$$

where:
AKT – methanogenic activity of sludge [$g\ COD_{CH4} \cdot g\ VSS^{-1} \cdot d^{-1}$],
R – parameter determined from the curve of the methane production [$CH_4 \cdot h^{-1}$],
W – conversion factor [$ml\ CH_4 \cdot COD^{-1}$] (assumed 418 $ml\ CH_4 \cdot COD^{-1}$ at 35°C),
V – volume of the sample [dm^3],

VSS – volatile suspended solids in the sample [$g\ VSS \cdot dm^{-3}$].

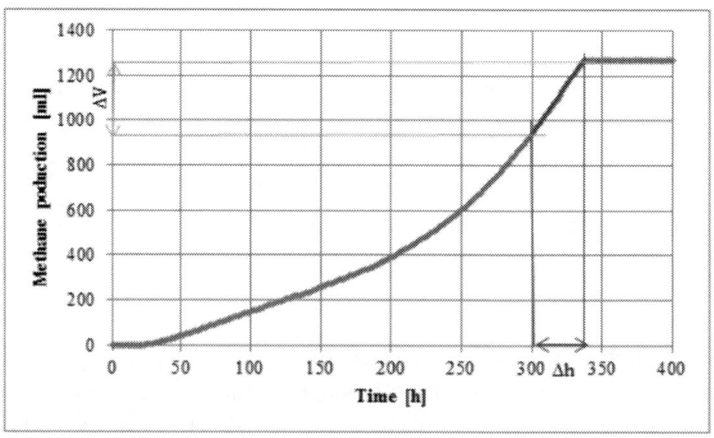

Fig. 1. Determination of the R factor from a methane production chart for digested sludge. The factor is determined as the ratio of the increase in methane production (ΔV) and time (Δh).

2.2 Filtration properties

A capillary suction time (CST) and specific resistance to filtration are the main sludge characterization techniques. CST measurements were conducted according to the Polish standards PN-EN 14701-1:2007 [19]. This parameter evaluates how easy moisture can be removed from the sludge cake; when a capillary suction time is small, the sludge easier (faster) releases liquid.

On the other hand, the specific resistance to filtration (SRF) is defined as the pressure required to make filtrate flow through the sludge cake, having unit mass of dry solids per unit area of filtration surface while a filtrate viscosity equals 1. The measurement of the specific resistance to filtration was carried out on the basis of the PN-EN 14701-2: 2013 [20]. The specific resistance to filtration (SRF) is calculation as follows:

$$SRF = (2 \cdot \Delta p \cdot A^2 \cdot b)/(\mu \cdot m) \qquad (2)$$

where:
SRF – specific resistance to filtration [$m \cdot kg^{-1}$],
Δp – pressure drop across the filter [Pa],
A – filtration area [m^2],
b – slope of the linear part of the curve obtained by plotting t/V vs. V [$s \cdot m^{-6}$], b factor determined as in Fig. 2,
μ – viscosity of filtrate at the sludge temperature [$Pa \cdot s$],
m– mass of solids deposited on the filtering medium per unit volume of filtrate [$kg \cdot m^{-3}$].

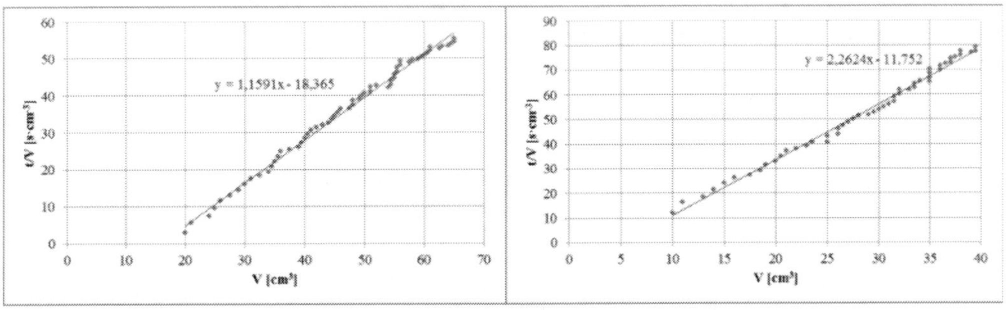

a) b)

Fig. 2. Results for b factor in digested sludge, a – before fermentation, b – after fermentation.

3 Results and discussion

The sludge samples (sewage sludge (D), water sludge (W)) were collected at the municipal water and wastewater treatment plants in Southern Poland. Water sludge was produced as a result of coagulation (PAX 19, XL 10, PAX 16), ozonation, dosing of powdered activated carbon and sand filters backwashing. The samples have been taken directly from sludge drying beds at the water treatment plant.

Table 2. Characteristic of sludge samples before and after digestion (preliminary research).

Parameters		Before digestion				After digestion			
		D	**W**	**1:0.5**	**1:0.3**	**D**	**W**	**1:0.5**	**1:0.3**
pH	[-]	7.15	7.02	7.05	7.40	6.94	6.82	7.11	7.25
Alkalinity	[$mgCaCO_3 \cdot dm^{-3}$]	511	186	279	349	1029	755	653	618
VS	[$gsm \cdot dm^{-3}$]	7.81	18.40	9.74	7.72	5.56	17.31	8.70	6.96
VSS	[$gsmo \cdot dm^{-3}$]	4.69	4.65	4.04	3.65	3.35	3.99	3.58	3.42
*tCOD	[$mgO_2 \cdot dm^{-3}$]	7910	4980	6470	6630	7498	3971	4655	4759
**sCOD	[$mgO_2 \cdot dm^{-3}$]	214	107	199	185	153	65	147	140

*tCOD – chemical oxygen demand of total organic suspensions, **sCOD – chemical oxygen demand of soluble organic compounds

Initially, only 3 series of preliminary studies were run to set right ratios of water sludge to sewage sludge; the proportion of sewage sludge to water sludge has been determined based on the organic content in the sludge. The results of physic-chemical analysis of the mixtures before and after digestion are shown in Table 2. The first test was performed for the sewage sludge (D), water sludge (W) and mixed sludge containing 50% (1:0.5) and 30% (1:0.3) of sewage and water sludge.

Biogas is produced as a result of anaerobic decomposition of organic compounds during methane digestion. Its volume and a production rate indicate the efficiency of the digestion process and signal any process limitations. Fig. 3 shows the biogas production in sewage sludge, water sludge and the samples that contained 50% (1:0.5) and 30% (1:0.3) of water sludge. In this study, biogas production varied depending on the ratios of sewage/water sludge though intensity of biogas production for both sludge (D) and (W) was very low at the beginning. After over 20 days of digestion, the produced biogas volumes were 0.26 $m^3 \cdot kgVSS^{-1}$ and - 0.17 $m^3 \cdot kgVSS^{-1}$, for sewage and water sludge, respectively. However, in the mixture with water sludge biogas production was much higher and reached 0.33 $m^3 \cdot kgVSS^{-1}$ for Sample 1:0.5 and 0.43 $m^3 \cdot kgVSS^{-1}$ for Sample 1:0.3. The sample mixed at a ratio of 1:0.3 produced the highest volume of biogas; it was about 65% higher than the value for sewage sludge only. It shows that water sludge has a stimulating effect on the biogas production. It should also be noted that a smaller amount of water sludge added to sewage sludge results in higher biogas volumes.

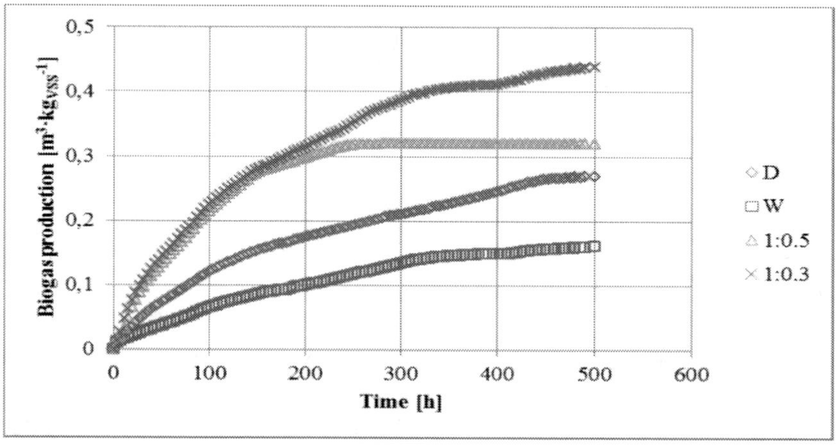

Fig. 3. The biogas production (preliminary research).

Another series of tests were performed for the mixed sludge containing 30% (1:0.3), 20% (1:0.2) and 10% (1:0.1) of water sludge and sewage sludge (D). The test results are shown in Table 3. After 20 days of digestion, the dry solid removals in the mixed sludge samples were as follows: 1:0.3–8.5%, 1:0.2 14% and 1:0.1–7.5%, while in the sewage sludge samples they exceeded 21 %. In the case of organic removal (COD), the mixed sludge showed up to 41% removal of sCOD, while in sewage sludge it was only 16%. Intensity of biogas production during the digestion process varied depending on the ratio of sewage sludge to water sludge. The highest production of biogas was observed, as in the preceding series, for the samples mixed at a ratio of 1:0.3 - 0.51 $m^3 \cdot kgVSS^{-1}$ (Fig. 4). In the sewage sludge samples (D) biogas production was the lowest (0.20 $m^3 \cdot kgVSS^{-1}$); it was about 60% lower than in Sample 1: 0.3.

During the study also some measurements of methanogenic activity (AKT) were made to investigate sludge digestion. The test results are shown in Tab.4. On the basis of the results it was stated that when water sludge was used to intensify the biogas production, the methanogenic activity increased from 5 to 122%, depending on the sludge mixing ratio. Methanogenic activity values to a large extent refer to the volume of biogas generated - the AKT value is the highest for Sample 1: 0.3, and the lowest for the samples D and 1: 0.1.

Table 3. Characteristic of sludge samples before and after digestion.

Parameters		Before digestion				After digestion			
		D	1:0.3	1:0.2	1:0.1	D	1:0.3	1:0.2	1:0.1
pH	[-]	7.51	7.16	7.22	7.21	6.89	6.89	7.04	6.98
Alkalinity	[$mgCaCO_3 \cdot dm^{-3}$]	1000	980	860	810	1100	1090	900	910
VS	[$gVS \cdot dm^{-3}$]	9.06	11.89	9.65	8.13	7.33	10.89	8.31	7.52
VSS	[$gVSS \cdot dm^{-3}$]	5.39	5.15	4.53	4.23	4.47	4.88	3.92	3.86
tCOD	[$mgO_2 \cdot dm^{-3}$]	7033	6227	4908	4469	6009	5740	3857	3677
sCOD	[$mgO_2 \cdot dm^{-3}$]	388	173	216	144	327	123	128	85
TP	[$mgP \cdot dm^{-3}$]	255	295	257	241	266	291	246	234
TN	[$mgN \cdot dm^{-3}$]	538	504	470	493	594	549	538	571
N_{NH4}	[$mgN_{NH4} \cdot dm^{-3}$]	266	154	126	126	291	224	258	224

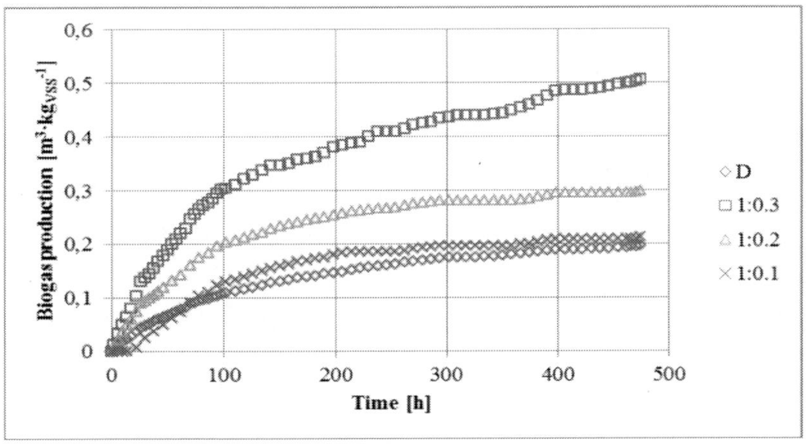

Fig. 4. The biogas production.

Table 4. Methanogenic activity test.

Parameters		D	1:0.3	1:0.2	1:0.1
Initial VSS	[$gVSS \cdot dm^{-3}$]	9.32	10.73	9.53	9.07
Final VSS	[$gVSS \cdot dm^{-3}$]	5.39	7.83	6.55	5.41
Initial sCOD	[$mgO_2 \cdot dm^{-3}$]	4911	4509	4018	3973
Final sCOD	[$mgO_2 \cdot dm^{-3}$]	1752	1581	1396	1731
R	[$mlCH_4 \cdot h^{-1}$]	7.43	12.07	9.88	7.21
AKT	**[$g\ ChZT_{CH4} \cdot g^{-1} \cdot VSS^{-1} \cdot d^{-1}$]**	**0.108**	**0.240**	**0.190**	**0.113**

Table 5. Filtration properties of sludge.

Parameters			D	1:0.3	1:0.2	1:0.1
Before digestion	CST	[s]	98	106	106	110
	SRF	[$m \cdot kg^{-1}$]	$2.43 \cdot 10^{13}$	$2.18 \cdot 10^{13}$	$2.21 \cdot 10^{13}$	$1.94 \cdot 10^{13}$
After digestion	CST	[s]	204	196	196	191
	SRF	[$m \cdot kg^{-1}$]	$2.85 \cdot 10^{13}$	$2.80 \cdot 10^{13}$	$3.13 \cdot 10^{13}$	$2.68 \cdot 10^{13}$

Corresponding to the tests, an analysis of capillary suction times (CST) and the specific resistance to filtration (SRF) were conducted. Since sludge generated during the treatment process has a very high moisture therefore, it must be thickened and dewatered. Measures such as CST and SRF determine whether sludge is susceptible to the above-mentioned processes. The value of SRF determines sludge permeability; the higher specific resistance, the lower efficiency of the dewatering process. On the other hand, CST determines a moisture release rate from the sludge; sludge that strongly holds water back has a higher CST.

Comparing the values of the filtration measures before and after digestion (Table 5) it can be concluded that anaerobic digestion significantly worsened dewaterability of sewage and mixed sludge. The value of CST after digestion were two times higher than before the process, while SRF increased by 17 - 42%. It is worth mentioning that both CST and SRF do not vary with the sludge mixing ratio, so if water sludge is used as a co-substrate in sewage sludge digestion the filtering characteristic of sewage sludge is not significantly affected.

4 Conclusions

1. The results of the research work showed that digestion of both water sludge and sewage sludge increased a biogas production. The biogas production for mixed samples was greater than for sewage sludge only by 15-60%, depending on the mixing ratio. The highest biogas production was observed for 1:0.3 sample – about 60% higher than for sewage sludge.
2. Methanogenic activity can help to evaluate the digestion process. The measurements confirmed the above conclusions. Analysis of this parameter showed the growth of methanogenic activity from 5 to 122% for mixed sludge, if compared to sewage sludge. The highest methanogenic activity was observed for Sample 1:0.3 - 0.240 g $ChZT_{CH4}·g^{-1}·VSS^{-1}·d^{-1}$, while for sewage sludge is was 0.108 g $ChZT_{CH4}·g^{-1}·VSS^{-1}·d^{-1}$.
3. Analysis of CST and SFR showed no effect of water sludge on the filtration characteristic of sewage sludge. However, digestion significantly worsened dewaterability of both mixed and sewage sludge.

References

1. Polish Act of 14 December 2012 about Waste (Journal of Laws from 2013, item 21)
2. W. Balcerzak, S. M. Rybicki, J. Kaszowski, Ochr. Środ. **29** , 3, p. 65-68 (2007)
3. W. Balcerzak, S. M. Rybicki, Ochr. Środ. **33**, 4, p. 67-69 (2011)
4. T. Ahmad, K. Ahmad, M. Alam, Proc. Environ. Sc. **35**, p. 950-955 (2016)
5. A. Nowacka, M. Włodarczyk-Makuła, Tech. Wod. **6**, p. 34-39 (2014)
6. M. Leszczyńska, M. Sozański, Ochr. Środ. i Zas. Nat. **40**, p. 575-585 (2009)
7. I. Płonka, B. Pieczykolan, M. Amalio-Kosel, K. Loska, Proceedings of ECOpole **6**, p. 337-342 (2012)
8. T. Ahmad, K. Ahmad, M. Alam, Journ. of Clean. Prod. **124**, p. 1-13 (2016)
9. K. Dassanayake, G. Jayasinghe, A. Surapaneni, C. Hetherington, Waste Man.**38**, p. 321-335 (2015)
10. M. Kyncl, S. Cihalova, M. Jurkova, S. Langarova, Inż. Min. **13**, 2, p. 11-20 (2012)
11. S. Szerzyna, Eko-Dok, p. 609-617 (2013)
12. I. Płonka, K. Barbusiński, Environ. Prot. Engineer. **3**, p. 59-67 (2010)
13. I. Płonka. Badania wpływu właściwości osadów z uzdatniania wody na ich przeróbkę, (Gliwce 2008)
14. J. Górka, M. Cimochowicz-Rybicka, Environ. Engineer. **5**, p. 123-129 (2016)
15. M. Cimochowicz-Rybicka, Aktywność metanogenna osadów ściekowych poddanych beztlenowej stabilizacji z zastosowaniem dezintegracji ultradźwiękowe, (Kraków 2013)

DOI: 10.1051/e3sconf/20171700025

16. G. Lettinga, L. Hulshoff Pol, Anaerobic reactor technology, Intern. Course on Anaer. Wast. Treat. (1991)
17. J. Young, R. Cowan, Respirometry for environmental science and engineering, (SJ Enterprises 2004)
18. I. Angelidaki, M. Alves, D. Boolznella, L. Borzacconi, L. Campos, A.J. Guwy, S. Kalyuzhnyi, P. Jenicek, J.B. van Lier, WS&T **59**, 5, p. 927-934 (2009)
19. PN-EN 14701-1:2007. Characterisation of sludges – Filtration properties – Part 1 – Capillary suction time (2007)
20. PN-EN 14701-2:2013. Characterisation of sludges – Filtration properties – Part 2 – Determination of the specific resistance to filtration (2013)

E3S Web of Conferences 17, 00026 (2017) DOI: 10.1051/e3sconf/20171700026
EKO-DOK 2017

Obtaining zeolites from slags and ashes from a waste combustion plant in an autoclave process

Agnieszka Grela[1,*], *Michał* Łach[2], *Dariusz* Mierzwiński[2], *Tomasz* Bajda[3], and *Janusz* Mikuła[2]

[1]Cracow University of Technology, Faculty of Environmental Engineering, Institute of Engineering and Water Management, Warszawska 24, 31-151 Cracow, Poland
[2]Cracow University of Technology, Faculty of Mechanical Engineering, Institute of Materials Engineering, 37 Jana Pawła II Av., 31-864 Cracow, Poland
[3]AGH University of Science and Technology in Kraków, Faculty of Geology, Geophysics and Environmental Protection, Department of Mineralogy, Petrography and Geochemistry, 30 Mickiewicza Av., 30-059 Cracow, Poland

Abstract. Waste combustion is associated with the generation of post-processing solid products – waste such as slag and ash. One of the promising technologies in waste management and processing is the synthesis of zeolites and other materials exhibiting sorption properties. The aim of this study was to characterise and assess the physicochemical properties of the waste and the products synthesised from it. This paper presents the possibility of synthesis zeolites from the slag and ash from two waste combustion plants. The investigated waste is classified as hazardous waste and denoted by the EWC code 190111*. The paper presents the results of physicochemical studies of these materials. As a result of synthesis in an autoclave at 140°C with the use of 2 M NaOH, and other compounds, such zeolite forms as chabazite and sodalite were obtained. Textural studies and ion-exchange capacity investigations carried out allowed characterisation of the sorption properties of the materials. It was found that the materials obtained are characterised by the BET specific surface areas of 25.45 m^2/g and 16.79 m^2/g.

1 Introduction

Waste combustion is a method that is a universally applied element of the waste management system in European Union countries. Incineration allows solid and liquid waste neutralisation that is safe for the environment and guarantees that all the requirements for the inertness of an ecological waste incineration plant are fulfilled [1]. Using the thermal neutralisation method, it is feasible to reduce the waste amount, but also to recover and use their intrinsic energy. At present, it is possible to reduce the waste volume up from to 80% (without slag processing) to 94% (with slag processing). Considering the weight of

* Corresponding author: agrela@pk.edu.pl

the waste, it is possible to reduce it from 60% to 70% [2]. The residues from waste incineration are fly ash and slag, or ash-slag mixtures and the like, so-called, waste. To correctly select the waste management and neutralisation method, above all their properties should be determined accurately. Most often, the following tests are performed: phase composition characterisation (XRD), chemical composition analysis, aqueous leaching testing, fineness and moisture content determination, morphology evaluation (SEM), pore size distribution and specific surface area measurements (BET, BJH), as well as, cation and anion exchange capacity determination (CEC and AEC) [3]. The results of studies carried out allow the basic ways of using the waste to be defined.

One of the promising technologies concerning the management and processing of the waste from waste incineration plants is the synthesis of zeolites and other materials exhibiting sorption properties [4]. Zeolites are crystalline aluminosilicate compounds with a highly developed inner structure [5, 6] with a characteristic channel structure, in which the channels are interlinked with SiO4 and AlO4 tetrahedra. This makes them very good sorbents, catalysts and ion-exchange materials [7]. In nature, molecular sieves that allow the formation of zeolites, are generally formed upon passing of geothermally heated water through silicate volcanic ash. Due to the high demand of the industry for synthetic zeolites, methods for producing these materials economically are still sought after. Initially, volcanic ash was used as starting material, and for the synthesis, hydrothermal reactions at elevated temperatures (50–105°C) were employed. In recent years, the feasibility of using fly ashes from hard coal and brown coal was successfully studied [8–14] as well as the ash from fluid boilers [15, 16] for the synthesis of zeolite materials. Initial approaches to the synthesis of zeolites focused on the replication of the high pressure and temperature conditions that occur in nature. Barrer (J. Chem. Soc., 1948, page 127) presented the first successful zeolite synthesis, and Milton (US Patent 2 882 243 (1959)) developed a large scale synthesis of zeolites under low temperature and pressure conditions, which made it possible to gain industrial importance. Owing to intense studies, it is possible to discover new zeolite structures and to obtain thorough knowledge of their properties.

All over the world, an increasing amount of scientific research is conducted on this subject. Ash from waste incinerators may be used as the starting material for the synthesis of zeolites such as: gismondine or gmelite [17, 18] zeolite A and zeolite P [18–21]. It is also possible to obtain zeolite materials from the waste incineration slag [22] in this way, e.g., tobermorite and hydrated sodium aluminium silicates were obtained that exhibited the BET specific surface area of 22.1 m^2/g (compared with 4.6 m^2/g for the starting material). The studies showed the suitability of the materials obtained for the adsorption of heavy metal cations. Zeolites produced from fly ash from waste incineration may be employed for the removal of cadmium, chromium and lead ions [23–26]. Other investigations [27] demonstrated that ash from wastewater sludge incineration may be used as a filter filler for the treatment of phosphorus containing wastewaters.

The aim of this study was to characterise and assess the physicochemical properties of the waste such as slag and ash from waste incineration and to characterise the synthesis products formed in the autoclaving process.

2 Experiments

2.1 Materials

Two waste types (waste from diagnosis, treatment and medical prevention and industrial waste produced in factories) with the EWC code 190111* – Bottom ash and slag containing dangerous substances, from two waste incineration plants in Poland, were subjected to study. Table 1 presents the designation of material samples used in the studies.

Table 1. Designations of samples of the materials used in the study.

Sample	Description
W1	190111* - Bottom ash and slag containing dangerous substances. Industrial and medical waste incineration plant
W1A	W1 after autoclaving process (2 M NaOH, 140°C, 6 h)
W2	190111* - Bottom ash and slag containing dangerous substances. Industrial and hazardous waste incineration plant
W2A	W2 after autoclaving process (2 M NaOH, 140°C, 6 h)

2.2 Syntheses in autoclave

The synthesis process was carried out in a PARR pressurised chemical reactor (autoclave). The post-processing waste from the thermal conversion of the waste denoted as 19 01 11* – Bottom ash and slag containing dangerous substances was charged into the reactor, then 2 M NaOH solution was added to an amount of 1000 ml per 250 g of waste. The mixture thus prepared was closed in the chemical reactor, in which the pressure was increased up to the level of 0.5 MPa. Next, after 5 minutes, the mixture was heated up to 140°C at a rate of 2°C/min, while the working pressure in the reactor was concurrently raised. The heating time at the reaction temperature was equal to 6 h, and the working pressure was 1.1 MPa. After heating, the reaction system was cooled down at a rate of 0.3°C/min. The total processing time was 24 h.

Afterwards, the NaOH solution used for the synthesis was removed (by decanting the liquid from above the sediment) into an appropriate container. The obtained reaction mixture was placed in a vessel with water warmed up to 30°C. To wash the mixture, a POLSONIC ultrasonic washer was employed. The washing was carried out for 30 min. After the washings were poured out, the reaction mixture was washed with water until the pH of the washings did not fall below 9. Next, the reaction mixture was dried at a temperature of 100°C for 12 h.

2.3 Examination methods

The phase composition of samples W1, W2, W1A and W2A were determined using powder X-ray diffraction Debye-Scherrer-Hull method. Diffractograms of samples W1, W2, W1A and W2A were recorded with the use of a Rigaku SmartLab X-ray diffractometer using the following parameters: CuKα radiation, a graphite reflection monochromator, tube voltage 45 kV, tube current 200, step scan mode: step size = $0.05°2\theta$, per 1 s. The values of interplanar distances were used for the identification of phases present in the samples based on data contained in the ICDD (International Centre for Diffraction Data 2014) Catalogue and XRAYAN software. For samples W1 and W2, their chemical composition was determined. The analyses were carried out with the use of a XRF WDX PANalytical Spectrometer.

The dioxin content was determined with gas chromatography coupled with mass spectrometry (GC-MS/MS) in the Laboratory of Trace Analysis at the Cracow University of Technology. The examination was carried out according to the P/01/03 procedure imposed on 11 03 2010. The results were expressed as the upper limit of TEQ (Toxic Equivalency) according to the I-TEF (1988).

Aqueous leaching tests were performed in the Mo-BRUK S.A. accredited laboratory in accordance with PN-EN 12457-4:2006. Samples for the determination of metal content were prepared according to PN-EN ISO 15587-2:2005. Depending on the type of substance analysed, one of the following methods was used: gravimetry, spectrophotometry, ion-chromatography, Inductively Coupled Plasma – Optical Emission Spectrometry (ICP-OES). The masses of the analysed samples were of ca. 2.5 kg each. Single-stage, batch test, at liquid to solid phase ratio of 10 l/kg in the case of below 10 mm grain-size materials (without size reduction or with size reduction).

Fineness testing was carried out in accordance with EN 451-2 by wet sieving. Moisture content was determined in delivered conditions by an analytical method of drying the ash in a moisture analyser (Radwag MAC50/1/NH) to constant mass.

The morphology of samples was studied with a JEOL JSM-820 scanning microscope. Samples were beforehand appropriately prepared. Small quantities of materials were dried to constant mass, and next, they were placed on a carbon substrate that ensures removing the charge from the sample. The materials were coated with a thin layer of gold by a JEOL JEE-4X Vacuum Evaporator.

Porosity and specific surface area measurements were performed with the use of a BET-ASAP 2020 Accelerated Surface Area and Porosimetry Analyser. Samples were degassed at a temperature of 373 K for 24 h. To determine the distribution of pore volume function in the materials analysed, experimental low-temperature (77 K) nitrogen adsorption isotherms were used. The Barrett-Joyner-Halenda (BJH) method was employed, and to determine the specific surface area, the Brunauer-Emmett-Taller (BET) method was applied.

Cation Exchange Capacity (CEC) was determined with the NH_4 sorption-desorption method. Anion Exchange Capacity (AEC) was determined with the phosphate method [28].

3 Results and discussion

3.1 Materials before autoclaving

Figure. 1 a and b present the recorded diffractograms of the waste samples, W1 and W2, respectively. Based on the analysis of the diffraction spectrum it was found that in sample W1 the following phases are present: plagioclase, wollastonite, goethite, maghemite, hematite, calcite, quartz (Fig. 1a), while in sample W2, the following phases were identified: clinochlore, calcite, plagioclase, wollastonite, gehlenite (Fig. 1b).

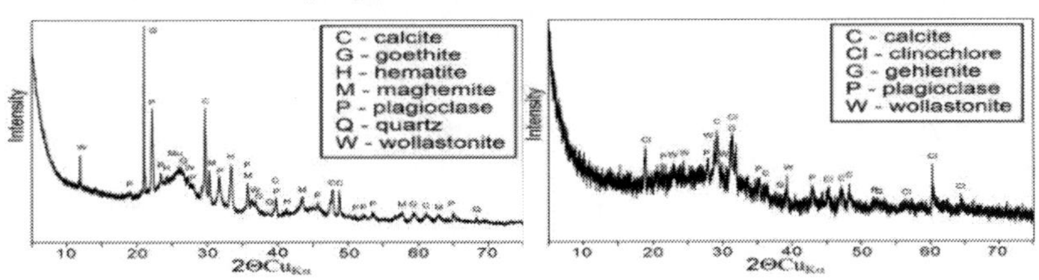

Fig. 1. X-ray diffraction pattern of samples: W1 (left) and W2 (right).

Table 2 presents the chemical composition analysis results for the waste samples. The dominating components are SiO_2 and Al_2O_3 that are of key importance for the processes of zeolite production. CaO (14-15 wt.%) adversely affects the process of zeolite phase formation. The SiO_2 and Al_2O_3 contents are similar in both waste samples (about 50%).

Table 3 presents the results of dioxin content analyses. The result is expressed as standardised toxic equivalency TEQ according to I-TEF od 1988 and presented as its upper limit. The total dioxin content of the W2 ash is almost fourfold higher compared to that of the W1 ash. Despite such differences, the dioxin content of both wastes is at a low level. In

Poland, there are no regulations concerning dioxin and furan emissions from waste incineration and their content of ashes and slags is not determined routinely.

Table 2. The chemical composition [wt. %] of samples W1 and W2 used in study.

Sample	SiO_2	TiO_2	Al_2O_3	Fe_2O_3	Na_2O	K_2O	CaO	MgO
W1	38.02	2.02	10.50	6.25	4.84	0.51	15.39	1.49
W2	36.61	3.29	8.95	16.45	11.08	0.61	14.22	1.80

Setting out dioxin emission standards would result in the necessity of carrying out appropriate controls. Dioxin emission testing is very costly and only a few research centres in Poland are able to take on the task of determining the dioxin concentration [29].

The toxicity of the ash and slag may be compared with the pollution limit values that are used for the ground or soil. A total dioxin content less than 0.005 ng/g allows the area to be used for agricultural purposes without any limitations. If the total dioxin content is in the range of 0.04-0.1 ng/g, then, it is permitted to use the area for fruit and shrub growing (however, with dioxin content control of other products), as well as for playing grounds for children. A total dioxin content in the range of 0.1-1.0 ng/g results in total unsuitability of the material for use in agriculture and horticulture, and it must be replaced [29] (Table 3).

Table 3 also shows selected physical properties of the waste samples. The test results are significant for their technological importance. They affect the alkaline activation process with respect to their water demand. Water demand depends on the moisture content level and their grain size (fineness). The drier and smaller the grains, the higher the absorptivity of the liquid activator. It was found that the studied waste samples differed slightly in their fineness, meant as the sieve residue (aperture size 0.045 μm) at wet sieving. The results obtained show that the differences in grain sizes of the tested waste samples are small.

Table 3. Results of the determination of dioxin content of studied wastes from incineration plants and results of physical properties testing of waste samples.

Sample	Total dioxin content (I-PCDD/F-TEQ) [ng/g]	Fineness [%]	Moisture [%]
W1	0.20 ± 0.053	69.74±1.29	28.30±2.74
W2	0.057± 0.015	58.73±0.48	0.42±0.04

For samples W1 and W2 aqueous leaching tests was performed. Based on the results obtained it was found that high amounts of chloride and barium ions undergo leaching from the W1 waste, whereas the W2 waste contains an elevated amount of fluorides. The content of Total Dissolved Solids (TDS) of about 7000 mg/kg is at a similar level for both wastes. Samples W1 and W2 contain the same amount of hexavalent chromium, Cr^{6+}.

Fig. 2 shows the morphology of the waste samples. Different grain and crystalline forms were observed that were characteristic of a given waste type. In sample W1, agglomerate forming plate-shaped grains and rods glued together were observed. Waste sample W2 is characterised by single, irregular particles of the size above 100-200 μm. Their edges are sharp and their surfaces non-uniform, with visible hollows.

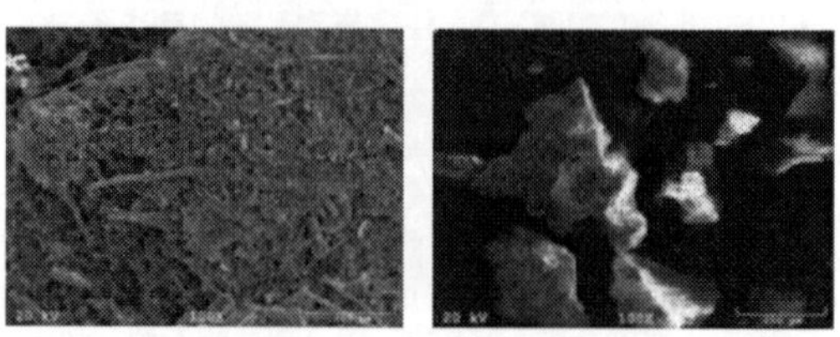

Fig. 2. SEM images of samples W1 (left) and W2 (right). Magnification 100x.

DOI: 10.1051/e3sconf/20171700026

Textural parameters were determined for samples W1 and W2 (Table 4). Both samples had a type II isotherm with a hysteresis loop of type H3. This corresponds to a mesoporous character with the formation of slit-shaped pores arising from the stacking of crystal particles. The specific surface area values (BET) of samples W1 and W2 were 22.30 m^2/g and 4.48 m^2/g. The total pore volume values were 0.035 cm^3/g and 0.006 cm^3/g, respectively (Table 4). The determined CEC and AEC values for samples W1 and W2 are also listed in Table 4. Theoretically, the higher the ion-exchange capacity of the sample, the larger the amount of anions or cations retained by ion-exchange, which means simpler to desorb. Comparing both samples, sample W1 was characterised by higher cation and anion exchange capacity values.

Table 4. Textural testing results and CEC, AEC results for samples W1 and W2.

Sample	BET $[m^2/g]$	BJH $[m^2/g]$	Single point adsorption $[cm^3/g]$	Adsorption average [nm]	BJH Adsorption [nm]	CEC [meq/ 100g]	AEC [meq/ 100g]
W1	22.30	18.47	0.0350	6.3	7.64	6.07	33.79
W2	4.48	3.81	0.0061	5.4	6.48	3.67	18.06

3.2 Materials after autoclaving

Figure 3 demonstrates the diffractograms of materials after synthesis. XRD analysis allowed identification of the following in the studied materials: in sample W1A: chabazite, gismondite, mullite, thomsonite, tobermorite, quartz; whereas in sample W2A: chabazite, melilite, mullite, sodalite, thomsonite and quartz. Chabazite, gismondite, thomsonite, sodalite belong to the zeolite group. These materials may be regarded as proper, expected synthesis products [30]. Other synthesis products are tobermorite, a hydrothermal alteration product of calcium carbonate occurring in the substrate of synthesis and melilite, a product of reaction between silica and alkali compounds from initial samples and 2 M NaOH, respectively. Other components, like mullite or quartz are unreacted reagents occurring in the starting samples.

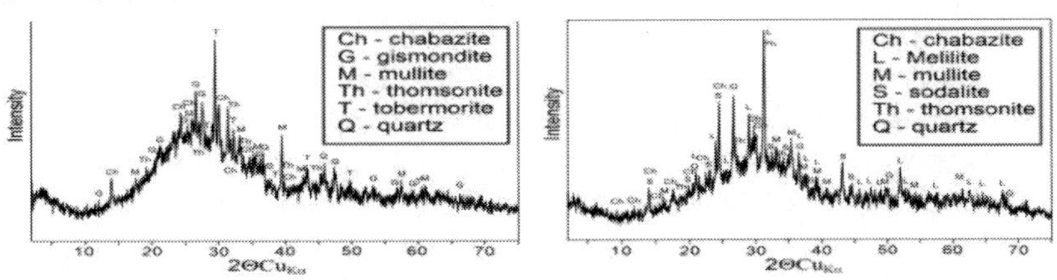

Fig. 3. X-ray diffraction pattern of samples: W1A (left) and W2A (right).

Figure. 4 (a-d) present SEM images of zeolite forms obtained as a result of hydrothermal synthesis from the studied waste samples in the autoclave. The morphology of these materials is totally different from that of the starting materials (Fig. 2). As a result of the synthesis, zeolite forms were obtained of chiefly spherical shape, placed on the surface of ash and slag particles. Large amounts of unreacted and amorphous fractions are also visible.

The synthesis products (W1A, W2A) show a similar type of isotherm and hysteresis loop of type H3 as in the case of the initial sample. However, we observed an increase in the SBET (25.42 m^2/g for W1A and 16.79 m^2/g for W2A), and the total pore volume (0.038, 0.027 cm^3/g respectively) indicates that the synthesis processes had a significant influence on the textural parameters (Table 5). In general, the synthesis products are characterised by an increase in the S_{BET} and V_{tot} compared with the starting samples. Comparison of CEC and AEC results for samples W1 and W1A (Tables 4 and 5), as well as

W2 and W2A (Tables 4 and 5) shows that the autoclaving process affected advantageously ion-exchange properties of samples W1A and W2A. The cation exchange capacity for sample W1A increased threefold in comparison with sample W1, and for sample W2A, it rose twice in comparison with sample W2. The anion exchange capacity did not change significantly in the case of samples W1A and W2A in comparison with samples W1 and W2.

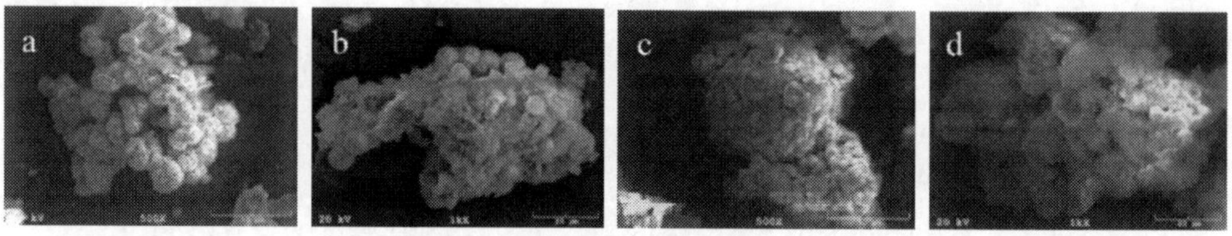

Fig. 4. SEM images of sample W1A (a ,b). Magnification 500x and 1000x. SEM images of sample W2A (c ,d). Magnification 500x and 1000x.

Table 5. Results of textural studies and CEC, AEC results for samples W1A and W2A.

Sample	BET [m²/g]	BJH [m²/g]	Single point adsorption [cm³/g]	Adsorption average [nm]	BJH Adsorption [nm]	CEC [meq/ 100g]	AEC [meq/ 100g]
W1A	25.42	17.93	0.0377	5.9	8.15	18.90	39.10
W2A	16.79	13.18	0.0273	6.5	8.27	6.25	27.28

4 Conclusions

The paper presents the results concern the change of physicochemical properties due to alkaline autoclaving. Based on the results obtained it was found that:

Post-processing waste from the waste incineration plants under study is useful for the production of zeolites in hydrothermal alkaline activation. Treatment of the waste from incineration plants in alkaline media leads to a change in the morphology of these materials and the formation of such zeolite forms as: chabazite, gismondite, thomsonite and sodalite.

The waste samples, with the EWC code 19 01 11* – Bottom ash and slag containing dangerous substances, subjected to autoclaving had different mineral composition, morphology and different cation and anion exchange capacities compared with the initial waste samples (raw material, starting/base material). In the case of the autoclaving process, based on the assessment of morphology and chemical composition, as well as the BET specific surface area, it was found that better results were obtained for waste sample W1A.

In the autoclaving process it is possible to obtain material with BET specific surface area values exceeding 25 m^2/g. Such materials may be used as sorbents.

XRD studies carried out demonstrated that in the autoclaving process, beside zeolite materials, the following were also obtained: tobermorite, a hydrothermal alteration product of calcium carbonate that occurs in the substrate of synthesis and melilite, a product of reaction between silica and alkali compounds from initial samples and 2 M NaOH, respectively.

The analysis of SEM images revealed that in the case of coarse grained slag, zeolite structures are formed on the surface of particles, which is why it is necessary to grind/mill them in order to improve the efficiency.

The authors would like to express their gratitude to the National Centre for Research and Development and the National Fund for Environmental Protection and Water Management for supporting this work with the grant GEKON1/05/213240/35/2015.

References

1. J. Łowkis, R. Jagiełło R., K. Kaczmarski, M. Surowiec, Ochr. Środ. **3-4** (1994)
2. W. Mokrosz, Inż. Chem. Proc. **25**, 1327 (2004)
3. M. Łach, J. Mikuła, M. Hebda, J. Therm. Anal. Calorim. **125**, 3 (2016)
4. Ch. H. K. Lam, A. W. M. Ip, J. P. Barford, G. McKay, Sustainability. **2**, 7 (2010)
5. T. Armbruster, M. E. Gunter, Mineral. and Geochem. **45** (2001)
6. A. Chica, Chem. Eng. **2013** (2013)
7. R. Aiello, G. Giordano, F. Testa, **142**, Studies in Surface Science and Catalysis (2002)
8. K. L. Chang, W. H. Shih, Ind. Eng. Chem. Res. **37** (1998)
9. G. C. Hollman, G. Steenbruggen, M. Janssen-Jurkovicova, Fuel. **78** (1999)
10. M. Inada, Y. Eguchi, N. Enomoto, J. Hojo, Fuel. **84** (2005)
11. A. Derkowski, W. Franus, E. Beran, A. Czimerova A. Powder Technol. **166** (2006)
12. F. Fotovat, H. Kazemian, M. Kazemeini M. Mat. Res. Bul. **44**, 2 (2009)
13. C. Belviso, F. Cavalcante, S. Fiore, Waste Manag. **30** (2010)
14. W. Franus, M. Wdowin, Environ. Monit. Assess. **186** (2014)
15. A. Grela, M. Hebda, M. Łach, J. Mikuła, Micro. Meso. Mat. **220** (2016)
16. A. Grela, M. Łach, J. Mikuła, M. Hebda, J. Therm. Anal. Calorim. **123**, 2 (2016)
17. G. Yang, T. Y. Yang, J. Hazard. Mat. **62** (1998)
18. M. Michihiro, T. Chikashi, M. Matsuda, J. Am. Ceram. Soc. **85**, 7 (2002)
19. R. P.˜a Penilla, A. G. Bustos, S. G.˜i Elizalde, J. Am. Ceram. Soc. **86**, 9 (2003)
20. T. Chikashi, M. Matsuda, Matsuda M., M. Michihiro, J. Ceram. Soc. **114**, 2 (2006)
21. M. Sallam, R. P. Carnahan, A. Zayed, S. Sunol, 16th Ann. North Americ. Waste Energ. Conf. (2008)
22. Y.W. Chiang, K. Ghyselbrecht, R. M. Santos, B. Meesschaert, J. A. Martens, Catal. Today. **190** (2012)
23. Y. S. Shim, Y. K. Kim, S. H. Kong, S. W. Rhee, W. K. Lee, Waste Manag. **23** (2003)
24. V.K. Gupta, I. Ali; V.K. Saini; T. Van Gerven, B. D. Van Bruggen, Ind. Eng. Chem. Res. **44** (2005)
25. Y. Tao, H. Kanoh, L. Abrams, K. Kaneko, Chem. Rev. **106** (2006)
26. J. Iqra, M. Faryal, R. Uzaira, T. Noshaba, Mat. Scien. Eng. **60** (2014)
27. J. Rodziewicz, A. Mielcarek, I. Kłodowska, W. Janczukowicz, E. Choińska-Żurek, A. Wolter, Inż. Eko. **48** (2016)
28. A. Grela, T. Bajda, Inż. Eko. **19** (2017) (to be published)
29. T. Pająk, Roczn. PZH. **47** (1996)
30. M. Łach, D. Mierzwiński, J. Mikuła, Inż. Eko. **18** (2017)

The influence of humic acids on desalination process with the use of electrodialysis

Martyna Grzegorzek[1,*], and *Katarzyna* Majewska-Nowak[1]

[1]Wrocław University of Science and Technology, Department of Environmental Engineering, Wybrzeże Stanisława Wyspiańskiego 27, 50-370 Wrocław, Poland

Abstract. With every year amount of drinking water in the world becomes smaller. One of the possible way to solve the problem with water scarcity is desalination. The aim of this paper is to evaluate the use of electrodialysis for salt removal in the presence of organic matter. During the tests installation PCCell BED-1-System was applied. Desalination was conducted with the use of standard ion-exchange membranes. The treated solutions contained 0.5 and 1 g $NaCl/dm^3$. The solution poured into a diluate chamber was also supplemented with humic acids (concentration amounted to 5, 10 and 15 mg/dm^3). The current intensity was equal to 0.11 A (current density equal to 1.72 mA/cm^2). The process was terminated when voltage reached 24 V. During the tests conductivity, colour and voltage was monitored. Also specific electrical energy demand (EC) was calculated. It has been found that electrodialysis can be used as an effective method for salt removal. In most cases conductivity was reduced by approximately 90%. The main factor which influenced EC and process duration was mineral salt content. Besides, it has been observed that organic matter had a slight impact on the process course.

1 Introduction

Humic acids (HA's) are organic matter common in soils, sediments and natural water. They are macromolecules which comprise humic substances. They can be described as polydispersed, colloidal and amorphous organic compounds [1]. Characteristic for HA's is that they are insoluble in acidic media, soluble in alkaline solutions and partially soluble in water. They cause deterioration of water colour and taste. The content of humic acids ranges from 20 $\mu g/dm^3$ (in groundwater) to 30 mg/dm^3 (in surface water) [2]. Also, it is worth to know that in natural water and soil they constitute a major part of DOC (dissolved organic carbon) [2–5].

Humic acids contain different functional groups. The type of these groups is determined by many factors, i.e. climate, origin, environmental conditions or age. The majority of this group are attributed to carboxylic acid or phenol functional groups [4].

It is worth to remember about that HA's are THM's (trihalomethanes) precursors. During the disinfection and chlorination, they may create carcinogenic and very toxic substances [5].

* Corresponding author: martyna.grzegorzek@pwr.edu.pl

Desalination means the removal of dissolved substances (like salts) from brackish, treated wastewater or seawater. The majority of Earth's surface (71%) is covered by water. But only 3% of total water content can be used as drinking water and 0.3–0.8% of this amount has a form of liquid, fresh water. Considering water salinity, we can divide water into three categories. Seawater is characterised by the content of total dissolved solids (TDS) equal to 35000 mg/dm^3. In brackish water (medium-salinity) TDS concentration is about 1000 – 15000 mg/dm^3, whereas in fresh water (low-salinity) TDS content is not higher than 500 mg/dm^3 [6–8].

It is important to remember that development of our civilisation resulted in reducing the water resources. It has been estimated that even 1.2 billion people may have problem with access to fresh water [5]. The current demand on potable water reached about 4500 billion m^3. It has been estimated that need for drinking water can increase about 53% till 2030 [9]. One of the possible way to overcome the problem with water scarcity is the salt water desalination. The significant growth of desalination market was noticed in the mid-1900s. The volume of desalinated water in 2016 reached value equal to 60 million m^3/day (the global scale) and in USA the amount of desalinated water reached 4.9 million m^3/day [10].

Various methods can be used for water desalination, e.g. thermal processes like distillation or evaporation. But they have disadvantages like high energy consumption and large gases emission what leads to greenhouse effect. Among others for these reasons membrane processes became more popular. High desalination efficiency, lack of necessity of chemicals use, space and energy saving and ease of operation are the main advantages of membrane techniques [11] Besides, when the salt content is lower than 5 g/dm^3 the membrane processes (i.e. electrodialysis, ED) are especially recommended for salt removal [12].

One of the advantages of electrodialysis is that it allows to achieve high feed recovery in comparison with distillation or other membrane techniques. It also should be noticed that ED does not need expanded pretreatment, because membrane scaling (one of the ED obstacles) can be minimized by using reversal electrodialysis. The first commercial installation consisted of electrodialysis was constructed in 1950s [12–14]. In this process ions migrate through the ion-exchange membranes and the potential difference is driving force. Anions migrate to the anode and cations are transported to the cathode. The cations can be transferred through the cation-exchange membranes but they cannot migrate through the anion-exchange membranes. The opposite situation occurs in case of anions. As the results two streams are formed - diluate and concentrate. ED is used for salt production, drinking water production, recovery of useful materials from effluents and industrial effluent treatment, whey demineralization, fruit juice deacidification or boiler water production. Also, it has been noticed that ED is more efficient economically when the salt content is low (not higher than 0.5%) in comparison with pressure-driven processes. However, when the content of dissolved solids is very high (about 30 g/dm^3) ED can also be successfully used for desalination [6, 9, 12, 15–16].

The aim of this paper was to evaluate the possibility of electrodialysis use for desalination process in the presence of organic matter.

2 Methods and materials

2.1 Installation and membranes

The experiments were conducted with the use of electrodialytic installation PC Cell Bed-1-System. The installation operated in a batch mode. The maximum value of current intensity which can be obtained with the use of this installation was equal to 5 A. The membrane stack was made of 11 cation-exchange membranes (PC-SK) and 10 anion-exchange (PC-SA) membranes. Each membrane had a surface area equal to 64 cm^2. The volume of diluate and concentrate amounted to 2 dm^3 (each). As an electrode solution 0.05 mol $NaCl/dm^3$ was used. Diluate, concentrate and electrode solution circulated in three independent circulation loops - the flow rate of each solution was equal to ±90 dm^3/h. The value of flow rate corresponded with the linear liquid velocity of 6.25 cm/s.

Current intensity during the ED process had a constant value - 0.11 A, what corresponded with current density equal to 1.72 mA/cm^2.

The process was terminated when the voltage value reached 24 V.

2.2 Reagents

Synthetic solutions were used during the experiments. They were prepared based on distilled water. They were made of mineral salt – sodium chloride (NaCl, molar mass 58.5 g/mol). The salt concentration amounted to 0.5 and 1 g $NaCl/dm^3$. Solution poured into diluate tank also contained organic matter – humic acids (HA) at dosages of 5, 10 and 15 mg/dm^3. Colour of solutions was equal to 25, 50 and 75 mg Pt/dm^3, respectively.

2.3 Methodology

Conductivity and colour were measured during the tests. The conductivity was monitored with the use of conductometer Elmetron CC-411. The measurement error was equal to ± 0.25%. Colour was determined with the use of spectrophotometer Hitachi U-1900. Wavelength during the measurements was equal to 350 nm and the calibration curve allowed to express the absorbance as colour (in mg Pt/dm^3) and the accuracy of measurement reached ±0.2%.

Also, in every 5 min. value of voltage was monitored. In the next step these voltage values allowed to calculate the specific electrical energy demand (EC). The specific electrical energy demand was estimated with the equation below:

$$EC = \frac{I\int_0^t U dt}{V_d} \qquad (1)$$

where: EC – specific electrical energy demand (kWh/m^3), I – current intensity (A), U – voltage (V), V_d – volume of the diluate (m^3), t – process duration (h).

3 Results

3.1 Conductivity

The variation of conductivity with time is presented in Fig. 1.

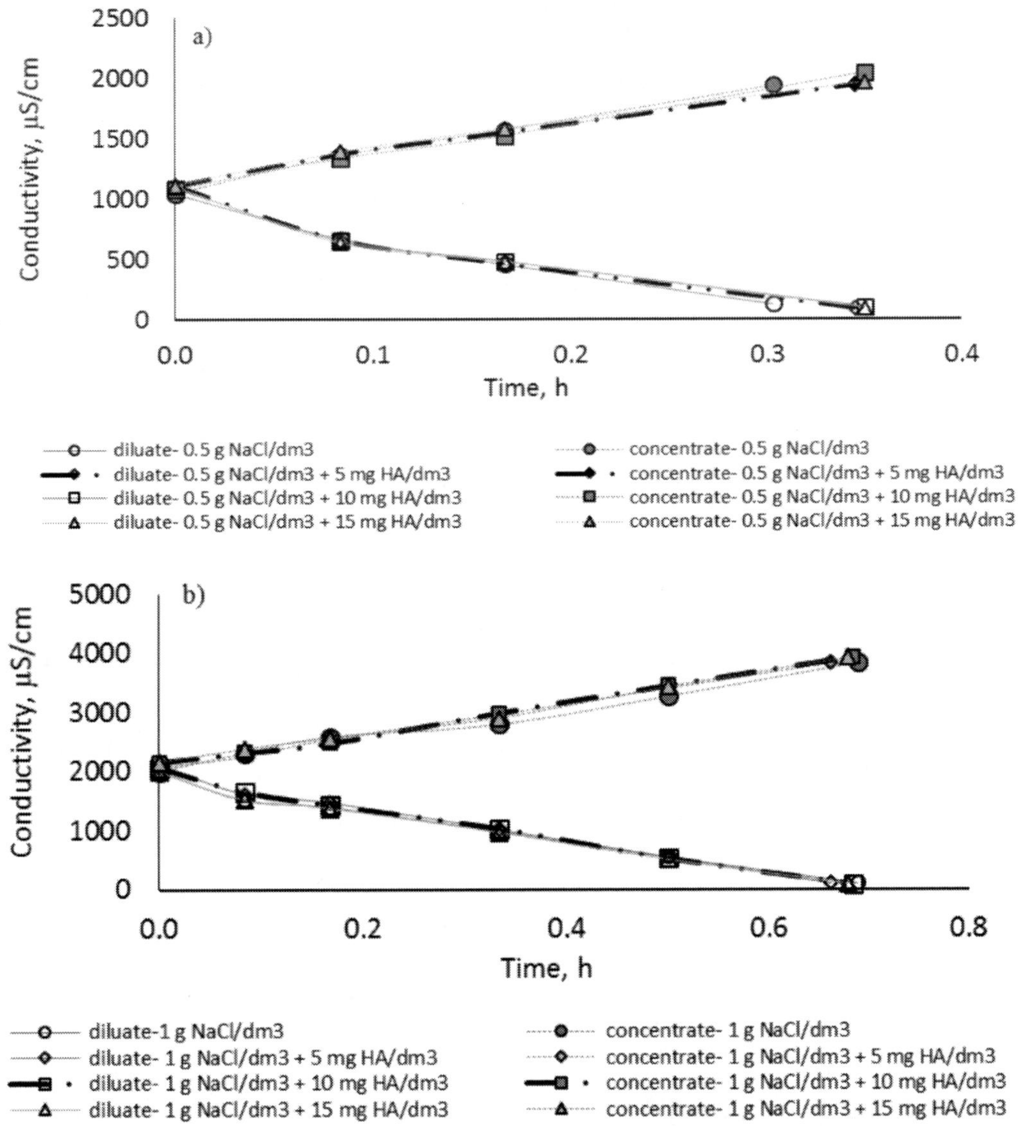

Fig. 1. Conductivity versus ED operational time for diluate and concentrate at various humic acids concentration: a) $C_0 = 0.5$ g NaCl/dm^3; b) $C_0 = 1.0$ g NaCl/dm^3; current density = 1.72 mA/cm^2.

It has been observed that the initial conductivity of solutions containing 0.5 g NaCl/dm^3 amounted to 1042 – 1110 µS/cm. The diluate conductivity decreased to 93.5 – 131 µS/cm at the end of ED process. The best results occurred when HA's content amounted to 5 mg/dm^3 and the worst effect was observed when solution without of organic matter dosage was treated.

The initial conductivity increased to 2000-2110 µS/cm when salt concentration reached 1 g NaCl/dm^3. The obtained results showed similar trend to that for solutions containing 0.5 g NaCl/dm^3 – the final conductivity was lowered to 87.7–104.7 µS/cm. When the HA's content was equal to 5 mg/dm^3, the results were least satisfying and the lowest conductivity was obtained when organic matter concentration reached 10 mg/dm^3.

 DOI: 10.1051/e3sconf/20171700027

The process duration was approximately 21 min when the salt content amounted to 0.5 g $NaCl/dm^3$ and it has been noticed that process was around two times longer when NaCl concentration was equal to 1 g $NaCl/dm^3$.

It has been observed that HA's did not have remarkable influence on process duration. The main factor which had impact on the length of the ED process was salt content. This is consistent with Faraday's law in which the amount of ionic equivalents which should be removed depends on time of the process and the current intensity.

3.2 Desalination efficiency

Desalination efficiency (expressed as the conductivity drop) was also calculated. The obtained results are presented in Fig. 2.

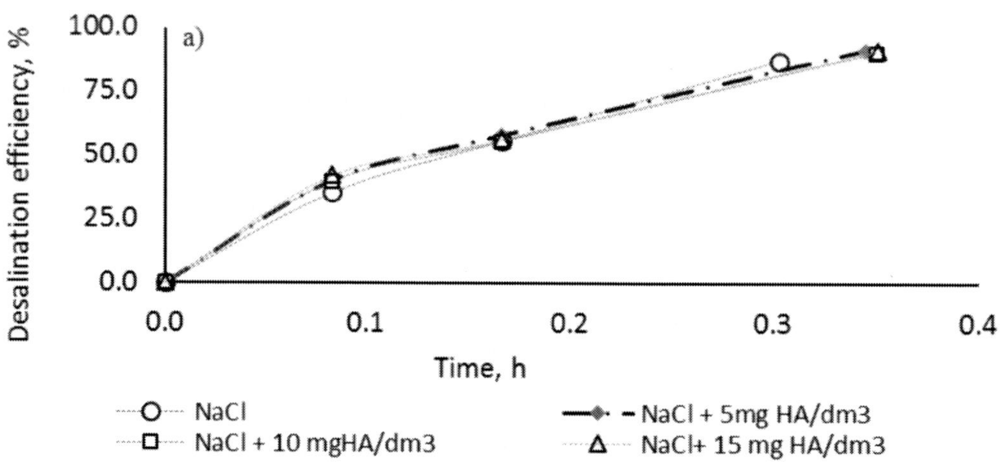

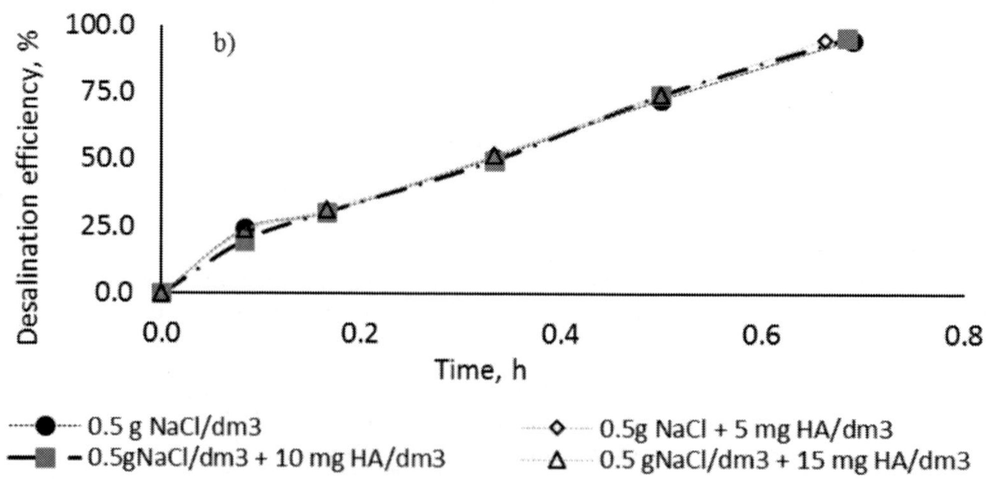

Fig. 2. Desalination efficiency versus ED operational time at various humic acids concentration: a) C_0 = 0.5 g $NaCl/dm^3$; b) C_0 = 1.0 g $NaCl/dm^3$; current density = 1.72 mA/cm^2.

When sodium chloride content amounted to 0.5 g NaCl/dm^3 the desalination efficiency was very satisfying and it varied from 87.4 to 91.6%. High desalination efficiency was also obtained for solutions containing 1 g NaCl/dm^3 – it reached about 95–95.7%.

It can be stated that organic matter has a minor impact on desalination efficiency. The presence of organic matter may insignificantly enhanced the desalination efficiency. It can be caused by the ions trapping inside the expanded structure of humic acids. The second one explanation is attracting the chloride ions by the Na^+ ions which had been bounded to negatively charged surface of HA macroparticles. As the result desalination efficiency is apparently enhanced, however this is due to the interactions between diluate components.

3.3 Colour

Colour was also measured during the ED tests. It was quite obvious that colour of solutions containing only sodium chloride (regardless of NaCl concentration) was close to 0 mg Pt/dm^3. It allows to claim that mineral salt does not contribute in the colour intensity of the experimental samples. As it has been mentioned previously, when HA content amounted to 5, 10 and 15 mg/dm^3, the initial colour of diluate was equal to around 25, 50 and 75 mg Pt /dm^3, respectively.

During the ED process depletion of diluate colour was observed. The final percentage colour reduction (at the end of electrodialysis) is shown in Fig. 3.

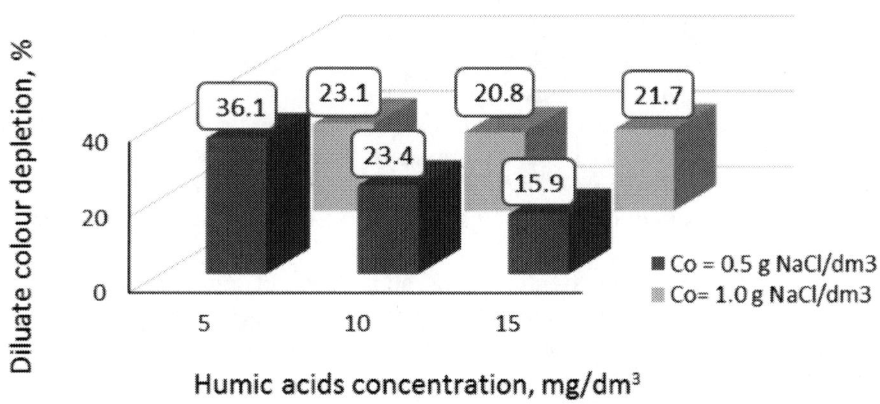

Fig. 3. Depletion of diluate colour versus HA concentration; current density = 1.72 mA/cm^2.

It has been observed that diluate colour decreased in the range of 15.9 – 36.1% (average 25.1%), when the NaCl content was equal to 0.5 g NaCl/dm^3 and in the range of 20.8 – 23.1% (average 21.9%), when salt concentration amounted to 1 g NaCl/dm^3. At the same time it has been noticed that concentrate was deprived of organic matter and its colour was close to 0 mg Pt/dm^3 in the course of ED process. This observation allows to state that HAs did not migrate from diluate to concentrate chamber. It can be assumed that colour depletion was caused by the deposition of humic acid macroparticles on membrane surface.

3.4 Specific electrical energy demand

Energy consumption is a crucial parameter for potential end-users, thus the specific electrical energy demand (EC) was calculated during the ED process. The EC was calculated with the use of equation (1).

The calculated energy consumption for various composition of treated solutions is presented in Fig. 4.

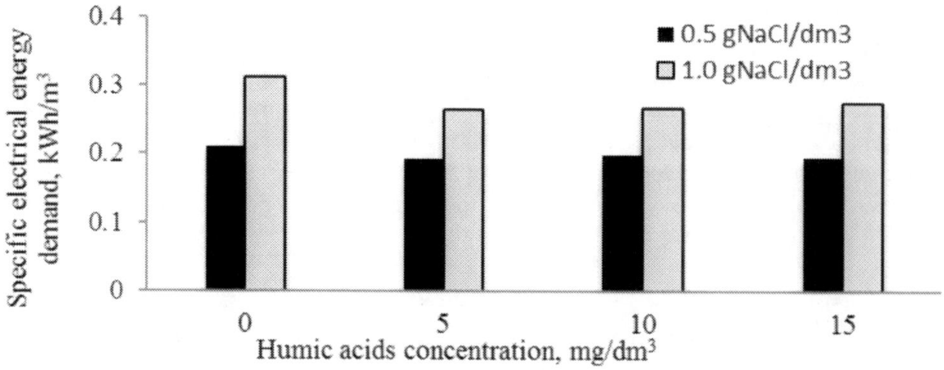

Fig. 4. Specific electrical energy demand versus humic acid and NaCl content; current density = 1.72 mA/cm^2.

It has been observed that the specific electrical energy demand was lower about 25.8–32.7% for solutions containing 0.5 g $NaCl/dm^3$ than for solutions with 1 g $NaCl/dm^3$. It has been noticed that the energy demand varied from 0.193 to 0.210 kWh/m^3. The highest value of EC occurred without HA's addition and the lowest when concentration of HA's was equal to 5 mg/dm^3.

ED experiments with solutions containing 1g $NaCl/dm^3$ arrived at the EC increased and it varied from 0.265 to 0.312 kWh/m^3. When organic matter content was equal to 5 mg/dm^3 EC was the lowest, whereas without HA's addition energy demand was the highest.

It has been found that in ED desalination plant with capacity equal to 2200 m^3/day and total dissolved solids content equal to 1.2 g/dm^3 energy demand for demineralisation was equal to 0.064 kWh/m^3. Whereas EC for pumping reached 0.4 kWh/m^3, the total energy consumed during the process was calculated as 0.464 kWh/m^3 [17]. This order of magnitude of energy consumption was comparable with EC calculated during experiments described in this paper.

It can be stated that humic acids had a slight impact on energy consumption. Macroparticles of humic acids deposited on the membranes surface may lead to the increase of membrane resistance, as well as the resistance of ED stack. As the result more energy is needed to conduct the desalination process.

4 Conclusions

It can be stated that electrodialysis can be used as an effective method for water desalination. The conductivity drop was close to 90%, regardless of the mineral salt content in experimental solutions. Besides, it has been observed that humic acids have a minor influence on desalination efficiency and process duration. It was worth to notice that the major factor which had impact on process duration was NaCl content.

The specific electrical energy demand (EC) was much higher for solutions containing g $NaCl/dm^3$ in comparison to solutions with low salt content (0.5 g $NaCl/dm^3$). The presence of organic matter (in a form of humic acids, HAs) in the treated solutions had a slight impact on EC value. It can be predicted that HAs are deposited on the membrane surface, thus increasing the resistance of ED stack. As a result more energy is needed to conduct the desalination process.

The deposition of HA particles on membrane surface was supported by diluate colour depletion. Concomitantly, the colour of concentrate remained relatively stable and it was close to 0 mg Pt/dm^3 during the ED process.

It is recommended to use the electrodialysis as final treatment preceded by other pre-treatment processes. Such approach allows to reduce costs of the electrodialysis and protect membranes against fouling (as a result membranes lifespan will be improved).

The financial support of the Faculty of Environmental Engineering (Wroclaw University of Science and Technology, grant no. 0402/0080/16) is greatly appreciated.

References

1. S. Zhang, L. Yuan, W. Li, Z. Lin, Y. Li, S. Hu, B. Zhao, Chemosphere **166**, 334–342 (2017)
2. M. Wang, L. Liao, X. Zhang, Z. Li, Appl Clay Sci. **67–68**, 164–168 (2012)
3. Klučáková, React. Funct. Polym. **109**, 9–14 (2016)
4. B.A. Gomes de Melo, F. Lopes Motta, M.H. Andrade Santana, Mater. Sci. Eng. C Mater. Biol. Appl. **62**, 967–964 (2016)
5. K. Szymański, A.W. Morawski, S. Mozia, Chem. Eng. J. **305**, 19–27 (2016)
6. M. Sadrzadeh, T. Mohhamadi, Desalination **221,** 440–447 (2008)
7. A.H. Galama, M. Saakes, H. Bruning, H.H.M. Rijnaarts, J.W. Post, Desalination **342**, 61–69 (2014)
8. M.A. Alghoul, P. Poovanaesvaran, K. Sopian, M.Y. Sulaiman, Renew. Sust. Energ. Rev. **13**, 2661–2667 (2009)
9. E.A. Abdel-Aal, M.E. Farid, F.S.M. Hassan, A.E. Mohamed, Egypt. J. Pet. **24**, 71–75 (2015)
10. J.R. Ziolkowska, R. Reyes, J. Environ. Manage. **167**, 15–22 (2016)
11. K. Wang, A.A. Abdalla, M.A. Khaleel, N. Hilal, M.K. Khraisheh, Desalination **401**, 190–205 (2017)
12. K. Ghyselbrecht, M. Huygebaert, B. Van der Bruggen, R. Ballet, B. Meesschaert
13. K. Tado, F. Sakai, Y. Sano, A. Nakayama, Desalinaiton **378**, 60–66 (2016)
14. B.A. Qureshi, S.M. Zubair, Desalination **394**, 195–203 (2016)
15. Ch. Xue, Q. Chen, Y.-Y. Liu, Y.-L. Yang, D. Xu, L. Xue, W.-M. Zhang, J. Membrane Sci. **493**, 28–36 (2015)
16. L.J. Banasiak, T.W. Kruttschnitt, A.I. Schäfer, Desalination **205**, 38–46 (2007)
17. S. Lahnid, M. Tahaikt, K. Elaroui, I. Idrissi, M. Hafsi, I. Laaziz, Z. Amor, F. Tiyal, A. Elmidaoui, Desalination **230**, 213–219 (2008)

The life cycle cost of a building from the point of view of environmental criteria of selecting the most beneficial offer in the area of competitive tendering

Beata Grzyl[1,*], *Emilia* Miszewska-Urbańska[1], and *Magdalena* Apollo[1]

[1]Faculty of Civil and Environmental Engineering, Gdansk University of Technology (GUT), Narutowicza 11/12, 80-233 Gdansk, Poland

Abstract. The article analyses environmental and ecological criteria of selecting the most beneficial offer in the aspect of LCC. Construction works contracts and the potential method of defining the above criteria, among others, is pondered on (for example by the recommendation of a material, which is supposed to be used, a ban on substances that are harmful for human health as well for the environment). In the relation to the above, it is necessary to define technical parameters that have an impact on the environment, for example the level of pollution and noise emission, electricity and water consumption, or stating the minimal involvement of a processed ingredient. In addition the article presents also an account of the life cycle cost of a building from the point of view of environmental criteria constituting an element of selecting the most beneficial offer in the area of competitive tendering.

1 Introduction

One of the key priorities of the EU implemented under the Europe 2020 strategy is sustainable development involving the support of the competitive economy, efficiently utilizing resources and environmentally friendly. In many EU documents promoting sustainable development, as a basis for making investment decisions, is shown a method of estimating the life cycle costs (LCC - Life Cycle Cost) of a building. The life cycle cost is now widely applied in various fields of life, it is also the subject of numerous studies and research carried out on behalf of private companies and public institutions. The new approach involves the integration of environmental considerations in the design of buildings at an early stage, and carrying out their evaluation in respect of environmental impact (in terms of energy consumption, the possibility of limiting the consumption of new raw materials, extending the period of operation, reducing the amount of waste) in all phases of the life cycle.

Public procurements accounts for a significant share of the market, and their value in the individual EU Member Countries was in 2015 10-20% of GDP [1]. The purpose of

* Corresponding author: beata.grzyl@wilis.pg.gda.pl

publicizing the environmentally-friendly public contracts, carried out on the basis of appropriate criteria, applicable to design, implementation and management of building structures, is to improve the environment by reducing the consumption of energy and natural resources, reducing CO_2 emissions, increasing the use of energy produced from renewable sources, prevention of a climate change. Making procurements, which are taking into account environmental criteria in assessing life-cycle costs, contributes to lowering the cost of use and maintenance of buildings but it is also part of a long-term development strategy for the region based on innovative technologies.

2 Building life cycle costs – the analysis and components

According to the idea of sustainable development, at the design stage of a building there is a need to apply LCC analysis in order to optimize the cost of its life cycle, taking into account safety, degree of functional failure, frequency of making maintenance, etc. According to [1] 80-90 % of costs incurred in the course of implementation stage, use and maintenance of the facility is determined at the design stage. In practice, in many cases the costs incurred in the full life cycle, however, are omitted, and investment decisions are made based on short-term criteria. Such activities cause negative consequences of a financial but also environmental nature (e.g. arises a need for earlier disposal of the obtained waste). Due to the fact that the lifetime of certain objects reaches 30-50 (70) years, and the costs associated with their operation and maintenance far outweigh the initial expenses (Table 1), the criterion of the lowest price cannot be the sole determinant in making an investment decision [2].

Table 1. Groups of costs and their percentage share in the life cycle costs of a building.

Stage	Cost share	Cost groups in terms of the scope of activities
Preparation and implementation	20%	- the concept and definition (including the cost of market research, design and analysis, defining requirements for object) - the design and development (including the costs of the documentation design, software) - the implementation (installation) of the object, installation of equipment (including production costs and delivery on market specific products and devices, environmental and social costs)
Use and maintenance	75%	- the use and handling (includes cost of repairs, maintenance, troubleshooting, spare parts, technical support found during the operation, the environmental and social costs) - the liquidation (includes the cost of demolition, dismantling, recycling and/or disposal, the environmental and social costs).
Demolition	5%	- liquidation, disposal, recycling

Source: own elaboration based on [3].

2.1 LCC analysis procedure

There are two main groups of methods of LCC analysis [4]:

- simple, uncomplicated serving for comparisons, allowing to select the optimal variant of the process without discounting,
- complex, including an analysis of discounted cash flows in the period of construction of the object to its demolition, integrating different elements of the costs obtained during the life cycle (including maintenance, operation, repair, inspection).

The selection of the appropriate LCC method depends on the nature, scope and complexity of the project. The methods that can be used to estimate the components of the building object life cycle cost (or its elements) are [4]:

- the engineering method for estimating the cost (direct testing of the product, component after component)
- the method of estimating the cost by analogy (an estimate based on experience gained from similar products or technologies)
- the parametric method of estimating the cost (the use of parameters and variables to develop correlation of estimating the cost).

In the LCC analysis one should take into account the following assumptions:

- the considered costs should relate to the same period of use,
- the comparison of cost should be subjected to only those projects (or their elements) which provide the projected period of use, functional and useful requirements.
- the analyzed costs in the proposed period of use should include all expenses related to the usage (e.g. the media) and the maintenance of facility (renovation, repair, inspection, troubleshooting), as well as liquidation,
- in the cost calculation one should take into account the technique of discounting. Methods of preparing the full life-cycle costing is included in [5].

2.2 Restrictions on the LCC analysis application

Despite many advantages, in practice LCC analysis faces difficulties and constraints characteristic of many tools for calculating costs [1].

1. The LCC analysis method is not strict, so as a result it can generate different results. This is due to fact that the costs included in the analysis are only approximate values (especially those assigned to operation, maintenance and demolition phase), and additionally a large number of reliable input data are required but they are difficult to obtain (e.g. from the property managers).
2. There is no accurate model of aging buildings, taking into account the development of civilization and the resulting changes in the cost calculation. There is also great difficulty in estimating at the design stage environmental and social costs.
3. In practice, the observed lack of broad knowledge of designers on the saving technologies and the associated costs.

3 The LCC analysis in public procurements

The basic legal acts adopted at EU level, regulating the policy regarding public procurements are Directives [6] and [7].

Under the current provisions of the Public Procurement Law [8], the evaluation criteria are "price or cost" or "price or cost and other criteria related to the object of the contract" (including environmental issues - energy efficiency of the contract, innovative aspects).

The aim of inclusion into the tendering environmental requirements is to obtain contracts with the best value for money, taking into account environmental issues. The Contracting Authority may take into account environmental issues in subsequent stages of the procurement procedure, at the stage of:

- description of the contract,
- qualification of contractors,
- selection of the best offer,
- determination of the conditions for the agreement implementation.

3.1 The environmental criteria for selecting the best offer

According to the Public Procurement Act being currently in force [9], in the description of the contract, the contracting authority may place requirements on the performance or functionality, including the environment, provided that the specified parameters are sufficiently precise to enable contractors to determine the purpose of the contract and the contracting authority giving the contract. In case of contracts for construction works, in the description of the contract, the contracting authority may require.: specific levels of impact on the environment and climate, specific packaging and labeling, use of specific processes and production methods at every stage of the object life cycle.

It should be noted, however, that the expectation by the customer to use too innovative solutions, may in practice result in the inability to implement the project because of too high offer prices.

The contracts for construction works in the public sector, including the cost of the life cycle and environmental criteria for selection of the best offer can be considered in a broad sense. These types of requirements, the implementation of a new object can be imposed on designers, contractors, subcontractors and managers. In the event when the subject of contract is the construction of a new facility or renovating an existing one, the contracting authority should examine the various options (providing at the same time in the tender documents the possibility of submission variant tenders), taking into account energy consumption and impact on the environment in terms of life cycle costs. The combination of such analyzes in the calculation of costs, i.e. the comparison of investment and long-term maintenance costs, just gives a complete picture of the total costs related to the acquisition, maintenance and use of the facility, it also allows for the indication of the most economically advantageous tender, which also provides high environmental efficiency.

By defining environmental criteria for public procurement, the contracting authority is obliged to apply the rules resulting from directives [6, 7].

1. The criteria may not give the contracting authority unrestricted freedom of choice, i.e. they must be formulated in such a way to allow all bidders to make their interpretation in the same way, they must provide an objective basis to differentiate offers.
2. The criteria must give the contracting authority the ability to verify the evidence, i.e. the orderer should clearly identify what evidence should present bidders within each criterion and how it will be judged.
3. The criteria should allow for an effective competition.
4. In the contract notice and contract documents, the contracting authority must specify the criteria that will be used to indicate the most economically advantageous tender and the relative weight, which he assign to each of them, in the form of a specific number or a range with an appropriate maximum spread, and sub-criteria to be applied and their weight.

Requirements for the best offer selection should vary depending on the scope of the future agreement. The subject of the contract may be a design object, the implementation of the object using "build" or "design and build" or "design, build and manage" investment methods.

3.2 The analysis of evaluation criteria used in practice

In order to determine how often in practice environmental criteria are applicable, the authors have analyzed 550 notices published in the Public Procurement Bulletin in the fourth quarter of 2016 by the contracting authorities operating in the province of Pomerania. The analysis allowed to draw up a statement of the most frequently used by the authorities evaluation criteria for construction works using "build" investment method / form. The results are presented in the Table 2.

Table 2. The frequency of use the criterion in tender procurement for works implemented in the form "build".

Type of criterion	Frequency of criterion use the [%]
Price	100
Period of quality guarantee	92.4
term of the contract	34.5
Date of payment	6.7
Employees experience	5.2
Environmental aspects	0.6
Social aspects	0.5
Technical parameters	0.4

Source: own data

The survey results indicate that the basic criterion for procurement is the price – it occurs in all analyzed procedures. Willingly used by the ordering authority criterion beyond price is warranty of quality (92.4%), term of the contract (34.5%), date of payment (6.7%) and experience of employees (5.2%). Other criteria (environmental, social, technical parameters) in practice are used very rarely. In each of the analyzed notices of procurement for works implemented in the form "build", the price criterion was in the importance of at least 60%. In 324 procedures (58.9%) occurred only 2 criteria. In none of the analyzed procedures were applied environmental criterion referred to the cost of the object life cycle.

3.3 Examples of evaluation criteria of the tender for construction works carried out by using "design and build" form

It is important to clearly define in the contract notice and tender documents, the expectations of the customer in meeting the environmental criteria. The ordering authority must determine the number of points giving for the fulfillment of each criterion. It is proposed that environmental award criteria were at least 15–20% of the total possible points. Such a mechanism will mobilize designers and contractors to present in the offers innovative ideas (e.g. in the field of design solutions, applied technologies and devices) reducing environmental impact. The ordering authority may just define his expectations (as a criterion for choosing the best offer) that their fulfillment assured to obtain better (i.e. lower, e.g. about 20–40%) indicators of demand for energy, water, waste, and emissions to the atmosphere of harmful substances. The size of the expected indicators can be applied to the value of indicators specific for the reference object that meets the current technical conditions. This information may also be attributed to the cost of the life cycle of the object (specifying them in the amounts or percentages in relation to the object that meets the current regulations). The description of criteria together with the weights and method of evaluation of the offer are given in Table 3.

3.4 The LCC example using cost impact on the environment

The life-cycle cost of a building is the basis for making long-term investment decisions, but also has a significant impact to increase their environmental performance (lower energy consumption, lower CO_2 emission). Increasing the initial capital costs typically results in lower running costs in the life cycle (Table 4) and an increase in the final value of the property.

Table 3. The description of criteria which the ordering entity will apply in selecting a tender together with weights of the criteria and method oftenders evaluation ("design and build" variant).

No.	Name of criterion	Weight of criterion	Method for assessing
1.	Price	30%	The offer price - up to 30 points The criterion 'offer price' the highest number of points will receive the offer with the lowest price (net). The following formula is applied: the number of points C1 = (Cmin / Cbad) x 30 where: Cmin - the lowest price net of all respondent tenders Cbad – the net price stated in the examined offer
2.	Compliance of architectural concept with the requirements and assumptions contained in PF-U and the Terms of Reference	30%	The compatibility of the designed surface will be the subject to the assessment, their layout and purpose of the assumptions contained in the PF-U and the Terms of Reference. In the criterion 'the compatibility of the concept' the highest number of points will receive the offer best corresponding to the requirements contained in the PF-U and the Terms of Reference. It will be applied the formula: number of points K2 = (Kodp / Kbad) x 30 where: Kodp - requirements mostly fulfilled of all tenders Kbad - the requirements contained in the examined offer
3.	Environmental aspects	20%	The energy consumption in the period of 1 year (weight criterion of 10%) and CO2 emissions in the period of 1 year (weight criterion: 10%) will be subjected to the assessment. In the criterion of 'environmental aspects' the contracting authority will award points according to the following formula: - the annual energy consumption/emissions of CO2 at the maximum level (according to the Terms of Reference - a reference to the reference building) - 5 points for each criterion. In the case to propose in the offer a better (lower) results in comparison with the maximum values specified in the Terms of Reference, there will be awarded points (according to timetabled intervals) in proportion to the improved performance - 10 points maximum for each criterion.
4.	Costs of life cycle (assuming lifetime = 30 years, the discount rate = 5% and the methodology to the LCC analysis specified in the Terms of Reference)	20%	In the criterion of 'life cycle costs of building' most points will receive the offer with the lowest cost. It will be applied the formula: the number of points K4 = (Kmin / Kbad) x 20 where: Kmin - the cost of the lowest among all the offers Kbad - the cost specified in the examined offer

Source: own study based on [10].

Table 4. The comparative analysis of investment cost and other parameters for a public building, carried out in Poland in the standard and energy-saving technology.

Description	Unit	Building (reference) in the standard technology	Building in the energy-saving technology
Demand for electricity	kWh/ year	69 474,00	45 331,00
Demand for heat	GJ/ year	2 719,00	184,00
Cost of heating and hot water (without lighting)	PLN/ year	210 808,50	43 002,00
The investment cost*	PLN	42 450 000,00	44 560 000,00
Avoided CO_2 emissions over 30 years	Mg/ year	-	9 262
Share of renewable energy in the energy balance	%	-	70,10
Building energy standard (heating, hot tap water, ventilation, air conditioning)	kWh/(m² year)	~ 90,00	19,24
Difference in investment costs	PLN	-	2 110 000,00
Simple payback period of additional investment	years	-	14,6

Source: [11]

* Investment cost does not include lifetime, discount rate, expenses related to conducting the activities of repair, maintenance and troubleshooting, etc., demolition and disposal of materials.

Results of the analysis indicate that taking into account investment cost value and additional benefits of reducing CO_2 emissions and share of energy from renewable sources in the energy balance, implementation of building in the energy-saving technology is better for the customer.

The energy costs in the course of the building use represent a substantial part of the total cost of ownership. The reduction of energy consumption is an effective solution both financially and environmentally. It is also important savings in terms of maintenance and replacement of equipment. It is worth considering a solution that provides an extended period of their exchange, as well as the selection of materials that reduce the frequency of carrying out maintenance operations, repair and overhaul. An important issue is the cost of object liquidation, including both the cost of removing and safe disposal of some materials.

4 Conclusions

The method of estimating the life cycle cost of a building is widely applied at the stage of decision-making in design, selection of technology, use or renovation. It also allows selection of the most cost-effective solution for planning and control of maintenance costs and use of the building. Analyzing a particular expense, taking into account the cost of acquisition, functionality maintenance, restoring the initial value and disposal, i.e. making a comprehensive analysis of the costs generated throughout the life cycle, it allows to determine the "real" value of the contract. Such approach should be applied, among others, by public authorities in the contracting process of construction works (in the variant "design and build" and "build") in order to maintain the highest efficiency of public spending.

Taking into account certain advantages of the analysis, it should be noted, however, that the life cycle cost of the building is a difficult issue for the widespread introduction and use in the area of public procurement, among others, due to the current practice, attitudes and habits of ordering authorities. The principal limitations in the broad application of environmental criteria to the public are:

- a small number of specific and clearly defined environmental criteria for the design and implementation stages of a work,

- concerns of ordering authorities regarding the correctness of assessment being in compliance with environmental criteria at the stage of the submitted tenders examination,
- lack of clear standards (guidelines) of the life cycle cost calculating method of an environmentally friendly object; a large number of methodology and data may result in the inability to compare results,
- lack of specialists in the field of determining the life-cycle costs,
- lack of broad dissemination of the applied environmental criteria,
- low awareness of the benefits arising from the use of environmentally friendly buildings,
- lack of a broad government support for the implementation and promotion of environmental criteria for selection of the best offer.

References

1. http://gpp-proca.eu/pl/wp-content/uploads/sites/6/2014/12/Green-ProcA-Guidelines-GPP-in-SEAPs-PL.pdf
2. B. Grzyl, A. Kristowski, K. Jamroz, A. Gobis, Methods of estimating the cost of the traffic safety equipment life cycle *Proceedings of GAMBIT* (to be published)
3. http://ec.europa.eu/DocsRoom/documents/5058/attachments/1/translations/en/renditions/pdfhttp://ec.europa.eu/environment/gpp/pdf/criteria/office_building_design/PL.pdf
4. PN-EN 63300-3-3 *Zarządzanie niezawodnością część 3-3: Przewodnik zastosowań. Szacowanie kosztu cyklu życia* (2006)
5. ISO 15686-5:2008 *Buildings and constructed assets: Service - life planning. Part 5: Life-cycle costing*
6. Dyrektywa Parlamentu Europejskiego i Rady 2014/24/UE z dn. 26 lutego 2014 r. w sprawie zamówień publicznych, uchylająca dyrektywę 2004/18/WE
7. Dyrektywa Parlamentu Europejskiego i Rady 2014/25/UE z dn. 26 lutego 2014 r. w sprawie udzielania zamówień przez podmioty działające w sektorach gospodarki wodnej, energetyki, transportu i usług pocztowych, uchylająca dyrektywę 2004/17/WE
8. https://www.uzp.gov.pl/__data/assets/pdf_file/0009/33102/UMl.pdf
9. Ustawa z dn. 22 czerwca 2016 r. o zmianie ustawy – Prawo zamówień publicznych oraz niektórych innych ustaw, Dz. U. z 2016, poz. 1020
10. Krajowa Agencja Poszanowania Energii S.A., Wytyczne realizacyjne dla budynku użyteczności publicznej w m. st. Warszawie, mające na celu zapewnienie optymalnego ekonomicznie poziomu wymagań dotyczących charakterystyki energetycznej budynku, B. Kacprzyk, Koszty cyklu życia - zastosowania praktyczne (w:) materiały konferencyjne, *Dokumenty kosztowe w procesie inwestycyjnym*, Konferencja Stowarzyszenia Kosztorysantów Budowlanych, Warszawa 17-18.03.2016, p. 60 (2016)
11. Krajowa Agencja Poszanowania Energii S.A., A. Węglarz, M. Pierzchalski, D. Koc, Koszty w cyklu życia budynku (w:) materiały konferencyjne, *Cena lub koszt cyklu życia. Nowe uwarunkowania w zamówieniach publicznych na roboty budowlane*, 22 Konferencja naukowo-techniczna, Ciechocinek 5-7.10.2016, p. 61 (2016)

Apparent water loss prevention using modern measurement tools

Joanna Gwoździej-Mazur[1,*]

[1]Department of Environmental of Engineering Systems, Bialystok University of Technology, 15-351 Bialystok, ul.Wiejska 45E, Poland

Abstract. The problem of discrepancies in balancing housing water meters indications in relation to the main water meters in multi-housing/family building is well known to both, property owners as well as to water supply and sewage companies. The article presents the results of research in the field of flows which are characteristic to the water supply connection in school complexes and the Conference and Training Center. using modern monitoring systems. Studies have shown that it is possible to eliminate the apparent loss occurring especially at night, resulting mainly due to improper selection/choice of the measuring device, which is the main water meter.

1 Introduction

Measuring instruments serve for measurement of a given physical quantity, processing of the information acquired from measurement and transferring this information to the observer. They are intended for performing a specific task under specific working conditions and in keeping with planned requirements. The most important requirement that must be met by measuring instruments is preservation of set metrological characteristics.

One of the reasons for apparent losses in water supply systems is the improper selection of the metering device on the water service line. The water meter should be selected on the basis of the hydraulic parameters in the pipe, i.e. nominal and maximum values of the volume stream of flowing water. For this purpose, the operating range of the water meter during the day must be determined precisely, preferably on the basis of actual measurements and service line monitoring.

The goal of this paper was to assess whether the installation of static, electromagnetic water meters will contribute to limiting apparent losses as well as to determine the structure of water consumption according to characteristic flow intervals describing the accuracy of water meters

2 Accuracy classes of water meters

Based on standard PN-ISO 4064, four accuracy classes of water meters have been introduced: A, B, C and D, and a characteristic flow rate for a given diameter has been

* Corresponding author: j.mazur@pb.edu.pl

 DOI: 10.1051/e3sconf/20171700029

determined for each of them. However, this classification proved to be insufficient, and large differences between water meters were present within the same class, so in practice, this data did not facilitate selection [1, 2].

Companies producing water meter attempted to solve this problem by providing additional parameters not required by the standard, but over time, the reliability of certain declarations was called into question.

Only a change of requirements for these devices that would account for technical progress and differences in design could solve this situation. For this purpose, standard PN-EN 14154 was used as the basis for requirements introduced into the Directive of the European Parliament and of the Council on measuring instruments (MID). Although companies producing water meter can still declare parameters characterizing water meters, this data is now verified.

The manufacturing process is also verified for repeatability of manufacturing quality equal to that of the item tested by the notified body so that a certificate can be issued not just for tested items.

Starting from 2017, Companies producing water meter should only apply water meter assessment in compliance with the MID directive. The MID directive also introduced different determinations of characteristic flows and terminology than before, i.e. [2]:

Q_1 – minimum flowrate – the lowest flowrate at which the water meter provides indications that satisfy the requirements concerning the maximum permissible errors (MPEs);

Q_2 – transitional flowrate – the transitional flowrate is the flowrate value occurring between the permanent and minimum flowrates, at which the flowrate range is divided into two zones, the 'upper zone' and the 'lower zone'. Each zone has a characteristic MPE;

Q_3 – permanent flowrate – the highest flowrate at which the water meter operates in a satisfactory manner under normal conditions of use, i.e. under steady or intermittent flow conditions. The numerical value of permanent flowrate Q_3 [m^3/h] is to be selected from among the following values: 1.0; 1.6; 2.5; 4.0; 6.3; 10; 16; 25; 40; 63; 100; 160; 250; 400; 630; 1000; 1600; 2500; 4000; 6300. In the transitional period, the following values are permitted: 1.5; 3.5; 6; 15; 20;

Q_4 – overload flowrate – The overload flowrate is the highest flowrate at which the meter operates in a satisfactory manner for a short period of time without deteriorating.

Water meter classes according to the MID directive are presented in the Figure below.

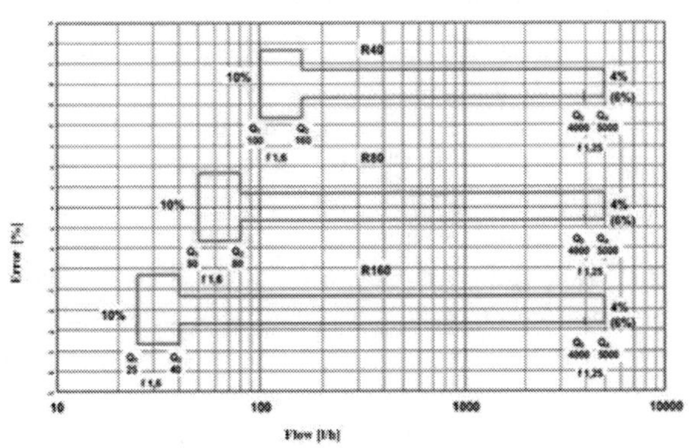

Fig. 1. Water meter classes according to MID [2].

According to legal regulations, an R value corresponding to different metrological classes of water meters is to be determined, where [2]:

$$Q_3=Q_4/1.25 \tag{1}$$

and:

$$R= Q_3/Q_1 \tag{2}$$

In relation to this, determining and selecting the proper metrological class of water meters of a given size is related to flowrate Q_1, which determines the boundary between measuring ranges characterized by different maximum permissible error values. Therefore, after determining the size of a water meter, the metrological class or minimum flowrate characterizing the water meter must be determined. Together, these two elements are to minimize apparent losses to below 1% (assumed at least during the initial period of measurement) [3–5].

In turn, metrological requirements according to the MID directive are as follows: the maximum permissible error in the lower range of flowrates from Q_1 inclusively to Q_2, not including Q_2, is ±5% (cold, warm and hot water), the maximum permissible error in the upper range of flowrates from Q_2 inclusively to Q_4, not including Q_4, is ±2% (cold water) and ±3% (warm and hot water).

3 Causes of losses and methods of preventing losses in a water supply system

When analyzing the causes of water losses in internal systems, water losses are to be divided into two categories – actual losses and apparent losses [6–9].

Actual losses are mostly the result of a poor technical condition of the water supply system. Failures of a system's pipes, joints and fittings have the greatest impact on the size of these losses.

Apparent losses are mainly the result of low accuracy of measurement of the volume of flowing water and non-simultaneity of readings of devices registering the amount of water supplied to the system (main water meter) and consumed by recipients (sub-meters on premises) [3, 7, 10, 11].

Making an accurate estimate of apparent losses is a very difficult task, and frankly speaking, practically impossible. This is due to the lack of a constant value of errors in measuring devices. According to legal regulations, metering devices must be within the margin of error, which is ±2% or ±5% for water meters (depending on flowrate) [12, 13].

When taking measures intended to reduce losses occurring in water supply networks and systems, both actual and apparent losses must be accounted for. Based on literature states, experience and direct contact with administrators and residents of housing cooperatives and communities as well as with recipients in public buildings, additional methods of preventing water losses in internal systems can be proposed [3, 6–8, 10]:

- metering of all premises and intake points. This should be the starting point of organizing water management in a building. As long as there is even one apartment billed on a lump-sum basis in a given building, balancing water consumption is out of the question,
- replacement of water meters (sub-meters) with meters of a higher metrological class, i.e. with greater accuracy of measurement and lower start-up threshold. Volumetric water meters that maintain their metrological class in every position of installation are particularly suitable for residential buildings. This is very significant, since it is often not possible to install a water meter in a horizontal position in buildings of this type due

to lack of space. A volumetric water meter also does not require straight segments before and after the water meter,

- keeping water meter replacement deadlines, i.e. making sure verification periods are observed. According to regulations, most water meters have a five-year verification period, after which they should be replaced. A water meter that has been in operation for a longer period does not necessarily have to give incorrect readings, but it is at greater risk of e.g. deposition of various contaminants or scale found in water, and thus, the probability of inaccuracy of measurement increases,
- mounting of radio transmitters on water meters. A frequent cause of differences in a building's water balance is the so-called "time-space" phenomenon, based on the fact that readings of individual sub-meters are given by recipients after a certain interval of time. One recipient may give a reading earlier because they are going on a trip at the end of the month, and another may give their reading later, simply because they forgot. Moreover, water meter readings are given with varying accuracy. Recipients may round the number of m^3 down or up. When there are several dozen water meters, a significant difference may arise in comparison to the main water meter. A radio system makes it possible to read all water meters at the same time and with the same, precise accuracy, as well as to save data e.g. at the end of every month,
- elimination of leaks at intake points. A large leak that floods a bathroom is removed immediately. However, small leaks are the most dangerous, since we either disregard them (leaking faucet) or cannot see them at all with the naked eye (leaking flush). Such leaks are also often not registered by sub-meters, but the main water meter is capable of metering them,
- overhauls of the internal water supply system. A building's water supply system should be modernized periodically. This makes it possible to eliminate old, frequently corroded, pipes and leaking joints, thus contributing to reduction of losses.
- maintenance of the proper water pressure in the internal system, particularly in the case of pressure spikes in the water supply network, and it is a good solution to install a pressure reducer on the service pipe to prevent an uncontrolled pressure increase in the internal system. Higher pressure will result in greater leakage.

4 Methodology of research

The object of conducted studies were characteristic flows in the water supply service pipe as well as analysis of the structure of water consumption in selected public buildings. Installation of the proper water meters and digital registration were possible thanks to the cooperation of the water company in Gołdap and the Sensus company. The process began with preliminary selection of buildings, after which a site survey was conducted for the purpose of detailed characterization of selected buildings in the scope of: building size (number of guests, students, services rendered, type and location of main water meter, etc.). Ultimately, two buildings were selected for study: pre-secondary school complexes and the Conference and Training Center. Flowrate measurements on the water service pipe covered a period of 46 days. A static, electromagnetic IPERL water mater from the Sensus company was used for tests. Residual magnetic field technology was applied in the water meter, ensuring a linear measuring range even at the lowest flowrate values. The magnetic field acting on water flowing through the measuring segment creates electrical voltage proportionally to its flowrate (principle of magnetic induction in flow measurement).

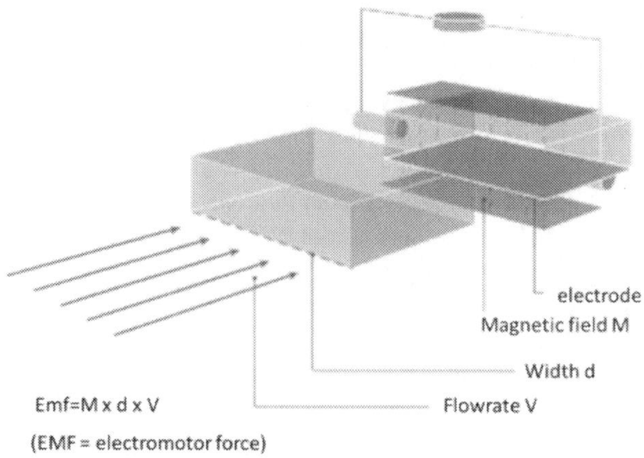

Fig. 2. Principle of magnetic induction in the IPERL water meter [14].

IPERL is equipped with an integrated low-power radio module operating at a frequency of 868 MHz or 433 MHz. Communication is provided by innovative "walk-by" or "drive-by" billing reading technology as well as by the capability of querying a device to obtain up to 2880 points with registered data. The principle of operation is based on reception of impulses transmitted by the transmitter installed on water meters by means of the Sensus Interface Radio Tool (SIRT), which is followed by registration and saving of these impulses using specialized software, which enables interpretation and analysis of read results later on.

5 Analysis of test results

The first studied case was a public building, namely a public school. Due to the nature of the buildings as well as large differences in water consumption over the course of a day, instant water flowrates were checked in the water service pipe. Among other things, minimum flowrates were registered, and an illustration of minimum flowrates is presented in Fig. 3.

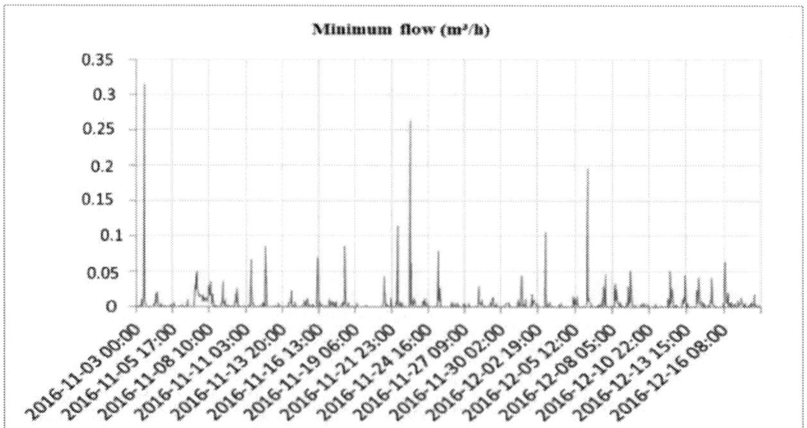

Fig. 3. Minimum water flowrate in the water service pipe to the complex of pre-secondary schools.

 DOI: 10.1051/e3sconf/20171700029

Minimum flowrates reflected the nature of the building, besides the fact that peaks were twice as high. Moreover, registration made it possible to state that the building's internal water system is serviceable, since the water period did not register any flows during the night hours. An addition site visit of the building was conducted and employees were interviewed in order to explain repeating elevated flow values on the water service line (03/11/2016 – time 2:00 P.M., 23/11/2016 – time 8:00 A.M., 06/12/2016 – time 10.00 A.M.). The repeating phenomenon is most probably related to maintenance of equipment installed in one of the building's workshops.

Characteristic intervals for different types and classes of water meters were selected for data analysis in order to illustrate water consumption in the studied building. Results based on monitoring are presented in table 1.

Table 1. Percentage of the minimum water flowrate consumed per hour in the complex of pre-secondary schools.

Flowrate m^3/h	Total %
< 0.01	90.64
0.01 - 0.02	4.61
0.02 – 0.03	2.23
> 0.03	2.53

The presented analysis of the hourly percentage of minimum flowrate consumed shows that flowrates from the interval from 0.00 to 0.02 had the greatest total share in minimum flowrates. It turned out that not all minimum flows were registered (apparent loss) until the time at which the water meter used for tests was replaced.

Another building which flowrates in the water service pipe were analyzed in detail was the Conference and Training Center. The distribution of minimum flowrates proved to be interesting. An example is shown in Fig. 4. Minimum flowrates are nearly continuously present in the studied building and make up a very large share of the structure of water consumption. The water meter registered an alarm signaling a water leak almost every day over the course of the period subject to analysis. It can be stated with high probability that the building contains intake points from which water is leaking in an uncontrolled manner.

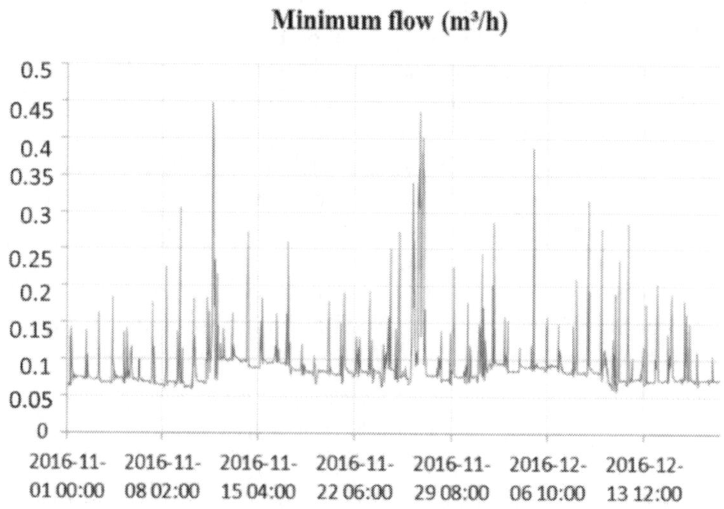

Fig. 4. Minimum water flowrate on water service pipe to the Conference and Training Center.

Due to the unserviceable water system in the building, the percentage distribution illustrating the class of water meters for both minimum and maximum flowrates on the water service pipe was investigated in detail (Table 2, Table 3).

Table 2. Percentage of the minimum water flowrate consumed per hour in the Conference and Training Center.

Flowrate m^3/h	<0.7	0.07-0.1	0.1-0.3	0.3-0.5	0.5-1.0	1.0-2.0	>2.0
Total %	3.27	29.57	15.90	13.37	26.60	9.21	2.08

Table 3. Percentage of the maximum water flowrate consumed per hour in the Conference and Training Center.

Flowrate m^3/h	<0.06	0.06-0.07	0.07-0.12	0.12-0.2	0.2-0.3	0.3-0.4	>0.4
Total %	0.15	20.51	68.35	7.28	1.49	1.78	0.45

Flowrates within the interval of 0.06 – 0.2 m^3/h had the greatest total share in minimum flowrates. In total, they make up 96.14% of all minimum flowrates. High values of maximum flowrates do not occur too frequently, are predictable, and it can be said that they are of lesser significance in the studied building than minimum flowrates. Flowrates registered in the interval from 0.07 to 1.0 m^3/h had the greatest share in the structure of maximum flowrates.

The example above shows that selecting a water meter is not an easy task. On one hand, a water meter must have a high metrological class in order to measure minimum flowrates, and on the other, it should also have the proper throughput, so that it can operate at maximum flowrates safely and without obstacle.

6 Summary

By registering and analyzing actual hydraulic conditions in a given water service line in public buildings, or the minimum stream of water volume, their total share in intervals defining the accuracy of metering devices was determined. Knowing the structure of water consumption in water supply systems, it can be stated whether the main water meter was properly selected. This is of enormous significance, because a water meter that is too small for the given working conditions will be at risk of hydraulic overload, which will result in premature wear of the mechanism, while a water meter that is too large provides lesser accuracy at minimum water flows (apparent loss).

One of possible ways to minimize apparent losses is to upgrade metering instruments to achieve better and more accurate measurements

Currently, the technical norms used to determine designed flows on the water service line do not reflect actual flows, which provides a basis for stating that installing water meters on the water service line, equipped with flow recorders and a high R factor, will make it possible to limit apparent losses.

Digital registration of the stream of water volume in the water service line provides the most reliable results. Information obtained on the subject of the water distribution profile makes it possible to precisely determine designed flow and values not exceeding it, which is an important factor in selection of main water meters.

Studies have been carried out in the framework of working No. S/WBiIŚ/02/2014 and financed from the funds for science MNiSW.

References

1. PN-EN 14154-1:2007 Water meters - Part 1: General requirements.
2. Directive 2004/22/EC of the European Parliament and of the Council on measuring instruments (OJ EU L 135 of 31/03/2004, 1-80)
3. T. Cichoń, J. Królikowska, A. Królikowski, Gaz, Woda I Technika Sanitarna, **6** (2013)
4. P. Tuz, Rynek Instalacyjny, **5** (2009)
5. P. Tuz, Rynek Instalacyjny, **6** (2009)
6. Water Audits and Loss Control Programs. Manual of Water Supply Practices M36. 3rd ed. AWWA. **283** (2009)
7. M. Rimeika, R. Albrektien, The 9th International Conference ENVIRONMENTAL ENGINEERING, Vilnius, Lithuania (2014)
8. G. L. Richards, M. C. Jonson, S. L. Barfuss, J. Am. Water. Works. Ass. **105(5)**, (2010)
9. T. Cichoń, J. Królikowska, Ecol. Eng. **48** (2016)
10. F. Arregui, E. Cabrera, R. Cobacho, J. Gracia-Serra, Water Practice and Technology **1**, 4 (2006)
11. W. Koral, Instal, **5** (2005)
12. J. Chudzicki, S. Sosnowski (Wydawnictwo Seidel-Przywecki, Warszawa 2005)
13. M. Kietniewski, W. Olszewski, E. Osuch-Pajdzińska (Oficyna Wydawnicza Politechniki Warszawskiej, 2016)
14. Materials company Sensus

Considering Risk Associated with the Realization of Waste Management Investment Plans in Cracow

Gabriela Hajduga[1,*], and *Agnieszka* Generowicz[1]

[1]Cracow University of Technology, Department of Environmental Engineering, Institute of Water Supply and Environmental Protection, 24 Warszawska, 30-155 Cracow, Poland

Abstract. To fulfill obligations to the European Union on waste management, the authorities of the city of Cracow, decided to build a waste incineration plant. Such investment involves considerable risks, not only financial but also social. The paper conducted a risk analysis based on the index net present value, identifies factors which are particularly exposed, and proposes solutions for reducing its level.

1 Introduction

Poland was obligated by the European Union, to implement regulations relating to environmental protection including solid waste management. The main legal act which controls the matter of municipal solid waste management is the Directive 2008/98/EC of the European Parliament and of the Council of 19 November 2008 on waste. It establishes major principles such as an obligation to handle waste in a way that doesn't have a negative impact on the environment or human health, an encouragement to apply the waste hierarchy and, implying application of the polluter-pays principle. In connection with new laws, every city in Poland, including Cracow, started intensified preparatory work for implementation of a modern waste management system. City government hired a group of specialist from the fields such as law, environmental protection and economics to create a few system solutions and recommend the best one. The result of their work was descriptive assessment titled "Evaluation of the Strategic Waste Management System of the City of Krakow", which is the main source of information in this article. Recommendation for implementation got the scenario, which includes an investment of a waste incineration plant. This article shows aspects which was taken into account in multi-criteria assessment conducted to emerge the best scenario and because this was the high-risk investment, also factors, that could have the highest impact, were identified and analyzed. For the purpose of this analysis the value of the net present value (NPV) was determined, which was used to carry out the sensitivity and scenario analysis, which culminated in a qualitative risk assessment [1, 2]. The aim of the article, was to present the assumptions of sensitivity analysis of the scenarios for management of waste system on the example of Cracow.

* Corresponding author: gabriela.hajduga@gmail.com

2 Scenarios for the development of waste management system in the city of Cracow

In created "Evaluation of…", four different scenarios of development of municipal waste management system were analyzed. Scenario S0 required expansion of the system for another landfill however, since 2013 it would not meet the legal requirements, therefore it was rejected. Scenario S1 not only required a new landfill, but also included creating subsequent lines in waste sorting plant, expansion of municipal solid waste composting plant, and new mechanical biological treatment plant of municipal waste. In addition, the implementation of this scenario required a large financial outlays on environmental education and the development of waste segregation "at source". Assumptions of scenario S2 largely are consistent with scenario S1, but in each of these variants, all above-mentioned investments have different capacities (can process different volumes of the waste stream). Scenario S3 was the most innovative one. Except of elements from other scenarios, it included a new investment, which was incineration plant with possibility of energy and heat recovery, a solution not commonly used in Poland [1, 3, 4].

Creating each well-functioning system of waste management, requires a detailed analyzes of factors such as: the expected amount of waste, used technology, economic opportunities and social attitudes, therefore, the final selection should never be guided by only one criterion, but every possible scenario should be evaluated. In Cracow, a multi-criteria analysis was conducted of all the proposed solutions, which includes 11 of the most important criteria divided into three groups (Table 1). For multi-criteria analysis, the compromise programming method was applied, which uses the concept of ordering individual strategies according to their distance from the established ideal point (Equation 1). The equation determines the value of the criterion that aggregates the distance measure of the studied strategy from the ideal point, while choosing the best strategy is based on the Equation 2. In calculations it was accepted, the limitation of the so-called acceptance threshold (Equation 3), which is a solution to the decision-making problem, ie, strategies that are ideally close to the ideal point marked in the result table by the symbol *. The method consists in finding the best strategy which takes into account all the mentioned above criteria. For the calculation it was necessary to create a hierarchy of all criteria defining the priorities of the participants in the decision making process. The criteria weights were adopted according to the authors of "Evaluation of…", as shown in Table 2. The criteria were organized into three groups, so subsequent numbers indicate the weight of individual criteria groups. In the first line, all the criteria are equivalent, in second one, the first group has assigned weight equals 5 while the others 1 and so on. The analysis results are shown in Table 2 [3, 5-7].

$$L_\alpha(s_n) = \sum_{m=1}^{M} w_m^\alpha \cdot (x'_m - r'_{NM})^\alpha \tag{1}$$

$$s_j = \bar{s} \Leftrightarrow L_\alpha(s_j) = minL_\alpha(s_j); n = 1,2,\ldots \tag{2}$$

$$s_n = 0{,}1 \cdot L_\alpha(s_n)_{min} \tag{3}$$

$L_\alpha(s_n)$ - distance measure of the studied strategy s_n from the ideal point, $\bar{s}$ - chosen strategy, w_m - weight factor of criterion m, x_m - m-coordinates of utopian point, r_{NM} - normalized value of the criterion, M- number of criteria, α- an exponent, measuring the deviation of the strategy from the utopian point.

Table 1. Classification and description of the criteria evaluated in choosing the best scenario [4].

CRITERIA	SYMBOL	NAME OF CRITERION
Minimization and recovery of waste	K1	Reduction of landfill waste.
	K2	Reducing the amount of landfilled biodegradable waste.
	K3	The recovery of secondary raw materials.
	K4	Energy recovery.
	K5	Exploitation time of the landfill.
Socio-political	K6	Compliance with the conditions set by the National Waste Management Plan, etc.
	K7	Compliance with EU and Polish law for the protection of the environment.
	K8	Regionality and prospectivity solutions.
	K9	Social acceptance. In Poland this is a highly controversial investment. People aren't aware of safety and benefits of building this investment.
Economic	K10	Monthly financial charge per capita.
	K11	The cost of 1 ton of waste disposal.

Table 2. The results of multi-criteria analysis to choose the best scenario for the waste management system for the City of Cracow [4].

Hierarchy of precedence (in order from Table 1)	**LINEUP OF STRATEGY**		
	alpha = 1	**alpha = 2**	**alpha = ∞**
1:1:1	S3*→S2→S1→S0	S3*→S1→S2→S0	S3*→S2→S1→S0
5:1:1	S3*→S1→S1→S0	S3*→S2→S1→S0	S3*
10:1:1	S3*→S2→S1→S0	S3*→S2→S1→S0	S3*
1:5:1	S3*→S2→S1→S0	S3*→S1→S2→S0	S3*
1:10:1	S3*→S2→S1→S0	S3*→S1→S2→S0	S3*
1:15:1	S3*→S2→S1→S0	S3*→S1→S2→S0	S3*
1:1:2	S3*→S1→S2→S0	S3*→S0→S1→S2	S3*
1:1:5	S3*→S0→S1→S2	S3*→S0→S1→S2	S3*→S1→S0→S2
1:1:6	S3*→S0→S1→S2	S3*→S0→S1→S2	S3*→S1→S0→S2
1:1:10	S3*→S0→S1→S2	**S0*→S3→S1→S2**	S3*→S1→S0→S2
5:1:5	S3*→S0→S2→S1	S3*→S0→S1→S2	S3*
1:5:5	S3*→S1→S2→S0	S3*→S1→S2→S0	S3*
5:5:1	S3*→S2→S1→S0	S3*→S2→S1→S0	S3*

3 Risk analysis

In the discipline of engineering and environmental protection, there are a variety of definitions of risk, but in the case of investments, the risks is associated with obtaining the benefits other than expected, ie. higher or lower, or in time and period other than planned. Making economic decisions it should be taken into account the risks associated with the danger of not achieving the intended purpose, in turn, failure to achieve the objective pursued may be associated not only with the occurrence of the loss, but also obtaining a result lower than expected [1, 8, 9].

3.1 Financial analysis

As it was already mentioned, the variant recommended for the implementation, was scenario in which, the investment of the incineration plant was planned therefore, only that scenario was analyzed. This object is a financial challenge and socially controversial. Investments of this type are classified, as large projects that could get financial support from the UE. However, they require analyzes, which is always evaluated by experts. This assessment concerns economic factors, environmental impacts (pollution reduction), social (creation of new jobs), and risk assessment. That is why it was possible to use the guidelines contained in the "Guide to Cost-Benefit Analysis of Investment Projects". According to the guide, the risk analysis should begin, by determining the net present value (NPV), which is used to estimate the cumulative net benefits of the investment project (Equation 4). It express the difference between the actual current cash inflows and outflows and it takes into account the changing value of money over time [8, 10, 11].

$$NPV = \sum_{t=1}^{n} \frac{B_t - C_t}{(1+i)^n} - I_0 \tag{4}$$

B_t - revenues in t year t [zł], C_t - costs in t years [zł], i - discount rate [%] - adopted at the level of 6%, n - the life of the investment object [years] - adopted 25 years, I_0 - investment costs [zł], B_t-N_t =NFC– net cash flow [zł].

According to the given information, the income earned by incineration plant include revenues for accepting waste, from the sale of heat, electricity and residents charges. Their value depends on the number of inhabitants, tariff rates and the quantity and quality of the treated waste, but such particulars data have not been available for analysis. Using the formula 4, it was possible to determine the values of NPV, which for this case amounted to -594 609 624 zł, and also the discounted cash inflows (Figure 1). Theoretically, investments for which NPV <0, should not be implemented, because they bring more losses than benefits, but in the case of investment of public benefit, qualifies the project to obtain subsidies from the EU [8, 9, 12]. In addition, Figure 1 shows that the difference between spending and revenues will decrease systematically and so after 25 years will be 75.5% lower than at the beginning, because even the costs will increase, revenues also obtain higher rate.

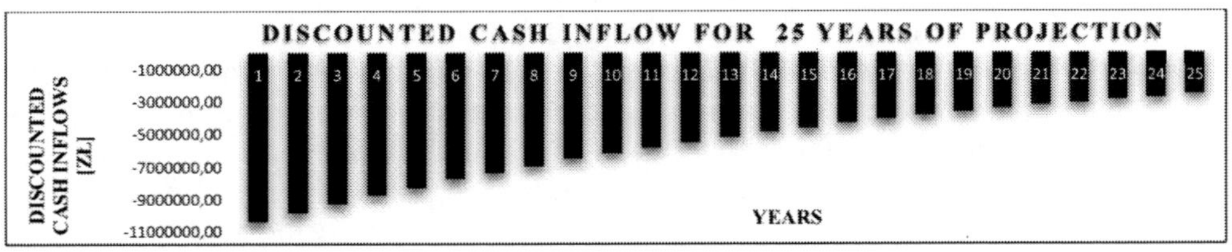

Fig. 1. Discounted cash inflow (source: own elaboration).

3.2 Sensitivity analysis

The sensitivity analysis consists in determining the impact of changes in individual variables, critical to the NPV value. The critical variables are identified as those, where the change in their value of +/- 1% compared to the baseline scenario, may change the value of the NPV index, according to the adopted criterion (+-/ 1% NPV). The analysis allows to determine the critical variables of incineration plant project. These changes can be positive or negative. The studied variables should be independent and deterministic as the most disaggregated. In the present case, such detailed data were not available, so changes were made in general values of discount rate, investment costs, operating costs and revenues (Table 3) [8, 10, 12, 13].

The analysis shows, that the investment is most sensitive to the change of the discount rate. Note however, its level was adopted on the basis of the recommendations of the EU. In fact, the adoption of the discount rate is highly subjective and moreover, it can change over time and depends on the rate of inflation. Therefore, in this case, should be consider the overall effect of changes in the discount rate, rather than specific values. In the case of manipulating the value of investment expenditures, changing their values has relatively little impact on the deviation of NPV. However, it should be noted, that the funding of this type of investment ranged from 30 to 60% of the cost, therefore so significant change of those costs would have much more significant impact on final result of NPV. Considering the increase/decrease of operating costs and revenues, in both cases it is shown that the change causes min 1.5% change in the NPV value. Therefore, in the later stages of risk analysis, should pay attention to the factors that affect their values.

Table 3. The results of the sensitivity analysis (source: own elaboration).

Variable		- 1% of the variable	+1% of the variable
Discount rate	**NPV**	- 592 779 908 zł	- 569 809 704 zł
	Change of NPV	2,07 %	1,84%
Investment costs	**NPV**	- 575 966 063 zł	- 585 030 496 zł
	Change of NPV	0,78 %	0,78 %
Operating costs	**NPV**	-570 563 438 zł	-590 433 106 zł
	Change of NPV	2,0 %	2,0 %
Revenues	**NPV**	-589 160 348 zł	-571 836 212 zł
	Change of NPV	1,49 %	1,49 %

3.3 Scenario analysis

In the first step of this method, authors had to specify three variants of the investment. The first of these - a base scenario, which does not introduce any changes. Due to the fact that the investment of a waste incineration plant is a project applying for obtaining grants (The EU has enabled obtaining grant funds only for combustion installations), other variants will vary in amount of subsidies. The amount of the grant could range from 30-60% of the cost. The second embodiment will refer to the possibility of obtaining 30% subsidy, and the third - 60%. For adopted variants, the values of NPV should be determined and then, each of the scenarios is credited with the probability of occurrence. Analyzing the EU subsidies in other countries, and the level of innovation introduced in the country by this project, it could be assumed, that the project will receive 30% subsidy of investment costs [7, 8, 13]. Therefore this variant should be treated as primary, which was credited with 50% the probability of occurrence. In the optimistic scenario, the grant will be awarded on the maximum level, and in pessimistic - that the project does not receive funding. These two last scenarios were credited with 25% probability of occurrence.The next step, the expected value needs to be determined (E_{NPV}), which must include the probability of occurrence of each scenario (Table 4). Then, the standard deviation (σ_{NPV}) of these values was determined and coefficient of variation *CV* on basis of Equation 5 [1].

$$C = \frac{\sigma_{NPV}}{E_{NPV}} \qquad (5)$$

Table 4. The results of calculations using scenario analysis (source: own elaboration).

	SCENARIO I	SCENARIO II	SCENARIO III
	Subsidy 30%	**without subsidies**	**Subsidy 60%**
NPV [zł]	- 444 531 774	-580 498 279	-308 565 268
Percentage change of NPV [%]	23	-	47
SCENARIO	**NPV [zł]**	**Probability**	**Probability·NPV**
Scenario I	- 444 531 774	0,5	-222 265 887
Scenario II	-580 498 279	0,25	-145 124 569
Scenario III	-308 565 268	0,25	-77 141 317

Expected value: - 444 531 774 zł; Standard deviation: 96 142 838 zł; Coefficient of variation: 0.22 [-].

Negative NPV after taking into account the support of the EU does not mean that the project is not desirable, and the project should be canceled. It means that it does not bring a proper financial return on national capital in accordance with the applicable reference value (6%). Such results are obtained often [7, 9]. In these cases, the particular importance of ensuring financial sustainability of the project. Deviation NPV amounted over 96 millions zł - therefore considered investment can bring the NPV of 96 millions higher or lower in relation to the expected value which represents more than 22% of its value.

3.4 Qualitative risk assessment

For the investment of waste incineration plant in Cracow, because of the scale of the project and because of the lack of detailed data, authors performed qualitative risk assessment, which performed a descriptive assessment of the probability (P), the materialization of a given factor, its severity (S) and overall, the assessment and reduction method using a risk matrix. Another object of this analysis was to determine adverse events, which may affect the project, the level of probability of their occurrence, severity of their effects and assess the level of risk, through the interpretation of a matrix consisting of an assessment of acceptable levels of risk and a description of measures to prevent or reduce risk in reference to its main types. The scale of severity has V stages, where I - is described as not impinging, and V - can have a disastrous impact on investment (Table 5). Events were credited with the probability of occurrence where [7, 9]:
A- Very improbable events - probability of the occurrence 0-10%,
B- Improbable events - probability of the occurrence 10-33%,
C- Events quite likely - probability of the occurrence 33-66%,
D- events likely - probability of the occurrence 66-90%,
E- events very likely - probability of the occurrence 90-100%.

Table 5. The classification of severity of risk factors [4, 7].

SCALE	MEANING
I	No impact on the social well-being, despite the lack of action.
II	A slight loss of social well-being generated by the project, with minimal impact on its long-term effects. However, the effects require remedial action.
III	Moderate. The project causes a loss of social welfare and/or financial loss, even in the medium and long term. Materialized effects can be addressed effectively.
IV	Critical. The project results in a significant loss of social welfare. Implementation of the risk of negating the basic functions of the project. Remedial action, even very complex, they are not able to compensate for the effects.
V	Catastrophic. Defectiveness of the project, which may result in serious or even total loss of the intended function.

On the basis of these two qualifications it was possible to create a risk matrix, which will characterize the level of risk of a given factor - P · S (Table 6) and further allowed to characterized those factors, that threaten the investment, rated their level and described actions to reduce them. Evaluating each of the factors was based on the experience of literature, as well as other investments in public utilities and infrastructure [7, 9].

Table 6. Assessment of the level of risk P·S [4, 7].

	S	I	II	III	IV	V
P	A	LOW	LOW	LOW	LOW	MODERATE
	B	LOW	LOW	MODERATE	MODERATE	HIGH
	C	LOW	MODERATE	MODERATE	HIGH	HIGH
	D	LOW	MODERATE	HIGH	VERY HIGH	VERY HIGH
	E	MODERATE	HIGH	VERY HIGH	VERY HIGH	VERY HIGH

The results of the qualitative risk assessment are presented below on the Table 7.

Table 7. The results of the qualitative risk assessment (source: own elaboration).

DESCRIPTION OF RISK	P	S	P·S	PREVENTION
RISKS AFFECTING THE DELAY IN PUTTING INTO OPERATION OF INVESTMENT				
Loss of financial liquidity	B	V	HIGH	The adoption of an appropriate time horizon for the project, and alternative funding sources.
Errors in the design documentation	C	III	MODERATE	Realization of the project documentation by a company with proper experience.
Delays in the schedule	D	V	VERY HIGH	Commissioned works and construction supervision by experienced engineering staff.
Unforeseen obstacles or weather conditions	C	IV	HIGH	Preparation of a new geodetic maps, site visit, the corresponding assumptions in the schedule.
Social discontent	D	IV	VERY HIGH	Conducting public consultation from an early stage of the project by the staff of specialists.
Delays in the signing of the contract for construction	C	III	MODERATE	Employment of professionals. Adoption of the time margin.
RISKS FROM THE DEMAND FO SERVICES				
Smaller stream and the calorific value of the waste than expected	C	III	MODERATE	The analysis should be carried out on the basis of cautious assumptions about the production of waste and a comparison with a similar area.
RISK OF FINANCIAL MANAGEMENT				
Exceeding the investment costs	C	III	MODERATE	Determining construction costs based on similar facilities and equipment on the basis of received tenders.
Exceeding the operating costs	B	V	HIGH	Operating costs are compared with the costs for similar projects in the similar period of time.
Problems with obtaining grants	C	IV	HIGH	Employment of specialists to prepare a motion
Revenue lower than expected	B	V	HIGH	Determining the initial price and the inclusion of increasing operating costs and potential revenue.

Risk analysis revealed two risk factors which have a very large impact on the investment. The first are delays in the schedule, which have impact on the subsidies from the EU, because any delay would cause a total loss of the funds and delays in putting the facility into service, thus, contributes to not meet environmental standards, and to impose financial penalties. Another threat are community protests and because Poland is country where environmental awareness is still being created, lack of information, can block the formation of incineration plant. Also five other factors of high risk were distinguished, which already were indicated at the stage of sensitivity analyzes. Other factors have been classified as moderate risk, which, should be monitored. In the table is not taken into account factors such as exceeding environmental standards or purchase of plot, as the project foresees the use of best available techniques, and the building plot, which was owned by the city, so the risk in their case would be sclassified as low, which does not mean that they should be completely ignored. Applied preventive measures should ensure decrease the risk to an acceptable level (moderate risk).

3 Conclusions

The analysis showed that the investment, which was a waste incineration plant in Cracow, from the very beginning was a complex investment burdened with considerable risks. This is the main reason why the decision on its implementation cannot be taken lightly, on the basis of taking into account only the economic or social issues. It was necessary to look at the investment comprehensively, considering not only the general benefits and also subsequent duties which are imposed on the owner of the object. Although the data used in the calculations were of a general nature, they allowed to illustrate the factors on which the project is particularly vulnerable, and qualitative risk assessment completed the financial analysis. Initially, taking into account only the economic aspects, the investment should not be taken into consideration at all. It should be remembered that this kind of investment, belongs to the municipal infrastructure, which do not generate enough income to be able to earn. In such cases, owner should ensure its financial sustainability. Due to the fact, that analysis showed that the factors of high risk are expected amount incurred operating costs and revenues, special attention should be paid to their forecasting eg. the rising costs of facility maintenance. But especially important aspect is the satisfaction of the public, who is pleased that the city will become more ecological, but it would be the best if the plant was built as far from their properties as possible, therefore, great importance should be given to public consultation, during which, in addition to the measurable benefits can be point out, these overlooked and overshadowed by a vicious stereotypes, such as the creation of new jobs and creating additional source of heat, extension of the operation time of the landfill Barycz, the aesthetics of the object itself as opposed to landfill, thriftiness of plot area - incinerator needs about 10 times less hectares than landfill, reducing methane emissions and a smaller decrease in property prices than in the case of construction of the landfill. Despite all possible aspects, despite the high risks and difficulties that could encounter project - decided to counteract the difficulties associated with the construction of incinerators.

References

1. B.Tchórzewska-Cieślak, J.R. Rak, K. Pietrucha, JPSRA SSARS, **1**, 185-195 (2011)
2. http://eur-lex.europa.eu/legal-content/en/TXT/?uri=CELEX:32008L0098
3. E. Georgopoulou, V. Hontou, N. Gakis, Y. Sarafidis, S. Mirasgedis, D.P. Lalas, A. Loukatos, N. Gargoulas, A. Mentzis, D. Economidis, T. Triantafilopoulos, K. Korizi, J Clean Prod, **6**(3), 359-373 (2008)
4. Ocena strategiczna Systemu Gospodarki Odpadami Miasta Krakowa wraz z wyborem wariantów lokalizacji zakładu termicznego przekształcania odpadów komunalnych, 2007 (*not published, released excerpts*)
5. K. Govindan, S. Rajendran, J. P.Sarkis, P. Murugesan, J Clean Prod, **98**(1), 66–83 (2015)
6. C. Kahraman, Springer Optim Appl, 16, 1-18 (2008)
7. A. Generowicz, Z. Kowalski, M. Banach, A. Makara, Waste Manage, **32**, 349-351(2012)
8. J. Rak, *Polska Inżynieria Środowiska - Prace*, 209-2017 (2012)
9. https://www.mos.gov.pl/fileadmin/user_upload/fundusze/Przewodnik_do_analizy_kosztow.pdf
10. H. Hu, X. Li, A. Dung Nguyen, P. Kavan, Int J Environ Res Publ Health, **12**, 7593-7614 (2015)
11. http://www.abcm.org.br/anais/encit/2010/PDF/ENC10-0663.pdf
12. S. Udomsri, A. R. Martin, T. H. Fransson, Waste Manage, **30**, 1414-1422 (2010)
13. A. Generowicz, J. Kulczycka, Z. Kowalski, M. Banach, J Environ Manage, **92**(4), 1314-1320 (2011).

Start-up analysis of partial nitrification process in SBR reactors for different initial conditions

Paula Iliaszewicz[1,*], *Stanisław* Miodoński[1]

[1]Faculty of Environmental Engineering, Wroclaw University of Science and Technology, Plac Grunwaldzki 9, 50-377 Wroclaw, Poland

Abstract. There are a lot of technical solutions that are designed to improve the efficiency of nitrogen removal from wastewater. One of the most interesting processes are autotrophic nitrogen removal process ANAMMOX. Essential for the ANAMMOX process is prepare the wastewater containing the correct proportion of ammonium and nitrite. The study tested strategies for start-up of the partial nitrification from wastewater streams from sludge dewatering after the methane fermentation process. The start-up was tested for partial nitrification process from full nitrification process in pilot scale. For the initial conditions was developed strategies. In this strategy was chosen initial parameters such as pH, temperature, and DO to the partial nitrification process will be well.

1 Introduction

Over the past several years, there has been intensive development of technology to nitrogen removal from the side stream, due to increased volume of wastewater containing high levels of nitrogen. New technologies provide efficiency of nitrogen removal of over 90% [1, 2]. These actions allow for reduction of the nitrogen load, which is directed to the main stream This causes an enhancement of the overall efficiency of nitrogen removal in WWTP. For the removal of nitrogen from the effluent can be trade on conventional processes nitrification and denitrification, which are often implemented in SBR reactors.

The Sequencing Batch Reactors (*SBR*) are often used for the purification of effluents. They have a lower cost of electricity consumption, a quick start-up period as compared to the flow reactors. Systems are very flexible. Operational parameters can be changed depending on the amount and composition of the influent wastewater [3]. Sewage effluent in the reactor allows to adjust the duration of the cycle, the length phases and time of aeration, depending on the pollutant load. It is relatively easy to keep them constant solid retention time (*SRT*), by withdrawing a certain amount of excess sludge.

One of many technologies for the removal of nitrogen from the effluent, that trade on this process is ANAMMOX (*Anaerobic Ammonium Oxidation*) [4]. This process takes place in "specific" anoxic conditions. They lead him bacteria from 5 types: *Bracadia, Kuenenia, Scaltndua, Anammoxglobus, Jettenia*. It is based on the autotrophic oxidation

* Corresponding author: p.iliaszewicz@gmail.com

of ammonium nitrogen ($N\text{-}NH_4$) to nitrogen gas, by means of an electron acceptor, which are nitrates ($N\text{-}NO_3$) [4, 5].

Stoichiometry reaction is given by Eq.1

$$NH_4^+ + 1.32NO_2^- + 0.066HCO_3^- + 0.13H^+ \rightarrow$$
$$1.02N_2 + 0.066CH_2O_{0.5}N_{0.15} + 0.26NO_3^- + 2.03H_2O \qquad (1)$$

For the correctness of the process, it is necessary to provide a substrate into a plant with a molar ratio of 1.3 $gN\text{-}NO_2/gN\text{-}NH_4$. This results from stoichiometry reaction (Eq.1). For each 1.32 mole of nitrite there is 1 mole of ammonium nitrogen [4, 5]. The benefits of the process include no need to provide a source of organic carbon, whereby operating costs are lower. Moreover this process does not cause as much biomass growth. The demand for oxygen is reduced to approximately 40 %. That way the energy consumption is reduced. The disadvantages of the ANAMMOX process are a long period of start-up process and the formation of nitrate in influent.

Necessary for the ANAMMOX process is the preparation of wastewater containing suitable proportions of ammonium and nitrite. As a result, it is very important to evaluate the possibility of starting a process of partial nitrification in pilot scale for the wastewater from the dewatering of digested sludge and preparation of the product of partial nitrification, which is a mixture of $N\text{-}NO_2$ and $N\text{-}NH_4$ in a suitable ratio to be able to use generated wastewater to process ANAMMOX.

Nitrification is a microbiological process, which happens in two stages. The oxidation process of ammonia ($N\text{-}NH_4$) to nitrites ($N\text{-}NO_2$), then to nitrate ($N\text{-}NO_3$). Aerobic conditions are very important for the process, oxygen acts as ultimate acceptor of electrons [6]. It takes place in the presence of unrelated groups of autotrophic nitrifying bacteria, which use the energy resulting from the oxidation of ammonia (*AOB - Ammonia Oxidizing Bacteria*) and nitrite (*NOB - Nitrite Oxidizing Bacteria*) [7].

The first step of nitrification bacteria carry AOB, which include among others *Nitorsomonas, Nitrosococcus, Nitrosospira.* It occurs with the participation of ammonia monooxygenase (AMO). Ammonia is oxidized to hydroxylamine (NH_2OH), which is the intermediate, then it is oxidized to nitrite ($N\text{-}NO_2$), in the presence of hydroxylamine oxidoreductase (HAO) [6].

2 Analysis strategy of partial nitrification process

There have been many studies about the start-up of partial nitrification with different parameters such as dissolved oxygen (*DO*) [8], high temperatures [9], low temperatures [10], free ammonia (*FA*) and free nitrous acid (*FNA*) [11].

FA determines the ammonia in the form of undissociated. The amount of FA depends on pH, temperature and concentration of ammonia nitrogen. FA is one of the major factors that inhibit NOB in the process of partial nitrification [12, 13]. However, the FNA, inhibits the activity of nitrifying bacteria. Inhibition of FNA is the delivery of bacteria into the interior of the additional proton that disrupts the pH of the both sides of the cell membrane, which then affects the synthesis of ATP [14].

The data presented in Table 1. refer to the start-up of partial nitrification process in the selected parameters. Tests were carried out on different reactors with different volumes. Depending on the solution it can be noticed, that most of the start-ups were below 1 gO_2/m^3, while pH oscillated between 6.5 ÷ 9.0. Such actions have been obtained for the inhibition of NOB bacteria. In the articles [15, 16, 17, 18, 19, 20, 8, 21, 22] which were focused only on the conditions, which allow to run the process, but composition of effluent was not controlled, which could be used in the two-stage deamonification. Ratio $N\text{-}NO_2/N\text{-}NH_4$ was significantly superior to the values of the optimum deamonification process.

Table 1. Characteristics of partial nitrification parameters operational in different experiments.

SYSTEMS	NLR[1]	SRT[2]	pH	Temp.	DO	AUR[3]
	$g\ N/m^3\cdot d$	d	-	°C	$g\ O_2/m^3$	mg N/gSS·h
SBR laboratory, volume 1.7 dm^3, synthetic wastewater, controlled DO using the variable mixing regime, not controlled ratio $N\text{-}NO_2/N\text{-}NH_4$, inhibition of FA and FNA, [15]	0.3÷1.2	-	7.8÷8.1	31	<1	-
Pilot-scale SBR, volume 10 dm^3, PN/A process, low-strength swage, not controlled ratio $N\text{-}NO_2/N\text{-}NH_4$, [16]	-	10	7.56÷7.71	32	<1	-
Pilot-scale SBR, volume 7m^3, controlled pH and BF, low temperature, analysis FISH, not controlled ratio $N\text{-}NO_2/N\text{-}NH_4$, [17]	-	15÷18	6.85÷7.85	11÷16	>2	-
SBR laboratory, volume 3 dm^3, wastewater treatment, not controlled ratio $N\text{-}NO_2/N\text{-}NH_4$, inhibition of nitrate – combination pH range and low DO, [18]	-	11	7.5÷9.0	32	<1	17÷22
SBR laboratory, wastewater treatment high $N\text{-}NH_4$, not controlled ratio $N\text{-}NO_2/N\text{-}NH_4$, different internal organic carbon sources was tested, operation of SBR with hydrolysed primary sludge, [19]	-	11	7.5÷9.0	32	<1	17
SBR laboratory, volume 3 dm^3, wastewater treatment, SHARON/DN and SBR, not controlled ratio $N\text{-}NO_2/N\text{-}NH_4$, [20]	-	11	7.3÷8.1	32	<1	22
SBR laboratory, the effect of regulation pH, DO and temperature on PN, not controlled ratio $N\text{-}NO_2/N\text{-}NH_4$, [8]	-	-	7.0÷7.5	<30	<2.5	-
Pilot-scale, reactor, inhibition of DO and FA, not controlled ratio $N\text{-}NO_2/N\text{-}NH_4$, [21]	0.16÷2.0	8	-	30	<2	-
Pilot-scale, Reactor AO, inhibition of FA and FNA, volume 36 dm^3, not controlled ratio $N\text{-}NO_2/N\text{-}NH_4$, [22]	0.08÷0.66	>35	6.5÷8.5	-	2÷4	-

[1] NLR - Nitrogen loading rate [2] AUR - Ammonia uptake rate [3] SRT - Solid retention time

2 Materials and methods

2.1 Reactor design

Pilot-scale SBR reactor with continuous stirring and aerating has active volume of 150 dm^3. In the reactor, there was temperature, pH and DO control. Sludge from excess sludge was injected into the reactor, to keep the appropriate suspended solid concentration and good sedimentation.

2.2 Start-up strategy

Leachate from the belt presses, which dehydrates fermented sludge, was the source of ammonia nitrogen to carry out the experiment.

Start-up of partial nitrification took place in the reactor in which the full nitrification process was run. Parameters were: DO 2 g O_2/m^3, SRT 8 day, temperature 25°C, pH 7.0, moreover alkalinity in influent was corrected.

Process was stared based on the analysis of literature. There were made the following steps.

The first and most important activity was turning off the correction of alkalinity in the influent. Then, operating parameters were changed in the reactor: DO 0.8 g O_2/m^3; sludge concentration in the reactor from 3.0 to 2.0 kg/m^3, while the target SRT was 2.5 day (it was not controlled in full nitrification process); the temperature was unchanged (25 °C); pH was changed to 7.1.

Analytical monitoring of influent and effluent was performed daily. Moreover concentration of sludge was determined in the reactor. On selected days AUR tests were carried out.

2.3 Analytical methods

Analytical monitoring was carried out using a cuvette test and the Hach DR 3900 spectrophotometer. Moreover alkalinity and the sludge concentration were controlled. In order to control the process, AUR tests were performed.

3 Results and discussions

The whole experiment lasted 73 days, including 13 days technological break, which was not plant at start of experiment. Nitrogen forms in effluent and NLR during the experiment was shown in Fig. 1.

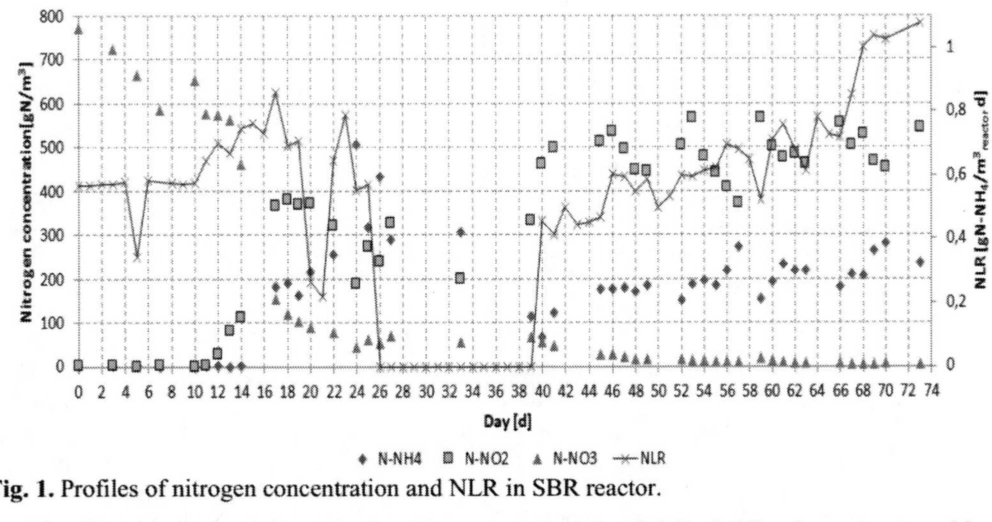

Fig. 1. Profiles of nitrogen concentration and NLR in SBR reactor.

The first 10 days of the experiment was an exploit of full nitrification process without modification of reactor cycle. Based on the strategies described in section 2.2., start-up was carried out on 10th day of experiment.

The reactor was operated according to the strategy to the 25th day.

In the days 25 ÷ 45, it was necessary to stop the reactor, becauseof technological break. The process of the partial nitrification was re-launched in the 45th day. The ratio of N-NO_2/N-NH_4 with reference to appropriate value for the ANAMMOX process (it has been accepted at 1.2 gN-NO_2/gN-NH_4.), was shown in Fig 2.

Increasing the NLR did not give the desired results. After each increase of NLR, adaptation occurred within 4÷6 days, and from then on it made the ratio N-NO_2/N-NH_4 worse to values of approx. 3.5 gN-NO_2/gN-NH_4, which cannot be considered as the base ratio. Due to the lack of effects of the target strategies (increasing NLR), there was a decision to lower pH value in the reactor. This action prevented full adaptation nitrifying bacteria to the new conditions. The effects of the activities has been noticed after 6 days. The value was approximately 2.3 gN-NO_2/gN-NH_4, instead of the primary 3.5 gN-NO_2/gN-NH_4.

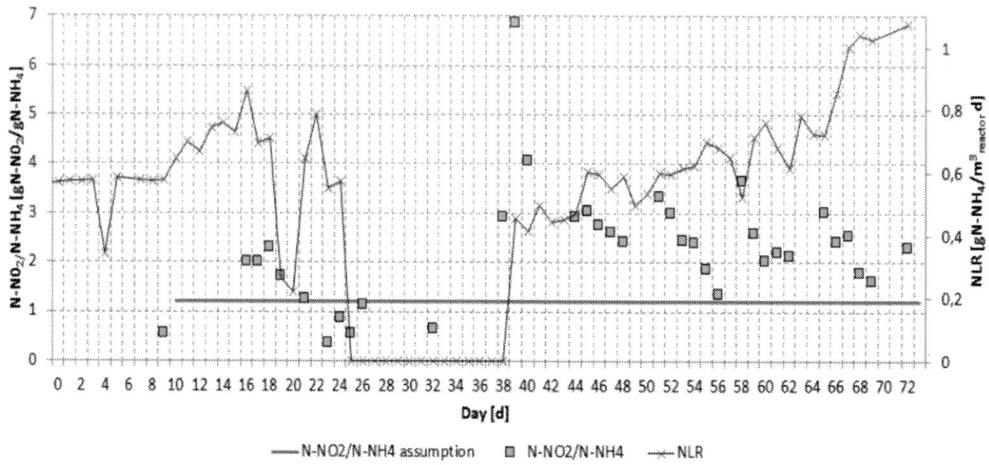

Fig. 2. Profiles of N-NO_2/N-NH_4 and NLR in SBR reactor.

In the days 46÷63 NLR was increased. Each increase in load caused momentary improvement ratio N-NO_2/N-NH_4, which after 4÷5 days returned to the initial values of approximately 3.5 gN-NO_2/gN-NH_4. The most probable cause of this situation was temporary nitrification overload, and then adaptation of nitrifying bacteria to new conditions. Adaptation of the nitrifying bacteria was noticeable, based on AUR, which was growing in the next days of the experiment (Table 2).

Table 2. Ammonia uptake rate.

		Days								
		12	14	17	19	47	54	61	67	68
AUR	gN/m^{3}min	0.66	0.70	0.19	1.17	0.14	0.21	0.34	1.29	1.08
	mgN/gSS d	349.7	485.8	203.5	1101.6	99.4	186.2	277.8	961.5	847.1

Value obtained after adaptation of nitrifying bacteria (at 3.5 gN-NO_2/gN-NH_4) meant that nitrification of about 75÷80 % of the nitrite was possible, which may be caused by higher alkalinity used in leaching or difficult to determine at this stage impact of pH adjustment (NaOH dosage). For this reason, further increasing of the load was pointless, because it probably would not allow to achieve satisfactory results as compared with N-NO_2/N-NH_4. Such action could lead to the crash of the process.

Because of that, since day 59 to 73 the strategy was modified, and pH was lowered and maintained in the reactor (change from 7.1 to 6.9). The effect of this action was improve, such as during the increasing of load in the reactor (approximately 1.6 gN-NO_2/gN-NH_4).

The process of adaptation to the new conditions has not been as effective, as in the previous part of the experiment. As the result of these actions, ratio did not return to value of approximately 3.0 $gN\text{-}NO_2/gN\text{-}NH_4$, but only to 2.3 $gN\text{-}NO_2/gN\text{-}NH_4$ in the last day of the experiment. Given the positive results further action were based on strategies for lowering pH.

Further actions will include gradual lowering of the set pH value in the reactor until the total shut down of adjustment of pH or obtaining of satisfactory ratio of $N\text{-}NO_2/N\text{-}NH_4$. Abrupt change is most likely impossible. Shut down of pH adjustment results in the significant ratio reduction. The high concentration of $N\text{-}NO_2$, which determines a high concentration of FNA, could also lead to inhibition of nitrification of the first phase. It would mean the crash of the process. To prevent lowering of the pH, value should be gradual.

4 Conclusions

- The transition from full nitrification process to partial nitrification was very rapid. At that time there was a washout of NOB. After 10 days, the participation of nitrite was 55% of total NO_x effluent. After the technological break it was possible to obtain up to 99% nitrite in total NO_x. The final obtained NLR was approx.1.05 gN/m^3d. Probably, there is potential load of SBR reactor, but at this stage it was not the subject of the experiment.
- Attempting to achieve the desired composition of the effluent by increasing the NLR, did not give the desired results. After each increase of NLR, adaptation occurred within 4÷6 days, and from then on it made the ratio $N\text{-}NO_2/N\text{-}NH_4$ worse to values of approx. 3.5 $gN\text{-}NO_2/gN\text{-}NH_4$, which cannot be considered as the base ratio. The most probable cause of this situation was temporary nitrification overload, and then adaptation of nitrifying bacteria to new conditions.

The authors gratefully acknowledge co-funding from the National Centre for Research and Development (grant no PBS2/B9/25/2014) and Municipal Water and Sewage Company Wroclaw.

References

1. E. Arnold, B. Böhm, P.A. Wilderer, Water Sci. Technol. **41**, 1 (2000)
2. C. Fux, S. Velten, V. Carozzi, D. Solley, J. Keller, Water Res. **40** (2006)
3. M. Mańczak, *Zalety i wady reaktorów okresowego działania –SBR, Seminarium nt.:"technologia oczyszczania ścieków w reaktorach SBR - teoria i praktyka* (Wyd. Seidel-Przywecki, Warszawa, 2007)
4. A. Mulder, A.A. van de Graaf, L.A. Robertson, J.G. Kuenen, Microbiology Ecology **16** (2006)
5. A.A.Van de Graaf, P. De Bruijn, L.A. Robertson, M.S.M Jetten, J.G. Keunen, Microbiology **142** (1996)
6. Z. Sadecka *Podstawy biologicznego oczyszczania ścieków,* (Seidel-Przywecki, Piaseczno, 2010)
7. J. Dosta, A. Gali, T.Benabdallah El-Hadj, S. Mace, J. Mata-Alvarez, Bioresource Technol. **98** (2007)
8. J. Wang, N. Yang, Process Biochemistry **39** (2004)
9. C. Hellinga, A. Schellen, J.W. Mulder, J.J Heijnen, Water Sci. Technol. **37** (1998)

10. S. Gu, S. Wang, Q. Yang, P. Yang, Y. Peng, Bioresoure Technology **112** (2012)
11. S. Park, W. Bae, B.E. Rittmann, Environ. Sci. Technol. **44** (2010)
12. Y. Jeanningros, S. E. Vlaemnick, A. Kaldate., Water Science & Technology -WST **61** (2010)
13. H. Li., S. Zhou., G. Huang, B. Xu, Process Safety and Environmental Protection **92** (2014)
14. C. Glass, J. Silverstein, J. Oh, Water Environment Research **69** (1997)
15. M. Soliman, A. Eldyasti, , Bioresource Technology **221** (2012)
16. Y. Miao, L. Zhang, Y. Yang, Y. Peng, B. Li, S. Wang, Bioresource Technology **218** (2016)
17. S. Gu, S. Wang, Q. Yang, P. Yang, Y., Bioresource Technology **112** (2012)
18. A. Gali, J. Dosta, J. Mata-Alvarez, .Environ. Technol. **28** (2007)
19. A. Gali, J. Dosta, J. Mata-Alvarez, Ind. Eng. Chem. Res. **45** (2006)
20. A. Gali, J. Dosta, M.C. van Loosdrecht, J. Mata-Alvarez, Ind. Eng. Chem. Res. **45** (2006)
21. J. A. Tora, J. Lafuente, J. Carrer, J. A. Baeza , Environmental Technology **33** (2012)
22. C. Wu, Y. Peng, L. Zhang, *Factors Influencing the Start-up and Stability of Partial Nitrification in a Continuous Flow Reactor,* (International Conference on Electronics, Communications and Control (ICECC), 2011)

Heat pumps as a way to Low or Zero Emission district heating systems

Piotr Jadwiszczak[1,*]

[1]Wroclaw University of Science and Technology, Faculty of Environmental Engineering, Wybrzeże Wyspiańskiego 27, 50-370 Wrocław, Poland

Abstract. In traditional district heating (DH) system heat is generated from fossil fuel (FF) combustion in heating only boilers (HOB) or in combined heat and power (CHP) plants. It results in greenhouse gases and other pollutants emission. The reduction of emission is one of the main target in EU climate policy. Among the alternative technologies in DH heat pumps (HP) play a crucial role and enable to decrease or even eliminate emission to create a low or zero emission (LZE) DH system. The emission reduction effect of integration the large scale HP units into DH systems can by defined by four groups of factors: the share of HP in the heat demand, the heat source for HP, the driving energy for HP and heat sink for HP. This paper illustrates the main options for large scale HP units application for LZE DH based on HP technology.

1 Introduction

District heating (DH) is an important option to supply the heat to the numerous consumers, especially in urban areas with high heat demand density. The DH system supplies heat generated in central source to the large number of consumers using double pipe distribution network. In traditional DH heat generation is based on the fossil fuels (FF) combustion, which results in greenhouse gases and other pollutants emission. The development in heat generation technologies open a wide opportunity to improve the DH systems in order to reduce or eliminate emission, including renewable energy sources (RES) and heat pumps (HP) technology. The significant reduction of emission in DH system creates the low or zero emission (LZE) DH system. The LZE DH meets the EU energy and climate targets in 2020, 2030 and 2050 and the challenges of sustainable energy future [1, 2].

The DH systems offer energy flexibility, which allows to create the fuel and energy mix system in large scale. The DH can integrate the different kinds of heat and energy generation technologies like conventional heat sources or power plants (coil, gas, oil etc.), combined heat and power (CHP) plants, thermal RES plants (thermal solar, geothermal, biomass etc.), HP units, thermal energy storage (TES), RES power plants (hydro power, wind power, PV ect.) [3]. The total emission in DH system is the sum of emissions from each used heat or power generation technology, including the technology share in energy

* Corresponding author: piotr.jadwiszczak@pwr.edu.pl

demand and the distribution networks losses. The integration of HP technology into a DH system is one of most effective ways to decarbonise the district energy sector, due to decreasing of FF, increasing of energy efficiency and increasing the share of RES through balancing. HP expands the range of renewable energy sources that can be used to supply DH including biomass, solar and geothermal heat, see water, river or lake, sewage, geothermal, industrial and waste heat.

Among the different kinds of developed HP technologies the electrically driven HP are the most widely used. Fig. 1 shows the principle of electrically driven heat pumps: the HP upgrades the low-grade heat (from heat source) to the useful heat (e.g. with the DH supply line as a sink) using electricity as a driving energy. The efficiency of HP describes the coefficient of performance (COP) and the seasonal performance factor (SPF). The COP is determined in the steady state and is defined as the ratio of the generated heat to the electricity needed to drive the HP. The SPF described the ratio of annual heat generation and annual electricity consumption of the HP.

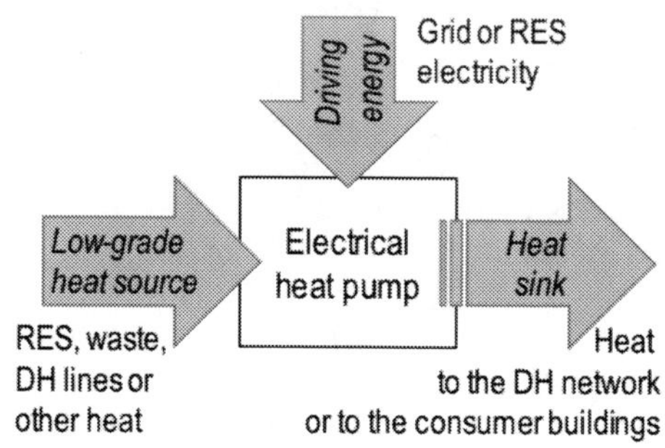

Fig. 1. The principle of electrically driven heat pumps.

There are many scenarios of HP integration into DH system. The HP technology in scale of DH systems includes both the large scale and the individual HP units. This article focuses on large scale electrically driven HP units, as a widely used technology in both the new and the existing DH systems across EU. This kind of HP units can be fitted into DH network as a part of central source or plant, or as a locally placed HP unit cooperating with central heat source. Centrally or locally placed the large scale HP unit can supply DH as a single heat source, in parallel with FF heat sources or in parallel with RES, with different share in covering the heat demand in DH. The large scale HP units can use the RES (e.g. ambient air, water or ground), waste (e.g. industrial), DH lines or other low-grade heat as a heat source; and the local or FF, CHP or RES grid electricity as a driving energy. The electrical large scale HP units enable higher share of RES in DH through balancing and utilizing the excess renewable electricity from wind and solar plants [5].

Each above mentioned factor influences on total emission value in DH. It is clear that the large share of HP, high COP and SPF, and low-carbon heat and electricity sources for large scale HP are preferred to create the LZE DH.

2 HP in LZE DH

Many lessons have been learnt since the first DH systems were installed. The HP technology has the large emission reduction potential, which can be realised in large scale in DH systems. The aim is a DH system that supplies the largest number of consumers with low or without any emissions. It can be achieved by integrating the large scale HP units into DH, with maximum share in heat demand covering, using a renewable heat sources and electricity from renewable energy sources.

Achieving a LZE DH system based on HP technology can be realised in many different option, and is defined by four groups of technical factors, as shown in Fig. 2. The groups are: 1) the HP share in heat demand covering, 2) the kind of RES heat source for HP, 3) the kind of HP driving energy, and 4) the kind of HP heat sink. The share of heat supplied from large scale HP can vary from 0 to 100% of yearly heat demand. The 0% share means a fully FF heat supply, and the 100% means heat supply only by HP as a single heat source. The middle shares can be realised by cooperation of large scale HP as a base heat source and a FF as a supporting (backup) heat source, or by cooperation of the large scale HP and other RES, and long term TES tank. The LZE DH requires to use only zero carbon renewables heat sources for HP, like a low-grade heat of water, ground, ambient air etc. It should be noted, that temperature and thermal capacity of heat source define the COP and SPF of large scale HP unit. The electrically driven HP units in DH can by powered by locally generated or grid electricity, both form FF (e.g. power plants or CHP) and RES power plants (e.g. hydro, wind or solar power plants). Only the RES electricity results in zero emissions, in each other case the emission is related with FF combustion and distribution grid losses. The needed temperature of HP heat sink (water to the DH supply line) determines the kind of HP technology and the value of achieved COP and SPF. The combination of all above mentioned factors affects on total emission in DH system. The local thermal conditions, economics, climate policy and legislation will form the structure of DH and its total emission.

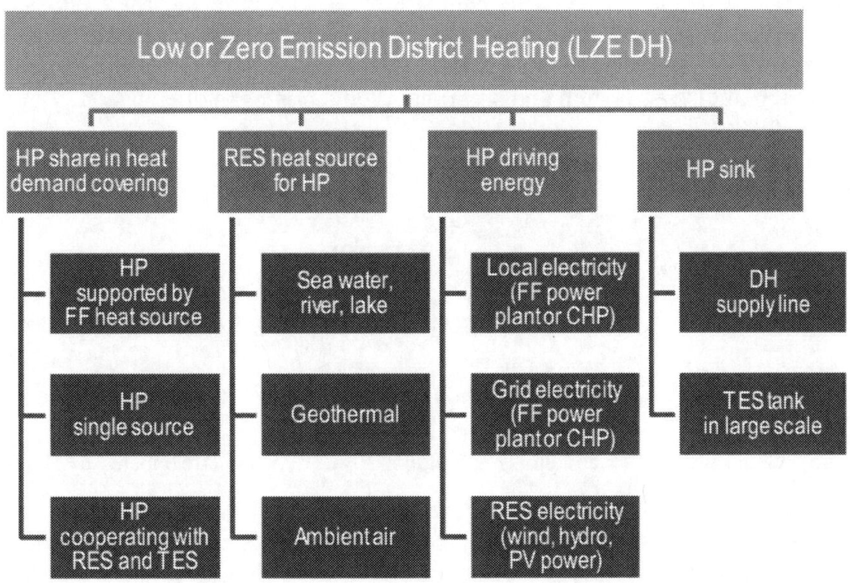

Fig. 2. The main factors in the pursuit of low or zero emission, heat pumps based district heating system.

3 Exemplary cases

The choice of proper technical solution for LZE DH based on large scale DH units is a complex engineering task and, as mentioned above, there are no one way to create the LZE DH using HP technology. According to the diagram in Fig. 2 the exemplary scenarios of LZE DH based on HP will be presented. All fuels and energy carriers, all energy sources and all energy losses are taken into account, in both generation and demand side.

3.1 Traditional FF based DH system

Figure 3 presents the fuel and energy flux, and emission placement and value in traditional FF based DH system. The FF heat source covers 100% of heat demand in DH. The FF combustion in central HOB or CHP plant affects with a high value of emission. The emission value is related to the heat demand in DH, the kind of FF, the heat source performance and the heat losses in distribution network.

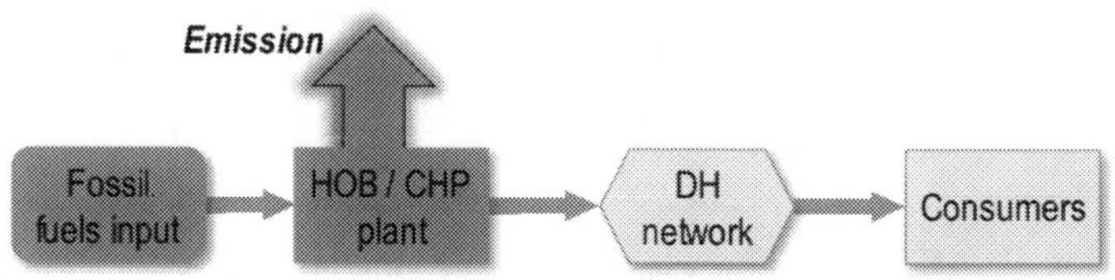

Fig. 3. The fuel and energy flux and emission value in traditional FF based DH.

3.2 Conventional FF based DH system boosted by HP in central plant

The conventional FF based DH can be boosted by large scale HP unit integration in central CHP plant, as is shown in Fig. 4. The emission is still placed in central FF CHP plant, but it is reduced. The large scale HP unit covers a part of heat demand using the heat from zero carbon RES and consuming the electricity generated in CHP. The integration of CHP and HP technology in central plant allows to decrease the total emission in DH, according to the CHP and HP ratio in heat supply. The adjusting of CHP heat and power generation, and HP heat supply ratio, taking into account the COP and SPF, will result in high efficiency of district system and significant reduction of total emission.

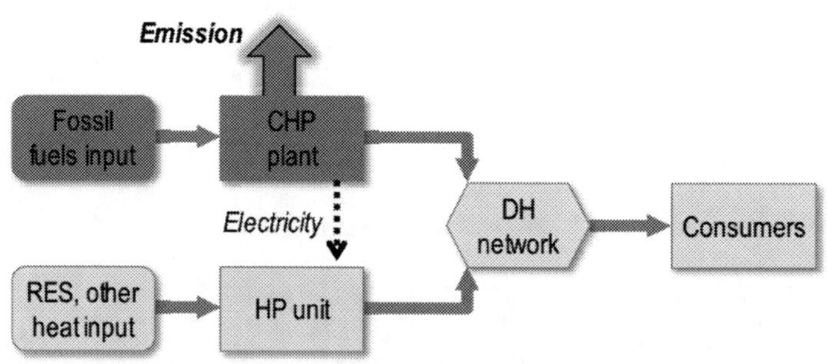

Fig. 4. The fuel and energy flux and emission value in FF CHP based DH boosted by HP.

The reduction of total emission in FF based DH boosted by large scale HP unit depends on the HP share in heat demand and achieved SPF. Fig. 5 shows the percentage reduction of total emission in DH system presented in Fig. 4, according to the share in heat demand and the SPF value of HP unit. Calculations was made for current Polish energy sector structure and emissions factors. It is evident, the higher share of the HP powered by CHP and the higher SPF, the greater reduction of total emission. Even by 99% share of HP the emission occurs, are reduced to 57 and 44% according to the SPF (e.g. SPF = 3.5 and SPF = 4.5), because HP is powered by electricity from FF CHP.

DOI: 10.1051/e3sconf/20171700032

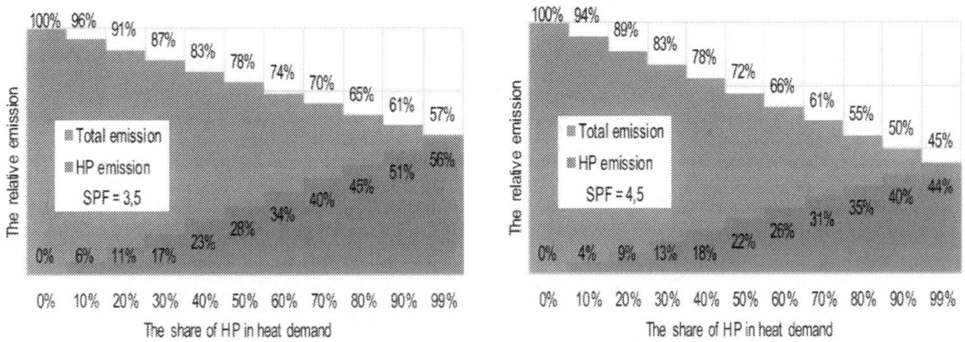

Fig. 5. The total emission and HP emission in FF CHP based DH boosted by HP, according to the HP share in heat demand and the SPF value.

3.3 Conventional FF based DH system boosted by locally placed HP

Figure 6 presents FF and HP based DH system, where the large scale HP unit is placed locally, near the favorable heat source (e.g. sea, river, lake water). In this case HP is located far from the central plant, but is still connected to the DH network. The HP unit consumes the grid electricity generated in FF CHP plant as a driving gear, so the distribution losses are included in total emission. Fig. 7 presents smaller reduction in emission compare to the previous case, 63 and 49% according to the SPF (e.g. SPF = 3.5 and SPF = 4.5).

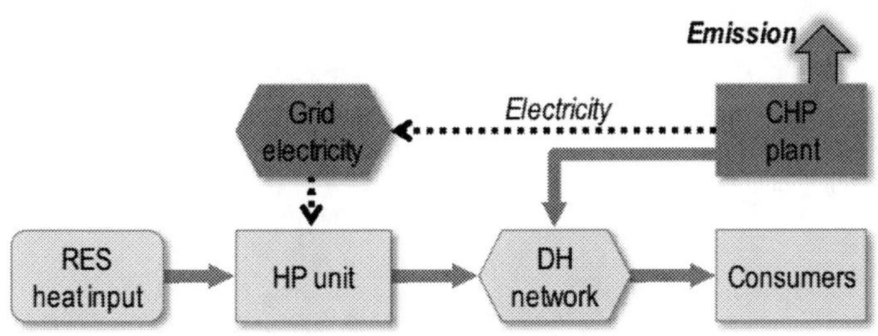

Fig. 6. The energy flux and emission value in FF CHP based DH with locally connected HP powered from grid electricity.

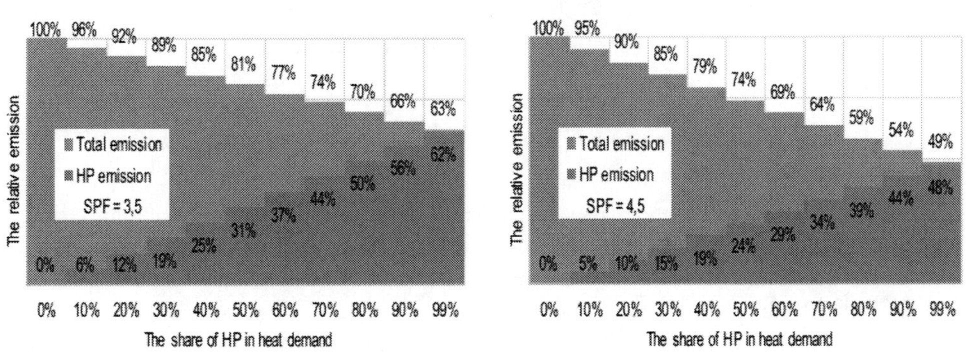

Fig. 7. The total emission and HP emission in FF CHP based DH boosted by HP, according to the HP share in heat demand and the SPF value.

 DOI: 10.1051/e3sconf/20171700032

3.4 DH system based on HP unit

The next options for LZE DH based on large scale HP is the HP single operation mode, where the large scale HP alone covers all heat demand in DH system. The HP unit can be powered by grid electricity from FF CHP or FF power plant (Fig. 8), by RES electricity from hydro, wind or solar power plant (Fig. 10), or simultaneously both by FF and RES electricity (Fig. 11).

The total emission in HP based DH shown in Fig. 8 includes the heat demand, the DH network and electrical grid losses and the SPF of HP. It should be mentioned, that the 100% share of large scale HP unit in heat demand requires the heat source for HP with stable and large thermal capacity, like geothermal, sea water, large lake or river, etc. The emission depends strongly on SPF of HP unit and on technology of electricity generation and distribution. It should be noted, that such a scheme strongly reduces, but does not eliminate the emission. By a high SPF and high power generation performance, the presented DH system may become the LZE DH. The lower SPF of large scale HP unit leads to the emission higher than in equivalent FF DH system, as shown in Fig. 9.

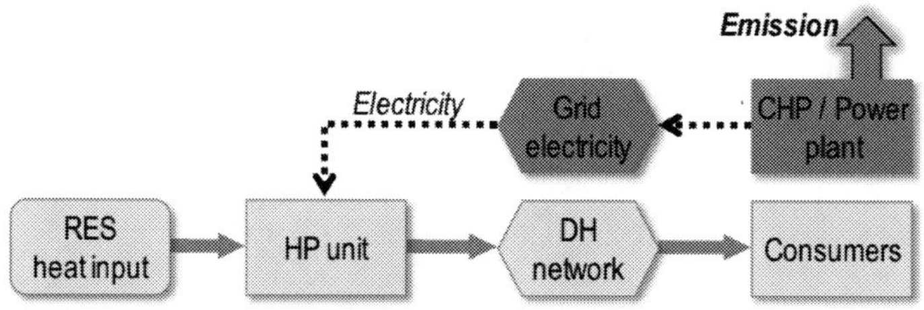

Fig. 8. The energy flux and emission value in LZE DH supplied by HP powered by grid electricity from FF CHP or power plant.

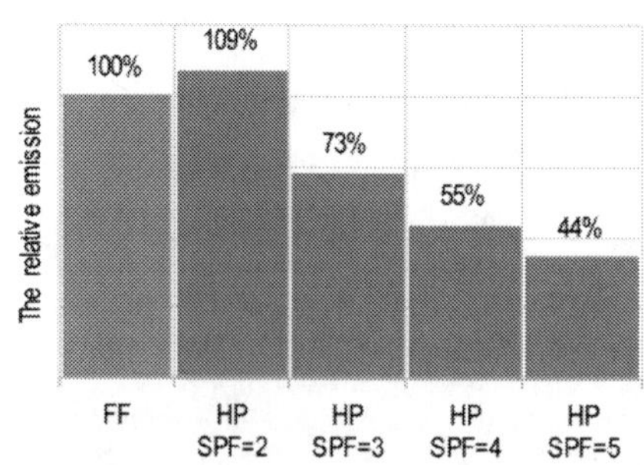

Fig. 9. The total emission depending on SPF of HP unit.

In HP based DH, were the large scale HP is powered only by RES electricity, the TES tank is recommended to balance the electricity and heat generation and demand side, as is shown in Fig. 10. The large scale HP unit powered by zero carbon electricity covers the 100% of heat demand without any emission. The needed electricity can by generated in stable RES like hydro power plant or in variable RES like wind or solar power plant, which influence on the needed TES scale. The thermal capacity of heat source for HP should be large and stable in scale of year, like geothermal, sea water, large lake or river, ect. to cover all heat demand, including the short or long term TES tank [4]. Due to using the zero carbon heat sources and the zero carbon driving energy for large scale HP unit the presented DH system certainly is a LZE DH.

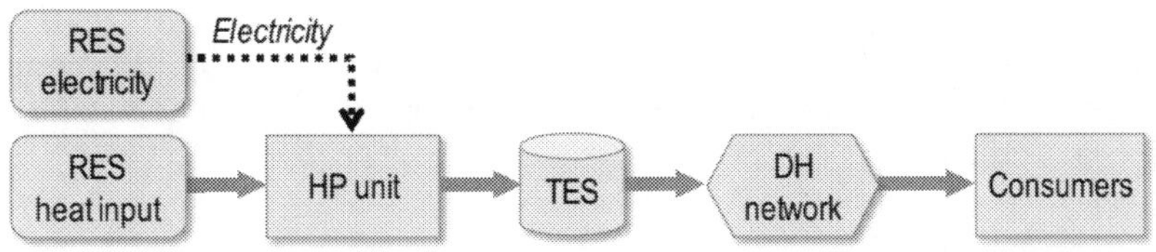

Fig. 10. The energy flux and emission value in LZE DH supplied by HP powered by RES electricity.

Fig. 11 shows the HP based DH in next option, were large scale HP unit is powered partially by RES and FF electricity generated in FF CHP or power plant, and distributed by electricity grid. The FF CHP or power plant can balance the variability of RES electricity generation or just complements RES power plants. It is clear, that large scale HP unit covers the 100% of heat demand, needs the zero carbon heat source with large and stable thermal capacity. The emission dependent on the achieved SPF and the ratio of consumed RES and FF electricity. By high SPF and large share of RES electricity, such DH system may become the LZE DH.

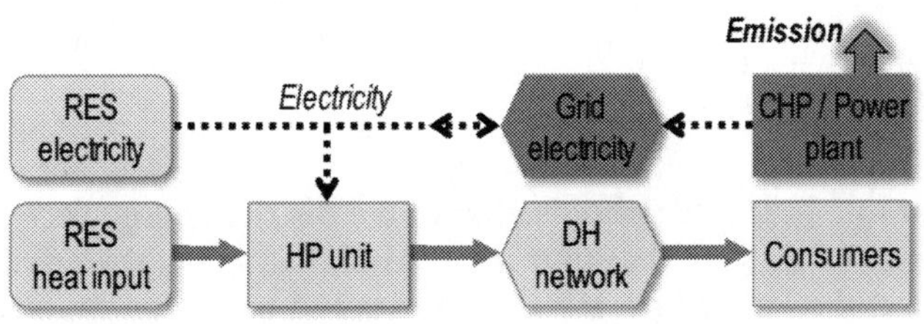

Fig. 11. The energy flux and emission value in LZE DH supplied by HP powered partially by RES and grid electricity.

3.5 PEB in DH system based on HP unit

The plus energy buildings (PEB) technology can be integrated into the DH network. Fig. 12 presents the large scale HP based DH system with TES, suppling both the PEB and traditional buildings. The energy fluxes between PEB and district energy systems are bidirectional, this means that the heat and the electricity is periodically consumed or produced in PEB. The large number of PEB in DH system has a significant influence on energy balance, and has a great potential to create the LZE DH system. The low or zero carbon heat generated in PEB decreases the needed thermal capacity of HP unit, but requires the long term heat accumulation in large scale TES tank. The large scale HP unit will integrate the electricity from RES and PEB, and storage it in form of heat in long term TES tank. It is evident, that the proper sizing of TES is crucial to the energy balance of the whole system. Each component of presented DH has no emission, so it certainly is a LZE DH.

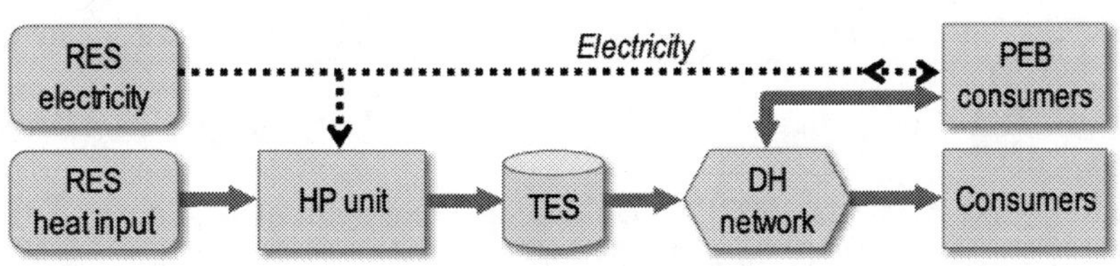

Fig. 12. The energy flux in LZE DH supplied by HP powered from RES and PEB.

The DH system in the next option contains thermal RES plant, e.g. large thermal solar plant. The large scale HP unit and thermal RES supply in parallel the large scale TES tank, as shown in Fig. 12. The thermal RES has priority in TES charging, and allows decreasing the needed thermal capacity of HP unit. The driving energy for both HP and thermal RES is generated in RES power plant and in PEB. The RES energy variability requires the long term and large scale energy storage. Each component of presented DH has no emission, so it certainly is a LZE DH.

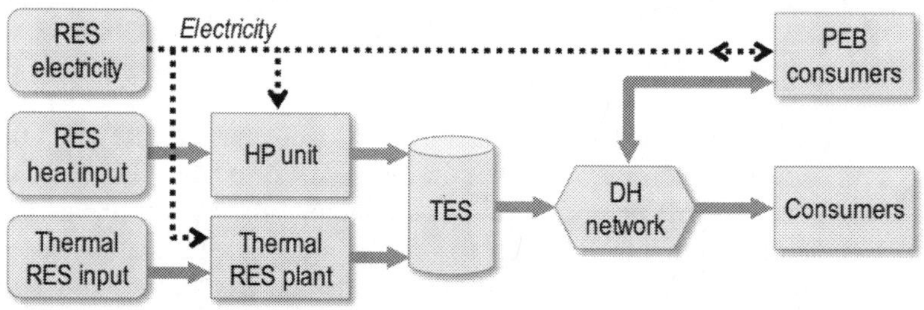

Fig. 13. The fuel and energy flux in DH with TES supplied partially by RES and HP powered from RES and PEB.

4 Conclusions

The emission saving potential of HP based DH is really significant. The HP technology plays an important role in the decarbonisation of the district energy sector. In every case the proper integration of large scale HP unit into DH should bring the environmental benefits. By reducing FF use, the large scale HP DH can lead to reduction in indoor and outdoor air pollution and the associated health impacts [5].

Depending on the kind of driving energy and heat source for HP unit, the DH may become the LZE. Low or zero carbon electricity and heat sources for HP are preferred. The LZE DH system based on large scale HP units allows to supply the low or zero carbon heat to the large number of consumer. The scenario of LZE DH based on HP technology can be replicated and scaled to many cities or residential areas.

The design of LZE DH must take into account the HP share in heat demand covering, the kind of RES heat source for HP, the kind of HP driving energy, and the kind of HP heat sink. The calculations of a total emission made for Polish energy sector structure prove the emission saving potential of HP technology in DH. All the emissions carried out along the whole energy flux were taken into account.

The total emission in HP boosted or HP based DH system strongly depends on the COP and SPF of HP unit, and the source of driving electricity. The possibility of emission savings was approved by LZE DH scenarios and calculations, which are investigated in this article.

References

1. EU Strategy on Heating and Cooling, SWD 24 final report (European Commission Brussel, 2016)
2. European Heat Pump Statistics, The European Heat Pump Association (EHPA), Brussels, 2009
3. DISTRICT ENERGY IN CITIES, Unlocking the Potential of Energy Efficiency and Renewable Energy (UNEP, Paris 2015)
4. V. Dagilis, L. Vaitkus, Combined heat pump and water-power plant at Kaunas Lagoon, The 9th International Conference ENVIRONMENTAL ENGINEERING, Vilnius, Lithuania 2014

 DOI: 10.1051/e3sconf/20171700033

Norfloxacin degradation by *Bacillus subtilis* strains able to produce biosurfactants on a bioreactor scale

Łukasz Jałowiecki[1*], *Joanna* Żur[2], and *Grażyna A.* Płaza[3]

[1]Department of Environmental Microbiology, Institute for Ecology of Industrial Areas,40-844 Kossutha 6 Katowice, Poland
[2]Department of Biochemistry, Faculty of Biology and Environmental Protection, University of Silesia in Katowice,40-032 Jagielońska 28 Katowice, Poland
[3]Department of Environmental Microbiology, Institute for Ecology of Industrial Areas, Katowice, Poland

Abstract. The discharge of antibiotics into the environment has become a major concern since this group of pharmaceuticals influence on microbial communities not only by its mode of action, but also because of the risk of a worldwide dispersal of antibiotic resistance genes (ARG). Antibiotics residues have been found in various environments such as waters, sediments, and soils. Moreover, most WWTPs are not designed to treat such kind of pollutants, which remain incompletely removed. Currently, biodegradation processes which involved bacterial strains with increased degradation capabilities is one of the most promising technique. The aim of this study was to evaluate the norfloxacin biodegradation potential of the three *Bacillus subtilis* strains named T-1, T'-1 and I'-1a on a bioreactor scale. The aerobic degradation was conducted in a 5-liter bioreactor on minimal salts medium in co-metabolic culture supplemented with glucose. The degradation rate of norfloxacin was determined with the HPLC technique. The surface tension was determined using ring method in order to observe the changes in biosurfactants production. Also, the biofilm formation abilities of the bacteria with two quantitative methods, crystal violet (CV) method and TTC-based test and enzymes production were evaluated.

1 Introduction

Currently, the annual production of antibiotics ranging from 100,000 to 200,000 tons per year [1, 2]. The increasing intake of pharmaceuticals results in uncontrolled pollution of the environment with these substances and their metabolites. Simultaneously, the popular aggressive chemical methods used in wastewater treatment plants (WWTPs) e.g. ozonation, Fenton processes or photocatalysis with TiO_2 are not able to complete removal of that kind of pollutants. Recently, biodegradation processes, which involved microorganisms with increased degradation capabilities is one of the most promising method used in WWTPs.

* Corresponding author: jalowiecki@ietu.katowice.pl

In this study microbial consortium composed of three *Bacillus subtilis* strains (T-1, T'-1 and I'-1a) was used for norfloxacin, the next-generation fluoroquinolone antibiotic degradation on a bioreactor scale. Fluoroquinolones, the synthetic class of antibiotics derived from nalidixic acid are widely used against both Gram-negative and Gram-positive microorganisms. All of the examined strains are able to lipopeptide (LPs) active surface compounds produce namely surfactins, iturins and fengycins [3, 4]. Biosurfactants, despite their ability to reduce the surface and interfacial tension are also classified as one of the chemical class of quorum sensing (QS) molecules. QS is defined as the regulation of gene expression in response to changes in cell-population density. Microorganisms use quorum sensing communications to regulate different physiological activities i.e. virulence, metabolites synthesis, sporulation and biofilm production. It is also widely known that microorganisms, which are able to surface-active compounds produced are characterized by good abilities to rapid colonization of different surfaces and initiating of biofilm formation [5]. In this study, two quantitative methods used for evaluation of biofilm formation diverse in terms of detection sensitivity were applied. Microorganisms with increased degradation potential may be used for construction of bacterial consortia and its subsequent immobilization. Besides this, microorganisms represents an alternative source of useful enzymes e.g. inulinases, cellulases, proteases or lipases, which have widespread uses in biotechnology, industry and medicine.

2 Materials and methods

Bacterial strains. Three *Bacillus subtilis* strains were used in the study named: T-1, T'-1 and I'-1a. They were isolated from sludge of an oil-refinery in Czechowice-Dziedzice in Poland [6]. In the previous study, the strains were identified and characterized by Płaza et al. [3] and Bernat et al. [4].

Degradation of norfloxacin. Degradation experiment was performed in a bioreactor (BioFlo 415, New Brunswick Scientific, US) containing 5 liters of mineral salts medium (MSM) with the following composition, (g L^{-1}): 3.78 $Na_2HPO_4 \cdot 12H_2O$; 0.5 KH_2PO4; 5 NH_4Cl; 0.2 $MgSO_4 \cdot 7H_2O$; 0.01 yeast extract. The cometabolic culture was supplemented with glucose. The bacteria were cultivated on LB medium up to 48 h, under shaking conditions (130 rpm) at 30°C. Then, the culture samples were centrifuged at 5 000 *g* for 10 min, and bacteria were resuspended in mineral salts medium. The suspensions of the each bacteria were added to the bioreactor. Norfloxacin was also added to the final concentration of 7.5 mg L^{-1}. The degradation rate of norfloxacin was determined with the HPLC technique using Merck Hitachi HPLC reversed-phase chromatograph equipped with a column Ascentis Express® C18, pre-column Opti-Solw® EXP, and UV/VIS DAD detector. The mobile phase was consisted of acetonitrile and 1% acetic acid (10:90 v/v; flow-rate 1 ml/min). The detection wavelength was set at 270 nm.

Determination of biosurfactants production. 50 ml of bacterial culture was centrifuged at 5 000 g for 15 min and the supernatant was used for the surface tension measurements. The surface tension was determined by DeNouya method using a ring with a tensiometer SIGMA 702 (Attension). Measurements were repeated at least three times, and an average value was used to express the surface activity of each sample.

Enzymes production.

Amylase activity. In order to detect the amylase activity, bacteria were plated on nutrient agar plates supplemented with 1% starch. The plates were then incubated at 28 ± 1°C for 48 h. After the appearance of the colonies on the starch-agar medium, the culture plates were flooded with 1% Lugol's iodine solution to identify amylase activity.

Cellulase activity. For the detection of cellulose activities, isolates were grown on plates with Carboxymethyl Cellulose agar (CMC) and flooded with 1% Lugol's iodine solution to identify the enzyme activity. The presence of cellulolytic activity was confirmed by the appearance of yellow colour around the colonies.

Inulinase activity. The bacteria were inoculated on MHI agar plates, which contained inulin as the sole carbon and energy source and incubated at 28 ± 1°C for 3 to 5 days. Inulin was sterilized separately and added before pouring the medium into the plates, with the final concentration of 2%. After incubation, plates were flooded with Lugol's iodine solution for 5 min. A clear halo appeared around the colonies, where the inulin has been degraded.

Lipase activity. Lipolytic activity of the bacterial strains was determined on blood agar plates and incubated at 28°C for 48 h. After autoclaving, the blood agar was cooled, then 5% (v/v) defibrinated sheep blood was added. The positive samples were then plated again on the Tween 80 agar plates with phenol red indicator and incubated at 28°C for 4 days. The lipase activity will bring about change in coloration from pink to lemon yellow.

Protease activity. The extracellular protease production was evaluated with two methods. In the first method the isolates were plated on enriched nutrient agar (2% starch) plates and incubated at 28 ± 1°C for 48 h. The appearance of a clear zone around the colony after flooding the plate with 1% Lugol's iodine solution confirmed the presence of proteolytic activity. In the second method, Skimmed Milk Agar (SMA) was used as a screening medium for the protease production. The plates were incubated at 28°C for 48 h. Development of clear zones around the bacterial strains confirmed the proteolytic activity of the strains.

Urease activity. In order to evaluate the urease activity, the bacteria were streaked on the Urea Agar Base (UAB) and incubated at 28°C for 48 h. The urease activity was determined by the appearance of the purple colour around the colonies.

Biofilm production. The abilities to biofilm production of the tested strains was performed according to procedures described by Stepanović et al. [7] for crystal violet (CV) and Brown et al. [8] for 2,3,5-triphenyltetrazolium chloride (TTC) reduction, with certain modifications. Each well of 96-well tissue culture plate were filled with 200 µl sterile LB medium and 20 µl of individual overnight culture strain dilution of 1:100 in fresh LB medium. The plates were incubated for 48 h at 30°C. Then, the wells were washed three times with 200 µl per well of sterile phosphate buffer saline (PBS, pH 7.2), emptied and left to dry. Next, the plates were fixed with 200 µl per well of methanol for 30 min, dried and then stained with 200 µl per well of 0.1% crystal violet dissolved in acetone:ethanol mixture (1:4, v/v) solution for 20 min. After staining followed by brief drying, 200 µl per well of 30% acetic acid was added into each well in order to resolubilized the dye bound to the adherent cells and incubated for 30 min. For TTC analysis, each well of the plate were filled with 200 µl sterile LB medium and 20 µl of overnight culture diluted to $OD_{600} = 0.05$. After incubation, the wells were washed with 200 µl per well of sterile phosphate buffer saline (PBS, pH 7.2), emptied and left to dry. Afterwards, 0.5% TTC solution were added to each well and incubated for 24 h. After the incubation period, the wells were emptied and stain with 200 µl of acetone/ethanol (1:4, v/v). Negative controls were obtained by incubating the wells only with 200 µl per well of LB medium, without bacteria. The optical density (OD) of the obtained solutions was measured at 600 nm and 490 nm for crystal violet and TTC assay, respectively, using a microtiter plate reader (Eppendorf).

3 Results and discussion

The concentration of antibiotic residues detected in the environment usually ranging from ng/L to µg/L level in natural water and wastewater and from µg/kg to mg/kg level in soil and sludge. Currently, antibiotics are considered to be emerging pollutants due to the risk of development, maintenance, transfer or spread of antibiotics resistant bacteria and antibiotics resistance genes (ARG) [2]. Norfloxacin is a synthetic fluoroquinolone antibiotic used for the treatment of conjunctivitis, gastrointestinal, urinary tract and skin infections with broad-spectrum antibacterial activity against most Gram-negative and Gram-positive bacteria, widely used in both human and veterinary medicine. Norfloxacin and other quinolones may enter into the environment mainly from hospital and municipal wastewaters. Simultaneously, the most intensively used techniques for wastewater treatment e.g. ozonation or UV/H_2O_2 and Fenton processes are not sufficient to complete removal of different class of pharmaceuticals e.g. antibiotics, non-steroidal anti-inflammatory drugs, hormones or lipid regulators [2, 9]. Moreover, it is widely known that antibiotics from fluoroquinolone class, i.e. ciprofloxacin and norfloxacin, are poorly biodegradable. Up to now, literature data on microbial degradation of norfloxacin remain scarce. Parshikov et al. [10] reported that species from *Escherichia*, *Pseudomonas*, *Microbacterium*, *Mycobacterium* and *Streptomyces* genera are able to transform norfloxacin by site-specific hydroxylation. Metabolism of norfloxacin *via N*-acetylation, oxidation, and breakdown of the piperazine ring has been reported for humans and fungi. *Microbacterium* sp. isolated from wastewater have been shown to transform the norfloxacin to four main metabolites, namely: *N*-acetylnorfloxacin, desethylene *N*-acetylnorfloxacin, 8-hydroxynorfloxacin and 6-hydroxynorfloxacin. Kim et al. [11] observed that during growth of the strain in medium supplemented with norfloxacin four main metabolites were produced. All of them were identified by liquid chromatography-tandem mass spectrometry (LC-MS/MS) and nuclear magnetic resonance (NMR) analyses. As results showed the metabolites were identified as: 8-hydroxynorfloxacin, 6-defluoro-6-hydroxynorfloxacin, desethylene norfloxacin, and *N*-acetylnorfloxacin [11]. As Lindberg et al. [12] and Santos et al. [2] reported more than 70% of fluoroquinolones present in the wastewater remained adsorbed to sludge, which may results in their releasing into the environment i.e. as biosolids, manures or wastewaters for agriculture. As many authors highlighted these antibiotic residues may affect microbial activity and structure in both soil and water environments. It is widely known that antibiotic residues in wastewater may results in occurrence of selective pressure on bacteria for development of resistance. However it is worth to note, that resistance to fluoroquinolones is multifactorial and include chromosome-mediated spontaneous mutation, production of multidrug-resistance (MDR) efflux pumps, modifying enzymes or target proteins [11, 13].

The high resistance of bacteria to fluoroquinolones results also from the multifunctionality of mechanisms responsible for this resistance, since MDR and RND (resistance-nodulation-division) pumps which are also involved in heavy metals or different agents i.e. ethidium bromide or hydrocarbons removal. Yang et al. [9] suggested that microorganisms play a crucial role in norfloxacin degradation in soil. Slightly changed nitrogen transformation and negligible effect of antibiotic on soil respiration suggested that the environmental concentration of norfloxacin may have a inconsiderable effect on microbial activity. Pan and Chu [14] suggested that the microbial transformation and degradation of norfloxain can be attributed to microbial resistance against fluoroquinolones. In this study, the bacterial consortium consist of the three examined strains from the *Bacillus* genus was able to degrade of 5.5 mg L^{-1} of norfloxacin during 26 days in cometabolic systems with glucose. (Fig. 1). The estimated reduction of norfloxacin was 75% during the experiment time. Species from the *Bacillus* genus are characterized by different industrial and biotechnological applications due to their good degradation abilities and production of various molecules, i.e. lipopeptide (LP) biological surface active compounds include surfactins, iturins, bacillomycins, fengycins and lichenysins. LP biosurfactants greatly reduce surface and interfacial tension and they are known as good emulsifiers, dispersing and foaming agents used in chemistry, food production, agriculture, cosmetics and in environmental biotechnology for soil and water treatment. One of the tested strain, *B. subtilis* I'1a synthetized iturin A C14 and C15, which antimicrobial effects on planktonic and sessile form were confirmed for *Escherichia coli, Serratia marcescens, Enterobacter cloacae, Proteus mirabilis, Citrobacter freundii* and *Enterococcus faecalis* [4]. During the time of the experiment the changes in surface tension were also measured. As results showed during norfloxacin degradation the surface tension of the bacterial culture was significantly decreased (Fig. 2).

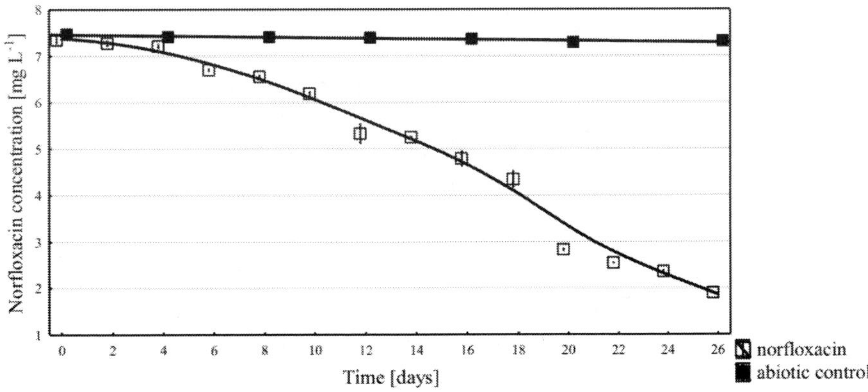

Fig. 1. Degradation of norfloxacin in the cometabolic culture by the bacterial consortium.

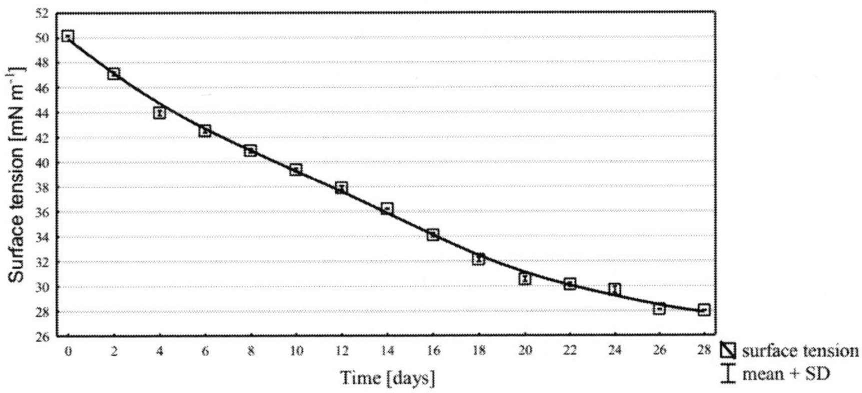

Fig. 2. Changes of the surface tension during the experiment period.

Biological surface active compounds and biosurfactants-producing microorganisms are widely used in hydrocarbon and metal-contaminated environments. Singh and Cameotra [15] used the lipopeptide biosurfactants, surfactin and fungycin, obtained from *Bacillus subtilis* A21 strain for heavy metals i.e. cadmium, cobalt, zinc, copper, lead, nickel and petroleum hydrocarbon removal from the soil samples collected from industrial landfill. As results showed soil washing technique with the use of biosurfactant solution significantly removed the petroleum hydrocarbon (64.5% of initial amount) and heavy metals (from 44.2% for cadmium to 32.07% for zinc) content. Barin et al. [16] investigated the impact of bacterial consortium consist of microorganisms able to biosurfactants produce and with increased biodegradation abilities on the improvement of biodecontamination processes. The results revealed that the consortium of *Bacillus subtilis* tb1 and *Pseudomonas aeruginosa* was able to kerosene hydrocarbons decontamination and efficiency of the removal was at least 25% higher than single species. Moreover, the biosurfactants showed potent antimicrobial activity for both Gram-positive, Gram-negative and fungi. The negative influence of surface-active compounds on *Bacillus megaterium* and

Saccharomyces cerevisiae [17], *Legionella pneumophila* [18], *Listeria monocytogenes* or *Bacillus cereus* [19], is well documented. It is worth to note, that up to now there is no scientific literature on the role of biosurfactants in the antibiotic degradation. One of the aim of this study was also to determine the biofilm formation abilities of the tested strains. Biofilm is composed of highly structured multispecies bacterial communities embedded in a self-produced polymeric matrix called extracellular polymeric substances (EPS) [20, 21]. Biofilms as a mode of the bacterial growth predominate in a different ecosystems. Microorganisms in biofilms are characterized by increased tolerance to xenobiotics and various environmental condition (e.g. chemical and shear stresses, acidification). Spatial organization, varied genes expression and different metabolic pathways exhibited by bacteria in biofilms allow to biodegradation, biotransformation or detoxification of various toxic compounds. In recent years biofilm-mediated bioremediation processes are widely used for environmental purposes. As Edwards and Kjellerup [22] noticed that indigenous bacterial communities are able to metabolize and/or degrade organic contaminants and oxidizing heavy metals, however their inadequate abundance and activity and limited access to nutrients results in insufficient rate of degradation. Thus, the introduction of bacteria with increased degradation abilities and biofilm formation capabilities seems to be reasonable. Biofilm formation was determined by two different quantitative methods, with the use of crystal violet (CV) and reduction of 2,3,5-triphenyltetrazolium chloride (TTC) (Tab. 1).

Table 1. Biofilm production by the *Bacillus* strains.

Strain	**CV**		**TTC (mg/ml)**
	Absorbance 600 nm	**Stepanović classification**	
I'-1a	2.761±0.16	strong biofilm producer	1.026±0.009
T'-1	0	no biofilm producer	0.0015±0,0008
T-1	0.972±0.09	weak biofilm producer	0,245719±0,001

Stepanović classification: nonbiofilm producer (OD less than or equal to ODc); weak biofilm producer (OD greater than ODc and less than or equal to 2xODc); moderate biofilm producer (OD greater than 2xODc and less than or equal to 4xODc) and strong biofilm producer (OD greater than 4xODc). This classification is based upon the cut-off value called ODc (ODc means average OD of negative control + (3 x standard deviations of negative control).

These two methods are different in terms of discrimination strength, crystal violet allow to detection the whole biofilm biomass include live and dead cells or acidic polysaccharides as well, while TTC staining allows to detect only metabolically active cells. On the basis of obtained results, for CV method, the examined strains were divided into the following categories using the classification of Stepanović et al. [7]. According to obtain results from CV method, the strain I'-1a was classified as strong biofilm producer, T'-1 as weak biofilm producer, while T-1 strain was unable to form the biofilm. The results from TTC staining were similar as obtained from CV methods, e.g. I'-1a and T'-1 produced 1.02 mg/ml and 0.24 mg/ml of formazan, respectively. The concentration of formazan produced by T-1 strain was estimated about 0.0015 mg/ml. Abilities of LPs produced by bacilli to affecting the bacterial adhesion properties, flagella development or to disturb the integrity of membranes and inhibition of biofilm formation is well documented [4]. The negative influence of mixture consist of fengycin and surfactin produced by *B. subtilis* and *B. licheniformis* on *E. coli* and *S. aureus* adhesion was observed by Rivardo et al. [23]. Wang et al. [24] suggested that bacteria coordinate i.e. biosurfactants and exopolysaccharide production which mediated biofilm formation by mechanism in which competition for common sugar precursors mannose-1-phosphate (Man-1-P) and glucose-1-phosphate (Glc-1-P) used for both expopolysaccharides and rhamnolipids synthesis serves as control of the bacterial motility, biofilm formation and surfactants production. Surface active agents secreted by Gram-positive bacteria often serves also as signaling molecules in quorum sensing process. Quorum sensing is a cell–cell communication mechanism that synchronizes gene expression in response to population cell density. Only when the population density is high, the accumulation of the signal in the extracellular environment is sufficient to activate the response. Bacteria which can self-produce biosurfactants possess some biological advantages related to colonization of hydrophobic surfaces and initiating of biofilm formation. Production of biosurfactants enable also rapid surface colonization which called swarming. This type of surface translocation is defined as coordinated movement across solid or semi-solid surface considered as the fastest type of bacterial motility [5]. Our results also confirmed that strain I'1a which produced the highest concentration of surfactants was classified as strong biofilm producer. Rhamnolipids produced by *Pseudomonas aeruginosa* have been shown to factor which influence of maintaining the channels between mushroom-shaped biofilm which allow the distribution of oxygen and nutrients and removal of metabolites from the biofilm structure. Boles et al. [25] demonstrated also that secretion of the right amount of rhamnolipids was critical for proper biofilm development. Solano et al. [5] highlighted that quorum sensing up-regulates the biosurfactants production. Further, the abilities of the particular strains to different enzymes production were also evaluated Obtained results showed that all of the examined strains produced inulinases, ureases, proteases, lipases and amylases. Celullases were produced by two strains, T-1 and T'-1. All of the tested enzymes are widely used for biotechnological and industrial purposes e.g. inulinases are used for ultra-high fructose syrup obtaining from inulin, bioethanol production and some chemicals production, like citric acid or lactic acid. Ureases are used for urea content analysis in urine, blood, wastewaters or heavy metals content in natural environments. Proteases, lipases and amylases have great commercial potential for detergents, pharmaceuticals, leather, dairy, food, chiral alcohols, carboxylic acids, amines or epoxides and dairy production. Microbial cellulases are of great interests in various industrial applications including pulp and paper, textile, biofuel and food industry, brewing, and agriculture[26- 28].

In conclusion, three *Bacillus subtilis* species have the specific properties, e.g. production of lipopeptide biosurfactants and various enzymes, biofilm formation, and their consortium was able to norfloxacin degrade in the cometabolic culture. The results obtained from the biodegradation experiment are promising. To our best knowledge, there are no available reports on the norfloxacin biodegradation by the strains which produced biosurfactants. Moreover, the role of biosurfactants in antibiotic degradation is still unknown.

This work resulted from the BONUS no 2112932-1 project was supported by BONUS (Art 185), funded jointly by the EU and The National Center for Research and Development (Poland).

References

1. W. Baran, E. Adamek, J. Ziemiańska, A. Sobczak. J. Hazard. Mater, **196**, 1 (2011)
2. L.V. Santos, A.M. Meireles, L.C. Lange, J. Environ. Manage.**154**, 8 (2015)
3. G.A. Płaza, M. Pacwa-Płucieniczak, Z. Piotrowska-Seget, R. Brigmon, E. Król *Environmental Sustainability*, 173 (2015)
4. P. Bernat, K. Paraszkiewicz, P. Siewiera, M. Moryl, G. Płaza, J. Chojniak, World J. Microbiol. Biotechnol, **32**, 157 (2016)

5. C. Solano, M. Echeverz, I. Lasa, Curr. Opin. Microbiol,**18**, 96 (2014)
6. C.J. Berry, S. Story, D.J. Altman, R. Upchurch, W. Whitman, D. Singleton, G. Płaza, R.L. Brigmon. Oxford University Press, Oxford, UK, pp 87 (2006)
7. S. Stepanović, D. Vuković, V. Hola, G. Di Bonaventura, S. Djukić, I. Ćiraković, F. Ruzicka. APMIS, **115**, 891 (2007)
8. H.L. Brown, A.H.Vliet, R.P. Betts, M. Reuter, J. Appl. Microbiol, **115**, 1212 (2013)
9. J. Yang, G. Ying, S. Liu, L. Zhou, J. Zhao, R. Tao, P. Peng, J. Environ. Sci. Health. Part B, **47**, 288 (2012)
10. I.A. Parshikov, J.B. Sutherland J Ind Microbiol Biotechnol, **39**,1731 (2012)
11. D.W. Kim, T.M. Heinze, B.S. Kim, L.K. Schnackenberg, K.A. Woodling, J.B. Sutherland. Appl. Environ. Microbiol, **77**, 6100 (2011)
12. R.H. Lindberg, U. Olofsson, P. Rendahl, M.I. Johansson, M. Tysklind, B.A.V. Andersson. Environ. Sci. Technol, **40**, 1042 (2006)
13. L.S. Redgrave, S.B. Sutton, M.A. Webber, L.J.V. Piddock, Trends Microbiol, **22**, 438 (2014)
14. M. Pan, L.M. Chu, Sci. Total Environ, **545**, 48 (2016)
15. A.K. Singh, S.S. Cameotra, Environ. Sci. Pollut. Res, **20**, 7367 (2013)
16. R. Barin, M. Talebi, D. Biria, M. Beheshti, Int. J. Environ. Sci. Technol, **11**, 1701 (2014)
17. M.I. Sriram, K. Kalishwaralal, V. Deepak, R. Gracerosepat, K. Srisakthi, S. Gurunathan, Colloids and Surfaces B: Biointerfaces, **85**, 174 (2011)
18. C. Loiseau, M. Schlusselhuber, R. Bigot, J. Bertaux, J.M. Berjeaud, J. Verdon. Appl. Microbiol. Biotechnol, **99**, 5083 (2015)
19. K.J. Perez, J.S. Viana, F.C. Lopes, J.Q. Pereira, D.M. Santos, J.S. Oliveira, R.V. Velho, S.M. Crispim, J.R. Nicoli, A. Brandelli, R.M Drummond Nardi, Front. Microbiol. **8**, 61 (2017)
20. T. More, J.S.S. Yadav, R.D. Tyagi, R.Y. Surampalli. J. Environ. Manage, **144**, 1 (2014)
21. A. Mitra, S. Mukhopadhyay. AIMS Bioengineering, **3**, 44-59 (2016)
22. S.K Edwards, B.V.Kjellerup, Appl.Microbiol.Biotechnol, **97**, 9909 (2013)
23. F. Rivardo, R.J. Turner, G. Allegrone, H. Ceri, M.G. Martinotti. App. Microbiol. Biotechnol, **83**, 541-553 (2009)
24. S. Wang, S. Yu, Z. Zhang, Q. Wei, L. Yan, G. Ai, H. Liu, L.Z. Ma, Appl. Environ. Microbiol, **80**, 6724 (2014)
25. B.R. Boles, M. Thoendel, P.K. Mol. Microbiol, **57**, 1210 (2005)
26. C. Neagu Bonciu, G. Bahrim, Innov. Rom. Food Biotechnol, **9**, 1 (2011)
27. R.C. Kuhad, R. Gupta, A. Singh, Enzyme Research, **2011**, 1 (2011)
28. J.L. Adrio, A.L. Demain. Biomolecules, **4**, 117-139 (2014)

The effect of thermohydrolysis on biological properties of selected chemical compounds

Monika Janas[1,*], *Alicja* Zawadzka[1]

[1]Lodz University of Technology, Faculty of Process and Environmental Engineering, Wolczanska 213, 90-924 Lodz, Poland

Abstract. One of the negative effects of the continuous development of civilization which is observed in recent decades is an increase in the number and variety of pollutants released into the environment. These substances have ecotoxic properties and pose a serious threat to humans and proper functioning of the ecosystem. Their presence in the environment forces the development of new pollution control techniques. One of the latest and very promising methods to destroy organic contaminants using the properties of water which exceeded the critical point is thermohydrolysis in supercritical water. The aim of this study was to investigate thermohydrolysis in supercritical water conditions and to check its suitability for removal of hardly biodegradable organic compounds present in wastewater. Tests for toxicity of these compounds to aquatic organisms such as sewage bacteria, crustaceans and guppy fish were also performed and the degree of their biological decomposition was determined. Biological studies were carried out on starting compounds and after the processes of thermohydrolysis. The research material included p-nitrotoluene sulfonic acid (PNTS), 4,4'-dinitrostilbene-2,2'-disulfonic acid (DNSDS) and 4,4'-diaminostilbene-2,2'-disulfonic acid (DASDS). The research leads to the conclusion that the reactions occurring during high-temperature thermohydrolysis of the three tested compounds are extremely complex. Results of biological studies indicate a significant decrease in toxicity and increase in the biodegradability of products obtained after the process of thermohydrolysis.

1. Introduction

Uncontrolled economic development proceeding for decades, in particular the intensification of industrial processes, contributed significantly to the pollution of natural environment. One of the industries that has had the greatest impact on the environment is chemical industry. This is due to the synthesis of thousands of new chemical compounds, often very toxic to the environment. Appearance of these substances in the environment forces the development of new techniques of their disposal [1–3].

In the last decade of 20th century considerable attention was paid to research and implementation of the methods which use supercritical fluids as a medium to conduct

* Corresponding author: monika.janas@dokt.p.lodz.pl

chemical processes. A particularly promising method of destroying organic pollutants is the process of thermohydrolysis in supercritical water which comprises thermal decomposition or transformation of the starting compounds. High effectiveness of this method results from the thermodynamic properties of water which change very rapidly, often by several orders, near the critical point (critical temperature T_{cr} = 374°C and critical pressure P_{cr} = 22.1 MPa) [4–6]. Viscosity and thermal conductivity of water in the near-critical region decrease rapidly to achieve a clear minimum at a pressure of 22.5—60 MPa, above which a further increase in temperature causes only a small increase in the parameters analyzed. Density and dielectric constant of water beyond the critical point decrease, and this in turn determines high solubility of a huge group of non-polar compounds in water, including gases and organic compounds. At the same time the loss of polarity causes that compounds with an ionic structure practically do not dissolve in it any longer [7–11].

Currently, the process of supercritical thermohydrolysis is not fully explained because of radical changes in the properties of water in the near-critical region. Accordingly, the aim of the study is to explore and present biological aspects of thermohydrolysis of organic compounds in supercritical water. These studies will be used to identify intermediate products formed in this process and to determine their potential use for other technologies. Only exact knowledge of the mechanism and kinetics of these reactions will allow us to control and optimize the process of thermohydrolysis in supercritical water.

2. Material and research methods

In the present work, experiments were carried out on model wastewater containing different concentrations of p-nitrotoluene sulfonic acid (PNTS), 4,4'-dinitrostilbene-2,2'-disulfonic acid (DNSDS) and 4,4'-diaminostilbene-2,2'-disulfonic acid (DASDS). The DASDS acid is a valuable intermediate used in the synthesis of stilbene dyes (replacing the carcinogenic benzidine dyes) and optical brighteners. It is a base for the production of several dyes including Brilliant Yellow, Chrysophenine G and Hessian Purple, Curcumin S, Mikado Yellow, Diamine Orange GR and others. Optical brighteners are widely used for bleaching cotton, viscose and paper, as well as to increase the whiteness of photo paper. The DASDS acid is obtained by oxidation of the PNTS acid by ambient air in a strongly alkaline medium, and then reducing the resulting nitro compound, i.e. the DNSDS acid, with the use of iron in a strongly acidic medium (Bechamp reduction). The factor that determines the quality of products and wastewater formed in the process is the efficiency of oxidation of the PNTS acid which ranges from 50 to 85%. The efficiency of reduction, which is the second step of the technological process, is approximately 90%. Because the efficiency of the two steps of DASDS production is relatively low, large amounts of raw material and intermediates get into the wastewater. These compounds are virtually non-biodegradable and toxic to aquatic organisms. The presence of these compounds in wastewater is a major problem for sewage treatment plants [12–15].

In the first stage of the study, solutions containing from 0.016 to 2.0 g/dm^3 for all compounds were subjected to thermohydrolysis. Experiments were carried out at temperature ranging from 20 to 600°C, at a pressure of 25.0 MPa and at different volumetric flow rates of the reaction mixture through the reactor. Figure 1 shows schematically the set-up for testing the process of thermohydrolysis in supercritical conditions.

 DOI: 10.1051/e3sconf/20171700034

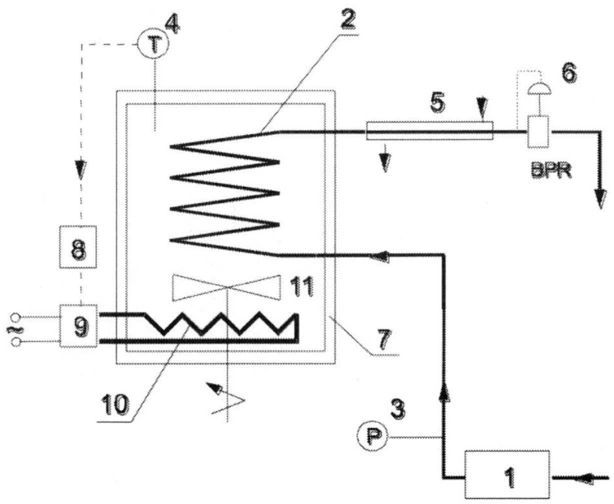

Fig. 1. Schematic of the set-up for testing thermohydrolysis in supercritical conditions.

1 – HPLC piston pump, 2 – flow reactor 3 – pressure measuring system, 4 – thermometer, 5 – cooler, 6 – back pressure regulator, 7 – electric oven, 8 – temperature regulator, 9 – transmitter, 10 – electric heaters.

Analytical methods were used in the study to follow changes in the concentrations of starting compounds and intermediate products during thermohydrolysis in supercritical conditions. Concentrations of the starting compounds and products of thermohydrolysis were determined by high performance liquid chromatography (HPLC). The analysis was performed on a Perkin Elmer liquid chromatograph with a DAD spectrophotometric detector which made it possible to sweep the spectrum from 195 to 360 nm. A standard C18 reverse-phase chromatographic column (Altech) 250 mm long and 4 mm in diameter with 5 μm particle diameter was used. Additionally, total organic carbon (TOC) (taken as a sum of all organic compounds, including intermediates, occurring in the reaction mixture) in the samples was determined by a coulometric method using Coulomat 702 Li/C (Ströhlein).

Biological tests of the compounds subjected to thermohydrolysis were performed in the final stage of the study. Toxicity tests were carried out on aquatic organisms: crustaceans – *Daphnia magna Straus* (water flea), fish – *Lebistes reticulatus Peters* (guppy fish), and three species of bacteria that are characteristic of activated sludge: *Bacillus subtilis*, *Proteus vulgaris* and *Pseudomonas fluorescens*.

Biodegradability of the tested compounds was evaluated using the test for determination of the ratio of biological oxygen demand over five days to chemical oxygen demand. The toxicity tests of the reaction mixtures obtained were compared with the toxicity tests of starting materials. The degree of biological decomposition of the reaction mixtures was investigated.

3. Results

Figures 2 to 4 show the dependence of reduction degree of the tested acid solutions on the reduction degree of TOC as a function of operating temperature of the reactor for the set flow rate of 2.5 cm^3/min. and constant initial concentration of 2 g/cm^3. With such process parameters the highest degree of pollutant reduction was achieved.

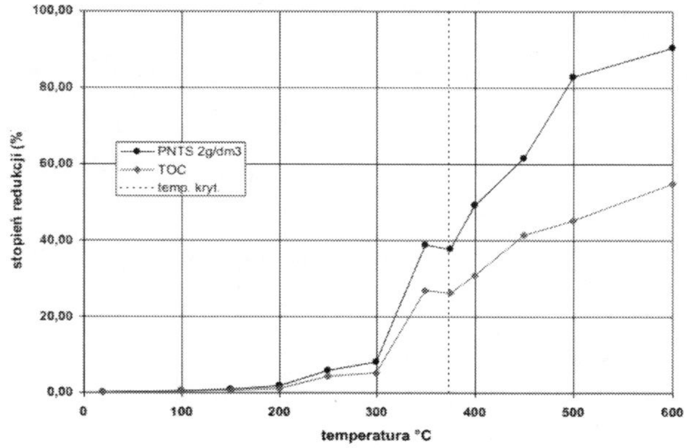

Fig. 2. The dependence of PNTS reduction and degree of TOC reduction on temperature. Volumetric flow rate of fluid through the reactor was 2.5 cm^3/min., inlet concentration was 2 g/dm^3, and pressure 25 MPa.

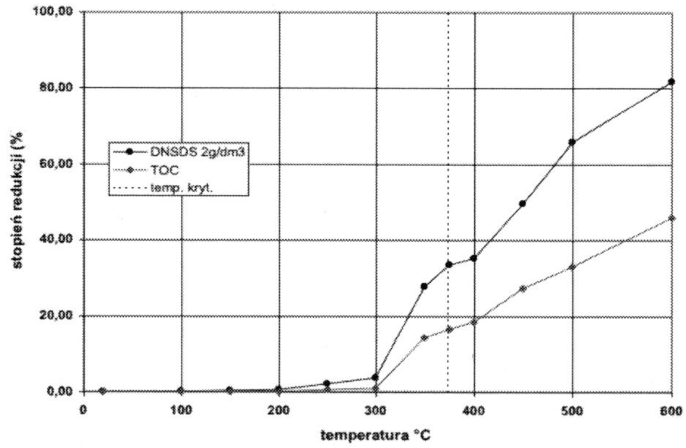

Fig. 3. The dependence of DNSDS reduction and degree of TOC reduction on temperature. Volumetric flow rate of fluid through the reactor was 2.5 cm^3/min., inlet concentration was 2 g/dm^3, and pressure 25 MPa.

The analysis of the curves which characterize the dependence of reduction degree of the compound on temperature leads to a conclusion that conversion occurs only in the near-critical region. At the initial stage, at relatively low temperatures ranging from 50 to 300°C, we can observe slight decomposition reaching about 2%. These results are similar for all tested compounds. A lower degree of TOC reduction in relation to the tested compounds can be explained by the fact that in the reaction mass due to the process of thermohydrolysis the compounds were transformed into simpler organic compounds whose structure includes carbon expressed as total organic carbon.

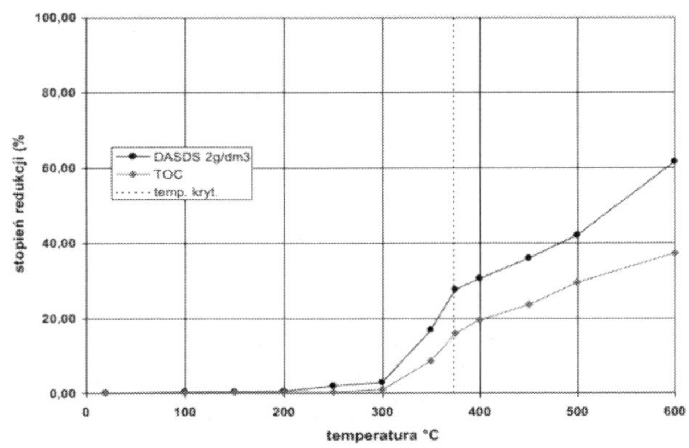

Fig. 4. The dependence of DASDS reduction and degree of TOC reduction on temperature. Volumetric flow rate of fluid through the reactor was 2.5 cm3/min, inlet concentration was 2 g/dm^3, and pressure 25 MPa.

Results of biological tests of the reactants subjected to the process of supercritical thermohydrolysis are given in Table 1.

Table 1. The results of PNTS, DNSDS and DASDS toxicity tests.

Determination	**Unit**	**PNTS**	**DNSDS**	**DASDS**
Toxicity of LC_{50} *Daphnia magna* – 96 hours	mg/dm^3	4.3	5.6	6.8
Toxicity of LC_{50} *Lebistes reticulatus* – 96 hours	mg/dm^3	68.4	154.6	178.4
Toxicity to sewage bacteria *Bacillus subtilis* *Proteus vulgaris* *Pseudomonas fluorescens*	 mg/dm^3 mg/dm^3 mg/dm^3	 5000 5000 5000	 5000 5000 5000	 5000 5000 5000
Biological decomposition	%	0	0	0

The LC_{50} indicators provide information on the compound concentration at which there is a 50% mortality of the tested species. When examining the toxicity of a compound against sewage bacteria, such concentration of the compound is taken at which no growth of bacteria after 96 h contact with the tested substance is observed, and a further lower concentration does not have a cidal effect on microorganisms.

There are big problems in carrying out toxicity tests of the biological reaction mixture after thermohydrolysis which can be fully compared with the results obtained for pure compounds. These problems stem from the fact that after the process of thermohydrolysis of the reaction mixture there is not only a single chemical compound but a solution of a mixture of several chemicals. Accordingly, it is impossible to determine the LC_{50} concentration, i.e. such a concentration at which half of the aquatic organisms survive. In the case of mixtures of chemical compounds we define the degree of dilution at which mortality of organisms is 50%.

Due to the fact that the toxicity of the tested compounds against sewage bacteria is higher than that of compounds used in the process, in a further step of the study no toxicity tests were carried out in relation to these organisms. In the first part of the study preliminary toxicity tests for the products of thermohydrolysis of PNTS, DNSDS and DASDS were performed at temperatures 200, 250 and 300°C, concentration 2 g/dm^3, pressure 25.0 MPa and flow rate through the reactor 2.5 cm^3/min in relation to daphnia. Results of the tests indicated that toxicity of these products of thermohydrolysis did not decrease in relation to this organism.

Therefore, for purposes of this study the toxicity tests were performed for three solutions of the tested compounds at a concentration of 2 g/dm^3 after thermohydrolysis at 400°C and 600°C, pressure 25.0 MPa and flow rate through the reactor 2.5 cm^3/min. Results of these tests are given in Table 2 below.

Table 2. Results of toxicity tests for PNTS, DNSDS and DASDS after thermohydrolysis.

Determination	Unit	Temp.	PNTS	DNSDS	DASDS
Toxicity of LR_{50} *Daphnia magna* – 96 hours	dilution	20°C	1+465	1+357	1+294
		400°C	1+480	1+367	1+302
		600°C	1+7	1+5	1+4
Toxicity of LR_{50} *Lebistes reticulatus* – 96 hours	dilution	20°C	1+29	1+13	1+11
		400°C	1+38	1+17	1+14
		600°C	1+2	1+1	1+1
Biological decomposition	%	20°C	0	0	0
		400°C	0	0.6	0.8
		600°C	95	91	86

Due to the inability to evaluate toxicity of the solution after thermohydrolysis, given in mg/dm^3, (because of the presence of a mixture of different reactants in different quantities) in relation to fish and daphnia, in this case the degree of dilution R_{50}, i.e. the number of dilutions of the reaction mixture, is determined to obtain a solution in which the organisms mortality reaches 50%.

To compare the two values – LC_{50} and LR_{50} for the tested solutions it was calculated that in order to obtain a solution with a concentration of 4.3 mg/dm^3 from the initial PNTS solution having a concentration of 2 g/dm^3 it must be diluted approx. 465 times. Then, the resulting solution would have toxicity of 50%. The toxicity of solution after the process of supercritical thermohydrolysis reached 50% at 8-fold dilution. Consequently, its toxicity to *Daphnia magna* at 600°C and pressure of 25.0 MPa decreased approx. 60 times. In relation to *Lebistes reticulatus* an about 15-fold decrease in PNTS toxicity is observed. Accordingly, for DNSDS a 60-fold decrease in toxicity to *Daphnia magna*, and a 7-fold decrease in the case of *Lebistes reticulatus* is observed. The toxicity of DASDS after thermohydrolysis to *Daphnia magna* decreased also about 60 times, and 5 times in relation to *Lebistes reticulatus*.

Slightly different results were observed for the tested compounds after thermo-hydrolysis at 400°C and pressure 25.0 MPa. At this temperature for all tested compounds a slight increase in toxicity to aquatic organisms in comparison with starting substances is observed. This is probably due to the maximum concentration of intermediate products observed at this temperature which are more toxic to aquatic organisms than the tested compounds. Properties of these reactants affect also the fact that we do not observe an increase in biodegradability of the reaction mixture at this temperature.

Results of the biological toxicity studies indicate that due to thermohydrolysis in supercritical conditions at a temperature of 600°C, pressure 25.0 MPa and flow rate through the reactor 2.5 cm^3/min of PNTS, DNSDS and DASDS the resulting products have altered properties in relation to aquatic biocenosis as compared to starting compounds. The products of supercritical thermohydrolysis are characterized by low toxicity to aquatic organisms and very high degree of biodegradation. A change in the toxicity of reactants subjected to thermohydrolysis results from a conversion of the tested compounds non-degradable by biological oxidation into simple compounds that are highly biodegradable. At the same time the mixture of these compounds exhibits reduced toxicity to fish and crustaceans.

4. Discussion and conclusions

In the study experiments were carried out on model wastewater containing different concentrations of p-nitrotoluenesulfonic acid (PNTS), 4,4'-dinitrostilbene-2,2'-disulfonic acid (DNSDS) and 4,4'-diaminostilbene-2,2'-disulfonic acid (DASDS). These compounds are used to produce a large group of dyes and optical brighteners. They are classified as non-degradable organic pollutants which are very difficult to remove from wastewater by conventional methods. This is due to high resistance of these compounds against biological oxidation and their relatively high toxicity to aquatic organisms.

Solutions containing from 0.016 to 2.0 g/dm^3 of all tested compounds were subjected to thermohydrolysis. Studies were carried out in the temperature range from 20 to 600°C, at a pressure of 25.0 MPa and different volumetric flow rates of the reaction mixture through the reactor.

The studies lead to the conclusion that destructive hydrolysis of the tested compounds is a very slow process. Examining the effect of certain factors on the rate of thermohydrolysis it was found that the process depended on the average residence time of the reaction mixture in the reactor, process temperature and concentration of the starting compound.

Investigation of the PNTS thermohydrolysis results in a conclusion that the maximum reduction of the compound of approx. 100% is achieved at a flow rate of the reaction mixture equal to 2.5 cm^3/min and at a temperature of 400°C for a starting concentration of 0.016 g/dm^3 PNTS (pressure 25 MPa). A higher starting concentration of this compound and a different flow rate through the reactor cause that 100% conversion of this compound is achieved at higher temperatures.

Results of the DNSDS conversion during thermohydrolysis suggest that at a flow rate of the reaction mixture equal to 2.5 cm^3/min and at a temperature of 600°C at a concentration of 0.016 g/dm^3 DNSDS (pressure 25 MPa) maximum 100% conversion of the compound is reached. Similarly to PNTS, higher starting concentrations of this compound and different flow rates through the reactor cause that 100% conversion of this compound is achieved at higher temperatures.

During thermohydrolysis of the tested compounds some amounts of intermediates are formed. This is confirmed by the value of TOC whose degree of conversion is lower than in the tested compounds. This is a result of the presence of other organic compounds in the reaction mixture. The structure of these compounds contains carbon determined as TOC.

The biological toxicity tests indicate that as a result of thermohydrolysis in supercritical conditions at a temperature of 600°C, pressure 25.0 MPa and flow rate through the reactor 2.5 cm^3/min of PNTS, DNSDS and DASDS we obtain products with changed properties in relation to aquatic biocenosis as compared to starting compounds. The products of supercritical thermohydrolysis are characterized by low toxicity to aquatic organisms and a very high degree of biodegradation. The change in toxicity of the reactants subjected to thermohydrolysis is a result of conversion of the tested compounds non-degradable by biological oxidation into simple compounds that are highly biodegradable. The mixture of these compounds exhibits reduced toxicity to fish and crustaceans.

References

1. F. Luck, Catal. Today **53**, 81-91 (1999) Wet air oxidation: past, present and future

2. L. Qian, S. Wang, D. Xu, Y. Guo, X. Tang, L. Wang, Water Res **89** (2016)
3. A. Zawadzka, G. Rogacki, Inżynieria i Aparatura Chemiczna **6** (2004)
4. P.E. Savage, Chem. Rev. **99** (1999)
5. H. Weingartner, E.U. Franck, Angew. Chem. Int. Ed. Engl. **44** (2005)
6. G. Brunner, J. Supercrit. Fluids **47** (2009)
7. H. Pińkowska, P. Wolak, A. Złocińska, Biomass Bioenergy **35**, 9 (2011)
8. Z. Shen, D. Yang, S. Wang, W. Wang, Y. Li, Int. J. Heat Fluid Flow **108** (2017)
9. B. Al-Duri, F. Alsoqyiani, I. Kings, J. Supercrit. Fluids **116** (2016)
10. K.M. Benjamin, P.E. Savage, Ind. Eng. Chem. Res. **44** (2005)
11. J. Chen, Y. Wang, S. Ding, J. Ding, M. Li, C. Zhang, M. Zou, JECE **4**, 3 (2016)
12. J. Abelleira, J. Sanchez-Oneto, J. R. Portela, E.J. Martinez de la Ossa, Fuel **111** (2013)
13. J.M. Prado, D. Lachos-Perez, T. Forster-Carneiro, M.A. Rostagno, Food Bioprod. Process. **98** (2016)
14. G. Brunner, J. Supercrit. Fluids **96** (2015)
15. A. Kruse, N. Dahmen, J. Supercrit. Fluids **96** (2015)

Concept of large scale PV-WT-PSH energy sources coupled with the national power system

Jakub Jurasz[1,*], *Jerzy* Mikulik[1], *Magdalena* Krzywda[1]

[1]AGH University of Science and Technology, 30 Mickiewicza Ave., 30-059 Kraków, Poland

Abstract. Intermittent/non-dispatchable energy sources are characterized by a significant variation of their energy yield over time. In majority of cases their role in energy systems is marginalized. However, even in Poland which is strongly dedicated to its hard and brown coal fired power plants, the wind generation in terms of installed capacity starts to play a significant role. This paper briefly introduces a concept of wind (WT) and solar (PV) powered pumped storage hydroelectricity (PSH) which seems to be a viable option for solving the problem of the variable nature of PV and WT generation. Additionally we summarize the results of our so far conducted research on the integration of variable renewable energy sources (VRES) to the energy systems and present conclusions which strictly refer to the prospects of large scale PV-WT-PSH operating as a part of the polish energy system.

1 Introduction

Accordingly to the recently published reports [1] and scientific analysis [2] the world has to urgently and rapidly transfer its energy system from the power plants fueled by finite energy resources to that based on renewable energy sources (RES) which are perceived as one of the solutions which can impede the climate change and its ominous consequences.

Historically, the RES where implemented as testing facilities for scientific purposes or were used in harsh terrain and conditions where the solutions based on conventional energy sources were not sufficient or completely not applicable. What is interesting, in the past large financial investments in the RES development were made mainly by the oil and gas industry [3, 4]. It is the oil and gas companies found photovoltaics and wind turbines as a viable energy sources in remote areas to power radio transmitters situated along the transmission line or for cathodic protection of pipes. Some even consider them as a suitable energy source for powering the autonomous pumping rigs.

On the large scale, the solar and wind energy sources firstly became popular on the islands. Especially those located far from the mainland and not on the main maritime transport routs. The main reasons for their popularity lie behind the: lack of natural energy resources on the islands which would be sufficient to cover the energy needs; usually favorable wind and/or solar resources for their development as well as necessary height

* Corresponding author: jakubkamiljurasz@gmail.com

differences to create pumped storage-hydroelectricity energy storing facilities; need to transport fuel from the mainland in from of usually heavy oil or coal to power conventional power plants what in consequence leads to a greater local pollution as well as decreases the society energy security as well as increase the energy costs.

However, the variable, intermittent, predictable to a limited extent and non-dispatchable nature of solar and wind energy sources makes their utilization an arduous task. Additionally their availability is not always (if at all) positively correlated with the energy demand [5]. But it is important to underlie that in case of Poland the changing patterns of the energy demand and their relation to the observed mean temperatures [7] which in turn are related to the irradiation values which affect determine the energy yield from PVs creates a market niche for solar powered air-conditioners [8].

An obvious solution to the problem of the intermittency of solar and wind resources is to couple PV and WT energy sources with energy storing device. But still even if energy storage will be added it is important to use both energy generating technologies because of their beneficial temporal complementarity which especially exhibits itself on an annual time scale [9,10]. The impact of using two or more complementary energy sources usually decreases the need for energy storage and makes the energy source more reliable. On a small scale such hybrids will be utilizing battery banks in from of e.g. lithium-ion batteries, whereas on a larger a pumped storage-hydroelectricity will be implemented. It is important to underlie that the PSH is the most mature and almost the only bulk energy storage technology available today. Some substitute technologies have been developed such as compressed air energy storage (CAES) [11,12] however, they not have been implemented on a large scale, two facilities are operating in Huntorf (Germany) and McIntosh (USA).

In this paper we will describe the basic concepts behind the operation of the wind and solar powered pumped storage-hydroelectricity operating as a part of the national power system. Drawn conclusions will be based on the calculations conducted in the so far conducted research available in [5,13].

2 Conceptual design of the PV-WT-PSH

A thorough analysis of the existing and proposed concepts of the PV/WT-PSH hybrids was made in [5, 13]. Therefore we will refrain ourselves from referencing those works once again and concentrate on describing the basic principles of such energy source operation and its impact on the national power system (NPS).

As already mentioned PV-WT-PSH hybrids consists of two non-dispatchable energy sources in form of photovoltaics which energy yield depends respectively on the current irradiation/irradiance and wind turbines which blades are moved by the flowing air masses and the propel the generator which in turns generate electricity. To some extent in case of PVs an assumption can be made that the energy yield is linear to the value of momentarily available irradiation, whereas in cases of WTs the energy output is strongly dependent on the wind speed to the power of three – naturally those characteristics are a significant simplification. Important aspects which should be also underlined include: the temperature impact of the PV cells efficiency and the fact that each wind turbine is represented by its specific power curve. In general the WT turbine power curve can be described by three characteristic wind speeds. Namely: $v^{cut\text{-}in}$ – is a wind speed at which the WT starts to generate electricity, v^{rated} – is a wind speed at which WT starts to generate electricity with its rated capacity, $v^{cut\text{-}off}$ – is a wind speed at which the WT must be shot down for safety reasons. Commercially available large scale wind turbines manufactured by Vestas, Gamesa or Enercon usually start to generate electricity when the wind speed is greater than 4 m/s and reach their rated power output for winds exceeding 12 m/s but no faster than

25 m/s. Those two energy sources cannot be controlled and their energy/power output can be predicted with a certain margin for error and for a limited time horizon [14, 15].

The second part of the considered hybrid energy source is the pumped storage-hydroelectricity. It is a relatively old concept. The first PSH went into the operation in 1909 at Schaffhausen in Switzerland and is still in operation [16]. Each PSH facility consists of two reservoirs. It is important that those reservoirs are close to each other but there remains a significant elevation difference between their water levels. This elevation difference is crucial because the potential (storing capacity and power output) of the PSH are mainly determined based on the basic equation determining the potential energy $E = mgh$, where: E – is the potential energy [J], m – is the mass [kg], g – is the gravitational acceleration [m/s^2], h – is the mentioned height difference [m]. The brief analysis of the above reminded equations indicates that the potential of given site for the PSH facility development will be to the same extent determined by the available water storing capacity of the upper and lower reservoir as well as the elevation difference between reservoirs. It is important to stress that both conditions must be met to given site be further considered for the PSH development. Naturally the subsequent analysis of the site predisposition for the PSH project will consider other criteria such as environmental impact, closeness to the transmission lines, amount and complexity of soil works etc.

Initially the PSH facilities were erected in order to compensate for the intra-day variability of energy demand and supply. It is clear that during night the energy demand is relatively lower than over daylight. Therefore the supply side must adjust to the occurring demand (please bear in mind that the concepts of the demand side management (DSM) are relatively novel and gained on popularity with the massive advent on VRES). So basically it is the conventional power plants that have to adjust their power output to meet the demand. But unfortunately, the conventional power plants are designed to operate within given range of parameters e.g. power output for which their efficiency is maximal and simultaneously the emissions are minimal. What is more sudden ramps up and down of the conventional power plants in terms of their power output does not positively impact their life span, reliability and economics.

The proposed almost two centuries ago solution is similar to that common on the other goods market. Meaning one has store the occurring energy surpluses and release them when the demand will occur. Naturally this solution is associated with additional questions like what is the optimal potential of the storage facility, how much energy can be freed per unit of time, what is the efficiency of the whole process, are there any loses if the storing period will elongate? We will address those questions in following paragraphs.

How this solution will work in case of the PSH facility? First of all the energy demand is smaller over the night. Some conventional power plants can curtail their power output but it is not possible in case of all of them. Therefore on the energy market occurs an energy surplus, meaning that the supply is greater than the demand and in consequence the energy price falls. Here comes the opportunity for the energy storing facilities in form of the PSH. They buy and then use that surplus energy to power the water pumps and force the flow of water from the lower reservoir to the upper one. Naturally some losses occur due to the non-perfect efficiency of the pumps. The volume of energy which can be pumped to the upper reservoir is determined by the pumps throughput and the reservoir available capacity. The night is over and a new day starts. People turn on lights, electric kettles and prepare for work. Offices and other businesses start their normal activities. The energy demand starts to pick up. It is followed by conventional power plants adjusting their power output. But they not always are able to keep up the pace. The situation on the energy market changes and now the energy demand is greater than the supply. In consequence the energy prices goes up. Here comes the opportunity for the PSH facilities which are characterized by the relatively fast reaction time. The during night pumped water to the upper reservoir can be now released back to the lower reservoir and its potential energy can be converted in water turbines into the electricity. The decisions whether the water from the upper reservoir should be discharged must be based on the difference between the cost at which it was pumped and the potential profit from discharging it and selling – especially in case when such facility operates on the energy market and should bring revenue.

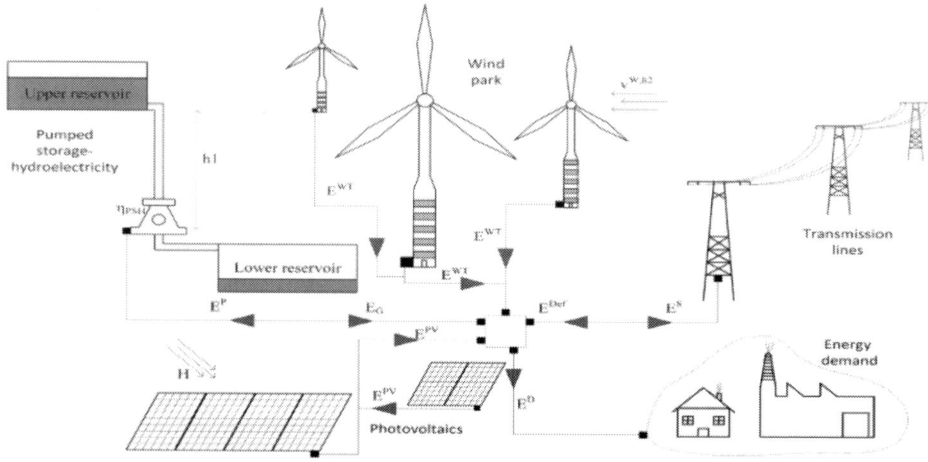

Fig. 1. Conceptual design of a grid connected PV-WT-PSH hybrid energy source.

Figure 1 visualizes the general concept of wind and solar powered PSH. As can be seen it is coupled to the energy demand source (here represented by a single house and a factory) and to the transmission lines. In the considered case the PSH consist of a single water turbine which can generate electricity or pump water from the lower to the upper reservoir (store electricity in a form of the potential energy). The water turbine is represented by its efficiency (η^{PSH}) which usually varies from 80 to 90% and its maximal power capacity. The capacity of the upper and lower reservoir is given in thousands of cubic meters but can be easily changed into energy units by considering the elevation difference between reservoirs (h1). In case of the non-dispatchable energy sources we have denoted to characteristic parameters which in the simulation models were used as hourly time series, namely: irradiation (H) and wind speed (v^{h2}) which is observed at the wind turbine rotor height (h2).

On the above presented figure, red dots indicate the possible energy flow directions. Naturally, from the energy sources (E^{WT} & E^{PV}) the energy flows only in one directions – they do not exhibit any self-consumption. An opposite situation can be observed in case of the energy consumers – they only exhibit the energy demand (E^{D}) – in reality some of the energy consumers may be also utilizing their own energy sources such as PV and WT and be so called *prosumers*, however for the simplicity of calculations we put aside this possibility. The energy flow in both directions is possible in case of the PSH and the transmission lines. In the first case we have a volume of energy which was used to pump the water to the upper reservoir (E^{P}) and the volume of energy generated (E^{G}) by discharging the water from the upper to the lower reservoir. In the best case scenario to PV, WT and the PSH should be able to cover the entire energy demand without generating energy surpluses. An energy surplus in case of such system which is connected to the transmission lines is understood as an energy which was generated from PVs and WTs and after using some part of it to cover the energy demand, the remaining part could not be stored in the upper reservoir, due to its limited capacity or the curtailed throughput of the water turbine. Naturally, an opposite situation may occur when then available energy from PVs and WTs and that stored in the upper reservoir will not be sufficient to cover the energy demand. In such circumstances there will be a need to cover this energy deficit from the grid (E^{Def}).

Energy deficits and surpluses have a random nature. It means that there will be a unpredictable energy flow between the PV-WT-PSH and the transmission lines (the energy system). Such situation is not desirable by the power system operators. They will have to adjust the power output of the conventional power plants with regard to the changing energy flow and direction. Therefore the PV-WT-PSH hybrid energy source should be sized in such a manner that the total sum of energy exchange with the power system is minimal. The best situation will be when the PV-WT-PSH would be able to guarantee the total energy self-sufficiency. This is theoretically possible but such system must be able to deal with extreme situations, like:

- Over prolonged time period the wind turbines and the photovoltaic installations are operating with their almost maximal power capacities. In consequence there occur a significant energy excess which must be stored in the upper reservoir. However, the energy storing capacity of the PSH is strongly limited and after several dozen hours or days there is no more possible to store there the energy from WT and PV which has not been used to cover the energy demand;
- A contrary case is when the energy from WT and PV is not sufficient to cover the energy demand. In consequence the volume of energy stored in the PSH is slowly declining. If the unfavorable weather conditions (low wind speed and irradiation) will prevail for next several hours or days the energy stored in the PSH will soon run out and the energy deficit will occur.

Even if such situation happen very rarely or on a lesser scale it will be still mandatory to oversize the whole system. By oversizing one should understood adding additional capacity installed in PVs and WTs. In general the energy generated by those both sources should be greater than the energy demand augmented by the losses occurring during the energy storing processes (not perfect efficiency of the water turbines). However, some additional capacity seems to be mandatory due to the long term variability of the mean annual wind speeds and general tendencies which indicate that the cumulative energy demands picks up. To sum up, the energy storage facility should be able to overcome the problem of the variable nature of the wind and solar energy sources. It is important to underlie that this problem may be overcome by coupling several such hybrid energy sources. It is however required that they must be relatively far apart if one wants to makes use of the spatial complementarity phenomena exhibited by some of the VRES. Spatial complementarity works similar to the temporal complementarity and it can be applied to the energy sources using the same energy resource e.g. wind parks located in two different areas. The impact of spatial complementarity is usually known as smoothing the energy generation curve by spatial distribution of variable energy sources [14,16]. The above described problem of the stochastic nature of the volumes and directions of the energy exchange with the grid has been addressed in [5] and the presented there conclusions indicate that to some extent it can be solved by applying adequate forecasts. Meaning that it will be hard to eliminate those energy flows by at least it is possible to predict them. Naturally some error will occur but they depend on the quality of the forecasting models.

3 PV-WT-PSH operation

We have already mentioned that the PV-WT-PSH hybrid energy source may aim at ensuring the energy autarky. This is especially a case in the isolated energy systems like those observed on the islands. However, here we consider them as a part of the larger energy system with all its positive and negative consequences. On the Fig. 2 we have visualized the operation on such system in situation when it should cover the whole energy demand. This is just an exemplar day which might have occurred. Please note the purple and squared area which indicate the unscheduled energy exchange with the grid. On the

Fig. 2–4 we have purposefully omitted the pumping operation of the PSH.

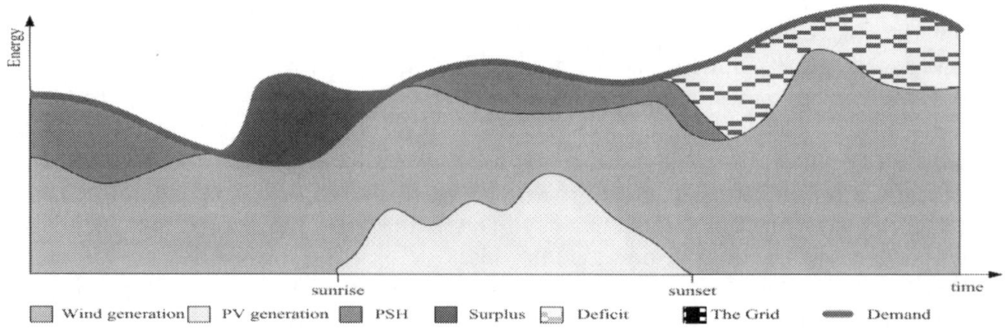

Fig. 2. The theoretical PV-WT-PSH operation assuming that some load is covered from the grid.

Because ensuring the energy self-sufficiency whilst minimizing the occurrence of the energy surpluses is a complicated and in some cases almost an impossible task we suggest two additional scenarios. On the Fig. 3 and 4 one can observe that some part (what is important a fixed value) of the energy demand is being covered from the grid (black squared area). The basic assumption which lies behind the scenario presented on the Fig. 3 is such that some part of the demand will be covered from the grid when there is no possibility to generate electricity from the PV installation. In general organizing hybrid energy system operation according to the theoretical energy yield from the PVs is a simpler task. This results from the fact that it is possible to exactly determine when they should generate energy. This is done based on the clear-sky models. Therefore, for the each day of the year one can establish precise hours when some volume of the energy should be delivered from the grid/energy system to cover the energy needs of the customers using the PV-WT-PSH hybrid. As shown in [13] such solution should not only ease the sizing procedure of the mentioned hybrid energy source but also minimize the volume of the unscheduled energy exchanges with the energy system. What is more, the values of the scheduled energy exchange (on the Fig. 3–4: The Grid) can vary for different parts of the year accordingly to the changing availability of the wind energy.

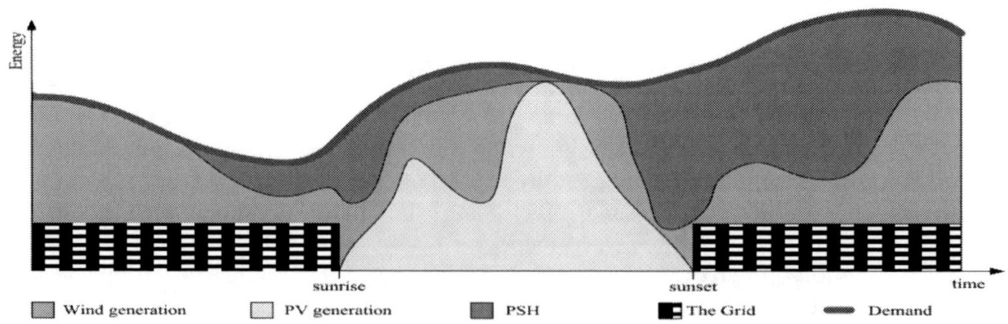

Fig. 3. The theoretical PV-WT-PSH operation assuming that some load is covered from the grid.

If the solution presented on Fig. 3 is not sufficient it is always possible to establish a fixed volume of the energy demand which should be covered from the energy system (see Fig. 4). This is also a beneficial situation because the power system will be dealing only with the so called base load (an energy demand which does not change and is observed over the whole year). In consequence the PV-WT-PSH will have to cover relatively smaller demand. The main benefit of such approach is such that the unscheduled energy exchanges can be significantly reduced and the from the perspective of the national power system operator the energy demand of the selected group of energy consumers is seen as a fixed value. Therefore, it is easy to schedule the operation of the conventional power plants. What is more such load can be covered from power plants which generate the electricity with the lowest costs.

Naturally, many other scenarios are possible. For example: the fixed value of energy delivered from the grid can vary for each hour of the day; the system operator would see it as very promising when the PV-WT-PSH will be used to perform so called *peak shaving* – what boils down to covering the energy demand it tends to exhibit the highest values (morning and evening energy demand peak).

With the advent of the variable renewable energy sources in the power systems researchers are searching for various approaches and solutions which can ease the process of their integration. Also from the perspective of the mentioned PV-WT-PSH hybrid a very promising tool is the so called demand side management (DSM) which may be able to slightly reduce the occurrence of energy surpluses and deficits [18]. An important aspect is also in the form of the privately owned energy sources like: small scale PV installations and wind turbines coupled with the battery banks. Such relatively small energy sources may also enable smoother and more beneficial for the power system operation of the PV-WT-PSH hybrid.

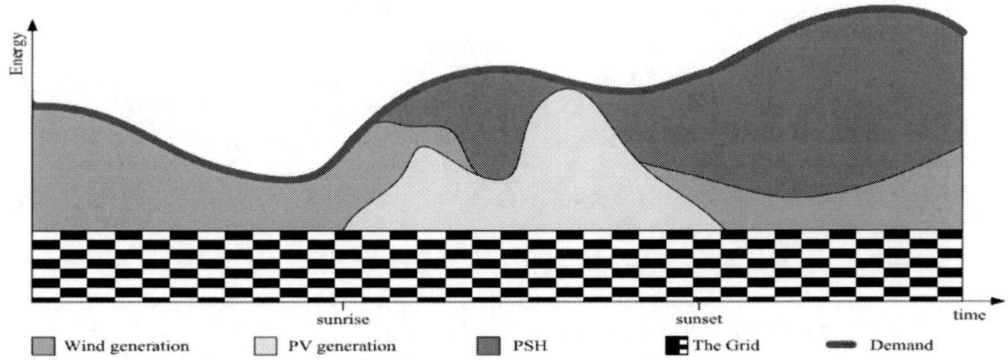

Fig. 4. The theoretical PV-WT-PSH operation assuming that some load is covered from the grid.

An open question remains about the possibility of creating such hybrid energy sources in the existing energy systems. As it has been already mentioned the PSH facilities were necessary to overcome the problem of the changing energy demand. In case of the PV-WT-PSH they would be also applied to compensate for the varying generation of non-dispatchable solar and wind energy sources.

The polish energy system has five operating PSH stations. The sixth which was commissioned in 1997 is currently operating as run-off-river power plant with pondage. One of the PSHs located in Żarnowiec (Northern Poland) was designed to cooperate with the first polish nuclear power plant but following governments delayed this project and its perspectives are murky. The biggest problem when it comes to the PSH facilities is finding a suitable location which will ensure sufficient height difference between reservoirs. Additional criteria which are usually used eliminate majority of the potential sites. The conducted in [13] literature review indicated that the cumulative capacity of the not-realized PSH projects in Poland amounts to 5.3 GW.

Summarizing above described potential of the PSH plants in Poland it is important to underlie that those projects are relatively old. It seems that due to the changing environmental regulations some of them could not be realized today. However, it is clear that Poland has significant potential for developing PSH facilities on a large scale. The described in this paper concept can be applied for much smaller PSH stations which combined with the PVs and WTs will be used to cover the energy needs of smaller communes: like villages or small towns.

4 Summary

The PSH stations coupled with the wind and solar energy sources are a common concepts known in the isolated energy systems located on islands. In this short post-conference publication we have briefly introduced this concept on the background on our recent works in area of the integration of variable renewable energy sources to the power system. Presented overview aims at: familiarizing the potential leader with those concepts; summarizing the general results of our works and to initiate the debate on the possibility of finding market niches for such solution in the existing and the future energy systems where RES will undoubtedly play a significant role.

References

1. S. A. Crate, & M. Nuttall, (Eds.). *Anthropology and climate change: from encounters to actions* (Routledge, 2016)
2. J. Hansen, P. Kharecha, M. Sato, F. Ackerman, D. Beerling, P Hearty. J., E. J. Rohling. PLoS One, 8(12), e81648 (2013)
3. J. Wang, J. O'Donnell, A. R. Brandt, Energy **118** (2017)
4. G. S. M. Teale, Journal of petroleum technology **36**, 05 (1984)
5. J. Jurasz. Energy Conversion and Management **136** (2017)
6. J. Jurasz, & J Mikulik. International Journal of Photoenergy (2016)
7. J. Jurasz, J. Mikulik, & A. Piasecki, *Przegląd Elektrotechniczny* **92** (2016)
8. J. Jurasz, M. Krzywda, & J. Mikulik. In *E3S Web of Conferences* (Vol. 10, p. 00059). EDP Sciences (2016)
9. J. Jurasz, A. Piasecki, & M, Wdowikowski. In *E3S Web of Conferences* (Vol. 10, p. 00032). EDP Sciences (2016)
10. J. Jurasz, & A. Piasecki. Acta Energetica **2,** 27 (2016)
11. F. Crotogino, K. U. Mohmeyer, R. Scharf, *Huntorf CAES: More than 20 years of successful operation* (Orlando, Florida, USA, 2001)
12. J. Taylor, & A Halnes. Analysis of compressed air energy storage. In *PCIC Europe 2010 conference record* (pp. 1–5). IEEE (2010)
13. J., Jurasz. *Optimization of the installed capacity in the solar-wind-pumped storage hydroelectricity power source.* (Unpublished PhD dissertation). AGH University, Cracow, Poland (2016)
14. J. Kleissl. *Solar energy forecasting and resource assessment*. Academic Press (2013)
15. M. Lei, L. Shiyan, J. Chuanwen, L. Hongling, & Z. Yan. *Renewable and Sustainable Energy Reviews* **13**, 4 (2009)
16. T. H. Douglas, (Ed.). *Pumped Storage: Proceedings of the Conference Organized by the Institution of Civil Engineers at Imperial College of Science, Technology and Medicine, London on 2-4 April 1990.* Thomas Telford (1990)
17. J. Jurasz, J. Mikulik, Rynek Instalacyjny **11** (2015)
18. M. R. Staats, , P. D. M. De Boer-Meulman, W. G. J. H. M. van Sark. *Sustainable Energy, Grids and Networks* (2016)

A concept development of an early warning system for toxic sewage detection

Anna Jurga[1,*], *Natalia* Gemza[2], and *Kamil* Janiak[1,2]

[1]Faculty of Environmental Engineering, Wroclaw University of Technology, Poland
[2] Wroclaw Municipal Water and Sewage Company, Poland

Abstract. The article presents a concept of an EWS (early warning system) for toxic sewage detection upstream from Wroclaw Wastewater Treatment Plant. It includes design and construction of three toxicity warning stations, situated in municipal sewage system. Early detection of toxic contaminants provides time for corrective actions. Wastewater toxicity level is based on activated sludge OUR (Oxygen Uptake Rate) measurement. The database of normal level of activated sludge OUR is collected during the measurement initial phase. Hence, if toxic event occurs, device will measure decreased OUR value and WWTP operators will be alarmed. Additionally, flow measurement is carried out. It allows to estimate the volume of a toxic stream. Development of the early warning system requires solving minimum six different scientific and technical issues and is a subject for a complex research project.

1 Introduction

Activated sludge poisoning by toxic constituents can occur in any wastewater treatment plant and can have catastrophic consequences for the treatment processes [1, 2] Toxic inflow can cause collapse of nitrification process and lead to significantly exceeded total nitrogen concentration in the effluent. Most of the times the inflow of toxic wastewater is noticeable after causing process failure and when it is too late for any corrective actions. Wastewater treatment plant can be equipped with an early warning system. It includes a few toxicity measurement points placed at selected locations of the sewage collection system. The toxic wastewater inflow is detected early enough for WWTP operators to react and introduce the appropriate set of corrective actions. Measurement points locations divide sewage system into main sectors. As a result, it is possible to search for the perpetrator of a toxic discharge effectively. Under normal circumstances, the extent of the sewerage network and a large number of sewage connections is practically impossible to locate toxic discharge.

The construction of an efficient system is a big challenge. Scientists all over the world make individual attempts to build such a system [3, 4]. There are also a number of commercially available devices for determining sewage toxicity. The first scientific challenge is to choose appropriate toxicity assessment method. Subsequent challenges

* Corresponding author: anna.jurga@pwr.edu.pl

include localization and calibration of the system in such way that its action would not give false positive and negative results. Aim of this paper is to design conceptual framework of the early warning system for Wroclaw Wastewater Treatment Plant for inhibition event prevention.

2 The concept of the system

2.1 General description of the EWS

Main components of the system are measurement units located in key points of sewer network. In order to achieve the most satisfying results, their location should provide sufficient time for reaction and minimize number of measurement units. Each measurement unit is composed of toxicity measurement apparatus and flow meter. Data from measuring points is continuously sent to PC where analysis is carried out according to an algorithm described in point 2.3. In case of toxicity detection, information is transferred to wastewater treatment control room as well as to sewer control room, since suggested corrective actions may include both wastewater treatment process and pumping station management.

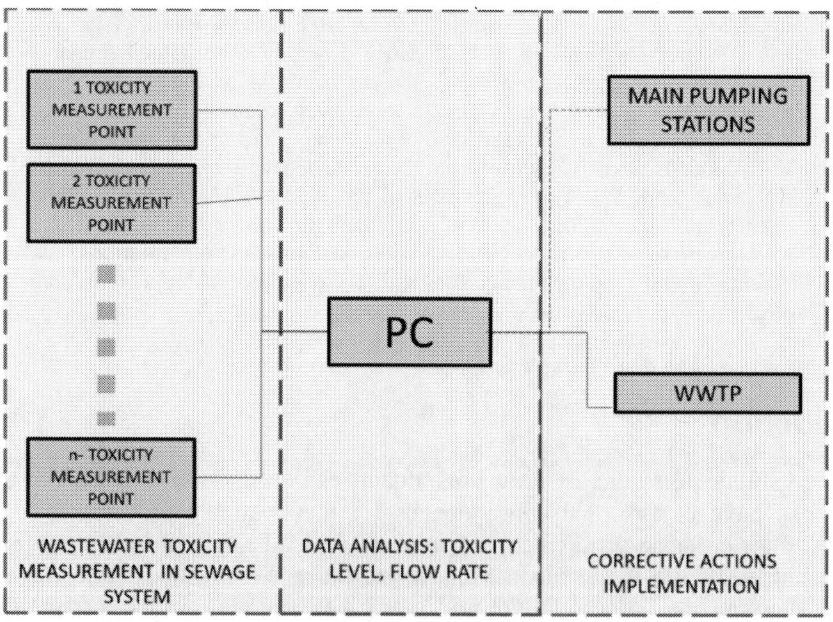

Fig. 1. Block diagram of toxicity warning system.

2.2 Location of measurement units

Figure 2. presents proposed locations of measurement units in Wrocław sewer system. In preliminary concept three measurement units are proposed. First one is located in the central part of Wroclaw, near Port Polnoc pumping station. This unit will monitor wastewater from the northern part of city. The second one is also located in the central part of city, near Port Poludnie pumping station. This station will monitor wastewater from the southern part of the city. There is *ca.* 13 km between those units and wastewater treatment plant. The third unit will be located at "Sleza" sewer outlet. This unit will monitor wastewater from the remaining part of city. There is circa 6,5 km from this unit to wastewater treatment plant.

With average wastewater velocity of 0,6 m/s, time for reaction will amount between 2,5 and 5,5 hours (0,5 hour is required for measurement). As this time is relatively short, the system will be expanded in the future with new units in further segments of sewer. More measurement units will allow to divide the sewer system into sectors which will increase chances for finding toxic wastewater sources.

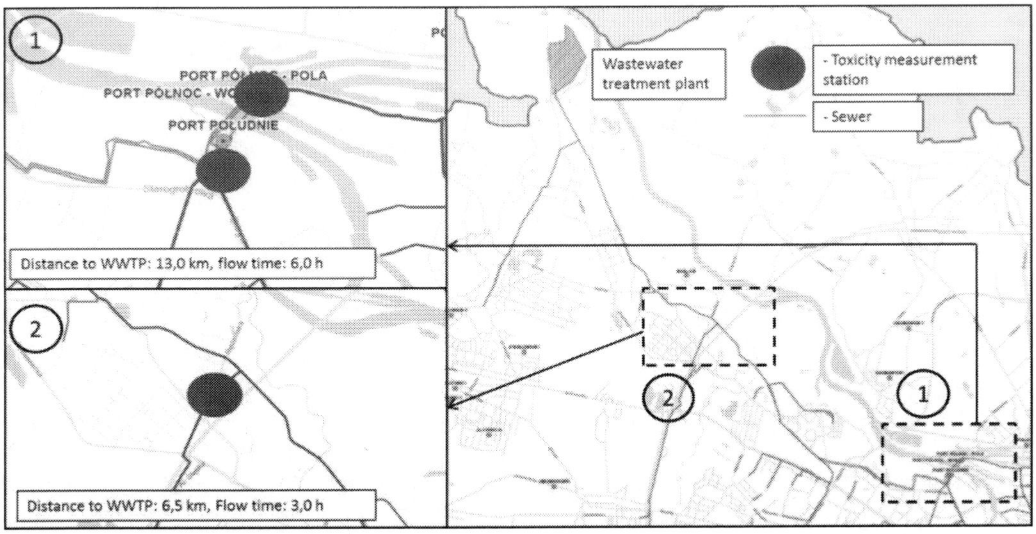

Fig. 2. Location of toxicity measurement points.

2.3 Measurement method

Toxicity of wastewater can be measured with many different methods [2, 3, 5–10]. Not every method can be used in toxicity warning system as they do not comply with its requirements which are: short time of response, easy handling, reliability and on-line data registration. In this concept of the system, a respirometric method has been chosen. Principle of this method is easy to understand, respirometric test can be conducted without need of very expensive and complicated equipment. It is also possible to on-line register measurement results. As nitrification inhibition will likely be the first result of toxic wastewater inflow to wastewater treatment plant, direct measurement of oxygen consumption by nitrifiers is a proper way to asses wastewater toxicity [11].

Respirometric measurement is conducted in respirometer. This device is composed of reactor filled with activated sludge, oxygen probe and stirrer. Measurement is conducted in controlled conditions with pH and temperature kept at given value. In order to carry respirometric measurement, sludge is aerated to oxygen saturation concentration. Afterwards aeration is switched off and sludge starts to use dissolved oxygen. Oxygen concentration changes are registered and converted to oxygen uptake rate. Healthy sludge fed with non-toxic wastewater will have high oxygen uptake rate, while sludge fed with toxic wastewater will have much lower oxygen uptake rate. The difference between those values can be converted into wastewater toxicity degree. Algorithm of an operation is described in point 2.3. Exemplary oxygen curve obtained during respirometric test is presented in Fig. 3.

 DOI: 10.1051/e3sconf/20171700036

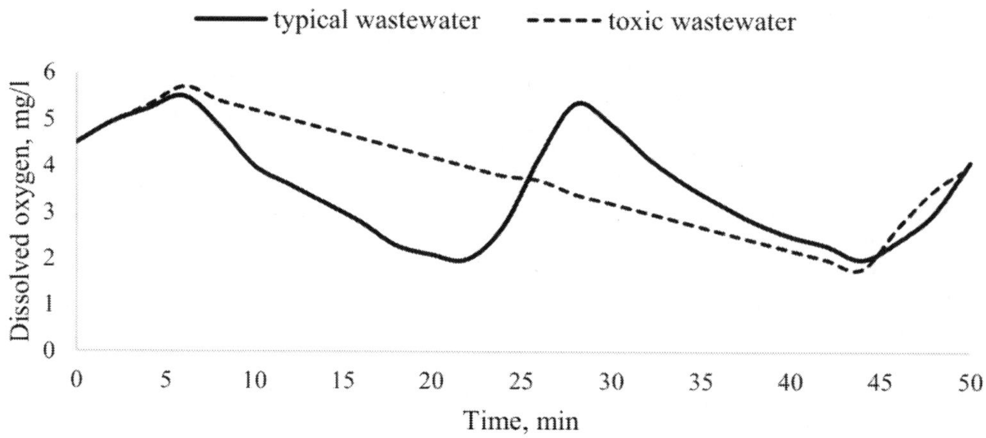

Fig. 3. Oxygen curve during respirometric test.

2.4 Algorithm of operation

Algorithm of operation is shown in Fig. 4. Every measurement unit conducts two subsequent respirometric tests every 30 minutes. Reference measurement (1) is used to check whether condition of sludge is not changing in time. In this measurement the sludge is fed with pure substrates such as ammonium salts. If oxygen uptake rate obtained with pure substrate is different than average value of previous measurements, correction coefficient need to be calculated (2) because condition of sludge has changed. For example: if reference value is 20% lower than the average one, this difference must be included in the analysis of the results obtained from test with wastewater. Without correction, false positive or false negative results may occur as change of sludge condition will be undetected. After reference measurement, one with sewage is conducted (3). Sewage dose is properly selected to ensure that sludge loading in respirometric reactor is the same as in wastewater treatment plant. Oxygen uptake rate from test with sewage is subsequently modified according to results from reference measurement and compared with typical value of oxygen uptake rate of sludge fed with non-toxic wastewater (4) As measurement points are placed on sewer transporting only fraction of WWTP influent volume it is necessary to calculate whether overall stream will be toxic in case of mixing toxic wastewater with non-toxic (5). In case when toxicity of overall stream is expected to be toxic, the red alert is announced. If toxicity of overall stream is not expected, the yellow alert is announced. Red alert should be followed with reaction of wastewater treatment plant crew while the purpose of yellow one is to increase attention.

DOI: 10.1051/e3sconf/20171700036

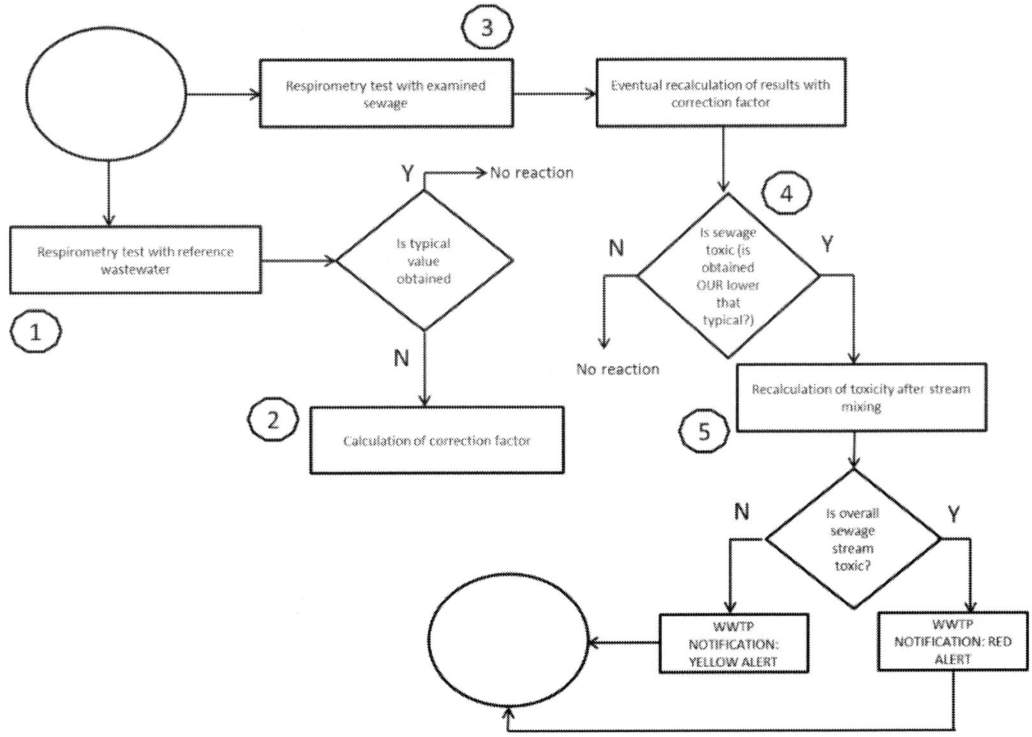

Fig. 4. Algorithm of operation.

3 Preventive actions after toxicity detection

Toxic sewage warning is sent to WWTP operators in advance due to right location of measuring points. It gives plant operators time necessary to prepare facility and introduce one of a few proposed preventive actions to minimize consequences of toxic stream entering biological chamber. Applicability of each of proposed preventive actions on a particular WWTP is limited by the specific plant configuration *i.e.* flow equalization basin or other wastewater storage tank, design and number of treatment lines and receiving body characteristic. Preventive actions accommodated to Wroclaw WWTP setup are as follows:

- lack of any additional flow equalizer ahead from inlet works
- primary settling tanks preceding biological chambers
- 5 parallel treatment lines
- effluent discharge into the river.

Preventive actions listed in the table below might be combined and implemented in a variety of configurations.

Table 1. List of the corrective actions.

No.	Corrective action	Description
1	Increasing/Decreasing main pumping station flow rate	Toxin concentration can be reduced by increasing sewage flow rate. In case of measuring points located upstream from pumping station reducing flow rate allows to dilute toxic compound with inflowing wastewater. Toxic stream identified downstream from pumping station requires corresponding action – increase in sewage pumped volume.
2	Chemically enhanced primary treatment	Chemical precipitation of heavy metal cations or sorption of soluble toxic compounds onto settling primary sludge flocs.
3	Hydraulically isolating one/a few treatment lines	A part of plant nitrifying biomass is protected from toxins casing inhibition which helps with faster subsequent process start-up.
4	Aeration shutting down/ reduction, shutting down agitators	Causing sludge settling helps with preventing toxins from penetrating easily though all sludge flocs. Mixing and/or aeration is recovered as soon as toxic stream leaves treatment line.

4 Scientific and technical challenges

4.1 Main research goals

In order to achieve full efficiency of the proposed system, an extensive research is needed. Main research goals are as follows:

a. determination of the number of OUR measurements required for reliable average value estimation,
b. determination of sample collection interval,
c. selection of substrates for reference measurement,
d. selection of reactor's parameters (SRT, pH, DO, T) in order to ensure that sludge in reactor is identically susceptible to toxic wastewater as activated sludge in wastewater treatment plant,
e. determination of typical oxygen uptake rate of sludge fed with non-toxic wastewater and pure substrates,
f. determination of correction factor of partial stream toxicity to WWTP influent toxicity.

4.2 How many repetitions of oxygen uptake rate measurement are necessary

This issue refers to measurements with pure substrate as well as with wastewater. Number of subsequent measurements of oxygen uptake rate is important due to confidence interval of obtained value. If confidence interval is large (small number of repetitions) one cannot be certain about true value of OUR and thus detection threshold is small. From this point of view more repetitions provide more reliable results. As each measurement takes *ca.* 10 minutes, with higher number of repetitions interval between samples gets too long and the system loses its ability of toxic wastewater fast detection. If the required number of repetitions is higher than 5–6 it will be necessary to shorten single measurement by changing aeration procedure. This can be achieved by alteration from aeration with air to artificial aeration with hydrogen peroxide as aeration is the most time consuming process during oxygen uptake rate measurement. This modification may change sludge behavior and cannot be implemented without investigation. It is highly possible that at least 3 repetitions of each measurement will be needed.

4.3 Sample collection interval

Minimal interval between samples depends on time of measurement and the number of repetitions of measurement. With 10 minutes for each measurement and 3 repetitions with pure substrate and 3 with wastewater, the minimal interval cannot be shorter than 60 minutes It is worth concerning whether interval can be longer without loss of ability for fast detection. Longer interval is beneficial as it lowers maintenance costs.

4.4 Substrates in reference measurement

It should be decided whether ammonium salts, organic compounds or mixture of both should be substrates in reference measurement. Nitrification is most sensitive process and its failure will be probably first symptom of sludge poisoning. From this point of view ammonium salts seems to be the most appropriate substrate for reference measurement as certainty that nitrification is stable will be most important for system detection threshold. Other substrates cannot be ruled out with this simple analysis, therefore this issue will be addressed.

4.5 Selection of parameters of the reactor in the monitoring station

This is probably the most significant research task. Ensuring the reliability of toxicity test at the monitoring station will require the calibration of operation parameters of reactor. Biomass sensitivity to toxic substances should coincide with the sensitivity of the sludge from WWTP. Otherwise, the system may respond with false results. As a consequence, the operator of WWTP may be exposed to adverse effects, e.g. financial. This task will require exposition of the biomass (from the reactor) and the activated sludge (from treatment plant) to the toxic contaminants. The answer will be examined and adjusted if necessary. Perhaps there is going to be a need for regular seeding reactors sludge from wastewater treatment, to ensure the validity of the test.

4.6 „Typical" values of the oxygen uptake rate

Attempts will be taken to precisely determine the typical oxygen uptake rate in the absence of toxic compounds in the wastewater. It should be taken into account that it is possible that factors other than the presence of toxic compound can cause drop in OUR value The adopted value of deviation from the typical parameters will influence the sensitivity of the system, as well as its ability to detect toxic waste. If the system is too sensitive it will cause a large number of false alarms. However, low sensitivity may result in oversight of toxic event.

A good calibration of the system will allow to obtain the most reliable response.

4.7 Estimation of the partial wastewater flow toxicity on the total stream

Another issue to be explored is the possibility of determining toxicity level downstream from measurement point – with a focus on WWTP inflow. A possible scenario is the dilution of the toxic substance on the way to the treatment plant. Attempts shall be made to assess whether it is possible to unambiguously estimate the toxicity after combining sewage streams. Next, examination of the factors on which it is dependent should be carried out. This constitutes another challenge which can significantly affect the reliability of the system.

5 Related works

The scientific community is aware of the problem, which is unforeseen discharge of toxic substances into the sewerage system [1, 2, 4, 9]. However, most of the scientific work focuses on the detection of inhibition only at the waste water treatment plant [2, 5, 7, 8]. As a result, processes for wastewater treatment are not protected and corrective actions cannot be undertaken. The concept of the EWS presented in this paper is convergent with the attitude presented by G.Black *et al.* [3]. Available analyses and research provide valuable information required for development and design of the device [1–5, 7–10]. Despite understanding the essence of the problem, EWS for toxic upsets prevention is not commonly used in sewage network. This constitutes the need of development and implementation of such technology with easy to use and reliable OUR measurements. Further work and development of the subject are advised.

References

1. K. Jönsson, C. Grunditz, G. Dalhammar, JLC. Jansen, Water Research **34** (2000)
2. P. Norman, P. Whalen, *A new method for upset prediction: ATP-based biomonitoring and statistical process analysis at two refinery wastewater plants*, Proceedings of the Water Environment Federation, WEFTEC 2009: Session 31 through Session 40, pp. 2506–2525(20) (2009)
3. G. Black, *Pollution prevention in wastewater networks: Development of biological early warning device*, (Cranfield University, 2016)
4. A. Shaw, N. Love, E. Roehl, *Feasibility testing of support systems to prevent upsets – WERF Final Report*, United States of America, IWA Publishing, ISBN: 1-84339-795-1 (2009)
5. W. Bourgeois, J. Burgess, R. Stuetz, Journal of Chemical Technology and Biotechnology **76** (2001)
6. M. Głodniok, K. Korczak, D. Zdebik, Prace Naukowe GIG. Górnictwo i Środowisko, **1** (2010)
7. J. Gorska-Surmacz, K. Gernaey, C. Demuynck, P. Vanrolleghem, W. Vestraete, Water Research **30** (1996)
8. S. Hong, I. Choi, BJ. Lim, H. Kim, **12** (2012)
9. A. Shaw, A. Fairey, J. Earle, N. Love, E. Roehl, R. Daamen, G. Olsson, J. Watts, *Online Instrumentation to Protect Wastewater Facilities from Toxic Chemicals*, Proceedings of the Water Environment Federation, WEFTEC 2006: Session 51 through Session 60, pp. 4553–4562(10) (2006)
10. Y. Xiao, C. De Araujo, CC. Sze, DC. Stuckey, Singapore, Journal of Hazardous Materials **286** (2015)
11. J. Alleman, K. Preston, *The biology of nitrifying bacteria, Second annual workshop on commercial aquaculture using water recirculating systems*, (Purdue University, 1991)

The analysis of mercury content in coals and ashes

Marta Kasprzyk[1], *Anna* Szpak[1,*], *Iwona* Jelonek[1], and *Magdalena* Wiktor[1]

[1]The University of Silesia, Faculty of Earth Sciences, 41-200 Sosnowiec, Będzińska 60 str. , Poland

Abstract. It is well known that mercury that enters the atmosphere is a serious threat to the natural environment. From the available literature, it can be stated that fossil fuels are among the main carriers of this fuel. The question arises whether it is the right approach. While the enrichment of bituminous coal produces limited amounts of mercury, they are being released into the atmosphere as a result of the combustion of coal. The situation is similar in the case of biomass, which releases large amounts of mercury into the atmosphere when subjected to the combustion process. While bituminous coal is the carrier of mercury, the application of currently known technology, rational production and the use of fossil fuels make it possible to maximally minimize mercury emissions into the atmosphere. This study evaluated the mercury content in the ashes resulting from the combustion of bituminous coal subjected to the enrichment process. The samples were collected from ten mines located in the Polish part of the Upper Silesian Coal Basin (USCB). In total, thirty coal samples and thirty products of the combustion process were examined. All sixty samples were subjected to physicochemical analysis, including: moisture content, ash content, sulfur content, carbon content, and Gross Calorific Value.

1 Introduction

The combustion of coal in both the power sector and private households is related with the emission of mercury compounds into the atmosphere. The reduction of the mentioned emissions is affected by the type of fuel burned, the quality, the type of installation, and the efficiency of flue gas purification systems.

According to the report prepared for the Convention of the United Nations (UN) made by the National Centre for Emissions Balancing and Management (KOBiZE), the Institute of Environmental Protection- National Research Institute, the value of mercury's emission in 2013 increased by 0.18% compared to 2012 [1]. Estimation of emissions of each air pollutant (Fig.1) was made based on the structure of emission sources included in the Guidance "EEA / EMEP Emission Inventory Guidebook" [2] and the Selected Nomenclature for reporting of Air Pollutants (SNAP). The SNAP 01 sector is among the largest emitters of mercury into the atmosphere [1, 2] meaning that the largest amounts of

* Corresponding author: anna.szpak@mixbox.pl

mercury are emitted as a result of the combustion processes carried out in the production and energy transformation sectors (Fig. 1).

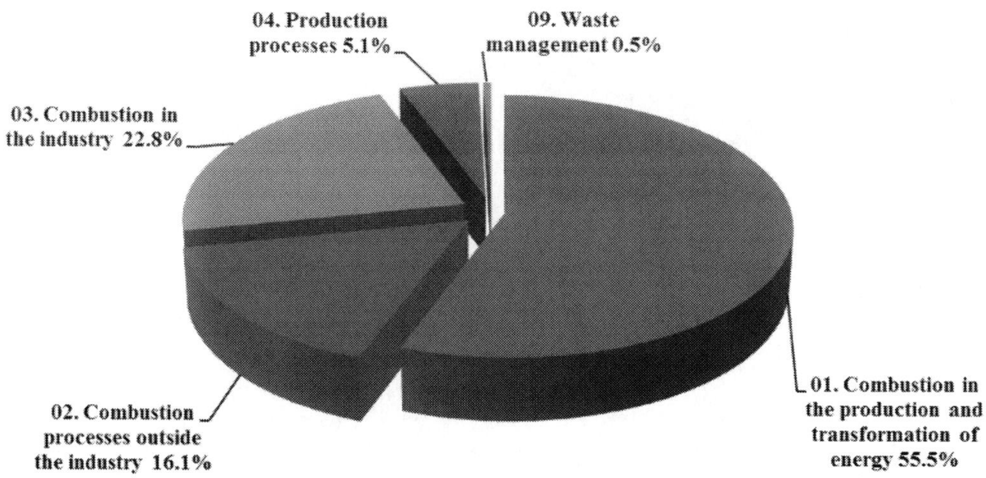

Fig. 1. The share of the largest sectors in mercury emissions in the year 2013 [1].

According to recent studies [3, 8], the average mercury content in coals from the Upper Silesian Coal Basin is 0.094 μg/g, with a minimum value of 0.020 μg/g and a maximum value of 0.235 μg/g. The authors of the mentioned studies pointed out the lack of association between the location of the sample collection and the mercury content [3].

Other studies reported a strong correlation between the sulphur content and the mercury contained in bituminous coal samples [3-8]. Such information is crucial for the preliminary assessment of the quality of coal and its suitability for a given technology.

2 Samples and analytical procedures

Bituminous coal samples from 10 mines were analyzed. The commercial samples of coal (i.e. subjected to the enrichment process in order to be used in power industry) were collected from different mines:

- Jastrzębska Spółka Węglowa S.A. (7 mines: Borynia, Jas-Mos, Zofiówka, Budryk, Knurów, Szczygłowice, Pniówek),
- Tauron Wydobycie S.A. (2 mines: Janina, Sobieski);
- Kompania Węglowa S.A. (8 mines: Piast, Ziemowit, Bolesław-Śmiały, Brzeszcze, Chwałowice, Jankowice, Marcel, Rydułtowy-Anna).

A total of 30 samples of bituminous coal, including 7 samples with low sulphur content, 10 with high sulphur content, and 13 mixed with coal slurries, were collected. After their combustion under controlled conditions in a muffle furnace in the laboratory at the Faculty of Earth Sciences of the University of Silesia the ash samples were collected. Coal and ash samples were subjected to a physico-chemical analysis. Samples for further examination were collected and prepared according to the current standards [9].

Each sample was divided into two equal parts, one of which was used to measure moisture content, ash content, sulphur content, carbon content, and Gross Calorific Value, and the other of which was used for the mercury content measurement. The samples of ash were also divided into two equal parts and subjected to ignition loss, sulphur dioxide, and mercury content measurements. All physico-chemical analyses were carried out according to applicable standards [9–15].

The measurement of total mercury content for all samples of coal and ash was performed using a Milestone DMA-80 direct mercury analyzer. Calibration of the analyzer was based on patterns provided by the manufacturer (NIST SRM). The analyzes were performed at the Laboratory of Water Analysis, Department of Earth Sciences, the University of Silesia.

3 Results and discussion

The results of mercury content and physico-chemical analysis of bituminous coal and ashes after the combustion process are presented in Table 1.

Table 1. Physico-chemical analysis, the mercury content in bituminous coal and ashes.

COAL								ASH				
Sample No.	Mar [%]	Mad [%]	Aar [%]	Star [%]	Car [%]	Qnet,ar [kJ/kg]	Hg [µg/kg]	Sample No.	Ignition loss [%]	St [%]	Ct [%]	Hg [µg/kg]
1	22.2	7.5	10.2	1.27	52.4	19 536	776.7	31	0.8	0.41	0.7	168.3
2	19.8	8.4	13.3	1.37	52.3	19 463	742.1	32	4.4	0.29	4.1	109.6
3	19.5	6.4	11.8	1.21	55.1	20 541	756.6	33	2.1	0.27	1.9	175.3
4	16.5	5.5	21.9	1.10	48.7	18 278	457.3	34	3.1	0.20	1.7	110.4
5	16.4	7.4	18.8	1.10	52.8	19 246	431.9	35	4.2	2.18	3.4	398.3
6	19.9	7.1	13.1	1.19	52.6	19 587	532.7	36	1.6	0.25	1.3	193.5
7	18.6	9.0	15.4	1.12	52.0	19 370	546.3	37	2.5	0.24	2.0	209.6
8	17.6	7.0	18.1	1.10	50.8	19 012	469.0	38	4.0	2.15	3.3	301.3
9	20.0	11.3	12.1	1.07	52.8	19 451	365.9	39	2.1	0.29	2.0	107.8
10	19.6	10.6	13.6	1.03	52.2	19 443	325.7	40	1.4	0.41	0.8	108.7
11	21.0	11.4	13.1	1.03	53.1	19 249	313.8	41	2.4	0.23	2.1	98.6
12	18.5	5.0	21.2	1.02	47.9	17 747	301.6	42	4.0	2.60	3.4	197.7
13	16.7	4.7	22.8	1.02	48.2	17 796	299.3	43	3.8	2.64	2.8	179.8
14	20.4	11.5	10.8	1.00	53.7	19 775	287.4	44	1.5	0.26	1.5	89.7
15	17.8	6.4	22.5	1.00	47.0	17 409	257.3	45	3.9	2.15	3.6	156.9
16	17.5	5.8	22.11	1.01	47.6	17 883	315.3	46	3.8	2.59	3.3	261.3
17	18.0	5.7	16.16	0.99	53.0	19 812	288.5	47	5.2	2.67	4.7	212.7
18	17.5	6.1	23.06	1.00	47.0	17 428	268.0	48	3.1	3.64	2.6	147.9
19	16.6	5.2	21.61	0.95	49.1	18 240	189.5	49	3.1	2.15	2.9	102.9
20	18.1	7.0	18.89	0.97	50.4	18 555	178.5	50	3.2	2.03	3.0	100.8
21	17.7	5.8	20.83	0.90	49.3	18 071	200.3	51	3.8	2.44	3.6	96.9
22	19.2	3.8	19.52	0.88	48.5	17 986	109.0	52	3.5	2.29	3.0	61.9
23	17.4	5.9	10.34	0.84	59.9	22 069	98.5	53	9.8	0.20	8.7	12.9
24	14.4	4.6	23.73	0.78	50.9	18 767	93.2	54	1.2	0.06	1.2	1.0
25	11.1	4.6	24.11	0.71	52.5	19 764	24.5	55	0.3	0.01	0.3	1.1
26	11.3	3.6	26.32	0.71	50.7	18 927	28.3	56	1.9	0.10	1.3	1.2

27	9.8	2.0	22.11	0.69	56.2	21 307	32.1	57	3.3	0.18	2.8	2.7
28	8.9	2.2	32.45	0.65	47.7	17 809	20.1	58	3.5	0.10	3.3	1.9
29	14.0	2.7	19.37	0.52	55.9	21 003	12.5	59	1.2	0.05	1.1	4.9
30	8.9	1.6	15.30	0.46	66.2	24 918	9.5	60	3.5	0.08	3.5	2.2

Mar – Moisture as received, Mad – Analytical moisture, Aar – Ash content as received, St ar – Total sulphur as received, Car – Carbon as received, Qnet,ar – Net Calorific Value as received, Hg – Mercury content, St – Total sulphur content, Ct – Total carbon content, Hg – Mercury content.

3.1 The results of coal analyses

The maximum sulphur content in 10 coal samples (sample No. 1-10) was in the range from 1.0% to 1.4%, while the maximum mercury content ranged from 325.7 mg/kg to 776.7 mg/kg. In the case of ash resulting from the combustion of coal, the mercury content was in the range from 107.8 μg /kg to 398.3 μg/kg. When it comes to coal samples (Samples No. 11-13) with the sulfur content from 1.0% to 0.8%, the mercury content ranges between 315.3 μg/kg and 98.5 μg/kg. For the samples which the lowest sulphur content (samples No. 24-30), the mercury content amounted to 9.5-93.2%. The average mercury content in all 30 samples is 291.0 μg/kg.

On the basis of physico-chemical analysis it has been shown that the tested coal is characterized by the following parameters: Total moisture as received is in the range from 8.9% to 2.2%, Ash content as received ranges from 10.2% to 32.6%, Total sulphur content is 0.5% to 1.44%, Carbon as received is 47% to 66.2%, and the Net Calorific Value as received is ranging between 17 409 kJ/kg to 24 918 kJ/kg.

3.2 The results of ash analyses

In the case of ash produced after the combustion of coal, the mercury content in these samples was highly variable. The highest concentrations of mercury, ranging from 398.34 μg/kg to 107.78 μg/kg were observed in samples of ash produced after the combustion of coal with the highest sulphur and mercury content (samples No. 31–40). In the case of ash samples resulting from the combustion of coal with the average sulphur and mercury content (sample No. 41–53), the mercury content ranges between 261.3 μg/kg and 12.9 μg/kg. When it comes to ash produced after the burning of coal with the lowest sulphur and mercury content (samples No. 54–60), the mercury content in all analyzed samples is 120.6 μg/kg.

The total sulfur content of ash is the range from 0.01% to 3.64%, while total carbon content and the ignition loss are the range of 0.3% to 8.7% and 0.34% to 9.8%, respectively.

3.3 Discussion of the results of statistical analyses

The correlation between all the parameters presented in Table 1 and the correlation of mercury content in coal and ashes was performed. The standard deviation R^2 value, which is a measure of the correlation strength between the variables, is quite strong in the case of coal parameters and amounts to 0.70 (total moisture as received vs. the sulfur content), 0.55 (total moisture as received vs. ash content), and 0.54 (total moisture as received vs. mercury content). A strong correlation between the total sulphur content and the total mercury content, with the standard deviation R^2 at 0.92, has also been observed (Fig. 2).

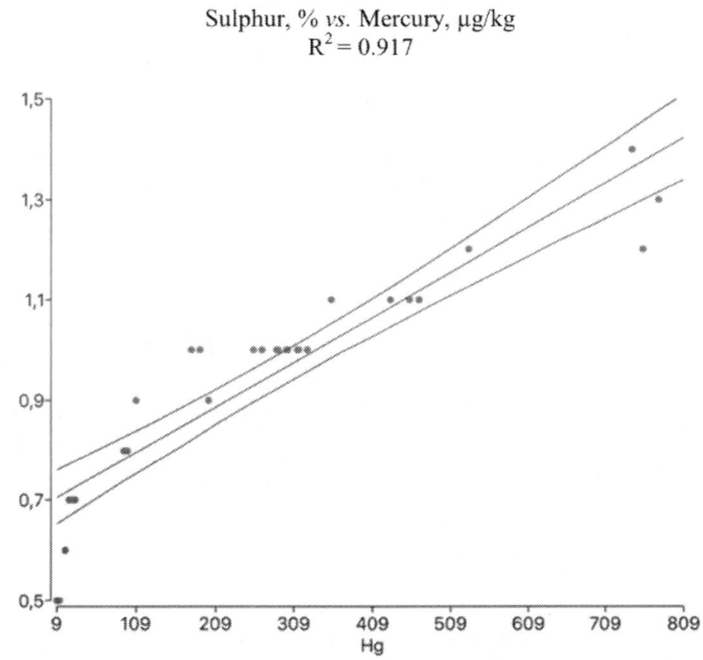

Fig. 2 The correlation between the total sulphur content and the mercury content in bituminous coal samples.

The analyzed correlation between the mercury content in coal and ash is significant, with the standard deviation R^2 value amounting to 0.56 (Fig. 3).

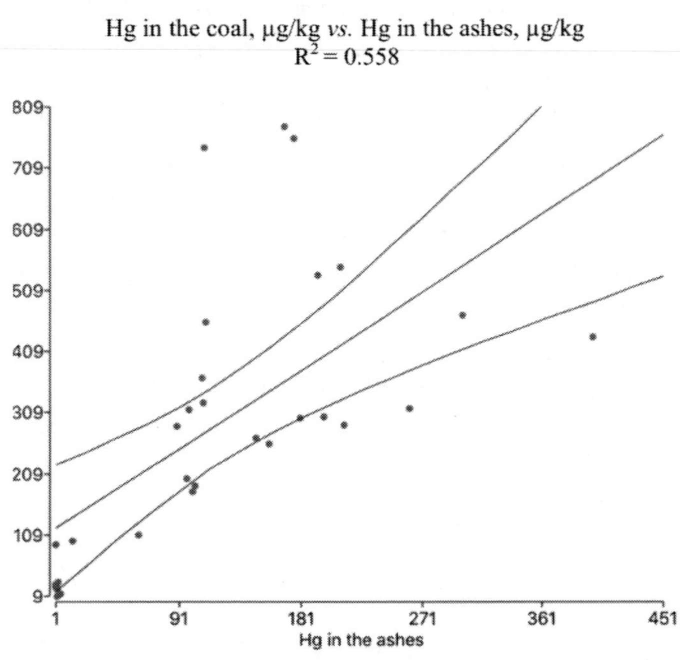

Fig. 3 The correlation between the mercury content present in the examined coal and in the ashes resulting from the combustion of coal.

In contrary, the relationship between the ash and mercury content in the analyzed coal and the mercury content in ash samples is much less pronounced. The coefficient of correlation R^2 is 0.32 and 0.14, respectively, while the analysis of variance has shown that the relationship between the variables is weak.

4 Conclusion

Based on the results of mercury determination in bituminous coal and furnace wastes, the following conclusions can be drawn:

1. The mercury content in the ash increases with increasing content of total sulfur in coal (Fig. 2).
2. The ash content of the examined coal is not related to the mercury content of coal and ash.
3. The mercury content observed in the coal decreases proportionally to the amount of mercury in ash (Tab. 1).
4. The mercury content in the fuel has a probably impact on mercury emissions and depends on the type of installation used in power plants or individual households.
5. The results have shown a high variability in the mercury content of bituminous coal, which is why a continuous control of this parameter is required.

References

1. B. Dębski i in., *Krajowy bilans emisji SO_2, NOx, CO, NH_3, NMLZO, pyłów, metali ciężkich i TZO w układzie klasyfikacji SNAP i NFR. Raport podstawowy*, **42** (2015)
2. *EMEP/EEA air pollutant emission inventory guidebook* 2009 (2013)
3. K. Bojarska, Z. Bzowski, Wiadomości górnicze, 657–662 (2015)
4. K. Wierzchowski, I. Pyka, Górnictwo i Geoinżynieria, 293-302 (2010)
5. B. Toole-O`Neil, S.J. Tewalt, R.B. Finkelman, D.J. Akers, FUEL, 47–54 (1999)
6. M. Wichliński, R. Kobyłecki, Z. Bis, Archives of Environmental Protection, 141–150 (2013)
7. K. Bojarska, MEC3 Third International Expert'sworkshop, 5–7 (2006)
8. T. Dziok, A. Strugała, A. Rozwadowski, M. Macherzyński, FUEL, 206–213 (2015)
9. PN-G-04502:2014-11 *Hard and brown coals – Sampling and preparation of samples for the laboratory tests – Primary methods*
10. PN-ISO 589:2006P *Hard coal - Determination of total moisture*
11. PN-ISO 1171:2002P *Solid mineral fuels - Determination of ash content*
12. PN-ISO 1928:2002P *Solid mineral fuels - Determination of gross calorific value by the bomb calorimetric method, and calculation of net calorific value*
13. PKN-ISO/TS 12902:2007P *Solid mineral fuels – Determination of total carbon, hydrogen and nitrogen – Instrumental methods*
14. PN-ISO 351:1999P *Solid mineral fuels - Determination of total sulfur – High temperature combustion method*
15. PN-EN 15935:2013-02 *Sludge, treated biowaste, soil and waste – Determination of loss on ignition*

DOI: 10.1051/e3sconf/20171700038

The impact of the time series resolution on the reliability of the maximum precipitation models

Bartosz Kaźmierczak[1,*], *Katarzyna* Wartalska[1], and *Marcin* Wdowikowski[2]

[1]Wrocław University of Science and Technology, Faculty of Environmental Engineering, Wybrzeże Wyspiańskiego 27, 50-370 Wrocław, Poland
[2]Institute of Meteorology and Water Management – National Research Institute, Podleśna Street 61, 01-673 Warsaw, Poland

Abstract. In the paper, two maximum precipitation models for Legnica were developed. For this purpose archival pluviographic records from the time span 1961–2010 were used. The first model was developed on the basis of rainfall data of durations ranging from 5 minutes to 6 days. In the second model, rainfall data with durations ranging from 10 min to 6 days were used, and precipitation amounts for 5 minutes duration were extrapolated. Generalized exponential distribution was used to develop the models. Both models were compared with measurement data using relative residual mean square error.

1 Introduction

The basic form of quantitative rainfall description are models of dependence: intensity I (in mm/min), unit intensity q (in dm^3/s·ha) or the height h (in mm) of precipitation on its duration t and the probability of exceedance p [1].

The relationship between precipitation intensity (unit intensity or height) and its duration is most often presented in terms of intensity-duration-frequency (IDF) or depth-duration-frequency (DDF) curves, for different probabilities p of precipitation occurrence. These curves represent the hyperbolic family given by the general equation [1, 2]:

$$I = \frac{a}{(t+b)^n} + c \tag{1}$$

in which: a, b, c and n are the empirical coefficients, dependent on the occurrence probability of a given rainfall and on the climatic and physiographic factors of the catchment. A series of homogeneous observation for decades is required for their determination [1].

Authoritative for designing of areas drainage systems are both short-term precipitation with a high-intensity and long-lasting precipitation with a significant territorial scope and high efficiency. Recommended by the European standard PN-EN 752:2008 modeling of existing or newly designed sewerage systems encounters in Poland a barrier of a lack of

* Corresponding author: bartosz.kazmierczak@pwr.edu.pl

access by designers to appropriate and reliable rainfall databases. At the entrance to hydrodynamic models are most commonly required storm hyetographs with a 5 minutes time resolution, such as Euler's type II model precipitation. The basis for the development of model precipitation are the models of maximum rainfall in the form of IDF or DDF curves [3–10].

In Poland, access to the precipitation source data is a matter for the Institute of Meteorology and Water Management (IMWM), the owner of the largest number of meteorological stations in the country.

By 2010, precipitation in the IMWM were recorded on paper pluviographs, which involved a very time-consuming development of measurement data. However, due to the number of the stations and the length of the measurement series (often decades) paper pluviographic strips are an extremely valuable research material on precipitation in Poland. With appropriate commitment, information about rainfall height can be read from the paper strips with a resolution time of 5 minutes. Based on these form of analogue but continous records of rainfall events, many models of maximum precipitation have been developed and currently applied in engineering practice [1, 11, 12].

Since 2007, the IMWM began to record precipitation by digital rain gauges, initially in parallel with traditional devices. Time series are currently recorded with a 10-minute time resolution. The assumptive resolution from the point of view of urban hydrology is a matter of concern because in the sewage systems designing and modeling it is needed information about shorter precipitation, lasting 5 minutes [2].

The aim of the study is to evaluate the impact of time series resolution on the reliability of the maximum precipitation models. The analysis, were conductedusing two maximum precipitation models based on archival pluviographs from Legnica from the time span 1961–2010. The first model was based on rainfall data of durations ranging from 5 minutes to 6 days. In the second model, rainfall data with durations ranging from 10 minutes to 6 days were used, and the amounts of precipitation with the duration of 5 minutes were extrapolated. Both models were compared with measurement data.

2 Materials and methods

Archived pluviographs from meteorological station of IMWM in Legnica from the time span 1961–2010 were used as research data. Measuring station in Legnica, as part of a national measurement and observation network at hydrological and meteorological service, is a synoptic station which is participating in the international weather monitoring program (Weather World Watch) as part of the World Meteorological Organization (WMO), of which Poland is a member. The station building is located on the south-eastern outskirts of the city of Legnica, at an elevation of 122 m above the sea level. The predominant land use in both the municipality and rural area around the station are fields and wasteland [12].

In order to determine the relationship between the amount of rainfall from duration and probability of exceedance $h(t,p)$, there must be done a selection of data on which the relationship will be developed. Elaborating archival pluviographs authors limited period of analysis to months from May to October (V–X) [1, 11, 12].

For the purpose of this paper, based on the Peak-Over-Threshold (POT) method [13–19] there were isolated from the tested 50-years period top 50 maximum amounts (h, mm) of rainfall for each of the following rainfall durations: t = 5, 10, 20, 30, 40, 50, 60, 90, 120, 180, 360, 720, 1080, 1440, 2160, 2880, 4320, 5760, 7200 and 8640 minutes.

In the first place, the top 50 amounts of rainfall were ordered decreasing. Then there were successively assigned to it the empirical probability of exceedance according to (1) from p = 0.020 (for the highest value) to p = 0.980 (for the lowest value) [1, 11]:

$$p(m,N)=\frac{m}{N+1} \tag{2}$$

where m is the sequence number within a decreasing ordered string of the number of N. The amount of rainfall recorded for selected values of empirical probability are shown in Table 1.

Table 1. The amount of rainfall (h, mm) recorded for selected values of empirical probability (m = 1, 5, 10, 25 and 50).

t, min	*p* 0.020	0.098	0.196	0.490	0.980	*t*, min	*p* 0.020	0.098	0.196	0.490	0.980
5	16.2	11.2	9.8	7.6	4.6	360	57.7	47.2	40.9	30.6	23.9
10	23.5	17.5	14.1	10.9	8.3	720	74.9	51.8	44.6	35.8	28.6
20	31.7	25.1	20.1	14.5	11.3	1080	77.3	57.7	51.6	41.3	32.4
30	32.0	26.9	24.5	16.0	12.5	1440	77.3	66.3	57.7	46.4	35.7
40	37.4	28.2	25.2	17.6	13.0	2160	114.8	77.3	61.1	49.4	38.7
50	39.8	31.0	25.2	18.2	13.7	2880	129.3	97.8	74.3	53.0	41.3
60	40.6	31.6	25.9	19.9	14.2	4320	143.1	97.8	77.6	57.4	41.3
90	41.0	36.2	29.4	21.6	16.5	5760	157.0	116.0	83.9	61.9	47.1
120	49.6	39.6	32.2	22.8	18.2	7200	158.5	121.1	86.0	69.9	51.7
180	57.6	40.7	34.5	26.3	20.0	8640	167.9	132.3	91.9	71.4	53.7

To describe the measurement data generalized exponential distribution (GED) was used. Likelihood function of this distribution describes an equation:

$$\ln L(\alpha,\lambda,\gamma)=n\ln\alpha+n\ln\lambda-\sum_{i=1}^{n}\lambda(x_i-\gamma)+(\alpha-1)\sum_{i=1}^{n}\ln\left(1-e^{-(x_i-\gamma)\lambda}\right) \tag{3}$$

where α, λ and μ are parameters of the distribution. Parameters can be determined by the maximization of the likelihood function or by solving the system of normal equations [20–24]:

$$\begin{cases}\dfrac{\partial L}{\partial\alpha}=\dfrac{n}{\alpha}+\sum\limits_{i=1}^{n}\ln\left(1-e^{-(x_i-\gamma)\lambda}\right)=0\\ \dfrac{\partial L}{\partial\lambda}=\dfrac{n}{\lambda}-\sum\limits_{i=1}^{n}(x_i-\gamma)+(\alpha-1)\sum\limits_{i=1}^{n}\dfrac{(x_i-\gamma)e^{-(x_i-\gamma)\lambda}}{1-e^{-(x_i-\gamma)\lambda}}=0\\ \dfrac{\partial L}{\partial\gamma}=n\lambda-\lambda(\alpha-1)\sum\limits_{i=1}^{n}\dfrac{e^{-(x_i-\gamma)\lambda}}{1-e^{-(x_i-\gamma)\lambda}}=0\end{cases} \tag{4}$$

Quantiles of a random variable for the GED described by the equation

$$h=\gamma-\frac{1}{\lambda}\ln\left(1-(1-p)^{\frac{1}{\alpha}}\right) \tag{5}$$

The coincidence of theoretical distributions with measured data was examined using the Anderson-Darling test for statistics [25, 26]:

$$A^2=-n-\frac{1}{n}\sum_{i=1}^{n}(2i-1)\left[\ln F(X_i)+\ln(1-F(X_{n-i+1}))\right] \tag{6}$$

where: x_i – i-th value in the decreasing ordered random sample, $F(x)$ – cumulative distribution function for the theoretical distribution. The null hypothesis H_0 (when the measurement data were suitable for tested theoretical distribution), were taken on a significance level of 0.05 if the A^2 test statistic was less than the critical value A_{kr}^2. The alternative hypothesis was taken otherwise. The critical values can be read from the statistical tables or obtained by Monte Carlo method [25, 27]. For GED distribution and $N = 50$ critical value $A_{kr}^2 = 0.723$.

Relative residual mean square error (R_{RMSE}) was used to evaluate the aptitude of investigated distributions and to describe the measurement data

$$R_{RMSE} = \sqrt{\frac{1}{n}\sum_{i=1}^{n}\left(\frac{h_{t,i} - h_{m,i}}{h_{m,i}}\right)^2} \cdot 100\% \qquad (7)$$

where: h_t – the theoretical amount of rainfall (mm), h_m – amount of rainfall from measurements (mm).

3 Results

Calculation results of particular parameters of GED were presented in Table 2. The parameter estimates were determined by numerical maximization of the log-likelihood function (3). The calculations were carried out for each of 20 durations of maximum precipitation amounts analyzed in the paper. GED distribution fulfils the compliance criterion A^2 for each of the 20 analyzed rainfall durations. There were also calculated relative residual mean square error statistics, covering the entire range of data – all 20 durations. In this case $R_{RMSE} = 3.3\%$.

Table 2. Calculation results of parameters of GED.

t, min	α	λ, 1/mm	γ, mm	A^2	R_{RMSE}
5	2.457	0.434	4.3	0.178	2.90
10	0.785	0.242	8.3	0.261	2.84
20	0.596	0.144	11.3	0.491	3.94
30	0.731	0.146	12.5	0.319	4.32
40	0.861	0.143	13.0	0.139	2.57
50	0.937	0.143	13.7	0.197	2.36
60	1.081	0.145	14.2	0.311	2.53
90	0.831	0.120	16.5	0.422	3.64
120	0.901	0.124	18.2	0.405	2.75
180	0.835	0.106	20.0	0.176	2.45

t, min	α	λ, 1/mm	γ, mm	A^2	R_{RMSE}
360	0.870	0.097	23.9	0.292	3.37
720	0.886	0.088	28.6	0.416	2.44
1080	0.910	0.084	32.4	0.255	2.35
1440	0.874	0.070	35.7	0.722	3.78
2160	0.879	0.059	38.7	0.474	3.13
2880	0.790	0.043	41.3	0.166	3.14
4320	0.924	0.044	41.3	0.424	3.53
5760	0.823	0.035	47.1	0.370	3.76
7200	0.633	0.031	51.7	0.396	3.94
8640	0.762	0.032	53.7	0.268	4.30

Following the equation (5) and the parameters listed in Table 2, the precipitation amount with any exceedance probability and selected rainfall duration ranging from 5 to 8640 minutes can be calculated. In this respect, it should be noted that the current 10-minute time step of rainfall registration will not allow in the future to estimate parameters for precipitation with the duration of 5 minutes. So these parameters will have to be extrapolated.

The parameters λ and γ have a clearly marked trend, while the parameter α is deprived of a trend. Since there is no dependency trend $\alpha(t)$, the mean value of $\bar{\alpha} = 0.837$ was assumed in the calculations.

Based on the calculated GED distribution parameters there were prepared plots (Fig. 1) showing their dependence on the rainfall duration (from 10 to 8640 minutes).

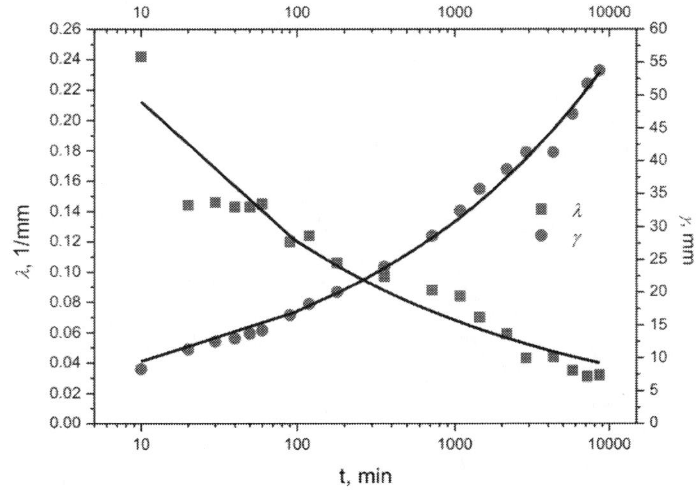

Fig. 1. The dependence of the parameters λ and γ of the rainfall duration.

The relationship of parameters λ and γ of the rainfall duration are described as functions:

$$\lambda = 0.376\, t^{-0.247} \tag{8}$$

$$\gamma = 5.261\, t^{0.256} \tag{9}$$

for the coefficient of determination of 0.929 and 0.994, respectively.

Finally, a model describing the dependence of the amount of rainfall on its duration and a specified exceedance probability, based on the quantiles of GED distribution (5), takes the form of:

$$h = 5.261\, t^{0.256} - 2{,}660\, t^{0.247} \ln\left(1-(1-p)^{1{,}195}\right) \tag{10}$$

Based on the obtained formula, precipitation amounts for the 5 min duration and the exceedance probability p = 0.020, 0.098, 0.196, 0.490 and 0.980 were computed. The received results were compared with the measurement data (Table 1) and the precipitation amount calculated using formula (5) with the parameters from Table 2. The results are shown in Table 3.

Table 3. Calculation results of parameters of GED.

p	h_m, mm	h_t, mm (5)	R_{RMSE}	h_t, mm (10)	R_{RMSE}
0.020	16.2	15.4	4.94%	22.8	40.74%
0.098	11.2	11.6	3.57%	16.5	47.32%
0.196	9.8	10.0	2.04%	13.8	40.82%
0.490	7.6	7.6	0.00%	10.3	35.53%
0.980	4.6	4.8	4.35%	8.0	73.91%

There were also calculated relative residual mean square error statistics, covering the entire range of data (t = 5 min). In this case R_{RMSE} = 2.9% for h calculated by (5) and R_{RMSE} = 41.5% for h calculated by (10). The fit quality of the equations for rainfall data from Legnica is shown in the *h-h* plot (Fig. 2).

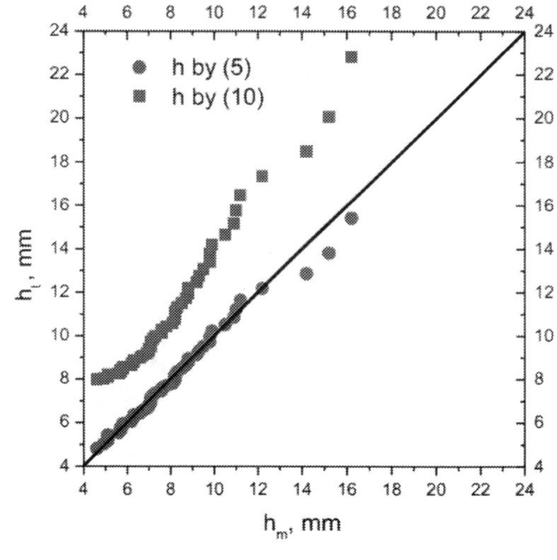

Fig. 2. The *h-h* plot for the GED distribution.

The analysis of the results clearly indicates that the extrapolation of the GED distribution parameters does not produce acceptable results. Calculated according to (10) rainfall heights of a $t = 5$ min duration do not correlate with the results of measurements.

4 Conclusions

In this paper, two maximum precipitation models for Legnica were developed. For this purpose archival pluviographic records from the time span 1961–2010 were used. The first model was developed on the basis of rainfall data of durations ranging from 5 minutes to 6 days. In the second model, rainfall data with durations ranging from 10 min to 6 days were used, and precipitation amounts for 5 minutes duration were extrapolated. Generalized exponential distribution (GED) was used to develop the models. Both models were compared with measurement data using relative residual mean square error. The conducted analysis allowed to draw the following conclusions:

- The maximum precipitation model (5) with the parameters shown in Table 2 accurately maps the measurement results - $R_{RMSE} = 3.3\%$. In particular, for rainfall with a duration of $t = 5$ min, the differences between the predicted and measured precipitation amounts at the level of $R_{RMSE} = 2.9\%$ were noted.
- In the case of using the model (10) to determine precipitation heights with a duration of $t = 5$ min (parameters extrapolation) obtained results were not correlated with the measurement results - $R_{RMSE} = 41.5\%$. Therefore, this method should be regarded as inappropriate for estimating precipitation amounts for the duration of $t = 5$ min.
- Precipitation amounts with a duration of $t = 5$ min can be estimated with some approximation based on the precipitation heights data for the duration of $t = 10$ min. The analysis of measurement data indicates that during the period considered, the maximum precipitation with a duration of $t = 5$ min represented on average 68% of maximum rainfall heights with a duration of $t = 10$ min. However, this estimation should be treated only indicatively, since this variability ranged in the case of the largest 50 precipitations from 55% to 75%. In order to draw up reliable conclusions, further research in this field involving other measurement stations is needed.
- In the light of the obtained results, it is requested to increase the temporal resolution of recorded by IMWM rainfall for 5-minute or shorter intervals.

The work was realized within the allocation No. 0401/0069/16 awarded for Faculty of Environmental Engineering Wroclaw University of Science and Technology by Ministry of Science and Higher Education in years 2016–2017.

References

1. A. Kotowski, B. Kaźmierczak, A. Dancewicz, *The modeling of precipitations for the dimensioning of sewer systems* (Polish Academy of Sciences, Warsaw, 2010)
2. A. Kotowski, *The principles of safe dimensioning of sewer systems* (Seidel-Przywecki, Warsaw, 2015)
3. E. Burszta-Adamiak, M. Mrowiec, Water Sci Technol **68**, 1 (2013)
4. A. Kotowski, K. Wartalska, M. Nowakowska, Ochrona Środowiska **38**, 1 (2016)
5. B. Kowalska, D. Kowalski, G. Łagód, M. Widomski, *Modelling of Hydraulics and Pollutants Transport in Sewer Systems with Exemplary Calculations in SWMM* (Lublin University of Technology, Lublin, 2013)
6. S. Y. Park, K. W. Lee, I. H. Park, S. R. Ha, Desalination **226**, 1–3 (2008)
7. T. G. Schmitt, *Kommentar zum Arbeitsblatt A 118 Hydraulische Bemessung und Nachweis von Entwässerungssystemen* (DWA, Hennef, 2000)
8. T. G. Schmitt, M. Thomas, KA – Wasserwirtschaft, Abwasser, Abfall **47**, 1 (2000)
9. D. Słyś, A. Stec, Environment Protection Engineering **38**, 4 (2012)
10. A. Wałęga, G. Kaczor, B. Stęplewski, Pol J Environ Stud **25**, 5 (2016)
11. E. Bogdanowicz, J. Stachý, *Maximum rainfall in Poland. Design characteristics* (The Publishing House of the Institute of Meteorology and Water Management, Warsaw, 1998)
12. B. Kaźmierczak, M. Wdowikowski, Periodica Polytechnica Civil Engineering **60**, 2 (2016)
13. W. Jakubowski, Meteorology Hydrology and Water Management **3**, 2 (2015)
14. A. Mailhota, S. Lachance-Cloutierb, G. Talbota, A.C. Favrec, J Hydrol **476**, 7 (2013)
15. S. Coles, *An introduction to statistical modeling of extreme values* (Springer Series in Statistics, Springer-Verlag, London, 2001)
16. C. Onyutha, Journal of Urban and Environmental Engineering **6**, 2 (2012)
17. W. L. Shinyie, N. Ismail, A. A. Jemain, Water Resour Manag **28**, 11 (2014)
18. S. Wi, B. J. Valdés, S. Steinschneider, T. W. Kim, Stoch Env Res Risk A **30**, 2 (2016)
19. A. G. Yilmaz, I. Hossain, B. J. C. Perera, Hydrol Earth Syst Sc **18** (2014)
20. R. D. Gupta, D. Kundu, J Stat Plan Infer **137**, 11 (2007)
21. R. D. Gupta, D. Kundu, J Stat Comput Sim, **69** (2001)
22. B. Kaźmierczak, A. Kotowski, J Hydrol **525** (2015)
23. F. Laio, G. Di Baldassarre, A. Montanari, Water Resour Res **45**, 7 (2009)
24. S. H. Lee, S. J. Maeng, Irrigat Drain **52** (2003)
25. R. D'Agostino, M. A. Stephens, *Goodness of Fit Techniques* (Marcel Dekker, New York, 1986)
26. S. Hongjoon, J. Younghun, J. Changsam, H. Jun-Haeng, Stoch Env Res Risk A **26**, 1 (2012)
27. A. S. Hassan, InterStat Electronic J (2005)

DOI: 10.1051/e3sconf/20171700039

Analysis of the possibility to cover energy demand from renewable sources on the motive power of the heat pump in low-energy building

Maciej Knapik[1,*]

[1]Institute of Thermal Engineering and Air Protection, Faculty of Environmental Engineering, Cracow University of Technology

Abstract. The article presents the problem of the demand for electricity for the heat pump and an analysis of the coverage of this demand by renewable energy sources such as wind turbines and photovoltaic cells, which generate electricity in low energy buildings. Low-energy and passive constructions are a result of introduction of new ideas in building design process. Their main objective is to achieve a significant reduction in demand for renewable primary energy, necessary to cover the needs of these buildings, mostly related to their heating, ventilation and domestic hot water This article presents the results of numerical analysis and calculations performed in MATLAB software, based on typical meteorological years. The results showed that renewable energy sources, can allow to cover a significant demand for electricity, that is required to power the heat pump it is economically justified.

1 Introduction

Low energy buildings and buildings modernized in terms of heat are characterized by lower energy demand for heating [1, 2]. In buildings of this type increased demand of energy for domestic hot water preparation in relation to the total energy demand can be observed. As a result, the proposed solutions allow to further lower energy demand by using the renewable energy sources. Research [3] shows that, to achieve the low demand for non-renewable primary energy, it is also necessary to use renewable energy sources, especially such as solar and wind power. Combining in one building, installations using these two types of renewable energy sources will provide benefits of mutual complementarity, when the energy generation in one of these installations will disappear or will be significantly reduced. In the autumn and winter period, when the possibilities of solar energy are significantly reduced, wind turbines will produce much more energy than photovoltaic cells due to usually occurring windy weather in this time. This situation is changing in the summer period, when the photovoltaic cells produce more energy, due to the longer day and much greater intensity of solar radiation. In the summer months, while the air masses are not moving too quickly, in result, it is observed lower wind speeds and therefore wind turbines produce limited amounts of power.

* Corresponding author: maciek.knapik@gmail.com

Application of the above described installations, which use renewable energy sources, can allow to cover a significant demand for electricity, that is required to power the heat pump [4,5] working for purposes of heating and domestic hot water. In case of excess electricity and the lack of heating needs, this energy can be used to heat the hot water in the tank. The use of heat pump in the above-mentioned system, will reduce the energy demand for primary fuels used to generate electricity.

2 Heat pump and renewable energy sources

In order to perform the calculations of electricity demand coverage for heat pump, the MATLAB software was used.

The prepared calculation script is based on typical meteorological years and data received from the manufacturers of heat pumps [6] and wind turbines [7]. The calculation script takes into account the variable energy demand for heating and variable production of renewable energy from photovoltaic cells and wind turbines. In a building with low energy demand, the largest energetic needs are generated by domestic hot water. In the model, a system with tank of water, based on calculation procedure given in Polish standard [8] is analysed. Based on the value of the thermal power Q_{tank} of the system with tank water - it can be calculated as follows:

$$Q_{tank} = 1{,}05\frac{Q_{without\_tank}}{[(N_h - 1)\Phi]} + 1 \tag{1}$$

where:

$Q_{without\_tank}$ – thermal power of system without a tank of water, based on the maximum hourly water flow, kW

N_h – hourly inequality rate of water consumption,

ϕ – the real rate of accumulation.

The analyzed system works according to the scheme presented in Fig.1. Renewable energy sources supply a heat pump with electricity, which provides heating needs of the building. In the summer months, the surplus of energy from renewable energy sources are routed directly to the heater in the domestic hot water.

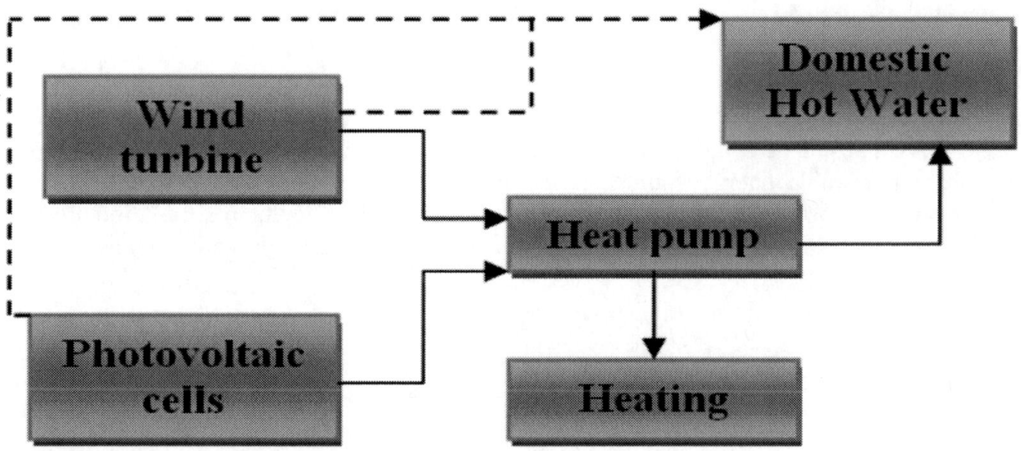

Fig. 1. Scheme of the analyzed system.

To carry out the analysis and calculations it is necessary to adopt a few assumptions – summarized in Table 1.

Table 1. The assumptions for the analysis and simulation.

Description	**Assumption**
Building type	Multifamily, modernized, low energy building, 18 kW heat losses
Location	Ustka, Poland (Polish first climatic zone[9])
Inhabitants	20 people
Hot water tank	1000 liters
Wind turbines	Selected 4 wind turbines with power 3 kW each
Photovoltaic cells	Selected 2 photovoltaic cells with power 300 W each (orientation S45^{0})
Heat pumps	Selected 3 heat pumps with power 11 kW each

Fig.2 presents the relation of turbine's power with wind speed, for wind turbine with a vertical axis of rotation with nominal power of **3 kW**.

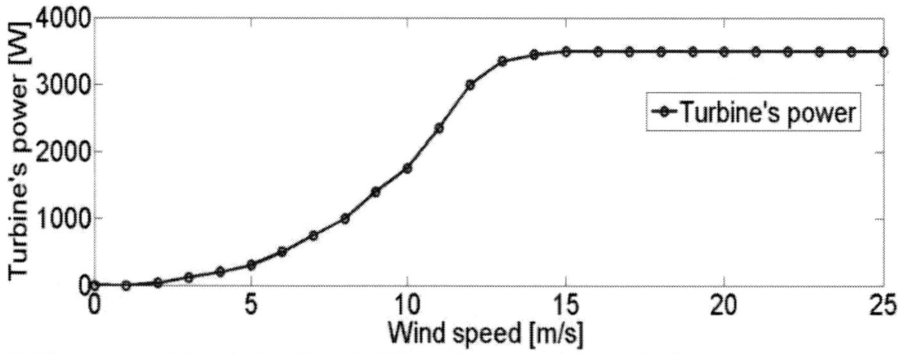

Fig. 2. The power of the wind turbine (3 kW) at the current wind velocity.

It can be seen that the selected wind turbine starts generating electricity at wind velocity equal to2 m/s and the nominal power is achieved at 12 m/s.

3 Results

As a result of simulation and analysis, was obtained a summary of energy produced from renewable energy sources (wind turbines and photovoltaic cells) in the specified location(according to the accepted assumption). Detailed results of simulation are presented in the Fig.3., which presents the potential electricity generated in within a year.

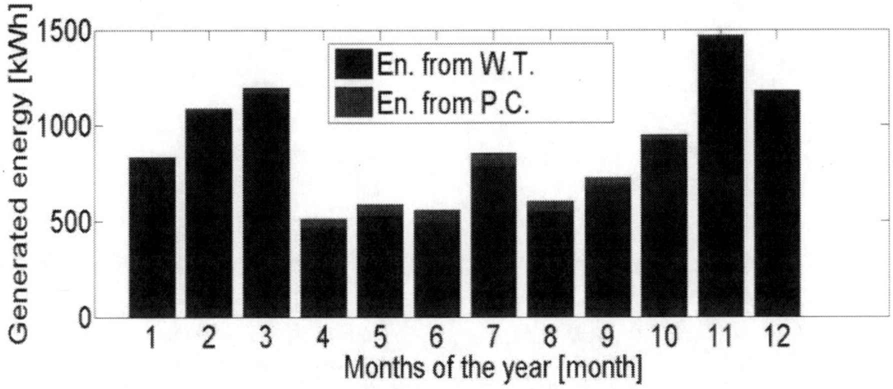

Fig. 3. Summary of energy generated by wind turbines and photovoltaic cells.

A significant difference between the energy produced by wind turbines and photovoltaic results from unequal selection. The selection is the result of adverse weather conditions (solar conditions) for specified location. From an economic point of view, it is reasonable to use more wind turbines. In another location, these conditions may be quite different.

Ustka is located in the Polish first climate zone. This means that, the design outdoor temperature amounts **-16** Celsius degrees. Outside temperature within a year is changing continuously. Therefore, it is also changing a temporary demand of heating energy in the heating season of the building. The heating season, this is the time, when building is not heated. In Poland, this period lasts on average from mid-April to October. As a result of the simulation, the detailed results of temporary of thermal load of the building was presented at Fig.4.

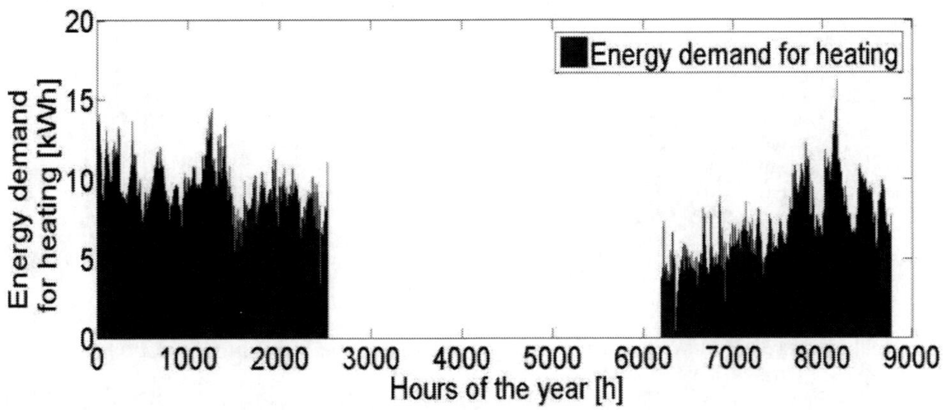

Fig. 4. Summary of energy demand for heating during a year, hour after hour.

Results presented in Fig.4. shows the variability of energy demand for heating purposes. In addition, it can be seen that the heat requirement for heating in a typical meteorological year was close to the predetermined value of the designed thermal load equal to 18 kW. For about half of the heating season, the temporary thermal load of the building oscillates around 50% of the design thermal load.

The energy efficiency of heat pump type air-water, depends mainly on the outside air temperature. The outside temperature affects on the thermal power of the heat pump and the amount of electricity consumed by it.

The Fig.5. presents the relation of the thermal power of the heat pump and also electrical energy consumed by it, depending on external conditions.

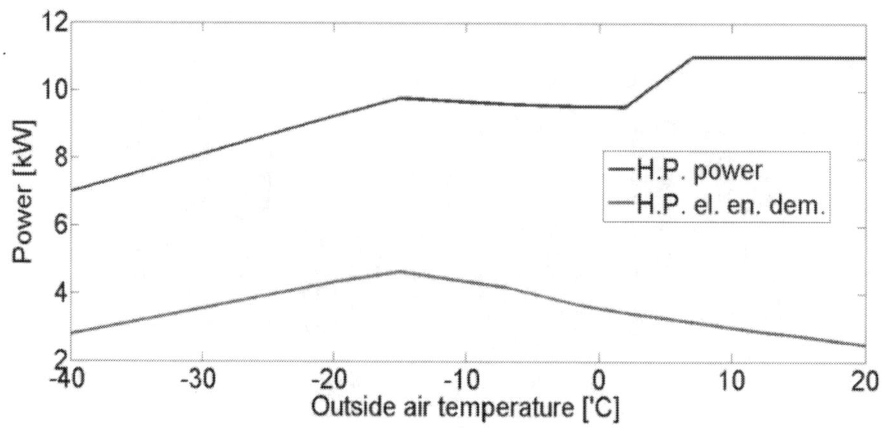

Fig. 5. Heat pump (11 kW) power and electrical energy demand for heat pump purposes in relation to outside temperature.

In presented simulation, it takes into account the evolving needs for electric energy to motive the heat pump and the thermal power of the selected heat pump. At low temperatures, the efficiency of the heat pump drops significantly, causing a drop in economic efficiency of investment [10].

Based on the data included in the Fig.5. and typical meteorological year was obtained a the following summary of demand for electricity to motive heat pumps in each hour of the year. It is presented on Fig.6.

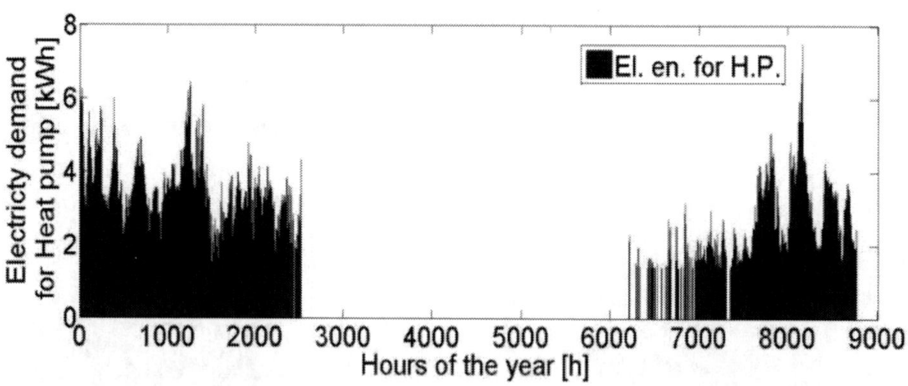

Fig. 6. Heat pumps(3 x 11 kW) electrical energy demand for heat pump purposes, hour after hour.

The demand of electricity to motive the heat pump coincides with the demand of thermal energy for heating of the building. After the heating season heat pump does not work and does not consume electricity. After the heating season electricity from renewable energy sources will be used to heat water in the hot water tank by the warmer.

The final result of the analysis and simulation is to present the coverage of energy demand for heat pumps by renewable energy sources. Fig.7. and Fig.8. presents respectively the coverage electricity demand for heat pumps used for heating during the heating season as well as during the summer months, renewable sources of energy are able to cover the energy needs for domestic hot water.

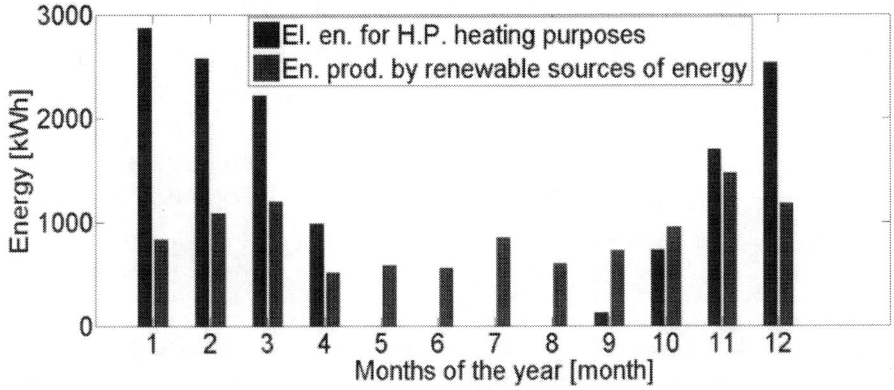

Fig. 7. Comparison of electricity demand for heat pumps with energy produced by renewable sources.

In the analyzed case, renewable energy sources, during the heating season are able to provide about 50% of the annual energy demand for heating purposes. Surplus of energy from the summer months will be used for domestic hot water preparation. From the other side, in the of low-energy building, higher energy demand is a process of the domestic hot water preparation, which is presented on Fig.8.

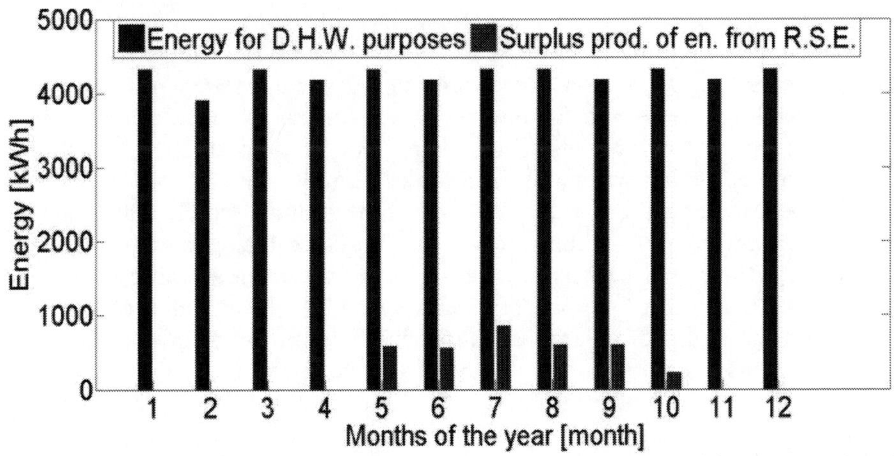

Fig. 8. Comparison of energy demand for domestic hot water purposes with energy produced by renewable energy sources.

By comparing the Fig.7. and Fig.8. can be observed the scale of the participation of energy demand for heating and domestic hot water preparation in the building of a modernized multi-family low energy demand.

The results of analysis and simulation do not present direct economic justification in the particular case.

Visible effects of the simulation can be observed when are compared the operating costs of the analyzed system with a system powered by non-renewable source of energy which is natural gas.

Assuming average prices for electricity, taking into account the day and night electricity tariffs and the average prices of natural gas, has been received a summary of operating costs, which is presented on Fig.9.

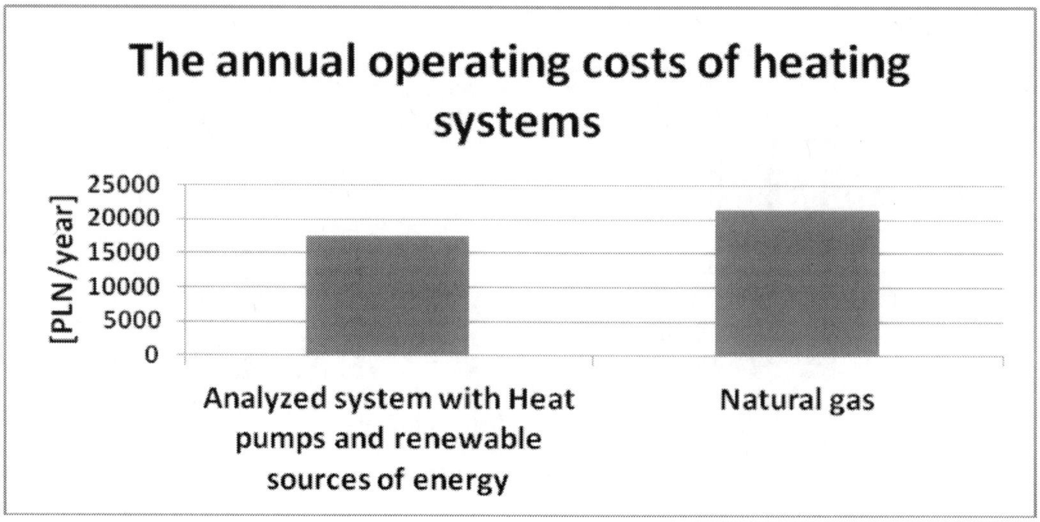

Fig. 9. Comparison of the operating costs of the analyzed system and heating system powered by natural gas.

4 Conclusions

The simulation and analysis in MATLAB software allowed to provide the coverage of the energy needs of the building for heating and domestic hot water purposes.

The effectiveness of the various renewable energy sources depends largely on the location and prevailing there advantageous or disadvantageous wind and solar conditions.

Experimental results show that in the low energy buildings about 20% of heat energy is used for heating domestic hot water. The aim of the study was an analysis of the results of installation cooperation using renewable energy with heat pump, which constitutes the sole source of heat for low energy building. It was also assessed the scope of coverage of the heat demand for heating and domestic hot water energy from renewable sources for the adopted, a typical example of low-energy building and accepted rules of dimensioning cooperating with the pump of solar installation and a small wind turbine. Based on the results of calculations and simulations was rated the coverage of the demand for heating and domestic hot water and was initially rated the economic efficiency of this solution source of energy for low energy building.

The simulation and analysis allow to conclude that, it is preferable to use renewable energy sources which produce electricity to motive heat pumps, which provide the energy needs of the building. The economic effect of analysis can be observed comparing the mentioned system with another system powered by other non-renewable source of energy – natural gas. In an article, for the analysis, it has been deliberately disregarded coal due to significant emissions to the atmosphere and considerable participation in the formation of smog in Polish cities.

References

1. W. Feist, U. Münzenberg, J. Thumulla, *Podstawy budownictwa pasywnego,* (2009, ISBN 83-923807-0-3)
2. E. Rylewski, *Energia własna*, (ISBN 83-917314-2-1)
3. M. Knapik, RI **9** (2016)

4. K. Wojtas, Polski Instalator **3** (2011)
5. K. Wojtas, Polski Instalator **4** (2011)
6. DAIKIN, technical materials and catalogs
7. Blitzmann, technical materials and catalogs
8. Polish standard, PN-90/B-01706: *Instalacje wodociągowe – Wymagania w projektowaniu* (1992)
9. Polish standard, PN-EN 12:831: *Instalacje ogrzewcze w budynkach – Metoda obliczania projektowego obciążenia cieplnego* (2006)
10. prEN 15316-4-2: *Heating systems in buildings – Method for calculation of system energy requirements and system efficiencies – Part 4-2: Space heating generation systems, heat pump systems* (2008)

Alternative device for ANAMMOX/denitrification rate assessment using manometric method

Łukasz Kokurewicz[1], and *Mateusz* Muszyński-Huhajło[1,*]

[1]Faculty of Environmental Engineering, Wroclaw University of Science and Technology, Place Grunwaldzki 9, 50-377 Wroclaw, Poland

Abstract. Anammox and denitrification processes play important role in nitrogen removal in wastewater treatment. In final stage of nitrogen compounds metabolism nitrogen gas is produced and released to atmosphere. Process rates can be estimated based on amount of produced nitrogen, but this kind of procedure requires special equipment that may be a significant obstacle in bringing this type of tests to common practice in wastewater treatment plants. A low-cost device was constructed to verify if such installation may be a reliable alternative for expensive commercially available products (e.g. OxiTop® system). Values of Specific Anammox Activity measured in different temperatures by constructed device and OxiTop® system were consistent in 86.3%. Average gas production based on pressure growth exceed 94.9% the values obtained by use of chemical analysis. Conducted tests confirmed that low-cost manometric device can be a useful tool in denitrification/Anammox sludge activity tests.

1 Introduction

Excess nitrogen loads directed with wastewater to lakes and rivers have always been an important issue for their water quality and aquatic biodiversity. Activated sludge process, known since the beginning of 20th century, in fact, was not able to solve this problem until the 50s, when Wuhrmann suggested oxic-anoxic (OA) reactor configuration to allow denitrification of nitrate previously produced in oxic reactor. However, the real breakthrough for biological nitrogen removal was noticed in 1962 when Ludzack and Ettinger developed an anoxic-oxic (AO) process. Reduction of external organic carbon requirement was a real boost for denitrification efficiency and allowed to meet more and more strict effluent quality standards. Efficient and sustainable N-removal has become standard on wastewater treatment plants (WWTPs) worldwide [1].

Much later, in 90s, another crucial for nitrogen removal discovery has been made when Mulder et al. [2] reported identification of new autotrophic microorganisms group able to anaerobic ammonium oxidation using nitrite, called Anammox (acronym for *ANaerobic AMMonium OXidation)*. From this moment, process was successfully introduced in side-stream installations for ammonia-rich streams treatment, with over 100 full-scale implementations by 2014 [3]. Nowadays, implementation of this process for the municipal

*Corresponding author: mateusz.muszynski-huhajlo@pwr.edu.pl

wastewater treatment (called as mainstream Anammox) is supposed to be the major step for energy self-sufficient and highly effective WWTPs [4].

Important role and increasing popularity of mentioned processes creates need for appropriate methods of process rate and efficiency evaluation, not only for research purposes, but also as self-control tools on full-scale plants. Due to dinitrogen gas production, as product of these reactions, manometric method is one of possible ways to evaluate their efficiency. In this paper, development and validation of a new, self-made installation for denitrification/Anammox process rate measurements is presented as a useful alternative for commercially available products.

1.1 Denitrification and Anammox –key elements in N-cycle

1.1.1 Denitrification

Due to its prevalence in wastewater treatment systems, in this study, only heterotrophic denitrification will be considered. It's biochemical process of nitrite and/or nitrate reduction under anoxic conditions with limited dissolved oxygen availability due to their facultative anoxic/aerobic metabolism. Denitrifying bacteria, which turn NO_x-N back into nitrogen gas for energy generation, play crucial role in natural nitrogen cycle and wastewater treatment. As a final stage of N-removal, denitrification is often limiting due to insufficient organic matter as an electron donor for full reduction of nitrogen oxides load. Metabolic pathway of this process in conventional activated sludge systems with full nitrification can be described by Eq. 1. As presented below, reaction increases the environment alkalinity due to the production of hydroxide ions [5].

$$NO^-_3 + 1.25CH_2O \rightarrow 0.5N_2 + OH^- + 0.75H_2 + 1.25CO_2 \quad (1)$$

1.1.2 Anammox

Discovery of anaerobic ammonium oxidation, performed by a group of bacteria with unique metabolism, shed new light on biological nitrogen cycle [6]. As it turned out, Anammox microorganisms are responsible for 30-70% of fixed nitrogen removal from marine environment and have ability to survive in a wide range of environments, also including activated sludge [7, 8]. In general, Anammox is an anaerobic multi-stage process: nitrite (NO_2^-) is reduced to nitric oxide (NO) that reacts with ammonium (NH_4^+) producing hydrazine (N_2H_4), which is oxidized to nitrogen gas (N_2) [9]. Reaction stoichiometry is given by Eq. 2 [10]:

$$NH_4^+ + 1.32NO_2^- + 0.066HCO_3^- + 0.13H^+ \rightarrow$$

$$1.02N_2 + 0.066CH_2O_{0.5}N_{0.15} + 0.26NO_3^- + 2.03H_2O \quad (2)$$

As mentioned above, Anammox process requires suitable nitrite/ammonium ratio in treated medium, usually achieved by combining Anammox with partial nitritation process (PN). Necessity of only partial oxidation of ammonium in this processes reduces net oxygen required for nitrogen removal to 1.71 mg O_2/mg NH_4^+-N converted into gas if influent biodegradable organic matter is used to denitrify residual NO_x-N [11]. Autotrophic metabolism of Anammox bacteria keeps this reaction independent from organic matter availability, as the inorganic carbon is used for biomass yield.

1.2 Manometric measurements technique

Rates of biological processes can be easily estimated based on their substrate utilization and products formation change in time. In some cases, such as denitrification and Anammox, when main product is a gas with low solubility, an indirect measurement of pressure growth due to gas production in air-tight vessel can provide information about process dynamics. Nitrogen gas, as main product of both reactions considered in this study, and its solubility change due to pressure growth can be considered as irrelevant in pressure range achieved in this type of tests. Otherwise than nitrogen, carbon dioxide has much higher solubility [12] and its production must be included in the gas production calculations or an CO_2 adsorbent should be used during denitrification rate measurements.

Described method has been proven to be adequate for activity and inhibition tests in number of studies, both for Anammox [13-17], as for denitrification [18]. Contrary to traditional measurements of bioprocess activity based on product/substrate profiles using chemical analysis, manometric method requires an air-tight vessel connected to a manometer for overpressure recording. Also, due to oxygen inhibition of described processes, an installation for flushing the sample and headspace with inert gas is needed. These facts make this type of measurements impossible without proper equipment. One of the most popular commercial devices used for this purpose is OxiTop® Control System (WTW, Germany) [13-17]. However, such specialized equipment can be considered as expensive a major obstacle for wide use of this method.

This paper presents a description of a self-made apparatus for manometric tests for denitrification/Anammox process rates assessment, including procedure description and validation test for mentioned installation based on comparison with analytical methods and parallel test performed in and OxiTop® device.

2 Materials and methods

2.1 Nitrogen gas production calculations based on recorded pressure change

Amount of nitrogen gas (N_2) produced during the test is calculated according to Clapeyron's ideal gas law, formulated in 1834 and describing relation between pressure, temperature and amount of gas in constant volume. Based on this relation, recorded pressure change (Δp, Pa) can be used for calculation of mass of produced gas (Δm, g) assuming that nitrogen molar volume equals 22.4 dm^3/mol and its density (ρ_{N_2}) equals 1.25 g/dm^3. A following equation (3) can be given:

$$\Delta m = \frac{\Delta p\, V_p}{RT} \cdot 22.4 \cdot \rho_{N_2} \tag{3}$$

where V_p - gas volume [m^3], R - gas constant [$\frac{J}{mol \cdot K}$] and T is temperature [K].

2.2 Measuring apparatus

2.2.1 Self-made reactor

A plastic vessel with volume of 390 cm^3 has been used for constructing an air-tight reactor for denitrification/Anammox batch test. The reactor is equipped with a diffusor for flushing the sludge sample with an inert gas, temperature sensor and 4 valves for sample collection, manometer connection and substrate dosing. Sample mixing is provided by using a magnetic stirrer. Gas headspace is connected through one of the valves with a manometer (Abatronic AB-8897) with data recording. Apparatus used during the tests has been presented in Fig. 1.

Fig. 1. Measuring apparatus on a magnetic stirrer (manometer not shown).

2.2.1 OxiTop®

A WTW's OxiTop® Control System was used to perform a parallel manometric test on the same Anammox sludge sample as in the self-made reactor. Device consisted of 650 cm^3 glass vessel with OxiTop® head for pressure control. All procedures, substrates and data interpretation was similar to one used in test performed in an alternative device described in this paper.

2.3 Samples and substrate source

Anammox sludge for performed batch tests was taken from a pilot-scale installation with a two-stage PN-Anammox in SBR with operational volume of $100dm^3$ treating real reject water from sludge dewatering at one of Polish WWTPs with capacity over 500000 p.e. Anammox process was stably operated for over 6 months before sludge samples were collected for batch tests presented in this study. Process was operated at temperature 23°C and pH=7.0. For batch tests, a synthetic medium was used containing: 1300 mg NH_4^+-N/dm^3, 1500 mg NO_2^--N/dm^3 and $NaCO_3$ to provide inorganic carbon source.

2.4 Test procedure

Sludge samples, taken from the SBR, were placed in an incubator set at temperature desired for the test. After the temperature stabilization, sample was put into batch test reactor. For performed tests headspace-to-sample volume ratio was set at about 1:3, resulting $290cm^3$ of sample of mixed liquor for single test. Sample concentration of nitrogen compounds was identified (NH_4^+, NO_2^-, NO_3^-) and, if required, sample was flushed with SBR effluent to remove present substrates which may cause gas production in the phase of gas-liquid equilibrium stabilization. To sustain anaerobic/anoxic conditions, sample was purged with nitrogen gas for 20-40 minutes with periodical dissolved oxygen and pH was adjusted to about 7.0 by dosing a 5% H_2SO_4 solution if needed. After that, reactor cover was sealed and headspace over the sample was purged for another 5 min to secure oxygen removal from the installation. When nitrogen flow was cut off, pressure was levelled with atmosphere and a 60-min phase of gas-liquid equilibrium stabilization begun, when sample was constantly mixed (see more in section 3.1.2). After this period, substrate was added to achieve initial concentrations of ammonium and nitrite at 60 and 50 mg N/dm^3, respectively. During substrate addition, headspace was also purged with N_2 to prevent any oxygen penetration into the vessel. After few minutes of mixing, a sample of mixed liquor was taken to measure initial test concentrations and pressure recording started. Test was conducted until pressure growth has slowed, typically for 60–120 minutes, depending on temperature and initial substrate concentrations.

2.5 Analytical methods

All samples were filtered with syringe filters, pore size 1.2μm and after that ammonium, nitrite and nitrate nitrogen concentrations were measured using Hach Lange's photometric cuvette tests: LCK303 (2 – 47 mg NH_4^+-N/dm^3), LCK342 (range 0,6 – 6 mg NO_2^--N/dm^3) and LCK340 (range 5 – 35 mg NO_3^--N/dm^3) and a DR3900 spectrophotometer. Total suspended solids (TSS) and volatile suspended solids (VSS) concentrations were determined using direct weight method according to Standard Methods guideline.

3 Results and discussion

3.1 Apparatus and test procedure validation

3.1.1 Batch reactor air-tightness tests

First step to verify if constructed installation can be used for manometric tests was to check its air-tightness by measuring potential pressure loss in time. To reduce impact of gas solubility in water, vessel was filled with a saturated saline solution (~26,5% NaCl). Firstly, low-pressure tests were conducted: overpressure in the vessel was kept at level around 3-5 kPa for 24 hours. No relevant pressure change was noticed that indicates proper air-tightness of the reactor.

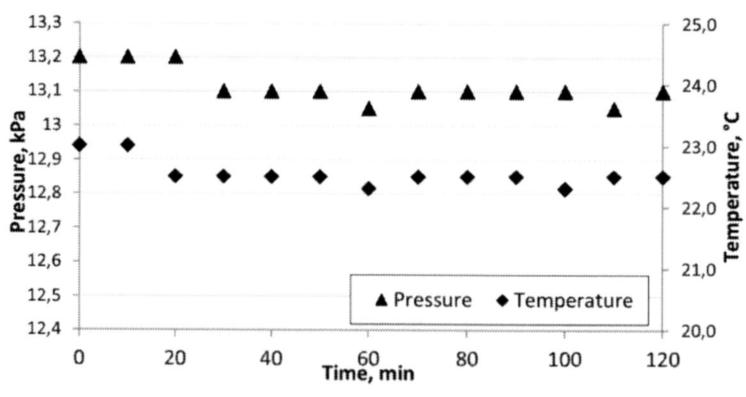

Fig. 2. Short-term apparatus air-tightness verification test results on saturated saline water (NaCl).

Range of pressure expected during manometric tests is about 5-20 kPa and takes around 2 hours which is strictly connected on the initial substrate concentration and process rate. Next step was to verify pressure change at higher pressure levels (~15 kPa) in a short-term (2 hour) tests. Results of long and short-term test, such as one presented in Figure 2, clearly indicates that constructed vessel provide sufficient tightness to conduct manometric tests.

Observed pressure fluctuation was strictly connected with temperature change of gas in the space over liquid phase.

3.1.2 Achieving liquid-gas phase balance

Due to nitrogen gas flushing, it's necessary to restore gas-liquid balance between sludge sample and gas space in the vessel. Continuous mixing provides proper contact between these phases but dynamics of this process was unknown. To decide how much time is needed to establish such balance after 40 minutes of purging the sample with N_2 a series of tests was conducted. Results are shown in Figure 3.

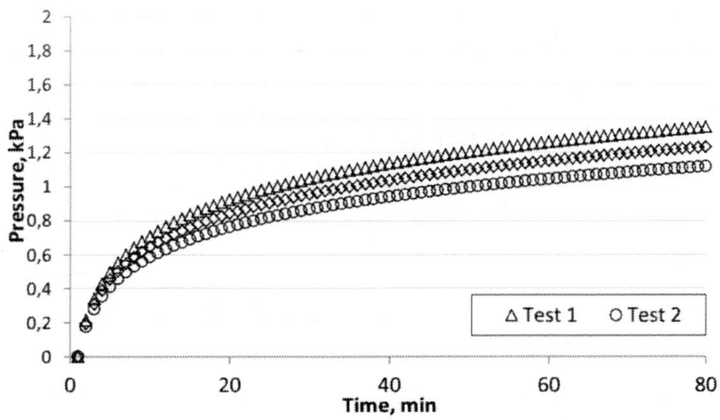

Fig. 3. Observed pressure change in headspace over a N_2-purged sample.

It can be clearly noticed that, after more than 60 minutes from cutting the flow of N_2, pressure growth slows down and further changes can be considered as insignificant for the final test results as average pressure growth in this period was equal 0.044 kPa/min. Such pressure stabilization period was present in methodology of all conducted tests.

3.2 Anammox process rate tests - Specific Anammox Activity (SAA)

3.2.1 Analytical methods vs produced nitrogen gas

Sample collection after substrate dosing and at the end of each test allows to verify amount of produced nitrogen gas with N-compounds left in the mixed liquor. NH_4^+-N, NO_2^--N and NO_3^--N concentrations were measured in all samples to calculate N mass balance which was compared with the mass of nitrogen in the headspace calculated from the pressure growth during the test (see point 2.1). Summary of gained results is presented in Table 1.

For the first test (ANMX 1*), real effluent from PN reactor was used as a batch test substrate (420 mg NH_4^+-N/dm^3 and 533 mg NO_2^--N/dm^3). Relatively low concentration of substrates resulted in big volume of added solution, which might have caused problems with abnormal overpressure in the vessel causing its possible unsealing. Another fact is, that such volume of substrate solution might not been measured precisely affecting calculations and lowering volume of headspace in the vessel. Based on this experience, only synthetic medium was used in further tests (see section 2.3).

Other tests revealed a good correlation between produced gas and depletion of nitrogen mass in the mixed liquor in wide range of temperatures (20-32°C) and maximal overpressure recorded during each test, equal to 7.3 kPa, 4.35 kPa and 9.37 kPa for ANMX1, ANMX2 and ANMX3, respectively.

Table 1. Summary of results obtained in performed tests.

Test no.	**Temp.**	**Nitrogen production / loss**		**SAA**		**Percentage ratio manometric / analytical**
		manometric	analytical	manometric	analytical	
-	°C	mg N	mg N	$g\ N\cdot gVSS^{-1}\ day^{-1}$		%
ANMX 1*	25	9.4	12.8	0.352	0.480	73.4
ANMX 1	26	8.0	7.9	0.320	0.315	101.3
ANMX 2	20	5.5	5.6	0.185	0.189	98.2
ANMX 3	32	12.2	13.2	0.812	0.876	92.4
ANMX 4	23	6.6	7.5	0.471	0.537	88.0

In Figure 4, typical pressure profile for performed manometric test is presented (ANMX1). It can be clearly noticed, that maximum process rate, observed as pressure growth due to gas production, was achieved between 5th and 60th minute of performed test when no substrate limiting occurred.

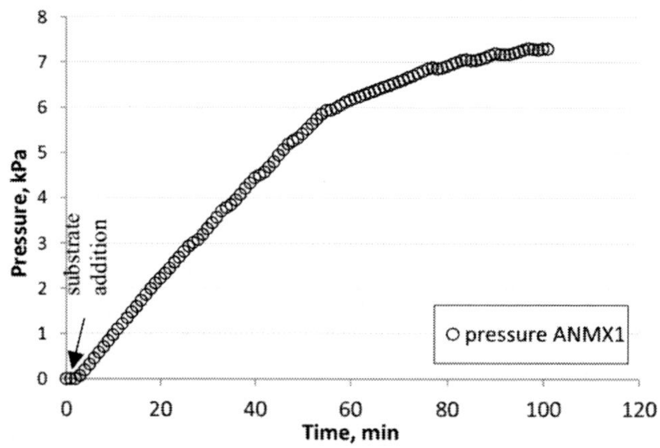

Fig. 4. Pressure profile during a manometric test (ANMX1).

3.2.2 Alternative device vs OxiTop®

Good conformity of the manometric and analytical results suggests that calculated SAA rates are highly reliable. To verify that assumption, a parallel test performed in a OxiTop®device was performed on the same Anammox sludge sample as in alternative device. Results gained from OxiTop® revealed that SAA was 0.94 $g\ N\cdot gVSS^{-1}\ day^{-1}$ for manometric calculations and 1.038 $g\ N\cdot gVSS^{-1}\ day^{-1}$ for analytical results.

Considering commercially available equipment as more reliable than constructed one, SAA obtained in alternative device was equal to 86.3% and 84.4% of values from the OxiTop® test for manometric and analytical results, respectively. This fact indicates need of further development and improvement of the construction and methodology of alternative apparatus to improve results quality.

4 Conclusions

- Manometric measurement technique is a very useful and reliable alternative for chemical tracking in case of denitrification/Anammox process rate. Lower number of samples than in analytical measurements and on-line control of gas production reduces required effort and provides a lot of useful information about process dynamics. Despite the obvious benefits of using this method, necessity of owning a specialized equipment may be an obstacle.
- A low-cost apparatus has been constructed as an alternative for currently available commercial products. Series of preliminary test has been conducted to verify air-tightness of constructed installation achieving positive results. Construction can be used as a potential batch test reactor for manometric tests.
- Air-liquid phase balance after purging the mixed liquor sample must be re-established, due to significant pressure changes that may affect final test results. Conducted tests revealed 60 minutes of mixing in sealed vessel as sufficient period for minimalizing impact of this effect.
- High-strength synthetic medium was more suitable as a batch test substrate due to lower volume added to the vessel.
- Tests performed in constructed reactor have shown good accuracy of achieved results in comparison with chemical tracking. Average gas production based on pressure growth was equal to 94.9% in 4 successfully performed test (standard deviation: 5.1%) in wide range of maximal recorded pressure and temperature of the test.
- SAA obtained in alternative device was equal to 86.3% and 84.4% of values from the OxiTop® test for manometric and analytical results, respectively.
- Despite promising results of manometric and analytical calculations in tests performed in the self-made apparatus, further tests must be performed to investigate causes of observed differences in comparison with OxiTop® test result.

References

1. E. Klaczyński, P. Ratajczak, Wodociągi i Kanalizacja **5** (2013)
2. A. Mulder, A.A. Van De Graaf, L.A. Robertson, J.G. Kuenen, Fems Microbiol. Ecol. **16** (1995)
3. S. Lackner, E.M. Gilbert, S.E.Vlaeminck, A. Joss, H. Horn, M.C.M. Van Loosdrecht, Water Res. **55** (2014)
4. M.C.M. Van Loosdrecht, D. Brdjanovic, Science **344** (2014)
5. V. Matějů, S. Čižinská, J. Krejčí , T. Janoch, Enzyme Microb. Tech. **14** (1992)
6. M.M.M. Kuypers, G. Lavik, D. Woebken, M. Schmid, B.M. Fuchs, R. Amann, B.B. Jørgensen, M.S.M. Jetten, P. Natl. Acad. Sci. USA. **102** (2005)
7. A.H. Devol, Nature **422** (2003)
8. B. Fu, J. Liu, H. Yang, T.C. Hsu, B. He, M. Dai, S.J. Kao, M. Zhao, X.-H. Zhang, J. Geophys. Res-Oceans **120** (2015)
9. B. Kartal, W.J. Maalcke, N.M. De Almeida, I. Cirpus, J. Gloerich, W. Geerts, H.J.M. Op Den Camp, H.R. Harhangi, E.M. Janssen-Megens, K.-J. Francoijs, H.G. Stunnenberg, J.T. Keltjens, M.S.M. Jetten, M. Strous, Nature **479** (2011)
10. M. Strous, J.J. Heijnen, J.G. Kuenen, M.S.M. Jetten, Appl. Microbiol. Biot. **50** (1998)
11. G.T. Daigger, Water Environ. Res. **86** (2014)
12. N.I. Kolev, *Multiphase Flow Dynamics 3: Turbulence, Gas Absorption and Release, Diesel Fuel Properties*, (Springer, 2002)

13. A. Dapena-Moraa, I. Fernándeza, J.L. Camposa, A. Mosquera-Corrala, R. Méndeza, M.S.M. Jettenb, Enzyme Microb. Tech. **40** (2007)
14. D. Scaglione, G. Buttiglieri, E. Ficara, S. Caffaz, C. Lubello, F. Malpei, Water Sci. Technol. **60** (2009)
15. E. Bettazzia, S. Caffazb, C. Vanninic, C. Lubelloa, Process Biochem. **45** (2010)
16. T. Lotti, W.R.L. Van Der Starc, R. Kleerebezema, C. Lubellob, M.C.M. Van Loosdrechta, Water Res. **46** (2012)
17. E.M. Gilbert, S. Agrawal, S.M. Karst, H. Horn, P.H. Nielsen, S. Lackner, Envir. Sci. Tech. **48** (2014)
18. M. Sánchez, A. Mosquera-Corral, R. Méndez, J.M. Lema, Bioresource Technol. **75** (2000)

Assessment of the impact of a parameter estimation method for the Nash Model on selected parameters of a catchment discharge hydrograph

Katarzyna Kołodziejczyk[1,*]

[1]Cracow University of Technology, Institute of Water Engineering and Water Management, 24 Warszawska Str 31-155 Cracow, Poland

Abstract. An analysis of the usefulness of two parameter calculation methods (*N* and *k* parameters) for the Nash Model was performed to transform effective rainfall into discharge based on two rainfall episodes gauged at the Kostrze gauging station as well as urban development data for the city of Cracow for 2014 and data obtained from a soil and agriculture map. The methods were the Rao et al. method and the Bajkiewicz-Grabowska method for regression relationships between instantaneous unit hydrograph model parameters and the physiographic parameters of a catchment. Effective rainfall was calculated for each rainfall episode using the SCS-CN method. A direct discharge hydrograph was calculated based on an effective rainfall hyetograph and using the Nash Model. Research has found that both studied methods yield comparable results, which indicates that both methods of effective rainfall transformation into discharge are useful. In addition, it has been shown that the impact of the Nash Model parameter estimation method on discharge hydrographs is minimal.

1 Introduction

Simple conceptual models find practical application in flood hydrograph modeling. Their main advantage consists in the small number of parameters, which makes them useful in ungauged catchments [1]. One of the frequently utilized models is the multi-reservoir instantaneous unit hydrograph (IUH) model created by Nash [2]. This model treats a catchment as a cascade of *N* linear reservoirs connected in a sequence, where the first reservoirs is recharged by effective rainfall, and subsequent reservoirs are recharged with water by previous reservoir. The instantaneous unit hydrograph [1, 3, 4, 5] is obtained at the mouth of the last reservoir in this sequence of reservoirs. The Nash Model is a two-parameter model whose parameters are *k* or storage parameter of each reservoir and *N* - the number of reservoirs.

This paper provides an analysis of the usefulness of two methods for calculating the *N* and *k* parameters of the Nash Model: (1) Rao et al. method [6], (2) relationships

[*] Corresponding author: katarzyna.kolodziejczyk@iiw.pk.edu.pl

 DOI: 10.1051/e3sconf/20171700041

developed by Bajkiewicz-Grabowska [3] for the transformation of effective rainfall into discharge.

2 Data

The study area consists of the Potok Kostrzecki catchment. It is a right-bank tributary of the Vistula River within the city limits of Cracow. The catchment is located in the southwestern part of the city in District VIII (Dębniki) (Fig. 1). It is a catchment characterized by an "average" degree of urbanization (16.1%), which is defined as the share of developed areas per total catchment surface area.

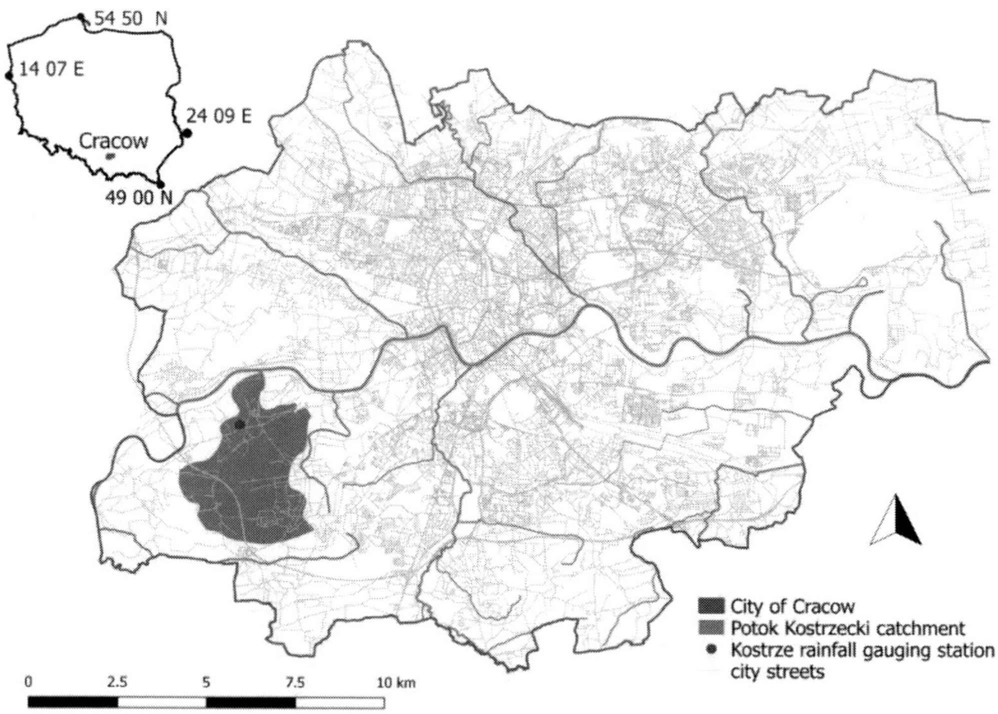

Fig. 1. Location of studied catchment.

Table 1 provides selected physiographic parameters for the studied catchment and stream.

Table 1. Selected catchment and stream parameters.

A (cat. surf. area)	B (cat. width)	H_{max} (max. elev.)	H_{min} (min. elev.)	ΔH (height diffs.)	J (cat. grad.)	L (stm. length)	i (stm. grad.)	U (share urban. area)
[km^2]	[km]	[m a.s.l.]	[m a.s.l.]	[m]	[m/km]	[km]	[‰]	[%]
10.32	1.66	260	200	60	9.68	5.96	6.3	16.11

Rainfall data in the study area includes rainfall measured at the Kostrze gauging station over the period May 27, 2014 to November 28, 2014 and made available by Municipal Waterworks and Sewer Enterprise in Cracow Joint Stock Company (MPWiK S.A.). Two rainfall episodes characterized by the highest totals (Fig. 2) were selected for study purposes. Other data used in the study included data on land development in Cracow in

DOI: 10.1051/e3sconf/20171700041

2014 obtained from the Land Survey Department of the City of Cracow as well as soil data from a soil and agriculture map made available for study by the Voivodeship Center for Geodetic and Cartographic Documentation in Cracow.

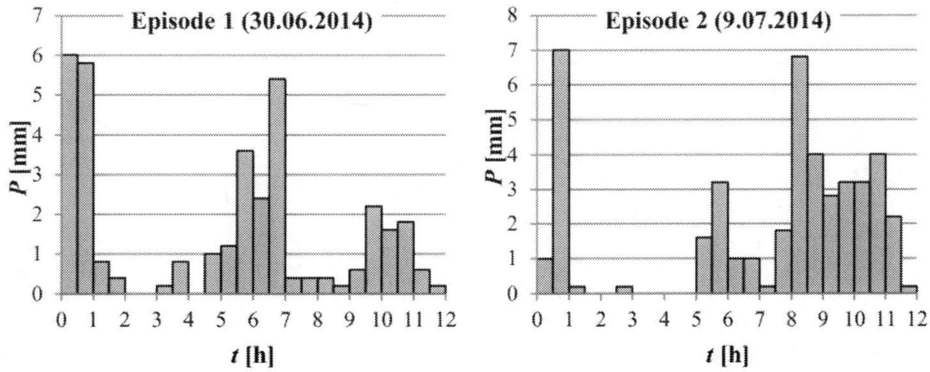

Fig. 2. Rainfall episodes.

3 Methods

The Nash Model treats the catchment as an imaginary retention reservoir recharged with effective rainfall P_e. The purpose of this type of model is to transform effective rainfall into discharge. Effective rainfall serves as the basic parameter in the Nash Model. In this study, this parameter was calculated for each rainfall episode using the SCS-CN method, which links effective rainfall with land use type, moisture levels in the catchment, soil type, and crop cultivation methods [7]. This method was created by the Soil Conservation Service of the United States, now known as the Natural Resources Conservation Service. It is used in the case of small and midsize catchments [8]. The SCS-CN method was adapted for use in Poland by Ignar [9] and other researchers. The details of this procedure are found in publication [10].

There exist a number of methods for estimating the N and k Nash Model parameters. The method proposed by Nash himself was that of moments, using initial moments of the first and second order of the gamma distribution. However, equations generated via this method turned out to be very sensitive to the quantity and accuracy of data, and also require accurate data on effective rainfall and discharge. Hence, this method cannot be used in many catchments. In order to make it possible to use the Nash Model for ungauged or inadequately gauged catchments, regression relationships were calculated between the N and k parameters and physiographic factors associated with each given catchment. Nash used data for a number of catchments in England to produce empirical formulas designed to determine his model's parameters [11].

Regression relationships between IUH model parameters and physiographic catchment parameters for Poland were generated by Bajkiewicz-Grabowska [3]. An attempt was also made to employ the geomorphologic instantaneous unit hydrograph (GIUH) to help estimate the N and k parameters for the Nash Model [12].

In this paper, two different estimation methods are used to determine the N and k Nash Model parameters: (1) Rao et al. method, (2) Bajkiewicz-Grabowska regression method. The parameters produced by each method were compared for resulting hydrographs. An analysis was then performed to assess the impact of each studied method of the estimation of the N and k parameters on the resulting discharge hydrograph.

According to the Rao et al. method [6], the runoff lag time LAG decreases with total depth of effective rainfall, and increases proportionately to the length of effective precipitation time [13]. In this method, the value of the k parameter is calculated using the following equation:

$$k = 0.56 \cdot A^{0.39} \cdot (1+U)^{-0.62} \cdot H^{-0.11} \cdot D^{0.22} \tag{1}$$

where: A – catchment surface area [km^2], U – share of developed land in catchment area [-], H – depth of effective rainfall [mm], D – effective rainfall duration [h].

The second parameter or the number of reservoirs N is calculated as follows:

$$N = \frac{LAG}{k} \tag{2}$$

where LAG or the discharge lag time equals [5, 13]:

$$LAG = 1.28 \cdot A^{0.46} \cdot (1+U)^{-1.66} \cdot H^{-0.27} \cdot D^{0.37} \tag{3}$$

According to Bajkiewicz-Grabowska [3], the N and k parameters of the Nash Model are calculated based on physiographic parameters. The equation was formulated based on a study of 48 catchments in Poland and covers six physiographic parameters of the stream and catchment. These parameters were identified based on their relevance in the formation of surface runoff. The parameters represent four basic groups of catchment characteristics: (1) catchment geometry, (2) relief, (3) river network and drainage conditions, (4) land use [3, 4].

The overall relationship can be expressed as follows:

$$N = \frac{5.66 \cdot T_u^{1.47} \cdot C_z^{3.62} \cdot J_r^{0.30}}{B_m^{1.02} \cdot D^{2.34}} \tag{4}$$

where: T_u – river network structure index [km^{-1}], C_z – catchment compactness index [-], J_r – gradient of main stream [m·km^{-1}], B_m – maximum catchment width [km], D – mean river network density [km/km^2];

$$k = \frac{76.2 \cdot \lambda^{0.24}}{N^{0.81} \cdot J_r^{0.64} \cdot R_L^{0.92}} \tag{5}$$

where: λ – forest cover [%], R_L – stream length index.

The river network structure index T_u is calculated using the following equation:

$$T_u = \frac{\sum N_i}{O} \tag{6}$$

where: $\sum N_i$ – number of streams (variable order) calculated on the basis of the river network classification by Strahler [-], O – catchment circumference [km].

The catchment compactness index C_z is defined as the relationship between the true circumference of the catchment P and the circumference of a circle with a surface area equal to that of the catchment (A):

$$C_z = 0.28 \cdot \frac{P}{\sqrt{A}} \tag{7}$$

The gradient of the main stream J_r is calculated as follows:

$$J_r = \frac{\Delta H}{L_m} \tag{8}$$

where: ΔH – height difference [m], L_m – maximum catchment length [km].

The mean river network density D is an important parameter that serves to describe the degree of drainage possible in a given catchment. It is calculated as the quotient of the total length of all streams L_i and the catchment surface area A:

$$D = \frac{\sum L_i}{A} \tag{9}$$

Forest cover λ is one of the most frequently used parameters characterizing land use. It is defined as the relationship between surface area covered by forests A_i and total surface area A of a given catchment:

$$\lambda = \frac{A_i}{A} \tag{100}$$

Stream length index R_L:

$$R_L = \frac{1}{M-1}\left(\frac{L_2}{L_1} + \frac{L_3}{L_2} + \cdots + \frac{L_M}{L_{M-1}}\right) \tag{11}$$

where : M – highest stream order in the catchment, L_i – mean length of sections of order i.

Once the N and k parameters are calculated via the two selected methods, and the Nash Model is applied, a direct discharge hydrograph may be calculated as follows:

$$Q_i = \sum_{j=1}^{i} h_{i-j+1} \cdot \Delta H_j, \qquad i = 1,2,\ldots, \tag{12}$$

where: h_i – unit hydrograph values [$m^3 \cdot s^{-1} \cdot mm^{-1}$], ΔH_j – depth of effective rainfall in the j-th time interval [mm].

4 Effective rainfall

The determination of effective rainfall is very important from the perspective of surface runoff analysis, as discharge hydrographs are calculated using this type of data, which constitutes that part of total precipitation that becomes transformed into surface runoff and eventually reaches the stream. Surface runoff begins at the point where precipitation exceeds the catchment's retention capacity [9, 11, 14–17].In the SCS-CN method, effective rainfall equals total rainfall less losses due to interception, infiltration, surface retention, and surface evaporation [9]. The *CN* parameter covers all factors affecting effective rainfall: soil type, land use, moisture conditions, other factors.

In this study, the *CN* parameter needed to be adjusted due to moisture conditions in the studied catchment (antecedent precipitation totals) in the case of both rainfall episodes. All rainfall episodes occurred in the vegetation season, where antecedent precipitation totals for the second (average) level of catchment moisture must remain within the 35 to 53 mm interval. Lower or higher values require a first or third moisture level designation and an appropriate adjustment of the *CN* parameter (Tab. 2) [15].

Table 2. Adjustment of the *CN* parameter to reflect catchment moisture levels.

Rainfall episode	1	2
Soil moisture level	II	II
Antecedent precipitation total [mm]	19.6	0.4
Adjusted level of soil moisture	I	I
CN [-] parameter	69	69
Adjusted *CN* [-] parameter	49.5	49.5

In the case of episode no. 1, effective rainfall appeared after 5.5 hours from the start of rainfall (Fig. 3). The effective rainfall total equaled 1 mm. The studied catchment is an area with a large share of forests, meadows, and agricultural land, which substantially increases its initial ability to retain rainwater. In addition, the data used in this study were collected at a time when the catchment was dry (low moisture level), which also caused a large part of the precipitation to infiltrate into the dry soil. As a result, effective rainfall was lower and also appeared later than normally.

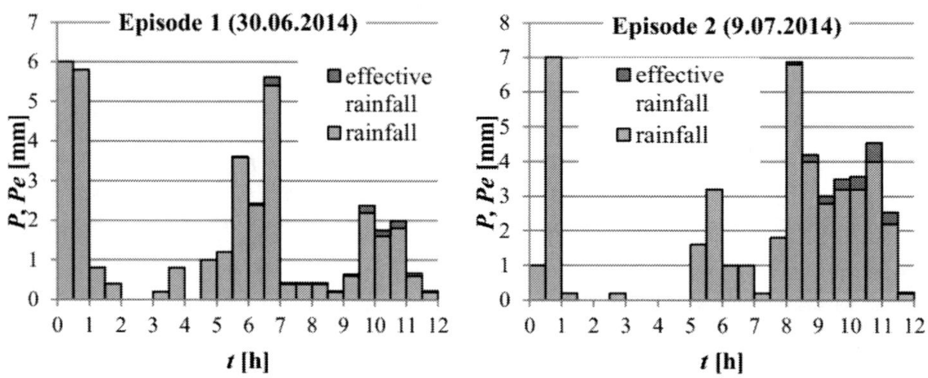

Fig. 3. Effective rainfall P_e.

Episode no. 2 occurred nine days after episode no. 1. The occurrence of episode no. 1 did not produce a significant impact on catchment moisture levels (also low moisture levels). The start of effective rainfall occurred eight hours after the start of rainfall, and its total equaled 2.06 mm (Fig. 3).

5 *N* and *k* parameters in the Nash Model

The following discharge hydrograph parameters were used to compare hydrographs that were produced based on model parameters calculated both using the Rao et al. method and the Bajkiewicz-Grabowska method: (1) time needed to reach a peak T_p, (2) height of peak wave q_p.

Figures 4 and 5 illustrate that the time needed to reach peak wave T_{p1} calculated using the Rao et al. method and the time needed to reach T_{p2} calculated via the Bajkiewicz-Grabowska method are equal for episodes no. 1 and no. 2. The value for episode no. 1 is 12 h, and 12.5 h for episode no. 2.

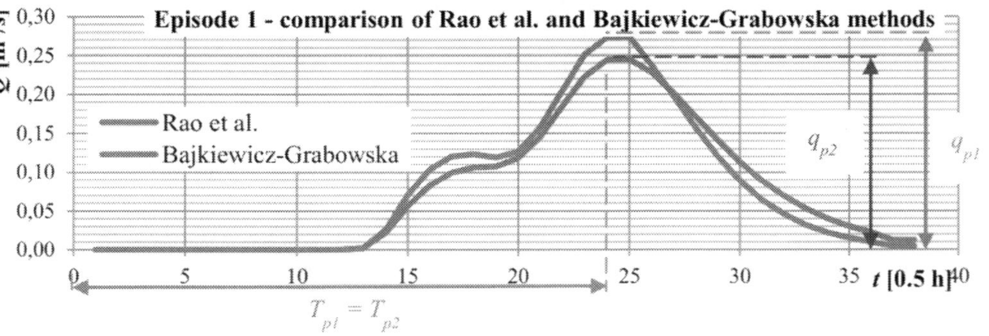

Fig. 4. Discharge hydrographs for episode no. 1.

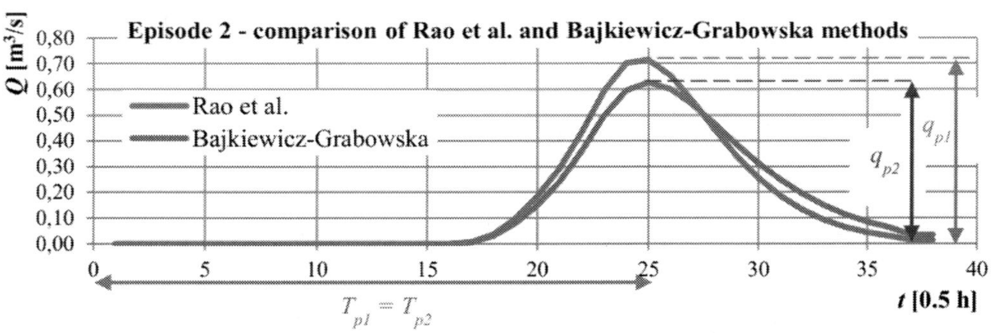

Fig. 5. Discharge hydrographs for episode no. 2.

The wave peak q_{p1} calculated using the Rao et al. method is slightly higher in both cases than the wave peak q_{p2} calculated for parameters estimated using the Bajkiewicz-Grabowska equation. For episode no. 1, the wave peak for the Rao et al. method equals 0.27 m^3/s, while for the Bajkiewicz-Grabowska method, it is 0.24 m^3/s. For episode no. 2, the corresponding values are as follows: 0.71 m^3/s, 0.63 m^3/s. The shape of the hydrograph for each examined episode is similar.

6 Summary and conclusions

The impact of the estimation method for parameters *N* and *k* in the Nash Model on several discharge hydrograph parameters was assessed using real rainfall data from 2014 along with land use data and soil type data for the study area. It has been shown that both compared discharge hydrograph parameters are either equal or close to equal, which further implies that the impact of estimation methods for the *N* and *k* parameters in the Nash Model is small. The Rao et al. method and the Bajkiewicz-Grabowska equation yield comparable results, which means that both methods can be used to transform effective rainfall into discharge.

The Rao et al. method requires little information about the catchment, which allows it to be used for hydrograph construction when complete observation data are unavailable [18]. It is successfully applied by many authors, among others Jovanovic [19], Banasik et al. [20]. Research has demonstrated that the hydrograph computed using Rao et al. method is approximate to the hydrograph calculated on the basis of rainfall-runoff data [18].

This study was part of a larger research endeavor: „Modern methods of design and management in engineering and water management", task 2: Application of the DPSIR analysis for assessment of state and prediction of flow dynamics and river-bed changes for the Vistula River and its tributaries in the Cracow agglomeration area.

References

1. A. Byczkowski, *Hydrologia (Hydrology)* (SGGW, **II,** 1999)
2. J. E. Nash, IASH – General Assembly of Toronto,**T. III.** (1958)
3. E. Bajkiewicz-Grabowska, Przegląd Geofizyczny, **Zeszyt 4**, Rocznik XXX (1985)
4. U. Soczyńska (red.), *Hydrologia dynamiczna (Dynamic hydrology)* (PWN, 1997)
5. K. Banasik, *Wyznaczanie wezbrań powodziowych w małych zlewniach zurbanizowanych (Assessment of flood events in small urban catchments)* (SGGW, 2009)
6. R. A. Rao, J. W. Delleur, B. S. P. Sarma, J HYDR ENG DIV-ASCE, ASCE **Vol. 98 (HY7)** (1972)
7. K. S. Mishra, V. P. Singh, *Soil Conservation Service Curve Number (SCS-CN) Methodology* (Water Science and Technology Library, **Vol. 42**, 2003)
8. D. H. Pilgrim, I. Cordery, *Flood runoff,* in: D. R. Maidment (ed.), (McGraw-Hill, Inc. *Handbook of hydrology,* 1993)
9. S. Ignar, *Metodyka obliczania przepływów wezbraniowych w zlewniach nieobserwowanych (Calculation methods for flood discharge in ungauged catchments)* (SGGW, 1993)
10. K. Kołodziejczyk, Acta Scientiarum Polonorum, seria Formatio Circumiectus, **15**, 4 (2016)
11. J. E. Nash, *Applied flood hydrology,* in: R. B. Thorn (ed.), (*River engineering and water conservation works,* Butterworths, 1966)
12. Z. Ziemońska, J. Żelaziński, Przegląd Geofizyczny, **T. XXLX, z. 3.** (1984)
13. A. Krajewski, K. Banasik, Acta Scientiarum Polonorum – Architectura, **12**, 4 (2013)
14. V. Mockus, *Soil Conservation Service (SCS) National Engineering Handbook,* **Section 4. Hydrology**, (Soil Conservation Service, US Department of Agriculture, Washington, D.C., 1964)
15. M. Ozga-Zielińska, J. Brzeziński, *Hydrologia stosowana (Applied hydrology)* (Wydawnictwo naukowe PWN, 1994)
16. J. Pociask-Karteczka, *Zlewnia. Właściwości i procesy (The catchment – properties and processes)* (Wyd. Uniwersytetu Jagiellońskiego, 2006)
17. E. Bajkiewicz-Grabowska, Z. Mikulski, Przegląd Geofizyczny, **T. XXLX, z. 3.** (1984)
18. A. Sikorska, K. Banasik, Acta Scientiarum Polonorum - Architectura, **7**, 4 (2008)
19. J. Jovanovic, *Hydrologic approches In Urban drainage system modeling,* Proceedings of the 3rd International Conference "Urban Storm Drainage", C. Maksimiwic, M. Radojkovic (eds.), Pergamon Press, Oxford, UK 185-208 (1986)
20. K. Banasik, M. Barszcz, L. Hejduk, *Current and perspective flood flow consequences of land use changes in Sluzew Creek (suburb of Warsaw),* Special aspects of urban flood management, Proceedings Coast Session Aquaterra Conference 2007, Technische Universitat Hamburg, 3-16 (2007)

SWOT analysis of wastewater heat recovery systems application

Sabina Kordana[1,*]

[1]Rzeszow University of Technology, Department of Infrastructure and Sustainable Development, al. Powstańców Warszawy 6, 35-959 Rzeszów, Poland

Abstract. Limited availability of fossil fuels and various issues connected with environment conservation lead to a growing interest in unconventional sources of energy. The latter include wastewater carried via sewage systems. Yet, the possibility to extract thermal energy from this source depends on a number of factors, location of heat recovery system being one of them. Viability of utilizing raw wastewater and effluent discharged to a receiving body after treatment process, as potential sources of energy, was assessed using SWOT/TOWS analysis. Strategies for developing systems of heat recovery from both types of wastewater have been defined, and recommendations related to their use in Poland have been formulated.

1 Introduction

In recent years there has been visible increase in the density of sewer networks [1]. In Poland this trend can be observed in both urban and rural areas, and one of its consequences is an increase in the volume of household and industrial wastewater carried by intercepting sewers. Such wastewater is recognized as one of the main point sources of freshwater pollution [2, 3], and its treatment to a required level is frequently a complicated and costly process [4]. Yet, wastewater produced by households and commercial establishments should not be seen exclusively as a problem, because it carries significant amounts of waste thermal energy which can effectively be recuperated and utilised in warming up water and heating indoor spaces. In the summer it can be used as a cold source for air-conditioning.

The need to utilize unconventional sources of energy, such as e.g. wastewater carried via intercepting sewers, is a consequence of excessive exploitation of fossil fuels. Of no small importance are also various issues connected with environment conservation as well as problems linked with climate change, whose results affect other domains of life [5, 6]. In Poland the proportional share of unconventional energy in the total energy consumption is growing, yet coal continues to be the main source of energy for district heating systems [7]. This situation is mainly caused by a lack of systematic knowledge related to the possible utilization of waste energy contained for instance in wastewater, as a result of which it is unproductively released into the environment [8].

* Corresponding author: sk@prz.edu.pl

In response to the above problem, the paper characterises raw and treated wastewater, from the viewpoint of their possible utilization as a source of thermal energy. The factors defined for this purpose include strengths and weaknesses of such heat recovery systems as well as potential opportunities and threats which may occur throughout the lifespan of such installations. SWOT analysis was applied as a tool, and its assumptions are described in the following section.

2 Research Methodology

The findings described in the paper have been acquired using SWOT analysis. This is one of the basic methods applied in strategic planning, and it can be used to assess any situation where strategic operations are required. It is assumed that this research method enables identification of factors characterising an entity or enterprise under consideration, in the context of a specified purpose, as well as classification of such factors into four groups. Two of these comprise positive, and the other two negative elements, as it was shown in Fig. 1 [9].

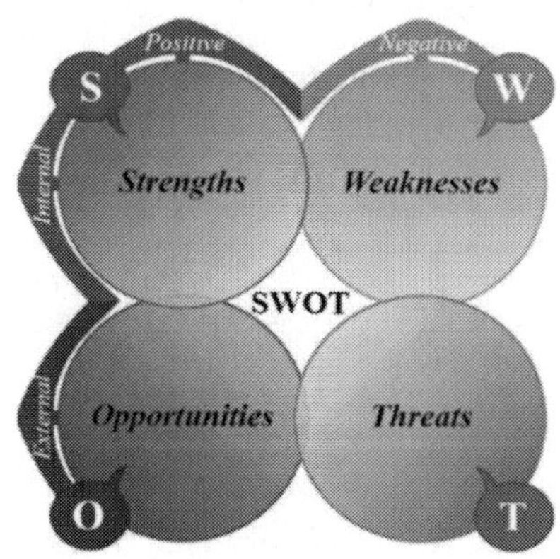

Fig. 1. SWOT analysis diagram.

The defined factors are assigned with weighting coefficients whose total within a given group should equal 1.0; additionally, interactions occurring between internal and external factors are determined. SWOT analysis is designed to examine relationships "from the inside towards the outside". Its purpose is to show, whether the identified strengths will make it possible to take advantage of the potential opportunities and to overcome the threats, and secondly to determine whether the specific weaknesses will magnify the threats and hinder the use of the opportunities [10].

SWOT analysis is supplemented with TOWS analysis, which is designed to identify interactions "from the outside towards the inside". As a result of its application it is possible to show whether the occurring opportunities will enhance the strengths and reduce the weaknesses, and secondly whether the possible threats will magnify the weaknesses and decrease impact of the strengths.

Based on the comprehensive findings of SWOT/TOWS analysis it is possible to determine development strategy for a given entity or enterprise and to define further recommendations [10].

3 Wastewater heat recovery

The idea of utilizing heat carried by wastewater was born a few decades ago, yet due to significant investment costs and poor efficiency of equipment in those times, the concept failed to gain recognition among investors and designers of energy systems. Today it is being investigated again, and the fact mainly results from the limited availability of fossil fuels. Other important reasons include the rapid developments in waste heat recovery technologies, as a result of which heat stored in wastewater may be recovered both in buildings [11] and during its transport via sewer systems [12].

In the first case, greywater from shower is usually the source of heat and the energy recovered from waste by heat exchanger is transferred to water carried by incoming water pipes [13]. On the other hand when energy is carried by wastewater transported via intercepting sewers, heat pumps or large exchangers installed inside or outside the sewer may be used; depending on the conditions in which the process is conducted the latter may function as an evaporator or an auxiliary device of heat pump.

4 Viability assessment of using wastewater as an energy source

In order to assess wastewater heat recovery systems and to increase environmental awareness of the public related to the viability of their implementation in Poland, this concept was subjected to SWOT analysis. Given the difference between the qualities of raw and treated wastewater, the assessment was performed separately for wastewater carried by intercepting sewers towards a treatment plant and for medium transported towards the receiving body after the treatment process.

4.1 Raw wastewater

Opinions provided by a team of experts representing academic community, business enterprises, and potential users of recovered heat, were examined to determine strengths and weaknesses of heat recovery from untreated wastewater. Potential opportunities and threats were also defined, considering the specific conditions of conducting this process in Poland. SWOT analysis took into account five most important factors classified in each group. The factors are presented in Fig. 2.

One of the most important parameters determining the possibility to recover heat from wastewater is its temperature, which depends mainly on the temperature of water used. Mean annual temperature of household wastewater carried via intercepting sewers is approximately 15°C, therefore it can be used as lower source of energy for heat pumps. Heat extracted from wastewater is most often applied for warming up water or for heating indoor spaces. In the summer it is possible to reset heat pumps into cooling operation mode, and to use untreated municipal wastewater as cold source for air-conditioning. Since energy contained in wastewater can be used in varied ways, these heat recovery systems can be applied in both small installations and in heating systems. Importantly, the use of this energy carrier makes it possible to significantly reduce dependence on availability of fossil fuels and on dominating weather conditions.

The strengths of the systems in question also include the general availability of raw wastewater, as it is generated practically in all residential or commercial buildings as well as industrial plants. Particularly favorable conditions occur in the case of trunk sewers which carry large quantities of warm wastewater. This is because flow rate is another parameter determining the possibility to utilize heat carried by wastewater.

Additional advantage of raw wastewater heat recovery systems lies in the fact that they can be located in direct proximity to heat consumers. Consequently, energy recuperated from wastewater can be utilized in the same buildings from which this medium is removed. In the most advantageous situation the installation in question is located in the vicinity of commercial facilities and shops. This is because buildings of this type are recipients of both heating and cooling energy [14], as a result of which energy potential of wastewater can be used most effectively.

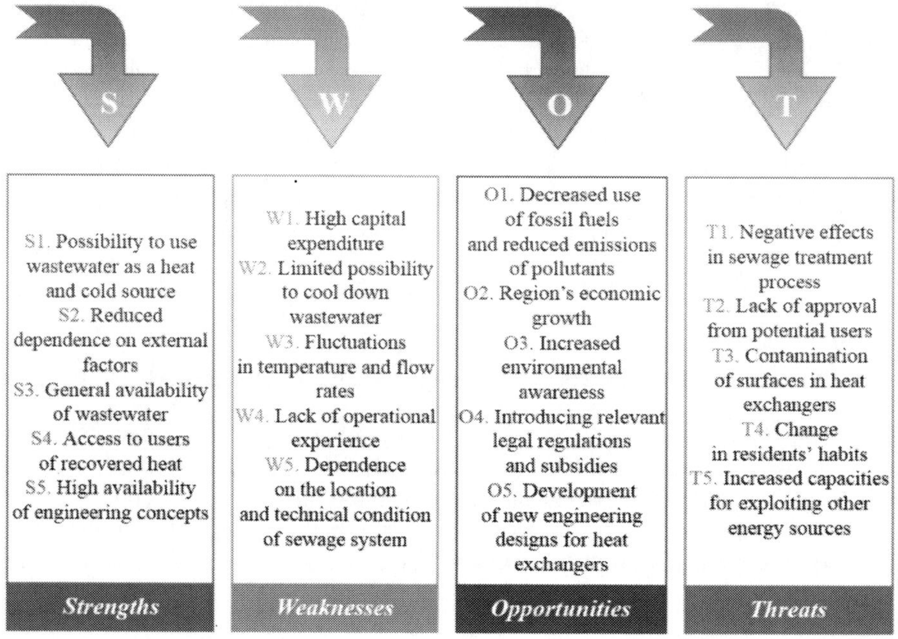

Fig. 2. SWOT analysis of exploiting raw wastewater heat recovery systems.

Another strength of such undertaking is linked with the large variety of available engineering concepts for heat exchangers, which means that they can be installed in both gravity and pressure sewers, and in specific cases also outside the sewage system. As a rule low capacity installations are applied in existing sewers. If a new sewage system is being built, or in a situation where worn out pipes are being replaced with new ones, it is possible to use heat exchangers embedded in sewage piping.

Wastewater heat recovery systems also have weaknesses. These include the significant capital expenditures linked with the cost of purchasing heat exchanger and pump, circulating pumps, and pipes of the heat exchange circuit. Notably, if a heat exchanger is situated upstream a treatment plant, the quantity of recovered thermal energy is reduced.

Weaknesses of the systems in question also include possible significant fluctuations in the temperature and flow rates of the warm medium. Such fluctuations usually occur within the diurnal and annual cycle. During a year differences in the temperature of wastewater may exceed ten degrees, which poses a technological challenge and may reduce effectiveness of an installation. In the case of systems designed for heat recovery from combined sewers there is another drawback connected with occurrence of precipitation which may significantly impact the system's effectiveness, just like in other facilities along sewage system [15, 16]. These problems are additionally compounded by the lack of experience in operating such systems, as in Poland they have not been used previously, as well as by the fact that feasibility of heat exchanger installation is linked with the technical condition and geometry of sewers.

The opportunities linked with extracting energy from raw wastewater include the potential positive environmental effects, resulting from reduced consumption of fossil fuels and decreased emissions of combustion products.

Construction of systems based on unconventional energy sources will also contribute to increased value of real properties located in their proximity and to creating new jobs necessary for the operation of such systems. Moreover, because of the modular structure of heat exchangers as well as the significant length of sewers, it will be possible to expand the systems with growing demand for energy. Consequently the above factors will contribute to region's economic growth.

Rapid expansion of the systems under consideration may also be facilitated by increased environmental awareness of the public and by newly implemented legal regulations enforcing utilization of energy carried by wastewater, particularly if such laws provide for an opportunity to acquire subsidies for construction of such installations. Additional stimulus promoting recovery of energy from wastewater will come from new developments in the technology; these will include engineering designs of heat exchangers, which will be dirt resistant and far more affordable than those available today.

Factors threatening implementation of these systems include their negative effects related to sewage treatment process. A decrease in the temperature of sewage, particularly if it is long lasting, results in visibly reduced effectiveness of biological treatment to which this medium is subjected [17]. Such operations may also be challenged by a lack of approval from users of recovered heat because ignorance related to this technology and the long time needed for return on investment are discouraging factors when it comes to building sewage heat recovery systems.

Another threat for the systems discussed here is connected with the possible contamination of surfaces in heat exchangers. This is linked with the significant amount of pollutants in raw wastewater; these may deposit on the surface of heat exchangers, impairing their effectiveness. This problem is particularly significant in the case of exchangers embedded in sewers, due to the fact that it is difficult to inspect such devices.

While designing systems for recovery of heat from raw wastewater it is necessary to remember about possible changes in people's preferences related to the temperature of water used for household needs, as well as temperature in indoor spaces. Other important issues are connected with new resources of fossil fuels, such as shale gas, which are being discovered, and with the fact that devices utilizing conventional fuels are more and more effective.

4.2 Treated wastewater

Factors taken into account in the analysis of systems designed for heat recovery from treated effluent are listed in Figure 3. Some elements assigned to the specific groups are identical as those considered in the analysis performed for raw wastewater. Given this, the following part of the paper discusses only those factors which are characteristic exclusively for treated wastewater.

Temperature of treated effluent is slightly lower than that of wastewater flowing into facilities of treatment plant, yet it is possible to induce its decrease by even a few degrees, enabling greater recovery of energy. Moreover, fluctuations in temperature and flow rate in effluent carried towards the receiving body are insignificant, which ensures stability of the system's operation. Additionally, because the effluent contains no contaminations, it can be carried directly into heat pump evaporator, without a need to use intermediate circuits.

A drawback of the systems in question, in addition to the required significant capital expenditures and lack of experience in operating such systems, is connected with the location of treatment plants, because these are generally situated far from potential users of recovered heat. Another challenge is the limited length of discharge pipes carrying the effluent towards the receiving body, and due to this in many cases location of heat exchanger outside the treatment plant is not feasible. One more weakness of purified wastewater heat recovery systems, taken into account in the analysis, is the lack of any activities promoting operations aimed at extracting energy from wastewater. This is because production of electric power from biogas is far more profitable for sewage treatment plant operators than recovery of energy carried by wastewater.

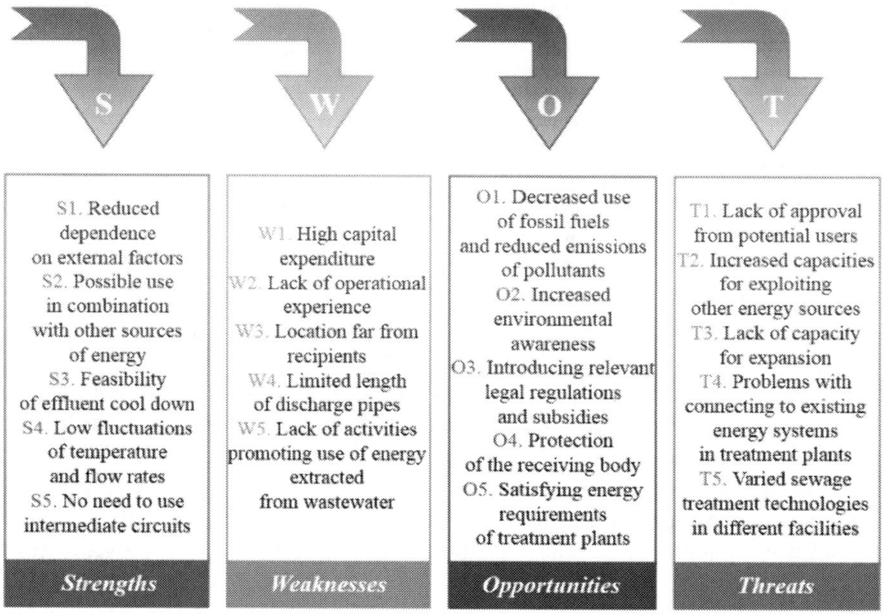

Fig. 3. SWOT analysis of exploiting treated wastewater heat recovery systems.

Opportunities identified as characteristic exclusively for treated wastewater include the possibility to protect the receiving body as a result of the fact that the effluent discharged has similar temperature as the water in the river. Furthermore, heat extracted from this medium may be applied for the needs of the treatment plant, for instance for warming up water used by its personnel, or for increasing the temperature of sewage flowing into the plant facilities, in order to improve effectiveness of biological treatment.

Factors recognized as threats to development of these systems include difficulties related to potential expansion due to the limited length of discharge pipes, which makes it impossible to connect additional modules of heat exchanger. Moreover, some sewage treatment plants may face problems in adjusting the designed heat recovery installation to the existing power system, particularly due to the fact that differences in sewage treatment capacity and technology in various facilities may be rather significant as a result of which standard solutions are not feasible.

4.3 Defining development strategy for wastewater heat recovery systems

During the next stage of the study, weights were assigned to the specific factors (separately for raw and treated wastewater), and interactions between internal and external factors were specified. Following the guidelines described in a study [10], eight supporting tables were drawn up for each type of wastewater, and based on these totals were determined for the interactions as well as products of weights and interactions. The obtained results of the analyses are shown in Tables 1 and 2. Yet, the presented findings related to both raw and treated wastewater do not provide an explicit recommendation regarding the choice of development strategy for heat recovery systems.

In the case of SWOT analysis, as well as the comprehensive SWOT/TOWS assessment of raw wastewater heat recovery systems, the highest scores were obtained for a combination of the positive factors, which is an evidence that these systems have a significant potential for development. This also shows a need for taking advantage of their strengths, in particular general availability of raw wastewater, in order to benefit from the potential opportunities, such as feasible improvement in the state of the natural environment. However, the results obtained for a combination of weaknesses and external factors are only slightly lower, which suggests the need for minimizing the drawbacks of these systems, most importantly the limitations related to cooling down wastewater. Furthermore, the findings presented in Table 1 show significant vulnerability of these systems to threats occurring in the future, mainly lack of approval from potential users and the possible contamination of surfaces of heat exchangers by raw wastewater. Due to this it is necessary to take action aimed at further development of wastewater heat recovery technologies and to promote sustainable management of waste energy among potential users, as this is the only approach making it possible to utilize the opportunities resulting from recovery of thermal energy from the medium transported towards treatment plant.

Table 1. The summary of analysis results for raw sewage.

Quadrant combinations	**SWOT analysis**		**TOWS analysis**		**SWOT/TOWS analysis**	
	Sum of interactions	Sum of products	Sum of interactions	Sum of products	Sum of interactions	Sum of products
S / O	**38/2**	**7.80**	22/2	4.75	**60/2**	**12.55**
S / T	16/2	2.95	**30/2**	**6.35**	46/2	9.30
W / O	36/2	7.05	18/2	3.65	54/2	10.70
W / T	36/2	6.55	24/2	4.90	**60/2**	11.45

Table 2. The summary of analysis results for treated sewage.

Quadrant combinations	**SWOT analysis**		**TOWS analysis**		**SWOT/TOWS analysis**	
	Sum of interactions	Sum of products	Sum of interactions	Sum of products	Sum of interactions	Sum of products
S / O	24/2	5.65	**28/2**	**6.45**	52/2	**12.10**
S / T	24/2	4.35	16/2	3.90	40/2	8.25
W / O	**32/2**	**6.20**	22/2	4.50	**54/2**	10.70
W / T	26/2	4.80	26/2	4.95	52/2	9.75

A different strategy should be applied with regard to systems designed for treated wastewater heat recovery. SWOT analysis showed the largest number of interactions for the combination of weaknesses and opportunities, which suggests that viability of heat recovery process is significantly affected by the weaknesses of the systems in question. In this case significant capital expenditures and limited length of discharge pipes downstream from treatment plant are of the greatest importance, yet benefitting from potential opportunities may be hindered by the other factors classified in this group as well. On the other hand, however, the high total of products obtained for the combination of the positive factors in SWOT analysis, as well as the results of TOWS analysis for the same combination of factors, suggest that opportunities existing in the external environment, mainly the possibility to satisfy energy related needs of treatment plants, may significantly strengthen development possibilities of these systems. Ambiguity of the findings is confirmed by the results of the comprehensive SWOT/TOWS analysis, and this shows the need for further research into utilization of thermal energy carried by treated wastewater.

5 Conclusion

The results of the analysis show that heat recovery systems have significant potential for development. However, in the case of raw or treated wastewater, the matter should not be approached without questioning. Utilization of heat carried by wastewater presents drawbacks, which will be difficult to overcome, given the existing legal regulations and the public's environmental awareness. Therefore, generalized recommendations with regard to utilization of energy contained in wastewater would be ill-advised or in fact impossible; consequently it is necessary to approach each project related to recovery of heat carried by wastewater on a case-by-case basis.

The conducted study has also shown that, contrary to the common opinion, raw wastewater presents more significant advantages, most notably the general availability of the resource. In Poland, however, the importance of this energy carrier is clearly underestimated, even though its utilization could increase the country's independence from external deliveries of fossil fuels and contribute to improving the state of the natural environment. Therefore it is necessary to conduct widespread promotion of this energy extraction method which will result in increased awareness related to sustainable energy systems and will contribute to gaining consumers' approval of energy extracted from wastewater. One must not forget, however, that the main purpose of sewer system is to remove and enable effective treatment of sewage, and these functions must not be neglected.

References

1. M. Starzec, J. Dziopak, M.I. Alexeev, Water and Ecology **1** (2015)
2. P. Koszelnik, Carpath. J. Earth Env. **9** (2014)
3. M. Zeleňáková, M. Čarnogurská, M. Šlezingr, D. Słyś, P. Purcz, Hydrol. Earth Syst. Sc. **17** (2013)
4. F. Hernandez-Sancho, M. Molinos-Senante, R. Sala-Garrido, Desalination **268** (2011)
5. B. Kaźmierczak, A. Kotowski, Theor. Appl. Climatol. **118** (2014)
6. K.B. Pochwat, JCEEA **63(2/II)** (2016)
7. J. Danielewicz, B. Śniechowska, M.A. Sayegh, N. Fidorów, H. Jouhara, Energy **108** (2014)
8. A. Stec, S. Kordana, Resour. Conserv. Recy. **105** (2015)
9. N. Pahl, A. Richter, Swot Analysis – Idea, Methodology and a Practical Approach (GRIN Verlag, 2009)
10. K. Obłój, Strategia organizacji (PWE, Warsaw, 2007)
11. M. Ramadan, T. Lemenand, M. Khaled, Energ. Buildings **128** (2016)
12. S.S. Cipolla, M. Maglionico, Energy Procedia **45** (2014)
13. S. Kordana, D. Słyś, J. Dziopak, J. Clean. Prod. **82** (2014)
14. D. Monfet, M. Corsi, D. Choinière, E. Arkhipova, Energ Buildings **81** (2014)
15. D. Słyś, J. Dziopak, Pol. J. Environ. Stud. **20** (2011)
16. D. Słyś, A. Stec, Environ. Prot. Eng. **38** (2012)
17. O. Wanner, V. Panagiotidis, P. Clavadetscher, H. Siegrist, Water Res. **39** (2004)

Hybrid ventilation in residential buildings – the proposal of research for the Polish climatic conditions

Maria Kostka[1,*]

[1]Wrocław University of Science and Technology, Faculty of Environmental Engineering, Wrocław, Poland

Abstract. Due to the geographic location and the influence of the polar maritime and polar continental air masses we observe a large variety and variability of external air parameters in Poland. Both during the whole year and also in shorter, even several-day periods. The parameters of outside air indicate that combination of mechanical and natural ventilation in detached buildings can be cost-effective and allows keeping the thermal comfort and good indoor air quality.

1 Climate

Poland is located in the transitional area between temperate oceanic climate on the west, a temperate continental on the east. The biggest influence on the climate are polar maritime air masses (ca. 60% of the days in a year) and polar continental air masses (ca. 30% of the days in a year). The first of these, causes cloudiness and high relative humidity in summer, while in winter - warming and the appearance of fog. The second of them carry a high temperature in summer and heavy frosts in winter. Moderate weather, warm, cloudy and without precipitation is the most common type. According to A. Woś [1] this type of weather occurs in central Poland for about 25% of time in summer, and for 17% of time in spring and autumn. Figure 1 shows the duration of the outside air temperature for Poland and Wrocław, developed on the basis of the statistical climate data from 58 weather stations, provided by the Ministry of Infrastructure and Construction [2]. The analysis does not include three stations (Hel, Śnieżka and Kasprowy Wierch), which, due to their location, are not representative of the whole country. The number of hours with the temperature below 0°C in Wrocław is 4% reduced in comparison with the national average. The total number of hours with the temperature between 0°C and 20°C, in spite of the differences for the individual values, is the same for the whole country. The differences in the outside temperature above 20°C can be observed clearly. In Wroclaw there is 12% of such hours more, compared with the national average. This commeasures with the data presented in the literature [3] where the southwest region of Poland, especially the Odra River Valley, is presented as the warmest in the country, with the average annual temperature of the outside air exceeding 8°C.

* Corresponding author: maria.kostka@pwr.edu.pl

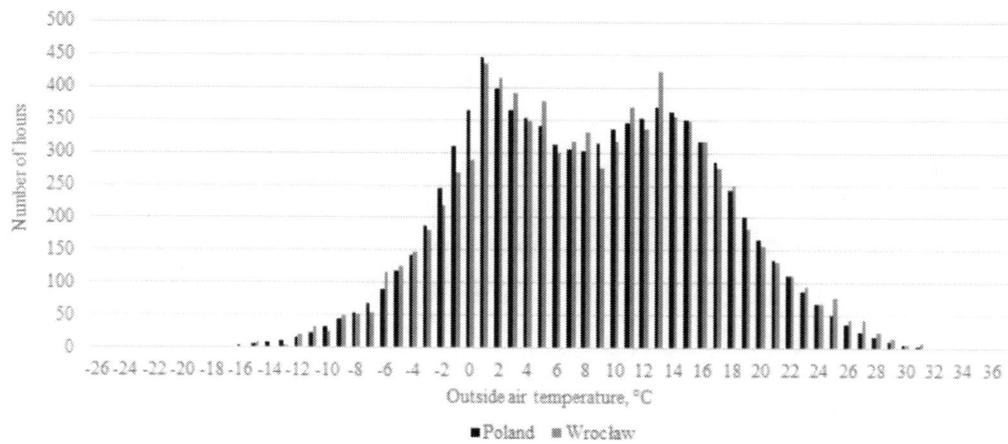

Fig. 1. The duration of the outside air temperature.

The percentage of outside air temperature ranges during the year for Poland and Wroclaw is shown in Figure 2. According to the drawing, for approximately 40% time of the year the temperature ranges between 1°C and 10°C. Clearly the dominant range is also 11°C - 20°C, which is almost 35%.

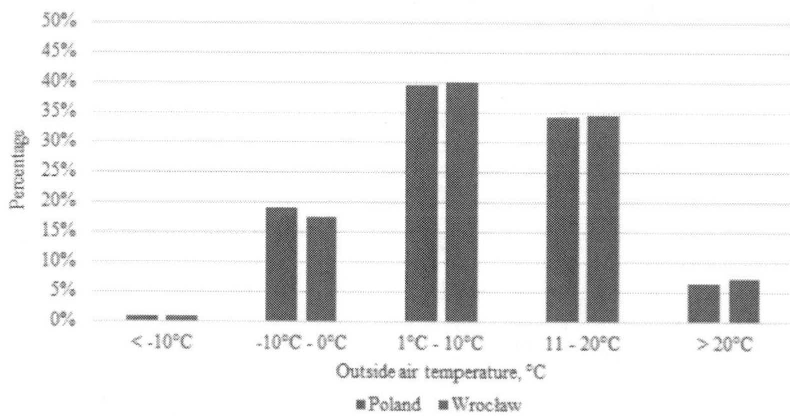

Fig. 2. The percentage of outside air temperature ranges during the year.

In Figure 3, the Mollier diagram describes the parameters of the outside air in Wrocław. The chart was developed on the basis of the Ministry data [2]. The highest outside air relative humidity in Poland is observed in winter – in January it ranges from 80 to 85%, and the lowest in summer – in June, about 55 - 60% [3]. According to the Ministry, the average value of relative humidity in January in Wrocław is approximately 85% and does not diverge from the value of the country median. In summer the relative humidity is much higher than the average for the entire country: June and July 74%, August 72%. One of the reasons of it is the presence of rivers and watercourses such as Odra, Bystrzyca, Ślęza, Widawa, Dobra and Zielona.

The winds from the west are prevalent in Poland (30 - 35%). South winds occurs during 20 - 25% of the year, and the east and north ones 15 - 20% [3]. East winds dominate in Wrocław only in March. In other months the west winds are prevalent [1]. The national average wind speed is 3 - 4 m/s [3] and coincides with the average for Wrocław, which is approximately 3 m/s [2].

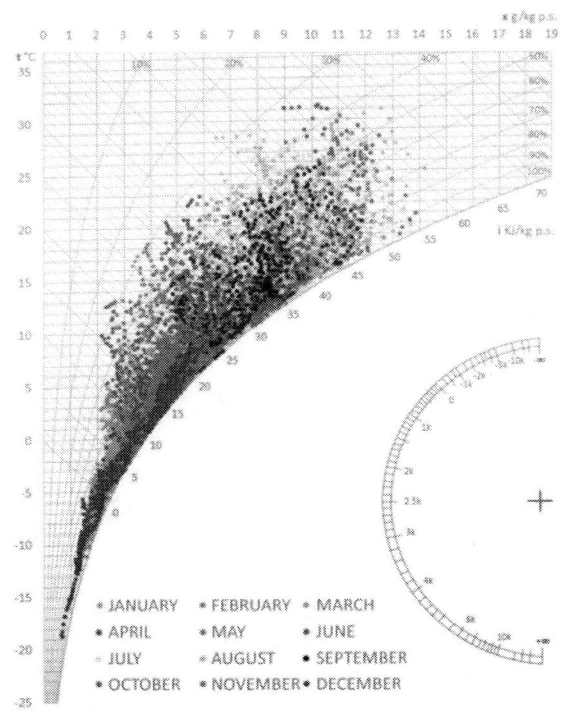

Fig. 3. The parameters of outside air – Wrocław.

2 Residential ventilation

The function of the ventilation in detached buildings is to ensure indoor air quality by removing the contaminants which appeared during regular exploitation of the building (e.g. laundry, cooking, metabolic contaminants) and replace it with fresh air. A typical ventilation in detached buildings in Poland is implemented in one of two ways – natural/gravity or mechanical. In the natural system the driving force of the air exchange is the pressure difference on the building partition, caused by the difference of internal and external temperature and wind. This is the basic system used over many years in the Polish construction industry. Along with the change of regulations related to the quality of buildings, their thermal insulation improvement and energy demand reduction the currently used mechanical systems are becoming popular. This type of ventilation in detached buildings reduces energy consumption, needed in the natural system for heating fresh air which is entering directly from the outside. Air exchange is provided by two fans placed in the air handling unit (AHU), equipped also with an exhaust air heat recovery exchanger, air filters and a heater. Typical heat recovery exchangers in the mechanical devices used in Poland are counterflow, cross–counterflow and rotary heat exchangers. Using of the heat recovery is crucial and justify the use of mechanical systems. In case of gravity ventilation used in low-energy building, financial expenses on outdoor air heating become a significant operating expense [4]. Mechanical ventilation allows reducing the cost of ventilation and ensures the constant air exchange efficiency, which is not possible in the case of natural ventilation in the Polish climate.

In detached buildings all sorts of natural systems assisted mechanically can be observed. They are called hybrid systems. P. Wouters, N. Heijmans, C. Delmotte and L. Vandaele [5] defined hybrid ventilation as a system that provides a comfortable internal environment using both, natural and mechanical systems at different times of the day and/or season of the year. Another solution can be mixed–mode ventilation (MMV) described by the University of California. It is designed and performed for exchanging the air in 3 ways: *concurrent mixed–mode operation*, *change–over design* and *zoned system* [6]. In *concurrent mixed–mode operation* mechanical device and operable windows operate in the same space and at the same time. In *change–over design* the building "changes-over" between natural ventilation and air conditioning unit on a seasonal or even daily basis. If we realized different ventilation strategies in different parts of building we can called this strategy *zoned system*.

3 Literature review

The mixed–mode ventilation is a relativity new subject area so there is no complete guide how to design and simulate such buildings [7]. There is also no direct literature sources evaluating the possibility of the use of this type of systems in the residential ventilation. But in the international professional press there are results of research on hybrid ventilation of this kind of buildings. The studies on the ventilation systems in residential construction, the natural ventilation potential and indoor air quality are currently conducting all over the world.

The assessment of the potential of the natural ventilation was made: for warm condition and temperate countries – I. Oropeza-Perez and P.A. Østergaad [8, 9], for warm and humid conditions – M. Haase, A. Amato [10], for United States – K. Hiyama, L. Glicksman [11] and Z. Cheng, L. Li, W.P. Bahnfleth [12]. Due to the lack of a comprehensive assessment of the possibilities for Poland, the closest example tests for a country with a temperate climate are the studies conducted by Ivan Oropeza-Perez and Poul Alberg Østergaad for Danmark house [9]. They showed that it is possible to maintain thermal comfort using the passive cooling with the 90% reduction of hours of the mechanical ventilation use. Because of the milder climate of Denmark we can expect smaller reductions in the use of mechanical ventilation in Poland.

The influence of window opening ventilation control on energy consumption and thermal comfort in residential building, naturally ventilated, located in Brazil presented by M.J. Sorgato, A.P. Melo and R. Lamberts [13]. They showed a significant impact of the residents' behavior on the thermal parameters of buildings. Research on optimization of opening dimensions for naturally ventilated buildings presented by L. Stephan, A. Bastide and E. Wurtz [14].

Studies of hybrid ventilation system for residential building with natural supply inlets and mechanical exhaust performed M. Kim and J. Hwang [15]. For an apartment located in Seoul they attempted to determine the airflow which provides comfortable thermal conditions and good air quality, also taking into account energy savings. Also hybrid systems studies described D. Jreijiry, A. Husaunndee and C. Inrad [16]. They compared two systems – mechanical extract and hybrid low-pressure in family building located in different locations (Greece, France and Sweden). The research showed that hybrid ventilation enables a decrease of more than 50% of the exposure of the occupants to high concentrations of CO_2 while reducing at the same time the electrical energy consumed by the fan by 90%.

Research of mixed–mode ventilation with natural ventilation and air conditioning system for office buildings presented M. Luo, B. Cao., J. Damiens, B. Lin and Y. Zhu [17], S. Ezzeldin, S.J. Rees [18], M. P. Deuble, R. J. de Dear [19] and A. Honnekeri, G. Brager, S. Dhaka and J. Mathur [20]. The authors focused on the building users thermal comfort and showed differences in their perception during the operation of natural and mechanical systems. Ezzeldin and Reeds also demonstrated the energy savings of mixed-mode ventilation.

The studies conducted all over the world showed that combination of natural and mechanical ventilation systems can reduce energy consumption and provide thermal comfort. However, none of the research shows the case of Poland. Due to the differences between the countries, including e.g. climatic conditions or typical building structures, the results of the research should not be directly transferred to Poland. For this reason, there is a need to assess the potential of the hybrid systems operating in the Polish conditions.

4 Ventilation system control strategy

Considering the ventilation systems, the parameters of outside air in winter cause that it is reasonable to use mechanical ventilation equipped with heat recovery from exhaust air. This saves the energy needed for building heating. Also during summer, in the case of high outside temperature and days of windless weather, the use of fans and heat recovery system is justified because it maintains a proper air exchange in the building and reduces overheating of the interiors. The time of spring, autumn, summer evenings and nights are often characterized by parameters of air, which allow ventilation of buildings in a natural way. As a result, the consumption of electricity is reduced.

The use of mechanical ventilation system with heat recovery is economically justified when energy recovery from exhaust air enable reducing the costs associated with the building utilization. This situation in single-family buildings in Poland concerns mainly the heating season. It generally means that this ventilation would be cost-effective as long as the supply air temperature after heat recovery exchanger is higher than the outside air temperature, which would be delivered to the building in the case of natural ventilation. In fact this situation is not obvious and requires analyzing the costs of obtaining energy from mechanical ventilation (fans work, heat exchanger freeze protection) and the cost of producing energy in the heat source used in the building. Also heat gains and the phenomenon of the interiors overheating during the summer should be considered.

The aim of the research is to elaborate the energy-efficient ventilation control algorithm for detached houses, allowing to minimize the heating and ventilation costs, maintain comfortable indoor conditions and air quality. The analysis includes the type of heat source, the efficiency of the heating system, the type and efficiency of the heat recovery exchanger and the object location in Poland. Algorithms for manual and fully automatic switching strategy will be considered. Figure 4–5 shows examples of ventilation control algorithms with unknowns marked, which will be defined on the basis of the carried out analysis.

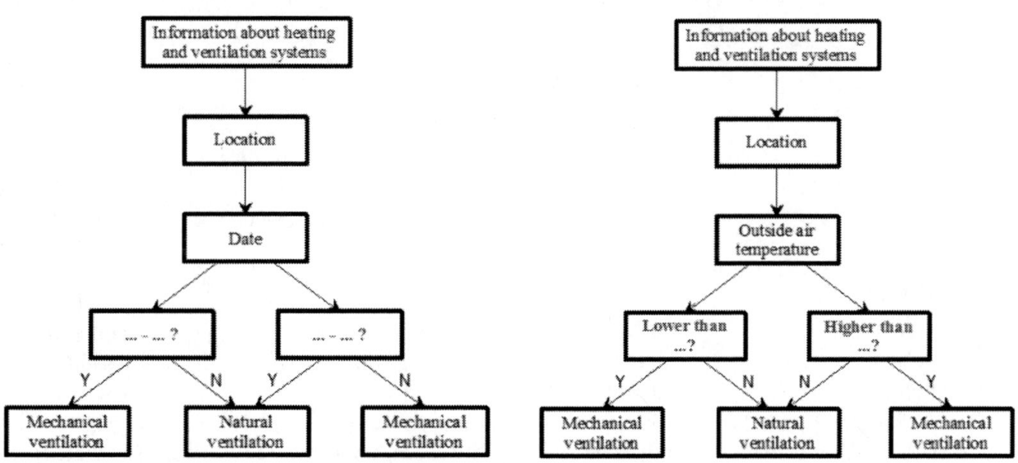

Fig. 4. Control algorithm for seasonal strategy (left) and outside air temperature strategy (right).

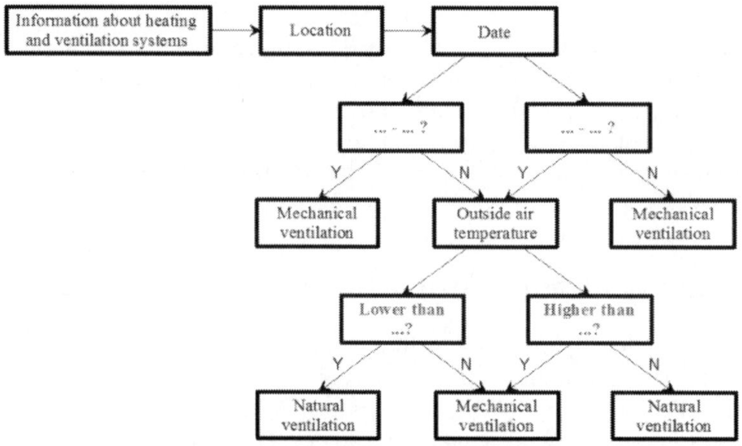

Fig. 5. Control algorithm for seasonal and outside air temperature strategy.

5 Proposal of the research in situ

For the proposed control strategies research on the existing detached house will be carried out. The hybrid change–over ventilation system was designed for building located near Wrocław.

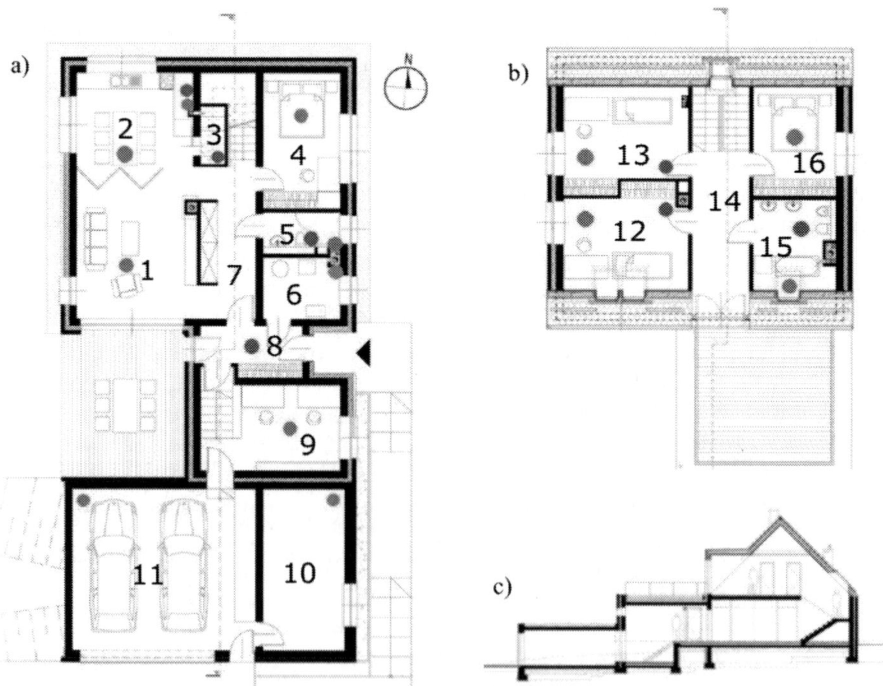

Fig. 6. Research object – detached house located near Wrocław: a) ground floor, b) attic, c) cross–section [21].

mechanical ventilation supply; mechanical ventilation exhaust; temporary natural ventilation exhaust; constant natural ventilation exhaust

1 –living room, 2 – kitchen, 3 – pantry, 4 – guest room, 5 – bathroom, 6 – technical room, 7 – corridor, 8 – vestibul, 9 – study room, 10 – utility room (unheated), 11 – garage (unheated), 12 – bedroom, 13 – bedroom, 14 – corridor, 15 – bathroom, 16 – bedroom.

The building area is 190 m^2, 136 m^2 of which is heated living space. It was built on a small hill and divided into 3 sections located at different heights (Figure 6c). The lowest part is the gravity ventilated garage and utility room, thermally isolated and separated by the airtight door. Above there is a study room, and the highest part is a two-storey living area. For those two parts the mechanical system and natural system was assumed and covered by the internal comfort and energy consumption research. The calculated heat demand of the building is 7.0 kW (51.5 W/m^2 for heated zone). The heat source for the building is an air–water heat pump.

Mechanical supply and exhaust air volume is 350 m^3/h and AHU is equipped with 85% efficiency rotary heat exchanger. In addition to the mechanical ventilation system for the ground floor rooms – the kitchen, the bathroom and the technical room (Figure 6a) the convective channels were constructed. The last of these rooms will be ventilated only in the natural way. For the rest of the rooms the separation of the systems are the spring return dampers. For the mechanical supply and exhaust installation the dampers closing by turning off the AHU were applied. The natural exhaust air from the bathroom in the attic (Figure 6b) will be provided through the opening of the roof window. The flow of air into the building during the action of natural ventilation is provided by manual unsealing/opening windows.

The real object research will consist of the measurements of indoor air parameters and parameters of air in the mechanical ventilation system (temperature, relative humidity, concentration of CO_2, VOC). The heating and ventilation systems energy consumption data will also be analysed.

6 Expected results

The scheduled research will allow to identify the areas of Poland and the hybrid ventilation systems control strategies with the best economic effects. The research on the real object will verify the effectiveness of the ventilation systems and will be a data source about the indoor air quality and the thermal comfort in the naturally and mechanically ventilated residential buildings.

Depending on the hybrid system control strategy, the expected reduction of energy demand for the ventilation is from about 15% to over 50%. It is necessary to include the energy consumption of all systems that provide thermal comfort in the building, so that the impact of hybrid ventilation on the final energy needed during the whole year will be assessed. The results of the final energy demand reduction depend mainly on the building energy standard, the climatic zone of Poland, the type of heat source and the ventilation system configuration. Expected reduction in final energy consumption for the purpose of ventilation and heating of buildings that meet current regulations in Poland is 10% - 20%. It should be noted that it is planned to strengthen regulations in the coming years. That will achieve better results, even exceeding 30%. For the building described above, the use of natural ventilation for example during the outdoor temperature of 15°C - 25°C will reduce the energy consumption for ventilation by about 25%. Seasonal use for example from May to the end of September will increase it to about 40%. The estimated reduction of final energy consumption for heating and ventilation of this building is approximately 20%.

7 Conclusions

Single-family buildings represent a large share of the residential market in Poland and they are impacting significantly the energy consumption. For this reason, it is necessary to explore solutions that will lead to a reduction of energy needs in this sector. The results of the world research and the initial analysis of the parameters of the outside air in Poland indicate the possibility of reduction of the energy consumption in buildings by using hybrid ventilation systems that combine mechanical and natural installations.

This paper presents the proposal of the research that will lead to the creation of hybrid ventilation strategies for Poland. With the calculation analysis and simulations, the potential of this systems to energy consumption reduction will be assessed. The research in the existing building will evaluate the indoor air quality, thermal comfort and efficiency of natural, mechanical and hybrid systems.

This work is supported by funds from the Ministry of Higher Education for research and development works of young scientists and doctoral students. Project number: 0402/0098/16.

References

1. A. Woś, *Struktura sezonowa klimatu Polski* (Bogucki Wydawnictwo Naukowe, Poznań, 1996)
2. http://mib.gov.pl/2-Wskazniki_emisji_wartosci_opalowe_paliwa.htm# [access: 26.01.2017]
3. U. Kossowska – Cezak, D. Martyn, K. Olszewski, M. Kopacz – Lembowicz, *Meteorologia i klimatologia, Pomiary, obserwacje, opracowania* (PWN, Warszawa, 2010)
4. M. Kostka, M. Szulgowska – Zgrzywa, Rynek Instalacyjny **3** (2016)
5. P. Wouters, N. Heijmans, C. Delmotte, L. Vandeale, *Classisication of hybrid ventilation cencepts*, The Second International Forum of Natural and Hybrid Ventilation, HybVent Forum 99, Sydney, Australia 1999
6. http://www.cbe.berkeley.edu/mixedmode/aboutmm.html [access: 27.01.2017]
7. J.C. Salcido, A.A. Raheem, R.R.A. Issa, Energ. Buildings **127**, 1008-1018 (2016)
8. I. Oropeza – Perez, P.A. Østergaard, Appl. Energ. **130**, 20-32 (2014)
9. I. Oropeza – Perez, P.A. Østergaard, Appl. Energ. **14**, 520-530 (2014)
10. M. Haase, A. Amato, Sol. Energy **83**, 389-399 (2009)
11. K. Hiyama, L. Glicksman, Energy **89**, 655-666 (2015)
12. Z. Cheng, L. Li, W.P. Bahnfleth, Build. Environ. **108**, 85-98 (2016)
13. M.J. Sorgato, A.P. Melo, R. Lamberts, Energ. Buildings, **133**, 1-13 (2016)
14. L. Stephan, A. Baside, E. Wurtz, Appl. Energ. **88**, 2791-2801 (2011)
15. M. Kim, J. Hwang, Energ. Buildings, **41**, 579-586 (2009)
16. D. Jreijiry, A. Husaunndee, C. Inard, Sol. Energy **81**, 227-239 (2007)
17. M. Luo, B. Cao, J. Damiens, B. Lin, Y. Zhu, Build. Environ. **88**, 46-54 (2015)
18. S. Ezzeldin, S.J. Rees, Energ. Buildings **65**, 368-381 (2013)
19. M.P. Deuble, R.J. Dear, Build. Environ. **54**, 53-60 (2012)
20. A. Honnekeri, G. Brager, S. Dhaka, J. Mathur, *Comfort and adaptation in mixed – mode buildings in a hot-dry climate*, 8th Windsor Conference: Counting the Cost of Comfort in a Changing World 2014
21. http://apeagroup.pl/ [access: 26.01.2017]

Application of nanofiltration membranes for removal of surfactants from water solutions

Izabela Kowalska[1], and *Aleksandra* Klimonda[1,*]

[1]Wroclaw University of Science and Technology, Faculty of Environmental Engineering, Wybrzeże S. Wyspiańskiego 27, 50-370 Wrocław

Abstract. The objective of the study was to compare effectiveness of nanofiltration (NF) process for treatment aqueous solutions containing various types of surfactants (anionic, cationic and non-ionic). The experiments were conducted with the use of Microdyn-Nadir® nanofiltration membranes (NP010 and NP030). The effect of surfactant type, its concentration and membrane cut-off on the process parameters (retention coefficient and permeate flux) was assessed. The experiments showed that separation of anionic and cationic surfactants depended on their concentration in the feed solutions. The retention coefficient of anionic surfactant ranged from 54 to 81% (NP010) and from 64 to 80% (NP030), while rejection of cationic surfactant varied from 48 to 85% (NP010) and from 51 to 88% (NP030). The values of retention coefficient of non-ionic surfactant were in the range of 69-77% and 79-88% for NP010 and NP030, respectively; and to a much lesser extent were depended on its concentration in the feed solutions in comparison with anionic and cationic compounds. Membrane characterized by smaller value of cut-off (NP030) allowed to obtain higher retention coefficients of surfactants. The membrane permeability deterioration was observed with the increase of surfactants concentration in the treated solutions. In all experiments, a sharp decline of the permeate flux was noticed for concentration below the CMC values.

1 Introduction

Surfactants are organic compounds that exhibit surface activity. Due to their amphiphilic structure, surface active agents lower the surface tension between two phases (two liquid or liquid and solid phases) and induct wetting, foaming, emulsifying and dispersing properties of the solutions [1]. A molecule of the surfactant contains both hydrophobic and hydrophilic unit. The hydrophilic part enables molecule to solubilize in polar liquids, while the hydrophobic portion allows the solubilisation in nonpolar liquids. Depending on the hydrophilic unit charge, surfactants are classified into four groups: non-ionic (0), anionic (-), cationic (+) and amphoteric (±) [2].

Due to the surface properties, surfactants are being used in numerous fields of the economy. The main applications of surfactants are: household detergents, toiletries,

* Corresponding author: aleksandra.klimonda@pwr.edu.pl

personal and body care products, cosmetics, industrial cleaning processes as well as textile, pulp and paper or construction industries [3]. In 2016, the surfactants market was estimated to USD 30.64 billion [4] and it is expected to reach USD 46.20 billion by 2022 [5]. The commercially most significant type of surfactants are anionic surfactants, which share more than 50% of the global market [6]. Value of the surfactant market in Poland in 2009 was estimated to USD 165 million [7].

Such wide range of application results in discharging surfactants into sewage system or directly into surface water [8]. Surface active agents occur both in municipal and industrial wastewater [2]. Large amounts of surfactants in sewage stream may result in disorder of biological treatment (activated sludge process) [9], hence surfactant concentration in industrial wastewater discharged into sewage systems in Poland cannot exceed 15 mg/dm^3 and 20 mg/dm^3 for anionic and non-ionic surfactants, respectively [10].

Insufficient wastewater treatment or direct releasing of contaminated sewage into water environment contribute to pollution of water bodies by surfactants. In the surface water, surfactants may reveal injurious effects to living organisms on account of their toxicity and their ability to solubilize of toxic organic compounds [11]. In surfactant-polluted water, oxygen permeation is limited [12]. Thus, the growth and development of the living organisms as well as self-cleaning processes are inhibited. When the surfactants concentration exceeds 0.1 mg/dm^3, the foaming of the water is observed [12].

In view of significant impact of surfactants on the environment, their removal from wastewater is indispensable. Additionally the economic factors lead to research on the recovery of surfactants in purpose of their reuse in manufacturing plants [13]. Pressure-driven membrane processes are employed in many fields of environmental engineering, mainly in water and wastewater treatment plants. Many studies relate to the usability of membrane techniques for removal of surface active agents from aqueous solutions [14–15]. Microfiltration (MF) is used as a pre-treatment method which enables removal of suspended solids before the main process which may be ultrafiltration (UF), recommended for the treatment of the micellar solutions (with critical micellar concentration, CMC) or nanofiltration (NF) which is regarded as an effective method in removal of surfactants in concentration below the CMC (surfactants monomers) [16–18].

The objective of experimental research was to compare effectiveness of nanofiltration (NF) process for purification of aqueous solutions containing various types of surfactants (anionic, cationic and non-ionic) at wide range of concentration.

2 Experimental

2.1 Solutions

Anionic (Triton QS-44, polyether phosphate ester), cationic (CTAB, cetyltrimethylammonium bromide) and non-ionic (Triton X-100, polyethylene glycol tert-octylphenyl ether) surfactants have been chosen for the experiments. The concentration of the surfactants in feed solutions was equal to 0.025, 0.1, 0.25, 0.5, 0.75, 1, 2 and 3 CMC values. The characteristics of surface active agents is shown in Table 1.

Table 1. Characteristics of surfactants.

Surfactant	Triton QS-44	CTAB	Triton X-100
Type	Anionic	Cationic	Non-ionic
CMC, mg/dm^3	250 (22°C)	335 (22°C)	165 (22°C)
Aggregation number	-	170	100-155
Diameter of the micelle, nm	-	3.5	8.5

The concentrations of the surfactants in aqueous solutions were determined using the potentiometric titration method on the apparatus 785 DMP Titrino.

2.2 Nanofiltration process

Flat-sheet nanofiltration Microdyn-Nadir® membranes (NP010 and NP030) made of polyethersulfone (PES) were used in the experiments. Membrane varied in molecular weight cut-off (MWCO) [17]. The parameters of investigated membranes are given in Table 2. The NF process was performed using a laboratory-scale system with an Amicon 8400 UF cell in a dead-end filtration regime. The effective surface area of the membrane was 45,4 cm^2. The volume of the filtration cell amounted to $350x10^{-6}$ m^3. The process was performed under the transmembrane pressure (TMP) of 0.20 MPa.

Table 2. Characteristics of Microdyn-Nadir® nanofiltration membranes.

Membrane type	Membrane polymer	Na_2SO_4 retention (%)	Pore radius [17] (nm)	MWCO [17] (Da)	Surface charge
NP010	polyethersulfone	25–40	0.80-1.29	1010-1400	negative
NP030	polyethersulfone	80–95	0.59-0.93	500-700	negative

Prior to the first filtration cycle, the brand-new membranes were soaked in distilled water for 24 hours and then the volume flux was measured under different transmembrane pressures in a range of 0.05-0.30 MPa. After each filtration cycle, the membranes were cleaned with the use of 0.1 M NaOH and treated with deionised water.

During the experiments, retention coefficient of surfactants and permeate volume flux were determined. To designate the membranes susceptibility to fouling, relative flux (J/J_0) was evaluated.

3 Results and discussion

3.1 Transport properties

Fig. 1 shows the effect of applied transmembrane pressure on membranes transport properties measured as a distilled water volume flux. Three brand-new membranes NP010 and three NP030 were examined.

It was noticed that membrane permeability increased with the applied pressure. The volume flux varied from 5.9 dm^3/m^2h (TMP = 0.05 MPa) to 42.2 dm^3/m^2h (TMP = 0.30 MPa) for membrane NP010 and from 1.1 dm^3/m^2h (TMP = 0.05 MPa) to 7.5 dm^3/m^2h (TMP = 0.30 MPa) for membrane NP030. The NP010 membranes obtained higher distilled water volume flux which is in accordance to membrane pore radius and MWCO reported by literature data [17].

The relative flux obtained after 60 min of the NF process is presented in Fig. 2. As can be seen, for all tested surfactants an increase of their concentration in the feed resulted in membrane permeability deterioration comparing with distilled water flux. During the filtration, a sharp decline of the membranes hydraulic properties was observed when the concentration of surfactants was below the CMC (J/J_0 for membrane NP010 and concentration equal to 1 CMC was 33%, 63% and 29% for Triton QS-44, CTAB and TRITON X-100, respectively; and for membrane NP030 and concentration equal to 1 CMC was 42%, 60% and 29% for Triton QS-44, CTAB and TRITON X-100, respectively). In concentration below the CMC value a phenomenon of surfactants monomer adsorption onto membrane pores occurs. Further increasing of surfactants concentration results in forming micelles which create polarization layer near the membrane surface. Thus, further decrease of membrane permeability is observed.

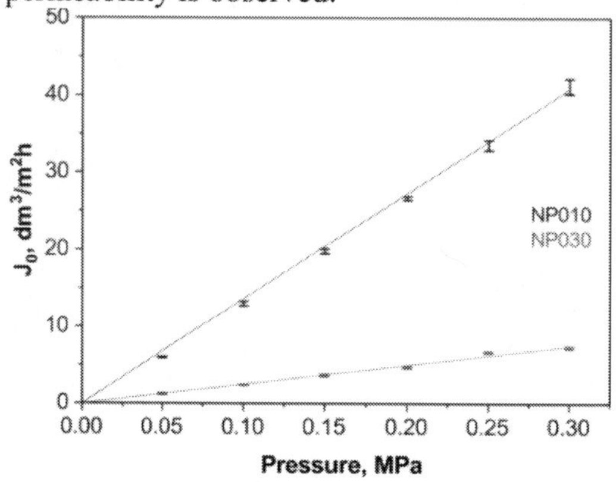

Fig. 1. Distilled water flux *vs* transmembrane pressure.

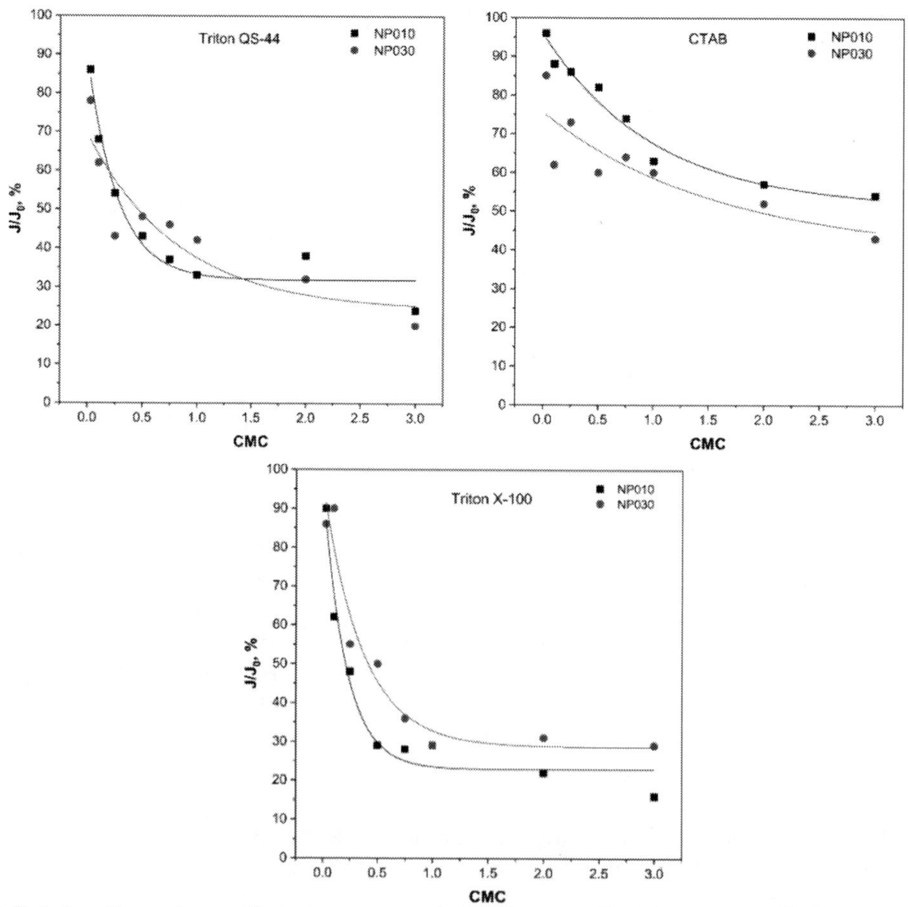

Fig. 2. Relative flux and *vs* surfactants concentration during nanofiltration process (TMP = 0.20 MPa).

During the filtration of non-ionic surfactant Triton X-100, membrane NP030 exhibited higher values of relative flux than membrane NP010 which corresponds with literature data. Kaya et. al [2] reported that membrane NF PES 10 (made of polyethersulfone, MWCO = 1000 Da) was more fouled than membrane N 30F (polyethersulfone, MWCO = 400 Da) during the non-ionic surfactant filtration. It should be mentioned that correlation between membrane pore size and micelle diameter/ monomer length affects process efficiency. Diameter of the membrane NP030 pore (0.59-0.93 nm [17]) is considerably smaller than surfactant's particle size (diameter of Triton X-100 micelle is 8.5 nm). As a result, adsorption of surfactants monomers in membrane pores is impeded. Membrane NP030 allowed to provide relative flux in the range from 86% (0.025 CMC) to 0.29% (3 CMC) when for membrane NP010 (pore radius 0.80-1.29 nm) values of J/J_0 varied from 90% (0.025 CMC) to 16% (3 CMC).

It was observed that presence of cationic surfactant (CTAB) in feed solutions exhibit the lowest influence on membranes transport properties. Obtained relative flux varied from 96% (0.025 CMC) to 54% (3 CMC) and from 85% (0.025 CMC) to 43% (3 CMC) for NP010 and NP030 membrane, respectively.

3.2 Separation properties

Fig. 3 shows the retention coefficient versus surfactant concentration during the filtration.

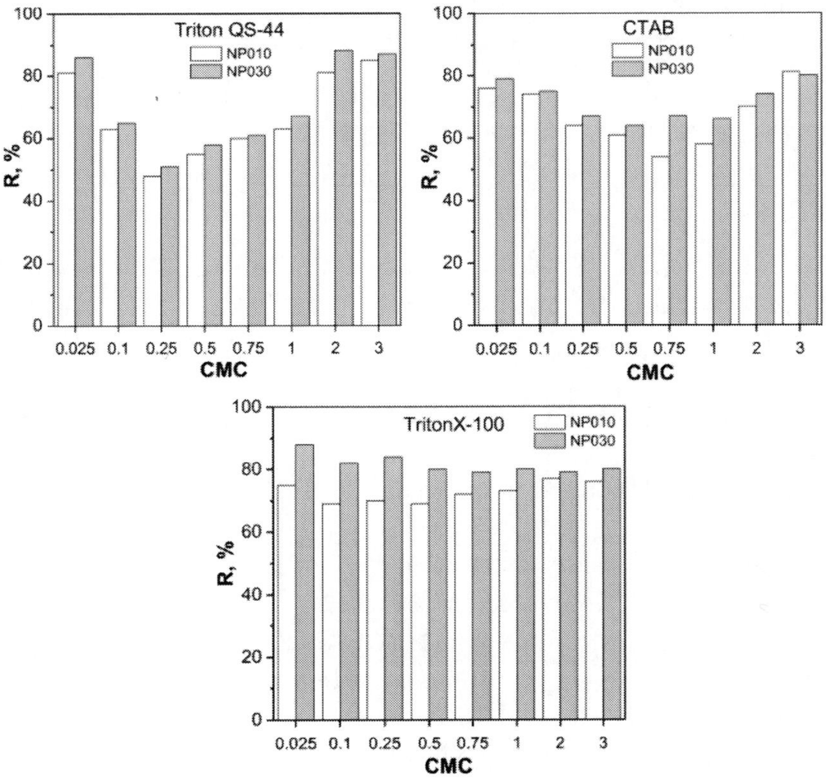

Fig. 3. Retention coefficient *vs* surfactants concentration during nanofiltration process (TMP = 0.20 MPa).

It should be mentioned that the best separation properties of anionic and cationic surfactant were obtained both for very low and the very high concentration in the feed.

Monomer adsorption on membrane surface and within the pores effected on good separation at low concentration [18]. When the concentration increased in the range below the CMC, membrane selectivity deterioration was observed. However, for the concentration above the CMC, the improvement of the retention due to forming of the micelles was noticed. Literature reports partially confirm the results obtained [19]. Membrane filtration of anionic surfactant alkyl-polyether-sulfate (at 20 °C) in a wide range of concentration (below CMC and far above CMC) resulted in retention coefficient from 92% (below the CMC) to >98,5% (~3 CMC).

3 Conclusions

The experimental research showed that nanofiltration process is a satisfactory method for removal of all tested types of surfactants from low-concentration solutions. In all experiments, membrane characterized by smaller MWCO allowed to obtain higher efficiency of surfactants separation.

The experimental results demonstrated that the concentration of surfactant in solution is a crucial parameter which significantly affects on the process efficiency. Increasing dosage of surface active agent in the feed resulted in deterioration of membranes' permeability and strongly affected on separation properties which differed greatly under various values of concentration.

There is a correlation between membrane pore diameter and surfactant molecule size. Removal of the non-ionic surfactant in the process with membrane NP030 exhibited the highest values of retention coefficient - in the rage of 79-88% – due to the significant differences in the monomer length and the membrane pore diameter.

This work was supported by a grant (No. 0401/0064/16) from the Department of Environmental Engineering, Wrocław University of Science and Technology.

References

1. K. Majewska-Nowak, I. Kowalska, M. Kabsch-Korbutowicz, Desalination **184** (2005)
2. Y. Kaya, C. Aydiner, H. Barlas, B. Keskinler, J. Membr. Sci. **282** (2006)
3. R.J. Farn, *Chemistry and Technology of Surfactants* (Blackwell Publishing Ltd, New Delhi, 2006)
4. http://www.researchandmarkets.com/reports/3927849/surfactants-market-by-type-anionic-non-ionic
5. http://www.researchandmarkets.com/reports/3640162/surfactants-market-analysis-by-product-by#relb0
6. Market Study: Surfactants (Ceresana, 2nd ed., 2015)
7. Prospectus of the PCC Exol dated June 27, 2012
8. T. Ivankovic, J. Hrenovic, Arh Hig Rada Toksikol **61** (2010)
9. A. Dereszewska, S. Cytawa, R. Tomczak-Wandzel, K. Medrzycka, Pol. J. Environ. Stud. **24**, 1 (2015)
10. Journal of Laws of 2016, item 1757 (*in Polish*)
11. R. Romczask-Wandzel, A. Dereszewska, S. Cytawa, K. Medrzycka, Research and application of new technologies in wastewater treatment and municipal solid waste disposal in Ukraine, Proceedings of a Polish-Swedish-Ukrainian seminar, Lviv Ukraine, October 26-28, 2006

12. C.L. Yuan, Z.Z. Xu, M.X. Fan, H.Y. Liu, Y.H. Xie, T.Zhu, J. Chem. Pharm. Res. **6**, 7 (2014)
13. H. Byhlin, A. Jonsson, Desalination **151** (2002)
14. L.A.T. Nguyen, M. Schwarze, R. Schomacker, J. Membr. Sci. **493** (2015)
15. I. Kowalska, Pol. J. Environ. Stud. **21**, 3 (2012)
16. L. Suarez, M.A. Diez, R. Garcia, F.A. Riera, J. Ind. Eng. Chem. **18** (2012)
17. Z. Kovács, W. Samhaber, Membrántechnika **12** (2008)
18. I. Kowalska, Desalination **221** (2008)
19. A.C. Archer, A.M. Mendes, R.A.R. Boaventura, Environ. Sci. Technol. **33** (1999)

Computational and the real energy performance of a single-family residential building in Poland – an attempt to compare: a case study

Piotr Kowalski[1,*], and *Paweł* Szałański[1]

[1]Department of Air Conditioning, Heating, Gas Engineering and Air Protection, Faculty of Environmental Engineering, Wrocław University of Science and Technology, 4/6 C.K. Norwida st. 50-373 Wrocław, Poland

Abstract. The paper presents energy use for heating and ventilation (one of the energy performance components) determined in three ways. A case of a single family building located near Wroclaw in Poland is analyzed. The first and the second variant are both computational and the third presents actual measured energy consumption. Computational variants are based on the Polish methodology for the EPC (the Energy Performance Certificate). This methodology is based on 'the Energy Performance of Buildings Directive 2010/31/EU'. Energy use for heating and ventilation is calculated using monthly method presented in EN ISO 13790. In the first computational option standard input data (parameters such as indoor and outdoor air temperature etc. are taken from standards and regulations) are implemented. In the second variant this input data are partially taken from measurements. The results of energy use from both computational variants are compared to the actual measured energy consumption. On the basis of this comparison the influence of three factors: solar radiation heat gains, building air tightness and the SCOP of the heat pump on energy use calculations are analyzed. Conclusions aim to point the differences between them and the actual energy consumption.

1 Introduction

In the common understanding of the buildings users the Energy Performance Certificate (the EPC), prepared in connection with the requirements of Directive [1], should allow to estimate the real energy consumption and the rough cost of heating. However, this idea is rather misleading and can cause some doubts. Thus, the question arises whether the calculation methods indicated by the Polish regulations on energy

* Corresponding author: piotr.kowalski@pwr.edu.pl

performance certificates [2] (mostly in line with European standards) can give a result close to the real energy consumption for the analyzed building.

It is obvious that regardless of calculation models, the implementation of the standard values (not real) of input data does not allow to obtain the result close to the real energy consumption. However, in the analyzed building a user has installed a self-designed measuring and logging system in order to observe the energy and water consumption and to manually control indoor air temperatures in the building. Therefore, the article attempts to answer the question: whether on the basis of measurements from this system supplemented by additional instantaneous measurements carried out by authors one can predict real energy consumption.

1.1 Description of the analyzed building

The subject of the study (Fig. 1 and 2) is located in Poland near Wroclaw. This is a two storey single-family house with a garage, an attic and without basement. Building was built in the traditional brick technology with two-layer walls of hollow bricks called Porotherm and insulated with an expanded polystyrene. The roof is insulated with a mineral wool. Heat transfer coefficients calculated due to [3,4] are as follows: external walls U_{wall} = 0,22 W/(m^2 K), the floor on the ground U_{equiv} = 0,18 W/(m^2 K), the ceiling of the attic $U_{ceilling}$ = 0,15 W/(m^2 K), the roof U_{roof} = 0,20 W/(m^2 K), windows U_{win} = 1,40 W/(m^2 K) – value assumed. The usable area (calculated using standardized methods) of the building is 228,3 m^2, plus garage area of 39,9 m^2, while the floor area with controlled temperature of the indoor air equals A_f = 308,7 m^2.

Fig. 1. The case study: a view of the single-family house.

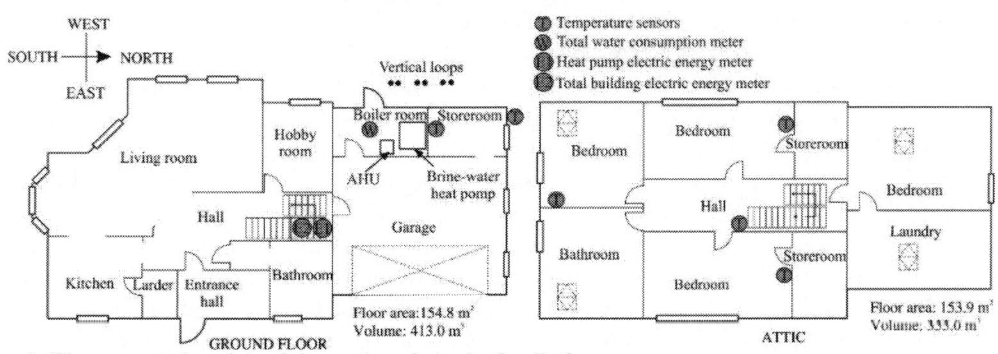

Fig. 2. The case study: plan of the analyzed single-family house.

The building external volume (calculated according to the external dimensions) equals 1175 m^3, ventilated-internal building volume is 768 m^3 and the ratio of external barriers area and external volume of the building A/V = 0,410 1/m. In 2013 the building was inhabited by two adults and one child and in 2015 by two adults and two children. The building structure was assumed as a semi-heavy. The facility is equipped with the installation of heating, ventilation, cooling and hot water supply (Fig. 3). The heat source is a water/brine heat pump (heating capacity 12 kW due to EN 14511). Ventilation in the building provides an air handling unit with heat recovery heat exchanger.

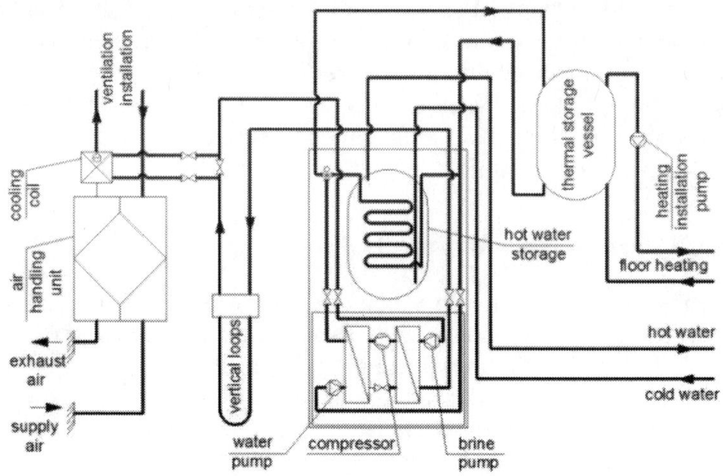

Fig. 3. The case study: A scheme of installations in analyzed building.

1.2 The measurements description

A measuring self-designed system (by P. Piotrowski) was installed in the building by the owner P. Piotrowski. Measurements were logged in 10 minute intervals. Measuring system consists of (Fig. 2):

- digital air temperature sensors type DS18B20 Maxim Integrated (accuracy ±0,5°C), where:
  - indoor air temperature sensors were located on the internal walls 30–40 cm above floor,
  - outdoor air temperature sensor was located on the northern wall 1,7 m above the ground,
- electric energy meter (accuracy class B) for the heat pump and heating installation pump,
- electric energy meter (accuracy class 1) for the whole house,
- typical water meter with pulse transmitter.

Additionally, authors carried out the below described measurements.

The mechanical ventilation supply and exhaust flows were measured once in two operating modes: low flow Monday – Sunday 24:00–6:00, Monday – Friday 9:00–14:00; normal flow Monday – Friday 6:00–9:00 and 14:00–24:00; Saturday – Sunday 6:00–24:00. Measured flows are assumed to be constant during analyzed period respectively for operating modes. The total flows were calculated as sum of flows measured on terminal elements of the ventilation installation. Flows on the terminal elements were measured using vane-type anemometer type Airflow LCA301 (accuracy ±1% of reading + 0.02 m/s). A heat recovery efficiency in air handling unit (AHU) was determined by measuring temperatures of supply, exhaust and outdoor air while outdoor air temperatures were close to the average air temperature of the heating season.

Temperatures were measured using an electronic instrument type TESTO 435-2 with air temperature probe (accuracy ±0,2 °C).

Electric power of heating installation pump was measured using wattmeter BRYMEN BM357s. This pump operates all the heating period.

2 Assumption to the calculations

The following calculations of energy demand and energy consumption apply to the part of the energy performance: for heating and ventilation purposes (without heating installation pump and fans in AHU). The comparison of the analyzed variants does not apply to the energy demand for hot water. In those calculations two computational variants were considered: with a standard input data (from standards and regulations) - variant I and with a selected real input data (mainly obtained from user self-designed measuring system and instantaneous measurements carried out by authors) – variant II. Moreover, in the variant III the real energy consumption for heating and ventilation was determined on the basis of the measurements.

2.1 Description of the computational model

Computational energy performance (variant I and II) was determined due to monthly method presented in Polish translation of EN ISO 13790 [5] with additional assumptions presented in Polish regulations [2]: calculations do not take into account heat flux of thermal radiation to the sky from building elements.

In particular, the calculation was carried out according to the models from the following standards:

- Heat transfer coefficients (U, $W/(m^2 K)$) on the basis of Polish translation of EN ISO 6946 [3].
- Linear heat transfer coefficients of thermal bridges (Ψ, W/(m K) assumed after Polish translation of EN 14683 [6].
- Heat loss via the ground with floor heating are based on Polish translation of EN ISO 13370 [4].
- Transmission and ventilation heat loss coefficient (H_T, H_V, W/K) on the basis of Polish translation of EN ISO 13789 [7].

2.2 Description of the input data

The final energy demand of the building depends on the number of input data, such as: outdoor and indoor air temperatures, solar and internal heat gains, air tightness of the building, exposure to the wind, the efficiency of the heating and ventilation systems and walls, roof, floor, windows, door heat transfer coefficients. The way of specifying them in particular variants is described below.

In variant I calculations were carried out with standard input data assumed on the basis of Polish regulations [2].

In variant II the following input data were obtained from the measurements or assumed on the basis of the measurements:

- measured outdoor air temperature was separately averaged for each month,
- indoor air temperature was averaged (for each month) from available sensors (Fig. 2) respectively for living space and utility area (garage, boiler room, store room on the ground floor),
- to estimate the internal heat gains:
  - total measured electric energy consumption for the heat pump,
  - electric energy consumption for exhaust fan in AHU,

- electric energy consumption for heating installation pump,

were subtracted from the total measured electric energy consumption of the building and the result was added to heat gains from residents,

- supply, exhaust flows and heat recovery efficiency of the ventilation installation.

Other input data in variant II were assumed on the basis of Polish regulations [2]. Seasonal control system efficiency was an exception, which was assumed as 1.0. It was due to basing on measured values of indoor temperatures in this variant. Temperatures were measured and logged every 10 minutes, and then averaged for every month. Therefore, it was assumed that the effect of temperature fluctuations [8] due to the control system efficiency was already taken into account in calculations. Variant I assumed seasonal control system efficiency differently - equal to 0.76 for the analyzed floor heating system with central and without local regulation [2]. It was due to the calculations based on the constant indoor temperature.

The results of the calculations in variant II were presented for two years: 2013 and 2015 except from June to September.

Variant III presents measured energy consumption for the same periods with the following assumptions:

- The final energy use for hot water was subtracted from the total measured heat pump energy consumption during the heating period to estimate the final energy consumption for heating and ventilation. In consequence, the final energy used by heating installation was estimated.
- The final energy consumption for hot water supply was determined on the basis of the heat pump energy consumption measurements in the summer months. The periods when energy was used by the heat pump only for hot water supply were taken into account. Average estimated value of final energy for hot water supply equals 51.7 kWh/month for the year 2013 and 74.4 kWh/month for the year 2015.

In all the variants energy consumption in June and September was omitted in order to avoid an error related to specifying the beginning and the end of the heating season.

Comparison of input data for all three variants are presented in the Table 1.

Table 1. Comparison of the assumptions for variants I, II and measurements for III.

Input data	Month / Variant	I	II	III	IV	V	X	XI	XII
Outdoor air temperature, °C	I [9]	-0.4	-0.7	2.8	7.3	12.7	8.9	3.8	-1.1
	II, III (2013)	0.3	0.0	-1.1	8.8	14.3	9.9	5.0	1.8
	II, III (2015)	2.0	1.2	5.1	8.8	13.6	7.8	5.5	4.8
Indoor air temperatures, °C: - living space - utility area (garage, boiler room, store room on the ground floor)	I	20.4 [10] 8.1 [10]							
	II, III (2013)	20.5 10.3	21.3 11.4	20.8 11.0	21.3 15.3	21.8 18.8	20.9 17.4	20.3 14.9	20.4 12.6
	II, III (2015)	21.7 12.9	22.0 12.4	22.7 15.4	22.2 17.0	23.2 20.6	23.7 17.9	22.6 16.1	22.8 15.1

Average supply, exhaust air flow and the (living space / utility area) m^3/h	I II, III	146 and 146 / 0 and 20 140 and 140 / 0 and 20
Heat recovery efficiency in ventilation system, %	I, II	60
Result of building air tightness test n_{50}, h^{-1}	I, II	4.0 [2]
Solar radiation heat gains	I,II	Calculated on the basis of climate data [9]
Internal heat gains (living space / utility area), W/m^2	I II (2013) II (2015)	6.8 / 6.8 [2] 1.16 / 2.51 1.41 / 2.51
SCOP (EN 14825 [12] - average climate, 35°C)	I, II	4.7 [11]
Seasonal control/heat storage/distribution efficiencies system	I	0.76 / 0.95 / 1.0 [2]
	II	1.0/ 0.95 / 1.0 [2]

3 Results

The results of the calculations in all the variants are presented in the Table 2. The final energy for heating and ventilation in variant I is much lower than in variant II. It is caused by different input data. Moreover, variant II gives much lower values than variant III. The difference can be due to the following reasons: the imperfection of the computational model and the discrepancies between the assumptive and the actual input data. It was estimated that the real air tightness test result, the real solar radiation gains and the real SCOP of the heat pump were the most unsure among the assumptive input data in view of the lack of information on their values.

Table 2. Final energy consumption for heating and ventilation.

Variant I	Variant IIe 2013	Variant III 2013	Variant IIe 2015	Variant III 2015
2723 kWh	3146 kWh	4547 kWh	3241 kWh	4648 kWh

The system created by the user does not give the possibility of determining this input data. Due to the inability to identify those values their impact on the result was analyzed. In the consequence six sub-options for variant II were defined in Table 3.

For all of the sub-options the SCOP of the heat pump was fitted so as to obtain the result of the energy demand equal to real energy consumption. In consequence the influence of solar radiation heat gains, building air tightness and efficiency of the heating system on energy use for heating and ventilation was presented (Figure 4).

Table 3. Sub-options for variant II.

Parameter	Sub-options of variant II					
	A	B	C	D	E	F
Solar radiation heat gains	50% of	100% of	150% of	50% of	100% of	150% of
	value calculated on the basis of climate data [9]					
Result of building air tightness test n_{50}	1.5 h^{-1} Maximum for building with mechanical ventilation due to Polish regulations [10]			4.0 h^{-1} Due to Polish regulations [2], value which should be assumed in calculations when test was not made		

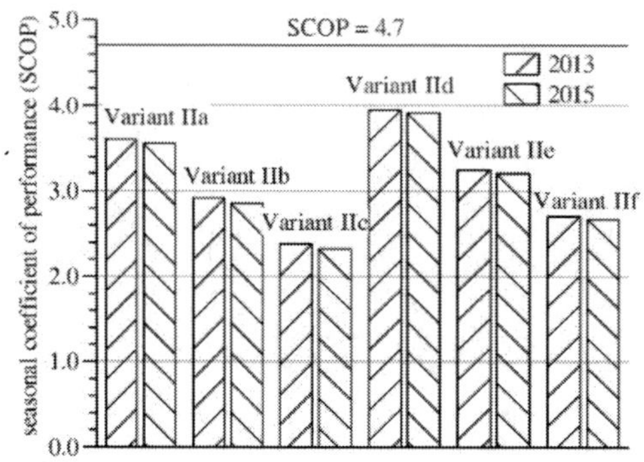

Fig. 4. Theoretical SCOP corresponding to the real final energy consumption for sub-options of variant II.

4 Conclusions

The presented analysis (especially comparison of variant I and III) shows that using monthly method (due to the Polish methodology for EPC) failed to predict the actual energy consumption. So it indicates that the EPC methodology is rather to assess the energy performance of the design of the building and not to predict the actual energy consumption. After changing a part of the standard input data on the real values (variant II) the result is closer to the real energy consumption (variant III). However, the difference still remains significant. It can be mainly caused by the imperfection of the model and the discrepancy between real values of the input data and values implemented in calculations:

- SCOP was assumed according to the EN 14825 standard and not measured,
- U-values were calculated or assumed and not measured,
- hot water energy consumption was assumed equal in winter-time and in summer-time,
- occupation patterns (internal gains) were declared by building user and not registered,
- solar gains were based on standard climatic data and not measured,
- the mechanical ventilation rate and heat recovery efficiency were measured once in each of the operating modes and not registered during all period.

The difference between the real SCOP of the heat pump and the standard SCOP (due to EN 14825) could be mainly caused by:

- the standard SCOP is determined for standard conditions and not for real,
- poor adjustment of the heat pump to the needs of the building,
- continuous operation of the buffer charging pump in the heating season.

Even if the input data (Variant IIa-IIf) is significantly changed the result does not coincide with the real energy use of the building (Figure 4). In order to reduce resulting uncertainty more real input data should be implemented to the calculations and the following actions should be undertaken in the future:

- expansion of the measuring and logging system with: a heat meter for the heat pump (to determine the real SCOP of the heat pump), solarimeter (to better estimation of solar gains), flow-meters and temperature sensors of ventilating air (for better estimation of ventilating flows and heat recovery efficiency),
- performance of an air tightness test of the building envelope and measurements of a real U-value.

Presented research and analysis are part of the research task financed from special-purpose subsidies of Ministry of Science and High Education of Poland earmarked for the development of young scientists and doctoral students.
The authors would like to thank Piotr Piotrowski for providing results of measurements.

References

1. Directive 2010/31/EU of the European Parliament and of the Council of 19 May 2010 *on the energy performance of buildings (recast),* Official Journal of the European Union, http://eur-lex.europa.eu/eli/dir/2010/31/oj
2. Regulation of Minister of Infrastructure and Economic Development (21 February, 2015) *on the methodology for determining the energy performance of building or part of the building and energy performance certificates*. (in Polish)
3. PN-EN ISO 6946 *Building components and building elements – Thermal resistance and thermal transmittance – Calculation methods* (in Polish)
4. PN-EN ISO 13370:2007 *Thermal performance of buildings – Heat transfer via the ground — Calculation methods* (in Polish)
5. PN-EN ISO 13790:2008 *Energy performance of buildings – Calculation of energy use for space heating and cooling* (in Polish)
6. PN-EN ISO 14683:2007 *Thermal bridges in building construction – Linear thermal transmittance – Simplified methods and default values* (in Polish)
7. PN-EN ISO 13789:2007 *Thermal performance of buildings — Transmission and ventilation heat transfer coefficients — Calculation method* (in Polish)
8. G. Dall'O', *Green Energy Audit of Buildings, A Guide for Sustainable Energy Audit of Buildings*, Springer-Verlag, London (2013)
9. Ministry of Infrastructure and Building, *Typical statistical climate data for the Polish area for building energy performance purposes*. (in Polish), Web Page: http://mib.gov.pl/2-Wskazniki_emisji_wartosci_opalowe_paliwa.htm), [accessed: 26 January, 2017]
10. Regulation of Minister of Infrastructure and Economic Development (12 April, 2002) *on the technical conditions, which are to be met by buildings and their location* (in Polish)
11. NIBE AB Sweden, *Installer manual NIBE F1245 Ground source heat pump*, IHB PL 1602-1 331498 (in Polish)
12. EN 14825 *Air conditioners, liquid chilling packages and heat pumps, with electrically driven compressors, for space heating and cooling – Testing and rating at part load conditions and calculation of seasonal performance*

E3S Web of Conferences 17, 00046 (2017) DOI: 10.1051/e3sconf/20171700046
EKO-DOK 2017

River Flow Prediction for Future Climate Using Long Series of Multi-Site Synthetic Data and MIKE SHE Model

Leszek Kuchar[1,*], *Slawomir* Iwanski[1], and *Leszek* Jelonek[2]

[1]Wroclaw University of Environmental and Life Sciences, Department of Mathematics, 50357 Wroclaw, Poland
[2]Institute of Meteorology and Water Management - Branch Wroclaw, 51618 Wroclaw, Poland

Abstract. A new simulation of daily flow for Kaczawa River, south-west Poland for extra long series of generated meteorological data (comparing to previous research) and selected climate change scenarios are presented. The Representative Concentration Pathways (RCPs) scenarios vs. SRES are introduced for simulations. The flow simulation in the river catchment is made using MIKE SHE hydrological model while the multisite data are generated by spatial weather generator SWGEN. Simulations are done for 2040 and 2060 while the simulations for the year 2000 are used as a background. The large number of new simulated series determined by the lead time, three climate change scenarios (RCP2.6 RCP4.5 and RCP6.0), and number of generated years (1000 for each case) is equal to 7000 for a single station. Finally, Pdf function for flow is presented as well probability of exceedance of maximum flow.

1 Introduction

The future climate change may have a significant impact on the forthcoming hydrology in the river catchment, runoff response, and predictions [1, 3, 5, 8, 24, 25, 28, 33]. The daily flow simulation, particularly seasonal extremes for future climate conditions given by different scenarios are important for the Polish rivers for several reasons [15, 17, 27]. The paper presents an example of new simulation of seasonal flow for the relative small catchment of Kaczawa River a left side tributary of the Odra River (Southwest region of Poland) for selected climate change scenarios is presented. Above simulations extended study are compared to the earlier research [15, 16, 17].

2 Methods

The idea of runoff simulation in the river catchment for future climate conditions given by different scenarios are presented in Figure 1 and the same philosophy comparing to earlier works is used [15, 16, 17]. This procedure combines a hydrological model, spatial

* Corresponding author: Leszek.Kuchar@gmail.com

weather generator producing synthetic weather data and climate change scenarios [12, 17]. Simulations give a knowledge for a decision support system.

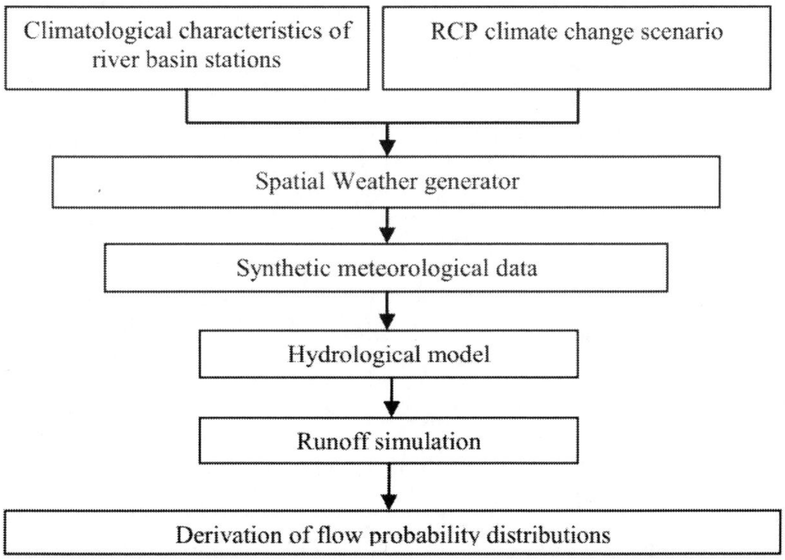

Fig. 1. Diagram of river flow simulation for future climate using synthetic meteorological data and climate change scenario.

2.1 Climate change scenario selection

Changes during the winter and summer season in different regions and scales of the country are symptomatic [3, 6, 12, 13, 24, 26]. The impact of climate change according to the approved set of emission scenarios described in the IPCC Special Report on Emission Scenarios (SRES) [10], and A1B (most used SRES scenario) were used as typical for Poland. Particularly, scenarios given by GISS Model E, HadCM3 and GFDL R15 assuming doubling the CO_2 concentration, which is expected for years 2050-2060 were applied. As the baseline conditions, year 2000 was chosen, according to reference period: 1990–2010. A new 5th IPCC Rapport 2014 [11] changes a philosophy of SRES to a new set of scenarios, the Representative Concentration Pathways (RCPs) and implies them for the new climate simulations. The RCP includes time series of emissions and concentrations of full suite of GHG gases and aerosols and chemically active gases, as well as land use cover. Predictions for surface temperature for Poland are more adequate comparing to previous SRES scenarios, and therefore, new RCP scenarios (particularly RCP2.6, RCP4.5 and RCP6.0) are introduced for simulations.

However, the experience with new RCPs scenarios are much lower comparing to previous emissions scenarios SRES, the most Earth System Model simulations were performed with prescribed CO2 concentrations reaching 538 ppm (RCP4.5), 670 ppm (RCP6.0) to the year 2100. Including also CH4 and N2O the combined CO2 equivalent concentration are 630 ppm (RCP4.5), 800 ppm (RCP6.0) [10, 11]. Therefore, in the simulations two main scenarios RCP4.5 and RCP6.0 (RCPs) similar to GISS Model E, HadCM3, GFDL R15 (previously used and based on SRES) and in addition RCP2.6 scenario as most probable for Poland are considered [15, 16, 17]. This is also a compromise between experience and recent knowledge for study on climate change impact on environment.

2.2 Synthetic weather data

Spatial weather generator SWGEN is used as best downscaling method to produce n-years of synthetic data on potentially possible weather course, k stations, the given time horizon and scenario [2, 4, 7, 12, 19, 32]. The year 2000 as the background of the potential changes in catchments flow is used together with 1000 years of synthetic data.

The SWGEN model generates precipitation by means of the first-order Markov chain to determine the occurrence of wet/dry days, and then for the amount of rainfall multidimensional two-parameter gamma distribution [12, 17]:

$$(\Gamma_m(\alpha_1,\beta_1), \dots, \Gamma_m(\alpha_k,\beta_k)) \tag{1}$$

where m is the number of months (m=1,...,12) and k is the number of locations, while daily values of solar radiation (SR), temperature maximum (T_{max}) and minimum (T_{min}) are considered as a multidimensional time series AR(1) in the following form:

$$\mathbf{X}_t = \mathbf{\Phi}_m \cdot \mathbf{X}_{t-1} + \boldsymbol{\varepsilon}_t \tag{2}$$

where $\mathbf{X}_t$ and $\mathbf{X}_{t-1}$ are vectors (m × 1) of normalized values for all three variables for day t and t-1, $\boldsymbol{\varepsilon}_t$ is a vector (m × 1) of independent random components normally distributed with vector of mean equal to zero and matrix of covariance $\mathbf{\Sigma}_m$, and $\mathbf{\Phi}_m$ (for m=1, ..., 12) is a matrix of parameters [12, 15, 17].

2.3 Flow simulation with rainfall-runoff model

In the next stage, generated data were applied to hydrological rainfall-runoff model MIKE SHE [9, 22] to simulate runoff for closing water-gauges. The catchment runoffs are evaluated with a different temporal step (one day, five days and ten days). The obtained outflows for various simulations are characterized by probability distribution functions. Three parameter gamma probability distribution is used as the best distribution fitting the monthly outflow, and Pareto pdf for daily maximum outflow within the year [12, 15, 29, 30].

3 Study Area

The above simulation scheme was applied for the Kaczawa basin that is one of the main left bank tributaries of the Odra River – the second biggest river in Poland. Daily data of solar radiation, maximum and minimum air temperature, and total precipitation of a 28-year data series (1981-2008) of meteorological network within or around the Kaczawa River catchment, were obtained for 16 stations of hydrological network from Institute of Meteorology and Water Management. In addition, daily data of flows from 6 closing water-gauges (partial catchments) were collected.

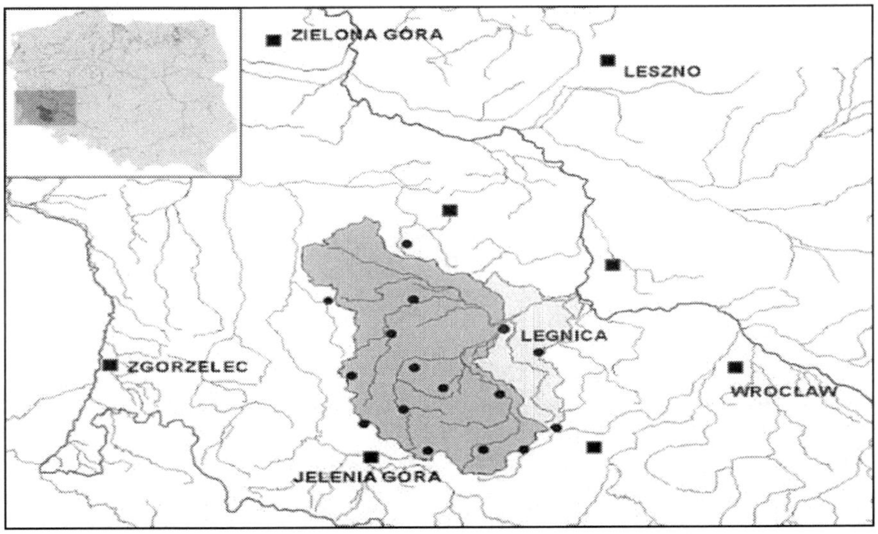

Fig. 2. The Kaczawa River catchment (left side tributary of the Odra River, area of 1807 km^2, main sit: Legnica 51°13'N, 16°14'E) with meteorological stations (•).

The basic climatological characteristics required by weather generator are computed [12, 18, 31], and spatial correlations between variables and stations are added to the characteristics [14, 18].

Next, on the basis of information coming from climate change scenarios RCP2.6, RCP4.5 and RCP6.0, averaged and rescaled for years 2040, 2060 basic climatologic characteristics are modified. Then, spatial weather generator SWGEN is used to produce new long series of 1000 years (comparing to previous study [16]) of synthetic data for 16 stations, given time horizon and scenario. The year 2000 as the background of potential changes in river flow is used together with 1000 years of synthetic data. Next, generated data is applied to hydrological model MIKE SHE to simulate daily flows for closing water-gauges. The flow is evaluated with different temporal step and characterized by Pdf functions.

4 Results

The simulations of daily runoff in the Kaczawa River catchment were done at discharge point in Piatnica. The number of simulations were determined by the time of horizon (2040, 2060 and 2000 as a background), three averaged climate change scenarios (RCP2.6, RCP4.5 and RCP6.0), and number of generated years (1000) for each case, with total of 7000 (2x3x1000+1000) years. The MIKE SHE model computed for a given year, a daily flow at discharge point, and maximum value was chosen to estimate parameters of density function.

It means that the parameters of Pareto probability distribution were estimated for 7 (2x3+1) combinations, based on long series of 1000 computed runoff each.

As an example, Figure 3 presents probability of exceedance of maximum flow at discharge point at Piątnica for present conditions (year 2000) and two RCPs scenarios for 2040 and 2060 within July, while Figure 4 shows graphs probability of exceedance of maximum flow for 2040 and 2060 within the period May – August.

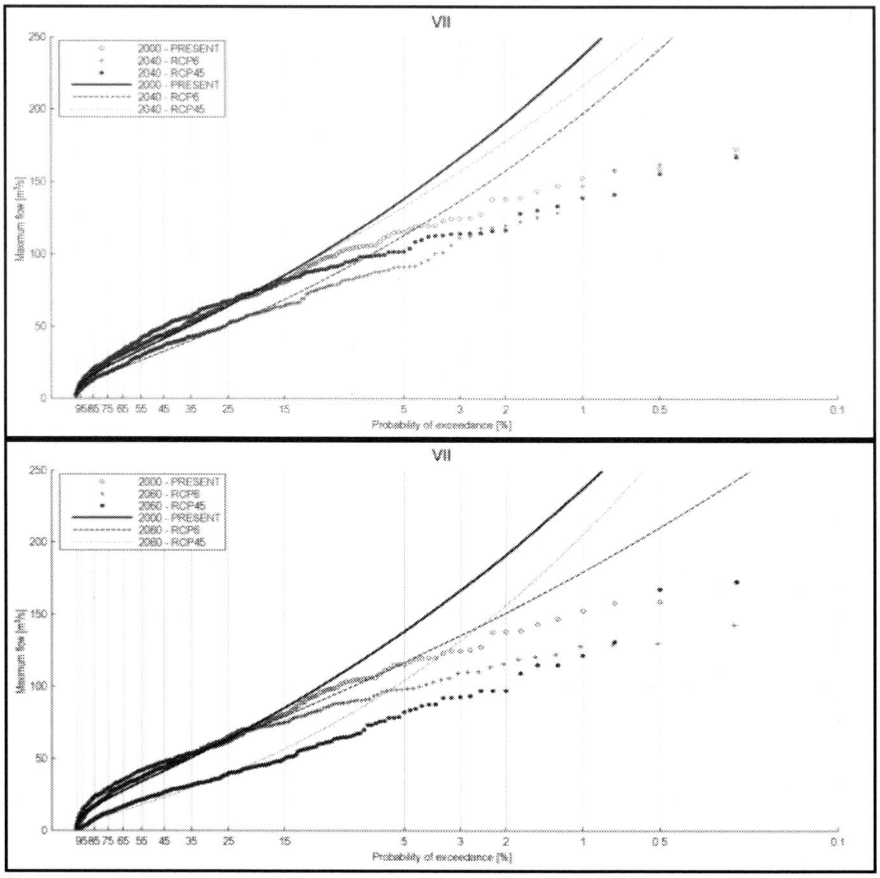

Fig. 3. Probability of exceedance of maximum flow within July at discharge point at Piatnica on the Kaczawa River, simulation for present conditions (year 2000) and two RCPs scenarios for 2040 and 2060.

The above simulations leads to the following remarks. The application of spatial weather generator SWGEN combined with hydrological rainfall-runoff model (MIKE SHE Ed. 2008) and climate change scenario, gives various possibilities to study changes in the river catchment coming up to 40-80 years [20, 21, 23]. The probability distribution of the extreme river flow gives detailed information on the moment characteristics, confidence intervals and critical values. It is an important tool for a decision support system. In case of extreme daily flow in the Kaczawa River for considered periods, the catchment shows changes depending on the climate change scenario and time to lead.

New simulations with RCPs scenarios RCP4.5 and RCP6.0 give similar results as previously used SRES A1B scenarios from GISS Model E, HadCM3 and GFDL R15 [17]. Application of Pareto probability distribution shows some uncertainties for tail fitting, and suggests future studies, particularly nonparametric estimation.

However, simulations show a growing risk for 1% (and lower) probability of exceedance of maximum flow, RCP6.0 and sixty-year lead time. Above risk is higher for a longer period, for example during the period May – August than July.

Application of long generated weather series (1000 vs. 500 years) gives similar results as recently obtained [16].

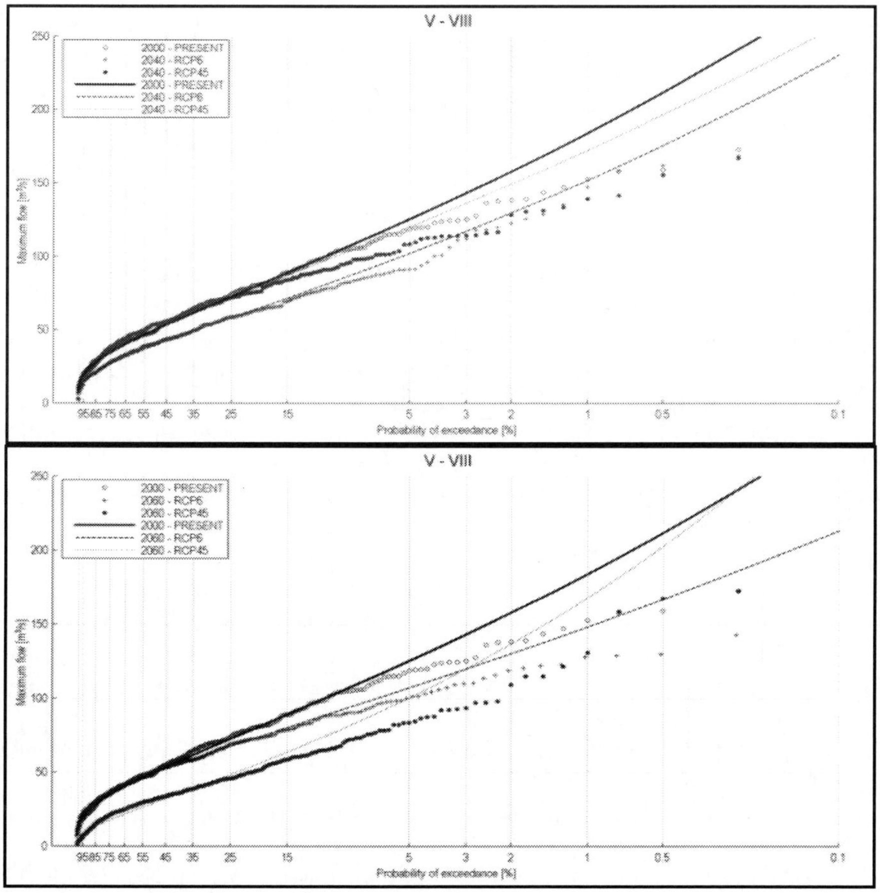

Fig. 4. Probability of exceedance of maximum flow during the period May – August at discharge point at Piatnica on the Kaczawa River, simulation for present conditions (year 2000) and two RCPs scenarios for 2040 and 2060.

References

1. T. Barnett, T. Malone, W. Pennell, D. Stammer, B. Semtner, W. Washington, Clim. Change, **62**, 1 (2004)
2. J.J. Beersma, T.A. Buishand, Clim. Res. **25**, 121 (2003)
3. S. Bergstrom, B. Carlsson, M. Gardelin, G. Lindstrom, A. Pettersson, M. Rummukainen, Clim. Res. **16**, 101 (2001)
4. F. Brissette, M. Khalili, R. Leconte, J. Hydrol. **345**, 121 (2007)
5. D.H. Burn, J. Hydrol. **160**, 53 (1994)
6. H. Chen, C.Y. Xu, S. Guo, J. Hydrol. **434-435**, 36 (2012)
7. J. Chen, F.P. Brissette, X.C. Zhang, Trans. ASABE, **57**, 1375 (2014)
8. N.S. Christensen, A.W. Wood, N. Voisin, D.P. Lettenmaier, R.N. Palmer, Clim. Change, **62**, 337 (2004)
9. D.N. Graham, M.B. Butts, *Watershed Models* (V.P. Singh, D.K. Frevert (Eds.), CRC Press, 2005)
10. IPCC-SRES-SPM WG III, *Emissions Scenarios, Summary for Policy-makers, Special Report* (IPCC, 2000)
11. IPCC WG I, II and III, *Climate Change, Synthesis Report* (IPCC, 2014)

12. S. Iwanski, L. Kuchar, Acta Scientiarum Polonorum - Formatio Circumiectus, **2**, 113 (2003)
13. R.N. Jones, F.H.S. Chiew, W.C. Boughton, L. Zhang, Adv. Water Resour. **29**, 1419 (2006)
14. M. Khalili, R. Leconte, F. Brissette, J. Hydrometeorol. **8**, 396 (2007)
15. L. Kuchar, S. Iwanski, L. Jelonek, W. Szalinska, Geografie, **119**, 1 (2013)
16. L. Kuchar, S. Iwański, L. Jelonek, *Water Resources, Hydraulics Facilities and Environment, The International Scientific-Practical Conference* (Azersu, Baku, Azerbaijan, 2017)
17. L. Kuchar, S. Iwanski, L. Jelonek, W. Szalinska, Meteorol. Hydrol. Water Manag. **2-2**, 49 (2014)
18. L. Kuchar, Math. Comp. Symul. **65**, 69 (2004)
19. W.A. Landman, S.J. Mason, P.D. Tyson, W.J. Tennant, J. Hydrol. **252**, 221 (2001)
20. L. Menzel, G. Burger, J. Hydrol. **267**, 53 (2002)
21. W.S. Merritt, Y. Alila, M. Barton, B. Taylor, S. Cohen, D. Nielsen, J. Hydrol. **326**, 79 (2006)
22. MIKE 11, *A modelling system for Rivers and Channels, User Guide*, (DHI Water and Environment, 2003)
23. M. Minville, F. Brissette, R. Leconte, J. Hydrol. **358**, 70 (2008)
24. D.I. Müller-Wohlfeil, G. Bürger, W. Lahmer, Clim. Change, **47**, 61 (2000)
25. C. Prudhomme, N. Reynard, S. Crooks, Hydrol. Process. **16**, 1137 (2002)
26. J.R. Thompson, Hydrol. Res. **43**, 507 (2012)
27. T. Tokarczyk, W. Szalinska, A. Tiukalo, J. Jelowicki, A. Chorazyczewski, Meteorol. Hydrol. Water Manag. **4-1**, 65 (2016)
28. T. Vansteenkiste, M. Tavakoli, V. Ntegeka, P. Willems, F. De Smedt, O. Batelaan, Hydrol. Process. **27**, 3649 (2012)
29. R.E. Walpole, R.H. Myers, S.L. Myers, K. Ye, *Probability and statistics for engineers and scientists* (Prentice Hall, NJ 2002)
30. M. Wdowikowski, B. Kazimierczak, O. Ledwinka, Meteorol. Hydrol. Water Manag. **4-1**, 53 (2016)
31. R.L. Wilby, Environ. Modell. Softw. **22**, 1705 (2007)
32. D.S. Wilks, Clim. Change, **1**, 898 (2010)
33. E. Zorita, H. von Storch, J. Climate, **12**, 2474 (1999)

The role of pH in the decomposition of organic micropollutants during the heterogeneous photocatalysis process

Edyta Kudlek[1,*], *Mariusz* Dudziak[1] , *Jolanta* Bohdziewicz[1], and *Gabriela* Kamińska[1]

[1]Silesian University of Technology, Faculty of Energy and Environmental Engineering, Konarskiego 18a, 44-100 Gliwice, Poland

Abstract. The study investigated that the change in the pH ranged from 5 to 7 affects the adsorption degree of selected organic micropollutants such as: polycyclic aromatic hydrocarbons (anthracene, benzo[a]pyrene), pharmaceutical compounds (diclofenac) and industrial additives (octylphenol and pentachlorophenol) on the photocatalyst surface (100 mg TiO_2/dm^3). For example, in suspension of pH 5 a 12% reduction in the concentration of octylphenol was noted, while in the suspension of pH 7 the concentration of this compound was reduced by 64%. In the case of anthracene an inverse relationship was observed. The concentration of this micropollutant decreased by 94% for pH 5 and only by 61% for pH 7 suspension. With the commencement of UV irradiation of reaction mixtures a gradual increase in the concentration of micropollutants reduction in aqueous solutions was observed. This indicates the decomposition of compounds as a result of their reaction with highly reactive $OH^•$ radicals or other reactive oxygen species generated during the process. For reaction mixtures of pH 7, already in the first 15 minutes, higher degrees of removal of micropollutants ware observed compared to suspension of pH 6. The obtained micropollutants removal degrees exceeded 91% for diclofenac, octylphenol, pentachlorophenol and amounted to 100% for anthracene and benzo(a)pyrene.

1. Introduction

Literature data indicate a continuous increase in the number of organic micropollutants in the water environment. These compounds belong to different chemical groups and have a differentiated impact on the water quality and also pose a serious threat to aquatic ecosystems [1–3]. The Water Frame European Directives (2000/60/EC) [4] listed some micropollutants including PAHs, EDCs, phenolic compounds and pharmaceuticals as "priority hazardous substances". It is therefore necessary to take any action to reduce their concentrations in the environment through the implementation of effective methods of their removal from aqueous streams. Advanced oxidations process, which include also the process of heterogeneous photocatalysis gives an opportunity to obtain satisfactory

* Corresponding author: edyta.kudlek@polsl.pl

effects of micropollutant decomposition [5]. The oxidation and reduction of compounds by means of highly reactive free hydroxyl radicals $OH^{\bullet}$ formed during the process depends on different operating parameters of the process, which include the wavelength emitted by the applied UV irradiation source [6], oxygen concentration in the irradiated water solution [7], catalyst dose and type [8] and also the physicochemical composition of the treated water [9, 10].

The paper presents the influence of the pH of water solution on the decomposition efficiency of chosen organic micropolutants from the group of polycyclic aromatic hydrocarbon – anthracene (ANT), benzo(a)pyrene (BaP), industrial admixtures – pentachlorophenol (PCP), octylphenol (OP) and pharmaceutical compounds – diclofenac (DCL) during heterogeneous photocatalysis. The photochemical process was conducted in the presence of TiO_2 as a catalyst. The kinetics of the process was described by means of Langmuir-Hisherwood equation. The concentrations of all compounds in water solution before and after their treatment were estimated by HPLC chromatography preceded by Solid Phase Extraction.

2 Materials and methods

2.1 Chemicals

All investigated organic micropollutants (octylophenol, pentachlorophenol, anthracene, benzo(a)pyrene, diclofenac,) were purchased from Sigma Aldrich (Poland). Their properties are shown in table 1.

Table 1. The structural and chemical properties of the studied micropollutants.

Structural formula	Chemical properties	
	Pentachlorophenol (PCP)	
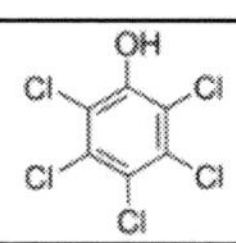	Molecular weight, g/mol	266.34
	Solubility in water, mg/L	80
	log K_{ow}	5.12
	4-tert-octylphenol (OP)	
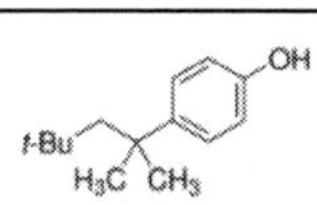	Molecular weight, g/mol	206.32
	Solubility in water, mg/L	7.0
	log K_{ow}	4.80
	Anthracene (ANT)	
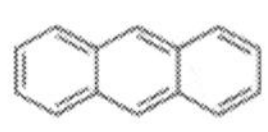	Molecular weight, g/mol	178.23
	Solubility in water, mg/L	0.044
	log K_{ow}	4.45
	Benzo[a]pyrene (BaP)	
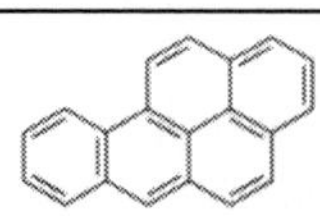	Molecular weight, g/mol	252.31
	Solubility in water, mg/L	0.00147
	log K_{ow}	5.97
	Diclofenak sodium salt (DCL)	
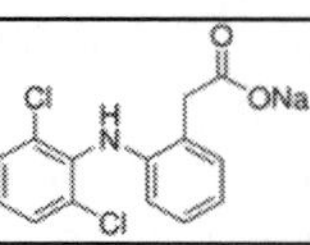	Molecular weight, g/mol	318.13
	Solubility in water, mg/L	50
	log K_{ow}	4.51

Due to the poor water solubility of the compounds from the group of polycyclic aromatic hydrocarbon (anthracene and benzo(a)pyrene) the stock solutions of all micropollutants were prepared with methanol (analytical standard), Avantor Performance Materials Poland S.A. (Poland). The study subjects, with a micropollutant concentration of 0.5 mg/L, were prepared by diluting the stock solution with deionized water. The pH of the water solutions were adjusted subsequently to 5.0, 6.0 or 7.0 using 0.1 mol/L HCl and 0.1 mol/L NaOH purchased from Avantor Performance Materials Poland S.A. (Poland).

TiO_2 applied as a catalyst into the process of heterogeneous photocatalysis was purchased from Evonik Degussa GmbH (Hanau, Germany) and labelled by the manufacturer with the symbol of P25.

2.2 Photocatalysis process

The process of heterogenic photocatalysis was conducted in a laboratory batch reactor with a volume of 700 mL by Heraeus (Germany) equipped with a medium-pressure mercury vapour UV lamp (150 W) placed in a cooling jacket. The temperature of the process did not exceed 20 ± 1 °C. The radiation emitted by the UV lamp have wavelength λ_{exc} equal to 313, 365, 405, 436, 546 and 578 nm. The scheme of the reactor was shown in figure 1.

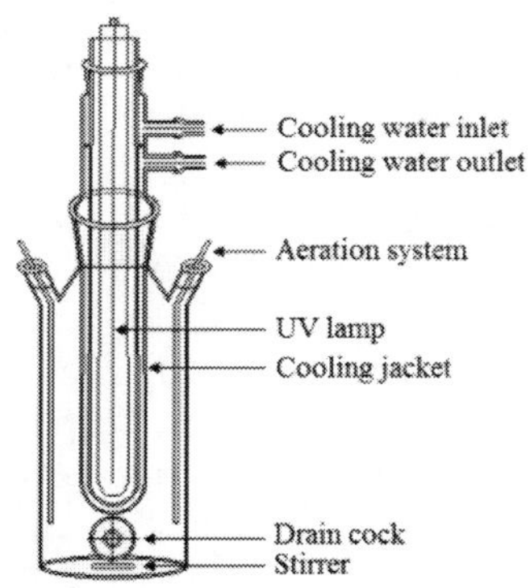

Figure 1. Scheme of the photocatalytic batch reactor [11].

The dose of catalyst was estimated during the preliminary stage of the research and was equal to 100 mg/L, whereas the contact time of the catalyst with the water solutions before UV irradiation process was set to 15 min. This dark mixing step ensured the adsorption of micropollutants on the surface of the catalyst. The reaction mixture was aerated by the use of an aeration pump of capacity of 4 L air per minute. Microfiltration membrane filters of 0.45 membrane pore sizes by Merck Milipore (Germany), were use for the separation of catalyst from the post-reaction mixture.

2.3 Kinetic of heterogeneous photocatalysis

Based on the Langmuir-Hinsherwood equation the kinetics of the photocatalysis process was calculated, as a correlation between the concentration of the micropollutant in the water solutions (C, mg/L) and the duration of the UV irradiation (t, min), according to the equation (1), where (K) is an equilibrium constant of the compound adsorption on the catalyst surface

$$r = \frac{dC}{dt} = k\left(\frac{KC}{1+KC}\right) \quad (1)$$

With the assumption, that the degradation process of micropollutants becomes a pseudo-first order reaction, the reaction rate constant (k, min^{-1}) may be established as a linear regression slope (Equation 2), where (C_0) and (C_t) are the compound concentrations in the water solutions before the commencement of the UV irradiation process ($t = 0$) and after a given time indicated as (t).

$$\ln\left(\frac{C_t}{C_0}\right) = -kt \quad (2)$$

The determined reaction rate constant (k) enables to calculate the half life ($t_{1/2}$) on the basis of the Equation (3):

$$t_{\frac{1}{2}} = \frac{\ln 2}{k} \quad (3)$$

2.4 Analytical procedure

The analytical procedure of investigated micropollutants was performed by the use of high performance liquid chromatography HPLC-UV by Varian (Poland) equipped with a Hypersil GOLD column by Thermo Scientific of length 25 cm, diameter 4.6 mm and graining 5 µm. The mobile phase was a mixture of acetonitrile and water in the volume ratio of 85:15 (v/v). The chromatographic analysis was preceded by solid phase extraction (SPE) of analyzed water samples.

For the SPE Supelclean™ ENVI-18 columns of a volume of 6 mL (1.0 g) by Supelco (Poland) filled with octadecylsilane C18 bed were use. For compounds from the group of polycyclic aromatic hydrocarbons – anthracene and benzo[a]pyrene and industrial admixtures - pentachlorophenol, octylphenol, together with pharmaceutical compounds – diclofenac separate extraction procedure were applied. The column bed before the extraction of PAH was conditioned with methanol and washed with deionized water. After the samples with the addition of an aqueous solution of hyamine reagent had completely passed, the SPE bed was dried under vacuum. Then the column bed was wetted with methanol and the analytes were eluted with dichloromethane. The extraction of pentachlorophenol, octylphenol and diclofenac was preceded by conditioning of the columns bed with acetonitrile and methanol. Afterwards, the column bad was rinsed with deionized water. The isolated compounds were eluted with acetonitrile/methanol mixture (60/40, v/v).

Obtained eluates were dried under high purity nitrogen flux and then the residues were re-dissolved in 50 µL of methanol and analyzed chromatographically. Details of the method are presented in the reference [12].

3 Results and Discussion

The first stage of the study was devoted to assess the adsorption process of the investigated organic micropollutants on the surface and/or inside of the photocatalyst particles. The absorption of compounds in the active centers of the semiconductor, constitutes the basis for a proper run of their photocatalytic oxidation and/or reduction. The adsorption degrees of the micropollutants were estimated on the basis of the changes in their concentration in water solution subjected to 15 minute contact time with the catalyst.

 DOI: 10.1051/e3sconf/20171700047

Figure 2 presents the rates of concentration decrease of OP, PCP, ANT, BaP and DCL present in deionized water solutions.

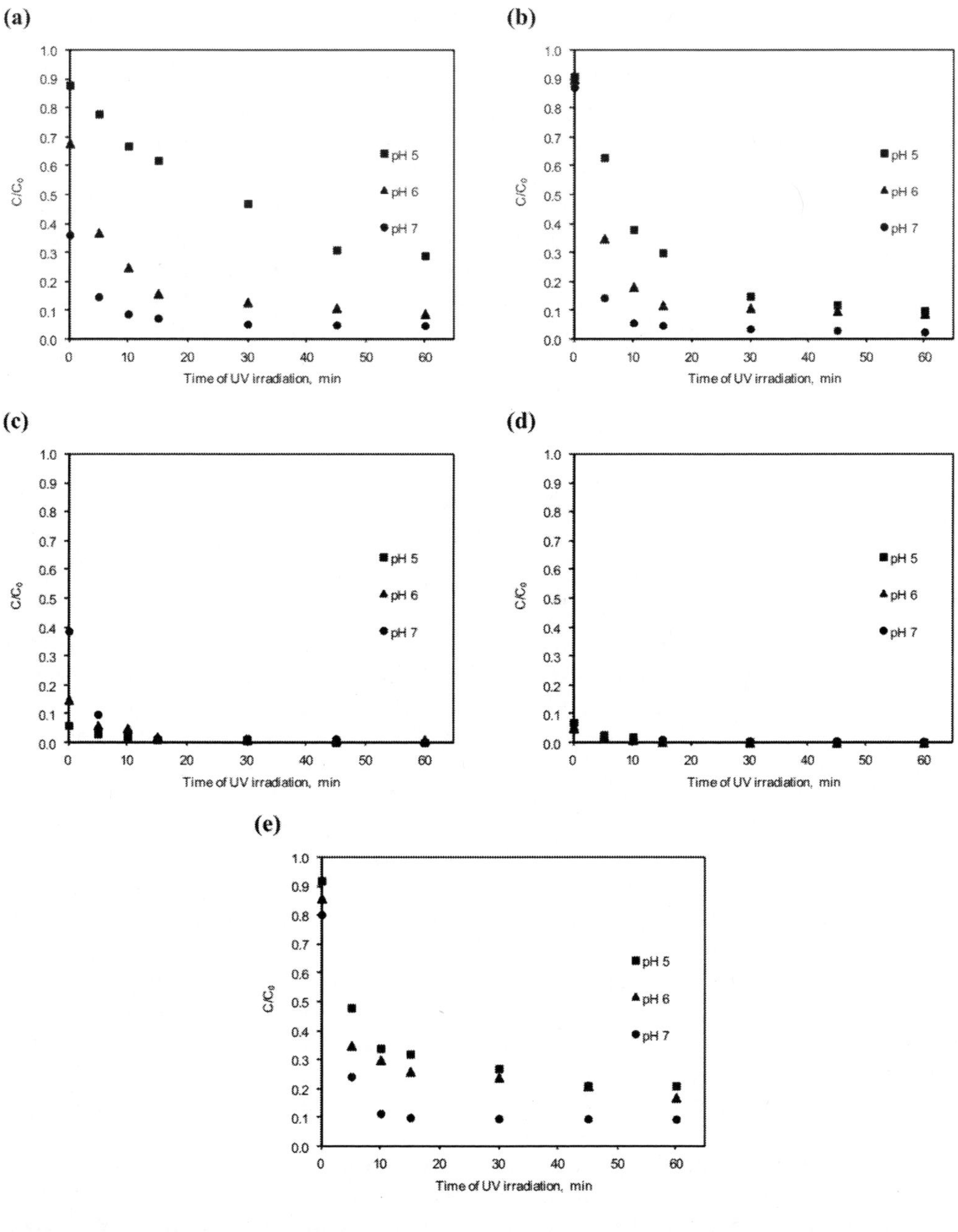

Fig. 2. Influence of solution pH on the degradation of **(a)** OP, **(b)** PCP, **(c)** ANT, **(d)** BaP and **(e)** DCL during the photocatalysis process.

It was observed that the adsorption degrees of OP and PCP increases with the increase of the pH of water solutions (Time of UV iradiation = 0). The 12% decrease of OP concentration, 9% PCP, 8% DCL was observed for pH 5. Whereas for solutions of pH 7 the concentration of micropollutants consecutively decreased by 64% for OP, 13% for PCP and about 20% for DCL. The adsorption degree of BaP was, regardless of the pH, at a constant level of 95%. Different correlation was observed in case of ANT. The adsorption degree of this compound decreased with the increase of pH from 94% obtained for water solutions of pH 5 to 61% for pH 7 solutions. It should be highlighted that for pH values above pH_{zpc} (the value at which the total charge of the catalyst surface shall be zero), the surface of the semiconductor assumes a negative charge and repel anionic compounds. Whereas for pH below pH_{zpc}, the surface assumes a positive charge, resulting in repulsion of cationic donors and electron acceptors [13].

The obtained results indicated that the absorption of micropollutants on the photocatalyst surface depends on the chemical structure of the compound and the change of the charge of semiconductor surface in different pH values of micropollutant water solutions.

The implementation of UV irradiation on the reaction mixtures initiates the photochemical decomposition of organic compounds adsorbed on the catalyst surface and located directly in the water solutions (photolysis process). The reaction of compounds with $OH^{\bullet}$ radicals, which are generated in accordance with reactions from (4) to (6), resulted in the decrease of the micropollutant removal.

$$Ti^{IV} + H_2O \rightarrow Ti^{IV} - H_2O \tag{4}$$

$$Ti^{IV} - OH^- + h^+ \rightarrow Ti^{IV} - OH^{\bullet} \tag{5}$$

$$Ti^{IV} - H_2O + h^+ \rightarrow Ti^{IV} - OH^{\bullet} + H^+ \tag{6}$$

The decomposition of compounds occur most intensively in the first 15 minutes of the process regardless of the pH of the water solutions. For example, the concentration of PCP decreased after the 15th minute of UV radiation by 70% for solutions of pH 5, 88% for pH 6 and 95% for pH 7. The removal degree of ANT and BaP obtained in the considered irradiation time of 15 min in all mixture independent of the pH value was very similar and exceed 98%.

After 60 min of UV irradiation a complete removal of ANT and BaP was obtained. In the case of water solutions containing OP, PCP and DCL a clearly effect of pH on the efficiency of the compound decomposition was observed. The concentration of OP decreases by 71% for pH 5, 91% for pH 6 and 95% for pH 7. The removal degree of PCP reached the highest value exceeding 98% in water solutions of pH 7. While the concentration of DCL in the solution of pH 5 after 60 minutes of UV exposure was reduced by more than 79% and in water solutions of pH 7 the removal degree exceeded 91%. This phenomenon may be a result of the increasing degree of adsorption of micropollutants on the surface of the catalyst, which was shown during the first stage of the study, or results from the increase in the amount of OH^- ions, which are reactants by the production of $OH^{\bullet}$ radicals.

The kinetics assessment revealed a significant difference between the rapidity of micropollutant decomposition, expressed on the basis of the value of the reaction rate constant (k), appointed during the first period of the process (up to 15 min) and the rapidity observed after the 15th minute of UV irradiation (Table 2). These differences may result from a competitive adsorption of by-products of micropollutants decomposition on the surface of catalyst, which can inhibit the oxidation and/or reduction of parent compounds. An inverse relationship was observed only in the case of ANT decomposition in water solution of pH 5, where the reaction rate constant was higher after the 15 minute of the process.

Table 2. Half-life of investigated organic micropollutants during photocatalysis.

Compound	pH	Reaction time, min	Reaction rate constant k, min^{-1}	R^2	Half-life $t_{\frac{1}{2}}$, min
OP	5	0 - 15	0.0241	0.98	34.29
		15 - 30	0.0180	0.94	51.65
	6	0 - 15	0.1053	0.99	17.95
		15 - 30	0.0126	0.99	185.86
	7	0 - 15	0.0947	0.92	11.96
		15 - 30	0.0089	0.79	364.80
PCP	5	0 - 15	0.0767	0.98	10.43
		15 - 30	0.0235	0.89	72.10
	6	0 - 15	0.1339	0.97	6.97
		15 - 30	0.0065	0.99	417.39
	7	0 - 15	0.1911	0.87	6.50
		15 - 30	0.0148	0.99	237.28
ANT	5	0 - 15	0.1156	0.99	30.58
		15 - 30	0.2103	0.92	4.94
	6	0 - 15	0.1245	0.95	19.31
		15 - 30	0.1213	0.96	21.39
	7	0 - 15	0,2117	0.99	8.41
		15 - 30	0.0333	0.87	124.66
BaP	5	0 - 15	0.1827	0.95	20.00
		15 - 30	0.1442	0.94	21.86
	6	0 - 15	0.1622	0.99	20.67
		15 - 30	0.1336	0.96	31.79
	7	0 - 15	0.0905	0.78	45.13
		15 - 30	0.0200	0.86	252.40
DCL	5	0 - 15	0.0749	0.88	14.28
		15 - 30	0.0101	0.90	169.00
	6	0 - 15	0.0703	0.80	13.16
		15 - 30	0.0094	0.96	198.71
	7	0 - 15	0.1395	0.89	8.40
		15 - 30	0.0013	0.93	2288.81

An interesting relationship was observed for the appointed compound half-lifes. The value of this parameter in the first 15 minutes of the process gradually decreases with the increasing pH of the water solutions. For example, the half-life of OP determined at pH 5 amounted 34 minutes, whereas at pH 7 the compound half-life was shorter than 12 minutes. Only in the case of BaP decomposition the half-live increased with the increasing pH of water solution and subsequently received values from 20 min in solutions of pH 5 to over 45 min in water solutions of pH 7. The half-live values of all investigated compounds estimated after the 15th minute of UV irradiation also increased with the increasing solution pH. An exception was only the half-live appointed for PCP in solutions of pH 7, with a value of 237 minutes, compared to solutions of pH 6, which half-live was longer than 417 minutes.

4 Conclusions

On the basis of the conducted investigations it can be observed that the absorption of micropollutants on the photocatalyst surface depends on the type of the compound and their chemical structure. The pH values of micropollutant water solutions influences the charge of the surface of the semiconductor, which alters the degree of adsorption of compounds. The adsorption degree of investigated micropollutants on the surface of the semiconductor reflected on the removal degrees observed during photocatalysis. It was noted that the concentration of micropollutants decreased with the increase in the pH of the water solution. Moreover the data analysis from the kinetics assessment of the photocatalysis process proved a significant difference between the rapidity of micropollutant decomposition observed during the first 15 min of the process and the rapidity of the process observed between the 15th and the 30th minute of UV irradiation.

The paper was prepared within the frame of the National Science Centre project based on decision no DEC-2013/11/B/ST8/04391.

References

1. B. Kuch, F. Kern, J. W. Metzger, K. T. von der Trenck, Environ. Sci. Pollut. Res. Int. **17**, 250 (2010)
2. Y. Sun, H. Huang, Y. Sun, C. Wang, X.-L. Shi, H.-Y. Hu, T. Kameya, K. Fujie, Environ. Pollut. **180**, 339 (2013)
3. H. I. Abdel-Shafy, M. S. M. Mansour, Egypt. J. Petrol. **25**, 107 (2016)
4. Water Frame European Directives 2000/60/EC of the European Parliament and of the Council of 23 October 2000 establishing a framework for Community action in the field of water Policy
5. M. J. Arlos, M. M. Hatat-Fraile, R. Liang, L. M. Bragg, N. Y. Zhou, S. A. Andrews, M. R. Servos, Water Res. **101**, 351 (2016)
6. I. Udom, P.D. Myers, M. K. Ram, A. F. Hepp, E.Archibong, E. K. Stefanakos, D. Y. Goswami, Am. J. Analyt. Chem. **5**, 743 (2014)
7. E. Kudlek, ACEE Archit. Civ. Eng. Environ. **8**, 93 (2015)
8. J. Bohdziewicz, E. Kudlek, M. Dudziak, Desalin. Water Treat. **57**, 1552 (2016)
9. T.A. Egerton, Catalysts **3**, 338 (2013)
10. N. Bouanimba, N. Laid, R. Zouaghi, T. Sehili, Desalin. Water Treat. **53**, 951 (2015)
11. E. Kudlek, M. Dudziak, J. Bohdziewicz, Water **8**, 1 (2016)
12. J. Bohdziewicz, M. Dudziak, G. Kamińska, E. Kudlek, Desalin. Water Treat. **57**, 1361 (2016)
13. B. Czech, K. Rubinowska, Adsorption, **19**, 619 (2013)

Earthworms as bio-indicators of chemical pollution in soils with drilling waste

Justyna Kujawska[1,*], and *Małgorzata* Pawłowska[1]

[1]Institute of Renewable Energy Engineering, Lublin University of Technology 40B Nadbystrzycka Street Lublin, Poland

Abstract. The deposited drilling waste may alter the functioning of soil ecosystems by disrupting the soil fauna. The wastes contain toxic compounds such as metals or hydrocarbons, which tend to accumulate in the living. In the research, earthworms were used u as bio-indicators of soil pollution resulting from the presence of pollution contained in drilling waste. The aim of this study is to evaluation of toxicity of the waste of spent oil and water drilling cuttings to earthworm *Eisenia fetida*. Oil-based and water-based drilling cuttings were added to soil, used six doses of drilling wastes (5%, 10%, 15%, 25%, 30%, 50%). The results showed that the 45 days LC50 of oil-based drilling cuttings ranged 18%, LC50 of water-based drilling cuttings ranged 33%. In general, mortality increases with increasing dosage waste.

1 Introduction

Depletion of crude oil and natural gas from conventional sources has caused an intensive search for alternative hydrocarbons. The global deposits of shale gas are estimated to 200 trillion m^3. However, it is only extracted by the United States and China [1]. Other countries, including Poland, have started to exploit the shale gas, but the drilling and hydraulic fracturing technologies raised social concerns. On the one hand, an increase in shale gas extraction could mitigate the emission of greenhouse gases. On the other hand, there is a risk of soil, ground and surface waters, as well as air becoming contaminated.

The produced drilling waste may pollute the environment, but their actual influence is not sufficiently known. Drilling wastes comprise drill cuttings and spent drilling fluid. The chemical composition of drilling wastes is diversified and depends on the type of drilling fluid utilized in the well. Two types of drilling fluids are commonly employed: water- and oil-based (mainly mineral oil). Apart from the basic substances, various additives such as loamy materials, tannins, surfactants,

* Corresponding author: j.kujawska@pollub.pl

glycol and metal salts are used. Hence, drilling wastes contain drilling fluid compounds and the compounds which percolate from the soil, such as zinc, lead, and hydrocarbons [2].

Horizontal drilling of Marcellus shale on the depth of 3700 m produced 500 t of solids, i.e. drill cuttings. These substances were characterized by elevated content of metal and xylenes [1].

Wastes produced in the course of shale gas extraction are stored at the drill site and subsequently transported to a landfill. In Poland, this is the main method of managing this type of wastes. Chemical compounds from the deposited wastes may infiltrate into the environment – mainly soil – and negatively impact the flora and fauna inhabiting it. The influence of drilling wastes on soil is not fully understood.

As far as studies on environmental pollution are concerned, in recent years there has been a growing interest in bioassays. These tests enable a quick assessment of toxicity of soil polluted by chemical compounds, as well as the analysis of pollutant impact on living organisms. The possibility of determining the pollutant dose causing toxicity is an advantage of this method [3].

Assays conducted on earthworms may be used for the evaluation of chemical pollution in soil. Earthworms constitute 60-80% of animal biomass in soils, having an influence on its quality. Earthworms bioaccumulate pollutants. A coherent relation between earthworms and the concentration of pollutants (mainly metals) in soils make them good biomonitoring organisms [4].

There are few works uses earthworms in monitoring of soil polluted with drilling wastes. There are a number of studies on the toxicity of drilling wastes on aquatic organisms; this is related to the fact that in the United States the drilling wastes were dumped into the ocean for many years. Studies indicate a toxic influence of drilling wastes on living organisms [5–9]. The knowledge about the toxic potential of chemical compounds found in drilling fluids may be used for the assessment of their impact on the environment.

The aim of the work was to evaluate the toxicity of water- and oil-based drilling fluids on the biomass and mortality of earthworms following 7, 14, 30, and 45 days.

2 Characteristics of soil and waste

The chernozem soil was collected in the vicinity of Hrubieszów (eastern Poland) from arable soil (up to the depth of 30 cm). The water-based drilling wastes were taken from Maćkowice well, where potassium/polymer drilling fluid was used. On the other hand, the oil-based drilling wastes were originated from the well which utilized oil-based drilling fluid. Seven mixtures of soil and drill cuttings were prepared, which differed in respect to the composition. The content of water- and oil-based drill cuttings equaled 5%, 10%, 15%, 25%, 30%, and 50% of weight.

3 Research methodology

The assay was carried out in glass containers with the volume of 250 cm^3, filled with soil containing an addition of water- and oil-based drilling cuttings.

Each sample contained 10 mature *Eisenia fetida* earthworms. Chernozem soil constituted the control sample. The water content amounted to 40÷60% (WHC). The assay was carried out over the period of 45 days. The earthworms were fed with finely ground manure. Following 7, 14, 30, 45 days since the earthworms were placed in containers, they were weighed and counted. The assay was conducted in triplicate for each mixture. The acute toxicity assay was performed in line with a PM-ISO 11269-1, 1998 standard. This method involves marking the mortality of mature *Eisenia fetida* earthworms after 7, 14, 30, and 45 days since they were placed in drill cuttings. The mortality coefficient of earthworms after n days was calculated with the following formula:

$$M_n = \frac{(A - B_n)}{A} 100\% \qquad (1)$$

where:
A – number of live earthworms in the control soil sample
B_n– number of live earthworms in polluted soil after n days
M_n – mortality coefficient after n days, %
n = 7, 14, 30, 45 days.

The lethal dose of LC50 waste was determined according to the linear scale. The mortality points were marked on the y-axis, while the dose of drilling cuttings was marked on the x-axis. The dose of LC50 is determined by the abscissa of the crossing of curve with the straight line corresponding to 50% mortality [10].

4 Results

The addition of drilling waste reduced the viability of *Eisenia fetida* earthworms when compared with the control sample, that was a chernozem soil (Fig. 1 and 2). However, the mortality of the tested organisms depended on the drill cuttings type and time of the earthworms exposure to the chemicals contained in the waste.

After 7 days of the experiment, the mixtures with the addition of 5, 10, 15, 25, and 30% of water-based drilling cuttings do not impact the number of earthworms. Mixtures with 25 and 30% addition of water-based drilling cuttings resulted in 30% mortality after 30 days. After 45 days, the mortality in mixtures grows along with the increase in the dose, resulting in the death of all earthworms. In the case of 50% dose of drilling cuttings, the mortality reached 100% (Fig. 1).

Mixtures with oil-based drilling cuttings are more toxic for earthworms than the water-based drilling cuttings. Already in the seventh day, the 25% oil-based drilling cuttings addition resulted in 10% mortality of earthworms; 30% addition caused 30% mortality, whereas 50% addition was lethal to 100% of specimens. After 14 days, the 25% addition of drilling cuttings brought about 25% mortality; 30% addition results in 90% mortality, while 50% causes all earthworms to die off. After 30 days, half of the specimens in the mixture with 25% oil-based drilling cuttings addition died, whereas with 30% dose, the mortality reached 100%. No earthworms survived after 45 days in the mixture with 25% oil-based addition (Fig. 2).

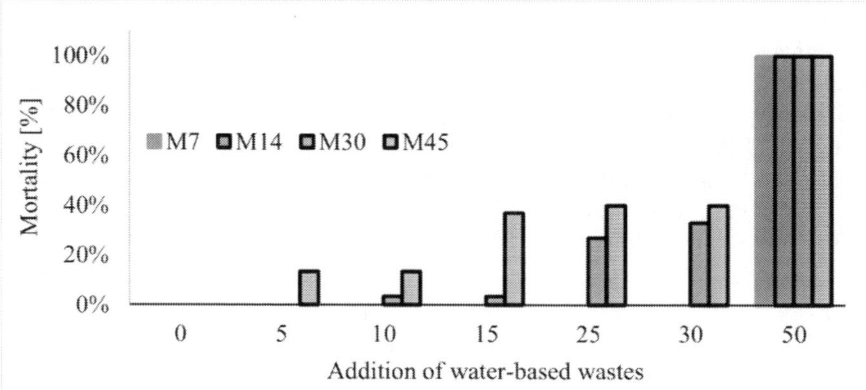

Fig. 1. The mortality coefficient of *Eisenia fetida* in mixtures containing water-based drilling cuttings after 7, 14, 30, and 45 days.

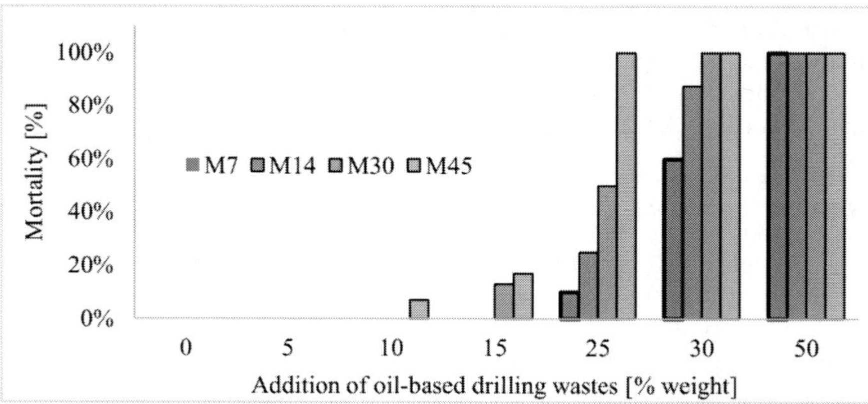

Fig. 2. The mortality coefficient of *Eisenia fetida* in mixtures containing oil-based drilling cuttings after 7, 14, 30, and 45 days.

The LC50 doses calculated on the basis of a linear relation between the waste dose and earthworms mortality after considered time of the experiment duration were presented in table 1. The dose of water-based drilling cuttings, which resulted in 50% of earthworms dying off in the considered periods of the time amounted to 33%. In the case of oil-based drilling cuttings, the dose was slightly lower and equalled 25% for 7, 14, and 30 days of the experiment; after 45 days, the value dropped even further, to 18%. The correlation coefficient for the mixture with water-based drilling cuttings after 7 and 14 days was moderate, while after 30 and 45 days it was very high. The correlation coefficients for mixtures with oil-based drilling cuttings addition are high. Such values of correlation coefficients indicate strong correlation between the dose and mortality coefficient.

Table 2 presents the impact of drilling cuttings on the biomass of earthworms, following 7, 14, 30, and 45 days. The biomass of earthworms declines greatly with the addition of drilling cuttings. In the case of mixtures with 5, 10, and 15% addition of water-based drilling cuttings, the biomass of earthworms decreases by the factor of 1.5 after 30 days, whereas after 45 days, it experiences a threefold decrease. With 25%, 30% addition, the biomass is reduced to one-third of the initial value.

Similar situation occurs in the mixture comprising oil-based drilling cuttings. In mixtures with 5, 10, and 15% addition, the biomass of earthworms after 30 days reduces by 1.5, while after 45 days – by 2.5 times in relation to the initial value. In a 25% oil-based drilling cuttings addition, the biomass is halved after 30 days.

Table 1. The correlation coefficient and the LC50 doses for the mixture with water-based and oil-based drilling wastes.

Time [days]	Mixtures with water-based drilling wastes		Mixtures with oil-based drilling wastes	
	Mortality coefficient R^2	LC50	Mortality coefficient R^2	LC50
7	0.64	33%	0.85	25%
14	0.64	33%	0.83	25%
30	0.91	33%	0.85	25%
45	0.92	33%	0.76	18%

Table 2. Comparison of earthworm biomass in mixtures of soil and drilling cuttings.

Dose of waste \ Time	Total biomass [g]				
	Start	After 7 days	After 14 days	After 30 days	After 45 days
The control sample					
0	2.21± 0.30	2.46±0.35	2.17± 0.23	2.78± 0.37	2.21± 0.25
Samples with water-based drilling cuttings addition					
+5%	2.27± 0.25	2.93±0.17	2.21 ± 0.07	1.41 ± 0.13	0.68± 0.27
+10%	2.22± 0.26	2.72 ± 0.49	2.37 ± 0.34	1.54 ± 0.33	0.94 ± 0.12
+15%	2.14±0.27	2.67 ± 0.59	2.38 ± 0.40	1.42 ± 0.27	0.70 ± 0.16
+25%	2.11 ± 0.37	2.46 ± 0.71	2.30 ± 0.70	0.67 ± 0.31	0.65 ± 0.21
+30%	2.09 ±0.46	2.40 ± 0.65	2.19 ± 0.41	0.42 ± 0.11	0.54 ±0.33
+50%	2.03 ± 0.35	0	0	0	0
Samples with oil-based drilling cuttings addition					
+5%	1.94 ± 0.16	2.51 ± 0.34	1.95± 0.21	1.27 ± 0.1	0.76 ± 0.03
+10%	1.96 ± 0.14	2.52 ± 0.22	2.21 ± 0.24	1.12 ± 0.24	0.61 ± 0.09
+15%	2.14 ± 0.41	2.16 ± 0.43	1.96 ± 0.18	1.18 ± 0.11	0.74± 0.06
+25%	1.95 ± 0.36	1.98 ±0.51	0.65 ± 0.09	0.91 ± 0.71	0
+30%	1.90 ± 0.54	0.80 ±0.34	0.33 ± 0.21	0	0
+50%	1.85 ±0.77	0	0	0	0

The obtained results show that the doses of oil- and water-based drilling cuttings which are non-toxic to soil amount to 5 and 10%. Drilling cuttings applied in the higher doses are toxic and when added in higher doses, result in the death of earthworms. This may be related to the presence of such chemical compounds as chromium salts, metals, and surfactants. Oils are compounds which disrupt the nervous system [11].

Studies on the impact of drilling waste on living aquatic and soil organisms (earthworms) were conducted. Ogeleka et al., Enuneku and Ayobahan, and Soegianto et al. investigated the influence of drilling fluids on living organisms [5–7]. In all studies, a negative impact of drilling wastes on tested organisms was observed.

The studies were carried out on the solutions decanted from the mixture of drilling mud-to-water in volumetric ratio of 1 to 9. This solution was defined to be 100% SPP (suspended particulate phase) equal to 1000000 ppm SPP. Any other concentration of SPP refers to a percentage of SPP that was obtained by volumetrically mixing of 100% SPP with the dilution water. The concentrations of the test solutions used in the studies were 3.125, 6.25, 12.5, 25 and 50% (equal to 31250, 62500, 125000, 250000, 500000 ppm SPP).

Enuneku et al. conducted a research on the influence of oil-based drilling fluid and drill cuttings to *Aporrectodea longa* earthworms. The biomass of earthworms increased in the range of 6.25% to 12.5% SPP and decreased when the dose reached 25% and 50% of SPP [6]. Soegianto et al. investigated the toxicity of drilling wastes on the tiger prawn larvae (*Penaeus monodon*). The mortality of the larvae reached 100% for the dose of 125000 ppm SPP, after 96h [7]. In the research of Ogeleka et al. [5] the dose of oil-based drilling wastes which was 100% lethal to the brackish water shrimp (*Palaemonetes africanus*) amounted to 25% after 72h and 50% SPP after 48h. The dose which resulted in the death of 50% of brackish water shrimp after 96h equaled 13, and 18%.

The obtained results and the review of literature indicate a diverse impact of drilling wastes on the biomass of living organisms. Determining the influence of dosage on the pollution of the environment is difficult and mainly dependent on the employed technology and drilling conditions.

In the research carried out by Soegianto et al., the dose of LC50 which resulted in half of the *P. monodon* larvae dying off amounted to 30000 ppm SPP [7]. Viega et al. investigated the impact of 11 spent drilling fluids on grass shrimp larvae (*Palaemonetes intermedius*). The doses which killed 50% of larvae after 96h ranged from 142 to 100000 ppm SPP. Other studies conducted by these authors showed that the dose of water-based drilling fluids that caused the death of 50% of *Mysidopsis bahia* after 96h, amounted to 27000 ppm SPP [8]. The results obtained by Mexican scientists showed that the dose of water-based drilling fluids which halved the population of *Litopenaeus setiferus* ranged from 475000 to 700000 ppm SPP.

The dose of wastes which kills 50% of living organisms varies, similarly to the impact of drilling wastes on their biomass. The lowest dose of drilling wastes which halves the population of organisms amounts to 1.5%. This indicates the necessary of producing drilling fluids characterized by the composition which is safe for the environment.

5 Conclusions

Along with the increase in the concentration of drilling cuttings in soil, the mortality grows, while the biomass of earthworms decreases. This means that the compounds in the drilling wastes are toxic to annelids. Mass storage of drilling wastes should be limited due to the negative impact on living organisms, especially earthworms. The studies and the review of literature indicate the diversity of toxic dose of drilling wastes, which stems from varied composition of fluids adjusted to the employed drilling technology. Drilling wastes should be treated prior to being stored or managed in other way. Undoubtedly, there is a need for a constant monitoring of the drilling site and landfills of drilling wastes.

References

1. M.P.J.A. Annevelink, J.A.J. Meesters, A.J. Hendriks, Environmental contamination due to shale gas development, Sci Total Environ **550** (2016)
2. E.O. Abbe, M.S. Grimes, D.G. Fowler, R.A. Boccaccini, J. Material Sci. **60**, 1 (2009)
3. A. A Al-Haleem, A. M. Khalid, RJPBCS **7**, 6 (2016)
4. T. Hirano, K. Tamae, Sensors **11** (2011)
5. D.F Ogeleka, L.E Tadararo-Aherobo, BEPLS **2** (2013)
6. A. A. Enuneku, S. U. Ayobahan, SU, J. Appl. Sci. Environ. Manage **18**, 4 (2014)
7. A. Soegianto, B. Irawan, M. Affand, GJER **2**, 1 (2008)
8. L.F. Viega, Z.T. Tostes, M.V. Reynier, G.F.R. Brandao, F.F. Oliveira, *Setac 22nd Annual Meeting. Changing Environmental Awareness: Societal Concerns and Scientific Responses* (Baltimore, Maryland, USA, 2001)
9. R., F. Nunez, A. Chiappa, M. Vasquez-Botello, C. De la Rosa-Duque, C. Vanegas, *Setac 22nd Annual Meeting. Changing Environmental Awareness: Societal Concerns and Scientific Responses* (Baltimore, Maryland, USA, 2001)
10. M. Łebkowska, M. Załęska-Radziwiłł, Bożena Słomczyńska, *Environmental toxicology* (Warsaw University of Technology, Warsaw, 2004) (in polish)
11. W.C. McMordie, *Oil base drilling fluids. Symposium on Research on Environmental Fate and Effect of Drilling Fluids and Cuttings* (Lake Buena Vista, Florida, USA, 1980)

E3S Web of Conferences **17**, 00049 (2017) DOI: 10.1051/e3sconf/20171700049
EKO-DOK 2017

Prediction of Availability Indicator of Water Pipes Using Artificial Intelligence

Małgorzata Kutyłowska[1,*]

[1]Wrocław University of Science and Technology, Faculty of Environmental Engineering, Wyb. Wyspiańskiego 27, 50-370 Wroclaw, Poland

Abstract. The paper presents the results of artificial neural networks application to the availability indicator prediction. The forecasted results indicate that artificial networks may be used to model the reliability level of the water supply systems. The network was trained using 147 and 173 operational data from one Polish medium-sized city (distribution pipes and house connections, respectively). 50% of all data was chosen for learning, 25% for testing and 25% for validation. In prognosis phase, the best created network used 100% of 114 and 133 values for testing. Following functions were used to activate neurons in hidden and output layers: linear, logistic, hyperbolic tangent, exponential. The learning of the artificial network was performed using following input parameters: material, total length, diameter. In the optimal models hyperbolic tangent was chosen to activate the hidden and output neurons in modeling the availability indicator of house connections during 68 epochs of training. Hidden and output neurons were activated (20 epochs of learning) respectively by hyperbolic tangent and linear function during the prediction of availability indicator of distribution pipes. The maximum relative errors in learning and prognosis step were equal to 0.10% and 1.20% as well as 0.27% and 1.15% for distribution pipes and house connections, respectively.

1 Introduction

The suitable management of water supply systems seems to be nowadays one of the crucial aspect influencing the proper maintenance of all parts of the distribution network. F. Ijjas [1] noticed that all problems related to water are connected with human being. Also the technical condition of water-pipe networks should be considered as important factor determining the way of management and maintenance of water utilities. Consequently, the reliability level of water-pipe networks should be estimated very precisely and should have an influence on the water utility management. There are a lot of investigations and studies carried out in Poland and abroad concerning the failure, safety and reliability analysis of water supply system which could be treated as a part of critical buried infrastructure [2–5].

Mathematical modeling of the reliability indicators of water-pipe network is a promising tool which seems to be used more often nowadays [6, 7]. There are a lot of mathematical approaches which are used to estimate the technical condition and

* Corresponding author: malgorzata.kutylowska@pwr.edu.pl

deterioration as well as the risk of failure of water pipes, e.g. statistical and physically based models or (very popular nowadays) fuzzy logic [8, 9]. Recently, also the model of water outflow from fault conduit was proposed [10].

In the literature there are a lot of artificial neural network modeling results related to failure rate, number of damages, water distribution profiles or time to failure [11–15], but there is lack of investigations concerning availability indicator. Till now this parameter was predicted only with applying typical mathematical models.

The aim of the presented paper was continuation, complement and development of investigations results of failure rate prediction using artificial networks [16]. Among other things, two parameters (failure rate and availability indicator) describe the reliability level of water-pipe network and should be estimated inclusively.

Availability indicator of house connections and distribution pipes in one selected medium-sized Polish city was forecasted using typical network structure – multilayer perceptron (MLP). The aim of this work was to construct relatively simple model (with basic information about the pipes given as the input signals) which would be used by water utility to predict in the future reliability level of water pipes without necessity of collecting more complicated data. The modeling results were compared with experimental data. Implementation of the model is simple which seems to be the main advantage.

2 Methodology and range of studies

2.1 Artificial neural networks

Artificial neural networks (ANN) are used to predict, classify, recognize, associate and analyze data, to filter signals and for optimization purposes. Neural networks enable to model non-linear and complex problems. The basis of ANN were described in literature [16-18]. In this paper only the main important information will be reminded. Today the most popular and best theoretically described are MLP networks [19]. This type of the network has one input layer, one or more hidden layers and one output layer.

To achieve good convergence between experimental and predicted values as well as to obtain generalization purposes, the ANN model should be learnt properly using learning methods [16, 18]. In the problem of availability indicator prediction following algorithms were used to learn the network: conjugated gradient algorithm, steepest descent method and quasi-Newton approach (BFGS algorithm). BFGS algorithm described in [16, 18, 20] seems to the best for solving the problem of availability indicator prediction.

The proper choice of activation functions, which are responsible for transferring the information between neurons and layers, is also very crucial problem during the ANN modeling. In the problem of availability indicator prediction following functions were used to activate neurons in hidden and output layers: linear (eq. (1)), logistic (eq. (2)), hyperbolic tangent (eq. (3)), exponential (eq. (4)).

$$f(x) = x \tag{1}$$

$$f(x) = \frac{1}{1 - e^{-x}} \tag{2}$$

$$f(x) = \frac{e^{x} - e^{-x}}{e^{x} + e^{-x}} \tag{3}$$

$$f(x) = e^{-x} \tag{4}$$

2.2 Artificial neural network structure for availability indicator prediction

The availability indicator (*AI*) prediction of distribution pipes and house connections was carried out on the basis of the operating data received from water utility in one selected Polish city [21]. The reliability level of the water-pipe network could be established during the analysis of selected reliability indicators, e.g. failure rate, average time of work and availability indicator [3, 4, 22].

$$AI = \frac{1}{1 + \lambda \cdot L \cdot T_o} \tag{5}$$

where
λ – failure rate in the time t, fail./(km·year);
L – average length of the pipes, km;
T_o – average time of pipes restoration, h.

The stationary availability indicator calculated by relationship (1) was treated in this paper as experimental one, in the contrary to predicted by ANN values of *AI*. The experimental values of failure rate (presented in [16]) were calculated according to the, described in the literature, equation (2) [3, 22–24]. Generally, the availability indicator means the probability that in every time t the facility is efficient (is working) [3, 22]. In practice, this indicator determines the probabilistic assessment of facility disposal. This is reliability indicator of renewal object which is damaged and then repaired. The average time of pipe restoration was assumed at the level of 8 hours. The detailed information about T_o were not registered in the operational documentation prepared by water utility.

$$\lambda = \frac{N(t)}{L \cdot \Delta t} \tag{6}$$

where:
$N(t)$ – the number of failures in the time t, unit;
Δt – the analyzed period of time, year.

The learning of the network was performed using the operational data from the time span 2001–2006. The realization of artificial neural network approach was carried out in the program Statistica 10.0. The table 1 shows the most important assumptions of ANN architecture for availability indicator prediction.

Table 1. Main assumptions of ANN architecture.

Type of conduit	**Number of data (total/learning/prognosis)**			**Max./min. number of hidden neurons**	**Number of hidden layers**	**Number of learning epochs**
House connections	306	173	133	14/1	1	5–886
Distribution pipes	261	147	114	12/1	1	5–442

The optimal ANN model, which predicted the availability indicator in the best way, was chosen in several steps. Firstly, tens ANN architectures were checked (created) using assumptions displayed in the table 1 and also "trial and error method" (because ANN modeling has features similar to a "black box"). Then the absolute and relative root mean squared errors were calculated for each model. Several models (characterized by the lowest errors) were chosen. The prognosis (prediction) step was carried out on the basis of these several models and then the optimal ANN model was selected. The prediction results (in learning and prognosis step) displayed in the next section were achieved using this one chosen model.

The learning of the artificial network was performed using following input parameters: material, total length, diameter. The availability indicator was an output signal predicted by ANN. Length, diameter and availability indicator were quantitative parameters. Each of mentioned above input signals was represented by one neuron. On the other hand, material was qualitative parameter and several neurons represented this signal (see table 2). The prognosis of availability indicator was performed using different data than in learning phase. This assumption is the main in ANN modeling. Firstly, the network should be learnt properly (show the proper patterns and learn the suitable generalization) and then it is necessary to check the model using new data which has not been known by the network previously.

3 Results and discussion

As it was mentioned above, the whole data set (306 and 261 data for house connections and distribution pipes, respectively) was divided into two subsets: one used for learning the artificial network, the second to make a prediction (prognosis) of availability indicator using selected before one optimal ANN model. The values of input and output parameters in learning and prediction phase are shown in the table 2.

Table 2. Input and output parameters in learning and prognosis step.

Type of conduit	Input parameters			Output parameter	
	Material	Length, km	Diameter, mm	Availability indicator	
				Experimental	Predicted by ANN
Learning step					
House connections	Grey cast iron, steel, galvanized steel, PE, PVC	23.4–29.9	25–100	0.962139–0.97864	0.962148–0.978639
Distribution pipes	Grey cast iron, steel, PE, PVC	57.3–59.7	80–200	0.970938–0.98183	0.970864–0.981838
Prognosis step					
House connections	Grey cast iron, steel, galvanized steel, PE, PVC	31.4–50.2	20–100	0.976071–0.989957	0.976926–0.978614
Distribution pipes	Grey cast iron, steel, PE, PVC	69.9–88.7	90–200	0.979353–0.992046	0.988310–0.991265

The ranges of the input parameters (material and diameter) were similar in learning and prognosis step because the water-pipe network structure has not changed significantly. The total length of house connections and distribution pipes was higher in prediction phase than in learning step because the water network has been extended year by year. In analyzed Polish city house connections were made of grey cast iron, steel, galvanized steel, PE and PVC. The distribution pipes were made of grey cast iron, steel, PE and PVC. Material is qualitative input signal and was represented by 5 and 4 neurons during availability indicator prediction of house connections and distribution pipes, respectively. The main features of the optimal ANN models are shown in the table 3.

Table 3. Optimal ANN model.

Type of conduit	Type of MLP	Activation function (hidden layer)	Activation function (output layer)	Relative mean squared errors, %
House connections	7-14-1	Hyperbolic tangent	Hyperbolic tangent	0.00–1.15
Distribution pipes	6-10-1	Hyperbolic tangent	Linear	0.00–1.20

It was necessary to use 68 and 20 epochs of learning to train the network properly and to obtain reasonable predictions of availability indicators of house connections and distribution pipes, respectively. In comparison with failure rate prediction (that was also modeled using ANN and BFGS learning algorithm) [16] the number of learning epochs was almost 1.7 times higher for house connections and six times lower for distribution pipes. We should remember that the simple comparison between ANN architecture used for failure rate and availability indicator prediction cannot be carried out because ANN structure is directly connected with the solved problem. ANN is like a "black box" modeling (we have not knowledge what is going on inside the network) and it is not possible to obtain one uniform (universal) architecture for every kind of problem. It is necessary to seek the proper ANN structure for specific problem which is currently solved.

The decision concerning the number of hidden layers and neurons is fundamental creating the ANN structure. It has been proven that one hidden layer is sufficient for modeling engineering problems [11, 12]. The proper assignation of number of hidden neurons also makes some difficulties because we should avoid the risk of overtraining. On the other hand, it is necessary to show the artificial network proper patter and to learn in sufficient way. In the problem of availability indicator prediction we have 7 or 6 input neurons (concerning house connections and distribution pipes). The optimal ANN structures consist of 14 or 10 hidden neurons, respectively.

Similar as in failure rate prediction [16], the bias value was very important relating to the training process. Hyperbolic tangent was the optimal function in hidden and output layer for availability indicator prediction of house connections. Availability indicator of distribution pipes was predicted with the lowest relative error using hyperbolic tangent and linear function to activate the neurons in hidden and output layer, respectively.

The results of ANN modeling of availability indicator are displayed in the figures 1 and 2. The predictions of availability indicator of distribution pipes were, from engineering point of view, satisfactory. In the learning step, Fig. 1a, (2001–2006) the relative errors varied between 0.00% and 0.10%. The correlation between experimental and predicted data was characterized by coefficients R=0.9918 and R^2=0.9836. Such convergence is very good. The prognosis step, Fig. 1b, was characterized by higher discrepancies between experimental and predicted data (R=0.3105). The values of predicted availability indicator were relatively higher than experimental what could be caused by the fact that the ANN model did not know (in prognosis phase) the input data previously. It is obvious that in prognosis the discrepancies between experimental and predicted values are higher than in learning step. Only in 2011 the predictions were almost exactly the same as experimental values of availability indicators.

a) b)

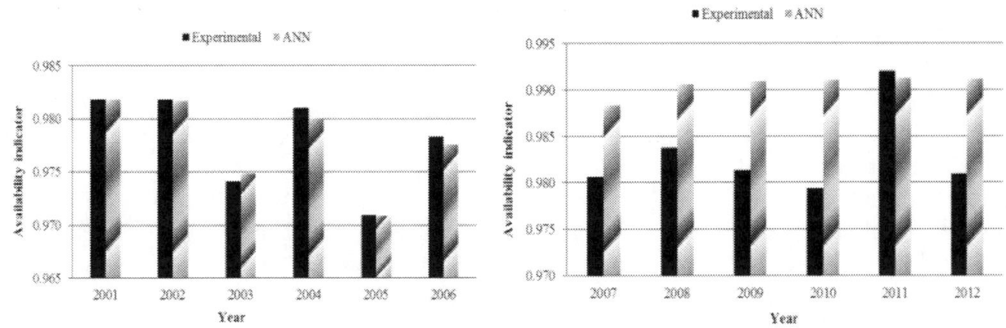

Fig. 1. Experimental and predicted availability indicator of distribution pipes, a) learning step, b) prognosis step.

The modeling results of availability indicator of house connections seem to be quite promising (Fig. 2). The predictions in learning phase (Fig. 2a) were characterized by coefficients R=0.9870 and R^2=0.9742. It means that the correlation between experimental values and prognosis is almost ideal. The relative errors were in the range 0.00 and 0.27%. Worse results were obtained in years 2007–2012 (prognosis step, Fig. 2b). Pearson correlation coefficient was equal to 0.3258, which is slightly better than in distribution pipes modeling. But still the relative error was not higher than 1.15%. In the prognosis phase one input signal (total length) was significantly higher than in learning step which could cause lower convergence. From engineering point of view the model MLP 7-14-1 could be recommended and implemented by water utility to forecast in the future the availability indicator of house connections because learning phase was characterized by relatively good correlation.

a) b)

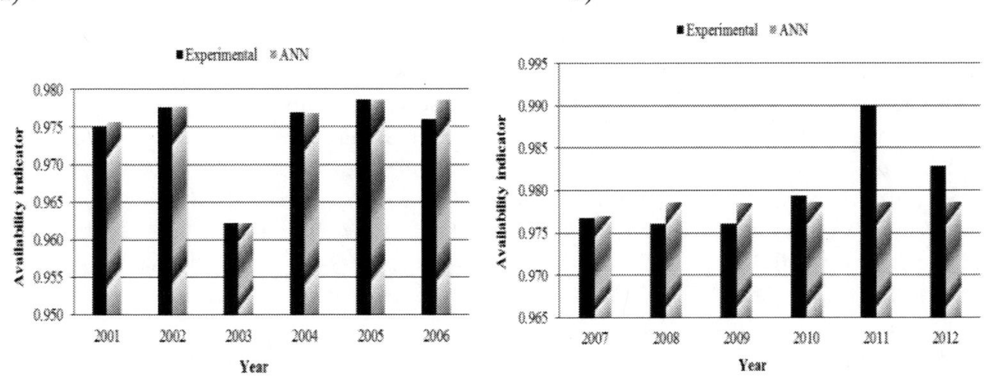

Fig. 2. Experimental and predicted availability indicator of house connections, a) learning step, b) prognosis step.

The results of availability indicator prediction of distribution pipes and house connections confirm that ANN modeling could be useful forecasting tool. Unfortunately, the above mentioned results show quite high discrepancies between experimental and predicted values in the prognosis step. This could be improved if bigger input vector is used. Generally, we should remember that the size of water-pipe network is increasing and also the diameters and materials are changing even year by year. In consequence of this, the input parameters could be slightly different in learning and prognosis step. The idea of, proposed in this paper, modeling is to adapt created models to changing conditions of water pipes. Next step could be training the network using input signals not from 6 but 12 years and the validation (prognosis) step will be done on next years (e.g. 2013, 2014). In such case the pattern will be recognized more precisely. The adaptation (changes in the network structure and in the ranges of input parameters) could not be performed to infinity and should be stopped when the reasonable level of the correlation between experimental and predicted values is achieved. Nevertheless, ANN modeling seems to be promising tool for prediction of the reliability level of water-pipe network. The main advantage of such modeling is relatively short time of calculation and possibilities of adaptation as well as the simple implementation by water utility for its own purposes. Moreover, it is necessary to remember that assessing the failure frequency of water pipes should be carried out together with the estimation of proper operation of sewerage systems, e.g. storm water system [25, 26]. These two municipal systems are very important for proper functioning of the whole buried infrastructure and should be considered together.

4 Conclusions

The results of availability indicator prediction in one Polish medium-sized city might be concluded as follows:

- The generalization phenomenon of ANN allows to describe the relationships (due to weight adaptation during the network training) between variables without having exact knowledge about these relations.
- It is needed to create the artificial network for each problem separately. The solved problem extorts the number of neurons, the kind of activation function and the training methods.
- There are some limitations of using ANN for prediction purposes. It is necessary to collect huge data base. This requirement is sometimes difficult to realize by water utilities. Some important, from forecasting point of view, variables are not registered. The situation is still improving because more and more water utilities attach importance to connect information about damages with GIS.
- It is also necessary to analyse deeply the data which are used as the training parameters in the process of neural network modelling. In some cases it is necessary to eliminate the outliers. If there is too much eliminated outlier data we should answer the question whether the modelling is rational. Unfortunately, real operational data are burden sometimes by huge mistakes due to e.g. improper or incomplete registration or lack of collecting all information in GIS database. In such cases it is necessary to eliminate or remain outliers reasonably. Before modelling approach it is required to cooperate with exploiters and to explain all inaccuracy as well as to complete data.
- The learning of the artificial neural network was performed using the operational data from the time span 2001–2006. The next years (2007–2012) were used to make a prognosis of availability indicator.
- BFGS method of learning was established as the best one in comparison with other training algorithms.
- MLP network structures (6-10-1 and 7-14-1) seem to be efficient for availability indicator predictions of distribution pipes and house connections, respectively.
- In the optimal models: hyperbolic tangent was chosen to activate the hidden and output neurons in modeling the availability indicator of house connections during 68 epochs of training. Hidden and output neurons were activated (20 epochs of learning) respectively by hyperbolic tangent and linear function during the prediction of availability indicator of distribution pipes.
- The maximum relative error in learning and prognosis step were equal to 0.10% and 1.20% as well as 0.27% and 1.15% for distribution pipes and house connections, respectively.

- The convergence between experimental and predicted values of availability indicator of distribution pipes and house connections was characterized by R^2=0.9836 and R^2=0.9742 (learning step), respectively. The prognosis phase gave worse results of modeling what could be improved by adaptation the model to the changing operational conditions.

The work was realized within the allocation No. 0401/0069/16 awarded for Faculty of Environmental Engineering Wroclaw University of Science and Technology by Ministry of Science and Higher Education in years 2016–2017.

References

1. F. Ijjas, Period. Polytech-Civ. **58**, 2 (2014)
2. I. Zimoch, E. Łobos, Environ. Prot. Eng. **38**, 3 (2012)
3. H. Hotloś, Quantitative assessment of the effect of some factors on the parameters and operating costs of water-pipe network (Wrocław University of Technology Publishing house, Wrocław, 2007) (in Polish)
4. M. Kwietniewski, J. Rak, Reliability of water supply and wastewater infrastructure in Poland (Polish Academy of Science, Warszawa, 2010) (in Polish)
5. J. M. Makar, Eng. Fail. Anal. **7**, 1 (2000)
6. M. Kutyłowska, Period. Polytech-Civ. **59**, 1 (2015)
7. D. Laucelli, B. Rajani, Y. Kleiner, O. Giustolisi, J. Hydroinform. **16**, 4 (2014)
8. Y. Kleiner, B. Rajani, Urban Water. **3**, 3 (2001)
9. B. Tchórzewska-Cieślak, Environ. Prot. Eng. **37**, 3 (2011)
10. M. Iwanek, D. Kowalski, M. Kwietniewski, Ochr. Sr. **37**, 4 (2015)
11. A. Shirzad, M. Tabesh, R. Farmani, KSCE J. Civ. Eng. **18**, 4 (2014)
12. R. Jafar, I. Shahrour, I. Juran, Math. Comput. Model. **51**, 9-10 (2010)
13. M. Tabesh, J. Soltani, R. Farmani, D. Savic, J. Hydroinform. **11**, 1 (2009)
14. W. Cieżak, J. Cieżak, Environ. Prot. Eng. **41**, 2 (2015)
15. R. Harwey, E. A. McBean, B. Gharabahgi, J. Water Res. Pl-ASCE. **140**, 4 (2014)
16. M. Kutyłowska, Eng. Fail. Anal. **47**, (2015)
17. M. Kabsch-Korbutowicz, M. Kutyłowska, Desalination Water Treat. **14**, 1-3 (2010)
18. R. Tadeusiewicz, Neural networks (Academic Publishing House RM, Warszawa, 1993) (in Polish)
19. X. Zeng, D. Yeung, Neurocomputing **69**, 7-9 (2006)
20. S. Osowski, Neural networks for information transferring (Warsaw University of Technology Publishing House, Warszawa, 2000) (in Polish)
21. Operational data given by Water Utility in one Polish city, 2013
22. M. Kwietniewski, M. Roman, H. Kłoss-Trębaczkiewicz, Reliability of water pipes and sewers (Arkady, Warszawa, 1993) (in Polish)
23. M. Kutyłowska, H. Hotloś, Eng. Fail. Anal. **41**, (2014)
24. D. Kowalski, K. Miszta-Kruk, Eng. Fail. Anal. **35**, (2013)
25. B. Kaźmierczak, M. Wdowikowski, Period. Polytech-Civ. **60**, 2 (2016)
26. A. Kotowski, K. Wartalska, M. Nowakowska, Ochr. Sr. **38**, 1 (2016)

Thermal comfort study of naturally ventilated office

Marta Laska[1,*]

[1]Wroclaw University of Science and Technology, Faculty of Environmental Engineering, Wybrz. Wyspianskiego 27, 50-370 Wroclaw, Poland

Abstract. It is well known that people spend about 90% of their lifetime in buildings where the indoor environment strongly influences their health and productivity. Therefore it is crucial to maintain proper internal conditions, and thus ensure comfort environment for building users. Due to the nature and complexity the thermal comfort is difficult to describe with physical parameters. However there is a number of available methods to assess the indoor conditions and define the comfort level - from simplified methods based on the measurement of basic parameters to complex approach – with the use of computational fluid dynamics engineering tools. The paper presents the outcome of the thermal comfort study for a naturally ventilated office located in Wroclaw (Poland). The chosen physical parameters has been registered in long- and short-term perspective, and later used for defining the boundary conditions and CFD model validation.

1 Introduction

People spend most of their time indoors. The building is a shelter from external conditions, so the interior has to be safe and comfortable for its users. The building environment strongly influences not only human comfort, but also the life, activity, health and productivity. Therefore the main concern of the engineers and researchers is to identify crucial factors and parameters influencing thermal comfort and maintain them on expected level. The combination of these physical and individual variables defines levels of thermal comfort and thus the user's satisfaction from the thermal environment. There is a certain number of methods to estimate the thermal comfort in rooms – from simplified, based on the variability of one parameter to complex approaches – based on the relationship between environmental and personal factors or implementation of numerical tools. These methods are still evaluating and are under improvement as they have limitations depending on outdoor conditions, building types, specific requirements of the users or the installations the building is equipped.

The indices and methods for estimating thermal comfort are evaluating from the 1930s [1]. Today the most common methods are based on Fanger's theory and operative temperature approach and are described in the European Standards 7730 and 15251 [2, 3]

* Corresponding author: marta.laska@pwr.edu.pl

respectively. The simple method based on relationship between the air temperature and relative humidity but under constant clothing insulation value (clo) of 0,6 and sedentary activity was developed by Rholes [4] His extended model dedicated for wider range of clo was described by Burati et all. in [5]. The adaptive predicted mean vote (aPMV), Human Thermal Model (HTM), adaptive comfort algorithm (ACA) in comparison with classic comfort approach represented by Fanger's theory are described in paper [6] as methods more flexible than classic adaptive approach. The example of the application of most complex approach in comfort study based on CFD modelling is described in [7]. The literature presents also examples of comfort study for mechanically ventilated enclosures [8, 9, 10] as well as naturally ventilated buildings [11–13].

The paper focuses on estimating comfort conditions in naturally ventilated office located in Wroclaw (Poland). The applied method is based on an approach described in standard EN 15251 [3]. It assumes that in naturally ventilated buildings working in free-running mode human body easier adapts to the surrounding due to his/her thermoregulation system. It is applicable for free-running buildings as well as periodically heated, cooled and running under a mixing mode, and is based on a theoretical value – an operative temperature. This method combines two parameters that need to be measured, namely indoor air temperature and mean radiant temperature. The literature [14] indicates that in well-insulated buildings, where the influence of exterior is small, the difference between air, mean radiant and thus operative temperature is negligible. To check this theory the relationship between these parameters is investigated and the statistical methods are applied. The investigated office was also subjected to a long-term assessment and CFD modelling. Therefore the last part of the article presents the outcome of these analysis.

The European Standards [2, 3] define also other methods for steady- and not-steady-state conditions based on temperature drifts, temperature steps, cycling temperatures or temperature ramps. However these are not under the interest of this paper.

2 Comfort study

2.1 Investigated room

The study was conducted in the naturally ventilated office room located on the first floor of the two-storey building. The main geometry data: room area equals 21,3m^2, room clear height is 3m, clear length of the external wall is 5,1m, two windows of 112x150cm. The room geometry is presented on the Fig. 1. The external wall and windows are directed towards south-west. The comfort conditions during heating season in the office are provided by classic radiator-based water system. In summer office works in a free-running mode. The European Standards [2, 3] defines the range of comfort operative temperatures depending on the class of the building. The investigated room can be classified as category II (B) – dedicated for renovated buildings or cat. III (C) – for existing buildings. In winter time the operative temperature range for cat. II (B) is defined as 20–24°C for cat. III (C) is defined as 19–25°C. The limit values for summertime are described by proper equations described in [3].

DOI: 10.1051/e3sconf/20171700050

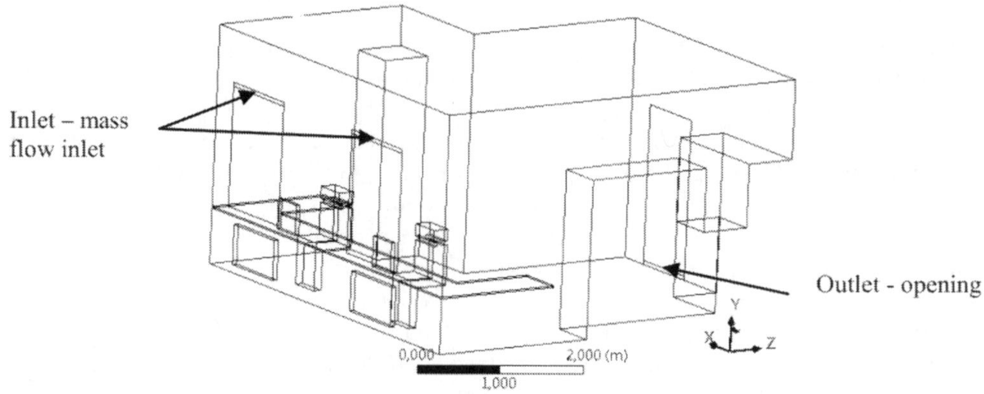

Fig. 1. Geometry of the investigated office with the description of main boundary conditions.

2.2 Indoor air temperature measurements

Measurement of air temperature in the room was performed continuously during the period from February to December 2014 in the 10-minute interval. The following data loggers have been used: Volcraft DL-141 TH with the accuracy of ±1°C for the temperature range from -10°C to +40°C and Rotronic CL11 with the accuracy of ±0,3°C for the temperature range from 0°C to +50°C. The variation of recorded parameter is presented on the Fig. 2. The surface temperatures were measured by the set of thermocouples connected to the logger.

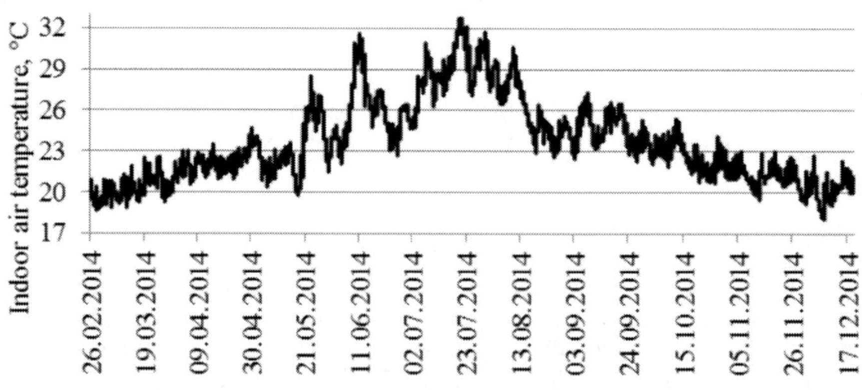

Fig. 2. The internal temperature variation during measurement period: 26.02.2014 to 19.12.2014.

The recording period started at 11:05 a.m. on 26th of February 2014. The last measurement took place at 14:15 p.m. on 19th of December 2014.

The highest temporary internal temperature of 32,7°C was measured during the night time on 21st of July when the office daily-average was 32,14°C. The lowest temporary internal temperature was recorded during morning hours (between 6 and 8 a.m.) on Monday the 8th of December and reached 18,1°C. The highest temperature at this day occurred between 6 and 7 p.m. when the mean external temperature was +1°C and reached 20,7°C.

2.3 Operative temperature

The air temperature and its variation are the main environmental parameters in comfort level estimations, however the most common simple indicator is an operative temperature.

It is a pure theoretical value, thus it cannot be obtained directly due to an empirical study. The operative temperature can be calculated from [14]:

$$T_o = H \cdot T_a + (1-H) \cdot T_r \tag{1}$$

where T_a is air temperature; T_r is mean radiant temperature; H is a value described by the relationship between convective heat transfer coefficient and linear radiation transfer coefficient. The H parameter is difficult to estimate and differs between researchers, therefore CIBSE (UK Charter Institution of Building Services Engineers) proposed to correlate the T_{op} with the air velocity [14]. For natural convection it can be assumed to be 0,1m/s thus the simplified equation describing operative temperature can be written as [14]:

$$T_o = 0{,}5T_a + 0{,}5T_r \tag{2}$$

To estimate the value of operative temperature and its variation for investigated office the inner surface temperatures of surrounding surfaces was recorded over the period of time: from 26th of September to 4th of December. The mean values of measured parameters and calculation outcomes are presented on the Fig. 3. The gap in data collection from 1st to 13th of November was caused by fluctuating problems with electrical network in the building and thus recording equipment.

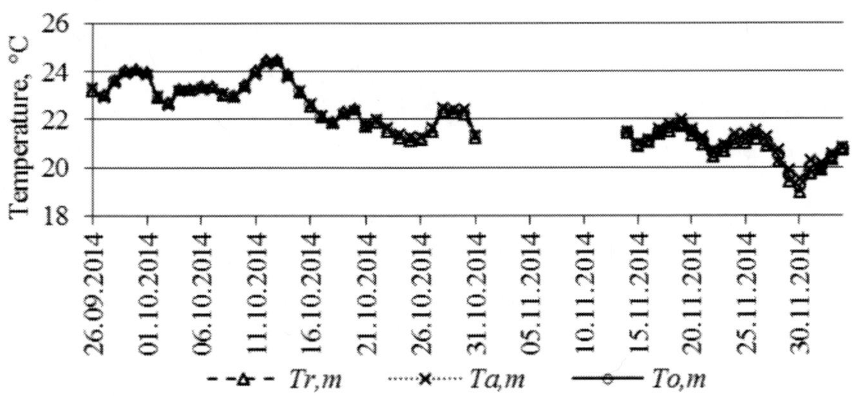

Fig. 3. The variation of mean values of radiant temperature ($T_{r,m}$), room air temperature ($T_{a,m}$) and operative temperature ($T_{o,m}$).

Data presented on the Fig. 3 clearly states that during the considered period, the discrepancies between mean indoor air and radiation temperature is negligible, and thus the variation of the operative temperature is close correlated with the room temperature.

2.4 Statistical relationship between air and radiant temperature

The statistical analysis is an important part of each scientific research. Therefore this paragraph presents the results of the statistical analysis of the relationship between indoor air temperature and the temperatures of surrounding surfaces, and their influence on the office air temperature. Persons correlation coefficient (r) defines strength of relationship between independent variables and p-value defines statistical significance of the relation for particular set of data points. The value of the coefficient can vary between -1 and 1, and thus the correlation is assumed to be strong if $|r|$ is bigger than 0,7 or medium to week if $|r|$ is in the range of <0; 0,7>.

The calculated Pearson correlation coefficients for the relationship between the indoor temperature and the temperature of inner surfaces of external wall, windows and flat roof were calculated and amount respectively 0.973; 0.855 and 0.989 what proves strong relation between analyzed data.

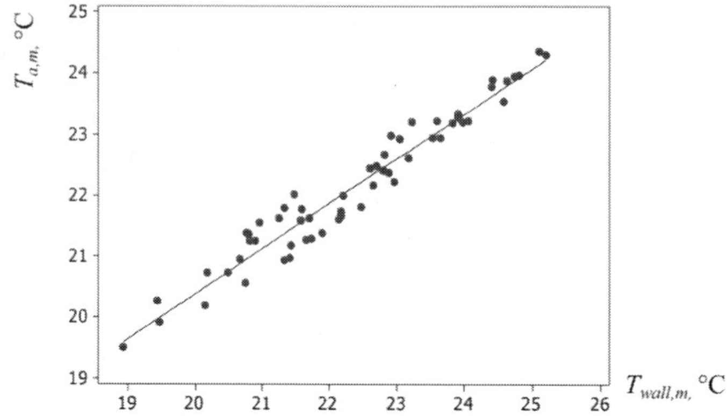

Fig. 4. Function 1 – describes the relationship between mean (daily-average) indoor temperature $T_{a,m}$ and mean temperature of the inner surface of external wall $T_{wall,m}$.

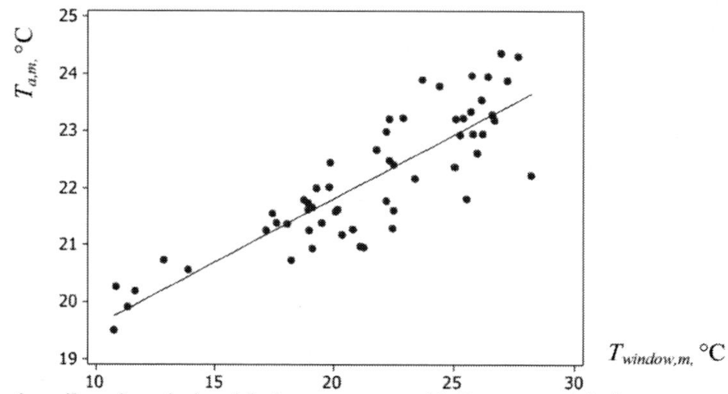

Fig. 5. Function 2 – describes the relationship between mean (daily-average) indoor temperature $T_{a,m}$ and mean temperature of the inner surface of windows $T_{window,m}$.

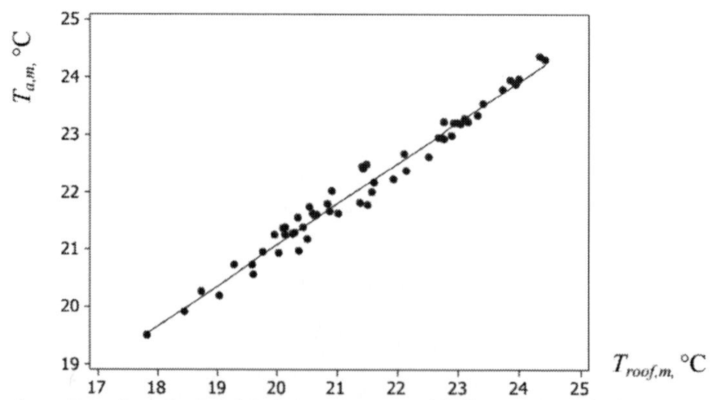

Fig. 6. Function 3 – describes the relationship between mean (daily-average) indoor temperature $T_{a,m}$ and mean temperature of the inner surface of flat roof $T_{roof,m}$.

To define functions describing the mathematical model of the relationship between air temperature and the temperatures of surrounding surfaces (and thus the room radiant temperature), the regression analysis has been applied.

The outcome of the analysis in graphic forms are presented on the Fig. 4., Fig. 5., Fig. 6. The mathematical equations describing these relationships are presented in Table 1. The residuals analysis which is out of scope for this paper.

To check the statistical significance of the correlation between the analyzed parameters, the significance level of 0,05 was assumed as for engineering applications. In all three cases the calculated P-value was lower than 0,05. Therefore, the null hypothesis H_0 that the data are independent is rejected and the alternative hypothesis H_a of a statistical significance is adopted. Present statistical significance (P-value) and strong variables correlations (r) are also confirmed by high value of determination coefficient R^2 for each function (Table 1), what is typically called "practical" significance and can be read as portion of real data variation explained or determined by the particular math model.

Table 1. The determination coefficients R^2 for the functions describing the relationship between the daily-average indoor temperature and the temperatures of internal surfaces of the building envelope.

Function no.	Mathematical model	Determination coefficient R^2
Function 1	y=5.612+0.7386x	94.7%
Function 2	y=17.35+0.2228x	73.0%
Function 3	y=6.859+0.7108x	97.7%

Presented in Table 1 models describe very well the relationship between discussed parameters therefore at this stage of the investigation there is no need for expanding the models for bigger number of freedom of the subsequent points or introducing new variables increasing the models complexity.

The proposed equations can be applied at the later stage of the research as a boundary conditions when considering CFD simulations of the office under dynamic state. However it is not the scope of this paper.

2.5 Long–term comfort assessment

The other approach of estimating comfort conditions assumes long-term measurements as described in [3]. The method classifies the enclosure to one of four categories of indoor conditions and is based on the daily average room temperature during the heating period.

Table 2. The percentage of time for a given temperature range (defined for a particular class of the building) in monthly cycle (on 24-hours basis).

Building category/ Month	I	II	III	IV
March	19.5%	45.4%	34.1%	1.0%
April	91.5%	6.8%	1.7%	0.0%
October	63.5%	29.4%	6.6%	0.5%
November	67.4%	23.9%	8.7%	0.0%
December (19 days)	19.0%	49.7%	21.7%	9.6%

Following [3] the room initially was classified as category III (with lower standards of comfort, as described in section 2.1), but the analysis of the temperature during office "working hours" indicated that the room can be classified as cat. II (with the percentage of people dissatisfied lower than 10% [2]).

Additional analysis of the percentage of time in a given month on a 24-hours basis when the office meets the class requirements has been performed. The time of occurrence of the temperature range corresponding to different classes of buildings is presented in Table 2.

2.6 CFD simulations

In the final stage of the investigation, the office was simulated in ANSYS (CFD tool) under steady – state conditions. The created geometry (Fig. 1) was divided on a certain number of control volumes. The defined boundary conditions are presented in Fig. 1. The other elements are defined as surfaces with constant temperature (taken from the measurements) or as adiabatic walls. For the simulation SST k-ω turbulence model was applied [8, 10]. The simulation outcomes are presented on Fig. 7 and Fig. 8.

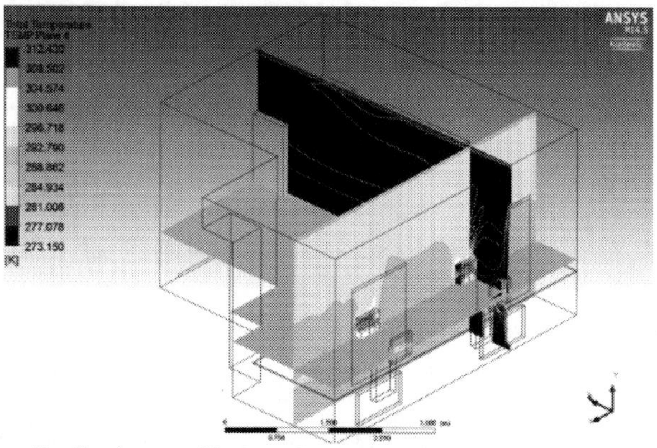

Fig. 7. The temperature distribution profile in a three control planes.

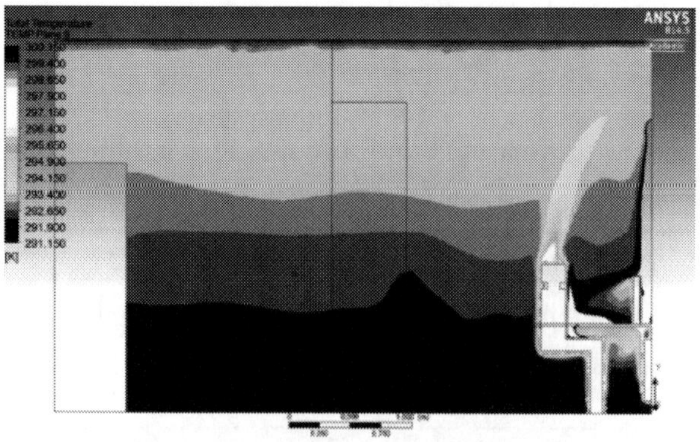

Fig. 8. The temperature distribution profile in a vertical plane. The temperature range: 18°C to 27°C.

The resulting maps of temperature in simulated office are in line with expectations. Measured room temperature for given boundary conditions was 20,5°C. According to the model, this level of temperature is achieved at about 1,7 m, which well corresponds with the sensor position and its readings within the tested room.

3. Conclusion

The paper presented chosen methods of assessing comfort conditions that can be applied for naturally ventilated rooms. The measurements of indoor temperature and surrounding surfaces were undertaken to calculate operative temperature, to perform long-term estimation of comfort, to define boundary conditions of the CFD office model and to compare with the simulation outcomes. The statistical analysis, as an important element of the scientific research, was applied. It was based on the regression method to establish and validate the relationship between measured parameters.

The paper indicated that there is a number of well described in the literature methods of assessing indoor comfort conditions. From the simple and common to the advanced and complex approaches, in short and long perspective. Each method has its own limitations and applicability. However, as thermal comfort is an important aspect in terms of building operation, satisfaction and productivity of building users, therefore the comfort investigation, even by application of simple methods, should be an important element of engineering practice.

References

1. M. Taleghani, M. Tenpierik, S. Kurvers, A. Dobbelsteen, Renew Sust Energ Rev, **26**, 201-215 (2013)
2. EN ISO 7730 *Analytical determination and interpretation of thermal comfort using calculation of the PMV and PPD indices and local thermal*
3. EN 15251 *Indoor environmental input parameters for design and assessment of energy performance of buildings- addressing indoor air quality, thermal environment, lighting and acoustics*
4. F.H. Rohles, Hum Factors, **13**, 553-60 (1971)
5. C. Buratti, P. Ricardi, M. Vergoni, Appl Energ, **104**, 117-127 (2013)
6. R. Holopainen, et all., Build Environ, **71**, 60-70 (2014)
7. P. Aryal, T. Leephakpreed, Energy Procedia, **79**, 183 – 188 (2015)
8. A.Stamou, I. Katsiris, Build Environ, **41**, 1171-1181 (2006)
9. M.Kavgic, D. Mumovic, Z. Stevanovic, A.Young, Energ Buildings, **40**, 1334-1343 (2008)
10. S. Hussain, P.H. Oosthuizen, Appl Therm Eng, **40**, 358-372 (2012)
11. J. Song, X. Men, Procedia Engineering, **121,** 1475 – 1481 (2015)
12. T. van Hooff, B. Blocken, Y. Tominaga, Build Environ, **114,** 148–165 (2017)
13. K.W.D. Cheong, et all., Build Environ, **38**, 63-73 (2003)
14. F. Nicol, M. Humphreys, S. Roaf, *Adaptive thermal Comfort. Principles and Practice* (Routledge, London, 2012)
15. M.A. Baharun, M. Surat, N.M. Tawil, A.I. Che-Ani, E3S Web of Conferences **3**, 01008 (2014)
16. P.O. Fanger, *Komfort cieplny* (Arkady 1974)

Freely available mean daily discharge series from Czechia: what can be inferred from them?

Ondrej Ledvinka[1,*]

[1]Czech Hydrometeorological Institute, Hydrology Database and Water Budget Department, Na Sabatce 2050/17, 143 06 Prague 412, Czechia

Abstract. Most hydrometeorological data from Czechia are still provided for a fee. This especially applies to time series with a finer step than monthly. The fact that the data must be paid with respect to an expert's appraisement and that the proper licence agreement ratification has to be performed causes a considerable delay in the data transfer to potential customers, unfortunately including scientists as well. Naturally, this time-consuming process is unpleasant to the experts on both sides. Due to a substantial rise of university students' requests for data, the Czech Hydrometeorological Institute's hydrologists launched a website from which long mean daily discharge series representing ten selected water-gauging stations can be downloaded. Besides other assessments, the series may play an important role when studying climate change impacts on water resources in Czechia. Therefore, the objective here was to extract from these series some preliminary information on long-term changes such as abrupt and gradual trends caused either by the construction of reservoirs or by climate variability itself. The main tool used was nonparametric trend analysis. Mainly different months were of interest so as to determine if there have been recorded some changes in seasonality. The results may be easily expanded by students.

1 Introduction

Although some basic information can be found on the internet, regime hydrometeorological data from Czechia usually are not free-of-charge. The main provider of these data here is the Czech Hydrometeorological Institute (CHMI) where the process of data transfer to customers is still somewhat weird, perhaps inheriting the practice of older employees who are not accustomed to working with modern technologies. The process namely consists in inappropriate exploiting the capacities of experts working with databases rather than in the supporting staff being activated. It applies also to necessary licence agreements where a requested expert initiates their ratification and, essentially, must be a part of the whole cycle rather than only a decision maker regarding data accessibility and quality. It is more than evident that this situation is not sustainable. On the one hand the number of experts is decreasing and, consequently, each individual is becoming more responsible, on the other hand the number of requests for data substantially rises. Naturally,

* Corresponding author: ledvinka@chmi.cz

the resulting considerable delays in data transfer adversely influence the scientific needs and capacity building development.

At the turn of 2014 and 2015, the deputy director for hydrology at the CHMI decided that selected hydrological series should be freely available exactly for scientific purposes and students so that the staff of the CHMI could invest their time into addressing other issues. In 2015, some negotiation with branch/regional offices as well as with the deputy director for meteorology and climatology took place so as to determine which water-gauging and climatological stations should be represented in such a set of free data. Finally, thanks to the cooperation with the Czech National Committee for Hydrology (CNCH), Czech and English websites (see http://cnvh.cz/) devoted to the presentation of the data set were launched at the beginning of 2016. Although initially dedicated to small experimental basins in the Jizera Mts. in the north of Bohemia, the data set includes also ten long mean daily discharge series representing wider territory of Czechia (see Table 1 and Fig. 1). Such data are particularly valuable when dealing with possible climate change (CC) impacts on water resources, which are intensively studied now.

The aim of this study was to draw the attention of scientific community to the mentioned data set and to initiate the preferable use of it so that the number of requests for the regime hydrological data directed to the CHMI could drop. At the same time, through a nonparametric trend analysis, it is illustrated here what can be inferred from this data set regarding CC. The results here should be considered preliminary, while mainly the students are encouraged to continue such research, either by adopting other, more sophisticated, methodologies or by suggesting other issues that can be addressed using this data set. Note that the data representing the experimental basins are overlooked intentionally here since they will be a subject of another study.

Table 1. Closer information on water-gauging stations whose discharge series were studied here (adapted from http://hydro.chmi.cz/ismnozstvi/). The order of numbers (from south to north) corresponds to the *y*-axis in Fig. 2.

Order (from south to north)	ID	Name	River	Catchment area [km^2]
10	240000	Děčín	Labe	51120.39
9	091000	Železný Brod	Jizera	791.27
8	210100	Stará Role	Rolava	126.35
7	058000	Úhřetice	Novohradka	458.91
6	294000	Bohumín	Odra	4663.77
5	167200	Nespeky	Sázava	4038.65
4	179000	Staňkov	Radbuza	701.54
3	151000	Písek	Otava	2913.70
2	135000	Modrava	Vydra	89.80
1	421500	Strážnice	Morava	9144.80

 DOI: 10.1051/e3sconf/20171700051

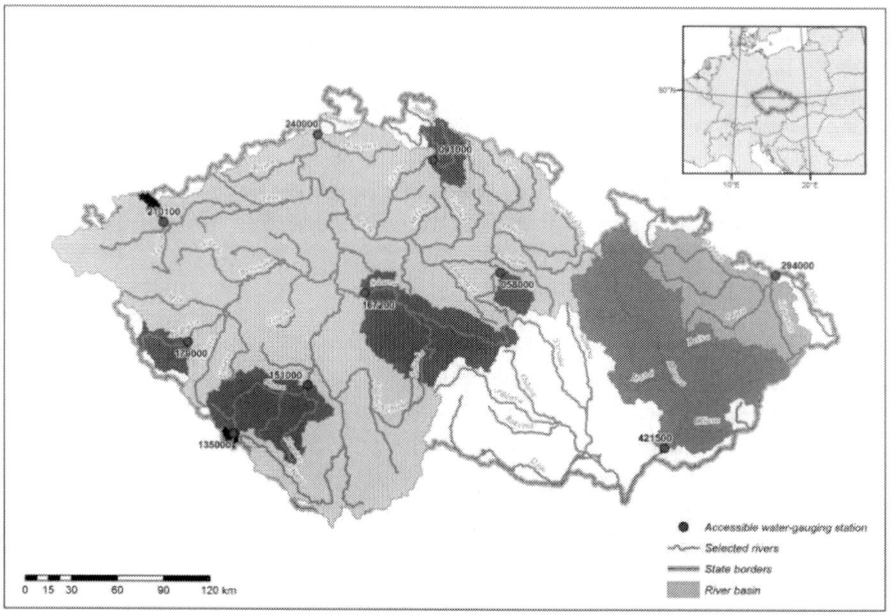

Fig. 1. Location of offered water-gauging stations with freely available long mean daily discharge series with respect to the borders of Czechia. All the geographical information, apart from the river network (source: Ministry of the Environment of the Czech Republic; other can be downloaded from http://www.dibavod.cz/), can be acquired at http://hydro.chmi.cz/ismnozstvi/ upon registration.

2 Brief exploratory analysis of selected discharge series

In time series analysis (TSA), it is necessary first to know what time periods are covered by the data. It is not sufficient to know when a time sires begins and ends, but also one has to be interested in the placement of missing values. Missing values (or gaps) are a common feature of time series and the daily discharge series offered online by the CHMI are not an exception. This fact is actually an advantage since the students can, at the very beginning of their career, face to this natural issue, where several reasons for the occurrence of gaps may be reflected (including political conflicts such as wars or issues connected to a momentary lack of employees at hydrological services needed for digitization of archived reports).

Figure 2 shows the availability of mean daily discharges at selected stations separately for all hydrological years (in Czechia defined as periods from 1 November of last year to 31 October) starting from 1888 (i.e. beginning of the longest series of station 240000) and proceeding to 2015. It may be deduced that the beginnings of the series differ. Irrespective of this fact, it may further be stated that stations 240000, 091000, 294000, 179000, 151000 and 421500 have uninterrupted time series composed of mean daily values. On the contrary, stations 210100, 058000 and 135000 have interrupted series, either somewhere close to the beginning or close to the end of the series. Only few daily values are missing in the series from station 058000. In the case of station 167200, it can be seen that it starts very late, which is caused by the fact that the measurement at this station succeeds the observations at another, closely located, station. It is questionable whether to merge this series with its predecessor. However, it is probable that it will happen in the future and the students will have another valuable longer series for their CC studies.

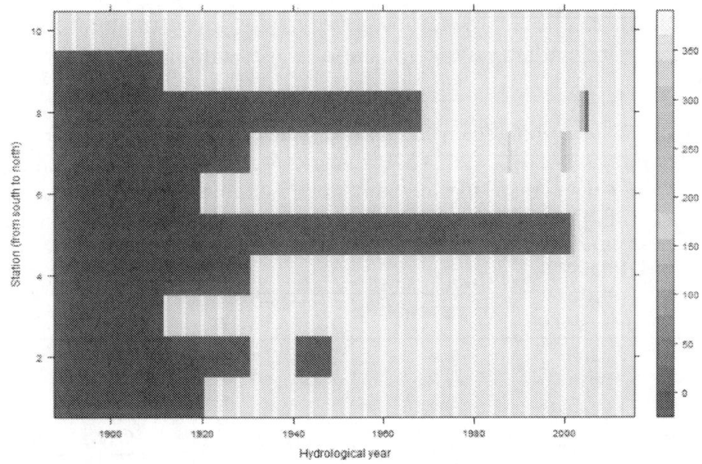

Fig. 2. An overview of the accessibility of mean daily discharges in every individual year from 1888 to 2015 representing ten water-gauging stations offered online by the CHMI. Leap years manifest themselves as the lighter stripes, meaning there are 366 instead of 365 values per year available.

Looking at Fig. 2, one may also notice several common periods that can be studied separately at the exploratory analysis stage. This can be useful, for instance, when investigating the influence of the presence of reservoirs. Thus, the whole period of hydrological years 1888–2015 was divided into four subperiods: entire period, 1931–2015, 1961–2015, and specifically 1981–2010 (i.e. current Czech hydrological reference period). It was not important how many stations measured at the beginning of the respective period. Rather it was aimed at possible changes in basic descriptive statistics. During the computation of the statistics, the missing values were ignored as well. Table 2 gives an overview of such pursuits.

Upstream reservoir operation may manifest itself in decreasing variation as well as in increasing lag-1 autocorrelation. Both may be apparent when assessing the series coming from station 294000 on the Odra River. At this station, also an increase in skewness can be observed, meaning that there have been less (or smaller) flood waves after the construction of reservoirs. On the other hand, rivers where the anthropogenic influence is believed to be very small experience a decrease in skewness (see e.g. station 135000 on the Vydra River). Of course, the descriptive statistics would change their values if monthly or annual series would be of interest. Particularly, apart from the long-term mean, they would decrease.

Furthermore, it is advisable, before performing a trend analysis, to decide whether the time series can be studied as a whole or it should rather be divided into several parts. Dividing the series is recommended especially when the entire series reveals inhomogeneity, which means that different parts of the record were collected under different (unequal) conditions. Therefore, a homogeneity check was carried out using an R tool prepared by climatologists in Canada. This tool is, to the author's knowledge, the only one that works with daily data as well as with missing values. The description of the method would not be trivial, and the readers are referred to [1] instead. Since the tool was designed for precipitation, discharge was transformed to runoff depth first.

Surprisingly, it was found that almost all the series are homogeneous regardless of the hypothesized change points connected to anthropogenic influences. The explanation may lie in the fact that the human impact on the investigated rivers started much earlier than the stations became operational. Anthropogenic influences may also be weaker than those of natural origin at larger rivers such as Labe. Nevertheless, there are two stations in the free data set where the null hypothesis of homogeneity had to be rejected at the level of significance $\alpha = 0.05$. While the series of station 179000 revealed only one change point at the beginning of the hydrological year 1965, the series of station 167200 showed three change points during the 2000s. Since the reasons for these change points are not known without any detailed metadata, the two stations were no longer included in subsequent analyses.

Table 2. Basic statistics describing ten mean daily discharge series offered online by the CHMI. Four periods (entire period, 1931–2015, 1961–2015 and *1981–2010*) are summarized in every single cell irrespective of missing values or the starting point of measurement. If the values are the same as those for the succeeding period (because there is no new information), they are not in bold.

ID	Long-term mean [$m^3 \cdot s^{-1}$]	Coefficient of variation	Skewness	Kurtosis	Lag-1 autocorrelation
240000	**312.00**	**0.87**	**3.23**	**18.07**	**0.968**
	310.00	**0.84**	**3.36**	**20.17**	**0.971**
	313.00	**0.80**	**3.51**	**24.74**	**0.974**
	315.00	*0.81*	*3.68*	*27.70*	*0.972*
091000	**17.00**	**1.09**	**4.21**	**31.87**	**0.831**
	16.70	**1.10**	**4.18**	**31.70**	**0.839**
	16.90	**1.10**	**4.22**	**33.78**	**0.841**
	17.20	*1.11*	*4.30*	*35.03*	*0.839*
210100	2.32	1.11	5.08	50.95	0.881
	2.32	1.11	5.08	50.95	0.881
	2.32	**1.11**	**5.08**	**50.95**	**0.881**
	2.45	*1.06*	*4.49*	*37.85*	*0.886*
058000	2.50	1.47	4.54	33.37	0.885
	2.50	**1.47**	**4.54**	**33.37**	**0.885**
	2.64	**1.41**	**4.68**	**36.75**	**0.890**
	2.54	*1.52*	*5.34*	*45.17*	*0.888*
294000	**43.10**	**1.29**	**6.58**	**100.27**	**0.895**
	43.00	**1.28**	**6.93**	**111.29**	**0.894**
	43.30	**1.25**	**8.08**	**155.57**	**0.903**
	41.60	*1.36*	*10.41*	*221.00*	*0.904*
167200	21.30	1.36	6.74	76.52	0.941
	21.30	1.36	6.74	76.52	0.941
	21.30	**1.36**	**6.74**	**76.52**	**0.941**
	23.00	*1.38*	*6.54*	*70.65*	*0.948*
179000	3.74	1.29	6.97	94.48	0.838
	3.74	**1.29**	**6.97**	**94.48**	**0.838**
	3.75	**1.27**	**7.16**	**106.35**	**0.826**
	3.84	*1.32*	*8.10*	*132.68*	*0.803*
151000	**23.60**	**0.98**	**7.16**	**158.68**	**0.864**
	23.40	**0.98**	**7.55**	**181.10**	**0.866**
	23.90	**0.99**	**8.47**	**219.63**	**0.873**
	24.40	*1.04*	*10.26*	*283.80*	*0.854*
135000	3.38	1.07	5.01	47.76	0.750
	3.38	**1.07**	**5.01**	**47.76**	**0.750**
	3.35	**1.06**	**4.56**	**36.91**	**0.760**
	3.46	*1.08*	*4.77*	*40.96*	*0.736*
421500	**58.60**	**1.09**	**3.53**	**18.57**	**0.936**
	59.00	**1.10**	**3.58**	**18.90**	**0.932**
	59.60	**1.05**	**3.72**	**21.64**	**0.931**
	59.30	*1.11*	*4.01*	*24.84*	*0.938*

DOI: 10.1051/e3sconf/20171700051

3 Trend analysis

Nowadays, investigators interested in trend analysis emphasize that the trend component having the lowest frequency [2] should be properly distinguished not only from short-term persistence (STP), but also from long-term persistence (LTP) effects. Both types of persistence cause most known trend tests to reject the null hypothesis of no trend too often. Therefore, where possible, so-called unit root tests were applied to find out if LTP could stand behind falsely detected trends. The possible effect of STP was considered directly in the trend test afterwards.

3.1 Testing for stationarity

There are several types of stationarity relating to time series. First, what is usually forgotten, one may look at the stationarity in various moments such as mean and variance. Hydrologists sometimes apply tests devoted to heteroscedasticity (i.e. nonstationarity in the variance) and apply models that account for it (e.g. [3]), but, traditionally, they study possible nonstationarity in the mean. Second, one may distinguish between deterministic and stochastic stationarity. While stochastic nonstationarity in the mean can be modelled by differencing (see e.g. [4]) and is related to persistence, deterministic nonstationarity can be modelled by mathematical curves (among which one may select a line as well) and is often exactly what investigators want to detect when trying to prove CC impacts. Here, deterministic nonstationarity in the mean (i.e. deterministic trend) was distinguished from stochastic nonstationarity in the mean (i.e. stochastic trend) that can be modelled by fractional differencing.

For this purpose, hydrologists and climatologists (e.g. [4]) jointly apply one of the unit root tests and a stationarity test because their null hypotheses are set oppositely. In particular, the Phillips–Perron (PP) test that is designed to confirm nonstationarity and the Kwiatkowski–Phillips–Schmidt–Shin (KPSS) stationarity test were applied here to decide what type of stochastic processes generates the time series.

Note that prior to this testing, the series have to be deseasonalized. Therefore, a method based on wavelets suggested in [5] was adopted here. Also, during stationarity testing and the associated Hurst exponent estimation, missing values are usually not permitted. Thus, only the complete series (stations 240000, 091000, 294000, 151000 and 421500) were subjected to this procedure.

3.2 Trend test employed

Trend analysis is a rapidly growing area in hydrology and there are a lot of tests that have been utilized by hydrologists. For their overview see, for example, [6]. In this paper, a methodolgy outlined in [7] was adopted, which means that the Yue–Wang (YW) modification of the nonparametric Mann–Kendall (MK) test was employed in particular. For more details the readers are referred to the original paper [8] or to [7]. Primarily, the test was applied to annual series, but also monthly series were of interest because they may reveal important changes in the seasonal course. Only the significance level $\alpha = 0.05$ was selected here. Note also, that, this time, the missing values were allowed, so, again, the above four different periods for the selected stations could be examined.

 DOI: 10.1051/e3sconf/20171700051

4 Results and discussion

The procedure aimed at checking the stationarity suggested that the deterministic trends at five selected water-gauging stations could not be rejected. It means that the trends detected via the subsequent YW–MK test may very likely be of the deterministic origin.

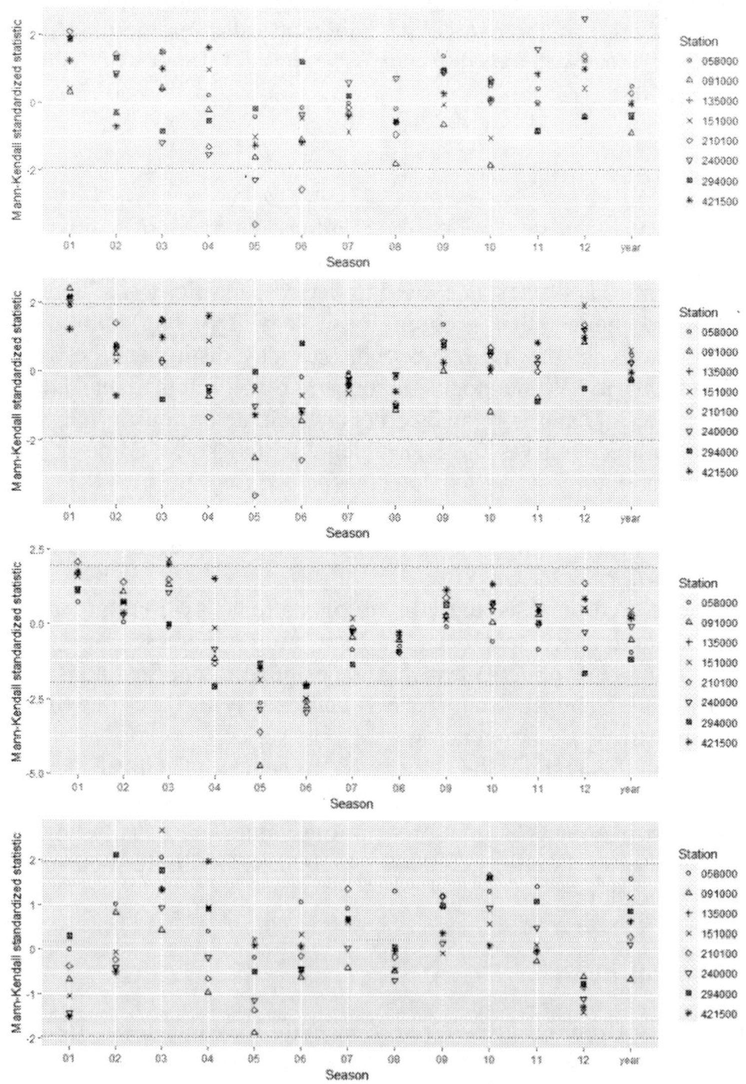

Fig. 3. Resulting standardized statistics of the MK test corresponding to eight selected water-gauging stations and four selected periods (from top to bottom): entire period, 1931–2015, 1961–2015 and 1981–2010. Dotted lines indicate the 95% confidence interval.

Figure 3 summarizes the results of trend analysis. It may be deduced that the trends are really rare here. Indeed, looking at the annual scale, there are hardly any trends. This clearly corresponds to the findings regarding precipitation in Czechia published in [7]. However, more interesting insights were provided by the monthly time scale. Mainly, a sinusoidal course in the standardized MK test statistics is apparent. It also seems that this course is becoming stronger when the starting point of investigated series is shifted ahead. This phenomenon may be associated with the behaviour of snow that seemingly melts out earlier now in Czechia and its mountains (e.g. [9]). Indeed, decreasing trends, if any, prevail at the turn of spring and summer, which is not so evident for the important period 1981–2010.

5 Conclusion and recommendations

In this paper, a freely available data set consisting of mean daily discharge series representing selected water-gauging stations located in Czechia was presented. Many of these series are relatively long, which enables students and other scientists, who are meant to be the main recipients of such data, to freely assess the possible response of Czech water resources to climate change. An example of trend analysis was suggested here, through which it was found that there are no changes in discharge when looking at the annual scale. However, when assessing monthly series, some trends were obvious. Notably, the strengthening decreases in discharge at the turn of spring and summer can pose some issues in water supply services in the future.

Nevertheless, not all aspects were studied here. The series that are now freely online may be assessed from other perspectives as well. For example, changes in seasonality should be evaluated more thoroughly. Also, a periodic behaviour of the time series could be studied, as well as the relationships among upstream and downstream places. Wavelet analysis and especially the methods dealing with the so-called wavelet coherence may be useful. From mean daily discharge series, one may also derive a lot of indicators in order to study hydrological drought patterns. Now, it depends on the students and other scientists if they will accept the offered chance to conduct some more rigorous work based on this data set.

Finally, it should be noted that Czechia is not the only country from Central Europe that currently provides at least some daily hydrometeorological data for free. In the case of Poland, similar data sets can be found at https://dane.imgw.pl/. German daily meteorological data can be downloaded from the official FTP server ftp://ftp-cdc.dwd.de/pub/CDC/. Probably, there will be more free data in the near future.

The author would like to thank the deputy director for hydrology at the Czech Hydrometeorological Institute. Mainly giving the chance to students from around the world to freely test their statistical skills on such a data set from Czechia is really appreciated.

References

1. X. L. Wang, H. Chen, Y. Wu, Y. Feng, and Q. Pu, J. Appl. Meteorol. Climatol. **49**, 2416 (2010)
2. Y.-F. Sang, Z. Wang, and C. Liu, Hydrol. Process. **27**, 2021 (2013)
3. R. Modarres and T. B. M. J. Ouarda, J. Hydrol. **499**, 1 (2013)
4. S. Fatichi, S. M. Barbosa, E. Caporali, and M. E. Silva, J. Geophys. Res. **114**, D18121 (2009)
5. E. Szolgayová, J. Arlt, G. Blöschl, and J. Szolgay, J. Hydrol. Hydromech. **62**, 24 (2014)
6. D. Machiwal and M. K. Jha, *Hydrologic Time Series Analysis: Theory and Practice* (Springer Netherlands, Dordrecht, 2012)
7. O. Ledvinka, Acta Hydrol. Slovaca **16**, 199 (2015)
8. S. Yue and C. Wang, Water Resour. Manag. **18**, 201 (2004)
9. Z. Kliment, M. Matoušková, O. Ledvinka, and V. Královec, J. Hydrol. Hydromech. **59**, 36 (2011)

Manometric measurements of temperature influence on Anammox activity

Maciej Leoniak[1], and *Piotr* Balbierz[1,*]

[1]Wroclaw University of Science and Technology, Faculty of Environmental Engineering, Wybrzeze Wyspianskiego27, 50-370 Wroclaw, Poland

Abstract. Wastewater treatment plants face more stringent effluent limits, especially with regard to total nitrogen, which force them to consider sidestream nitrogen removal from digested sludge dewatering liquor which can contain up to 20% of incoming nitrogen load. One of the most promising technologies for sidestream nitrogen removal is the deammonification process which is based on partial short-cut nitrification and the Anammox process. Anammox process is an anaerobic ammonia oxidation with nitrite conducted by slowly growing autotrophic bacteria. Ensuring adequate activity of sensitive Anammox bacteria is a key condition for effective deammonification. As the main product of Anammox bacteria is the nitrogen gas, the assessment of Anammox activity can be based on manometric measurements which are a relatively simple and cost-effective alternative to traditional tests involving determination of soluble nitrogen forms. This paper presents the principles of manometric method as well as results of temperature influence on Anammox activity obtained in a commercially available lab-scale set-up.

1 Introduction

The potential existence of chemolithotrophic bacteria anaerobically oxidizing ammonia with nitrite/nitrate to dinitrogen gas was predicted, based on the theoretical thermodynamic calculations, already in the 1970s. However, it was not until 1990s, that the effects of such a process were observed in reality, first in the technical systems, and later widespread in the natural environment. Later it was discovered that this process is important at global scale as it considerably affects the global nitrogen cycle, especially in the oceans, where it outcompetes denitrification [1, 2].

The bacteria responsible for this process, usually called anammox bacteria (abbreviation for ANaerobic AMMonium OXidation), belong to the order of *Planctomycetes*, and three candidate genera has been identified so far, including *Candidatus Brocadia, Candidatus Kuenenia* and *Candidatus Scalindua*. Different species rarely exist together, suggesting different requirements and optimal conditions for growth [1, 2].

In engineering practice anammox process becomes more commonly applied as wastewater treatment plants face more stringent effluent limits, especially with regard to

* Corresponding author: piotr.balbierz@pwr.edu.pl

total nitrogen, which force them to consider sidestream nitrogen removal from digested sludge dewatering liquor which can contain up to 20% of incoming nitrogen load [3].

The most innovative and promising biological technology for sidestream nitrogen removal is deammonification which is based on partial short-cut nitrification and the anammox process [4]. As raw reject water contains mainly ammonium, the partial nitrification step, conducted by strictly aerobic ammonium oxidizing bacteria, is required to oxidize approximately half of the initial ammonium to nitrite. In the anammox step ammonium (electron donor) and nitrite (electron acceptor) are converted by strictly anaerobic anammox bacteria into dinitrogen gas and small amounts (~10%) of nitrate without external organic carbon source, according to equation [3, 5]:

$$NH_4^+ + 1.32NO_2^- + 0.066HCO_3^- + 0.13H^+ \rightarrow \\ \rightarrow 1.02N_2 + 0.256NO_3^- + 0.066CH_2O_{0.5}N_{0.15} + 2.03H_2O \quad (1)$$

The advantage of deammonification is that it allows to decrease oxygen demand by 60%, and organic carbon demand and sludge production by 90%, compared to conventional nitrification/denitrification systems [1, 3, 5].

In full-scale installations, deammonification is usually operated as a one-stage system, where both steps with different requirements take place in one reactor, which results in higher volumetric efficiency and lower investment costs, but makes the process control more complicated. An alternative are two-stage systems, where partial nitritation and anammox are split in two reactors in series, which allows to optimize process condition for each of the steps separately [3].

The weakness of deammonification is that both steps, and especially anammox, are conducted by extremely slowly growing autotrophic bacteria with a long doubling time of 11 days [6]. This means that in case of washout or toxicity incidents, the recovery of the system will take a long time, during which the untreated reject water will increase the loading of the mainstream system. Therefore, it is crucial to operate the deammonification installations at optimum conditions (temperature, pH, substrate ratio) to support the maximum growth rate of anammox bacteria [7].

1.1 Temperature effect on Anammox activity

Temperature is one of the most important factors affecting microbial growth in general, and biological wastewater treatment in particular. In engineering applications, the temperature dependency of biological processes is usually expressed using the following equation [7]:

$$\mu_T = \mu_{T_r} \theta^{(T-T_r)} \quad (2)$$

where μ_T and μ_{Tr} are the maximum specific growth rate at process temperature T and reference temperature T_r, respectively, and θ is the temperature correction factor.

It has been generally reported that the optimum temperature for growth of anammox bacteria was in the range of 30–40°C [8], whereas other authors claimed to observe the highest activity even at 43°C [7]. Therefore, most of up to date anammox applications were on warm (T > 30°C) and high strength (>0.1 g N/L) streams, mostly reject water from digested sludge dewatering. Moreover it was shown that temperature decline leads to severe anammox activity suppression [9].

However, recently it was proved that successful anammox operation at temperatures around 20°C is possible, especially with immobilized biomass [10, 11]. Moreover, works conducted on marine anammox samples from cold regions reported significant activity even at temperatures below 10°C [5].

On the other hand, the upper limit for the anammox operation has been reported in the range of 45–55°C, where significant irreversible loss of activity due to biomass lysis was observed [5, 7].

Recently, Lotti *et al.* [12] and Sobotka *et al.* [7] showed results suggesting that, contrary to other biological process, the temperature dependency of anammox activity cannot be expressed by a single Arrhenius coefficient, as the temperature induced decline in activity is stronger at lower temperatures. Both studies concluded that for the higher temperature range (15–40°C) single characteristics can be assumed with temperature coefficient θ of 1.07 and a corresponding activation energy of 51–66 kJ mol^{-1}, which is close to value of 70 kJ mol^{-1} postulated by Strous *et al.* [13]. Whereas in the low temperature range (10–15°C) a much higher temperature coefficient θ of 1.65 and a corresponding activation energy of 231-437 kJ mol^{-1} was determined.

Moreover, Sobotka *et al.* [7] observed a change in the molar conversion ratio of substrates NO_2-N/NH_4-N, which in the temperature range 15–30°C was close to a typical value of 1.3, but decreased to a value of 1.0 below 15°C – this suggests a significant change in the anammox metabolism at low temperature.

Additional factor that needs to be considered is the potential adaptation of anammox bacteria to lower temperatures. Long term studies showed that adapted anammox biomass has a higher specific rate at lower temperatures, but the extent of adaptation is difficult to predict and likely to depend on the duration of operation [12, 14]. Some researchers suggested that complete adaptation may take months or even years of operation [15].

Therefore, the influence of temperature change on anammox activity, especially in the lower temperature range, and the ability of anammox bacteria to adapt to lower temperatures is still a valid scientific question.

The aim of this study was to determine the short-term influence of temperature on the activity of anammox biomass cultivated at relatively low temperature (23°C) using manometric measurements in batch tests, in order to estimate the values of temperature coefficient and activation energy.

2 Materials and methods

This study was conducted at the pilot plant used for assessment of different reject water treatment processes, equipped with four 150 dm^3 SBR-type reactors and fed with real reject water from full-scale sludge dewatering station at Wroclaw WWTP in south-west Poland.

Sludge samples were taken from a pilot-scale Anammox SBR fed with partially nitrified real sludge dewatering reject water and operated at temperature of 23°C, pH 7.0, an SRT of 50 d and MLSS of 2.0 kg TSS m^{-3}. At start-up this SBR was seeded with granular sludge from an upflow sludge blanket Anammox reactor, and operated in stable conditions for 180 days at the volumetric loading of >0.7 g N L^{-1} d^{-1} with treatment efficiency of >85%.

As the main product of the anammox reaction (Eq. 1) is the poorly soluble nitrogen gas, manometric measurements may be applied to track the rate of the process as it is proportional to the pressure increase rate in a gas-tight reactor.

2.1 Batch test procedure

To determine the short-term influence of temperature on the activity of anammox biomass cultivated at relatively low temperature (23°C) a series of manometric batch activity tests in temperatures between 17 and 32°C were conducted according to the methodology described in the literature [2, 16].

The tests were performed in a commercially available WTW OxiTop Control AN6 set-up (Figure 1.) equipped with three-necked glass bottles with a total volume of 650 mL, closed with a manometric measurement head with pressure sensor and puncturable gas-tight septum for substrate injection and sampling. Anammox biomass sampled from the pilot SBR was used for the test, 480 mL of sludge was transferred to the bottle (liquid/total volume ratio of 0.75). The headspace and liquid volume were sparged with nitrogen to remove residual oxygen and saturate the liquid with nitrogen. The initial pH value was set to 7.0. The bottles were placed on magnetic stirrer located in a thermostatic chamber at a set temperature until stable conditions were reached. Then the substrates in form of a stock solution (NH_4Cl 1.3 g N/L, $NaNO_2$ 1.56 g N/L, $NaHCO_3$ 5.46 g/L) were injected (initial ammonium and nitrite concentrations 50 and 60 mg N/L, respectively), the initial sample taken and pressure equalized to the ambient one. Nitrogen gas production was tracked by continuous measurement and recording of the overpressure in the headspace until decrease in the process rate was observed. The final pH value was measured in order to check that it was in the optimal range. At the end of the test a sample was taken and analysed for TSS, VSS, ammonium, nitrite and nitrate.

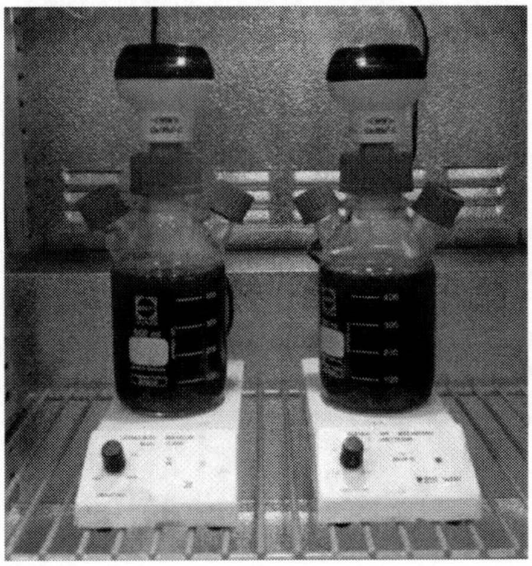

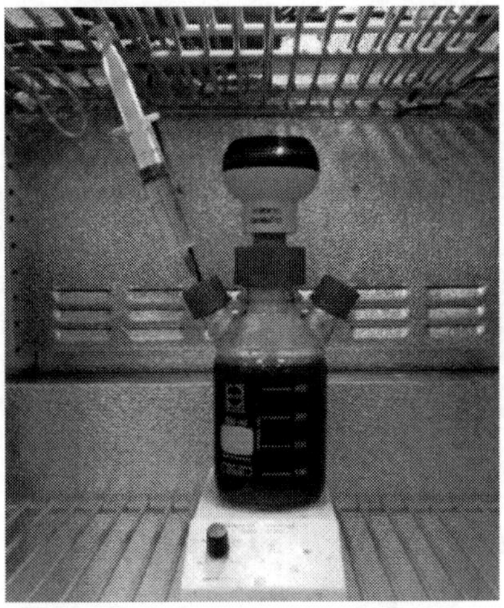

Fig. 1. Experimental set-up in a thermostatic chamber (left) and dosing of substrate (right).

2.2 Calculations

The overpressure increase in the headspace of the bottle during the test was used to calculate the total amount of nitrogen gas produced using the ideal gas law equation. The amount of nitrogen removed from the liquid phase was also calculated based on the initial and final concentrations of ammonium, nitrite and nitrate. The difference between both amounts was used to estimate the relative error of the method.

The nitrogen gas production rate (n, mol $N_2 \cdot h^{-1}$) was calculated from the maximum slope of the curve of the pressure increase in the headspace along the time (a, $Pa \cdot h^{-1}$):

$$n = a \frac{V_G}{RT} \tag{3}$$

where V_G is the volume of the headspace (m^3), R the ideal gas constant (8,314 $J \cdot mol^{-1} K^{-1}$), and T the temperature (K).

The specific anammox activity (SAA, g N_2-N gVSS^{-1} h^{-1}) was calculated from the gas production rate n divided by the biomass content in the bottle (XV_L, g VSS)

$$SAA = \frac{n}{XV_L} \frac{28g\,N}{mol\ N_2} \tag{4}$$

where X is biomass concentration (g VSS m^{-3}) and V_L is the volume of the liquid phase (m^3)

The activation energy (E_a, J mol^{-1}) reflects the reaction temperature dependence and can be estimated graphically from the plot of the natural logarithm of the Arrhenius equation:

$$\ln r = -\frac{E_a}{R} \cdot \frac{1}{T} + \ln A \tag{5}$$

where r is the specific anammox reaction rate (mol gVSS^{-1} d^{-1}) and A is the reaction frequency factor.

The relationship between activation energy and temperature correction factor θ (Eq. 2) is given by the equation:

$$\theta = \exp\left(\frac{E_a}{RT_1T_2}\right) \tag{6}$$

2.2 Analytical methods

Samples were filtered through standard 1.2 μm glass fibre filters. Ammonium, nitrite and nitrate concentrations were determined by spectrophotometry using Hach cuvette tests.

Dissolved oxygen and pH were checked at the start and end of the test using Hach HQD Intellical LDO and pH electrodes connected to a HQ40D meter.

Biomass concentration was measured as total (TSS) and volatile suspended solids (VSS) at the end each test according to EN-872 – standard direct method for suspended solids on glass fibre filters with 1.2 m pore size.

3 Results and discussion

The short-term temperature influence on specific anammox activity (*SAA*) was determined in a series of batch manometric tests in the temperature range between 17 and 32°C conducted at least in duplicates for each temperature.

The obtained *SAA* values as a function of process temperature are presented in Figure 2., whereas the Arrhenius plot used to estimate the activation energy is shown in Figure 3.

The observed values of *SAA* varied between 0.39 and 1.48 g N_2-N gVSS^{-1} d^{-1} for the temperature of 17 and 32°C, respectively, and they were 56% lower and 69% higher than the *SAA* value of 0.88 g N_2-N gVSS^{-1} d^{-1} obtained for the temperature of 23°C at which the pilot-scale SBR was operated. The measured *SAA* values fall into the typical range reported in the literature [7, 12].

The obtained relative errors between the amount of nitrogen gas produced based on manometric measurements and the amount of nitrogen removed from the liquid based on analytical determination were smaller than 6±10% indicating a good consistence.

The temperature correction factor $\theta = 1.096$ determined for the temperature range of 17 to 32°C with a corresponding activation energy of 68 kJ mol^{-1} are in accordance with the literature which reports values of 1.065-1.1 for the temperature correction factor [7, 17] and values of 51-70 kJ mol^{-1} for the anammox activation energy [7, 12].

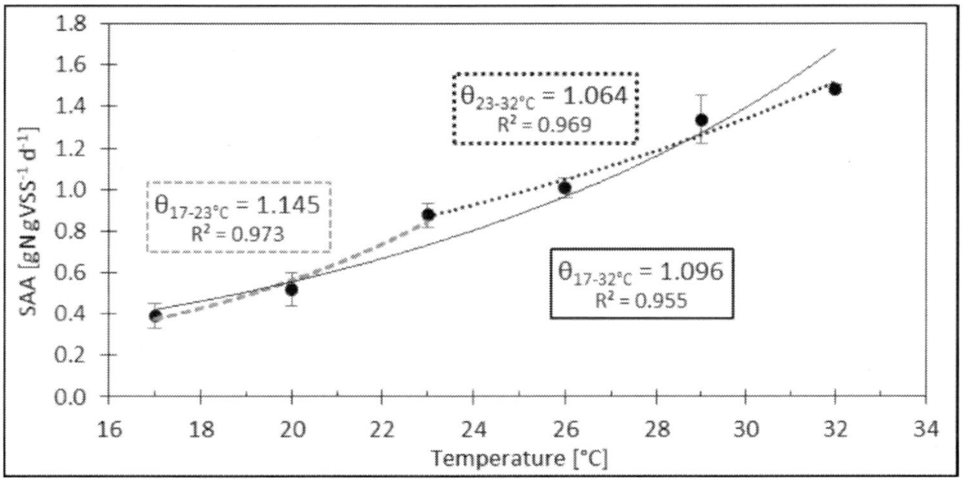

Fig. 2. Specific anammox activity as a function of temperature with the temperature correction factor θ for different temperature ranges.

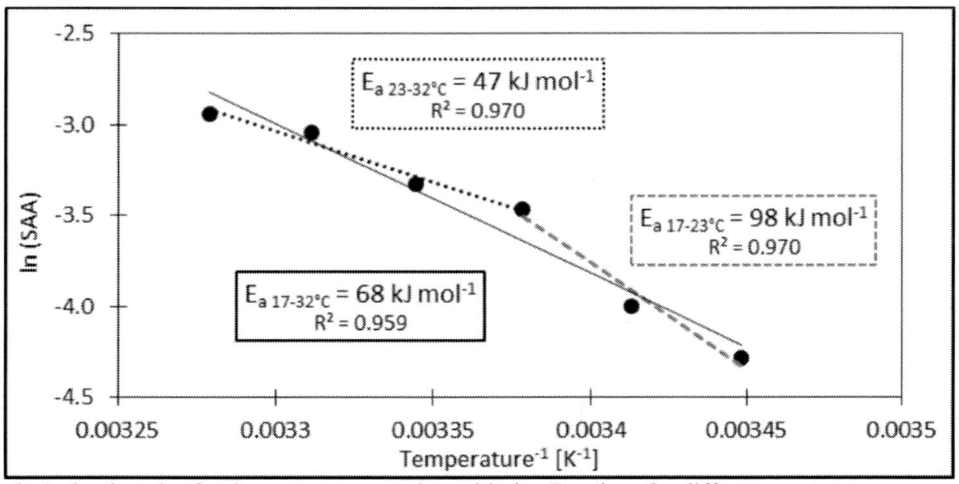

Fig. 3. Arrhenius plot for the anammox reaction with the E_a values for different temperature ranges.

However, closer inspection reveals that actually two temperature ranges with different characteristics can be distinguished: a lower range 17–23°C and a higher range 23–32°C, which is well visible at Figure 3.

For the lower temperature range (17–23°C), the estimated temperature correction factor θ was equal to 1.145 with a corresponding activation energy of 98 kJ mol^{-1}, whereas for the higher temperature range (23–32°C), the estimated temperature correction factor θ was equal to 1.064 with a corresponding activation energy of 47 kJ mol^{-1}. These values are still relatively close to the literature values mentioned above.

It is worth mentioning that the two distinguished temperature ranges were below and above the temperature of 23°C at which the pilot-scale SBR was operated. Considering the fact that the tests were performed without biomass acclimatization, this may suggest that a short-term temperature effect due to a shock caused by sudden temperature change may have contributed to the observed differences [18]. Lotti *et al.* [12] recommended an acclimatization period of 24–48 h to allow biomass to recover from the temperature shock effect. This, however, considerably extends the duration of the tests, but still does not account for the possible long-term temperature effect due to biomass adaptation to new conditions.

5 Conclusions

Manometric batch tests allowed to characterize short-term influence of temperature in the range of 17-32°C on the activity of anammox biomass cultivated at relatively low temperature (23°C), and estimate the temperature correction factor θ equal to 1.096 and a corresponding activation energy of 68 kJ mol^{-1}. The potential temperature shock effect and long-term bacterial adaptation require further investigation.

Acknowledgements – The authors gratefully acknowledge co-funding from the National Centre for Research and Development (grant no. PBS2/B9/25/2014) and Municipal Water and Sewage Company Wroclaw.

References

1. J. Makinia, *Mathematical modelling and computer simulation of activated sludge systems*, London, New York, IWA Publishing (2010)
2. A. Dapena-Mora, I. Fernandez, J.L. Campos, A. Mosquera-Corral, R. Mendez, M.S.M. Jetten, Enzyme Microb. Tech. **40** (2007)
3. S. Lackner, E.M. Gilbert, S.E. Vlaeminck, A. Joss, H. Horn, M.C.M. van Loosdrecht, Water Res. **55** (2014)
4. T. Liu, D. Li, J. Zhang, Y. Lv, X. Quan, Biochem. Eng. J. **105** (2016)
5. J. Dosta, I. Fernandez, J.R. Vazquez-Padin , A. Mosquera-Corral, J.L. Campos, J. Mata-Alvarez, R. Mendez, J. Hazard. Mater. **154** (2008)
6. M. Strous, J.J. Heijnen, J.G. Kuenen, M.S.M. Jetten, Appl. Environ. Microbiol. **50** (1998)
7. D. Sobotka, K. Czerwionka, J. Makinia, Water Sci. Technol. **73**, 10 (2016)
8. A. Daverey, P.C. Chei, K. Dutta, J.-G. Lin, Int. Biodeter. Biodegr. **102** (2015)
9. S.W.H. vanHulle, H.J.P. Vandeweyer, B.D Meesschaert, P.A. Vanrolleghem, P. Dejans, A. Dumoulin, Chem. Eng. J. **162**, 1 (2010)
10. G. Cema, J. Wiszniowski, S. Zabczynski, E. Zabłocka-Godlewska, A. Raszka, J. Surmacz-Gorska, Water Sci. Technol. **55**, 8/9 (2007)
11. K. Isaka, T. Sumino, S. Tsuneda, J. Biosci. Bioeng. **103**, 5 (2007)
12. T. Lotti, R. Kleerebezem, M.C.M. van Loosdrecht, Biotechnol. Bioeng. **112**, 1 (2015)
13. M. Strous, J.G. Kuenen, M.S.M. Jetten, Appl. Environ. Microbiol. **65** (1999)
14. J. Vazquez-Padin, I. Fernadez, M. Figueroa, A. Mosquera-Corral, J-L. Campos, R. Mendez, Bioresource Technol. **100** (2009)
15. S. Wu, A.S. Bhattacharjee, D.G. Weissbrodt, E. Morgenroth, R. Goel, Bioresource Technol. **219** (2016)
16. M.C.M. van Loosdrecht, P.H. Nielsen, C.M. Lopez-Vazquez, D. Brdjanovic, *Experimental Methods in Wastewater Treatment*, London, IWA Publishing (2016)
17. G. Sin, D. Kaelin, M.J. Kampschreur, I. Takacs, B. Wett, K.V. Gernaey, L. Rieger, H. Siegrist, M.C.M. van Loosdrecht, Water Sci. Technol. **58**, 6 (2008)
18. J.H. Hwang, J.A. Oleszkiewicz, Water Environ. Res **79**, 9 (2007)

Assessment of the possibility of using flocculation to improve properties of ultrafiltration membranes used in the purification of swimming pool water system washings

Edyta Łaskawiec[1,*], *Mariusz* Dudziak[1] , and *Joanna* Wyczarska-Kokot[1]

[1]Institute of Water and Wastewater Engineering, Silesian University of Technology, Konarskiego 18, 44-100 Gliwice, Poland

Abstract. The paper presents the possibility of using an in-line mode flocculation–ultrafiltration system for the purification of backwash water from flushing the beds of pressure filters used in the indoor swimming pool water treatment circulation. The effect of flocculation operational conditions (the flocculant dose in the range from 16 to 240 mg/dm^3 and the process temperature of 8–30°C) on the transport-separation properties of ultrafiltration membranes has been examined. To establish the effectiveness of the conducted processes, the contaminant retention coefficients, among others, were determined (based on the measurements of turbidity and UV254 ultraviolet absorbance). A significant influence of the flocculation process operational parameters on the performance of ultrafiltration was noted. With increasing flocculant dose, the efficiency of contaminant removal in the ultrafiltration process increased simultaneously. Moreover, the change in the temperature conditions of the flocculation process had a fundamental effect on the transport properties of the ultrafiltration membranes. The most advantageous temperature for conducting the processes was considered to be 21°C. Under those conditions, all flocculant doses contributed to a significant improvement in the transport properties of the ultrafiltration membranes. High values of the contamination retention coefficients and the permeate volumetric flux were obtained already at a flocculant dose of 80,160 and 240 mg/dm^3.

1 Introduction

The filtration process constitutes a basic element of water treatment systems in swimming pool facilities. During the flow of water through the bed, contaminant particles are retained in pores, which results in an increase in hydraulic losses. To prevent the bed breakthrough phenomenon and the contaminants passing on to the treated water, it is necessary to conduct bed washing with the reverse water and/or air stream at the

* Corresponding author: edyta.laskawiec@polsl.pl

appropriate frequency. In order to conduct the process in a correct manner, it is necessary to use 4–6 m^3 of water per each m^2, as a result of which a waste water stream, called backwash water, is formed. In the facility examined within the present study, the swimming pool circulation includes two 1800 mm-diameter filters which are washed every 48 hours, which generates a monthly water consumption from 245 to 365 m^3. The backwash water is discharged directly to the sewage system, which involves both wastage and high water drawing and waste water discharge costs [1–5]. Tests on the use of membrane pressure ultrafiltration for backwash water purification are being presently conducted [7]. It has been determined, though, that the chemical compounds and suspension present in the backwash water intensify membrane pore clogging. Hence, methods for supporting these processes are being searched for, which will increase the magnitude of the permeate volumetric flux, while maintaining the high separation abilities of the membranes.

The present investigation comprised the exploration of the possibility of using flocculation as an operation aiding the membrane ultrafiltration process. The purification system operated in an in-line batch mode, whereby the flocculant was introduced directly to the filtration cell. During the tests, an attempt was made to establish the operational parameters of the flocculation process (the flocculant dose and the process temperature), which contribute to an improvement in membrane transport–separation conditions, such as the permeate volumetric flow and the contaminant retention coefficient.

2 Materials and test methods

2.1 Characterization of backwash water tested

Backwash water used in the tests was a stream of waste water from washing multi-layered pressure filters constituting an element of an indoor swimming pool water treatment system. The facility under investigation is characterized by a high load with bathers, being the place of swimming lessons taken by children and young people of nearby schools and a form of recreation for adult city dwellers. A mixture of backwash water originating from the swimming pool circulation and a hot tub was used in the tests. The basic physicochemical parameters of the backwash water are given in Table 1.

Table 1. The physicochemical parameters of the backwash water.

Parameter	Unit	Value
Reaction (pH)	-	7.14
Conductivity (PWC)	µS/cm	960.00
Turbidity	NTU	26.60
Colour (C_{Pt})	mgPt-Co/l	306.00
Ultraviolet absorbance (UV254)	m^{-1}	23.50

2.2 Analytical procedures

The quality assessment of the test backwash water before and after the purification processes was made based on selected physicochemical properties. The measurement of the conductivity (PWC) and reaction (pH) of tested samples was done with an inoLab® 740 multi-parameter meter (WTW, Measuring and Analytical Technical Equipment). The ultraviolet absorbance at a wavelength of 254 nm was measured using a UV VIS Cecil 1000 supplied by Analytik Jena AG, with a cuvette optical pathlength of 1 cm. The UV_{254} value was determined based on the measurement method presented by US EPA [6], and the final analysis result was expressed in m^{-1}. For the determination of the turbidity of samples, an EUTECH Instruments Turbidimeter, Model TN-100, was employed. The measurement of the colour was performed using a UV VIS Spectroquant® Pharo 300 spectrophotometer (Merck) using a wavelength of 340 nm.

2.3 Operational parameters of the purification process in the in-line flocculation–ultrafiltration system

For carrying out backwash water flocculation, a product under the commercial name of Flockfix (supplied by Chemoform AG), being a solution of dialuminium pentahydroxychloride, commonly used in swimming pool facilities, was used. The colourless liquid was characterized by density at a level of 1.024 g/cm^3 and active substance concentration in the range from 2.5% to 10%, as stated by the manufacturer. To assess the effectiveness of a single flocculation operation within preliminary testing, typical beaker tests were performed, which included: A one-minute fast mixing process (at 250 rpm), followed by 25 minutes' flocculation at a mixer rotational speed of 20 rpm conducted in a four-stand laboratory coagulator (Velp Scientifica). The process was conducted at different backwash water temperatures, namely 8, 21 and 30°C, for each of the assumed doses. The temperature conditions were assumed based the considered backwash water storage variants – in an external tank in spring–summer conditions or in a decanter located in the swimming pool understructure (21°C), in an external tank in winter conditions (8°C) and in a flow-through tank with the filtration process being conducted directly after washing the filters (30°C). Directly after the process of flocculation started the membrane filtration.

The tests of the in-line mode flocculation–membrane ultrafiltration process involved the determination of the relationship between the conditions of the conducted flocculation process and the transport–separation properties of the ultrafiltration membranes in a wide range of flocculant doses (16, 80, 160, 240 mg/dm^3) and the above-mentioned temperatures. In this process, the flocculant was fed directly to the membrane cell which, at the same time, played the role of a reaction chamber. Filtration was started after the fast mixing and flocculation processes.

For membrane filtration, an ultrafiltration membrane with the symbol MW supplied by GE Osmonics Inc. (USA), whose characteristics, along with the operational parameters of the ultrafiltration process, are given in Table 2.

Table 2. Characteristics of the ultrafiltration membrane and the operational parameters of the process.

Membrane-forming material	Cut-off molar mass, Da	Membrane active filtration surface area, m^2	Trans-membrane process pressure, MPa	Deionized water permeate volumetric flux, $Jw \cdot 10^{-5}$, m^3/m^2s
Polyacrylonitrile	50000	0.00385	0.2	2.60

The membrane was placed in a 380 cm^3 steel filtration cell equipped with a magnetic stirrer. Before starting filtration, the new membrane was conditioned by filtrating deionized water until the permeate volumetric flux stabilized. The proper filtration process was conducted in a dead-end unidirectional filtration system until the 50% of the feed volume was received. At the same time, selected physicochemical parameters were verified in successively received 25 cm^3-volume permeates and the magnitudes of the permeate volumetric flux were determined. Six measurements were taken at each cycle. After each cycle comprising the filtration of the post-flocculation solution under specified conditions, the membrane was washed with deionized water. This was aimed at removing any residues of the suspension deposited on the membrane surface. After that stage, the permeate volumetric flux was determined again for deionized water, which was used for determining the degree of membrane pore clogging.

2.4 Parameters of the assessment of the ultrafiltration membrane transport–separation properties

The assessment of the transport properties of membranes in filtration processes was made based on the volumetric flux of deionized water, J_v (during conditioning the membrane with water), and the permeate, J_w (the proper filtration process), from the following equation:

$$J_w = \frac{V}{F \cdot t}, \frac{m^3}{m^2 \cdot s} \tag{1}$$

where: V - volume of water of permeate, m^3; F- membrane active surface area, m^2; t - filtration time, s.

To assess the separation properties of membranes, the retention (R) was determined, whose value was determined based on the reduction in the values of respective contamination indices:

$$R = \left(1 - \frac{c_p}{c_n}\right), \% \tag{2}$$

where: c_p - concentration (index value) of contaminants in the permeate stream, c_n - concentration (index value) of contaminants in the feed.

The intensity of the reduction in membrane transport abilities (membrane pore clogging degree) was determined by calculating the value of the relative permeate volumetric flux:

$$\alpha = \frac{J_w}{J_v} \tag{3}$$

where: J_w – value of the permeate volumetric flux at the sixth measurement point, $m^3/m^2 \cdot s$; J_v – value of the deionized water volumetric flux, $m^3/m^2 \cdot s$.

3 Test results

Figure 1 shows the transport–separation characteristics of membranes for the processes of filtering, respectively, raw backwash water and backwash water subjected to flocculation at a temperature of 21°C with a flocculant dose of 6, 160 and 240 mg/dm^3. A significant influence of the flocculation process operational parameters on the performance of ultrafiltration was noted. During the filtration of raw backwash water (with no flocculant), the average relative permeate volumetric flux was 0.74, which indicates a reduction in membrane efficiency by 26%. Whereas, the values of the contaminant retention coefficients, as determined by the measurement of turbidity or UV254 ultraviolet absorbance, amounted to approx. 98% and approx. 97%, respectively. The obtained results confirm the high effectiveness of the ultrafiltration process in removing the suspension present in the backwash water tested and part of the contaminants determined by the UV254 ultraviolet absorbance measurements. These results also document the fact of the occurrence of the membrane pore clogging phenomenon that reduces the values of the permeate volumetric flux. In turn, with increasing flocculant dose, a distinct increase in contaminant retention degree value was observed. By contrast, the relative permeate volumetric flux decreased in those conditions. With the increase in flocculant dose from 16 to 240 mg/dm^3, the average relative permeability of the membrane decreased from 1.26 to 0.97. Under the temperature conditions tested, all of the applied flocculant doses increased the membrane transport conditions. Moreover, already a flocculant dose of 80 mg/dm^3 allowed higher contaminant retention coefficient values to be obtained, compared to the results of raw backwash water filtration.

a)

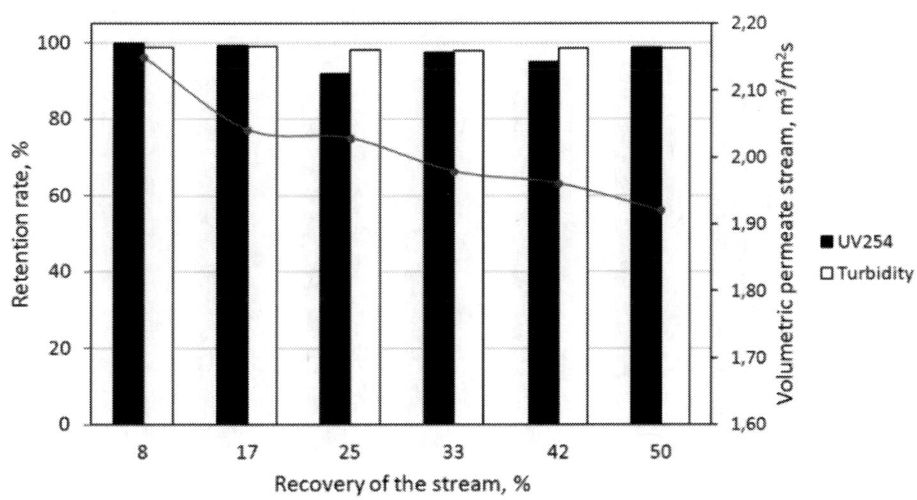

b)

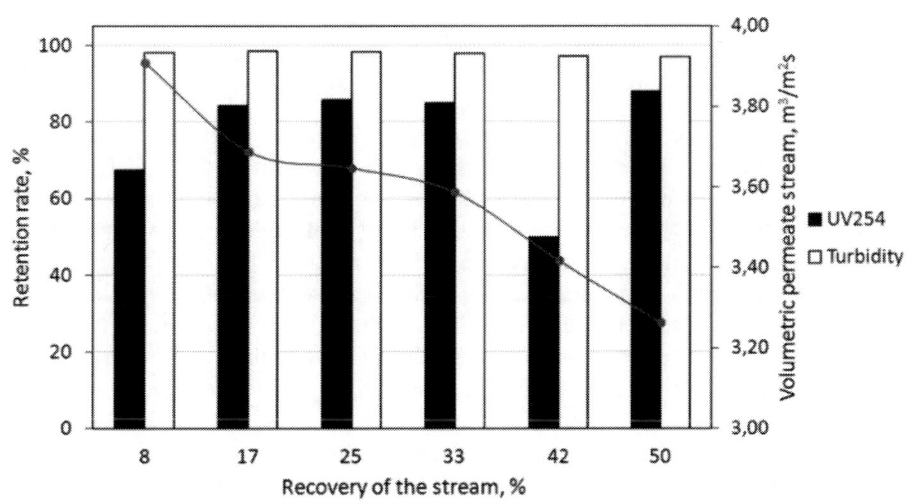

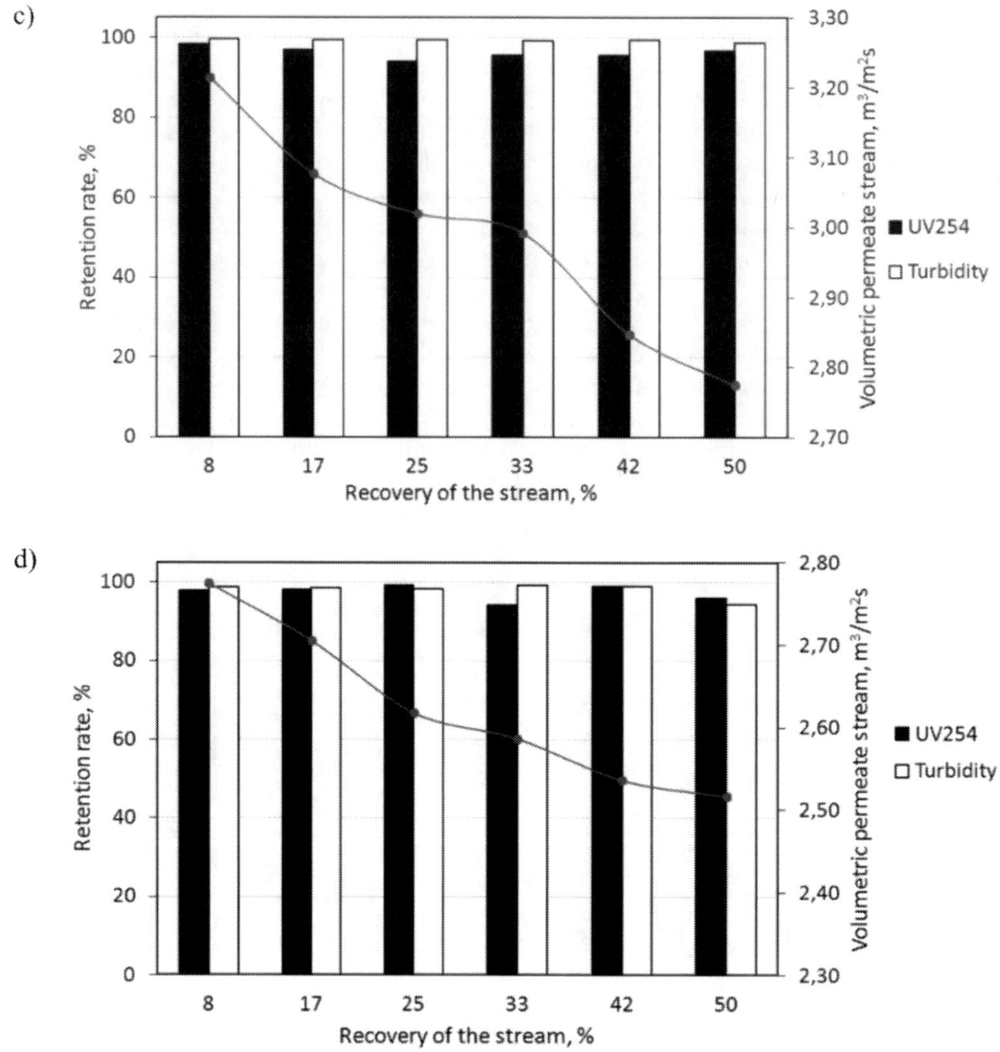

Fig. 1. The transport and separation properties of the ultrafiltration membrane during the filtration of a) raw backwash water, and backwash water with a flocculant dose of: b) 16 mg/dm^3 c) 160 mg/dm^3 d) 240 mg/dm^3, at a temperature of 21°C.

Figure 2 shows a comparison of the transport–separation abilities of the membrane for processes conducted at a temperature of 8 and 30°C with a flocculant dose of 16 mg/dm^3. Also for backwash water that underwent flocculation at temperatures of 8 and 30°C, an increase in the separation abilities of the ultrafiltration membrane was observed with increasing flocculant dose. Whereas, the values obtained at lower flocculant doses were higher for flocculation–ultrafiltration conducted at a lower temperature.

The change in the temperature conditions of the flocculation process had a fundamental effect on the transport properties of the ultrafiltration membrane. At a temperature of 8°C, an increase in the average relative permeate volumetric flux was noted only in the case of the filtration of backwash water with a flocculant dose of 160 mg/dm^3 (α = 0.76). In the remaining cases, the relative permeate volumetric flux value α was lower than that for the purification of raw backwash water. Furthermore, in spite of the high values of contamination retention coefficients for the flocculation–ultrafiltration process conducted at

30°C with a flocculant dose of 160 and 240 mg/dm^3, much lower values of permeate volumetric fluxes were obtained (α at a level of approx. 0.50).

a)

b)

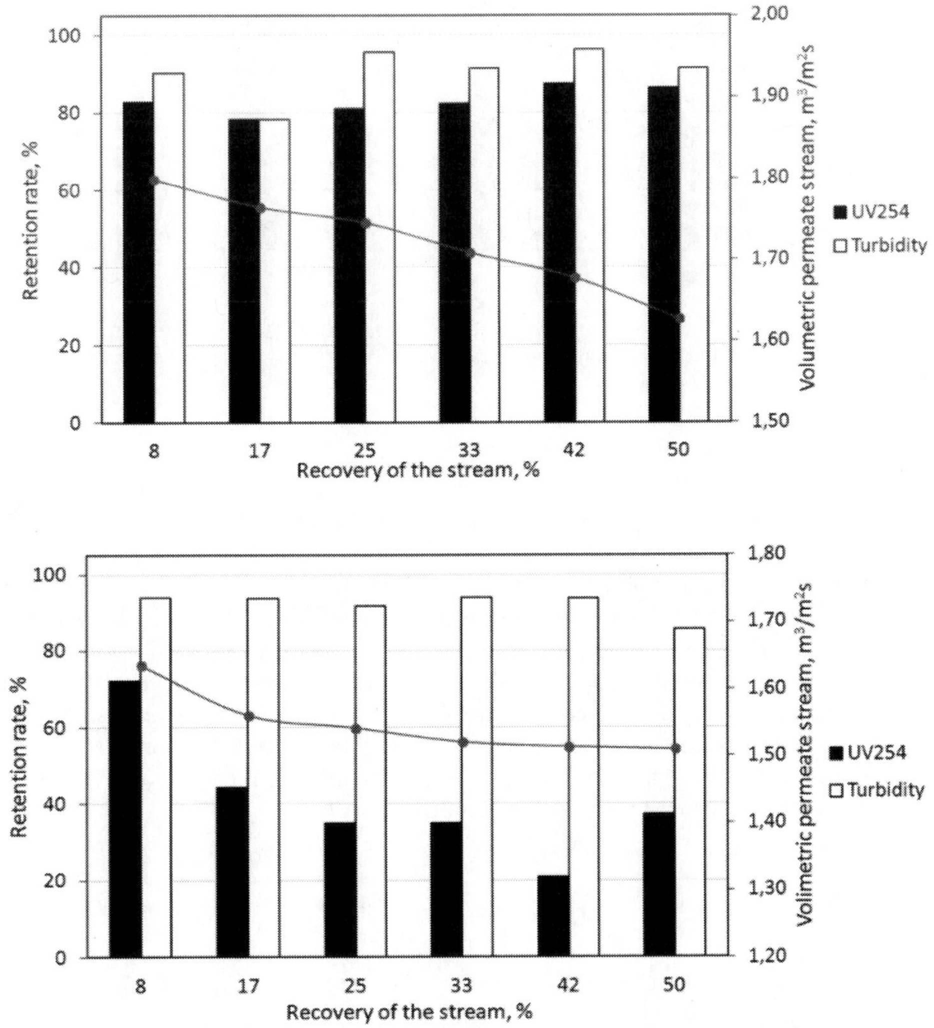

Fig. 2. The transport and separation properties of the ultrafiltration membrane – backwash water after flocculation with a flocculant dose of 16 mg/dm^3 at a process temperature of 8 (Fig. 2a) and 30°C (Fig 2b), respectively.

Among the presented backwash water purification systems, the most advantageous transport–separation conditions were shown by the variant with the flocculation process conducted at 21°C and with a flocculant dose of 80 mg/dm^3. Fig. 3 compares the effects of using the above-mentioned dose for three temperature variants.

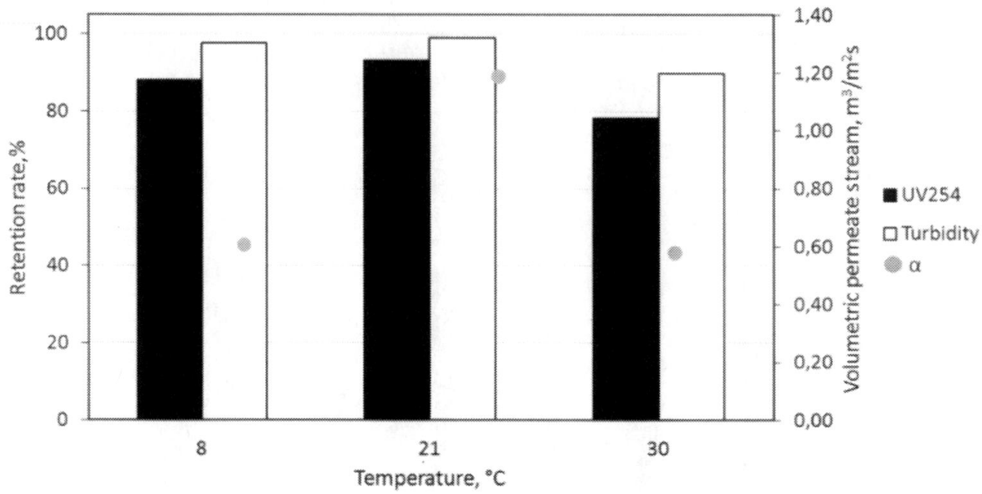

Fig. 3. The transport and separation properties of the ultrafiltration membrane – backwash water with 80 mg/dm^3 flocculant doses.

4 Summary

The investigation has confirmed the high effectiveness of pressure membrane systems in removing contaminations from backwash water coming from the swimming pool water circulation. The use of hybrid (flocculation–ultrafiltration) processes could improves the transport–separation properties of ultrafiltration membranes. The flocculation process has a particular importance in this case, as it increases the degree of removal of organic substances from the waste stream. It should be borne in mind, however, that the flocculant dose depend on the waste stream composition and concentration and on the nature of backwash water obtained in a given facility, including its physicochemical parameters. It is proposed that tests in a purification system with capillary immersed membranes should be carried out within further investigations on this subject.

References

1. P. Wiercik, M. Domańska, Sci. Rev. Eng. Env. Sci., **54**, 333–343 (2011)
2. I. Zimoch, Ochr Sr, **4**, 17–22 (2013)
3. M. Meggeneder, J. Łomotowski, P. Wiercik, Ochr Sr, **4**, 53–56 (2011)
4. M. Leszczyńska, M. M. Sozański, Environ, **40**, 575–585 (2009)
5. J. Wyczarska-Kokot, Proceedings of ECOpole, **2**, 9, 803–813 (2015)
6. B. B. Poter B.B., EPA Document, Method 415.3 (2009)
7. E. Łaskawiec, M. Dudziak, J. Wyczarska-Kokot, Apar. Bad. Dydakt., **21**, 238–245 (2016)

The content of fulvic acids in the primary effluent at the Płaszów WWTP in Kraków

Dominika Łomińska[1,*], and *Anna M. Anielak*[1]

[1]Department of Environmental Engineering, Institute of Water Supply and Environmental Protection, Cracow University of Technology, 24 Warszawska, 31-155 Cracow, Poland

Abstract. Municipal wastewater contains significant amounts of organic substances of both natural and anthropogenic origin. Wastewater treatment removes most of contaminants, however new substances are formed as a result of biochemical changes taking part during the treatment process; some of them remain in the final effluent and may pose a threat to life and health of living organisms. Humic substances (HS), which include soluble fulvic acids (FA) are contaminants commonly found in wastewater. This is why the research attempted to examine the content of fulvic acids in mechanically treated wastewater at the Płaszów Wastewater Treatment Plant (WWTP) in Kraków. The study was conducted to assess the quantities of the pollution load coming to the plant. FA were separated (isolated) at a hydrophobic ion exchanger, and then subjected to qualitative analysis to determine their elemental composition and degree of contamination with metals and other substances. Studies were also conducted in the infrared spectrum to determine the effects of internal structure of fulvic acids. Humic acids (HA) are precursors of chloroorganic compounds and other carcinogenic, mutagenic and teratogenic substances in oxidation and disinfection processes. Therefore the problem of FA in the final effluent is important, because of their amount and location of a wastewater discharge point.

1 Introduction

According to a lot of authors humic substances (HS) are the most widely distributed organic materials in the environment [1]. HS are natural products resulting from decomposition of organic matter (plants and animals) found in soils, sediments and natural waters. Their colour depends on their origin, concentrations and pH of the solution; it may change from yellow to dark brown. The molecular weight of HS ranges from a few hundred to a few thousand Daltons [2, 3]. In general HS are a mixture of different organic molecules. Although they have been known for over 200 years, their permanent chemical structure have not been established yet, and many models of their structure are known [4]. Their chemical form depends on: conditions under which process of peat runs, hydration of the local environment, type and duration of humification [5]. This is closely connected with a variety of their physicochemical properties.

* Corresponding author: dominika.lominska@pk.edu.pl

Humic substances can be classified according to their solubility as: humic acids (HA), fulvic acids (FA), hymatomelanic acids and humin. HA dissolve in alkaline solution and precipitate in a strongly acidic conditions (pH < 2). FA dissolve in both alkaline and acid solution while humin remain insoluble in acids and alkalis. Hymatomelanic acids (ulmine) dissolve in ethanol [2, 4]. Quite often HA predominate in dissolved organic matter (DOM), as represented mainly by FA, which constitute approx. 50% of the DOM in aquatic ecosystems. The most important fraction, natural organic matter (NOM), comprises humic and fulvic acids, due to their solubility [6, 7].

The subject of the research - FA are a fraction of humus compounds soluble in water, NaOH solution and diluted acids. Their colour is from yellow to yellowish brown. Like HA their construction is not exactly known yet. It is believed that fulvic acid molecules create benzene rings linked in a polymer structure using a visa hydrogen [8, 9]. Because of the fact that FA are smaller and have more carboxyl groups, they are much more chemically reactive than HA [10]. The elementary composition of FA is as follows: carbon 40 – 50%, oxygen 44 – 50%, and nitrogen 1 – 3% [11]. According to Ukrainian researchers, „FA are major constituents of humic substances; their amount may reach 81 - 95% of the total HS"[6, 7]. Since HS have a high content of functional groups and a high resistance to biodegradation they play an important role in both aquatic and land environments. Additionally, HS as natural organic acids are source of organic carbon for plants and microorganisms as well as play an important role in a biogenic cycle. They also take part in biochemical and nutrient involving processes [6, 7, 12]. In 1974 the researchers showed that HS may serve as precursors of carcinogenic and mutagenic compounds produced during oxidation and disinfection processes. They include: trihalomethanes (THM), haloacetic acids, halogenonitriles, halogenoketones, trichlorophenol, trichlorobenzenes, hydroxi-furanes and others. HS easily create soluble metal complexes which are hard to remove. During oxidation that is often used at the beginning of technological treatment system, they create by-products of oxidation which are much harder to remove that just HS. Therefore it should be aim to minimalize their concentration in water which is subjected to purification processes. This is the reason why HS should be studied in the environment and eliminated from water before entering a water treatment plant, otherwise they could pose a serious threat to the whole environment [13]. As it was reported by Pempkowiak [14] HA found in the effluent from the municipal wastewater treatment plant are different from HA isolated from the Vistula River. It means that municipal wastewater treatment plants are one of the sources of HS formation. Concentrations of HA in surface waters range from 1 to 9 mg/dm^3 and the wastewater treatment plant effluent discharge may have a significant impact on a surface water quality [14]. The next very important thing is a fact that FA are introduced with raw sewage to WWTP as not biodegradable compounds, so they are largely discharged with effluent. Previous studies other authors have shown that some amounts of FA are discharged with treated sewage. Taking into account in the above facts the authors examined the content of FA in a mechanically treated wastewater (primary effluent) from the Płaszów WWTP in Kraków in order to obtain information whether these substances are introduced to wastewater treatment plants and to receive knowledge in which quantities.

2 Materials and methods

FA were extracted according to the method described in [15]. Then FA were examined by qualitative analysis to determine the elemental composition of acids and the degree of contamination with heavy metals and other substances. Studies were also carried in the infrared spectrum to determine an internal structure of FA. The analysis were performed in the certified laboratory EkotechLAB in Gdańsk.

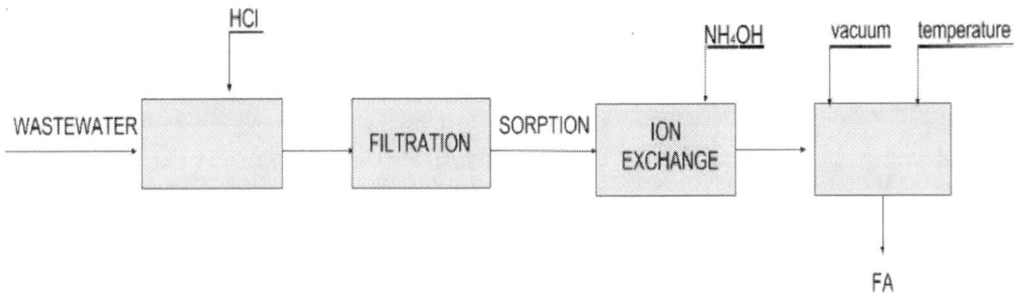

Fig. 1. Extraction of FA from wastewater at the Płaszów WWTP.

2.1 Analytical methods

The analysis of an elemental composition (C, H, N and P) of FA was performed by the EkotechLab laboratory according to own research procedures. The analysis of carbon, hydrogen and nitrogen was carried out using a combustion technology with chromatography detection (elemental analysis) and samples of 5-10 mg. The analysis of other elements was performed using X-ray fluorescence technique (XRF). The studies were carried out in the vessel for powder analysis, on a Prolen film of a thickness of 4 µm. The research was conducted on the WD-XRF spectrometer, model S8 Tiger manufactured by Bruker and elemental analyzer Flash EA manufactured by Thermo.

EkotechLab laboratory performed thermogravimetric method according to the own research procedure. An ash content was determined using a thermogravimetric method at a heating rate of 50°C/min up to 700°C, then 700°C by 5 minutes at the air flow of 100 mL/min. The analysis was conducted on the thermogravimetr SDT Q600 manufactured by TA Instruments.

Identification of the compounds was performed using a FTIR technique with the ATR accessory. Spectral range was as follows: 650 - 4000 cm^{-1} with resolution of 1 cm^{-1}, the scan number: 32. The study was conducted on the FT-IR IS10 spectrometer manufactured by Thermo Nicolet.

2.2 Wastewater treatment plant characteristic

The Płaszów WWTP is located in the south-eastern part of Kraków in the Podgórze district, and it covers the area of 50 ha.

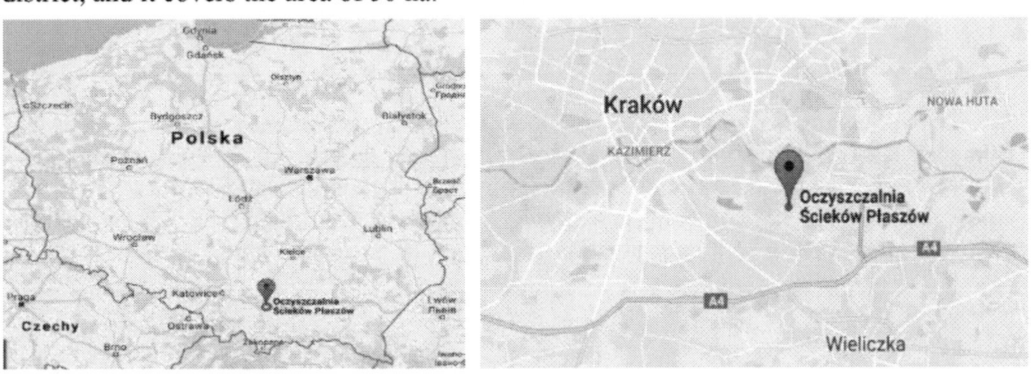

Fig. 2. Location of the Płaszów WWTP (Source: https://maps.google.com).

The wastewater treatment plant treats mainly wastewater from industrial plants as well as from individual users. The Płaszów WWTP comprises mechanical and biological wastewater trains as well as sludge processing and biogas trains. The mechanical stage consists of: coarse bar screens, pumping station, fine bars screens, grit chambers, preliminary clarifiers and a second pumping station. The biological stage consists of biological reactors and secondary clarifiers. The sludge processing train includes: mechanical thickeners, sludge disintegration unit, gravity thickeners, belt presses, anaerobic digesters and a biogas generation train (biogas desulphurization, biogas tanks and boilers). The wastewater treatment plant effluent is discharged through the outlet cascade into the Drwina river. The plant works as a three stage Bardenpho system and operates five identical biological reactors. The wastewater treatment process comprises the following unit processes: pre-denitrification, biological phosphorus removal, nitrification and denitrification. The average wastewater flow is 165 thousand m^3/day while a capacity of a biological stage is 320 thousand m^3/day. The concentration of activated sludge ranges from 3.5 to 5.5 kg of dry solids/ m^3. Excess sludge together with the primary sludge, after thickening to a 5% solid content, is pumped to the anaerobic digesters (information acquired at the Płaszów WWTP), [16].

3 Results and discussions

On the basis of data obtained from the Płaszów WWTP the authors [17] determined the parameters of wastewater entering the biological stage (primary effluent). The concentrations of BOD_5 in the primary effluent remained within the range 212 ± 72 mg O_2/L while the COD concentrations were 422 ± 170 mg O_2/L. A high content of nitrogen was observed, N_{tot} 56 ± 12 mg N/L while the concentration of P_{tot} was 5.7 ± 1.7 mg P/L [17].

The study involved analysis of FA extracted from primary effluent at the Płaszów WWTP with the characteristics as follows: COD =240.72 mg O2/L, N_{tot} = 64.96 mg N/L and P_{tot} = 6.37 mg P/L.

Table 1. Characteristic of the primary effluent (with extracted FA) from two separate studies.

Parameter	Primary effluent (authors data)	Primary effluent [17]
COD mg O_2/L	240.72	422 ± 170
BOD_5 mg O2/L	-	212 ± 72
N_{tot} mg N/L	64.96	56 ± 12
P_{tot} mg P/L	6.37	5.7 ± 1.7

The wastewater samples had a volume of 50 L (mechanically treated wastewater). Characteristics of fulvic acids (elemental composition) are shown in Table 2 and 3. The acids isolated at a concentration of 14.8 mg/L had a dark brown colour. They were contaminated with inorganic substances (Cl, Si, Na, Ca, K, P, Mg, and Fe, Br). The ash content was 11.55%, which indicated a large amount of inorganic impurities.

DOI: 10.1051/e3sconf/20171700054

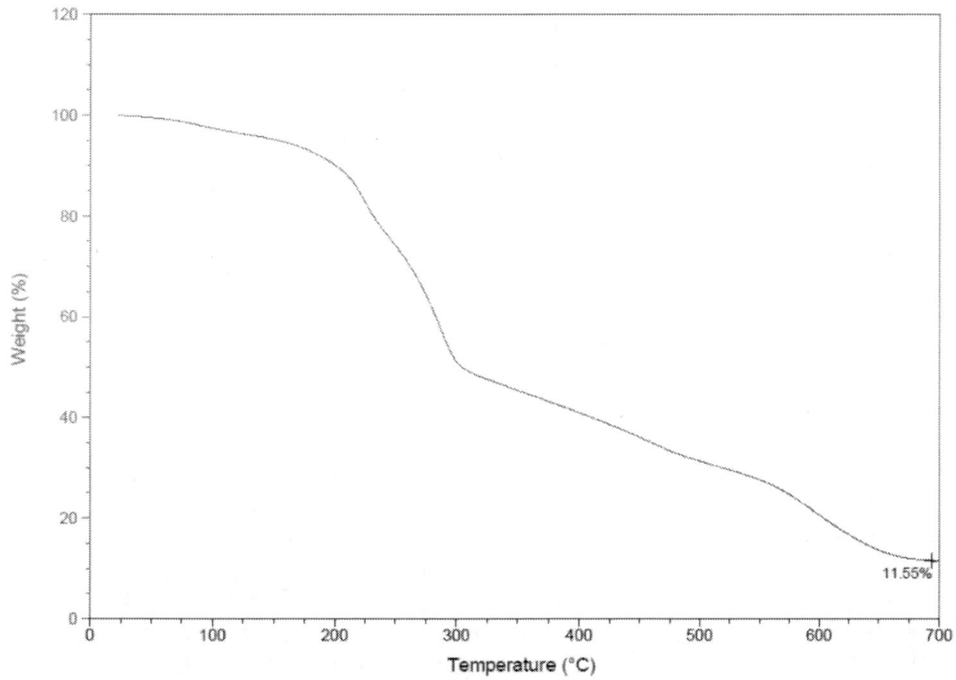

Fig. 3. Mass reduction in wastewater sample (after primary treatment) as a function of temperature.

Looking at the curve it can be seen that the sample mass decreases rapidly along with a temperature rise. At the ash content of 11.55% the sample weight remained unchanged, despite the temperature rise up to 700°C. The mass remained constant at a temperature of approx. 680°C, which was slightly higher than the temperature specified in the applicable standards (550°C). The elemental composition of FA extracted from the effluent is shown in Table 1.

Table 2. Elemental composition of FA extracted from the primary effluent.

Element (in ash-free mass)	Content [%]
C	33.24
H	4.60
N	10.03
S	1.87
O	50.26

Table 3. Ash content and atomic ratios in FA extracted from the primary effluent.

Sample	C/O	C/H	C/N	Ash	FA
	[-]	[-]	[-]	[%]	mg/L
Primary effluent	0.66	7.23	3.31	11.55	14.81

Table 4. Trace elements content in FA extracted from the primary effluent.

	Elements							
	Cl	Si	Na	Ca	K	Mg	P	Fe
	[%]	[%]	[%]	[%]	[%]	[%]	[%]	[%]
Primary effluent	13.62	2.58	1.74	0.93	0.37	0.29	0.28	0.04

The main components of extracted FA are: oxygen which content in ash-free mass is 50.26% and carbon in an amount of 33.24%. According to [18] a high oxygen content in the structure of FA may indicate the presence of carboxyl, carbonyl, methoxy, hydroxyl, alcohol, ester and steering groups [18]. According to table 2 hydrogen content is 4.60%, sulfur content is 1.87 while a nitrogen content is 10.03%. The high content of nitrogen may indicate the presence of amino acids in their structure. According to table 3 the C/N ratio is relatively small and equals 3.31 due to a high nitrogen content in FA (10.03%). The C/O ratio provides information on acids hydrophobicity. The more carbon the greater the ratio and the greater hydrophobicity of the acids. The C/O ratio for the FA from the wastewater was 0.66, so it can be concluded that the acids extracted from the mechanically treated wastewater were hydrophilic and had a lot of functional groups. Another issue is the C/H ratio, which served as an indicator of aromatization and condensation of HA. According to Pempkowiak the C/H ratio in HA isolated from the marine environment was below 10, while for a regular land humus it was above 10 [14]. The C/H ratio in FA from the primary effluent was 7.23 (below 10), so they seem to be less aromatized and condensed than the ones from the regular land humus. According to Table 4 a trace element of the largest content in FA extracted from the wastewater was chlorine (13.62%), next was silica (2.58%) and then sodium (1.74%).

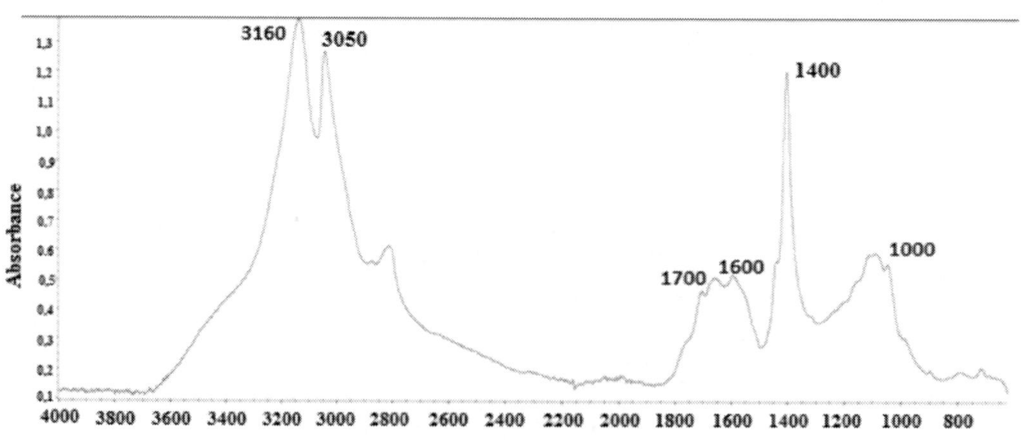

Fig. 4. The IR absorption spectra of FA separated from the primary effluent.

The IR absorption spectra of FA were conducted to analyse their structure. The analysis was performed based on the work [19, 20, 21, 22]. A spectral range was characteristic for this kind of substances: 650 - 4000 cm^{-1}. A quite broad peak appeared around 3160 cm^{-1} that was usually attributed to O-H stretching from alcohols, phenols and carboxylic related groups and also N-H stretching from amide and amince functional groups. The next peak around 3050 cm^{-1} was caused by C-H asymmetric and symmetric stretching aromatic compounds. The peak around 1700 cm^{-1} and 1600 cm^{-1} can be attributed to the C=C and

C=O stretching vibrations, coming from aromatic rings and carboxylic, ketone or ester groups repsectively. The peaks from 1400 cm^{-1} to 1000 cm^{-1} are probably due to C-O stretching and $-CH_3$ deformation vibartions. According to authors [19, 20, 21, 22] the complex heterogeneous nature of FA samples give rise to rather broad peaks observed [19, 20, 21, 22], what exactly the same has been observed in analysis of the primary effluent.

4 Conclusions

The primary effluent from the Płaszów WWTP was analysed for a presence of fulvic acids. FA isolated at a concentration of 14.8 mg /L had a deep brown colour and a high ash content (11.55%), which indicates their considerable contamination with inorganic substances. FA extracted from the primary effluent at the wastewater treatment plant comprise carbon, nitrogen, hydrogen, and oxygen, which is symptomatic for FA.

Based on the IR spectrum analysis it was concluded that FA extracted from the primary effluent have a characteristic comparable with FA extracted from surface waters (Oyster River and Suwannee River) from soil [19, 20, 21] and from Kołobrzeg mud [11, 13].

The plant effluent complies with the applicable legal standards. According to the data published in the Statistical Yearbook of the Republic of Poland (2016), 8 827,8 million m^3 of wastewater was discharged directly from wastewater treatment plants in 2015, including 514,3 million m^3 of wastewater treated only in mechanical processes [23]. FA that are introduced with raw sewage to wastewater treatment plants, as not biodegradable compounds, are largely discharged with the effluent. Assuming that the FA concentrations are like in the primary effluent, their load discharged to the receiving water can be easily calculated. In 2015 in Poland there were 20 mechanical sewage treatment plants and their total capacity was 900 000 m^3/d [23]. The average concentration of FA in the primary effluent is approx. 15 g/m^3 so the daily load calculated based on the data from the Statistical Yearbook is significant and amounts to 13.5 tons/day. While the average wastewater flow in Płaszów WWTP at Kraków is 165 thousand m^3/day and the concentration of FA is 14.8 g/m^3, so the daily load can be calculated approx. 2.4 tons/day. FA are the most mobile components and that is why we usually have to deal with them and they become the most problematic issue.

The main goal of the study was to complete a balance and assess the amount of fulvic acids in the primary effluent at the Płaszów WWTP in Kraków. The studies have shown high concentrations of FA in primary effluent; they can penetrate into the surface waters and serve as precursors of oxidation and disinfection by-products at surface water treatment plants. The obtained results indicate necessity to carry out detailed analysis of the amount of discharged HS in specific biological reactor processes.

References

1. M. Kalina, M., Klučáková., P. Sedláček, Geoderma 207-208, 92-98 (2013)
2. Noel E. Palmer Ray von Wandruszka, Fresenius J Anal Chem **371**, 951–954 (2001)
3. S.S. Gonet, B. Debska, A. Zaujec, In: Clapp CE, Hayes MHB, Senesi N, Griffith SM (eds), International Humic Substances Society, St Paul, MN, 95–101 (1996)
4. A.M. Anielak, *Wysokoefektywne metody oczyszczania wody*. PWN, Warszawa, 28-32 (2015)

5. M. Drobnik, T. Latour, *The Influence of Thermal Aand Chemical Processes on The Content and Structure of Humic Acids - The Basic Ingredient of Peat Preparations,* ROCZN. PZH, 62, **4**, 453 – 462 (2011)
6. P. N. Linnik, Ya. S. Ivanechko, R. P. Linnik., V. A. Zhezherya, Russ. J Gen. Chem., **83**,13, 2715–2730 (2013)
7. P. N. Linnik, Ya. S. Ivanechko, R. P. Linnik., V. A. Zhezherya, J. Water Chem. Techno+, **35**,6, 295–304 (2013)
8. Gonet S.: Struktura substancji humusowych. Zesz. Probl. Post. Nauk Roln., **411**, 184-194 (1993)
9. M. Huculak-Mączka, K. Hoffmann, J. Skut, J. Hoffmann, Proceedings of ECOpole, **4**, 2, 383-387 (2010)
10. S. Amir, M. Hafidi, L. Lemee, J. R. Bailly, G. Merlina, M. Kaemmerer., J. C. Revel, A. Ambles, J. Anal. Appl. Pyrolysis, **77**, 149-158 (2006)
11. A.M. Anielak, M. Grzegorczuk-Nowacka, Pol. Environ. Stud., **20**, 6, 1381-1386 (2011)
12. D. Łomińska, Czasopismo Techniczne, **1**-Ś/2016, 80-81 (2016)
13. R. Świderska, A. M. Anielak, *The significance of electrokinetic potential in the adsorption process of humic substances,* Rocznik Ochrona Środowiska, **6**, 31–49 (2004)
14. J. Pempkowiak, H. Obarska-Pempkowiak, M. Gajewska, D. Ruta, Przem. Chem., **87**, 5, 542–54 (2008)
15. A. M. Anielak, M. Polus, D. Łomińska, T. Żaba, Przem. Chem., **95**,1, 110- 113 (2016).
16. A. M. Anielak, T. Żaba, M. Polus, P. Beńko., T. Bochnia, Łuszczek B., Przem. Chem., **95** (2), 314-319 (2016)
17. D. Łomińska, A. M. Anielak, *Płaszów WWTP as a source of humic acids,* Leaving the Ivory Tower : bridging the gap between academia, industry, services and public sector : 8 th Eastern European Young Water Professionals Conference, Poland, Gdańsk (2016)
18. M. Grzegorczuk-Nowacka, *Adsorpcja kwasów fulwowych z wodnych roztworów*, praca doktorska, Kraków, 52-53 (2012)
19. B. Zhu, D. K. Ryan, J Environ. Radioactiv., **153**, 97-103 (2016)
20. X. He, B. Xi, Z. Wei, X. Guo, M. Li, D. An, H. Liu, Chemosphere, **82**, 541-548 (2011)
21. S. Huo, B. Xi, H. Yu, L. He, S. Fan, H. Liu, J. Eviron. Sci., **20**, 492-498 (2008).
22. W. Zieliński, A. Rajca, *Metody spektroskopowe i ich zastosowanie do identyfikacji związków organicznych.* Praca zbiorowa, WNT, Warszawa (2000)
23. Rocznik Statystyczny Rzeczypospolitej Polskiej (2016)

Method of floc classification after the coagulation process

Ewelina Łukasiewicz[1,*], and *Mariusz* Rząsa[1]

[1]Opole University of Technology, Department of Thermal Engineering and Industrial Facilities, 5 Mikołajczyka Street, 45-271 Opole, Poland

Abstract. Coagulation is one of the processes of water and wastewater treatment. During this process, flocs are formed and are characterized by different shapes, structures and dimensions. Due to this fact, removing of these particles may produce some problems for researchers because the sedimentation way of this particles is disturbed. In this paper, the method of flocs classification is presented where coefficients k_1, k_2 and k_3 were calculated. An appropriate value of coefficient k enables the classification of the considered particle to a specific group. The idea of the classification method consists in calculating surface areas for a particle image registered using a microscope. After that, the images are analyzed (computer image processing). The analyzed images will be used for future studies, in which the settling phenomenon will be described.

1 Introduction

Safe drinking water is essential to the health and welfare of the community, and water from all sources must undergo some form of purification before consumption. A very important step in the process of purifying is a coagulation and flocculation process, which is widely used due to its simplicity and cost-effectiveness. Regardless of the nature of the treated sample (e.g. various types of water or wastewater) and the overall applied treatment scheme, coagulation-flocculation is usually included, either as a pre-, or as post-treatment step [1–3]. Properly carried coagulation not only provides a high degree of removal of colloidal suspensions and settling, but is alsoassociated with these other pollutants, including micropollutants. The impact of coagulation involves an effective reduction of turbidity, color intensity, indicating the organic impurities in these precursors of disinfection by-products and chemical oxidation, and the content of a number of micropollutants, mainly present in the purified water in the form of sparingly soluble form or associates with solid particles. The suspended particles vary considerably in source, composition charge, particle size, shape, and density. Correct application of coagulation and flocculation processes and selection of the coagulants depend upon the understanding of the interaction between these factors [4,5,6]. In practice, the coagulation is carried out by coagulants, where the hydrolysis products cause electrolyte coagulation and the coagulation by the colloid with an opposite sign. Coagulation occurs in two phases: in the first - after

[*] Corresponding author: e.lukasiewicz@po.opole.pl

the addition of a coagulant, lasting seconds - chemical and physical reactions take place leading to the destabilization of colloidal particles, and in the second phase – lasting much longer – flocculation – the flocs are created as a result of transport and collisions of destabilized particles (Fig. 1). Flocs can be efficiently removed by the process of sedimentation/flotation and filtration [7, 8]. After mixing, the processed water is directed to clarifiers. Settling tanks are devices in which a process that removes the suspension from purified water occurs. They are used to retain the settleable suspension easily, either natural or produced by the coagulation process and chemical precipitation of sparingly soluble compounds. Contact settlers with a suspended sediment are used to treat water by coagulation and are usually called clarifiers [9, 10]. Due to the fact that the drag coefficient depends on both the shape and density of particles, there are some difficulties in determining it. The determination of this coefficient is significantly hindered due to different forms of flocs as a result of the coagulation process. Sediment structures formed during this process are diverse in terms of their shapes and degree of density [11, 12]. Taking the above into consideration, it can be observed that the modelling of this process is extremely difficult. It is possible to group flocs, taking their characteristics into account, and then formulate general equations describing the movement of particles belonging to a given group.

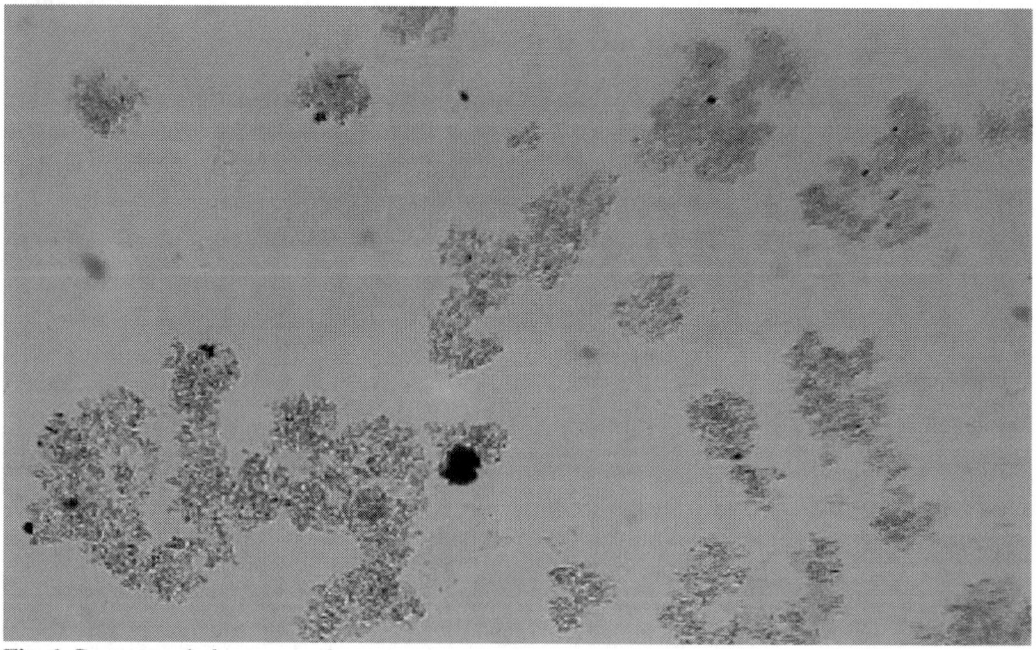

Fig. 1. Post-coagulation suspension.

2 Classification

This paper presents the classification of flocs. The classification concerns particle distribution, taking into account their movement during the settling process. Due to technological reasons, the settling time of the sediment is the most important parameter. Therefore, the classification proposed by the authors mainly focuses on the separation of groups of flocs according to their characteristic features affecting the settling velocity.

In order to form a suspension, model water based on distilled water was prepared. 24 g of sodium chloride, 12 g of sodium sulfate, and 0.05 g of iron sulfate were dissolved in 1 liter of water. 60 g of aluminum hydroxychloride per 1 cubic meter were used as a coagulant. After the coagulation process, suspension samples were taken and examined under a microscope.

While examining the microscope images, the authors decided to classify individual particles statistically. Flocs were divided into three characteristic groups. The classification takes into account the structure of the floc in terms of mass concentration (presence of macropores) and its shape. Flocs were divided into: compact particles, particles having a distinct mass concentration nucleus around which sediment crystallization networks are formed (flocks without mass concentration), and particles having a porous sponge structure (porous conglomerates).

Based on the floc analysis, three separate types of particles characterised by a different settling way were distinguished (Fig. 2). Compact particles are characterised a homogeneous structure. Their shape is similar to a sphere filled with the same microporous structure. Micropores are specific features of a post-coagulation material. Particles of this type settle in a straight line with a slight zigzag motion. Due to their small size and mass, their movement is repeatedly disrupted by fluid movement triggered by the movement of other particles, considerably higher in size, in their vicinity. The second group of particles (star-shaped particles) has a distinct mass concentration nucleus around which sediment crystallisation networks are formed (flocs without mass concentration) (Fig.3). Protruding arms cause the particle to move in a circular motion around its own axis while settling. The circular motion causes the particle, whose direction is constantly changing, to be influenced by the Magnus effect. This effect causes the particles of this type to settle in a spiral motion. Over time, star – shaped particles connect to create bigger agglomerates. In case of big agglomerates, the circular motion around their own axis gradually disappears, and particles of this type settle in a straight line. They are characterised by considerably big macropores, through which fluid may flow during their motion. This type of structure is similar to fluid flow through a porous medium [Fig. 4].

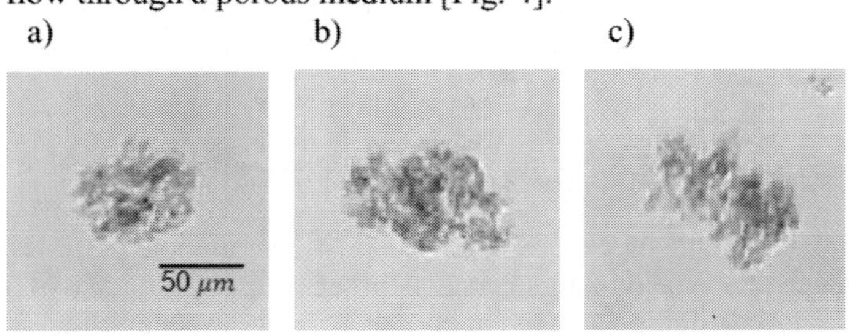

Fig. 2. Compact structure of flocs.

Therefore, the flocs presented above (Fig. 2.) are flocs with a compact structure containing a large number of pores. These particles are characterized by the compact structure and relatively regular shape of a floc. Pores occurring in it are a natural feature of this type of suspensions. Settling of such particles can be described using the Stokes' law [7]. Due to various shapes, it is necessary to determine an appropriate drag coefficient experimentally.

The star-shaped particles have a nucleus with concentrated mass surrounded by a number of sediment crystallization arms (Fig. 3.). In case of such particles, the weight of a nucleus mainly affects the settling velocity. However, irregular crystallization arms change the direction in which particles settle. A non-linear trajectory significantly extends both settling path and time. The mathematical description of such particles requires much more complex mathematical models.

The third group includes porous conglomerates consisting of the types of particles present in a given suspension, combined into clusters. Their structure resembles a sponge, they are very unstable and they can easily change their shape while settling.

Figure 4 presents the mass that is not concentrated in one place. The structure developed a porous structure (like a sponge). In addition, it was observed that the porous structure of flocs enables them to combine into larger conglomerates. According to the phenomenon of settling conglomerates, this type of structure is classified as a separate group in the classification of particles. The occurrence of a distinct porous structure will result in an additional flow of liquid through open pores.

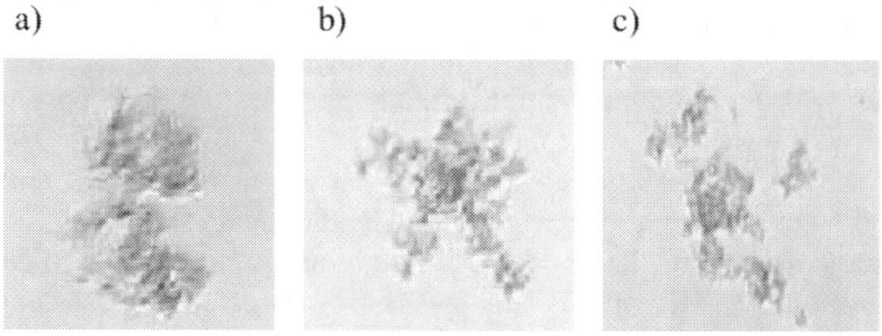

Fig. 3. Flocks without mass concentration.

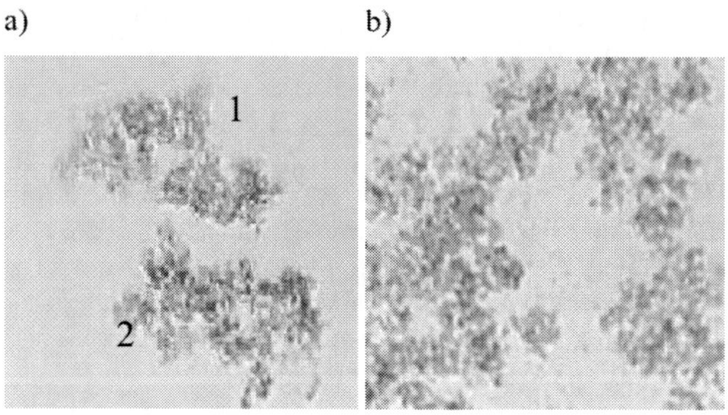

Fig. 4. Porous conglomerates.

3 Idea of the method of particle classification

The presented method of classification of post-coagulation particles is based on the process of image analysis (computer image processing). The idea of the classification method consists in calculating surface areas for a image registered using the microscope. Due to a great amount of data (microscope images), it is necessary to apply an automated classification of the flocs of the suspension. In order to carry out such a classification, reference levels in the form of numeral parameters are vital.

The microscope image of a particle (Fig. 5a) is subjected to a binearization process (Fig. 5b). Subsequently, for this image the surface area is calculated for particle A_1. In the second stage, the computer image analysis is carried out to close all open spaces (Fig. 5c). On this basis, surface area A_2 is calculated. A similar operation can be performed on the image of Figure 5b to make all surfaces closed (Fig. 5d). This operation enables the determination of surface A_3 constituting the surface of the particles inside the outline.

The calculation of areas A_1, A_2 and A_3 is the basis for determining the classification criterion. Based on the relationships of selected surface areas, a characteristic factor k_1, k_2, and k_3 is determined.

$$k_1 = \frac{2\cdot|A_2 - A_1|}{A_2 + A_1}100\% \qquad k_2 = \frac{2\cdot|A_3 - A_1|}{A_1 + A_3}100\% \qquad k_3 = \frac{2\cdot|A_2 - A_3|}{A_2 + A_3}100\% \qquad (1)$$

Based on the values of coefficients k_1, k_2, and k_3, the classification is carried out to assign the selected particles to specific groups, in line with the proposed classification. The assignment algorithm assumes appropriate threshold values for coefficients k_1, k_2 and k_3. The assignment to the specific groups was based on the criteria listed in Table 1.

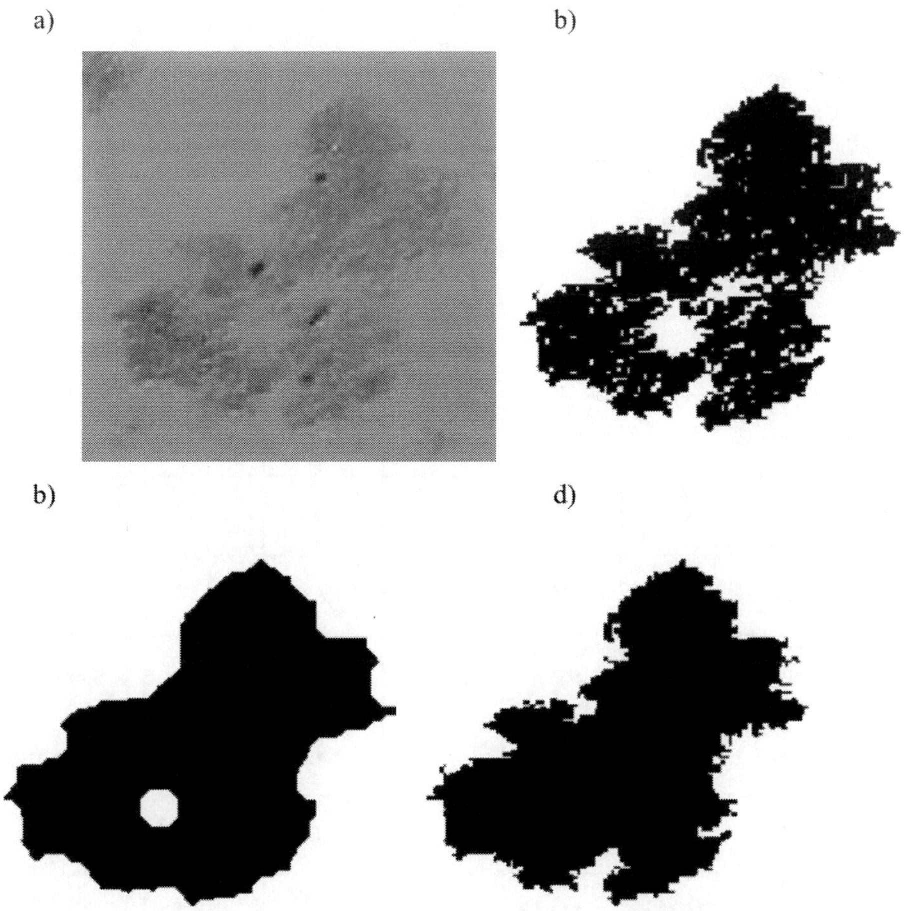

Fig. 5. Image processing: a) microscope image of a particle b) image after binearization c) image after the closure of open spaces d) image after filling in confined spaces.

Table 1. Criterion for the allocation of particles to specific groups.

Group name	k_1	k_2	k_3
Compact	$k_1 < 25\%$	$k_2 < 10\%$	$k_3 < 15\%$
Star-shaped	$k_1 > 25$	$k_2 < 10\%$	$k_3 > 15\%$
Porous conglomerates	$k_1 > 25\ \%$	$k_2 > 10\%$	$k_3 > 15\%$

For the purposes of computer classification of particles, it is necessary to determine the appropriate thresholds below which the individual particles are clearly classified. The particles which will be located on the border of threshold numbers are assigned to the group for which the results are most similar to the thresholds. In case of not fulfilling one of the three criteria, the particles were classified by meeting as many criteria as possible. The threshold values were determined experimentally.

4 Experimental studies

Experimental studies were conducted involving the determination of surface A_1, A_2 and A_3 for the particle images (Figures 2, 3 and 4). The results of the calculations are presented in Table 2. The values of the areas were calculated on the basis of the number of pixels comprising the image of the floc. Due to the fact that coefficients k_1, k_2, and k_3 are the quotients of the areas, it is not necessary to convert the metric units of the surface area.

Table 2. Criterion for the allocation of particles to specific groups.

Particle name	A_1	A_2	A_3	k_1 [%]	k_2 [%]	k_3 [%]	Group
Fig. 2a	1174	1459	1336	21.65	12.91	8.80	compact
Fig. 2b	1851	2221	1981	18.17	6.78	11.42	compact
Fig. 2c	1607	1999	1740	21.74	7.95	13.85	compact
Fig. 3a	2516	3455	2721	31.45	7.83	23.77	Star shape
Fig. 3b	1847	2743	2052	39.04	10.52	28.82	Star shape
Fig. 3c	1322	1897	1388	35.73	4.87	30.99	Star shape
Fig. 4a-1	2840	3894	3268	31.30	14.01	17.48	Porous conglomerates
Fig. 4a-2	2868	3948	3233	31.69	11.97	19.91	Porous conglomerates
Fig. 4b	9878	14224	11340	36.06	13.78	22.56	Porous conglomerates

During the analysis of the results in Table 2, it was noted that in two cases, for particles with Figure 3b and Fig. 2a, k_2 criterion was not fulfilled. This proves the validity of adding third k_3 criterion, since in such cases the classification of the particles to a specific group will be met if the other two criteria are fulfilled. This solution improves the reliability of the method of classification.

5 Summary

The proposed computer method for the classification of a polydisperse post-coagulation suspension allows the satisfactory classification of its particles. The automated classification method is used for the analysis of microscopic images of such a suspension. This issue is important in view of the fact that studies concerning the settling process that contains such suspensions are necessary to learn about the concentration of each fraction suspended in a sedimentation process. In order to apply this method to other suspensions, the appropriate values of coefficients k_1, k_2 and k_3 must be experimentally determined.

References

1. R.M. Rząsa, E. Podgórni, Pol. J. Environ. Stud. **23**, 6, 2157-2151 (2014)
2. N.D. Tzoupanos, I.A. Zouboulis, 6th IASME/WSEAS International Conference On Heat Transfer, Thermal Engineering And Environment (HTE'08) Rhodes, Greece, August 20–22 (2008)
3. M. Ariffin Abu Hassan, T. Pei Li, Z. Zainon Noor, Journal of Chemical and Natural Resources Engineering, **4**, 1, 43–53 (2009)
4. H. Zemmouri, M. Drouiche, A. Sayeh, H. Lounici, N. Mameri, Procedia Engineering **33**, 254–260 (2012)
5. R. Bergamasco, L.C. Konradt-Moraes, M.F. Vieira, M.R. Fagundes-Klen, A.M. Salcedo Vieira, Chemical Engineering Journal **2** (2011)
6. Y. Xu, T. Chen, F. Ciu, W. Shi, Chemical Engineering Journal 287, 225–232 (2016)
7. J. Gumińska, M. Kłos, Gaz, woda i technika sanitarna, **5**, 195 (2011)
8. M. Świderska-Bróż, J. Krupińska, Ochrona Środowiska, **3**, 86, 9–13 (2002)
9. M. Kłos, J. Gumińska, Ochrona Środowiska, **4**, 29, 73–76 (2007)
10. G.H. Ganser, Powder Technology, 77, 145 (1993)
11. M. Malakootian, A. Fatehizadeh, Iran. J. Environ. Health. Sci. Eng., **7,** 3, 267 (2010)
12. D. Ghernaout, A. Boucherit, Journal of Research & Developments in Chemistry, **2** (2015)

E3S Web of Conferences 17, 00056 (2017) DOI: 10.1051/e3sconf/20171700056
EKO-DOK 2017

Preliminary results of thermal comfort analysis in selected buildings

Grzegorz Majewski[1,*], *Marek* Telejko[1], and *Łukasz J.* Orman[1]

[1]Kielce University of Technology, al. Tysiąclecia Państwa Polskiego 7, Kielce 25-314, Poland

Abstract. The paper analyses thermal comfort in an intelligent building. Intelligent buildings are now widely seen both in Poland and abroad as a result of developments in engineering sciences. However, data on the thermal comfort provided in the buildings are limited. The research described in this paper was carried out in ENERGIS intelligent building of the Kielce University of Technology. The parameters measured included air temperature, relative air humidity, air flow rate and carbon dioxide concentration. Average radiation temperature was determined. The key element was the study of thermal sensation in occupants present at the time of the measurements. The studies were performed in classrooms during the class. The students completed a questionnaire at the beginning and end of the class. The persons present in the classrooms evaluated their thermal sensations using a seven-level thermal comfort scale. The questionnaire included such questions as the type of clothing and the degree of physical activity of the people sitting in the classroom. On the basis of the measurements of classroom microclimate parameters, PMV (Predicted Mean Vote) and PPD (Predicted Percentage Dissatisfied) indices were determined. The responses from the questionnaires provided actual average assessment of thermal sensation and actual percentage of dissatisfied people.

1 Introduction

Thermal comfort is one of the climatic elements with variable parameters such as air temperature, air humidity, relative air flow rate, mean radiant temperature, thermal insulation of clothing and physical activity. The study of thermal comfort began in the 20th century when microclimate in buildings could be controlled. This would not have been possible without the development of thermodynamics in the 19th century, which markedly contributed to the growth of thermal control technology. The discoveries of two scientists, Richard Mollier, regarded as the precursor of thermodynamic calculations, and Willis Carrier, the creator of modern air conditioning, gave rise to a new discipline referred to as thermal comfort. The interdisciplinary aspect of thermal comfort is reflected in physical, psychological, physiological, social and cultural factors that have to be considered, along with clothing type, eating habits, climate and building types. One of the first scientists to study thermal comfort was Bedford, who in 1936 proposed a seven-point scale of thermal

[*] Corresponding author: majewskigrzegorz@wp.pl

sensations [1]. Other researchers of that time, such as Winslow, Herrington and Gagge [2], also focused their professional interests on thermal comfort issues. In 1957 Yaglou and Minard [3] developed the Wet Bulb Globe Temperature (WBGT) indicator to evaluate average effect of temperature on a human being in the period representative of the nature of his or her work. In the sixties and seventies of the 20th century, L. O. Fanger published the thermal comfort equation and PMV (Predicted Mean Vote) and PPD (Predicted Percentage Dissatisfied) indices [4, 5]. These indices continue to apply as recommended by, for example, PN-EN ISO 7730 [6].

The current standards defining the mean vote (PMV) and the percentage of dissatisfied persons (PPD) include the European standard PN-EN 15251 [7] and the American ASHRAE Standard 55-2013 [8]. A very large amount of research is being conducted worldwide on thermal comfort of occupants of various buildings in which many people are gathered at the same room at the same time. In [9] Ricciardi and Buratti reported the result from their investigations carried out in a theatre building. Fabbri [10] studied thermal comfort in a kindergarten. This paper focuses on studying thermal comfort of occupants of teaching rooms in the building of the Faculty of Environmental, Geomatic and Energy Engineering at the Kielce University of Technology. Actual thermal sensations obtained from the occupants were compared with the PMV and PPD indicators. The studies of thermal comfort in classrooms have been conducted in many parts of the world. M. A. Nico, S. Liuzzi and P. Stefanizzi [11] investigated thermal comfort of students in university classrooms and compared the sensations reported by men and women. The results indicate that female students are more sensitive to cold and tend to wear thicker clothes. Cao et al. [12] studied thermal comfort in classrooms at universities in three Chinese cities representing different climatic zones: Harbin as the coldest city, Beijing with a little higher temperature and Shanghai with the highest temperatures. In Hrabin, more than 70% of data fell in comfortable zone. In Shanghai, more than 80% of data were in the winter comfortable zone. The percentage of people satisfied with the cold conditions in Beijing was lower than those in Hrabin and Shanghai. Abdallach [13] demonstrated that at the indoor temperature higher than 28°C, the PMV indicator is lower than 90%, and the PPD indicator is greater than 10% in particular in Muslim female students with covered heads. The study revealed that 83% of students surveyed prefers cooler air temperatures. Similar investigations were conducted by O'Donovan et al [14] and Katafygiotou and Serghides [15].

2. Research procedure

The studies were carried out in the ENERGIS building occupied by the Faculty of Environmental, Geomatic and Energy Engineering at the Kielce University of Technology. An indoor air quality monitor, an anemometer and a pyrometer were used to measure microclimate parameters such as air temperature, air flow rate, temperature of the walls surrounding the rooms and carbon dioxide concentration. Measurement results were recorded at 5 minute intervals. The microclimate parameters were measured at 130 cm, that is at the height of a sitting person's head. The temperature of the walls was measured at the beginning and at the end of the classes. A 1m spacing was maintained for wall temperature measurements carried out at the height of 130 cm. Temperatures of the floor and ceiling were also recorded. The assessment of thermal comfort was performed through anonymous questionnaires in which the students evaluated their thermal sensations on a seven-point thermal comfort scale. A total of 101 students in three classrooms participated in the survey by filling in the questionnaires at the beginning and at the end of the classes. The studies were conducted in summer. Test no. 1 was performed in a lecture hall with a net area of 93.13 m^2 on 20 June 2016 between 13:35 and 14:20. Test no. 2 was performed in the lecture hall with a net area of 157.03 m^2 located on the western side on 28 June 2016 between 10:00 and 11:20. Test no. 3 was performed in the lecture hall located on the eastern side, of 53.14 m^2 net area on 4 July 2016 between 13:00 and 14:45. Mechanical ventilation and air conditioning were on during the tests.

3 Test results and comparison to existing models

3.1 Test results

The indoor parameters are shown in Table 1. All the occupants were assumed to have metabolic rate at 1.2 met (70 W/m²). The questionnaire results indicated that thermal insulation of clothing in all the respondents was I_{cl} = 0.50 clo (0.080 m^2KW). Test no. 1 showed carbon dioxide concentration twice as high as that recorded during tests no. 2 and no. 3. This result is associated with different numbers of occupants present during the tests. Sixty people were present in the room during the Test no. 1, with 25 people occupying the room in Test no. 2 and 16 people in Test no. 3. Indoor air temperatures at the beginning of the tests were higher than those at the end of the tests. During the class the air conditioner was switched on, which lowered the air temperature. In the break between the classes, the air conditioner was switched off, allowing the temperature to rise. The value of the air temperature at the beginning of the class was recorded prior to switching on the air conditioner.

Table 1. Summary of microclimate parameters in three university classrooms.

At the beginning of class/ At the end of class	Outdoor temperature [°C]	Indoor temperature [°C]	Indoor temperature relative humidity [%]	Indoor air flow rate [m/s]	Indoor carbon dioxide concentration [ppm]
TEST NO 1 (Classroom 1)					
At the beginning of class	23.0	25.9	65.8	0.15	1194
At the end of class	23.0	25.1	56.7	0.15	2128
TEST NO 2 (Classroom 2)					
At the beginning of class	22.0	29.2	51.3	0.12	477
At the end of class	22.0	28.3	49.9	0.08	678
TEST NO 3 (Classroom 3)					
At the beginning of class	23.0	29.6	39.2	0.09	479
At the end of class	23.0	28.4	43.2	0.12	731

Figure 1 shows actual mean thermal sensations of the students at the beginning and end of the classes. Assessments of their thermal sensations varied with the time of assessment.

In test 1, at the air temperature above 25°C thermal sensations decreased at the end of the class, whereas in tests 2 and 3, at 28-29°C, the sensations increased.

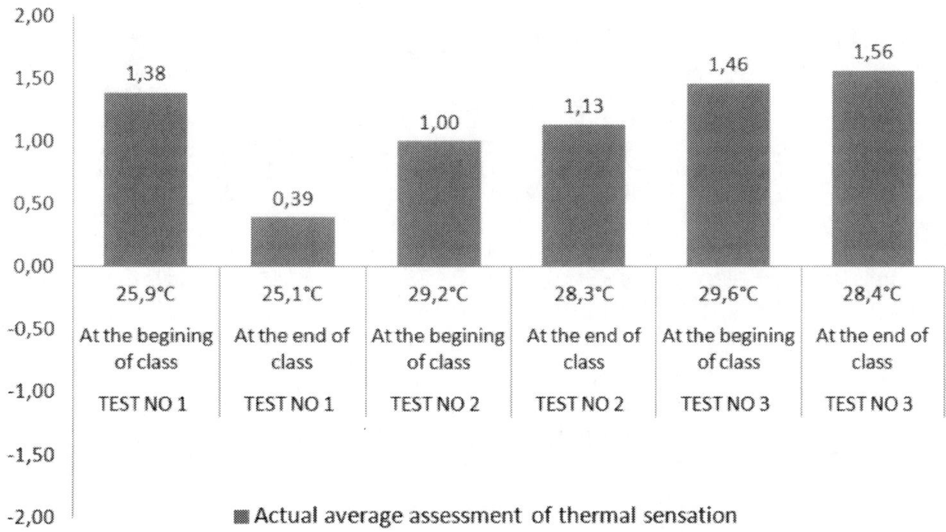

Fig. 1. Real mean assessment of thermal sensations of respondents at the beginning and at the end of class.

Figure 2 shows the change in the percentage of dissatisfied students at the end of class compared with their sensations at the beginning of class. As in chart 1, at air temperature above 25°C, the percentage of students dissatisfied at the end of class decreased. In tests 2, and 3, at 28-29°C the percentage of dissatisfied students increased relative to the beginning of class.

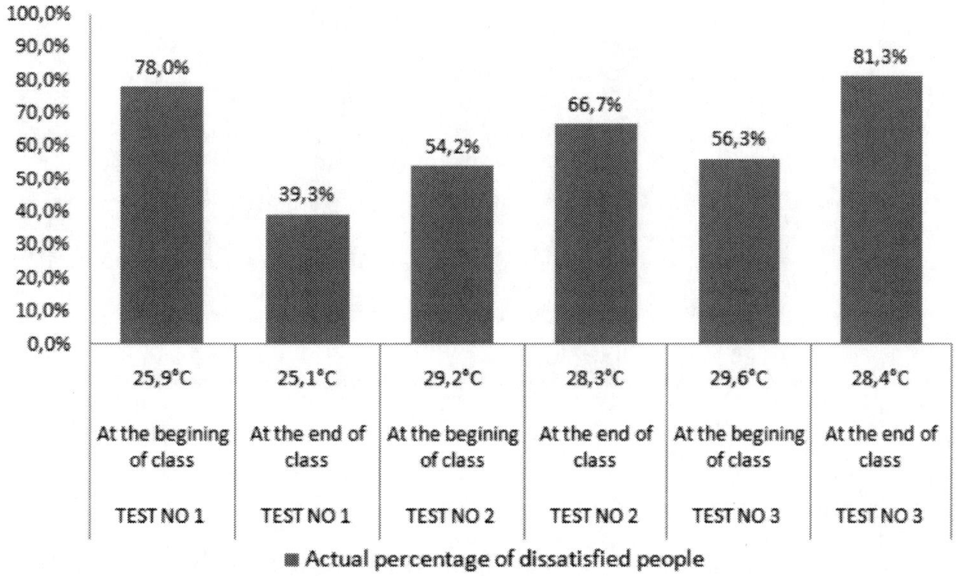

Fig. 2. Real percentage of respondents dissatisfied with their thermal sensations at the beginning and end of class.

Figure 3 compares mean sensations of female and male students. In most cases, the female students were more sensitive to temperature than men. The reverse situation occurred in one case only.

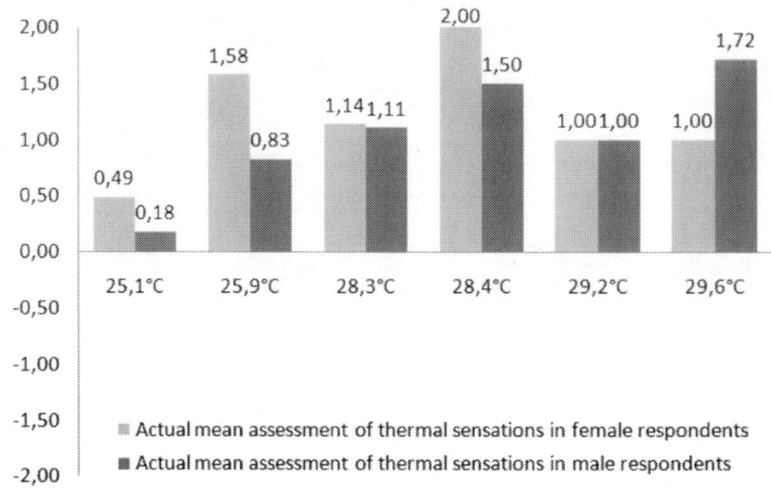

Fig. 3. Actual sensations of female respondents versus those of male respondents.

Figure 4 compares the percentage of dissatisfied female and male students. In the majority of cases, the female students were more dissatisfied with thermal conditions in the rooms.

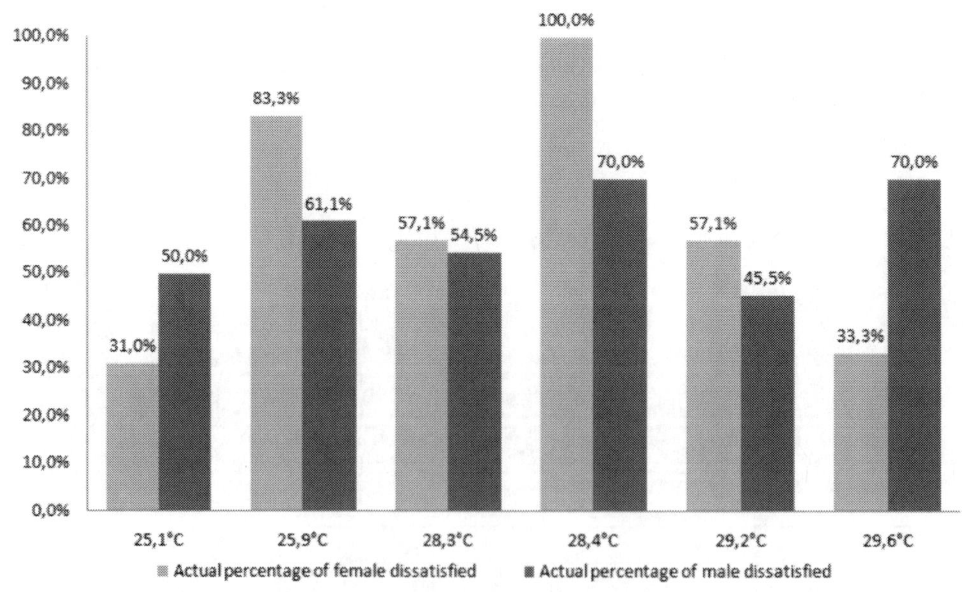

Fig. 4. Percentage of dissatisfied female respondents versus dissatisfied male respondents.

3.2 Comparison of the results to available models

Figure 5 compares the PMV indicator according to ASHRAE-55-2013 and EN-15251 with the actual mean assessment of thermal sensations from the study. It follows from the figure that in most cases the actual mean assessment of thermal sensations is markedly higher than that obtained from the calculation of PMV in compliance with the two standards. Only in one case the assessment of thermal sensations was lower than the PMV.

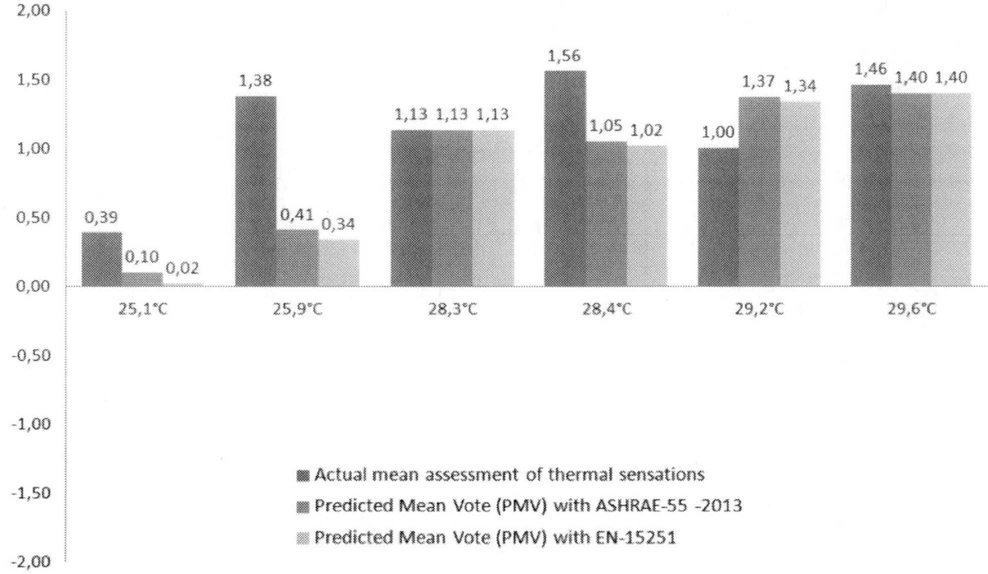

Fig. 5. The PMV depending on ambient temperature according to ASHRAE-55-2013 and EN-15251 versus actual mean assessment of thermal sensations.

Figure 6 shows a comparison of the PPD according to ASHRAE-55-2013 and EN-15251 with the actual percentage of dissatisfied people found during the study. A markedly higher actual percentage of dissatisfied people was found than that shown by the PPD indicator calculated according to the American and European standards.

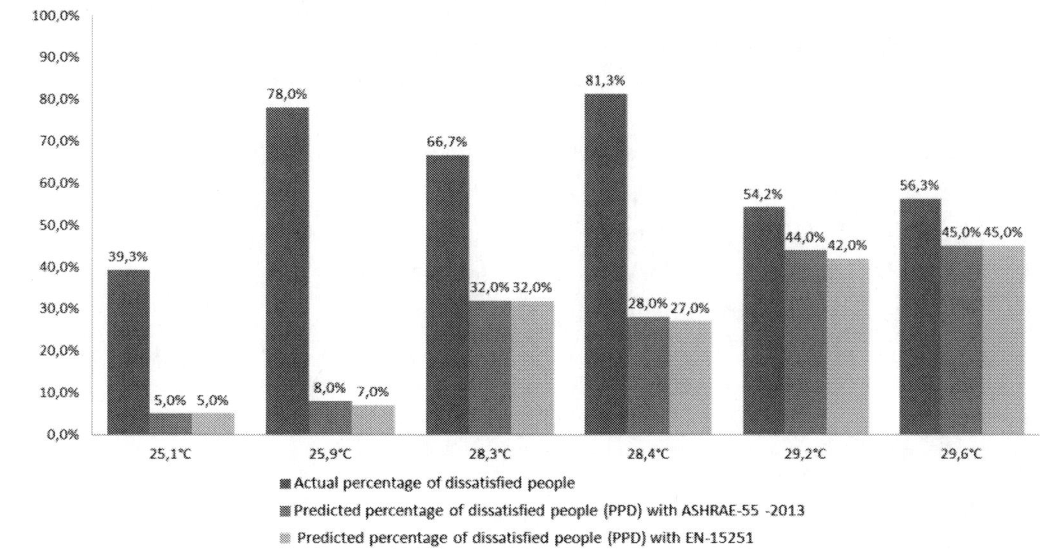

Fig. 6. The PPD depending on ambient temperature according to ASHRAE-55-2013 and EN-15251 versus the actual percentage of dissatisfied respondents.

4 Conclusions

The study indicates that most occupants of the classrooms (39-81%) are dissatisfied with their thermal sensations. According to the results obtained, female respondent were more than male respondents dissatisfied with the thermal conditions in the classrooms. The experimental results were compared to the PMV and PPD indices commonly used in compliance with ASHRAE-55-2013 and EN-15251. Substantial differences were observed between the thermal comfort indices and the actual assessment derived from the people present in the classrooms. Despite technological advancement of intelligent buildings, a high percentage of people was dissatisfied with the thermal conditions in the building. In view of the above, to provide optimal thermal comfort conditions, it seems expedient to modify the thermal sensation indices in buildings and to develop operating guidelines for the control of heating, ventilation and air conditioning systems.

References

1. T. Bedford, Ind. Heal. Res. Board, **76** (1936)
2. C.-E. A. Winslow, L. P. Herrington, A. P. Gagge, Pierce J. B., Laboratory of Hygiene, New Haven, **16,** 288–299 (1937)
3. C. P. Yaglou, D. Minard, Am. Med. Ass. Arch. Ind. Hlt., **16,** 302–316 (1957)
4. P. O. Fanger, ASHRAE Trans, **73,** 1–4 (1967)
5. P. O. Fanger, *Thermal comfort-analysis and applications in environmental engineering,* (Danish Technical Press, Copenhagen 1970)
6. EN ISO 7730, 2005. Ergonomics of the thermal environment - Analytical determination and interpretation of thermal comfort using calculation of the PMV and PPD indices and local thermal comfort criteria, Inter. Stand. Org. (2005)
7. PN-EN 15251, 2007. Indoor Environmental input parameters for design and assessment of energy performance of buildings addressing indoor air quality, thermal environment, lighting and acoustics, Inter. Stand. Org. (2007)
8. ASHRAE Standard 55-2013 Thermal Environmental Conditions for Human Occupancy (2013)
9. P. Ricciardi, C. Buratti, Energ. and Build. **99,** 243–252 (2015)
10. K. Fabbri, Build. and Envir. **68,** 202–214 (2013)
11. M. A. Nico, S. Liuzzi, P. Stefanizzi, Appl. Ergon., **48,** 111–120 (2015)
12. B. Cao, M. Luo, M. Li, Y. Zhu, Energ. and Build. **133,** 469–477 (2016)
13. A. S. H. Abdallach, Proc. Engin. **121,** 1674–1681 (2015)
14. A. O'Donovan, P. D. O'Sullivan, M. D. Murphy, Energ. and Biuld. **135,** 312–323 (2017)
15. M. C. Katafygiotou, D. K. Serghides, Sustain. Cit. and Soc., **13,** 303–312 (2014)

E3S Web of Conferences 17, 00057 (2017) DOI: 10.1051/e3sconf/20171700057
EKO-DOK 2017

Contamination of water in Oliwski Stream after the flood in 2016

Karolina Matej-Łukowicz[1*], and *Ewa* Wojciechowska[1]

[1]Department of Sanitary Engineering, Faculty of Civil and Environmental Engineering, Gdansk University of Technology, Narutowicza 11/12, 80-233 Gdańsk, Poland

Abstract. In the article pollution of stream waters with surface runoff from an urbanized area caused by an extremely high rainfall is discussed. The analyzes were carried out after the rainfall of the depth 152 mm which took place in Gdańsk on 14th and 15th July 2016. This extreme rainfall caused urban flooding, damage of several retention ponds and pollution of surface waters. In the article the results of physical and chemical analyzes of the water samples from Oliwski Stream, inflowing to the Gulf of Gdańsk at the beach in Jelitkowo, are presented. The samples were collected at six points along the Stream in order to evaluate potential pollution sources. The results of the study indicated elevated concentrations of phosphorus compounds and nitrates (V). Additionally, the concentrations of total suspended solids (TSS), solids granulometry and grain size distribution along the stream was investigated.

1 Introduction

Both quality and quantity of storm water runoff is varying in a wide range, depending on factors associated with the rain event, catchment characteristics and development etc. Every year is different in terms of annual precipitation, monthly precipitation or number of rainy days. The usual fluctuations of rainfall depth and intensity are becoming even more variable due to climate change. At the same time, due to intensive urbanization and the increase of paved areas, the runoff volume has grown rapidly over the last few decades [1]. The existing urban drainage systems are no longer capable of safe collection of the increased amount of runoff, quite often resulting in urban flooding. The phenomenon of rapid urban flood has been repeated so often in different locations of the biggest world cities that a special term *flash flood* (FF) has been created to name it [2].

In addition to the increased amount of run-off, another hot environmental problem is runoff quality, often being the cause of pollution of receiving surface waters. Storm water is delivered to the surface waters by several pathways, including direct precipitation, infiltration followed by groundwater outflow, urban storm water sewers discharge and surface runoff. According to [1] the atmospheric pollution accounts for only 10-20% of storm water contamination, while the actual pollution is created when the contaminants from the catchment are washed out by the runoff [3–5].

Total suspended solids (TSS) is considered to be the major pollutant of storm water runoff [6], not only due to its large contents in the runoff, but also due to the fact that it carries a lot of adsorbed pollutants to mention just heavy metals [7,8], total nitrogen, COD and BOD_5. The loads of pollutants discharged to surface receivers with the runoff depend to

* Corresponding author: karolina.matej@gmail.com

some extent on atmospheric pollution but mainly on characteristic of watershed, including the intensity of road transportation, vicinity of industry and type of land usage.

The flooding events are known to have significant impact on aquatic ecosystems due to major changes of physical, chemical and microbiological properties of surface waters. This is especially true for urban flooding events when many anthropogenic substances, toxic to biota, are washed away from the urbanized area and discharged to the surface waters. The analyzes of pollutants concentrations (TSS, COD, BOD_5 and dissolved organic carbon DOC) were carried out by Hidehikoi et al [9]. These authors examined the pollutants concentrations in the vertical profile, concluding that there was no statistically significant relationship between depth and pollutant concentrations.

The solution to the problem of too high amounts of impurities could be wetland systems. The first systems were already built in Gdansk, in ZOO (Catchment Rynarzewski Stream, which flows into the Oliwski Stream) [10].

In this article the concentrations of pollutants in Oliwski Stream in Gdańsk after the urban flash flood event on 14th-15th July 2016 are presented. Oliwski Stream inflows directly to the Gulf of Gdańsk in Jelitkowo, where one of the most attractive Tricity beaches and bathing places is located. Thus, pollution of the stream has a direct impact on coastal waters. Moreover, the loads of nutrients discharged by streams and rivers contribute to eutrophication of the Baltic Sea, which is the high priority environmental issue.

2 Materials and Methods

2.1 Study site

Stormwater runoff from the city of Gdańsk is discharged to more than 20 streams. The largest of them, regarding flow rate (0.29 m^3/s) and the second longest (9.57 km) is Oliwski Stream. The average hydraulic slope of the stream is 14.1‰ [11]. The stream takes its origin near Złota Karczma and its catchment area is approximately 3050 ha [12]. The major inflows are Graniczny Channel, Rynarzewski Stream, Czysta Woda Stream, Bernard's Stream and Zajączkowski Stream. The Oliwski Stream receives runoff from diverse catchment including forests, streets, parks, residential districts as well as the city zoo in Gdańsk, which is located directly at Rynarzewski Stream.

In the catchment area of the Oliwski Stream there are thirteen retention ponds [12] of the total retention volume equal to 70 000 m^3 and the total area of 13.5 ha. Apart of retention, they also have recreational function.

2.2 Collecting of samples

The analyzes of water of Oliwski Stream were carried out after the extremely intensive rainfall episode in July 2016. The samples of water were taken from the mainstream using a scoop and poured into 1 liter glass bottles.

The sampling points location was selected after analysis of catchment characteristics and development (Fig. 1). Six representative points were chosen:

1. The Valley of Joy – sampling point located in the forest with limited access of mechanical vehicles. Potential pollution sources at the distance from the stream source: the motorway S6, run-off from forest area.
2. The Valley of Joy – sampling point located in the forest with limited access of mechanical vehicles. Potential pollution sources between sampling points 1 and 2: a single residential building, the motorway S6, run-off from forest area.

 DOI: 10.1051/e3sconf/20171700057

3. Below retention pond no. 11 – crossroads of Bytowska and Kwietna streets, below the bridge. Potential pollution sources between sampling points 2 and 3: allotments, fishery ponds.
4. Before retention pond no. 8 near Spacerowa street. Potential pollution sources between sampling points 3 and 4: roads, residential buildings, zoo.
5. Residential District Przymorze Małe, Rzepicha Str. Potential pollution sources between sampling points 4 and 5: roads, residential buildings, animal faeces, city park in Oliwa.
6. The discharge in Jelitkowo, Jantarowa Str. Potential pollution sources between sampling points 5 and 6: roads, residential buildings.

The samples from each point were taken in triplicate.

The samples of water were immediately transported to the laboratory in cooling conditions were physical and chemical analyzes of pollutants concentrations were performed.

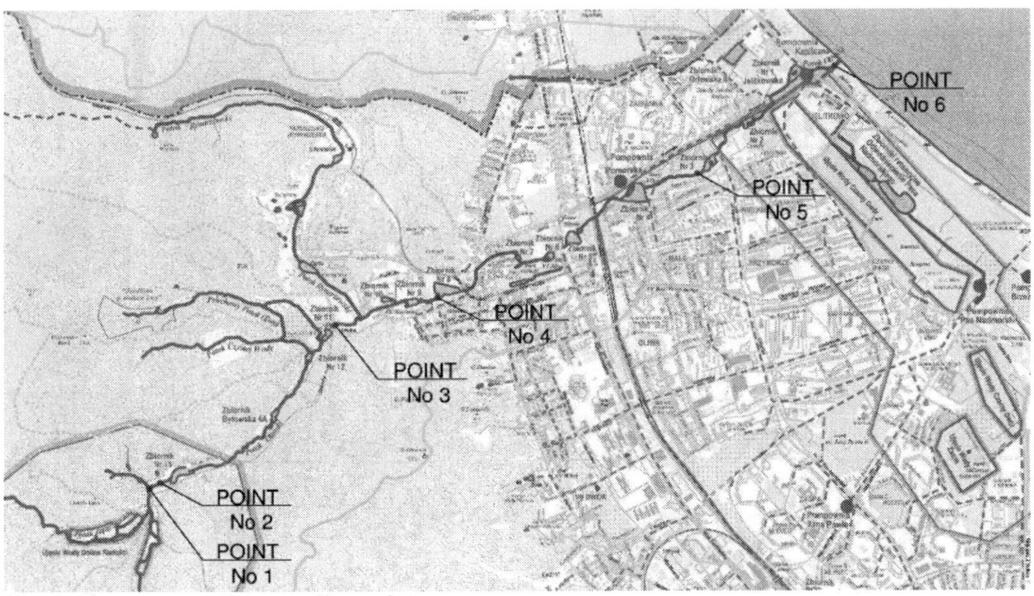

Fig. 1. Six representative sampling points.

2.3 Analytical methods

In the collected samples of water from Oliwski Stream the concentrations of the following pollutants were measured: TSS, COD, ammonia NH_4^+, NO_2^-, NO_3^-, PO_4^{3-} and total phosphorus (Ptot). The chemical analyses were carried out using cuvette tests LCK Hach and spectrophotometer VIS DR3900 Hach, according to the following standards:

NH_4^+ - ISO 7150-1, DIN 38406 E5-1,
NO_2^- - EN ISO 26777, DIN 38405 D10,
NO_3^- - ISO 7890-1-2-1986, DIN 38405 D9-2,
PO_4^{3-} - ISO 6878-1-1986, DIN 38405 D11-4,
Ptot - ISO 6878-1-1986, DIN 38405 D11-4),
COD - ISO 6060-1989, DIN 38409-H41-H44)
TSS concentration was measured according to PN-EN 872:2007.

Granulometric composition of suspended solids was measured using laser granulometer produced by Malvern Instruments Ltd 2000 allowing for volumetric analysis of solids size during diffraction of laser beam. The instrument allowed for measurement of solids size in the range from 0.02 to 2000 μm. The volume of analyzed samples was equal to 0.6 dm^3.

3 Results and Discussion

3.1 Hydrological characteristics of the catchment area

The average precipitation in Oliwa in the period 2007-2014 was equal to: annual – 670.8 mm, monthly (July) – 91.0 mm. The rainfall depth on 14th July 2016 was equal to 152 mm (Fig. 1), being two times higher than in the whole July 2013. The annual and monthly (July) precipitation in Oliwa in the years 2007-2014 are presented in Table 1.

Table 1. July and annual precipitation [mm] in the years 2007-2014 measured at meteo station located in Gdańsk Oliwa [12].

Date	2007	2008	2009	2010	2011	2012	2013	2014	average
July	92.4	65.2	124.4	119.7	77.6	148.5	77.5	22.7	91.0
annual	771.6	669.9	659.8	863.2	445.0	826.9	647.7	482.2	670.8

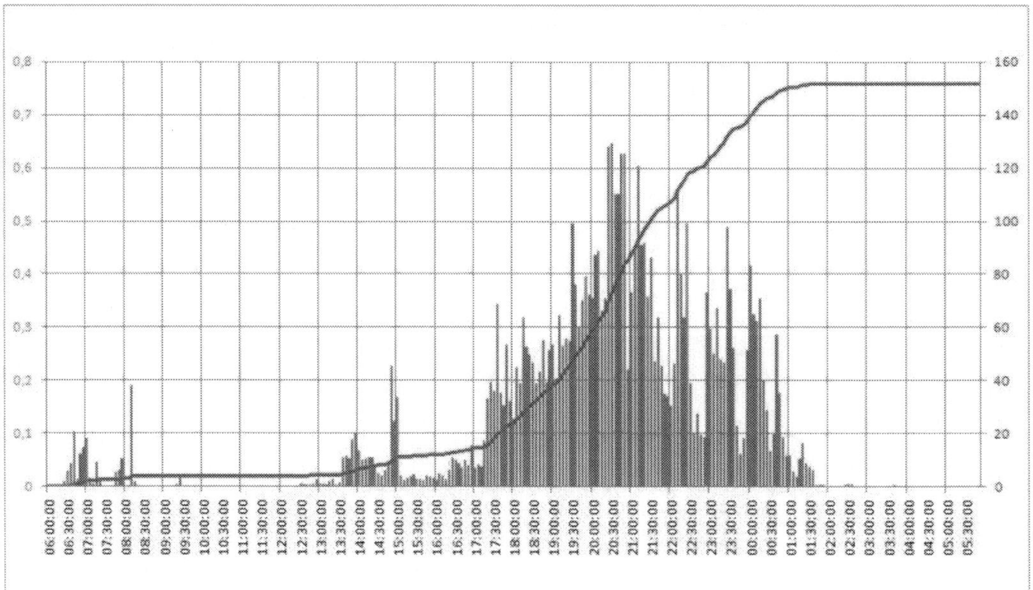

Fig. 2. Rainfall intensity [mm/min] and mass curve of rainfall in Oliwa between 6:00 14.07 and 6:00 15.07 [13].

Due to intensive extreme rainfall and large surface runoff discharged to receiving surface waters numerous locations in the city of Gdańsk were flooded. The catchment area of Oliwski Stream was one of most affected. Water of the Stream overflowed several retention ponds, flooding residential areas and the zoo. At one of the retention ponds the floodgate was damaged, which caused further flooding of the nearby area.

3.2 Concentrations of nutrients and organic matter

Results of chemical analyzes of collected samples of water are presented in Table 2. The analyzes of nutrients concentrations included ammonia, nitrates (III), nitrates (V), orthophosphates and total phosphorus. Ammonia originates during biochemical decay of organic matter, It is also released with domestic wastewater. High concentrations of ammonia in surface water indicate contamination with sanitary sewage or animal faeces.

Nitrates (III) are formed during decay of nitrogen present in organic matter or reduction of nitrates (V) in anaerobic conditions. Concentrations of nitrates (III) in surface waters are usually low due to oxidation of NO_2^- to NO_3^-. Presence of nitrates (III) indicate contamination with fish faeces or animal decaying remains. Since nitrates (III) are transitory form between NH_4^+ and NO_3^-, they should be determined together with nitrates (V). Nitrates (V) are the final product of nitrogen transformation, thus their presence in surface waters at low concentrations are quite typical. Additional sources of nitrates (V) are municipal and industrial wastewater as well as artificial fertilizers washed away by surface runoff. Nitrates (V) present at higher concentrations are responsible for eutrophication of surface waters. Phosphorus compounds undergo biochemical changes in the environment as nitrogen species and the final form are orthophosphates. Orthophosphates in surface waters originate from sanitary sewage, surface runoff and fish breeding. Otherwise than nitrogen, phosphorus mobility is low, thus large quantities are deposited in the sediments.

Organic matter in the surface water is formed during natural decay processes of animal and plant detritus. Allochtonic organic matter is delivered with wastewater inflow. Organics in higher concentrations disturb self-cleaning process and cause deterioration of organoleptic properties of water. Within this study COD was measured as an indicator of organic matter concentration.

Table 2. Concentrations of organic matter and nutrients in Oliwski Stream after the extreme rainfall on 14-15.07.2016.

Analyzed parameter		**POINT 1**	**POINT 2**	**POINT 3**	**POINT 4**	**POINT 5**	**POINT 6**	**Rainless period (average)**
NITRATES III (NO_2^-)	[mg/dm^3]	0.04	0.04	0.08	0.09	0.11	0.07	0.02
NITRATES V (NO_3^-)	[mg/dm^3]	1.61	1.83	1.75	2.18	2.29	1.37	0.57
AMMONIA (NH_4^+)	[mg/dm^3]	0.19	0.50	0.20	0.18	0.13	0.14	0.05
ORTOPHOSPHATES(PO_4^{3-})	[mg/dm^3]	0.42	2.25	0.68	2.16	0.88	0.55	0.45
COD	[mg/dm^3]	23.00	24.30	30.00	30.90	31.70	22.50	8.68
TOTAL PHOSPHORUS P_{TOT}	[mg/dm^3]	0.63	0.27	0.79	0.24	0.18	0.15	0.08

The concentrations presented in Table 2 were compared with the values given in Regulation of Environment Minister from 21st July 2016 regarding classification of surface waters and environmental quality standards for priority pollutants. Concentration of ammonia at point 2 and 3 exceeded the boundary value for I class. Concentrations of nitrates (III) corresponded to the II class at all sampling points. Nitrates (V) concentrations at sampling points 1,2,3 and 6 was within the limits for II class, while at points 4 and 5 exceeded the boundary value for II class. During floods sanitary sewage are often discharged to surface waters (for instance from pumping stations) which may be the reason of deterioration of the stream quality. Concentrations of orthophosphates exceeded the values for II class. Total phosphorus was out-of-class at point 1 and 3, while at sampling points 2 and 4 it corresponded to the II class limits. At sampling points 5 and 6 the I class limits for total phosphorus concentrations were fullfilled. The significant increase of orthophosphates concentration at sampling point 4 was caused by flooding of the city zoo.

Another reason for the increase of total phosphorus concentration could be resuspension of phosphorus deposited in the bottom sediments during turbulent flow of water.

Concentrations of COD at sampling points 1, 2 and 6 corresponded to the I class limits, while at sampling points 3, 4 and 5 the limits for II class were exceeded. This was obviously caused by flooding of animal cages at the zoo as well as damage of retention tank and the stream bed below it which causes additional inflow of organic matter from the catchment area.

3.3 Concentrations and granulometric analyzes of suspended solids

Total suspended solids concentrations, measured as a weight of solids in the volume of analyzed sample, is a standard determination bringing just quantitative information. Qualitative analysis of solids can bring more information about solids granulometry that can be useful for instance in designing of storm water treatment devices and has already been performed for some areas in Poland [14, 15].

Fig. 4 presents concentrations of TSS in water samples from Oliwski Stream measured according to PN-EN 872:2007. Fig. 4 presents characteristic diameters of solids: d 0.1 (solids of this diameter and smaller represent 10% of the total solids number), d 0.5 and d 0.9 (solids of this diameter and smaller represent 50% and 90% of the sample, respectively). Fig. 6 presents granulometric spectrum of suspended solids, measured with laser granulometer.

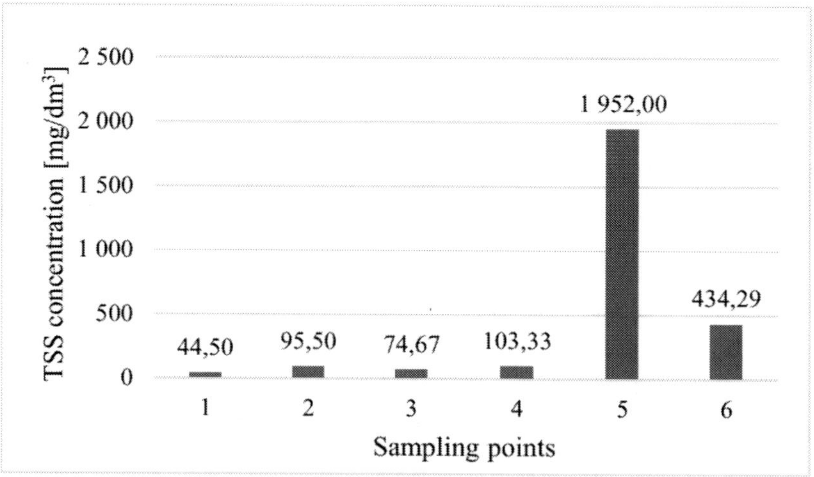

Fig. 3. TSS concentration in sampling points 1 – 6, 15.07.2016.

TSS concentrations at sampling points 1-3 were below the limits defined by Regulation of Environment Minister [16]. At sampling points 4-6 the increased surface runoff and, most of all, damage of the overflow of one of retention tanks, resulted in the increase of TSS concentrations, with the maximum at sampling point 5 (1 952 mg/dm^3 = 1,95 g/L). Similarly, in sampling points 1-3 the bigger solids were dominating (Figs. 4 and 5), while at sampling points 4-6 size of the solids was variable. Granulometric analyzes (Fig.6) shows that the sizes of solids were decreasing downstream, which is opposite to the situation during dry weather, when solids sizes are increasing in sampling points downstream.

DOI: 10.1051/e3sconf/20171700057

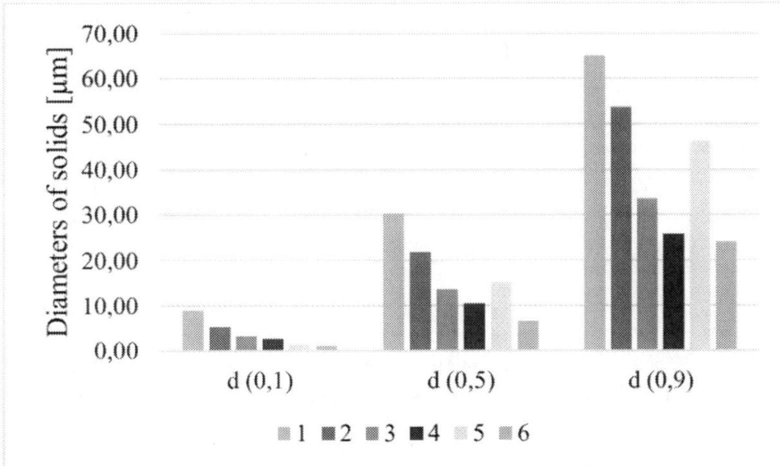

Fig. 4. Characteristic diameters of solids in sampling points 1 – 6, 15.07.2016.

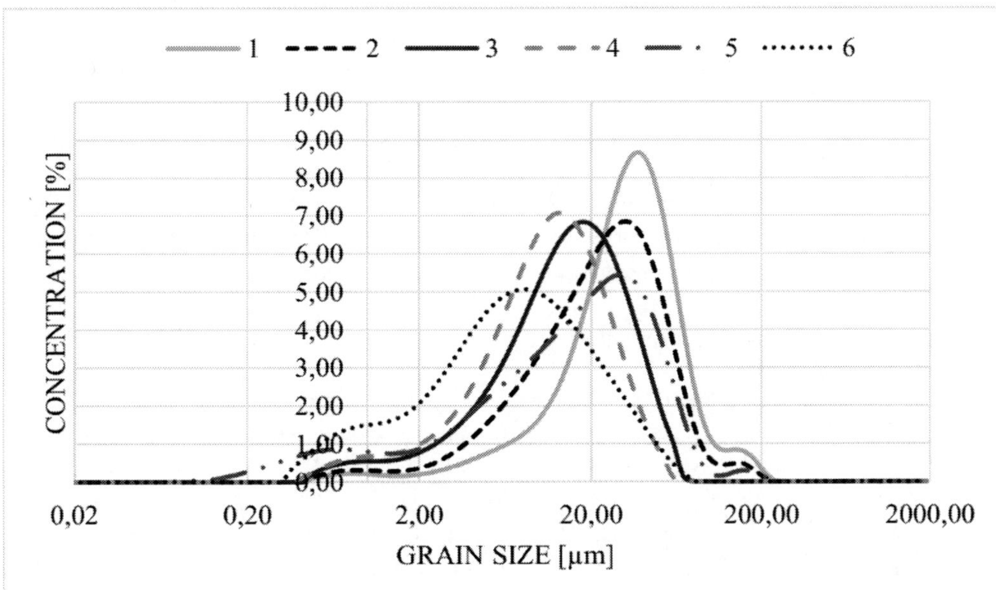

Fig. 5. Granulometric spectrum of suspended solids in sampling points 1 – 6, 15.07.2016.

4 Conclusions

1. Urban storm water system cannot safely drain the runoff produced by rainfall of the depth of 152 mm during 16 hours, which results in urban flood.
2. As a result of urban flood on 15 July 2016 the waters of Oliwski Stream got polluted and could be defined as out-of-class since some of the analyzed parameters exceeded the limits of II class water.
3. Pollution of Oliwski Stream resulted mostly from surface runoff from the zoo area, resuspension of bottom sediments and damages of retention ponds.
4. The decrease of concentrations at subsequent sampling point downstream can result from retention of polluted runoff in the retention tanks and dilution.

References

1. E. Wojciechowska, M. Gajewska, N. Żurkowska, M. Surówka, H. Obarska-Pempkowiak *Zrównoważone systemy gospodarowania wodą opadową*, Wyd. Politechniki Gdańskiej, 147 ss. (2015)
2. A. Magnuszewski, *Procesy korytowe rzek nizinnych a bezpieczeństwo powodziowe*. Warszawa: Uniwersytet Warszawski, Wydział Geografii i Studiów Regionalnych (2013)
3. M. Gajewska, A. Wargin, *Identyfikacja zanieczyszczeń potencjalnie występujących w ściekach opadowych*, http://www.innowrota.pl/sites/default/files/images/A.Wargin_M.Gajewska_1.pdf, (2010)
4. E. Bandowska, D. Bandzierz, *The possibilities of local stormwater management in the context of its quality and quantity*, Proceedings of ECOpole **8** (2014)
5. E. Ociepa, A. Kiesiel, J. Lach, *Zanieczyszczenia wód opadowych spływających do systemów kanalizacyjnych*, Proceedings of ECOpole 2010, Vol. **4**, No **2**, 465–469 (2010)
6. A. S. Sánchez, E. Cohim, R.A. Kalid, *A review on physicochemical and microbiological contamination of roof-harvested rainwater in urban areas*, Sustainability of Water Quality and Ecology, Vol. **6**, Pages 119–137 (September 2015)
7. K. Barbusiński, W. Nocoń, K. Nocoń, J. Kernert, *Rola zawiesin w transporcie metali ciężkich w wodach powierzchniowych na przykładzie Kłodnicy*, Ochrona środowiska **2**, vol. **34**, 33–38 (2012)
8. M. Zawilski, G. Sakson, *Ocena emisji zawiesin odprowadzanych kanalizacją deszczową z terenów zurbanizowanych*, Ochrona środowiska **2**, vol. 35, 33–40 (2013)
9. H. Hidehiko, T. Masaharu, U. Noburu, M. Masaru, *Water quality and pollution load during flood and non-flood periods in an urban tidal river*, Journal of Japan Society of Civil Engineers, Ser. B1 (Hydraulic Engineering) Vol. **69** No. 4 p. I_1723-I_1728 (2013)
10. H. Obarska-Pempkowiak, M. Gajewska, E. Wojciechowska, M. Stosik. Constructed wetland systems for aerial runoff treatment in the Gulf of Gdansk region. Rocznik Ochrona Środowiska, Tom **13**, 173-185 (2011)
11. http://www.gedanopedia.pl/?title=POTOK_OLIWSKI
12. http://www.gdmel.pl/
13. http://www.klimat.ug.edu.pl/?page_id=5092
14. E. Burszta-Adamiak, M. Kęszycka, J. Szwed, *Ocena przydatności granulometru laserowego do analizy zawiesin zanieczyszczających wody opadowe odprowadzane z terenów zurbanizowanych*, Woda-Środowisko-Obszary Wiejskie, t. **7**. (2007)
15. E. Burszta-Adamiak, R. Stodolak, *Ocena składu granulometrycznego zawiesiny w mokrym opadzie atmosferycznym na tle jego właściwości fizykochemicznych*, Woda-Środowisko-Obszary Wiejskie, t.**7**. (2007)
16. Rozporządzenie Ministra Środowiska z dnia 18 listopada 2014 r. w sprawie warunków, jakie należy spełnić przy wprowadzaniu ścieków do wód lub do ziemi, oraz w sprawie substancji szczególnie szkodliwych dla środowiska wodnego, Dz.U. 2014, poz. 1800 (2014)

E3S Web of Conferences **17**, 00058 (2017) DOI: 10.1051/e3sconf/20171700058
EKO-DOK 2017

The analysis of unit costs of preparing hot water for various sources of heat

Aleksandra Mazur[1,*], and *Daniel* Słyś[1]

[1]Rzeszów University of Technology, Department of Infrastructure and Sustainable Development, al. Powstańców Warszawy 12, 35-959 Rzeszów, Poland

Abstract. The article presents an analysis of energy consumption and the cost of its production by the systems of supply of hot water depending on the applied heat sources and destination of a building. The calculations have taken into account the efficiency of the adopted installation solutions that have a direct impact on energy consumption. The following parameters have been determined: the size of the seasonal demand for hot water, an annual amount of fuel needed to its production and annual operating costs. The calculations have been performed for residential buildings and public buildings, whilst maintaining the same usable space. The aim of the article is to determine the rate of the amount of fuel, needed to heat to the desired temperature, a unit volume of water depending on the type of heat source and destination of a premises. For the adopted variants of installation, the ratio of production costs of unit volume of hot water has been also determined. The indicators of the fuel amount and the cost of hot water pointed out in the article may be one of the criteria for selection of an effective system for the preparation of hot water.

1 Introduction

Due to rising fuel prices, as well as constantly decreasing natural resources, it is essential to manage our deposits consciously and responsibly, as well as use them effectively. For this reason, the choice of a proper heating system is extremely important, including a system for the preparation of hot water. Today, there is a significant improvement in the thermal insulation of construction bulkheads, resulting in a decrease in energy demand for central heating purposes, and consequently there is an increase in the system for the preparation of hot water. The selection of an appropriate system to supply a building with hot water is not an easy decision and it should be based on a multi-criteria analysis [1]. There are certain attempts that aim at reducing water consumption, as well as reducing the energy required to heat the hot water through recycling greywater and using heat exchangers DWHR [2, 3]. A reasonable idea seems to be the replacement of inefficient heat sources, with those of higher efficiency; or replacement of the conventional energy carriers with their alternatives. Hence, numerous analyses of installations are being performed, in which renewable energy sources, such as solar collectors, photovoltaic panels, or heat pumps are used [4–7].

* Corresponding author: a.mazur@prz.edu.pl

The system for the supply of hot water for a building should be matched to the characteristics of water consumption by users. The size of the daily consumption of hot water depends on many factors, such as: sanitary facilities; time of day and year, and the number of users and their habits. Most of these factors are closely connected with the purpose of the building.

The aim of the article is to analyse the energy consumption and the cost of its production in the systems which supply hot water, depending on the applied heat source and the purpose of the building. The scope of the article covers the calculation of the seasonal demand for hot water, the efficiency of the analysed system, the annual amount of fuel necessary for the production of hot water, and the annual operating costs. These values have been calculated with the use of the ArCADia-TERMO 6.5 computer software. The indicator of the amount of fuel needed to heat a unit volume of hot water to the desired temperature, as well as the rate of production costs per unit volume of hot water, depending on the type of heat source and the premise's functions, have been determined.

2 Installation variants

For the analysis, six variants of systems which supply a building with hot water have been adopted:

- variant 1: capacitive electric heater,
- variant 2: coal-fired boiler for coal with a storage tank,
- variant 3: biomass boiler with a storage tank,
- variant 4: condensing gas boiler with a storage tank,
- variant 5: oil boiler with a storage tank,
- variant 6: heat pump air / water with a storage tank.

Calculations have been carried out for residential buildings and public buildings such as: an office building, an apartment building, a restaurant, a shopping centre, a hotel, a warehouse, a school, a hospital, and a sports hall.

3 Analysis of the results

The legal basis for the calculation is the Ministry of Infrastructure and Development regulation, dated 27 February 2015, on the methodology for determining the energy performance of a building, or part of a building, and energy performance certificates [8], according to which the seasonal demand for hot water is calculated using the following formula [8]:

$$Q_{w,nd} = V_{Wi} \cdot A_f \cdot c_w \cdot \rho_w \cdot (\theta_w - \theta_0) \cdot k_R \cdot t_R / 3600 \ [kWh/year] \quad (1)$$

where:

V_{Wi} – unitary, daily demand for hot water [$dm^3/(m^2 \cdot day)$],
A_f – heated floor space [m^2],
c_w – specific heat of water [kJ/(kg·K)],
ρ_w – density of water [kg/dm^3],
θ_w – calculative temperature of the hot water in the draw-off tap [°C],
θ_0 – calculative water temperature before its heating [°C],
k_R – correction factor, due to an interruption in use of hot water. [-],
t_R – number of days in the year [day].

The values of the unitary, daily demand for hot water, and the values for the correction factor have been adopted in accordance with Table 27 of the Ordinance [8]. For all of the analysed buildings, an equal heated area of 1600 m^2 has been assumed. The calculative temperature of hot water has been assumed at the level of 55°C, in accordance with [9], and the cold water at the level of 10°C. A summary, of seasonal calculations for hot water demand $Q_{w,nd}$ are presented in Table 1.

Table 1. Seasonal demand for hot water $Q_{w,nd}$.

Type of building	Correction factor k_R [-]	Unitary, daily demand for hot water V_{Wi} [$dm^3/(m^2 \cdot day)$]	Seasonal demand for hot water $Q_{w,nd}$ [kWh/year]
Office building	0.70	0.35	7494
Apartment building	0.90	1.60	44045
Restaurant	0.80	2.50	61174
Shopping centre	0.78	0.60	14315
Hotel	0.60	3.75	68821
Warehouse	0.70	0.10	2141
School	0.55	0.80	13458
Hospital	1.00	6.50	198816
Sports hall	0.42	0.25	3173

The efficiencies of the calculation systems used in determining the demand for hot water has been set in the study. The total system efficiency is calculated as the quotient of the following elements: the average annual efficiency of the heat production from energy carrier, the average annual efficiency of heat transfer from the heat source to the draw-off taps, the average annual efficiency of the heat accumulation in the capacitive elements, and the average annual efficiency of heat utilization, whose values have been set at 1,0 [8]. The volume of each efficiency adopted on the basis of tables 9, 12 and 14 of the Regulation [8] are summarized in Table 2.

Table 2. Summary of the efficiency of systems which supply hot water.

Variant	Energy carrier	Efficiency of the heat production	Efficiency of the heat transfer	Efficiency of the heat accumulation	Total system efficiency
1	Electric energy	0.96		1.00	0.67
2	Coal	0.83			0.49
3	Beech wood	0.83	0.70		0.49
4	Natural gas	0.85		0.85	0.51
5	Heating oil	0.85			0.51
6	Electric energy	2.60			1.55

The efficiency of the system is a parameter of fundamental importance as it directly affects the amount of energy used to produce hot water. The highest efficiency is observed in the system with a heat pump which uses a renewable source of heat. While taking into account the conventional systems, the highest efficiency is noted in the an electric heater installation. In this case, not only is the efficiency of the equipment high, but also the efficiency of accumulation, due to the absence of additional equipment installed in the hot water tank. The higher the efficiency of the system of supplying the building in hot water, the lower the operating costs are, borne by an investor and the installation is more economical.

The calculative, annual amount of electricity required in the system for preparation of hot, tap water is calculated by the formula (2) and in other energy sources, the formula (3) [8] is used.

$$C_W = Q_{k,W}/A_f \ [kWh/(m^2 \cdot year)] \tag{2}$$

$$C_W = \frac{Q_{k,W} \cdot 3{,}6}{A_f \cdot W_o} \ [reference\ unit/(m^2 \cdot year)\,] \tag{3}$$

where:
$Q_{k,W}$ – annual demand for an end use energy delivered to the building system for the preparation of hot water [kWh/year)],
A_f – heated floor space [m^2],
W_o – calorific value of the fuel [MJ/reference unit].

The total annual amount of fuel is calculated using the formula (4), then the results are given in Table 3.

$$R = C_W \cdot A_f \ [reference\ unit/(year)] \tag{4}$$

Table 3. Summary of the total annual amount of fuel depending on building types.

Type of building	The total annual amount of fuel					
	Variant 1 [kWh/year]	Variant 2 [kg/ year]	Variant 3 [kg/ year]	Variant 4 [m^3/ year]	Variant 5 [l/ year]	Variant 6 [kWh/ year]
Office building	11152	1971	3545	1486	1470	4844
Apartment building	65544	11583	20838	8735	8640	28471
Restaurant	91033	16087	28942	12132	12000	39544
Shopping centre	21302	3764	6772	2839	2808	9253
Hotel	102412	18098	32560	13649	13500	44487
Warehouse	3186	563	1013	425	420	1384
School	20027	3539	6367	2669	2640	8700
Hospital	295865	52284	94061	39429	38999	128517
Sports hall	4722	835	1501	629	623	2051

Estimated annual operating costs, which, excluding the price of the energy source, consist of the cost of the operation of auxiliary equipment, including circulating pumps and the hot water tank charging pumps. Demand for end use energy for the drive of the pumps for the analysed systems for the preparation the hot water is equal to 653 kWh/year, which amounts to 87 EUR annually. Only in the case of the system with an electric heater, the value is lower, since there is no need to mount a charging pump on the tank, a fixed demand for end use electric energy is equal to 467 kWh/year, which is 62,3 EUR annually. The unit cost of heat for specific fuels are assumed on the basis of available data from the utilities suppliers or their price lists, and they are summarized in Table 4. The results of the calculation of annual operating costs are presented in Table 5.

Table 4. The unit cost of heat for specific fuels.

Type of fuel	Electric energy	Coal	Beech wood	Natural gas	Heating oil
Price	0,13 EUR/kWh	216 EUR/t	60,8 EUR/m^3	0,47 EUR/m^3	0,66 EUR/l

Table 5. Annual operating costs.

Type of building	**Annual operating costs [EUR/year]**					
	Variant 1	**Variant 2**	**Variant 3**	**Variant 4**	**Variant 5**	**Variant 6**
Office building	1549	513	617	787	1050	733
Apartment building	8801	2590	3201	4204	5748	3883
Restaurant	12200	3563	4412	5804	7949	5360
Shopping centre	2903	900	1099	1425	1927	1321
Hotel	13717	3998	4952	6519	8932	6019
Warehouse	487	209	238	287	362	272
School	2733	852	1038	1345	1817	1247
Hospital	39511	11385	14142	18669	25638	17223
Sports hall	692	267	311	384	495	361

The calculated indicators for the amount of energy necessary to heat a unit volume of hot water to a desired temperature depending on the type of fuel and the purpose of the building. They are summarized in Table 6.

The determined indicators of production costs per unit volume of hot water of the analysed hot water preparation systems and building types are given in Table 7 and graphically in Figure 1.

Table 8 summarizes the average levels of energy required and the cost of production per unit volume of hot water depending on the heat source.

Analyzing the results of the calculations presented in the article, it is noticeable that the preparation cost of domestic hot, tap water is affected not only by the type of energy source, but also by the function of the premises.

Table 6. Indicators for the amount of energy necessary to heat a unit volume of hot water.

Type of building	The indicator of the amount of fuel needed to heat a unit volume of hot water					
	Variant 1 [kWh/m³]	Variant 2 [kg/m³]	Variant 3 [kg/m³]	Variant 4 [m³/m³]	Variant 5 [l/m³]	Variant 6 [kWh/m³]
Office building	19.91	3.52	6.33	2.65	2.62	8.65
Apartment building	25.60	4.52	8.14	3.41	3.37	11.12
Restaurant	22.76	4.02	7.24	3.03	3.00	9.89
Shopping centre	22.19	3.92	7.05	2.96	2.92	9.64
Hotel	17.07	3.02	5.43	2.27	2.25	7.41
Warehouse	19.91	3.52	6.33	2.65	2.62	8.65
School	15.65	2.76	4.97	2.09	2.06	6.80
Hospital	28.45	5.03	9.04	3.79	3.75	12.36
Sports hall	11.81	2.09	3.75	1.57	1.56	5.13

Table 7. Indicators of production costs per unit volume of hot water.

Type of building	Indicators of production costs per unit volume of hot water [EUR/m³]					
	Variant 1	Variant 2	Variant 3	Variant 4	Variant 5	Variant 6
Office building	2.77	0.92	1.10	1.41	1.88	1.31
Apartment building	3.44	1.01	1.25	1.64	2.25	1.52
Restaurant	3.05	0.89	1.10	1.45	1.99	1.34
Shopping centre	3.02	0.94	1.14	1.48	2.01	1.38
Hotel	2.29	0.67	0.83	1.09	1.49	1.00
Warehouse	3.04	1.30	1.49	1.79	2.26	1.70
School	2.13	0.67	0.81	1.05	1.42	0.97
Hospital	3.80	1.09	1.36	1.80	2.47	1.66
Sports hall	1.73	0.67	0.78	0.96	1.24	0.90

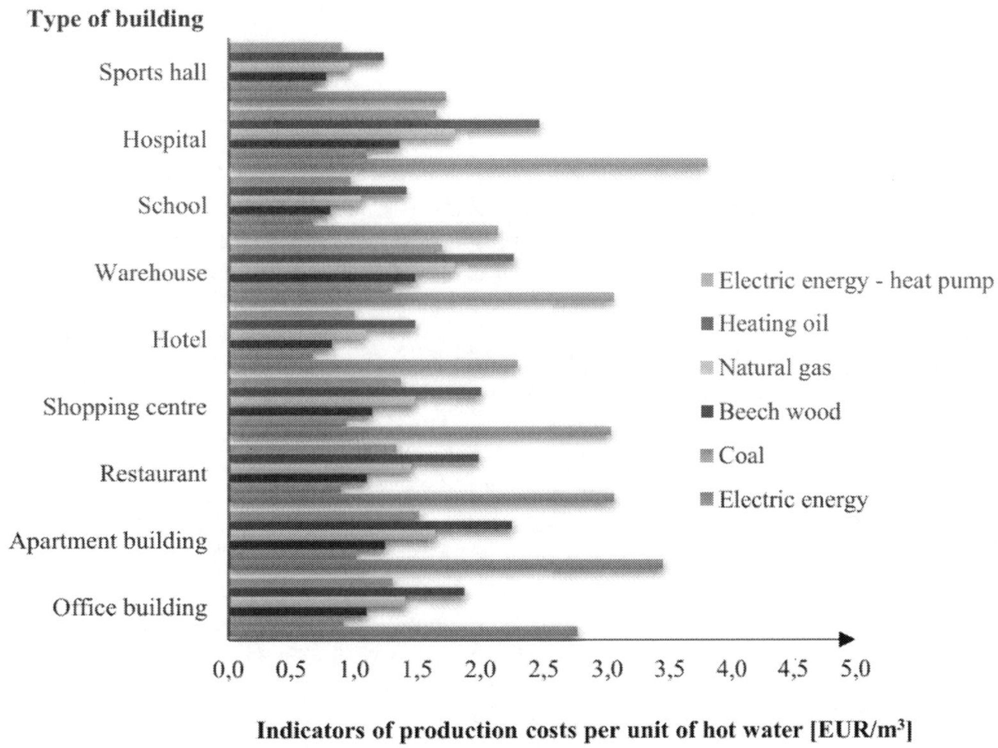

Fig. 1. Indicators of production costs per unit of hot water depending on the type of heat source and destination of a building.

Table 8. Average levels of energy required and the cost of production per unit volume of hot water.

Average levels of energy required to production unit volume of hot water					
Electric energy [kWh/m^3]	Coal [kg/m^3]	Beech wood [kg/m^3]	Natural gas [m^3/m^3]	Heating oil [l/m^3]	Electric energy - heat pump [kWh/m^3]
20,37	3.60	6.48	2.71	2.69	8.85
Average levels of the cost of production per unit volume of hot water [EUR/m^3]					
Electric energy	Coal	Beech wood	Natural gas	Heating oil	Electric energy - heat pump
2,81	0.91	1.10	1.41	1.89	1.31

The cheapest source of heat is a coal-fired boiler, hence its popularity is at a high level, especially in households. This solution does not seem to be optimal, given the user's comfort and the impact on the natural environment. In terms of the economical aspect of a system for the supply of hot water supply, are in second and third place respectively: a biomass boiler (beech wood) and a heat pump (air/water). It turns out that investment in alternative energy sources is not always the most cost-effective option [10]. In the case of heat pumps, it is due to the fact that the transfer of "free" energy from the heat source to the installation involves the consumption of electricity. Clearly, the more expensive source of heat, from the rest, is a boiler using heating oil. The most expensive option is the electric heater.

Due to the characteristics of water partitioning, as well as the specifics of the building, the greatest amount of energy is required for the preparation of hot water in a hospital, and the least in a sports hall.

4 Summary

The selection of the right energy source, which powers the hot, tap water supply system in a building is one of the most important decisions to make for an investor, especially if it relates to large facilities where water consumption is significant. Besides the comfort of the user, or sanitary security, the fundamental role is played by the economical criterion.

A comparative analysis of systems for the preparation of hot, tap water, carried out in the article, showed that the cost of production of one cubic meter of hot water ranges from 0,67 EUR to as much as 3,80 EUR, depending on the energy source and the function of the premises. The cheapest source of heat is a coal-fired boiler, and the most expensive is a capacitive electric heater. For some types of buildings, e.g. a school or a hotel, it is more than three times more expensive than a coal boiler, and more than two times more expensive in comparison with a heat pump.

Every investment must be considered individually, and the selection of a suitable system for hot, tap water should be based on multi-criteria analysis. The indicators of the amount of energy and production costs of the hot water, set out in the article, provide information which is helpful in making the right choice.

References

1. S. Kordana, D.Słyś, JCEEA, **XXXIII**, **63 (2/II/16)**, 225–240 (2016)
2. D.Słyś, S. Kordana, ENERG BUILDINGS, **71**, 1–11 (2014)
3. A. Stec, S. Kordana, RESOUR CONSERV RECY, **105**, 84–94 (2015)
4. C. Naldi, M. Dongellini, G. L. Morini, Energy Procedia, **78**, 1117-1122 (2015)
5. X. Sun, Y. Dai, V. Novakovic, J. Wu, R. Wang, Energy Procedia, **70**, 394-401 (2015)
6. G. Comodi, M. Bevilacqua, F. Caresana, C. Paciarotti, L.Pelagalli, P.Venella, APPL ENERG, **164**, 944–955 (2016)
7. T. Szul, Journal of Research and Applications in Agricultural Engineering, **56(2)**, 161–164 (2011)
8. Journal of Laws of 2015, item 376, Rozporządzenie Ministra Infrastruktury i Rozwoju z dnia 27 lutego 2015 r. *w sprawie metodologii wyznaczania charakterystyki energetycznej budynku lub części budynku oraz świadectw charakterystyki energetycznej* (Regulation of the Minister of Infrastructure and Development of 27 February 2015 *on the methodology for determining the energy performance of a building or part of a building and energy performance certificates*)
9. Journal of Laws of 2002, No. 75, item 690, as amended, Rozporządzenie Ministra Infrastruktury z dnia 12 kwietnia 2002 r. *w sprawie warunków technicznych, jakim powinny odpowiadać budynki i ich usytuowanie* (Regulation of the Minister of Infrastructure of 12 April 2002 *on technical specifications that buildings and their location should comply with*)
10. T. Szul, Journal of Research and Applications in Agricultural Engineering, **58(2)**, 158–162 (2013)

Hydrographic and navigational analysis of Lower Oder shipping conditions – case study of the backflow occurrence

Piotr Medyna[1,*], and *Paulina* Sobkowicz[2]

[1]Instytut Nawigacji Morskiej, Akademia Morska w Szczecinie, Wały Chrobrego 1-2, 70-500 Szczecin, Polska
[2]Centrum Naukowo – Badawcze Analizy Ryzyka Eksploatacji Statków, Akademia Morska w Szczecinie, Wały Chrobrego 1-2, 70-500 Szczecin, Polska

Abstract. One of the significant limitations of inland waterway transport on the Oder River is the water level. Its variability determines the minimum clearance under the bridges and thus the maximum drought of moving vessels. An important factor affecting the water level in the Oder River is the phenomenon of wind backwater. Certain pressure field distribution situations may affect in an increase of water level on the coast and thereby block the outflow of river waters. In this paper an example of the limitations of inland navigation in the estuarine part of the Lower Oder under the weather conditions on the southern Baltic Sea in autumn 2016 is presented. For the analysis area of Szczecin Waterway Node and section of Oder River from Szczecin to Gryfino was chosen. Based on data the weather and navigational conditions of the area of the Lower Oder were assessed. Occurring winds, currents and changes in water level has been analyzed. In the situations of unfavorable raise of the water level in the river, inland navigation is redirected to the section of Regalica River, and a section of the port of Szczecin region is closed. This causes obvious difficulties and results in prolonged ships routes. Additional obstacle that take place here is an arbitrarily determined time of railway bridge span lifting in Szczecin Podjuchy.

1 Introduction

Sailing conditions on the inland waterways are determined by navigational and hydro-meteorological factors. The navigational factors include, among others parameters of inland waterways, state of buoyage and navigational infrastructure.

The group of hydro-meteorological factors include the impact of winds and currents, ice phenomena, high-water stage and low-water stage. The water level in the river determines the transit depth and the minimum clearances under bridges and thus the transport capacities of inland vessels in the area.

In the case of conditions of exceeding the highest navigable water, navigation on the section marked out by water gauges indicating exceeding is closed.

* Corresponding author: p.medyna@am.szczecin.pl

The impact of factors mentioned above varies depending on the location of the river section in the river basin. A special phenomenon affecting only a mouth section of the river is the phenomenon of backwater caused by storm surges, which can be seen on the waters of the Lower Oder.

2. Hydrographic and navigational analysis of Lower Oder

Oder River at km 704.1 divides into Western Oder and Eastern Oder. In the region of the river division on the Western Oder there is located valvular weir that prevents shipping directly from the Oder River to Western Oder. On the 3.0 km of Western Oder counting from this weir, HFM Canal debouches to the Oder. Canal since Hohensaaten, runs laterally in relation to the Oder. Western Oder is the original riverbed.

Western Oder has a total length of 36.55 kilometers counting from the weir in Widuchowa to the bridge on the Castle Route (Trasa Zamkowa) in Szczecin. Western Oder is divided into three main sections:

- Western Oder: km 0,0 – km 3,0 (HFW Canal)
- Western Oder: km 3,0 – km 17,15 (end of frontier section)
- Western Oder: km 17,15 – km 36,55 (Szczecin, Castle Route (Trasa Zamkowa) bridge.

Council of Ministers of 07. May 2002 regulation on the classification of waterways defines the parameters required for shipping lanes, together with the assignment to the class [1]. Western Oder has the class Vb. Tab. 1 shows the operating parameters for the river holding class Vb.

Table 1. Operating parameters of inland waterway with the Vb class [1].

Parameter	Value
Minimum dimensions of the shipping lane	
Width of the shipping lane	50.0 m
Transit depth	2.8 m
Radius of the arc of axis of the shipping lane	800.0 m
Minimum dimensions of the canal	
Width of the waterway	45.0 m
Smallest water depth in the canal	3.5 m
Radius of the arc of axis of the shipping lane	800.0 m
Minimum dimensions of the locks	
Width of the lock	12.0 m
Length of the lock	187.0 m
Depth at the low threshold	4.0 m

The width of Western Oder shipping lane is primarily limited by bridges. There is a railway bridge in Szczecin (35.59 km) and road bridge in Gryfino (km 14.64). Location of the bridge spans reduces the accessibility of the shipping lane.

In the case of the railway bridge in Szczecin width of the shipping lane at the right and left bridge span is 10.0 m, while the entire width of the span arch of road bridge in Szczecin the width is only 17.5 m. Under the road bridge in Gryfino navigable route is limited to a width of 30 m [9].

The problem is also the minimum clearances under bridges. For the class Vb waterway minimum clearance (above the highest navigable water) should be 5.25 m.

Currently, the vertical clearance under the road bridge in Szczecin at the lowest point of arc span is a total of width of 3.40 m. Under the railway bridge in Szczecin, this value is 3.79 m.

Transit depth is determined based on measuring the water level and announced daily in shipping information. Therefore this parameter is variable and depends on the part of the river:

- Western Oder: km 0.0 – km 3.0 – transit depth variable
- Western Oder: km 3.0 – km 17.15 – transit depth variable
- Western Oder: km 17.15 – km 36.55 – transit depth constant

Information of transit depth can be obtained from web site, using Transit Depth Calculator. An example is shown in Table 2.

Table 2. Transit depth calculator of RZGW (Regional Water Management) [6].

Section of the Oder river	km	Water gauges	The current measurement			Current state	
			Measurement date	Water level	the transit depth	enter the water level	the transit depth
Western Oder	km 0,0 -0,3	Gryfino O.Z (Water gauges in Mescherin)	19.12.2016	529	212		
Western Oder	km 0,3 – 17,1	Gryfino O.Z (Water gauges in Mescherin)	19.12.2016	529	401		

Current water level is measured by water gauges. On the waters of Western Oder there are two water gauges administered by Poland: Gryfino (km 14.4) and Szczecin (35.95 km) and two stream gauges administered by Germany: Gartz (8.06 km) and Mescherin (km 14.08).

Exceeding the states of highest navigable water (WWŻ) is associated with reduced clearances under bridges, increased waving (especially on the waters of Szczecin Waterway Node – Szczeciński Węzeł Wodny) and restrictions to navigation. One of the limitations occurring on the Oder River is closing for navigation the lower sections of the Western Oder due to reduced clearance under the road/rail bridge over Western Oder.

In such cases, the shipping goes through Regalica River under the railway drawbridge. Railway drawbridge opens at fixed hours that make impossible the smooth flow of inland ships. Tab.3. shows the actual bridge opening hours.

Table 3. Rail drawbridge on the river 733.7km opening schedule [9].

Bridge opening period	**Time of real daily bridge opening by administrator (PKP PLK S.A.)**
11.12.2016 – 11.03.2017	08:20 – 08:40
	10:10 – 10:30
	12:20 – 12:40
	13:56 – 14:16
	16:05 – 16:25
	18:00 – 18:20

3 An example of weather situation causing backwater and rising water levels in the estuary part of Oder River

Fig.1. shows the situation of rapidly changing pressure field distribution over Europe, which consequently caused a short-term storm surge, backwater on the Oder and exceeding alarm levels on the river. Storm surge is understood here as a sudden rise of water level in coast area caused by passing low-pressure system, which would not occur without the accompanying strong winds [3].

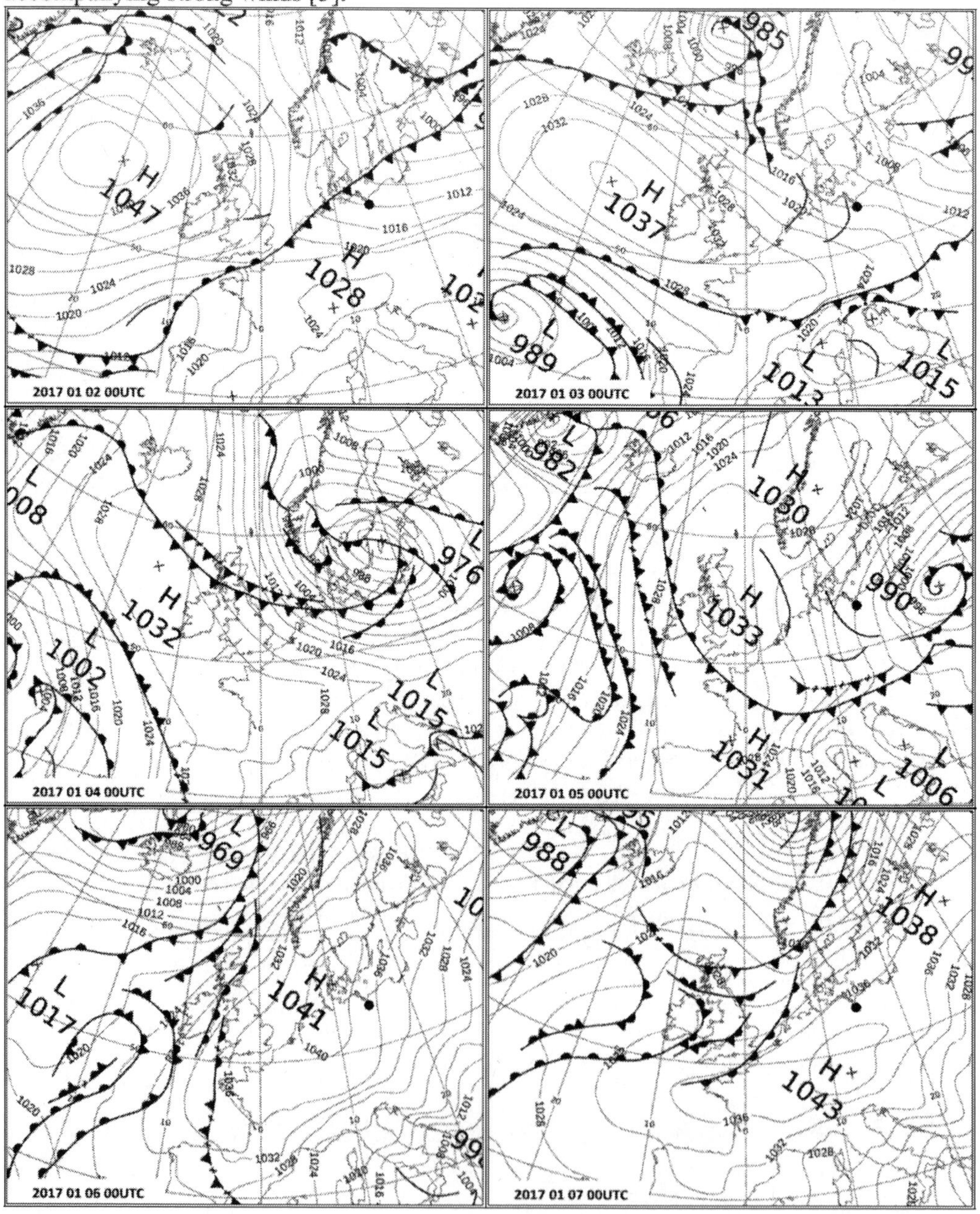

Fig. 1. Pressure field distribution on 02-07 January 2017 (dot indicates estuary section of the Oder River) [7].

January 2, 2017 00UTC weather situation in this region is causing by high pressure system with the center on the west of Ireland with a maximum pressure in the center 1047hPa, and by low pressure 993hPa with the center south of Lake Onega in Russia. Lower Oder area is ahead of the advancing cold front. This causes the advection of air masses from the west sector, so the situation conducive to pour ocean waters to the Baltic Sea through the Danish Straits (Fig. 1.).

On 3. January the pressure in the center of high decreases to 1037hPa, while the area occupied by it decreases assuming the stretched shape on the latitudinal axis WNW-ESE. Over Norwegian Sea low-pressure system with the pressure of 985hPa is marked clearly. Air circulation is still causing the occurrence of westerly winds in the relevant area.

January 4. – the low-pressure center is located in the area of central Baltic over Gotland. The minimum pressure at the center is 976hPa. In conjunction with the passing of the next cold front the winds begin to change their direction from west to north-west. After 1PM warning levels are exceeded at gauges in Gryfino and Szczecin Most Długi (Long Bridge). After 1PM warning levels are exceeded at gages in Gryfino and Szczecin Most Długi (Long Bridge). Regional Water Management (Regionalny Zarząd Gospodarki Wodnej) in Szczecin in the Communication 1/2017 warns about the possibility of reaching or exceeding the state of great navigable water (WWŻ) on the gauges Szczecin Most Długi (Long Bridge), Gryfino Western Oder, Szczecin Podjuchy, Gryfino Eastern Oder and Widuchowa [9].

On 5th January low-pressure system moves away a little to the east, the pressure increases to the value 991hPa, decreases the value of barometric gradient, while the winds take north direction. Away from south-west Europe a high pressure area is organising and high pressure is stretching from Spain (1031hPa) through the British Isles (1033hPa) and then to Scandinavia (1030hPa). Alarm conditions on certain sections of the estuary of the Oder and Vistula are exceeded (fig. 2.). After 5AM an alarm level is exceeded at the water gauge in Gryfino and after 7AM the same situation is at the gauge on the Most Długi (Long Bridge) in Szczecin.

On 6th January low-pressure system starts to disappear moving further east. The high pressure system is expanding and its center is located over the Jutland Peninsula (1041hPa). As a result, the winds diminish. After 6AM the water falls below an alarm level on gauge in Szczecin Most Długi (Long Bridge) and after 3PM at the water gauge in Gryfino.

On 7th January the center of High moves slightly south and the pressure system extends covering most of the Europe. The wind directions over the area of Lower Oder begins to be mainly from south-west sector.

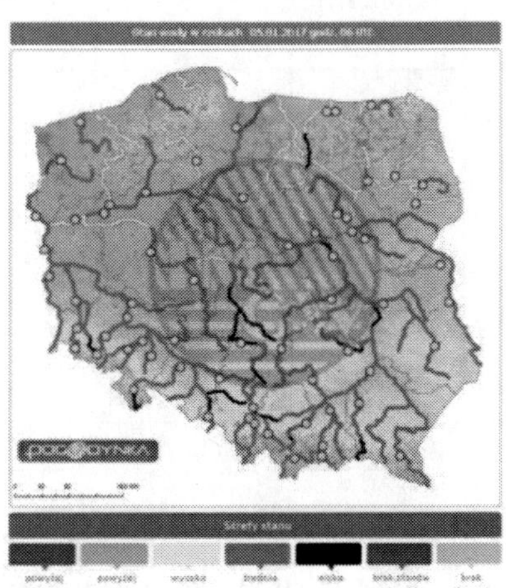

Fig. 2. Water levels in rivers on 5th January 2017, 0600UTC [08].

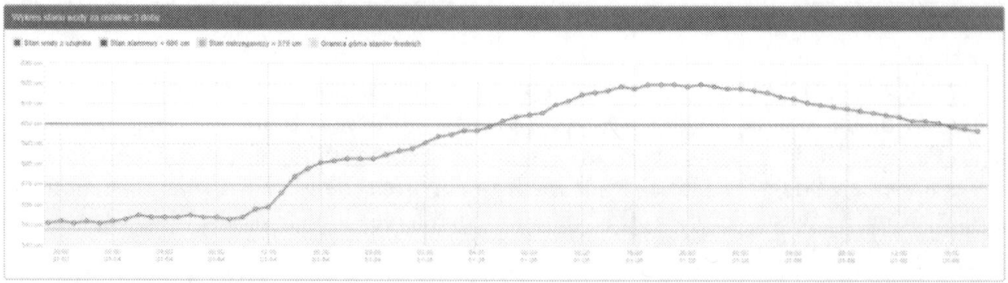

Fig. 3. Water level for last 3 days on water gauge in Gryfino, 6. January 2017, 06UTC [08].

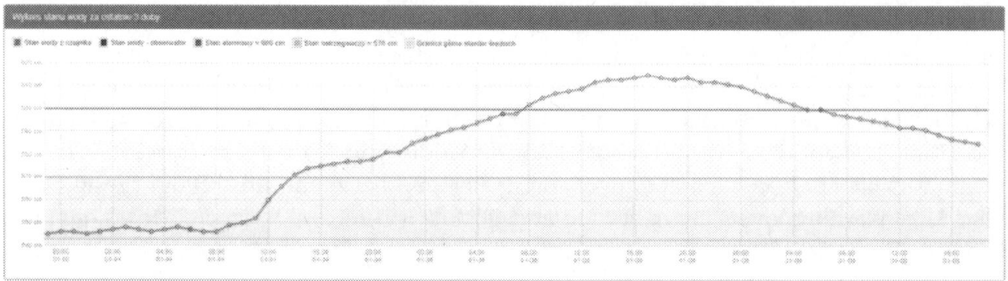

Fig. 4. Water level for last 3 days on water gauge in Szczecin Most Długi (Long Bridge), 6. January 2017, 06UTC [08].

Raising of the water level has marked clearly on 5. January, 2017 and the next day the situation began to return to normal (Fig. 3, 4). Hydrological situation in the morning on January 5. (06 UTC) is presented in Table 4. Warning levels were exceeded all the way from Świnoujście up to Gryfino. The highest overdraft of warning level was recorded in Świnoujście, at 0600UTC it was 62cm.In Trzebież warning level has been exceeded by 49cm, about 32cm in Gryfino, in Szczecin Most Długi (Long Bridge) about 28cm, and in Szczecin Podjuchy about 27cm. Exceeding of the alarm level disposed as follows: Świnoujscie - 42cm, Trzebież - 29cm, Gryfino - 2cm. In Szczecin Most Długi (Long Bridge) and in Szczecin Podjuchy exceeding of the alarm level was not recorded. At the same time, in Szczecin wind speed of the magnitude 17m/s was recorded and in Świnoujście measured wind speed was 21m/s.

Table 4. Hydrological situation on January 5th, 2017 at 0600 UTC on selected water gauges [8].

Gauge position	Water level (cm)	Level change (cm/24h)	Warning level (cm)	Alarm level (cm)	Alarm level exceeding (cm)	Warning level exceeding (cm)
Świnoujście	622	94	560	580	42	62
Trzebież	589	49	540	560	29	49
Szczecin Most Długi	598	51	570	600	-	28
Szczecin Podjuchy	607	51	580	610	-	27
Gryfino	602	47	570	600	2	32

4 Conclusions

Several factors contribute to the occurrence of extreme sea levels on the Polish coast, leading consequently to the phenomenon of backwater and exceeding the alarm conditions in the estuary sections of rivers [5]. One of them is the pouring the southern Baltic by wind generated flow of water through the Danish Straits, and thus the higher state of the sea reached even before the surge [2]. The second is the effect of wind, which consists of direction towards the shore, its speed and duration. The third is a deformation of the water surface level caused by the movement of the low pressure system over the central or southern Baltic Sea, generating waves of great length - baric wave and fluctuations in the water level of a quasi-seiche nature [4].

The situation described above was therefore typical for this part of the coast region. It was not a particularly strong surge, 94cm in Świnoujście, it was also not a particularly long - storm surges generated by lows typically persist through a period of 2 to 5 days. The weather situation, however, caused exceeding of the alarm levels on the river and temporarily preclude shipping on some of its sections.

The area of Lower Oder region is difficult to navigate. There are a number of restrictions imposed by the width of the shipping lane or clearance under bridges. Weather conditions affect the level of water in the river, which can interrupt or exclude shipping. Proper regulation of the river provides correct flow of the river. Navigational infrastructure does not act as regulatory and does not minimize the impact of backflow, which greatly affects the value of the vertical clearances under bridges, minimizing the possibility of transport. This shows the particular importance of current information relevant to navigation and weather conditions in Lower Oder area.

This research outcome has been achieved under the grant – Modelowanie ryzyka eksploatacji statku manewrującego na akwenie ograniczonym No 4/S/INM/15 financed from a subsidy of the Ministry of Science and Higher Education for statutory activities.

References

1. Rozporządzenie Rady Ministrów z dnia 07 maja 2002r. w sprawie klasyfikacji dróg wodnych (Council of Ministers of 07. May 2002 regulation on the classification of waterways)
2. I. Stanisławczyk, M. Sztobryn, *Zmiany napełnienia Bałtyku jako wskaźnik oceanicznych wlewów powierzchniowych.*, XII Międzynarodowa Konferencja Naukowo-Techniczna – Rola Nawigacji w Zabezpieczeniu Działalności Ludzkiej na Morzu, Wydawnictwa AMW, Gdynia (2000)
3. M. Sztobryn i inni, *Storm Surges in the Southern Baltic Sea (Western and Central Parts).*, Berichte des Bundesamtes fur Seeschifffahrt und Hydrographie, **Nr 39** (2005)
4. B. Wiśniewski, *Oddziaływanie niżów barycznych na stany wód w estuarium Odry.*, (w) Geotechnika w warunkach oddziaływanie morza, Politechnika Szczecińska, Szczecin (2002)
5. B. Wiśniewski, T. Wolski, *Katalogi wezbrań i obniżeń sztormowych poziomów morza oraz ekstremalne poziomy wód na polskim wybrzeżu.*, Akademia Morska, Szczecin (2009)
6. http://kgt.szczecin.rzgw.gov.pl
7. http://www.metoffice.gov.uk
8. http://www.pogodynka.pl
9. http://www.szczecin.rzgw.gov.pl

DOI: 10.1051/e3sconf/20171700060

One-stage deammonification start-up control using ion-selective electrodes (ISE) sensor

Stanisław Miodoński[1,*], and *Mateusz* Muszyński-Huhajło

[1]Faculty of Environmental Engineering, Wroclaw University of Science and Technology, Plac Grunwaldzki 9, 50-377 Wroclaw, Poland

Abstract. Dynamic development of side-stream ammonia-rich streams treatment technologies creates need of new methods of control techniques. So far, most of available technologies are based on relatively simple sensors, such as pH probes. This article presents the results of an ion selective electrode used for controlling a one-stage deammonification process in the start-up period. Sensor applicability under alternating aerobic/anaerobic conditions and large fluctuations of measured substances concentrations was investigated. ISE probe was also used to verify Anammox bacteria response on elevated dissolved oxygen exposition at the beginning of the anaerobic phases preceded by aerobic phases.

1 Introduction

Discovery of anaerobic ammonium oxidation (Anammox) had great impact on biological nitrogen removal (BNR) process development [1]. An important trend in recent years is to focus on turning away from traditional nitrification-denitrification path and try to make deammonification a main tool for nitrogen removal from wastewater [2]. Current state of knowledge about this process limits its application to high-strength ammonium streams (i.e. centrate water from sludge dewatering), named as side-stream treatment. Robust deammonification in cold streams with low ammonium concentration, such as municipal wastewater (mainstream treatment), is still subject of intense research and should be considered as an infancy stage so far [3-6]. Until 2014, there was over 100 full-scale side-stream installations, where 88% of them was in one-stage reactor configuration, which means that partial nitritation and Anammox (PN/Anammox) processes occur in a single reactor [7].

Considering fact, that characteristics of nitritation and Anammox process are completely different, operating one-stage installation for such processes is a great challenge. Especially in context of dissolved oxygen (DO) control, which, as necessary for nitrifiers and, on the other hand, critical for Anammox biomass, must be kept at certain level to avoid affecting both processes. It was reported, that Anammox bacteria was reversibly inhibited by low DO concentrations (<1-2% air saturation) and irreversibly by higher DO concentrations (>18% air saturation) [8]. In general, Anammox bacteria are

* Corresponding author: stanislaw.miodonski@pwr.edu.pl

considered as difficult to cultivate due to low growth rate and biomass yield and high sensitivity to environmental conditions [9].

This study describes the use of an ion-selective electrode (ISE) ammonium and nitrate probe during a start-up of a one-stage PN/Anammox process in a pilot-scale SBR treating real centrate water from sludge dewatering at a Polish large (over 500000p.e.) wastewater treatment plant (WWTP). Aim of this paper was to highlight potential benefits/threats derived from using this type of measuring devices in such advanced process.

1.1 PN/Anammox process

1.1.1 Anammox

Discovery of anaerobic ammonium oxidation, was a revolutionary step towards more efficient N-removal in wastewater treatment [10]. In general, Anammox is an anaerobic multi-stage process which stoichiometry is given by Eq. 1 [11]. Described group of microorganisms are autotrophic bacteria using inorganic carbon for biomass yield.

$$NH_4^+ + 1{,}32NO_2^- + 0{,}066HCO_3^- + 0{,}13H^+ \rightarrow$$

$$1{,}02N_2 + 0{,}066CH_2O_{0,5}N_{0,15} + 0{,}26NO_3^- + 2{,}03H_2O \quad (1)$$

1.1.2 Partial nitritation

Contrary to traditional nitrification-denitrification pathway of nitrogen removal, for Anammox purposes, only partial oxidation of ammonium to nitrite is required (see Eq. 1) To achieve this state, constant suppression of nitrite oxidizing bacteria (NOBs) is required along with stable ammonium/nitrite ratio around 1.3, which is suitable for Anammox process. Selective inhibition of NOBs can be achieved by creating proper operating conditions in the reactor, such as: high free ammonia/nitrous acid concentrations [12], low dissolved oxygen concentrations [13] or elevated temperature [14]. As nitritation consumes alkalinity due to two protons production per mole of oxidized ammonium, final NO_2/NH_4 ratio depends on influent alkalinity and may be controlled by pH maintained in the reactor [15]. Necessity of only partial oxidation of ammonium in this processes reduces net oxygen required for nitrogen removal to 1.71 mg O_2/mg N-NH_4 converted into gas if influent biodegradable organic matter is used to denitrify residual N-NO_x [16].

1.2 On-line monitoring for wastewater treatment

Development of advanced and more complex technologies used in wastewater treatment would not be possible without development in the field of products used to gather and process information about conditions of ongoing treatment processes. The end of 90s of 20th century brought a distinct change in process control at WWTPs as number of new chemical and physical on-line sensors became popular along, as well as the earlier spread of SCADA (Supervisory Control and Data Acquisition) systems. It became possible to measure practically all significant parameters in real-time at each stage of wastewater treatment and allowed to create new control strategies based on so called "safety points". Such control allows to run predefined preventive or repair procedure in case of violation one of set points values, enhancing process stability [17].

DOI: 10.1051/e3sconf/20171700060

To optimize side-stream treatment nitrogen removal, ISE sensors for on-line ammonium and nitrate concentrations measurement demonstrated their usefulness as well as luminescent dissolved oxygen sensors. These advanced tools in combination with control algorithms allowed to optimize aeration process for nitrification purposes [18, 19]. Direct measurement of nitrogen compounds concentration allows to precisely control biological nitrogen removal processes, which is commonly used in mainstream treatment (i.e. aeration control based on N-NH_4 concentration profile [20]) and can be easily adopt for side-stream treatment purposes (i.e. control of aeration phases duration for achieving appropriate N-NH_4/N-NO_2 ratio in deammonification process). The obvious benefit from the use of such in-situ sensors is quick and automatic collection of information on key parameters for installation operation. Ex-situ meters require additional sampling device and proper sample preparation (usually filtration) before measurement itself, which increases the level of complexity of the measuring system and extends the time from sampling to obtain information. The specificity of mentioned measurement methods makes in-situ sensors more favourable due to lower operational and maintenance costs, however they require regular calibration and verification of their indications [21, 22].

2 Materials and methods

2.1 Experimental set-up

2.1.1 Pilot scale SBR

Start-up was performed in a 150L SBR treating real reject-liquor from digested sludge dewatering. Reactor was operated at predefined time cycles, consisting of filling, mixing with intermittent aeration, sedimentation, decantation phases. Reactor was equipped with number of sensors connected to SCADA system providing parental control of following parameters: temperature, pH, DO level, filling/decanting volume and mixing intensity. N-NH_4/N-NO_3 ISE probe was installed in the reactor without connection to SCADA and was used to modify SBR operating parameters manually by the operator. Reactor scheme was shown in the Figure 1.

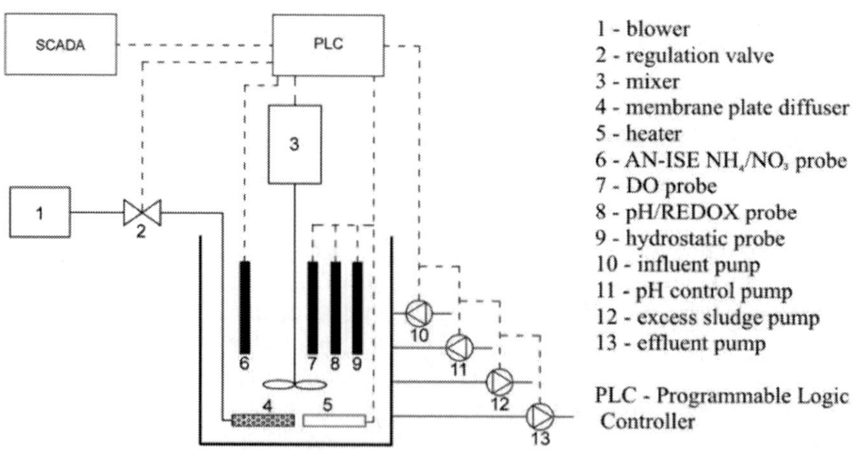

Fig. 1. Pilot scale SBR set-up.

2.1.2 Treated medium characteristics

Pilot installation treated real centrate from sludge dewatering collected from a belt press without flushing water used for belt cleaning to avoid excessive dilution. Sludge processed in dewatering unit was after mesophilic digestion process (HRT about 40d). Due to intermittent operation of belt press, centrate was collected periodically in a storage tank to provide supply of influent medium for SBR. After each filling, medium composition was determined. During start-up period, typical influent ammonium concentration, and alkalinity were equal 660 g N-NH4/m^3 and 56 g $CaCO_3/m^3$, respectively.

2.1.3 Combined N-NH_4 and N-NO_3 AN-ISE Sensor

Due to major role in presented study, more specific details about Hach's (Germany) AN-ISE combined ammonium and nitrate probe are presented in this section. In pilot SBR, a single probe of this type is installed and connected to Hach's Universal Controller Display Module SC1000, providing on-line measurements and data recording. AN-ISE probe contains ion-selective electrodes with special membranes, that only specific type of ion can adhere. Potential difference between selective electrode and reference electrode occurs, proportionally to measured NH_4^+/NO_3^- ion concentration. Interference of potassium and chloride when measuring ammonium and nitrate, respectively, are compensated by built-in electrodes. To prevent electrodes contamination, a compressed air cleaning unit is periodically used. Except that, probe was manually cleaned once a week during presented experiment. Sensor calibration was performed based on regular verification of its indications with a reference method (Hach's cuvette tests – indophenol blue method for ammonium and 2,6–dimethylphenol method for nitrate). Each time, full matrix correction protocol was used (NH_4^++K^+, NO_3^-+Cl^-) for calibration. Table 1 summarizes selected information about AN-ISE sensor provided by the manufacturer [23].

Table 1. AN-ISE sensor specification [23].

Parameter	Value	Comments
Measurement range	0 – 1000 mg/dm^3	N-NH_4, N-NO_3, K+, CL-
Accuracy	±5%+0.2mg/dm^3	% measured value
Response time	<3 min	-
Sample pH range	5 – 9	-
Sample temperature range	2 – 40°C	-

2.1.4 Analytical methods.

Collected samples were filtered with syringe filters, pore size 1.2μm and after that ammonia, nitrite and nitrate nitrogen concentrations were measured using Hach photometric cuvette tests: LCK303 (2 – 47 mg N-NH_4^+/dm^3), LCK342 (range 0.6 – 6 mg N-NO_2^-/dm^3) and LCK340 (range 5 – 35 mg N-NO_3^-/dm^3) and a DR3900 spectrophotometer. Total suspended solids (TSS) and volatile suspended solids (VSS) concentrations were determined using direct weight method according to Standard Methods guideline.

2.1 One-stage deammonification start-up

Start-up of one-stage deammonification process was performed in a SBR treating centrate from sludge dewatering. Sludge inoculum for the process was taken from full-scale deammonification installation at another WWTP, where, due to technical issues, side-stream installation was not operated for over two months. In this case, described actions can be considered as attempt to re-start the whole process. As the main objective of the experiment was to achieve stable PN/Anammox process operating at nitrogen loading rate (NLR) over 0.2kg $N\text{-}NH_4\cdot(m^3\cdot d)^{-1}$.

Start-up was preceded by a series of lab-scale tests to evaluate PN and Anammox process rates in the inoculum. Based on gained results, initial NLR has been determined and process started. SBR was operated in 3 hour cycles. Single cycle consisted of: one filling phase at the beginning (time adjusted depending on NLR), 7 aeration phases (5 to 7 minutes, adjusted based on process efficiency) separated by anoxic phases. Total reaction phase took 126 minutes with continuous mixing. At the end of each cycle 20-40 minutes were provided for sludge settling and 5-10 minutes for effluent decantation. Cycle scheme during described period is presented in Figure 1.

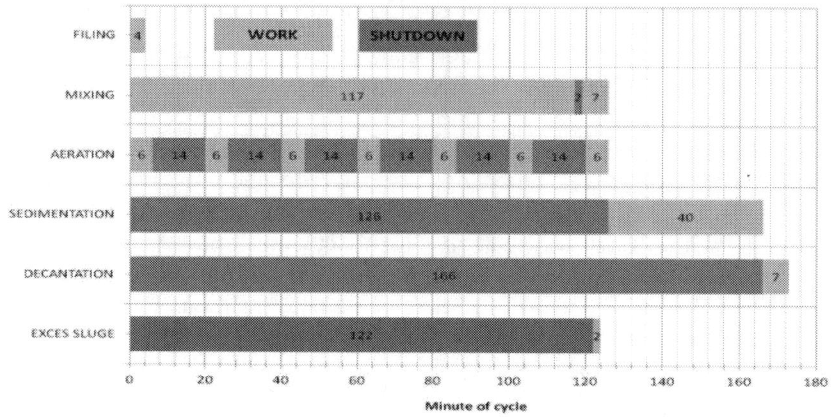

Fig. 1. SBR cycle scheme during the start-up period.

Process was operated under the similar conditions to the full-scale installation: temperature 27°C, pH= 7.1, DO in the aeration phase 0.4-1.0 mg O_2/dm^3. NLR and aeration time was adjusted daily by the operator based on N-compounds concentration in the effluent (analytical method) and $N\text{-}NH_4^+/N\text{-}NO_3^-$ profiles from the AN-ISE probe. NLR and effluent quality is presented in Figure 2.

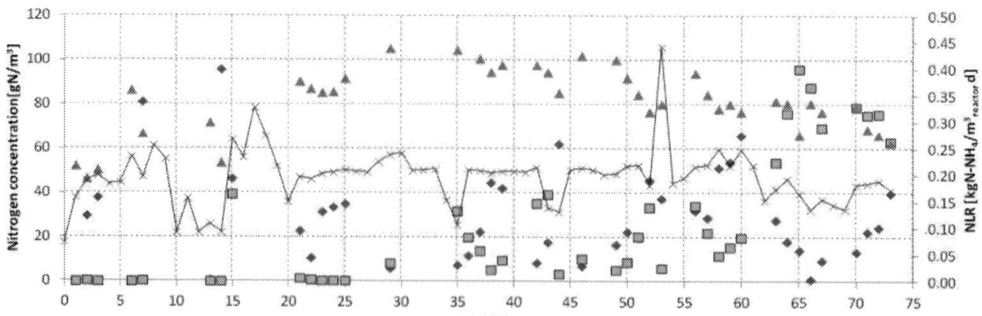

Fig. 2. SBR effluent composition (daily average) and NLR during the start-up period.

3 Results

3.1 Ammonium uptake rate (AUR) in oxic / anaerobic conditions

Ammonium concentration was continuously measured using the AN-ISE probe. Each day of start-up period, an average AUR for aerobic and anaerobic conditions was calculated for single SBR cycle. AUR under aerobic conditions mostly represented reduction of ammonium concentration due to activity of ammonium oxidizing bacteria (AOBs). Notable decrease of NH_4^+ level during anaerobic conditions can be explained by the Anammox bacteria activity.

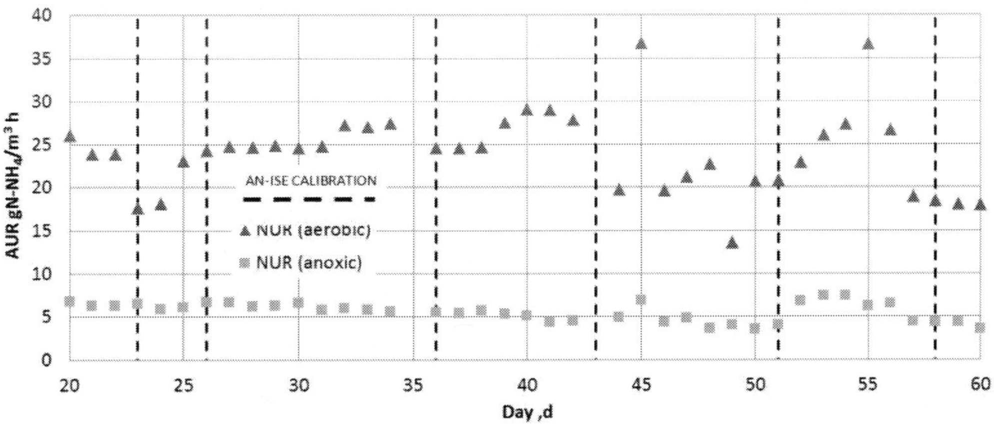

Fig. 3. Average anoxic and aerobic AUR rates during the start-up period. Dashed lines indicates sensor calibration.

Figure 3 presents daily average AUR values for anaerobic and aerobic conditions calculated from AN-ISE measurements in selected single cycle each day. Dashed vertical lines indicates days when sensor calibration was performed. During first 43 days of experiment, average AUR value for aerobic and anaerobic conditions showed no significant variations and were, on average, equal 24.44 g $N\text{-}NH4\cdot(m^3\cdot h)^{-1}$ (SD: 4.56 g $N\text{-}NH4\ (m^3\cdot h)^{-1}$) and 5.88 g $N\text{-}NH4\ (m^3\cdot h)^{-1}$ (SD: 1.25 g $N\text{-}NH4\cdot(m^3\cdot h)^{-1}$), respectively. After that period, slight reduction of these values occurred, probably due to periodic NH_4 and NO_2 accumulation (see Fig. 2) resulting in elevated free ammonia / free nitrous acid concentrations causing bacterial activity inhibition. In this period, aerobic AUR was 22.77 g $N\text{-}NH4\cdot(m^3\cdot h)^{-1}$ (SD: 5.56 g $N\text{-}NH4\cdot(m^3\cdot h)^{-1}$) and anaerobic AUR was 4.73 g $N\text{-}NH4\cdot(m^3\cdot h)^{-1}$ (SD: 1.48 g $N\text{-}NH4\cdot(m^3\cdot h)^{-1}$). Despite these observations, it can be clearly noticed that significant impact on measured AUR values has time elapsed since sensor calibration. Changes in sensor parameters did not occurred linearly (whole indication curve did not shift up or down), which may cause big differences in AUR values before and right after re-calibration.

3.2 Lag phase in Anammox activity after aeration phase.

Return of Anammox bacteria to their normal activity at the beginning of anaerobic phases was investigated. After end of each aeration phase, DO level in the reactor dropped below 0.1 mg O_2/dm^3 in less than 40 seconds due to presence of ammonium and nitrifiers activity. After that short time, a stagnation of ammonium concentration decrease was still noticeable due to Anammox lag phase after period of inhibition due to elevated dissolved oxygen presence. Using the AN-ISE probe length of the lag phase was measured based on the ammonium concentration at different stages of the start-up. Results are presented in Figure 4. It can be clearly noticed, that a clear suppression of Anammox activity occurred for several minutes (5-7 min) in the first days of the experiment. After 25 days, such phenomenon was reduced to less than one minute, what suggests possible Anammox adaptation to periodic aerobic conditions and reduction of the time needed to restore its normal metabolism.

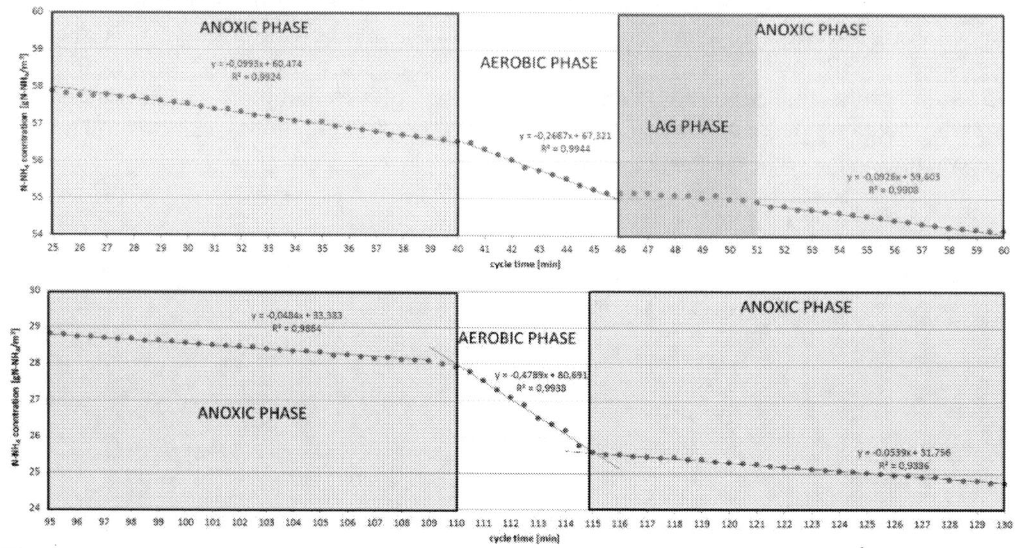

Fig. 4. Anammox lag phase after aerobic phase at the beginning of start-up day 22th, (upper) and at 63th day of experiment (lower).

4 Conclusions

- Ion Selective Electrode for ammonium and nitrate measurements can be a useful tool during a one-stage deammonification process start-up and operation. Quick and reliable information about process condition can be helpful in daily control and detection of stability disturbances connected with ammonium accumulation or unwanted activity of NOBs resulting nitrite concentration growth.
- AUR determination using AN-ISE probe during aerobic and anaerobic conditions is possible only when time between calibrations is short enough. In reactor where large fluctuations of measured values occur, sensor indications quality are quickly decreasing. To keep gained data at certain level of accuracy frequent calibrations are needed, more than once a week.
- Observed Anammox inhibition due to alternating aerobic/anoxic conditions was only periodical despite relatively high DO concentrations for such process (up to 1g O_2/m^3). During described process start-up, adaptation of Anammox bacteria to periodic aerobic conditions was identified as a reduction of the lag phase in the ammonium decrease after the aerobic phase. Remarkable reduction of the lag phase was observed, from 5-7 minutes at the beginning of the experiment to less than
  a minute after 25 days. fact indicates that AN-ISE probe can be successfully used in tests of factors affecting processes which result in ammonium concentration change.

The authors gratefully acknowledge co-funding from the National Centre for Research and Development (grant no PBS2/B9/25/2014) and Municipal Water and Sewage Company Wroclaw.

References

1. A. Mulder, A.A. van de Graaf, L.A. Robertson, J.G. Kuenen, Fems Microbiol Ecol, **16**, 177-183 (1995)
2. M.C. van Loosdrecht, D. Brdjanovic, Science, **344**, 1452-1453 (2014)
3. G. Xu, Y. Zhou, Q. Yang, ZM. Lee, J. Gu, W. Lay, Y. Cao, Y. Liu, Appl Microbiol Biot, **99**, 2485-2490 (2015)
4. B. Wett, S.M. Podmirseg ,M. Gomez-Brandon, M. Hell, G. Nyhuis, C. Bott, S. Murthy, Water Environ Res, **87**, 2084-2089 (2014)
5. P. Regmi, M.W. Miller, R. Bunce, D.M. Hingley, D. Kinnear, B. Wett, S. Murthy, C.B. Bott, WEFTEC. **15,** 1994-2017 (2012)
6. R. Lemaire, H. Zhao, C. Thomson, M. Christensson, S. Piveteau, S. Hemmingsen, F. Veuillet, P. Zozor, J. Ochoa, WEFTEC. **6,** 2183-2197 (2014)
7. S. Lackner, E.M. Gilbert, S.E.Vlaeminck, A. Joss, H. Horn, M.C.M. Van Loosdrecht, Water Research, **55**, 292–303 (2014)
8. R.C. Jin, G.-F. Yang, J.J. Yu, P. Zheng, Chem Eng J **197**, 67–79 (2012)
9. M. Strous, J.A. Fuerst, E.H.M. Kramer, S. Logemann, G. Muyzer, K.T. van de Pas-Schoonen, R. Webb, J.G. Kuenen, M.S.M. Jetten, Nature **400** 446–449 (1999)
10. M.M.M. Kuypers, G. Lavik, D. Woebken, M. Schmid, B.M. Fuchs, R. Amann, B.B. Jørgensen, M.S.M. Jetten., P Natl Acad Sci USA, **102,** 6478-6483 (2005)
11. M. Strous, J.J. Heijnen, J.G. Kuenen, M.S.M. Jetten, Microbiol Biot, **50**, 589-596 (1998)
12. A.C. Anthonisen, R.C. Loehr, T.B.S. Prakasam, E.G. Srinath, J Water Pollut Con F, **48**, 835-852 (1976)
13. R. Blackburne, Z. Yuan, J. Keller, Biodegradation, **19**, 303-312 (2008)
14. M.S.M. Jetten, S. Logemann, G. Muyzer, L.A. Robertson, S. de Vries, M.C.M. van Loosrecht, J.G. Kuenen, A Van Leeuw J Microb, **71**, 75-93 (1997)
15. C. Fux, M. Boehler, P. Huber, I. Brunner, H. Siegrist, Water Res, 99, 295-306 (2002)
16. G.T. Daigger, Water Environ Res **86**, 204-209 (2014)
17. K.P. Tsagarakis, K. Stamatelatou, *Sewage Treatment Plants: Economic Evaluation of Innovative Technologies for Energy Efficiency* , IWA Publishing (2015).
18. D. Jenkins, J. Wanner, *Activated sludge: 100 years and counting*, IWA Publishing, 2014
19. L. Åmand, G. Olsson, B. Carlsson., Water Sci Technol, **67**, 2374–2398 (2013)
20. H. Siegrist, L. Rieger, C, Fux, M. Wehrli, Water Sci Technol, **50**, 35–43 (2004)
21. S. Winkler, L. Rieger, E. Saracevic, A. Pressl, G. Gruber, Water Sci Technol, **50**, 105–114 (2004)
22. D. Kaelin, J. Eugster, K. Rottermann, C. Banninger, H. Siegrist, Water Sci Technol, **58**, 6-29 (2008)
23. *AN-ISE sc: Combination sensor for ammonium and nitrate – Data Sheet*, HACH (2012)

The influence of prefabricated pipe cement coatings and those made during pipe renovation on drinking water quality

Anna Młyńska[1,*], and *Michał* Zielina[1]

[1]Cracow University of Technology, Faculty of Environmental Engineering, Institute of Water Supply and Environmental Protection, ul. Warszawska 24, 31-155 Cracow, Poland

Abstract. Nowadays, cement coatings are often used as an anticorrosion protection of the internal surfaces of manufactured ductile iron water pipes. The protective cement linings are also commonly used for old water pipe renovation. In both cases, the cement lining is an excellent anticorrosion protection of the pipelines, effectively separating the pipe wall from the flowing water. Moreover, cement linings protect the pipelines not only by a mechanical barrier, but also by a chemical barrier creating a highly alkaline environment in water contact with the metal pipe wall. In addition, cement coatings have an ability for so-called self-regeneration and provide the improvement of hydraulic conditions inside the pipelines. In turn, the differences between the analysed cement coatings mainly depend on the types of cements used and techniques of cement mortar spraying. As was expected, they influence the quality of water having contact with the coating. A comparison of the impact of cement coatings manufactured in factories and sprayed on building sites during the renovation on drinking water quality parameters was performed in the study. The experiments were conducted in laboratory conditions, using the test stands prepared for this purpose. The results include analysis of selected water quality parameters for the samples contacting with cement mortar and collected during the investigation.

1 Introduction

1.1 Protective cement mortar coatings

Pipes made of different materials are used in water supply systems. Water pipes manufacturers offer plastic pipes made of polyethylene (PE), polyvinyl chloride (PVC) and more recently also glass reinforced plastic (GRP). However, traditional materials such as steel, iron and cast iron are still commonly used, especially for large diameter pipes working under high pressure. The limitation of the application of traditional materials for water supply system is corrosion. It can be reduced by using internal protective coatings, traditionally made of cement, but presently also often made of resins [1].

[*] Corresponding author: a.mlynska13@gmail.com

The protective cement coatings are made during the manufacturing process in factories but also on building sites during the renovation of corroded water pipes [2]. Cement mortar lining is the oldest and still the most popular renovation trenchless technique for old water pipes [3]. Pipe renovation by cement mortar lining was first time performed in the 1930s [2–4]. In turn, the first water supply system consisting of iron pipes with internal cement coatings was built in 1922 [4–6].

Among many others, the most important advantages of protective cement coatings are mechanical and chemical barriers separating the metal pipe wall from the water, creation a high alkaline environment in which the corrosion process does not occur, the ability to self-regeneration [5, 6] and a long-term improvement of hydraulic conditions inside the pipelines [3, 6].

The differences between cement coatings made during manufacturing processes in factories and cement coatings prepared on building sites during renovation are mainly related to the different cement types and techniques. Blast furnace slag cement is most often used in first case [7, 8], whereas it is noted, that Portland cement is mainly used for renovation purposes. Prefabricated cement coatings are produced in two ways. In the first, cement mortar is sprayed inside the pipe rotating with a high speed; in the second, cement mortar is sprayed inside the stationary pipe by a situated centrally rotating head. In both cases, after cementing, the pipe is spun with a high velocity and then vibrated to achieve smooth and dense cement coating surface [5]. During the renovation process, the cement mortar is lined by using a centrally situated rotating head, which is pulled through the pipeline with a constant speed. The rotating velocity is also constant in order to achieve a homogenous cement coating with the same thickness along the whole water pipe cross section [2]. By using different techniques of cement coating preparation, the finally obtained protective coatings have a different microstructure and thus durability and also the probability of leaching pollutants from the lining into the water is greater or lesser. In general, cement coatings sprayed on a building site during renovation have a rather homogenous structure, whereas prefabricated coatings, lined in factories, as a consequence of pipe spinning, are characterised by the sand grains located close to the metal pipe wall inside the coating. It causes dense, smooth and less porous cement coating from the water side and the cement coating is more resistant to aggressive waters flowing through the pipe [9–11].

1.2 The risk of water contamination resulting from contact with the cement mortar coating

The risk of the leaching chemical elements from the cement coatings into the water is determined by the type of cement used. For example, the greatest amount of aluminium is leached from a high-alumina cement coating. In turn, the greatest amount of calcium is leached from a Portland cement coating [9, 10]. In the case of using blast furnace slag cement for coating, the risk of leaching a greater amount of aluminium and calcium is rather small. Many experiments proved that the greatest impact of cement coating on water quality initially is mainly related to the significant growth of pH [4, 10, 12–15] and aluminium concentration [4, 10, 13–15]. It is also probable that some heavy metals will be leached from the cement coating into the water. Numerous investigations on leaching trace elements from cements into the water were conducted and described in the literature [4, 13–17].

Even a small amount of heavy metal in the human body can contribute to many different health problems. In the Netherlands, in 1996, a great amount of aluminium leaching from a new cement coating was noticed. This leached aluminium contributed to the death of some patients in a haemodialysis centre as a consequence of aluminium poisoning. Aluminium was leached to the water from a newly-installed water system covered inside by a protective cement coating. The concentration of aluminium was forty times higher than the specified limit [18]. Thus, the concentration of aluminium in water making contact with a cement coating in the initial period of time should be controlled.

Since both pipe cementing techniques and type of used cements significantly influence the dissolution and leaching of chemical components from the cement coating, a comparison of cement coatings manufactured in factories and prepared on a building site during renovation on water quality was performed in this study. The paper presents the change of chemical elements such as aluminium, chromium, lead and cadmium in water samples coming into contact with two different cement coatings - manufactured in a factory and prepared in a laboratory.

2 Materials and methods

2.1 Tested water pipes with internal cement coatings

In the performed experimental research, two different types of water pipes with internal cement coatings were used.

The first of them (Fig. 1a) was a piece of ductile iron water pipe with a prefabricated cement coating manufactured by one of the water pipe producers. The length of tested water pipe was about 16.0 cm with a nominal diameter equal to 100 mm. The external surface of this pipe was coated with a metallic zinc-aluminium alloy and a blue epoxy layer. The internal surface of the pipe was protected by a cement coating made of blast furnace slag cement (CEM III) during the manufacturing process, with a thickness of 0.65 cm.

In turn, the second tested water pipe sample (Fig. 1b), with a length of 25.0 cm and a nominal diameter of 80 mm, was made of steel and manually coated inside with cement mortar with a thickness of 0.70 cm. The cement mortar was prepared by mixing Portland cement (CEM I) with quartz sand in a 1:1 ratio and with water, maintaining a water to cement ratio equal to 0.35. The curing time of the fresh cement mortar coating was 24 h. Both the properties of the manually prepared cement coating and the curing time were the same as in the case of the cement coatings sprayed on building sites during the renovation process.

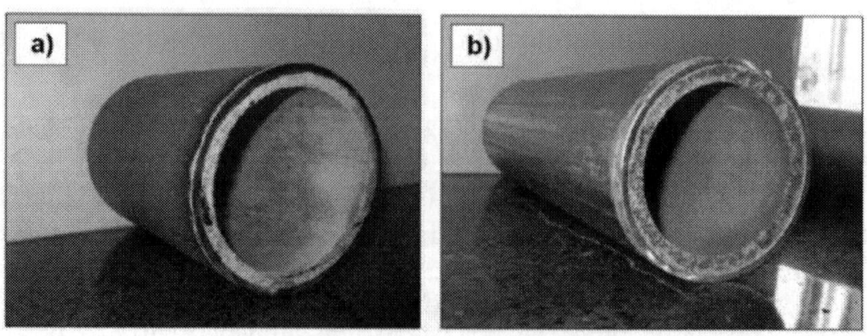

Fig. 1. Water pipes with internal cement coatings.

a) Ductile iron water pipe with prefabricated cement coating.

b) Steel water pipe with manually prepared cement coating.

2.2 Test stands

Each of the two tested water pipe samples was installed on a separate test stand in the laboratory. The construction details of the two test stands are presented in Fig. 2. All of the experiments were conducted parallel in static conditions. The interiors of both samples were filled with the same water collected from the outflow of the largest water treatment plant for Cracow. The water was in contact with the cement coatings for the same periods of time. The raw water used in the experiments was relatively soft (consisting of small amount of ions), becoming rather aggressive to the cement mortar linings.

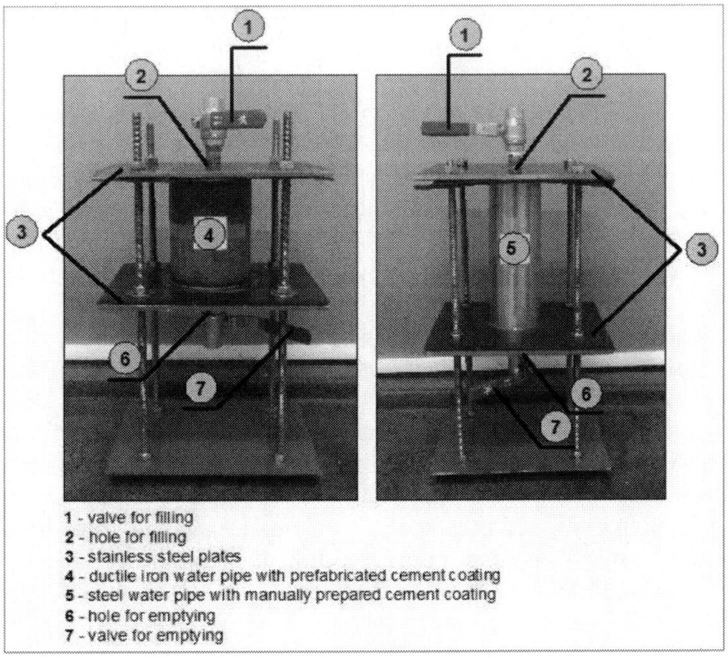

Fig. 2. Test stands for the static experiments.

2.3 Water sampling

The tested water samples inside the two pieces of pipes were in contact with the internal cement coatings for exactly the same periods of time. The periods were timetabled in accordance with the Dutch Standard (EA NEN 7375:2004) [19] and Polish Standard (PN-EN 14944-3:2008) [20]. During sampling, the whole amount of water from both pipe interiors was replaced with a new raw water. The waters were replaced and collected parallel soon after the following periods of time: 0.25, 1, 2.25, 5, 6, 9, 13, 16, 33, 36, 52 and 56 days.

2.4 Tested water quality parameters

Several physicochemical water quality indices were tested for all of the water samples collected from the ductile iron pipe with prefabricated cement coating and from the interior of steel pipe with manually covered cement mortar lining and also for the raw water samples. The paper presents the results of the analysis of the aluminium, chromium, cadmium and lead concentration changes. Based on the laboratory measurements of chemical elements concentrations and also based on the methods presented in the Dutch

Standard [19], the expected cumulative leaching of aluminium and chromium from 1 m^2 of the cement mortar coating surface into the water was counted and presented below.

3 Leaching contaminations from cement coatings into the water – results and discussion

3.1 Aluminium leaching

It is clearly evident from Fig. 3 that the aluminium was leached much more intensively from the manually prepared cement coating into the water than from prefabricated cement coating for the whole period of the investigation.

The greatest amount of the leached aluminium from the manually prepared cement coating, exceeding even 20.0 mg/m^2, was noticed during the first 5 days of its contact with water. After this time, the amount of leached aluminium was gradually decreased and on day 56 of the experiment, the leaching had totally stopped. In turn, for the prefabricated cement coating, a small leaching of aluminium, equal to 0.14 mg/m^2, was noticed just after 2 days of contact with water. In the 33-day experiment, again some increase of leaching (0.34 mg/m^2) was observed, whereas almost a total lack of any leaching was noticed till the end.

For the whole investigation period, about 160 mg of aluminium in total was leached from 1 m^2 of the manually prepared cement coating surface, whereas only about 0.50 mg/m^2 was leached from the prefabricated cement lining. It shows that the total amount of aluminium leached from the cement coating made during renovation can be even more than three hundred times more than that from a prefabricated cement coating.

Since aluminium is important component of Portland cement, leaching of aluminium from manually prepared coating is much more intensive than from prefabricated cement coating covering inner side of the pipe. In turn, the amount of aluminium contained in prefabricated mortar lining made of blast furnace slag cement is much smaller.

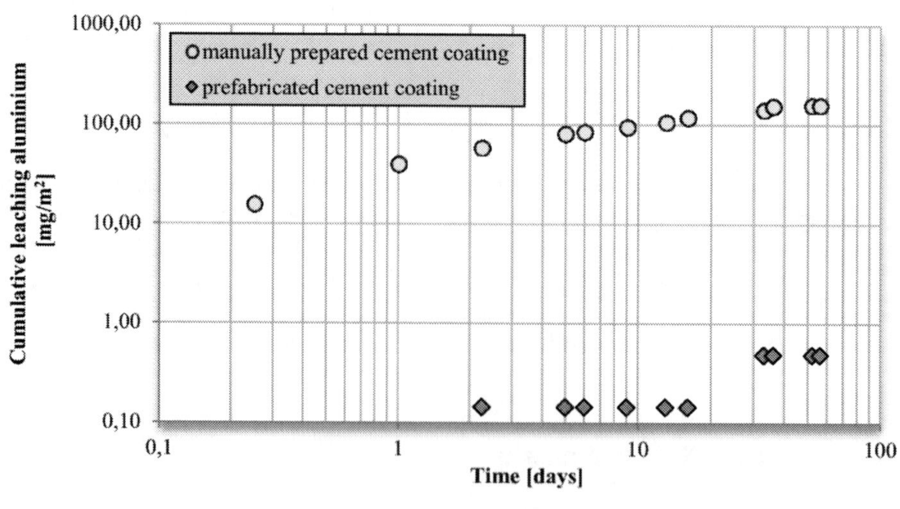

Fig. 3. Cumulative leaching of aluminium from manually prepared and prefabricated cement coatings.

3.2 Chromium leaching

In comparison to aluminium, chromium was leached from the both analysed cement coatings much less intensively (Fig. 4).

In the case of the manually prepared cement coating, the most intensive chromium leaching was noted during the first day of contact with water. In the following days, the intensity of the leaching process was smaller and at the end of the study period chromium leaching had stopped. The total, cumulative amount of leached chromium from cement coating made of Portland cement was a little over 4.5 mg/m^2 and thus, was about thirty-five times smaller than the total cumulative amount of leached aluminium. In the case of prefabricated cement coating, only during the first 6 h of contact with water, 0.024 mg of chromium from 1 m^2 of cement coating was leached. As can be seen, the risk of leaching chromium from cement coatings made during the renovation process is much greater than in the case of prefabricated coatings. Nevertheless, in none of the two cases is there any serious risk of exceeding concentration limits by the leaching of chromium from the inner cement linings. Only high concentration of chromium in raw water could be threatening to human health.

The same as in case of the aluminium, it was observed much more intensive leaching of chromium to water contacting with manually prepared cement coating than to water contacting with prefabricated coating. It was due to different types of cement used in preparation of both specimens and also different methods of preparation.

However, since chromium is one of the trace elements in cement, chromium was leached from both kinds of coating less intensive than aluminium.

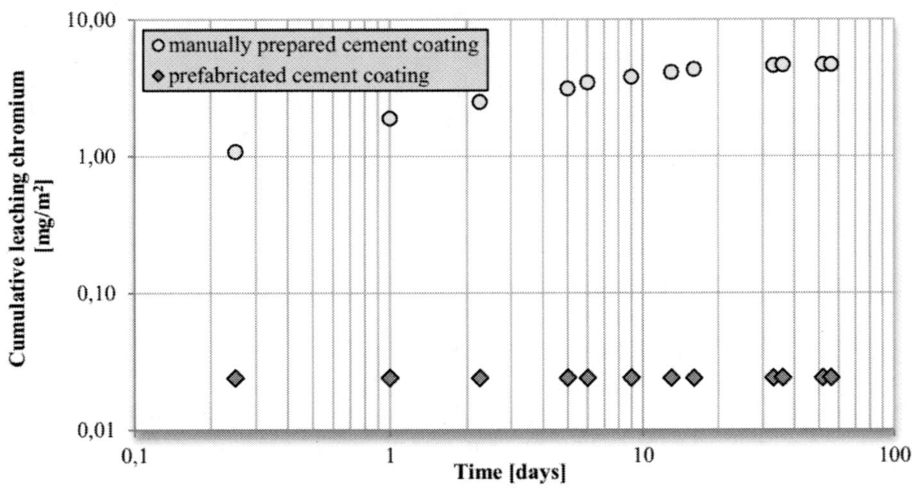

Fig. 4. Cumulative leaching of chromium from manually prepared and prefabricated cement coatings.

3.3 Lead and cadmium leaching

In all collected water samples, the concentrations of lead and cadmium were very low from contact with the prefabricated and manually prepared cement coatings. The concentration of lead was five times smaller than its limitation value according to drinking water standards [21], while the cadmium concentration was as much as ten times smaller than the maximum value regulated by [21]. It confirms that the risk of leaching of toxic heavy metals from both analysed cement coatings into the water was negligible because of low heavy metal contents in both cements.

4 Conclusions

Both kinds of internal cement coating, manufactured in factories and made during the renovation of old pipes, guarantee relatively good protection against corrosion. However, due to different techniques of lining, the coatings are characterised by different microstructures. The structure made during renovation is quite homogenous and porous in the whole cross section, whereas a prefabricated coating consists of two parts. The first part, which attaches to pipe wall, consists of a cumulated amount of sand grains, whereas the second part, which makes contact with water, is quite smooth and dense, without many sand grains. It results in the prefabricated cement coating to be more durable and resistant to dissolution and leaching out of chemical components.

The conducted experiments suggest that some risk of leaching contaminations from the both analysed cement coatings is only related to aluminium. Leaching of dangerous heavy metals such as chromium, lead or cadmium was revealed to be very low or negligible. As was expected, the most significant leaching of all pollutants was observed during the initial period.

Based on the experiments, it was found that the leaching of aluminium and chromium from coatings made during renovation is much more intensive than from prefabricated coatings made in factories.

The risk of drinking water contamination from the fresh cement coating soon after renovation process seem to be relatively low. However, it is still important to determine an appropriate contact time of fresh cement mortar lining with flowing water, as well as the composition of the mortar, flushing velocity and time.

Prefabricated cement coatings seem to be quite safe for drinking water. Even soon after the pipe installation, any exceeding of drinking water quality parameters is not expected.

However, some other experiments concerned water pipe lining should be done in the future, e.g. investigating the influence of different cement types for pipe renovation or hydraulic conditions on drinking water quality. The results of such experiments will help to determine the maximum allowable length of pipe put under renovation depending on pipe diameter, hydraulic conditions or cement types.

References

1. G. Tsakiris, V. Tsakiris, *Pipe technologies for urban water conveyance distribution systems*, Water Utility Journal, **3**, 29–36 (2012)
2. American Water Works Association, *Rehabilitation of water mains*, Manual of water supply practices–M28, Second Edition (AWWA, Denver 2001)
3. N. Damodaran, J. Pratt, J. Cromwell, J. Lazo, E. David., R. Raucher, Ch. Herrick, E. Rambo, A.K. Deb, J. Snyder, *Customer acceptance of water main structural reliability* (AwwaRF, Denver 2005).
4. A.K. Deb, S.B. McCammon, J. Snyder, A. Dietrich, *Impacts of lining materials on water quality* (Water Research Foundation, Denver 2010)
5. American Water Works Association, *Ductile-iron pipe and fittings*, Manual of water supply practices–M41, Second Edition (AWWA, Denver 2002)
6. R.W. Bonds, *Cement-mortar linings for ductile iron pipe* (Ductile Iron Pipe Research Association, Alabama 2005)
7. Information materials of Duktus Company
8. Information materials of Saint-Gobain PAM UK
9. I.S. Meland, *Durability of mortar linings in ductile iron pipes*, Durability of Building Materials and Components 8: Proceedings of the Eighth International Conference on Durability of Building Materials and Components, 170-179 (Vancouver 1999)

10. M. Zielina, W. Dąbrowski, E. Radziszewska-Zielina, *Cement mortar lining as a potential source of water contamination,* World Academy of Science, Engineering and Technology, **8(10)**, 636-639 (2014)
11. A. Wassilkowska, K. Kuc, M. Zielina, *A study of the composition of cement mortar lining in water and sewage pipelines,* Technical Transactions, Civil Engineering, **4-B**, 131–136 (2015)
12. A.J. Whelton, M. Salehi, M. Tabor, B. Donaldson, J. Estaba, *Impact of infrastructure coating materials on storm-water quality: Review and experimental study*, Journal of Environmental Engineering, **139(5)**, 746–756 (2013)
13. M. Zielina, A. Młyńska, T. Żaba, *Experimental research on deterioration of drinking water quality after cement mortar pipe lining,* Technical Transactions, Civil Engineering, **4-B**, 145–152 (2015)
14. A. Młyńska, M. Zielina, *Experimental research on the impact of different hardness waters on their contamination by protective cement mortar linings after pipe renovation*, Technical Transactions, Environment Engineering, **1-Ś**, 97–104 (2016)
15. M. Zielina, A. Młyńska, T. Żaba, *Przenikanie zanieczyszczeń z wyprawy cementowej do wody pitnej w przewodach wodociągowych po cementowaniu,* Zaopatrzenie w wodę, jakość i ochrona wód, 535-544 (PZIiTS, Poznań, Toruń 2014)
16. B. Kowalska, D. Kowalski, M. Kwietniewski, K. Miszta-Kruk, J. Chudzicki, *Badania zawartości jonów metali ciężkich w wodzie kontaktującej się z wykładziną cementową rur wodociągowych*, Ochrona Środowiska, **33(4)**, 41–45 (2011)
17. Q. Guo, P.J. Toomuluri, J.O. Eckert, *Leachability of regulated metals from cement mortar linings,* Journal AWWA, **90(3)**, 62–73 (1998)
18. K. Berend, T. Trouwborst, *Cement-mortar pipes as a source of aluminum*, Journal AWWA, **91(7)**, 91–100 (1999)
19. EA NEN 7375:2004, Leaching characteristics of moulded or monolithic building and waste materials/Determination of leaching of inorganic components with the diffusion test – "The tank test", Netherlands Normalisation Institute Standard, Aprol 2005
20. PN-EN 14944-3:2008, Wpływ wyrobów cementowych na wodę przeznaczoną do spożycia przez ludzi – Metody badań – Część 3: Migracja substancji z produkowanych fabrycznie wyrobów cementowych, Luty 2008
21. Rozporządzenie Ministra Zdrowia z dnia 13 listopada 2015 roku w sprawie jakości wody przeznaczonej do spożycia przez ludzi (Dz.U. 2015 poz. 1989)

E3S Web of Conferences **17**, 00062 (2017) DOI: 10.1051/e3sconf/20171700062
EKO-DOK 2017

Analysis of water losses in two selected water distribution systems

Anna Musz-Pomorska[*1], *Małgorzata* Iwanek[1], *Katarzyna* Parafian[2], and *Katarzyna* Wójcik[2]

[1]Faculty of Environmental Engineering, Lublin University of Technology, ul. Nadbystrzycka 40B, 20-618 Lublin, Poland, phone: +48 81 538 44 81
[2]Graduate of the Lublin University of Technology

Abstract. The analysis of water losses should precede the decisions on repairing or modernizing a water network. Water balance and water losses indicators established by the International Water Association (IWA) standards can constitute the basis for the analysis. The methods recommended by IWA are gaining increasing popularity in many countries, including Poland. The aim of the paper is the analysis and comparison of water losses in two middle-sized water distribution systems during the period of 10 years. The compared networks are similar in respect to many parameters, including water intensity indicator value (circa 48 $m^3/d/km$). Analyses were conducted on the basis of water losses indices recommended by IWA, such as Real Losses Level per connection per day (*RLL*), Non-Revenue Water Level (*NRWL*) and Infrastructure Leakage Index (*ILI*). The results indicated that water losses in both systems are lower than in the literature data for other similar networks.

1 Introduction

The problem of water losses in distribution systems is one of the main concerns of the water systems managers both in Poland and around the world. It is commonly known that real losses are caused mainly by leakages as a result of water network breakages and failures. Therefore, it is not surprising that many scientists use the most recent technological and scientific developments to work out efficient methods for limitation of failures and their results [e.g. 1–8]. However, it should be emphasized that real losses of water cannot be eliminated totally [9].

International Water Association (IWA) proposes four methods of leakage management [9]: Active Leakage Control, Pipeline and Assets Management, Speed and Quality of Repairs and Pressure Management. An analysis of water losses in a distribution system on the basis of the IWA balance should be an integral part of the management and should precede the decisions on repairing or modernizing a water network.

The aim of the paper is to analyze of water losses in two middle-sized water distribution systems during the period of 10 years (2005-2014). The obtained values of the performance indicators enabled to evaluate the condition of the systems.

[*] Corresponding author: a.musz@wis.pol.lublin.pl

2 Materials and methods

The first stage of our investigations involved acquiring operating data from the water companies in two towns – A and B. Part of the data that were not measured by a company and could not be determined exactly, were estimated on the basis of literature or on the basis of a network operator's experience. The acquired data were used to create a simplified IWA water balance and to calculate selected performance indicators of water losses. An analysis of the calculation results enabled to evaluate the condition of the water networks.

2.1 Description of the networks

Two analysed water distribution systems are located in middle-sized towns (A and B) in the eastern part of Poland. The parameters shown in Table 1 bear a strong resemblance to both systems. However, the system in town A is slightly bigger, which is reflected in the volume of water pumped from intakes and sold to customers as well as in the values of Water Network Intensity Indicator calculated as:

$$WNII = \frac{SIV}{L_m} \tag{1}$$

where: *SIV* – System Input Volume expressed as m^3/day, L_m – length of mains [km].

Table 1. Characteristics of the water distribution systems at the end of 2014.

Parameter	Town A	Town B
Length of mines [km]	83.04	70.20
Number of connections [pcs.]	3 996	3 834
Length of connections [km]	77.68	54.38
Population [person]	29 018	22 463
Average operating pressure [MPa]	0.39	0.30
Water Network Intensity Indicator (*WNII*) [m^3/day/km]	49.18	47.02

The material structure of the networks is different. In the first network (town A), plastic pipes (PVC-U – 46% and PE-HD – 12%) and cast iron (31%) predominate; 11% of pipes is made of asbestos-cement (A-C). The network in town B is made of cast iron (34%), A-C (31%), plastics (PVC-U –23% and PE-HD – 4%) and steel (8%).

2.2 Calculations of water loss performance indicators

A simplified IWA water balance constituted the basis for calculating the indicators of water losses, as shown in Table 2. The values of *SIV* and *BAC* for both systems, as well as *UAC* for the system in town A were obtained from the water companies. *UAC* for the system in town B was assumed as 6000 m^3/yr, according to the information obtained from the network operator. *AL* were calculated as a sum of Unauthorised Consumption (*UC*) and Customer Metering Inaccuracies (*CMI*). *UC* and *CMI* were assumed according to literature [10–11] as 2% and 3% of *SIV*, respectively. *RL* were calculated as a difference between *SIV* and the other three components of the balance shown in Table 2.

Table 2. Components of a simplified IWA water balance.

System Input Volume (*SIV*)			
Billed Authorised Consumption (*BAC*)	Unbilled Authorised Consumption (*UAC*)	Apparent Losses (*AL*)	Real Losses (*RL*)

To evaluate water losses, performance indicators recommended by IWA and commonly used in Poland were calculated, including Real Losses Level per connection per day (*RLL*), Infrastructure Leakage Index (*ILI*) and Non-Revenue Water Level (*NRWL*). The first indicator was determined according to the formula:

$$RLL = \frac{CARL}{N_c} \tag{2}$$

where *CARL* – Current Annual Real Losses corresponding to *RL* from an IWA water balance expressed as dm^3/day, N_c – number of service connections.

Infrastructure Leakage Index (*ILI*) is defined as:

$$ILI = \frac{CARL}{UARL} \tag{3}$$

where *UARL* – Unavoidable Annual Real Losses [dm^3/day], calculated according to the equation:

$$UARL = [18 \cdot L_m + 25 \cdot L_c + 0.8 \cdot N_c] \cdot P \tag{4}$$

where L_c – average distance from property line to a customer meter [km], in Poland usually corresponding to the length of connections, *P* – average operating pressure head [m H_2O].

Non-Revenue Water Level (*NRWL*) was calculated as percent of *SIV* according to formula:

$$NRWL = \frac{SIV - BAC}{SIV} \cdot 100\% \tag{5}$$

The obtained performance indicators results for two systems in question were analysed and used to evaluate the condition of systems according to World Bank Institute Physical Loss Assessment Matrix (Tab. 3) [12]. The values of indicators were also compred to literature data for similar systems in Poland.

Table 3. Components of a simplified IWA water balance.

Technical Performance Category	***ILI***	***RLL* [dm^3/connection/day] at an average pressure of:**			
		20m	**30m**	**40m**	**50m**
A	1-2	<50	<75	<100	<125
B	2-4	50-100	75-150	100-200	125-250
C	4-8	100-200	150-300	200-400	250-500
D	>8	>200	>300	>400	>500

3 Results and discussion

Tables 4 and 5 show the components of simplified IWA water balances for towns A and B, necessary to calculate the selected water loss performance indicators.

The values of *SIV* in both systems remained stable during the analyzed period. The difference between the highest and the lowest value for the first system (town A) equaled 144 628 m^3/yr, which is 10.8% of average *SIV* in the period. The corresponding values for the second system (town B) are 113 200 m^3/yr and 9.4%. The average values of *SIV* for both systems are comparable – the relative difference (related to the first system) equaled less than 16%.

Table 4. Components of the water balance for the system in town A over the period 2005-2014, expressed in m^3/yr.

Year	*SIV*	*BAC*	*UAC*	*AL*	*RL*
2005	1 516 038	1 275 684	68 158.0	75 801.9	96 394.1
2006	1 481 882	1 304 421	72 566.0	74 094.1	30 800.9
2007	1 43 8227	1 268 594	54 307.7	71 911.4	43 414.0
2008	1 438 180	1 230 961	73 702.7	71 909.0	61 607.3
2009	1 459 074	1 166 664	118 491.7	72 953.7	100 964.6
2010	1 408 449	1 204 796	76 734.6	70 422.5	56 496.0
2011	1 445 893	1 227 401	80 564.7	72 294.7	65 632.7
2012	1 397 690	1 217 792	69 171.9	69 884.5	40 841.6
2013	1 392 010	1 190 144	63 078.2	69 600.5	69 187.3
2014	1 361 410	1 133 267	62 297.3	68 070.5	97 775.2

The values of *RLL* for systems in towns A and B are shown in Fig. 1. In each year over the period 2005-2014, *RLL* for the first system was higher than for the second one, with a slight difference in 2007 and the highest discrepancy in 2009. The first system was characterized by *RLL* varying over a wide range (from 78.17 dm^3/connection/day to 126.59 dm^3/connection/day) during the whole analyzed period. For the second system, after 2007 *RLL* appreciable reduced from more than 80 dm^3/connection/day to less than 40 dm^3/connection/day and in the second half of the period in question, it remained stable (about 60 dm^3/connection/day). The values of *RLL* indicate Category A according to WBI Target Matrix (Tab. 3) over the majority of years in the analyzed period (excluding 2005, 2009 and 2014) for the first system and after 2007 for the second system, which means that further loss reduction may be uneconomic in the systems.

Table 5. Components of the water balance for the system in town B over the period 2005–2014, expressed in m^3/yr.

Year	*SIV*	*BAC*	*UAC*	*AL*	*RL*
2005	1 271 800	1 023 164	6 000	50 872	171 067
2006	1 255 700	1 092 502	6 000	50 228	88 505
2007	1 234 500	1 047 422	6 000	49 380	118 124
2008	1 182 700	1 066 423	6 000	47 308	49 546
2009	1 165 900	1 043 028	6 000	46 636	58 476
2010	1 193 600	1 053 277	6 000	47 744	74 690
2011	1 197 100	1 057 341	6 000	47 884	76 980
2012	1 206 700	1 065 872	6 000	48 268	80 486
2013	1 194 500	1 062 548	6 000	47 780	69 372
2014	1 158 600	1 017 788	6 000	46 344	81 677

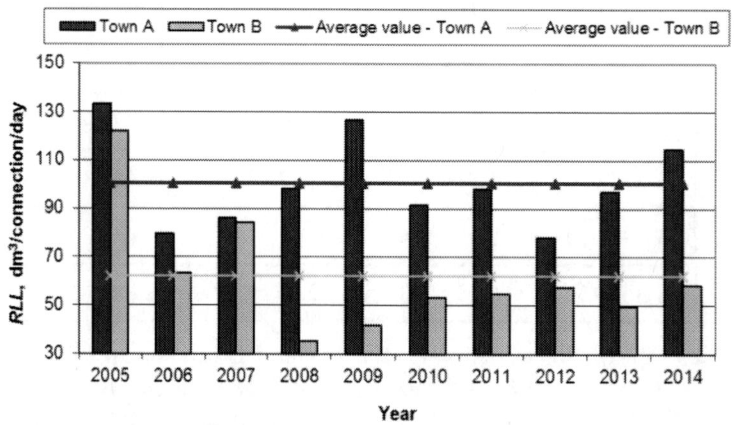

Fig. 1. Operational indicator *RLL*.

The values of *ILI* shown in Fig. 2 correspond to the values of *RLL*. For the first system, *ILI* exceeded 1.5 in 2005, 2009 and 2014, whereas for the rest of the period in question it was in the range of 1-1.5. For the second system, in the first 3 years of the period, *ILI* was clearly higher than in the first system, reaching the values between 1.8 and 3.6. In 2008 for the second system, *ILI* fell to the value of 1.0 and since 2010 it has remained stable (about 1.5) and comparable with the values for the first system. The values of *ILI* lower than 2 indicate Category A, according to WBI Target Matrix (Tab. 3), over the whole period in question for the first system (unlike for *RLL*) and after 2007 for the second system (like for *RLL*) as well. It should be emphasized that *UARL* used in the calculation of *ILI*, is a reliable predictor for a system with more than 5000 service connections, density of connections greater than 20 per km of mains and average operating pressure greater than 25 m H_2O [13-15], and the first of these conditions is not met by any system in question. However, on

 DOI: 10.1051/e3sconf/20171700062

the basis of tests and the analysis, the guidelines for New Zealand [16] recommend reducing this limitation by replacing 3 conditions by a single one: *UARL* calculation should be reliable, if $(L_m \cdot 20 + N_c)$ exceeds 3000. This condition is met by both systems in question.

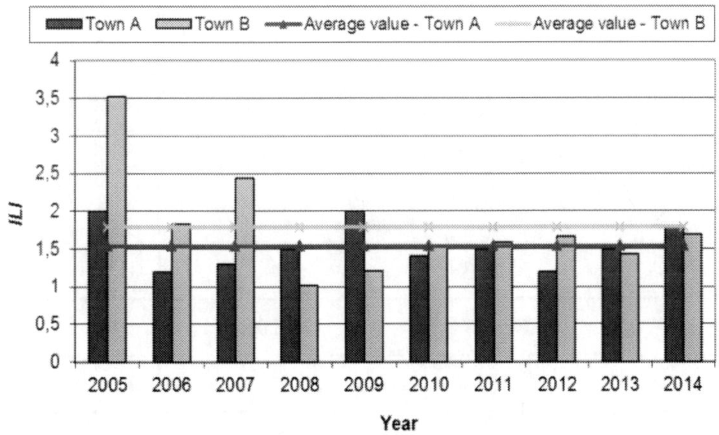

Fig. 2. Operational indicator *ILI*.

The values of *NRWL* (Fig. 3) are higher for the first system in all but 2 years (2005 and 2009) of the period in question. Similarly to the previous indicators, *NRWL* varied in a wide range in the whole period as far as the first system is concerned, whereas for the second system, it remained stable after 2007.

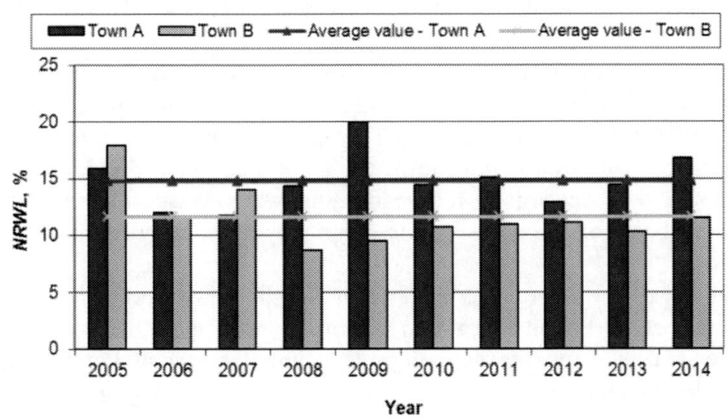

Fig. 3. Financial indicator *NRWL*.

An increased awareness of the significance related to losses management, which has occurred in Poland over the last decade, made calculating water losses performance indicators a popular practice [e.g. 17-22]. Table 6 presents the published average values of *RLL*, *ILI* and *NRWL* for Polish water distribution systems, similar to the examined systems according to *WNII* (formula (1)) or/and population. The values given in Table 6 indicate that water losses in the water systems of towns A and B are definitely the lowest in comparison with other towns. Moreover, *NRWL* for the city of Kos in Greece (population of ca. 17 350 from September to February, and ca. 40 000 from March to August) ranges from 10.7 to 63.7 during the period 1999-2008 [23], significantly exceeding the values calculated for the towns A and B. The results of comparison between the considered Polish towns, other Polish towns described in the literature, and the Greek city can suggest a very good condition of the examined systems, but may also be a cause of doubt pertaining to the accuracy of the estimated components of water balances, necessary to determine the water losses indicators.

Table 6. Water losses in water distribution systems in selected Polish towns.

Town	*WNII*	Population	*RLL*	*ILI*	*NRWL*	Literature
-	[m^3/day/km]	-	[dm^3/(connection/day)]	[-]	[%]	-
Sanok	40.35	39 569	625.83	10.28	40.2	[24]
Myszków	30.68	32 499	159.78	2.26	27.2	[25]
Krosno	38.30	47 307	889.3	7.3	-	[26]
Jasło	47.01	36 363	362.9	5.0	20.3	[27]
Mielec	47.17	60 827	124.20	1.83	24.30	[28]
Town A	**49.18**	**29 018**	**100.34**	**1.54**	**14.7**	-
Town B	**47.02**	**22 463**	**62.09**	**1.79**	**11.65**	-

4 Conclusions

The analysis of water losses in two middle-sized water distribution systems in towns A and B during the period of 10 years suggests their very good condition. The results indicate an improvement in the managing the system of town B after 2007, which was confirmed by the system operator and involved, i.a., reduction of the operating pressure. Average values of the calculated performance indicators were higher in the case of town A, but both can be classified as systems which do not require a further loss reduction according to WBI Loss Target Matrix. The comparison with other water distribution systems also indicates a good condition of the examined networks. However, the ultimate conclusion should be preceded by reinvestigating water losses in the networks in towns A and B on the basis of the IWA water balances with a reduced number of estimated components.

Real water losses are strongly connected with leakages occurring during breakages or failures of a network. In Polish water distribution systems, including systems in tows A and B, they results mainly from the age of pipes, annual range of temperature, freezing of ground, unstable foundation and human mistakes during designing and construction of a network. The activities undertaken to reduce water losses include renewal or replacement of old pipes, optimization procedures for pressure control, detection and location of leakages, operational repairs. However, the activities results are not always sufficient, so it is recommended to supplement the investigation of water losses on the basis of the IWA water balances in the examined systems by failures analysis, to facilitate assessment of the systems conditions and formulating the ultimate conclusion. Thus, reinvestigating water losses in the networks in towns A and B with a reduced number of estimated water balances components, as well as the networks failure analysis will be the subject of our future investigations.

References

1. D. Kowalski, K. Miszta-Kruk, Eng. Fail. Anal. **35**, 736-742 (2003)
2. M. Kutyłowska, H. Hotloś, Eng. Fail. Anal. **41**, 23-29 (2014)
3. M. Iwanek, D. Kowalski, M. Kwietniewski, Ochrona Środowiska **37**(4), 13-17 (2015)
4. M. Kutyłowska, Period. Polytech. Civil. Eng. **59**(1), 37-43 (2015)
5. M. Kutyłowska, Eng. Fail. Anal. **47**, 41-48 (2015)
6. P. Suchorab, B. Kowalska, D. Kowalski, Rocznik Ochrona Środowiska **18**(2), 416-427 (2016)
7. M. Iwanek, B. Kowalska, E. Hawryluk, K. Kondraciuk, Eksploatacja i Niezawodnosc - Maitenance and Reliability **18** (2), 278-284 (2016)
8. M. Iwanek, P. Suchorab, M. Karpińska-Kiełbasa, Period. Polytech. Civil. Eng. Paper No 9728 (2017) (to be published)
9. A. Lambert, Water **21**, 50-51 (2003)
10. G. Merlo, *IWSA Workshop* (Warszawa 1992)
11. T. Bergel, Gaz, Woda i Technika Sanitarna **8**, 322-325 (2012)
12. R. Liemberger, K. Brothers, A. Lambert, R. McKenzie, A. Rizzo and T. Waldron, *Water Loss Conference* **1** (Bucharest, Romania, 2007)
13. A. Lambert, R. McKenzie, *Proceedings of IWA Conference–Leakage Management: A Practical Approach* (Lemesos, Cyprus 2002)
14. W. Winarni, Civil. Eng. Dimension **11**(2), 126-134 (2009)
15. C. Lenzi, C. Bragalli, A. Bolognesi, M. Fortini, Procedia. Engin. **70**, 1017-1026 (2014)
16. R. McKenzie, A. Lambert *Benchmarking of Water Losses in New Zealand. Manual.* (2008)
17. M. Kwietniewski, Ochrona Środowiska **35**(4), 9-16 (2013)
18. A. Choma, M. Iwanek, B. Kowalska, D. Kowalski, Instal **10**, 61-65 (2014)
19. J. Rak, D. Trojnar, Czasopismo Inżynierii Lądowej, Środowiska i Architektury XXXI, 61 **1**(14), 245-256 (2014)
20. K. Pietrucha-Urbanik, B. Tchórzewska-Cieślak: Journal of KONBiN. **1**(33), 233-242 (2015)
21. M. Iwanek, A. Musz-Pomorska, B. Kowalska, D. Kowalski, M. Chołody, Instal **1**(369), 40-43 (2016)
22. A. Musz-Pomorska, M. Iwanek, P. Suchorab, A. Brodaczewska, JCEEA. **1(**I/16), 179-189 (2016)
23. V. Kanakoudis, S. Tsitsifli, Urban Water **7**(5), 267-285 (2010)
24. I. Piegdoń, B. Tchórzewska-Cieślak, Gaz, Woda i Technika Sanitarna **10**, 450-452 (2012)
25. W. Kędzia, E. Ociepa, Inżynieria i Ochrona Środowiska **18(**4), 525-535 (2015)
26. K. Pietrucha-Urbanik, A. Studziński, Gaz, Woda i Technika Sanitarna **10**, 452-454 (2012)
27. J. Rak, Ł. Sypień, JCEEA. **3(**13), 5-18 (2013)
28. K. Kujawska, J. Rak, JCEEA. **2**(I/16), 445-454 (2016)

The changes in the microstructure of ion-exchanged clays

Edyta Nartowska[1,*], *Tomasz* Kozłowski[1], and *Marta* Kolankowska[1]

[1]Kielce University of Technology, Faculty of Environmental, Geomatic and Energy Engineering, Division of Geotechnical and Hydraulic Engineering, av. 1000-lecia Państwa Polskiego 7, 25-314 Kielce, Poland

Abstract. Three natural bentonites and their monoionic forms were investigated by the X-Ray Diffractometry (XRD) and Scanning Electron Microscopy (SEM). SEM photographs were analysed using the Photoshop CS4 software and the pore parameters were calculated. The parameters were compared to a interplanar spacing d001 value changes between different homoionic forms. The results suggest the connection between the type of exchangeable cation with the structural parameters of the pores and physicomechanical parameters of the soil, therefore the ion exchange method could be used for controlling soil properties corresponding to their suitability for use in the industry.

1 Introduction

Recently, there has been growing interest in the application of clays to environmental engineering, environmental protection engineering and industry. Bentonites are a special group of clays, consisting essentially of the crystalline structure montmorillonite, which is responsible for a plasticity, thixotropy, high degree of swelling, particular expansive properties of the soil and a high potential adsorptivity to water. Thanks to the unique properties, bentonites are used e.g. for stabilizing boreholes and deep earthworks, in building guardrails while storing radioactive waste and for removing the toxic and radioactive contaminants from the environment and also for halting the spread of the pollution in the soil, water and air. Some characteristics of the soil, such e.g. clay fraction affect to their engineering parameters [1]. Also, connections between the type of exchangeable cation and soil mechanical properties [2,3] and changes in microstructural parameters [4,5] are confirmed. Despite this, looking for the connections between the microstructure features with the soil properties is very essential.

Homoionic forms of the bentonites were examined with respect to thermal properties [6] and unfrozen water content [7]. In both cases a significant impact of the type of exchangeable cation on the studied parameters was confirmed.

The aim of the study was to determine the changes in the microstructure of clays after the ion exchange, which may affect the properties of the clay soils and thus their suitability for use in a specific areas of engineering and industry.

* Corresponding author: enartowska@tu.kielce.pl

2 Materials

Research material consisted of three model soils: two clays of the Clay Minerals Society SWy-2 from Wyoming and Stx-1b from Texas and one clay BSvk from Jeslovy Potok, Slovakia deposits (imported to Poland by the ZGM Zebiec in Zebiec near Starachowice) and their homoionic forms ($Na^+/K^+/Fe^{3+}/Ca^{2+}$), obtained from natural soils by the repeated saturation of the fraction less than 0.063 mm and subsequent purifying from the solutes by diffusion. The natural forms of clays are characterized by a high montmorillonite content of 85%. Such soil are called 'bentonites'.

A properties of the thirteen tested soils are characterized in Table 1 and Table 2.

A preliminary analysis revealed that external surface area and specific surface area depends on the type of exchangeable cation. For Ca^{2+} form of the bentonites there was an increase in external and specific surface area whereas for Na^+ and K^+ forms a decrease in external and specific surface area was observed. A plastic limit was higher after the ion exchange, regardless of the cation type. Clay fraction content after the ion exchange was dependent on the soil type (regardless of the cation type): it decreased for SWy-2 bentonite and increased for Stx-1b and BSvk bentonites. The reason for similar behaviour of the bentonites from Texas and Slovakia could be a similar content of the major metals after the ion exchange (Table 2). bentonites from Texas and Slovakia are characterized by several times lower content of the Na^+ , three times greater content of the and about two times greater content of the Ca $^{2+}$ than SWy-2 bentonite. In turn, Bentonite from Wyoming has a two times greater content of the Ca^{2+} and five and seven times lower content of the Na^+ and K^+, respectively.

It appears that a decrease in the clay fraction content is associated with an increase in Ca^{2+} cation content in bentonites, what is surprising compared to an increase in external and specific surface area in these soils. This may indicate that a montmorillonite content, being a part of the clay and silt fraction contents, determines a specific surface area of the soil. Adictionally, a probably reason for changes in soil properties could be the differences of the planar-assign distance in a phase of montmorillonite 15A and crystallinity between bentonites with different leading cation.

This observation may be relevant for engineering calculations of the colloidal activity according to Skempton, which depends only on clay fraction content. There is a need to further research towards the method of the determination of the parameters forming the part of formula for colloidal activity (mainly liquid limit and clay fraction content).

3 Methodology

3.1 X-Ray Diffraction (XRD)

X-ray diffraction analysis was conducted by using a Bruker D8 advance powder diffractometer in Debye- Sherrer geometry, Cu-K $_{\alpha 1}$ radiation (λ = 1,5406Å) from a Johansson-type monochromator and LynEye position-sensitive detector. The measurements were carried out at 2θ from 4.51° to 70°. The applied voltage 3540 kV with the 530 mA current.

Phase identification was made based on the database PDF 4+ whereas the estimation of the individual phases was made using a semi-quantitative method in Diffrac Eva program. The main purpose of the XRD study was to determine 001 planes i.e., the planes parallel to the surface of the sample.

DOI: 10.1051/e3sconf/20171700063

Table 1. Properties of the clays investigated in this study.

Property / Clays	SWy-2	Swy-2 Ca^{2+}	Swy-2 Na^{+}	SWy-2 K^{+}	SWy-2 Fe^{3+}	Stx-1b	Stx-1b Ca^{2+}	Stx-1b Na^{+}	Stx-1b K^{+}	B Svk*	B Svk Ca^{2+}	B Svk Na^{+}	B Svk K^{+}
Plastic limit PL [%]	35	44	45	43		55	64	60	67	46	48	57	56
Clay fraction Cl** [%]	45	40	40	40	41	15	25	29	19	15	19	26	18
External surface area*** S'[m^2/g]	55	92	55	50	79	99	104	66	68	105	130	77	52
Specific surface area*** S [m^2/g]	333	552	330	302	474	593	622	394	408	633	780	462	310

*bentonite from Jelsovy Potok, Slovakia **Laser diffraction method ***Water sorption test

Table 2. Exchangeable major cation content in dry soil matrix (mg/kg $_{dry\ soil}$).

Cation content / Clays	SWy-2	Swy-2 Ca^{2+}	Swy-2 Na^{+}	SWy-2 K^{+}	SWy-2 Fe^{3+}	Stx-1b	Stx-1b Ca^{2+}	Stx-1b Na^{+}	Stx-1b K^{+}	B Svk*	B Svk Ca^{2+}	B Svk Na^{+}	B Svk K^{+}
Ca^{2+}	4890	23301	2699	2657	976	10016	14595	730.4	687.5	5083	9694	2941	442
Na^{+}	5009	345	6160	1500	478	1054	211	8353	590.9	587	138	8244	598
K^{+}	1306	912	731	6654	1352	972.5	448	1167	20892	1017	589	1513	17900
Fe^{3+}	3063	3453	2347	2014	86198	3412	3069	2618	3011	5496	5686	4403	4279

 DOI: 10.1051/e3sconf/20171700063

3.2 Scanning Electron Microscopy (SEM)

The experiments were conducted with a Quanta 200 scanning electron microscope at 10 kV voltage applied. The air-dried samples (8-10 mm diameter) of the monoionic bentonites were observed and their microstructures were analyzed.
The surfaces of the samples were covered with a 40 nm layer of gold to prevent electrisation and photographed at two magnifications: x1000 and x5000. The photographs taken at the lower magnification were used to determine the quantitative pore space parameters, whereas the qualitative analysis of the soil microstructure and the types of contacts were based on the photographs taken at x5000 magnification.

The SEM photographs were analysed with Photoshop CS4 – based Numerical Image Analysis (NIA) method. Since the lighter areas on the SEM images of the soils correspond to the mineral particles, while darker areas – to the spaces between the particles, it is possible to identify pores and soil particles and to define the total surface area, height, perimeter, width and the total number of pores. Additional values required for the quantitative analysis, such as the number of ultra/micro pores, were calculated using Excel.

Pores are very important elements of the soil structure: e.g. they store water and affect its transport between the soil particles.[8]. Pores can be classified by size according to Grabowska-Olszewska [9]:

- ultrapores < 0.1 μm- completely filled with the bound water
- micropores 0.1-10 μm- possible capillarity, filtration
- mezopores 10-1000 μm.

4 Results

4.1 XDR results

The peak positions at 2θ and X-ray counts for the natural bentonites and their homoionic forms are shown in figures 1-3. Table 3 shows interplanar spacing d001 values i.e. planes parallel to the sample surface.

For homoionic forms of the SWy-2 bentonite, there was a reduction in the value of the 2θ angle compared with the base material and, according to the Bragg equation, an increase of d001 value.

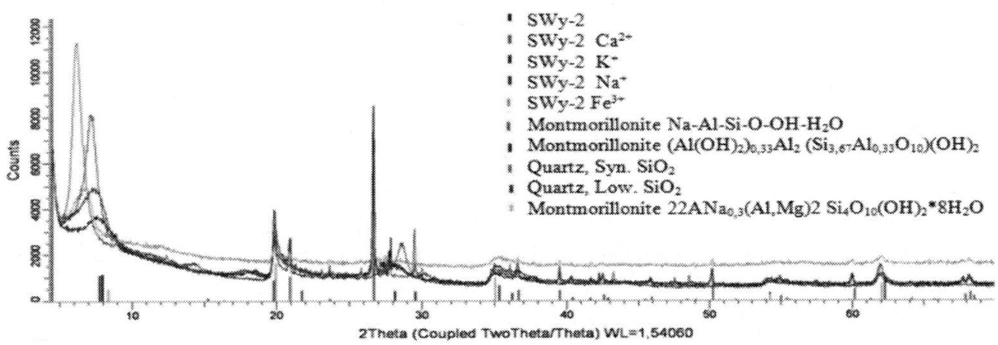

Fig. 1. The peak positions at 2θ and X-ray counts for the natural SWy-2 bentonite and its' homoionic forms.

In the case of homoionic forms of the STx-1b bentonite, there was an increase in the 2θ angle relative to the base material and a reduction of d001 values.

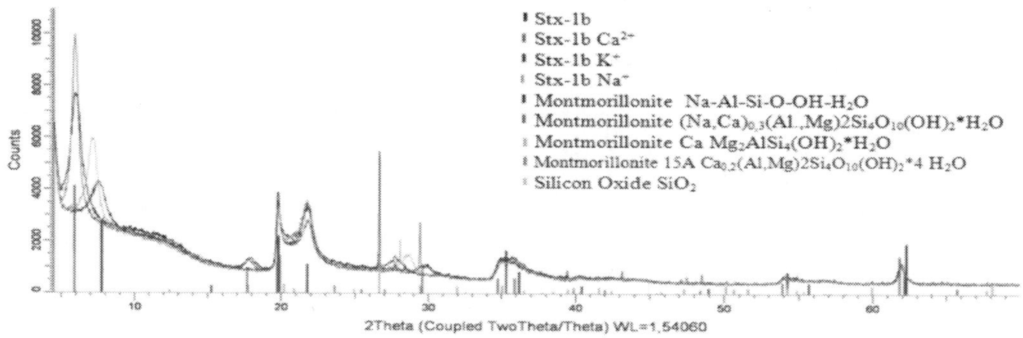

Fig. 2. The peak positions at 2θ and X-ray counts for the natural STx-1b bentonite and its' homoionic forms.

For homoionic forms of the BSvk bentonite there was an increase in the 2θ angle compared with the base material and reduction the interplanar spacing d001 values.

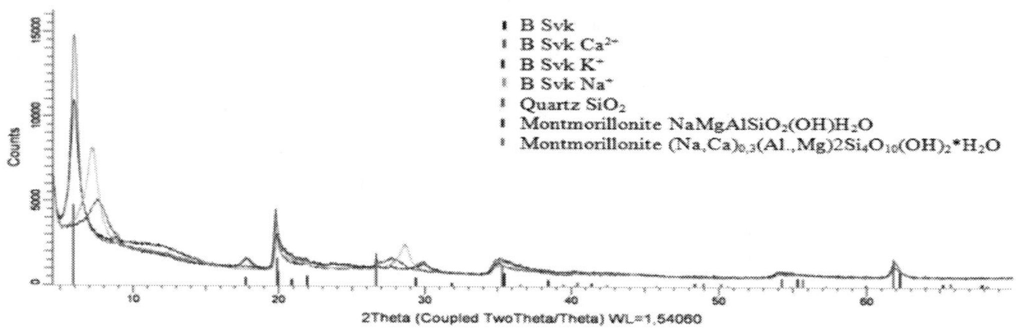

Fig. 3. The peak positions at 2θ and X-ray counts for the natural BSvk bentonite and its' homoionic forms.

Table 3. Collected interplanar spacing d001 values (parallel to the sample surface).

Type of form / Clays	Stx-1b	SWy-2	BSvk
	d001 [Å]		
base	14.88	11.44	14.87
Na^+	12.37	11.95	12.27
K^+	11.66	11.86	11.56
Ca^{2+}	14.84	14.33	14.88
Fe^{3+}	-	13.26	-

4.2 SEM photographs' analysis

The photographs of the bentonite microstructures are shown in Fig. 4–6.

Microstructural analysis [10] of the 5 ion-varieties of the SWy-2 bentonite reveals an almost two-fold increase in the total surface area of the pores for Na^+ modification and two-fold and three-fold decrease in the total area of the pores in/for Ca^{2+} and Fe^{3+} varieties compared to the natural bentonite. Also, a more than four-fold decrease in the number of ultrapores was observed for all the monoionic forms of the bentonite SWy-2.

Ion exchange of STx-1b bentonite caused the significant decrease in the total area of the pores. The largest decrease (4.5-fold) was obtained For the Ca^{2+} form. Aditionally, K^+ and Ca^{2+} bentonite forms were characterized by fewer ultrapores and micropores, whereas in the case of the Na^+ form, an increase in the numbers of ultrapores and a decrease in the number of micropores were observed.

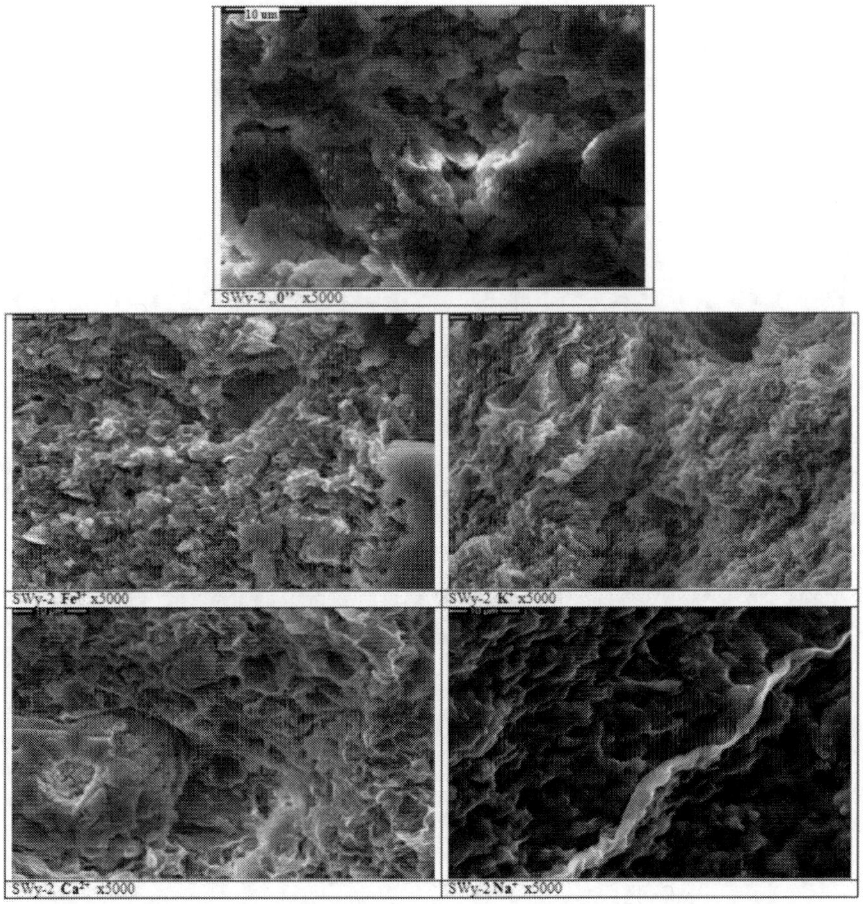

Fig. 4. SEM photographs of the natural SWy-2 bentonite and its' homoionic forms.

Fig. 5. SEM photographs of the natural Stx-1b bentonite and its' homoionic forms.

The K^{+} form of the BSvk bentonite was characterized by almost two-fold decrease of the total pore surface area and the number of micropores and 1.5-fold increase in the number of ultrapores. In the case of Ca^{2+} form, there also was a two-fold decrease in the total pore surface area, but the number of ultrapores and micropores decreased respectively 2 and 3 times. For the Na^{+} form, the following changes were observed: a small increase in the total pore surface area, a two-fold increase in the number of ultrapores, which probably occurred at the expense of reduced number of micropores.

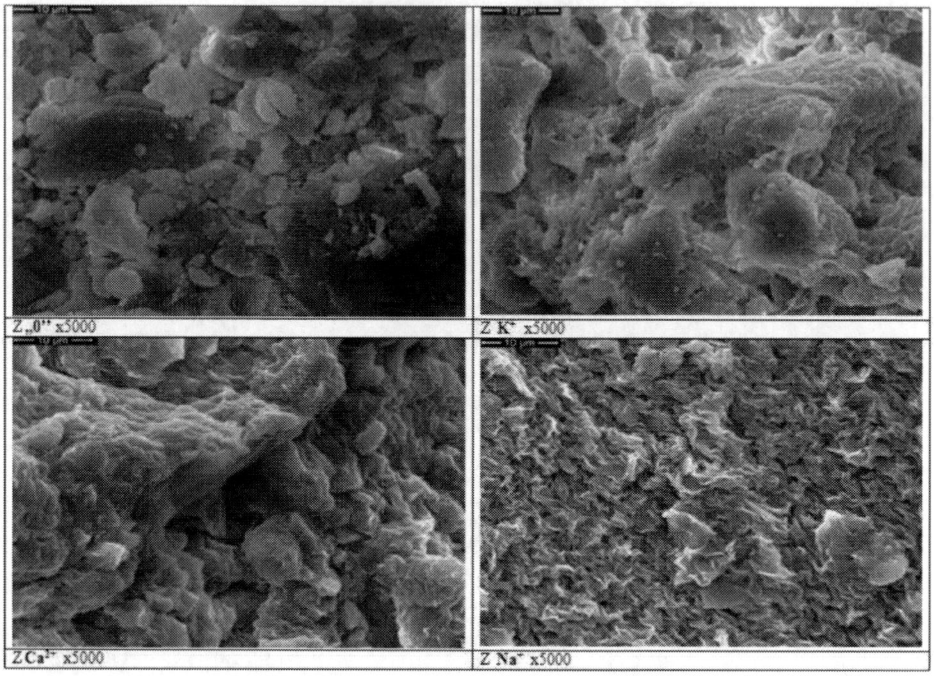

Fig. 6. SEM photographs of the natural BSvk bentonite and its' homoionic forms.

Furthermore, dependencies between pore structural parameters and the major cation (Na^{+}, K^{+}, Ca^{2+}, Fe^{3+}) were found, regardless on the soil type. The observations are gathered in Table 4.

Table 4. Changes in structural parameters of the pores.

Parameters	The major cation in clays (SWy-2; Stx-1b, BSvk*) after ion exchange			
	Na^{+}	K^{+}	Ca^{2+}	Fe^{3+}
Total surface area of pores [μm^2]	1.5-fold decrease for Stx-1b and 2-fold increase for SWy-2(higher Cl)	2-3 fold decrease for BSvk and Stx-1b **	2-4 fold decrease	3–fold decrease
Total circuit of pores [μm]	1.5-3 fold increase	decrease for BSvk and Stx-1b** increase in SWy-2 (higher Cl)	3-4 fold decrease for BSvk and Stx-1b (higher S, less Cl)	decrease

ultrapores	1.5-4 fold increase in quantity	2-4 fold increase in quantity in SWy-2 and BSvk *** and 2 fold decrease for Stx-1b (higher S)	2-2.5 fold decrease in quantity in BSvk and Stx-1b 4 fold increase in quantity	4-fold increase
micropores	2-3 fold decrease in the average area	decrease in quantity in BSvk and Stx-1b** decrease in the average area for BSvk and SWy-2**	1.5-3 fold increase in quantity and the average area for BSvk and Stx-1b 2.5- fold decrease in the average area for SWy-2	2-fold decrease in quantity and 3-fold the average area

* bentonite from Jelsovy Potok, Slovakia ** soils have a similar the amount of clay fraction (Cl) *** soils have a similar the specific surface area (S)

5 Conclusions

On the basis of the soil microstructure analysis, XRD results and physical properties comparison it can be assumed that the specific surface area of bentonite is related with the type of exchangeable cation. Because of that the specific surface area is connected with sorption capacity of the soil, the ion exchange in soils may be considered as a tool to control their properties.

In addition, the type of exchanged ion affects the d001 plane value. The XRD and SEM results indicate that the d001 planes have a close relationship with the number of pores in the soil. The increase in interplanar spacing d001 value causes an overall increase in the number of pores, in particular micropores. Decrease in d001 value leads to a reduced number of pores, in particular micropores.

A type of exchangeable cation has the influence on the pore structural parameters and physicomechanical parameters of the soil, but the type of the soil is not without importance and determines the intensity of changes after the ion exchange.

References

1. K. Clare, Road Research Laboratory Harmondsworth, Middlesex
2. A. R. Dexter and, K. Y Chant., Journal of Soil Science, **42**, 219–226 (1991)
3. Solanki, P., and Zaman, M., *Scanning Electron Microscope*, V., In-Tech Publisher, Chapter **38**, 771-798 (2012)
4. Y. M. Sergeyev, B. Grabowska-Olszewska, V. Opisov et al., Journal of Microscopy (1980)
5. B. Grabowska-Olszewska, Prz. Geol. **49**, 299–302
6. T. Kozlowski, A. Rusin, E. Nartowska, Appl.Clay.Sci. (2014)
7. T. Kozlowski, E. Nartowska, Vadose Zone Journal (2013)
8. B. Grabowska-Olszewska, , Przegląd Geologiczny, **31**, 162-165 (1983)
9. W. Chesworth, *Encyclopedia od Soil Science*, **476** (2007)
10. B. Grabowska-Olszewska, V. Opisov, V. Sokolov, *Atlas of Microstructure of Clay Soils* (1984)

Review on the quality of sediments from the stormwater drainage system in the urban area.

Nicole Nawrot[1,*], and *Ewa* Wojciechowska[1]

[1]Faculty of Civil and Environmental Engineering, Gdańsk University of Technology, 11/12 Narutowicza Str, 80-233 Gdansk, Poland

Abstract. The main task of the stormwater drainage system (SDS) is a safe drainage of rainwater and snowmelt from the urban area to the receiver. The flow of rain water in the drainage pipes is directly related with the formation of sediments in the whole stormwater system. In addition, pollutants from land runoff get adsorbed to the sediments. The sludge is mainly formed in those elements of SDS, wherein the flow conditions allow for sedimentation. This article provides an overview of the literature concerning the characteristics of sediments from SDS, with a particular focus on heavy metals in sediments deposited in the urban catchment area.

1 Introduction

Stormwater drainage systems (SDS) discharge rainwater or snowmelt runoff generated within the catchment to receivers – rivers, lakes, reservoirs [5]. SDS is equipped with devices to separate pollutants [4], which should be removed before rainwater is discharged to the receivers. These are stormwater inlets, sedimentation and retention tanks. In urban areas the amount of sludge produced in these devices is variable and usually difficult to estimate. This is primarily due to the lack of requirements for monitoring of the sediments. The most frequently analyzed and controlled elements are the concentrations of pollutants in stormwater which have limit concentrations while discharged to the receiver – soil or surface water [20]. The major sources of stormwater pollution in urban area are traffic, atmospheric dust, products of fuel combustion and to a lesser extent fertilizers and herbicides [1]. The suspended solids (SS) contained in runoff is the most important in terms of quantitative and qualitative evaluation of sediments. The SS fulfill an important role in the transportation of other pollutants. In storm sewage from urban area mineral SS dominate the organic SS. Mineral suspension constitutes about 80-99% of the total suspension. Characteristics of rainwater greatly influences the quality of deposits set aside in the SDS. Being aware of the presence of sediments in SDS it should be taken into account that the sediments must be periodically removed and properly managed or disposed in order to maintain proper operation of the system. Thus, it should be assessed to what extent the sediments from SDS are contaminated and whether they constitute a threat to the environment and human health. Unfortunately, in Poland there are neither quality standards

* Corresponding author: nawrotnicole@wp.pl

for sediments from SDS nor technical guidelines for the treatment and disposal of sludge. Helpful in assessing the quality of sludge from SDS may be Geochemical criteria developed by the Polish Geological Institute PIB [7], ecotoxicological criteria (PEC Probable Effects Concentration) [6] and German LAWA classification 1997 (Lander-Arbeitsgemeinschaft Wasser) [19]. In this study, the data available for sediments from SDS and bottom sediments deposited in the retention tanks, receiving runoff from the urban area was collected. The speciation analysis of heavy metals (HM) in sediments, which is essential for assessing the scale of the potential hazard to the environment is also discussed. The article presents guidelines that can be useful for the study of sediments from SDS.

2 Quantity and quality of sediments from SDS devices and retention tanks on the basis of literature data

2.1 Elements of SDS, where the sludge is formed

The initial part of the SDS, which accumulates sediments are storm inlets with sediments tanks and local devices used to treat runoff – settling tanks and separators of petroleum substances [1]. A different part of the system are retention reservoirs, whose primaly function is flood protection. Sediments in reservoirs are formed as a result of the accumulation of autochthonous and allochthonous substances. While deposits from local storm sewage treatment facilities don't pose a direct danger to man, the retention tanks are public and often available to the fishery. Since most of the potentially hazardous contaminants (like HM) are deposited in sediments, it is important to assess the bioavailability of HM. Thus, apart from the total HM contents in sediments, speciation analysis is often used that allows for the quantitative determination of the various chemical forms in which the metals are bound in the bottom sediments [16].

2.2 Quantitative assessment of sediments

The amount of sludge produced in SDS depends on many factors, such as water content, particle size distribution, the efficiency of pretreatment. Quantification can often be affected by a material error, which creates a problem with designing of sedimentation tanks. The most effective way to assess the quantity of sludge is by far the observation of existing devices for at least one year time [1]. The data published in the literature is very divergent. Environnement Groupe Saint Dizier [1] estimated the quantity of SS removed annually from the surface areas of various development. The amount of SS discharged from urban centers during one year is at least 490 [kg/ha·a], a maximum of 4280 [kg/ha·a], an average of 1590 [kg/ha·a]. For industrial areas, these values are at minimum of 450 [kg/ha·a], maximum 1700 [kg/ha·a], an average of 720 [kg/ha·a]. Dąbrowski [15] gives the number of SS discharged from one hectare of sealed surface by runoff for several catchment areas of the basin from 347 to 2340 kg suspension, on average 487 [kg/ha·a]. Still, load of SS discharged per year per one ha of sealed area doesn't reflect directly the amount of sludge in elements of SDS. Herold P. Guy already in 1970' presented the problems with sediments deposited in Urban areas (Washington 1970). Even then, it was recognized that the problem of control of the amount of sediments should be respected in the same level of importance as other environmental problems [10].

2.3 Characteristics of sediment according to previous studies

One of the few information about sediments deposited in the storm inlets comes from studies carried out in Bialystok [1, 8]. Research was conducted with street inlets in areas with significant traffic. The analysis showed low usefulness of these deposits for recultivation due to high content of HM. Another study was performed in 1996–1998 [1]. Four inlets were selected, of different exposure to anthropogenic pollution. Three of them were at the streets of varying intensity of traffic and one at the bicycle path. The results were described by Grabarczyk in 1999 [8], who assessed that the sediments from street inlets have higher concentrations of substances of anthropogenic origin (eg. four times higher value of hydrocarbons in comparison to the inlet located on the bicycle path, similar disparities in case of HM). The most contaminated sediments were recorded after snowmelt. Sludges from storm inlets predominantly consistied of mineral substances [1,8]. The largest share of the organic substances was recorded in the inlets, which accumulated a large amount of leaves and plant debris. Most oily deposits came from the grooves serving the drainage areas with significant traffic. Most total Kjeldahl nitrogen was observed in the inlets located on roadways with high traffic. In studies done by Krolikowski et al. [1] it was showed that Zn constituted of more than 68% of the total HM content. The analysis indicated that inlets can be divided into two groups which can be referred to as "low" or "highly loaded". Inlets highly loaded serve the catchments with significant traffic, city centers. The low loaded grooves are located within the residential districts of detached houses, serving partly pedestrian and cycling paths and streets with little traffic [1]. In reference to the cited classification (Tables 2-4) only the concentration of Cd was exceeded in sediments from storm inlets [1]. The study of sediments from sewage pretreatment devices was also performed by researchers from Bialystok [1]. Sludges from these devices are characterized by a higher water content in comparison to the inlets sediments. The concentration of total Kjeldahl nitrogen is also related to the nature of the basin. Like in the case of the sediments from storm inlets, Zn was dominating HM (representing close to 70% of the investigated metals). Krolikowski et al. (2005) determined the relevant statistical relationship between the concentration of Pb, Zn, Cr and Cu. The contents of individual HM in the sediments was closely linked to each other (correlation coefficients within 0,85-0,97) [1]. Szyprowska et al. [9] studied physico-chemical properties and grain size of sludge produced in elements of SDS from the streets in the agglomeration of Warsaw in 2010-2011. They chose several devices located at the streets with high average daily traffic ADT over 25000 vehicles per day. In all samples the dominance of mineral SS above the organics was found (73,1-97,1% dw). For the sample collected from the separator on the Siekierkowska route of ADT 100000 vehicles per day, exceeded concentrations of Zn and Pb were observed according to the geochemical classification (Table 2) and LAWA (Table 4). Szyprowska et al. (2012) confirmed that the same sediments accumulated in sediment-flotation devices contain approximately 80-90% of the hydrocarbons present in the runoff from roads. The evaluation of sediments coming from sedimentation tank located at Piekoszowska street in Kielce was conducted by Bak et al. (2012) [11]. Sludges were examined in the context of environmental hazards and the possibilities for their disposal. Sediment grain composition showed a high content of dust and clay fractions with a grain diameter less than 0,063mm (65-72%), which is responsible for the sorption capacity of pollutants. 42% is the fraction of grain diameter of less than 0,025 mm, the remaining fractions were sand and organic debris. Sediments showed the diverse content of HM. Large disparities between the samples taken at the inlet and outlet of the tank were observed for Zn (inlet 70,8 mgZn/kg dw, outlet 750,9 mgZn/kg dw). The source of Zn are the products of weathered surface streets, broken discs and tires. The presence of carbonates was revealed, which in combination with a weakly alkaline pH of sediments suggests that metals are present in the forms of sparingly soluble compounds and, therefore weakly mobile in the environment. Salata and Bak (2015) [12] also performed an analysis of sediments of two retention tanks within the city of Kielce – open reservoir next to Witosa street (in use since 2003, W_{IT} – urbanized area) and closed reservoir next to Jesionowa street (in use since 2001, J_{ES} – industrial area). The analysis of pollutant concentrations in the sediments of these tanks has shown that, in relation to the geochemical criteria (Table 2) the limit concentrations of Zn were exceeded in the sediment tank W_{IT}, thereby classifying sludge from this tank in group II. For the sludges from tank J_{ES} indicator metals concentrations were not exceeded, so the overall assessment of sediment ranks as the second class according to geochemical criteria. The analysis of Ecotoxicological PEL indicator (see Table 3) showed that the sediments W_{IT} and J_{ES} were harmful to the environment and had a negative impact on living organisms, since the concentrations of Pb and Zn were exceeded. In 2012 Bak, Gorski et al. conducted a survey for the bottom sediments of the Suchedniow tank [13] – dam reservoir. The forests and arable lands constituted the greatest part of the catchment (45% and 18%, respectively). The share of residential areas did not exceed 5%. In the period 2009-2011 in the surrounding area the works related to the construction of the expressway S7 were performed. In the course of this work there was a significant increase in the volume of sediments deposited in the reservoir and the deterioration of water quality (high share of mineral SS). Concentrations of HM showed large spatial heterogeneity in the studied sediments. According to the geochemical classification (Table 2) Suchedniow tank sediments could be classified in class II for the content of Cr, Cu, Ni, Zn, class III because of Pb and class IV due to the Cd content. The share of Pb and Cd in the general contamination of sediments was greatest. Dabrowska (2009) studied speciation of HM in sediments from Kozlowa Gora tank. The highest concentrations of HM were found for Zn and according to geochemical criteria (Table 2) were classified as contaminated sediments. In addition, the value of the ecotoxicological indicator PEL (see Table 4) for Cd (3,5mg/kg) was exceeded in two samples, for Zn (315mg/kg) for one sample and for Pb (91mg/kg) in sediments from all measurement points. Kozlowa Gora sediments were characterized by a low share of the ion-exchangeable fraction of metals. Most Zn and Cr were found in the fraction of practically insoluble compounds. Cu occurred mainly in the organic fraction, and Pb and nickel in conjunction with the carbonates. Cd was held in a fraction of hydrated iron oxides and manganese in the carbonate fraction depending on the measurement point. In 2006 K.V.Heal et al. characterized the quality and quantity of sediments deposited in three ponds and one wetland receiving urban surface run-off in Dunfermline, Scotland, UK. The aims of the study were to determine the correlations between the concentrations of HM and provide recommendations for the design and maintenance of sustainable urban drainage system ponds (with regards to sediments issues). The concentration of Cr, Cu, Ni, Zn increased significantly in the sediments in all retention tanks from 1999 to 2002, though concentrations of most metals returned to 1999 values in 2003. The highest concentrations of HM were found for Ni (average value AV from 63,6 to 83,9mg/kg) for all ponds and according to geochemical criteria (Table 2) the sediments could be classified as II cleanlines class. Elevated concentrations of metals were also observed for Cr (AV 118 mg/kg) for one of the analyzed ponds in Duloch Park. Sediment depths varied within each pond, with sediment initially accumulating near inlets and in the primary basins [24]. Nine retention tanks receiving highway run-off in urbanized areas throughout the central and southern part of Florida were selected to investigate HM accumulation in bottom sediments. Sediment cores were sampled from each studied retention tank. The top layer of sediment core was the most hydrated. The moisture content in the top layer varied from a maximum of 76.4% at the North Miami Beach pond to a minimum of 49.6% at the New Smyrna pond. The greater part of metals (particularly Pb, Zn, Cu) showed significantly higher concentrations in the top loose sediments. These metals amounted to more than 76% of total metals detected in urban and highway runoff, excluding iron. High content of metal concentrations reported for Fe (from 236mg/kg (Clearwater, firm sediment) to 8367mg/kg (Ocala, loose sediment), Pb (max concentration 1047mg/kg – North Miami Beach), Zn (max concentration 538mg/kg – Orlando) and Cd (max concentration 53mg/kg – Fort Myers). According to the mentioned classification, the sediments should be classified as contaminated [25]. Comparison of heavy metal content of sludge from tanks shows Table 1.

Table 1 Summary of the contents of heavy metals in retention tanks [mg/kg].

Spot	Type of treatment system	Location/form of sediment		Sediments metals concentration						References
				Zn	Pb	Cu	Ni	Cr	Cd	
Poland	Inlets in Białystok	Inlet low loaded		28.1-167.0	0-32.5	2.5-22.2	-	3.2-9.4	-	[1]
	Inlets in Białystok	Inlet highly loaded		30.0-429.0	0-76.6	2.6-44.6	-	4.1-20.2	-	
	Elements of SDS Warsaw	Siekierkowska route		1142	116.4	-	-	-	-	[9]
	Retention tank (open) Kielce	Inlet		70.8	1.0	9.5	-	-	0.1	[11]
		Outlet		750.9	4.4	2.6	-	-	0.2	
	Suchedniow tank	Retention tank		177.5-216.9	59.4-110.0	26.4-42.9	27.3-31.4	37.4-63.9	5.0-10.7	[13]
	Kozlowa Gora tank	Tank		84-627	41-121	2.8-11.2	3.1-16.1	1.7-15.4	0.8-7.6	[14]
Dunfermline, Scotland	Halbeatth Pond	Retention tank	1 inlet. 1 outlet	5.5-151.3	0-57.8	9.6-28	14.9-111.7	4.9-136.5	0-0.75	[25]
	Linburn Pond		4 inlets. 1 outlet	20.6-199.4	5.8-45	5.6-36.2	28.6-108.2	0-165.2	0-0.64	
	Pond 7		3 inlets. 1 outlet	52.2-101.8	8.7-27.6	9.9-22.7	27.9-145.3	8.0-128.0	0-0.81	
	Wetland pond		2 inlets. 1 outlet	50-136.2	5.4-39.8	10.0-24.8	6.1-121.1	0-178.7	0-1.33	
Florida	Orlando	Retention tank	loose	538	1025	73	29	68	3	[26]
			firm	43	171	14	19	37	4	
	Tampa		loose	56	94	13	12	30	2	
			firm	12	18	2	9	17	2	
	New Smyrna		loose	20	124	7	44	19	31	
			firm	2	36	3	17	5	13	
	Fort Myers		loose	13	192	7	82	28	53	
			firm	2	64	3	30	17	16	
	Ocala		loose	286	371	31	56	112	2	
			firm	22	33	5	17	15	1	
	Palm Bay		loose	79	237	22	52	128	34	
			firm	4	26	2	6	11	3	
	Melbourne		loose	71	159	46	72	55	1	
			firm	5	40	4	59	7	1	
	Clearwater		loose	91	119	16	37	41	1	
			firm	8	105	5	39	11	1	
	North Miami Beach		loose	291	1047	33	82	71	4	
			firm	10	113	6	57	38	1	
	Average		loose	161	374	28	52	61	15	
			firm	12	67	5	28	18	5	
	S.D.		loose	176	384	21	24	38	20	
			Firm	13	52	4	20	12	6	

2.4 Speciation analysis of HM

The greater part of HM that end up in water is bound and transported with the SS, which leads to deposition of metals with sediments in retention tanks. Contaminants in sediments have a negative impact on the aquatic ecosystem and, indirectly, on the components of the biotope and biocenosis as well as humans. The concentrations of harmful substances in the sediments are much higher than in water. HM are not permanently fixed in bottom sediments. Often metals show susceptibility to migrate from bottom sediments into the water column, causing recontamination. This phenomenon is called the mobility of HM and depends on the form in which the metal is present and physicochemical conditions. Some of the factors which determine the forms of metals speciation in sediments are pH, redox potential value or ionic strength [16]. In acidic soils and sediments (pH = 5,5) labile ionic form dominate, while in neutral (pH = 6,6-7,2) and alkaline (pH above 7,2) soils and sediments less mobile, difficult soluble hydroxy oxygen forms are dominating [16]. Changing environmental conditions, acidity, redox potential, the concentrations of inorganic and organic ligands affect the mobility and bioavailability of the metal [17]. Speciation analysis is specifying the forms of the presence of HM and their quantitative determination, using selective extraction method, and (often) sequential extraction. The results of speciation analysis lead to the conclusion about the origin of the trace elements, their occurrence, as well as the possibility of migration in the environment.

3 The sediments classifications

Based on the analysis of the literature reports several criteria have been presented that may be used for the classification of sediment from SDS.

3.1 Geochemical criteria developed by the Polish Geological Institute PIB

The geochemical classification for aquatic sediment was first presented by Bojakowska and Sokołowska (1998) [7]. This classification presents three classes of sediment cleanliness. The limit value for the quality class is determined using the principle that the contents of anomalous element in the environment takes concentration higher than the sum of the average content of this element and two standard deviations laid down for the population studied. Additionally, geochemical classification takes into account the geochemical background for individual elements.

Table 2 Classification of sediments on the basis of geochemical criteria ppm [mg/kg] [7].

Metal	Geochemical background	Cleanliness class		
		I	II	III
Zinc (Zn)	48	<200	<1000	<2000
Lead (Pb)	10	<50	<200	<500
Copper (Cu)	6	<20	<100	<200
Nickel (Ni)	5	<30	<50	<100
Chrome (Cr)	5	<20	<100	<500
Cadmium (Cd)	<0.5	<1	<5	<20
Mercury (Hg)	<0.05	<0.1	<0.5	<1.0

3.2 Ecotoxicological classification

Ecotoxicological studies allow for assessment of the impact of pollutants in sediments (heavy metal compounds, radionuclides, hydrophobic organic compounds) on elimination of ecologically important groups of benthic organisms. Ecotoxicological classifications are used to estimate the exposure of biocenoses of bottom sediments in water reservoirs to the presence of substances accumulating in sediments, (also undergoing desorption to the aqueous phase). Worldwide there have been many methods developed to determine quality standards (SQG - Sediment Quality Guidelines) [6], which are used to assess the risk caused by the presence of organic compounds and HM in sediments. Some of the indicators are based on empirical methods – these are indicators ERL and ERM (Effects Range-Low, Effects Range-Median [21]), TEL and PEL (Threshold Effects Level, Probable Effects Level [23]), SLC (Screening Level Concentration [22]). Examples of the described indicators for PAHs are presented in the Table below.

Table 3 SQG values for selected polycyclic aromatic hydrocarbons [6].

Compound	SQG µkg				
	Level of threshold effects			Level of adverse effects	
	SLC	TEL	ERL	PEL	ERM
Anthracene	160	50	90	240	1100
Naphthalene	410	30	160	390	2100
Benzo[a]pyrene	400	90	430	760	1600

3.3 LAWA classification from 1997 (Lander-Arbeitsgemeinschaft Wasser)

It was developed in cooperation with the Federal Environment Agency (Umweltbundesamt) in Germany in 1997 [19]. This classification divides the water, bottom sediments and SS into seven classes of purity, depending on the degree of contamination of these components of aquatic ecosystems with HM. Class I refers to uncontaminated sediment, without anthropogenic interference. Class I-II - sediments unpolluted or with very small anthropogenic interference. Class II corresponds to moderately polluted sediments, and it is a reference for the other levels of pollution. Class II-III - twice the Class II - defines moderate to significant contamination of sediments, Class III - four times Class II - defines significant pollution, class III-IV - eight times Class II - refers to a very strong pollution, Class IV - means more than eight times exceeding class II - is an ultimate pollution. Comparative value is the average contents of HM expressed in mg/kg.

Table 4 The concentration of various HM in mg/kg for each class of sediments classification LAWA [19].

Metal	Purity class						
	I	I-II	II*	II-III	III	III-IV	IV
Zinc (Zn)	≤100	≤150	≤200	≤400	≤800	≤1600	>1600
Lead (Pb)	≤25	≤50	≤100	≤200	≤400	≤800	>800
Copper (Cu)	≤20	≤40	≤60	≤120	≤240	≤480	>480
Nickel (Ni)	≤30	≤40	≤50	≤100	≤200	≤400	>400
Chrome (Cr)	≤80	≤90	≤100	≤200	≤400	≤800	>800
Cadmium (Cd)	≤0.3	≤0.6	≤1.2	≤2.4	≤4.8	≤9.6	>9.6
Mercury (Hg)	≤0.2	≤0.4	≤0.8	≤1.6	≤3.2	≤6.4	>6.4

4 Conclusions

1. Quality of sediments deposited in SDS should be monitored and subjected to environmental classification. It is of relevance to develop standards dealing specifically with this type of sludge. The priority should be to determine the exact directions of management and disposal.
2. The most restrictive of the existing sediments quality criteria appear to be eco-defined indicators PEL and TEL.
3. Speciation analysis of HM in sediments from retention tanks allows to determine metals mobility and lability and to assess their susceptibility to migration from the sediments into the water column, and vice versa.

References

1. A.Królikowski, K.Grabarczyk, J. Gwoździej-Mazur, A.Butarewicz, *Osady powstające w obiektach systemu kanalizacji deszczowej* (Monografia PAN, nr 35, Lublin, 2005)
2. H.Obarska-Pempkowiak, M.Gajewska, E.Wojciechowska, *Hydrofitowe oczyszczanie wód i ścieków* (Wydawnictwo Naukowe PWN, Warszawa 2010)
3. A.Kabata-Pendias, H.Pendias, *Biogeochemia pierwiastków śladowych* (Wydawnictwo Naukowe PWN, Warszawa, 1993)
4. E.Wojciechowska, M.Gajewska, N.Żurkowska, M.Surówka, H.Obarska-Pempkowiak, *Zrównoważone systemy gospodarowania wodą deszczową* (Wydawnictwo Politechniki Gdańskiej, Gdańsk, 2015)
5. Z.Suligowski, A.Tuszyńska, Problemy zagospodarowania wód opadowych w Polsce. Kanalizacja ogólnospławna, http://retencja.pl
6. R.Kalinowski, M.Załęska-Radziwiłł, Ochrona Środowiska i Zasobów Naturalnych **40** (2009)
7. I.Bojakowska, G.Sokołowska, Przegląd Geologiczny **46**, 1 (1998)
8. K.Grabarczyk, *Wpływ ścieków deszczowych na zawartość zanieczyszczeń w osadach zatrzymywanych w urządzeniach kanalizacji deszczowej systemu rozdzielczego, rozprawa doktorska* (Politechnika Białostocka, Białystok, 1999)
9. E.Szyporowska, H. Sawicka-Siarkiewicz, A. Nechay, Ochrona Środowiska i Zasobów Naturalnych **54** (2012)
10. H.P. Guy, *Sediment Problems in Urban Areas* (Water in the Urban Environment, Washington, 1970)
11. Ł.Bąk, L.Dąbek, E.Ozimina, A.Sałata, *Ocena jakości osadów pochodzących ze zbiornika otwartego miejskiej kanalizacji deszczowej w kontekście zagrożenia dla środowiska oraz możliwości ich zagospodarowania* (Monografia PAN, nr 100, Lublin, 2012)
12. A.Sałata, Ł.Bąk, Proceedings of ECOpole **9**, 1 (2015)
13. Ł.Bąk, J.Górski, A.Rabajczyk, M.Szwed, Proceedings of ECOpole (2015)
14. L.Dąbrowska, Ochrona Środowiska i Zasobów Naturalnych **49** (2011)
15. W. Dąbrowski, Gaz, Woda i Technika Sanitarna **6** (2001)
16. J.Pawłowski, M.Rozental, A.Drzewińska, S.Neffe, Biuletyn WAT, vol. LXIII, nr **4** (2014)
17. E.Bezak-Mazur, *Specjacja w Ochronie i Inżynierii Środowiska* (Komitet Inżynierii Środowiska PAN, *Monografie*, nr 20, Kielce, 2004)

18. A.Hulanicki, *Współczesna chemia analityczna. Wybrane zagadnienia* (Wydawnictwo Naukowe PWN, Warszawa, 2001)
19. LAWA-Arbeitskreis, „Zielvorgaben" in Zusammenarbeit mit LAWA-Arbeitskreis „Qualitative Hydrologie der Fließgewässer", Beurteilung der Wasserbeschaffenheit von Fließgewässern in der Bundesrepublik Deutschland – Chemische Gewässergüteklassifikation, Berlin (1998)
20. Rozporządzenie Ministra Środowiska z dnia 18 listopada 2014r. w sprawie warunków, jakie należy spełnić przy wprowadzaniu ścieków do wód lub do ziemi, oraz w sprawie substancji szczególnie szkodliwych dla środowiska wodnego (Dz.U. 2014 poz. 1800).
21. E.R.Long, D.D.Macdonald, S.I.Smith, F.D.Calder, Environmental Management **19**, 1 (1995)
22. D.D.Macdonald, Approach to the assessment of sediment quality in Florida coastal waters, vol.1 – Development and evaluation of the sediment quality assessment guidelines, Report prepared for the Florida Department of Environmental Protection, *Tallahasse*, FL : 1–140 (1994)
23. Australian and New Zealand Environment and Conservation Council, ANZECC interim sediment quality guidelines, *ANZECC Report for the Environmental Research Institute of the Supervising Scientist*, Sydney, Australia (1997)
24. K.V. Heal, D.A. Hepburn, R.J. Lunn, Sediment management in sustainable urban drainage system ponds, *Water Science & Technology*, Glasgow (2006)
25. Y.A. Yousef, L'Yu Lin, Win Lindeman, T. Hvitved-Jacobsen, The Science of the Total Environment **146/147** (1994)

The bioaccumulation of heavy metals in *Brassica napus* L. in the area around Turów Power Station, Poland

Alicja Niedźwiecka[1], and *Dorota* Zamorska-Wojdyła[1,*]

[1]Wrocław University of Science and Technology, Faculty of Environmental Engineering, Wybrzeże Wyspianskiego 27, 50-70 Wrocław

Abstract. *Brassica napus* L. is a known bioaccumulator of copper, zinc, cadmium, lead, chromium, nickel and arsenic from soils. The metal ions are accumulated in the roots, stems, leaves and seeds of the oilseed rape. The samples of soils and plants were collected in the area around the Turów power station (Bogatynia city, Lower Silesia). The soil samples were collected from the surface layer of 0-25 cm. Roots, stems and pods of the oilseed rape were used in the study. The environmental samples were digested in HNO_3, 60%, using the Microwave Digestion System. Metal concentrations have been dermined through the FAAS method. Three heavy metals - zinc, copper and lead – have been analyzed. The content of zinc was higher than the content of copper in all samples (plants and soils). In the roots of the oilseed rape higher concentration of metals compared to other parts of the plant was observed. In the soil samples, there was no correlation between the concentration of pollution and the distance from the power plant. Permissible concentrations of heavy metals relative to the standard according to the Polish Ministry of Environment Regulation from September 1st, 2016, have not been exceeded. The permissible pollution indexes (Wn) in soils were exceeded compared to the geochemical background in uncontaminated soils of Poland. The bioaccumulation coefficients of heavy metals indicate lower metal concentrations in plants than in soils

1 Introduction

The main factor for the contamination of soil with metals is the industrial activity. In Poland, soils contaminated with heavy metals are usually observed in heavily industrialized areas. The processes of mining, also the extraction of metal ores, as well transporting, remaking and finally utilizing the products, are the leading sources of anthropogenic contamination of soils. Large amounts of pollutants are brought into the soil and the land from the metallurgical waste on landfills, and also as a result of gas and dust emissions created during the processing of fossil fuels in power and heating plants. Heavy metals can also get into the soil from the exhaust of road and railway transport vessels [1–3].

Heavy metals are are the persistent contamination and are accumulated in biotic elements of the ecosystem. Traditional methods of soil remediation are based on the immobilization of metals or, optionally, immobilization and removal from the environment

* Corresponding author: dorota.zamorska-wojdyla@pwr.edu.pl

(*in situ* and *ex situ* methods). The phytoremediation process can be used in industrial areas and along the roads, utilizing plants to purify the soil-water environment [4-6]. Plants have a natural ability to dispose, collect or remove contaminants from the soil. In phytoremediation specific species of plants are used, which can have lesser ability to acquire the pollutant, but their rate of growth is very high. These include e.g. the oilseed rape (*Brassica napus* L.), mustard greens (*Brassica juncea* L.), maize (*Zea mays* L.). The oilseed rape (*Brassica napus* L.) is known to accumulate copper, zinc, cadmium, lead, chromium, nickel and arsenic from the soil. Metals accumulate in the roots, stem, leaves and seeds [7].

2 Materials and methods

Samples of soil and oilseed rape were collected in June 2016 from the farmlands along road no. 352 in Lower Silesia (district of Zgorzelec, municipality of Bogatynia), in the immediate neighborhood of the Turów power station. In each location three soil samples were collected from ca. 4 m^2 soil surface. Soils from a given location were sieved and averaged after drying, giving one laboratory sample. The soil samples have been collected from the surface layer of the soil profile (up to 25 cm of depth).

Fig. 1 Locations of sampling.

Samples with location no. 2 were collected the nearest to the Turów power station. The remaining sampling places were at a greater distance from the source of the pollution emissions (Fig. 1, Table 1).

Table 1. The origin of environmental samples.

No.	Location
1	Laboratory of the Sewage Research and Sludge Processing Station.
2	Turów Power Station
3	Działoszyn
4	Działoszyn – Bratków Road
5	Bratków
6	Lutkogniewice
7	Krzewina

Analogically, three plants were collected at the same soil surface. Roots, stems and pods of the oilseed rape after separation were dryed and homogenized, giving an averaged laboratory samples. Soil samples of approx. 0.2 g of dry matter and plant samples of approx. 0.15 g of dry matter (the same amount of root, stem and pods) were digested in 8 cm^3 of HNO_3, 60%. Microwave mineralization process was conducted in a closed system in a condition of linear temperature rise (up to about 220 °C) and microwave radiation (power 500 W). The measurments of the concentration of metals were conducted through the FAAS method, with the use of the iCE 3500 Thermo Solaar. The spectroscopy analysis of three heavy metals (zinc, copper and lead) was carried out. The measurements in solutions were made in six replicates (six readings on the apparatus) and then arithmetic mean were calculated.

The accumulation index (Wn) of metal in soil was calculated as the ratio of the average content of chemical element in the contaminated soil versus to the geochemical background. The bioaccumulation coefficient (WB) of metal in a plant (separately for roots, stems or pods the oilseed rape) was calculated as the ratio of the average content of chemical element in a part of a plant to the metal concentration in the soil.

Chemical analyses have been carried out in the certified Laboratory of Toxycology and Environmental Research in the Faculty of Environmental Engineering at the Wroclaw University of Science and Technology.

3 Results

3.1 The content of zinc, copper and lead in soils

The highest content of copper and zinc was identified in soil samples collected from the location closest to the source of emissions, the Turów power station (Fig. 2).

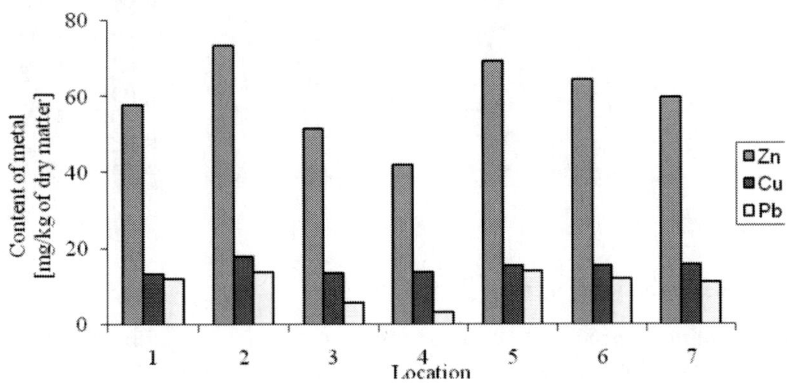

Fig. 2. The content of zinc, copper and lead in soils.

The content of zinc in soil sample no. 2 was 73.3 mg/kg of dry matter. This value is higher than the average zinc content in the Lower Silesia soils (59.8 mg/kg of dry matter [8]). In the area of Bogatynia, from where the samples were collected, the average content of zinc is even higher, 207.7 mg/kg of dry matter. The values of zinc content in soils are in the limit of the average value occurring in the region. The content of copper in soil sample no. 2 was 17.98 mg/kg of dry matter (in the Lower Silesia soils it is 21.5 mg Zn/kg of dry matter and in the area of Bogatynia it is in the range of 7.3 - 19.2 mg Zn/kg of dry matter) [8]. High concentrations of zinc and copper in the soil were also observed in samples no. 5, 6 and 7, collected a few kilometers away from the source of pollution emissions. The highest concentration of lead in sample no. 5 is 14.10 mg/kg of dry matter. Comparable concentrations of Pb were observed in soil samples, in descending order, no. 2, 1 and 6.

Copper is present in soil samples within the range of 13.37 to 17.98 mg/kg of dry matter. In sample no. 1 the lowest copper content in soil was observed. The concentration of zinc is within the range of 42.11 to 73.34 mg/kg of dry matter and the lowest content of Zn is in sample no. 4. Concentrations of zinc and copper exceeded the geochemical background level in surface soils for the whole Poland (background values: Cu – 6.5 mg/kg, Zn – 33 mg/kg, Pb – 18 mg/kg [9]). The concentration of lead in all measurement points was below the geochemical background and ranges from 3.35 to 14.10 mg Pb/kg of dry matter. Its content is the lowest in location no. 4.

The concentration of metals in soils is lower than the limit values according to the standard (the Polish Ministry of Environment Regulation from September 1st, 2016, regarding the manner of conducting the assessment of the contamination of ground [10]) and indicates lack of soil contamination of the investigated area

3.2 The pollution indexes of soils for zinc, copper and lead

Pollution indexes (Wn) were calculated for soil samples, informing about the enrichment of the soil with heavy metals. The Wn factor is expressed as the ratio of the metal in soil to the geochemical background in Poland [9]. The highest, above 2-fold, was the enrichment of all soil samples with copper (fig. 3), in relation to the average value of the geochemical background in Poland (Cu of 6.5 mg/kg [9]). The pollution indexes ranged from 2.1 to 2.8 in the case of copper. The maximum value Wn of 2.8 for copper occured in location no. 2.

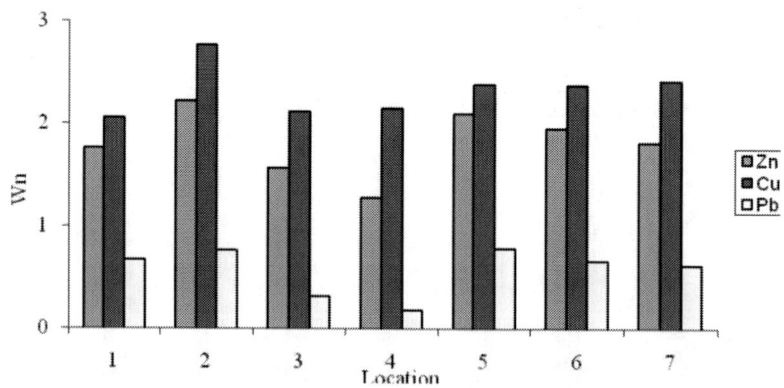

Fig. 3. Wn indexes for heavy metals.

The Wn values for zinc were in the range of 1.3 – 2.2. The maximum accumulation of zinc occured in sample no. 2 (Wn of 2.2). The lowest Wn value of 1.3 was recorded in location no. 4. The concentrations of zinc in soils, as in the case of copper, in all measurement samples were higher than the geochemical background in Poland (Zn of 33 mg/kg [9]).

The Wn values obtained for lead (0.20-0.80) indicated a small accumulation of metal in the soil samples. The pollution indexes do not show the excess of lead relative to the geochemical background (Pb of 18 mg/kg [9]). The highest enrichment of soils with this element were found in samples no. 2 and 5 (Wn 0.80). The minimum value was observed in soil no. 4 (Wn 0.19).

 DOI: 10.1051/e3sconf/20171700065

3.3 The content of zinc and copper in the oilseed rape (*Brassica napus* L.)

In the roots, stems and pods of the oilseed rape (*Brassica napus* L.) the content of zinc, copper and lead was assayed through the FAAS method. The concentrations of lead in the biomass were below the quantification limit for the flame technique of the atomic absorption spectroscopy.

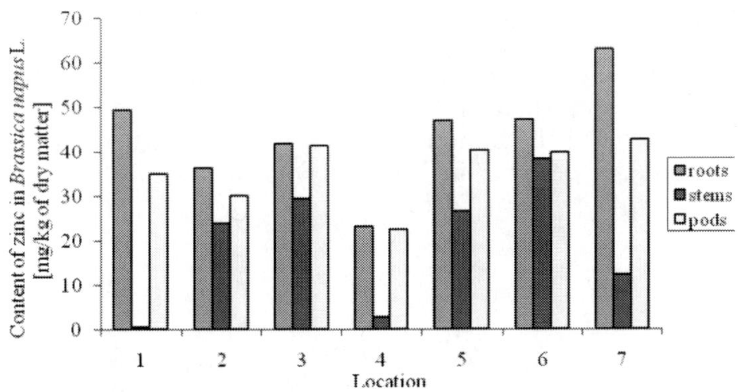

Fig. 4. The content of zinc in *Brassica napus* L.

The highest concentrations of zinc were observed in the roots of the oilseed rape (Fig. 4). The content of Zn ranged from 23.2 to 63.0 mg/kg of dry matter. Decreasing amount of metal was observed sequentially in samples no. 7, 1, 6, 5, 3, 2 and 4. The content of zinc in the pods of the oilseed rape were in the range of 22.5 to 42.8 mg/kg of dry matter.

The stems of the oilseed rape in samples no. 2 and 4 had the content of zinc below 3 mg/kg of dry matter. In other samples content of 12.0 to 38.4 mg Zn/kg of dry matter was observed. Maximum concentration of zinc in the stem was identified in sample no. 6. Comparison of the results with the allowable concentration of Zn in the forage plants from the grassland (50–100 mg Zn/kg of dry matter [11]) showed no exceedance of the 100 mg/kg of dry matter value in the oilseed rape. In sample no. 7 the highest value of 63.0 mg Zn/kg of dry matter was recorded.

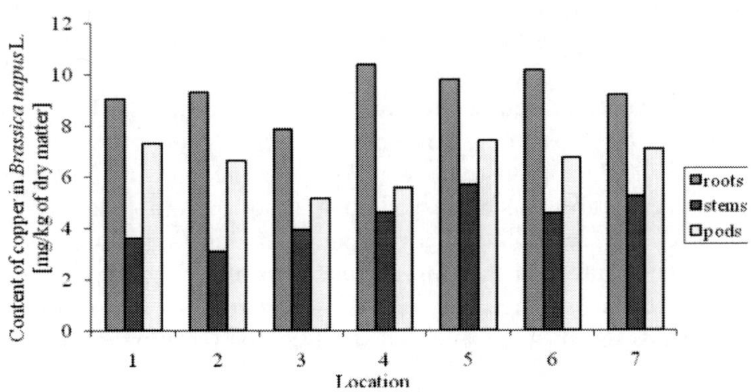

Fig. 5. The content of copper in *Brassica napus* L.

The concentrations of copper in the oilseed rape (*Brassica napus* L.) were significantly lower than those of zinc. Similarly to zinc, the highest concentration of copper was observed in the roots of the oilseed rape from 7.8 to 10.4 mg/kg of dry matter (Fig. 5). Accumulation of copper in the pods was in the range of 5.2 - 7.4 mg/kg of dry matter. The lowest concentration of copper, ranging from 3.1 to 5.7 mg/kg of dry matter, was observed in the stems. Essentially, the copper content in the plants did not exceed the permissible concentration for forage plants (10 mg Cu/kg of dry matter [11]). The exceptions are samples from locations no. 4 and 6, because the values in roots of the oilseed rape were slightly exceeding the limit value for Cu (ca. 0.5 mg/kg of dry matter).

3.4 The bioaccumulation coefficients of zinc and copper in *Brassica napus* L.

In order to check the possibility of accumulation of heavy metals by the oilseed rape (*Brassica napus* L.) the bioaccumulation coefficient, expressed as content of metal in the plant organs to the content of metal in the soil, was calculated. The highest bioaccumulation of zinc in the oilseed rape occurred in the roots, with smaller bioaccumulation observed in pods and stems (Fig. 6).

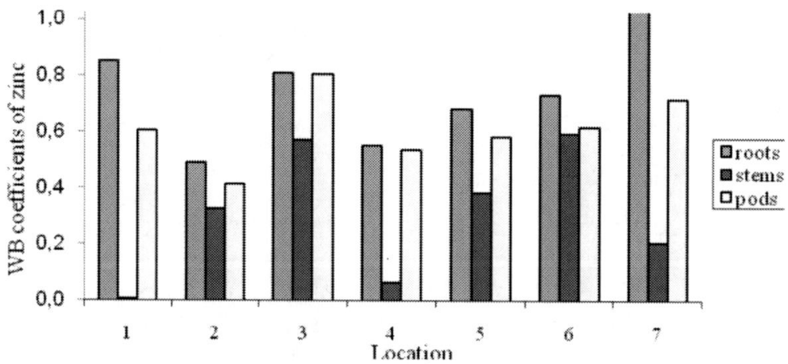

Fig. 6. WB coefficients for zinc.

The WB coefficients for zinc in the roots of the oilseed rape ranged from 0.50 to 1.05, in the pods from 0.41 to 0.80 and in the stems up to 0.60. In sample no. 7 the highest value of the bioaccumulation coefficient of 1.05 for zinc was recorded in the roots. High value of the WB for zinc was also identified in samples of the roots in locations no. 1 and 3: 0.83 and 0.81, respectively. Average values of the bioaccumulation coefficient for zinc indicate smaller bioaccumulation in the tissues of the oilseed rape relative to the concentration of metal in the soil.

Copper is the element bioaccumulated by the oilseed rape to a lesser extent than zinc (Fig.7). The values of the bioaccumulation coefficients in roots are in the range of 0.52 to 0.74, in the pods 0.37–0.55 and in stems of 0,17–0,37. The majority of copper accumulated in roots of the oilseed rape was observed in a sample from location no. 4. The WB coefficient values of bioaccumulation for copper, similarly to those for zinc, indicated lower concentrations of metal in the oilseed rape than the concentration of copper in the soil. Due to the excess of copper and zinc in soil samples relative to the geochemical background in Poland, the oilseed rape can be used for the phytoremediation process of soils enriched with these metals.

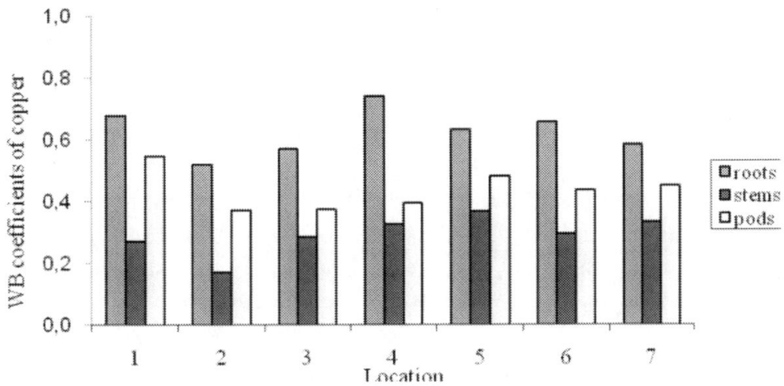

Fig. 7. WB coefficients for copper.

Phytoremediation of soils contaminated with heavy metals conducted under greenhouse conditions and field conditions in the region of Apulia (Italy) confirms the ability of metal bioaccumulation in shoots and roots of the oilseed rape in the following order of Cr > Zn > Cu > Pb. Additionally, the presence of compost and strains of *Bacillus licheniformis* stimulates the removal of metals from the soil and accumulation in the tissues of the oilseed rape. A higher contents of Cr, Pb, Zn, Cd, Ni, and Cu were found in roots than in stems [12]. In biomass of *Brassica napus* L. from agricultural fields irrigated with sewage was accumulated about twice more of Zn and Cu in compared with control [13].

In process of phytoremediation, an important are not only the number of heavy metals and their contents in soil but also interaction between trace elements. Increase of Pb, Cd, Zn contents in the soil cause an increased concentration of these metals in the roots, stems and seeds of *Brassica napus* L. Lead is accumulated mostly in roots, while zinc is present in aerial parts of the oilseed rape. Lead and zinc together in the soil affect the increasing of lead concentration in stems and much lower content of lead in the roots of the oilseed rape, compared with the process conducted in the soil contaminated only through one heavy metal [14]. The oilseed rape shows moderate tolerance to heavy metals and can be used in land remediation moderately polluted with heavy metals [15-16].

In presented studies confirmed the ability to bioaccumulation of heavy metals in *Brassica napus* L. collected from moderately polluted soils around of the Turów power station. The oilseed rape collected from fields can be used as feed. Because contents of zinc and copper in biomass do not exceed of the limit values: 100 mg Zn/kg and 30 mg Cu/kg [11].

4 Conclusion

1. The content of metals in soil samples does not exceed permissible concentrations set by the Polish Ministry of Environment Regulation from September 1st, 2016 (the soil depth of 0 to 0.25 m).
2. In all soil samples enrichment with copper and zinc (average of 2-fold, with respect to the geochemical background in Poland) was observed.
3. Comparison of the content of copper and zinc in soils with average concentrations of these metals in the soils of Lower Silesia (21.5 mg Cu/kg, 59.8 mg Zn/kg) proved that the results are within the average values for the region. In four samples (location no. 2, 5, 6, 7) the content of metals exceeded 1.5-times the average content of metals for the province of Lower Silesia.
4. The pollution indexes (Wn) for copper in soil are higher than for zinc and lead.

5. Sample no. 2 collected from a location nearest to the Turów power station has the highest Wn values for copper (2.8) and zinc (2.2).
6. There has been little excess of copper content in the roots of the oilseed rape (sample no. 4:10.36 mg Cu/kg of dry matter, sample no. 6: 10.14 mg Cu/kg of dry matter) compared with the permissible concentration of these metals for the forage plants from grassland (10 mg Cu/kg of dry matter). There were no exceedances of zinc content in *Brassica napus* L.
7. The values of the bioaccumulation coefficient (WB) for zinc in the oilseed rape are higher than for copper and are as follows: roots - 0.74 (Zn) and 0.63 (Cu), pods – 0.61 (Zn) and 0.44 (Cu), stems - 0.31 (Zn) and 0.29 (Cu). The most substantial accumulation of metals in *Brassica napus* L. was observed in the roots, slighter in the pods and then in the stems.
8. The remediation process of moderately polluted soil with heavy metals around the Turów power station can be carried out with the oilseed rape using.
9. The oilseed rape harvest from the studied areas around the Turów power station can be used as feed.

References

1. A. Karczewska, C. Kabała, Scientific Journal of Wroclaw University of Environmental and Life Sciences - Agronomy **XCVI**, 576 (2010)
2. R. Czubaszek, K. Bartoszuk, CEE, **2** (2011)
3. A. Hołtra, D. Zamorska-Wojdyła, EPE **42**, 3 (2016)
4. A. Astel, A. Cyżyk, A. Parzych, Lab **19**, 4 (2014)
5. S. Gawronski, Agricola **79**, 5 (2011)
6. A. Hołtra D. Zamorska-Wojdyła, EPE **40**, 4 (2014)
7. G. Dabrowska, K. Hrynkiewicz, A. Mierek-Adamska, A. Goc, OilseedCrops **XXXIII**, 2 (2012)
8. *Report on the state of Lower Silesia in 2014, Poland* (http://www.wroclaw.pios.gov.pl)
9. A. Kabata-Pendias, H. Pendias, *Trace elements in soil and plants* (CRC Press, Inc., USA, 2011)
10. The Polish Ministry of Environment Regulation from September 1st 2016, regarding the manner of conducting the assessment of the contamination of ground, Journal of Laws 2016 pos. 1395
11. The Polish Ministry of Agriculture and Rural Development Regulation for January 23rd, 2007 on permissible concentrations of undesirable substances in feed, Journal of Laws 2007 no. 20 pos. 119
12. G. Brunetti, K. Farrag, Geoderma, **160** (2011)
13. T. Cakmakci, Y. Ucar, Pol. J. Environ. Stud. **23**, 6 (2014)
14. V. Angelova, R. Ivanova, G. Todorov, K. Ivanov, Commun. Soil Sci. Plan., **39** (2008)
15. L. Marchiol, S. Assolari, P. Sacco, G. Zerbi, Environ. Pollut., **132** (2004)
16. V.M.J. Grispen, H.J.M. Nelissen, J.A.C. Verkleij, Environ. Pollut., **144** (2006)

Preliminary assessment of the possibility of supporting the decomposition of biodegradable packaging

Lidia Niekraś[1,*], and *Ewa* Moliszewska[1]

[1]Opole University, Department of Natural Sciences and Technology, ul. Kominka 6A, 45-032 Opole, Poland

Abstract. This article presents a preliminary evaluation of the possibility of using grass biomass from a sports field as a compost ingredient which positively affects the degree of decomposition of the biodegradable wrappings. For 5 months the biodegradable bags were stored, both empty and filled with organic waste in the heap of grass clippings. After that period, fragments of the bags were observed under the microscope and then assessed the state of their decomposition. The results indicate that the biomass used favourably affected the process of bag degradation, however the speed of decomposition of the empty bags was quicker than the bags filled with the organic waste.

1 Introduction

Growing amount of the non-biodegradable waste materials obtained from the non-renewable fossil sources, as well as their excessive storage in natural environment, induced materials industry to the search of the alternative solutions that shall be equally beneficial for both environment and consumer. As a result of the mentioned search, in 1995, first biomaterials - materials obtained from polymers which derive both from fossil materials and from renewable materials (biomass), appeared on the market. To the first group of the polymers belong, for instance, PBAT, PCL and PBS, whereas to the second group belong i.a. PLA (poly(lactic acid), polylactide), PHA (polyhydroxyalkanoates), TPS (thermoplastic starch). At the moment, it is worth pointing out that not every biomaterial will succumb to biodegradation in natural environment, as the biodegradation is not polymers' derivation-dependent but its chemical structure-dependent [1]. The examples are, available on the market, renewable sources polymers with the Eko or Bio prefix, for instance, Bio-PE (polyethylene terephthalate), Bio-PP or Bio-PET [2]. The name, that sounds positively, relates only to the technique of their production during which polymers from petrochemical materials are not being used, what contributes, as well, to the reduce of greenhouse gas emissions, however, not to the decrease of the amount of stored waste. Waste that will be made of, in fact, will not decompose in natural environment but are going to be mixed into "plastic" stream of waste.

* Corresponding author: lniekras@wp.pl

In 2015 global production of the plastic materials totalled 300 million tonnes of which biomaterials represented only 1% [3]. European Bioplastics estimates that until 2018 their share in polymer materials will increase from 1.6 million tonnes in 2013 to 6.7 million tonnes per year. However, optimism of this forecast is being dampened down by the fact that the maximum growth of the production is predicted for non-biodegradable biomaterials such as Bio-Pe or Bio-Pet [2]. On the other hand, in case of biodegradable biomaterials, production of the PLA, i.e., polylactide, is going to be the fastest growing one [3].

Packing industry is the biggest customer of the polymer materials. This sector consumes nearly 40% of all polymer materials [4], therefore, it is going to be the main costumer of the biomaterials. What is more, considering the fact that the packets are of short duration parts of the products, it can be expected that in the short run, accumulation of waste from biodegradable packets will rise in natural environment. Even if we assumed that the packets would succumb to natural degradation, the fact that this process needs both an appropriate period of time and proper conditions, must be remembered.

Organic waste is an appropriate place of lodging of the worn biodegradable packets. During process of their liquidation, two basic biochemical treatment methods of waste are being used, i.e., composting or anaerobic digestion [1]. Composting, as the natural utilization, is the less expensive method of developing of the biodegradable waste. It consists in transformation of organic substance with presence of oxygen and microorganisms into CO_2, water and biomass [5]. The appropriate conditions are necessary for the proper process of composting: efficient supply of oxygen, high level of humidity, pH, and appropriately high temperature [5]. It is thought that the temperature is the main factor that decides about the species number and diversity of the composting microorganisms. Mesophilic, termophilic and temperature decreasing phases can be distinguished during the maturation of the compost [6]. Moreover, every phase is characterised by distinct microbiological composition. For instance, in termophilic phase, bacteria are dominant [7], while yeasts are not detected, whereas in temperature decreasing phase, fungi of the *Aspergillus* spp. and *Mucor* spp.,[6] are being observed , and the presence of actinobacteria is noticed in mature compost [8]. The above information shows that the composting must take place in controlled conditions what is possible only during industrial composting. In case of home composting, even the range of temperatures is significantly lower, what makes the process harder and much slower [9].

In purpose of minimising of organic waste storing, the method of home composting is increasingly popular. More and more frequent appearance of biodegradable biomaterials in the stream of waste should be expected. On the market, there are even biodegradable plastic bags, used for household waste bins, which were declared by the manufacturer [10]. According to the above information, if the conditions in compost bin are not appropriate enough, the amount of the plastic film with organic waste will be very large in a short time. Moreover, closing of waste in a bag, may, for example, cause a limit of access of oxygen to the composting components and dieback of aerobic organisms, which can be replaced by anaerobic organisms [11]. This, as the result, may cause a complete inhibition of the process.

Considering, household waste bins' beddings are important for the process of composting [12], so the research of the influence of grass cuttings biomass's potential on household waste films was realized. Grass cuttings were used as the factor stimulating decomposition of the biodegradable plastic bags used for filling in home compost bins. In the present article, a characteristic of grass cuttings' biomass from a pitch was made. In the last part, initial results of the observation of the range of researched plastic film's decomposition, conducted with use of light microscope, are presented.

DOI: 10.1051/e3sconf/20171700066

1.1 A brief characteristics of the lawn grasses creating turf of a pitch

The following three species of the turff grasses are predominant on the Polish stadiums: common meadow-grass *Poa pratensis* L., English ryegrass *Lolium perenne* L. [13] and two subspecies forms of red fescue [14]: Chewing's fescue (*Festuca rubra* ssp. *commutata)* and Creeping red fescue (*Festuca rubra* ssp. *rubra*) [15]. 50-70% of the total mixture of grasses intended for pitches falls for common meadow-grass. Maximally 20% of English ryegrass and about 10% of Red fescue. The selection of such and no other grasses results from their high resistance to mechanical damages (trampling and frequent mowing) as well as high resistance to diseases, including the most frequent in our geographical area - stem rust - caused by the fungus Puccinia graminis and leaf spot, caused by Drechslera sp. Fungi. Furthermore, these are short grasses with firm root system that are tillering strongly[16–20]

1.2 Maintanance that improves the quality of biomass derived from the grass mown on a pitch

The high quality of turf is provided by complex maintenance that can be distinguished to basic, supporting and regenerating maintenance [22]. Providing proper humidity of grass substratum guarantees good growth from planting to the moment of grasses tillering. Average daily need of turf for water is about 4 litres of water/m2 and during spring and summer period it increases to about 7.5-15 litres of water/m^2. As the grasses use nutrients present in the shallow soil, the sterilisation of the layer high in nutrients appears very quickly. That is why the turf of the pitch should be fertilised regularly with nitrogen, phosphorus and potassium (in following proportion: 4:1:2) and regularly aerated what improves gas exchange and increases water absorption [23].

In view of the foregoing, one can expect that the care of how the pitch turf looks will cause that the quality of waste biomass created as a result of frequent mowing and verticulation (i. e. clearing the necrosing parts of grasses) is going to be better than, for example, the quality of biomass received as a result of mowing the city park lawn.

Biomass of the grass species that create turf, in contrast with fodder grasses, is not tall, therefore incineration seems as energy unprofitable. Moreover, in accordance with Turski, organic waste with high humidity and lack of structure, as for example, grasses, essentially qualify for methanogenesis [26]. However, in our country, the development of biogas plants is slow, so other solutions shall be found [27]. Undoubtedly, the less expensive method of organic waste treatment is composting, meanwhile in Polish conditions the dominant is still uncontrolled storage of organic waste. Due to the above, searching for solutions that will change the situation, is justified.

The aim of the experiment was the initial evaluation of the potential of mown grass's biomass from the turf of a pitch to help with decomposition of the biodegradable packets used for bedding of the home compost bin.

2 Experimental materials

Prisms of mown grass from the 8000 square meter pitch of Centre of Sport and Tourism in Dobrzyń (Kuyavian-Pomeranian Voivodeship) were used.

The composition of the mixture of grasses for the turf included: 70% of English ryegrass, 15% of common meadow-grass and 15% of red fescue. Before the planting, bedding of the turf was analysed in laboratory. In accordance with the analysis, professional firm that builds and renovates pitches, matched the proper plan of fertilising and a schedule of basic, supporting and regenerating maintenance. Frequency of mowing and the amount of sourced biomass was not constant and was dependent on growth rate of the grass. From

March to the first half of April the grass was mowed once a week and every mowing was giving on average 1.8m3. Then, from April to June and afterwards, from July to August, the frequency of mowing increased to two times a week. However, in the first case, there was 3.9m3 of biomass per mowing, while in the second case 2.1m^3. In September and in October the turf was mowed once a week and 1.3m3 of biomass was obtained.

Biodegradable bags (according to manufacturer's declaration) (Fig. 1a) were stretched on a wooden frame and in such a form, they were placed in the prism of grass from the pitch (Fig. 1b) in July. In order to compare the results of observation, the prism of grass from the city park lawn, was prepared (the species composition was the same). The city park lawn was not maintained as well as the pitch turf. What is more, it was exposed to excrements of animal origin (Fig. 1c). It was also decided to examine the biodegradable bags according to the producer instruction, and for this purpose they were filled with organic household waste (selected eggshells, peelings from vegetables and fruits). This part of the experiment should show if the substances present in the household waste will really help to increase the process of their decomposition (Fig. 1d). The experiment took 5 months, during this time the prisms were regularly dampened. After this period of time, the range of the examined bags' decomposition was evaluated and the microscopic observation was made.

3 Results

In the experiment the general condition of the composted bags was assessed as well as their structure, colour, flexibility, tearing and cracking and the degree of transparency (Tab. 1). The results clearly showed that the producer' declaration concerning the possibility to biodegradation in the six week period of time is not possible, although some symptoms of biodegradation were noticed. The best degree of biodegradation was observed for the clean bags composted in the prism of the pitch grass and for the bags filled with household waste and buried in the prism of the pitch grass (Tab. 1).

Fig. 1. A biodegradable film, a) a new, not used film – control, b) composted with football pitch grass, c) composted with grass, d) composted with household.

Table 1. Influence of the way of composting on the degree of decomposition of biodegradable bags.

Sample / Feature	1- Control (a new bag)	2 - A bag composted in a prism of municipal grass	3 - A bag composted in a prism of pitch grass	4 - A bag with household waste composted in a prism of cut pitch grass
General condition	*In touch – silky, with no visible delamination Fig 1a*	*Unpolished with no delamination, in the experimental frame partly not changed Fig 1b*	*Unpolished with numerous points of delamination, in the experimental frame partly lack of the film Fig 1c*	*Unpolished but very comparable to the control one (new one) Fig 1d*
Structure	*Guttiferous*	*As a thin paper*	*Parchment*	*As a thin parchment*
Colour	*Light-green colour Fig 3a*	*Light-green colour Fig 3b*	*Khaki colour Fig 3c*	*Khaki colour Fig 3d*
Flexibility	*Has inherent flexibility*	*Lack of flexibility, fragile*	*Lack of flexibility, fragile*	*Lack of flexibility, fragile*
Susceptibility to tearing and cracking	*Stretchable, bursting with adequate strength along the fibers*	*Susceptible to cracking and tearing with any effort*	*Very susceptible to cracking and tearing with any effort*	*Very susceptible to cracking and tearing with any effort*
Clarity	*A low transparency*	*A low transparency*	*A great transparency*	*A very high degree of transparency*

The results of the observation of the 3 samples correlated in table 1, show that in comparison with control sample, the only visible feature in all tested samples is lack of the bag's flexibility. The other features are not visible in all tested samples. The look of a bag composted in a municipal grass is the most similar to the original bag. Its light-green colour and level of clarity (Fig. 3b) is exactly the same as in control sample (Fig. 3a). The differences in susceptibility to tearing and cracking are the smallest there as well. This may suggest that the unmaintained municipal grass additionally exposed to, for example, excrements of animal origin, does not contain substances supporting decomposition (e.g. enzymes) of the biodegradable bag declared by the manufacturer. On the other hand, bags stored in a prism of pitch grass, both the clean ones and filled with organic waste, show the significant differences in comparison with control sample. Their khaki colour (Fig. 3c and 3d), susceptibility to tearing and cracking with any effort, finally the structure of a bag changed from guttiferous to parchment-like or even to structure of a thin tissue paper, may show the presence of factors beneficially influencing the level of biodegradation of examined packets in biomass of cut pitch grass. In order to countercheck this assumption, the further research of biomass of cut pitch grass are necessary. Nevertheless, one may assume that the way of maintenance may have a beneficial influence on the biodegradable potential of biomass obtained in that way (e.g. the amount and quality of enzymes). This can be showed by the fact that both municipal lawn and pitch turf had the same species composition in the research.

However, in the conducted microscopic observations (optical microscope magnifying), such a vast changes in the look of experimental packets (Fig. 2b, 2c, 2d) like in case of macroscopic evaluation (Fig. 1a, 1b, 1c, 1d) were not possible to be observed. It was hard to see even the smallest cracks or perforations in bags that after 5 months confirm progressive process of their biodegradation.

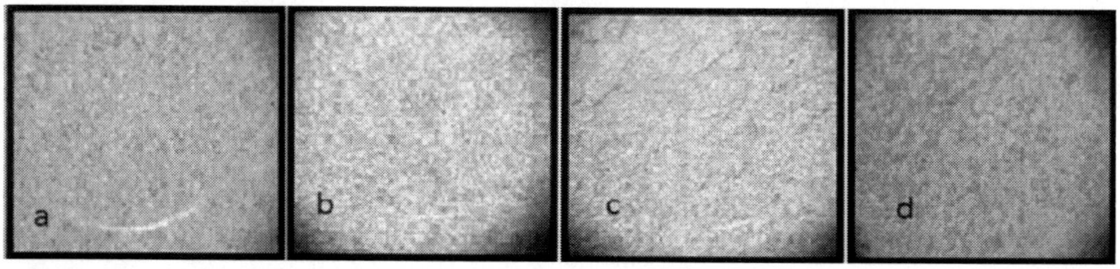

Fig. 2 Microphotograph of biodegradable film, a) a new, not used film – control, b) composted with grass, c) composted with football pitch grass, d) composted with household.

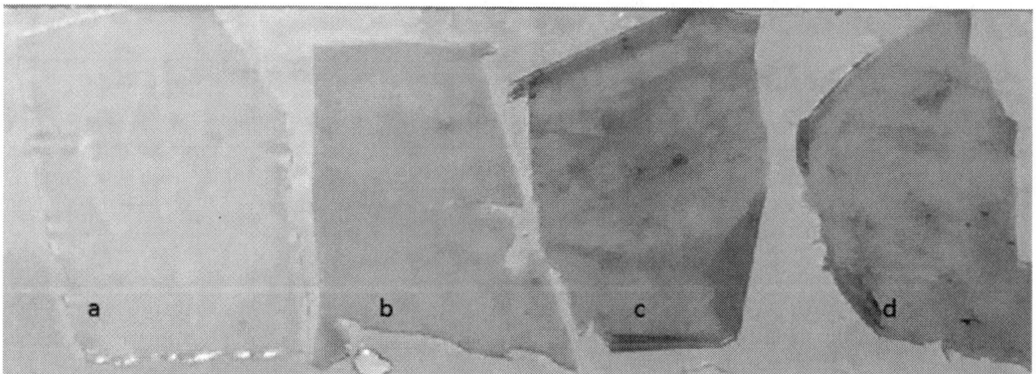

Fig. 3 Macrophotograph of biodegradable film, a) a new, not used film – control, b) composted with grass, c) composted with football pitch grass, d) composted with household.

4 Conclusions

1. Different ways of composting biodegradable bags for household waste proceed in with varying results of the degree and the rate of decomposition.
2. None of the methods of composting of biodegradable bags for household waste was compatible with the manufacturer's declaration.
3. Further research is needed to investigate the conditions of biodegradation of biodegradable bags, including studies using household waste and grass cut from pitch.

References

1. http://www.plastice.org/fileadmin/files/Polish_WEB.pdf
2. R. Malinowski, Inżynieria i Ochrona Środowiska, **18** (2), 215—231 (2015)
3. H. von Pegnelli, Biotworzywa, **1**, 7—12 (2015)
4. http://www.plasticseurope.org/documents/document/20150127125822-tworzywa-fakty_2014_final.pdf
5. A. Jędrczak, *Biologiczne przetwarzanie odpadów* (PWN, Warszawa, 2007)

6. J. Ryckeboer, J. Mergaert, K. Vaes, S. Klammer, D. De Clercq, J. Coosemans, H. Insam, J. Swings, Ann. Microbiol., **53**, 349–410 (2003)
7. P. Manczarski, *Kompostowanie odpadów komunalnych. Referat na Forum Technologii Ochrony Środowiska POLEKO 2007* (2007)
8. C. G. Golueke, Bio Cycle, **33**, 55—57 (1992)
9. http://www.cobro.org.pl/nip/images/stories/PORADNIK/bioplastics pl_internet.
10. L. Nikeraś, Interdyscyplinarne zagadnienia w inżynierii i ochronie środowiska, **7**, 129-136 (2016)
11. K. Pilarski, A. Pilarski, Technika Ogrodnicza, Rolnicza, Leśna, **1**, 16—17 (2009)
12. K. Ishii, S. Takii, J. Appl. Microbiol., **95**, 109–119 (2003)
13. P. Dąbrowski, B. Pawluśkiewicz, Przegląd Naukowy Inżynieria i Kształtowanie Środowiska, **51**, 27–35 (2011)
14. K. Wolski, M. Talar-Krasa, A. Dradrach, M. Szymura, M. Bierniacik, S. Świerszcz, Łąkarstwo w Polsce, **18**, 241—254 (2015)
15. B. Rutkowski, A. Hempel A., *Trawniki* (PWRiL, Warszawa, 1986)
16. S. Prończuk, Biul. IHAR, **186**, 127–132 (1993)
17. https://biologiaroslin.wordpress.com/2012/06/16/trawy-na-murawy/
18. http://www.ihar.edu.pl/gazony.php
19. D. Martyniak, Biul. IHAR, **228**, 335–344 (2003)
20. M. Prończuk, *Choroby traw – występowanie i szkodliwość w uprawie na nasiona i użytkowaniu trawnikowym* (Monografie i Rozprawy Naukowe, IHAR, Radzików, 2000)
21. http://www.trawnik.com/?aktyw_kat=4#.WJzFU3-XeM8
22. http://www.trawasportowa.pl/pdf/pielegnacja.pdf
23. www.bip.lubniewice.pl/bip_download.php?id=852
24. http://www.mrgreengrass.pl/uslugi/zabiegi_pielegnacji_boisk.html
25. http://www.victus.pl/porady-eksperckie/pielegnacja_trawnika/
26. S. Baran, R. Turski, *Wybrane zagadnienia z utylizacji i unieszkodliwiania odpadów* (Wydawnictwo AR, Lublin, 1999)
27. http://magazynbiomasa.pl/szanse-dla-biogazu-po-1-stycznia-2016-r/

The national scenarios for emissions reduction in residential heating sector in Poland

Elżbieta Niemierka[1,*], and *Piotr* Jadwiszczak[1]

[1]Wroclaw University of Science and Technology, Faculty of Environmental Engineering, Wybrzeże Wyspiańskiego 27, 50-370 Wrocław, Poland

Abstract. Today's residential heating sector (RHS) in Poland is based mainly on fossil fuels combustion, both in individual and large-scale heat sources. The RHS emissions significantly contribute to the outdoor air quality deterioration, especially by low-stack emissions in the heating season. Following the example of other EU countries, the national scenarios for emissions reduction in RHS are an effective tool for outdoor air quality improvement. The scenarios contain a replacement of the existing heat sources in RHS with heat generation technologies in the scale of the country, like high-efficiency gas boilers, biomass boilers or heat pumps. The study takes into account today's RHS structure in Poland, the scenarios feasibility and achieved emissions reduction.

1 The anthropogenic emission in residential heating sector

In Poland, the residential heating sector (RHS) is based on fossil fuel combustion, both in individual heat sources (HS) and large-scale district HS, including heat-only boilers and combined heat and power plants (CHP). The average age of individual HS in Poland is almost 10 years and the technical condition of boilers, quality of fuels, combustion processes and emissions are here completely out of control [1]. The large scale district HS and CHP can supply a large number of consumers and offer lower emissions due to high-efficiency and effective exhaust treatments.

Today's RHS in Poland consumes 550 PJ/year, representing 21% of total heat demand in Poland, and is responsible for emissions of 36% of sulfur oxides (SOx), 64% of carbon monoxide (CO), 40% of total suspended particles (TSP) and 87% of polycyclic aromatic hydrocarbons [2-3].

The poor outdoor air quality problem is associated not only with emissions of well-known CO_2 and other greenhouse gases but also with the emission of SOx, nitrogen oxides (NOx), CO, TSP, benzo(α)pyrene (BαP) etc. A particular threat is the low-stack emissions from individual HS in the RHS, which result in smog during the heating seasons. The high risk of smog occurs in urban areas with a high heat demand density and a large number of small fossil fuels HS. In 2014, 92% of the world population was living in places where the World Health Organization air quality guidelines levels were not met [4].

* Corresponding author: elzbieta.niemierka@pwr.edu.pl

Poor outdoor air quality concerns not only Polish cities. The occurrence of the Big Smoke of London in 1952 led to the creation the Clean Air Act 1956 as a tool to combat smog. Since then, by prohibiting from using coal-fired boilers, replacing power plant outside urban areas, increasing the height of chimneys of local emitters, introducing fees and restricting the movement of high emissions vehicles in the cities centre, it has been able to improve ambient air quality [5]. Analogous regulations are in force and benefit in many European cities, for example in Vienna. It promotes buildings energy performance improvement and extends district heating system. It resulted in a reduction of TSP (PM10) by up to quarter and elimination exceed values [6]. These examples indicate that central, comprehensive scenarios to combat smog are an effective tool and bring more benefits than voluntary programs directed to individuals [7]. It is evident, that the common solutions are needful in Poland, in the scale of the region and even the whole country, to improve the outdoor air quality.

2 Emissions in Polish RHS

As mentioned above the heat demand in RHS in Poland amounts to 550 PJ/year. The heat is generated in individual and central HS using different fuels (coal, lignite, coke, wood, natural gas, LPG and oil) and energy carriers (district heating and electricity). Fig. 1. shows the share structure of fuels and energy carriers in the Polish RHS [8-10] and the emissions factors respectively [1, 11-12]. Fuels like natural gas, LPG and fuel oil have low emissions ratios and fuels like coal, lignite, coke and wood are characterised by higher factors, especially in SOx, CO, TSP and BαP. It should be noted that the wood can be regarded as a renewable energy source, but it also causes local emissions, despite a zero balance CO_2, which was included.

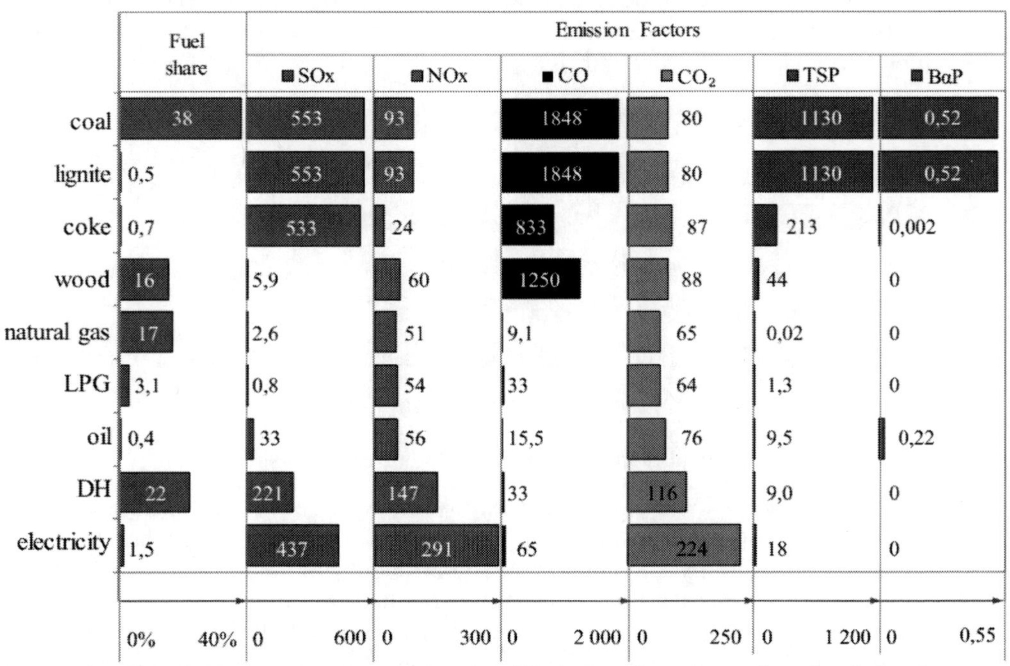

Fig. 1. The shares of fuels and energy carriers in RHS (in % of heat demand) and emission factors resulting from the fuel combustion or energy carrier consumption by the final customer (in Mg/PJ and for CO_2 in Gg/PJ) (based on [1, 8-12]).

Emissions in the RHS may be limited in two main ways - by reducing heat demand (building refurbishment or thermo-modernization) [1] and by replacing HS with high efficient devices [13-14], that consume low-emission fuels or with individual renewable energy sources. It should be noted that there is no one universal scenario to improve outdoor air quality by the emissions abatement at RHS. It is required to analyse the current state and to identify the applicable scenarios of changes. As mentioned above, the national scenarios are the best solution for the effective reduction of emissions in RHS. The development of proper national scale scenarios necessitates better knowledge of the energy sector structure, the heat demand profiles, the heat generation technologies, the fuels and energy carriers mix, the emissions factors etc. The paper presents the 7 developed exemplary scenarios of national scale changes in the structure of heat supply in RHS in Poland, to eliminate the smog risk and improve the outdoor air quality. The target is to decrease both the low-stack emission (LE) and total emission (TE) in RHS, mainly in individual HS. The LE means the significant amount of contaminants emitted directly into the neighbourhood and has a considerable impact on people's lives and it contributes to their health status. The TE affects outdoor air quality in the scale of the country and above borders, and to changes in the global environment.

3 The national scenarios

The national scenarios have the greatest impact on the choice of strategy changes leading to the emissions reductions. Therefore 7 national scenarios were developed and compared to the current emission in Polish RHS.

Nowadays in Poland heat-only boiler plants, power plants, CHP, district heating systems and individual boilers are mainly based on coal, lignite and wood combustion, accompanied by high emissions factors. Fig. 2. shows the shares of fuels combustion or energy carriers consumption in particular pollutants emission. Almost entire emissions of cancerogenic BαP and TSP is connected with LE, caused by coal combustion. A large share of pollution takes wood (CO_2, CO and NOx) and coal-based HS (CO_2, NOx and SOx).

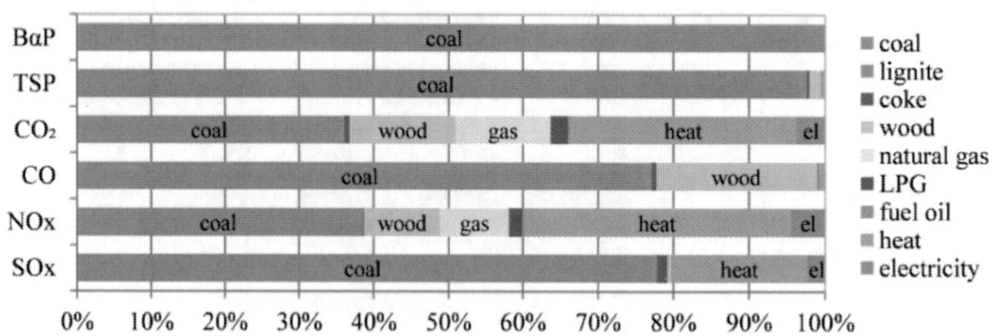

Fig. 2. Current shares of fuels and energy carriers in emissions from Polish RHS.

All the proposed in national scenarios modifications will be a concern in the individual HS sector as causing the greatest smog risk. They do not include the upgrading in gas, oil, district heating and electricity HS, due to its good efficiency or small share in current RHS emission. All national scenarios are presented in detail in Table 1 and contain a replacement of all individual HS based on fossil fuels combustion with other, more effective heat generation technologies like high-efficiency coal boilers, biomass wood boilers, gas boilers, heat pumps and district heating plus heat pump mix (Fig. 3.). The study takes into account today's structure of energy sector and RHS sector in Poland (Fig. 2.), the scenarios feasibility and achieved reductions of emissions. The results of each scenario will be shown as the calculated LE and TE reductions, referenced to the current state.

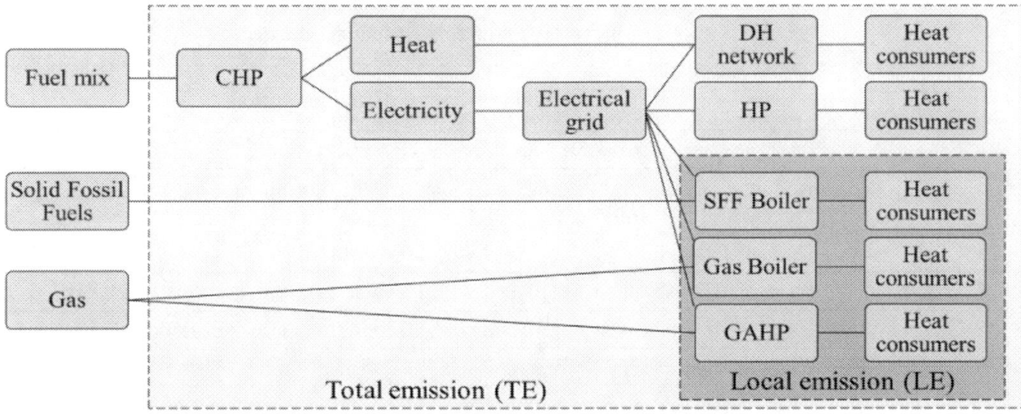

Fig. 3. The energy and heat flow diagram for LE and TE calculation in all scenarios.

In each scenario, the LE and TE were calculated as described below, according to Fig. 3., Eq. 1 and Eq. 2. The calculation diagram as energy and heat flow in all scenarios is shown in Fig. 3. The scheme takes into account the all technical factors related to RHS, energy sector, fuels and energy carriers mix, LE and TE in Poland.

The LE was calculated as the sum of LE_k of pollutants emitted from all kind of local HS predicted in national scenario. The LE of the k-th contaminants (LE_k) are associated with individual HS, and takes into account the share of the i-th fuel ($m_{F,i}$, %) [8-10], the space heating participation ($n_{heating,i}$, %) [3] in residential sector heat demand (HD, PJ/annum), the average efficiency of HS (η_i, %) [1, 13-14] and the emission factor from i-th fuel of k-th pollutant ($EF_{i,k}$, ton/PJ) [11] - Eq.1. The LE does not include emissions associated with grid electricity and district heating consumption, due to transferring of emissions to the central plants away from the consumers.

$$LE_k = HD \cdot \sum (m_{F,i} \cdot n_{heating,i} \cdot \eta_i \cdot EF_{i,k}) \quad (1)$$

The TE is related to both local and national emission, and additionally includes the share of the j-th energy carrier ($m_{EC,j}$, %) [8-10], the space heating participation ($n_{heating,i}$, %) [3] in residential sector heat demand (HD, PJ/annum), the average efficiency of HS (η_j, %) [1,13-14], the efficiency of energy carrier generation ($\eta_{j,g}$, %) and transmission ($\eta_{j,t}$, %) [15-16], and the emission factor j-th energy carrier of k-th pollutant ($EF_{j,k}$, ton/PJ) [12]. The $EF_{i,k}$ value is dependent on energy generation technology and transmission efficiency as well as on usable and excess energy ratio in CHP plant. Contemporary HS, both conventional and renewable, require an auxiliary power supply ($AP_{el.,k}$, PJ) [17]. In Polish RHS it is typically grid electricity, which causes an additional amount of contaminants. Consumption of grid electricity moves the emissions to the central power or CHP plants, which increases the national TE without increasing the risk of smog. According to the above dependencies was created equation that was used in the calculations of emissions for all national scenarios - Eq. 2.

$$TE_k = LE_k + \sum (AP_{el.,k} \cdot EF_{j,k}) + HD \cdot \sum (m_{EC,j} \cdot n_{heating,j} \cdot \eta_j \cdot \eta_{j,g} \cdot \eta_{j,t} \cdot EF_{j,k}) \quad (2)$$

The proposed modifications in 7 analysed scenarios are described in Table 1. The names of scenarios come from the main change applied in RHS. All scenarios were compared in terms of emissions, calculated on the basis of Eq. 1 and Eq. 2.

Table 1. Characteristic national changes scenarios.

National scenario	Characteristic of national scenario
1. Base	Current state - current HD, and current HS structure in RHS
2. BR	Reduced HD due to buildings refurbishment or thermo-modernization, and current HS structure
3. Coal	Current HD and all SFF replaced with HE coal boilers (class V)
4. Wood	Current HD and all SFF replaced with HE wood biomass boilers (class V)
5. Gas	Current HD and all SFF replaced with HE condensing gas boilers
6. GAHP	Current HD and all SFF replaced with gas absorption heat pumps
7. HP	Current HD and all SFF replaced with electrically driven heat pumps
8. HP&DH	Current HD, all SFF replaced by mix of electrically driven heat pumps and district heating system according to CHP heat and electricity generation ratio
	Where: HE – high-efficient, SFF – solid fossil fuels

The effects of the national scenarios, calculated according to Eq. 1 and Eq. 2, are shown in the Fig. 4-7. Fig. 4. presents the reductions of fuels and energy carries consumption in RHS due to the improvement of individual HS efficiency and changes of fuels and energy carriers mix in following scenarios. The lowest consumption as compared to Base scenario corresponding to HP and HP&DH, while the largest to Coal and Wood scenario. The fuels and energy carriers consumption is the base for LE and TE calculations in all national scenarios.

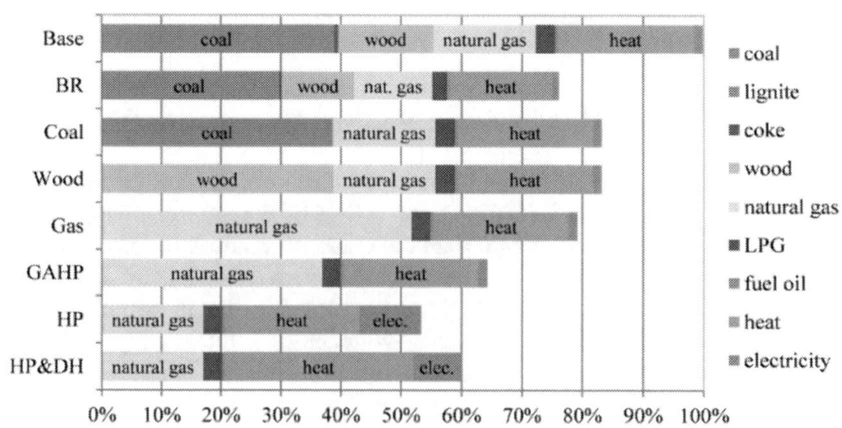

Fig. 4. The fuels and energy carriers consumption for all scenarios in relation to the Base.

Fig. 5. shows the relative reductions of pollutants calculated both in LE and TE for all national scenarios. In every scenario, the LE and TE will be reduced, which will be contributed to the outdoor air quality improvement. The scenario BR based only on the old buildings refurbishment, with no changes in fuels and energy carriers mix, results in 24% reduction of emissions (reduced to 76,3% in relation to the Base scenario). The calculation takes into account the statistical data on the age, the size of dwellings and the heat demand for all existing buildings in the Polish sector. The BR scenario predicts refurbishment of all buildings constructed before World War II, which constitute about 22% buildings in Poland, to the current energy standards [1].

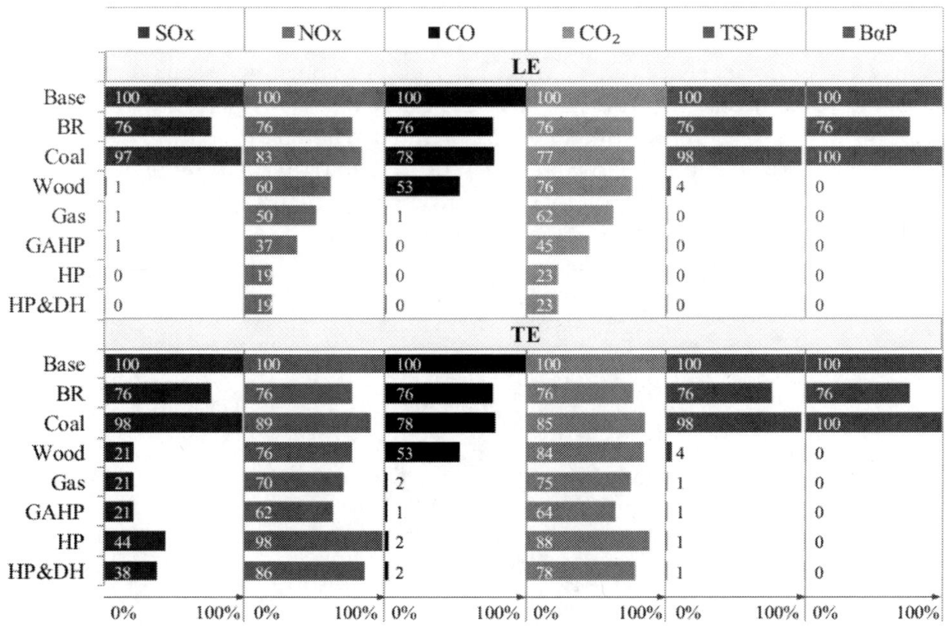

Fig. 5. Relative emissions of different contaminants in individual scenarios according to the Base.

The further scenarios predict replacement of all low efficient individual heat sources with high efficient heating technologies in the scale of the country. The Coal scenario comprises replacement of all individual HS based on solid fossil fuels with high-efficiency coal-burned boilers (V class) and results in only a limited decrease in LE pollutants such as NOx (83%), CO (78%), CO_2 (77%). The changes are incommensurate to the scale of national changes in RHS.

The Wood scenario results in the lowest LE among solid fossil fuels individual HS. Reduction of low-stack emission of CO_2 is similar to Coal, however, we notice significantly lower relative emissions of TSP (4%), SOx (1%) and BaP (<1%).

Application of natural gas condensing boilers (scenario Gas) eliminates the same compounds as burning of wood, while reducing of CO_2 (62%) and NOx (50%), and does not emit CO (1%).

The best results can be achieved in GAHP scenario. Low emissions factors for natural gas and high efficiency of GAHP result in avoiding of CO, TSP, BaP, SOx and incomparable reductions of CO_2 (45%) and NOx (37%) emissions.

The HP scenario is one of the most effective in reducing LE of CO_2 and NOx up to about 20% and eliminates other contaminants. Replace the fossil fuels with electricity generated in central plants will transfer emissions, what is evident in LE and TE changes. Apply a large amount of electrically driven heat pumps causes an increase the generation of excess heat in CHP units and raise the emissions up to 12% (for NOx). The use of a large amount of heat pump units in Polish energy sector should be associated with the increase of district heating consumers.

The HP&DH scenario predicts a mix of individual heat pumps and district heating supply in best proportion for current heat and electricity generation sector in Poland (heat-only boilers, power plants and CHP units). The coal-based CHP in HP&DH scenario will

DOI: 10.1051/e3sconf/20171700067

slightly diminish NOx (86%), CO_2 (78%) and eliminate the CO, TSP, and BaP (<2%) emissions.

The relative reductions of TE for all scenarios (with equal weights of all contaminants) are shown in Fig. 6. The large worth means the significant decrease of TE in each scenario.

The LE and TE reductions in five most effective scenarios (Wood, Gas, GAHP, HP and HP&DH) are present in Fig. 7. The great value means the meaningful reduction of emissions. It should be mentioned that the LE and TE reductions occur on the different levels in the same scenario. From LE and TE point of view, there are two groups of scenarios with the best emissions reductions. The heat pumps based scenarios (HP and HP&DH) are characterised by a large reduction of LE and only small reduction of TE. The gas based scenarios are characterised by good decreasing both in LE and TE.

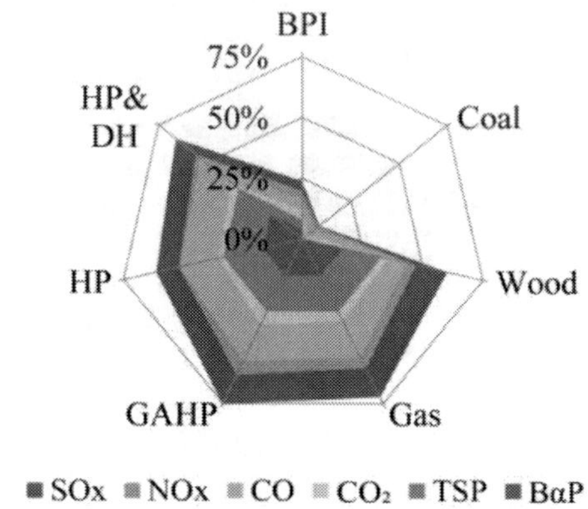

Fig. 6. The relative reductions in total emission for all scenarios (with equal weights of all contaminants – the total absence of emissions of one pollutant means a change from 12,5% to 0%).

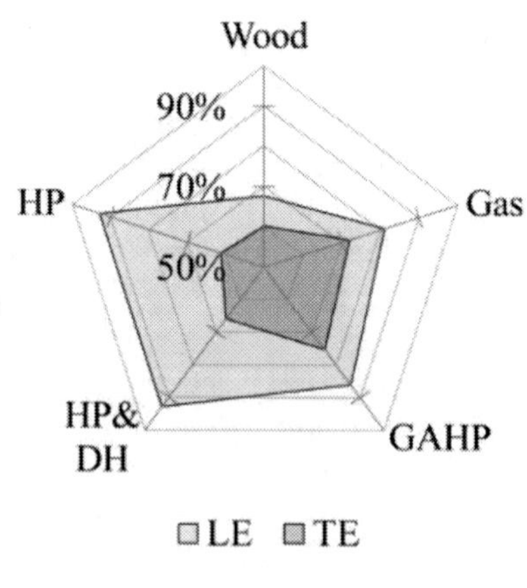

Fig. 7. The LE and TE reductions in the most effective scenarios.

4 Conclusions

In Poland, air quality problem is caused mainly by low-stack emission. The proposed changes in individual heat sources sector will improve local air quality, both in local and country scale. Following the example of other countries, it was noted that national scenarios of changes in residential heating sector are an effective way to combat smog and poor air quality. The presented article can be used to design development path and created obligatory regulations that will reduce the emissions of pollutants and improve outdoor air quality. The presented national scenarios are monogenic, thus hardly realistic to implement. The real scenario will be a mix of fuels and energy carriers with different proportions dictated by their availability, pricing, government policy, etc. The analysis showed that the fuels with the greatest impact on emissions in Poland are coal and wood. Four of the seven presented scenarios are worthy of mention because they incomparably reduce the LE and TE. The highest TE reductions are achieved for GAHP (75%), gas boilers (72%) and by balancing the number of users electric heat pumps and district heating network in HP&DH scenario (66%). Exploit energy carriers can reduce LE by 93%, however, a small share of renewable energy sources in Polish CHP negatively affects the overall balance - TE of contaminants allowing their limitation of only 66%.

It is difficult to identify the best national scenario of reducing the LE and TE in Poland, but observation of change trends demonstrate that in the near future it will be HP&DH based scenarios.

References

1. Instytut Ekonomii Środowiska, Efektywność energetyczna w Polsce. Przegląd (2015)
2. IOŚ-PIB, KOBiZE Krajowy bilans emisji SO2, NOx, CO, NH3, NMLZO, pyłów, metali ciężkich i TZO w układzie klasyfikacji SNAP i NFR (2015)
3. Central Statistical Office, Energy consumption in households 2012 (2015)
4. http://www.who.int/mediacentre/factsheets/fs313/en/
5. http://cleanair.london/wp-content/uploads/CAL-217-Great-Smog-by-GLA-20021.pdf
6. http://krakow.wyborcza.pl/krakow/1,44425,17728684,Wiedenskie_nauki_dla_Krakowa__Jak_zrobic_miasto_skrojone.html
7. https://www.nfosigw.gov.pl/download/gfx/nfosigw/pl/nfoopisy/1207/1/17/poprawa_jakosci_powietrza_20151109_100411.pdf
8. Central Statistical Office, Energy efficiency in Poland in years 2004-2014 (2016)
9. https://bdl.stat.gov.pl/BDL/
10. Central Statistical Office, Zużycie paliw i nośników energii w 2014r (2015)
11. https://krajowabaza.kobize.pl/docs/male_kotly.pdf
12. http://kobize.pl/pl/fileCategory/id/28/wskazniki-emisyjnosci
13. Dz.U. 2015 poz. 376
14. PN-EN 303
15. Urząd Regulacji Energetyki, Energetyka cieplna w liczbach – 2014 (2015)
16. https://www.export.gov/article?id=Poland-Electrical-Power-Systems
17. Dz.U. 2015 poz. 376

Numerical analysis of the heat exchanger energy efficiency depending on location from the floor

Magdalena Orłowska[1,*]

[1]Technical University of Koszalin, Faculty of Civil Engineering, Environmental and Geodetic Sciences, al. Śniadeckich 2, PL-75-453 Koszalin, Poland

Abstract. The article is a continuation of a series of articles the author 1 on convective heat transfer of heat exchangers working in different building conditions. Location of the window sill, housing enclosing the distance of the radiator from the floor is not without significance, as it turns out. The article presents the results of numerical analysis regarding installation of the heater flat plate in three different distances from the floor.

1 Introduction

The phenomenon of convection has been known for a long time and widely used technique [1, 2]. Natural convection heat transfer is associated with the movement of fluid, due to the gravitational field. Fluid with a higher temperature increases its volume, thus reducing its density and rise upwards. Fluid having a temperature lower its density increases, which causes it to fall down [3]. Natural convection is characterized by the extraordinary complexity which hinders detailed and accurate analysis. They are however, very intense fields of studies, both theoretical and experimental or numerical. These studies provide more interesting elements which could be in use. Besides the importance of cognition tend to have the results of increasing the intensity of the convection heat transfer [4].

Heater Installation is usually proposed by the manufacturer. There is always assembly instructions with it connectors attached to the heater determine the distance from the wall, sill or floor. Leaving aesthetic issue aside, it`s interesting how parameters such as distance from a building structures has impact on the thermal effect – performance of the heater.

* Corresponding author: magdalena.orlowska@tu.koszalin.pl

Table 1. Distances proposed by different data sources [5, 6].

Distance	cm
From the floor	10
From the wall	Depending on the type of handles for hanging the radiator
From the windowsill	10

2 Methodology of Numerical Research

All simulations were performed using the simulation code Ansys Mechanical APDL / CFD-Flo Preferences for GUI Filtering -Flotran CFD 2D Flotran 141 – Library of Elements Types [7]. The program is based on solving the balance equations of energy, momentum and mass used in CFD- Computational Fluid Dynamic model. The standard version of the simulation code allows you to conduct modeling based on traditional dependence phenomenological equations and additional closures, describing the previously mentioned turbulent momentum and energy streams. The test system was modeled in two-dimensional space. Program calculates after entering the boundary conditions: temperature of the walls of the heat exchanger, room air velocity and gravity [8].

The article presents couple of selected systems with different geometry Fig. 1 and attempted to explain its effects on the intensity of heat exchange (heat transfer coefficient).

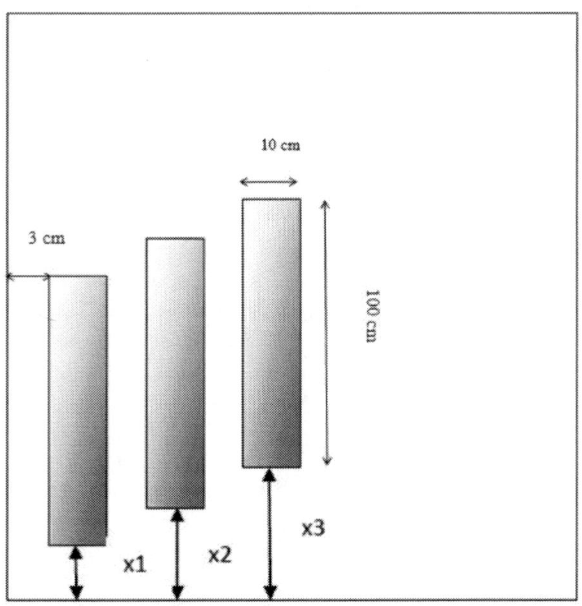

Fig. 1. Scheme subjected to numerical analysis.

3 Calculation Results

The study used simulation flat plate heat exchanger with a thickness of 10 cm and a height of 100 cm with 3 cm distance between the wall and the radiator. The heater was tested at three different distances from the floor x1 =5 cm x2=10 cm and x3 =15 cm. The dimensions of the room in which the heater was in a width of 3 m, 2,5 m height. The third dimension - length - due to the two-dimensional character of analysis were omitted. Heater wall temperature of 70°C, the walls of a room of 20°C.

The results of numerical analysis gave the following graphs: Figures 2, 3, 4, 5:

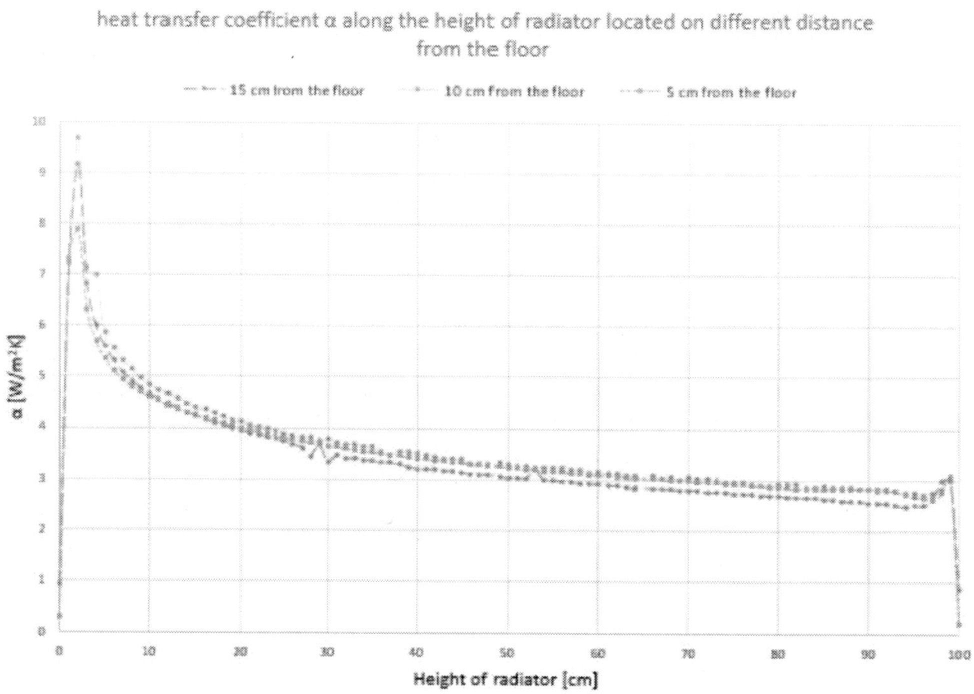

Fig. 2. Heat transfer coefficient α [W/m^{2}K]along the height of radiator [cm].

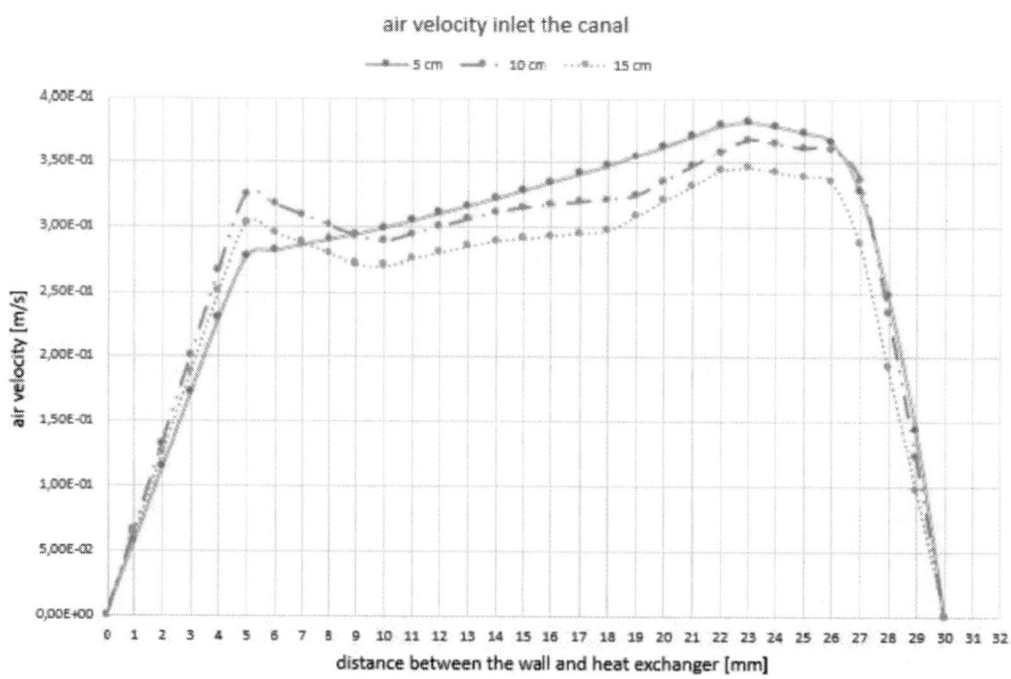

Fig. 3. Air velocity w [m/s] along width at inlet the canal between the wall and heat exchanger [mm].

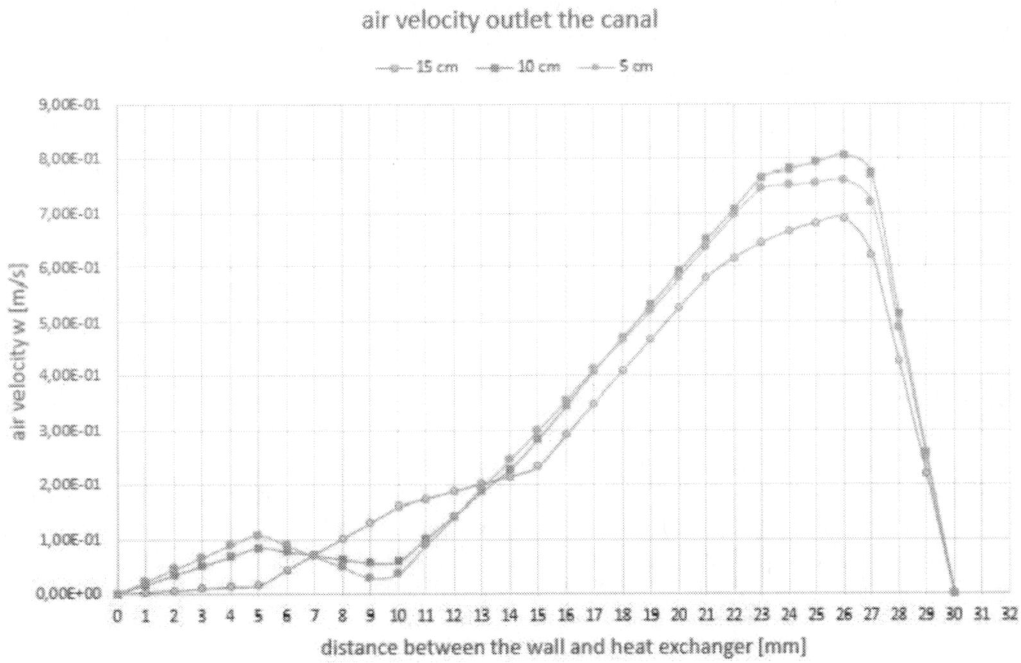

Fig. 4. Air velocity w [m/s] along width at outlet the canal between the wall and heat exchanger [mm].

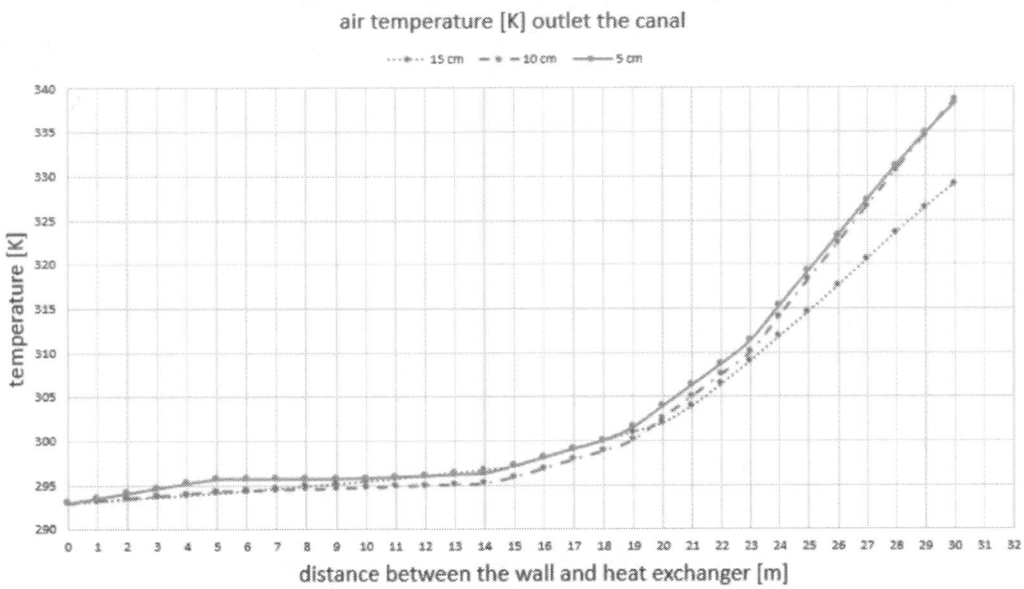

Fig. 5. Air temperature [K] along width at outlet the canal between the wall and heat exchanger [mm].

The highest average value of heat transfer coefficient $\alpha = 3{,}58$ [W/m^{2}K] obtained in the case of a heater located at a distance of 10 cm from the floor, then 5 cm $\alpha = 3{,}46$ [W/m^{2}K] and 15 cm $\alpha = 3{,}37$ [W/m^{2}K].

The highest average value of the air velocity in the inlet to the channel obtained in in the case of a heater located at a distance of 5 cm from the floor w = 2,77 [m/s], then 10 cm w = 2,74 [m/s] and the lowest 15 cm w = 2,54 [m/s].

The highest average value of the air velocity in the outlet to the channel obtained in in the case of a heater located at a distance of 10 cm from the floor w = 3,21 [m/s], then 5 cm w = 3,13 [m/s] and the lowest 15 cm w = 2,83 [m/s].

The highest average value of the air temperature at the outlet of the T = 304,68 K was prepared in a 5 cm distance between the floor and the radiator, the average value temperature in the system from a distance of 10 cm T =303,7 K and lowest temperature T = 302,6 K system of 15 cm distance.

If the heater is too low from the floor it causes the speed of the incoming airflow to the inlet of the channel between the wall and the radiator is large but it is reduced later. System with 15 cm distance performed the worst in terms of received speed and air temperature and the ratio of the heat transfer α.

4 Conlusions

Finally the best system in terms of heat exchange results are achieved with a heater positioned 10 cm from the floor. It allows air to circulate freely and is considered optimal from a distance. The results of the numerical analysis can be a valuable source of data. We`ve received important informations on the flat radiators. This informations can be useful for companies involved in installation of radiators or repairing the apartment and struggling with the proper placement of heat exchangers.

References

1. M. Orłowska, M. Czapp, Rocznik Ochrony Środowiska **XIV** (2012)
2. S. Czapp, M. Czapp, M. Orłowska, Numerical and experimental investigation of thermal convection near electric devices with vertical channels, 2016 International Conference on Information and Digital Technologies (IDT), Rzeszow, 2016, pp. 54–58 (2016)
3. M. Szpakowska, Badanie wpływu warunków brzegowych na intensyfikację konwekcyjnej wymiany ciepła od płyty pionowej w przestrzeni częściowo ograniczonej, PhD thesis (2010)
4. M. Czapp, H. Charun , J. Baran, Chłodnictwo **6** (1980).
5. www.instalreporter.pl/ogolna/montaz-grzejnikow-zaleznie-od-ich-rodzaju-na-roznych-scianach/
6. www.regulus.com.pl/faq/jaka-jest-prawidlowa-odleglosc-grzejnika-od-parapetu-i-podlogi
7. Ansys, User Instruction
8. M. Orłowska, Numerical analysis at a plate heat exchanger with covers- selected variations, Heat Transfer and Heat Exchangers – Heat Transfer and Renewable Sources of Energy HTRSE 2016, pp. 225–228 (2016)

The optimization of calibration of the computer model describing the flows in the water distribution network; Study of the example

Maria Orłowska-Szostak[1,*]

[1]Faculty of Civil and Environmental Engineering, Gdansk University of Technology, ul. G. Narutowicza 11/12, 80-233 Gdansk, Poland

Abstract. The paper describes the issues of optimization of calculations performed during the calibration of a computer model describing the flow in the water distribution network. Optimization concerns the efficiency of these calculations which means not only the volume and speed of calculations, but above all, the optimal use of the measurements made during calibration to obtain the best, i.e. the most real value of the searched parameters/characteristics of the system components. The paper includes a short introduction which orders the range of methods of calibration and the area of calculation methods used in this class of issues, and then the description of an example where in the calibration process was used a computational tool called EPANET Calibrator developed by scientists – engineers of cooperating Brazilian universities. This tool is based on genetic algorithm.

1 Introduction

Calibration of the model describing the phenomenon of flows occurring in the system of transport and distribution of water intends to define the characteristics of system components (parameters appearing in their analytic descriptions), and so to ensure that the model shall describe this phenomenon with required accuracy. The aim of calibration is not to measure the whole system, because in this case we would know everything we want about the system and computer modeling could lose its sense because it would often be enough to monitor the results of measurements. During the calibration process one carries out measurements only at the selected/possible optimal points/ nodes of the system (see, e.g.: [1]), and then the results of these measurements are used to determine the possible optimal characteristics/parameters (the methods of performance at these step of creation of a computer model are described later in the paper) and thanks to that it is created a calibrated model, which is a tool for simulating of all possible real behaviors of the system.

The calibration process is characterized by indeterminacy resulting from the following facts:

- mentioned, rarely distributed measurement points,

* Corresponding author: maria.o@pg.gda.pl

- the fact that they are not known either searched characteristics of the system components (especially the roughness of pipes) and flows in the network (or water partitions) occurring during performing these measurements.

There are used different calibration methods, and the aim of each of them is the removing (reducing the value) the said indeterminacy of the task. Each method requires, for example, performing a variety of measurements used in the calibration. Frequently used method of reducing the number of unknown roughness and partitions [2] is a grouping of pipes of virtually identical roughness coefficients and grouping of junction nodes with virtually identical timing patterns of water partitions. The criteria for the creation of such groups are not always convincing, although this method is widely used.

The efficiency of the calibration also depends on the method of performance (intuitive ones, methods/algorithms implemented numerically) for the optimal use of the above mentioned measurements in the calibration of the model, i.e. as mentioned above, in the determination of characteristics (parameters appearing in the analytical descriptions of these characteristics) of the system components.

Among the above-mentioned calibration methods, the most popular are the following ones:

(1) method, which can be called "direct"

This method uses measurements of pressure and flow at selected points of the system. Generally, the pressures are measured for calibration purposes at pipe junction selected nodes (not at all nodes) of the water distribution network. In small and medium-sized systems the junction nodes for measuring are selected intuitively (as described later in this paper, in the description of the calibration carried out for the example water distribution system). For the larger systems, it is reasonable to optimize the number and location of junction nodes at which we perform the measurements (see e.g. the theoretical basis of *sampling design* [1]). In each case, the flows are usually measured at all objects, such as pumping stations of water intakes, the zonal pumping stations, storage tanks and finally in selected pipes. The weakness of the described calibration method is an unknown water partitions patterns occurring at the time of the measurements. This is the reason of the strong indeterminacy of the whole task of calibration. We try to “arrange“ the partitions of water as reliably as possible between the junction nodes, so that their sum was compatible in the each zone with the measured supplies in these zones, and at the same time to use all available data and information on the consumption of water by individual customers. The determination of simultaneous partitions of water for all customers during performing the aforementioned measures of pressure and flows, without taking (practically impossible to perform) continuous measurements at those customers, was not possible. Therefore, the accuracy of the calibration was limited and difficult to estimate. However, during long use of so calibrated model it was possible further verification of its parameters, which caused that the designated roughness of the pipes was treated with more and more confidence. From the other hand, applying genetic algorithms for the measuring data made with the 'direct' method is very much prospering what was shown in the article.

(2) the fire hydrant flow test

This method better cope with the described indeterminacy of calibration. It is tried to bring to the situation in which we would be practically very close to determination of water consumption in the supplied area, with simultaneous values of pressures and flows measured as in the previous method. This method is based on taking identical measurements of pressures and flows and the estimation of consumption of water in the supplied area, as in the previous method, except that:

- it is considered and measured the specific night situation at the hour of minimum consumption, close as possible to zero consumption,

- when making the above-mentioned measurements, one or more, usually several fire hydrants are fully open, and at the same time their discharges and pressures are measured.

The discharges of opened hydrants should be significant/dominant water distributions in the system during measurements needed for calibration; other , "night" distributions are low. In this way we are approaching a situation where we know the consumption occurring during measurements, and therefore we can substantially reduce the indeterminacy of calibration which is characteristic in the previous method.

(3) the tracer test

Tracer test methods of the calibration of models are the most developed in hydrology, in modeling of groundwater flow. The use of tracers method in the calibration of models describing flows in the water distribution systems is based, in addition to measurements performed as in "direct" method, on introducing a specific quantity of tracer to the network, especially at its points of supply. By measuring the concentration of this tracer in many points on the network, using the proper equipment, and assuming generally correct assumption that the distribution of tracer in the network can be described with the piston model and the amount of tracer flowing in and flowing out in the junction node are equal, it is measured the speed of flow in the water network, considerably reducing the indeterminacy which is characteristic for the "direct" method (see above). Calibration accuracy depends here on the density and location of the measurement points, whereas in the "direct" method characterized by a lack of real, current data describing the water consumption during calibration, described indeterminacy of the task occurs.

It should be emphasized that very good results of the calibration of model considering the roughness of pipes is ensured by fire hydrant flow test, because during the calibration process using this method it is not necessary to determine the accurate values of consumption of water or to determine the timing patterns of such consumption. Therefore, sometimes presented approach to the calibration of the model [2] involves the use of the fire hydrant flow test (where it is simulated a single technical situation in the system using SPA - *Single Period Analysis*), and additionally, if we are interested in consumptions in the supplied areas (their values along with timing pattern), then at already calibrated roughnesses it is performed a calibration of the said consumption patterns using e.g. the tracer method or using the results of continuous monitoring of pressures and flows. It is applied here the space-time simulation EPA – *Extended Period Analysis.*

Let's consider these methods applied for the optimal use of the above-mentioned measurements in the calibration process. The task involve the solving of nonlinear systems of equations describing the behavior of the system (implicit constraints in the model) and searching of the roughness vector of the pipes (group of pipes) and sometimes the consumption vector i.e. decision vectors ensuring the global minimum of the objective function. That objective function is constructed so that by using the method of the least squares of differences between the measured and simulated values it could be possible to determine the actual values of searched roughness and, if necessary , consumption [3]. Solving of the mentioned system of equations of conservation is a known issue. They are applied known, classical numerical methods (Newton-Raphson method for nodal pressures corrections, gradient method) and heuristic methods, which include genetic algorithms [4].
Taking into account the methods of searching for the vector minimizing the value of the said objective function, it must be emphasized that in comparison to the classical numerical methods, computer heuristic methods protect against the determination of a local minimum treated as a global minimum. Therefore (but not only therefore), they are more and more often used by engineers. Regarding the issues and their resolving, which are the essence of the calibration of the model of water distribution network the best are genetic algorithms [2, 5, 6].

This paper will describe a computer model of flows occurring in the sample water distribution network. The model was created using the EPANET program. It was described the calibration of the model using the above-mentioned direct method, based on the measured pressures and flows in designated points of the water distribution system.

2 The analyzed exemplary water distribution system

The water supply system consists of the water intake and treatment station working in a two-stage pumping system. The water intake is located in the central part of the city and consists of eight deep-water wells. Water is drawn from the decks of quaternary aquifers [7]. The total yield of the wells is 440 m^3/h with depression from 7.5 to 16 m. Water is taken alternately from individual groups of wells to ensure the required time-varying capacity of the intake. Water from drilled wells is pumped by the first stage submersible pumps to the treatment plant. Once treated, the filtered water flows into reservoirs of clean water. Two field tanks, with a volume of 2,100 m^3 each, serve as storage tanks in the water supply system [8]. The dispatcher maintains a constant water level in the tank at a height of 3.7 m, measured from the bottom of the tank. The second stage pumps draw water from the tanks and pump it to the network. These pumps are controlled by VSD to provide the desired pressure at the outlet of the pump station: 4.5 bar during the day and 4.0 bar overnight. These values were appointed by the dispatcher of the system basing on his exploitation experience, however they fulfill all of the major constraints given in professional literature, see e.g. [9].

The water supply network is a loop-type with few branched network elements. The outline of the analyzed water supply system, executed in the EPANET program, was shown in Fig. 1. The structure of the network, including the materials that the pipes are made of, is characteristic of typical urban networks in Poland [10]. Cast iron pipes account for more than half of all urban network pipes. A significant part of the network was also made of PVC [8]. The age structure of the network has been carefully worked out by the authors because it is an important factor in determining the initial roughness of the pipes during model calibration. Cast iron pipes are the oldest types here. Information was also gathered on ongoing flushes and other regenerative pipe processes as they have a significant impact on the pipes roughness [11].

3 Basic data for the model of water distribution system

Carefully gathered data and standard software based on EPANET were used for construction of the computer model. Specifically, data describing the consumption of water and the network topology were collected, as well as any necessary parameters characterizing it.

When determining reliable water consumption, the results of continuous measurements of the second-stage pump station delivery, pumping water directly to the network were set as the starting point. Base consumptions and time patterns of inequality of hourly water consumption for individual connections were determined with taking into consideration some specific information about water consumption of individual objects or more often the sizes of the objects, their character, and sometimes the diameter of the connections themselves.

4 Applied method of calibration of model, applying a genetic algorithm with respect to the analyzed water distribution system

Taking into account the assortment of measurements, the applied method of the calibration of model can be classified as the direct method, described in chapter 1, i.e. the method which uses the measurements of pressure and flows at selected points of the water distribution system.

The drawing up of the results of above-mentioned measurements , i.e. estimation of roughness of pipes forming a water distribution network as well as emitter coefficients was carried out using a genetic algorithm. It was applied EPANET Calibrator software, developed by Resan Labs from Brazil [3,6,12].

The authors of the EPANET Calibrator software give the form of the objective function in optimization task of calibration of model of the system of transport and distribution of water. Therefore, the criterion of the optimization task is to minimize the said objective function [12]

$$MinFO = \sum_{i=1}^{N} \left[\frac{\Delta P_i}{\frac{\sum_{i=1}^{N} P_i^{'}}{N}} \right]^2 + \sum_{j=1}^{M} \left[\frac{\Delta Q_j}{\frac{\sum_{j=1}^{M} Q_j^{'}}{M}} \right]^2 \tag{1}$$

where: $P'i$ - set of pressure values, $Q'i$ - set of flow values, ΔPi - difference of pressure between simulated and observed data, ΔQi - difference of flow between simulated and observed data, N - number of monitored nodes, M - number of monitored pipes.

The coordinates of the decision vector can be roughnesses of pipes and emitter coefficients, which in the objective function occur in the implicit form. In turn, implicit restrictions in this optimization task are conservation equations describing the flows in the system. Due to formal reasons, mentioned coordinates of the decision vector are under restrictions with explicit character. The optimal solution is the decision vector meeting the above criterion.

The consumption of water in time and space of the supplied area and also the initial roughnesses of pipes have been determined as precisely as possible due to their high importance for the convergence and the stability of the algorithm. Localization of measuring points was made following the selection of sites with the highest pressure fluctuations (primarily terminal nodes of the network branches) and in the areas with the largest water consumptions (around the "centre of gravity of the system").

Schematic visualization of the water supply system modeled by us was shown on the Fig. 1. In this picture groups of pipes were marked; how we recalled in chapter 1, we are creating the groups of pipes of similar roughness in order to reduce the number of unknowns / searched roughnesses.

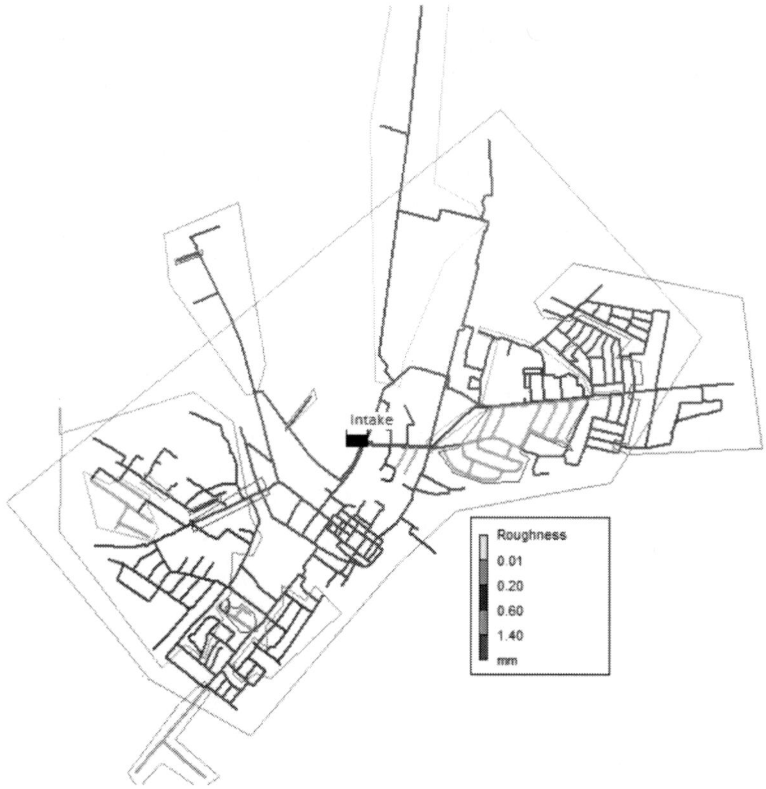

Fig. 1. Scheme of the modeled water supply system with groups of pipes marked by their roughnesses.

Essential findings and remarks about calibration are given in the last chapter devoted to conclusions. Only brief calibration results are provided in this chapter, which ensure that the model adequately describes the operation of the water supply system.

Fig. 2 shows the comparisons of observed 24 h pressures with simulated ones before and after calibration, at two exemplary nodes.

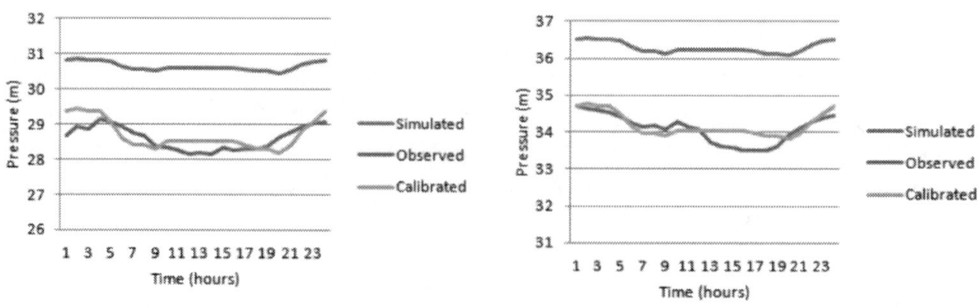

Fig. 2. Comparison of observed pressures with simulated ones before and after calibration, at exemplary nodes; Initial population: 200, Generations: 400, Mutation probability: 0,01, Crossover probability: 0,9.

As can be seen, as a result of the performed calibration, i.e. estimation of the pipes roughnesses and the determination of the emitters coefficients, the accuracy of the system operation simulation is very good, what was not possible to be stated with the reference to not-calibrated model, although starting values of the roughnesses of pipes and the partitions of water were assumed with the great care. Differences between pressures simulated with the help of calibrated model and pressures that were measured do not exceed permissible values given at the work [13], and they even are definitely smaller.

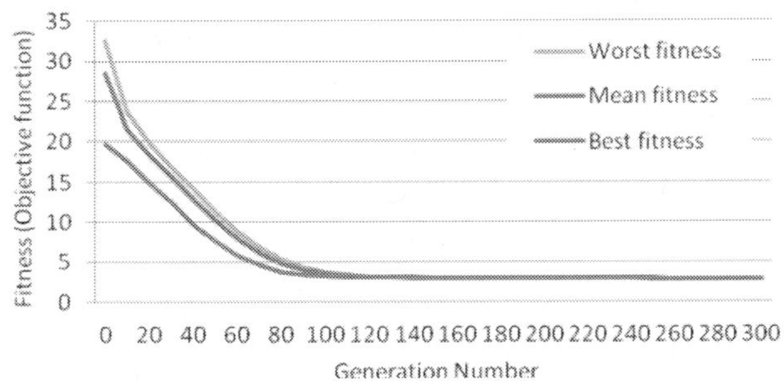

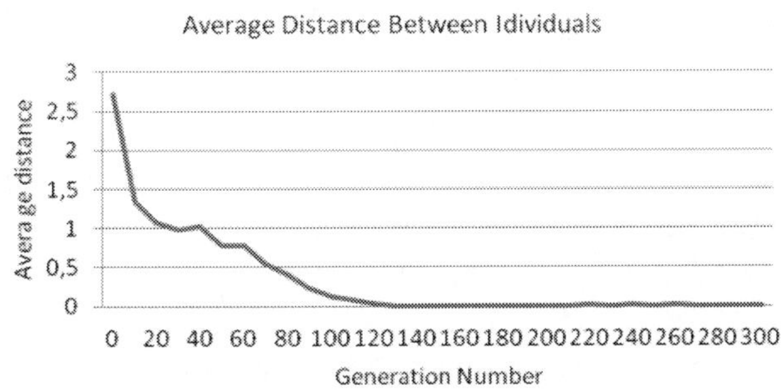

Fig. 3. Characteristics of the convergence of the applied genetic algorithm.

As can be seen, calculations conducted with the EPANET Calibrator software were not only stable, but also were characterized by a very good convergence. In contrast with often applied "manual" roughnesses adjustments while making the calibration, these calculations are proceeding automatically and quickly, not requiring the interference of the engineer drawing the computer model up. It also helps to avoid the tendency of setting the local minimum of the objective function.

5 Conclusions

1. The author carried out a calibration of the model using the “direct” method, i.e. using measurements of pressure and flow in selected locations of the system. Thanks to applying presently standard methods of reducing the number of unknown roughnesses , as well as thanks to advantages of the genetic algorithm, the direct method of calibration turned out to be completely sufficient.
2. It is important to choose the proper localization of the measuring points. The method of locating these points described in the article was sufficient. More sophisticated methods of sampling design are desirable only for large and complex water supply systems.
3. In the process of the calibration we are measuring both pressures and flows and at set partitions of water the result of the calibration exploiting the EPANET Calibrator software is a vector of roughnesses of pipes and emitters coefficients.
4. Significant elements affecting the quality (convergence) of the calibration of the model are: carefully determined water consumptions of the historical period used for the calibration as well as carefully determined demand patterns. It is also the condition for estimating possibly accurate leakages.

5. The graph presented in the Fig. 3 is demonstrating a very good, better than assumed while starting the calibration, convergence of the optimization process, which is a contribution of both the genetic algorithm implemented in the EPANET Calibrator software and carefully prepared starting data recalled in the previous chapter.
6. Applying described method of calibration and software based on EPANET, i.e. EPANET Calibrator utilizing the genetic algorithm, the computer model providing very good accuracy of the system operation simulation was obtained. Differences between pressures simulated with the help of calibrated model and pressures that were measured do not exceed permissible values given at the work [13], and they even are definitely smaller.

References

1. C. A. Bush, and J. G. Uber, J Water Res Pl-Asce, **1249**,6, 334-344 (1998) (http://dx.doi.org/10.1061/(ASCE)0733-9496(1998)124:6(334))
2. S. Lingireddy and L. E. Ormsbee, Civ Eng Environ Syst , **19**, 1, 13-39 (2002) (http://dx.doi.org/10.1080/10286600212161)
3. N. Abe, P. B. Cheung, *EPANET Calibrator. User Guide.* (2008)
4. E. Keedwell, Soon-Thiam Khu, Eng Appl Artif Intel, **18**, 461–472 (2005) (http://dx.doi.org/10.1016/j.engappai.2004.10.001)
5. Di Nardo, M. Di Natale, C. Gisonni, M. Iervolino, J Water Supply Res T **64**,1, 36-46 (2015) (http://dx.doi.org/10.2166/aqua.2014.004)
6. W. Stanisławski, M. Zmarzły, Pomiary Automatyka Robotyka,**12**, 61-64 (2010)
7. A.Wróbel, *Comprehensive analysis of the operation of water supply network, for example for the town of Działdowo, using a computer model of the network created in Epanet 2.0,* Master's thesis done under the direction of M. Szostak-Orłowska, Department of Sanitary Engineering, Faculty of Civil and Environmental Engineering, Gdańsk University of Technology, Poland (2015), in Polish
8. D. Sarnowski, *Gospodarka wodno-kanalizacyjna w Działdowie,* Działdowo, Poland (2012), in Polis
9. V. Ghorbanian, B. Karney and Y. Guo, J Water Res Pl-Asce,**142**,8 (2016) (http://dx.doi.org/10.1061/(ASCE)WR.1943-5452.0000665)
10. M. Kwietniewski, M. Tłoczek, L. Wysocki et al., *Rules for selection of materials and design solutions for the construction of water pipes,*(Publishing House Izba Gospodarcza Wodociągi Polskie, Bydgoszcz, Poland, 2011), in Polish
11. Cz. Grabarczyk, *Liquid flow in pipes: calculation methods*, (Publishing House Envirotech Sp. z o.o., 1997), in Polish
12. Narumi Abe, P. B. Cheung, *Epanet Calibrator – An integrated computational tool to calibrate hydraulic models*,(Integrating Water Systems. Boxall & Maksimovic (eds) Taylor & Francis Group, London, ISBN 978-0-415-54851-9, 2010)
13. T. M. Walski, D. V. Chase, D. A. Savic et al., *Advanced water distribution modeling and management*, (Haestad Press , Waterbury, CT USA, 2003)
14. L.A. Rossman, *EPANET 2 Users Manual*, (EPA, United States Environmental Protection Agency, Cincinnati, 2000)

Impact of type of wastewater treatment process on the antibiotic resistance of bacterial populations

Adriana Osińska[1,*], *Ewa* Korzeniewska[1], *Monika* Harnisz[1], and *Sebastian* Niestępski[1]

[1]Department of Environmental Microbiology, Faculty of Environmental Sciences, University of Warmia and Mazury in Olsztyn, Poland

Abstract. Antibiotic resistant bacteria reach the environment directly with faeces, and indirectly with sewage discharged from wastewater treatment plants (WWTPs). The aim of the study was to determine the level of removal of bacteria during wastewater treatment. Samples of untreated and treated sewage were collected from 13 WWTPs with different capacity, modification of treatment and type of inflowing wastewater. Microbiological characterization of the samples included determination of the total number of bacteria resistant (ARB) to β-lactams and tetracyclines as well as the number of *Escherichia coli* resistant to the same drugs. The counts of ARB and *E. coli* were determined on TSA and mFc media with/without antibiotic supplementation, respectively. The highest percent of reduction in number of ARB (at least 99.9%) and *E. coli* (above 99.3%) was obtained for WWTPs with A_2O system. The lowest percentage reduction of ARB and *E. coli* was observed for WWTPs operating with SBR system. The lowest number of microorganisms resistant to analyzed antibiotics was observed for ARB and *E. coli* resistant to cefotaxime and doxycycline. The results indicate a large variation in the removal of antibiotic-resistant bacteria in WWTPs depending on modifications of treatment system and type of inflowing wastewater.

1 Introduction

The overuse of antibiotics in medicine and agriculture caused a reduction of their efficacy against most of infectious diseases. According to the World Health Organization (WHO), the increase of antibiotic resistance among bacteria is one of the most important global problems [1]. In addition, according to the report prepared by O'Neill [2] due to infections caused by bacteria resistant to antimicrobial agents, 700,000 people die every year, and by 2050 this number is expected to grow to over 10 million. The definition of drug-resistant bacteria given by ECDC [3] is based on the limit values relating to the clinical strains. That is why it seems necessary to have an alternative approach to the microbial characteristics of environmental strains [4]. However, so far there is no

* Corresponding author: adriana.osinska@uwm.edu.pl

exhaustive data allowing to assess the consequences of the occurence of antibiotic resistance in environmental bacteria.

Another global problem is the access to safe and clean water [5]. It should be remembered that water resources are decreasing dramatically all over the world, especially due to global warming and climate change [6]. The discharge of treated sewage with a high burden of drug-resistant microbes into the surface water reservoirs may prevent their further use for economical purposes. The adverse effect of the phenomenon of antibiotic resistance on the health of both humans and animals, as well as its economic impact, enforces taking actions which objective is to monitor the spread of the antibiotic resistance in microorganisms of aqueous environments [7].

The occurrence and growth of antibiotic resistance in microorganisms populations are inevitable and conditioned by changes in the way of evolution, because bacteria can be quickly adapted to changing environmental conditions [8]. Antibiotic resistance may be intrinsic in the bacteria with a natural predisposition for recombination and genetic mutation or acquired as a result of selective pressure of antimicrobial substances present in the environment [9]. It may also be the result of a process of horizontal gene transfer between microorganisms using of mobile genetic elements (plasmids, transposons, integrons) [10, 11].

Wastewater treatment plants are considered as an important reservoir (hot-spots) of antibiotic-resistant bacteria (ARB), and antibiotic resistance genes (ARGs) [12-14] representing a major source of their dissemination in the environment [15, 16]. The wastewater flowing into WWTPs create favorable conditions for the survival of ARB and transfer of ARGs, due to the high content of both microbial biomass and the nutrients [13].

Safe and simultaneously economically reasonable methods of discharges of the treated wastewater to the environment are a major challenge in the management of wastewater. Nowadays, technology of treatments is primarily focused on the removal of contaminations such as organic matter, nitrogen and phosphorus compounds, which can cause oxygen depletion and eutrophication of waters receiving treated wastewater. However, they do not focus on the removal of ARB and ARGs [16]. Quantitative determination of the presence of indicators of fecal contamination such as *Escherichia coli* and *Enterococcus faecalis*, taken into consideration in the evaluation of surface water, is also recommended by many authors to assess the microbiological quality of wastewater [17, 18].

Although there has been a reduction in the number of bacteria at the level 99% in the process of treatment and disinfection of wastewater, together with the WWTPs' effluents the significant amounts of ARB enter into the environment. These ARB are characterized by higher resistance to antibiotics due to the transfer of genes between bacteria present in the wastewater [19]. Their higher number is still observed in the treated sewage [20, 21], which can be transferred to subsequent environments, posing a potential threat to their safety and public health. Therefore, in this article we compare the use of various modifications of the wastewater treatment technologies based on activated sludge in the removal of both the total number of antibiotic-resistant bacteria and *E. coli* resistant to the most commonly used class of antibiotics which are beta-lactams and tetracyclines.

2 Materials and methods

2.1 Study sites and sampling

Samples of untreated and treated sewage was collected from 13 WWTPs located in Warmia and Mazury District. WWTPs treat the wastewater based on activated sludge with different modification (biological, biological with elevated nutrient removal) and type of inflowing wastewater (domestic sewage, hospital sewage, wastewater from food industry), various capacity and number of population served. Wastewater treatment plant were divided into 4 categories, due to the applied modification of sewage treatment system: A - WWTPs with A_2O system, B – WWTPs with mechanical-biological system, C - WWTPs with SBR reactors, D – WWTPs with mechanical-biological system with elevated removal of nutrients. Samples of wastewater were collected in summer (July) and in winter (February) into the sterile bottles, transported to the laboratory at the temperature of 4°C and processed on the day of collection.

2.2 Physicochemical parameters

The level of physicochemical parameters value of wastewater samples, including biochemical oxygen demand over 5 days (BOD) and chemical oxygen demand (COD) of sewage were evaluated in parallel to microbial analysis. The methodology to assess those parameters was used in accordance with the APHA [22]standard methods.

2.3. Number of bacteria and bacteria resistant to antibiotics

To obtain 8–80 colony forming units (CFU) per plate, untreated wastewater (UWW) and treated wastewater (TWW) samples were decimal diluted with saline water and passed through a cellulose filter (pore diameter 0.45 uM, Millipore). Greater accuracy was achieved by plating triplicates. The total number of bacteria and bacteria resistant to β-lactams (amoxicillin, cefotaxime) and tetracyclines (oxytetracycline, doxycycline) as well as the total number of *Escherichia coli* and *Escherichia coli* resistant to the same drugs was determined on plates containing the TSA medium (Oxoid) and the mFc Agar medium (Merck) with/without amoxicillin (2 μg/mL), cefotaxime (2 μg/mL), oxytetracycline (2 μg/mL) and doxycycline (2 μg/mL) supplementation. Antimicrobial dose was determined in accordance to EUCAST [23] guidelines. The total number of microorganisms were incubated at 30°C for 48 h. *E. coli* were cultured at 44.5±0.2°C for 24 h and after incubation, *E. coli* colonies were counted based on the number of dark blue colonies formed on the mFc Agar medium.

2.7. Data analyses

Statistical analyses were carried out using the STATISTICA 10 software package (StatSoft Inc., 1984–2011). The data were not normally distributed, and the Kruskal–Wallis test, a non-parametric version of the classical one-way analysis of variance (ANOVA), was used to determine variations in the abundance of the studied bacterial groups and the physicochemical parameters of samples from different groups of WWTPs. The Spearman's rank correlation was used to detect the correlations between physicochemical parameters and microbial counts.

3 Results and discussion

3.1. Physicochemical parameters concentration

The highest average values of BOD in UWW were found in samples from WWTPs in group B (606 mg/L), while the lowest were found in samples from WWTPs belonging to group A (401 mg/L). During the process of wastewater treatment in all analyzed WWTPs, BOD values of were reduced in 97-99.8%. In contrast, the highest values of COD were observed among UWW samples from WWTPs belonging to group D (1540 mg/L), whereas the lowest were found in UWW samples from WWTPs of group C (1199 mg/L). COD values in the treatment process were diminished at percentage reduction of 92–98%. Both of those parameters were directly positively correlated with the number of microorganisms analyzed in this study ($p<0.05$).

3.2. The total number of bacteria and antibiotic-resistant bacteria to beta-lactams and tetracyclines

The average total number of bacteria in untreated wastewater ranged from 1.47×10^7 to 6.05×10^7 CFU/mL, while in the treated wastewater from 8.25×10^4 to 2.77×10^5 CFU/mL. After the treatment process, the total number of bacteria was decreased by the level of 91.12 to 99.99% (Table 1). The lowest percentage of reduction was observed in WWTPs from group C, while the highest one was noticed in WWTPs from group A.

Table 1. Total counts of bacteria, counts of bacteria resistant to beta-lactams (amoxicillin, cefotaxime) and tetracycline (oxytetracycline, doxycycline).

	A		B		C		D	
	UWW *	TWW **	UWW *	TWW **	UWW *	TWW **	UWW *	TWW **
TB	60.5±41	27±26 (99.7-99.9) ***	27±10	8.3±3.7 (68-99.6)	15±8	22±19 (91-99.9)	15±8	9±2.8 (98.8-99.5)
TBA	16±8.5	2.6±2 (99.96)	16± 4	8±3 (0-99.8)	8.4±1.7	11±8.9 (95-99.9)	6±1	8±3 (96.7-99.39)
TBC	1.98±0.9	2.9±2 (94-95.7)	1.4±0.6	1.3±0.7 (25-99.8)	1.3±0.7	0.4±0.2 (96.7-99.9)	0.6±0.01	0.7±0.4 (98.7-99)
TBO	2.8±1.9	1.1±1 (99.6-99.9)	1.9±0.5	1.5±0.7 (47.5-99.5)	0.9±0.3	3.6±3 (82-99.9)	0.7±0.3	0.5±0.1 (98-99.5)
TBD	2.7±2	0.01±0.002 (94-99.3)	0.03±0.01	0.1±0.005 (94-99.9)	0.03±0.01	0.14±0.1 (79-99.8)	0.06 ±0.01	0.07±0.003 (98-99)

* CFU/mL$\times 10^6$ ** CFU/mL$\times 10^4$ ***% percent of reduction in the brackets
CFU – colony forming unit, UWW – untreated wastewater, TWW – treated wastewater, TB – counts of total bacteria, TBA – counts of bacteria resistant to amoxicillin, TBC – counts of bacteria resistant to cefotaxime, TBO – counts of bacteria resistant to oxytetracycline, TBD – counts of bacteria resistant to doxycycline, A, B, C, D - modification of sewage treatment system.

The average number of bacteria resistant to beta-lactams and tetracycline in the untreated wastewater ranged from 7.04×10^5 to 1.6×10^7 CFU/mL and from 3.32×10^4 to 2.75×10^6 CFU/mL, respectively. In both, the untreated and treated wastewater, the bacteria resistant to amoxicillin (up to 80% of all bacteria), which belongs to an older generation antibiotics, were the most frequently occurring (Fig. 1).

The lowest percentage (3%) were bacteria resistant to doxycycline which belongs to a new generation of drugs from the class of tetracyclines. Alexander et al. [4] in their study showed that WWTPs contribute to increasing of antibiotic resistance in bacterial populations, although their number was decreased. The conventional processes of wastewater treatment based on activated sludge and anaerobic digestion, create ideal conditions for horizontal gene transfer of mobile elements presented in the bacterial cells, contributing to an increase in the prevalence of antibiotic-resistant bacteria [8, 10]. Many researchers [16, 24-26] indicate that WWTPs are not sufficiently effective in removing contaminants such as ARB or ARGs. A high concentrations of nutrients and a high concentrations of bacteria in the wastewater (particularly in the activated sludge), which is essential for effective treatment, support the spread of antibiotic resistance among bacteria.

The most common antibiotics used in the treatment of people all over the world are penicillins, cephalosporins and carbapenems belonging to the group of beta-lactam antibiotics. The second ones are tetracyclines, which are the most popular antibiotics in animal treatment. This accounts for approximately 95% of total antibiotics used in the world. The reduction of number of bacteria resistant to beta-lactams after treatment processes ranged from 95.10 to 99.98%. The highest percent of reduction in the number of bacteria resistant to both amoxicillin and cefotaxime was observed in WWTPs of the group A. These plants also most effectively reduced the number of bacteria resistant to oxytetracycline and doxycycline at the level of 99.99 and 100%, respectively. The lowest percentage of reduction of the number of bacteria resistant to amoxicillin (95.1%), oxytetracycline (82.3%) and doxycycline (79.4%) was observed in WWTPs belonging to group C (Table 1). Although WWTPs usually reached quite a high level of reduction in the number of antibiotic-resistant bacteria, WWTPs' effluents discharged into the environment on average from 1.05×10^2 to 2.77×10^5 CFU/mL (Table 1). It should be noted that the percentage of antibiotic-resistant bacteria in the total number of microorganisms increased after the treatment process. This was particularly noticeable in the case of WWTPs from group B, where an increase of the percentage of bacteria resistant to the older as well as newer generations of antibiotics was observed. Also Yuan et al. [19] and Korzeniewska et al. [27] reported that, despite of the high percentage of ARB reduction, discharge of wastewater causes the threat of aquatic environments.

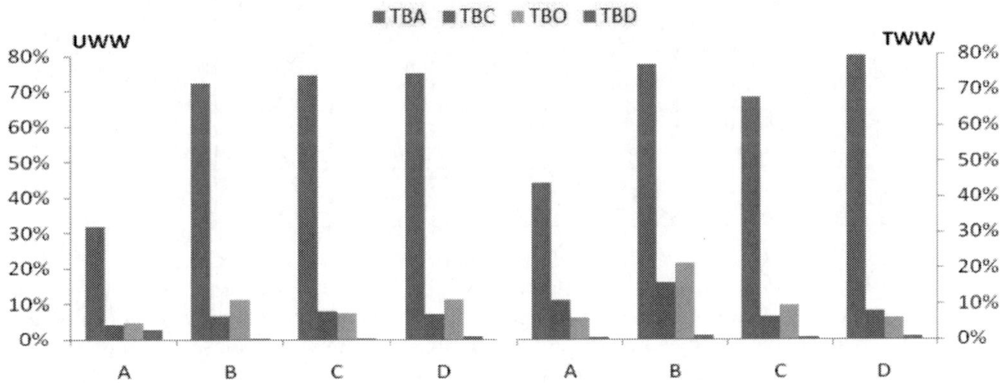

Fig. 1. The share of antibiotic resistant bacteria in total counts of microorganisms (explanation of abbreviations under Table 1).

3.3 The total number of *E. coli* and *E. coli* resistant to beta-lactams and tertracyclines

The excrement of animals or humans, incoming into sewage treatment plant, affected mainly on a high bacterial load of wastewater [18; 28]. *Escherichia coli*, which is a natural microbiota of gastrointestinal tract of humans and animals, can get into the environment directly with faeces, and indirectly with sewage effluents discharged from wastewater treatment plants. Therefore, the quantitative determinations of *E. coli* as an indicator of faecal contamination is recommended by many authors to assess the microbiological quality of the wastewater [18], as well as the contamination of aquatic ecosystems [29, 30]. The highest average number of *E. coli* were found in samples of untreated sewage from WWTPs which belongs to a group A, the lowest ones were found among WWTPs of the group B. Their number reached the amount of 6.55×10^5 and 9.76×10^4 CFU/mL, respectively (Table 2). Regardless of the type of treatments, *E. coli* resistant to antibiotics of older generation - amoxicillin and oxytetracycline were predominant group (15–58% and 7–12%, respectively) in samples of untreated sewage. After the wastewater treatment, apart from *E. coli* resistant to amoxicillin, the dominance of *E. coli* bacteria resistant to a new generation drugs – cefotaxime and doxycycline have been also observed. Similar results were obtained by Osińska et al. [31], who indicate an increase of percentage of antibiotic-resistant *E. coli* in the the total number of *E. coli* in wastewater after the treatment process as well as growth of multi-drug resistance of the microorganisms.

The reduction of the number of antibiotic-resistant *E. coli* ranged from 56.7 to 99.9%. The WWTPs which belongs to the group C were characterized by the lowest percentage of reduction. Among *E. coli* bacteria resistant to analyzed antibiotics in untreated wastewater, *E. coli* resistant to amoxicillin were the most frequently observed, while the most rarely were isolated *E. coli* resistant to cefotaxime ($4.45x10^5$ and $3.67x10^2$ CFU/mL, respectively). The reduction of the number of bacteria resistant to beta-lactams ranged from 0 to 99.9% (Table 2). No reduction was observed for both of bacteria resistant to amoxicillin (WWTPs belonging to groups B and C) and cefotaxime (WWTPs belonging to groups C and D).

Table 2. Total counts of *Escherichia coli*, counts of *Escherichia coli* resistant to beta-lactams (amoxicillin, cefotaxime) and tetracycline (oxytetracycline, doxycycline).

	A		B		C		D	
	UWW *	TWW **	UWW *	TWW **	UWW *	TWW **	UWW *	TWW **
EC	65.5±50	2.3±1 (99.7-99.9)***	9.8±3	18.7±1 (68-99.6)	21.2±11	52.8±51 (56-99.9)	23.7±1.3	13.8±1.6 (99.3-99.5)
ECA	44.5±38	2.1±2 (99.96)	2.7±0.8	5.2±1.6 (0-99.8)	4.8±2	25.30±24 (0-99.8)	3.6±1.6	7.5±1.4 (94-98.6)
ECC	0.2±0.02	1.3±0.2 (94.7-95.7)	0.1±0.03	1.4±0.7 (25-99.7)	0.04±0.02	0.43±0.3 (0-99.8)	0.09±0.03	4.9±2.6 (0-97.5)
ECO	2.1±1	0.3±0.07 (99.6-99.9)	1.3±0.4	3.5±1 (47.5-99)	1.2±0.7	8.9±2 (72-99.6)	2.1±1	1.4±0.1 (99.4-99.7)
ECD	0.3±0.04	2.2±2 (94-99)	0.3±0.1	0.74±0.2 (5-99.51)	0.2±0.1	12.7±11 (0-96.8)	0.7±0.5	1.3±0.7 (87-98.5)

* CFU/mL x 10^4 ** CFU/mL x 10^2 ***% percent of reduction in the brackets
UWW – untreated wastewater, TWW – treated wastewater, CFU – colony forming unit, EC – counts of total *E. coli*, ECA – counts of amoxicillin resistant *E. coli*, ECC – counts of cefotaxime resistant *E. coli*, ECO – counts of oxytetracycline resistant *E. coli*, ECA – counts of doxycycline resistant *E. coli*
A, B, C, D - modification of sewage treatment system.

Similar trends were also observed in the reduction of the number of bacteria resistant to tetracycline. There was also no reduction in the number of doxycycline resistant *E. coli* in WWTPs which belong to group C (Fig. 2). The lowest percentage of reduction of the number of *E. coli* resistant to oxytetracycline was observed in WWTPs belonging to the group B (47.5%). The highest efficiency in the removal of bacteria resistant to both beta-lactams and tetracyclines, which reached up to 94-99.9%, was observed for WWTPs of group A. The results of our study, which indicate an evidence of insufficient removal of ARB in WWTPs using conventional wastewater treatment systems were confirmed also by the study conducted by many authors [24, 32].

Despite the significant reduction of the number of antibiotic-resistant *E. coli*, we noted that after the sewage treatment process the percentages of *E. coli* bacteria resistant to all tested antibiotics increased in wastewater treated in WWTPs belonging to groups B, C and D. However, in wastewater treated in WWTPs of group A, the percentage of bacteria resistant to antibiotics from old generation was decreased. The percentage of bacteria resistant to cefotaxime and doxycycline, which are antibiotics from new generation groups, in the total number of *E. coli* has been increased. Some researchers [28, 33] suggest that WWTPs are effective in removal of bacteria, including indicator bacteria like *E. coli*.

However, there are also studies which suggest that the number of bacteria did not decrease after the wastewater treatment processes, regardless of the type of treatment [34].

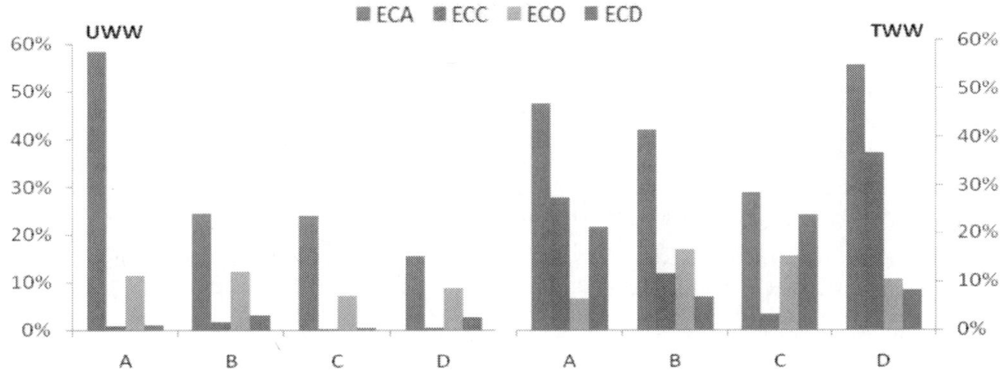

Fig. 2. The share of antibiotic resistant *Escherichia coli* in total counts of *E. coli* (explanation of abbreviations under Table 2).

4 Conclusions

The presented results indicate that wastewater treatment plants are an important source of antibiotic-resistant bacteria. They also indicate different level of removal of antibiotic-resistant bacteria, depending on the technology used for wastewater treatment. The highest percent of reduction in number of ARB (at least 99.9%) and *E. coli* (above 99.3%) was obtained for WWTPs with A_2O system. Although most of WWTPs showed a high efficiency of reduction of the number of antibiotic-resistant bacteria during the treatment process, they continue discharge with treated wastewater the massive amounts of microorganisms into the natural environment. Futhermore, the studies have shown that during the wastewater treatment process, the percentage of antibiotic-resistant bacteria in the total number of bacteria underwent increase. The presented results indicate that some technological solutions used by WWTPs insufficiently reduced the number of antibiotic-resistant *E. coli*, which is a serious threat to public health. This indicates the need of improving of new modification of wastewater treatment processes and continuous monitoring of treated sewage discharged into the environment.

References

1. A. F. Leonard, L. Zhang, A.J. Balfour, R. Garside, W. H. Gaze, Environ. Int. **82** (2015)
2. J. O'Neill, http://amrreview.org/sites/default/files/160525_Final%20paper_with %20cover.pdf (2014)
3. ECDC, European Centre for Disease Prevention and Control. Antimicrobial consumption interactive database (ESAC-Net). http://ecdc.europa.eu/en/healthtopics/antimicrobial-resistance-and-consumption (2015)
4. J. Alexander, A. Bollmann ,W. Seitz, T. Schwartz, Sci. Total Environ. **512–513** (2015)
5. M. A. Shannon, P. W. Bohn, M. Elimelech, J. G. Georgiadis, Mariñas, B. J., Mayes, A. M., Nature **452** (2008)
6. R. M. Carr, U. J. Blumenthal, D.D. Water Sci. Technol. **50** (2004)

7. J. L. Martinez, T.M. Coque, F. Baquero, Nat. Rev. Microbiol. **13** (2015)
8. L. S. Frost, R. Leplae, A. O. Summers, A. Toussaint, Nat. Rev. Microbiol. **3** (2005)
9. J. Davies, Are antibiotics naturally antibiotics? J. Ind. Microbiol. Biotechnol. **33**, 7 (2006)
10. E. Marti, B. Huerta, S. Rodríguez-Mozaz , D. Barceló, J. Jofre, J. L. Balcázar, Water Res. **61** (2014)
11. J. L. Martinez, Science **321** (2008)
12. B. Berglund, J. Fick, P. E. Lindgren, Environ. Toxicol. Chem. **34** (2015).
13. L. Rizzo, C. Manaia, C. Merlin, T. Schwartz, C. Dagot, M. C. Ploy, I. Michael, D. Fatta- Kassinos, Sci. Total Environ.**447** (2013)
14. T. Yang , Z. Zeng, L. Rao , X. Chen, D. He, L. Lv, J. Wang, L. Zeng, M. Feng, J. H. Liu, Vet. Microbiol. **170** (2014)
15. E. Kotlarska, A. Łuczkiewicz, M. Pisowacka, A. Burzyński, Environ. Sci. Pollut. Res. **22** (2015)
16. A. Pruden, Environ. Sci. Technol. **48** (2014)
17. L. K. Dick, E. A. Stelzer, E. E. Bertke, D. L. Fong, D. M. Stoeckel, Appl. Environ. Microbiol. **76** (2010)
18. J. D. Dubreuil, Curr. Issues Mol. Biol. **14** (2012)
19. Q. B. Yuan, M. T. Guo, J. Yang , Environ. Sci.: Processes Impacts, **16** (2014)
20. F. F. Reinthaler, J. Posch , G. Feierl, G. Wust , D. Haas , G. Ruckenbauer , F. Mascher, E. Marth, Water Res. **37** (2003)
21. S. Borjesson, S. Melin, A. Matussek and P. E. Lindgren, Water Res., **43**, 4 (2009)
22. APHA, American Public Health Association, Standard Methods for the Examination of Water and Wastewater, 20th ed. Washington D.C. (1998)
23. EUCAST, European Committee on antimicrobial susceptibility testing. Breakpoints tables for interpretation of MICs and zones diameters. Version 4.0, http://www.eucast.org (2014)
24. D. Rafraf, I. Lekunberri, A.Sànchez-Melsió, M. Aouni, C. M. Borrego, J. L. Balcázar, Environ. Pollut. **219** (2016)
25. E. Korzeniewska, M. Harnisz, J. Environ. Manage., **128** (2013)
26. M. Harnisz, E. Korzeniewska, I. Gołaś, Chemosphere **128** (2015)
27. E. Korzeniewska, A. Korzeniewska, M. Harnisz, Ecotox. Environ. Safe., **91** (2013)
28. R. Hendricks, E. J. Pool, J. Environ. Sci. Health A Tox. Hazard Subst. Environ. Eng. **47** (2012)
29. S. Li, S. Wengang, Z. Yufa, T. Yujing, G Yanxia, M. Zengmin, Environ. Sci. Pollut. Res. **22** (2015)
30. Y. Titilawo, L. Obi, A. Okoh, Sci. Total Environ. **523** (2015)
31. A. Osińska, E. Korzeniewska, M. Harnisz, S. Niestępski, Sci. Total Environ. **15** (2017)
32. S. Rodriguez-Mozaz, S. Chamorro, E. Marti, B. Huerta, M. Gros, A. Sànchez-Melsió, C. M. Borrego, D. Barceló, J.L. Balcázar, Water Res. **69** (2015)
33. M. Oliveira, I. Serrano, S. Van Harten, L.J. Bessa, F. Bernardo, P.M. da Costa, Environ. Sci.Pollut. Res. **23** (2016)
34. V. K. Sharma, N. Johnson , L. Cizmas, T. J. McDonald, H. Kim Chemosphere **150**, (2016)

Exposure of urban agglomeration population to the selected components of PM_1 emitted from low emission sources

Łukasz Pachurka[1,*], *Wioletta* Rogula-Kozłowska[2,3], *Anna* Chlebowska-Styś[1], *Alicja* Nych[1], and *Anna* Zwoździak[1]

[1]Wroclaw University of Science and Technology, Faculty of Environmental Engineering, Unit of Ecologistics and Environmental Risk Management, Wybrzeże Wyspiańskiego 27 St., 50-370, Wroclaw, Poland
[2]Institute of Environmental Engineering, Polish Academy of Sciences, M. Skłodowskiej-Curie 34 St., 41-819, Zabrze, Poland
[3]The Main School of Fire Service, Faculty of Fire Safety Engineering, Słowackiego 52/54 St., 01-629 Warsaw, Poland

Abstract. The sources of gaseous and particulate (PM) pollutants in Polish cities are mainly: municipal sector, industry, emissions from the road transport and the upstream emission (pollution "flowing" to the cities, derived from emission sources located outside of cities). The residents of the cities are mainly exposed to air pollutants from low-emission sources (i.e. municipal sector and road traffic). In the paper, the results of the study from field campaign, conducted in January of 2016 in Wroclaw will be presented. During the field campaign the 24-h concentrations of submicron particulate matter (PM_1) and 24-h concentrations of selected PM_1-bound heavy metals were determined. The cancer risk associated with inhalation exposure to arsenic, nickel, and cadmium to the city's residents, based on the US Environmental Protection Agency (US EPA) standards methodology including the so-called lifelong chronic exposure of adult and child, was calculated. Measurements results showed that in Wroclaw during the winter season in 2016 the 24-h concentrations of PM_1-bound arsenic, nickel and cadmium ranged from 0.51 to 4.26 ng/m^3, 0.21-52.89 ng/m^3 and 0.08-1.01 ng/m^3, respectively. Obtained calculations results of cancer risk values for inhalation exposure to arsenic were: for men: $6.11 \cdot 10^{-6}$, women: $7.30 \cdot 10^{-6}$, children: $14.90 \cdot 10^{-6}$, to nickel: for men: $1.91 \cdot 10^{-6}$, women: $2.29 \cdot 10^{-6}$, children: $4.67 \cdot 10^{-6}$, to cadmium: for men: $0.37 \cdot 10^{-6}$, women: $0.44 \cdot 10^{-6}$, children: $0.91 \cdot 10^{-6}$. The values obtained for inhalation exposure among children indicated the high potential risk of cancer, mainly for arsenic exposure.

[*] Corresponding author: lukasz.pachurka@pwr.edu.pl

1 Introduction

Air pollution contributes to the deterioration of the environment through the introduction of gaseous, liquid and solid substances into the atmosphere [1, 2]. Cardiovascular disease is considered to be the main cause of death in the European Union, accounting for about 40% of all deaths. In terms of a number of inhabitants, it constitutes about 2 million people per year [3]. In Poland, the premature death caused by the cardiovascular disease is twice as high compared to other Western European countries [4]. The most often cardiovascular disease that occurs in society is hypertension. Beyond the incorrect lifestyle, the factor responsible for the occurrence of cardiovascular disease is air pollution caused by particulate matter, $PM_{2.5}$ (fine particulate matter; a fraction of atmospheric particles with aerodynamic diameter Dp not greater than 2.5 microns) and PM_{10} (Dp $\leq$ 10 microns) [5]. The emissions data from 2013 show that combustion processes outside the industry (51%), road transport (13%), combustion processes in the transformation of energy (10%), combustion processes in industry (7%), other vehicles and equipment (7%), waste management (5%), production processes (5%,) distribution of fossil fuels and their distribution (1%), the use of solvents and other products (1%) and agriculture (0.2%) are mainly responsible for the emission of $PM_{2.5}$ [6].

The conducted worldwide studies have shown an association between increased hospitalization due to cardiovascular disease and the number of daily deaths from cardiovascular and respiratory diseases due to environmental exposure to a fine fraction of particulate matter, $PM_{2.5}$ [5].

Heavy metals such as nickel and cadmium cause a threat to human health and life, due to their good absorption through the respiratory system. Formation of complexes with proteins causes depositing cadmium in the liver and kidney. The toxic effects of cadmium deposition are hypertensive disease, impaired reproductive function, kidney disorder, cancer (kidney, prostate) [7].

Inhalation exposure to nickel contributes to allergic reactions and damage to mucous membranes. Other disorders related to long-term exposure to metal include malignancy, bone marrow changes and chromosomes disturbances in the metabolism of plasma proteins [7]. Exposure to arsenic presence contributes to the incidence of lung cancer [8, 9].

The aim of the study was to analyse the variability of 24-h concentrations of PM_1 (Dp $\leq$1 microns) and PM_1-bound heavy metals, i.e. cadmium, arsenic and nickel and assessment of inhalation exposure of the residents of the selected area of the city of Wroclaw associated with the presence of the above heavy metals in PM_1.

2 Materials and methods

The 24-h concentrations of PM_1 in the selected measuring site (Biskupin - Kosiby St.) were determined in the study carried out during a two-week campaign in the heating (winter) and non-heating (spring) season in 2016 in Wroclaw. The measuring site was located in the area covering both high and low buildings.

The sampling and weighing of PM_1 were performed according to the methods described in [10, 11]. The arsenic (As), cadmium (Cd) and nickel (Ni) concentrations were identified in all 24-h samples of PM_1. Metals quantitative analysis was preceded by the samples mineralization in a mixture of spectrally pure (Merck, Ultrapur) acids. Before mineralization, samples were flooded with 5 ml of nitric acid HNO_3 (65%), 1 ml of hydrofluoric HF (40%) and 1 ml of perchloric acid $HClO_4$ (70%). Mineralization was carried out in a microwave reaction system Multiwave 3000 from Anton Paar, in the sealed polytetrafluoroethylene PTFE vessels under pressure (mineralization furnace parameters: a pressure rise of 0.3 bar/s; 240°C; pressure 60 bar; mineralization time 50 minutes).

Chemical analysis was carried out using ICP spectrometer MS Elan DRCE produced by Perkin Elmer. Certified multielement standard stock solutions Periodic table mix 1 and Transition metal mix 2 (Fluka) were used as calibration solutions. The limits of quantification of elements were as follows: Ni -<0.01 μg/sample, As <0.025 μg/sample, Cd <0.003 μg/sample. The analytical method was validated by determination of As, Cd, and Ni in the Standard Reference Material, SRM 1649a, (Urban Particulate Matter Dust), obtained from the National Institute of Standard and Technology (NIST). Recovery rates were as follows: As (80%); Cd (83%); Ni (118%).

In order to determine the risk of cancer through inhalation on PM_1-bound Ni, As and Cd the methodology of risk assessment developed by US Environmental Protection Agency (US EPA) was applied. This methodology includes the so-called lifelong chronic adults and children exposure to selected PM-bound heavy metal [12, 13]. Based on the literature [7, 14, 15], in the scenario calculation, it was assumed that both adults and children may be exposed 24 hours a day and 365 days a year to the impact of polluted air. Among the data of the age of the population, the average time of exposure for children and adults were assumed as 6 and 70 years respectively [7, 14, 15, 16].

The use of methods of health risk assessment described by the US EPA allows of estimation the expected risks and health at a certain concentration of air pollution - in this case of inhalation. In this scenario, the daily intake depends on the concentration level of Ni, As and Cd in the study area.

The sampled amount (estimated daily intake) of the chosen heavy metal, defining the amount of substance entering the body throughout the day, based on 1 kg of body weight, was calculated from the formula (1).

$$EDI = (C \cdot IR \cdot AF \cdot F \cdot ED)/BW \cdot AT \qquad (1)$$

where:
EDI - estimated daily intake [mg/d·kg]
C - the average concentration in the air [mg/m^3]
IR - the daily lung ventilation (intake rate) [m^3/d]
AF - bioavailability factor [unitless]
F- the frequency of exposure [d/year]
ED - the duration of exposure [year]
BW - the average body weight [kg]
AT - the averaging time [d]

The following parameters were assumed in the study for calculation of average weight for women: 65.4 kg, men: 78.1 kg and children: 16 kg. Daily ventilation was assumed as following: for adults 20 m^3/d, for children - 10 m^3/d. Other parameters were: 24 hours a day, the bioavailability factor – 1, the frequency of exposure - 365 d/year, the duration of exposure - 365 days a year. These figures are based on a 50-percentile distribution of body weight in anthropometric data [7, 14, 15]. The average concentration of the heavy metal for the resident was defined as the average value of the metal concentrations determined in the 24-hour samples of PM_1 collected during the heating and non-heating season. The assessment included the carcinogenic effects of As, Cd, and Ni on the basis of the formula (2).

$$CR = EDI \cdot SF \qquad (2)$$

where:
CR - cancer risk
EDI - estimated daily intake [mg/d·kg]
SF - slope factor

Slope factor was assumed from the toxicological database (IRIS): inhalation exposure to As 15.1, Cd 6.3 and Ni: 0.84 [kg·d·mg^{-1}] [17]. The values obtained were compared with an acceptable level of risk of cancer of $1 \cdot 10^{-6}$ to $1 \cdot 10^{-4}$ [7, 14, 15].

3 Results and Discussion

Average concentration levels of PM_1 in selected European cities recorded in the summer and winter were listed in Table 1. In Wroclaw for both winter and spring season, the average concentrations of PM_1 are lower than those obtained in studies conducted in Zabrze (Poland) and selected European cities (Table 2). Average concentration of PM_1 in Wroclaw in winter is approximately 63 and 41% lower than the levels recorded in Milan and Zabrze, while for the spring season, the average concentration of PM_1 in Wroclaw is about 27; 54, 58 and 38% lower than the PM_1 concentrations in Melplitz, Genoa, Milan and Zabrze.

Table 1. Concentrations of PM_1 [μg /m^3] in selected European cities [18, 19].

City (country)	Sampling point	The average concentration in winter [μg/m^3]	The average concentration in summer [μg/m^3]
Zabrze (Poland) [18]	Urban background	30.8	13.1
Melplitz (Germany) [19]	Rural background	10.0	11.0
Genoa (Italy) [19]	Urban background	11.5	17.4
Milan (Italy) [19]	Urban background	48.8	19.4

Table 2. Descriptive statistics for 24-hour concentrations of PM_1.

Parameter	Concentration [μg/m^3]
the average concentration in heating season – winter (09.01- 26.01.2016) (the range of 24-h concentrations)	18.16 (2.84-38.20)
median (09.01- 26.01.2016)	15.55
standard deviation (09.01- 26.01.2016)	10.66
the average concentration in non-heating season – spring (09.05- 23.05.2016) (the range of 24-h concentrations)	8.05 (3.21-11.53)
median (09.05- 23.05.2016)	9.31
standard deviation (09.05- 23.05.2016)	2.88
the average concentration in heating and non-heating season – winter (09.01- 26.01.2016); spring (09.05- 23.05.2016) (the range of 24-h concentrations)	13.76 (2.84-38.20)
median (09.01- 26.01.2016) and (09.05- 23.05.2016)	10.64
standard deviation (09.01- 26.01.2016) and (09.05- 23.05.2016)	9.73

In accordance with European and Polish regulation, the levels of some PM_{10}-bound heavy metals in the framework of the State Environmental Monitoring (SEM) should not exceed nickel - 20 ng/m^3, arsenic - 6 ng/m^3 and cadmium - 5 ng/m^3 [20, 21].

Table 3 presents the average concentrations of PM_1-bound Ni, As and Cd in selected European cities. Based on the literature data in other European cities and own studies, there were no exceedances of selected metals in a fraction PM_1 (in relation to the value of the permissible level set for a fraction of PM_{10}).

Table 3. Concentrations of heavy metals indicated in particulate matter PM_1 in various European cities [7, 22, 23 and own studies].

Place of sampling, country	Site location	Nickel [ng/m^3]	Arsenic [ng/m^3]	Cadmium [ng/m^3]
Wrocław (Poland): Kosiby St., heating season (09.01- 26.01.2016)	Urban background	8.90 (0.21-52.89)	1.58 (0.51-4.26)	0.23 (0.08-1.01)
Wrocław (Poland): Kosiby St., non-heating season (09.05- 23.05.2016) [22]	Urban background	2.72 (0.34-7.35)	1.52 (0.71-4,19)	0.27 (0.07-0.91)
Zabrze (Poland), January-March 2008 [23]	Urban background	0.17	2.14	0.75
Barcelona (Spain), 2005-2006 [23]	Urban background	-	0.6	0.3
Milan (Italy), winter 2002 [23]	Residential-commercial area	7.0	2.0	-
Upper Silesian Agglomeration (2007-2009) [7]	-	3.29	2.15	1.52

Table 4 summarizes the calculated values of the estimated daily intake by inhalation of PM_1-bound Ni, As and Cd in Wroclaw. The highest daily intake of Ni taken by inhalation may be received by children ($5.56\cdot10^{-6}$ mg/d·kg) - due to the fact of 50% more inspired air than adults per kilogram of body weight. The respiratory system of children is also not fully developed compared with one of the adults. In a study conducted in the Upper Silesian Agglomeration, part of the Upper Silesian Industrial Region, on the basis of data from the Regional Inspectorate of Environment Protection (RIOP) from 2007-2009, dosage amount of Ni was calculated which amounted to: for children: $2.06\cdot10^{-6}$ mg/d·kg; women: $1.01\cdot10^{-6}$ mg/d·k; men: $0.84\cdot10^{-6}$ mg/d·kg [7]. The values obtained for daily intakes in non-heating season in Wroclaw are 17 (children), 18 (women) and 17 (men)% lower than those obtained in the Upper Silesian Agglomeration. In the case of the heating season, the values obtained were 270 (children), 269 (women) and 271 (men)% higher than those obtained in the Upper Silesian Agglomeration. The higher dosage amounts of Ni collected by the human body in the Silesian Agglomeration could depend on the industrial character of the region.

Table 4. Daily intakes by inhalation for selected heavy metals.

Place of sampling	Nickel Daily intake [$\cdot10^{-6}$ mg/d·kg]			Arsenic Daily intake [$\cdot10^{-6}$ mg/d·kg]			Cadmium Daily intake [$\cdot10^{-6}$ mg/d·kg]		
	children	adults		children	adults		children	adults	
		women	men		women	men		women	men
Wrocław: Kosiby St., heating season (09.01- 26.01.2016)	5.56	2.72	2.28	1.00	0.48	0.40	0.14	0.07	0.06
Wrocław: Kosiby St., non-heating season (09.05- 23.05.2016)	1.70	0.83	0.70	0.95	0.47	0.39	0.17	0.08	0.07

Based on the daily intakes, the value of cancer risk (Table 5) for the analysed heavy metals was calculated. The highest potential inhalation cancer risks associated with exposure to As and for the group of children, which is associated with the value of the slope factor were found for As: 15.1; cadmium 6.3 and nickel: 0.84 [kg·d·mg^{-1}] [17].

Table 5. Inhalation risk of cancerous on selected heavy metals in each group.

Place of sampling	Nickel CR [·10^{-6}]			Arsenic CR [·10^{-6}]			Cadmium CR [·10^{-6}]		
	children	adults		children	adults		children	adults	
		women	men		women	men		women	men
Wrocław: Kosiby St., heating season	4.67	2.29	1.91	14.90	7.30	6.11	0.91	0.44	0.37
Wrocław: Kosiby St., non-heating season [14]	1.43	0.70	0.59	1.43	0.70	0.59	1.06	0.52	0.44

Risk considered insignificant or acceptable by the US EPA is at a value of $1 \cdot 10^{-6}$ to $1 \cdot 10^{-4}$. The high risk requiring protective action is $1 \cdot 10^{-3}$ [7, 14, 15]. The calculations showed the highest risk values for the residents of the study area in Wroclaw during the heating season, respectively for children: $14.90 \cdot 10^{-6}$, women: $7.30 \cdot 10^{-6}$ and men: $6.11 \cdot 10^{-6}$. The values of the risk of cancer associated with exposure to As for the selected area of the city of Wroclaw were higher in winter than the values obtained on the basis of studies conducted in the years 2007-2009, where the potential cancer risks of exposure to arsenic for the residents of Wroclaw Agglomeration amounted to $0.93 \cdot 10^{-6}$ for men, $1.11 \cdot 10^{-6}$ for women and $2.27 \cdot 10^{-6}$ for children [7].

Determined cancer risk values for inhalation exposure in the measurement campaign in the winter season for individual metals amounted to As - men: $6.11 \cdot 10^{-6}$, women: $7.30 \cdot 10^{-6}$, children: $14.90 \cdot 10^{-6}$, Ni - men: $1.91 \cdot 10^{-6}$, women: $2.29 \cdot 10^{-6}$, children: $4.67 \cdot 10^{-6}$, Cd - men: $0.37 \cdot 10^{-6}$, women: $0.44 \cdot 10^{-6}$ children: $0.91 \cdot 10^{-6}$

The comparison of the obtained value of daily intakes taken was possible by the assumption of the same input data for the calculation, as in the literature [7]. The values of As daily intakes for the residents of Wroclaw in both the heating (children: 1.0; women: 0.48, men 0.40 [$\cdot 10^{-6}$ mg/d·kg]) as well as non-heating season (children: 0.95, women: 0.47, men 0.39 [$\cdot 10^{-6}$ mg/d·kg]) were lower due to the less contribution of industrial sources than in other city (Legnica: children: 4.53, women: 2.22, men: 1.86 [$\cdot 10^{-6}$ mg/d·kg] located in the industrial region [7].

The calculations in this study showed that the results obtained in the cancer risk are at an acceptable level for the selected site in Wroclaw.

4 Summary

Exposure to heavy metals such as cadmium, arsenic, and nickel may in the future be a potential cause of cancerous changes. The increase in the concentrations of heavy metals in the air is mostly related to human activities, including a large proportion of low-emissions from fuels in households.

In Wroclaw, children are the most exposed to the inhalation cancer risks due to the presence of arsenic in the air, and obtained results may indicate the potential risk of cancer ($14.90 \cdot 10^{-6}$). Beyond the level of concentration of analysed heavy metals, the risk of cancer is affected by the size of the slope factor (for arsenic: 15.1 [kg·d·mg^{-1}]). To fully determine the scale of the health risks for selected heavy metals, other means of exposure to contamination, i.e. the oral route and through the skin, should be considered in the future.

Current regulations regarding concentration levels of some heavy metals refer to the average annual values for particulate matter PM_{10}. Components of the sub-micron fraction can be transported to the alveoli in the human body. Therefore, steps should be taken to identify targets of annual and daily average concentrations of PM_1. None of the heavy metals show average concentrations higher than acceptable levels (Ni - 20 ng/m^3, As - 6 ng/m^3 and Ca - 5 ng/m^3).

The investigations were co-financed within the framework of the order No. B50601 with the specific subsidy granted for the Faculty of Environmental Engineering Wroclaw University of Science and Technology (W-7) by the Minister of Science and Higher Education to conduct research and development work and related tasks contributing to the development of young scientists and doctoral students in 2015/2016 and within the project No. 2012/07/D/ST10/02895 (ID 202319) financed by the National Science Centre, Poland (NCN).

References

1. A. Malec, A. Borowski, Inżynieria Ekologiczna = Ecological Engineering, **50**, 161-170, Dec. (2016)
2. J. Krzeszowiak, A. Michalak, K. Pawlas, Medycyna Środowiskowa = Environmental Medicine, **18 (2)**, 66-73 (2015)
3. L. Ryden, J. Martin, S. Volqvartz, Eur J Cardiov Prev R, **14 (3)**, 355-356 (2007)
4. M. Marek, A. Chłoń Damińczak, Kaleta, J. Mazur, P. Miśkiewicz, D. Poznańska, J. Stokwiszewski, B. Wojtyniak, T. Zdrojewski, *Social inequalities in health in Poland,* The European Office of the World Health Organization, Copenhagen (2012)
5. G. Dziubanek, E. Marchwińska-snatched, A. Piekut, M. Rusin, I. Hajok, Hygeia Public Health, **49 (1)**, 75-80 (2014)
6. Chief Inspectorate of Environmental Protection (CIEP), *Fine dust in the atmosphere. A compendium of knowledge about air pollution particulate matter in Poland,* http://www.gios.gov.pl/images/aktualnosci/Pyly_drobne_w_atmosferze.Kompendium_wiedzy.pdf, (2016), [Access: 22.1.2017]
7. M. Trojanowska, R. Świetlik, Medycyna Środowiskowa = Environmental Medicine, **15 (2)**, 33-41 (2012)
8. M.N. Bates A.H. Smith, C. Hopenhaym-Rich, Am J Epidemiol, **135 (5)**, 462-476 (1992)
9. K. Widziewicz, W. Rogula-Kozłowska, K. Loska, Atmospheric Pollution Research, **7 (5)**, 884–894 (2016)
10. Ł. Pachurka, I. Sówka, W. Rogula-Kozłowska, B. Mathews, A. Chlebowska-Styś, A. Zwoździak, [in:] Health determinants/ editors J. Chmielewski, M. Florek-Łuszczki, M. Szpringer, Warsaw: Environmental Institute: National Research Institute, 133-141 (2016)
11. I. Sówka, Ł. Pachurka, M. Przepiórka, W. Rogula-Kozłowska, A. Zwoździak, Rocznik Ochrona Środowiska = Annual Set The Environment Protection, **18 (2)**, 603-614 (2016)
12. US EPA: Guidelines for Human Health Risk Assessment of Chemical Mixtures, Federal Register (51 FR 34014-34025), Washington, D.C. (1986).
13. US EPA: Risk Assessment Guidance for Superfund, Vol. I, Human Health Evaluation Manual (Part A), EPA/540/1-89 /002, Office of Emergency and Remedial Response, Washington, D.C. (1989)
14. M. Biesiada, *Health risk assessment of Wiślinka inhabitants of the effects heaps of phosphogypsum,* Institute of Occupational Medicine and Environmental Health (2006)
15. M. Trojanowska, R. Świetlik, Medycyna Środowiskowa= Environmental Medicine, **16 (2)**, 14-22 (2013)
16. W. Szymczak, N. Szeszenia-Dabrowska, *Estimating health risks associated with environmental pollution*, Library Environmental Monitoring (1995)

17. https://www.epa.gov/iris, [Access: 22.1.2017]
18. A. Zwoździak, M.I. Gini, L. Samek, W. Rogula-Kozłowska, I. Sówka, K. Eleftheriadis, J Aerosol Sci, **103**, 38-52 (2017)
19. W. Rogula-Kozłowska, K. Klejnowski, P. Rogula-Kopiec, L. Ośródka, E. Krajny, B. Błaszczak, B. Mathews, Air Quality, Atmosphere and Health, **7**, 41-58 (2014)
20. Directive 2004/107/EC of the European Parliament and of the Council of 15 December 2004 relating to arsenic, cadmium, mercury, nickel and polycyclic aromatic hydrocarbons in ambient air. (2004)
21. Regulation of the Minister of Environment of 24 August 2012 on the levels of certain substances in the air, OJ 2012 item. 1031. (2012)
22. Ł. Pachurka, I. Sówka, W. Rogula-Kozłowska, A.Chlebowska-Styś, A. Zwoździak, [in:] Ukrainian- Polish Conference: the problems of air pollution and purification: control, monitoring, catalytic, photocatalytic and Sorption methods of treatment: proceedings, November 6-8, 2016 Kyiv/ [editors: NV Vlasenko, E.V.Gubarenì]. Dnipro: Serednâk T.K., 51 (2016)
23. W. Rogula-Kozłowska, B. Kozielska, K. Klejnowski, S. Szopa, Arch Environ Prot, **39** (1), 53-65 (2013)

The impact of the work under partial load on the energy efficiency of an air-to-water heat pump

Krzysztof Piechurski[1,*], *Małgorzata* Szulgowska-Zgrzywa[1], and *Jan* Danielewicz[1]

[1]Wrocław University of Science and Technology, Wybrzeże Stanisława Wyspiańskiego 27, 50-370 Wrocław, Poland

Abstract. The energy efficiency of air-to-water heat pumps used for central heating depends significantly on the correctness of the device choice. Under given climate conditions, the temperature of the bivalent point defines the amount of energy supplied from an additional source, usually an electric heater, in periods when the power of the heat pump is too low. Commonly overlooked fact is that excess capacity is also unfavourable and has significant, negative impact on the energy efficiency of such a system. In the article, basing on the measurement data of the real installation of the air-to-water heat pump, a significant decrease in the energy efficiency of the device during periods of low energy demand was shown. The average monthly value of SCOP in the warm months was similar or even lower than in the winter months. The reason of this is the operation of the device under partial load. Although the values of COP were very high in short periods of operation, frequent switching on and off of the heat pump reduced its energy efficiency. Also, the length of operation in stand-by mode, i.e. the length of the periods between successive heating cycles, had negative influence on energy efficiency.

1 Introduction

Air-to-water heat pumps can be successfully used as a single source of heat in a building, even in the Polish climate conditions. Good sizing of the unit and conscious choice of the bivalent point is crucial to obtain good seasonal performances [1]. The problem of undersizing of air-to-water heat pumps is often analyzed. However, equally important but overlooked in the discussion is the problem of oversizing the unit and its operation in part-load conditions. Consequences of the work under partial load are dependent on the regulation system of the device used and the type of heating system. Many devices adjust the power generated by cyclic switching on and off (on/off controlled systems). This type of regulation can cause unstable work of a heat pump in dynamic conditions. This is particularly visible in case of a large disproportion between the thermal load of the building and the maximum capacity of the heat pump. In autumn and spring, the heat demand of the building is usually small. At the same time, the capacity of the air-to-water heat pump is high. The greater the disproportion, the shorter the compressor cycles and the higher the number of on-off cycles. In consequence, a significant degradation of the

* Corresponding author: krzysztof.piechurski@pwr.edu.pl

energy efficiency compared to manufacturer data based on EN 14511 [2] standard is observed [3]. According to this standard, the tests are carried out in steady state conditions at full capacity, hence many aspects of the actual work in dynamic conditions are not considered. In order to reduce the efficiency losses caused by the short cycling, the capacity control based on changing the rotational speed of the compressor by a frequency inverter is commonly used. This kind of regulation affects the achieved COP of the heat pump and is limited to the maximum power reduction (usually approx. 30% of the nominal power). It also reduces the range of the on/off regulation, but does not eliminate it completely. Below the minimum power (maximally reduced), this type of regulation will also occur in devices equipped with a frequency inverter [4]. In this article, the analysis of the air-to-water heat pump operation under partial load has been described. Additionally, the difference between the heat pump's efficiency level declared by the manufacturer and the actual measured values has been shown. Finally, the methods of calculation of the device's efficiency under partial load have been discussed, together with their accuracy comparing to the actual measured values.

2 Description of research and analysis

The analyzes presented in this paper were based on measurement data of the air-to-water heat pump, supplying heat to a low temperature installation. The heat is distributed to the offices of approx. 300 m^2 situated on the 1st floor and to the air handling unit room in which the heat source – an air-to-water heat pump with a compressor (split type) of the nominal power of 10.6 kW (A2/W35°C) – is installed. The heat pump performance data – the heating capacity (Q_{HP}) and the declared coefficient of performance ($COP_{decl.}$) measured according to EN 14511 standard are shown in Table 1. The unit works in the monoenergetic mode. In the periods of insufficient heating capacity or the ambient temperature below –15°C, the built-in electric heater delivers auxiliary energy. The installation does not have any heat buffer and is not used for domestic hot water. For further analysis, it is also important that the unit has a frequency inverter installed. It allows the capacity control of the heat pump up to the value of 5.0 kW.

Table 1. Declared heating capacity (Q_{HP}) and $COP_{decl.}$ of the unit, according to EN 14511.

	T_a, °C	-15	-7	2	7	10	12	20
Q_{HP} (kW)	T_{in} = 35°C	5.80	8.47	10.60	14.60	14.80	15.82	17.90
	T_{in} = 45°C	5.20	7.50	10.00	13.10	14.10	14.70	16.80
$COP_{decl.}$	T_{in} = 35°C	1.89	2.62	3.25	4.29	4.40	4.63	5.29
	T_{in} = 45°C	1.50	2.10	2.80	3.10	3.40	3.50	4.10

For the purpose of this paper, analysis of the system behavior over the heating season (September 2013 – May 2014) has been made. Energy consumption (Q_H) and the monthly average ambient temperature ($T_{a,avg}$) is shown in Fig. 1. The monthly average inlet temperature (t_{in}) shows an insignificant change in the range of 30.1–31.5°C. The flow, the inlet (t_{in}) and return temperatures (t_{out}), the ambient temperature (t_a) and the electric power consumption (Q_{el}) are measured. The data were recorded every 1 minute. The data collected allowed to analyze the monthly energy efficiency of the installation and the factors affecting this parameter. It was examined how the cycle length of the compressor and the energy loss caused by the cycling on/off impacts the COP comparing to the one declared by the manufacturer under these conditions ($COP_{decl.}$). In addition, measurements of the electricity consumption between the compressor work periods allowed to determine the energy consumption of the heat pump in stand-by mode. One of the observation was the constant work of the circulation pump.

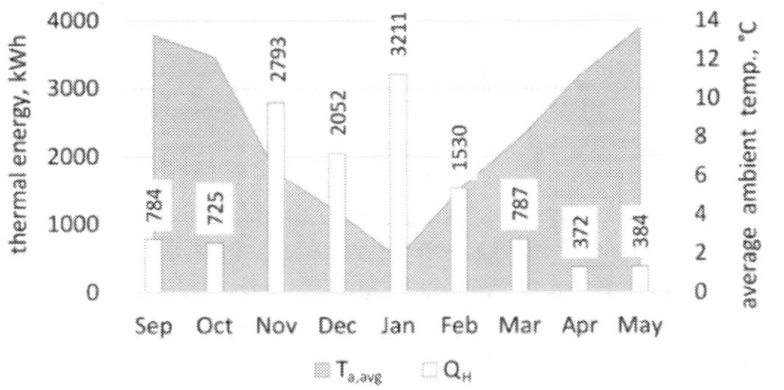

Fig. 1. The thermal energy for the building and ambient temperature in the analyzed months.

3 Data analysis

3.1 Analysis of the heat pump's monthly energy efficiency

The measurements allowed to determine the monthly values of the energy efficiency of the air-to-water heat pump. Two monthly coefficients of performance have been calculated: $SCOP_{month}$ (1) and $SCOP_{cycle}$ (2).

$$SCOP_{month} = \sum Q_H / \sum Q_{el,month} \tag{1}$$

$$SCOP_{cycle} = \sum Q_H / \sum Q_{el,cycle} \tag{2}$$

To determine the energy efficiency of the device without the losses resulting from the compressor off-mode energy consumption, the value of $SCOP_{cycle}$ has been calculated. It only includes the electrical energy consumed during the work of the compressor $Q_{el,cycle}$. The $SCOP_{month}$ includes all the electricity supplied to the heat pump during the month $Q_{el,month}$. Thus, the difference between the $SCOP_{month}$ and the $SCOP_{cycle}$ is the energy used in the stand-by mode.

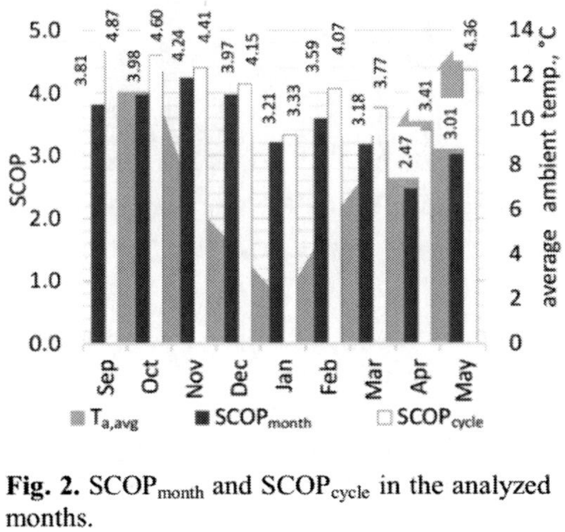

Fig. 2. $SCOP_{month}$ and $SCOP_{cycle}$ in the analyzed months.

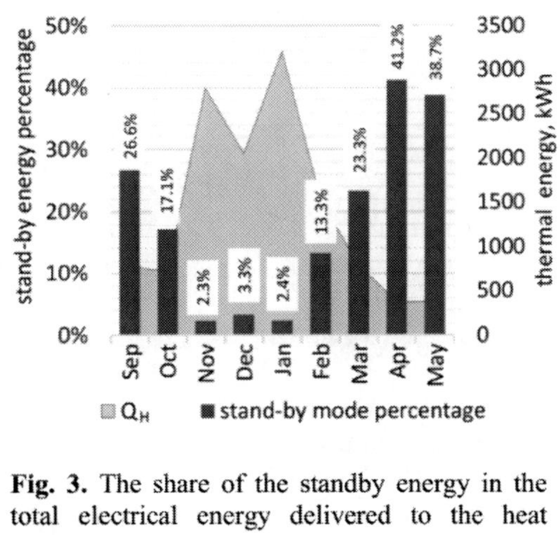

Fig. 3. The share of the standby energy in the total electrical energy delivered to the heat pump.

The results of this calculation with the average ambient temperature are summarized in Fig. 2. The SCOP value depends on the ambient temperature. However, this is not the only factor influencing its value. The decrease in SCOP of warmest months (March, April, May) is in fact clearly visible. In April, the value of $SCOP_{month}$ was only 2.47, even though the average ambient temperature was 11.1°C. For comparison, in January ($T_{a,avg}$ = 1.8°C) the value was 3.21. This reflects the significant impact of factors other than the lower and upper source temperatures. Estimating the SCOP, it is necessary to take into account that the heat pump does not operate only in heating mode (active mode). Especially when the ambient temperature is high, there is also significant share of the stand-by mode operation (which is the time when the compressor is not working). The electrical energy is consumed by the automatic control system, the circulation pump and by the fan (during starts and stops). In Fig. 2 it is clearly visible that the differences between the $SCOP_{month}$ and the $SCOP_{cycle}$ are higher in the warmer months. This is consistent with the calculated share of energy consumed during the operation in this mode compared to the total electrical energy consumption, as shown in Fig. 3.

Despite of the small power consumption in the stand-by mode (in the presented case, according to the manufacturer's data approx. 25 W, without taking into account the circulation pump), as a consequence of low heating demand of the building, the share of the energy consumed in this mode in the warmer months is very large, e.g. in April and May about 40%. This means that most of the time the heat pump was operating without the compressor switched on, consuming electrical energy without generating heat energy. The situation is quite different in the colder months (November – January), where the share of the energy consumed in stand-by mode ranges from 2.3% to 3.3% and only slightly affects the $SCOP_{month}$ in relation to the $SCOP_{cycle}$ (for example 3.21 to 3.33 for January). The electrical energy consumed in this mode lowers the average monthly value of $SCOP_{month}$, however it has no effect on the energy efficiency of the device when it is operating in the heating mode. Still, the achieved $SCOP_{cycle}$ is in many months lower than the estimates based on the relatively high ambient temperature and the working curves from the manufacturer (the COP_{decl} given in Table 1). For the monthly data this has been shown in Fig. 4. The value of this decrease can by expressed as the ratio of $SCOP_{month}/SCOP_{cycle}$. For the analyzed system it has been shown against the compressor run time in Fig. 5.

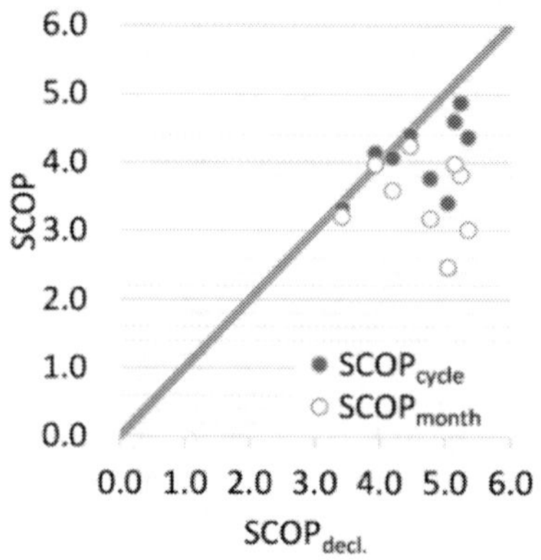

Fig. 4. $SCOP_{cycle}$ and $SCOP_{month}$ versus $SCOP_{decl}$.

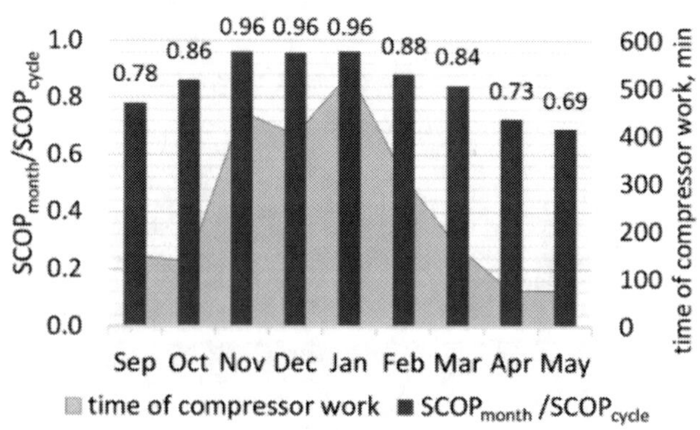

Fig. 5. The share of $SCOP_{month}$ in $SCOP_{cycle}$ in the analyzed months.

In the months of low ambient temperature, the heating demand of the building is higher, the cycles of active compressor work are usually longer and the intervals between them shorter. This results in a low share of operation time spent in the stand-by mode and a high average cycle length. The opposite situation occurs in the warm months, where short cycling and a significant time between the compressor operation lowers the average monthly value of SCOP. The impact of the average cycle length on the ratio of an actual

$SCOP_{cycle}$ to the $SCOP_{decl.}$ declared by the manufacturer is shown in Fig. 6. In particular, for very short cycles (below 10 min) a significant decrease in the energy efficiency is observed, reaching over 30%. Fig. 7 shows how does the ratio of $SCOP_{month}$ to $SCOP_{cycle}$ changes depending on the stand-by operation time as a percentage of the total time. The change is significant, e.g. for 60% heat load (which implies 40% of time in the stand-by mode) it equals about 25%. Similar results were observed in other studies [4, 5].

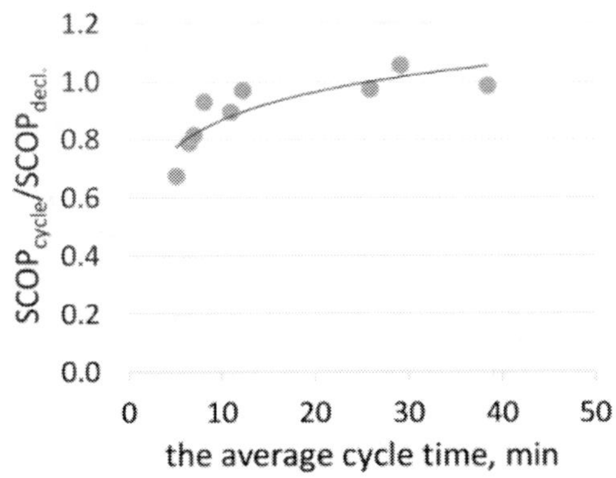

Fig. 6. The share of $SCOP_{cycle}$ in $SCOP_{decl}$ versus the average cycle time.

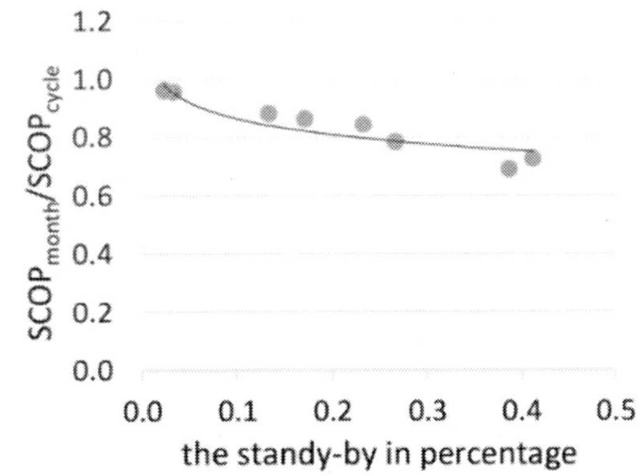

Fig. 7. The share of $SCOP_{month}$ in $SCOP_{cycle}$ versus the stand-by time as a percentage of total time.

3.2 Analysis of the heat pump operation

Fig. 8 shows the profile of the average hourly heat demand of the building and the heat pump's capacity during the entire month (January). Due to the various periods of ambient temperature conditions in this month, a wide range of the heat pump's operating parameters was observed. Most of the time, the capacity of the heat pump corresponds to the heat demand of the building. This is when the ambient temperature is below 5°C. However, there are also periods of very low heat demand of the building, outside of the regulation range of the heat pump (ambient temperatures above 5°C).

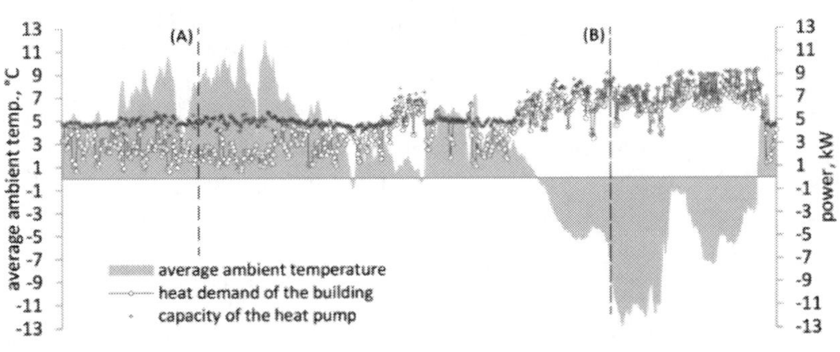

Fig. 8. The profile of the hourly heat demand of the building and the heat pump's capacity.

The difference in operation of the device in these extreme conditions is shown in Fig. 9, illustrating a period of 5 hours of constant work of the device. Picture (A) shows the operation of the heat pump in part-load conditions, with the capacity regulated by the frequency inverter and the on/off cycling. One can notice frequent compressor starts and the work with the minimum power for a short time. Part (B) illustrates the operation of the device with a high heating demand of the building. Compared to (A), longer cycles are observed, interrupted periodically due to the freezing of the heat exchanger. The heat pump performs the process of defrosting by reversing the refrigerant cycle, using heat accumulated in the heating system. The calculations demonstrating the effect of these factors on the COP of the device are shown for both periods in Section 3.3.

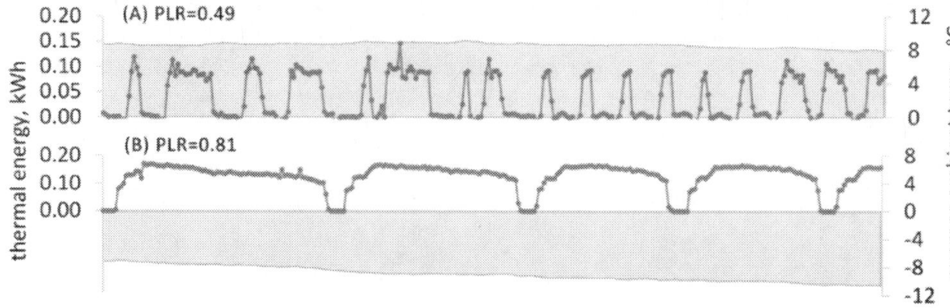

Fig. 9. The operation of the heat pump in two different part-load conditions.

3.3 Operation under partly load in the current European standards

The problem of operation under partly load is discussed in the current European standards for the calculation of seasonal energy efficiency – EN 14825 [6] and EN 15316-4-2 [7]. These documents have different approaches to the problem.

The standard EN 14825 defines the COP_{pl} taking into account the impact of the partial load factor as a function of PLR and Cc, described by equation (3). The PLR is the ratio of the building's heat demand (Q_H) to the current capacity of the heat pump (Q_{HP}). The Cc factor takes into account the decrease of the energy efficiency of the device due to the cyclic operation (on/off control) and power consumption in the stand-by mode. These factors are not considered when testing the device according to EN 14511 and during the calculation of $COP_{decl.}$. The Cc value should be determined individually for each heat pump. Unfortunately, it is not measured by the manufacturers. This often results in accepting the value suggested by the standard – Cc=0.9.

$$COP_{pl} = COP_{decl.} \cdot PLR \,/\, [PLR \cdot Cc+(1\text{-}Cc)] \tag{3}$$

The standard EN 15316, on the other hand, defines the COP_{pl} basing on the value of heat demand of a building (Q_H), $COP_{decl.}$ and the amount of electrical energy consumed in the stand-by mode ($Q_{el,stand\text{-}by}$), as described by equation (4). Power of the device in the stand-by mode is usually contained in the manufacturer's data. However, it does not cover the power of the circulation pump, which often occurs in installations. In addition, any imprecise calculation of the working time in the stand-by mode may result in significant inaccuracies in estimating the COP_{pl}.

$$COP_{pl} = Q_H \,/\, (Q_H/COP_{decl.}+Q_{el,stand\text{-}by}) \tag{4}$$

A detailed analysis of calculating COP_{pl} has been made basing on the measurements of the periods A and B illustrated in Fig. 9. Table 2 shows the measurement results from both periods. The PLR and COP_{pl} were calculated. On this basis, the actual values of the reduction factor PLF and the parameter Cc have been determined. For the tested heat pump, the Cc factor is defined as 0.82. Table 3 shows the actual values of working time (τ), compressor run time (τ_{on}), compressor off time (τ_{off}) and the electric energy consumption during the compressor work ($Q_{el,cycle}$), stand-by ($Q_{el,stan\text{-}by}$) and defrosting ($Q_{el,defr.}$).

Table 2. The measurement results for periods A and B (two different part-load conditions).

	T_a	T_{in}	$COP_{decl.}$	Q_H	Q_{el}	Q_H	Q_{HP}	PLR	COP_{pl}	PLF	Cc
	°C	°C	-	kWh	kWh	kW	kW	-	-	-	-
period (A)	8.6	30.9	4.80	11.22	2.77	2.24	4.58	0.49	4.05	0.84	0.82
period (B)	-9.0	31.6	2.60	37.19	14.91	6.69	8.29	0.81	2.49	0.96	0.82

Table 3. The electrical energy used in periods A and B (two different part-load conditions).

	$Q_{el,cycle}$	$Q_{el,stand\text{-}by}$	$Q_{el,defr.}$	τ	τ_{on}	τ_{off}	$Q_{el,stand\text{-}by}$
	kWh	kWh	kWh	min	min	min	W
period (A)	2.4	0.4	0.0	300	141	159	136
period (B)	14.3	0.0	1.04	300	249	51	-

The measured data were used to calculate the value of hypothetical COP_{pl} according to the standards [6, 7]. In the case of the standard EN 14825, basing on the actual PLR value and the assumed in the standard parameter Cc=0.9, the discrepancy between the calculated and the actual COP_{pl} was significant, as shown in Table 4. The inaccuracies were at the level of 2% and 7%, for period B and A respectively. In the case of the standard EN 15316, basing on the manufacturer's information about the stand-by energy and the actual operation time in this mode, the discrepancies between the calculated and the actual COP_{pl} were 4% and 13% respectively.

Table 4. The discrepancy between the calculated and the actual COP_{pl} for periods A and B.

	PLR	Cc	COP	Error
	-	-	-	-
EN 14825, A	0.49	0.9	4.34	7%
EN 14825, B	0.81	0.9	2.54	2%

	$Q_{el,stand\text{-}by}$	$Q_{el,stand\text{-}by}$	COP	Error
	W	kWh	-	-
EN 15316, A	25	0.066	4.67	13%
EN 15316, B	25	0.021	2.60	4%

Fig. 10 summarizes the value of PLF calculated basing on the Cc according to the standard EN 14825 and the approximate value of this parameter according to EN 15316 including the PLF for the actual C_C based on the measurements. This is noticeable that the actual value of this parameter is much lower than the one assumed in the calculations. The problem of estimating the degradation coefficient is also analyzed in the work [3] where the importance of this issue was emphasized.

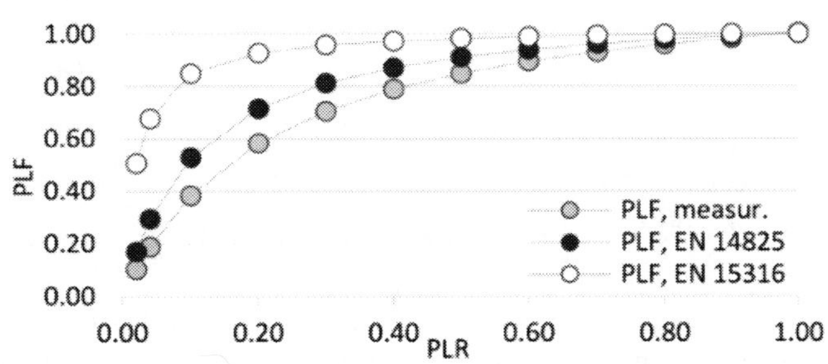

Fig. 10. The PLF factor: measured and according to the standards EN 14825, EN 15316.

Fig. 11 shows how the estimated values of COP_{pl} fit the actual COP of the unit in January 2014. Analyzing the measurements themselves, the decrease of the unit's energy efficiency due to the on/off control is noticeable. It occurs when the heat demand is outside of the range of the frequency inverter operation, i.e. in ambient temperatures above 5°C. Furthermore, the problem of correct estimation of the COP_{pl} is clearly visible in these methods. The values of COP_{pl} calculated according to the standards (C_C=0.9 and the stand-by mode power 25 W) are shown on the left side of Fig. 10. Better precision is obtained when calculating according to the standard EN 14825. The standard EN 15316-4-2 does not

DOI: 10.1051/e3sconf/20171700072

represent the actual work of the tested heat pump. Better accuracy is achieved when the actual measured data is used in the calculations for both standards: Cc=0.82 and the stand-by power consumption of 136 W (including the circulation pump). Although the accuracy is improved, still none of the methods reflects correctly the trend of COP_{pl} of the unit.

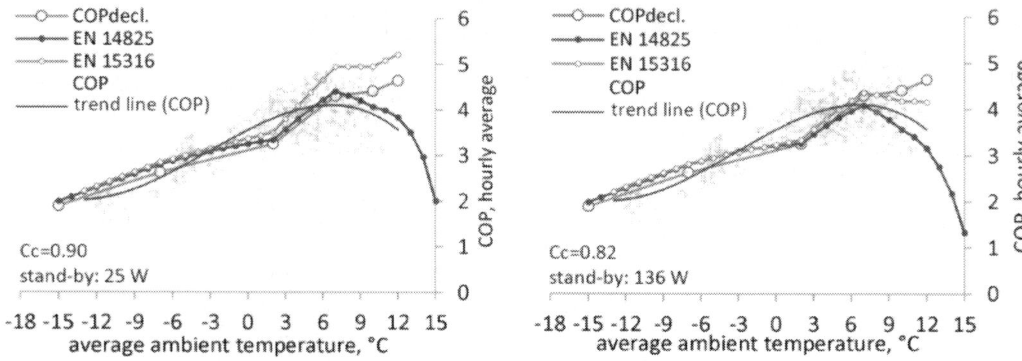

Fig. 11. The estimated values of COP_{pl} and the actual COP in January 2014.

4 Conclusions

Basing on the analysis of the SCOP in multiple following months it was found that there is a significant difference between the operational energy efficiency of the heat pump and the efficiency declared by the manufacturer. The analysis of the hourly data allowed for the calculation of the actual value of Cc, which describes the level of reduction of COP_{pl} in relation to the declared COP_{decl}. For the analyzed device, the value of Cc is 0.82 and it is lower than the standards EN 14825 and EN 15316 suggest. The COP calculation methods described in these standards ignore the impact of the defrosting process and the circulation pump's energy consumption when the compressor is turned off. Probably both methods assumed that the influence of these parameters is included in the COP obtained during the tests according to EN 14511. In fact, these tests are performed in the steady-state conditions and do not properly reflect these issues. It is necessary to either change the test procedure of the COP or the calculation procedure of the COP_{pl}.

References

1. K. Piechurski, M. Szulgowska-Zgrzywa, Rynek Instalacyjny **6**, s. 35–38, 40 (2016)
2. Air conditioners, liquid chilling packages and heat pumps with electrically driven compressors for space heating and cooling, EN 14511:2013 (2013)
3. D. A. Waddicor, E. Fuentes, M. Azar, J. Salom, Appl Therm Eng **106**, 275-285 (2016)
4. L. Cecchinato, Energy Conversion and Management **51**, 1500–1509 (2010)
5. J. M. Corberán, D. Donadello, I. Martínez-Galván, C. Montagud, International Journal of Refrigeration **36**, 2251–2261 (2013)
6. Air conditioners, liquid chilling packages and heat pumps, with electrically driven compressors, for space heating and cooling — Testing and rating at part load conditions and calculation of seasonal performance, EN 14825:2013 (2013)
7. Heating systems in buildings - Method for calculation of system energy requirements and system efficiencies – Part 4-2: Space heating generation systems, heat pump systems, EN 15316-4-2 (2008)

Ability to Determine the Quality of Indoor Air in Classrooms without Sensors

Tomasz Pietrucha[1,*]

[1]Laboratory of Sensor Technique and Indoor Air Quality Studies, Faculty of Environmental Engineering, Wrocław University of Science and Technology, Wybrzeże Wyspiańskiego 27, 50-370 Wrocław, Poland

Abstract. The search for new possibilities for assessing indoor air quality is one of the major challenges of our times. Results of many present investigations organized around the world indicates, that we are still not able to cope with inadequate air quality in many types of building. This fact results in a deterioration of occupants comfort and in the long time damage to health. For this reason indoor air quality in educational institutions deserves special attention. Children are more sensitive to contamination of indoor air and their ability to concentrate is dependent on the state of the air in the room, which affect later examination results. Considering the fact, that classrooms are used in a schematic way, in this paper it is proposed to use currently available calculation formula for determining the concertation of carbon dioxide, which is an indicator of indoor air quality. Results of calculations were compared with measurement data of carbon dioxide concertation collected in classrooms with natural ventilation. With appropriate data and assumptions calculation formula allow to determine the quality of indoor air in classrooms.

1 Introduction

Nowadays, keeping good indoor air quality in many types of buildings is the major challenge. Indoor air influence building occupant health, comfort and work productivity. Many studies are carried out in schools, because children are more sensitive for inadequate indoor air quality and they spend most of their day-time in classrooms. Children are particularly susceptible to indoor air pollution when compared to adults due to their underdeveloped immune and respiratory systems and high inhalation rates per body mass [1]. One of the first extensive studies to investigate indoor air quality in schools was carried out in Stockholm (Sweden) by Elias Heyman in the XIX century and the results were not satisfactory [2]. Now, many research in schools indicates that the problem of inappropriate indoor air quality still exists and it is spread around the world in developing and developed countries. For example in schools in Portugal [1], Japan [3], and Denmark [4]. In all of the works, authors claims that one of the problems is inefficient natural ventilation, which consists buoyancy-driven ventilation and ability of opening windows. Insufficient rates of fresh air supply to the classrooms are connected with inappropriate

* Corresponding author: tomasz.pietrucha@pwr.edu.pl

indoor air quality, occurrence of SBS symptoms among pupils [3] and weaker results in the exam pass rates [5]. Natural ventilation types are common among schools. In practice they are very cheap in maintain, but they can't provide good indoor air quality when pupils are in the classroom. Because of all these evidences researches are looking for possibilities to provide good indoor quality in schools with natural ventilation. The best solution for this problem probably would be a mechanical ventilation, but it is very expensive. Moreover in many cases it could be impossible to mount mechanical ventilation, because of the structure of the building. Some curious idea was presented in publication [6]. Authors used carbon dioxide concertation as visual feedback for pupils. Carbon dioxide is usually used to investigate indoor air quality and rate of ventilation [7]. This compound is emitted from people during exhalation, and the emission rate depend on their activity [8]. Moreover, currently technology provides easy measurement of this gas [9]. Connection between carbon dioxide concentration and description of indoor air quality is presented on Table 1.

Table 1. Description of Indoor Air Quality on the basis of carbon dioxide concentration [10].

Carbon dioxide concentration	Description of indoor air quality
350-400 ppm	Fresh air, perfect conditions
< 600 ppm	Acceptable conditions of indoor air quality in rooms
1 000 ppm	The upper limit of fresh air
1 500 ppm	Air perceived as stuffy and not fresh
2 000 ppm	People with respiratory illnesses may receive cough, weakened people may faint
> 10 000 ppm	Bad air quality causes increased breathing rates, problems with respiration, headaches, nausea

Researches constructed special device, which indicates state of indoor air quality by green, yellow and red lamps on the basis of carbon dioxide concertation. Pupils were instructed how the device work and when they should open windows. Conclusions of this investigation shows, that visual information about indoor air quality for occupants results in improvement of indoor air quality in classroom. This method seems very promising, but still it needs founds and abilities, because this kind of device should be in every classroom in school. In this paper it is claimed, that to determine indoor air quality on the basis of carbon dioxide concentration there is no need of sensor. Concentration of carbon dioxide could be calculated using common known equation, but for that there are needed appropriate data. The problem of the data was considered in this paper.

2 Experimental section

Key to approach presented in the article are assumptions: I. Lessons are held in a schematic way. All pupils enter and leave the classroom at the same time. II. Fresh air in classroom is provided only by buoyancy-driven ventilation. III. Windows and doors of the classroom are closed during classes (which is normal situation, especially in cold months). IV. Before of each lesson classroom is properly ventilated by opening windows (carbon dioxide concentration in classroom is at outdoor level). Real-field measurements were collected in six classrooms in different schools. Schools were named from A to F. In every school was buoyancy-driven ventilation and openable windows. It was the only way to ventilate the room. Measurement were conducted in cold months, between November and March. Classes were held in the following manner: I. Before lesson classroom is properly ventilated by opening windows. II. Children enter the classroom – lesson begins. During class room is only ventilated by buoyancy-driven ventilation. III. Children leave the classroom. Classroom is ventilated by opening windows before next lesson (Figure 1.). Carbon dioxide was measured during all day, but for the later analysis were taken 45-minutes parts between entering and leaving the classroom by pupils. It was measured by nondispersive infrared sensor. The devices range was 0 – 5000 ppm, accuracy was ± 50 ppm ± 3% of measurement value and resolution was 1 ppm. The value was recorded every 1 minute. In three classroom (A, B and C) was measured amount of air exhausted by buoyancy-driven ventilation. On the basis of the data collected air exchange was calculated. Ventilation rate was measured by ventilation meter (anemometer) with flow hood two times per lesson. The devices range was 0.3 – 20.0 m/s, accuracy was ± 0.1 m/s ± 1.5% of measurement value and resolution was 0.1 m/s. Information about number of pupils was delivered from the teacher. Examples of equation to calculate concentration of carbon dioxide could be found in many publications [11, 12, 13]. In this work carbon dioxide was calculated using equation (1).

$$C = [q/(nV)](1 - 1/e^{nt}) + (C_0 - C_i)1/e^{nt} + C_i \qquad (1)$$

C is the calculated concentration of carbon dioxide [ppm], q is a emission of carbon dioxide [l/min], V is a volume of the room [m^3], e is a mathematical constant (2.718), n is an air exchange per hour [h^{-1}], t is a time [min], C_i is a carbon dioxide concentration in outdoor air [ppm], C_0 is a carbon dioxide concentration in the room at start, t = 0 [ppm].

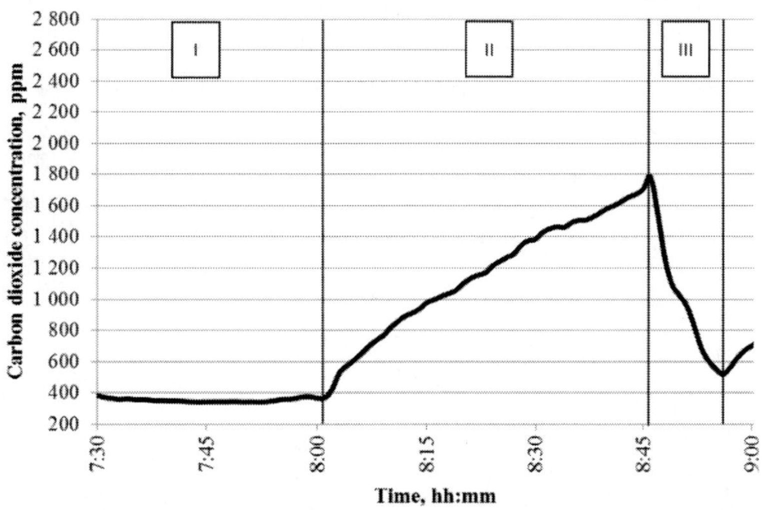

Fig. 1. Changes of carbon dioxide concentration and schedule of the measurement procedure during lessons: I. Before lesson classroom is properly ventilated by opening windows. II. Children enter the classroom – lesson begins. During lesson room is only ventilated by buoyancy-driven ventilation. III. Children leave the classroom. Classroom is ventilated by opening windows before next lesson.

3 Results and discussion

Before calculating carbon dioxide concentration using equation (1) multiple data should be considered. Concentration of carbon dioxide in outdoor air varies from 350 to 450 ppm [14]. For the calculation value of 400 ppm was assumed. Information about emission of carbon dioxide from people could be find in publication [8]. Emission depends on people activity, so there was assumed that pupils' activity is between seated in quiet and office work. For calculations 0,27 l/min was adopted. Air exchange was calculated on the basis of measurement of air exhausted by buoyancy driven ventilation and volume of the classroom using equation (2), where n is an air exchange [h^{-1}], V is a volume of the room [m^3] and Q is a mean rate of air exhausted by buoyancy driven ventilation in classroom [m^3/h]. Measurement of ventilation rates was performed 2 times per lesson (part II. from schedule). Mean value of ventilation rate for classroom A, B and C, calculated air exchange and air exchange assumed for calculation is presented on Table 2. Air exchange used in the formula for each classroom is a mean value of air exchange from school A, B and C.

$$n = Q/V \tag{2}$$

Table 2. Mean value of ventilation rate for classroom A, B and C, calculated air exchange and air exchange assumed for calculation.

Classroom	Volume [m^3]	Mean rate of ventilation [m^3/h]	Air exchange [h^{-1}]	Air exchange applied to calculation formula [h^{-1}]
A	157,1	135.1	0.86	0.75
B	215,1	204.3	0.95	
C	143,3	61.6	0.43	
D	141,9	-	-	
E	348,1	-	-	
F	154,2	-	-	

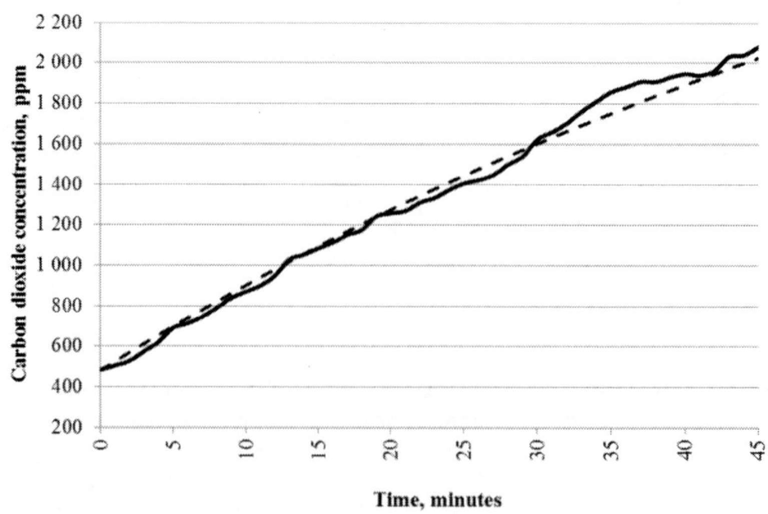

Fig. 2. Example of measured and calculated values of carbon dioxide in time during one lesson. Solid line – measured, dashed line – calculated.

On the basis of upper assumptions carbon dioxide concentration was calculated and compared with 45 minutes carbon dioxide measurement parts (lessons) selected from daily measurement. There were selected 9 parts form school A, B – 5, C – 6, D – 8, E – 4 and F – 7. Example of measured and calculated values of carbon dioxide in time during one lesson (45 minutes) is presented on Figure 2. Calculated and measured concentration was compared in two points: first point was in a minute, when measured value exceed 1000 ppm and the second point was the forty-fifth minute of the lesson (end of the lesson). First point was usually achieved after 10-20 minutes of a lesson. Table 3. presents values measured and calculated in these two points in school A. In the Table 4 there was checked, that the calculated concentration is in the range of measurement error of the measured value which corresponds. It is showed in percent. Moreover in the Table 5 there are mean values of the differences between calculated and measured concentration in ppm.

Table 3. Measured and calculated concentration of carbon dioxide in two comparative points (school A).

Part	1000 ppm exceeded		End of the lesson	
	Measured	Calculated	Measured	Calculated
1	1029	1019	2080	2023
2	1036	978	2022	2071
3	1013	946	1761	1715
4	1021	1038	1785	1959
5	1010	1071	1895	2103
6	1001	1063	1952	2244
7	1014	956	1404	1394
8	1009	914	1434	1322
9	1019	1001	1635	1607

Table 4. Number of cases when calculated concentration was in the error range of measured concentration.

School	1000 ppm exceeded	End of the lesson
A	89%	56%
B	80%	60%
C	100%	33%
D	88%	75%
E	75%	25%
F	71%	71%

Key feature to assessing indoor air quality using this calculation formula are data and assumptions. Data like number of people, emission of CO_2 from occupants or outdoor concentration of carbon dioxide are easy to get. The main problem could be the data about air exchange provided by buoyancy-driven ventilation in classrooms. On the basis of measurement from classrooms A, B and C mean value was calculated and it was 0,75 h^{-1}. This data was applied and results were very positive. Most of the calculations was very similar to Figure 2. and these presented in Table 3., even in classroom D, E and F where was no data about air exchange. Measured and calculated concentrations were more similar in point when measured value exceed 1000 ppm than in the end of the lesson which can be seen in Table 4. and Table 5.

Table 5. Mean difference between calculated and measured concentration in ppm.

School	1000 ppm exceeded	End of the lesson
A	50	108
B	29	111
C	51	144
D	48	61
E	38	316
F	71	85

This is very promising, because when concentration of CO_2 raise 1000 ppm proper actions should be taken to improve indoor air quality. Classroom E is designed for 90 occupants and the rest of the classes for 20-30 pupils. Calculated concentration in 3 of 4 cases in this classroom was in the error range of measured concentration and the mean difference was only 38 ppm. It can be said that this calculation of carbon dioxide could be applied in different size classes. Moreover that rate of exhausted air provided by buoyancy-driven ventilation in schools was very similar and lack of knowledge about rate of ventilation is not a problem to determine indoor air quality on the basis of calculated carbon dioxide concentration. But this method is only usable under assumptions given in Experimental section. Without this foundations it is impossible to use this method to determine indoor air quality. Comparing this method with the method described in the publication [6], for determining the concentration of CO_2 with appropriate assumptions and data do not need a CO_2 sensor. This is very huge advantage of this method and because of that it could be applied in every school with natural ventilation, for example an application for a smartphone.

4 Conclusions

There could be drawn the following conclusions: (1) If lessons take place in accordance with the assumptions it is possible to determine the concentration of CO_2 in the air which will provide information about indoor air quality. (2) In most cases at the point when 1000 ppm was exceeded, differences between calculated and measured CO_2 concentration do not exceed the error of measurement device (3) The differences was higher at the end of the lessons but classroom should be ventilated before carbon dioxide concertation exceed 1500 ppm (Figure 2. and Table 3.).

The conference participation was financed from the statutory funds of Faculty of Environmental Engineering, Wrocław University of Science and Technology from the project 0401/0066/16.

References

1. J. Madureira, I. Pacienecia, C. Pereira, J. P. Teixieira, E. de O. Fernandes, *Indoor air quality in Portuguese schools: levels and sources of pollutants*, Indoor Air **26**, 526–537 (2016)

2. J. Sundell, *On the history of indoor air quality and health*, Indoor Air **14**, 51–58 (2014)
3. M. Takaoka, K. Suzuki, D. Norback, *Sick Building Syndrome among Junior High School Students in Japan in Relation to the Home and School Environment*, Global Journal of Health Science, Vol. **8** No. 2 (2016)
4. D. Wyon, P. Wargocki, J. Toftum, G. Clausen, *Classroom ventilation must be improved for better health and learning*, REHVA. Eur. HVAC. J., **3**, 12–16 (2010)
5. M. J. Mendell, E. A. Eliseeva, M. M. Davies, A. Lobschied, *Do classroom ventilation rates in California elementary schools influence standardized test scores? Results from a prospective study*, Indoor Air **26**, 546–557 (2016)
6. P. Wargocki, N. A. F. Da Silva, *Use of visual CO2 feedback as a retrofit solution for improving classroom air quality*, Indoor Air **25**, 105–114 (2015)
7. A. Szczurek, M. Maciejewska, T. Pietrucha, *CO2 and volatile organic compounds as indicators of IAQ*, 6th AIVC Conference, 5th TightVent Conference, 3rd venticool Conference, Madrid, Spain 23–24 September 2015
8. D. S. Dougan, L. Damiano, *CO2-based demand control ventilation do risks outweigh potential rewards?*, ASHRAE Journal **46** (2004)
9. N. Mahyuddin, H. B. Awbi, *A review of CO2 measurement procedures in ventilation research*, International Journal of Ventilation **10**, 353–370 (2012)
10. T. Pietrucha, *Measurement of carbon dioxide concentration for assessment of indoor air quality in the lecture hall*, 13th Students' Science Conference, Polanica-Zdrój Poland, 17–20 September 2015
11. http://www.engineeringtoolbox.com/pollution-concentration-rooms-d_692.html [on-line: 25 January 2017]
12. P. Kapalo, S. Vileckova, F. Domnita, O. Voznyak, *Determine a methodology for calculating the needed fresh air*, The 9th International Conference "ENVIRONMENTAL ENGINEERING", Vilnius Lithuania, 22–23 May 2014
13. D. Kosicanova, R. Nagy, I. Senitkova, *Transmission and distribution of pollutants in relation to the environment*, Technical University of Kosice (2011), ISBN 978-80-553-0829-6
14. A. P. Jones, *Indoor air quality and health*, Atmospheric Environment **33**, 4535–4564 (1999)

Removal of organic compounds from natural underground water in sorption and sono-sorption processes on selected activated carbons

Andżelika Pietrzyk[1,*], and *Dorota* Papciak[1]

[1] Department of Water Purification and Protection, Rzeszow University of Technology, Faculty of Civil and Environmental Engineering and Architecture, Aleja Powstańców Warszawy 6, 35-959 Rzeszow, Poland

Abstract. The article rated removal efficiency of organic matter in the processes of sorption and sono-sorption of underground water grasped for municipal purposes. The studies were conducted in laboratory scale and verified in pilot scale at the Water Treatment Plant Tarnobrzeg-Jeziórko. In the research used granular activated carbons, ie. WD-Extra, WG-12, Norit Row 0.8 and Filtrasorb 300. The processes efficiency was evaluated on the basis of changes in the following parameters, ie.: total organic carbon (TOC), permanganate index, UV absorbance, turbidity and colour. The ultrasounds were generated by means of disintegrator Sonics&Materials VCX 130, using the sonication time of 1 and 5 minutes. The results obtained for the batch tests allowed to observe a beneficial effect of ultrasound on the efficiency of the removal of organic material in the sorption process. The combination of sonication and sorption on activated carbon increased the efficiency of the removal of organic matter by 6–37% for TOC, and 18.6–27.9% for permanganate index, depending on the sorbent used. The positive laboratory results were not confirmed in a pilot scale. In the flow conditions the sonication process did not affect the efficiency of removal of organic matter on the filter model with a bed of activated carbon.

1 Introduction

Effective elimination of organic contaminants is now one of the major problems encountered in water purification plants. The presence of natural organic matter (NOM), particularly humic substances in the drinking water, is a negative phenomenon causing deterioration of organoleptic properties of water as well as contributing to the formation of by-products of disinfection and oxidation [1, 2]. A considerable content of organic substances in water entering the distribution system is one of the main causes of the instability of water conditioning the secondary microbial growth on the network [3, 4]. The content of organic matter in natural waters is very variable and depends mainly on the type

* Corresponding author: d332@stud.prz.edu.pl

of water, geographic location, temperature, salinity, pH, microbial activity and the nature of the surrounding catchment [5, 6].

The unit processes used in the process of water purification systems, i.e. coagulation, adsorption, ion exchange, chemical oxidation and membrane processes, often do not provide sufficient removal of organic matter, and their efficiency largely depends on the composition of purified water, the properties of the existing NOM fractions and the conditions of the process [3].

The need to reduce the content of organic compounds in intaken waters caused an increase in interest in the methods of connecting the unit processes with chemical agents (Fenton reaction) or physical (ultrasound, UV radiation), which are designed not only to effectively eliminate organic pollutants, but also to reduce the amount used at reagent stations [7–11].

In the last decade there has been more and more interest in ultrasound process, i.e. sono-sorption. Their effectiveness is confirmed by many studies. The efficiency of sono-sorption is influenced, among others, by the degree of contamination of the solution, sonification time and its intensity, and kind of activated carbon. It is worth noting that the majority of experiments conducted to date, relating to the sono-sorption process were performed on the basis of model solutions prepared on the basis of commercial preparations of humic acids, while a much smaller amount of them applies to natural waters [2, 12–14].

The aim of the research was to evaluate the possibility of using ultrasonic energy to intensify the removal of NOM from natural groundwater. The effectiveness of lowering the content of organic substances in the combined sorption and sonification processes using granular activated carbons was determined in the article.

2 Materials and methods

2.1 The subject of research

The study was carried out on the natural underground water purified at the Water Treatment Plant (WTP) Tarnobrzeg-Jeziórko. The WTP is supplied from an unconfined quaternary aquifer with a depth of about 15 m bgl using 27 wells. The physicochemical composition of water drawn and directed to the treatment station varies greatly, because it depends essentially on the well included in the operation and the productivity. Drawn water does not meet sanitary requirements for drinking water in terms of turbidity (8-14 NTU), colour (40–100 mg Pt/l), the permanganate index (11–18.1 mg O_2/l), ammonium ion (1.20–1.98 mg NH_4^+/l), iron (14–44 mg Fe/l) and manganese (0.74-2.58 mg Mn/l). High water colour, which correlates with the elevated permanganate index value and total organic carbon (TOC; 14–20 mg C/l), indicates the presence of water of natural organic matter that may be present in combinations of complex compounds of iron and manganese.

Although the quality of the treated water which is directed to the water supply meets the requirements of the Regulation of the Minister of Health of 13 November 2015 relating to the quality of water intended for human consumption (Journal of Law 2015, pos. 1989), research has been carried out in order to further reduce the content of organic substances in the treated water of the following composition: TOC 7–11 mg C/l, with a permanganate index of 4.7–5 mg O_2/l and colour of 10-16 mg Pt/l.

Water after filtration on anthracite-sand deposits was used in the study prior to the final process of disinfection (Fig.1) at the parameters set out in Table 1.

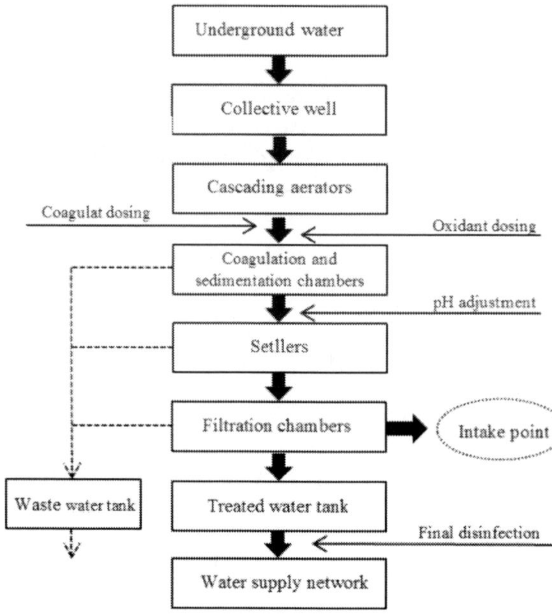

Fig. 1. Technological diagram of Water Treatment Plant.

Table 1. Physicochemical parameters of water used in the test.

Parameters	Unit	Value
TOC	mg C/l	8.36
Absorbance UV	254nm,1m^{-1}	16.46
Permanganate index	mg O_2/l	4.3
Colour	mg Pt/l	12
Turbidity	NTU	1.08

2.2 The selection of activated carbon and determining the time of sonification

The studies were carried out on the basis of batch tests according to the methods described in the work [14]. Adsorbent dose amounted 4g/l. The prepared samples were shaken on a laboratory shaker for two hours at a speed of 140 revolutions/min, after which the samples were filtered through the coarse filter and analyzed physicochemical. The following parameters were monitored: TOC (using a TOC Sievers 5310 C analyzer), permanganate index (by the $KMnO_4$ method),UV absorbance and colour (using a Merck spectrophotometer) and turbidity (by nephelometry, with a Hach-Lange turbidity meter). The parameter designations were made in accordance with the test procedures.

The sonification process was carried out before process sorption using a Sonics & Materials VCX 130 disintegrator with an amplitude of 100%. The study used two exposure times: 1 and 5 minutes, the volume of sonified solution was 250 ml. The amount of energy input during the one minute sonification was about 2000 J, whereas at 5 minutes, this value increased to approximately 11500 J. After subjecting the water sample to ultrasound action, the sorption process was carried out.

The study used the following sorbents, i.e. WD-Extra, WG-12, Norit Row 0.8 and Filtrasorb 300. A detailed technical specification of activated carbons is presented in Table 2. The sorbents were washed with distilled water (in order to remove ash) prior to commencing the tests, and then dried at 145°C.

Table 2. Technical specifications of activated carbons (according to the manufacturer).

Parameter	WG-12	WD-Extra	Norit Row 0.8	Filtrasorb 300
Ball-pan hardness	96.4 %	97 %	98 %	95 %
Methylene blue adsorption	33 cm^3	30 cm^3	24 cm^3	28 cm^3
Iodine number	1074 mg/g	998 mg/g	≥1100 mg/g	1065 mg/g
Apparent density	450 g/dm^3	478 g/dm^3	400 g/dm^3	480 g/dm^3
pH	-	11.0	alkaline	alkaline
Moisture	0.7 %	-	≤ 5 %	2 %
Ash content	6 %	-	7 %	7 %

2.3 Determination of the effect of sonification on NOM removal efficiency in the sorption and bio-sorption processes

The tests were carried out in a semi-technical scale, in the flow conditions using filtration columns with the parameters specified in Table 3. The model filters were filled with two types of activated carbon, i.e. WD-Extra and WG-12.

The model filter filled with WD-Extra granular activated carbon was incorporated to work in October 2015. Following a year of operation, a biofilm was formed on the bed which allowed the removal of organic pollutants in the bio-sorption process. The operating time of the model filter at the time of the research on the process of sono-sorption was 430 days.

A filter column with WG-12 deposit has been operating since July 2016 - (150 days), and the adsorption processes still dominate on the bed. The test water was introduced over both beds by gravity via a peristaltic pump at a rate of 1 m/h.

The sonfication process was carried out before biofiltration process, i.e. a water sample having a volume of 250 ml was sonified for 5 minutes and collected in the reservoir; after 2 hours it was placed onto the filter beds. In this way, the toxicity of free radicals on the micro-organisms which form a biofilm was reduced.

Table 3. Parameters of the filter model.

Operating parameters	Unit	Model filters
Height of the carbon bed	m	1.40
Height of the supporting layer	m	0.22
Diameter	m	0.055
Filtration velocity	m/h	1
Contact time	h	0.5

3 Evaluation of research results

3.1 Batch tests

The analysis of the results obtained shows that the degree of removal of organic contaminants measured with the parameters i.e. TOC, UV absorbance, permanganate index, and the colour was different depending on the sorbent used and sonification time.

Analyzing the changes in TOC, it was found that the efficiency of removal of organic compounds in the classical sorption ranged from 19.5% (WG-12) to 45.3% (Filtrasorb 300). As a result of sonification, the efficiency of NOM removal from the tested water increased. As a result of 5 minutes exposure of the sample to ultrasound, we observed a 55% reduction in organic content in the tested water. The exception was only the

WD-Extra activated carbon for which the removal efficiency reached 33.9%. The results indicate thereby, that the ultrasound disintegration had no significant effect on the sorption process using WD-Extra (there was a slight increase in the efficiency by only 2% (1minute) and 6% (5 minutes). However, in the case of WG-12 employing the process of sonification, the efficiency increased by 27.3% (1minute) and 37.1% (5 minutes) (Fig. 2).

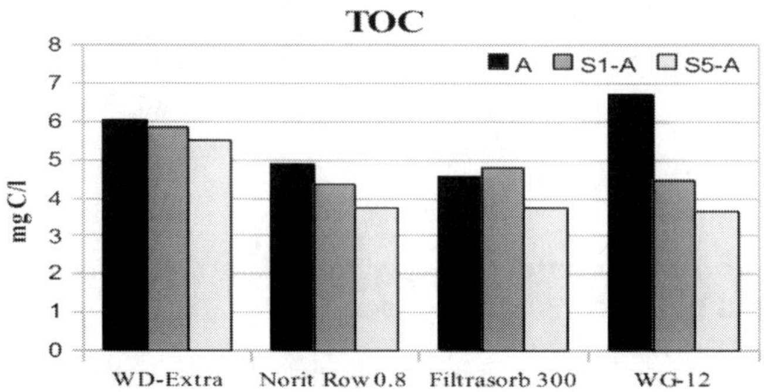

Fig. 2. Change of the TOC value in the tested water (A - adsorption, sono-sorption: S1-A – sonification (1minute) + adsorption, S5-A – sonification (5 minute) + adsorption).

The greatest reduction in permanganate index was in the case of WG-12 activated carbon. As a result of classical sorption, the content of organic matter decreased to a value of 2.4 mg O_2/l, whereas the sonification process made it possible to reach the level of 1.2 mg O_2/l (5 minutes). The effectiveness of the processes amounted to 44.2% and 72.1% respectively. The lowest efficiency of NOM removal expressed by permanganate index was reported for WD-Extra. In this case, the use of 1 minute sonification did not yield any results and the prolonged sonification time allowed the value of permanganate index only to be reduced to 2.6 mg O_2/l. The Filtrasorb 300 and Norit Row 0.8 were characterized by comparable efficiency to remove organic matter from the tested water; the value of permanganate index during 5-minute sonification was reduced to 1.9 - 2.1 mg O_2/l (Fig. 3).

The UV absorbance is another parameter closely related to the structure of organic compounds. In analyzing changes in this parameter, it was found that activated carbons such as WG-12, Norit Row 0.8 and Filtrasorb 300 are characterized by similar sorption properties. The efficiency of the processes carried out for these sorbents oscillated between 62.2 - 67.7%, while for the WD-Extra was lower and amounted to 43.5 - 51.3%. In the case of absorbance, the sonification of the tested water did not contribute to the intensification of sorption (Fig. 3).

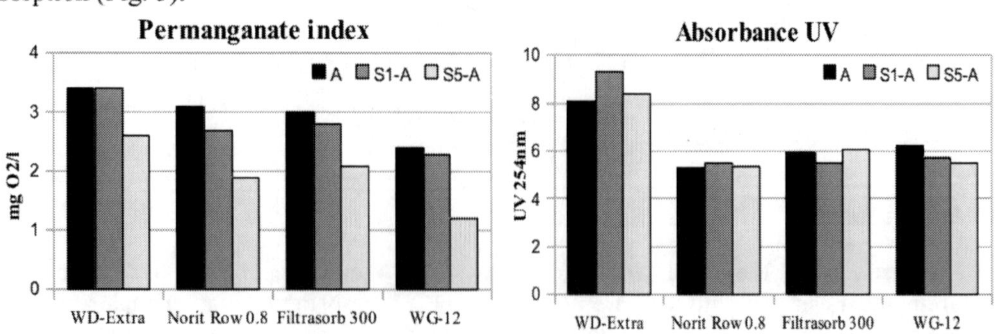

Fig. 3. Change of the permanganate index and UV absorbance value in the tested water (A - adsorption, sono-sorption: S1-A – sonification (1minute) + adsorption, S5-A – sonification (5minute) + adsorption).

The literature data shows that in the majority of studies carried out so far, there was no increase in sono-sorption effectiveness expressed by absorbance observed either. Subjecting the adsorbate to ultrasonic modification results in forming compounds with a specific structure, which additionally absorbs radiation thereby increasing the value of this parameter [2, 14]. This phenomenon is identified with a simultaneous increase in turbidity, which was also confirmed in the study.

As a result of the sorption process in all the analyzed samples, reduction in the colour of the test water occurred. Most preferred changes were observed for WG-12 (efficiency reached 50%), whereas the degree of colour reduction ranged from 16.7 to 33% in the other sorbents. Assisting the sorption process with ultrasound adversely affects the colour of water (Fig. 4).

The sorption and sono-sorption process, regardless of the adsorbent, contributed to the increase in turbidity of the tested water. The increased turbidity of solutions which have been subjected to ultrasound action may be caused by aggregation of cleaved fragments of organic compounds [2]. Whilst shaking the samples, abrasion of grains of activated carbons occurs and as a result the solution a certain amount of organic carbon, ash or other components of the sorbent can pass to solution.

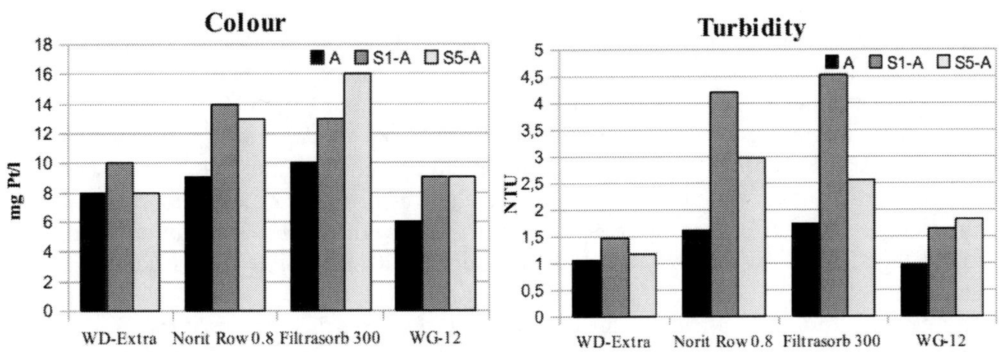

Fig. 4. Change of the colour and turbidity value in the tested water (A – adsorption, sono-sorption: S1-A – sonification (1minute) +adsorption, S5-A – sonification (5minute)+adsorption).

The research on the effectiveness of sonification shows that the ultrasonic field has a beneficial effect on the course of sorption process under laboratory conditions. The best efficiency of removal of organic compounds, expressed as TOC and permanganate index, was observed for the sono-sorption process with a sonification time of 5 minutes. A prolonged exposition time of the sample to ultrasound greatly improved the efficiency of the sorption process (e.g. for WG-12 the value of TOC and permanganate index additionally decreased by 37 and 28% respectively). The negative effect of the sonification process is visible while the increase in turbidity and colour of the tested water, and the highest values of these two parameters were recorded for Norit Row 0.8 and Filtrasorb 300.

The activated carbons used in the study were characterized by different sorption properties, as evidenced by the varied effectiveness of conducted of sorption and sono-sorption processes. The ability of the sorbent to adsorb pollutants is conditioned by, among others, pore structure (the number of micro-, meso- and macropores), the size of the surface area and the amount and type of impurities present.

In the test conducted, among all analyzed activated carbons, the best removal efficiency of NOM was marked by WG-12 characterized by a high proportion of macropores. As a result of the sonification process alongside with a significant reduction in the TOC, permanganate index and UV absorbance, the lowest increase in colour and turbidity also occurred.

Studies carried out on model humic solutions indicate that the use of ultrasonic disintegration is particularly advantageous in the case of sorbents with a large share of microspores. The literature data indicate that activated carbons with a predominant amount of pores, the size of less than 0.2 μm, can have lower removal efficiency of macromolecular organic compounds, because they are not able to adsorb impurities present in the water [14].

3.2 Flow conditions

The model filters filled with activated carbon have a great efficacy in the removal of organic matter.

As a result of filtration of the water sample through a bed of WD-Extra (biologically activated deposit) the TOC values decreased by 25.4%, permanganate index by 35.7%, the UV absorbance of 39.2% and colour by 70%. However, in the case of WG-12 (the bed is dominated by adsorption processes) the reduction of the analyzed indicators was 62% – TOC, 59.5% – permanganate index, 67% – UV absorbance and 100% – colour respectively.

Earlier studies conducted at a water treatment plant show also that the expansion of the technological process of second degree filtering on biologically activated carbon filters greatly increases the efficiency of removal of organic matter [15].

The tests performed in a semi-technical scale on model carbon filters using the sonification process before biofiltration process did not confirm the results obtained in the batch tests - there was no increase in the efficiency of removal of organic compounds. In further studies there is a need to verify whether the lack of results is not conditional upon, among others, too low intensity of the ultrasound field, insufficient sonification time or contact time (Tab. 4).

Literature reports indicate that the use of the sonification process prior to the introduction of treated water on the filter bed can also contribute to the reduction of organic matter by creating adsorbed compounds more easily, biodegradable and better absorbed by the microorganisms forming the biological membrane [12]. The efficacy of the process may significantly be determined by the susceptibility of organic compounds to ultrasound, predominantly present in the analyzed waters. Therefore, further research will be focused on the determination of specific NOM fractions contained in the groundwater.

Table 4. The efficiency of removal of organic substances on model carbon filters.

Parameter	Unit	The water supply to the filter	The water at the outlet of the model filters			
			WD-Extra		WG-12	
			Biofiltration process	Sonication and biofiltration process	Biofiltration process	Sonication and biofiltration process
TOC	mg C/l	7.93	5.92	5.88	3.01	2.92
Perman. index	mg O_2/l	4.2	2.7	2.6	1.7	1.6
Absorbance UV	254nm, 1m^{-1}	15.52	9.44	10.08	5.12	5.16
Colour	mg Pt/l	10	3	6	0	0

4 Conclusions

1. The organic matter present in natural underground water was removed with varying efficiency depending upon the specific type of granular activated carbon, and the process (sorption, sono-sorption).
2. The application of the phenomenon of ultrasonic cavitation positively influenced the course of the sorption process in laboratory conditions. The best efficiency of removal of organic compounds, expressed as TOC and permanganate index was observed for the sono-sorption process with a sonification time of 5 minutes. A prolonged exposure time of the sample to ultrasound increased the removal efficiency of NOM removal by 6–37% for TOC, and 18.6–27.9% for the permanganate index depending on the sorbent. The WG-12 was activated carbon for which the greatest intensification of the sorption process was found.
3. A negative effect of the use of sonification process was an increase in turbidity and colour of the water sample. The highest values of these parameters were recorded for the Norit Row 0.8 and Filtrasorb 300 during 1 minute sonification time.
4. The research carried out in a semi-technical scale did not confirm the results obtained in the laboratory conditions. The application of the sonification process prior to filtration on granular activated carbons did not affect the efficiency of removal of organic contaminants. No visible results may be conditioned, among others, by insufficient intensity of ultrasound field sonification time, or contact time.
5. The effectiveness of the sono-sorption process depends on the susceptibility of organic compounds to ultrasound action; so further research will be focused on the determination of the NOM fraction contained in the groundwater.

References

1. A. Urbanowska, M., Kabsch-Korbutowicz, Environ. Prot. Eng.,vol.**42**, no. 2 (2016)
2. V. Naddeo, V. Belgiorno, R.M.A. Napoli, Desalination **210** (2007)
3. M. Wolska, Environ. Prot. Eng., vol. **41,** no. 2 (2015)
4. W. Liu, H. Wu, Z. Wang, S.L.Ong, J.Y. Hu, W.J. Ng, Water Res. **36** (2002)
5. A. Matilainen, E. T. Gjessing, T. Lahtinen, L. Hed, A. Bhatnagar, M. Sillanpää, Chemosphere **83** (2011)
6. A. Pietrzyk, D. Papciak, JCEEA, vol. **33**, no. 63 (2016) (In Polish)
7. A. Durán, J. M. Monteagudo, A. J. Expósito, V. Monsalve, Chem. Eng J. **284** (2016)
8. L.J. Xu, W. Chu, N. Graham, Water Res. **47** (2013)
9. F. Chemat, P.G.M. Teunissen, S. Chemat, P. V. Bartels, Ultrason. Sonochem. **8** (2001)
10. C. G. Joseph, G. L. Puma, A. Bono, D. Krishnaiah, Ultrason. Sonochem. **16** (2009)
11. M. Sillanpaa, *Natural Organic Matter in Water* (IWA Publishing, 2015)
12. P. Olesiak, L. Stępniak, Acta. Biochim. Pol., vol. **61**, no. 4 (2014)
13. L. Stępniak, M. Kusiak, E. Okoniewska, Rocz Ochr Sr, vol. **15** (2013) (In Polish)
14. M. Kusiak, L. Stępniak, *Inżynieria Środowiska-Młodym Okiem*, vol. **1** (2012) (In Polish)
15. D. Papciak, J. Kaleta, A. Puszkarewicz, B. Tchórzewska-Cieślak, J. Eco. Eng., vol. **17**, no. 3 (2016)

Hydraulic analysis of functioning of the drainage channel with increased retention capacity

Kamil Pochwat[1,*]

[1]Department of Infrastructure and Sustainable Development, Faculty of Civil and Environmental Engineering and Architecture, Rzeszow University of Technology, al. Powstańców Warszawy 6, 35-959 Rzeszów, Poland

Abstract. Hydraulic overload of networks and its related objects is a problem that is more and more common in drainage systems. The most popular way to prevent it from its occurring is to increase the storage capacity of the system. At present, there are numerous solutions allowing to increase retention in drainage systems, with the selection of the most beneficial one being dependent on numerous factors; it must also be preceded by a technical and financial analysis. Financial analysis is of particular importance here, since due to significant dimensions and complicated structure, these objects require high investment outlays for its construction. As a consequence, it is a priority to undertake actions aimed at developing such solutions that will allow for significant increases of retention in the drainage system, and cost reductions at the same time. One of such solutions is the *Underground tank for combined sewage and stormwater* that is now patent-pending. The application of this solution allows to multiply retention capacity of the drainage system, without the necessity of modernizing the existing sewer system. This paper presents the application potential of the subject method for increasing sewer retention, along with its hydraulic function analysis. The research carried out and presented was based on the analysis of individual phases of functioning of the subject solution.

1 Introduction

A phenomenon more and more commonly encountered in traditional drainage systems involves hydraulic overloading of the network and the related facilities [1, 2]. This is an effect of replacing biologically active surfaces with water-tight surfaces [3, 4]. Of no small importance are issues associated with climate change [5, 6], in particular intensified precipitation [7, 8]. Therefore, in order to enable their proper functioning, it would be advisable to transform them into advanced sustainable drainage systems based on applying retention and control of wastewater flow within the system [9–11]. As a rule there are a few options for increasing retention capacity of the system. The choice of the most beneficial

* Corresponding author: kp@prz.edu.pl

possibility should be based on an analysis of various criteria, most importantly including technical and economic factors. Due to their considerable dimensions and their complicated structure, retention facilities require high capital expenditures related to their construction [12]. Hence, it is advisable to apply such solutions which allow for minimizing their necessary cubic capacity, and consequently the capital expenditure, while ensuring the assumed level of operational security for sewage systems.

The purpose of the study is to perform functional hydraulic analysis of an innovative concept of Retention reservoir for rainwater and combined sewage during the specific phases of its functioning.

2 Description of the concept of Retention reservoir for rainwater and combined sewage

Retention reservoir for rainwater and combined sewage is an innovative solution which enables significant increase in retention capacity of sewage system without a necessity to change the positioning levels of sewers transporting wastewater along their flow route. The device in question has been presented and described in axonometric scheme (Fig. 1).

Fig. 1. Visualization of retention reservoir for rainwater and combined sewage: 1 – intercepting sewer, 2 – retention chamber, 3 – transit sewers.

Retention reservoir for rainwater and combined sewage has a retention chamber hydraulically connected with intercepting sewer via two-way transit sewers, whose bottom is located above the bottom of the intercepting sewer.

Designing a retention reservoir, which does not require varied levels of intercepting sewer positioning, increases the possibilities for staging of investment process and for expanding drainage systems as a given catchment area grows, without a need to reconstruct drain interceptors.

By increasing retention capacity of a sewage system it is possible to significantly reduce capital expenditures required by construction or modernization of such sewage system. Additionally, the invention may be applied both in existing and in newly designed retention facilities.

In specific cases, it is possible to increase volumetric flow rate of sewage passing to the reservoir chamber, by applying suitably shaped impounding structures.

An important aspect of the presented solution is the bottom of retention chamber which should be positioned above the bottom of the intercepting sewer, because this enables redirecting wastewater back to the sewage system. However, if part of rainwater were to be

 DOI: 10.1051/e3sconf/20171700075

subjected to a process of infiltration through the soil, it would be possible to lower positioning of the reservoir bottom in relation to the intercepting sewer.

3 Dimensioning of Retention reservoir for rainwater and combined sewage

The use of this solution in engineering practice is associated with the adoption of appropriate procedures for its dimensioning. Figure 2 shows the dimensioning algorithm of retention reservoir for rainwater and combined sewage.

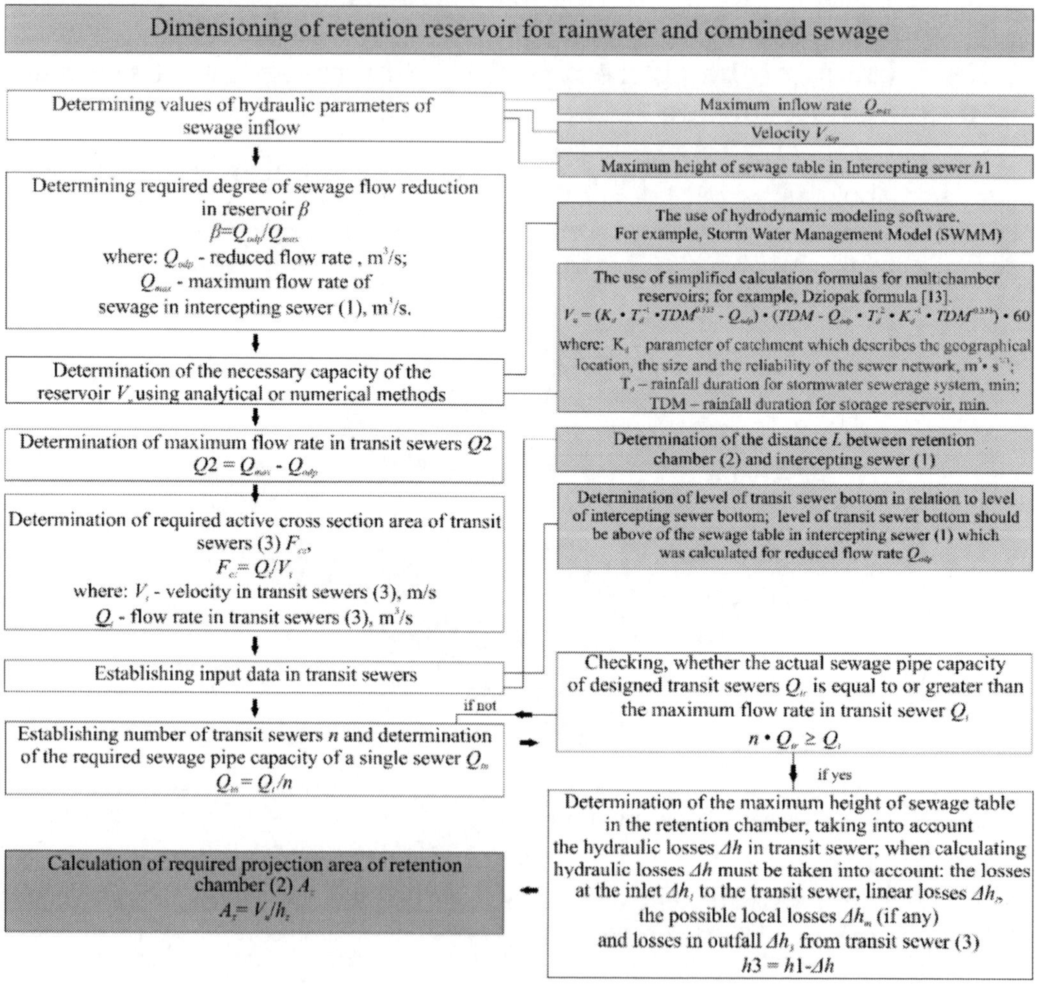

Fig. 2. The dimensioning algorithm of retention reservoir for rainwater and combined sewage.

The developed algorithm (Fig. 2) contains the procedure for dimensioning the individual components of retention reservoir for rainwater and combined sewage. A Particularly important element of this procedure is to determine the input parameters for the calculation i.e. flow rate in intercepting sewer, reduced flow rate, distances between intercepting sewer and the retention chamber, and the height of sewage table in the intercepting sewer for reduced flow rate. The issue of determining the hydraulic losses in the transit sewer is also important. They affect the maximum height of sewage table in the retention chamber. Use of the presented algorithm allow to design a retention tank for rainwater and sewage in combination.

4 Functional phases of Retention reservoir for rainwater and combined sewage

Hydraulic analysis performed during the study examined operation of Retention reservoir for rainwater and combined sewage at the specific stages of its filling during rainfall. Schematic section of the reservoir is shown in Figure 3, which also presents outline of wastewater flow.

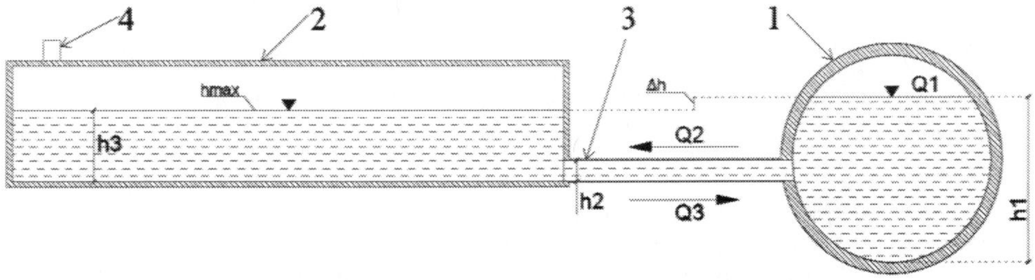

Fig. 3. Longitudinal section of retention reservoir with highly efficient hydraulic system: 1 – intercepting sewer, 2 – retention chamber, 3 – transit sewers, 4 - ventilation
h1 – level of wastewater in intercepting sewer, h2 – level of wastewater in transit sewer, h3 – level of wastewater in retention chamber of the reservoir, Δh – hydraulic losses in transit sewers,
Q1 – volumetric flow rate of wastewater in intercepting sewer, Q2 – volumetric flow rate of wastewater reaching retention chamber, Q3 – volumetric flow rate of wastewater leaving retention chamber.

4.1 Functional Phase I of the solution

The first functional phase of the installation, at the start of rainfall, is shown in Fig. 4. Because of the low flow rate in intercepting sewer (1), rainwater is not distributed between transit sewers (3). The system operates like conventional storm water drainage.

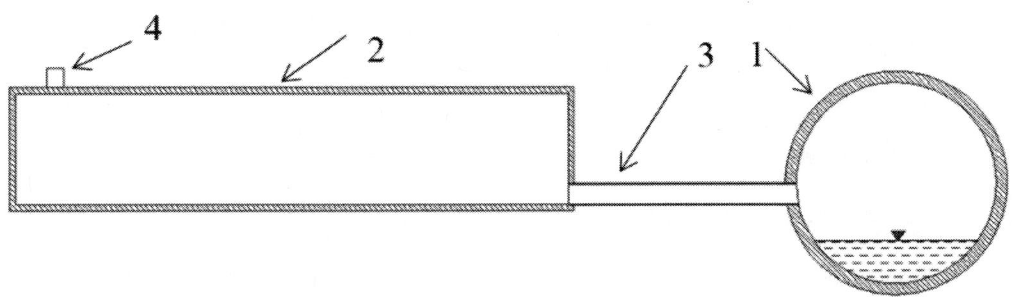

Fig. 4. Phase I of the reservoir operation.

- Boundary conditions in terms of high of fluid: $h1.1 > 0$, $h2.1 = 0$, $h3.1 = 0$.
- Boundary conditions in terms of flow rate: $Q1.1 > 0$, $Q2.1 = 0$, $Q3.1 = 0$.

4.2 Functional Phase II of the solution

The second phase, shown in Fig. 5, reflects the situation when increased volumetric flow rate leads to accumulation of wastewater in sewer (1), up to the bottom of transit sewers (3), yet still does not induce their operation.

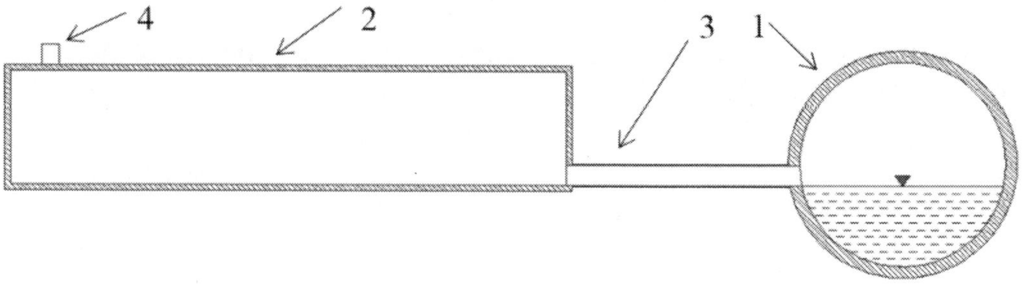

Fig. 5. Phase II of the reservoir operation.

- Boundary conditions in terms of high of fluid: h1.2 > 0, h1.2 > h1.1, h2.2 = 0, h3.2 = 0.
- Boundary conditions in terms of flow rate: Q1.2 > 0, Q1.2 > Q1.1, Q2.2 = 0, Q3.2 = 0.

4.3 Functional Phase III of the solution

During the third phase, shown in Fig. 6, the level of wastewater is higher than the level of the base of transit sewers (3). During this phase, the volume of wastewater above the bottom of transit sewers (3) is transported via sewers to the reservoir accumulation chamber (2). Here, the volumetric flow rate is divided into rainwater directed via intercepting sewer (1) to the receiving body and rainwater carried via transit sewers (3) into retention reservoir (2). This is when the solution in question starts functioning; this results in increased sewer retention capacity of the section along which reservoir (2) is located and it enables significant reduction in the rate of flow along the intercepting sewer 1 further on to the system located downstream, or to the receiving body.

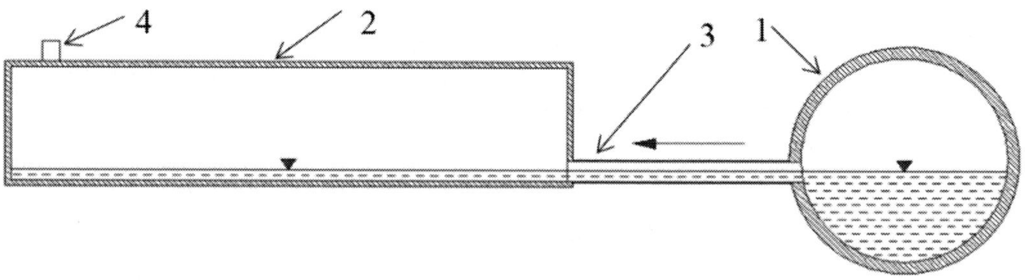

Fig. 6. Phase III of the reservoir operation.

- Boundary conditions in terms of high of fluid: h1.3 > 0, h1.3 > h1.2> h1.1, h2.3 > 0, h3.3 > 0.
- Boundary conditions in terms of flow rate: Q1.3 > 0, Q1.3 > Q1.2 > Q1.1, Q2.3> 0, Q3.3 = 0.

4.4 Functional Phase IV of the solution

The fourth functional phase of the invention, shown in Fig. 7, is the next and final stage of filling in the reservoir. During this phase, rainwater transported via transit sewer (3) leads to maximum accumulation resulting in its pressure-based functioning. Further filling of reservoir (2), and alignment of liquid levels in sewer (1) and reservoir (2) are enabled by efficiently operating ventilation (9).

 DOI: 10.1051/e3sconf/20171700075

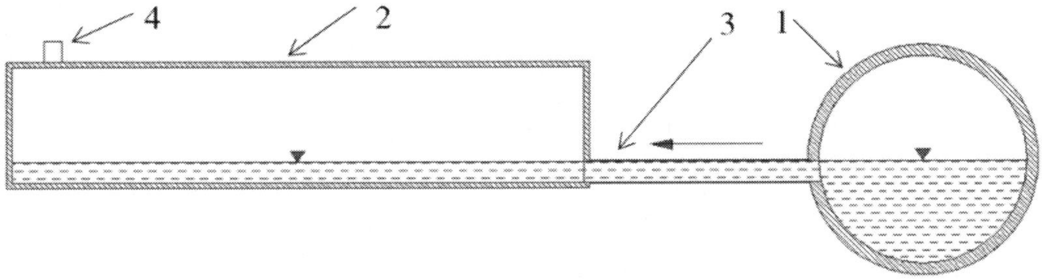

Fig. 7. Phase IV of the reservoir operation.

- Boundary conditions in terms of high of fluid: h1.4 > 0, h1.4 > h1.3> h1.2> h1.1, h2.4 > 0, h2.4 > h2.3, h3.4 > 0, h3.4 > h3.3.
- Boundary conditions in terms of flow rate: Q1.4 > 0, Q1.4 > Q1.3 > Q1.2 > Q1.1, Q2.4> 0, Q2.4 > Q2.3, Q3.3 = 0.

4.5 Functional Phase V of the solution

The fifth phase of the invention operation, as shown in Fig. 8, is the final stage of the process. As a result of the increased flow rate in transit sewer (3), the reservoir is filled in and the level of wastewater accumulated in it reaches the maximum design capacity h_{max}. In the case of rainfalls exceeding critical parameters, for which the system is designed, rainwater may be released from the system in a controlled manner in retention reservoir (2), rather than at the location of intercepting sewer (1).

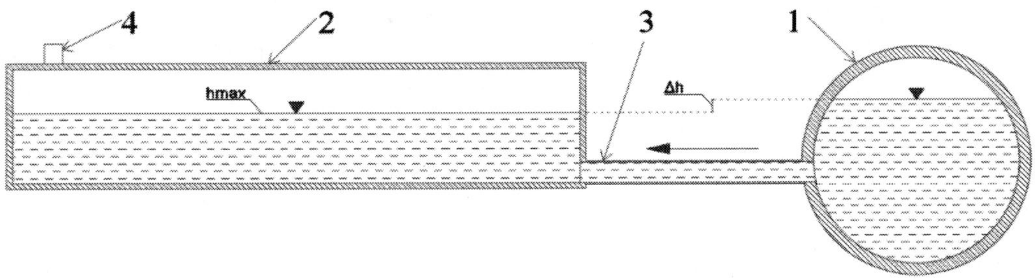

Fig. 8. Phase V of the reservoir operation.

- Boundary conditions in terms of high of fluid: h1.5 > 0, h1.5 > h1.4 > h1.3> h1.2> h1.1, h2.5 > 0, h2.5 = h2.4, h3.5 > 0, h3.5 > h3.4 > h3.3.
- Boundary conditions in terms of flow rate: Q1.5 > 0, Q1.5 > Q1.4 > Q1.3 > Q1.2 > Q1.1, Q2.5> 0, Q2.5 > Q2.4 > Q2.3, Q3.5 = 0.

4.6 Functional Phase VI of the solution

During phase 6, shown in Fig. 9, reservoir (2) is emptied. After rainfall stops, the level of rainwater in sewer system (1) gradually decreases. As a result, the same transit sewers 3, which during the previous phases led rainwater towards reservoir (2), now carry the rainwater back to intercepting sewer (1). This is enabled by the construction of transit sewers (3) running downward to intercepting sewer (1).

DOI: 10.1051/e3sconf/20171700075

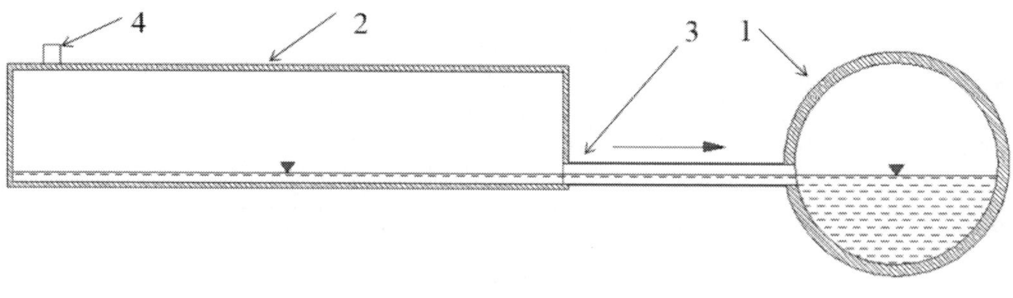

Fig. 9. Phase VI of the reservoir operation.

- Boundary conditions in terms of high of fluid: h1.6 > 0, h1.6 < h1.4 < h1.5, h2.6 > 0, h2.6 < h2.5, h3.6 > 0, h3.6 < h3.4< h3.5.
- Boundary conditions in terms of flow rate: Q1.6 > 0, Q1.6 < Q1.3 < Q1.4 < Q1.5, Q2.6=0, Q3.6 > 0.

4.7 Functional Phase VII of the solution

The final seventh phase of reservoir operation is presented in Fig. 10. At this stage, level of rainwater in intercepting sewer (1) is lower than the level of the bottom supports in reservoir (2), as well as the lowest positioning of transit sewer (3). During this phase the reservoir is completely emptied.

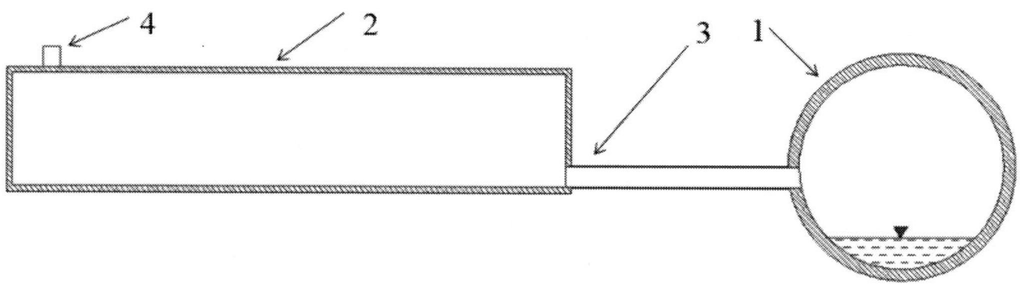

Fig. 10. Phase VII of the reservoir operation.

- Boundary conditions in terms of high of fluid: h1.7 > 0, h2.7 = 0, h3.7 = 0.
- Boundary conditions in terms of flow rate: Q1.1 > 7, Q2.7 = 0, Q3.7 = 0.

5 Summary and final conclusions

The hydraulic model of *Retention reservoir for rainwater and combined sewage* shows that by applying this solution it is possible to significantly increase retention capacity of a sewage system. Its strengths most notably include the possibility to deposit excess rainwater or combined sewage in the reservoir retention chamber, without a necessity to lower its bottom in relation to the bottom of intercepting sewer. This is particularly important in existing drainage systems where due to the construction of conventional retention reservoirs it is frequently necessary to use pumping devices in order to increase the level of wastewater. Application of the solution in question in existing sewage systems makes it possible to ensure high standard of drainage and to improve hydraulic security of the system.

Research carried out in the publication led to the following conclusions:

- filling of retention chamber occurs after obtaining the appropriate level of sewage table in the intercepting sewer. Thus, the solution can be used in existing sewage systems, in which there are periodic hydraulic overload,
- the procedure for reservoir dimensioning should include issue of determine the hydraulic losses in the transit sewers, which have an influence on maximum level of sewage table in retention chamber,
- in case of a small flow rate of sewage in intercepting sewer the retention chamber does not function. Therefore installing additional flow regulators in the intercepting sewer between transit sewers allow to force the inflow to retention chamber.

Further research related to the findings discussed in this article will include physical modeling and CFD comparison. Development of these models enable further analysis of the reservoir in a different conditions of sewage inflow.

References

1. M. Starzec, J. Dziopak, M.I. Alexeev, Water and Ecology **1**, 41-50 (2015)
2. A. Stec, S. Kordana, Resour. Conserv. Recy. **105**, 84–94 (2015)
3. M. Zeleňáková, G. Markovič, D. Kaposztásová, Z. Vranayová, Procedia Engineering **89**, 1515–1521 (2014)
4. T.D. Fletcher, H. Andrieu, P. Hamel, Adv. Water Resour. **51**, 261–279 (2013)
5. D. Słyś, S. Kordana, Energ. Buildings **71**, 1–11 (2014)
6. S. Kordana, D. Słyś, J. Dziopak, J. Clean. Prod. **82,** 58–69 (2014)
7. K. Pochwat, D. Słyś, J. Dziopak, Gaz Woda i Technika Sanitarna **7**, 269-272 (2013)
8. A. Kotowski, B. Kaźmierczak, J Hydrometeorol **14**, 1958–1965 (2013)
9. K.B. Pochwat, JCEEA **63**(2/II), 351–358 (2016)
10. Y.J. Chen, P. Cowling, F. Polack, S. Remde, P. Mourdjis, Eur. J. Oper. Res. **257**, 494–510 (2017)
11. B.R.T. Vilane, E.J. Mwendera, Afr. J. Agric. Res. **6**, 1313–1321 (2011)
12. Y. Kim, T. Kim, H. Park, M. Han, KSCE J. Civ. Eng. **19**, 1585–1590 (2015)
13. Dziopak J. *Multi-chamber storage reservoirs in the sewerage system* (Czestochowa, 1997)

E3S Web of Conferences 17, 00076 (2017) DOI: 10.1051/e3sconf/20171700076
EKO-DOK 2017

Toxicity assessment of water at different stages of treatment using Microtox assay

Marta Pogorzelec[1,*], and *Katarzyna* Piekarska[1]

[1]Wroclaw University of Science and Technology, Faculty of Environmental Engineering, Wybrzeże Wyspiańskiego 27, 50-370 Wrocław, Poland

Abstract. Number of potentially toxic hydrophobic organic contaminants e.g. polycyclic aromatic hydrocarbons, pesticides, polychlorinated biphenyls and dioxins having entered aquatic environment, including potential sources of drinking water. Unfortunately, not all micropollutants can be removed during water treatment processes. What is more, disinfectants can react with some organic compounds already present in the water, and form disinfection by-products which also can be toxic. The aim of this study was to assess toxicity of water at different stages of water treatment and to verify usefulness semipermeable membrane devices in monitoring of drinking water. For this purpose, semipermeable membrane devices (SPMDs) were deployed in a surface water treatment plant. To determine the effect of water treatment on the presence of toxic micropollutants, study was conducted for a period of 5 months. Three sampling places were chosen: raw water input, stream of water just before disinfection and treated water output. After sampling dialysis in organic solvent was carried out and extracts were then analyzed with the Microtox acute toxicity test. The study has indicated the utility as well as some limitations of combining SPMDs with bioluminescence assay in the monitoring of biological effects of bioavailable hydrophobic pollutants in drinking water.

1 Introduction

Surface water, in comparison with the groundwater, contains an increased number of microbiological and chemical contaminants [1, 2]. Many of organic micropollutants, including polycyclic aromatic hydrocarbons (PAHs), pesticides, dioxins and polichlorinated biphenyles, could be a hazard to the living organisms, because of their toxic potential [3, 4]. Therefore, it is necessary to carry out water treatment processes in order to enable it to human consumption. One of the main stages of water tratment is disinfection, which allows to eliminate or minimize the bacteriological risk. The purpose of disinfection is to destroy the living and spore forms of pathogenic organisms and preventing their re-development in the water distribution system. Insufficiently effective disinfection or lack thereof are the cause of waterborne diseases. Disinfection of water by chemical methods is the addition of strong oxidants. However, disinfectants (most frequently chlorine,

* Corresponding author: marta.pogorzelec@pwr.edu.pl

chlorine dioxide and ozone) can react with chemicals already present in water, that can lead to a number of other hazardous substances, referred as disinfection by-products (DBPs), which also may be toxic, genotoxic and carcinogenic [1–5].

Detection of micropollutants at low concentrations is extremely difficult. Accordingly, the water sample containing contaminants should be concentrated prior to analysis. Classical methods require the transport of large quantities of water from sampling places to laboratory, which creates logistical problems. To avoid this, in recent years popularity of passive samplers, including semipermeable membrane devices (SPMDs) has increased [6-10]. Semipermeable membrane devices are designed for concentration of the hydrophobic organic compounds present in water and air environments. SPMD is consisting of a neutral lipid with high molecular weight (> 600 Da), the most often triolein, sealed within a thin wall of low density polyethylene (LDPE) tube. The polyethylene membrane is considered to be non-porous, but the random thermal motions of the polymer chain leads to the formation of instantaneous cavities with maximum diameter of 10 Å. These membrane mimicking the biological membranes, thus allow selective diffusion of dissolved low molecular weight organic compounds, and their condensation in triolein [11, 12].

The combination of semipermeable membrane devices with bioassays allows the assessment of toxic or genotoxic potential of environmental samples. The effectiveness of the method has been demonstrated for short-term bioassays, including Microtox™ and *Daphnia pulex* toxicity analysis [7, 8,13–15].

2 Materials and methods

2.1 SPMDs deployment

Semipermeable membrane devices of the standard configuration (91.4 cm long, 2.5 cm wide, membrane thickness 70-95 µm, containing 1 ml of 99% purity triolein) from ExposMeter AB (Tavelsjő, Sweden) were deployed in in a surface water treatment plant called "Mokry Dwór" located in Lower Silesia, nearby Wrocław (Poland). Three places were chosen for sampling: A- pumping station "Czechnica" as a raw water input, B- pipeline with a stream of water just before ozone disinfection and C- clean water tank as a treated water output. Characteristics of the sampling points is described below.

In the study, samples were taken monthly over a period of about five months (II–VI : February 2016-July 2016), which means that after every month membranes were exchanged. Exceptions were membrane labeled "III–IV" (March 2016- April 2016), which, due to technological problems at the water treatment plant, remained in water for about eight weeks. Each membrane represents the state of water in a given month.

2.2 Characteristics of sampling places

2.2.1 A- Pumping station "Czechnica"- raw water input

The water that flows into the pumping station "Czechnica" is surface water without any treatment from the Oława River. The river has a length of 91.7 km and a catchment area of 1167.4 square kilometers. The source of Oława is in the Sudeten Foreland at an altitude of about 315 meters above sea level. Its composition is heavily dependent on the seasons. This water is characterized by high content of suspensions, colloids and dissolved organic compounds. In comparison with infiltrated or groundwater is characterized by a low content of minerals such as manganese and iron.

2.2.2 B- Pipeline- stream of water just before ozonation

The water stream leaving the sand filter is free of suspensions, both those present in the raw water input, as well as those formed during the coagulation process. Organic compounds affecting the taste and odor are still present. To remove them effectively on a carbon filters, ozone is used to breaking them into smaller molecules, which increases the biodegradability of organic compounds. Generated during ozonation disinfection by-products are adsorbed on a bed of coal or decomposed by a biomass in the bed.

2.2.3 C- Clean water tank- stream of water after disinfection

In the clean water tank there is a water after complete treatment and disinfection with chlorine dioxide. The water from this tank is collected and pumped directly to the water supply network. This water meets the requirements of water intended for human consumption in accordance with the Regulation of the Minister of Health in Poland. The dose of disinfectant dispensed before the clean water tank must be large enough to prevent secondary bacterial contamination of the water supply system.

2.3 Samples preparation

After one month of sampling SPMDs were transported to the laboratory where mechanical and chemical cleaning was carried out according to good SPMDs practice [16]. Then membranes were dialyzed separately in 130 cm^3 of n-hexane to release compounds bounded in membrane. After 18 hours of dialysis in 18°C the solvent was replaced with fresh one (130 cm^3) and dialysis was continued for another 8 hours. Then n-hexane has been completely removed using the rotary evaporation. The extracts intended for the Microtox™ analysis were dissolved in 5 cm^3 of dimethyl sulfoxide (DMSO).

2.4 Microtox™ toxicity analysis

The Microtox™ toxicity analysis is based on bioluminescence of the marine bacteria *Vibrio fisheri*. Luminescent bacteria emit light when they are in optimal environmental conditions. Presence of toxic substance causes change of cellular state, which is rapidly reflected in a decrease of light emission [17]. The analysis with the Microtox™ Basic Test ware carried out following the standard procedures of the manufacturer. Fourteen double dilutions of each sample (A, B, C), solvent (hexane) and diluent (DMSO) were prepared. The luminescence intensity was measured with the Microtox™ Model 500 Analyzer at 0-, 5- and 15-min exposure time. For the calculations of percent inhibition of light emission, the following equations were used:

$$R_t = \frac{I_{ct}}{I_{c0}} \quad (1)$$

$$G_t = \frac{R_t \times I_{s0}}{I_{st}} - 1 \quad (2)$$

$$\%inhibition = \frac{G_t}{1+G_t} \cdot 100 \quad (3)$$

where R_t is a correct for any inhibition induced by a negative control sample (blank cuvette) and I_{c0} and I_{ct} are the absolute light intensities produced by the negative control at time *t* and at initial time *0*, respectively. I_{s0} and I_{st} are the light intensities produced by water sample at time *t* and at initial time *0*, respectively. G_t is the ratio of the light lost at time *t* to the light remaining at time t, which can be convert to *% inhibition*. The final expression of the samples toxic response is presented as EC_{50} which is the Effective Concentration of the samples which reduces the light emission by 50%, and was estimated from the linear regression of the log of each concentration level of the contaminant versus the percent inhibition. It was expressed as the amount of SPMD triolein [mg] per 1 cm^3 of sample [18-19]. The toxicity equivalents, calculated as $1/EC_{50}$, ware used for graphical representation of the data.

3 Results and discussion

The analysis results of acute toxicity test Microtox ™ are shown in Table 1 and in Figures 1-6, as EC_{50} values and $1/EC_{50}$ values after an exposure time of 5 and 15 min.

Table 1. EC_{50} and $1/EC_{50}$ parameter values after 5 and 15 min exposition for water samples from different stages of treatment (A: pumping station- raw water input; B: pipeline- stream of water just before ozonation; C: clean water tank), in particular months (II-VI).

Sample	EC50		1/EC50	
	t=5	t=15	t=5	t=15
diluent	3.583	2.659	0.279	0.376
hex	1.382	3.065	0.724	0.326
II A	0.668	0.467	1.496	2.143
II B	1.049	0.702	0.953	1.424
II C	0.808	0.509	1.238	1.963
III-IV A	0.334	0.199	2.994	5.015
III-IV B	0.422	0.300	2.372	3.338
III-IV C	0.355	0.190	2.816	5.252
V A	0.366	0.240	2.730	4.163
V B	0.422'	0.264	2.371	3.791
V C	0.474	0.272	2.109	3.681
VI A	0.367	0.247	2.721	4.055
VI B	0.466	0.362	2.144	2.765
VI C	0.568	0.403	1.760	2.483

In the analysis of the toxicity test Microtox ™ a positive effect of treatment processes at the level of toxicity of water has been observed. In some cases, after disinfection toxicity has grown, but still it was lower value than value for the raw water. Toxicity was changing with each month, however, to be able to draw accurate conclusions related to the impact of weather conditions on the level of toxicity of water, observation should be conducted for a longer time, at least for one year.

The results obtained for the solvents (n-hexan, DMSO) used to prepare samples for analysis (cleaning, dialysis, dissolving material) indicates that these did not have any additional contaminants and their impact on the results was negligible.

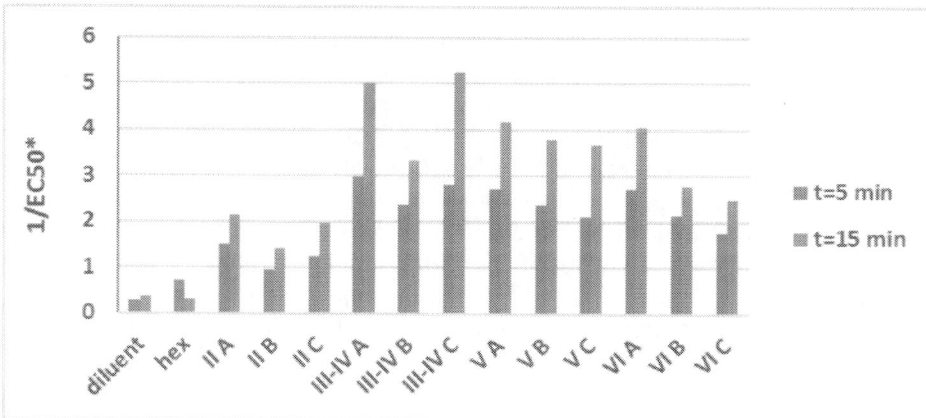

Fig. 1. $1/EC_{50}$ parameter values after 5 and 15 min exposition for water samples from different stages of treatment (A: pumping station- raw water input; B: pipeline- stream of water just before ozonation; C: clean water tank), in particular months (II–VI).

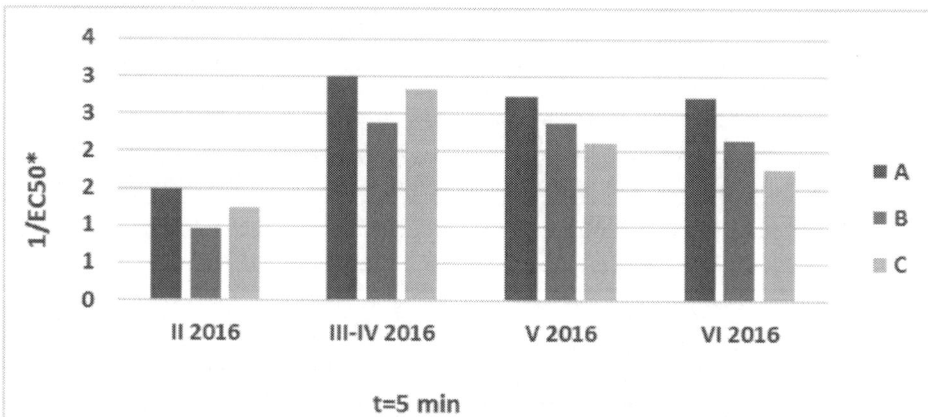

Fig. 2. $1/EC_{50}$ parameter values after 5 min exposition for water samples from different stages of treatment (A: pumping station- raw water input; B: pipeline- stream of water just before ozonation; C: clean water tank), in particular months (II–VI).

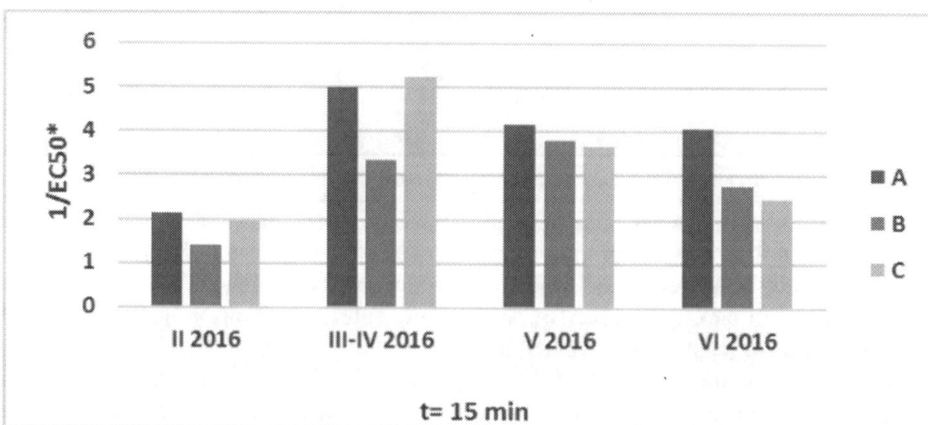

Fig. 3. $1/EC_{50}$ parameter values after 15 min exposition for water samples from different stages of treatment (A: pumping station- raw water input; B: pipeline- stream of water just before ozonation; C: clean water tank), in particular months (II–VI).

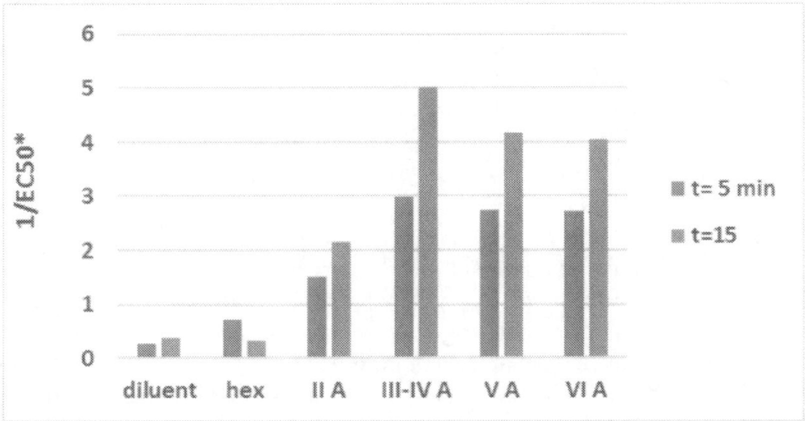

Fig. 4. 1/EC_{50} parameter values after 5 and 15 min exposition for water samples from sampling place A (pumping station- raw water input), in particular months (II–IV).

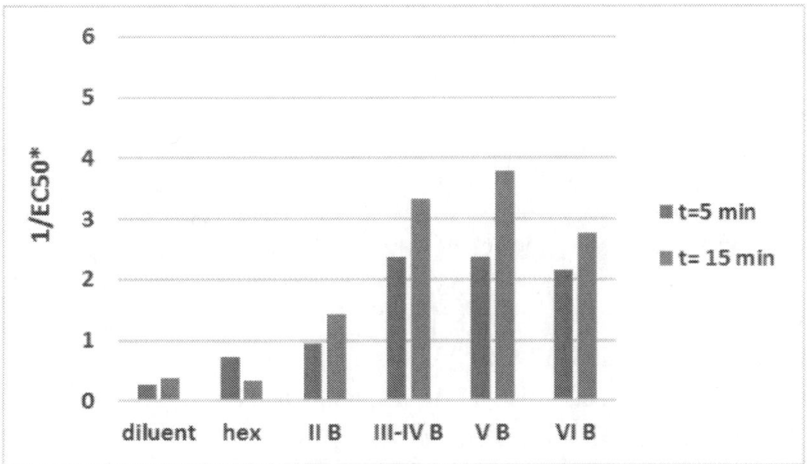

Fig. 5. 1/EC_{50} parameter values after 5 and 15 min exposition for water samples from sampling place B (pipeline- stream of water just before ozonation), in particular months (II–IV).

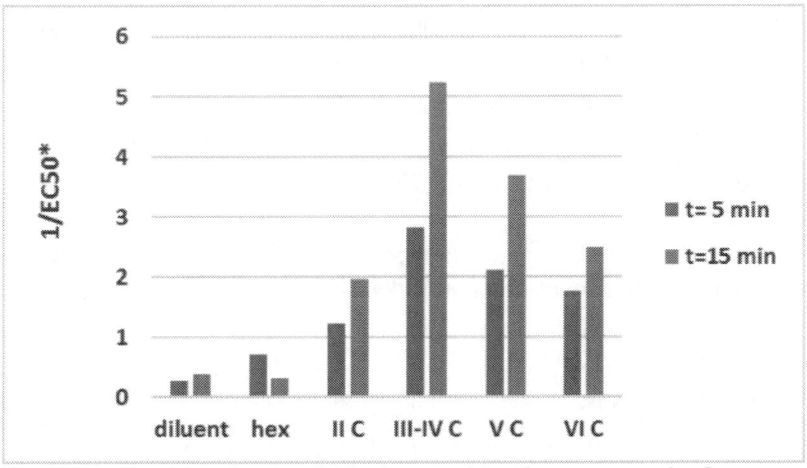

Fig. 6. 1/EC_{50} parameter values after 5 and 15 min exposition for water samples from sampling place C (clean water tank), in particular months (II–IV).

3 Conclusions

The current "bottleneck" of biological assays is problem with concentration of investigated compounds. Due to low concentration of micropollutants in water, it is necessary to fit appropriate technique for concentration of them in order to be detectable during the chemical and biological analysis. The proposed method uses passive samplers with semipermeable membrane, which are placed directly at the place of sampling, what eliminates logistical problems.

The study indicates that the use of semipermeable membrane devices can be an effective tool for the analysis of the aquatic environment, in which organic micropollutants are present at very low concentrations. Nevertheless, analysis of the toxicity of water at different stages of treatment should be conducted for a longer term, to take into account the impact of changing weather conditions on obtained results. Moreover, this analysis should also be supplemented by chemical analyzes and genotoxicity assays, to determine the health quality of water and its impact on living organisms.

The research financed by the funds of the Ministry of Science and Higher Education for young scientist and doctoral students order number: B50557.

References

1. J. Fawell, M. J. Nieuwenhuijsen, Brit. Med. Bull. **68**, 199-208 (2003)
2. M. Świderska-Bróż, *Mikrozanieczyszczenia w środowisku wodnym*, (Wydawnictwo Politechniki Wrocławskiej, Wrocław, 1993)
3. L. Kowal, M. Świderska-Bróż, *Oczyszczanie wody* (PWN, Warszawa, 2007)
4. C. Zani, D. Feretti, A. Buschini, P. Poli, C. Rossi, L. Guzella, F. Di Caterino, S. Monarca, Res. **587**, 26–37 (2005)
5. S. D. Richardoson, Trend. Anal. Chem. **22**, 666–684 (2003)
6. W.L. Cranor, D.A. Alvarez, S.D. Perkins, G.A. Tegerdine, R.C. Clark, J. N. Huckins, *Use of semipermeable membrane device (SPMD)technology for a probabilistic assessment of hydrophobic organic contaminants in selected reaches of Virginia rivers* (U. S. Geological Survey, 2005)
7. G. Gilli, T. Schiliro, Pignata, D. Traversi, E. Carraro, C. Baiocchi, R. Aigotti, D. Giacosa, E. Fea, Chemosphere **61**, 1691–1699 (2005)
8. Y. Lu, Z. Wang, J. Huckins, Aquat. Toxicol. **60**, 139–153 (2002)
9. A. Marrucci, B. Marras, S.S. Campisi, M. Schintu, Mar. Pollut. Bull. **75**, 69–75 (2013)
10. J.F. Mueller, M.R. Mortimer, J. O'Brein, T. Komarova, S. Carter, Mar. Pollut. Bull. **63**, 73–76 (2011)
11. J.N. Huckins, J.A. Lebo, M.W. Tubergen, G.K. Manuweera, V.L. Gibson, J.D. Perry, *United States Patent, Patent Number 5,098,573* (1992)
12. J.N. Huckins, *Fundamentals of SMPDs. Monitors of organic chemicals in the environment (*Springer, 2006)
13. J.D. Petty, J.N. Huckins, S.B. Jones, T.P. Boyle, Chemosphere **41**, 311–321 (2000)
14. D. Sabaliūnas, J.R. Lazutka, I. Sabaliuniene, Environ. Pollut. **109**, 251–265 (2000)
15. D. Sabaliūnas, A. Södergren, Environ. Pollut. **96**, 195-205 (1997)

16. D. Alvarez, *Guidelines for the use of the semipermeable membrane device (SPMD) and the polar organic chemical integrative sampler (POCIS) in environmental monitoring studies*, (U.S. Geological Survey, Techniques and Methods 1–D4 2010)
17. G. Węgrzyn G., A. Czyż, Oceanologia **44**, 291-305 (2002)
18. I.D. Gaudet, *Standard Procedure for MICROTOX Analysis. WCMUC* (1994)
19. R. James, *Strategic Diagnostics Inc. Microtox® Rapid Toxicity Testing System. Environmental Technology Verification Report* (U.S. EPA, 16–17, 2003)

E3S Web of Conferences **17**, 00077 (2017) DOI: 10.1051/e3sconf/20171700077
EKO-DOK 2017

Comparison of the composition of wastewater from fruit and vegetables as well as dairy industry

Monika Puchlik[1,*], and *Joanna* Struk-Sokołowska[1]

[1]Bialystok Technical University, Department of Technology in Engineering and Environmental Protection, 45E Wiejska Street, 15-351, Poland

Abstract. The aim of the study was to compare the composition of wastewater from the facility producing fruit and vegetable juices and dairy processing plant in Podlasie province. Their composition is also influenced by processes of washing and disinfecting of production lines. The washing process moves solid, colloidal, and dissolved impurities to the wastewater, depending on the type of raw material processed and the technological process used. At the same time, cleaning and disinfecting components at the amounts difficult to determine, penetrate the wastewater. The seasonal oscillations in the quantity and quality of sewage composition originating from fruit-vegetable and dairy processing became a problem for many conventional wastewater treatment plants, to where they are discharged. Based on the survey, it was found that wastewater from the fruit and vegetable industry as well as dairy industry contained large amounts of organic matter expressed in BOD_5 (fruits and vegetables processing from 860 to 3 200mg O_2/dm^3, dairy from 1 410 to 3 850 mgO_2/dm^3) and COD (fruits and vegetables processing from 919 to 3 700 mgO_2/dm^3, dairy from 1 680 to 5 420 mgO_2/dm^3). Significant differences were found in concentrations of nitrogen and phosphorus in compared raw wastewater.

1 Introduction

Food industry companies with the specifics of individual branches can be a source of many threats for all elements of the environment - soil, water, air, plants, animals, and humans. The food industry is part of the national economy department, which is engaged in the acquisition and processing of natural resources, and is one of the links in the food economy composed of many members closely related to one another [1]. It includes fruit and vegetable sector as well as dairy facilities producing a wide range of products, which occupy a central place in the food economy. Its aim is to satisfy people's basic needs, which are determined by the needs of a consumer and the needs and opportunities, especially in agriculture. Their common feature is the strong differentiation: there can be both pre-processing of agricultural products, and obtaining products with high degree of processing. The chemical composition of wastewater is very diverse, in its deposit mass consists of

[*] Corresponding author: m.puchlik@pb.edu.pl

organic and inorganic compounds. The aim of the paper is to compare the composition of the raw sewage from dairy as well as fruit and vegetable industry. It was observed that the seasonal variations in quality of wastewater from the food industry have become a problem for many conventional sewage treatment plants, where they are discharged.

2 Characteristics of raw sewage from the fruit and vegetable as well as dairy industry

Fruit and vegetable production is characterized by seasonal campaigns, that fall mainly from June to October [2, 3]. Activities in the processing and preservation of fruits and vegetables was shown by 88% of entities, while the remaining 12% was the production of juices and beverages [4]. Due to the nature and variety of processed raw material, wastewater from fruit and vegetable industry are difficult to characterize [5]. Most wastewater is produced in the first stage of manufacture. Wastewater generated from washing the fruits and vegetables are typically used once. In industry, water consumed in the first stage is preliminarily purified on a sieve and discharged to drain after washing the whole lot of fruits and vegetables, and then replaced with new water portions. In opinion by Nawirska [5], wastewater from fruit and vegetables plants are characterized by a high content of carbohydrates and minerals as well as variable composition depending on the raw material processed or time of the year [5]. Their composition is also affected by the processes of cleaning and disinfection of production lines. The washing process moves solid, colloidal and dissolved contamination to the wastewater depending on the type of raw material processed as well as on the process used. At the same time the sewage receives difficult to determine amounts of various components of detergents and disinfectants. Depending on wastewater generation place, following can be distinguished: raw material washing, cleaning, peeling, blanching, cooling (water after cooling), filling. The composition and amount of wastewater also depend on: the type, quality, settings, and size of dishwashers, type and origin of the raw material [6, 7]. Wastewater generated during production of fruit concentrates are characterized by a pH values at the range from 5.8 to 9.4 and the COD value from 1 030 up to 5 630 mgO_2/dm^3 [8]. In plants producing varied assortment, e.g. salads, pulps, or pickles, the pH can range from 4.9 to 7.7, while COD value from 5 260 to 270 mgO_2/dm^3. These values depend on the production technology and currently processed assortment [8]. Wastewater from the fruit and vegetable production is low in nitrogen and its main impurity is an organic matter. According to the literature, BOD_5 value ranges from 500 to 5 000 mgO_2/dm^3, and during the intensive production, it even exceeds 5 000 mgO_2/dm^3 [2, 9–11]. These levels are much higher than in household sewage [2, 12, 13]. Composition of wastewater produced in dairy processing plants depends primarily on the production profile. Other factors influencing the load of wastewater are raw materials used, technological level of the facility, processes of cleaning and disinfection, and the amount of water used [14, 15]. Dairy wastewater are characterized by high loads of COD and BOD_5 (fats, sugars, proteins), great qualitative variability (production line washing stage) [16,17,18,19]. According to literature data, dairy waste from all production lines can be purified altogether [20]. Composition of wastewater produced in the dairy processing plants depends mainly on the type of production and seasonality. There has been nearly 20% increase in supplied material in relation to the annual average in the summer and decrease in winter. Other factors influencing the composition and load of wastewater are raw materials used, technological level of the plant, processes of cleaning and disinfection, and the amount of water consumed. The second type of wastewater produced in dairy facilities are non-production sewage, among others, sanitary, post-cooling, storm, which are usually combined with the production wastewater and purified together [19,23]. Among plants that process milk, following should be distinguished: milk receiving points, small dairy plants (producing pasteurized milk, sour cream, kefir, yogurt, cottage cheese), as well as medium and large dairies with extended or full production profile (additionally producing melted and long maturing cheese, butter, powdered milk, ice cream) [21]. Typical dairy plant in Poland discharges contaminations of an average BOD_5 about 1167 mgO_2/dm^3 and COD 2077 mgO_2/dm^3 [22,23]. Among dairy processing facilities, there are also objects discharging over 5000 m^3/d wastewater, in which BOD_5 value reaches 6000 mgO_2/dm^3, and COD 9000 mgO_2/dm^3 [22,23]. Literature data show that the pH of wastewater depends on the type of production in a plant, and ranges from 6.4 to 7.6 [13]. According to Dąbrowski, nitrogen and phosphorus concentrations in dairy wastewater mainly depends on the quantity of chemicals used during the cleaning of devices in dairy plants [24].

3 Materials and Methods

The study was conducted in the laboratories of the Department of Technology in Engineering and Environmental Protection. Samples of raw sewage were collected 4 times a month from May to July in 2016 from two facilities located in Podlasie province. Facility A - processing of fruit and vegetables, producing mainly juice of fruit and vegetable purée, as well as facility B - dairy, in which the main products were hard cheese, yogurt and cottage cheese. The collected samples of raw sewage were subject to determinations of the following items in accordance with the current methodology:

- COD_{Cr} – dichromate method according to PN-74/C-04578.03;
- BOD_5 – manometric method using Oxi-Top Standard system;
- $P_{tot.}$ – spectrophotometric method according to PN-EN ISO 6878:2006;
- $N_{tot.}$ – spectrophotometric method using UV-VIS Pharo 300 device;
- acidity – potentiometric method PN-EN ISO 10523:2012;
- total suspended solids – gravimetric method PN-EN 872:2007+Ap1:2007.

4 Results

Wastewater from the fruits and vegetables and dairy processing (Table 1) were characterized, in comparison to municipal wastewater from collecting sewage, by many times higher values of BOD_5 and COD_{Cr} and comparable levels of nitrogen and phosphorus. Figure 1 shows variable volume fraction of wastewater from the fruit and vegetable as well as dairy industry since May to July 2016. The smallest volume of wastewater at fruit and vegetable processing plant, amounting to 6 322 m^3, was recorded in May, while the largest in July (16 527m^3). In dairy processing facility, the smallest volume of wastewater was recorded in May – 20 091m^3, while the largest in July – 22 800m^3. A significant increase in the volume of wastewater (more than 2-fold) was observed during the summer season for the plant producing fruit and vegetable juices (Figure 1). This is due to seasonal work in this type of production plant. The scope of changes in BOD_5 and COD values from May to July is presented in Table 1. Calculations show that the ratio of COD to BOD_5 for raw sewage in analyzed facilities is less than 2, which indicates that the wastewater is easily biodegradable. Figure 2 shows the range of variation of the average monthly values of COD and BOD_5 values in raw sewage in 2016 for the plant producing fruit and vegetable juices and other products as well as for dairy processing plant. The highest load of COD in raw sewage was reported in the dairy in July, amounting to 123 000 kg, while the lowest in May, 5 809 kg, for fruits and vegetables processing plant. The maximum monthly load for BOD_5 in raw sewage was observed in July with a value of 88 920 kg for dairy, whereas the lowest in May 5 436 kg in the fruit and vegetable facility.

Figure 3 shows the average monthly change in the load of total nitrogen and total phosphorus in raw sewage. The lowest value of the total nitrogen load was recorded for fruit and vegetable processing plant in May, 252 kg, while the maximum of 991 kg in July. In the dairy facility, the lowest value of total nitrogen load was recorded in June – 1 487 kg and the highest, 1 983 kg, in July. Given the load of total phosphorus, it increased from 59 kg in May to 264 kg in July in fruit and vegetable processing plant. In the dairy, the lowest value of total phosphorus load was observed in May – 252 kg and the highest in July – 501 kg. The highest monthly average value of COD concentrations for processing plants were recorded in July: dairy – 5 420 mg O_2/dm^3, fruit and vegetable processing plant – 3 700 mg O_2/dm^3; the lowest in May, 1 680 mg O_2/dm^3 in the dairy, and 919 mg O_2/dm^3 for processing fruits and vegetables (Table 1). The highest average monthly amount of organic matter expressed in BOD_5 in raw wastewater was 3 200 mg O_2/dm^3 in July in fruit and vegetable processing facility, while 3 900 mg O_2/dm^3 in the dairy. The lowest concentration of total suspended solids in raw sewage was 200 mg/dm^3 for dairy in May, while the highest was recorded in July with a value of 420 mg/dm^3 for the processing of fruits and vegetables (Table 1). An analysis of monthly average concentrations of total nitrogen and total phosphorus in the raw wastewater from the fruit and vegetable as well as dairy processing, reveals that the highest value of total nitrogen, 87 mgN/dm^3 was recorded in July for dairy plant, while the lowest value, 40.0 mgN/dm^3, in May for juice production. The highest concentration of total phosphorus in raw sewage was 22.0 mg P/dm^3 in July in the dairy, and in May, more than 2 times lower amounting to 9.4 mg P/dm^3 in the plant processing fruits and vegetables. It has been observed that wastewater directly related to the production of food has a variable pH value from 4.3 up to 9.4 (Table 1).

Table 1. Average monthly values for raw sewage.

Parameter	**Unit**	**Raw wastewater**					
		Fruits and vegetables processing			**Dairy**		
		V	VI	VII	V	VI	VII
BOD	mgO$_2$/dm^3	860	2 200	3 200	1 220	2 800	3 900
COD	mgO$_2$/dm^3	919	3 350	3 700	1 680	3 300	5 420
$N_{tot.}$	mgN/dm^3	40.0	48.0	60.0	76.0	69.0	87.0
$P_{tot.}$	mgP/dm^3	9.4	14.8	16.0	12.6	16.0	22.0
TSS	mg/dm^3	249	356	420	200	245	286
pH	-	5.5-7.2	4.3-7.1	4.6-7.9	6.1-9.3	6.6-9.0	6.8-9.4

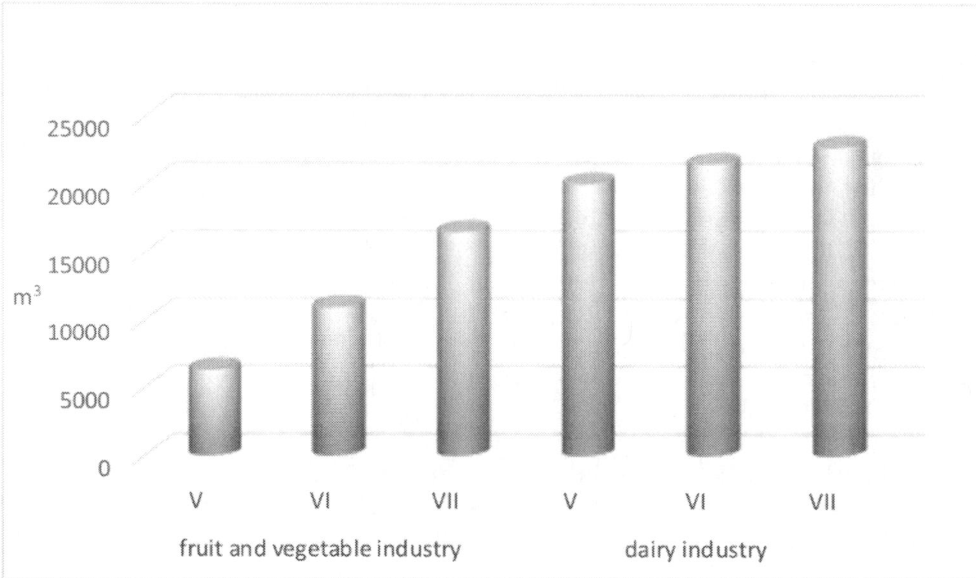

Fig. 1. The average monthly volume of raw sewage from the fruit and vegetable as well as dairy industry since May to July 2016.

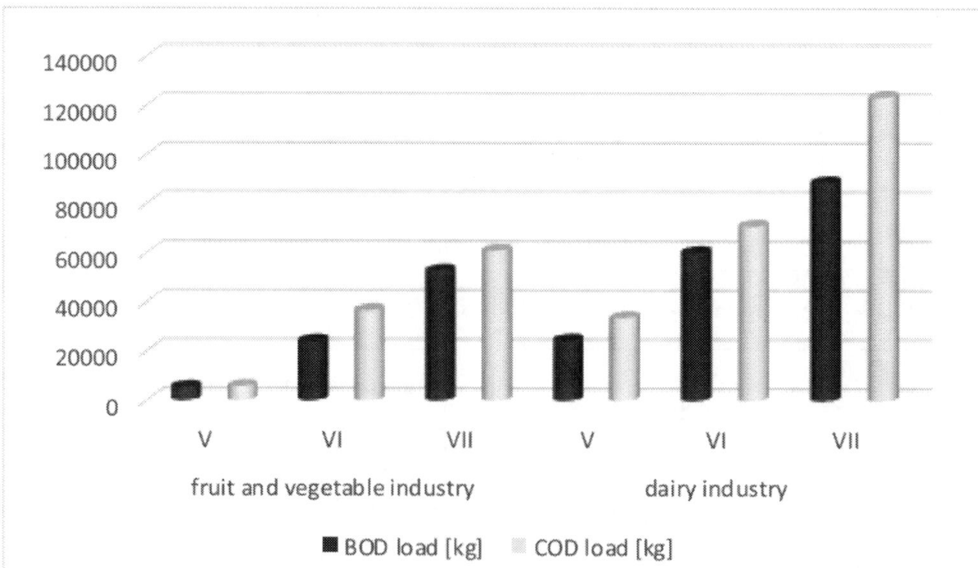

Fig. 2. The average monthly COD and BOD_5 loads in raw sewage from the fruit and vegetable as well as dairy industry since May to July 2016.

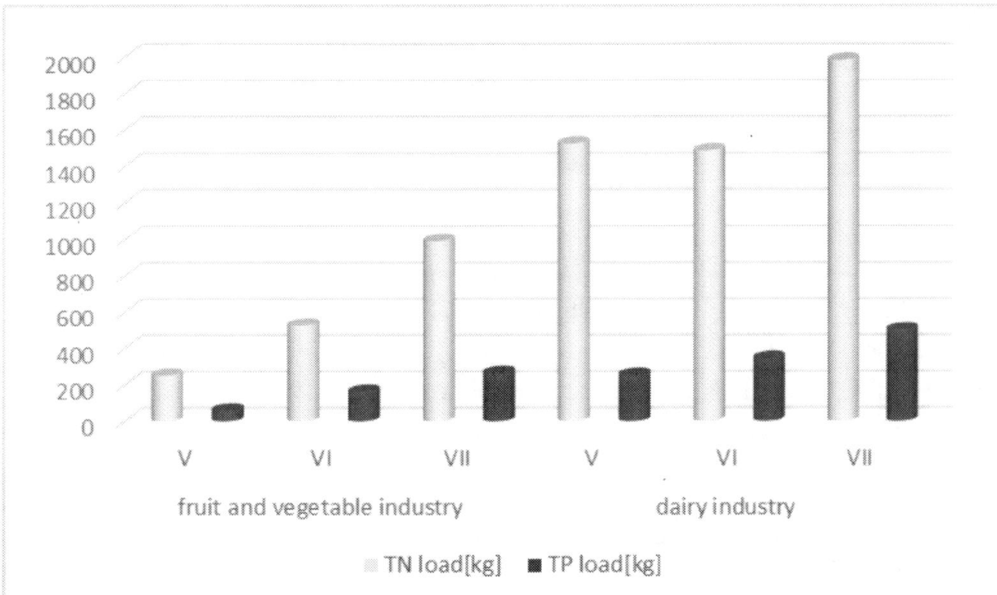

Fig. 3. The average monthly $N_{tot.}$ and $P_{tot.}$ loads in raw sewage from the fruit and vegetable as well as dairy industry since May to July 2016.

5 Conclusions

Analysis of determinations allow for the following statements:

1. In the analyzed facilities, the main contaminant was organic matter expressed in COD and BOD_5.
2. The ratio of COD to BOD_5 for raw wastewater is smaller than 2, which indicates that this sewage is readily biodegradable.
3. There were small variations in the concentrations of total nitrogen and phosphorus resulting from the production profile and chemicals used for the cleaning of equipment in facilities.
4. Significant fluctuations in the quantity and quality of wastewater from the food industry during the production campaigns may represent a problem for many conventional sewage treatment plants, to which they are discharged.

The research has been carried out in the framework of project No. MB/WBiIŚ/13/2015 and financed from the funds for science MNiSW.

References

1. M. Puchlik, J. Struk-Sokołowska, E. Wołejko, U. Wydro, Problem oczyszczania ścieków z przemysłu spożywczego w małych i średnich przedsiębiorstwach, Interdyscyplinarne zagadnienia w inżynierii i ochronie środowiska, T**7** (2016)
2. M. Soroko, Oczyszczanie ścieków z małych przetwórni owocowo-warzywnych w oczyszczalniach hydrofitowych, Woda Środowisko Obszary Wiejskie, T **3** (2003)
3. M. Soroko, Skuteczność oczyszczania ścieków z małych ubojni i przetwórni owocowo-warzywnych w systemach hydrofitowych, PAN, Oddział w Krakowie **3**, 1 (2006)

4. S. Kierczyńska, Produkcja warzyw i owoców oraz ich przetwórstwo w Polsce, Roczniki Seria, T **X**, 4 (2008)
5. A. Nawirska, Gospodarka wodno-ściekowa w przemyśle owocowo-warzywnym., Agro Przemysł **3**/2007 (2007)
6. A. Nawirska, L. Szymański, Gospodarka wodno-ściekowa w zakładach przemysłu spożywczego, Wydawnictwo Akademii Rolniczej we Wrocławiu (2002)
7. M. Puchlik, Application of constructed wetlands for treatment of wastewater from fruit and vegetable industry, J. Ecol. Eng. - Vol. **17,** 1 (2016)
8. D. Kopiec, Ścieki z zakładów spożywczych– doświadczenia z Miejskiej Oczyszczalni Ściekow w Błoniu, Przegląd Komunalny **2** (2007)
9. B. Koziorowski, Oczyszczanie ścieków przemysłowych. Warszawa: WTN (1980)
10. F. Meinck, H. Stoof , Kohlschutter, Ścieki Przemysłowe (1975)
11. B. Talik, B. Chudzik, Oczyszczanie ścieków z małych przetwórni owocowo-warzywnych zlokalizowanych na wsi. IMUZ T **20,** 3 (2000)
12. M. Puchlik, K. Ignatowicz, W. Dąbrowski ,Influence of bio-preparation on wastewater purification process in constructed wetlands, J. Ecol. Eng. - Vol. **16,** 1 (2015)
13. B. Bartkiewicz, K. Umiejewska ,Oczyszczanie ścieków przemysłowych, Wyd PWN (2010)
14. J. Struk-Sokołowska, Wpływ ścieków mleczarskich na frakcje ChZT ścieków komunalnych, IE **24** (2011)
15. M. Wojnicz, A. M. Anielak, Badania nad możliwością oczyszczania ścieków przemysłu mleczarskiego w systemie SBR z wykorzystaniem zeolitu naturalnego, Oczyszczanie ścieków i przeróbka osadów ściekowych, Oficyna Wyd. Uniwersytetu Zielonogórskiego, **2** (2008)
16. B. Buczak, Oczyszczanie ścieków mleczarskich z uwzględnieniem usuwania azotu-mity i rzeczywistość, III Konferencja Naukowo-Techniczna. Woda i ścieki w przemyśle spożywczym. Białystok (2010)
17. B. Demirel, O. Yenigun, T.T. Onay, Anaerobic treatment of dairy wastewater: a review. Proc. Biochem. **40** (2005)
18. E. Neczaj, M. Kacprzak, T. Kamizela, J. Lach, E. Okoniewska, Sequencing batch reactor system for the co-treatment of landfill leachate and dairy wastewater. Desalination, **222** (2008)
19. J. Struk-Sokołowska, K. Ignatowicz, Współoczyszczanie ścieków komunalnych i mleczarskich w oczyszczalniach typu SBR. Rocznik Ochrony Środowiska (Annual Set The Environment Protection), **15** (2013)
20. W. Janczukowicz, M. Zieliński, M. Dębowski, Biodegradability evaluation of dairy effluents originated in selected sections of dairy production. Biores. Techn., **99** (2008)
21. A. M. Anielak, Gospodarka wodno-ściekowa przemysłu mleczarskiego, Agro Przemysł **2** (2008)
22. J. Struk-Sokołowska, Wpływ ścieków mleczarskich na frakcje ChZT ścieków komunalnych, IE **24** (2011)
23. J. Struk-Sokołowska, Ocena efektywności pracy oczyszczalni ścieków
24. w Siemiatyczach, Interdyscyplinarne zagadnienia w inżynierii i ochronie środowiska, T **4** (2014)
25. W. Dąbrowski, Oczyszczanie odcieków z oczyszczalni mleczarskich w systemach hydrofitowych, Monografia (2014)

E3S Web of Conferences **17**, 00078 (2017) DOI: 10.1051/e3sconf/20171700078
EKO-DOK 2017

Numerical modelling of sulphate ion concentration in wastewater from a closed cooling system

Paweł Regucki[1,*], and *Barbara* Janowska[1]

[1]Wrocław University of Science and Technology, Faculty of Mechanical and Power Engineering, Wybrzeże Wyspiańskiego 27, 50-370 Wrocław, Poland

Abstract. The paper presents a mathematical model that allows for predicting daily changes of sulphate ion concentration in water circulating in a closed cooling system consisting of condensers and cooling towers. This is an important issue because an excessive and uncontrolled increase in the concentration of SO_4^{2-} in the circulating water may cause the corrosion of concrete parts of the channels and cooling towers, as well as an increase in the concentration of calcium salts, which can accelerate the process of its deposition on the exchange surfaces inside condensers. The goal of the paper is to propose an original mathematical model, which under certain simplifications poses an analytical solution to this problem. The approach enables calculating the asymptotic value of the sulphate ion concentration in circulating water, as well as the minimal volumetric flow rate of wastewater allowing the SO_4^{2-} ion concentration to be preserved below legal limits. The simplified analytical solution enables indicating the optimal volumetric flow rate of wastewater fitting the actual number of working power units. This is especially important because the daily amount of water released from the cooling system as wastewater usually reaches a level of several thousand m^3 and its proper management could significantly reduce system maintenance costs.

1 Introduction

In conventional and thermal power plants, water is widely used in many technological processes, including, e.g.: the water-steam cycle in power boilers, the cooling installations in boiler auxiliary equipment or the cooling systems cooperating with condensers. Each of these processes requires constant access to fresh water sources with appropriate physical and chemical properties in order to maintain the continuity of electricity and heat production [1]. On the other hand, power plants are also significant producers of pollutants and wastewater [2, 3]. The remarkable increase in the production of wastewater and its influence on the environment has resulted in the latest industrial waste regulations introduced by the Polish Ministry of Environment dated to 18 November 2014 and the EU directive [4], which impose, as of the beginning of 2016, new limits on the physical and

* Corresponding author: pawel.regucki@pwr.edu.pl

chemical parameters of wastewater released into natural reservoirs. The increase in environmental responsibility together with high penalties for exceeding allowable wastewater limits has forced companies to look for optimal methods of industrial waste management.

One of the largest fresh water consumers in a power unit is the cooling system supporting the condensers, which are responsible for the condensation process of spent steam from the last stage of the turbine. The latent heat released in the condensation process is transferred to the water circulating in the cooling installation. Due to the fact that the amount of released heat is often calculated in hundreds of MWs, the power units must either be located close to natural water reservoirs (lakes, rivers, seas) or possess cooling equipment like fan coolers or cooling towers. An efficient cooling system is especially important as the thermodynamic conditions inside a condenser have a direct influence on the electricity production efficiency of the whole power unit. The cooling system's demand for large amounts of water on the one hand, and the recent legal regulations connected with the chemical parameters of wastewater on the other, mean that power plant operators are forced to introduce optimal methods of water resource management.

Efficient wastewater management needs scientific backgrounds and is based on mathematical models [5]. Focusing on the optimization of fresh water consumption, the Authors present an original mathematical model which allows for predicting daily changes of sulphate ion concentration in water circulating in a closed cooling system consisting of condensers and cooling towers. This is an important issue because an excessive and uncontrolled increase in the concentration of SO_4^{2-} in the circulating water may cause the corrosion of concrete parts of the channels and cooling towers [6, 7], as well as an increase in the concentration of calcium salts, which can accelerate the process of its deposition on the exchange surfaces inside the condensers.

2 Closed cooling system

The closed cooling system is one of the largest fresh water consumers in a power plant. As a result, knowledge about the scientific processes influencing the changes in the sulphate ion concentration in the installation is essential for any attempts at establishing a mathematical model. The scheme of a closed cooling water system is presented in Fig. 1.

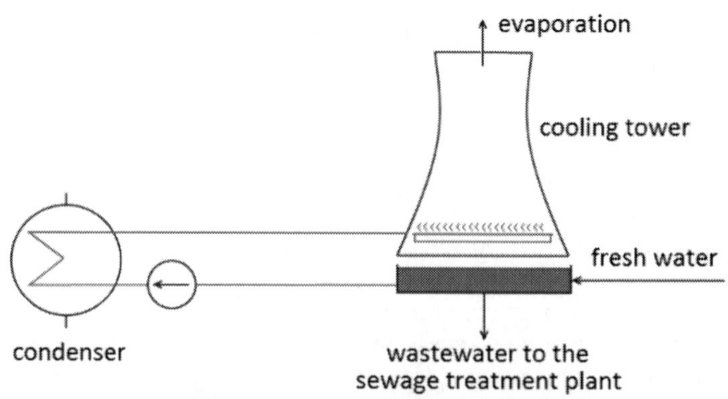

Fig. 1. Scheme of a closed system of cooling water in a power plant.

The water circulating in this system is heated in a condenser by condensing steam from the last stage of a turbine, and is then cooled down in a cooling tower. The main mechanism of water temperature decrease inside the cooling tower is connected with its partial evaporation in contact with counterflowing air [8]. This phenomenon causes a continuous decrease of the mass of the water in the installation and has a direct influence on the increase of the concentration of chemical compounds (among others, the abovementioned sulphate (VI) ions) in the circulating water. Due to the fact that the volume of water circulated in the system is usually at the level of tens of thousands m^3, it is impossible to apply any chemical methods enabling the reduction of concentration of sulphate ions [9].

The only reasonable method to avoid the uncontrolled increase of the undesirable SO_4^{2-} ion concentration in cooling water involves the periodic discharge of wastewater to a sewage treatment plant. In order to preserve a constant amount of water in the cooling installation, the missing mass of water is replenished from natural reservoirs. This action leads to the reduction of the total concentration of SO_4^{2-} ions in the closed system because the fresh water usually has a lower concentration of sulphate ions in comparison to its actual concentration in the closed cooling system.

Measurements carried out in a professional power plant indicate seasonal changes in the concentration of SO_4^{2-} ions (denoted by x, [g/m^3]). Sample results are presented in Table 1.

Table 1. Monthly changes of mean concentration of sulphate ions x and mean temperature of cooling water T at the reservoir of a cooling tower.

Month	I	II	III	IV	V	VI	VII	VIII	IX	X	XI	XII
T, °C	15.7	14.0	14.5	17.4	19.1	19.9	21.2	22.0	20.6	20.4	16.7	16.9
x, g/m^3	175.0	160.0	190.2	206.4	252.0	303.7	313.0	316.2	253.6	276.4	190.7	195.4

The presented data indicate that the mean monthly concentration of SO_4^{2-} ions changes over the course of the year. There are two factors which significantly influence this situation: the high intensity of water evaporation in the cooling tower during the summer and the reduction of the operation parameters of the cooling tower in winter. It is observed that in summer about 90% of the transported heat is released through the evaporation process, which causes approximately 1% of the mass of circulating water to evaporate in the cooling towers [8]. On the other hand, in winter the intensity of water evaporation is limited not only by the low ambient temperature but also for technological reasons forcing the temperature of the cooling water in the installation to amount to no less than 12°C. In order to preserve the temperature of the circulating water at above 12°C in winter only part of it is cooled down in the cooling towers. The intensity of water loss is reflected in the highest value of SO_4^{2-} ion concentration in July and August (up to ~320 g/m^3) in opposition to the lowest concentration in February (approx. 160 g/m^3). Keeping in mind that legal regulations allow for a much higher SO_4^{2-} ion concentration in wastewater (maximally x_{limit} = 500 g/m^3 [5]), the operator of the closed cooling system releases more sewage than required. This situation provides an opportunity to optimize wastewater management in order to reduce the operating costs of the system.

3 Results and discussion

3.1 Mathematical model

Analysing the thermal–flow processes in a closed cooling system, the Authors indicate three main physical mechanisms leading to changes in the sulphate ion concentration in circulating water:

- evaporation of water in cooling towers, $q_{v,ct}$,
- replenishment of fresh water, $q_{v,fw}$,
- discharge of wastewater to a sewage treatment plant, $q_{v,dw}$.

The relationship between these three abovementioned processes is schematically presented in Figure 2.

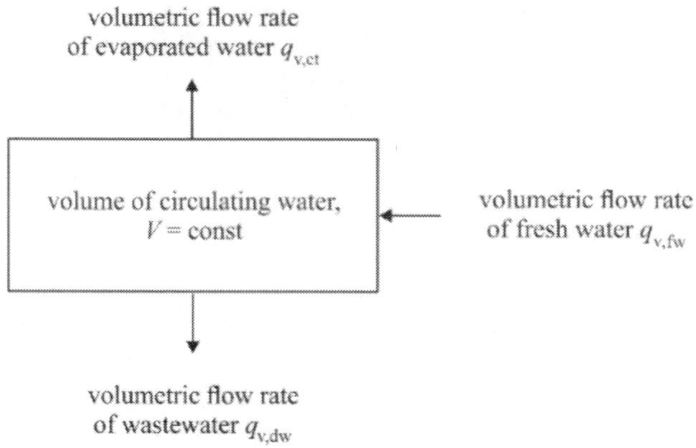

Fig. 2. Schematic balance of circulating water in a closed cooling system.

Due to the fact that the total volume of circulating water in the closed system V is conserved, the mass conservation law for volumetric flow rates must be satisfied:

$$q_{v,fw} = q_{v,ct} + q_{v,dw} \tag{1}$$

The analysis of losses and replenishment of water in the closed system has allowed the Authors to formulate a differential equation describing the changes of SO_4^{2-} concentration in circulating water (t – time in days):

$$V\frac{d\,x(t)}{dt} = -q_{v,dw}(t)\cdot x(t) - q_{v,ct}(t)\cdot x_{ct} + q_{v,fw}(t)\cdot x_{fw}(t) \tag{2}$$

where:
$x(t)$ – current concentration of sulphate ions in circulating water,
x_{ct} – sulphate ion concentration in vapour released in the cooling tower,
$x_{fw}(t)$ – concentration of sulphate ions in fresh water.
All the concentrations are in g/m^3, the volumetric flow rates in m^3/d, and the total volume V in m^3.

Equation (2) can be analytically solved under the assumptions that volumetric flow rates $q_{v,dw}$ and $q_{v,ct}$ are constant over time, sulphate ion concentration in fresh water x_{fw} is fixed and vapour leaving the cooling tower is chemically clean (x_{ct} = 0 g/m^3). This reduces (2) to the form of a first order ordinary differential equation with constant coefficients a and b:

$$\frac{d\,x(t)}{dt} + a\cdot x(t) - b = 0 \tag{3}$$

where: $a = q_{v,dw}/V$ and $b = q_{v,fw}\cdot x_{fw}/V$. The solution of (3) is the function $x(t)$, which describes change in SO_4^{2-} concentration over time:

$$x(t) = \frac{b}{a} + \left(x_0 - \frac{b}{a}\right)\cdot e^{-at} \tag{4}$$

where x_0 is the initial SO_4^{2-} ion concentration in the closed cooling system.

It is worth noting that the model does not have physical sense for coefficient $b = 0$ because in this case either the mass conservation law (2) is not satisfied ($q_{v,fw}$ is zero) or the concentration of sulphate ions in fresh water x_{fw} amounts to zero. On the other hand, the coefficient a must also have a nonzero value (which means that $q_{v,dw} \neq 0$), because otherwise the equation (2) is reduced to the form:

$$\frac{d\,x(t)}{dt} - b = 0 \tag{5}$$

which has the linear solution $x(t) = b \cdot t + x_0$ and the concentration $x(t)$ tends to infinity over time.

The SO_4^{2-} ion concentration in fresh water varies between 20 - 40 g/m^3 depending on local chemical conditions; thus, in the model, the Authors have used $x_{fw} = 40$ g/m^3. Figure 3 presents an example of solutions (4) obtained for $q_{v,ct} = 45600$ m^3/d and the total volume of the circulating water equalling $V = 90000$ m^3.

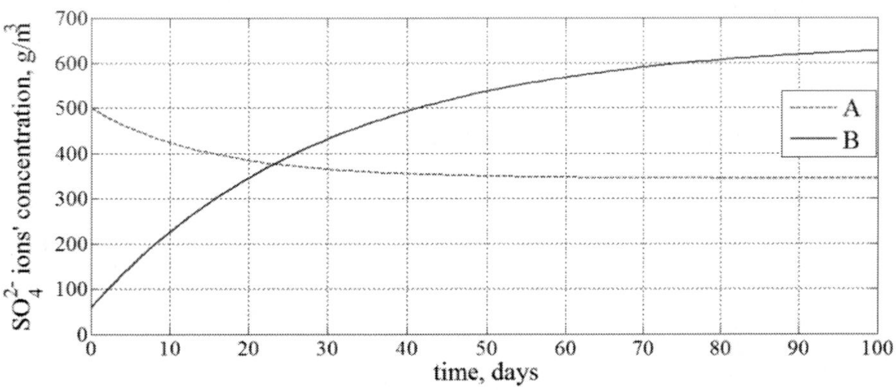

Fig. 3. Example of analytical solutions of (4) for different initial conditions: A (red dashed line) - $x_0 = 500$ g/m^3 and $q_{v,dw} = 6000$ m^3/d; B (blue solid line) - $x_0 = 60$ g/m^3, $q_{v,dw} = 3000$ m^3/d.

The curves A and B in Figure 3 present two possible solutions of (4) depending on the initial condition x_0 and values of the a and b coefficients. For a fixed volumetric flow rate of evaporated water inside cooling towers $q_{v,ct}$ and a fixed total volume of water circulating in installation V, the value of the discharge of wastewater to a sewage treatment plant $q_{v,dw}$ has a decisive influence on the shape of the presented plots. In case A, the flow rate $q_{v,dw} = 6000$ m^3/d causes a decrease in the initial SO_4^{2-} ion concentration over time to the level ~350 g/m^3, while in case B the low value of $q_{v,dw} = 3000$ m^3/d causes the increase of $x(t)$ over time up to ~650 g/m^3. It is worth noting that both curves have the asymptotic solutions x_∞ if time tends to infinity. The exact value of x_∞ can be calculated for both cases from the limit:

$$x_\infty = \lim_{t\to\infty} x(t) = \lim_{t\to\infty}\left[\frac{b}{a} + \left(x_0 - \frac{b}{a}\right)\cdot e^{-at}\right] = \frac{b}{a} \tag{6}$$

For the presented examples, the asymptotic value x_∞, calculated for formula (6), is equal to 344 g/m^3 and 648 g/m^3 respectively. From the mathematical point of view, the fixed value of x is reached also when the bracket $(x_0 - b/a)$ in formula (4) amounts to zero. In this case, $x(t)$ and x_0 are simply equal to x_∞.

3.2 Numerical study

It is worth comparing the simplified analytical solution (4) of the presented mathematical model with its numerical study. To do so, numerical methods were used to solve the time-dependent equation (2) satisfying mass conservation law (1). Equation (2) was discretized using the updated Euler method [10]:

$$p_{i+1} = x_i + \Delta t \cdot \left(-\frac{q_{v,dw}}{V} \cdot x_i + \frac{(q_{v,dw} + q_{v,ct})}{V} x_{fw} \right)$$
$$x_{i+1} = x_i + \frac{\Delta t}{2} \cdot \left(-\frac{q_{v,dw}}{V} \cdot (x_i + p_{i+1}) + 2\frac{(q_{v,dw} + q_{v,ct})}{V} x_{fw} \right) \quad (7)$$

where $x_i = x(t_i)$ is the numerical solution for $t_i = t_0 + \Delta t \cdot i, (i = 0,1, \ldots)$ and Δt is a time step. In order to estimate the accuracy of the numerical solutions obtained from formula (7), they were compared with analytical functions (4) derived for the set of parameters $\{x_0, x_{fw}, q_{v,dw}, q_{v,ct}, V\}$ analysed in cases A and B (presented in Fig. 3). The relative error E_{rel} was calculated according to the formula:

$$E_{rel} = \frac{|x_{anal}(t_i) - x_{num}(t_i)|}{x_{anal}(t_i)} \cdot 100\% \quad (8)$$

where x_{anal} and x_{num} denote the analytical and numerical solution respectively. The comparison of E_{rel} calculated for cases A and B is presented in Fig. 4. In both analysed examples, the relative error is less than 0.012% and decreases at the ends of the interval.

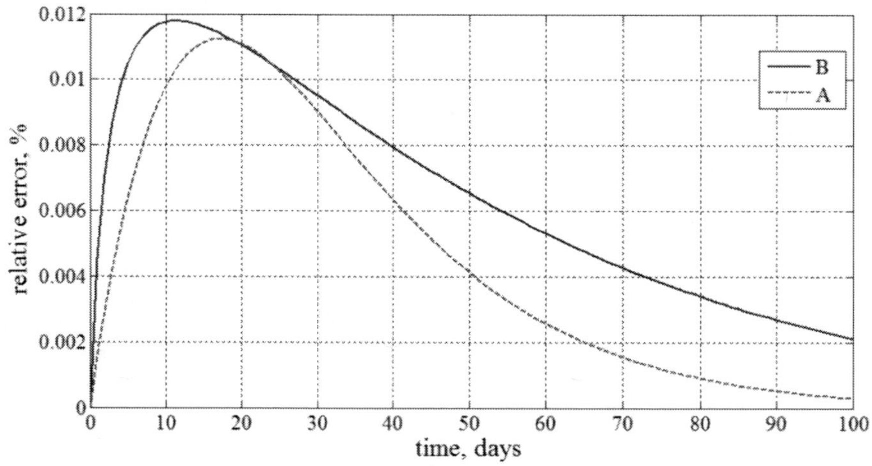

Fig. 4. Relative error E_{rel} of analytical and numerical solutions calculated from formula (8) for cases A and B presented in Fig. 3.

The measurements conducted on the cooling water installation have shown that despite the fact that mean values of volumetric flow rates are conserved, its daily values can vary by even ±15%. It is worth mentioning that the concentration of sulphate ions in fresh water x_{fw} is rather stable and depends on the chemical properties of local water reservoirs. In order to assess the suitability of the analytical solution (4) to predict the change of SO_4^{2-} ion concentration in a real installation, the numerical scheme (7) was modified to take into account the possibility of the daily changes of the $q_{v,dw}$ and $q_{v,ct}$ parameters:

$$p_{i+1} = x_i + \Delta t \cdot \left(-\frac{q_{v,dw}(t_i)}{V} \cdot x_i + \frac{(q_{v,dw}(t_i) + q_{v,ct}(t_i))}{V} x_{fw} \right)$$
$$x_{i+1} = x_i + \frac{\Delta t}{2} \cdot \left(-\frac{(q_{v,dw}(t_i) \cdot x_i + q_{v,dw}(t_{i+1}) \cdot p_{i+1})}{V} + \ldots \right. \quad (9)$$
$$\left. \ldots + \frac{(q_{v,dw}(t_i) + q_{v,dw}(t_{i+1}) + q_{v,ct}(t_i) + q_{v,ct}(t_{i+1}))}{V} x_{fw} \right)$$

An example of a numerical solution including the fluctuation of $q_{v,dw}$ and $q_{v,ct}$ at a level of ±15% is presented in Figure 5. The calculations were done for the parameters used in case A and compared with the analytical solution (4) for mean values of $q_{v,ct} = 45600$ m^3/d and $q_{v,dw} = 6000$ m^3/d.

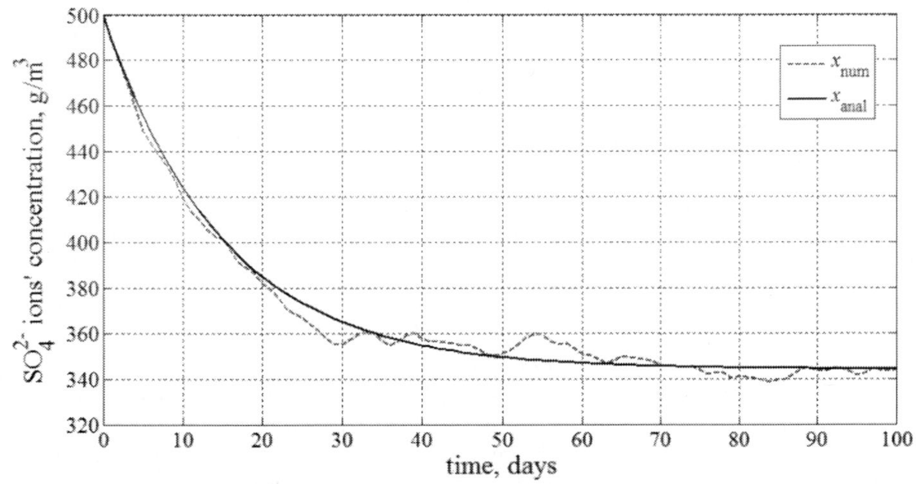

Fig. 5. Comparison of analytical solution x_{anal} (black solid curve) and numerical results x_{num} (red dashed curve) obtained from formula (9) for $x_0 = 500$ g/m^3, $q_{v,dw} = 6000$ m^3/d, $q_{v,ct} = 45600$ m^3/d and $x_{fw} = 40$ g/m^3.

The relative error for the abovementioned example is presented in Figure 6. The presented data shows that the maximum value of E_{rel} is approximately 3.3%, which indicates very high agreement of the analytical solution with the numerical data even if fluctuations of the volumetric flow rates $\{q_{v,dw}, q_{v,ct}\}$ are taken into account.

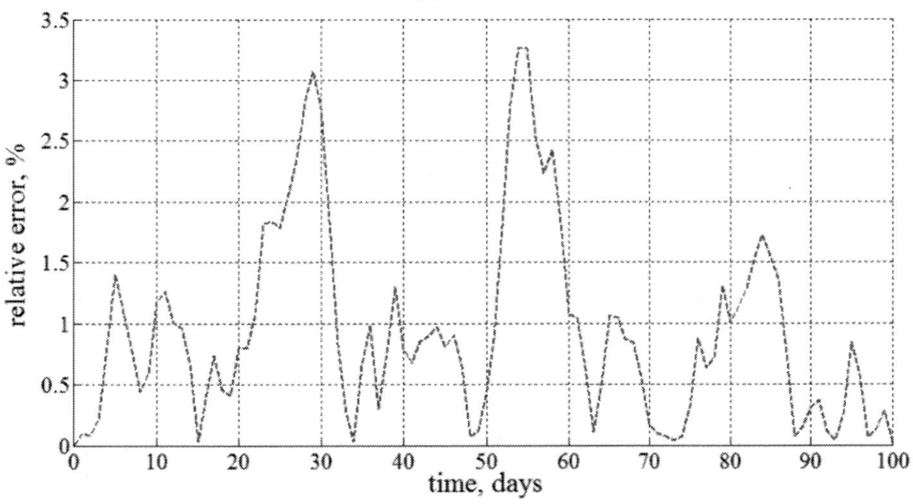

Fig. 6. Relative error calculated from formula (8) for solutions presented in Figure 5.

4 Closing remarks

The mathematical model created by the Authors is the first in literature which describes the changes in the sulphate ion concentration in a closed cooling system. The goal of the model lies in the fact that its analytical solution (4) enables estimating the minimum requirement of the wastewater volumetric flow rate $q_{v,dw}$ in order to satisfy the current legal regulations of x_{limit} = 500 g/m^3. Example values of minimal wastewater volumetric flow rate, parametrized by SO_4^{2-} ion concentration in fresh water x_{fw} and the volumetric flow rate of vapour from cooling towers $q_{v,ct}$, are presented in Table 2.

Table 2. Minimal values of volumetric flow rates $q_{v,dw}$ (in m^3/d) required to satisfy legal regulations imposed on sulphate ion concentration x in wastewater from a power plant x_{limit} = 500 g/m^3.

SO_4^{2-} ion concentration x_{fw}, [g/m^3] / volumetric flow rate $q_{v,ct}$, [m^3/d]	20	30	40
35000	1459	2235	3044
40000	2554	2554	3479
45600	1900	2911	3966

This knowledge could help operators and decision makers to introduce planned and cost-effective wastewater management in power plants, reducing the consumption of fresh water as well as the amount of industrial waste discharged into the environment. In order to automate the wastewater management process, it is possible to implement the mathematical model into a computer program that could be included in the power plant's monitoring system.

References

1. J. Stańda, *Woda do kotłów parowych i obiegów chłodzących siłowni cieplnych* (WNT, Warszawa, 1999)
2. S. Fleischli, Blog: Message to EPA: Time to Modernize America's Power Plants - Cooling Systems Included. https://www.nrdc.org/experts/steve-fleischli/message-epa-time-modernize-americas-power-plants-cooling-systems-included (Last cited 4 January 2016)
3. W. Hermanowicz *et al.*, *Fizyczno – chemiczne badanie wody i ścieków* (Arkady, Warszawa, 2010)
4. Council Directive 91/271 / EEC of 21 May 1991. Urban Waste Water Treatment (Acts. Office. EC L 135, 05.30.1991, p. 40, as amended. d.; Acts. Office. Polish special edition, ch. 15, v. 2, p. 26), the Directive of the European Parliament and of the Council 2010/75/EU of 24 November 2010. on industrial emissions (integrated pollution prevention and control) (recast) (OJ. office . EC L 334, 17.12.2010, p. 17, as amended. d)
5. I. Skoczko, P. Ofman, E. Szatyłowicz, R. Och. Środ. **18** (2016)
6. F. Bellmann, W. Erfurt, H-M. Ludwig, Cement & Concrete Comp. **34** 1 (2012)
7. J. Condor, K. Asghari, D. Unatrakarn, **DOI**: 10.1016/j.egypro.2011.02.506 (2011)
8. L. D. Berman, *Evaporative cooling of circulating water* (Pergamon Press, New York, 1961)
9. R. Silva, L. Luciana, J. Rubio, Minerals Eng. **23,** 15 (2010)
10. J. H. Mathews, K. D. Fink, *Numerical methods using Matlab* (Prentice Hall, New York, 2004)

An evaluation of the possibility of water chemical quality restoration at the user by filtration through an activated silica bed

Katarzyna Rucka[1,*], and *Małgorzata* Wolska[1]

[1]Wrocław University of Science and Technology, Faculty of Environmental Engineering, Wybrzeże S. Wyspiańskiego 27, 50-370 Wrocław, Poland

Abstract. The usefulness of activated silica-filled CACF filters from EURO FILTR'EAUX for restoring chemical quality of water delivered to the end user, and therefore limiting scale formation on the surface of devices, and consequently their damage was studied. The study was performed for three different water types, which were characterized by differing hardness and carbo-acidic aggressiveness levels, which allowed an evaluation of the usefulness of the given filtration bed for treating different water types. The study encompassed changes in water chemical stability levels as well as determining the bed useful life. The effect of water-activated silica bed contact on water chemical stability was studied. It was found that filtrates were characterized by greater chemical stability levels than the supplied water, which decreased the risk of scale formation of water installations and decreased water aggressiveness. A decrease in free chloride and dissolved oxygen concentrations was observed, which resulted in a decrease in aggressive water action on water installations.

1 Introduction

1.1 Water quality for human consumption

The requirements for water destined for human consumption in Poland are defined by the Ordinance of the Minister of Health of 13 November 2015 [1] and in generally do not differ from WHO guidelines [2]. Therefore exceeding the limits defined in this ordinance may cause health hazards for consumer. From the moment water leaves the treatment plant until it reaches the consumer, many factors have an effect on its quality, including chemical and biological stability of the transported water [3]. During transport, recontamination occurs, which, among others, causes a deterioration in water taste and smell [4]. The development of biofilms in water pipes and the resultant microorganism diffusion into water results in increased disinfectant consumption, which may yield a significant deterioration in water quality in the water system and the release of corrosion products. On the over hand the disinfectant is necessary to limit of microorganism develop in water. Consequently, the user receives aggressive water, and more rarely with a tendency to deposit calcium carbonate. A lack of water stability delivered to the user and its changed sensory qualities are frequent complaints from customers concerning water quality.

* Corresponding author: katarzyna.rucka@pwr.edu.pl

Ensuring chemical water stability limits calcium carbonate precipitation, which increases installation and equipment lifetimes, including heating elements in home appliances.

Due to this, home-based water treatment systems are becoming more common, as they may limit the adverse effects of secondary contamination and ensure chemical and biological stability. The object of this study was an evaluation of the CACF filter with an activated silica bed.

1.2 Water chemical stability

Due to the large amount of factors affecting water chemical stability, it is very difficult to unambiguously and precisely determine the effect of water on materials it contacts. Water chemical stability is described by saturation, indicator of the aggressiveness (I), Langelier (IL) or Ryzner (IR) indices, which characterize the calcium-carbonate equilibrium [5]. Values of water indicators which define water as corrosive is presented in the table 1.

Table 1. Water corrosiveness indicator values.

Indicator	Value
IL	< 1
IR	> 6.8
I	> 1

Besides the calcium-carbonate equilibrium, the water aggressiveness and tendency to release precipitates inside water pipes and installation is decided by many factors. Low pH and a high water alkalinity may cause an increase in corrosion. The presence of dissolved oxygen and free chlorine in water accelerate corrosion reactions. Dissolved substances cause an increase in conductivity and increase the rate of corrosion [6].

1.3 CACF Filters

Many water consumers use home based water treatment systems. From a customer's viewpoint, high chlorine concentrations are important, as they influence the taste of water, along with water hardness, which causes scale deposits on water installations and household appliances. Filters, due to simple installation and the possibility of eliminating chemical substances and bacteria in one device, are most commonly used in households. Different types of filters are in use, making use of reverse osmosis or activated silica beds [7].

Due to its sorbent properties, silica has been used for water treatment for many years. Activated silica is commonly used, and its material properties depend on the activation method and the activating factors [8]. Water-silica contact results in oxygen accumulation around metaloorganic complex, which determines the disinfectant properties of silica [9]. The adsorbing properties of silica are used for softening water, since its surface absorbs calcium and magnesium ions [10]. Activated silica is used in filters mounted close to the water intake, especially for limit water aggressiveness and scale forming potential. For this study a CACF filter was used, produced and commonly used in France. The aim of this study was determining and confirming the usefulness of the used filter for treatment of water of varying hardness.

2. Methods and course of study

The study was conducted for three internal installations, supplied with water of varying quality and properties. Two of these installations were supplied with surface water supplied to two different cities, while the third installation was supplied with infiltration water. From each installation, two samples were taken: supply and filtrate. The evaluation of water quality included parameters concerning calcium-carbonate equilibrium and oxidant concentrations, which affect the water aggressiveness level. A physio-chemical analysis of water concluded pH measurement, specific conductivity, general alkalinity, general hardness, calcium, magnesium, iron, free chlorine, diffused oxygen and carbon dioxide concentrations as well as water turbidity. Based on these measurements, aggressive carbon dioxide content was determined, along with water aggressiveness and corrosivity indicators.

Table 2 presents a list of Polish Standards used in the analysis of water quality parameters.

Table 2. List of Polish Standards used in the analysis of water quality parameters.

Indicator	**Unit**	**Method of determination**	**Standard**
pH		Potentiometric	ISO 5969
Alkalinity	val/m^3	Titration	PN-EN ISO 9963-1:2001/Ap1:2004
CO_{2F}	gCO_2/m^3	Metotitration	PN-C-04547-01:1974 PN-C-04547-03:1974
Turbidity	NTU	Nephelometric	PN-EN ISO 7027:2003
Dissolved Oxygen	g O_2/m^3	Potentiometric	PN-EN 25814:1999
Free chlorine	g Cl/m^3	DPD	PN-EN 937:2009
Specific conductivity	µS/cm	Direct measurement	PN-EN 27888:1999
Hardness	g Pt/m^3	Titration with EDTA	PN-ISO 6059:1999
Calcium	g Ca/m^3	Titration with EDTA	PN-ISO 6058:1999
Magnesium	g Mg/m^3	Titration with EDTA	PN-C-04554-04:1999 Ap.A
Iron	g Fe/m^3	FAAS	PN-ISO 8288:2002

An evaluation of the effects of water-filtration bed contact were performed based on the differences in the calculated indicators and carbon dioxide content in the filtrate. The calcium carbonate (scale) precipitation potential was determined based on changes in the values of pH, turbidity, water hardness and carbon dioxide concentration.

For the installation supplied with infiltration water, the bed useful life was also determined, taking the end of bed capacity as the moment when filtrate quality parameters deteriorated. The flow characteristic determination was performed based on the differences in pressure measurements before and after the filter and the amount of water that flowed through the filter. This was possible due to a flow meter installed on the installation and manometers before and after the filter.

The study was conducted for one and a half months, with water samples taken once a week. For the installation supplied with infiltration water, the study was conducted up to the moment of end of bed capacity.

3. Results

All the water supplies had an average hardness, though the largest values were exhibited by surface water from water supply 1 (Table 3).

In the majority of samples from all three water supplies, aggressive carbon dioxide was present, which testified to aggressive character of water. Surface water from supply 1 was characterized by the highest concentration of this gas and the greatest water alkalinity.

All three water sources fulfilled the requirements for water for human consumption, apart from turbidity (Fig.1), whose exceedances were probably connected with the presence of deposits inside water pipes and the flushing of these deposits during non-uniform water usage periods. Such a dependency was also found by Barbot et al. [11].

In all the water supply samples, dissolved oxygen was present, and in water from sources 2 and 3, free chlorine was also present, increasing water aggressiveness.

The greatest iron content was found from infiltration water from supply 3. In general the water being supplied to each installation differed in aggressiveness levels and insoluble contaminant content, which decided water turbidity levels. A lack of water chemical stability is testified to by the values of the water aggressiveness indicators (Table 3) and the presence of oxidants and aggressive carbon dioxide. Therefore, it was reasonable to use these water supplies for studies of changes in water chemical stability levels due to flow through activated silica filtration beds.

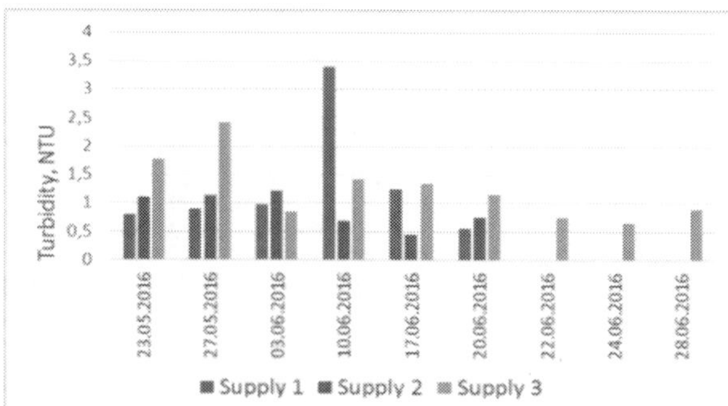

Fig. 1. Water supply turbidity.

Due to water-filtration bed contact an increase in pH values was found in all filtered water samples from all three installations. The increase in pH values was accompanied by an increase in alkalinity for water from supply 1, which was in the range of 0.05–0.5 val/m^3, and it increased with filter bed usage time and at the final period of the study was equal to 0.1 val/m^3. However, no change or an increase in this parameter was found for water from the two other installations. The difference in changes in this parameter probably results from different aggressiveness of the supply waters.

It is significant from a chemical stability standpoint that no change in water hardness was found in any of the filtrates, which leads to the hypothesis the increase in water stability results from a change in contaminant form and not content. This hypothesis is confirmed by the increase in water specific conductivity in all filtrate samples, which differed depending on water type (Fig. 2).

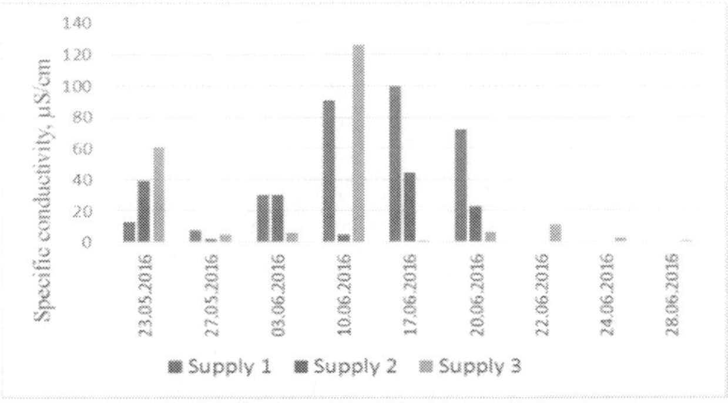

Fig. 2. Increased in water specific conductivity caused by contact with activated silica bed.

Furthermore, an increase in water chemical stability is testified to by the decrease in aggressive carbon dioxide that was found and amounted to 134 g CO_{2agrr}/m^3, 3.5 g CO_{2agrr}/m^3 and 5.8 g CO_{2agrr}/m^3 respectively for water supplies 1, 2 and 3. Consequently, the concentration of this gas found in filtrates was much smaller, and in the majority of samples no presence of it was detected.

Contact with a activated silica also ensured a decrease in chlorine content in all water samples, however its concentration in filtrates from supplies 2 and 3 was in the range of 0.07-0.43 g Cl_2/m^3 and 0.22-0.36 g Cl_2/m^3 respectively. A decrease in chlorine concentration was accompanied by decreases in oxygen content (fig.3). Probably the oxidants were absorbed by active silica which is good adsorbent.

A consequence of the decrease in aggressive carbon dioxide and oxidant content was a decrease in water aggressiveness and a restoration of calcium-carbonate equilibrium, which is testified by the calculated water corrosiveness indicators. In all filtrate samples, lower values of IR and I indices were found, along with IL changes indicating a decrease in water aggressiveness levels.

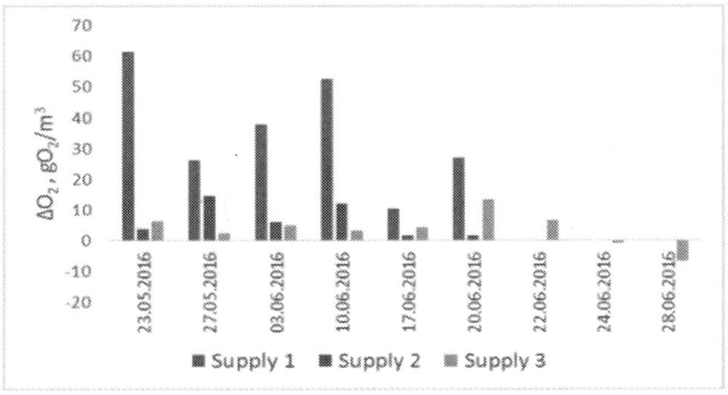

Fig. 3. Effectiveness of dissolved oxygen removal during water filtration.

Apart from a decrease in water aggressiveness, a decrease in non-soluble contaminant content was observed, measured as turbidity. The effectiveness in decreasing this parameter was in the range of: 2.2-80.9%, 19.7-58.1% and 8.2–64.3% respectively for supplies

DOI: 10.1051/e3sconf/20171700079

1, 2 and 3. Among the suspended contaminants that were removed were iron compounds, whose concentrations were effectively lowered in all waters.

For water supply 3, the filter useful life was determined. Studies were conducted until the end of filtration bed capacity. The end of bed capacity caused a lack of increase in specific conductivity and a smaller increase in pH and an increase in calcium ion content, which testified to flushing of contaminants from the bed. An increase in dissolved oxygen was also observed along with a lack of change in free chlorine concentrations. The end of bed capacity occurred after obtaining 147 m^3 of filtrate. In order to confirm the end of bed capacity, tests of increase in flow resistance at three filtrate flow volumes through the bed were conducted (Fig.4).

Based on this analysis, it may be stated that despite a large volume of filtrate that had flowed through bed, the increase in resistance and therefore the decrease in supply valve efficiency, was insignificant. It may be concluded that the limiting factor in determining the end of bed useful life is not water flow hydraulics, but a changes in filtrate quality.

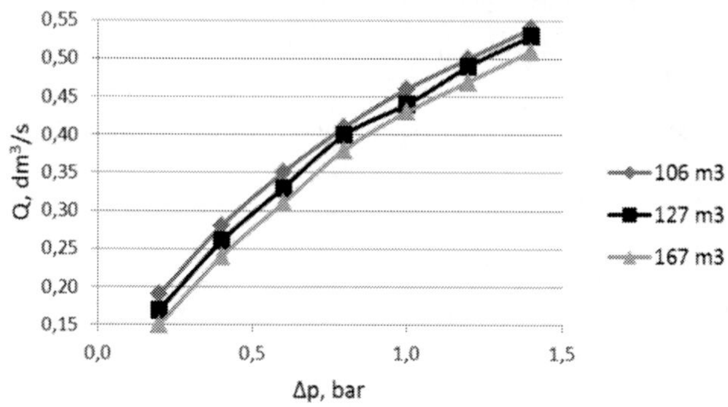

Fig. 4. Effect of volume of water filtered through the bed on changes in supply valve efficiency.

Table 3. Ranges of values of analyzed water quality indicators.

Indicator	Unit	Supply 1						Supply 2						Supply 3					
		Before			After			Before			After			Before			After		
		min	max	avg	min	max	avg	min	max	avg	min	max	avg	min	max	avg	min	max	avg
pH	-	6.23	7.68	-	7.08	7.68	-	7.63	8.06	-	7.76	8.08	-	7.21	8.08	-	7.33	8.09	-
Alkalinity	val/m^3	4.4	4.9	4.7	4.5	5.0	4.8	2.5	3.1	2.9	2.4	3.1	2.9	3.4	3.7	3.5	3.0	3.6	3.4
CO_{2F}	g/m^3	0.0	187	43.4	9.8	52	21.5	4.4	8.8	6.6	3.5	6.6	4.7	4.2	25.0	8.1	3.2	15.0	6.3
CO_{2agrr}	g/m^3	0.0	152.0	27.7	0.0	18.0	3.83	0.0	3.5	0.67	0.0	0.0	0.0	0.0	13.0	1.67	0.0	0.0	0.0
Turbidity	NTU	0.56	3.40	1.31	0.53	0.88	0.69	0.45	1.21	0.89	0.28	0.92	0.61	0.65	2.42	1.25	0.65	1.29	0.93
Dissolved Oxygen	g O_2/m^3	4.19	12.05	6.90	2.67	5.60	4.10	5.87	10.80	7.46	5.52	10.40	6.99	6.36	9.32	7.22	6.22	8.86	6.94
Free chlorine	g Cl/m^3	0.00	0.00	0.00	0.00	0.00	0.00	0.18	0.53	0.33	0.11	0.25	0.19	0.38	1.42	0.81	0.13	1.06	0.56
Specific conductivity	µS/cm	482	667	611	582	693	663	415	597	511	438	636	534	449	644	606	575	650	627
Hardness	°tw	20.0	21.5	20.4	19.8	21.5	20.3	9.6	16.0	12.6	9.5	16.0	12.5	14.0	15.0	14.6	14.0	15.0	14.63
Calcium	g Ca/m^3	100.0	104.0	102.0	98.6	104.0	101.1	56.4	90.7	73.43	55.7	90.7	70.6	78.5	85.7	82.2	80.7	85.7	83.2
Magnesium	g Mg/m^3	25.7	30.0	26.5	25.7	30.0	26.5	7.3	14.7	10.23	7.3	14.7	10.2	1.7	22.1	13.0	11.8	22.1	14.27
Iron	g Fe/m^3	0.00	0.03	0.014	0.00	0.002	0.005	0.00	0.04	0.01	0.00	0.01	0.002	0.02	0.31	0.11	0.00	0.19	0.07
IL	-	-1.07	0.51	0.11	-0.02	0.43	0.23	0.06	0.37	0.22	0.14	0.48	0.30	-0.37	0.58	0.33	-0.27	0.59	0.29
IR	-	6.59	8.37	7.07	6.78	7.19	7.07	7.23	7.74	7.44	7.74	7.74	7.37	6.72	7.95	7.16	6.91	7.87	7.27
I	-	0.00	68.2	12.8	0.00	4.63	0.89	0.00	1.55	0.27	0.00	0.00	0.00	0.00	4.45	0.54	0.00	2.34	0.26

4 Conclusions

An evaluation of water quality before and after filtration allowed for posing the following conclusions

1. The quality of all filtrate water samples fulfilled the requirements for water for human consumption, with the exception of turbidity.
2. Filtrates were characterized with a greater chemical stability level than the supply water, which decreased the potential for scale formation in water pipes and household appliances, and limited the water aggressiveness level.
3. A decrease in dissolved oxygen concentration and free chlorine contributed to a lowering of the aggressive effects on water pipes.
4. Lowering of free chlorine concentration causes an improvement in the taste and olfactory qualities of water.
5. Contact with the filtration bed caused a decrease in turbidity and iron concentration, which allows for fulfilling the requirements for water for human consumption.
6. Studies have confirmed the manufacturer's data concerning the volume of water that can flow before bed capacity is reached. The end of bed capacity is decided by worsening filtrate quality and not increasing flow resistance
7. The filtrate quality worsens gradually after the end of bed capacity, and not suddenly as is the case if there is intense contaminant flushing from the filtration bed.

References

1. Ordinance of the Minister of Health of 13 November 2015, Dz.U.2015.1989 date 2015.11.27
2. World Health Organization. Guidelines for drinking-water quality [electronic resource]: incorporating first addendum. Vol. 1, Recommendations – 3rd edition 2006.
3. M. Świderska-Bróż, Ochr. Sr. 25, 4 (2003)
4. M.J. Lethola, M. Laxander, I.T. Miettinen, A. Hirvonen, T. Vartiainen, P.J. Martikainen, Water Res. **40**, 1 (2006)
5. M. Esmaeili-Vardanjani, I. Rasa, V. Amiri, M. Yazdi, K. Pazand, Environ. Monit. Assess. 187, 53 (2015)
6. N.C. Burtch, H. Jasuja, K.S. Walton, Chem. Rev. **114**, 20 (2014)
7. http://www.woda.com.pl
8. Surface Active Silica Sols Effect of PEG-Silica Interactions, Department of Chemical and Biological Engineering, *Division of Applied Surface Chemistry,* Chalmers University of Technology Goteborg, Sweden (2011)
9. M. Laskowska, Ł. Laskowski, K. Fijałkowski, M. Kacprzak, Inżynieria Materiałowa **35**, 6 (2014) (in Polish)
10. A. Szkatula, M. Balanda, M. Kopeć, The European Physical Journal of Applied Physics **18**, 1 (2002)
11. E. Barbot, N.S. Vidic, K.B. Gregory, R.D. Vidic, Environ. Sci. Tech. **47**, 6 (2013)

Methods of assessment of stormwater sediments quality

Aleksandra Sałata[1,*], and *Lidia* Dąbek[1]

[1]Kielce University of Technology, Department of Environmental, Geomatics and Energy Engineering, aleja Tysiąclecia Państwa Polskiego 7, 25-314 Kielce, Poland

Abstract. Concentration of heavy metal (cadmium, copper, chromium, nickel, lead and zinc) in sediments collected from the stormwater treatment plant located in the urbanised catchment were investigated using geo-accumulation index and enrichment factor to determine metal accumulation and pollution status. Total metal concentrations varied widely in studied materials and the mean values were higher than their background values. The I_{geo} results indicate that tested sediments were uncontaminated with respect to Cd. The study area is moderately to strongly contaminated with Zn, Pb and Cu. The other elements are within the scope moderate contamination.

1 Introduction

The sediments found in urban areas are composed of a diversity materials that range from mineral compounds and biogenic materials to organic and inorganic materials of human origin [1]. Atmospheric particles deposited on impermeable surfaces of cities are the main sources of urban sediments. This type of sediment may take various routes, including atmospheric, soil, or bodies of water, depending on some specific conditions both natural and anthropogenic processes thus, it carries associated pollutants along with it, such as metals, especially zinc (Zn), lead (Pb), copper (Cu), nickel (Ni), cadmium (Cd), and chromium (Cr), which are frequently used in industrial activities and discharged without any control into the rainwater or sewage system [2, 3].

Metals are a group of pollutants of high ecological significance. They are not removed from water by self-purification, but accumulate in suspended particulates and sediment [4–6]. Sources of metals in aquatic sediments are natural or anthropogenic sources. Main natural sources are weathering of soils and rocks and atmospheric deposition. Discharging agricultural, municipal, residential or industrial waste products into water bodies are anthropogenic sources [7]. Moreover, the long-term environmental monitoring requires an evaluation of the degree of contamination in urban areas with particular emphasis upon understanding the environmental processes controlling sediment distribution in the assessment of trace elemental studies [8–9].

* Corresponding author: asalata@tu.kielce.pl

2 Materials and methods

2.1 Study site

Analyzed catchment, with a total area of 82 hectares, from the east and the northern part is limited by the open ditch running along the city boundaries and receiving rainwater flowing down from the dense forest complex. The sedimentation tank, which collected sediments for testing, is a open horizontal settling chamber of stormwater treatment plan named OWD Witosa. This property was built in the seventies. Currently, after reconstruction realized in 2009, it consists of separation chamber, open horizontal settling chamber length of 50 m, coalescence separator of ϕ 3.0 m, water pumping station and drainage system. Technical parameters of the settler shown in Table 1. Wastewater is delivered to the separation chamber with storm overflow by the ferroconcrete collector ϕ 1400 mm from the northern part of the city of Kielce. Sewage system consists of a main collector with a diameter ϕ 1200 and 1400 mm (L = 762 m) and lateral channels (ϕ 300–800 mm) with a total length of approx. 7.0 km. It receives wastewater from estate area, in which is located 400 single and multi-family residential. At present residential buildings covers an area of 36.5 ha, which represents approx. 44.5% of dewatered area, the 13.6 ha of these terrains are sealed surfaces (roads, parking lots, squares, roofs). The remaining part of the catchment are green areas – 45.5 ha (55.5%), of which the majority are meadows and wasteland (31.5 ha), forests – 12.2 ha and 1.8 ha are orchards. Treated wastewater is discharged into the Silnica river via channel having the diameter ϕ 1400 mm at km 11 + 580 [10].

Table 1. Technical parameters of sedimentation tank OWD Witosa.

Parameter	Value	Unit
Bottom surface	1268.0	m^2
Water surface area at normal water mirror	1387.0	m^2
Total depth of settling chamber	3.45	m
Depth at the maximum water mirror	3.00	m
Depth at normal water mirror	1.50	m
Active capacity of settling chamber	2080	m^3

2.2 Sampling

The sediment samples were collected during two months of spring season (May – June) 2014, in accordance with the PN-EN ISO 5667-15:2009 standard method [11]. Within sedimentation tank, sediments were collected from two sites in this open reservoir, namely inlets and outlets. Three sampling points in inlet and another three in outlet – 18 sediment samples. Samples were collected using a stainless steel standard bottom grab sampler, or Eijkelkamp equipment for sites with difficult access, and placed in acid-washed plastic containers (~1L). Between the series of sample collecting, all equipment was rinsed with water from the reservoir. Then, the samples were transported to the laboratory and homogenized using a solvent-cleansed stainless steel bucket and a spoon. The physical and chemical parameters were measured immediately afterwards.

2.3 Determination of heavy metal

The sediment samples for heavy metal determination were oven-dried at 105°C on glass dishes until the constant weight, homogenized with a pestle and mortar, and each of the weighed samples (approximately 0,2 g) were transferred into Teflon vessels, digested with 7 ml HNO_3 in a microwave oven (Multiwave 3000, Anton Paar). The digestates were left to cool at room temperature and then filtered through nitrocellulose membrane filter. The filtered digestates were diluted with distilled water to 100 ml in a volumetric flask. Total metal concentration of lead, chromium, cadmium, copper, nickel and zinc was determined using by atomic emission spectrometry with inductively coupled plasma ICP Optima 8000 (Perkin Elmer) with certified multi element standards [12]. All the reagents were of high quality and analytical grade. All solutions were prepared using double distilled water. All plastic, quartz and glassware were soaked in 10% nitric acid for at least 24 h hours and rinsed with ultra pure water. Total heavy metal concentrations were expressed in mg per kg of dry mass of sediments.

2.4 Geoaccumulation index (I_{geo})

Geoaccumulation index (I_{geo}) was developed by Műller [13] and had widely been used in trace elements studies of sediments [3,7,14]. To quantify the degree of heavy metal pollution Igeo was calculated using Eq. (1):

$$I_{geo} = \log_2\left(\frac{C_n}{1.5B_n}\right) \qquad (1)$$

where: C_n is the measured concentration of the metal n in the urban sediment and B_n is the geochemical background value of given metal in the shale [15,16] and the 1.5 factor was introduced for the purpose of assessing the natural fluctuations in the content of a given substance in the environment with minimum anthropogenic influence. I_{geo} classes are given in table 2.

2.5 Enrichment factor (*EF*)

Normalized enrichment factor is applied to differentiate metal source originating from anthropogenic and natural means. This involves normalization of the sediment with respect to reference elements such as Al, Fe, Mn, Ti, Sc, Li and Cs [14]. Normalized EF of metals in urban sediments from OWD Witosa of each site was calculated using Eq. (2).

$$EF = \left(\frac{C_n}{C_{Fe}}\right)_{sample} \Big/ \left(\frac{C_n}{C_{Fe}}\right)_{background} \qquad (2)$$

where (C_n/C_{Fe})sample is the ratio of concentration of the element of concern to that of Fe in the sediment sample (mg/kg dry weight) and (C_n/C_{Fe})background is the same ratio in an unpolluted reference material [17]. According to Chen et al. [18] the EF values were categorized into seven classes where EF < 1 indicates no enrichment, EF < 3 is minor enrichment, EF = 3–5 is moderate enrichment, EF = 5–10 is moderately severe enrichment, EF = 10–25 is severe enrichment, EF = 25–50 is very severe enrichment, and EF > 50 is extremely severe enrichment (Table 2). Enrichment factor values between 0.5 and 1.5 indicate the metal is entirely from natural processes, whereas values greater than 1.5 suggest that the sources are more likely to by anthropogenic [19].

Table 2. Enrichment factor (EF) and I_{geo} classes in relation to sediment quality.

EF	Sediment quality	I_{geo}	I_{geo} classes	Sediment quality
< 1	No enrichment	≤ 0	0	Unpolluted
< 3	Minor enrichment	0 – 1	1	Unpolluted to moderately polluted
3 – 5	Moderate enrichment	1 – 2	2	Moderately polluted
5 – 10	Moderately severe enrichment	2 – 3	3	Moderately to highly polluted
10 – 25	Severe enrichment	3 – 4	4	Highly polluted
25 – 50	Very severe enrichment	4 – 5	5	Highly to very highly polluted
> 50	Extremely severe enrichment	5 – 6	6	Very highly polluted

3 Results and discussion

3.1 Heavy metal concentrations

Total metal concentrations varied widely in studied materials. The minimum, maximum, average, standard deviations (SD) and background concentrations for each element are presented in Table 3. Mean values were higher than their background values, suggesting that these metals in stormwater sediments from OWD Witosa were influenced by anthropogenic sources, except cadmium that was no detected in all sediment samples. The results showed that zinc (Zn), copper (Cu) and lead (Pb) with the highest concentrations (983.43±103.09 mg/kg d.w., 69.39±12.46 mg/kg d.w., 124.87±5,87 mg/kg d.w.) and these values exceeds the values of background several times. Mean concentration of Cr is 36.31 mg/kg d.w. and of Ni 23.75 mg/kg d.w., respectively. Values exceeding the standards presented in the next decreasing order Zn>Cu>Pb>Cr>Ni>Cd. It was also observed differences in the metal concentrations on the inlet and outlet of the sedimentation tank. For each element the concentration values were less than a few to several percentage points at the inlet than at the outlet of the reservoir. Elevated concentrations of zinc and lead may result from many processes that occur in the aquatic environment; to a large extent, they depend on sediment composition, thus on the character of the catchment.

Table 3. Heavy metal concentrations in sediments compared with local background values.

Study site	Parameter	Concentration, mg/kg dry weight					
		Cd	Cu	Cr	Zn	Pb	Ni
Inlet	Average	n.d.	60.07	27.97	894.55	119.38	19.70
	Min	n.d.	51.27	25.00	887.38	117.63	18.57
	Max	n.d.	70.46	31.25	902.46	121.04	21.10
	SD	n.d.	7.14	2.40	6.64	1.43	0.80
Outlet	Average	n.d.	78.71	44.64	1072.31	130.36	27.79
	Min	n.d.	66.99	42.37	998.55	127.37	26.07
	Max	n.d.	90.37	47.03	1196.37	132.69	30.00
	SD	n.d.	9.15	1.51	69.02	1.84	1.67
Total	Average	n.d.	69.39	36.31	983.43	124.87	23.75
	Min	n.d.	51.27	25.00	887.38	117.63	18.57
	Max	n.d.	90.37	47.03	1196.37	132.69	30.00
	SD	n.d.	12.46	8.79	103.09	5.87	4.36
Background		0.5	7	6	73	15	5

 DOI: 10.1051/e3sconf/20171700080

3.2 Index of geoaccumulation I_{geo}. Pollution status

The geoaccumulation index includes seven grades ranging from 0 to 6 (unpolluted to very strongly polluted). The geoaccumulation factor results indicate that tested sediments were uncontaminated with respect to Cd (cadmium was not detected in all samples), therefore not even placed on bloxplot diagram (Fig. 1). The Zn accumulation level is 3.16 which indicates to strong contamination. Tested sediments are moderately to strongly contaminated with Pb (2.47) and Cu (2.70). The other elements are within the scope moderate contamination ($1<Igeo<2$). The high contents of lead, zinc and copper are probably a result of anthropogenic activity in watershed area. Many of these terrains are sealed surfaces (roads, parking lots, squares, roofs) and more than half of the area (55.5%) are the green areas where can be used plant protection products as a chemical fertilizers and pesticides - a potential source of heavy metal contamination.

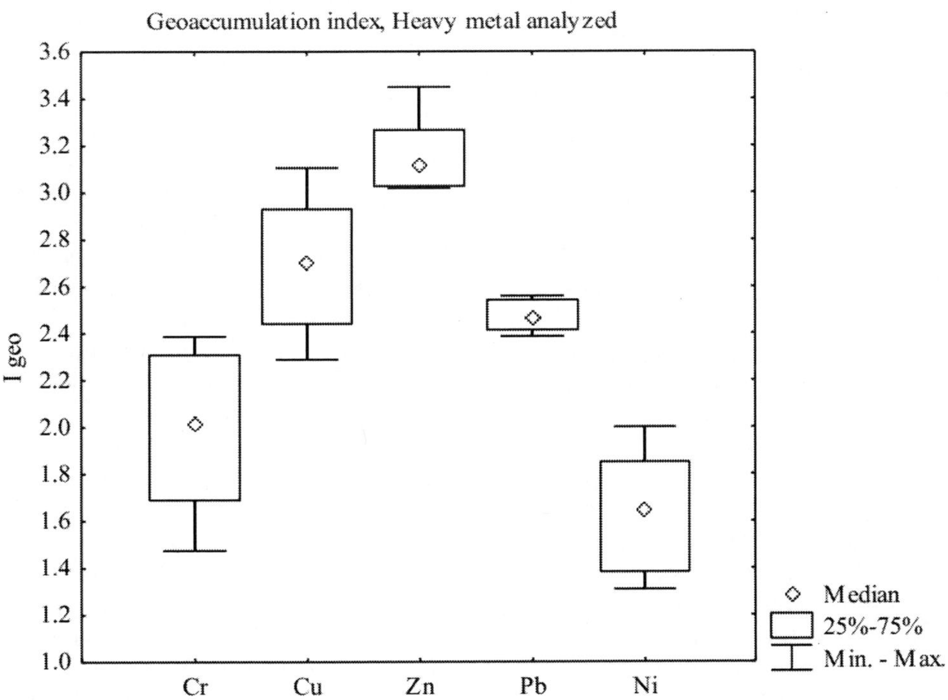

Fig. 1. Boxplot diagram of the geoaccumulation index (I_{geo}) values.

3.3 Enrichment factor (EF). Pollution status

In present study, the enrichment factor (EF) was used to assess the level of contamination and the possible anthropogenic impact in the sediments collected from urbanised stormwater treatment plant. To identify anomalous trace metal concentrations, the normalization of element is required. Several authors used the iron (Fe) to normalize contaminants [17,20] therefore in the present study concentration of Fe has been used as a normalization standard. Table 4 shows enrichment factor (EF) values for Cd, Cu, Cr, Zn, Pb and Ni. The values of EF suggest that all trace elements except cadmium originate from anthropogenic inputs and tested sediments are very enriched with them. The EF values for Zn are the highest among the metals and it has severe enrichment (EF from 13.54 to 23.76). Slightly lower values but still within the scope of this class have been obtained for Cu and

Pb (13.08±3.82 and 10.88±1.91, respectively). Results for Ni and Cr indicate to moderately severe enrichment (6.35±1.91 and 8.21±2.85, respectively). The enrichment of heavy metals in this study has been observed to be relatively high. The differences in individual EF values may result in the nature of heavy metals, their pollution loads, as well as speciation forms of trace elements occurrence in sediment-water complex. Therefore so high content of heavy metals due to ability to sorption on fine particle which are forming deposits.

Table 4. Enrichment factor (EF) values of selected heavy metals in stormwater sediments.

Parameter	Enrichment factor (EF)					
	Cd	Cu	Cr	Zn	Pb	Ni
Average	<1	13.08	8.21	17.75	10.88	6.35
Min	<1	8.38	4.56	13.54	8.62	4.09
Max	<1	18.68	11.28	23.76	12.98	8.81
SD	<1	3.82	2.85	4.01	1.91	1.91

4 Conclusion

The present study is a assessment of the stormwater sediment contamination by geochemical indicators – geoaccumulation index Igeo and the enrichment factor (EF) for six selected heavy metals (cadmium, copper, chromium, zinc, lead and nickel). Results showed that the heavy metal concentration are very high and it has been reported a significant excess of a geochemical shale. The results indicate that zinc (Zn), copper (Cu) and lead (Pb) had the highest means concentrations exceeds the value of background several times. Cadmium was no detected in all sediment samples. It was also observed differences in the metal concentrations on the inlet and outlet of the sedimentation tank. For each element the concentration values were less than a few to several percentage points at the inlet than at the outlet of the reservoir.The Igeo results indicate that tested sediments were uncontaminated with respect to Cd. The Zn accumulation level is 3.16 which indicates to strong contamination. The study area is moderately to strongly contaminated with Pb (2.47) and Cu (2.70). The other elements are within the scope moderate contamination (1<Igeo<2). The high contents of lead, zinc and copper are probably a result of anthropogenic activity in watershed area. Many of these terrains are sealed surfaces (roads, parking lots, squares, roofs) and more than half of the area (55.5%) are the green areas where can be used plant protection products as a chemical fertilizers and pesticides – a potential source of heavy metal contamination. Additionally, the enrichment of heavy metals in this study has been observed to be relatively high – from moderately severe enrichment (Cr, Ni) to severe enrichment (Zn, Pb, Cu). The differences in individual EF values may result in the nature of heavy metals, their pollution loads, as well as speciation forms of trace elements occurrence in sediment-water complex. These findings indicate that more attention should be paid to metal contamination in urban sediments and its regular monitoring. Particularly, with the use of geochemical tools.

The work reported herein was undertaken as part of the statutory project No. 05.0.10.00/2.01.01.01.0026 (MNSP.IKTW.15.001).

References

1. X. Yuan, L. Zhang, L. Li, C.Wang, J. Ji. Sediment properties and metal pollution assessment in the river, eustary and lake environments of fluvial plain, China. *Catena,* **119**, 52–60 (2014)

2. S. Charlesworth, M. Everett, R. Mcarthy,A. Ordoñez, A. DeMiguel. A comparative study of heavy metal concentration and distribution in deposited street dusts in a large and a small urban area: Birmingham and Coventry West Midlands, UK. *Environment International*, **29**, 563–57 (2003)
3. L.L. García Martínez, C. Poleto. Assessment of diffuse pollution associated with metals in urban sediments using the geoaccumulation index (Igeo), *Journal of Soils and Sediments* **14**,1251–1257 (2014)
4. K. Karlsson, M. Viklander, L. Scholes, M. Revitt. Heavy metal concentrations and toxicity in water and sediment from stormwater ponds and sedimentation tanks. *Journal of Hazardous Materials*, **178**, 612–618 (2010)
5. S. Caeiro, M.H. Costa, T.B. Ramos, F. Fernandes, N. Silveira, Coimbra A., Medeiros G., Painho M. Assessing heavy metal contamination in Sado Eustary sediment: An index analysis approach. *Ecological Indicators*, **5**, 151 – 169 (2005)
6. Ł. Bąk., J. Górski, K. Górska, B. Szeląg. Suspended solids and heavy metals content of selected rainwater waves in an urban catchment area: A case study, *Environmental Protection*, **34(2)**, 49-52 (2012)
7. F. Cevik, M.Z. Lugal Göksu, O.B. Derici, O. Findik. An assessment of metal pollution in surface sediments of Seyhan dam by using enrichment factor, geoaccumulation index and statistical analyses, *Environmental Monitoring and Assessment*, **152**, 309–317 (2009)
8. A. Kumar, Singhal R.K., Rout S., Ravi P.M. Spatial geochemical variation of major and trace elements in the marine sediments of Mumbai Harbor Bay, *Environmental Earth Sciences*, **70**, 3057–3066 (2013)
9. B. Gałka. Assessment of the degree of contamination bottom sediments of small water objects at the park in Wroclaw – Pawlowice, *Environment and Natural Resources Protection*, **42**, 233–239 (2010)
10. Discharge Consent for Rainwater and Thaw-Water Treated at the Witosa Treatment Facility, Kielce (2008)
11. PN-EN ISO 5667–15:2009. Water Quality. Sampling. Part 15: Guidance on the Preservation and Handling of Sludge and Sediment Samples (2009)
12. PN-EN ISO 11885:2009. Water Quality. Determination of selected elements using optical emission spectrometry with inductively coupled plasma (ICP-OES, 2009)
13. G. Müller. Schwermetalle in den Sedimenten des Rheins—Veränderungen seit. Umschau, **24**, 778–783 (1971)
14. A. Zahra, M.Z. Hashmi, Malik R.N., Z. Ahmed. Enrichment and geo-accumulation of heavy metals and risk assessment of sediments of the Kurang Nallah-Feeding tributary of the Rawal Lake Reservoir, Pakistan. *Science of the Total Environment*, **470-471**, 925–933 (2014)
15. K.K. Turekian, K.H. Wedepohl. Distribution of the elements in some major units of the earth's crust. *Geological Society of America Bulletin*,72(2),175–192 (1961)
16. I. Bojakowska, G. Sokołowska. Geochemical classess of water sediments. *Polish Geological Review*, **46(1)**, 49–54 (1998)
17. H.A. Ghrefat, Y. Abu-Rukah, Rosen M.A. Application of geoaccumulation index and enrichment factor for assessing metal contamination in the sediments of Kafrain Dam, Jordan. *Environmental Monitoring and Assessment*, **178**, 95–109 (2011)

18. C.W. Chen, C.M. Kao, Chen C.F., Dong C.D. Distribution and accumulation of heavy metals in the sediments of Kaohsiung Harbor, Taiwan. *Chemosphere*, **66(8)**, 1431–1440 (2007)
19. J. Zhang, C.L. Liu. Riverine composition and estuarine geochemistry of particulate metals in China: Weathering features, anthropogenic impact and chemical fluxes. *Estuarine Coastal and Shelf Science*, **54**, 1051–1070 (2002)
20. E. Esen, F. Kucuksezgin, E. Uluturhan. Assessment of trace metal pollution in surface sediments of Nemrut Bay, Aegean Sea. *Environmental Monitoring and Assessment*, **160**, 257–266 (2010)

 DOI: 10.1051/e3sconf/20171700081

The evaluation of the classical Kirby-Bauer antibiogram method for the determination of antibiotic resistant microorganisms in tap water

Agata Siedlecka[1,*], and *Katarzyna* Piekarska[1]

[1]Wroclaw University of Science and Technology, Faculty of Environmental Engineering, Wroclaw, Poland

Abstract. Water supply systems are the main source of potable water for many inhabitants. The European Union and Polish laws demand high quality of tap water, but the aspect of antibiotic resistant bacteria (ARB) is not mentioned. The special emphasis is placed on the presence of pathogenic microorganisms in water, determining by the presence of chosen indicators. Although these regulations provide quite good quality of drinking water, the neglect of resistance vectors could lead to a serious threat to human health. In this study the randomly chosen sample of water from the Wroclaw water supply system was tested in order to evaluate the usefulness of Kirby-Bauer antibiogram method for the assessment of the presence of ARB. The susceptibility testing with 6 chosen antibiotics was performed on bacterial strains isolated from the sample. The diameters of growth inhibition zones were measured and the preliminary identification of strains was carried out, including Gram staining and bacterial cellular morphologies assessments. Although full recognition of the drug resistance phenomenon demands the identification of bacterial species for the comparison with susceptibility testing guidelines, this preliminary research could answer the question, whether the Kirby-Bauer method is sufficient for investigations of ARB dwelling in Wroclaw tap water or the other solution should be applied.

1 Introduction

The safety of potable water is one of the most important factors affecting human health. It is obvious, that the intake of poorly treated drinking water could lead to the outbreak of an epidemic caused by pathogenic and opportunistic microorganisms dwelling in contaminated water, as it was in Peru, 1991, as the result of negligence the disinfection of drinking water in water supply systems [1]. According to the Polish law, the quality of tap water in terms of microbiological contamination is assessed by regarding the presence of the chosen indicators: coliform bacteria, thermotolerant coliform bacteria, *Escherichia coli* species, *Streptococcus faecalis* species and *Clostridium spp.* (in extended research), as the presence of these bacteria in water samples could be easily identified and they fulfil

* Corresponding author: agata.siedlecka@pwr.edu.pl

DOI: 10.1051/e3sconf/20171700081

the conditions of faecal contamination indicators, *i.e.* they are the typical human intestinal pathogens, do not form the spores (allowing to detect the recent contamination), their lifespan is longer than the lifespan of pathogens, *etc.* [2]. Though these regulations provide the suitable quality of treated tap water, the antibiotic resistance factor is completely neglected. There is no guideline to identify antibiotic resistant bacteria (ARB) or antibiotic resistance genes (ARGs) in drinking water and there is no research applied regarding this matter. However, the literature review indicates, that the aspect of antibiotic resistance of bacteria dwelling in tap water is present all over the world, as, for example, tetracycline and chloramphenicol resistant *E. coli* revealed in a public rural water supply system in France [3], and could be perceived as the emerging threat, since the presence of ARB or ARGs leads to the further prevalence of antibiotic resistance in the environment and, if consumed with potable water, in human organisms [4, 5]. From this point of view, resistant species dwelling in tap water (as well as free ARGs), even if harmless, could be considered as potential resistance vectors, since they could transfer the resistance features to pathogenic strains, hereby contributing to the appearance of pathogenic, resistant strains.

The presence of ARB in tap water may be due to water treatment processes, especially biological activated carbon, ozonation and chlorine disinfection units [6], as in these processes bacteria could acquire new features because of their genomic flexibility [7]. Users of drinking water contaminated with ARB could be carriers of resistant microorganisms, as proven in Canada with resistant *E. coli* strains [8]. From antibiotic resistance point of view, the safety of drinking water is not provided even if resources of water are groundwater, especially in karst terrain area [3].

The aim of this study is to evaluate the classical, Kirby-Bauer antibiogram method as a tool for the determination of ARB dwelling in tap water. At the same time, the authors made the attempt to preliminary assess whether there is an antibiotic resistance phenomenon among the bacteria isolated from the Wroclaw tap water sample, although with the presented data only rough assumption may be implemented regarding this matter.

2 Materials and methods

2.1 Sample collection and isolation of microorganisms

In order to isolate bacterial strains dwelling in tap water, 1 L of the water sample was collected to sterile, 1 L blue-cap bottle. The sampling point was located in the laboratory at the Faculty of Environmental Engineering at Wroclaw University of Science and Technology. The sampling was carried out according to the procedure described in Kańska *et. al.* [9]: the tap was chemically disinfected and the sample of water was collected after 10 minutes of not disturbed water flow to eliminate the risk of the intake of bacteria dwelling near the entrance of the tap. Immediately afterwards, 1 L of drinking water was filtrated through the sterile mixed cellulose ester membrane filters, pore size 0.2 μm, diameter 50 mm (Whatman™) in the sterile filter unit (Nalgene®). The filters were placed on R2A agar plates (BTL) and incubated at 22°C and 37°C for 5 days. Several randomly chosen colonies of each filter were subjected to the further testing and isolated on R2A agar plates [10]. After 4 days of incubation at the optimal temperatures (*i.e.* 22°C and 37°C) pure bacterial strains were transferred to R2A agar slants and further incubated. All the isolated strains were stained according to the procedure described as Gram staining [10] with the purpose of the preliminary differentiation.

2.2 Antibiotic susceptibility testing

The isolated strains were subjected to antibiotic susceptibility testing. In this approach, the classical Kirby-Bauer antibiogram method, as well as modified antibiogram methods were applied, as strains dwelling in tap water are known to have different features (*i.a.* nutrient and temperature requirements) than pathogenic strains, for which the standard antibiogram was elaborated [11].

6 widely known and frequently applied antibiotics of different antibiotic classes [12] were chosen, as presented in the table 1. Beyond tetracycline, all these antibiotics were applied in ARB research in surface and drinking water distribution systems in South Africa [13].

Table 1. Antibiotics selected for the disk diffusion method.

Antibiotic	Class of antibiotics	Assay [mcg/disc]	Symbol
tetracycline	tetracyclines	32.9	TE 30
kanamycin	aminoglycosides	31.6	K 30
erythromycin	macrolides	16.1	E 15
amoxicillin	aminopenicillins	11.7	AX 10
chloramphenicol	chloramphenicols	32.8	C 30
trimethoprim	synthetic chemotherapeutic	5.9	TMP 5

According to Kirby-Bauer antibiogram procedures [11], density of bacterial strains solutions was adjusted to 0.5 McFarland standard (bioMerieux) by suspending the inoculum in sterile saline (0,9 % NaCl). Subsequently, Mueller-Hinton agar (BioMaxima) plates were swabbed and within 15 minutes the sterile cotton antimicrobial susceptibility test discs (BioMaxima) were placed onto plates. Each strain was subjected the incubation at recommended 35°C and its optimal temperature, *i.e.* 22°C or 37°C, for the comparison of their growth under the different temperature conditions on Mueller-Hinton agar. The incubation period was 5 days.

3 Results

3.1 Characterization of microorganisms

The membrane filter incubated at 37°C overgrew with white, yellow and orange colonies of different sizes (varying from < 0.5 mm to 5 mm), while the membrane filter incubated at 22°C overgrew with white and yellow colonies of different sizes (varying from < 0.5 mm to 3 mm). Except the differences in pigmentation, size and height, the colonies shape (punctiform / round), surface (smooth), texture (butyrous) and edge (entire) were similar. The results of the cultivation of microorganisms concentrated on membrane filters are shown in the Fig. 1.

Fig. 1. Membrane filters incubated at 37°C (left) and 22°C (right) after 5 days of incubation.

As the results of the isolation procedure [10], 26 pure strains were obtained, including 13 mesophilic (37°C, no 1–13) and 13 psychrophilic (22°C, no 14–26) strains. Most of the strains turned out to be gram-negative coccobacilli, but in some cases also gram-positive cocci were revealed, indicating the pleomorphism phenomena. Moreover, gram-positive bacilli were observed. Nevertheless, further investigations must be implemented in order to identify the obtained strains.

2.2 Antibiotic susceptibility testing

The diameters of zones of inhibited bacterial growth were measured for each strain incubated on Mueller-Hinton agar at 35 °C, as well as for the strains incubated at their optimal temperatures (*i.e.* 37°C for the strains numbered as 1÷13 and 22°C for the strains numbered as 14–26). Not every strain was exposed to every drug and few of the strains were incubated without any antibiotic disc, in order to observe its non-disturbed growth on Mueller-Hinton agar. The table 2 presents the values of diameters [mm] obtained after incubation at 35°C (T_{35}) and 37°C or 22°C (T_{opt}) around the applied antibiotic disks, marked as their symbols (look at the Table 1). In cases where there was no inhibition, '0' was measured, neglecting the diameter of the cotton disk.

Table 2. The diameters of inhibited growth. '– ' means that there was not any growth on the plate, so the diameters could not be read. '0' means that there was no inhibition of growth (despite the presence of the disk). 'CL' means 'control' – the strain cultivated without any antibiotic disk. 'N/D' means 'no data' – the given antibiotic was not applied.

Strain no	TE 30		K 30		E 15		AX 10		C 30		TMP 5	
	T_{35}	T_{opt}	T_{35}	T_{opt}	T_{35}	T_{opt}	T_{35}	T_{opt}	T_{35}	T_{opt}	T_{35}	T_{opt}
1	-	-	-	-	-	-	-	-	-	-	-	-
2	-	-	-	-	-	-	-	-	-	-	-	-
3	-	-	-	-	-	-	-	-	-	-	-	-
4	-	-	-	-	-	-	-	-	-	-	-	-
5	-	-	-	-	-	-	-	-	-	-	-	-
6	-	-	-	-	-	-	-	-	-	-	-	-
7	40	36	N/D	28	39	36	20	20	26	14	38	34
8	-	-	-	-	-	-	-	-	-	-	-	-
9	-	-	-	-	-	-	-	-	-	-	-	-
10	-	-	-	-	-	-	-	-	-	-	-	-
11	CL	CL	CL	CL	CL	CL	CL	CL	CL	CL	CL	CL
12	55	60	N/D	N/D	N/D	N/D	26	N/D	40	N/D	0	0
13	CL	CL	CL	CL	CL	CL	CL	CL	CL	CL	CL	CL
14	CL	CL	CL	CL	CL	CL	CL	CL	CL	CL	CL	CL
15	-	60	-	N/D	-	23	-	0	-	36	24	0
16	-	-	-	-	-	-	-	-	-	-	-	-
17	-	56	-	26	-	19	-	0	-	36	-	16
18	45	48	N/D	36	N/D	28	20	22	44	44	0	14
19	46	40	48	34	36	24	20	21	50	44	0	12
20	58	60	N/D	N/D	22	18	N/D	N/D	N/D	N/D	N/D	N/D
21	44	34	42	34	30	30	15	16	45	40	0	0
22	-	44	-	50	-	20	-	0	-	36	-	27
23	-	58	-	44	-	N/D	-	12	-	34	-	34
24	-	-	-	50	-	N/D	-	30	-	-	-	-
25	50	44	N/D	42	38	N/D	16	24	50	44	0	16
26	-	50	-	N/D	-	41	-	20	-	-	-	-

3 Discussion and conclusions

According to antibiotic susceptibility testing guidelines, drug resistance could be determined only when the species of tested microorganisms are established [14]. Seemingly, without these data, only the rough assumption might be implemented. However, these elaborated databases are the perfect tool mainly for antimicrobial susceptibility assessments of clinically relevant strains, as pathogenic strains, and it is likely, that species isolated from tap water samples are not included in these guidelines, as apparently harmless ones. This shortage must be overcome by assumptions based on the given diameters of zones of inhibited bacterial growth for known, pathogenic species, for preliminary determination of the potential drug resistance. Thereby, the cut-offs of diameters were accepted as following:

- resistant, when the diameter is < 10 mm
- intermediate, when the diameter is 10÷20 mm
- susceptible, when the diameter is > 20 mm.

Although these assumptions might be regarded as too lenient and contribute to the omitting of truly resistant species, the results show that even under these circumstances many strains could be considered resistant. Indeed, the cut-off values differ for various antibiotics, as well as for various species.

According to the assumption presented above, only amoxycillin and trimethoprim could be regarded as potential non-effective drugs and resistance to erythromycin, amoxycillin, chloramphenicol (just in one case) and trimethoprim could be regarded as intermediate. Tetracycline and kanamycin appear to be efficient against the obtained strains. For the mesophilic strains (no 1–13), the diameters measured in 35°C and 37°C did not differ significantly, as the difference of temperature conditions was irrelevant. However, for the psychrophilic strains (no 14–26), these differences tend to be essential in many cases. Some strains appear to be susceptible, when incubated at 35°C but resistant, when incubated at 22°C. This lead to the conclusion, that classical Kirby-Bauer antibiogram method is not the appropriate tool for psychrophilic microorganisms.

Most of the selected antibiotics used in this study are known to have a broad-spectrum against gram-positive and gram-negative strains (except erythromycin, which is effective mainly against gram-positive strains, although in this study gram-negative coccobacilli revealed to be susceptible to erythromycin). To confirm the obtained results, further investigations must be applied, including the classification of the strains to species.

This preliminary research is the evidence, that the classical Kirby-Bauer antibiogram method is the insufficient tool for the investigations of ARB dwelling in tap water. Water supply systems are the specific environment for microorganisms, because of the following features [15]:

- concentration of nutrients (*i.a.* organic carbon), which is strongly limited
- temperature conditions, which are not suitable for many bacterial species
- occasionally inconvenient hydraulic conditions (*i.a.* water stagnation)
- lack of light, lack of many competitive species

These features make that water supply systems could be regarded as the niche, where only the minor part of whole plethora of microorganisms are able to survive, but still, the biodiversity of biofilm and planktonic forms of microorganisms occurring in pipes is wide, as it is shown in the numerous studies [15–17]. It is worth to mention that the guidelines for susceptibility testing are still subjected to changes and improvements, as new data are taken into account. This leads to many critical evaluations of CLSI (NCCSL) and EUCAST handbooks [18–21], but to date, as indicated in this study, there is a lack of knowledge and outlines for many species.

The main disadvantages, that make the Kirby-Bauer antibiorgam method inappropriate for bacteria of water-pipelines origin are the following:

- tap water microorganisms generally grow slowly, so the cultivation period need to be extended. It is almost impossible to use the fresh, 18-hour culture for testing, as recommended [11]
- adjusting the inoculum to 0.5 McFarland standard should provide the even growth on Mueller-Hinton agar plates – however the results showed, that many isolated strains created separated colonies
- the recommended height of Mueller-Hinton agar on a 100-mm plate (petri dish) is 4 mm [11], but during the incubation at 35°C more than 3-days agar of that height may dry up
- ingredients of Mueller-Hinton agar make it rich in nutrients medium, and because of this, bacteria cultivated previously on poorer R2A agar could be unable to adapt to new conditions
- tap water microorganisms transferred from their natural environment to the laboratory and cultivated on agar medium could have changed their biochemical and physiological features, lost their characteristics and even DNA sequences (plasmid loss), what could make the attempts to identify the strains unsuccessful, hereby precluding the eventual susceptibility testing confirmation. From this point of view, the authors suggest, that in respect of the investigations of antibiotic resistance phenomena in tap water, research focused on DNA (*i.e.* ARGs), rather than ARB cultivation, should be applied
- it is obvious, that plenty of microorganisms of each environment cannot be cultivated on agar medium in laboratory conditions. There is the assumption, that only several percentage of bacterial biodiversity are able to be obtained and the main part remains elusive and could be tested only by metagenomics tools [22]. These considerations are important also for the anaerobic forms of bacteria and strains requiring extremely incubation conditions. Instead of the Kirby-Bauer antibiogram method, ATB panels (Antibiotic Susceptibility Testing Strips, bioMérieux) could be applied [16], nevertheless this tool also requires cultivable, oxidant microbial colony to be tested. Other studies revealed that in water supply systems the alterations in the composition of the bacterial community were higher among cultivable communities than 16S rRNA sequences [17], indicating that some part of biomass could be lost during the classical isolation procedure. That could lead to the omission of many ARB.

In view of the fact that the isolated strains were selected randomly from all colonies obtained from the tap water sample, it is possible, that majority of them belongs to the same genus or even species, what in turn could explain the similar antibiotic resistance profiles of several investigated strains. Therefore, the classification of these strains is needed and should be implemented in further research. In fact, it is worth to mention that the obtained strains could be naturally resistant, *i.e.* biochemically invulnerable to selected antibiotics applied in this study.

There are many studies which contribute to broadening the knowledge of ARB in tap water, but to date, the majority of them are focused on pathogenic or opportunistic isolates. Therefore, it is crucial to underline, that resistant tap water strains, even if harmless, could act simply as the resistance vectors (*e.g* ARGs) and from this point of view could be regarded as a potential threat and should be monitored.

The conference participation was financed from the statutory funds of Faculty of Environmental Engineering, Wroclaw University of Science and Technology from the project 0401/0066/16.

References

1. https://www.cdc.gov/Mmwr/preview/mmwrhtml/00001912.htm [24/01/2017]
2. A. Grabińska-Łoniewska, E. Siński, *Mikroorganizmy chorobotwórcze i potencjalnie chorobotwórcze w ekosystemach wodnych i sieciach wodociągowych* (Seidel-Przywecki, Warszawa, 2010)
3. E. Laroche, F. Petit, M. Fournier, B. Pawlak, J. Hydrol. **392**, 12–21 (2010)
4. V. Figueira, E.A. Serra, I. Vaz-Moreira, T.R.S. Brandâo, C.M. Manaia, J. Water Health **10**, 1 (2012)
5. S.G. Mulamattathil, C. Bezuidenhout, M. Mbewe, C.N. Ateba, J. Pathogens **2014**, ID 371208 (2014)
6. X. Bai, X. Ma, F. Xu, J. Li, H. Zhang, X. Xiao, Sci. Total Environ. **533**, 24–31 (2015)
7. C. Narciso-da-Rocha, I. Vaz-Moreira, C.M. Manaia, Sci. Total Environ. **466–467**, 127–135 (2014)
8. B.L. Coleman, M.I. Salvadori, A.J. McGeer, K.A. Sibley, N.F. Neumann, S.J. Bondy, I.A. Gutmanis, S.A. McEwen, M. Lavoie, D. Strong, I. Johnson, F.B. Jamieson, M. Louie, ARO Water Study Group, Epidemiol. Infect. **140**, 633–642 (2012)
9. Z. Kańska, A. Grabińska-Łoniewska, M. Łebkowska, E. Rzechowska, *Ćwiczenia laboratoryjne z biologii sanitarnej, część I* (Oficyna Wydawnicza Politechniki Warszawskiej, Warszawa, 2006)
10. E. Kocwowa, *Ćwiczenia z mikrobiologii ogólnej* (Państwowe Wydawnictwo Naukowe, Warszawa, 1984)
11. R. Schwalbe, L. Steele-Moore, A.C. Goodwin (ed.), *Antimicrobial Susceptibility Testing Protocols* (CRC Press, Boca Raton, 2007)
12. Z. Markiewicz, Z.A. Kwiatkowski, *Bakterie antybiotyki lekooporność* (Wydawnictwo Naukowe PWN, Warszawa, 2001)
13. S.G. Mulamattathil, C. Bezuidenhout, M. Mbewe, S. Afr. J. Sci. **110**, 11/12 (2014)
14. *CLSI. Performance Standards for Antimicrobial Susceptibility Testing*, (Clinical and Laboratory Standards Institute, Wayne, 2017)
15. T. Traczewska, M. Sitarska, I. Biedroń, *Ekologiczne i techniczne aspekty powstawania biofilmu w wodzie* (Oficyna Wydawnicza Politechniki Wrocławskiej, Wrocław, 2014)
16. I. Vaz-Moreira, O.C. Nunes, C.M. Manaia, Appl. Environ. Microbiol. **77**, 16 (2011)
17. I. Vaz-Moreira, C. Egas, O.C. Nunes, C.M. Manaia, FEMS Microbiol. Ecol. **83**, 361–374 (2013)
18. M. Hombach, P. Courvalin, E.C. Böttgera, Antimicrob. Agents Ch. **58**, 7 (2014)
19. J.A. Kiehlbauch, G.E. Hannett, M. Salfinger, W. Archinal, C. Monserrat, C. Carlyn, J. Clin. Microbiol. **38**, 9 (2000)
20. A. Rodloff, T. Bauer, S. Ewig, P. Kujath, E. Müller, Dtsch. Arztebl. Int. **105**, 39 (2008)
21. L.N. Mutanda, A.M. Omari, I.A. Wamola, East Afr. Med. J. **66**, 7 (1989)
22. T.A. Brown, *Genomy* (Wydawnictwo Naukowe PWN, Warszawa, 2012)

E3S Web of Conferences **17**, 00082 (2017) DOI: 10.1051/e3sconf/20171700082
EKO-DOK 2017

Impact of damper opening time on work of storage matrix regenerative heat exchanger

Maciej Skrzycki[1,*], and *Maciej* Besler[1]

[1]Department of Air Conditioning, Heating, Natural Gas Engineering and Air Protection, Faculty of Environmental Engineering, Wroclaw University of Technology, Wyb. Wyspianskiego 27, 50-370 Wroclaw, Poland

Abstract. Devices that by heat exchange take heat from exhaust air stream and transfer it to the external air stream in order to pre heat are called heat recovery heat exchangers. The heat exchange may take place through the wall (diaphragm) or using heat storage liquid (these devices are called recuperators) or by direct contact of exhaust and external air stream with the heat transfer surface - these devices are called regenerators. The publication presents the work characteristics of the storage matrix heat exchanger based on laboratory tests. In order to achieve the objectives the research station was designed and built.

1 Introduction

Currently in Poland described heat exchanger is not very well known and widely used, even though its construction was known and developed long before the rotary heat exchanger. This heat exchanger is an attractive alternative to other heat exchangers, both recuperators and regenerators. The purpose of this article is a closer representation of the storage matrix heat exchanger, its construction, high efficiency, but occasionally reported in the literature.

The storage matrix heat exchangers are often also called accumulative heat exchangers or reversible heat exchangers and operate on the principle of alternating washing out of accumulative storage matrixes within a specified period of time by an exhaust air stream and external air stream [1]. Air distribution through the chambers of storage blocks is arranged by using dampers, mechanically coupled with each other. The following is a flowchart of storage matrix heat exchanger.

The exchanger consists of two independent storage blocks and a dampers system with actuators for changing the direction of air flow. Storage matrix is similar to the matrix used in rotary heat exchangers. The main difference lies in the layered arrangement of plates and the undulating arrangement between the layers of sheets of metal. Storage matrix chambers are separated from each other by a vertical partition. After a predetermined time, the air flow is reversed by dampers. As a result of matrix temperature difference and air stream temperature, outlet air stream temperatures changes. Time of one cycle of air flow through the storage matrix chambers are variable and depend on the manufacturer of approx.

* Corresponding author: maciej.skrzycki@pwr.edu.pl

 DOI: 10.1051/e3sconf/20171700082

20 seconds [2] to even 100 seconds [3]. The graph in Fig. 1 shows an operation of storage matrix heat exchanger [4].

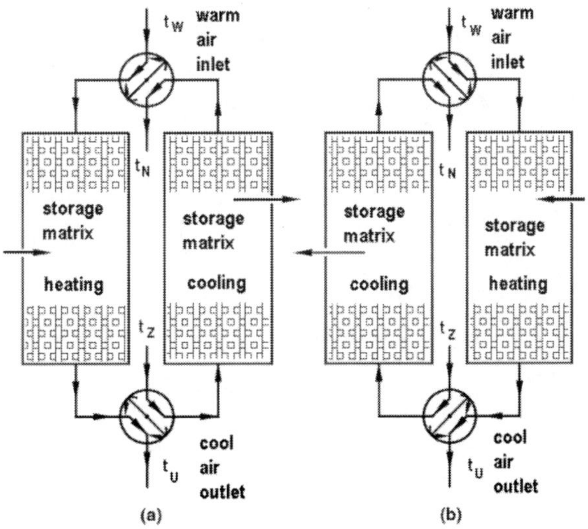

Fig. 1. Principle of operation of storage matrix heat exchanger: a) charging phase, b) discharging phase [4].

Storage blocks chambers have a responsive mechanism to change the direction of air flow during even less than 1 second [2] - it is shown in Fig. 1 and 2.

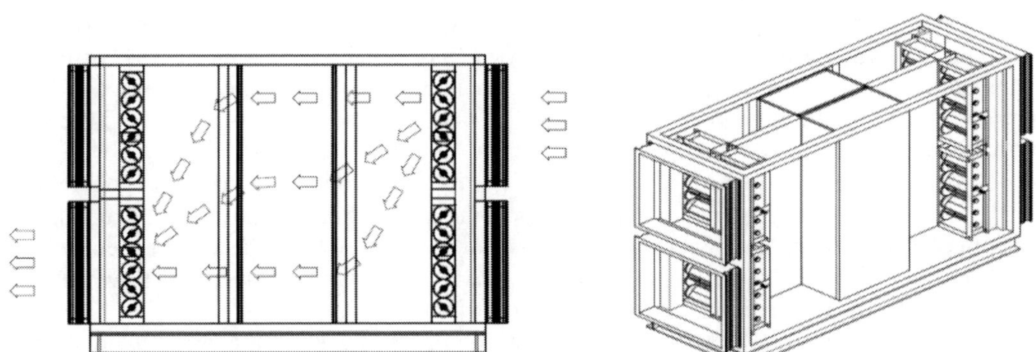

Fig. 2. Airflow through the single heat exchanger chamber in one cycle of work of the device and the three dimensional view of the device (fig. authors).

The average heat flux transferred in a storage matrix heat exchanger can be assumed as follows [5]:

$$Q^{sr} = \frac{Q_{ak}}{\tau^{ok}} = \bar{\kappa} c \rho V \left(t^{GRZ} - t^{CH}\right) \tag{1}$$

Q_{ak} – heat accumulated in the matrix of a regenerator, kW
τ^{ok} – regenerator cycle, s

$$\bar{\kappa} = \frac{\vartheta^{g}_{sr}\left(\tau^{GRZ}\right), \vartheta^{g}_{sr}(o)}{t^{GRZ} - t^{CH}} \frac{1}{\tau^{ok}} \tag{2}$$

t^{GRZ}, t^{CH} – air stream temperatures in the charging and discharging phase, °C
$\upsilon^{g}_{śr}$ – average matrix temperatures at the end and beginning of charging phase, °C
c, ρ, V – specific heat, density, air stream,
κ – heat flux reduced in the regenerator , kW

In analogy to the rotary heat exchangers, also in the storage matrix heat exchangers the cycle is split into two equal phases: heat transfer from the air to the matrix and from the matrix to the air. This process (shown in Fig. 3) can be described as $\tau_0 = \tau_1 + \tau_2$ [5].

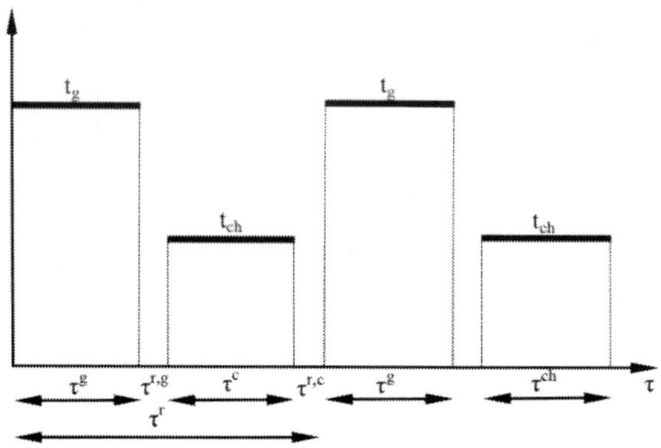

Fig. 3. The temperatures of air streams entering the regenerator in accordance of time: t_g – warm air stream temperature, t_{ch} – cool air stream temperature, τ^r – regenerator work time, τ^g – time of charging phase, τ^{ch} – time of discharging phase, $\tau^{r,g}$ – reversion time after charging phase, $\tau^{r,g}$ – reversion time after discharging phase [5].

The work of storage matrix regenerators is symmetrical – time of flow of warm air stream through a first chamber of heat exchanger is equal to the time of flow of cool air stream through the second chamber of heat exchanger. The sum of these times is the time of cycle of air flow through the heat exchanger [2]. The time needed to change the direction of the flow of air streams is assumed as negligible.

At the time of switching of dampers at the inlet and outlet of heat exchanger the air streams contacts with each other. Accordingly, the use of storage matrix heat exchangers is similar to the rotary heat exchangers and is not recommended if there is an emission of harmful substances, aromatic substances, etc. because of the risk of flow of these substances in the heat exchanger to the external air stream [2].

Storage matrix heat exchangers respective to the flow of the air streams can be compared to the counterflow heat exchangers. In the heat exchanger cross-section the temperature fluctuations are recurrent in each cycle of work of exchanger – in charging and discharging phase [6]. These temperatures decrease along with decreasing of cycle time or the length of the air flow in the heat exchanger, as well as with the change of the heat capacity of the storage matrix.

2 Research station and description of tests

Tests were performed on storage matrix heat exchanger type BA-19-500-0600x1350-V-G1. The main element is storage matrix made of aluminum plates having a thickness of 0,1 mm placed at a distance of 2,0 mm from each other, between which are placed further layers of corrugated sheets of aluminum. The matrix length is 500mm and it is placed in housing with 585mm width and 1350mm height (Fig. 5). In each duct in the inlet and in the outlet are arranged two dampers (total 8 dampers), each other connected with the actuators.

 DOI: 10.1051/e3sconf/20171700082

Fig. 4 shows the appearance of the tested heat exchanger at the research station and Fig. 5 shows the constructional dimensions of the tested heat exchanger.

Fig. 4. Construction of storage matrix heat exchanger (photog. authors).

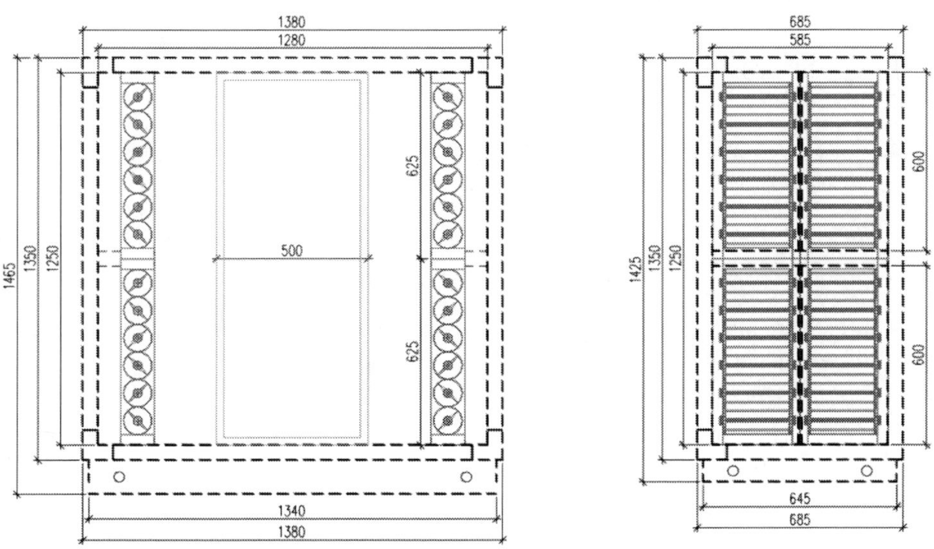

Fig. 5. The constructional dimensions of the tested heat exchanger (fig. authors).

DOI: 10.1051/e3sconf/20171700082

Air flow value was determined at $V=1200m^3/h$. The variable value was the air flow through the matrix chambers of heat exchanger. The initial period of air flow was set at 30 seconds with 30 second step of damper opening/close time. During the measurements exhaust air temperature was taken, as a constant value throughout the measurement period at $t_w=25°C \pm 1,0°C$.

The following charts show the results of measurements of efficiency of the tested heat exchanger (Fig. 6 and Fig. 8), and the temperature of the air streams (Fig. 7 and Fig. 9).

The temperature measurements were based on the guidelines in the standard [7], the air flow measurements were based on the guidelines in the standard [8].

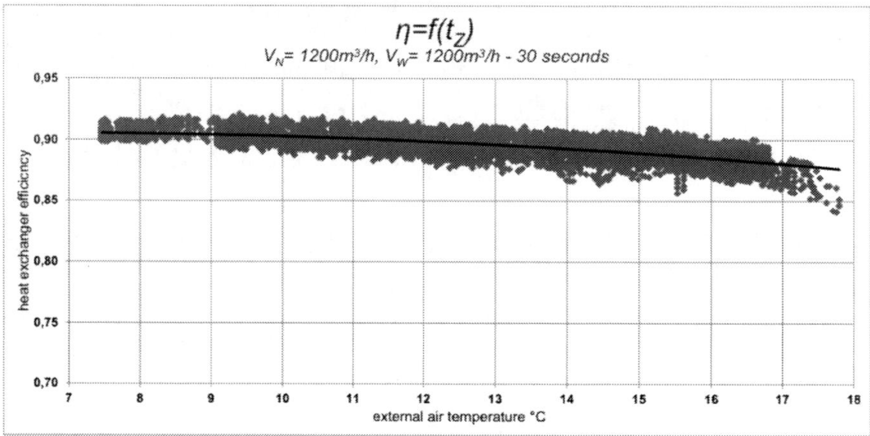

Fig. 6. Efficiency chart of tested heat exchanger in an exemplary measurement period by 30-second airflow through the storage matrix chambers.

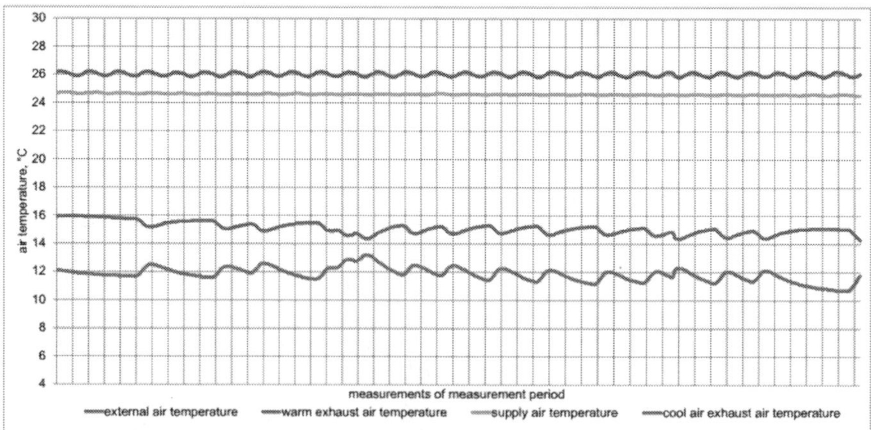

Fig. 7. Chart of variability of air temperatures of tested heat exchanger in an exemplary measurement period by 30-second airflow through the storage matrix chambers.

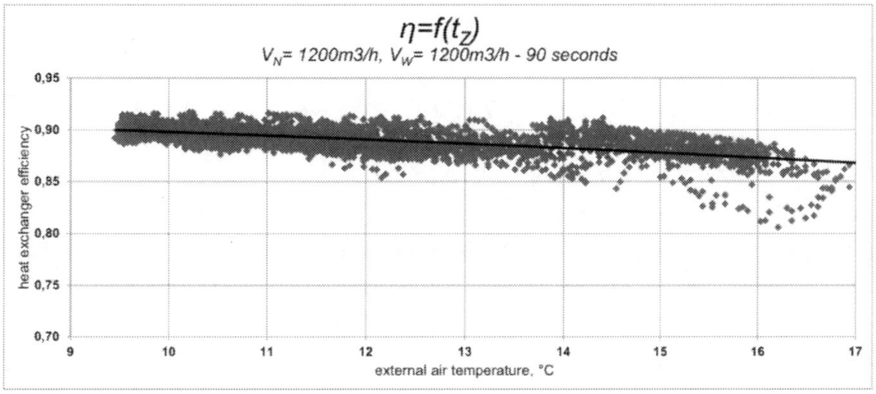

Fig. 8. Efficiency chart of tested heat exchanger in an exemplary measurement period by 90-second airflow through the storage matrix chambers.

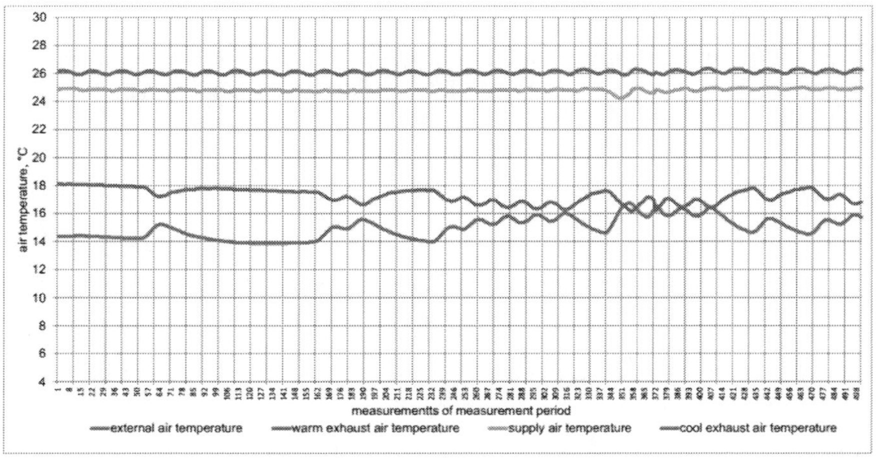

Fig. 9. Chart of variability of air temperatures of tested heat exchanger in an exemplary measurement period by 90-second airflow through the storage matrix chambers.

3 Summary

Regardless to the time between the dampers changeover high coefficient of heat recovery efficiency was achieved. The values obtained the requirements of the current legislations and the measurements confirmed the efficiency of this type of heat exchangers.

As it is shown in the graphs, heat transfer efficiency is higher for low external air temperatures. Longer warm-up time (damper changeover time) of storage matrix by the exhaust air stream will not always result in more energy storage. It is due to the thermal capacity of the storage matrix and heat transfer coefficient which is also dependent on the speed of the stream. For longer periods, the excess of the available energy is not absorbed by the storage matrix. Determination of optimum length of time between the dampers changeover must take into account the air flow velocity and accumulative capacity of the material used.

Considered storage matrix heat exchanger in the measuring period worked with an average efficiency of approx. 90% over the entire external air temperatures.

DOI: 10.1051/e3sconf/20171700082

References

1. T. Kuppan, *Heat exchanger design handbook*, Marcel Dekker Inc., New York-Basel 2000
2. *Fenix air handling unit with Superblok heat exchanger* (in Polish), Web page: http://climaprodukt.com/produkt/fenix/
3. *Resolair air handling unit* (in Polish), Web page: http://www.menerga.pl/produkty,2,centrale-menerga,11,resolair-6264666865,72
4. T. Skiepko, R. K. Shah, Int. J. Heat Mass Tran. 48 (2005) 1608–1632, Web page: http://www.sciencedirect.com/science/article/pii/S0017931004005095
5. E. Kostkowski, *Heat flow*, Silesian University of Technology Publishing House, 2006 (in Polish)
6. E. Kalinowski, *Heat transfer and heat exchangers*, Warsaw University of Technology Printing House, Warszawa 1995 (in Polish)
7. PN–EN 308:2001 *Heat exchangers. Test procedures for establishing the performance of air to air and flue gases heat recovery devices*, Polish Committee for Standardization, Warsaw 2001 (in Polish)
8. PN-ISO 5221 *Air distribution and air diffusion -- Rules to methods of measuring air flow rate in an air handling duct*, Polish Committee for Standardization, Warsaw 1994 (in Polish)

The Content of Heavy Metals (Cu, Zn, Cr, Ni, Pb) in The Soil Near The Arterial Roads in Wroclaw (Poland)

Karolina Sobczyk[1], *Anna* Hołtra[1,*]

[1]Wroclaw University of Science and Technology, Faculty of Environmental Engineering, Wybrzeze Wyspianskiego 27, 50-370 Wroclaw, Poland

Abstract. The concentrations of heavy metals in soils along the motorway bypass of Wroclaw (AOW) and the Eastern Ring Road of Wroclaw (WOW), Poland, have been determined. The soil samples were collected from the levels of 0-25 cm within 2 m from the edge of the road. The mineralizates were prepared in HNO_3, 60%, using the Microwave Digestion System. The content of Cu, Zn, Cr, Ni and Pb in soils were determined using FAAS method. The physicochemical parameters, the conductivity and pH of the soil solutions were measured to evaluate the salinity of the soils and their active and exchangeable acidity. The pollution indexes (W_N) showing the enrichment of soils in metals have been determined. Excess of metal concentrations in soils compared to the geochemical background in uncontaminated soils of Poland has been observed. Permissible concentrations of heavy metals relative to the standard for soils, according to the Polish Ministry of Environment Regulation from September 1st, 2016, have not been exceeded.

1 Introduction

The number of cars on Polish roads, both private and from the transport sector (mainly transit), has been steadily increasing. The large majority of vehicles is powered by petrol (approx. 54%) and diesel fuel (approx. 41%). Pollution of the atmosphere and the soil is often caused by the exploitation of vehicles constructed with out-of-date production technologies. Anthropogenic sources of heavy metals in soils along the roads most commonly include: liquid spills of transported cargo, leakages of fluid operating systems of the car (Pb: liquid in the battery, formerly a component of gasoline), exploitation of wear parts of vehicles (Pb – in the bearing and solder alloys, covers of the electric cables, Pb and Ni in anti-corrosion agents, Cu in the brake system and radiator), abrasion of car tires (Zn), accidents and collisions (Cr in paints and varnishes) and dusts appearing as a result of incomplete combustion of diesel fuels and abrasion of brake pads and the clutch [1–4]. The concentrations of heavy metals in soils of Poland were shown in Table 1.

* Corresponding author: anna.holtra@pwr.edu.pl

Table 1. The content of heavy metals in soils along the Polish roads.

Location	Zn	Pb	Cu	Ni	Cr
	[mg/kg]	[mg/kg]	[mg/kg]	[mg/kg]	[mg/kg]
Soils along the Warsaw roads [5]	329.0 – 352.0	186.0 – 246.0	37.4 – 70.3	-	-
Soil along the arteries of Olsztyn [6]	19.6 – 112.4	19.7 – 196.0	7.2 – 26.9	7.3 – 23.6	5.4 – 21.6
Soils along the road no. 7 Olsztynek – Płońsk [7]	19.7 – 75.6	6.8 – 17.4	3.3 – 11.3	-	-
Dust road in Lublin [8]	202.0	23.0	65.0	27.0	53.0
Drainage ditches along the roads of Roztocze National Park and Przemysl Foothills Landscape Park [9]	40.0 – 256.0	10.2 – 29.0	4.0 – 23.0	4.0 – 37.3	-

Heavy metals accumulate in soils and then in the living organisms. In the organism, there are not decomposed, only deposited in tissues and organs. As a rule, they are toxic, especially if present in excess. They may interfere with biochemical and biological processes, endangering health. Over time, the concentration of heavy metals increases and they begin to react with the enzymes, causing the disease. The activation mechanism of metals is also based on the impact of the stress factor on the organism [10].

2 Materials and methods

The soil samples have been collected in June 2016 from the surface layer of the soil profile (up to 25 cm of depth), approx. 2 m from the edge of the motorway bypass of Wroclaw (AOW, no. 1-5) and the Eastern Ring Road of Wroclaw (WOW, no. 6-10) – map.1.

No.	Location
1	AOW Długołęka (intersection of road no. 98 and Widawska Street)
2	AOW Node of North Wroclaw (exit from the roundabout into road no. 5)
3	AOW Node of Stadium Wroclaw (Kosmonautów Street)
4	AOW Node of Airport Wroclaw (exit from the roundabout into Graniczna Street)
5	AOW Node of South Wroclaw (exit from road no. E40 into road no. A8)
6	WOW Jagodno (exit from the roundabout into road no. 395)
7	WOW Siechnice (exit from the roundabout into Opolska Street)
8	WOW Blizanowice (exit direction Blizanowice)
9	WOW Kamieniec Wrocławski (exit from the roundabout into Wrocławska Street)
10	WOW Dobrzykowice (exit from the roundabout into Strachocińska Street)

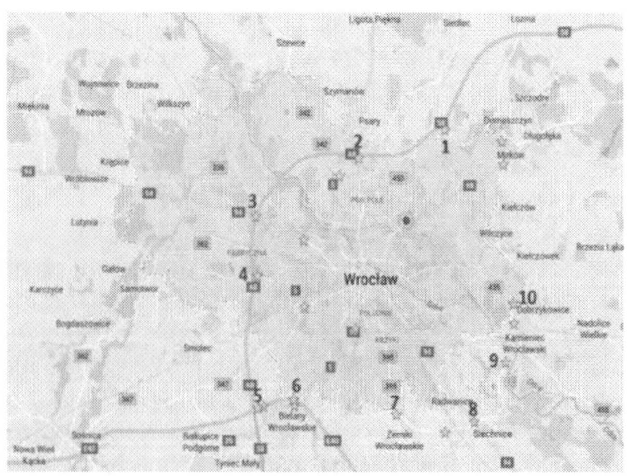

Map 1. Descriptions of test stands.

In a given location, the soil samples were collected from three points located within the distance of about 50-100 cm from each other. Three soil samples from a given location were screened and averaged after drying. The measurements of pH in water and 1M KCl and the electrical conductivity EC were made according to PN-ISO 10390:1997. The soil solutions were made in a stoichiometric ratio of soil to a solution 1:2.5 (m:v). Soils used for the mineralization process were prepared in accordance to PN-ISO 11465:1999. The soil sample weighing about 0.2 grams of dry matter digested in 8 ml of 60%-nitric acid. The microwave mineralization process was conducted with the use of Start D device (Milestone). The concentrations of zinc, copper, nickel, chromium and lead have been determined through the Flame Atomic Absorption Spectroscopy (FAAS) method. In the spectroscopy analysis the iCE 3500 Solaar Thermo device (Thermo Scientific) [PN-ISO 11047:2001] was used. The arithmetic mean and the standard deviation were calculated with the three replicates of the sample.

Chemical analyses have been carried out in the certified Laboratory of Toxycology and Environmental Research in the Faculty of Environmental Engineering at the Wroclaw Uniwersity of Science and Technology.

3 Results and discussion

Measurements of pH indicated active acidity (pH in H_2O) and exchangeable acidity (pH in 1 M KCl solution) in the soil solutions. The pH parameter allows to evaluate the risk of migration of heavy metals deep into the soil profile. The measurement of proper conductivity indicated the content of ions in the soil. The EC parameter allowed to evaluate the salinity of soils. The result below 25 mS/m indicated low salinity and above 75 mS/m showed high levels of salinity [11]. Spectroscopic analysis informs about the content of Cu, Zn, Pb, Cr and Ni in soils. On that basis the indexes of the enrichment of soils in heavy metals compared to the uncontaminated geochemical background in Poland can be calculated [12].

3.1 pH and EC measurements in soil solutions

The most alkaline character of the soil (pH of 8.08 in H_2O and pH of 7.66 in 1 M KCl) was identified in sample no. 7 WOW Siechnice (Fig. 1). It may be due to the ongoing work at the metallurgical waste landfill in Siechnice, located approx. 200 m from the road. Transportation of waste contributes to the pollination of the road and surrounding area. pH in the waste solution is 8.92 in H_2O and 8.52 in 1 M KCl. In a slightly alkaline environment most of the heavy metals are immobilized (hydroxide forms), which causes the low ability of metals to migrate deep into the soil. Soil sample no. 10 (WOW Dobrzykowice) demonstrated acidic character (pH 6.67 in H_2O and pH 5.71 in 1 M KCl). In the medium with pH below 6.5 metals have mostly cation forms. The ability of metals to migrate is much higher, which may pose a threat to the environment.

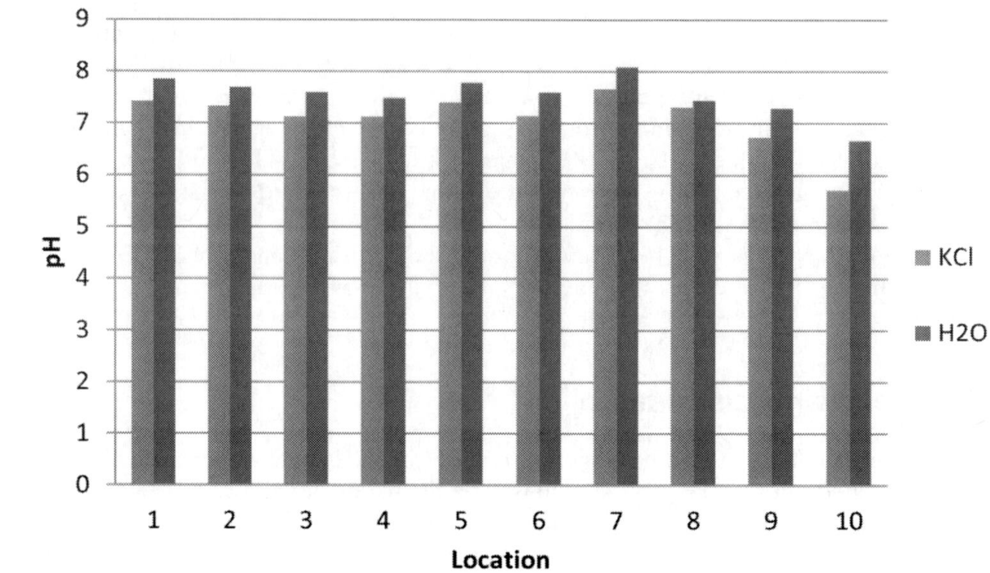

Fig. 1. pH measurement in soil solutions.

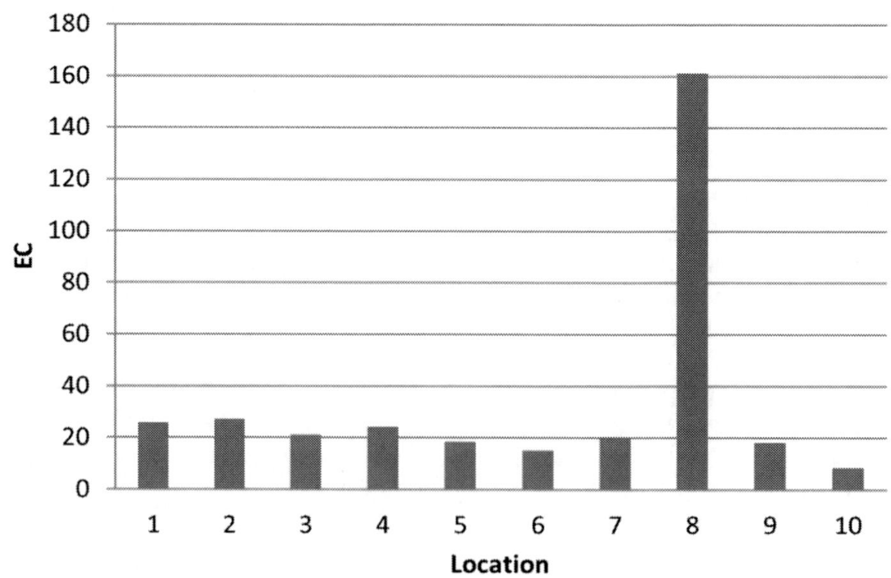

Fig. 2. EC measurement in soil solutions.

The results of the measurement of proper conductivity show that essentially the soils are slightly salted, the EC values are below 25 mS/m (Fig. 2). In sample no. 2 (AOW Node of North Wroclaw) exceeds the EC limit by about 2.16 mS/m, therefore the soil is of medium salinity. The highest conductivity of 161.35 mS/m is in sample no. 8 (WOW Blizanowice). The high salinity of the soil is probably the effect of the accumulation of the road salt during winter. Anions have lesser sorption ability than cations, therefore only irrational human activities can result in such a high soil salinity.

3.2 The content of Cu, Zn, Pb, Ni and Cr in soils

None of soil samples exceeds the permissible concentrations of heavy metals, relative to standard for soils, in accordance with the Polish Ministry of Environment Regulation from September 1st, 2016 (Table 2, Fig. 3) [13].

Table 2. The content of heavy metals in soils.

No.	Cu [mg/kg]		Zn [mg/kg]		Pb [mg/kg]		Ni [mg/kg]		Cr [mg/kg]	
	x	s	x	s	x	s	x	s	x*	s
1	27.79	0.93	72.75	0.17	17.18	1.80	34.08	0.88	13.54	1.29
2	41.21	0.20	298.27	1.12	30.17	0.11	18.53	0.28	28.60	1.18
3	29.71	1.13	116.71	0.67	23.01	0.64	6.86	0.46	11.54	1.38
4	17.70	0.46	83.57	1.19	13.79	1.60	9.41	0.53	3.58	0.78
5	11.52	0.88	48.97	1.11	6.50	0.93	7.38	0.43	12.43	0.36
6	14.76	0.16	68.04	1.25	10.91	1.54	5.74	0.99	15.14	1.33
7	8.73	0.22	52.53	1.10	3.59	0.50	17.46	0.53	79.99	0.70
8	21.32	1.00	93.43	0.48	11.57	0.81	12.02	0.99	114.54	2.44
9	10.45	0.67	71.23	1.08	5.93	0.51	24.26	0.65	12.54	0.81
10	6.84	0.30	29.96	0.97	no	no	4.39	0.94	6.13	0.29

x - arithmetic mean; s – standard deviation for 3 measurements, no - below the limit of Pb quantification by FAAS method.

The highest concentrations of copper, zinc and lead were recorded in sample no. 2 (AOW Node of North Wroclaw). The content of heavy metals is respectively 41.21 mg Cu/kg of dry matter, 298.27 mg Zn/kg of dry matter and 30.17 mg Ni/kg of dry matter. In general, the lowest concentration of metals are observed in soil sample no. 10 (WOW Dobrzykowice): 6.84 mg Cu/kg of dry matter, 29.96 mg Zn/kg of dry matter. The concentration of lead is below the limit of quantification of Pb by Flame AAS method.

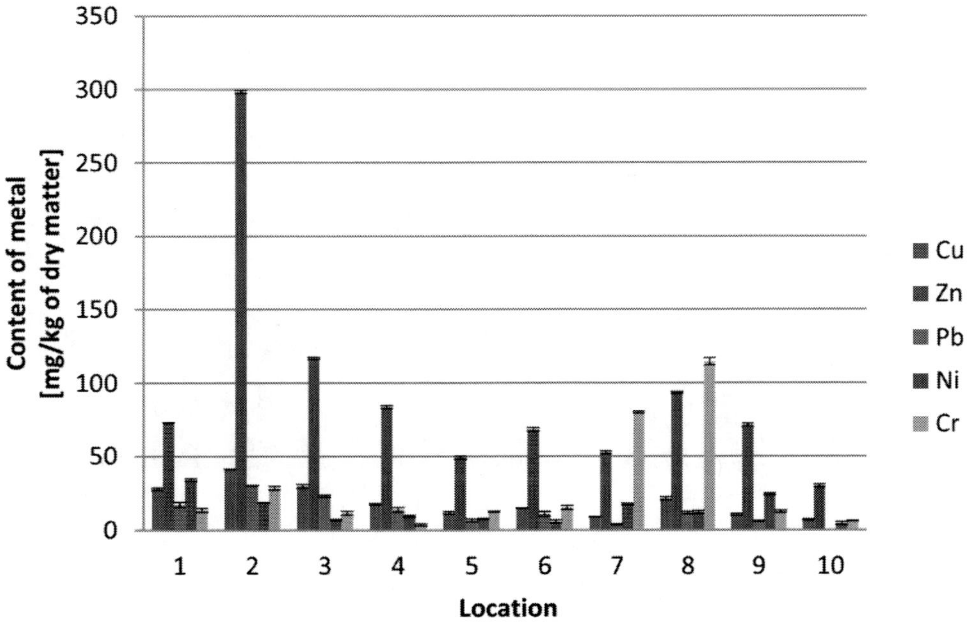

Fig. 3. The content of trace elements in soils.

The highest concentration of nickel was 34.08 mg/kg of dry matter and it was observed in sample no. 1 (AOW Długołęka). Again, the lowest concentration of heavy metal (4.39 mg Ni/kg of dry matter) was recorded in soil sample no.10 (WOW Dobrzykowice). Very high content of chromium (114.54 mg/kg of dry matter) was observed in sample no. 8 (WOW Blizanowice). The contamination of soil with chromium may be connected with the transport of metallurgical waste (compounds of iron and chromates) from the heap in Siechnice. The lowest content of chromium is 3.58 mg/kg of dry matter in soil sample no. 4 (AOW Node of Airport Wroclaw).

3.3 The pollution index

The pollution index (W_N) determines the enrichment of soils with heavy metals in relation to the geochemical background in the soil surface levels for the entire surface of Poland. The geochemical background in Poland is as follows: 6.5 mg Cu/kg; 33 mg Zn/kg; 18 mg Pb/kg; 11 mg Ni/kg; 40 mg Cr/kg [12].

Table 3. The pollution indexes of heavy metals in soils compared to the geochemical background.

	Pollution indexes (W_N)				
No.	Cu	Zn	Pb	Ni	Cr
1	4.28±0.14	2.20±0.01	0.95±0.10	3.10±0.08	0.34±0.03
2	6.34±0.03	9.04±0.03	1.68±0.01	1.68±0.03	0.72±0.03
3	4.57±0.17	3.54±0.02	1.28±0.04	0.62±0.04	0.29±0.03
4	2.72±0.07	2.53±0.04	0.77±0.09	0.86±0.05	0.09±0.02
5	1.77±0.13	1.48±0.03	0.36±0.05	0.67±0.04	0.31±0.01
6	2.27±0.02	2.06±0.04	0.61±0.09	0.52±0.09	0.38±0.03
7	1.34±0.03	1.59±0.03	0.20±0.03	1.59±0.05	2.00±0.02
8	3.28±0.15	2.83±0.01	0.64±0.05	1.09±0.09	2.86±0.06
9	1.61±0.10	2.16±0.03	0.33±0.03	2.21±0.06	0.31±0.02
10	1.05±0.05	0.91±0.03	no	0.40±0.09	0.15±0.01

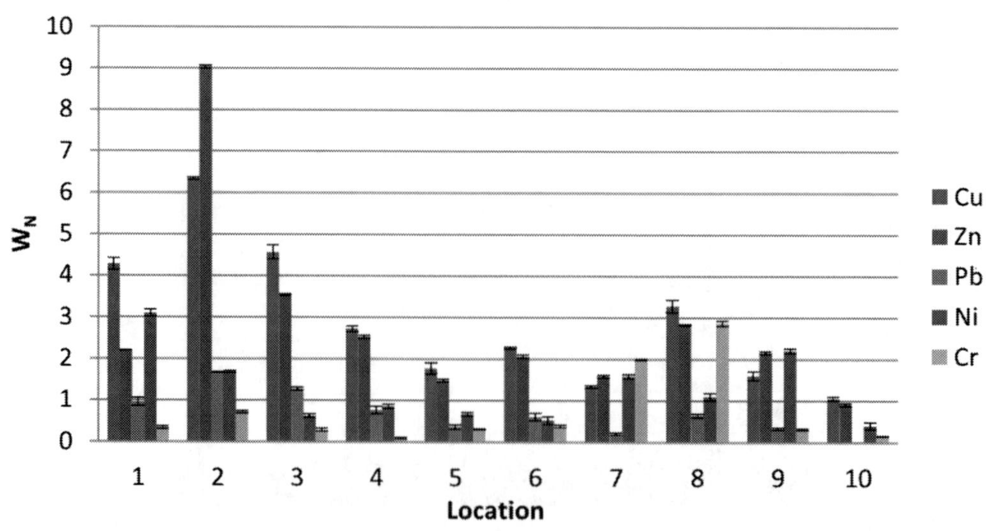

Fig. 4. The pollution indexes (W_N) of trace elements in soils.

The test soils are enriched with copper and zinc (table 3, fig. 4), except for sample no. 10 (WOW Dobrzykowice), in which there is no excess of zinc (W_N 0.91). The highest values of pollution index (W_N) of copper and zinc are in sample no. 2 (AOW Node of North

Wroclaw), 6.34 and 9.04 respectively. The W_N indexes in the remaining soils are in the range of 1.05-4.57 for Cu and 2.06 - 3.54 for Zn. Exceeding of lead content in soil is observed again in the sample no. 2 (W_N 1.68, AOW Node of North Wroclaw) and also in the sample no. 3 (W_N 1.28, AOW Node of Stadium Wroclaw). Two soil samples: no. 7 (WOW Siechnice) and no. 8 (WOW Blizanowice) are enriched with chromium. The values of W_N are respectively 2.00 and 2.86. Five locations demonstrate excess of nickel concentration in compared to the geochemical background of Poland. The W_N indexes range from 1.09 to 3.10.

4. Conclusions

- Soil sample no. 7 (WOW Siechnice) is of alkaline character (pH 8.08 in H_2O and pH 7.66 in 1M KCl). The reason may be the current ongoing work on the metallurgical waste landfill in Siechnice commune. pH of the waste sample from the heap is of 8.92 in H_2O and 8.52 in 1 M KCl. In this environmental conditions, the ability of heavy metals to migrate into the soil is inhibited.
- Soil sample no. 10 (WOW Dobrzykowice) is of faintly acidic character (pH in H_2O – 6.67, pH in 1M KCl – 5.71). In soil environments, at pH below 6.5, most elements appear in the cation forms and can migrate into the soil profile depth, causing the environmental hazard.
- Soil samples are generally faintly salified, with EC below 25 mS/m. Only sample no. 8 (WOW Blizanowice) has a very high EC value of 161.35 mS/m. High salinity of the soil may be caused by a large accumulation of salts after winter thaw in the hollow area.
- None of the soil samples exceeds the permissible concentrations of heavy metals, in accordance with the Polish Ministry of Environment Regulation from September 1st, 2016, regarding the manner of conducting the assessment of the contamination of ground.
- All soil samples are enriched with copper with reference to the geochemical background. The content of zinc is not exceeded only in soil sample no. 10. The enrichment of the soils with copper and zinc is on average three-fold. Sample no. 2 (AOW Node of North Wroclaw) has the highest Wn indexes of copper, zinc and lead, 6.34, 9.04 and 1.68 respectively.The accumulation of lead was observed in two soil samples. The enrichment of soil with nickel was observed in five samples.
- Average values of the pollution indexes (W_N) are combined into the following order: Cu > Zn > Ni > Pb > Cr.

References

1. K. J. Aslam, S.A. Khan, S.H. Khan, J. Saudi Chem. Soc., 17 (2013)
2. R. Czubaszek, K. Bartoszuk, CEE, **2** (2011)
3. A. Chrzan, G. Formicki, PECO, **6**, 2 (2012)
4. A. Hołtra, D. Zamorska-Wojdyła, EPE **42**, 3 (2016)
5. K. Czarnowska, T. Kozanecka, Soil Sci. Ann., **54**, 4 (2001)
6. A. Bieniek, B. Bieniek, Soil Sci. Ann., **59**, 3 (2008)
7. Z. Ciećko, A. Żołnowski, M. Wyszkowski, *Integrated monitoring of the natural environment* (Turpress, Toruń, 2004)
8. A. Kiebała, M. Kozieł, W. Zgłobicki, Eng. and Protection of Environ. **18**, 3 (2015)

9. J. Szyszlak-Bargłowicz, T. Słowik, G. Zając., W. Piekarski, Annual Set The Environment Protection, **15** (2013)
10. A. Kabata-Pendias, H. Pendias, *Trace elements in soil and plants* (CRC Press, Inc., USA, 2011)
11. J. Minczewski, Z. Marczenko, *Analytical chemistry* (PWN, Warsaw, 2011)
12. A. Kabata-Pendias, H. Pendias, *Biogeochemistry of trace elements* (PWN, Warsaw, 1999)
13. The Polish Ministry of Environment Regulation from September 1st, 2016, regarding the manner of conducting the assessment of the contamination of ground, Journal of Laws 2016 pos. 1395

Quantifying and reporting greenhouse gas emissions at local level

Izabela Sówka[1,*], and *Yaroslav* Bezyk[1]

[1]Unit of Ecologistics and Environmental Risk Management, Faculty of Environmental Engineering, Wroclaw University of Science and Technology, Wyb. Wyspiańskiego 27 St., 50-370, Wroclaw, Poland

Abstract. Cities as global centers of consumption and production often are a significant and growing source of greenhouse gas (GHG) emissions. At the same time, local authorities are increasingly taking action on climate change by focusing on reducing GHG emissions and efficiency improvement opportunities. To assess and reduce the overall greenhouse gas emission level from an urban area, it is necessary to identify all the activities and processes which generate these emissions. GHG inventory gives an opportunity to get wider knowledge for city's community about spatial emission processes and emissions contribution of key sources categories at the local scale. Inventory is being used for decision-making purposes and strategic planning in emission reduction policy. The goal of this paper was to clarify the major methodological challenges of GHG monitoring at the urban level. The paper is based on the discussion of different methods and approaches to assessing GHG emissions at the local level. It is presented sectoral GHGs emission trends in selected urban areas and compared CO_2 emission level in different countries and metropolises and variable European cities guidance. The study determines the inventory tools of GHGs emission taking into account the characteristics of main sources at local levels.

1 Introduction

Strong ability to take effective action on climate mitigation and adaptation strategies, monitoring and verifying progress in GHGs reduction depends on reliable GHG emission inventory. Under the United Nations Framework Convention on Climate Change, the GHG inventory is reporting emissions and sinks of anthropogenic greenhouse gases within a region and period [1]. Based on the Kyoto Protocol classification system, the most important direct greenhouse gases emitted by humans include carbon dioxide (CO_2), methane (CH_4), nitrous oxide (N_2O), and several fluorine-containing halogenated substances (hydrofluorocarbons, perfluorocarbons and sulfur hexafluoride) [2].

The internationally-accepted methods used to calculate GHG emissions and sinks are provided by Intergovernmental Panel on Climate Change (IPCC). Applying defined by the IPCC guidelines uses standardized field data from commonly determined processes in sectoral categories (energy, industrial processes and product use, agriculture, waste

* Corresponding author: izabela.sowka@pwr.edu.pl

DOI: 10.1051/e3sconf/20171700084

management) [3], but not detailed enough for local inventory system of different countries and regions to show the overall carbon footprint of city activities. The newest 21st yearly session of the Conference of the Parties (COP) to the UNFCCC 1992, was held in Paris from 30 November to 12 December 2015, proclaimed the important role of city's level actions in the field of climate change [4].

The GHG emissions reporting at the local scale is a precondition for creating a city's mitigation mechanisms and removal of GHGs [5]. Recently, in practice of developing the methodological framework for calculating urban GHGs emission were involved local authorities from different countries. As a result, a network of local initiatives such as ICLEI – Local Governments for Sustainability, C40 Cities Climate Leadership Group, the Covenant of Mayors is represented in international climate negotiations.

The comparability of inventory data between different cities is limited due to varying methodological approaches and frequency of reporting process. Since regions and countries have contrasting economic and social structures, CO_2 emissions can be expressed in CO_2-eq emissions by Gross Domestic Product (CO_2-eq /GDP) and per capita (CO_2-eq /cap). These indicators show the relationship between the size of city's population and the amount of GHGs emitted and improvements in the emissions associated with energy generation [6]. The CO_2-eq/GDP and CO_2-eq/capita measurement help to investigate the relationship between the structure of the economy, the composition of local community and emissions and used as a benchmark for making an allocation of emission sources.

2 Variability of CO_2 emissions in European cities

The numeric per capita GHG emission values are represented at the city-scale while incorporating the scope of related emissions versus the larger national scale emission levels (Fig. 1). In general, wealthier countries as well as cities having higher emissions per capita, due to higher rates of consumption and more energy-intensive lifestyles, likewise urbanization brings higher energy usage, although other factors such as population density and geography also have an influence [7].

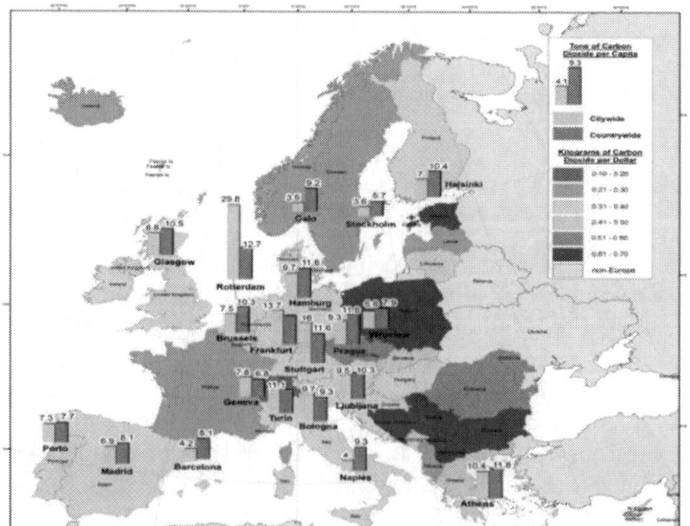

Fig. 1. Europe-wide CO_2 emissions per capita and unit of GDP in 2005. Source: Base map source: OpenStreetMap.

The created map below on the data from the Word Bank [7] shows the European countrywide and citywide CO_2 emissions per capita (tons of CO_2-eq per person) and per

Gross Domestic Product (kg CO_2 per PPP $1 GDP). The expressed greenhouse gas emission per GDP from selected countries represents how green are the different European economies. Poland's per capita emissions with 7.9 CO_2-eq per person, relative values to 2005, are similar to the European Union (EU) average too, but at the same time the amount of GHGs emitted per GDP in Poland are much higher – 0.6 kg CO_2 per PPP $1 GDP. The main reason of this is lower income level and less emissions-efficient of the Polish economy.

Given the emission level in Poland and the share of fossil fuels in energy sector, the critical difference of Poland's energy production and emissions structure from EU in 2005 was due to the dominance of coal and lignite (above 90%) in total energy production [7]. To summarize, the reflection of GHG emissions from city's and national level is not only in absolute form but in the manner of per inhabitant, which helps to develop a more holistic GHG emissions footprint that shows the overall impact of a city's activities on local and national GHG emissions reduction targets.

3 Approaches to GHG accounting and analysis of existing methodologies

3.1 GHG inventory tools used at the local level

Main approaches that can be applied for estimating GHG emissions are emission factors (quantifying the emissions per activity); continuous emissions monitoring (accurate and real-time data); source testing (periodic sampling and analysis of emitted species); material balance (Input = Output + Emissions). In the range of gases and GHG emission sources, the emission factors method and mass balance method are commonly used for the inventory process [8].

Emission factors specific to the urban area are often based on the recommended by Intergovernmental Panel on Climate Change (IPCC) for country-level "tiered" (referred Tier 1, 2, 3 data source) measuring rules [3]. The tier 1 approach reflects default activity data, e.g. readily available energy statistics, production statistics and corresponding emission factors for typical/familiar process conditions. Tier 2 approach is more specific to regions, which includes country-specific information on process conditions and technological details. Tier 3 approach is the most detailed method and can use facility level data, including sophisticated models [3].

The concept of 'scope' allocations with regards to GHG emissions for communities allows accounting for direct and indirect emissions. Three "scopes" categories depending on where they physically occur (within and without city boundaries) are defined in the GHG Protocol standard for cities known as 'Global Protocol for Community-Scale Greenhouse Gas Emission Inventories' (GPC) [9]. The GPC Protocol was created in cooperation with World Resources Institute, C40 Cities Climate Leadership Group and Local Governments for Sustainability (ICLEI) [9].

The several of programs for improving the process of quantification of greenhouse gases are designed by ICLEI. The major of them are: Clean Air and Climate Protection (CACP) Software a management tool for accounting emission, associated with energy and power consumption, fuel use, and waste disposal [10]; Harmonized Emissions Analysis Tool (HEAT +) supporting program for creating GHG inventory and cost-effective impact on emissions reductions from energy consumption, transport and waste sources [11].

In 2008 was established the Covenant of Mayors (CoM) ¬ the largest cities initiative for combating climate change. The CoM association has taken commitments regarding the EU CO_2 reduction objectives by more than 20% by 2020 compared to 1990 [12]. For the reduction targets, the participated cities have to provide the Baseline Emission Inventory (BEI) quantifies the amount of CO_2 emitted due to energy consumption in their territory through the implementation of Sustainable Energy Action Plans [13].

The Greenhouse Gas Regional Inventory Process (GRIP) for city-scale emissions accounting was developed by the University of Manchester. The GRIP Programme is based on three calculations methods: level 1, level 2 and level 3, similar to the IPCC's tiered approaches. Level 1 is associated with the high standard of accuracy; level 2 includes individually collected or cumulative dataset with average accuracy records; level 3 is determined as a category with the highest level of uncertainty [14].

As follows from the above, the inventory tools and methodologies furthermore differ with how cities allocate emissions to the territory where they were emitted. For instance, in city's emissions accounting and reporting practices it should be clearly distinguished whether only the direct (sources located in cities) and/or also indirect (upstream) emissions are calculated (clear measurement frameworks), what can have major consequences for the outcome of the inventory process.

3.2 Measurement boundaries, scopes and sources of GHG emission

Depending on the level of approach in estimating GHG emissions, there is top-down approach (used in national GHG inventories) and bottom-up approach (applied at the regional or municipal level). Data collection and processing in bottom-up approach guarantees a relatively accurate inventory at the local level, but accounting procedures can have varieties from site to site. Data aggregating and scaling in top-down approach reflects the national average statistics for an emission source but due to the little interaction between different levels not accurately reflect local conditions of emissions [15]. Finally, local level inventories are often based on a mix of bottom-up and top-down approach.

The area where emissions are measured relates to the boundaries of the measurement. It includes different aspects such as activity data (data on the fuel consumption or the relevant electricity/heat production in or outside the territory); scaling data principle (GHG are allocated within the geographic boundaries of a city or out-of-boundary); allocation of emission loads on the respective sources and consumption sectors (allocation methods and references for emissions related to energy sources used to production electricity and heat).

Electricity using activities carried out by cities are the cause of GHG emissions elsewhere, which are taken into account when allocating CO_2 emissions and fossil fuel used to electricity/heat produced. An allocation - is the distribution of emissions to the main product and co-products [16]. The allocation mechanism of CO_2 emission allowances differs significantly depending on: the variety of method used; the efficiencies of dedicated electricity/heat plants; thermodynamic medium temperature; other reference efficiencies assumed; energy policy reasons in the country (different taxation to CHP) [17]. There are two widely used categories of allocation methods [18, 19]:

Thermodynamic Allocation Methods based on:

- Energy content of heat and electricity: emissions are allocated to the produced heat and electricity is based on the primary energy content of the produced products;
- Exergy content of heat and electricity as the main product (exergy - energy that is available to be used) covers the main part of the CO_2 emissions, and the heat is the by-product (anergy - the balance from converted the energy, unavailable part).

Economic Allocation Methods based on:

- The Benefit Distribution Method: all emissions are allocating to heat production;
- Alterative Power Method: electricity covers all emissions of CHP production;
- The equal mix of both energy and heat side: total emissions is allocated to fixed sharing 50% - 50% between power and heat products.

To provide accuracy and to improve transparency of GHG inventory process, it is necessary to identify scope and sources of GHG emissions. Based on the latest tools of ICLEI, the GPC Protocol is counting the direct and indirect emissions, grouped it into three boundaries (scopes 1, 2, 3) [9]. Scope 1 including all direct GHG emissions within the city geographic boundary; Scope 2 including all indirect emissions occurred inside or and outside local authority (consumption of electricity, heat or steam); Scope 3 estimates all other indirect emissions that occur outside local authority (electricity transportation related activities, waste generation) [9]. The major sources of GHG emissions from city activities covered by GPC Protocol are following: stationary energy; transportation, industrial processes and product use, waste, agriculture, forestry, and other land use (AFOLU), other indirect emissions and sources of fugitive emissions [9].

Attributing anthropogenic carbon concentrations to a specific source is challenging because emissions vary depending on type and location, technology type and operational practices, meteorological conditions, time framework, etc. [20]. Then, the exact amount of emitted GHG emissions from selected sources at urban area can be hard to determine, and some sources are better known than others. With pure data of actual field emission measurements, individual inventories can use data for a specific sector from national statistics if the relevant data are difficult to obtain at the local level and could be scaled. The rule is that scaling data from national GHG inventories can be applicable within local authority if data for a selected sector are broken down from the national GHG inventory and sector share is not considerable in overall emissions from the city [21].

3.3 Structure of GHG emission inventory process

The process of preparing the GHG inventory includes following stages: identifying geospatial features of territory on inventory, clarification of sources and terms of GHG emissions, selection of an emissions quantification methodology, determination emission factors for the data collected, evaluating uncertainties, verification and improvement inventory, interpretation and communicating of inventory results, performing a life-cycle inventory and monitoring of future greenhouse gas emissions [22]. The overall architecture of the accounting procedure is based on the value chain concept: the primary activities data are transformed in the form of a value-added model of future GHG emissions scenarios (Fig. 2).

Within the territorial framework identification is required detailed knowledge about the area of emission inventory and/or about geospatial features; identifying key categories and term of conditions for the emergence and fate of greenhouse gas emissions [23]. In the next step, it is important to separate existing assessment approaches and methodologies and planning possibilities of implementation in local conditions and/or identified sources. The step of selecting methods is provided due to the realization of the two important tasks: first - analysis of inventory scope and city-induced frameworks and checking impacts of selected level of detail on the final result of GHG inventory.

During second objective it should be planned the core features of adaptation of chosen inventory methods to the local level. Using this option, municipal administrators can choose for evaluation either the direct impact of GHG emissions from a different economy and the residential sectors within the city operational boundary and/or indirect and embodied GHG emissions, which occur inside or outside urban geographical/ political boundary. The calculation beyond selection of the detail standards and allocation of emissions also required the determination of the corresponding territory-specific emission factors [24]

In a process for quantifying GHG emissions one of the most important parts is a clear estimation of the level of uncertainties during information processing and verification for

 DOI: 10.1051/e3sconf/20171700084

inventory data. The consideration of the possibility to use developed GHG inventory materials (data sources, inventory plan, the scale of the inventory, calculation tools, presenting results) and specific emission factors for selected activity data should be provided by monitoring, reporting, and verification (MRV) process at regular time intervals [25].

Monitoring covers the scientific part of MRV process (accuracy and precision of the estimates); by reporting is provided data harmonizing (better access to information and high level of transparency); verification is an assessment of the completeness and accuracy of the collected database [26] In this step, to monitor the emission loads of different production and consuming sectors in terms of improving inventory system and GHG reduction strategy the Life Cycle Assessment (LCA) method could be applied. A reliable life-cycle inventory (LCI) performance as a value-added methodological structure is crucial to achieving inventory goals [27].

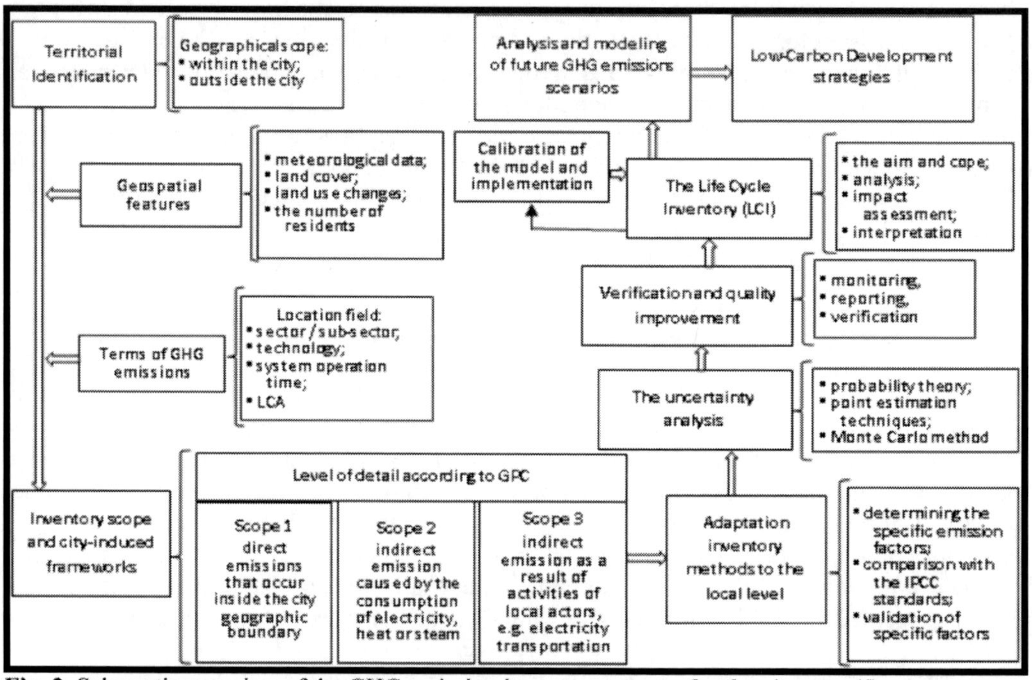

Fig. 2. Schematic overview of the GHG emission inventory process for the city area (Source: Author's construct).

The following step is the interpretation of the GHG inventory results. The identified emissions trends in research area are used to impact on activities where emissions occur. The aim is to reduce given GHGs amount, improve energy efficiency and implement innovative sectoral and cross-sectoral low carbon initiatives. Further to predict future change in GHGs, it could be prepared geostatistical forecasting model of the spatial distribution of greenhouse gas emissions that helps to estimate risks (health, environmental, etc.) associated with a wide variety of global warming emissions, as well as to manage emissions reduction efforts and allocate resources.

4 Conclusions

Cities play a dominant role in global consumption and production. They are associated with serious problems like air pollution, greenhouse gas emissions, waste, and poverty. But urban areas also offer opportunities for sustainable development. The world's municipalities experience shows that the building process of GHG emissions quantification and reporting could enable cities to set realistic reduction targets and manage their mitigation and adaptation efforts. The local inventory data gives an opportunity to get wider knowledge about spatial emission processes in the sectors of the economy not only at the national level but also at the level of administrative units.

Accordingly, the first stage of this research was devoted to studying GHG emission accounting and inventory techniques at national and local level. The study has shown that many initiatives on the GHG inventories at a local level are being applied in different European countries and worldwide. Some of the most advanced GHG accounting tools are analyzed, and the most important differences between these tools are highlighted.

Finally, it was shown that there are existing meaningful differences across the respective accounting tools and inventories of compared cities in Europe. Depending on the purpose of the GHG calculation different methods may be used. For instance, the choice whether only direct or also indirect emissions are taken into account (scope of the measurement), analytical framework and approaches to allocating emissions allowances have major consequences for the outcome of the inventory process. Territorial framework and timing variability, nature of future data releases have also influence on progress towards updating emission factors and identifying activity data across all sectors.

The investigations were co-financed within the framework of the project No. 0402/0077/16 with the specific subsidy granted for the Faculty of Environmental Engineering Wroclaw University of Science and Technology (W-7) by the Minister of Science and Higher Education to conduct research and development work and related tasks contributing to the development of young scientists and doctoral students in 2016/2017.

References

1. UNFCCC: United Nations Framework Convention on Climate Change. United Nations, FCCC/INFORMAL/84 GE.05-62220 (E) 200705, *Secretariat of the United Nations Framework Convention on Climate Change*, Bonn, Germany, **15** pp., (1992).
2. J. Depledge, United Nations Framework Convention on Climate Change (UNFCCC) Technical paper: Tracing the Origins of the Kyoto Protocol: An Article-by-Article Textual History (PDF), UNFCCC, (2000).
3. J.T. Houghton, L.G. Meira Filho, B.A. Callander, N. Harris, A. Kattenberg, and K. Maskell (eds.). IPCC (1996) Climate Change (1995): The Science of Climate Change. Intergovernmental Panel on Climate Change. Cambridge University Press. Cambridge, United Kingdom.
4. World Resources Institute. Aniruddha Dasgupta. COP21 Highlights Importance of City Actions in the Climate Fight, (2015). Available at http://www.wri.org/blog/2015/12/cop21-highlights-importance-city-actions-climate-fight
5. M. Sippel, Cities in Germany and their climate commitments: More hype than substance? Discussion Paper. University of Stuttgart. IER – Institute of Energy Economics and Rational Energy Use. June (2010).
6. B. Dawson, M. Spannagle. The complete guide to climate change, Routledge, Oxon, United Kingdom, (2009).
7. WRI (World Resources Institute)/World Bank. World Resources: The Wealth of the Poor - Managing Ecosystems to Fight Poverty. Washington, DC: WRI, (2005).
8. K. Ritter, M. Lev-On, and T. Shires. Understanding Uncertainty in Greenhouse Gas Emission Estimates: Technical Considerations and Statistical Calculation Methods."

American Petroleum Institute, Washington DC, Background paper for EPA 19th International Emissions Inventory Conference, San Antonio, Texas, (2010).

9. The Global Protocol for Community-Scale Greenhouse Gas Emission Inventories (GPC). An Accounting and Reporting Standard for Cities. November (2014).
10. International Council for Local Environmental Initiatives (ICLEI) - Clean Air and Climate Protection (CACP) Software. Available at http://en.openei.org/wiki/International_Council_for_Local_Environmental_Initiatives_(ICLEI)_Clean_Air_and_Climate_Protection_Software_Tools
11. International Council for Local Environmental Initiatives (ICLEI) - HEATplus Harmonized Emissions Analysis Tool. Available at http://heat.iclei.org/heatplusgpc/indexnew.aspx
12. Covenant of Mayors for Climate & Energy, (2016). Available from: http://www.covenantofmayors.eu/The-Covenant-of-Mayors-for-Climate.html
13. How to develop a sustainable energy action plan (SEAP) – guidebook, Luxembourg: Publications Office of the European Union, (2010).
14. S. Carney, S. Shackley. The greenhouse gas regional inventory project (GRIP): Designing and employing a regional greenhouse gas measurement tool for stakeholder use. Energy Policy 37, (2009).
15. C. Nick Hewitt (Editor), Andrea V. Jackson (Editor). *Handbook of Atmospheric Science*: Principles and Applications. ISBN: 978-0-632-05286-8. 648 pages, (2009).
16. The impact of modified EU ETS allocation principles on the economics of CHP-Based district heating networks. Günther Westner and Reinhard Madlener. FCN Working Paper No. 4/2011, (2011).
17. WBCSD/WRI. Allocation of Emissions from a Combined Heat and Power (CHP) Plant – Excel tool. (2006). Available at http://www.ghgprotocol.org/calculation-tools/
18. M. Harmelink, Assessment of CO_2 emissions of electricity and heat used at industrial plants. Harmelink Consulting, The Netherlands Lex Bosselaar, (2014).
19. Arto Nuorkivi. Allocation of Fuel Energy and Emissions to Heat and Power in CHP. Energy-AN Consulting. FINAL REPORT. Sep 7, (2010).
20. U. S. Environmental Protection Agency. Office of Policy, Planning, and Evaluation. Inventory of U.S. Greenhouse gas emissions and sinks, 1990-1994. EPA 230-R-96-006. Washington, DC, November, (1995).
21. N. Bader, R. Bleischwitz, Measuring Urban Greenhouse Gas Emissions: The Challenge of Comparability. Cities and Climate Change, (2009).
22. B. Vigon, D. Tolle, B. Cornaby, H. Latham, I. Batte. Life-cycle assessment inventory guidelines and principles. Columbus, (1993).
23. German Informative Inventory Report. Chapter 2.6 - QA/QC and Verification Methods. Updates, (2016). Available at http://iir-de.wikidot.com/qa-qc-and-verification-methods
24. American Association of State Highway and Transportation Officials (AASHTO). Greenhouse gas emission inventory methodologies for state transportation departments, (2011).
25. J. Wintergreen, T. Delaney, ISO 14064, International Standard for GHG Emissions Inventories and Verification. Available at https://www3.epa.gov/ttnchie1/conference/ei16/session13/wintergreen.pdf
26. V. Bellassen, N. Stephan, M. Afriat, E. Alberola, A. Barker, J. Chang, P. Chiquet, C. Cochran, I. Deheza, M. Dimopoulos, C. Foucherot, C. Jacquier, G. Morel, R. Robinson, R. Shishlov. Monitoring, Reporting and verifying emissions in the climate economy. Nature Climate Change, (2015), 5, (4), 319-328.
27. B.V. Babu, Life Cycle Inventory Analysis. International Conference on Life cycle assessment & Life cycle management methodologies. Malaysia. December 4-5, (2006).

GIS-based modelling of odour emitted from the waste processing plant: case study

Izabela Sówka[1,*], *Agnieszka* Grzelka[1], *Yaroslav* Bezyk[1], and *Urszula* Miller[1]

[1]Wroclaw University of Science and Technology, Faculty of Environmental Engineering, Unit of Ecologistics and Environmental Risk Management, Pl. Grunwaldzki 9, 50- 377 Wroclaw

Abstract. The emission of odours into the atmospheric air from the municipal economy and industrial plants, especially in urbanized areas, causes a serious problem, which the mankind has been struggling with for years. The excessive exposure of people to odours may result in many negative health effects, including, for example, headaches and vomiting. There are many different methods that are used in order to evaluate the odour nuisance. The results obtained through those methods can then be used to carry out a visualization and an analysis of a distribution of the odour concentrations in a given area by using the GIS (Geographic Information System). By their application to the spatial analysis of the impact of odours, we can enable the assessment of the magnitude and likelihood of the occurrence of odour nuisance. Modelling using GIS tools and spatial interpolation like IDW method and kriging can provide an alternative to the standard modelling tools, which generally use the emission values from sources that are identified as major emitters of odours. The work presents the result, based on the odour measurements data from waste processing plant, of the attempt to connect two different tools – the reference model OPERAT FB and GIS-based dispersion modelling performed using IDW method and ordinary kriging to analyse their behaviour in terms of limited observation values.

1 Introduction

Geographic Information System (GIS) has been widely used in many fields related to the environmental protection, including for example hydrology, meteorology and also to estimate a dispersion of air pollutants [1, 2]. GIS as an information processing technique, helps to collect, store, display, process and analyse spatial geographic data [3, 4]. That is why GIS is a suitable tool to also analyse and visualize environmental models based on spatial data. As an example, it can visually show pollution condition and environmental quality [3, 5].

Odours are a specific type of pollution in terms of the nuisance that is caused by them is mostly a derivative of their subjective perception [6]. Therefore, prediction and modelling odour dispersion is relatively difficult. There are several air pollution dispersion models, which simultaneously use GIS tools for instance to visualize the achieved data. The most

* Corresponding author: izabela.sowka@pwr.edu.pl

commonly used ones are CALPUFF and AERMOD. Both of them are designed generally for air pollutants, but they also meet the requirements for the use of data about the odour concentrations or emissions. AERMOD and GIS were employed to model odour dispersion from biosolids reuse sites in Orange County in Virginia [8]. Some model based on odour dispersion calculations using a mathematical model CALPUFF and geostatistical analysis – ordinary kriging was used to characterize the odour nuisance caused by a distillery located in the south-west of Poland [7]. There are also other applications that integrate GIS and dispersion models – DUSTRAN use EPA atmospheric dispersion models and ArcGIS in order to stimulate the dust dispersion [9]. In India CALINE-4 model and TransCAD software, based on GIS was used to evaluate concentrations of vehicular pollutants [10]. In Poland the reference model based on Pasquill formula and adapted to the description of methodology for performing such calculations in accordance with Regulation of the Minister of Environment dated 26 January 2010 on reference values for some substances in the air is used [11].

Among the methods that are used in the world for the purpose of determining the olfactory impact of emission sources, there can be distinguished methods such as the dynamic olfactometry, field studies and questionnaire surveys [3]. All of them can be used to obtain a discrete data for subsequent modelling the impact of odours using GIS tools. By measuring the geographical coordinates at the measuring points it is possible to obtain a spatial nature of data and subsequently process them and visualize in GIS.

The goal of the study was to attempt to connect two independent but consistent tools - the reference model (OPERAT FB software) and a GIS-based odour dispersion modelling on selected area performed using Geographic Information System and to analyse their behaviour in terms of limited observations values.

2. Characteristics of selected methods used in the study

2.1 Dynamic olfactometry and reference model

The dynamic olfactometry is one of the odour concentration measurement techniques that can be used to obtain a discrete data for subsequent modelling in GIS. Procedures for determination of odour concentration by this method are described in the standard *PN-EN 13725:2007 Air quality. Determination of odour concentration, with the use of dynamic olfactometry method* [12]. In this sensory method a team of evaluators assess air samples containing odorants in various concentrations. The evaluators are selected in accordance with guidelines given in the standard, with the use of certified reference material (n-butanol in nitrogen). In the simplest version of the method – "yes – no" - the evaluators are about to signal, whether they sense the smell in the presented gas stream. The gases are diluted with an odourless air dynamically. The initial dilution is chosen in the way, it is not possible to determine the presence of odour. The dilutions are decreasing. Among presented samples there are so called "blind samples", in which instead of an odour, a clean, odourless air is presented. One measurement consists of four series. The results are calculated as the result of one measurement team – the average of all individual measurements. It is also the odour concentration in a given sample given in ou_E/m^3 units [6, 12, 13]. On the base of the determined odour concentrations it is possible to calculate the odour emission from individual sources and overall odour emission from the studied facility as well as to calculate odour concentrations at reception points by using available dispersion models according [11].

2.2 GIS-based modelling

Spatial interpolation is one of the most advanced procedures in GIS systems. Using this method it is possible to estimate the analysed magnitude for areas where measurements were not performed. Interpolation is a part of the science of geostatistics, which combines geography and statistics. The basis of the geostatistics is the Tobler law that says that objects that are adjacent each other in time and / or in space are more alike than objects that are farther apart [1, 14, 15]. Geosatistics can convert the point data into the spatial data. Geostatistics can estimate the value of the data at the points where, due to time, cost, or technical limitations performance of the measurements is impossible. It assumes that the spatial variability of natural phenomena can be modelled using the spatial autocorrelation, so it is possible to get the results as a continuous surface. It is important to properly select appropriate interpolation functions and choose how to determine the weight to be attributed to the surrounding areas [15]. We can distinguish such methods of interpolation like deterministic and probabilistic. The first includes Inverse Distance Weighted (IDW) method, radial basis function, trend surface analysis, second consists of different sorts of kriging – simple kriging, ordinary kriging, block kriging and point kriging [14]. Since the kriging method is based on the statistical properties of the measured points it provides an accuracy analysis of the predictions while IDW interpolation (based on a linear combination of nearby observations) does not [16].

2.2.1 Inverse-distance weighting method (IDW)

The main assumption of the IDW (Inverse-distance weighting) method is that each sample has an impact on their surroundings. In IDW method the value of the variable in the interpolation point is determined as a weighted average of the surrounding sampling points. The weight of the variables decreases with distance. The use of higher value of the power increases the influence of closer measurement points. This makes it possible to calculate the variable at any point in the field on the basis of the measurement data [15, 17, 18]. The effect of the calculation depends on such interpolation parameters as the value of the power, the number of neighbours and a radius of search [17].

The advantages of the IDW are simplicity and speed of execution. It gives reliable results whilst maintaining the values of samples. It characteristic is not to exceed the interpolated values. This is important, for example, for measurements that use a strict numerical scale [17].

2.2.2 Kriging

Kriging is a geostatistical interpolation method of moving weighted average. Kriging technique is applied to the local estimation. In the process of estimation there are included only data located near the area of estimation. In this method a set of the weightings assigned to the samples minimalizes the variance of the estimation. The variance of the estimation is calculated as a function of the adopted model of an empirical variogram and mutual location of samples and reciprocally relative to a point or a block which is the subject of estimation [14, 15]. Kriging is a stepwise process. In the first stage the empirical variogram should be calculated. Then the theoretical model, also referred as the geostatistical must be matched. The hardest step is the spatial autocorrelation that fit the theoretical function to the course of the variogram. In the kriging a set of kriging linear equations is solved, which results in receive a kriging weights (coefficients). The coefficients in the kriging equation represent the covariance between samples in the area of estimation and between samples and the area of estimation [14, 15].

3. Research methodology and results discussion

Odour concentrations measurements were taken in the area of the waste processing plant, located on the suburbs of a chosen city in Poland. The plant is located in distance approximately 1200 m to the west to nearest residential buildings and surrounded by fields and bands of green. Volume flow rates (m^3/s) and odour concentrations (ou_E/m^3) were measured in four regular measurement series at a certain time (according to VDI 3883) [19, 20]. Modelling parameters in the software OPERAT FB based on parameters of the gases (Table 1), odour emission rate; meteorological conditions and topography; pollution transport in the atmosphere. The determined coefficient of aerodynamic roughness of $z_0 = 0{,}55$ m was assumed.

Table 1. Characteristic of the emitters and basic parameters of the emitted gases.

Emitter No.	Emitter high [m]	Emitter diameter [m]	Speed of the gaseses [m/s]	Volumetric heat capacity [kJ/m^3/K]	Emitter location X [m]	
					X [m]	Y [m]
E1	10.5	0.345	2.29	1.30	463.5	240.9
E2	11	0.345	2.29	1.30	463	232.8
E3	12.5	0.345	2.29	1.30	462.5	224.7
E4	13	0.345	2.29	1.30	463.5	218.1
E5	13	0.345	2.29	1.30	465.1	203.5
E6	11	0.345	2.29	1.30	466.1	193.9
E7	13	0.345	2.29	1.30	465.6	187.8
E8	11	0.345	2.29	1.30	466.6	182.8
E9	10.5	0.345	2.29	1.30	467.1	174.7
E10	1.5	0.18	0.04	1.30	366.6	371.6
E11	3	0.18	0.04	1.30	382.8	291.9

In the next step, for verification used tool (OPERAT FB software), the Geographic Information System (GIS) – based model of the odour plume and potential impacts on the offsite receptors was developed in the present study. Integrated modelling approach involving GIS to highlight odour concentrations estimation by physically connection of two rasters - results of measured concentrations simulated in OPERAT FB and a raster datasets covering an area under research – has been used. Results of odour dispersion modelling (displayed as concentration predicted map) are obtained from an application of ArcGIS spatial analyst tools such as ordinary kriging and inverse-distance weighting method (IDW). Measurement data were used to estimate the annual average odour concentrations (modeled by OPERAT FB) and graphically represented in GIS software as a raster (Fig. 1). Calculations are made in grid format receptors (size 1800 m × 1800 m and step 50 meters. The model was developed by physical connection in GIS environment the OPERAT FB modeling output and a visualizing odour impact area. The geographic coordinate system WGS 84 / UTM zone 33N (for Poland) was assigned to the data frame.

Fig. 1. Annual average odour concentration calculated with reference model (OPERAT FB software). Map source: OpenStreetMap.

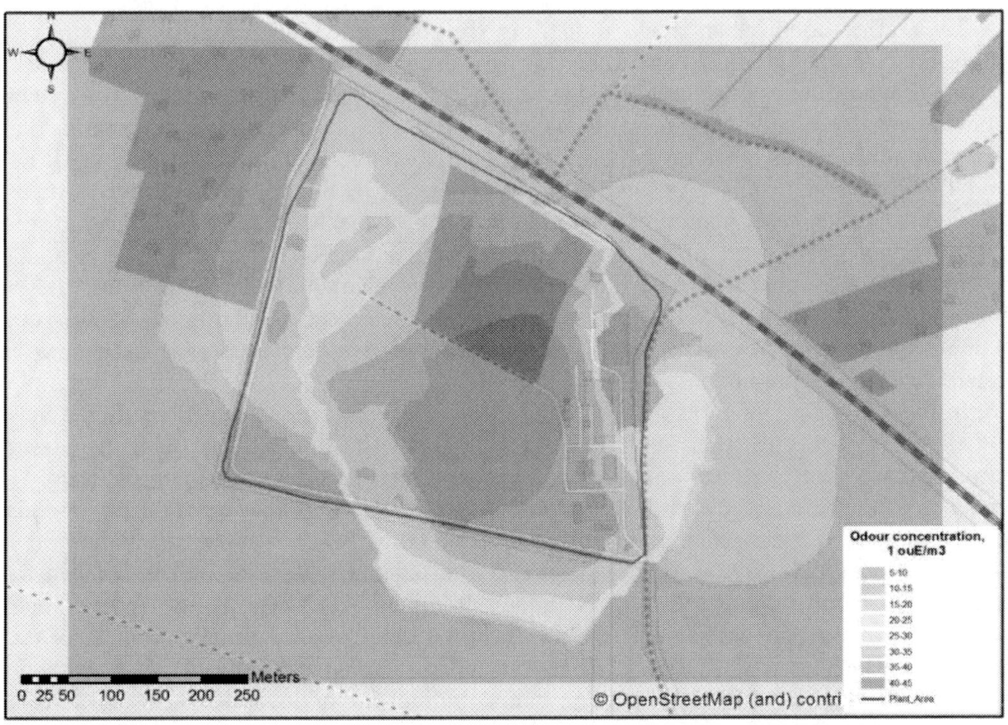

Fig. 2. Odour dispersion model built with ordinary kriging in GIS software. Map source: OpenStreetMap.

Fig. 3. Odour dispersion model built with Inverse-distance weighting method in GIS software. Map source: OpenStreetMap.

Then, the odour dispersion modelling results were incorporated into the GIS environment. The geostatistical analysis was used to predict concentration values at locations with odour concentrations determinated and calculated on the basis of dynamic olfactometry measurements at a limited number of points. The atmospheric odour dispersion model was obtained by performing spatial analysis by use of the ordinary kriging and IDW method from GIS (ArcMap10.3.1) (Fig. 2, 3). Suitable areas were created as a weighted raster were minimum error variances (root mean square error, mean absolute error, variance of errors) of kriging interpolation have been adopted. The geostatistical interpolation was provided with the prediction error characteristics as follows: the root mean square standardized of the value is 0.6454; the mean standardized error is equal 0.0386; the average normal value is 0.215. All the errors were used to measure the best accuracy of model predictions.

In the study contour plots of modelled odour concentrations are well predicted by the OPERAT FB and GIS tool. The OPERAT FB model shows a similar trend in measured grid distance but provides a significant underprediction effect in sight distance. The analysis of the results showed that interpolated concentration data points on the boundary isolines obtained from calculations by using the OPERAT FB software, mainly in the south-east polygon, have a more widespread interpolated surface i.e. 1200–1600 m from the studied facility compared to GIS model. It might be caused by the overestimation despite there was no objective basis for the predicting in the measurement scales and meteorological characteristics (e.g. direction and maximum wind speed). In the hot-spot area in the framework of plant the odour concentrations are on level of 20-30 ou_E/m^3 (using OPERAT FB software) and above average amount in the GIS interpolation (30–45 ou_E/m^3). In the GIS interpolation prediction map the lower concentration points are very close (200-–300 m) to the waste processing plant area. The differences in interpolation polygon areas between the kriging and IDW method are not significant (visualization effect). The

contour (isoline) differences were especially in the bottom central quadrant of the map where the odours from emitters were sampled most rarely. In general, kriging as Gaussian process regression works better when input data is close to Gaussian distribution. Interpolated
a raster surface from odour sampling points using a kriging/IDW technique is calculated with high precision in the range of applied measurement data, but it may be a down fall in prediction variables because depends on the number of points and size of the search area. In all of the interpolation cases, the higher the density of sampled cells is, the more accurate will be the interpolated value.

4. Conclusions

The success of the odour dispersion model prediction depends on the comprehensive input data including odor emission rates, meteorological data, sources and surface characteristics, as well as the quality and relevance of the dispersion model. The comparison and performance of two different, but consistent tools for application in spatial analysis and modeling odour dispersion, the OPERAT FB software and ordinary kriging and IDW in the GIS environment, was evaluated in this study. Interpolated a raster surface from odour sampling points using a kriging/IDW technique is calculated with high precision in the range of applied measurement data, but it may be a down fall in prediction variables because depends on the number of points and size of the search area. Consequently, the available interpolation techniques (kriging/IDW) would work better depending on the statistical properties of the data.

This work was supported by a statutory research No. 0401/0065/16.

References

1. P.A. Longley, M.f. Goodchild, D.J. Maguire, D.W. Rhind, *Geographic Information Systems and Science* (Wydawnictwo Naukowe PWN SA, Warszawa, 2006)
2. J. Urbański, *GIS w badaniach przyrodniczych* (Wydawnictwo Uniwersytetu Gdańskiego, Gdańsk, 2011)
3. X. Li, Z.Z. Xu, Y.T. Qiu, J.Y. Qi, S.C. Tang, Adv Mat Res. **610**, 831-835 (2013)
4. N. Bozon, B. Mohammadi, B. Appl Geomat, **1**, 3, 59-74 (2009)
5. K. Fedra, Comput Environ Urban Syst., **23**, 443-457 (1999)
6. I. Sówka, *Metody identyfikacji dorotwórczych gazów emitowanych z obiektów przemysłowych* (Oficyna Wydawnicza Politechniki Wroclawskiej, Wrocław, 2011)
7. M. Skrętowicz, J. Zwoździak, *Dynamic-statistical odour dispersion model using the CALPUFF model and geostatistical analysis* (Proceedings of the 17th International Confernce on Harmonisation within Atmospheric Dispersion Modelling for Regulatory Purposes 9-12 May 2016, Budapest, Hungary, 2016)
8. E. Intarakosit, M. Ramirez, *GIS-Based Modeling of Odor Dispersion from Biosolids Reuse Sites* (Proceedings of the ESRI International User Conference, San Diego, California, 2007)
9. F.C. Rutz, B.L. Hoopes, D.W. Crandall, K.J. Allwine, *Development of a GIS Based Dust Dispersion Modeling System,* (Pacific Northwest National Laboratory (PNNL), Richland, WA (US), 2004)

10. N. Sharma, K. Bhandari, P. Rao, A. Shukla, *GIS applications in air pollution modelling* (Proceedings of the 6th International Conference 'Map India 2003', 2003)
11. Regulatin of the Minister of Environment dated 26 January 2010 on reference values for some substances in the air (Journal of Laws 2010 no. 16, item 87)
12. PN-EN 13725:2007 Air quality. Determination of odour concentration, with the use of dynamic olfactometry method
13. I. Sówka, U. Miller, P. Sobczyński, Environ. Protect. Eng., **42**, 3 (2016)
14. B. Namysłowska-Wilczyńska, *Geostatystyka. Teoria I zastosowania* (Oficyna Wydawnicza Politechniki Wrocławskiej, Wrocław, 2006)
15. I. Sówka, M. Skrętowicz, A. Nych, P. Zwoździak, J. Zwoździak, Przem. Chem., **91**, 5, (2012)
16. E. Intarakosit, M. Ramirez. *GIS-Based modelling of odour dispersion from Biosolids Reuse Sites*. Wilfrid Laurier University, e-thesis, (2010)
17. I. Sówka, Z. Kowalski, M. Skrętowicz, A. Makara, P. Sobczyński, K. Stokłosa, Przem. Chem., **92**, 6 (2013)
18. I. Sówka, P. Sobczyński, M. Skrętowicz, U. Kita, J. Zwoździak, *Zastosowanie wybranych narzędzi matematycznych oraz geostatystycznych do oceny zapachowej oczyszczalni ścieków* (Polska Inżynieria Środowiska; IV Kongres Inżynierii Środowiska, Monografia T.1. pod red. Marzenny R. Dudzińskiej, Artura Pawowskiego, Lublin: Komitet Inżnierii Środowiska PAN, 261-268, 2012)
19. VDI 3883: Olfactometry. Static sampling. Verein Deutscher Ingenieure, 2011
20. The Wroclaw University of Science and Technology, Department of Environmental Engineering, Olfactometric Research Laboratory, Technical Report, June 2013
21. P. P. Siska, and I - Kuai Hung. *Assessment of Kriging Accuracy in GIS Environment*. (Proceedings of ESRI Conference). CD Publication, San Diego, CA, (2001)
22. M. Brattoli, G. De Gennaro, V. De Pinto, A. Demarinis Loiotile, M. Lovascio s., penza, *Odour detection methods. Olfactometry and chemical sensors, Sensors*, (2011), 11, 5290

 DOI: 10.1051/e3sconf/20171700086

Financial efficiency of rainwater utilization system in single-family house

Agnieszka Stec[1,*]

[1]Department of Infrastructure and Sustainable Development, Rzeszow University of Technology, al. Powstańców Warszawy 6, 35-959 Rzeszów

Abstract. Designing of sustainable water systems should be aimed at reducing the consumption of tap water and the use of alternative water sources, such as rainwater and graywater. Therefore, the aim of the researches conducted was to determine the cost-effectiveness of the economic exploitation of rainwater utilization system in single-family house. As a tool for the analysis, the methodology Life Cycle Cost was used. It provides a comparison of different investment options and the opportunity to choose the one that is characterized by the lowest costs over the entire period of exploitation of the object. The research was conducted for four installation variants: the traditional solution and a solution in which rainwater was used for flushing toilets, washing and watering the garden. Variable parameters for calculations applied in the model were, among other things, different number of occupants and different length of exploitation of the installations. Additionally, the study included co-financing for the initial investment, which could be an incentive for residents to undertake this type of installation. The analysis conducted has shown that the systems with the use of rainwater enable significant reductions in the consumption of drinking water, while the variant with the traditional system was a most cost-effective solution only in few cases.

1 Introduction

A growth of the world's population, urbanization and climate changes, adversely affect water resources. According to numerous forecasts, the world's population will increase from the present level of 7.3 billion to 9.8 billion by 2050 [1] thus contributing to an increased global demand for water by 55% [2]. Urban dwelling populations will, during the same period, also increase to about 66% [1]. These demographic changes and the resulting increase in consumption, combined with climate changes, will have adverse impacts on the functioning of municipal water and sewerage infrastructure and water resources management [3]. This creates the necessity to look for new technologies and alternative solutions that will result in total or partial withdrawal from the traditional, centralized water and wastewater management.

Climate changes observed in recent years also affect the amount and intensity of rainfall and they often cause the occurrence of floods and hydraulic overload of sewer systems [4]. To counteract these phenomena various devices and objects for retention and infiltration of

* Corresponding author: stec_aga@prz.edu.pl

rainwater are used in catchments, as well as solutions which allow using rainwater installations in buildings [5, 6].

Designing of sustainable water systems should be aimed at reducing the consumption of tap water, among others, through the use of dredging batteries and devices to reduce the amount of water consumed, economic use of rainwater and recycling of gray water [7–9].

Economical use of rainwater can affect not only the protection of fresh water resources, but also reduce the financial outlays for the construction and maintenance of existing and new water supply networks [10, 11]. Thanks to the fact that the collected rainwater from roofs of buildings has a low degree of contamination, it is most commonly used for purposes where quality of potable water is not required, and that its treatment will not require the use of sophisticated chemical and biological processes. Rainwater Harvesting Systems (RWHS) are primarily used for toilet flushing, watering green areas, washing, car washing and irrigation of farmland [12–15]. The applicability of RWHSs and water savings achieved depend on many factors which include: the amount of rainfall and the frequency of their occurrence and the demand for non-potable water. These parameters also affect the profitability of the financial implementation of these systems in buildings [16–18].

Taking this into consideration the aim of the research conducted was to determine the cost-effectiveness of the economical exploitation of rainwater utilization system in a single-family house located in Poland. As a tool for the analysis, the Life Cycle Cost methodology was used. It provides a comparison of different investment options and the opportunity to choose the one that is characterized by the lowest costs over the entire period of exploitation of the object. The research was conducted for four installation variants.

2 Methodology

In order to determine the cost-effectiveness of the application of the RWHS system and the possibility of saving potable water a financial analysis for four different variants of sanitary installations in the building was carried out:

- Variant 0 - traditional systems (Fig. 1),
- Variant 1 - installation with the use of rainwater for toilet flushing (Fig. 2),
- Variant 2 - Installation with the use of rainwater to toilet flushing and washing (Fig. 3),
- Variant 3 - Installation with the use of rainwater for toilet flushing, washing and watering the garden (Fig. 4).

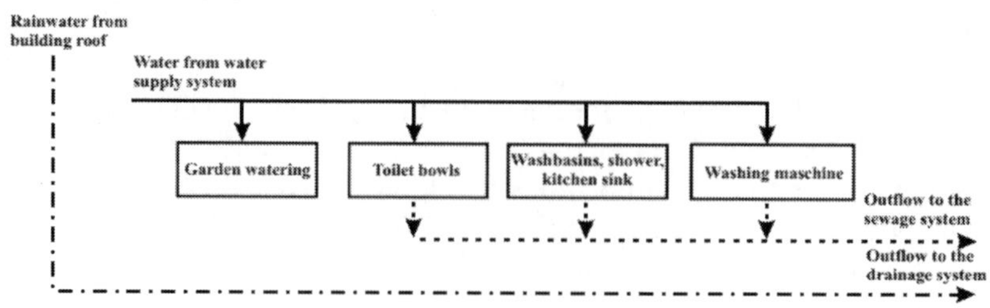

Fig. 1. The solution of installation in Variant 0.

In the study the Life Cycle Cost methodology was used which allowed to take into account all the costs connected with investment (*INV*) and use (*OCt*) of a system and the residual value (*RV*), which is the remaining value at the end of the study period [19]. *LCC* costs for the variants of installation analyzed are based on the formula (1). However, according to the guidelines in the work [20] in justified cases when the life of the system exceeds the length of the period of analysis, the residual value can be omitted. Similar assumptions were made by other researchers who analyzed the financial efficiency of RWHSs [21]. Taking this into account and the assumed lengths of the *LCC* analysis, the *RV* value was not considered. It was also assumed that T parameter in the financial model was a variable parameter, thanks to which it was possible to determine the impact of the length of the analysis period on the profitability of the investment.

$$LCC = INV + \sum_{t=1}^{T} (1+r)^{-t} \cdot OC_t + RV \quad (1)$$

where:
INV – investments, €;
OCt – operating costs in the year t, €;
RV – residual value, €;
T – duration of the *LCC* analysis, years;
r – constant discount rate;
t – another year of the system use.

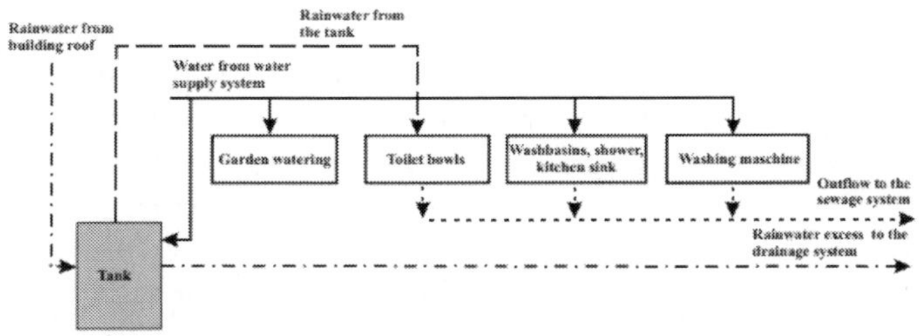

Fig. 2. The solution of installation in Variant 1.

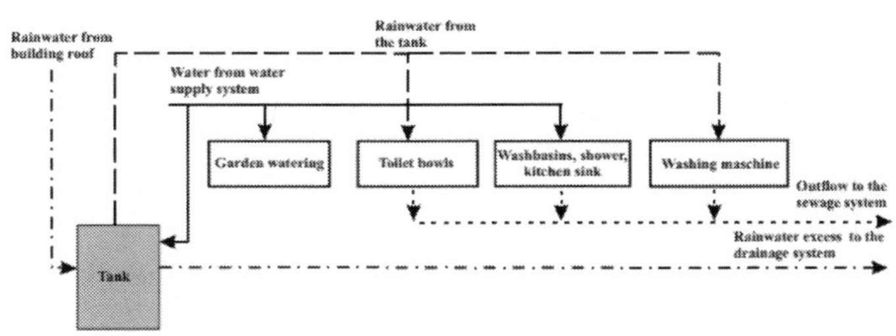

Fig. 3. The solution of installation in Variant 2.

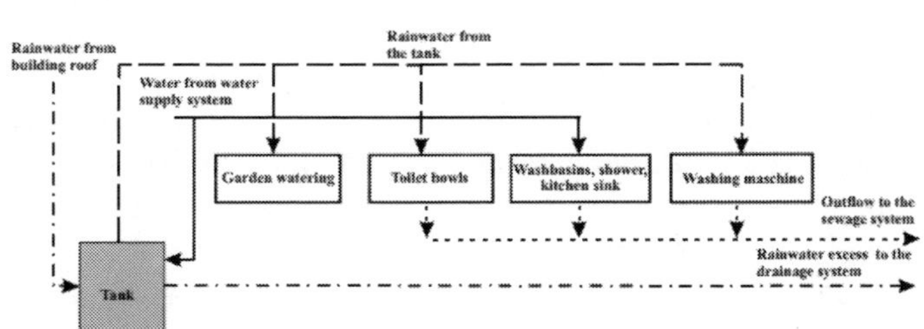

Fig. 4. The solution of installation in Variant 3.

The functioning of the system of rainwater harvesting was analyzed using a simulation model developed by Słyś [22]. This made it possible to determine the amount of rainwater that could be used or discharged into the sewerage system in each of the variants of the installation. Calculation algorithm of a simulation model applied was based on a daily water balance. A variable parameter of the simulation model was the number of users of the system (3, 4 and 5 people), which determined the size of the daily demand for water in the building, and this in turn affected the required capacity of the reservoir in the RWHS system.

3 Case study

Research to determine the financial efficiency of the use of RWHs was performed for a single-family building located in Rzeszów (the south-east part of Poland). A daily structure of water consumption in the building per capita is as follows: toilet flushing – 35 dm^3/d, washing – 18 dm^3/d, drinking and cooking – 4 dm^3/d, washing up – 47 dm^3/d, washing dishes – 12 dm^3/d, cleaning and other needs – 8 dm^3/d [23]. It was also assumed that in the period from May to September, 3 times a week a backyard garden will be watered (area 500 m^2), in the amount of 2,5 dm^3/m^2 [24].

Rainwater from the roof of the building with an area of 150 m^2 and a coefficient of runoff ratio of 0,9 will be discharged into an underground tank located near the building. In contrast, from the reservoir via the pump it will be transported to the inner system. On the basis of the demand for water for non-potable uses and guidelines for producers, rainwater tanks of 1,5 m^3 for Variant 1 and of 2 m^3 for Variants 2 and 3 were chosen. The exception for Variant 2 is the case when the system is used by 3 residents, where the volume of the reservoir is also 1,5 m^3. In the calculations a 7-day period of storage of water in the tank is assumed. Because of the possibility of bacterial growth after this time the tank is emptied. In periods of shortage of rainwater the reservoir will be fed with water from the water supply network. However, during intense and prolonged rainfall an excess of water is drained from the tank to the sewage system.

In the simulation model actual rainfall data for Rzeszów from a 10-year period (2003–2012) was recorded. The average annual precipitation H in the analyzed period amounted to 695 mm.

4 Input data of the financial model

The data characterizing the building considered, and the inputs of the financial model are summarized in Table 1. Investments and unit costs resulting from the operation of the system were established on the basis of producer prices and the applicable tariffs for water supply and sewage disposal.

In each of the investment variants, in the *LCC* analysis the initial investments INV_0 arising from the implementation of internal plumbing and sewage were considered. In addition, in variants 1, 2 and 3 capital expenditure associated with the use of rainwater harvesting system were also considered. These costs for the installation of the tank with a capacity of 1,5 m^3 amounted to 1448 euro while for the tank of 2 m^3 the cost was 1728 euro. In view of the fact that the rainwater will be used for washing, the RWHS system in Variants 2 and 3 is additionally equipped with a filter-pipe.

In turn, the operating costs of each variant include the costs of purchasing water from the water supply and sanitary sewage to the sewage network and fees for draining rainwater to the sewage system (the whole water for Variant 0, excess of water for Variant 1, 2 and 3). In Variant 1, 2 and 3 energy costs resulting from the transport of pumping water from the tank into the building and the cost of replacing a pump every 10 years of operation in the RWHS system were also taken into account.

Table 1. Data used in the calculation of LCC costs.

Parameter	Parameter value
Analysis period T	15, 20, 25 years
The annual increase in electricity prices	4%
The annual increase in the prices of purchase of water from the water-pipe network	6%
The annual increase in the prices of rainwater discharge to the sewage network	4%
The annual increase in the prices of sanitary sewage discharge to the sewage system	6%
The annual cost of filter cartridges	€10
The cost of purchasing electricity in the year 0	0.139 €/kWh
The cost of purchasing water from the water-pipe network in the year 0	0.880 €/m^3
The cost of sanitary sewage discharge to sewage network in the year 0	0.894 €/m^3
The cost of discharge of rainwater to the sewage network in the year 0	0.719 €/m^3
The cost of purchasing and installing the RWHS - tank 1,5 m^3 $INV_{RWHS\,1,5}$	€1448
The cost of purchasing and installing the RWHS - tank 2 m^3 INV_{RWHS2}	€1728
The cost of purchasing and installing the additionally filter INV_F	€72
The cost of purchasing and installing the sanitary systems INV_0	€1920
The discount rate r	5%

Increased initial investments resulting from the use of rainwater harvesting system can be a barrier for potential investors, for whom the financial criterion is the decisive one. In this connection, the studies were also conducted which included some subsidies to make variants of installation of the system of economic use of rainwater. The subsidies amounted to 15, 30 and 45%. Such aid for initial investment could come from the state budget or non-governmental organizations, as in case of many countries around the world.

5 Result and discussion

The studies conducted on a simulation model of rainwater harvesting system functioning in the analyzed building showed that rainwater was not able to completely replace tap water required to meet the demand for water for non-potable uses. Depending on the installation variant and number of residents an decrease in total tap water per year ranged from 14,6% to 30,3%. The high proportion of water from the sewage system in the water demand for non-potable uses is due to insufficient irregularity of the surface of the roof and the occurrence of precipitation during the year.

The results of simulation tests allowed to assess the effectiveness of financial investment on the utilization of rainwater in the building. The financial analysis showed that the use of Life Cycle Cost methodology in the evaluation of the profitability of the investment was the right choice, because the decision only on the basis of the initial investment might lead to the selection of a solution that in the long term might make very high operating costs. Based on the results obtained, which are listed in Table 2, it was found that only in 4 calculation cases the variant with conventional installations (Variant 0) was a solution with the lowest ratio of LCC. This was mainly in shorter periods of analysis when T was 15 years. At the same time it should be noted that for these cases the differences in the values of this index between Variant 0 and Variant 3 were insignificant. It also turned out that during the longest period T of 25 years, irrespective of the number of installation users, the best solution financially was always Variant 3, in which the rainwater was used for toilet flushing, washing and watering the garden. A similar situation with one exception was observed for the 20-year analysis period. The exception is the case when the installation is utilized by 3 people. This is caused by less need for rainwater and, consequently, lower savings which result from water purchase from the network for 20 years. The least cost-effective option was the concept of the system where rainwater was used only for toilet flushing (Variant 1). Regardless of the number of inhabitants and the length of operation of the installation the highest values of *LCC* were obtained.

Table 2. Summary of the results of Life Cycle Cost analysis for different investment variants.

Variant	The number of occupants								
	3 persons			4 persons			5 persons		
	$T = 15$ years	$T = 20$ years	$T = 25$ years	$T = 15$ years	$T = 20$ years	$T = 25$ years	$T = 15$ years	$T = 20$ years	$T = 25$ years
Variant 0	7 426	9 384	11 410	8 694	11 115	13 628	9 961	12 847	15 846
Variant 1	8 977	10 339	12 218	10 138	11 811	14 105	11 335	13 360	16 090
Variant 2	8 850	9 978	11 741	10 100	11 447	13 573	11 286	12 973	15 530
Variant 3	8 670	9 393	10 949	9 879	10 969	12 968	11 120	12 613	15 074

Tests results presented in Table 2 show that the differences between the values of the *LCC* for all analyzed variants and calculation cases are not great. This is an influence of additional investment costs in alternative installations supplied with rainwater, which are not fully compensated by lower operating costs obtained through the use of rainwater for non-potable uses. In addition the lower profitability of variants of RWHS is affected by the operating costs associated with the replacement of the pump every 10 years and the annual cost of replacement filter cartridges. In this regard the research was also carried out and it aimed at determining the effect of financing to make rainwater harvesting system in the single-family building. It was assumed that funding for the initial investment will amount to 15%, 30% or 45%. The results of these tests are summarized in Tables 3, 4 and 5.

Based on the results obtained one can notice that a subsidy of up to 45% will not change the cost-effectiveness of the use of Variant 0 for the analysis period T of 15 years. That means that in a shorter analysis period the reduction of the amount of potable water consumption, and the resulting from it cost savings due to RWHS application, does not fully compensate for the increased initial investment and the cost of replacing parts of the system. This will only reduce the differences in values of *LCC* ratio between Variant 0 and other variants of investment. On the other hand, when the installation is utilized by 3 residents and $T = 20$ years, the subsidy for initial investments in the amount of already 15% will change the most beneficial variant so far (Variant 0) in favor of Variant 3.

Table 3. Results of the *LCC* analysis for different installation variants in the building, taking into account 15% of the funding for capital expenditure.

Variant	The number of occupants								
	3 persons			4 persons			5 persons		
	$T = 15$ years	$T = 20$ years	$T = 25$ years	$T = 15$ years	$T = 20$ years	$T = 25$ years	$T = 15$ years	$T = 20$ years	$T = 25$ years
Variant 0	7 426	9 384	11 410	8 694	11 115	13 628	9 961	12 847	15 846
Variant 1	8 744	10 105	11 984	9 904	11 577	13 871	11 101	13 126	15 856
Variant 2	8 605	9 734	11 497	9 830	11 177	13 303	11 016	12 704	15 260
Variant 3	8 400	9 123	10 679	9 609	10 699	12 698	10 850	12 343	14 804

Table 4. Results of the *LCC* analysis for different installation variants in the building, taking into account 30% of the funding for capital expenditure.

Variant	The number of occupants								
	3 persons			4 persons			5 persons		
	$T = 15$ years	$T = 20$ years	$T = 25$ years	$T = 15$ years	$T = 20$ years	$T = 25$ years	$T = 15$ years	$T = 20$ years	$T = 25$ years
Variant 0	7 426	9 384	11 410	8 694	11 115	13 628	9 961	12 847	15 846
Variant 1	8 510	9 871	11 750	9 671	11 343	13 637	10 867	12 892	15 622
Variant 2	8 361	9 489	11 252	9 560	10 907	13 034	10 746	12 434	14 990
Variant 3	8 131	8 853	10 410	9 339	10 429	12 429	10 580	12 074	14 534

Table 5. Results of the *LCC* analysis for different installation variants in the building, taking into account 45% of the funding for capital expenditure.

Variant	The number of occupants								
	3 persons			4 persons			5 persons		
	T = 15 years	*T* = 20 years	*T* = 25 years	*T* = 15 years	*T* = 20 years	*T* = 25 years	*T* = 15 years	*T* = 20 years	*T* = 25 years
Variant 0	7 426	9 384	11 410	8 694	11 115	13 628	9 961	12 847	15 846
Variant 1	8 276	9 638	11 516	9 437	11 109	13 403	10 634	12 658	15 388
Variant 2	8 116	9 244	11 008	9 290	10 637	12 764	10 476	12 164	14 720
Variant 3	7 861	8 584	10 140	9 070	10 159	12 159	10 310	11 804	14 264

6 Conclusions

Economical use of rainwater for indoor and outdoor non-potable purposes makes it possible to reduce the consumption of tap water, and thus protect fresh water resources. It also reduces the amount of rainwater discharged into the sewerage system so improves their functioning. The search for alternative water sources is becoming increasingly important, since a climate change and the growing world population favor the occurrence of water deficits. In addition, fees associated with the operation of water supply and sewage systems in residential buildings represent a significant part of costs paid during the year for the use of the entire facility. Taking these problems into account, the analysis of Life Cycle Cost of the use of rainwater harvesting system in a residential building located in Poland was conducted.

The studies showed that rainwater can be a valuable source of water in the residential building. The use of rainwater for toilet flushing, washing and watering the garden can reduce the consumption of water from the water supply, depending on the variant of the installation, from almost 15% to over 30%. The lowest investment costs are in Variant 0, but after taking into account the operating costs, it turned out that only at the shortest period of analysis of 15 years, and low demand for rainwater, this option was still the most cost-effective. In all other cases, the lowest LCC value obtained for the variant where the rainwater was used to flush toilets, wash and water the garden (Variant 3). The use of rainwater harvesting system for toilet flushing only (Variant 1), or toilet flushing and washing (Variant 2) has a higher index values of *LCC*, but still for longer periods of analysis these options are more cost effective than the traditional solution for the installation.

The results of the financial analysis have a practical aspect and can be a valuable source of information for potential investors to use rainwater harvesting systems already at the stage of investment planning and decision-making. In addition, as confirmed also in studies, funding for initial investment will increase financial efficiency of the use of rainwater harvesting systems. This may be especially important in the case of single-family houses, for which, due to low demand for water resulting from the small number of residents, an implementation of rainwater harvesting system incurs very high investment.

References

1. The United Nations World Water Development Report 2015 Water For a Sustainable World. https://www.unesco-ihe.org/sites/default/files/wwdr_2015.pdf. Accessed 30 December 2016
2. OECD Environmental Outlook to 2050: The Consequences of Inaction, OECD Publishing, Paris. DOI: http://dx.doi.org/10.1787/9789264122246-en
3. A.S. Silva, E. Ghisi. Journal of Environmental Management **180**, 82-93 (2016)

4. B. Kaźmierczak, A. Kotowski. Theor. Appl. Climatol. **118**, 285-296 (2014)
5. K. Pochwat, D. Słyś. Journal of Civil Engineering, Environment and Architecture **63**, 359-369 (2016)
6. D. Słyś, A. Stec. Ecol. Chem. Eng. S **21**, 623-635 (2014)
7. J.Y. Cheng Leong, K.S. Oh, P.E. Poh, M.N. Chong. J. Clean. Prod. **142**, 3014-3027 (2017)
8. S. Kordana, D. Słyś, J. Dziopak. J. Clean. Prod. **82,** 58-69 (2014)
9. A. Stec, S. Kordana. Resour. Conserv. Recy. **105**, 84-94 (2015)
10. T.R. Gurung, A. Sharma. J. Clean. Prod. **67**, 26-36 (2014)
11. C.M. Silva, V. Sousa, N.V. Carvalho. Resour. Conserv. Recycl. **94**, 21-34 (2015)
12. E. Ghisi, P.N. Schondermark. Water Resour. Manag. **27**, 2555-2576 (2013)
13. C. Matos, C. Santos, I. Bentes, M.A. Imteaz, S. Pereira. Water Resour. Manag. http://dx.doi.org/10.1007/s11269-015-1040-9.
14. R. Karima, M. Bashara, M.A. Imteaz. Resour. Conserv. Recy. **104**, 61–67 (2015)
15. G. Markovič, D. Káposztásová, Z. Vranayová. Transactions on Environment and Development **10**, 242-249 (2014)
16. S.R. Ghimire, J.M. Johnston, W.W. Ingwersen, T.R. Hawkins. Environ. Sci. Technol . **48**, 4069-4077 (2014)
17. X. Liang, M.P. Van Dijk. Resour. Conserv. Recy. **55**,1100-1108 (2011)
18. T. Morales-Pinzón, R. Lurueña, J. Rieradevall, C.M. Gasol, X. Gabarrell. Resour. Conserv. Recy. **69**, 130-140 (2012)
19. S. Fuller, S. Petersen. Life Cycle Costing Manual for the Federal Energy Management Program/National Institute of Standards and Technology. NIST Handbook 135, the U.S. Department of Energy. Available: www.fire.nist.gov/bfrlpubs/build96/PDF/b96121.pdf (accessed November 2016)
20. Life Cycle Cost Handbook. Guidance for Life Cycle Cost Estimation and Analysis. Office of Acquisition and Project Management, U.S. Department of Energy, Washington. Available: www.energy.gov/sites/prod/files/2014/10/f18/LCC%20Handbook%20Final%20Version%209-30-14.pdf (accessed November 2016)
21. A. Rahman, J. Keane, A.M. Imteaz. Resour. Conserv. Recy. **61**, 16-21 (2012)
22. D. Słyś. Water Environ. J. **23**, 318-325 (2009)
23. J. Chudzicki, S. Sosnowski. Instalacje wodociągowe: projektowanie, wykonanie, eksploatacja, Wydawnictwo „Seidel-Przywecki" Sp. z o.o., Warszawa 2009.
24. Rozporządzenie Ministra Infrastruktury z dnia 14 stycznia 2002 r. w sprawie przeciętnych norm zużycia wody (Dz. U. 2002 Nr 8, poz. 70)

The influence of the ground parameters' assumptions on the low enthalpy heat pump's energy source simulation's results

Ewelina Stefanowicz[1], and *Natalia* Fidorów-Kaprawy[1,*]

[1]Wrocław University of Science and Technology, Faculty of Environmental Engineering, 27 Wybrzeże Wyspiańskiego St., 50-370 Wrocław, Poland

Abstract. In the paper the simulations of the low enthalpy energy source for brine-to-water heat pump using EED software had been presented. Authors indicated the influence of the ground parameters' assumptions on the low enthalpy heat pump's energy source simulation's results. Based on the real borehole geological profile and data available in literature, the ground thermal conductivity coefficients have been calculated making different assumptions concerning the ground thermal properties. The calculated ground thermal conductivity coefficients along with the calculated source sizes, heat pump parameters and the heated building thermal performance served as the input data for the simulation software. The outcomes of the analyses include the temperatures of brine entering the heat pump as well as the brine initial temperature. Outcomes served for the low enthalpy energy source performance evaluation.

1 Introduction

Renewable sources of energy gained the popularity back in the 80s, when the global energy crisis began. Some new buildings' heat supply systems were introduced and ground source heat pumps have been among them. Also already in 80s the research on the introduction of the computer software, enabling analysis of the low enthalpy energy sources for heat pumps, have been performed in the USA and Sweden [1]. The analysis of the simulation outcomes let designers optimize the low enthalpy energy source performance in the subsequent years of the heat pump operation. Earth Energy Designer – the vertical ground heat exchangers' design software is the example of such simulation computer program. Along with the development of the computer software, the need for the analysis input data determination methods, particularly the ground thermal conduction, have increased. There are several methods, like the pilot drilling or Thermal Response Test (TRT), that let determine the ground properties. The pilot drillings let the designers preliminarily asses the ground thermal conduction coefficients. The parameters of the particular layers can be found, based on the created ground profile. The determination of the ground moisture is the greatest issue. The thermal conductivity values of the ground with various level of saturation can be found in the literature. For example the thermal

* Corresponding author: natalia.fidorow@pwr.edu.pl

conductivity of the dry sand equals 0,4 W/(m·K), while for the saturated one it reaches 2,4 W/(m·K) [2]. The difference of the assumed specific extraction rate of the dry and saturated sand equals 20 W/m. The implementation of wrong assumptions leads to the design faults, that may result with low enthalpy energy source extensive cool down in the first years of the system operation. The second method that let determine the ground heat conduction, the TRT test, is carried out primarily for the bigger installations, because of the small availability and high costs. It is estimated that the TRT test establishment starts to be economically feasible for the systems with 10 or more boreholes, because it enables the proper low enthalpy energy source sizing [3]. The analysis methods of ground parameters are employed only for big installations due to high costs. The common practice is the low enthalpy source sizing on the basis of the assumed unit thermal performance of the ground. This procedure, although consistent with guidelines, may also lead to the low enthalpy energy source overcooling, the heat pump operation in emergency mode and substantial operational costs increase. The system efficiency decrease may be visible already in the first years of the system operation and depend on the difference between the assumptions made and the real ground properties.

2 Analysis method and system description

In the paper the design process and the thermal performance analysis of brine-to-water low enthalpy energy source had been performed based on the geological profile of the ground and taking into account various assumptions concerning the ground thermal properties and design guidelines. The purpose of the analysis was to indicate how those assumptions influence the system sizing and the simulation outcomes.

2.1. The geological profile and assumptions concerning ground properties

In the Fig. 1 the geological profile of the ground on the depth of 100 meters had been presented. Based on the layers description, the thermal conductivity (λ, W/(m·K)) and the volumetric heat capacity (s, MJ/(m^3·K)) of each layer had been assumed in two variants and the weighted average of the thermal conductivity and volumetric heat capacity of the ground serving as low enthalpy energy source had been calculated. The ground thermal conductivity and volumetric heat capacity in the first variant for all ground types excluding granite (clay, gravel and sand) had been assumed as an average between dry and saturated ones and the sand and granite content in the clay had been assumed as 30%. The ground thermal conductivity and volumetric heat capacity in the second variant for all ground types excluding granite had been assumed as for the saturated ground, the sand and granite content in the clay had been assumed the same as in first variant.

Based on the thermal conductivity coefficients it is possible to establish the specific extraction rate of the ground (q, W/m) used for the heat pumps' low enthalpy energy source sizing. There is a number of guidelines that may be used for that purpose [2,4,5] and they do not give the same outcomes. For the purpose of analysis authors decided to use the values recommended by Polish Organisation of the Heat Pump Technology Development (PORT PC – (in polish) Polska Organizacja Rozwoju Technologii Pomp Ciepła) [2] and the Swiss Standard SIA 384/6 (Erdwärmesonden) [5]. The obtained data have been used for low enthalpy energy source sizing and as the input data for the simulation software EED (Earth Energy Designer).

DOI: 10.1051/e3sconf/20171700087

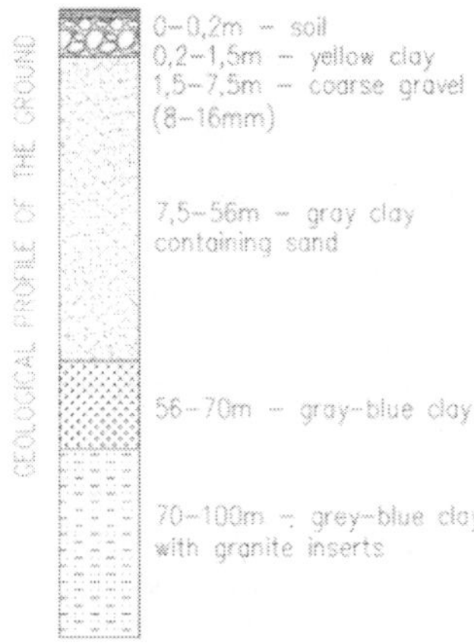

Fig. 1. The ground vertical profile.

2.2. The heat pump system and the building thermal properties description

The energy efficient small size commercial building had been assumed for the purposes of the analysis. The building heat load equals to 16,5 kW and its yearly energy demand is 24 000 kWh, while the heated area is 600 m^2. The building is equipped with the underfloor water heating system with flow temperature of 35°C. The energy demand for each month is given in the Table 1. The energy demand for hot tap water preparation purposes is covered by electric heaters and is not the purpose of this analysis.

Table 1. Monthly energy demand of the office building.

No	Month	Energy demand, MWh	No	Month	Energy demand, MWh
1	January	3.72	7	July	-
2	February	3.55	8	August	-
3	March	3.00	9	September	1.46
4	April	2.38	10	October	2.09
5	May	1.54	11	November	2.81
6	June	-	12	December	3.46

The energy demand for the spatial heating purposes is fully covered by the brine-to-water heat pump, which performance parameters, essential for the purpose of the system analysis, are given in table 2. The low enthalpy energy source for the heat pump consists of the single U-type (PE 32 x 3,0) boreholes, 85-100 m deep. The number and final depth of the boreholes have been calculated assuming four different unit thermal performance values obtained using two thermal conductivity values and two different guidelines. Additionally two other boreholes' numbers and lengths have been assumed using the values of the specific ground extraction rates from tables contained in design guidelines for engineers, without any previous calculations in order to find out how such source would act under heat load conditions for both thermal conductivity values described above. The heat transferring agent had been assumed as monoethylenglycole 25% solution in water. The flow rate had been assumed as 0,24 l/s per borehole and it ensures the turbulent flow in the boreholes and the minimum flow rate for the heat pump.

Table 2. Heat pump technical data (according to EN 14511).

T_H	T_B	-5°C	0°C	5°C	10°C	15°C
35°C	COP_{HP}	3.94	4.44	4.80	5.14	5.56
	Q_{HP}, kW	14.3	16.7	19.2	21.6	24.2
	Q_e, kW	11.0	13.8	16.0	18.0	20.4

2.3. The low enthalpy energy source performance simulation

In order to indicate the influence of the ground parameters' assumptions on the low enthalpy heat pumps' energy source simulation results authors performed the simulations in various variants. The assumptions for the simulations are gathered in Table 3.

In variants from 1.1 to 1.4, first the thermal conductivity had been calculated, according to the assumptions, and then the specific extraction rates of the ground (q, W/m) have been established according to different standards. Based on the established specific extraction rates and the heat pump cooling capacity, the demanded number and depth of the boreholes have been calculated, assuming that the average thermal conductivity of the ground does not change if the borehole length is in range of 85 to 100 m. Afterwards the systems performance have been simulated in EED (Earth Energy Designer) software, which is proven to give the results being in good accordance with the measurements [6].

In variants from 2.1 to 2.4 the specific extraction rates have been assumed (range used most often by designers), and then the sources were sized for two assumed values, without considering the ground thermal conductivity. The simulations in EED software have been carried out assuming two different values of thermal conductivity and volumetric heat capacity for both source sizes.

Table 3. The assumptions made for the simulation variants.

Variant no	Thermal conductivity λ, W/(m·K) and volumetric heat capacity s, MJ/(m^3·K)	Specific extraction rate of the ground, (q, W/m) assumptions
1.1	dry and saturated soil average	according to PORT PC
1.2	dry and saturated soil average	according to SIA 384/6
1.3	saturated soil	according to PORT PC
1.4	saturated soil	according to SIA 384/6
2.1	as in variants 1.1 and 1.2	20 W/m
2.2	as in variants 1.1 and 1.2	40 W/m
2.3	as in variants 1.3 and 1.4	20 W/m
2.4	as in variants 1.3 and 1.4	40 W/m

The input data for the simulations were: thermal conductivity of the ground, volumetric heat capacity of the ground, ground surface temperature, geothermal heat flow, boreholes' characteristics (number, configuration, depth, distance, diameter, heat transfer agent flow and U-pipe diameter and material), heat transferring agent properties, the thermal load of the building (as given in table 1) as well as the number of simulation years and the month of the heat pump system operation begin. The simulation outcomes are the average brine temperatures at the end of each month in 1st, 2nd, 5th, 10th and 25th year of system operation.

Thermal conductivity and volumetric heat capacity and the source size were implemented according to assumptions and calculations. The other properties are common to all system variants. The ground surface temperature assumed for the area of town Jelenia Góra $t_g = 6{,}9°C$; the geothermal heat flow $q_{gr} = 0{,}06$ W/m^2; boreholes distance $b_{ds} = 6{,}0$ m; ground heat exchanger material and diameter PE DN32 PN10 (32 x 3,0 mm). Heat transferring agent properties are: thermal conductivity $\lambda_{br} = 0{,}48$ W/(m·K), specific heat capacity $c_p = 3795{,}0$ J/(kg·K), density $\rho_{br} = 1052$ kg/m^3, viscosity $\eta_{br} = 0{,}0052$ kg/(m·s) and freezing point $t_f = -14°C$. All simulations have been performed for 25 years and the system starts in September.

3 Analysis, results and discussion

In following section the results of performed calculations, analyses and simulations are discussed in details. First the design process undergoes the basic analysis, than the low enthalpy energy sources' simulation results are analysed and discussed.

3.1. Thermal properties of the ground and boreholes sizing

Calculated thermal conductivity and established unit thermal performance of the ground is presented in table 4 along with the assumed thermal conductivity coefficients of the particular layers. The values described with subscript 1 concern the variant where ground parameters were assumed as an average for dry and saturated ones and the values described with subscript 2 concern the variant where ground parameters were assumed as for saturated ones. The values of the thermal conductivity for particular ground types (gravel, clay and sand both saturated and dry, as well as granite) have been taken as recommended values form the EED software database.

Table 4. Thermal conductivity coefficients and unit thermal performance of the ground.

No.	Depth, m	Ground type	Share in the structure	λ_1, W/(m·K)	λ_2, W/(m·K)
1	2,0 – 7,5	Coarse gravel (8-16 mm)	5.5/100.0	1.10	1.80
2	7,5 – 56,0	Gray clay with 30% of sand	48.5/100.0	1.12	1.84
3	56,0 – 70,0	Gray-blue clay	14.0/100.0	1.00	1.60
3	70,0 – 102,0	Gray-blue clay with 30% of granite	32.0/100.0	1.72	2.14
Weighted average λ coefficient, W/(m·K):				**1.3**	**1.9**
Specific extraction rate of the ground (PORT PC guidelines), W/m [2]:				35	41
Specific extraction rate of the ground (SIA 386/4 standard), W/m [5]:				23	32

The simulation of heat pumps' energy source in EED software requires the volumetric heat capacity of the ground as an input data. They have been calculated using the same assumptions and method as for thermal conductivity and are equal to: $s_1 = 2{,}0$ MJ/(m^3·K) and $s_2 = 2{,}4$ MJ/(m^3·K).

It can be seen that the designers, who have the data concerning the geological profile of the ground, but make different assumptions concerning the ground moisture content will calculate the thermal conductivity coefficients that vary a lot (around 46% in analysed case). Moreover if they use different standards concerning the relationship between the thermal conductivity coefficient and specific extraction rate of the ground the established values show even greater difference (in analysed case values differ almost twice).

The boreholes number and depth have been calculated for established and assumed ground specific extraction rates. The heat pump cooling capacity have been taken from table 2 for the brine temperature $T_B = 0$°C and the heating agent flow temperature $T_H = 35$°C. The calculation outcomes are given in Table 5.

Table 5. The demanded boreholes depth and quantity.

Variant no	Unit thermal performance of the ground, *q*, W/m	Demanded length of the ground heat exchanger	Calculated number and depth of boreholes
1.1	35	394 m	4 · 99 m
1.2	23	600 m	6 · 100 m
1.3	41	337 m	4 · 85 m
1.4	32	431 m	5 · 87 m
2.1 and 2.3	20	690 m	7 · 99 m
2.2 and 2.4	40	345 m	4 · 87 m

The length and quantity of boreholes have been sized in the way that fits most precisely to the demanded summary length of the heat exchanger. As expected, the sizes of the low enthalpy energy source also vary greatly, depending on the assumptions made. Those values have direct impact on the low enthalpy heat source investment cost, but also on the source performance during long years of operation.

3.2. The low enthalpy energy source simulation outcomes

As mentioned before the simulations of the heat pump's low enthalpy heat source performance have been executed for two design approaches. In first one ground heat exchanger sizes are the effect of geological profile analysis and specific extraction rates established based on the approved standards. In second one the heat source sizing is based on the assumed specific extraction rates (taken from design guidelines) without considering the geological profile. The outcomes are gathered and discussed in following subsections.

3.2.1. The simulation variants 1.1 to 1.4

The outcomes of the low enthalpy energy source simulation are presented in the Fig. 2 and Table 6. The differences among the variants are only due to the assumptions made by authors while calculating the thermal conductivity coefficient of the ground and while assuming the specific extraction rates of the ground, which affected the source size.

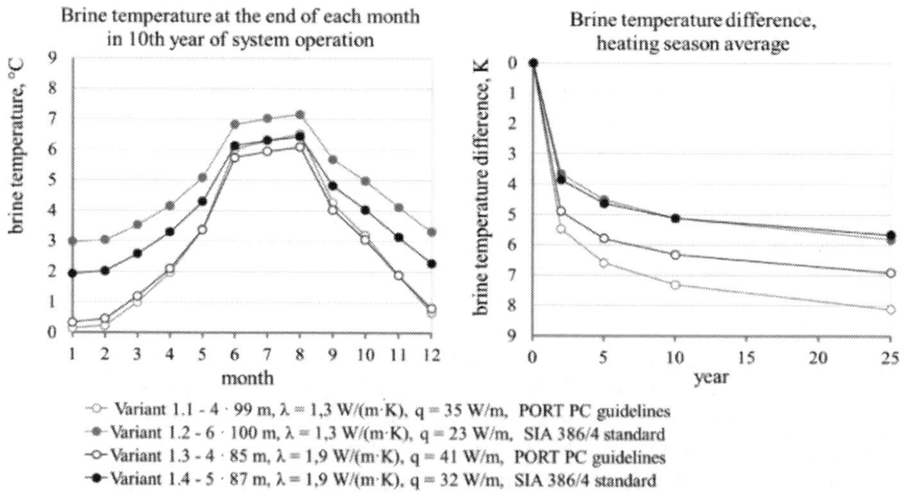

Fig. 2. The outcomes of the system variants 1.1 to 1.4 simulation in EED software.

On the left side of the Fig. 2 the brine temperature values at the end of each month in 10th year of the heat pump system operation for each simulation variant are presented. The difference in brine temperature, depending on the variant, is about 3 K in winter and about 1 K in summer. Those temperature changes are due to the various boreholes thermal load (various source sizes) bud also due to different brine initial temperatures among variants. On the right side of Fig. 2 the difference between the brine initial temperature (before the installation commissioning) and the seasonal average brine temperature are shown. The temperature differences for variants 1.2 and 1.4 (SIA 386/4 standard) are aligned, and those for variants 1.1 and 1.3 (PORT PC guidelines) are divergent. This means that the relationship between the thermal conduction coefficient of the ground and its specific extraction rate proposed in SIA 386/4 standard, which is not linear [5], assigns such specific extraction rates to the different thermal conductivity coefficients that cause similar boreholes performance (similar cool down). The relationship proposed in PORT PC (linear one [2]) assigns such values that for smaller thermal conductivity coefficients give greater brine cool down. Table 6 brings the additional information about the brine initial temperature for each variant, the temperature of the brine in the end of January (the middle of heating season) and in the end of August (just before the heating season, after source self-regeneration during summer) in 25th year of heat pump operation as well as the difference between the initial and operational values. The difference in the initial brine temperatures are due to different boreholes' depth and thermal conductivity coefficients of the ground. The brine temperature in the heating season determines one of the proper sizing conditions, and the temperature difference between the initial and end of summer temperature (which approximately corresponds to the global ground cool down) determines other proper boreholes' sizing condition.

Table 6 Brine temperatures and temperature changes after 25 years of operation.

Simulation	V 1.1	V 1.2	V 1.3	V 1.4
t_{begin}	9.2	9.2	8.2	8.3
t_{25} (January)	-0.7	2.2	-0.3	1.4
Δt_{25} (January)	9.9	7.0	8.5	6.9
t_{25} (August)	6.5	6.5	5.5	5.9
Δt_{25} (August)	2.8	2.8	2.7	2.4

Although all sources are designed according to the approved standards, the outcomes differ one from another. First of all it should be noticed that despite the different assumptions concerning ground moisture, and obtaining different ground thermal conductivity coefficients, as well as despite concerning different relationships between thermal conductivity coefficients and specific extraction rates, all outcomes confirm the source proper sizing. The temperature difference between the initial and the end of summer value do not exceed 3 K. The temperature of the brine in the end of January, which is the lowest value, in case of applying SIA 386/4 standard for specific extraction rates determination, is above 0°C, and in case of applying PORT PC guidelines the brine temperature only slightly drops below 0°C. PORT PC guidelines allow for the smaller (cheaper) source sizing, so the source slightly worse performance was to be expected, because the source is more heavily loaded (especially when thermal conductivity is lower).

Assuming that the calculated thermal conductivity coefficient is the same as real one, the effects described above will be probably true not only while simulating the heat pump's energy source, but also for real installations, because the EED software results proved to be close to the measurements [6]. It is also interesting to find out what will happen if the thermal conductivity is unknown, and the source is designed based on rough assumptions.

3.2.1. The simulation variants 2.1 to 2.4

The outcomes of the simulations are given in in Fig. 3 and Table 7. It has been assumed that both designed sources (for 20 W/m and 40 W/m specific extraction rates) performance will be simulated for the ground with thermal conductivity coefficients of 1,3 W/(m·K) and 1,9 W/(m·K).

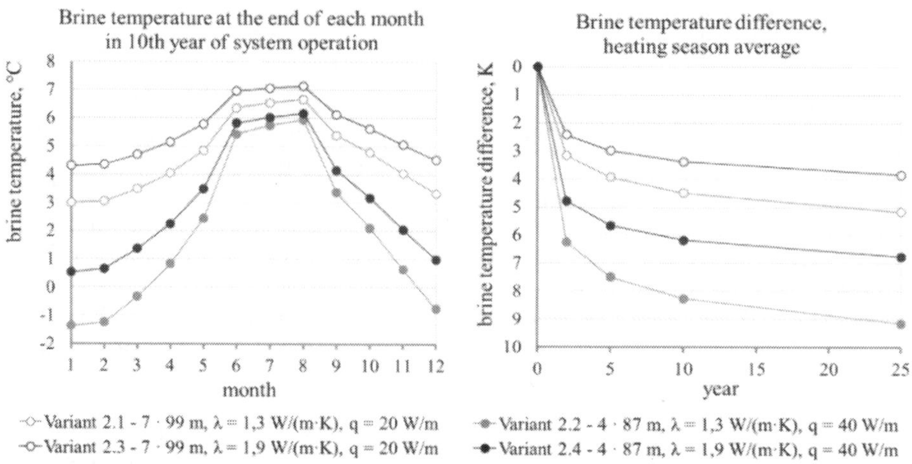

Fig. 3. The outcomes of the system variants 2.1 to 2.4 simulation in EED software.

Table 7 Brine temperatures and temperature changes after 25 years of operation.

Simulation	V 2.1	V 2.2	V 2.3	V 2.4
tbegin	9.2	8.9	8.5	8.3
t25 (January)	3.0	-2.3	3.8	-0.1
Δt25 (Januray)	6.2	11.2	4.6	8.3
t25 (August)	6.7	5.1	6.7	5.6
Δt25 (August)	2.5	3.8	1.8	2.7

In Fig. 3 the brine temperature values at the end of each month in 10th year of the system operation are presented for each variant, as well as the difference between the brine initial temperature (before the installation commissioning) and the seasonal average brine temperature. It can be noticed, that for variants 2.1 and 2.4, which somehow corresponds to the design guidelines, the installations work properly. As it was expected, in case where the designer assumes high extraction rate (40 W/m) and the ground conditions are poor (1,3 W/(m·K)), like in variant 2.2, the low enthalpy energy source will get overloaded, and may freeze. Variant 2.3 is interesting, because one could expect much better performance of the system, in which the low enthalpy energy source is oversized. The simulation results indicate however that the brine temperatures are only slightly higher that for the variant 2.1, and the ground cool down is only a bit smaller.

4 Conclusion

The assumptions made while sizing the brine-to-water heat pump's low enthalpy energy source influence the simulation outcomes, which vary in wide range. Those differences may influence further decisions, including one concerning the installation realization.

While sizing the low enthalpy energy source size, it is important to combine various methods of the data concerning the ground properties gathering; pilot drilling, TRT test and computer simulations; as each of them may indicate some inaccuracy. The knowledge of soil parameters and the appropriate design guidelines selection enables the correct operation of the installation in subsequent years. The analysis based on the computer simulations is the tool that let verify the proper low enthalpy energy source performance.

The change of ground's thermal conduction value and the boreholes depth influence initial, calculated (EED software) brine temperature, strictly related to the unloaded ground temperature. The disproportion between assumed ground specific extraction rate and one resulting from the ground thermal conductivity value is visible in simulation results.

References

1. J.D. Spitler, S.E.A. Gehlin, Renew. Sust. Energ. Rev., **50**, 1125–1137 (2015)
2. *Guidelines for the design, completion and handover of installations with heat pumps. Part 1. Energy sources for heat pumps.(in Polish: Wytyczne projektowania, wykonania i odbioru instalacji z pompami ciepła. Część 1. Dolne Źródła do pomp ciepła.)* PORT PC (2013)
3. M. Wajman, Tech. Chłodn. I Klimat. **12**, 546–557 (2011)
4. B. Sanner, Geothermische Energie, **26/27**, 1–4 (1999)
5. *SIA 384/6 Erdwärmesonden,* Swiss Society of Engineers and Architects (2010)
6. G. Hellström, B. Sanner, M. Klugescheid, T. Gonka, S. Martensson, *Proceedings of Megastock Conference 1997,* Sapporo, Japan, 247–252 (1997)

Rate of change in concentration of organic compounds and nutrients in municipal and dairy wastewater treated in full technical SBR scale

Joanna Struk-Sokolowska[1,*], and *Monika* Puchlik[1]

[1]Technical University Of Bialystok, Department of Technology in Engineering and Environmental Protection, Wiejska St. 45E, Bialystok 15-351, Poland

Abstract. Municipal WWTP often receive the industrial wastewater, including dairy sewage. The problem of sewage treatment from this sector of economy is not completely solved, especially in plants of medium and small milk processing capacity. The aim of the study was to determine the rate of change in organic compounds and nutrients in municipal wastewater with the participation of dairy products during the processing phases of SBR-type reactor in a technical scale. The study was conducted in WWTP in Sokolka. Rate of changes in the concentrations of organic compounds, nitrogen and phosphorus in wastewater during the processing phases of SBR reactor was determined. Presence of dairy sewage (13% volume participation) affected the decrease in the rate of changes in organic compounds concentrations defined as BOD_5 during all processing phases of SBR reactor. Significant influence of dairy sewage on rates of change in nitrogen and phosphorus compounds concentrations during filling, aeration, stirring, sedimentation and decantation phases, was found.

1 Introduction

In municipal wastewater treatment technology, the activated sludge in a batch SBR system is often used. Wastewater treatment is carried out in a single-chamber bioreactor, in which during the alternating phases, processes of integrated removal of carbon, nitrogen and phosphorus occur, as well as the distribution of treated wastewater from the process medium, takes place. There are few reports [1–3] in the literature on the rate of change in organic compounds and nutrients during SBR processing phases.

Municipal sewage treatment plants often receive the industrial wastewater, including dairy sewage. The problem of sewage treatment from this sector of economy is not completely solved, especially in plants of medium and small milk processing capacity. Dairy wastewater is characterized by high load of COD and BOD_5. This includes dissolved and crystallized fats (glycerol, triglycerides), sugars (lactose), and protein (casein) in colloidal form and as a clot. Shares of these substances can vary. Quantity of dairy sewage

* Corresponding author: j.struk@pb.edu.pl

is characterized by high seasonal variations (20% increase in spring and summer). Additional oscillations relate to the daily and hourly shift work dependent on the size, production profile, and shifts at milk processing plant [4–12]. Biological treatment of dairy wastewater is high-efficient due to the significant share of easily biodegradable substances [13–16]. It is possible to co-treat the dairy and municipal wastewater to achieve the desired C:N:P ratio in order to carry out highly effective denitrification and biological phosphorus removal. However, analysis of the up-to-date knowledge showed that there is a lack of data on the effects of dairy sewage on the efficiency of municipal wastewater treatment in sequential chambers.

The aim of the study was to determine the rate of change in organic compounds and nutrients in municipal wastewater with the participation of dairy products during the processing phases of SBR-type reactor in a technical scale.

2 Material and methods

2.1 Area of study

The study was conducted in mechanical-biological treatment plant in Sokolka. The facility consists of mechanical sewage treatment station (blocked device of type Rotomat Huber Ro5/780/2 with capacity of 120 $l{\cdot}s^{-1}$ equipped with the cylindrical sieve with a diameter of 780 mm and 2 mm clearance holes, sand traps purged from the scrubber sand RoSF4tC, fat separator, and press for dewatering of remains). Mechanically treated waste-water flows into the intermediate pumping station. They are pumped into the averaging two-chamber retention tank with an active volume of 1800 m^3. This tank is equipped with 2 hyperboidal stirrers of type HCMA with aeration grate and in mixer 2 hyperboidal stirrers of type HCM. There is the possibility of dosing the coagulants. Wastewater is pumped from the tank to the sequential biological reactors (SBR). The process of biological treatment is carried out in the cyclic system, in four cylindrical chambers of the active capacity of 1500 m^3, equipped with hyperboidal stirrers of type HCMA. Duration of a single cycle is 8 hours 10 minutes (filling phase – 0.7 h, aeration phase – 4 h, the stirring phase – 1 h, sedimentation phase - 1.5 h, decantation phase - 1 h). Oxygen conditions vary over a range from 1.8 to 4 g $O_2{\cdot}m^{-3}$. The maximum daily capacity of the plant is 6000 m^3, and the equivalent population (PE) 25070. The object receives domestic and industrial wastewater from the milk processing plant. Dairy Cooperative in Sokolka discharges sewage to municipal sewage treatment plant after the pretreatment process. Table 1 presents types of wastewater and sludge subjected to physicochemical determination.

Table 1. Types of municipal wastewater and sludge subjected to physicochemical determination.

Sample	Hour	Phase of treatment process
S1	6.00 a.m.	Raw wastewater
S2	6.20 a.m.	Wastewater after mechanical treatment collected from the retention chamber*
S3	7.00 a.m.	Wastewater from SBR (end of filling phase, before aeration phase)
S4	8.10 a.m.	Wastewater from SBR, after 60 minutes of aeration (total duration 240 min.)
S5	9.10 a.m.	Wastewater from SBR, after 120 minutes of aeration
S6	10.10 a.m.	Wastewater from SBR, after 180 minutes of aeration
S7	11.10 a.m.	Wastewater from SBR, after 240 minutes of aeration
S8	12.10 a.m.	Wastewater from SBR (end of stirring phase)
S9	1.40 p.m.	Wastewater from SBR (end of sedimentation phase – 90 minutes)
S10	2.40 p.m.	Wastewater from outlet (end of decantation phase – 60 minutes)
$S_{a.slud.}$	11.00 a.m.	Activated sludge from SBR chamber
		* Periodic supply of over-sediment water from decanter centrifuge

Table 2 shows the average flows and loads of municipal and dairy wastewater flowing into the sewage treatment plant according to operational data from 2010-2014.

Table 2. Quantities and loads of contaminants in dairy and municipal wastewater.

Qd_{max} (d-daily; max – maximum)	$m^3 \cdot d^{-1}$	6000
Qd_{av} (d-daily; av – average)	$m^3 \cdot d^{-1}$	3350
$Qd_{av\ liquid\ wastes}$ (d-daily; av – average)	$m^3 \cdot d^{-1}$	27
PE (equivalent population)		25070
$Qd_{av\ dairy}$ (d-daily; av – average)	$m^3 \cdot d^{-1}$	430
Volume share of dairy sewage in municipal wastewater	%	13
Load of BOD_5 (Biochemical Oxygen demand)	$kg \cdot d^{-1}$	1504
Load of $BOD_{5\ dairy}$	$kg \cdot d^{-1}$	103
Share of BOD_5 load of dairy sewage in municipal wastewater	%	7
Load of COD (Chemical Oxygen Demand)	$kg \cdot d^{-1}$	2811
Load of COD_{dairy}	$kg \cdot d^{-1}$	201
Share of COD load of dairy sewage in municipal wastewater	%	7

Table 3 presents selected data on the production profile of the dairy processing plant.

Table 3. Production characteristics of dairy processing plant.

Amount of processed milk [$m^3 \cdot d^{-1}$]	80.00
The ratio of wastewater produced [$m^3 \cdot d^{-1}$]	4.81
Leading products	cottage cheese, sour cream, kefir, yogurt

2.2 Analytical Methods

The scope of analytical determinations made in samples included analysis of the following indicators: BOD_5, COD, TOC, N_{tot}, P_{tot} - in wastewater; V, MLSS, SVI – in activated sludge. Analyses were carried out in accordance with APHA (2012) [17]. Based on the study results, calculations of carbon, nitrogen and phosphorus removal efficiency (total and partial), as well as the rate of change in parameters during the processing phases of SBR reactor, were carried out. Total removal efficiency was calculated from the difference of a given parameter concentration at the inlet (C_O) and outlet (C). Efficiency of the partial of organic carbon, nitrogen and phosphorus removal from wastewater was calculated from the difference of a given parameter concentration at the start (C_X) and end (C_Y) of the treatment phase.

Rate of changes in the concentrations of organic compounds, nitrogen and phosphorus in wastewater (r_V) during the processing phases of SBR reactor was determined from the formula:

$$r_V = (A - B) / T \tag{1}$$

r_V – rate of the parameter change in wastewater, $mg \cdot dm^{-3} \cdot h^{-1}$
A – concentration at the start of processing phase, $mg \cdot dm^{-3}$
B – concentration at the end of processing phase, $mg \cdot dm^{-3}$
T – durability of processing phase, h.

To determine the rate of change in parameters in wastewater relative to the concentration of activated sludge in SBR reactor (r_X) for particular processing phases, following dependence was applied:

$$r_X = [(A / X_p) - (B / X_k)] / T \tag{2}$$

r_X – rate of the parameter change in wastewater, $mg \cdot g_{DM}^{-1} \cdot h^{-1}$
X_p – concentration of activated sludge in SBR at the start of processing phase, $g_{DM} \cdot dm^{-3}$
X_k – concentration of activated sludge in SBR at the end of processing phase, $g_{DM} \cdot dm^{-3}$.
To calculate the rates (r_V, r_X), a simplification was assumed that concentrations of organic compounds and nutrients at the end of the processing phase are equal to the concentrations at the start of the subsequent SBR processing phase. It was also assumed that changes in biomass of the activated sludge during an 8-hour SBR cycle are not significant [1, 2, 14]. Accordingly, the equation takes the following form:

$$r_X = [(A - B) / X] / T \quad (3)$$

X – concentration of activated sludge in SBR, $g_{DM} \cdot dm^{-3}$.
In order to determine the influence of dairy sewage on the rate of change in the concentrations of organic compounds and nutrients, municipal wastewater (collected before connecting with a stream of dairy sewage) was purified in a model SBR reactor presented in paper by Struk-Sokołowska and Ignatowicz [18], according to the methodology described in the publications by Struk-Sokołowska [2, 18]. Municipal wastewater treatment in a model SBR reactor was performed in this study. Processing parameters and the activated sludge were the same as in the wastewater treatment plant in Sokolka.

3 Results

Table 4 presents characteristics of activated sludge in Sokolka.

Table 4. Characteristics of activated sludge ($S_{a.slud}$) in Sokolka.

Settled sludge volume	V	$cm^3 \cdot dm^{-3}$	510.0
Concentration of solids in the sludge suspension	MLSS	$kg_{DM} \cdot m^{-3}$	3.8
Sludge Volume Index	SVI	$cm^3 \cdot g_{DM}^{-1}$	134.2

3.1 Rates of parameters change in wastewater in various stages of treatment

The rates of changes (r_V and r_X) in particular parameters of municipal wastewater with 13% share of dairy sewage are presented in Table 5.

In mixed filling phase in SBR in Sokolka wastewater is mixed with the biomass. It initiates biological reactions, degrades the organics by using residual oxygen. De-nitrification may occur in this fill. Average rate of total nitrogen removal during 40-minute filling phase in SBR content was 64.3 $mg \cdot dm^{-3} \cdot h^{-1}$ (in reference to the activated sludge biomass concentration was 16.92 $mg \cdot g_{DM} \cdot h^{-1}$). During this phase, some decrease in the total phosphorus concentration was recorded. Rate of phosphorus removal was 6.4 $mg \cdot dm^{-3} \cdot h^{-1}$ (1,68 $mg \cdot g_{DM}^{-1} \cdot h^{-1}$). Average rate of organic matter removal expresses as BOD_5 during filling phase amounted to 668.6 $mg \cdot dm^{-3} \cdot h^{-1}$ (in relation to the biomass -176 $mg \cdot g_{DM} \cdot h^{-1}$). Average rate ($r_V$) of organic matter removal determined by parameter COD during dephosphatation in SBR was 922.1 $mg \cdot dm^{-3} \cdot h^{-1}$ (242.7 $mg \cdot g_{DM} \cdot h^{-1}$).

Aeration phase in SBR in Sokolka may be mixed react or aerated react type. Mixed react allows anoxic condition to achieve de-nitrification and aerated react allows aerobic condition during aerated fill to achieve complete nitrification. The aeration phase lasting 240 minutes cause the removal of the total nitrogen from wastewater at mean rate of 1.5 $mg \cdot dm^{-3} \cdot h^{-1}$ (0.39 $mg \cdot g_{DM} \cdot h^{-1}$). It should be underlined that this rate during initial 60 minutes amounted to 7.8 $mg \cdot dm^{-3} \cdot h^{-1}$ (2.05 $mg \cdot g_{DM} \cdot h^{-1}$), while during the subsequent 180 minutes, it decreased up to 0.5 $mg \cdot dm^{-3} \cdot h^{-1}$ (0.13 $mg \cdot g_{DM} \cdot h^{-1}$) or the increase of the total nitrogen concentration at the rate of 1.2 $mg \cdot dm^{-3} \cdot h^{-1}$ (0.32 $mg \cdot g_{DM} \cdot h^{-1}$), was recorded.

With the beginning of the aeration phase, the concentration of nitrogen increased as a result of nitrification, before the settling phase. Before the decantation phase the concentration of nitrogen decreased, which may result from denitrification effected by denitrificators which contain organic matter in their cells [3]. This phase caused the total phosphorus removal from wastewater at the average rate of 1.1 $mg \cdot dm^{-3} \cdot h^{-1}$ (over 2 $mg \cdot dm^{-3} \cdot h^{-1}$ during initial 120 minutes). It should be emphasized that between 120 and 180 minute of the aeration phase, sudden release of total phosphorus at the rate of 3.2 $mg \cdot dm^{-3} \cdot h^{-1}$ was observed, and then between 180 and 240 minute of that phase, total phosphorus concentration decrease at the rate of 3 $mg \cdot dm^{-3} \cdot h^{-1}$. In order to remove phosphorus biologically, alternate aerobic and anaerobic conditions have to be ensured. During the anoxic phase, phosphorus concentration in the wastewater under treatment increased as a result of introducing raw wastewater and secondary phosphorus release to the liquid under anaerobic conditions. As the time of aeration increased, the phosphorus concentration decreased, as they were assimilated by certain bacteria strains (e.g., Acinetobacter sp., Arthrobacterglobiformis) in as much as 25% of dry cell weight [3]. Aeration of wastewater and activated sludge during 240 minutes resulted in the removal of organic matter (BOD_5) with an average rate of 1.5 $mg \cdot dm^{-3} \cdot h^{-1}$ (0.39 $mg \cdot g_{DM} \cdot h^{-1}$). During the initial and final 60 minutes of aeration, the removal rate was 2 $mg \cdot dm^{-3} \cdot h^{-1}$ (0.53 $mg \cdot g_{DM} \cdot h^{-1}$), while during the other 120 minutes, it dropped to 1 $mg \cdot dm^{-3} \cdot h^{-1}$ (0.26 $mg \cdot g_{DM} \cdot h^{-1}$). During this phase removal of organic matter referred to as COD occurred at an average rate of 1.6 $mg \cdot dm^{-3} \cdot h^{-1}$. It should be underlined that during the first 60 minutes, the rate of impurities removal was 4.7 $mg \cdot dm^{-3} \cdot h^{-1}$ (1.24 $mg \cdot g_{DM} \cdot h^{-1}$), while during the next 180 minutes of aeration, it decreased in the range from 0.9 to 0.1 $mg \cdot dm^{-3} \cdot h^{-1}$. An increase in concentration TOC was caused by carbon as similation by heterotrophic organisms. Organic matter can also be precipitated and adsorbed in activated sludge [3].

Table 5. Rates of changes in nutrients and organic compounds during processing SBR phases.

Processing phase	N_{tot}		P_{tot}		BOD_5		COD		TOC	
	r_V	r_X	r_V	r_X	r_V	r_X	r_V	r_X	r_V	r_X
filling	64.3	16.92	6.4	1.68	175.95	668.6	242.66	922.1	34.11	129.6
aeration	1.5	0.39	1.1	0.29	0.39	1.5	0.42	1.6	-0.05↑	-0.2↑
after 1/4	7.8	2.05	2.3	0.61	0.53	2.0	1.24	4.7	6.29	23.9
after 1/2	-1.2 ↑	-0.32↑	2.1	0.55	0.26	1.0	0.24	0.9	-0.11↑	-0.4↑
after 3/4	0.5	0.13	-3.2↑	-0.84↑	0.26	1.0	0.21	0.8	-0.45↑	-1.7↑
end	-1.2 ↑	-0.32↑	3.0	0.79	0.53	2.0	0.03	0.1	0.37	1.4
stirring	0.5	0.13	1.5	0.39	-	-	0.32	1.2	1.21	4.6
sedimentation	1.1	0.29	-0.1↑	-0.03↑	0.18	0.7	-1.08↑	-4.1↑	1.47	5.6
decantation	0.2	0.05	-	-	-	-	0.13	0.5	-	-

r_V - $mg \cdot dm^{-3} \cdot h^{-1}$; r_X – $mg \cdot g_{DM}^{-1} \cdot h^{-1}$, (-) – no change recorded; (↑) – increase

In sedimentation phase in SBR in Sokolka separation of solids as settled sludge is allowed to provide clarified supernatant under quiescent condition. No influent or effluent currents interfere with settling process. In some cases gentle mixing during initial stages of settling allows clearer effluent and more concentrated sludge. Phase of sedimentation of the activated sludge in SBR caused the re-release of total phosphorus in wastewater, however, it was much lower (rate 0.1 $mg \cdot dm^{-3} \cdot h^{-1}$), than during aeration phase. The rate of changes in organic compounds defined by BOD_5 during this phase amounted to 0.18 $mg \cdot g_{DM} \cdot h^{-1}$ and the increased amounts of organic substances (as COD) in wastewater was recorded (4.1 $mg \cdot dm^{-3} \cdot h^{-1}$). The concentration of TOC decreased during the settling phase, which may have been caused by carbon consumption by denitrificators [3].

In decantation phase in SBR in Sokolka clarified treated supernatant is allowed to remove within the predetermined cycle time. Decanters are used in this process which may be of floating type or fixed type. During decantation phase total phosphorus removal and changes in organic compounds expressed as BOD_5 were not recorded in wastewater. During this phase the concentration of total phosphorus was not recorded to increase as a result of the phosphorus being released to the liquid from the precipitate [3]. Organic compounds (as COD) were removed at the rate of 0.5 $mg \cdot dm^{-3} \cdot h^{-1}$ (0.13 $mg \cdot g_{DM} \cdot h^{-1}$). Rate of total nitrogen removal was 0.2 $mg \cdot dm^{-3} \cdot h^{-1}$ (0.05 $mg \cdot g_{DM} \cdot h^{-1}$). When the accumulated organic matter was exhausted, the nitrification process took place until the end of the decantation process, in the presence of oxygen [3].

Number of literature references on the municipal wastewater treatment with a specific and strictly controlled share of dairy sewage, including biodegradation kinetics in SBR, is insufficient [1-3]. Data on qualitative and quantitative characteristics of raw dairy sewage are relatively well-documented [4-16], but at the same time there are serious gaps in relation to the municipal sewage with a specific share of industrial (dairy) wastewater. For this reason, it is impossible to compare the results with literature data.

Table 6 illustrate rates recorded in SBR-type reactors in wastewater treatment plant in Sokolka, where municipal wastewater containing dairy sewage is processed. These rates were compared to values achieved in reactors treating municipal wastewater not containing dairy sewage.

Table 6. Comparing the rate of changes (r_V, r_X) in the concentrations of compounds in wastewater and wastewater enriched with dairy sewage (mixed wastewater) during phases of the SBR cycle.

Waste-water	Parameter				
	BOD_5	COD	TOC	N_{tot}	P_{tot}
	$r_V - r_X$	$r_V - r_X$	$r_V - r_X$	$r_V - r_X$	$r_V - r_X$
Filling – 40 minutes					
mixed	668.6 – 179.95	922.1 – 242.66	129.6 – 34.11	64.3 – 16.92	6.4 – 1.68
municipal	754.3 – 198.50	831.1 – 218.71	117.4 – 30.89	62.7 – 16.5	-5.1↑ – -1.34↑
Aeration – 240 minutes					
mixed	1.5 – 0.39	1.6 – 0.42	-0.2↑ – -0.05↑	1.5 – 0.39	1.1 – 0.29
municipal	2.4 – 0.63	4.8 – 1.26	3.1 – 0.82	3.2 – 0.84	2.4 – 0.63
Stirring – 60 minutes					
mixed	-	1.2 – 0.32	4.6 – 1.21	0.5 – 0.13	1.5 – 0.39
municipal	0.1 – 0.03	0.9 – 0.24	4.0 – 1.05	0.9 – 0.24	2.2 – 0.58
Sedimentation – 90 minutes					
mixed	0.7 – 0.18	-4.1 ↑ – -1.08↑	5.6 – 1.47	1.1 – 0.29	-0.1↑ – -0.03↑
municipal	1.4 – 0.37	4.0 – 1.05	2.1 – 0.55	1.2 – 0.32	-0.8↑ – -0.21↑
Decantation – 60 minutes					
mixed	-	0.5 – 0.13	-	0.2 – 0.05	-
municipal	1.0 – 0.26	0.1 – 0.03	-0.4↑ – -0.11↑	-0.2↑ – -0.05↑	-0.2↑ – -0.05↑
r_V – $mg \cdot dm^{-3} \cdot h^{-1}$; r_X – $mg \cdot g_{DM}{}^{-1} \cdot h^{-1}$; (-) – no change recorded; (↑) – increase					

The presence of dairy sewage contributed to a reduction in the rate of change of organic compounds concentrations defined by BOD_5 during all processing phases of SBR reactor. Taking into account concentration of organic compounds (as COD), the 13% share of dairy sewage contributed to the decrease in the change rate only during the aeration phase. A significant influence of dairy sewage on the rate of the nitrogen and phosphorus compounds concentration change during the filling, aeration and stirring phases, was recorded. The rate of nitrogen removal during the filling phase was lower in the case of municipal wastewater treatment only. In the case of phosphorus, presence of dairy wastewater caused phosphorus binding, while for municipal wastewater, its release was observed. During the phases of aeration and stirring, reduction in the rate of phosphorus and nitrogen changes was recorded due to the presence of dairy wastewater. At the same time, during the sedimentation phase, there were comparable rate values with respect to nitrogen compounds and significantly higher rates of phosphorus release in the SBR reactor treating the municipal with no dairy wastewater. During decantation phase, impact of the dairy sewage on the rate of change in nitrogen and phosphorus compounds was smaller. Differences in the rates of change in nitrogen and phosphorus concentrations were derived from the rate of organic compounds exhausting.

The results reported in literature on biodegradation kinetics in SBR of wastewater contaminated with phenol, pesticides, herbicides as well as tannery and slaughterhouse wastewater [19-23].The rates of changes in organic compounds and nutrients of municipal wastewater with 13% share of dairy sewage can not be compared with the results of other authors.

4 Summary

The 8-hour work cycle of the SBR reactor has ensured high effectiveness of contaminants removal from municipal wastewater. Presence of dairy sewage affected the decrease in the rate of changes in organic compounds concentrations defined as BOD_5 during all processing phases of SBR reactor and the increase in organic compounds concentrations (COD) during filling, the stirring and decantation phases of SBR reactor. Exploring the kinetics of the wastewater treatment process in SBR reactor reveals characteristic changes in nitrogen concentrations for the nitrification and denitrification processes. Moreover, changes typical of phosphorus removal by phosphate microorganisms are also observable. Significant influence of dairy sewage on rates of change in nitrogen and phosphorus compounds concentrations during filling, aeration ,and stirring phases, was found. During the sedimentation phase in SBR-type reactor was found a decrease in the value of organic compounds (BOD_5, TOC) and N_{tot} and an increase of COD and P_{tot}. During decantation phase only COD and N_{tot} value decreased.

The research has been carried out in the framework of project No. S/WBiIS/3/2014 and financed from the funds for science MNiSW.

References

1. W. Janczukowicz, S. Grabowski, J. Pesta, R. Brzozowska, Kinetics of orthophosphate release and uptake in the volatile fatty acid-fed sequencing batch reactor treating dairy wastewater, Pol. J. Natur. Sc. **22**, 2, 256–270 (2007)
2. J. Struk-Sokołowska, Zmiany frakcji ChZT w procesie oczyszczania ścieków komunalnych i mleczarskich w oczyszczalni typu SBR, PB, (2015) (in Polish)
3. M. Wojnicz, A. M. Anielak, Kinetics of Dairy Wastewater Treatment in the SBR System. Arch. Env. Prot. **36**, 3, 27–37 (2010)
4. F. M. E. Emerald, D. T. S. A. Prasad, M. R. Ravindra, H. A. Pushpadass, Performance and biomass kinetic of activated sludge system treating dairy wastewater. Intern. Journ. Dairy Techn. **65**, 4, 609–615 (2012)
5. Y. W. Gong, H. X. Hang, X. N. Cheng, Treatment of dairy wastewater by two – stage membrane operation with ultrafiltration and nanofiltration. Wat. Scie. Tech. **65**, 5, 915–919 (2012)
6. S. J. Lim S, P. Fox, A kinetic analysis and experimental validation of an integrated system of anaerobic filter and biological aerated filter. Bios. Tech. **102**, 10371-10376 (2011)
7. N. Mehrdadi, G. R. Nabi Bidhendi, M. Shokouhi, Determination of dairy wastewater treatability by bio-trickling filter packed with lava rocks – case study PEGAH dairy factory. Wat. Scie. Tech. **65**, 8, 1441–1447 (2012)

8. E. Neczaj, M. Kacprzak, T. Kamizela, J. Lach, E. Okoniewska, Sequencing batch reactor system for the co-treatment of landfill leachate and dairy wastewater. Desali. **222**, 404–409 (2008)
9. M. Passeggi, I. Lopez, L. Borzacconi, Integrated anaerobic treatment of dairy industrial wastewater and sludge. Wat. Scie. Tech. **59**, 3, 501–506 (2009)
10. T. Z. Pentado, R. S. S. Santana, A. L. B. Dibiazi, S. C. Pinho, R. Ribeiro, G. Tommaso, Effect of agitation on the performance of an anaerobic sequencing batch biofilm reactor in the treatment of dairy effluents. Wat. Scie. Tech. **63**, 5 (2011)
11. A. Tawfik, M. Sobhey, M. Badawy, Treatment of a combined diary and domestic wastewater in an up-flow anaerobic sludge blanket (UASB) reactor followed by activated sludge (AS system). Desali. **227**, 167–177 (2008)
12. A. A. L. Zinatizadeh, Y. Mansouri, A. Akhbari, S. Pashaei, Biological treatment of a synthetic dairy wastewater in a sequencing batch biofilm reactor: statistical modeling using optimization using response surface methodology. Chem. Indust. Chem. Engine. Quart. **17**, 4, 485–495 (2011)
13. R. Ayeche, Treatment by coagulation-flocculation of dairy wastewater with the residua lime of National Algerian Industrial Gases Company (NIGC-Annaba). Energy Proc. **18**, 147–156 (2012)
14. W. Janczukowicz, M. Zieliński, M. Dębowski, Biodegradability evaluation of dairy effluents originated in selected sections of dairy production. Biores. Techn. **99**, 4199–4205 (2008)
15. G. R. Munavalli, P. S. Saler, Treatment of dairy wastewater by water hyacinth. Wat. Scie.&Tech. **59**, 4, 713–722 (2009)
16. N. S. A. Mutamim, Z. Z. Noor, M. A. A. Hassan, A. Yuniarto, G. Olsson, Membrane bioreactor: Applications and limitations in treating high strength industrial wastewater. Chem. Engine. Journ. **225**, 109–119 (2013)
17. E. W. Rice, R. B. Baird, A. D. Eaton, L. S. Clesceri, Standard Methods for the Examination of Water and Wastewater, Ame. Pub. Hea. Ass. (APHA), Ame. Wat. Wor. Ass. (AWWA), Wat. Envir. Fed. (WEF), 22 nd Edition (2012)
18. J. Struk-Sokołowska, K. Ignatowicz, Municipal and Dairy Wastewater Co-treatment Using SBR Technology. Ann. Set of Environ. Protect. **15**, 1881-1898 (2013)
19. Z. Duan, Microbial degradation of phenol by activated sludge in a batch reactor. Envi. Protec. Enginee. **37**, 2, 53–63 (2011)
20. Ch. Deepak, Biodegradation of pesticide-contaminated wastewaters in denitrifying sequencing batch reactors. University of Canterbury, New Zealand (2016)
21. E. Celis, P. Elfsiniotis, N. Singhal, Biodegradation of Agricultural herbicides in seqencing batch reactors under aerobic or anaerobic conditions. Wat. Res. **42**, 3218–3224 (2008)
22. G. Durai, M. Rajasimman, N. Rajamohan, Kinetic studies on biodegradation of tannery wastewater in a sequential batch reactor. J. Biote. Res., **3**, 19–26 (2011)
23. P. Kundu, A. Debsarkar, S. Mukherjee, Treatment of slaughter house wastewater in a sequencing batch reactor: performance evaluation and biodegradation kinetics, Hind. Publis. Corpor. BioMed Res. Intern., 1–11 (2013)

E3S Web of Conferences **17**, 00089 (2017) DOI: 10.1051/e3sconf/20171700089
EKO-DOK 2017

Application of the selected classification models to the analysis of the settling capacity of the activated sludge – case study

Bartosz Szeląg[1,*], and *Piotr* Siwicki[2]

[1]Kielce University of Technology, Faculty of Environmental, Geomatic and Energy Engineering, Tysiąclecia Państwa Polskiego Ave. 7, Poland
[2]Warsaw University of Life Sciences - SGGW, Faculty of Civil and Environmental Engineering, Nowoursynowska St.166, Poland

Abstract. The study presents the development of classification models for sedimentation of activated sludge using the artificial neural networks (ANN), logistic-regression (RL), and linear discrimination model (LDM). The input consisted of indicators of wastewater quantity and quality (biochemical oxygen demand, chemical oxygen demand, total suspended solids, total nitrogen and total phosphorus at the inflow to the wastewater treatment plant) and operational characteristic of the bioreactor (pH, temperature of activated sludge, mixed liquor suspended solids, concentration of oxygen in the nitrification chamber, amount of PIX dosing).The prediction quality of the developed models was measured with: sensitivity, specificity, and computed errors. The calculations of the sedimentation were performed for sludge volume index (SVI). The results indicate that successful predictions were obtained using ANN, RL and LDM methods, which is supported by the fit of computations to measurement results. The study shows that for the wastewater treatment plant of concern, sedimentation properties can be obtained using only the loads of organic compounds, mixed liquor suspended solids, temperature, pH of activated sludge, concentration of oxygen in the nitrification chamber and amount of PIX dozing. Other analysed variables appear to be statistically insignificant for the sludge volume index.

1 Introduction

Engineering practice shows that the activated sludge method is one of the most frequently used and most effective technologies of domestic wastewater treatment. Sludge volume index (SVI) is one of the parameters employed to assess the performance of the wastewater treatment plant [1-3]. Sludge volume index makes it possible to predict metastable states, the occurrence of which is related to the excessive growth of filamentous bacteria. During wastewater clarifications in secondary settlement tanks, the presence of those bacteria can lead to an increase in the content of total suspended solids and of organic carbon in the outflow from the settler. Therefore, to be able to predict metastable states in

* Corresponding author: bszelag@tu.kielce.pl

good advance, and to make it possible for the treatment plant staff to select optimum settings of the bioreactor parameters to ensure effective wastewater treatment, it is necessary to develop mathematical models that are capable of determining continuous value or the linguistic value. In the former case, the discrete values of the measured dependent variable are predicted [4–6]. In the latter case, the range of variation of the measurement results provides a basis for a division into classes. The ranges of variation of the parameter under analysis, set by the researcher, correspond to those classes [7–8].

Numerous regressive methods are used to model the discrete value of the sludge volume index. The methods include Artificial Neural Networks and their modifications, Support Vector Machines, Random Forests, Boosted Trees, and k-Nearest Neighbour. For the analysis of the settling capacities, the classification models based on the methods listed above are also used. In those models, the SVI value is described by a binary (zero-one) variable, or linguistic values are assigned to SVI value ranges. In the classification models, the number of independent variables describing a given phenomenon can be smaller than in the regression models, because it is not the discrete value that is modelled but exclusively the class membership. That fact is of considerable importance in practical applications, especially those concerning wastewater treatment facilities. That means it is possible to reduce the number of variables needed to describe a given phenomenon. With the classification approach, sludge settling capacity can be estimated relying on the extreme SVI values. Additionally, SVI values can be corrected by means of altering the facility settings [9]. In the regression and classification black box methods, which include the ones mentioned above, the model structure being a basis for a given dependent variable prediction is generated at the training stage. However, to develop the model, it is necessary to implement complex numerical algorithms. Occasionally, model development would require designing specialist computer programs. Although such models show satisfactory predictive abilities in estimating SVI, their practical applications to technological areas will involve the use of additional software and make it necessary to integrate the computational and measurement modules, which can be a difficult task. The literature review shows that the logistic regression model was used to analyse the settling capacity. However, the correlation produced by [7] and also [8] did not allow the possibility of correcting the bioreactor technological parameters in order to improve the settling capacity of the activated sludge. In economic, and also medical and social sciences, to analyse binary data the so-called linear discriminant model is used. The results of computations demonstrate that the model has better predictive abilities than logit regression models or black box methods of ANN-type, and others.

The objective of this paper is to compare predictive abilities of the selected classification methods for the control of the settling capacity of the activated sludge. The paper analyses the possibility of modelling sludge settleability with the use of the black box models and dichotomy models, such as the logistic regression model and linear discriminant model. Additionally, the aim of the analyses is to develop a simple regression model for the continuous control and monitoring of the settling capacity of the activated sludge. That could be based on the quality indexes of wastewater influent and operational parameters of aeration tanks. Thus, it would not be necessary to implement complex numerical algorithms, or install additional measurement systems at the wastewater treatment plant.

2 The object of investigation

The analyses were conducted for the urban wastewater treatment plan (WWTP) having the design capacity of 72.000 m^3/d, located in the terrain of the commune of Sitkówka - Nowiny. The plant collects wastewater from the city of Kielce, the commune of Sitkówka -

Nowiny, and partially also from the commune of Masłów. The wastewater delivered to WWTP is mechanically pretreated using step screens and aerated grit chambers with grease separators. Next, wastewater is pumped to primary clarifiers, from which it is delivered to the biological unit, i.e. bioreactor with dephosphatation, denitrification and nitrification tanks. Then, together with activated sludge, wastewater is transferred to four secondary clarifiers, and after clarification it flows to the receiving water, i.e. the river Bobrza.

Continuous monitoring conducted by the company Wodociągi Kieleckie Sp. z o.o. at the treatment plant provides measurements of influent wastewater quantity, and also of the operational parameters of the bioreactor (Table 1).

3 Methodology

In this paper, the linear discriminant model, the logit model and the artificial neural network were employed to analyse sludge settling capacity. The results of measurements of influent wastewater quality and the reactor operational parameters were utilised to this end. The former included the following: biochemical oxygen demand, the content of ammonia, nitrate and nitrite nitrogen, total suspended solids, and total phosphorus. The bioreactor parameters were as follows: mixed liquor suspended solids (MLSS), activated sludge temperature and pH, amount of dosage of ferric coagulants (PIX), oxygen concentration in the nitrification tank. To identify the dependence between the variables of concern and activated sludge index, a correlation matrix was constructed based on the measurements of wastewater quantity and quality and the bioreactor parameters. In the analyses, the criterion for the assessment of the impact of the bioreactor operational parameters and influent wastewater quality was the value of the sludge volume index (SVI), equal to SVI_{lim} = 150 cm^3/g [7]. When the measured value of SVI was greater than SVI_{lim}, it was assigned the value of 1, otherwise it was given the value of 0.

Linear discriminant model (LDM) and logistic regression are used to analyse binary data and they offer the possibility of predicting the occurrence or non-occurrence of an event. In addition, those models are frequently applied to economical, and also medical and social sciences. The logit model is also often employed in river engineering, geomorphology, environmental protection, microbiology, wastewater treatment, and in analyses of the operation of components of sewage systems [8,10,11]. In addition to the applications above, the linear discriminant model was also used to assess the condition of stormwater drainage pipe and to analyse landslide occurrence [12,13]. Although linear discriminant model (LDM) has much better predictive abilities than the logit model, it has not been used for the analysis of the settling capacity of the activated sludge. The dichotomy models mentioned above can be described by the following dependences:

- the logit model:

$$p(\mathbf{X}) = \frac{exp(\mathbf{X})}{1+exp(\mathbf{X})} \tag{1}$$

- the linear discriminant model:

$$Z(\mathbf{X})_k = \mathbf{X}_k \tag{2}$$

in which: k – number of classes to be separated, and thus the number of discrimination functions; in the problem under consideration k = 2; in the first case, function Z_1 describes membership in a group, in which SVI < 150 cm^3/g, whereas function Z_2 expresses combinations of linear parameters x_i, for which SVI > 150cm^3/g, p – probability of exceeding the limiting value SVI_{lim} that is the basis for sludge volume index prediction, which correspond to the value p=0.50; X – vector made from linear combination of independent variables, expressed by the formula:

$$\mathbf{X} = \sum_{i=1}^{j} \beta_i \cdot x_i + \beta_0 \tag{3}$$

where: x_i – variables describing sludge volume index, which include the following: MLSS, sludge temperature and pH, amount of dosage of ferric coagulants (PIX), the content of ammonia nitrogen and that of organic substances in influent wastewater, and others; β_0, β_i – coefficients determined with the maximum likelihood method.

The advantage offered by the logit model is the possibility of interpreting the value of $\exp(\beta_i)$. For that purpose, the notion of chance value is used. It is defined as the ratio of the probability of an event occurrence to the probability of its non-occurrence (1 – p). The effect of the increment of independent variables by the value of Δx_i ($i = 1, 2, \ldots, j$) on the chance of the event occurrence can be computed on the basis of the so-called chance ratio expressed as $OR=\exp(\Sigma\beta_i \cdot \Delta x_i)$.

Neural networks can be used to simulate linear and non-linear processes, optimisation, classification and control [5, 6, 14]. For the sake of modelling the processes taking place in the environment, the network termed the multilayer perceptron (MLP) is used most frequently. In MLP networks, input signals are multiplied by the values of weights and then transferred to the neurons of the hidden layer, where summation occurs in individual neurons. The sums received are transformed using a nonlinear activation function and conveyed to the output neurons. The optimal structure of the neural network for the prediction of SVI binary value was sought by computing the measures of fit between simulation results and measurements (SENS, SPEC, R_z^2). It was assumed that number of neurons in the hidden layer was (3–10) and the activation function was linear, expotential, sine, sigmoidal and hyperbolic tangent. For the output layer, the linear activation function was adopted.

The accuracy of predictions obtained with the statistical models was established based on the values of sensitivity (SENS), specificity (SPEC) and computed error (R_z^2), described in detailed in the study by Harrell [15], and other papers.

4 Results

Based on the results of measurements of wastewater quantity and quality and the bioreactor technological parameters, the range of parameter variation was determined (Table 1) in order to establish the applicability of the developed classification models.

The data provided in Table 1 indicate that daily wastewater inflows to the WWTP and the wastewater quality indexes varied substantially. That considerably affected the operational parameters of aeration tanks; the value of the mixed liquor suspended solids (MLSS) ranged 1.91–6.59 kg/m^3 and oxygen concentration in the nitrification tank changed 0.55–5.78 mg/dm^3, whereas the mean value of dissolved oxygen (DO) was 2.56 mg/dm^3. The data in Table 1 demonstrate that in the period of concern, the settling capacity of the activated sludge deteriorated, which is indicated by both the mean value of SVI = 186 cm^3/g and the maximum one that was 320 cm^3/g.

Based on the measurement results (Table 1), the values of the coefficient of correlation (r) between the considered variables describing influent quality and the bioreactor technological parameters were determined (Table 2).

Table 1. Range of variation of values of parameters describing wastewater quantity (Q) and quality (BOD_5, TSS, NH_4^+-N, TN) and aeration tank operational parameters (T_{sl}, pH, MLSS, RAS, PIX, DO).

Variable	Minimum	Mean	Maximum
Q, m^3/d	32564	40698	86592
T_{sl}, ^0C	10.0	15.9	23.0
pH	7.2	7.7	7.8
MLSS, kg/m^3	1.98	4.26	6.59
PIX, m^3/d	0.0	0.8	1.93
DO, mg/dm^3	0.55	2.56	5.78
SVI, cm^3/g	95	186	320
BOD_5, mg/dm^3	127	309	557
TSS, mg/dm^3	126	329	572
NH_4^+-N, mg/dm^3	24.4	49.4	65.9
TN, mg/dm^3	39.9	77.7	124.1

Table 2. Values of the coefficient of correlation (r) between the variables considered and the sludge volume index.

Variables	BOD_5	COD	TSS	NH_4^+-N	TN	pH	T_{sl}	MLSS	DO	PIX	Q
r	0.21	0.15	0.17	0.25	0.19	-0.21	-0.46	-0.61	-0.17	-0.12	-0.10

On the basis of the data in Table 2, it can be concluded that weak correlations hold between the value of sludge volume index and BOD_5 (r=0.21), COD (r=0.15), TSS (r=0.17), NH_4^+-N (r=0.25), and TN (r=0.19) in the influent. In addition, weak correlation was observed between SVI value and daily inflow into the wastewater treatment plant (r=-0.10). As regards the bioreactor parameters, such as DO, pH, and PIX, a weak dependence between SVI and the parameters of concern was found. Additionally, Table 2 indicates high and medium correlation between sludge volume index and activated sludge temperature (r =-0.46) and MLSS (r=-0.61). The results obtained (Table 1) are confirmed by investigations conducted by Lou and Zhao [4], Bayo et al.[7]. Due to the fact that the value of the sludge volume index alone does not give direct information on the aeration tank loading, in further analyses pollutant loads were introduced (L_{BOD5}). Based on the measurement data (Table 1) and the results of analyses (Table 2), binary variables were developed to describe sludge volume index. By applying the step regression method, variables x_i and coefficients β_i were determined in the model described by Eqs. (1) and (2). In the case under consideration, the equations can be expressed by the following dependences:

- the logistic regression:

$$p(\mathbf{X}) = \frac{\exp(15.79-0.39\cdot T_{sl}-1.33\cdot MLSS-1.83\cdot PIX-1.18\cdot DO+0.00008\cdot L_{BOD5})}{1+\exp(15.79-0.39\cdot T_{sl}-1.33\cdot MLSS-1.83\cdot PIX-1.18\cdot DO+0.00008\cdot L_{BOD})} \quad (4)$$

- the linear discriminant model:

$$Z = \begin{cases} 18{,}66\cdot MLSS - 11{,}36\cdot T_{sl} - 367{,}50\cdot pH - 7{,}48\cdot PIX + 0{,}001L_{BOD5} - 0{,}14\cdot DO - 1528{,}19 \\ 20{,}24\cdot MLSS + 11{,}92\cdot T_{sl} + 370{,}19\cdot pH + 9{,}18\cdot PIX - 0{,}001L_{BOD5} + 1{,}44\cdot DO - 1567{,}04 \end{cases} \quad (5)$$

As regards neural networks, the best classification abilities are demonstrated by the model that has 7 neurons in the hidden layer and the activation function takes the form of the hyperbolic tangent. The comparison of the parameters describing the fit of computational results to measurements, namely SPEC, SENS, and R_z^2 are presented in Table 3. In the logistic regression and LDA models, the remaining variables, i.e. pH, recirculation ratio, total suspended solids, total and ammonia nitrogen turned out to be statistically insignificant at the adopted confidence level p = 0.05. In the classification model based on artificial neural network method, the independent variables of SVI value, given in Eqs. (4) and (5), were adopted. The data listed in Table 3 show that the classification models based on the logit model and artificial neural networks have identical predictive abilities.

Table 3. Parameters of fit of computational results to measurements (SPEC, SENS, R_z^2) in models for sludge volume index classification.

Parameter	Methods		
	logit	LDM	ANN
SPEC	0.827	0.804	0.875
SENS	0.873	0.841	0.826
R_z^2	0.854	0.826	0.854

The lowest fit of computational results to measurements was obtained for the model based on the linear discriminant analysis (LDM). For instance, with the logit model, out of 55 events analysed, during which values of SVI < 150 cm^3/g, 46 events were classified correctly. Out of 55 events, during which sludge settling problems occurred (SVI > 150 cm^3/g), the model classified 49 events correctly. With the model based on the LDA method, out of 55 events considered, during which SVI < 150 cm^3/g, 44 cases were classified correctly, so were 47 events in 55 when SVI > 150 cm^3/g. Due to the fact that the ANN-based statistical model is an implicit dependence SVI = f(x_i) and has identical predictive abilities as the logistic regression model described by Eq.(4), binomial logit model was utilised to identify the regression dependence in further analyses.

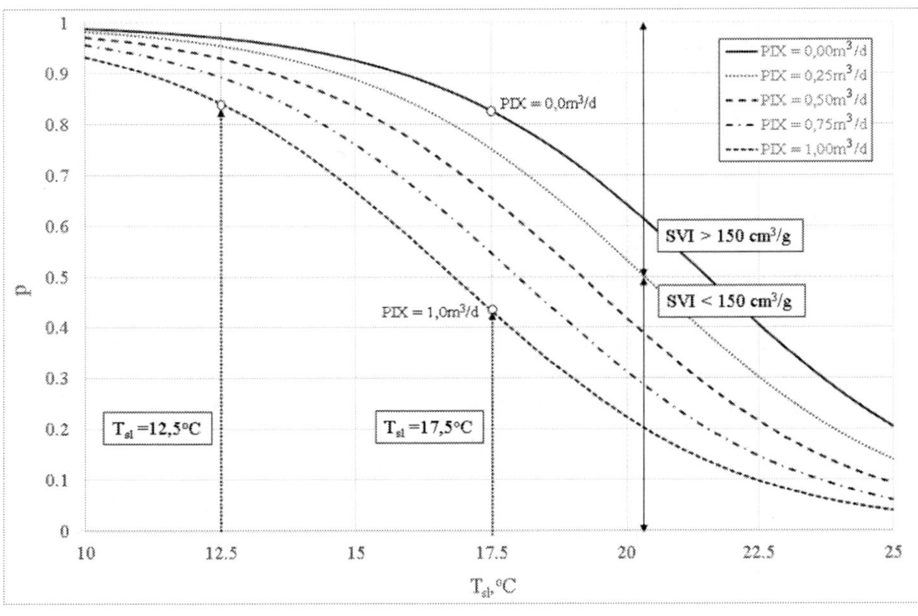

Fig. 1. The effect of activated sludge temperature (T_{sl}) and the amount of dosage of PIX on the probability of the SVI value exceeding SVI_{lim} = 150 cm^3/g.

The logistic regression model, described by formula (2), is an example of a simple regression dependence that makes it possible to find the value of sludge volume index. On the basis of this dependence, curves can be plotted which illustrate the impact of individual variables of the model on the activated sludge settling (Fig. 1).

The curves show the effect of sludge temperature and the amount of PIX dosed on the probability that SVI value will exceed SVI_{lim}=150 cm^3/g for mean values of MLSS, pH, BOD_5 load and oxygen concentration. Additionally, based on the results, chance ratios (OR) were computed. They allow the possibility of assessing the influence of increase in individual independent variables by the value of Δx_i (Table 4) on the risk of the event occurrence (exceeding of SVI_{lim}).

For instance, on the basis of the curves (Fig.1), it can be stated that temperature increase from T_{sl} = 12.5°C to T_{sl} = 17.5°C results in the reduced probability of SVI_{lim} value being exceeded from p = 0.85 to p = 0.42. In the latter case, that means the computed value of the sludge volume index will be lower than 150 cm^3/g. Additionally, the analysis of curves shows that when sludge temperature is T_{sl} = 17.5°C and PIX is not dosed, the probability of SVI_{lim} value being exceeded is equal to p = 0.83.

Table 4. Results of computations of chance ratios (OR) for individual independent variables of SVI.

ΔMLSS, kg/m^3	OR	ΔT_{sl}, ^0C	OR	ΔPIX, m^3/d	OR	ΔL_{BOD5}, kg/d	OR	ΔDO, mg/dm^3	OR
0.25	0.72	0.5	0.82	0.25	0.63	2000	1.17	0.1	0.89
0.50	0.51	1.0	0.68	0.50	0.4	3000	1.27	0.2	0.79
0.75	0.37	1.5	0.56	0.75	0.25	4000	1.38	0.3	0.7
1.00	0.26	2.0	0.46	1.00	0.16	5000	1.49	0.4	0.62

When the amount of PIX delivered into the aeration tank is $1m^3/d$, the value p computed with formula (2) is 0.40. That means the value of sludge volume index determined for these conditions will be lower than 150 cm^3/g. The values of chance ratios (OR) in Table 4 show that the variables accounted for in the model substantially affect the probability that SVI_{lim} value will be exceeded. The data in Table 4 indicate that an increase in BOD_5 load in the influent by the value of 2000 kg/d leads to increased probability of SVI_{lim} being exceeded by 37%. However, e.g. increase in MLSS or T_{sl} by 0.50 kg/m^3 and 1°C, respectively, results in reduced probability of SVI_{lim} being exceeded by 49% and 32%.

5 Conclusions

On the basis of analyses conducted for the study, it is possible to state that the linear discriminant model, the logit model and artificial neural networks can be applied to the assessment of the settling capacity of the activated sludge. The best classification abilities with respect to sludge volume index were demonstrated by the logit model and artificial neural networks. That was confirmed by the determined values of parameters of simulation fit to the measurement results. Because the ANN model is a complex and implicit dependence that requires the implementation of complex numerical algorithms, its practical application under the treatment facility service conditions is limited. Therefore, for the control and monitoring of the sludge settling capacity, the logistic regression model proposed in the study is readily applicable. It has the form of a simple dependence and it does not require the implementation of complex algorithms as most software packages provide special modules for the computations of the model parameters.

Additionally, the analyses performed for the study demonstrate that the value of sludge volume index (SVI_{lim} = 150 cm^3/g) could be classified with satisfactory accuracy on the basis of the measurement results of the MLSS and activated sludge temperature, oxygen concentration in the nitrification tank and BOD_5 load in the influent wastewater. Taking into account that settling process of activated sludge is extremely complex, it is necessary to conduct further analyses. They should primarily rely on the logit model that would ultimately aim at taking into account the interaction of variables describing wastewater quantity and quality, and also the bioreactor parameters and the sludge biocenosis.

References

1. J. Łomotowski, A. Szpindor, *Nowoczesne systemy oczyszczania ścieków* (2002)
2. A.M.P. Martins, K.R. Pagilla, J.J. Heijnen, M.C.M. Van Loosdrecht, Bulking filamentous sludge - a critical review, Wat. Res. **38**, 793 (2004)
3. Z. Dymaczewski, J.A. Oleszkiewicz, M. M. Sozański, *Poradnik eksploatatora oczyszczalni ścieków*. Wydanie II (1997)
4. I. Lou, Y. Zhao , Sludge Bulking Prediction Using Principle Component Regression and Artificial Neural Network, Math Probl Eng **2012**, 1 (2012)
5. B. Szeląg, J. Gawdzik, Application of Selected Methods of Artificial Intelligence to Activated Sludge Settleability Predictions, Pol J Environ Stud **25**, 1709 (2016)
6. F. Li, J. Qiao J., Y. Han, C. Yang, A self - organizing cascade neural network with random weights for nonlinear system modeling, Appl Soft Comput **42**, 184 (2016)
7. J. Bayo, J.M. Angosto, J. M, J. Serrano-Aniorte, Evaluation of physicochemical parameters influencing bulking episodes in a municipal wastewater treatment plant. *Water Pollution VIII: Modelling, Monitoring and Management,* 531 (2006)
8. E. Bezak-Mazur, R. Stoińska, B. Szeląg, Ocena wpływu parametrów operacyjnych i występowania bakterii nitkowatych na objętościowy indeks osadu czynnego – studium przypadku, Rocznik Ochrona Środowiska **18**, 487 (2016)
9. L. Belanche, J. Valdes, J. Comas, I. Roda, M. Poch, Prediction of the bulking phenomenon in wastewater treatment plants, Artificial Intelligence in Engineering **14**, 307 (2000)
10. J. Łomotowski, M. Dańczuk, Application of the microwave energy to the hygienization of sewage sludge, EPE J **36**, 77 (2010)
11. T. Heyer, J. Stamm, Levee reliability analysis using logistic regression models – abilities, limitations and practical considerations, Georisk Assessment and Management of Risk for Engineered Systems and Geohazards **7**, 77 (2013)
12. D.H. Tran, A.W.M. Ng, B.J.C. Perera, S. Burn, P. Davis, Application of probabilistic neural networks in modelling structural deterioration of stormwater pipes, Urban Water J **3**, 175 (2006)
13. A.M. Ramos-Cañón, L.F. Prada-Sarmiento, M.G. Trujillo-Vela, J.P. Macías, A.C. Santos-R, Linear discriminant analysis to describe the relationship between rainfall and landslides in Bogotá, Colombia, Landslides **13**, 671 (2016)
14. E. Gatnar, *Podejście wielomodelowe w zagadnieniach dyskryminacji i regresji* (2012)
15. F. Harrell, *Regression Modeling Strategies with Application to Linear Models, Logistic Regression, and Survival Analysis* (2001)

E3S Web of Conferences **17**, 00090 (2017) DOI: 10.1051/e3sconf/20171700090
EKO-DOK 2017

Performance comparison of a silica gel-water and activated carbon-methanol two beds adsorption chillers

Adam Szelągowski[1,*], and *Andrzej* Grzebielec[1]

[1]Institute of Heat Engineering, Faculty of Power and Aeronautical Engineering, Warsaw University of Technology, Poland

Abstract. The aim of the study is to compare the efficiency of adsorption refrigerating equipment working with different working pairs. Adsorption cooling devices can operate with a relatively low temperature of heat sources while consuming only a small amount of electricity for the operation of auxiliary equipment. Refrigerants used in adsorption devices are substances that do not have a negative impact on the environment. All that makes that adsorption refrigeration seems to be a good solution for utilizing renewable and waste heat sources for cold production. To carry out the experiment the adsorption cooling device has been developed and researched in Institute of Heat Engineering at Warsaw University of Technology. The test bench consisted of two cylindrical adsorbers, condenser, evaporator, oil heater and two oil coolers. In order to perform the correct action it has been developed and implemented special control algorithm device, allowed to keep the temperature in the evaporator at a preset level. The unit tested for two sorption pairs: activated carbon – methanol, and silica gel – water. For activated carbon - methanol working pair it was obtained energy efficiency rating (EER) equals to 0.14 and specific cooling power (SPC) of 16 W/kg. For silica gel - water EER of refrigeration unit was 0.25 and SPC was equal to 208 W/kg.

1 Introduction

Solid sorption is used in many applications including thermally driven refrigeration technologies. Solid sorption or adsorption refers to processes where a vapor is taken up by a solid when both the gas and the solid are in contact. Vapour molecules are accumulated on the surface of the solid and remain attached to it [1]. This process can be reversed and the adsorbed phase can be released from the adsorbent by applying heat to the compound. Adsorption refrigeration covers two processes, which are the heating-desorbing process and the cooling-adsorbing process. Because of that the simple traditional cycle is a type of intermittent refrigeration cycle [2]. If the heat source can be provided continually and the continuous refrigeration effect is required, two adsorber bed or multi adsorber beds device need to be designed for an adsorption refrigeration system, for which the heating and

* Corresponding author: adam.szelagowski@itc.pw.edu.pl

cooling processes of multi adsorbers will be complementarily arranged [3]. Adding every extra adsorber to device makes the cooling effect of the device is more stable, but it is associated with more complex and more difficult controlling devices work. Because of those reasons the most common adsorption units with continuous cooling effect contain two beds [4]. Figure 1 shows thermodynamics states of adsorbers for two bed unit during the operation. During the first half-cycle adsorber 1 is being heated, while adsorber 2 is being cooled, during the second half-cycle adsorber 1 is being cooled, while adsorber 2 is being heated [5]. During the first phase of working process (line A-B in Fig. 1), adsorber absorbs applied heat thus increasing the adsorbent's temperature which induces a pressure increase from evaporation pressure to condensation pressure. This phase is also known as isosteric heating [5]. The temperature of adsorbent bed increases from t_A to T_B. After this vapours of refrigerant are released from the bed, and they move towards condenser where they cool down and condense. While adsorber receiving heat continues but being connected to condenser, desorption of vapor get induced and liquefaction of desorbed vapor in condenser further takes place (line B-C in Fig. 1). The temperature of the adsorbent bed increases from t_B to t_C, but pressure stays at condensing pressure level. This phase is also known as isobaric desorption [5]. The liquid refrigerant flows through expansion value and evaporated in the evaporator providing cooling power [6]. Meanwhile, temperature of second adsorbent reduces due to release of heat (line C-D in Fig. 1) which induces pressure decrease from condensation pressure to evaporation pressure. This phase is also known as isosteric cooling [5]. The pressure get reduced from condensation pressure p_c to evaporation pressure p_e and temperature also decreases to t_D. The adsorbent temperature continues to reduce due to further release in heat which causes adsorption of vapor due to connection with evaporator. The adsorbent is cooled from t_D to t_A, pressure stays at evaporation level. This phase is also known as isobaric adsorption [5]. The whole process of adsorption cooling requires only heat energy which can be supplied through gas or oil fired or solar energy [7]. Unlike absorption processes, adsorption cooling systems have the advantage, that can be driven by lower temperature of heat source [8]. Driving heat source, can be at such low temperature as 50°C. Therefore, solar radiation, waste heat, as well as geothermal energy can be used to power refrigeration unit systems. Using solar energy, as a free and renewable type of energy, to drive adsorption cooling systems is considered an attractive option as well as a focal point of interest. Adsorption cooling systems can be also used in cogeneration systems [9]. Refrigerant used as a working fluid in adsorption cooling devices usually do not have a negative impact on the environment. In conjunction with the use of waste and renewable energy sources, application of these devices use introduce a good environmental and social factor [10]. Selection of the working pair is essential to adjust the application. In application where cooling temperature can be above 0°C water can be used as a refrigerant with any adsorbents. For applications below 0°C methanol, ethanol or ammonia with activated carbon can be used [11]. Multi bed devices allow to improve efficiency of the unit by recovery the heat between the adsorbers. Such solution can improve the efficiency of the whole system by about 30%. Adsorption units that work in basic continuous cycle have large possibility to recover heat from one adsorber to another. In the beginning phase of the process the first adsorber has to be cooled, when second one has to be heated. Temperature difference, between point C and A (Fig. 1) is large enough to transport heat from one bed to another. Creating such heat recovery system in the beginning phase of process the external heat source and external cooling system is not needed.

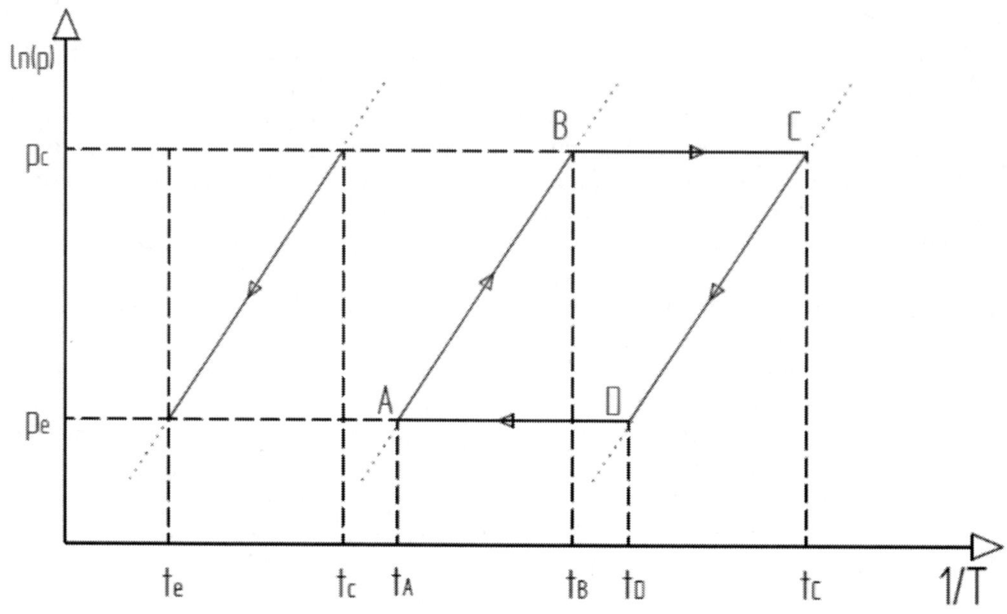

Fig. 1. Thermophysics states of adsorpbers during the operation.

2 Construction of the apparatus

The thermal wave adsorption refrigeration device was designed and built at the laboratory of the Division of Refrigeration and Energy in Buildings (formerly Division of Processes Equipment and Cooling), Faculty of Power and Aeronautical Engineering, Warsaw University of Technology. The thermal wave device allows to recover heat between adsorber beds during the process [12]. When the direction of oil circulation is reversed, the secondary fluid recovers the heat from one adsorber and transfers it to the second one. The installation is designed to examine this device comprises three circulation systems: heat transfer fluid, refrigerant and heat sink. Figure 2 shows an overview of the apparatus. The adsorber consists of an inner brazen tube with 12 mm outer diameter, through which the heat transfer medium flows, supplying the heat to the bed or taking the heat back from it. The inner tube is mounted inside an outer tube with 28 mm out diameter. Between the tubes there is a sorption material bed. In the centre of the inner tube a pipe is soldered, through which the refrigerant flows to and from the bed. The whole assembly is housed in a tight foam insulation. The heat flux should occur between the adsorbent bed and the heat transfer medium rather than between the bed and the air surrounding the adsorber. Device was tested with two working pairs. In one configuration, the methanol as refrigerant and activated carbon as sorption material were selected as the working pair. In the second configuration, the water as refrigerant and silicagel as sorption material were selected.

DOI: 10.1051/e3sconf/20171700090

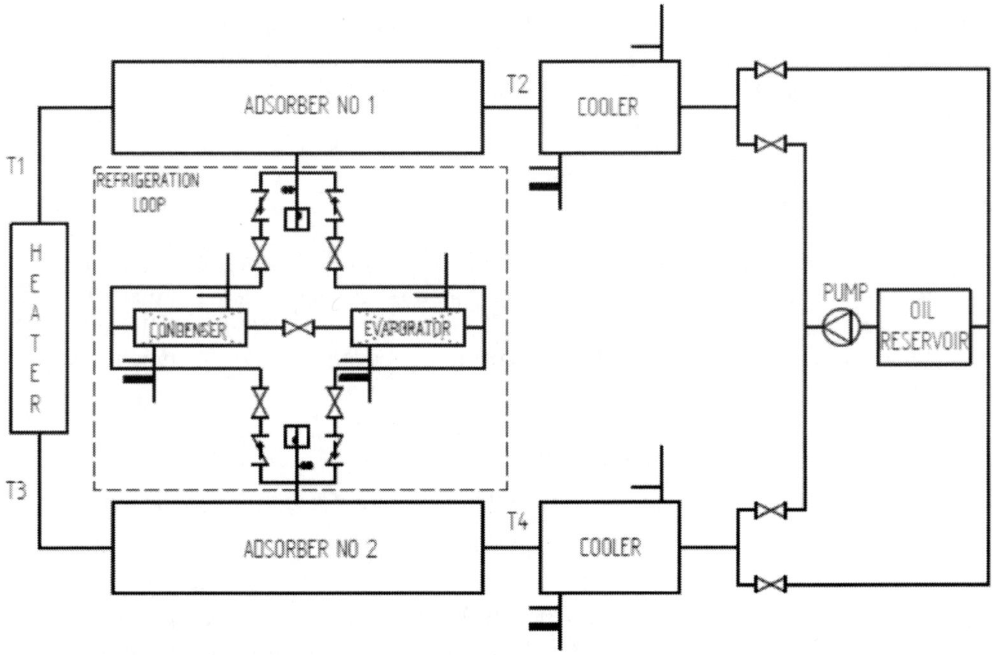

Fig. 2. Thermal wave adsorption refrigeration device.

2.1 Secondary fluid circulation

The heat is transferred to and from the adsorber beds by the mineral oil circulation system. Oil system comprises a circulating pump, oil surge tank, five heat exchangers consisting of two coolers, two adsorber beds and one electric heater (Fig. 2). An assembly of four solenoid valves was mounted to control the direction of oil flow. In the first half cycle oil is pumped from surge tank through first cooler, first adsorber bed, than through the electric heater to the second adsorber bed. Downstream oil flows through second cooler to the surge tank. In the second half cycle, the oil flows in the opposite direction. The open surge tank protects the system against an uncontrolled pressure rise, and compensates for thermal expansion of the working fluid.

2.2 Refrigerant circulation

The refrigerant circulates in the system between adsorber beds, condenser and evaporator. The refrigerant circulation system comprises a set of solenoid valves and in case of methanol non-return valves which provide a connection between the adsorber beds on one side and the condenser and evaporator on the other. Each adsorber bed is connected with condenser and evaporator (Figure 2). This set and proper control of the opening of the valves makes the stream of refrigerant from the desorbing bed always flow to the condenser, and then, after flowing through the electronic expansion valve, to the evaporator. Finally, the vapor refrigerant is adsorbed by the second adsorber bed. Initialization and opening time for valves are determined on the pressure values in the adsorbers.

2.3 Heat sink circulation

To dissipate heat from the system heat sink circulation is required. In order to determine heat exchanger capacities, a water circulation system was introduced. The water flows through condenser, evaporator and both heat sink exchangers. All capacities in mentioned heat exchangers are calculated according to Equation 1.

3. Methodology

In the context of experimental studies a continuous oil temperature were measured at inlet and outlet of adsorbers (Fig. 2). There was also measured temperature of the water before and after all heat exchangers through which water flows. Based on these parameters instantaneous heat capacities have been determined according to the equation 1:

$$\dot{Q} = \dot{m} \cdot c_p \cdot (T_{out} - T_{in}) \tag{1}$$

Mass flow $\dot{m}$ was determined on the volumetric flow rate $\dot{V}$ according to the equation 2:

$$\dot{m} = \dot{V} \cdot \rho \tag{2}$$

The oil volumetric flow rate $\dot{V}$ was determined by measuring the known volume filling time. Volume flow rate was measured during flow in both direction. In both cases, the measured value eqals $\dot{V}$=1.7 l/min. For the temperature measuring oil density was ρ =880 kg/m^3.

For temperature measurements were used NTC5 temperature sensors with measurement uncertainty of 1%. Filling time was measured with accuracy of ±1s and the tank volume was 2 liters. During experiment, there was measured pressure of refrigerant in the adsorbers with piezoelectric quartz pressure transducer having an accuracy of measurement of ±0.3%. Based on the pressure P, there was determined saturation temperature T for evaporation and condensation in accordance to the saturation line equation [13]:

$$\ln P = a + \frac{b}{T} \tag{3}$$

Where the coefficients a and b for methanol are a=20.84 and b = -4694. For water a=20.5896 and b = -5098.26. In the equation 3, the temperature is given in Kelvin degrees and the pressure is calculated in mbar. The temperature determined from the measured pressure determines the heat source requirements the adsorption device can co-operate.
The energy efficiency rate *EER* of adsorption refrigeration devices represents amount of coolness generated in evaporator to amount of heat generated in heat source ratio. It can be calculated from the equation (4)

$$\mathrm{EER} = \frac{Q_C}{Q_{Hs}} \tag{4}$$

wherein Q_C is the amount of heat obtained in evaporator during one half cycle according to equation 5. Q_{HS} is the amount of heat supplied to the system at the same time by heat source according to the equation 1 and trapezoidal integration method (equation 6).

$$Q_C = m_{ads}(w_{max} - w_{min})h_{fg} \tag{5}$$

In equation (5) m_{ads} is mass of adsorbent in a single bed [$kg_{adsorbent}$], w_{max} is maximum uptake of refrigerant in adsorbent [$kg_{refrigerant}/kg_{adsorbent}$], w_{min} is minimum uptake of refrigerant in adsorbent, h_{fg}[J/ $kg_{refrigerant}$] is specific latent heat of vaporization of refrigerant.

$$Q_{HS} = \int_{t_0}^{t_1} \dot{Q}_{HS}(t)\, dt \tag{6}$$

wherein t_0 is half cycle start time, and t_1 is time of half cycle period. $\dot{Q}_{HS}$ is the heat flux supplied to the system during selected time step.

4. Research results

The pressure value in each adsorber was measured and it was used to operate the electromagnetic valves that control the refrigerant and oil flow direction. A pulse type electronic expansion valve was used, so that the pressure and evaporation temperature could be controlled using researchers' software. The boiling point of both tested refrigerants (methanol, water) was set to be 10°C. Fig. 3 and 4 show heat fluxes during four full half cycle, during the further course of the experiment the results are reproducible and characteristics are like those presented.

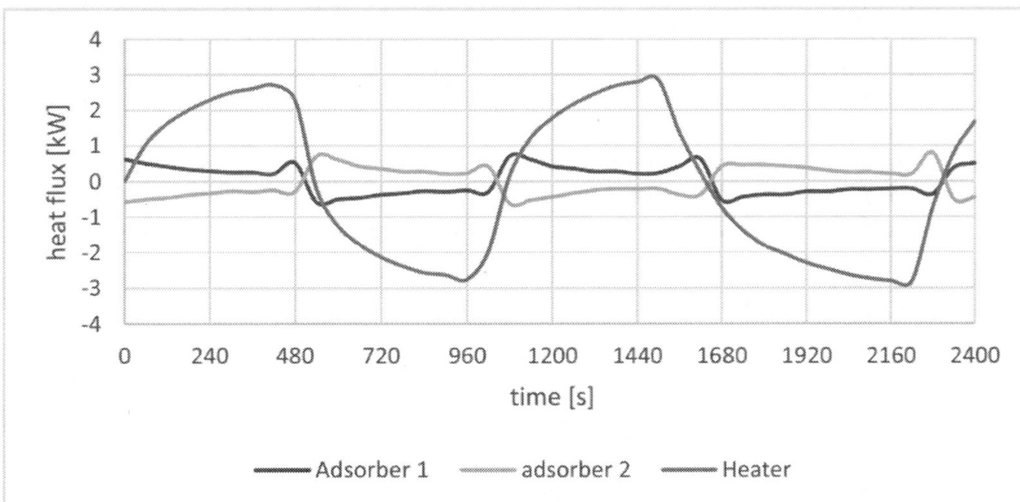

Fig. 3. Heat fluxes provided / received for component for activated carbon - methanol working pair.

 DOI: 10.1051/e3sconf/20171700090

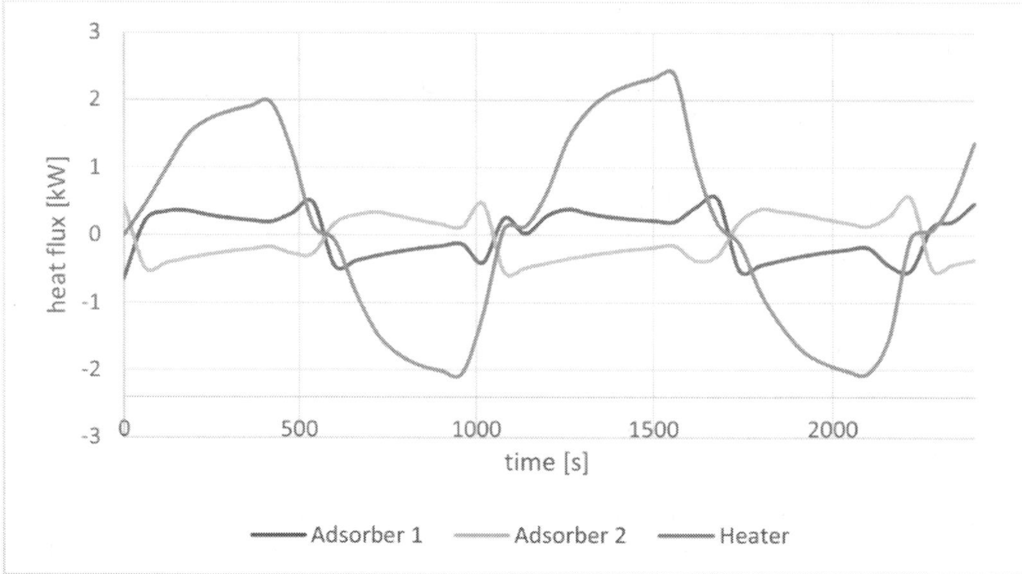

Fig. 4. Heat fluxes provided / received for component for silica gel - water working pair.

In the first half-cycle the temperature difference between T1 and T2 represents heat of desorption while the temperature difference between T3 and T4 corresponds to heat of adsorption in the second adsorber. Based on the measurements, real instantaneous powers of each component of the system could be determined. The energy efficiency rating EER of adsorption devices is determined based on Eq. (4). With measurements of instantaneous thermal power, the amount of heat can be determined from the integrals of the power over the time. What has been done here by trapezoids. For investigated case the average value of EER was obtained at level 0.14 for activated carbon - methanol and 0.25 for silica gel – water. Specific cooling power (SPC) was 16 W/kg for activated carbon - methanol and 208 W/kg for silica gel – water.

3 Summary

In this article the results of the experimental research concerning a state-of-the-art thermal wave adsorption refrigeration device are presented. Due to different working temperatures of the adsorbers, the change in the thermal power of each system component could be determined. The measurement results are presented in Figures 3 and 4 show four half cycles of the adsorption device operating under steady state conditions. In the following experiment, the cycle is repeated, so that these results can be regarded as representative for the calculation of the energy efficiency rating of the adsorption refrigeration apparatus.

There was energy efficiency rating (EER) obtained at level 0.14 for activated carbon – methanol and 0.25 for silica gel – water working pair. Specific cooling power (SPC) was 16 W/kg for activated carbon - methanol and 208 W/kg for silica gel – water. Low energy efficiency was due to the excessive duration of the half cycles representing approximately 6 minutes.

References

1. A. Ruciński, A. Rusowicz, K. Rucińska Przem. Chem. **1**, 7 (2016)

2. H. Z. Hassan, A. A. Mohamad, Y. Alyousef, H. A. Al-Ansary, Sustain. Energy Rev. **45** (2015)
3. H. Deshmukh, M. P. Maiya, S. Srinivasa Murthy, Appl. Therm. Eng. **82** (2015)
4. T. X. Li, R. Z. Wang, H. Li, Prog. Energy Combust. Sci., **40**, 1 (2014)
5. P. Goyal, P. Baredar, A. Mittal, A. R. Siddiqui, Renew. Sustain. Energy Rev. **53** (2016)
6. W. Kuczyński, T. Bohdal, H. Charun, Exp. Heat Transf. **26**, 1 (2013)
7. I. I. El-Sharkawy, H. AbdelMeguid, B. B. Saha, Appl. Energy **126** (2014)
8. A. Rusowicz, A. Ruciński, Environ. Eng. **1-3** (2011)
9. M. Jaworski, M. Bednarczyk, M. Czachor, Appl. Therm. Eng. **96** (2016)
10. P. Cyklis and K. Janisz, Chem. Process Eng- Inz, **36**, 3 (2015)
11. A. Grzebielec, A. Rusowicz, A. Ruciński Przem. Chem. **1**, 6 (2015)
12. A. Grzebielec, A. Rusowicz, R. Laskowski, Chem. Process Eng-Inz. **36**, 2 (2015)
13. G. Cacciola, G. Restuccia, Int. J. Refrig. **18**, 2 (1995)

E3S Web of Conferences 17, 00091 (2017) DOI: 10.1051/e3sconf/20171700091
EKO-DOK 2017

Absorbance based water quality indicators as parameters for treatment process control with respect to organic substance removal

Sławomir Szerzyna[1,*], *Marek* Mołczan[1,2], *Małgorzata* Wolska[1,2], *Wojciech* Adamski[1], and *Jacek* Wiśniewski[1]

[1]Faculty of Environmental Engineering, Wrocław University of Science and Technology, Wyb. Wyspiańskiego 27, 50-370 Wrocław
[2]Wroclaw Municipal Water and Sewage Company, New Technologies Center, Na Grobli 14/16, 50-421 Wrocław

Abstract. The dependence between individual water quality indicators and the UV254 was presented in the paper. Derived dependences allow to calculate values for these indicators using only the absorbance measurement, which could be easily on-line monitored. This approach make possible to reduce number of performed laboratory analyzes, used for the efficient controlling of water purification process. Identified relations are not universally applicable even for individual specific water source. These could be described with sufficient precise only for water samples from particular treatment step. Presented methodology can be helpful for exploitation of real water treatment systems to quick and efficient evaluation of individual purifying processes.

1 Introduction

The presence of organic substances in water for human consumption and the possibility of removing organic matter in unit water treatment processes has been the object of studies worldwide for many years [1–5]. Due to possibility of generation of organic substances harmful to human health in unit processes [6–7], especially during the disinfection process, it is necessary to ensure a most effective elimination of organic substances during water treatment. The amount and type of generated disinfection by-products (DBPs) depends on the types and properties of natural organic matter present in treated water. One of key parameters deciding the potential for forming chlorinated organic compounds is the molecular mass of the particle [8–11]. Studies [8] have shown that DBPs generation is determined also by dissolved organic carbon (DOC) content and UV absorbance at 254 nm wavelength values (UV_{254}) as a measure of aromatic substance content. UV absorbance measurements are used to evaluate chlorinated organic substance precursor content [12]. According to Korshin et al. [13], these precursors exhibit a much greater correlation with

* Corresponding author: slawomir.szerzyna@pwr.edu.pl

absorbance at a wavelength of 272 nm (UV_{272}) than at 254 nm. Uyak and Demirbas [14] have also shown that the amount of trihalomethanes (THM) and choloroorganic acids generated are dependent on absorbance at a wavelength of 272 nm, and specifically its decrease during unit water treatment processes. Lobanga et al. [15] utilized UV absorbance measurements at various wavelengths for an evaluation of susceptibility of organic substances contained in surface waters in South Africa to adsorption on granulated activated carbon. Apart from chlorinated organic substance precursors, UV absorbance values correlated with the content of substances, mainly humic and fulvic acids, which cause natural yellow-brown color of water [16].

UV absorbance measurements are very significant with regards to operating water treatment process systems. They yield knowledge concerning aromatic substance content and their susceptibility to removal using commonly-used unit processes. The wide spread nature of indicators based on UV absorbance stem from the simplicity of this measurement and the fact that frequently it supplements or replaces more advanced analytical methods, i.e. chromatography or DOC analysis.

This paper presents the variability in indicators based on UV and visible light absorbance measurements in surface water as supplied and after water treatment unit processes, and their relationship with other water contamination indices. The aim of these studies was evaluating the possibility of controlling a water treatment system by absorbance measurements, therefore avoiding the time lag in the case of a decrease in raw water quality.

2 Study methodology

The study was conducted at a research station operating two water treatment lines, each with a throughput of 3 m^3/h that filter water from the Oława river. Water samples for analysis were taken during 16 months of treatment line operation, at the inlet to both lines and after each unit process (raw water– R, water after coagulation and sedimentation– CS, water after sand filtration– SF, water after absorption on activated carbon – AC). The samples were taken once a day five days a week. The organic substance content was determined for the samples by the use of the following indicators: UV absorbance at wavelengths of 254 and 272 nm, visible light color intensity at wavelengths of 340 (B340) and 410 nm (B410), total organic carbon (TOC) and dissolved organic carbon. Additionally, for all water samples, pH, conductivity, temperature, alkalinity, turbidity, ammonia, nitrite and nitrate nitrogen concentrations, phosphate, sulfate, chloride, iron, manganese, aluminum and dissolved oxygen concentrations were measured, the results of which are not presented in this paper. All analysis was performed according to current applicable standards.

Total and dissolved organic carbon analysis was performed with a Shimadzu TOC-L analyzer, while absorbance and color intensity measurements were performed with a Shimadzu UV-1800 spectrophotometer.

3 Discussion of results

During the study period, the input water and the water after each of the unit processes was characterized by a large variability in composition (Table 1), which allowed for a certain generalization of the relationships that have been obtained. Because the study was conducted with various input water temperatures (4.4–25.7 °C), this factor, which influences the intensity and type of transformations during the water treatment process, was also taken into account. The study results have confirmed the existence of a correlation between the UV_{254} absorbance values and other indicators of organic substance water contamination. Fig. 1. presents the dependence of UV_{272} absorbance on UV_{254} absorbance for water taken from different locations in the water treatment line.

Table 1. Value ranges analysed water quality parameters.

	R	CS	SF	AD
Temperature, °C	4.4 ÷ 25.7	4.8 ÷ 25.6	5.2 ÷ 25.8	6.1 ÷ 25.8
pH	6.68 ÷ 8.41	6.63 ÷ 8.13	6.43 ÷ 8.06	6.32 ÷ 10.1
Conductivity, μS/cm	288 ÷ 738	292 ÷ 746	293 ÷ 795	299 ÷ 750
Alkalinity, val/m^3	2.16 ÷ 3.58	1.79 ÷ 3.49	1.76 ÷ 3.40	1.86 ÷ 3.34
Turbidity, NTU	2.6 ÷ 28.6	0.3 ÷ 14.3	0.0 ÷ 3.4	0.0 ÷ 4.2
Oxygen, gO_2/m^3	1.31 ÷ 12.21	3.24 ÷ 12.32	2.65 ÷ 12.10	2.37 ÷ 11.53
UV_{254}, m^{-1}	6.08 ÷ 14.87	3.86 ÷ 13.55	3.76 ÷ 11.98	0.44 ÷ 8.04
UV_{272}, m^{-1}	5.19 ÷ 11.95	3.28 ÷ 10.95	3.06 ÷ 9.68	0.36 ÷ 6.39
Color at 340nm, g/m^3	5.70 ÷ 14.46	3.02 ÷ 12.75	2.73 ÷ 11.08	0.37 ÷ 6.39
Color at 410nm, g/m^3	7.82 ÷ 24.62	2.52 ÷20.24	2.36 ÷ 18.31	0.32 ÷ 8.18
TOC, gC/m^3	3.23 ÷ 5.90	2.71 ÷ 5.69	2.30 ÷ 5.22	0.98 ÷ 4.05
DOC, gC/m^3	3.21 ÷ 5.68	2.42 ÷ 5.45	2.47 ÷ 5.03	0.93 ÷ 4.01

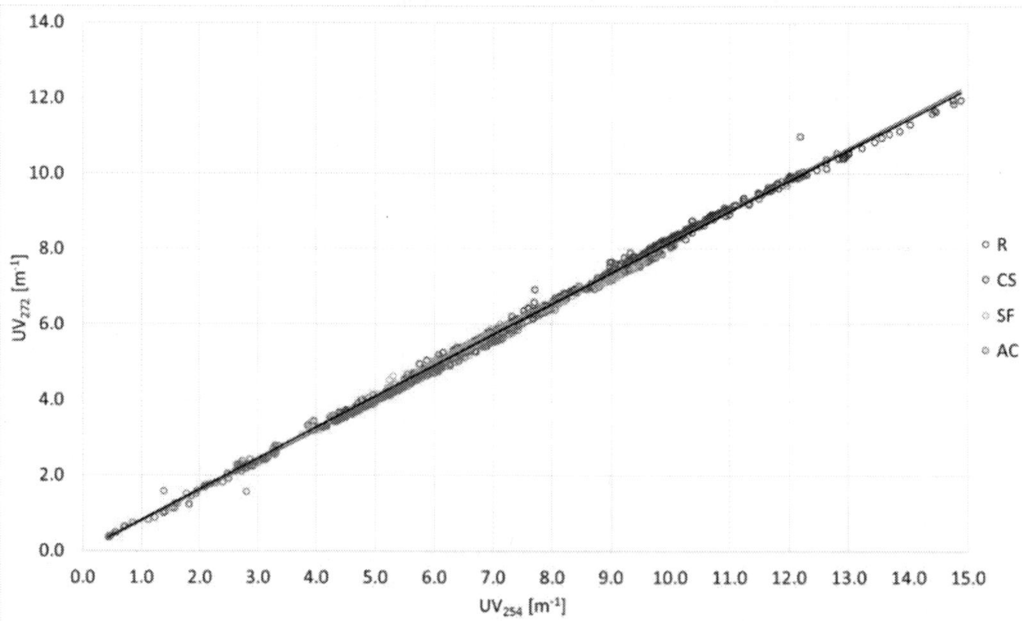

Fig. 1. Correlation between UV_{272} absorbance and UV_{254} absorbance for water after different unit treatment processes.

Analyzing the relationships that were obtained it may be concluded that the correlation between the two types of UV absorbance is linear regardless of the measured water origin. The directional coefficients of equations (1–4) obtained for waters after the different unit processes do not show significant differences:

$$\text{Raw water: } y=0.822x \quad (1)$$

$$\text{Water after coagulation and sedimentation: } y=0.8185x \quad (2)$$

$$\text{Water after rapid filtration : } y=0.8158x \quad (3)$$

Water after adsorption: $y=0.7941x$ (4)

The coefficients of determination R^2 for the presented equations are in the range of 0.9923 to 0.9964. The individual equations may be replaced by one general equation (5), which describes the relationship between UV_{272} absorbance and UV_{254} absorbance in the entire range studied of these indicators, for all types of water

$$y=0.8173x \quad (5)$$

with also very good fit to the linear model (R^2=0.9976).

As reported in literature [16], both absorbance at 272 nm as well as at 254 nm are a measure of chlorinated disinfection by-product precursor content. This is also confirmed by the linear correlation found independent of water type, and similar directional coefficient values indicate a measurement of organic substances of similar properties. The obtained results (Fig.1) indicate a large utility of the coagulation and sedimentation process (removal effectiveness of UV_{254} 11–47% and UV_{272} 11–44%) and adsorption (removal effectiveness of UV_{254} 18–99% and UV_{272} 18–99%) in removing these contaminants. Low values of the UV_{254} and UV_{272} reduction effectiveness results from adsorption with saturated active carbon and high values results from adsorption with virgin active carbon.

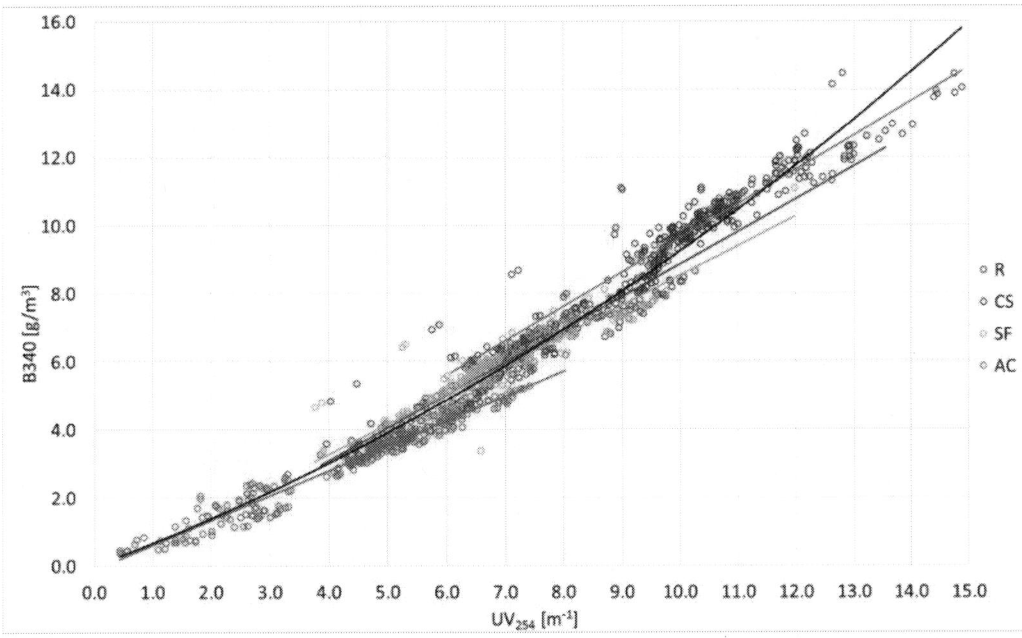

Fig. 2. Correlation between 340 nm values color and UV_{254} absorbance for water after different unit processes.

The correlation between color at 340 nm and absorption at 254 nm is shown in Fig. 2. This indicates a relationship between refractive substance content (evaluated via absorbance at 254 nm) and the concentrations organic substances responsible for the color of water. From the presented correlation after subsequent unit processes it may be concluded that the greatest lowering of the values of these indicators took place during coagulation and, to a lesser degree, during adsorption. In the range of individual water types, the dependence between these two water quality indicators may be described by linear equations (6–9) with directional coefficients differing more than in the case of UV_{254} and UV_{272}. R^2 for these equation is in the range of 0.8611–0.9579.

$$\text{Raw water: } y=1.0045x-0.4164 \quad (6)$$

$$\text{Water after coagulation and sedimentation: } y=0.9608x-0.7501 \quad (7)$$

$$\text{Water after rapid filtration : } y=0.875x-0.2315 \quad (8)$$

$$\text{Water after adsorption: } y=0.7271x-0.1113 \quad (9)$$

Knowing the origin of studied water, with the use of these equations it is possible to relatively precisely calculate the color value at 340 nm based on UV_{254} absorbance. If the origin of the water is not known, a correlation in polynomial form may be used (see comment above):

$$y=0.0282x^2+0.6421x \quad (10)$$

The dependency of color at 410 nm on UV_{254} absorbance (fig. 3.), similarly as in the case of color at 340 nm and UV_{254} may be described by a linear equation only in the range of values obtained for the individual water types:

$$\text{Raw water: } y=1.7633x-1.7494 \quad (11)$$

$$\text{Water after coagulation and sedimentation: } y=1.6222x-3.0607 \quad (12)$$

$$\text{Water after rapid filtration : } y=1.3174x-1.1816 \quad (13)$$

$$\text{Water after adsorption: } y=0.7514x+0.0551 \quad (14)$$

In this case, these equations are less precise than in the case of the relationship between color at 340 nm and UV_{254} absorbance (R^2 in the range of 0.5822 to 0.6899) with the best correlation found for water taken after the last treatment step (AC – R^2 equal to 0.6899).

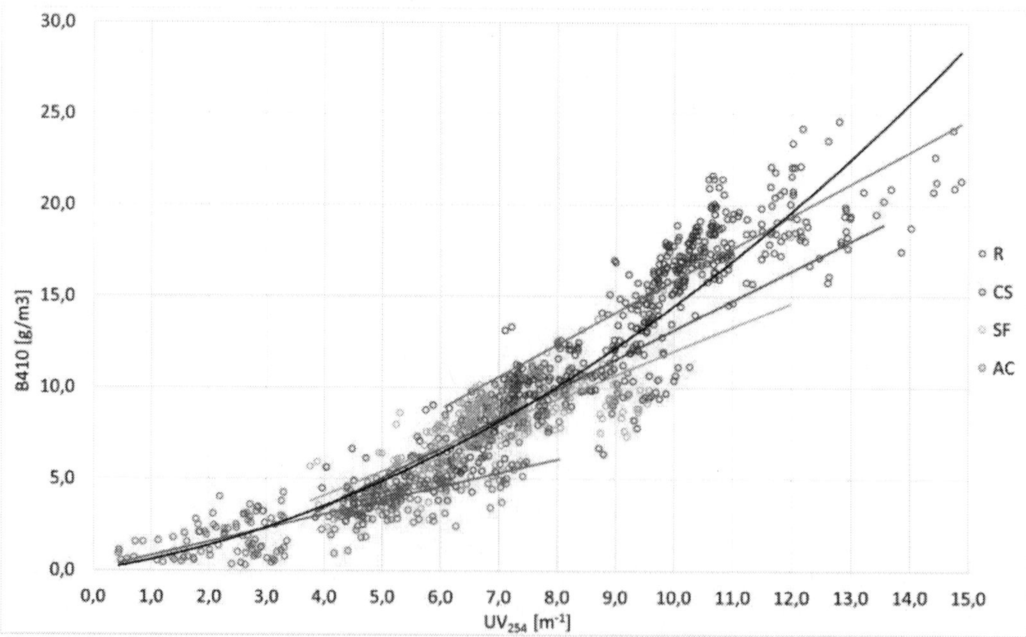

Fig. 3. Correlation between 410 nm values of color and UV_{254} absorbance for waters after different unit treatment processes.

The differences found between the values of determination coefficients may testify to slightly different properties of organic substances as measured by absorbance at 340 nm and 410 nm. A general equation describing the relationship of color at 410 nm and UV_{254}, absorbance, regardless on the measured water type, which describes this relationship more precisely (R2 equal to 0.8801) is presented below.

$$y=0.0947x^2+0.5009x \quad (15)$$

Correlations of total organic carbon and dissolved organic carbon to UV_{254} absorbance are presented in two subsequent figures (figs. 4 and 5). Both cases are best described by general equations which do not take into account the type of water measured. The correlation between total organic carbon and UV_{254} absorbance is described by equation 16, while equation 17 describes this relationship for dissolved organic carbon:

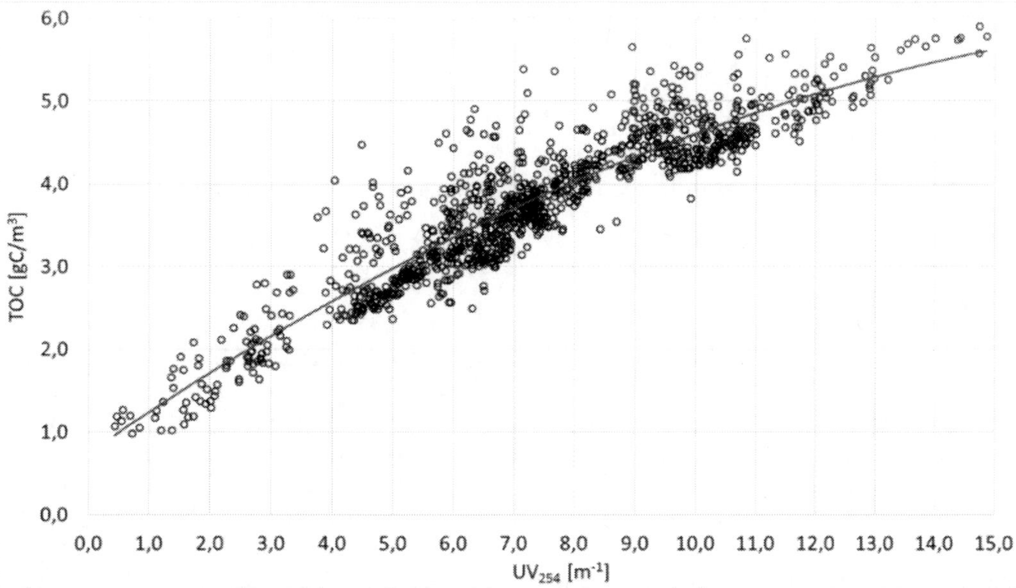

Fig. 4. Relationship between total organic carbon and UV_{254} absorbance at all water treatment stages.

$$y=-0.0122x^2+0.5082x+0.745 \quad (16)$$

$$y=-0.0074x^2+0.4258x+0.9228 \quad (17)$$

Both equations describe these relationship relatively precisely (R^2 determination coefficients of 0.8482 and 0.8718 respectively). These equations allow for replacing time-consuming organic carbon analysis in treated water by values obtained through calculation. Consequently, these calculation may be utilized for an initial evaluation of the potential of organic substances to be removed by individual unit processes and the choice of process parameters, eg. reagent dosages. However, this requires an analysis of the relationships between changes in these parameters in unit processes.

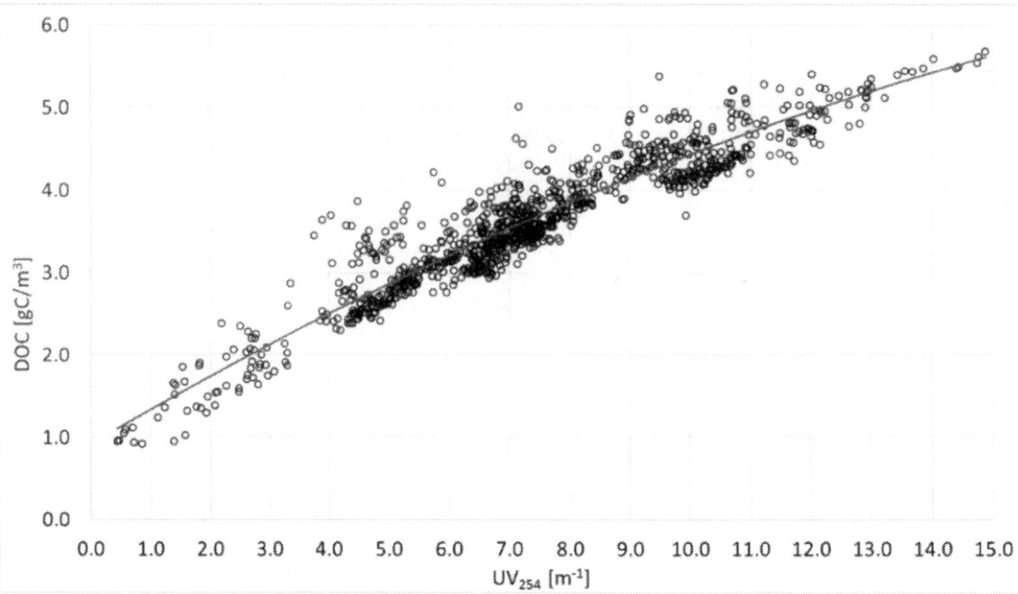

Fig. 5. Relationship between dissolved organic carbon and UV_{254} absorbance at all water treatment stages.

4 Conclusion

All the presented indicators of organic substance content correlate with ultraviolet absorbance at 254 nm. A linear correlation was found for the relationship between absorbance at UV_{272} and UV_{254}. In the case of the correlation between colors at 340 nm and 410 nm, total organic carbon and dissolved organic carbon the general equations describing the correlation were of polynomial form. The equations determined in this study may serve to calculate values of remaining organic substance content indicators based only on UV_{254} measurements. This allows to obtain information about water quality very quickly, without the need to perform more time-consuming and specialized analysis, which is particularly useful for personnel controlling flow-type water treatment processes. Furthermore, if the treatment system has on-line type absorbance sensors, then the values from these sensors may be utilized for real-time calculations of remaining organic contamination indicators. This is particularly important if total organic carbon or dissolved organic carbon values, whose measurement are time-consuming, are used for controlling or on-going evaluation of treatment effectiveness. It must be remembered, however, that it is not possible to completely forego laboratory analysis, and the results from such analysis should be used for a continuous verification of the equations obtained.

This publication was made possible by the National Center for Research and Development grant (PBS3/B9/44/2015) "Research on effectiveness of new water treatment technology as a step towards a shift in thinking about water utility sector" (WODTECH), and thanks to involvement of both project consortium members (Wrocław Municipal Waterworks and Drainage Company and Wrocław University of Science and Technology).

References

1. Z. Hasan, S. H. Jhung, J. Hazard. Mater. **283** (2015)
2. W. W. Tang, G. M. Zeng, J. L. Gong, J. Liang, P. Xu, C. Zhang, B. B. Huang, Sci. Total Environ. **468** (2014)
3. S. Metsämuuronen, M. Sillanpää, A. Bhatnagar, M. Mänttäri, Sep. Purif. Rev. **43** (1) (2014)
4. M. Inyang, E. Dickenson, Chemosphere **134** (2015)
5. A. Alsbaiee, B. J. Smith, L. Xiao, Y. Ling, D. E. Helbling, W. R. Dichtel, Nature **529** (2016)
6. R. Y. Yeh, M. J. Farré, D. Stalter, J. Y. Tang, J. Molendijk, B. I. Escher, Water Res. **59** (2014)
7. B. K. Mishra, S. K. Gupta, A. Sinha, Journal of Environmental Health Science and Engineering **12** (1), 73 (2014)
8. E. E. Chang, P. C. Chiang, H. J. Hsing, S. Y. Yeh, Toxic and Radioactive Waste Management (2007)
9. E.E. Chang, P.C. Chiang, W.Y.Tang, S.H. Chao,H.J. Hsing, Chemosphere **58** (2004)
10. J.D. Plummer, J.K. Edzwald, Water Sci. Technol. **35** (2001)
11. P.C. Chiang, E.E. Chang, C.H. Liang, Chemosphere **46** (2002)
12. J-B. Serodes, M. J. Rodriguez, H. Li, Ch. Bouchard, Chemosphere **51** (2003)
13. G. V. Korshin, W. W. Wu, M. M. Benjamin, O. Hemingway, Water Res. **36**, 13 (2002)
14. V. Uyak, K. D. Demirbas, Environ. Forensics **15**, 1 (2014)
15. K. P. Lobanga, J. Haarhoff, S. J. van Staden, Water SA **39**, 3 (2013)
16. T. Thebel, C. D. Swartz, I. R. Morrison, W. J. Engelbrecht, R. E. Loewenthal, P. Krüger, WISA 2000 Biennial Conference, Sun City, South Africa (2000)

METHOD OF WATER CONSUMERS SAFETY ANALYSIS AND ASSESSMENT

Dawid Szpak[1,*]

[1]Rzeszow University of Technology, Department of Water Supply and Sewage Systems, Al. Powstańców Warszawy 6, 35-959 Rzeszów, Poland

Abstract. The subject of the publication is to present the method of water consumers safety analysis and assessment to implement the so-called Water Safety Plans (WSP), recommended by the World Health Organization (WHO) in Poland. The paper presents the method of water consumers safety analysis and assessment, which is based on failure mode and effects analysis (FMEA). The failure mode and effects analysis is widely used to analyse the reliability and safety of products and processes in technical systems. The procedure in the FMEA is based on the identification of potential undesirable events, assessment of effects and determining the causes of these events, assessment of frequency, the possibility of detection, estimating validity of the actions and estimating how to prevent undesirable events, re-examining and assessment of these events. The developed method of water consumers safety analysis and assessment is a complement and extension of recommended for WSP matrix methods, which according to the author's opinion is not sufficient for a proper analysis of undesirable events that may happen in collective water supply system (CWSS). The use of the FMEA method will allow for better identification of existing hazards and to determine the cause–effect relationships between them. Reliable access to drinking water is one of the basic conditions for the development of civilization.

1 Introduction

According to the act [1] collective water supply system (CWSS) is part of the critical infrastructure. This system is crucial for the functioning of society and the state. Wrong functioning of CWSS may be a threat to people health or lives. Therefore CWSS must be characterized by a high degree of reliability and safety [2, 3, 4].

Water consumers safety is defined as the probability of avoiding threats associated with the consumption of poor quality water or lack of water supply. Water consumers safety is affected by a number of factors, which include, among others:
- the state of water supply infrastructure,
- the state of monitoring,
- exposure to internal and external threats,
- way to protect consumers against the consequences of undesirable events,

* Corresponding author: dsz@prz.edu.pl

- specialized services for renovation and repair.

The need to constantly improve the level of water consumers safety has already been noticed by the World Health Organization (WHO) in the sixties of the twentieth century. Since then at first International Standards for Drinking-Water and later the Guidelines for Drinking Water Quality appeared regularly. Currently the fourth edition of the Guidelines for Drinking Water Quality issued in 2011 [5] is valid. One of the achievements of the Guidelines for Drinking Water Quality is the introduction of concept and assumptions for the so-called Water Safety Plans (WSP). The main purpose of the WSP is to ensure safe access to drinking water.

In 2009 the WHO published the Water safety plan manual: step-by-step risk management for drinking-water suppliers [6] which is a kind of instruction for water and sewage companies. This manual contains 11 stages of the development and implementation of the WSP. Currently in Poland there is no obligation to implement the WSP, while many Western European countries have entered this model in force, which contributed to improving the quality of functioning of CWSS and reducing the number of undesirable events. Against the countries of 'old' European Union we have in this regard a lot to do, so one should look for new opportunities that will allow for proper implementation of the WSP. The extension of the WSP is Water Cycle Safety Plan (WCSP) [7].

The main aim of this work is to present the methodology of analysis and assessment of water consumers safety for the purpose of implementing the WSP.

2 Characteristics of undesirable events in CWSS

The membership of CWSS to critical infrastructure requires from water companies to ensure the continuity of operation and a fast recovery of the performance in case of failure. There are three groups of factors that may cause undesirable events in CWSS: technical, human and environmental factors [8, 9].

Threats in water supply can occur at any stage of the production of water intended for human consumption: in water source, during treatment and storage, as well as in water distribution subsystem.

The main threats to water consumers safety are [6]:

- unfavourable weather conditions, e.g. floods, droughts,
- contamination of raw water which cannot be removed by conventional treatment process and the use of an alternative treatment technology is not possible,
- bad water and sewage management in the city/municipality,
- ineffective water treatment process,
- lack or reduction of the supply of electricity to the water treatment plant and zone pumping stations,
- secondary water contamination in water supply network or water tanks,
- failures of water supply network,
- intentional harmful action of the third parties,
- lack of adequate protection of key objects for the reliable CWSS functioning.

The frequency of undesirable events can be reduced by:

- diversification of water sources,
- extension of the monitoring system,
- conducting regular inspections of the technical condition of the objects,
- maintaining efficient and suitably equipped emergency water and sewerage brigade,
- introduction of multibarrier system,
- development and implementation of WSP.

 DOI: 10.1051/e3sconf/20171700092

3 Materials and methods

In the study we used the method of failure mode and effects analysis FMEA which is widely used to analyse the reliability and safety of products and processes in technical systems [10]. The standard [11] describes the FMEA method and provides guidelines for its use for different purposes.

The method of failure mode and effects analysis FMEA allows to analyse the product or process, to determine weak points and then to introduce improvements and new solutions to eliminate the source of undesirable events. For this reason, this method can be used to analyse water consumers safety in the implementation of WSP. The method of FMEA in CWSS was shown in Fig. 1 [12].

The assessment of water consumers safety is carried out based on risk priority number (RPN) according to the equation [10, 12–14]:

$$RPN = S \cdot O \cdot D \tag{1}$$

where:
S – Severity – point weight associated with the importance of undesirable event,
O – Occurrence – point weight associated with the frequency of undesirable event,
D – Detection – point weight associated with the ability to detect undesirable event.

The individual parameters are described by an integer from 1 to 10, which is assumed on the basis of expert knowledge and experience of CWSS users. The result of the analysis is the RPN, taking values ranging from 1 to 1000. The higher RPN value, the lower level of water consumers safety. The input criteria for the individual parameters are assumed on the basis of the information contained in Tables 1, 2 and 3 [8, 12, 13].

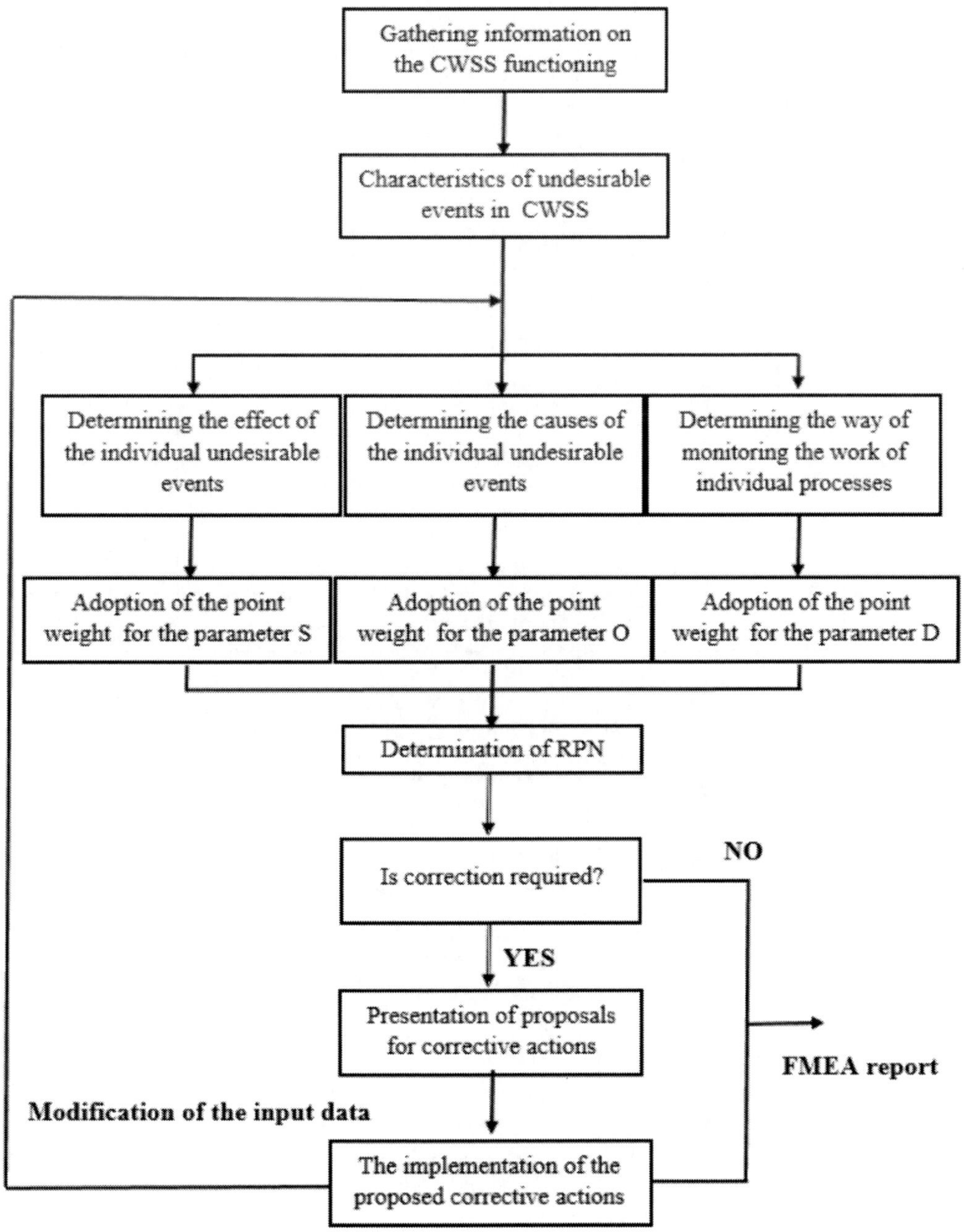

Fig. 1. The process of FMEA to assess water consumers safety.

Table 1. Criteria and point weights for the parameter S.

Severity	Description	Point weight
remote	There is no discernible effects, failure does not affect functioning of the system, water is proper for human consumption according to Act [15]	1
low	Disruptions in the operation of individual subsystems are not felt by consumers, water is proper for human consumption according to Act [15]	2
		3
moderate	Disruptions in the operation of individual subsystems cause a high degree of water consumers dissatisfaction, conditional suitability of water for consumption according to Act [15]	4
		5
		6
high	The system does not work, there may be a threat to the water consumers health, water is not proper for human consumption according to Act [15]	7
		8
very high	Undesirable event is a threat to water consumers health and lives, water is not proper for human consumption according to Act [15]	9
		10

Table 2. Criteria and point weights for the parameter O.

Occurrence	Frequency of appearance	Point weight
remote	< for 1 000 000 days	1
low	1 for 20 000 days	2
	1 for 4000 days	3
moderate	1 for 1000 days	4
	1 for 400 days	5
	1 for 80 days	6
high	1 for 40 days	7
	1 for 20 days	8
very high	1 for 8 days	9
	1 for 2 days	10

Table 3. Criteria and point weights for the parameter D.

Detection	Description	Point weight
very high	Complete monitoring, on-line equipment, potential undesirable event almost certainly will be detected	1
high	Potential undesirable event is detected by automated checks, which lead to error detection and protection from its development	2
		3
moderate	Undesirable event will not be detected until the reduction in the quality of water intended for human consumption or performance loss	4
		5
		6
low	Undesirable event will not be detected until inspection	7
		8
remote	There is no monitoring system, there is no chance to detect potential undesirable event	9
		10

The safety assessment is based on comparison of the obtained value with the adopted scale. The paper proposes three levels of water consumers safety:

- tolerable level of safety (TLS) - $RPN \leq 40$, in which as a result of the CWSS operation there is not any threat to water consumers lives or health,
- controlled level of safety (CLS) - $40 < RPN \leq 100$, in which as a result of the CWSS operation, mainly due to the consumption of contaminated water, there may be a threat to consumers' health, but there are sufficient safety barriers,
- unacceptable level of safety (ULS) - $RPN > 100$, beyond which as a result of the CWSS operation consumers are at risk of loss of health or lives.

4 Results of analysis

Table 4 shows a fragment of the FMEA analysis carried out for the CWSS located in the south-eastern Poland. Due to the limited volume of the publication the whole analysis of water consumers safety, covering tens of undesirable events, was not shown. Every undesirable event, its effects and causes is listed in the subsequent lines.

If the value of the RPN is higher than 100 (ULS), one must submit a proposal for remedial actions, indicate the person responsible for their implementation and the planned completion date. After the introduction of corrective actions the analysis must be made again, the value of RPN must be determined and the level of water consumers safety must be assessed.

DOI: 10.1051/e3sconf/20171700092

Table 4. A fragment of the FMEA analysis for the assessment of water consumers safety.

Phase of the process	Requirement/ Function	Potential type of failure	Potential effect of failure	S	Potential cause of failure	Current state				
						Prevention	O	Control	D	RPN
Water intake	Intake of good quality water	Excessive contamination of river water (bacterium, viruses, oils, petrol, detergents, pesticides, polycyclic aromatic hydrocarbons)	Contaminated water flows into Water Treatment Plant	5	Flood	The establishment of water intake protection zones, the ability to close water intake in case of water source contamination, alternative water sources	6	Automatic water quality monitoring at protective station and at water intake, biomonitoring	2	60
					Major industrial failure		2		2	20
					Harmful action of the third party		1		2	10
Water treatment	Maintaining qualification of the water treatment plant	Water at the output of WTP is not proper for human consumption according to Act [15]	Contaminated water is pumped to the WDS or there is no water supply	7	Failure of WTP equipment	Continuous supervision of the water treatment process, duplication of equipment	5	Automatic water quality monitoring at several points and at the output of WTP	2	70
					Failure of control system		4		2	56
Water distribution	Maintaining proper quality of supplied water	Secondary biological water contamination in plumbing or water supply network	Contaminated water is located in WDS	9	Lack of biological water stability	Observance of procedures, continuous modernization of plumbing and water supply network	5	Study of water quality made by county sanitary-epidemiological station	2	90
					Inappropriate content of disinfectant		3		2	54
					Inappropriate hydraulic conditions in network		3		2	54

5 Conclusions

The FMEA method can be an important tool for analysis and assessment of water consumers safety for the purpose of implementing the WSP. It allows to identify the weak points of the system and determine actions which must be taken in order to raise the level of water consumers safety. The presented analysis showed that if undesirable events occur, water consumers safety is at an acceptable or controlled level.

Further studies on this topic should be aimed at seeking solutions allowing to develop the conventional FMEA method by, among others, introducing to the FMEA method the fuzzy set theory. It is also necessary to continue the discussion on the criteria values for the parameter RPN and the assessment of its value.

References

1. Ustawa z dnia 26 kwietnia 2007 r. o zarządzaniu kryzysowym (Dz.U. 2007 Nr 89 poz. 590) z późniejszymi zmianami
2. K. Boryczko., B. Tchórzewska-Cieślak: Environment Protection Engineering, **40(4)**, 77–92 (2014)
3. B. Tchórzewska-Cieślak, J. Rak: Method of identification of operational states of water supply system. ENVIRONMENTAL ENGINEERING III, Ed. Pawlowski, L; Dudzinska, MR; Pawlowski, A. Taylor & Francis Group, 521–526 (2010)
4. K. Pietrucha-Urbanik: Engineering Failure Analysis, **57**, 137–142 (2015)
5. Guidelines for Drinking-water Quality. Fourth Edition. WHO, Geneva (2011)
6. Water safety plan manual: step-by-step risk management for drinking-water suppliers. WHO, Geneva (2009)
7. M.C. Almeida, P. Vieira, P. Smeets: Water cycle safety plan framework. Final. Prepared 2013.025 (2013)
8. B. Tchórzewska – Cieślak, D. Szpak: Ochrona Środowiska , **37(3)**, 43–47 (2015)
9. B. Tchórzewska-Cieślak, K. Boryczko, I. Piegdoń: Possibilictic risk analysis of failure in water supply network. Safety and Reliability: Safety and Applications, Taylor & Francis Group, 1473–1480, London (2015)
10. DH. Stamatis : Failure mode and effect analysis—FMEA from theory to execution. New York: ASQC Quality Press (1995)
11. PN-EN 60812:2009. Analysis techniques for system reliability: procedure for failure mode and effects analysis (FMEA)
12. A. Pillay, J. Wang: Reliability Engineering and System Safety, **79**, 69–85 (2003)
13. Y.M. Wang, K.S. Chin, G.K.K. Poon, J.B. Yang: Expert Systems with Applications, **36**, 1195–1207 (2009)
14. M. Kumru, P.Y. Kumru: Applied Soft Computing, **13**, 721–733 (2013)
15. Rozporządzenie Ministra Zdrowia z dnia 13 listopada 2015 r. w sprawie jakości wody przeznaczonej do spożycia przez ludzi (Dz.U. 2015 poz. 1989)

E3S Web of Conferences **17**, 00093 (2017) DOI: 10.1051/e3sconf/20171700093
EKO-DOK 2017

Improvement of energy efficiency of natural gas combustion by applying a homogeneous combustion

Jacek Szymczyk[1], *Ilona* Cieślak[1], *Agnieszka* Jach[1,*], *Rafał* Pyszczek[1], *Andrzej* Teodorczyk[1], and *Weiyu* Cao[2]

[1]Warsaw University of Technology, Institute of Heat Engineering, Nowowiejska 21/25, 00-665, Warsaw, Poland
[2]University of Michigan, Department of Mechanical Engineering, Ann Arbor, Michigan 48109

Abstract. In many heat devices designers and operators meet the problem of low efficiency of combustion and restricted emission standards. This process should be improved to maximize its efficiency and satisfy additional requirements as, for example, uniform temperature fieldin combustion chamber, low noise level or very low NOx emission. These requirements are satisfied by homogeneous combustion. Such combustion method is particularly attractive for the steel or glass industry or power industry based in particular on natural gas. In this paper factors, which have the biggest influence on performance of flameless combustion, are discussed, among others: momentum of fuel and oxidizer, composition of the mixture, the temperature of the inlet gases. Additionally, blind simulations of combustion in a combustion chamber of a furnace are run to assess how high is the influence of these factors individually. Numerical simulations are performed in a CFD code AVL Fire. The detailed chemical kinetics mechanism GRI-mech 3.0 is used for combustion calculations. Calculations results are correlated with experimental data. Blind simulations and experiment provide similar level of NO_X emission (~6-8 ppm). Experiments showed that the effect of the addition of ethylene to fuel on emissions of NO_X, CO, THC is not significant. Similarly, numerical simulations predict that influence of ethylene is negligible. CO, THC and CO_2 were on a stable level across all cases. NO_X emissions increases when mass flow of air and fuel increases due to higher heat release in the same volume, what results in higher temperature of combustion products. When temperature of fuel increases NO_X level decreases.

1 Introduction

Restricted emission standards and improvement of efficiency are one of the key issues in combustion. One of the methods which help to fulfil these requirements is homogeneous combustion, often called flameless combustion because of not visible flame front

* Corresponding author: ajach@itc.pw.edu.pl

in combustion area. Such combustion method is particularly attractive for the steel or glass industry or power industry based in particular on natural gas.

In the literature, flameless combustion (FLOX) has some other names: MILD (Moderate and Intense Low Oxygen Dilution) combustion [1], [2], HPAC combustion (Highly Preheated Air Combustion) [3], HTAC (High Temperature Air Combustion) [4], FDI (Fuel Direct Injection) [5]. Several reviews have also been published [1, 6, 7].

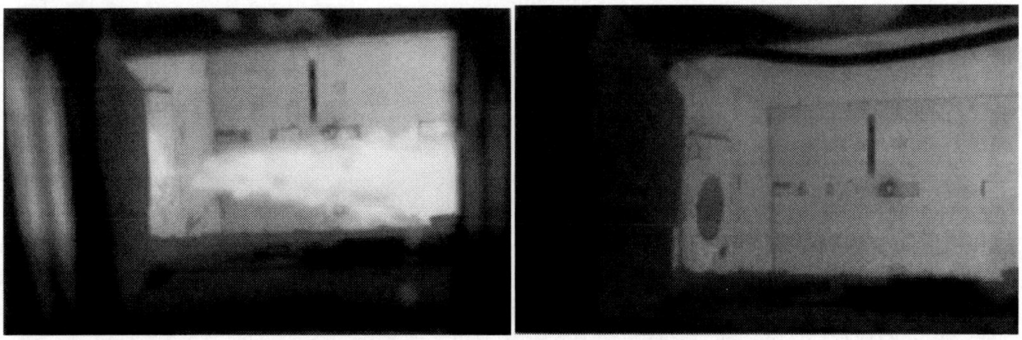

Fig. 1. Comparison flame (left) and flameless (right) firing natural gas [8].

The comparison of flame and flameless combustion is presented in Fig. 1. The difference between conventional flame and flameless combustion is recirculation ratio [9]:

$$K_v = \frac{M_E}{M_F+M_A} \quad (1)$$

where: M_E - mass of exhaust gas, M_F - mass of fuel, M_A - mass of air

For conventional combustion K_v is about 0.3, but in flameless combustion K_v should be more than 2 [7]. Fuel-air equivalence ratio is more than 1 in stable flameless combustion. Flameless combustion occurs when very dilute reactants are preheated to very high temperatures [1].

The main advantage of flameless combustion is reduction of NO_X emission. Moreover, the researchers noted more homogenous temperature distribution, reduced thermal stress for the burner or less burner faults [10]. Also typical combustion roar disappears due to not visible flame front.

In this paper an experiment performed at University of Michigan in Ann Arbor is described. The influence of doping of ethylene to natural gas on emissions is analyzed. Than blind simulations in a commercial CFD code AVL Fire were performed using the same geometry and boundary conditions. The influence of temperature of fuel, doping of ethylene and mass flow of fuel and air on emissions level (NO_X, CO, THC) was analyzed.

2 Experimental investigation

The experimental furnace is installed at University of Michigan in Ann Arbor. It is used mainly for simulation of flameless combustion.

All experiments were run in the vertical, 120 cm height furnace (Fig. 2a and b). The furnace was built as reversed bucket, where the fuel and oxidizer nozzles were installed vertically on the top of combustion chamber and exhaust gases left the furnace through windows in the lower part. The lowest element of the furnace was a moveable tub with simple heat exchanger made of the copper pipes. The role of heat exchanger was cooling down and - more important - measuring of the radiation level. The water flow was kept on a stable level of 0.5 kg/s. On the top of the combustion chamber one fuel vertical nozzle (in the axis) and four vertical oxidizer nozzles (symmetrical on diameter 560 mm) were installed. Each pipeline went through the hood and there was possibility to preheat each of them.

The aim of experiments, among others, was to achieve the temperature field and the content of O_2 and pollutants (NO_X, CO, THC) in each measuring port (probe) on 4 depths - on the fuel axis (r = 0 mm), 75, 160, 230, 280 mm (oxidizer nozzle axis). These probes were installed on the sidewall in three sets, where each set has 5 probes. Fuel flow in the nozzle was ~0.11 kg/min and oxidizer ~1.56 kg/min, fuel and oxidizer nozzles diameter was 16 mm.

During experiments there was a possibility to measure all parameters in 4 levels: in the highest – 220 mm, 430 mm, 640 mm and 850 mm from the top. The lowest probe (1060 mm from the top) showed on fig. 2a was not used.

The process of flameless combustion demands under stoichiometric conditions, where the mixture is rich and excess air ratio $\lambda < 1$ (equivalence ratio $\varphi = 1/\lambda > 1$). Starting procedure (about 1 hour) did not allow to keep fixed rich mixture because of safety and the probability of explosion but when temperature reached more than 850°C conditions could be slowly changed from lean toward reach mixture. When temperature was higher than minimal (850°C) flameless combustion could occur.

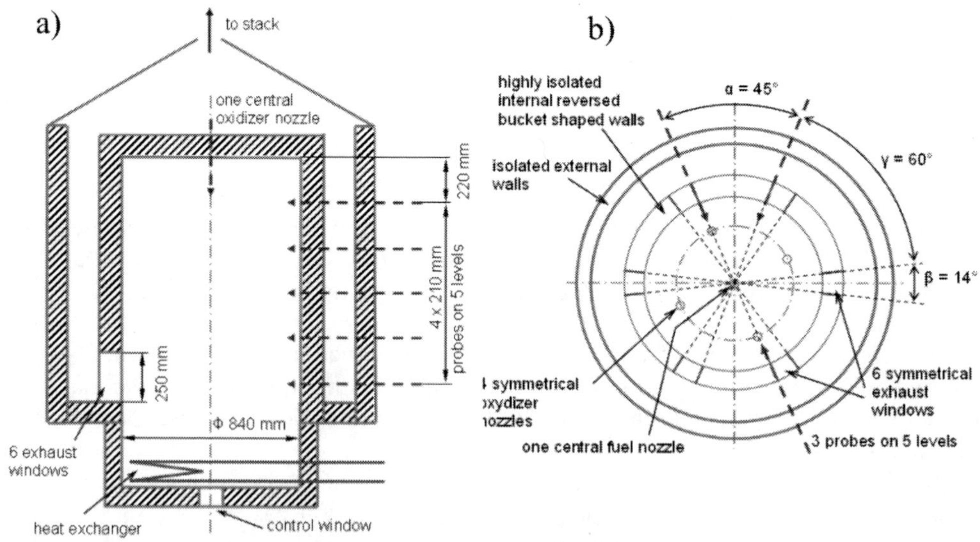

Fig. 2. Testing furnace: a – vertical view, b - horizontal view.

The literature [7, 11, 12] shows that combustion in enhanced oxygen can be more effective and radiation can be stronger compared to combustion in normal air (21% oxygen and 79% nitrogen). But the radiation could be enlarged more after adding heavier hydrocarbons to fuel. In this experiment the influence of ethylene doping was investigated. Fuel was preheated up to 473°C for radiation enhancement. In Table 1. cases measured during experiment are summarised.

Table 1. Cases measured during experiment.

Name	Oxygen/nitrogen (vol)	Ethylene [%]
35O2NG	35/65	0
35O2NG15E	35/65	15

In Fig. 3 (NO_X) the influence of ethylene on NO_X emissions is shown. Measuring points were located not so close to the top (to fuel and oxidizer nozzles) and one set of points was located straight below fuel nozzle and the second set – straight below oxidizer nozzle. Presented experimental results and results described in [7] show, that for some oxygen/ethylene ratios ethylene can lower NO_X emission by about 3 ppm in comparison to flameless combustion in pure air. In this case adding ethylene in some levels decreases NO_X content but on the bottom of the furnace this influence is much smaller probably because temperatures are smaller and the temperature field is more flat. The highest values of NO_X after ethylene adding were in fuel nozzle stream and they were between around 5 ppm and 6.8 ppm and were similar to the case where ethylene was not added [7]. The highest concentration of NO_X is on the top and drops when distance from the top H rises. Overall NO_X generation during flameless combustion compared to flaming combustion was low (about 1–3 ppm in comparison to 110–130 ppm [7]).

In this case the increase of NO_X concentration after adding of ethylene can be observed for H1060 mm, R0 mm case so this is quite opposite than overall tendency shown in Fig. 3. During the experiment semi-waves of exhaust gases were observed in the combustion chamber what suggests that it was not flameless homogenous combustion. On the other hand, small gradients of temperature and low values of NO_X show, that it was also not flaming combustion.

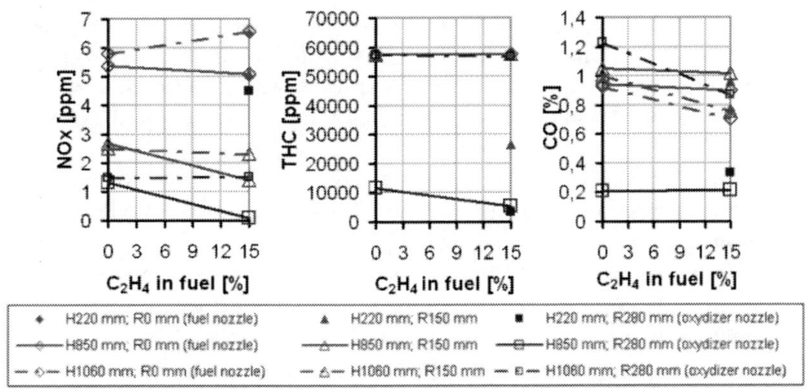

Fig. 3. Concentration of NOx, THC and CO in exhaust gases in dependence on the content of ethylene in the fuel mixture based on natural gas in three distances from the top H (220 mm, 850 mm and 1060 mm) and three radii R (0 mm, 150 mm and 280 mm).

In Fig. 3 (CO) the influence of ethylene on CO emissions along the furnace radius is also shown. It is observed that the CO concentration in measuring points is on the low level.

In all measuring points (except for point straight below the oxidizer nozzle) adding of ethylene to fuel decreases CO concentration. In the oxidizer nozzle stream CO concentration first is small (near the top) and going down rises to reach nearly similar values independently on radius. Adding of ethylene causes lowering of CO in one height, but has no influence on the bottom near the outflow windows.

Analysis of THC showed that the concentration profile is flat in almost the whole volume. The only difference is in two points – heights 220 mm and 850 mm in oxidizer nozzle stream, 10000 ppm. There is observed very large gradient of concentration between H = 850 mm and the bottom, and is equal to 125,2 ppm/mm.

It is interesting that there is increase of radiation and efficiency of combustion (in the meaning of transporting heat to heated water) after adding of ethylene. It was observed that the temperature of exit water in heat exchanger rises at a higher value in case 35O2NG15E unlike case 35O2NG from 26°C to 28.5°C. This increase is caused by ethylene but also by preheating of fuel and oxygen enhancement. Each of these elements can increase intensity of heat generation and radiation but not always causes drop of NO_X emission.

3 Simulation

Numerical simulations of combustion were performed using the AVL Fire code. AVL Fire is the leading CFD simulation tool in the field of combustion analysis. It specializes in the accurate prediction of processes as injection nozzle flow, fuel injection, combustion, emission and exhaust gas aftertreatment.

In this study real geometry of furnace (described in previous section) was investigated. Number of mesh elements and nodes is 368993 and 373305 respectively.

The furnace runs with equivalence ratio of EQR = 1.15 (air excess number of $\lambda = 0.87$). In total 14 simulations were run. Table 2. presents composition of fuel, mass flow of fuel and air (35% O_2), temperature of fuel for each case. Temperature of air was the same in every case and was of 300 K. Temperature of combustion chamber wall was 1100 K, with exception of the bottom wall (a location of a heat exchanger in experiment) - 298 K. Pressure was of 1 bar. The cases: C_a (base case without ethylene) and C_a15_C2H4 (15% of ethylene in fuel) are the most similar to the experimental ones.

The compressibility is solved for a weak compressible flow where density is a function of temperature only. Wall treatment is set to Hybrid Wall Treatment. In this model the "fitting" function between laminar and turbulent part of the boundary layer is derived from the relation of the friction velocity. This wall treatment should ensure a gradual change between viscous sublayer formulations and the wall functions. The default Standard Wall Function heat-transfer model is chosen, as recommended by the AVL Fire manual [15].

Turbulence is a key issue in CFD modeling. In this work turbulence is modeled by using k-ζ-f turbulence model. The k-ζ-f model is recommended by the AVL Fire manual for application in combustion simulations [15].

The General Species Transport Module and the General Gas Phase Reaction Module are used to calculate combustion. In this study the GRI-Mech 3.0 was used [13]. This the latest version and most commonly used detailed reaction model for natural gas combustion with 325 reactions and 53 species. The homogeneous chemical reactions are solved in the whole computational domain. The enthalpy sources resulting from the homogeneous reactions are taken into account in the gas-phase enthalpy equation.

Table 2. Analyzed cases (green color informs on a changed value in comparison to the base case).

Case	%CH_4 in fuel	%C_2H_4 in fuel	Mass flow fuel [kg/s]	Mass flow air [kg/s]	Temperature fuel [K]
C_a (base)	100	0	0.00183062	0.0166613	760
C_a05	100	0	0.00091531	0.0083307	760
C_a08	100	0	0.0014645	0.0133290	760
C_a12	100	0	0.00219674	0.0199936	760
C_a15	100	0	0.00274593	0.0249919	760
C_a20	100	0	0.00366124	0.0333226	760
C_a360	100	0	0.00183062	0.0166613	360
C_a560	100	0	0.00183062	0.0166613	560
C_a860	100	0	0.00183062	0.0166613	860
C_a960	100	0	0.00183062	0.0166613	960
C_a05_C2H4	95	5	0.00184765	0.0166613	760
C_a10_C2H4	90	10	0.00186880	0.0166613	760
C_a15_C2H4	85	15	0.00198900	0.0166613	760
C_a20_C2H4	80	20	0.00190833	0.0166613	760

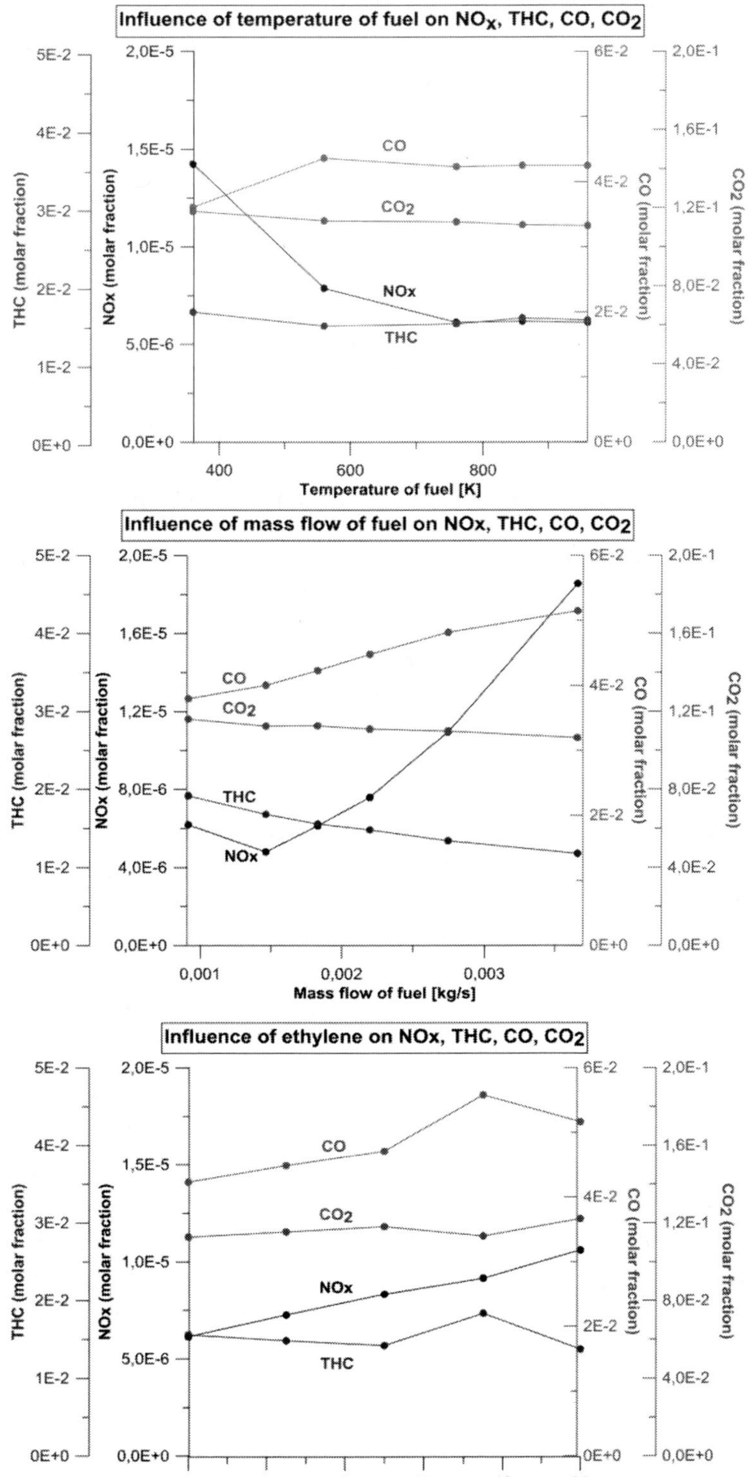

Fig. 4. Influence of temperature of fuel (top), mass flow of fuel (centre) and ethylene doping (bottom) on NO_X, THC, CO and CO_2 emissions.

Results from simulation are presented in Fig. 4 and 5. In Fig. 4. the influences of temperature of fuel, doping of ethylene and mass of fuel on NO_X, THC (Total Hydrocarbon), CO and CO_2 are presented. Emission of NO_X increases with the increase of mass flow of fuel and ethylene doping. Emission of NO_X decreases with the increase temperature of fuel, and then stabilizing for temperature of fuel higher than 760K. Emission of CO_2 is almost constant, only slightly decreases for higher mass flow of fuel. Emission of CO is the lowest for temperature of fuel of 360 K and the highest for temperature of fuel of 560 K. Further increase of temperature of fuel leads to a stabilization of emission of CO. Emission of CO increases with the increase of mass flow of fuel and ethylene doping. Emission of THC decreases with the increase of mass flow of fuel. The influence of temperature of fuel and ethylene doping on THC emission is negligible.

Figure 5. (left) presents NO_X in standard unit (mg/Nm^3) to describe emission of furnaces and boilers. Emission limit values for existing medium combustion plants with a rated thermal input equal to or greater than 1 MW and less than or equal to 5 MW, other than engines and gas turbines for natural gas is 250 mg/Nm^3 [14]. NO_X emissions in the illustrated furnace is much lower than the limit.

The high influence on mean temperature in combustion chamber (fig. 5. right) has mass flow of fuel. Mean temperature in combustion chamber increases by 174 K when mass flow of fuel increases fourfold. The effect of temperature of fuel and ethylene doping is negligible.

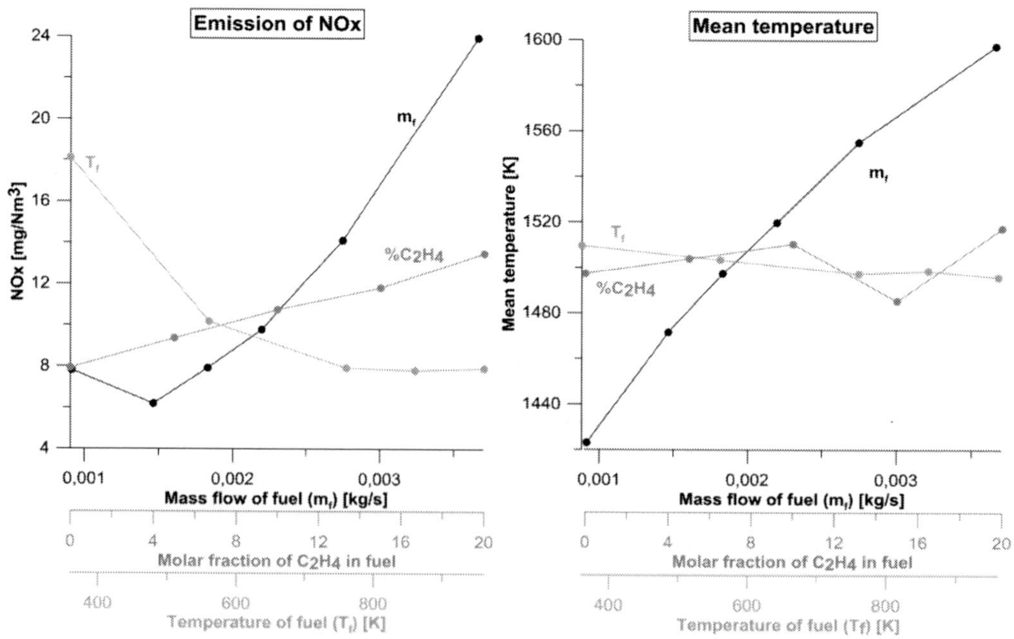

Fig. 5. Influence of temperature of fuel (pastel blue), mass flow of fuel (black) and ethylene doping (orange) on NO_X emission (left) and mean temperature in combustion chamber (right).

4 Conclusion

In the paper experimental and numerical investigations of natural gas combustion in a furnace were performed. The most important conclusion is that blind simulations and experiment provide similar level of NO_X emission (~6-8 ppm). Experiments showed that the effect of the addition of ethylene to fuel on emissions of NO_X, CO, THC is not significant. Similarly, numerical simulations predict that influence of ethylene is negligible.

CO, THC and CO_2 were on a stable level across all cases. NO_X emissions increases when mass flow of air and fuel increases due to higher heat release in the same volume, what results in higher temperature of combustion products. When temperature of fuel increases NO_X level decreases.

In the next step of numerical investigation influence of equivalence ratio, radiation and recirculation ratio will be assessed.

The authors acknowledge professor Arvind Atreya for support in collecting the data needed to write this article.

References

1. A. Cavaliere and M. De Joannon, Prog. Energy Combust. Sci. **30** (2004)
2. M. De Joannon, A. Cavaliere, T. Faravelli, E. Ranzi, Proc. Combust. Inst. **30** (2005)
3. M. Katsuki and T. Hasegawa, Symp. Combust. **27** (1998)
4. H. Tsuji, A. K. Gupta, T. Hasegawa, M. Katsuki, K. Kishimoto, M. Morita, High temperature air combustion: From Energy Conservation to Pollution Reduction (CRC Press LLC, 2003)
5. B. A. Fleck, A. Sobiesiak, H. A. Becker, Combust. Sci. Technol. **161** (2000)
6. J. A.Wünning, J. G.Wünning, Prog. Energy Combust. Sci. **23** (1997)
7. J. Szymczyk, P. Olszewski, Arch. Combust. **36** (2016)
8. A. Milani, A. Saponaro, IFRF Combust. J., Article Number 200101 (2001)
9. H. Yu, “Flameless Combustion of Natural Gas in the SJ/WJ Furnace” (2008)
10. J. G.Wünning, Flameless combustion and its applications, 14th IFRF Members Conference (2004)
11. A. Atreya, “Ultra-High Efficiency and Ultra-low Emissions Combustion Tech for Manufact Industries” (2013)
12. N. Krishnamurthy, W. Blasiak, A. Lugnet, Development of High Temperature Air and Oxy-Fuel combustion technologies for minimized CO2 and NOx emissions in Industrial Heatin, The Joint International Conference on “Sustainable Energy and Environment (SEE)”, (2004)
13. G. P. Smith, D. M. Golden, M. Frenklach, N. W. Moriarty, B. Eiteneer, M. Goldenberg, C. T. Bowman, R. K. Hanson, S. Song, W. C. Gardiner, V. V. Lissianski, and Z. Qin, “GRI-mech 3.0 mechanism.” http://www.me.berkeley.edu/gri_mech/
14. European Commission, “Directive of the European Parliament and of the Council on the limitation of emissions of certain pollutants into the air from medium combustion plants”, Off. J. Eur. Union 442, 19 (2013)
15. AVL FIRE Manual, version 2014.2

Influence of the air phase on water flow in dikes

Witold Tisler[1,*], and *Adam* Szymkiewicz[1]

[1]Gdansk University of Technology, Faculty of Civil and Environmental Engineering, Department of Geotechnics, Geology and Marine Civil Engineering, ul. Narutowicza 11/12, 80-233 Gdańsk, Poland

Abstract. Numerical models are often used to describe flow and deformation processes occurring in dikes during flood events. Modeling of such phenomena is a challenging task, due to the complexity of the system, consisting of three material phases: soil skeleton, pore water and pore air. Additional difficulties are transient loading caused by variable in time water levels, heterogeneity of the soil or air trapping. This paper presents a brief review of the influence of the air phase in soil on water flow and pore pressure generation, with focus on applications related to stability of dikes, earth dams and similar structures. Numerical simulations are carried out to investigate the differences between the Richards equation and the two-phase flow model, using an in-house code based on the finite volume method. A variety of boundary problems are considered, including seepage through flood dikes, dike overtopping and water level fluctuations. Special attention is paid to the problem of air trapping, which occurs when water flows over the top of a dike. Such a phenomenon occurred during experiments on model dikes reported in the literature, ultimately leading to development of cracks and damages in dike structure.

1 Introduction

When considering water movement in soil, it is often assumed that the air present in the soil is connected with the atmospheric air, which has approximately constant pressure. It means that the air in pores can escape freely before advancing water and thus has no influence on the infiltration rate. In most flow scenarios considered in civil and environmental engineering the flow of the air phase is neglected. Consequently, most of the available software for modeling unsaturated zone flow is based on the Richards equation, which does not account for the air flow. However, there is an increasing evidence that in some situations air flow must be taken into account: heterogeneous soils [1–3], rapid downward infiltration [4], water table fluctuations [5], seismic events [6] or overtopping of dikes [7–10]. In such cases the two phase flow model should be taken into account. In this type of analysis two equations (air and water) are coupled together. In general, each of the phases consists of multiple chemical components, which can move between phases. Pore air, for instance, is a mixture of gases, including water vapor, while pore water contains many dissolved substances, including gases. The number of phases and components included in the mathematical model depends on the problem under consideration. In many

[*] Corresponding author: wittisle@pg.gda.pl

applications focusing on the water flow, a sufficient accuracy can be achieved with a simplified model, where both air and water are considered as immiscible and the deformation of the solid skeleton is treated in a simplified manner [2, 11]. The main goal of the paper is to present numerical simulations of water and air flow in dikes carried out using an in-house code developed by the authors, based on the control volume finite element approach (CVFE) [2, 11]. We compare solutions obtained using the full two phase model with the solutions of the Richards equation, which neglects the air flow.

2 Mathematical model

2.1 Equations

Mathematical model used in this article is based on the following assumptions: fluids are immiscible, soil particles are incompressible, process is isothermal and all deformations are caused only by changes in the pore pressure and can be described by a linear isotropic elastic law. Under these assumptions, the governing equation can be written in the following form:

$$\frac{\partial(\rho_\alpha S_\alpha n)}{\partial t} - \nabla \cdot \left[\frac{\rho_\alpha \kappa k_{r\alpha}}{\mu_\alpha} (p_\alpha + \rho_\alpha g z) \right] = 0 \tag{1}$$

where α means air or water phase respectively, ρ_α – density, S_α – saturation, n – porosity, t – time, κ – intrinsic permeability, μ_α – dynamic viscosity, $k_{r\alpha}$ – relative permeability, p_α – pressure in fluid, g – gravity, z – depth. In the two phase flow model, we have two separate equations, where in the first one index α is replaced by 'w' (water) and in the second one by 'a' (air). The flow model proposed by Richards is obtained by retaining only one equation, where the considered phase is water. Although one of equation is omitted, equation (1) retains its nonlinearity. However it is easier to solve than the two phase flow model. The governing equations contain several dependent variables, so they have to be completed by additional constitutive relationships. The values of the two saturations sum up to one:

$$S_w + S_a = 1 \tag{2}$$

Pressure difference between the non-wetting fluid (air) and the wetting fluid (water) is called capillary pressure or suction:

$$p_c = p_a - p_w \tag{3}$$

The density of each fluid is assumed to depend only on the fluid pressure:

$$\rho_w = \rho_{w0} + \beta_w (p_w - p_{w0}) \tag{4}$$

$$\rho_a = \rho_{a0} + \beta_a (p_a - p_{a0}) \tag{5}$$

The value of capillary pressure is related to the water saturation. This relationship is called SWRC (soil water characteristic curve) or SWRC (soil water retention curve) and can be expressed by various analytical formulae. The most popular models are proposed by Brooks-Corey, van Genuchten or Gardner. In this article we used van Genuchten formula:

$$\frac{S_w - S_{wr}}{1 - S_{wr} - S_{ar}} = S_{ew} = \left[1 - \left(p_c / p_g\right)^{n_g}\right]^{-m_g} \tag{6}$$

where S_{ar} is residual saturation of fluid, S_{ew} - effective water saturation, p_g - scaling pressure in the retention function, n_g and m_g are the exponents in the retention function. Permeability of the medium describing the Mualem- van Genuchten model:

$$k_{rw} = \sqrt{S_{ew}}\left[1 - \left(1 - \left(S_{ew}\right)^{1/m_g}\right)^{m_g}\right]^2 \tag{7}$$

$$k_{ra} = \sqrt{1 - S_{ew}}\left[1 - \left(S_{ew}\right)^{1/m_g}\right]^{2m_g} \tag{8}$$

Volumetric deformations of the soil skeleton are accounted for in a simplified way, by assuming linear relationship between the porosity and the average pore pressure:

$$n = n_0 + \beta_s\left(p' - p'_0\right) \tag{9}$$

where n_0 is the initial porosity and β_s is compressibility coefficient, related to soil Young modulus E and Poisson ratio ν :

$$\beta_s = \left(1 - n_0\right)\frac{2(1 + \nu)}{E} \tag{10}$$

p' is the average pore pressure and can be defined as:

$$p' = S_{ew}\, p_w + \left(1 - S_{ew}\right) p_a \tag{11}$$

Nonlinearity of main equation or equations results from the relationship between saturation and capillary pressure or permeability and type of the soil (clay, sand, silt). Besides definition of the constitutive relationship it is also necessary to select appropriate par of unknowns. We distinguish three most common solutions: pressure formulation, pressure-saturation formulation and saturation formulation. In our previous study on this subject [10] we used a commercial code FlexPDE, where pressure-based formulation was implemented. However, the mixed formulation (pressure-saturation) seems to be better suited for the cases where one of the phases may disappear (such as the air phase during variably saturated flow in soils) [2, 11]. Therefore the mixed formulation was used in this paper.

3 Numerical implementation

The governing equations for unsaturated and two-phase flow were solved using in-house code implemented in Fortran. The algorithm is based on the control-volume finite-element approach, which combines features of the finite volume method (FVM) and the finite element method (FEM) - for details see e.g. [2]. Spatial discretization of the governing equations is performed on a dual grid, where finite volumes (control volumes) are built around nodes of the primary finite element grid. In two dimensions either triangular or quadrilateral elements can be used, while the resulting control volumes are irregular polygons, see Fig. 1. The fluxes of water and air through each segment of the control volume boundary are approximated using shape functions for the corresponding finite element. We use a vertex-centered scheme, where the primary unknowns (water pressure and water saturation) are associated with nodes of the primary finite element grid, which become centers of the finite volume of the dual grid. In such a way the values of unknowns at the outer boundaries of the domain are obtained directly from the solution. For the discretization in time the fully implicit first-order method was used. The nonlinear algebraic system arising at each time step are solved using Newton-Raphson iterative scheme.

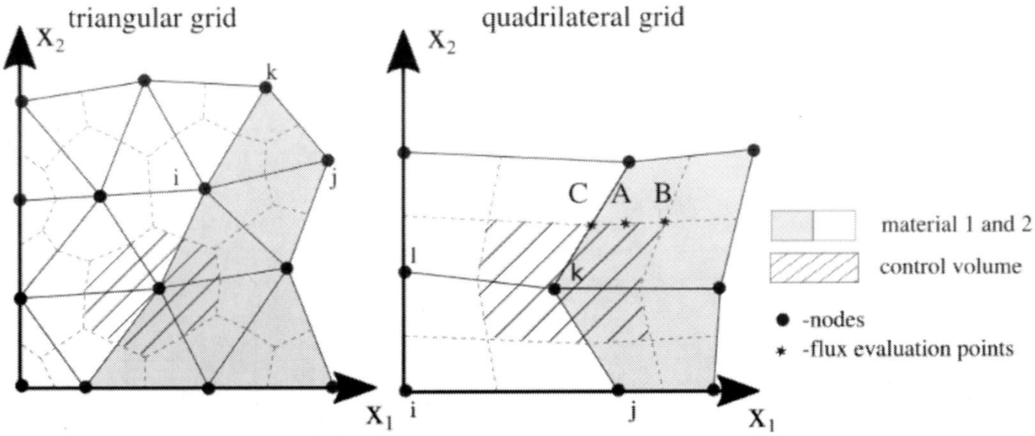

Fig. 1. Numerical grid for vertex centred finite volume method.

One of the main advantages of using an in-house code is a greater flexibility with respect to the choice of boundary conditions and material constitutive models, as compared to many commercial software packages. In unsaturated flow simulations it is often necessary to switch between Neumann and Dirichlet boundary condition, for example when modelling the infiltration or evaporation processes at soil surface, or seepage faces on slopes, such as the one presented in the example below. In this article this solution was used in modelling the seepage face of the dam. When water flow out only from some part of the boundary we assume a Dirichlet but for the rest which is unsaturated we assume Neumann. Position of the saturated-unsaturated interface is not known a priori and must be obtained iteratively during simulation.

4 Numerical examples

4.1 Seepage through a dike

4.1.1 Geometry, material parameters and boundary conditions

The influence of air phase on the water flow in dikes is investigated for two different patterns of seepage. The first test represents seepage through an embankment caused by an increase of the water level at the outer slope (Fig. 2). We consider homogeneous dike placed on an impermeable base. Soil parameters correspond to loamy sand are listed in Table 1 [13].

DOI: 10.1051/e3sconf/20171700094

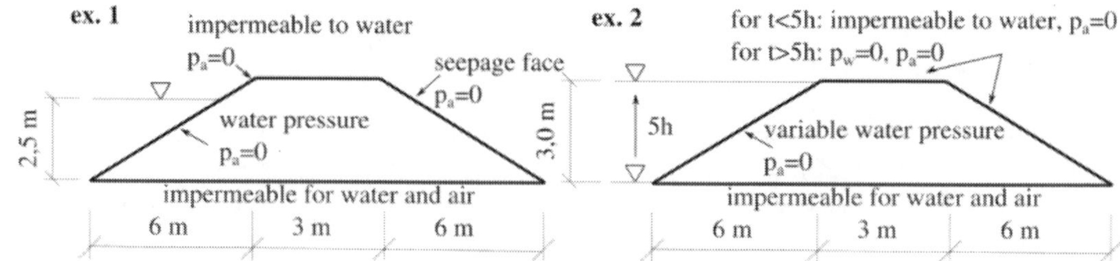

Fig. 2. Dike geometry and boundary conditions used in simulations.

Table 1. Soil parameters.

S_{wr}	S_{ar}	p_g	n_g	m_g	n_0	ν	E	p'_0
0,16	0,02	1308 Pa	1,89	0,471	0,41	0,25	$5x10^7$ Pa	0

For both fluid phases parameters are listed below:

Table 2. Fluid parameters.

Water				air			
ρ_{w0}	β_w	p_{w0}	μ_w	ρ_{a0}	β_a	p_{a0}	μ_a
1000 kg/m^3	$4{,}9x10^{-10}$ 1/Pa	0	10^{-3} Pa·s	1,22 kg/m^3	$1{,}2x10^{-5}$ 1/Pa	0	$1{,}57x10^{-5}$ Pa·s

Initially the dike is in unsaturated state with water pressure equal to 0 (atmospheric pressure) at the base and decreasing hydrostatically to -29,43 kPa at the top of the dike. In the whole domain air pressure was assumed constant and equal to 0 (atmospheric pressure). The time of rising of the water table is assumed to be negligibly small, so the water table level was given at the height of 2,5 m right at the beginning of the calculations. At the submerged part of the slope the boundary condition is given as hydrostatic and rising hydrostatically from 0 at the water table to 24,53 kPa at the bottom of the dam. For the rest of the left slope and the crest is assumed that gas phase can freely escape. The liquid phase can only flow out from the right sight of the dam. This part of the boundary is considered as seepage face. It means that it is impermeable for water as long as the water pressure at the boundary is negative. If the water pressure reaches 0, water starts to flow out from the soil, but no pressure built up is allow due to surface runoff. Boundary condition changes to pressure-type, p_w=0. For two phase flow model simulations, a constant value of air pressure p_a=0 is assumed along the slopes and top of the dike.

Numerical grid was performed using quadrilateral elements with computational points in the grid nodes (vertex centred method Fig. 1). Here we show results for an 800 elements and 626 nodes because for results for denser grid were qualitatively the same. Time step varied from 10^{-12} to 1000 s.

4.1.2 Results

Results of the simulations are shown in fig. 3–4. Fig. 3 show the distribution of water saturation in dike at an early stage of the simulation ($2 \cdot 10^5$ s). Shape of the saturation front is qualitatively similar but it is definitely longer compared to the 2PH model. At the end of the simulation this difference is a little smaller (steady state has not been reached).

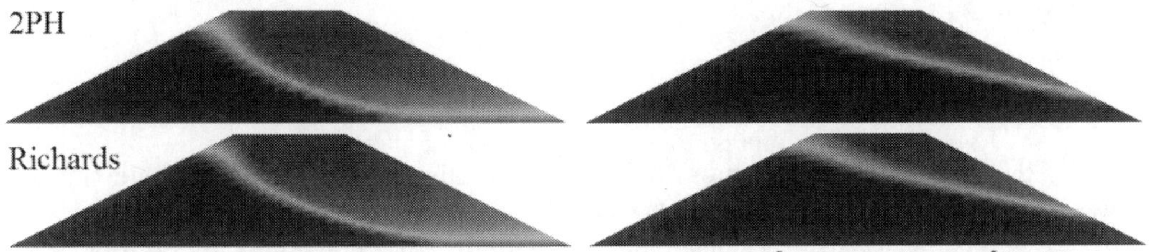

Fig. 3. Example 1: distribution of water saturation in dike for $t=2\cdot10^5$ s (left) and $t=7\cdot10^5$ s (right).

Figre 4 shows the total outflow of water from the right slope of the dike (seepage face) as a function of time. Initially value was nearly the same for both models, greater differences can be observed in the middle of the simulated period. It is associated with reaching the seepage face by the water. From this time differences are more visible. This is due to later start of the outflow for two phase model and consequently about 10% smaller discharges at the end. Received results are consistent with expected behaviour of both solutions. For Richards we expected that water should propagate faster than for 2PH. It is caused by the lack of presence the air in the domain, so water can flow out without any difficulties. While in 2PH water needs to displaced air during the whole process therefore infiltration takes a little longer.

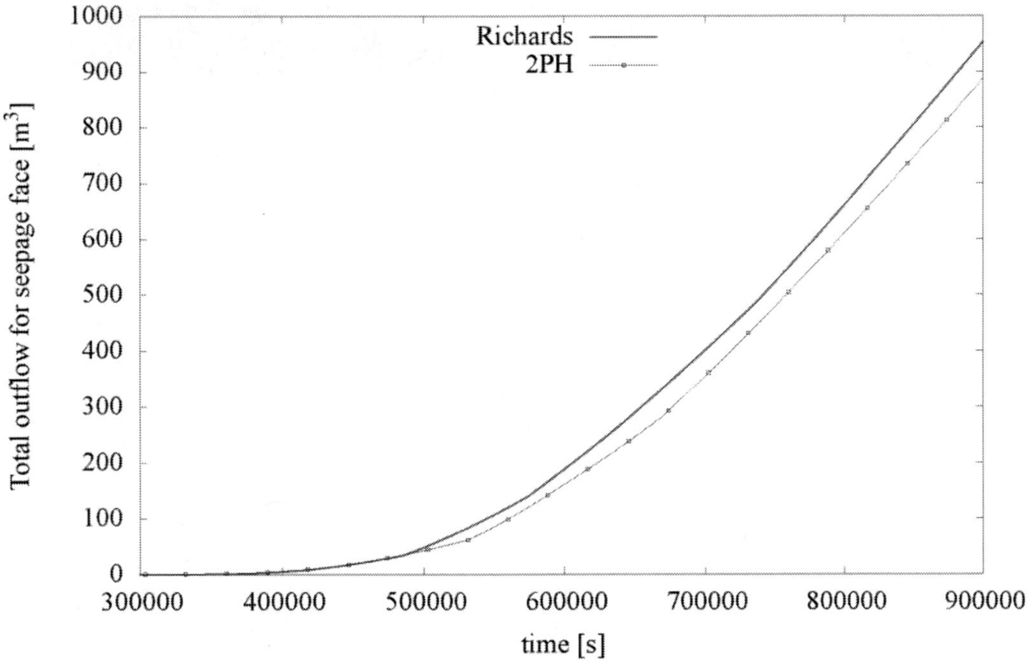

Fig. 4. Example 1: distribution of seepage flux at the inner slope of the dike.

4.2 Overtopping of a dike

Second simulation shows overtopping of a dike (fig. 2). This situation occurs when water table rise over the crest of the dam eg. during heavy rainfall. Geometry, material parameters and initial conditions are the same as in example 1. However, this time we assumed that water rise during 5 hours to the top of the dike and remained there till the end of the simulation. Initially crest and inner slope were treated as impermeable for water. When the overtopping started, the boundary condition was changed to $p_w=0$. As in the first example air pressure is $p_a=0$ for all of the boundaries with the exception of the bottom. Numerical grid and range of time steps were the same as in example 1.

4.2.2 Results

Figures 5–6 show distribution of water saturation shortly before achieving fully saturation according to Richards equation. We can observed that in 2PH model at the bottom right corner is still large unsaturated or partially saturated area while in Richards the dike is almost fully saturated. In the two-phase solution inflow of water is hampered by the necessity to displace the air through the water saturated medium surrounding the unsaturated zone. In Richards equation water infiltrates with the same velocity into an unsaturated area, regardless the possibility of air escaping, thus region of entrapped air is definitely larger in 2PH. Process of saturation during time can be also see on figure 5. We can observe that the fully saturated state occurs about 20 hours later in 2 phase flow model than in Richards.

Fig. 5. Example 2: distribution of water saturation in dike for $t=0{,}5\cdot10^5$ s (left) and $t=1{,}7\cdot10^5$ s (right).

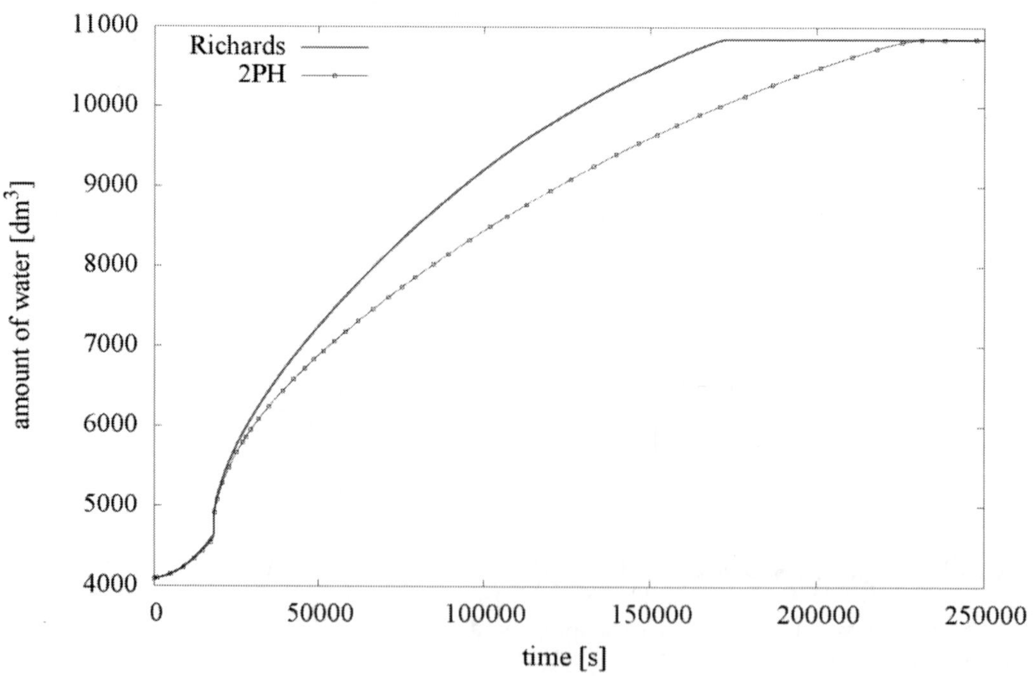

Fig. 6. Example 2: evolution of total volume of water in dike.

5 Conclusions

Numerical simulation presented in the article showed differences between the unsaturated flow model (Richards equation) and the two-phase model. Infiltration process occurring in dike is slower because water needs to displace air which is typically not accounted for in geotechnical and hydrological modelling software [14]. In addition when possibility of outflow for air phase is further limited, the impact is even greater. This can happen when we are dealing with overtopping of dike or heavy rainfall during infiltration [15–17]. The results confirm the findings from our previous study [10] performed using a commercial code FlexPDE based on the standard finite element formulation. The code developed by the authors, described in this paper, offers greater flexibility with respect to the boundary conditions and material constitutive models and therefore is suitable for further extensions.

Problem of air trapping in large soil structures has a significant impact on their stability and surface structure. In articles [8, 9] showed that entrapped or escaping air cause cracking near the surface which results in a damage of the structure. Cracks appear between the entrapped air areas and outer boundaries. While such phenomena were not investigated in this contribution it is believed that an explicit taking into account of the air flow in the simulation is a first step to their modelling.

References

1. P. Szymańska, W. Tisler, C. Schutz, A. Szymkiewicz, I. Neuweiler, R. Helmig, Acta Geophys, **6** (2016)
2. A. Szymkiewicz, Springer, *Modelling water flow in unsaturated porous media: accounting for nonlinear permeability and material heterogeneity* (2013)
3. G. Oettl, R. Stark, G. Hofstetter, Elsevier, *Numerical simulation of geotechnical problems based on a multi-phase finite element approach* (2004)
4. J.-O. Delfs, W. Wang, T. Kalbacher, A. K. Singh, O. Kolditz, Conference Paper (2013)
5. X. Kuang, J. Jiao, H. Li, Water Resour Res, **49** (2013)
6. A. R. Khoei, T. Mohammadnejad, Comput Geotech, **38** (2011)
7. J. Jiao, X. Kuang, L. Wan, X. Wang, D. Mao, Water Resour Res, **47** (2011)
8. D. Leśniewska, *Integrated flood risk analysis and management methodologies*, Technical Report (2007)
9. P. Bogacz, J. Kaczmarek, D. Leśniewska, Technological Sciences, **11** (1999)
10. W. Tisler, A. Szymkiewicz, Conference Paper (2014)
11. R. Helmig, Springer, *Multiphase flow and transport process in the subsurface: a contribution to the modeling of hydrosystems* (1997)
12. M. Marinas, J. W. Roy and J. E. Smith, Ground Water, **51(1)** (2013)
13. R. Carsel, R. Parrish, Water Resources Res, **5**, 24 (1988)
14. D. G. Fredlund, H. Rahardjo, M. D. Fredlund, Wiley, *Unsaturated soil mechanics in engineering practise* (2012)
15. A. Szymkiewicz, A. Kryczałło, IMiG, **2** (2011)
16. S.-H. Yu, K.-J.S. Kung, Comput Geotech, **23** (1997)
17. X. Zhang, Y. ZHU, Ch. Fang, Science Direct, **21**, 5 (2009)

E3S Web of Conferences **17**, 00095 (2017) DOI: 10.1051/e3sconf/20171700095
EKO-DOK 2017

Impact of the duct static pressure reset control strategy on the energy consumption by the HVAC system

Juliusz Walaszczyk[1,*], and *Aleksandra* Cichoń[1]

[1]Wroclaw University of Science and Technology, Faculty of Environmental Engineering, Wybrzeze S. Wyspianskiego 27, Wrocław, Poland

Abstract. This article addresses different duct static pressure control strategies which could be implemented in variable air volume air-conditioning systems (VAV). Two pressure reset control strategies are compared to the commonly used control solution based on the "Constant static pressure" method. First pressure reset control strategy, known as PID Control, uses signals from VAV boxes controllers to reset duct static pressure in a way that one of the VAV dampers is maintained almost entirely open. Second strategy decreases static pressure setpoint until an adjustable number of pressure requests occur. As a response to the certain amount of requests, static pressure setpoint is increased. This strategy is called Trim & Respond. Both static pressure reset control strategies described in this paper are considered to have more significant potential for energy savings than the "Constant static pressure" method. In order to validate this potential, several simulations for different control strategies were carried out and the obtained results are compared and analysed. The theoretical limit of the energy savings - set of the optimal control actions, was estimated with Nelder-Mead algorithm and also presented in this article. General description of the static pressure control strategies "Constant static pressure", PID Control and Trim & Respond is given.

1 Introduction

Variable air volume systems have been applied since the end of the World War II [1], while the term VAV (variable air volume), was formulated and first used almost two decades later in the sixties of the last century [2]. VAV systems became widely popular in the middle of the seventies, as a result of the 1970 energy and oil crisis. In continental Europe, VAVs gained their popularity in the eighties. Currently they are commonly applied and implemented in the modern HVAC installations.

A typical VAV system consists of variable air volume zone boxes with dampers and variable speed fans. In response to varying loads in the ventilating zones, VAV damper's position changes in order to maintain required airflow. The changing damper's position affects on a static pressure value in a duct system. As static pressure changes,

* Corresponding author: juliusz.walaszczyk@pwr.edu.pl

a fan responds by adjusting its speed to the point which allows to maintain static pressure at a given setpoint. For example, when zone's load decreases, less air volume is required and the zone's damper closes. At the same time fan reduces its speed to maintain constant duct static pressure. Lowered fan speed results in reduced total energy consumption. "Constant static pressure" control strategy is an efficient solution for energy savings but it is considered that the static pressure reset strategy is much more effective [3, 4].

In order to investigate the impact of the duct static pressure reset control strategy on the energy consumption by HVAC system, PID Control and Trim & Respond strategies were described and compared to the "Constant static pressure" method. In order to present the energy savings limit and potential for future research, optimal control actions was investigated. Optimal control actions were estimated using Nelder-Mead algorithm and then used as a benchmark for other presented strategies (Fig. 1).

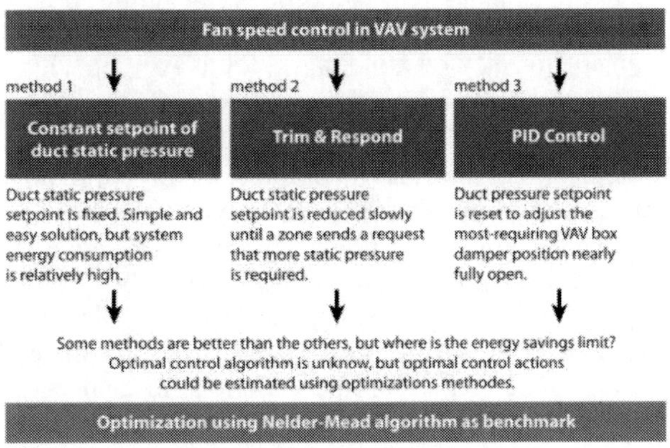

Fig.1. Graphical presentation of the research methodology.

2 Duct Static Pressure Reset Control Strategies

Since 1999, ASHRAE Standard 90.1 has required that static pressure setpoint should be reset for each system with direct digital control (DDC) at the zone level, instead of maintaining constant value of the duct static pressure. Static pressure reset control strategies saved fan energy and at the same time all zone's demands are met and required amount of air is supplied [5]. It is considered that the demand-based static pressure setpoint reset has potential to save fan energy up to 50%, compared with maintaining the constant pressure setpoint [6]. In the next two subsections, the most popular pressure reset control strategies PID Control and Trim & Respond are described.

2.1 PID Control

PID Control and its simpler set, the PI control, have been used to regulate a wide range of individual devices in complex HVAC systems. In the past few decades with increasing use of Direct Digital Control (DDC) units, PI and PID Control methods have been widely implemented in the pressure reset control strategies [5]. PID Control approach involves a standard proportional-integral-derivate (PID) control loop to determine the static pressure setpoint. Under this strategy, as the zone's demand decreases and the zone's damper begin to close, the duct static pressure setpoint is reduced until the most requiring VAV box damper is nearly wide open (it is recommended to not use the 100% value, because the PID loop requires the error margin on both sides of the setpoint [7]). In the recent studies, PID

Control method is pointed out as a simple way for energy savings, but is not recommended while the other method, Trim & Respond is available [6].

2.2 Trim & Respond

With Trim and Respond (T&R) approach, the duct static pressure setpoint is reduced slowly until any zone sends a request that more static pressure is required. As a response to the certain amount of pressure requests, system controller increases the pressure setpoint. The importance of each zone's request could be adjusted in order to ensure that the critical zone's demand is always satisfied. On the other hand, the T&R method allows identifying the zones that are too demanding (rogue zones) so that they could be ignored and excluded from the control loop. In many research, T&R method has been found to be the most effective approach to reduce energy costs. The main advantage of T&R method over the PID Control strategy is that the T&R method does not require knowledge about damper position, while PID logic does.

The important advantage of the T&R control strategy is that it is easier to deal with rogue zones. A rogue zone could occur in any VAV system for a number of different reasons hence the "rogue zone" definition is not unified as well. In general, a rogue zone is defined as a VAV box that generates the disproportionately large amount of requests over time. Static pressure reset strategy is a demand-based sequence so that it works properly only if there is a good agreement between installation sizing and actual loads. If any VAV box is undersized, a damper position would be constantly open and send requests to increase static pressure. In this case, while most of the other damper positions are around 40% open, this zone may be identified as a rogue zone and therefore no reliable static pressure reset is possible. One of the possible solutions to deal with the rogue zone is to omit it in the control loop. A recent version of ASHRAE Standard from 2013 has required that DDC system control logic should allow to identify rogue zone and the operator should be able to exclude this zone from the control algorithm. With the T&R logic, it is relatively easy to ignore one or more rogue zones with adjusting a certain number of pressure requests that must be received before the pressure is increased. It is much more difficult to do it with the standard PID logic so that the T&R method is more recommended [8].

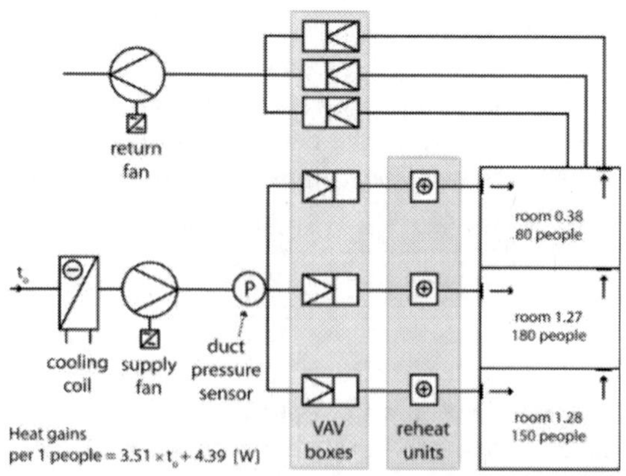

Fig. 2. Diagram of the modelled VAV system.

3 Model and simulations

In this article different pressure control strategies were compared based on the simulation results. The simulation of 1 hour period was carried out for different outdoor temperatures from range 13–19°C with 0,5°C step. Constant heat gains at given outdoor temperature were assumed. In this simulation heat gains depend only on the number of people in the room and change linearly in the function of the external temperature. Schematic diagram of the modelled installation is presented on the Fig. 2.

3.1 Total energy demand

All pressure control algorithms presented in this article, were compared on the basis of the total energy demand calculated over the period of 1 hour. "Constant static pressure", PID Control, and optimal control actions were investigated under the system's steady state. As the steady state is once established it is possible to calculate the total energy consumption directly from the power demand (Eq. (1)).

$$E = P \cdot t \tag{1}$$

where: E - total energy demand [Wh], P - steady state power demand [W], t - time [h].

Total power demand is the sum of the power demands of each part of the VAV system. Fan power P_F is calculated using Eq. (2).

$$P_F = \Delta p \cdot V / \eta \tag{2}$$

where: Δp - fan pressure [Pa], V - volumetric airflow [m^3/s], η - fan efficiency.

Thermal power of each air treatment process in the air-conditioning unit is calculated using Eq. (3).

$$P_{term} = V \cdot \rho \cdot c_v \cdot \Delta T / \eta_{term} \tag{3}$$

where: ρ - air density [kg/m^3], c_v - specific heat [J/(kg·K)], ΔT - temperature difference at the inlet and outlet of the thermal unit [K], η_{term} - normalized thermal power coefficient.

Thermal power in the air treatment process is calculated using the coefficient η_{term} (Eq. (4)). The coefficient value is estimated with the thermal energy cost and the cost of equivalent electric energy. Normalization was made in order to compare an electric and thermal energy at the same basis.

$$\eta_{term} = \textit{electric energy cost} \,/\, \textit{thermal energy cost} \tag{4}$$

3.2 Constant static pressure simulation

"Constant static pressure" strategy maintains the constant value of the static pressure in the point of the installation, where the main duct divides into the individual VAV branches (see Fig. 2). Static pressure is set at the level that provides the designed airflow at nominal heat gains. In this strategy fan speed is adjusted to maintain constant static pressure level (P_{max} = 534,5 Pa) under every operational conditions.

3.3 PID Control simulation

PID Control simulation was carried out with the assumption that at least one of the VAV dampers was almost open. The air resistance factor S (Eq. (5)) [9] was determined for every branch in the VAV system (with damper fully open). Every increase in the S indicates that the VAV damper closes. To described the situation when the damper is "almost" entirely open, S (determined while damper fully open), should not increase more than 4.

$$S = \Delta p_n / w_n^2 \tag{5}$$

where: S – air resistance factor, Δp_n - pressure drop over duct branch [Pa], w_n – airflow speed in n branch [m/s].

3.4 Trim & Respond simulation

The static pressure setpoint under T&R algorithm is determined as a response for requests from the zone level. In this research, requests are established into different thresholds. While the difference between the set and the measured value of the single VAV airflow volume is less than $0{,}02 m^3/s$, then the first type of request is generated. While the same difference is less than $0{,}1 m^3/s$, the request of the second type is send. As a result of this approach, every second type request forces the first type request to generate as well, so that two requests may be send from the single zone. This strategy results in increasing system flexibility, so that every zone's demand is met, even if only a single zone requires higher static pressure.

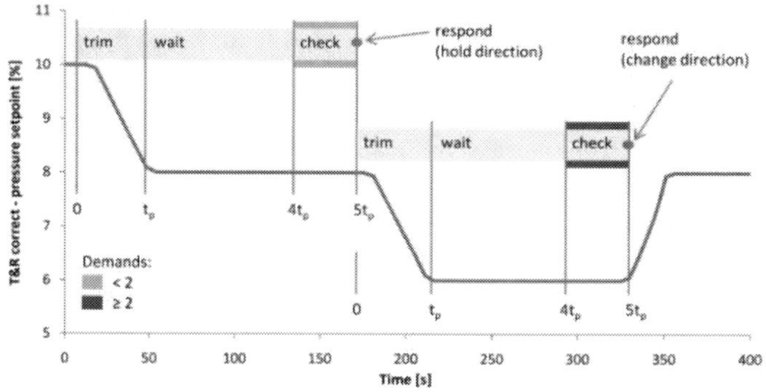

Fig. 3. Pressure setpoint value as a maximum static pressure percentage, indicates by the T&R algorithm over time.

The T&R strategy implemented into simulation proceeds in a following order (Fig. 3): **Step 1**: Started at 0 until t_p, pressure setpoint value is reset downwards by 2% (where 0% is P_{min}=133.4 Pa and 100% is P_{max}=534.5 Pa). **Step 2**: Pressure setpoint reset causes decrease in fan speed (due to fan speed control loop) and hence VAV boxes become more open (due to VAV dampers control loops). These events take time, therefore from time t_p to $4t_p$, control algorithm stands by, because all control loops should stabilize. **Step 3**: From time $4t_p$ to $5t_p$, control algorithm receives the requests from the zones. **Step 4**: At the time $5t_p$, control algorithm decides whether to increase or keep decreasing the pressure setpoint. The decision depends on the number of request received from the VAV boxes. In this study, the number of 2 requests was set as a limiting value.

While the T&R method is characterized by the varying pressure setpoint, fan pressure and fan power varies as well. In case of the non-steady state, the Eq. (1) cannot be used to calculate total energy demand. In case of T&R strategy, instantaneous power demand must be integrated with respect to time.

3.5 Optimal control simulation

It is possible to identify a set of the control actions for which a HVAC system consumes minimum energy. This set of optimal control actions could be delivered from the Nelder-Mead simplex optimization method. In this research the optimization process was carried out multiply for different start points. The optimization results were used as a benchmark for "Constant static pressure" and pressure reset control strategies: PID Control and T&R. Matlab built-in Nelder-Mead simplex method, implemented in the function *fminsearch*, was used. Optimization was made on the function P_{HVAC} (Eq. (6)).

$$P_{HVAC}(\Delta p_{SD}; \Delta p_{RD}; t_o; V_1; V_2; V_3) = P_{termCC} + P_{termR1} + P_{termR2} + P_{termR3} + P_{SF} + P_{RF} \quad (6)$$

where: P_{HVAC} - total system power [W], Δp_{SD} (Δp_{RD}) – supply (return) duct static pressure [Pa], t_o - outdoor air temperature [°C], V_1 (V_2, V_3)- supply airflow to the room 0.38 (1.27, 1.28) [m^3/s], P_{termCC} – cooling power [W], P_{termR1} (P_{termR2}, P_{termR3})– reheat power for room 0.38 (1.27, 1.28) [W], P_{SF} (P_{RF}) – supply (return) fan power [W].

4 Results

In this section, the results of the conducted simulations are presented. On the Fig. 4 and 5, fan power under "Constant static pressure", PID Control, T&R and optimal control actions is presented with regard to the static pressure setpoint signal in the T&R algorithm. The graph on Fig. 4 presents the simulation results in 13°C outdoor temperature, and graph on Fig. 5 in 16°C.

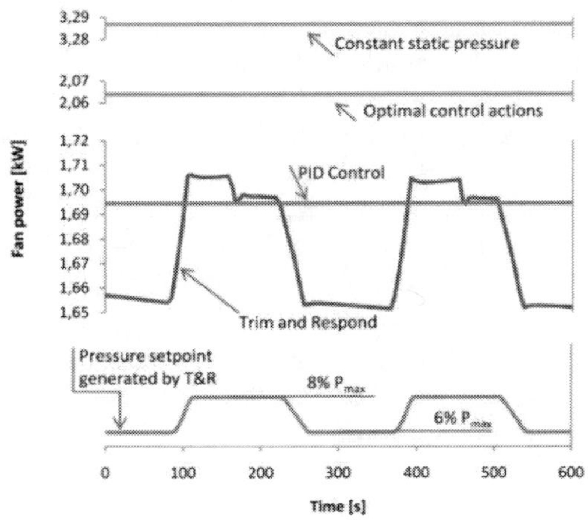

Fig. 4. The simulation results of the fan power for 13°C outdoor temperature.

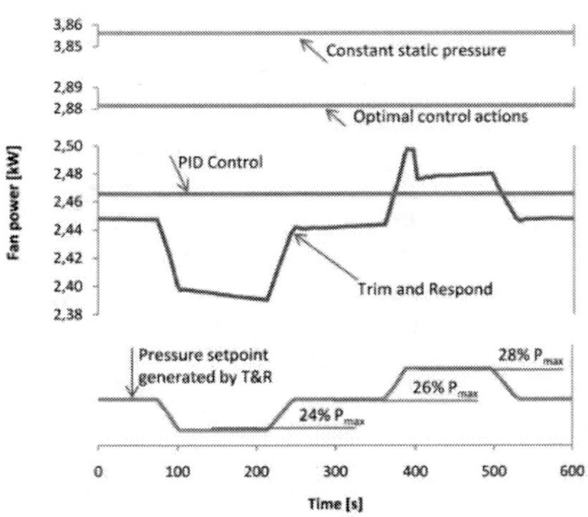

Fig. 5. The simulation results of the fan power for 16°C outdoor temperature.

Besides T&R, all analyzed methods are characterized by the constant fan power. This results from the system's steady state. Under T&R fan power changes periodically. While the pressure setpoint increases (in other word a required amount of requests occurs), the fan speed controller reacts and hence the fan power increases (under-regulations are registered). If there are no further requests from the zones' VAV boxes, T&R strategy starts to decrease the static pressure setpoint. As in the case of 13°C outdoor temperature, the T&R control loop proceeds in one step (single pressure setpoint increase of 2%) or in the two or more steps sequence like in the case of 16°C outdoor temperature (two pressure setpoint increase of 4% summarily).

The highest fan power demand is noticed for the "Constant static pressure" method. Fan power under PID Control was found to be almost 50% less than the one under constant static pressure regulation. Optimal control actions results in decreasing fan power demand while comparing to the "Constant static pressure". At the same time, fan power demand under optimal control actions is higher than under PID Control and T&R method, which seems to contradict with the word "optimal" itself. Under these control actions, the "optimal" effect is achieved by the fact that these actions employ not only pressure control strategy, but temperature control strategy as well. In this case, whenever possible, heat gains from the zones are removed by increased airflow volume rather than increasing cooling power. Thus, the supply airflow temperature required to meet zones demands, could be higher than under other pressure control strategies, and so the cooling energy consumption decreases. At the same time, increased airflow volume increases the fan power demand, which could be observed on the Fig. 6.

Any difference in fan power demand results not only from the different airflow volume, but also from the pressure loss on the VAV boxes dampers. Basically, the airflow volume setpoint for a specific outdoor temperature is identical in every analyzed control methods (under T&R method, the airflow volume fluctuates periodically around the constant setpoint). This does not concern the optimal control algorithm, under which the airflow volume is set based on overall (fan and air treatment) energy demand. In case of the same airflow volume under "Constant static pressure", PID Control and T&R method, any difference in fan power derives from different damper's position. While dampers are more open, pressure loss is low and hence less energy is needed to transport required amount of the air. Under the PID Control and T&R method, at least one damper is always almost open, which results in lower pressure loss in the duct system in comparison with "Constant static pressure" method.

5 Analysis and conclusions

In this article three possible static pressure control strategies were described and compared with regard to the energy consumption. "Constant static pressure", PID Control and Trim & Respond control strategy were investigated and referenced to the optimal control actions. Optimal control actions were used to present a hypothetical limit of the energy savings.

On the Fig. 6, total energy consumption, which is the sum of the energy consumption by thermal treatment and fan energy (on the left) and fan energy consumption alone (on the right), for different pressure control strategies, is presented. The highest total and fan energy consumption is noticed for the "Constant static pressure" method. While total energy consumption is taken into consideration, pressure reset control methods (PID Control and Trim & Respond) have a potential to reduce energy consumption from 5 (for 19°C) to 26% (for 13°C) compared to the "Constant static pressure" method. Furthermore, fan energy consumption under pressure reset control methods was found to be from 22 (for 19°C) up to 48% (for 13°C) lower.

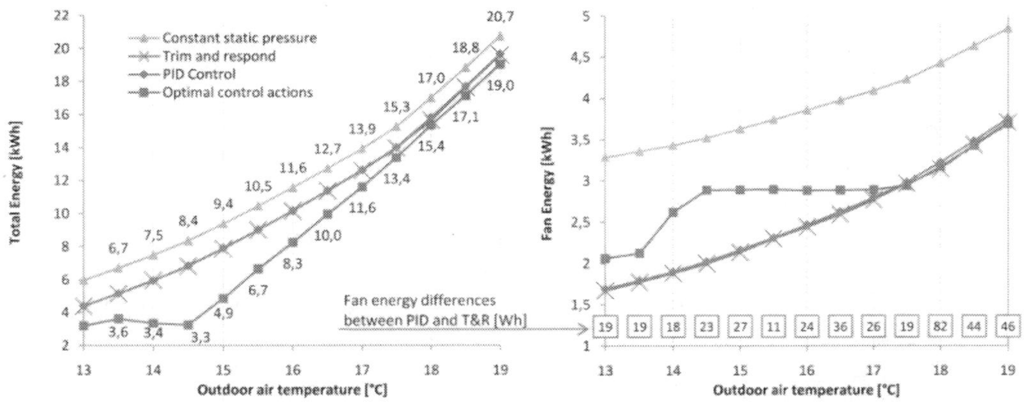

Fig. 6. Total energy consumption (on the left) and fan energy consumption (on the right) for different pressure control strategies.

Energy consumption in the thermal air treatment process is the same under PID Control, T&R and "Constant static pressure". This results from the fact that the parameters of supply air depend only on the current demand. The pressure control strategies different from each other only in the matter of approach in decreasing pressure setpoint and, in the final effect, fan energy consumption.

In case of the pressure reset control strategies (PID Control and T&R), the abovementioned differences are minimal as it could be seen on the Fig. 6. Total energy consumption, which is the sum of the fan and thermal air treatment energy consumption, shows no significant differences under PID Control and T&R control strategy.

The slightly higher fan energy consumption under PID Control method compared with the T&R strategy is noted for every outdoor temperature. However, these differences were found to increase with an increasing outdoor temperature. This could results from the fact that in the PID Control strategy, VAV damper position is not fully open. This additional minor air resistance on the VAV damper has the major effect on the pressure loss. It is caused by the fact that pressure loss increases with increasing airflow volume (pressure loss rises with the squared airflow speed). Under the T&R control strategy, at least one VAV damper is periodically fully open for a while, so that the fan energy consumption is lower than under PID Control (Fig. 4 and 5). At the same time, this could lead to a situation when airflow is lower than required setpoint, which is the main disadvantage of the T&R method. In conclusion, total and fan energy consumption under PID Control and T&R strategy are similar, but with minor advantage of the T&R method. Energy consumption is reduced but at the expense of meeting zones' demand.

Above all, it is clearly noticeable that pressure reset control methods (PID Control and Trim & Respond) have a great potential to reduce energy consumption compared with the "Constant static pressure" control method. Since it is not possible to reduced static pressure setpoint under a certain critical value, it is assumed that no further energy savings are possible while using PID Control or T&R control strategy. However, the optimization process showed a promising potential in reducing energy consumption. At the outdoor temperature range from 13 to 17°C, airflow volume under optimal control actions is relatively high, so that the fan energy consumption increases (Fig. 6 on the right). At the same time, this approach allows to lower the cooling/reheating demand, so that the increased air volume is able to meet the zones' requirements without additional energy consumption in the thermal air treatment process. This issue could be also addressed otherwise. While supply air temperature increases, cooling/reheating energy consumption decreases and fan energy consumption rises at the same time. Supply air temperature optimization process is usually called Supply Air Temperature (SAT) reset and could be investigated in the future researches.

References

1. A. Pełech, *Wentylacja i klimatyzacja - podstawy* (2008)
2. G.S. Okochi, Y. Yao, *A review of recent developments and technological advancements of variable-air-volume (vav) air-conditioning systems*, Renewable and Sustainable Energy Reviews, **59** (2016)
3. R.W. Haines, *Control Systems for Heating, Ventilating and Air Conditioning*, Van Nostrand Company, **E 4** (1987)
4. H. Recknagel, E.-R. Schramek, *Kompendium Ogrzewnictwa i Klimatyzacji* (2008)
5. B.W. Householder, *A study on static pressure reset and instability in variable air volume HVAC systems* (2011), Graduate Theses and Dissertations
6. Energy Design Resources, *The advanced variable air volume (VAV) system design guide* (2009)
7. S.T. Taylor, *Resetting Setpoints Using Trim & Respond Logic*, ASHRAE JOURNAL, **November** (2015)
8. T. Hartman, Terminal regulated air volume (trav) systems, **99** (1993)
9. G. Liu, M. Liu, *Supply Fan Control Methods for VAV Systems Using a Fan Airflow Station*, ASHRAE Transactions **NY-08-055** (2008)

Probabilistic approach of the Upper and Middle Odra basin daily rainfall modeling

Marcin Wdowikowski[1,*], *Andrzej* Kotowski[2], *Paweł B.* Dąbek[3], and *Bartosz* Kaźmierczak[2]

[1]Institute of Meteorology and Water Management – National Research Institute, 01-673 Warsaw, Podlesna Street 61, Poland
[2]Wrocław University of Science and Technology, Faculty of Environmental Engineering, Wybrzeze Wyspianskiego 27, 50-370 Wroclaw, Poland
[3]Institute of Environmental Protection and Development, Wrocław University of Environmental and Life Sciences, Poland pl. Grunwaldzki 24, 50-363 Wroclaw, Poland

Abstract. The aim of this study was to obtain the maximum daily rainfall descriptions based on 9 probability distributions in 12 meteorological stations located in Upper and Middle Odra river basin. Analysis included long-term period from 1961 to 2010. Regarding to AIC and BIC informational criterions Gamma distribution that appeared to be best fitted probability distribution to measurement rainfall data series. For several stations, Pareto, Weibull and generalized exponential distributions were also possible to use. For practical purposes it is necessary to carry out a similar analysis for much shorter rainfall intervals. The final evaluation of the suitability of individual probability distributions for constructing maximum rainfall probabilistic models requires further research.

1 Introduction

The reliability of urban drainage systems is not fully achievable due to the random nature of precipitation. However, it must be designed in the way that fulfill the contemporary standards of drainage, which is defined as the adaptation of the system to receive the maximum (predicted) storm water streams with a frequency of appearance equal to the acceptable (socially acceptable) frequency of spillage into the terrain [1]. Recent extreme natural phenomena, such as intense or long-lasting rainfall, as well as more frequent river floods or urban floods, caused with significant economic losses. That is why it is so important to systematically obtain rainfall analysis, including the determination of the statistical incidence of their maxima's [2].

The Blaszczyk model, based on rainfall measurements from the turn of the 19th and 20th centuries, is formally recommended to design drainage systems of urbanized areas in Poland, which is criticized by the scientific community. More and more often new models of maximum rainfall are used, such as Kupczyk-Suligowski or Bogdanowicz-Stachy regional models. New models, for various reasons, gives much higher rainfall values compared to the Blaszczyk formula. Due to practical reasons, in Poland, it is urgent to

* Corresponding author: marcin.wdowikowski@imgw.pl

replace the rainfall model of Blaszczyk's, in the dimensioning of drainage systems of terrains, with more accurate models (local impact models). The identification of differences in the results of papers published in the last two decades is also needed.

Both the synthetic description of the maximum rainfall structure as well as the probabilistic modelling of maximum precipitation based on the long term period used in climate analysis (eg 50 years) for the Upper and Middle Odra river basin, is currently not referenced in the problematic literature. At the same time, studies that presents a comparative analysis of a large number of probability distribution models is also insufficient. The scientific papers published in Poland in the case of spatio-temporal variability includes shorter periods of analyzed data or smaller number of measurement points (eg Polish Atlas of Climate for the 1971-2000 period [3], Maximum rainfall in Poland [4], Heavy and torrential rainfall during 1954-1981 in Poland [5] or Projections of changes in precipitation extremes in Poland [6]). Likewise, other parameters (eg flood indicators [7]) were analyzed and the maximum precipitation was limited to selected catchment areas or geographic regions (eg Widawa or Nysa Klodzka [8, 9]) or even individual measurement points (eg Agrometeorological Observatory in Swojec [10]). For the maximum precipitation modelling purposes, the literature of the subject presents the application of only selected probability distributions that can be used to characterize heavy rainfall in very general way (eg Blaszczyk's model) or only with very local impact (eg. in Cracow [11], Lodz [12] Czestochowa [13], Kielce and Kielce Uplands [14], Warsaw [15] or Wroclaw [1, 2, 16]). The main reason for this was the limited access of researchers to IMWM-NRI database or analysis conducted on their own measurement network.

The main aim of this study is to present the possibility of using the different probability distributions for maximum precipitation modeling in the Upper and Middle Oder basin, ie in the area in Poland that is often affected by the occurrence of extreme values. As the research material were used daily precipitation recorded in between 1961 and 2010 at 12 IMWM-NRI meteorological stations (from Cieszyn to Slubice). Maximum daily precipitation were described by the probability distributions most commonly recommended in hydrology, such as Gamma and Weibull distributions, and the GED distribution recommended by Kotowski and Kaźmierczak [1, 2, 16, 17, 18]. The quality of fitting the theoretical to empirical rainfall distributions will be assessed on the basis of the *AIC* and *BIC* information criteria.

2 Study area and methods

2.1 Odra river basin

The Upper and Middle Oder Basin was the research area. Odra is one of the largest rivers in the catchment of the Baltic Sea, and after the Vistula it is the second largest river in Poland.

The basin covers the administrative regions of the Silesian, Opole, Lower Silesia, Wielkopolska and Lubusz voivodeships. The length of the river is 854 km in total, including 742 km in Poland, and 204 km in the Lower Silesian voivodeship (from Olawa km 216 + 500 to Glogow km 392 + 900). Oder springs are located in the Czech Republic in the Odreckie Vrchachy, at an altitude of 634 m a.s.l. On the whole length of the Odra river can be divided into two parts: the mountain, the first 50 km (Upper Oder), and the lowland - the remaining part (Middle and Border Oder). From Kedzierzyn-Kozle to Brzeg Dolny Odra it is channeled at 186 km, in details it is built with 24 stages of waterlogging, including 23 permanently navigable in the navigation season (usually from 15th March to 15th December). The difference between the water level between the beginning and the end of the channel is about 64 m. In the distal Odra flows freely and its current is regulated with a spike. The extreme precipitation values occurrence, especially the maximum values, significantly determine the water relations of the Oder River Valley and directly contribute to the emergence of dangers from violent floods and floods in urban areas [1, 7, 19, 20]. Long-term regional rainfall causes over-retention of the soil and the formation of increased surface runoff in the catchments. According to the authors of the study [19], more than 75% of the floods in the Oder River are recorded after prolonged and intense rain events. Occurrence of water from river beds causes local flooding (spring period) or widespread floods (in 1997) or larger floods in the Oder River (2001, 2006, 2010 or 2013), which is a consequence of the increasing of heavy rainfall in this region. The increasing tendency of intensive rainfall occurrence in the Upper and Middle Oder basin, as well as the potential effects of urban floods (Wroclaw, Legnica, Klodzko, Głogow, etc.) and in the main catchment areas (Kwisa, Bobr, Kaczawa) became the main reason of authors interest upon this subject. Odra flooding has also significant impact for hydrogeological droughts in lowland part of basin [20].

2.2 Measurement stations

Research material used in the work were rainfall daily data series from 50 years (1961–2010), derived from 12 IMWM-NRI measurement stations. The daily totals from the characteristic points of the Upper and Middle Oder basin, represented by regional and climatological stations, also are used to prepare meteorological forecasts and statements, including short media releases (press, radio and television). The location of meteorological stations in the analyzed river basin, taking into account the terrain, presented in Figure 1 and their characteristics are given in Table 1.

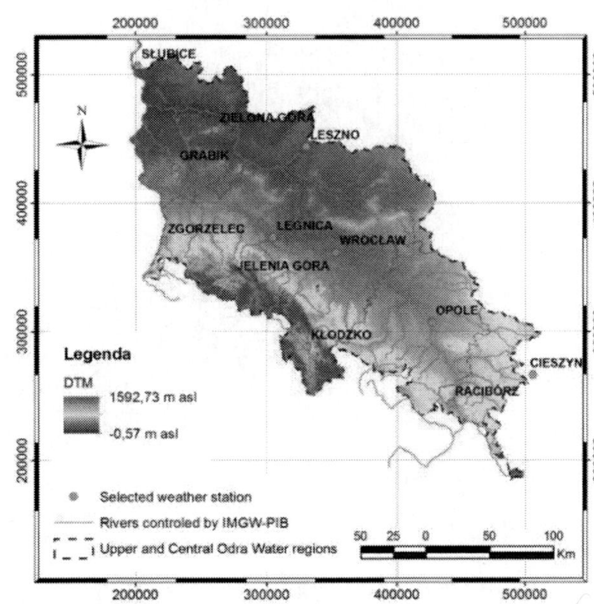

Fig. 1. Location of meteorological stations IMGW-PIB in Upper and Middle Oder (own elaboration, based on PUWG-92 coordinate system).

The digital terrain model (DTM) was used to visualize the terrain (Fig. 1). The free terrain data was obtained from the Central Office of Geodesy and Cartography (CODGIK).

Figure 2 shows the spatial distribution of the highest precipitation levels recorded at particular meteorological stations between 1961 and 2010 (based on Table 1). Spatial

 DOI: 10.1051/e3sconf/20171700096

distribution uses the Inverse Distance Weighting (IDW) estimation method conducted in geostatic ArcGIS software.

Table 1. Characteristics of IMWM-NRI meteorological stations in Upper and Middle Oder basin Indications: H - elevation, $h_{d,max}$ - maximum daily precipitation $h_{R,min}$ - minimum annual precipitation $h_{R,max}$ - maximum annual precipitation $h_{R,av}$ - average annual precipitation in years 1961–2010.

Station	Attitude	Latitude	H [m a.s.l.]	$h_{d,max}$ [mm]	$h_{R,min}$ [mm]	$h_{R,max}$ [mm]	$h_{R,av}$ [mm]
Raciborz	181127	500340	206	92.9	443	1001	642
Cieszyn	186492	497550	295	129.5	640	1348	934
Opole	175812	503740	165	99.0	374	868	624
Klodzko	163651	502613	356	101.2	369	586	602
Wroclaw	165357	510614	120	74.4	381	776	596
Legnica	161227	511133	122	85.9	361	795	537
Jelenia Gora	154720	505401	342	119.3	462	1005	698
Zgorzelec	150131	510810	203	81.1	346	809	616
Leszno	163205	515009	91	79.2	349	733	547
Zielona Gora	153128	515549	192	80.0	385	770	587
Grabik	150641	513909	165	90.2	380	902	627
Slubice	143711	522055	53	132.5	361	861	566

Analyzing the characteristics of maximum daily and annual precipitation (min and max) and annual averages from given long-term period (Table 2), it is clear that the maximum daily precipitation typically account for about 10% of the largest annual precipitation and almost 50% of annual minimum precipitation.

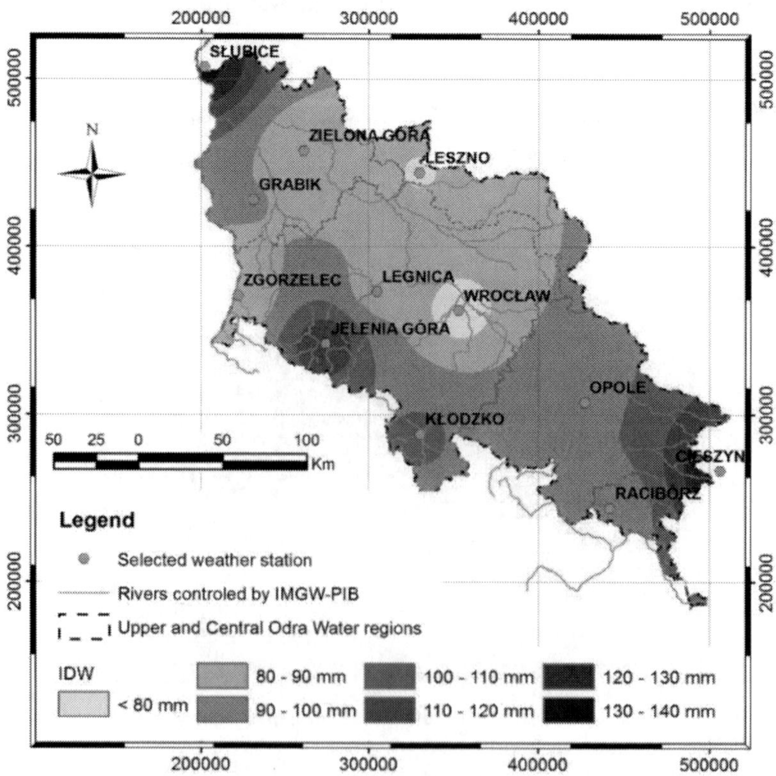

Fig. 2. Spatial variability of the daily precipitation in the years 1961–2010 (own elaboration, based on PUWG-92 coordinate system).

2.3 Probability distributions and its fitting to measurement data

According to the principles of modeling of maximum precipitation patterns, widely described by Kotowski [1, 2, 16, 17], the highest values of daily precipitation (h) for each measurement station were determined for the purpose of measuring data series. This procedure is not limited to the criterion of the highest value in a given year (AMS annual maximum series), because it may result that a significant precipitation value that could have occurred at least twice in the same year, may be excluded. So prepared in that way elements of the random variable were sorted decreasing, and then they were assigned empirical probability of exceedance, from p = 0.02 (for the highest value) to p = 0.98 (for the smallest value), according to the formula:

$$p\ (m,N) = m/(N+1) \tag{1}$$

where: m – The position (number) of the random sample element in the descending order, N – number of random samples.

In literature [1, 2], the concept of precipitation in years C is often used in the quantitative description of precipitation, interchangeably with the probability of occurrence of precipitation, which is described by a simple relation:

$$C = 100/p \tag{2}$$

The theoretical probability distributions used in hydrology were used to describe measurement data. Table 2 lists the distributions and the abbreviations used in the study.

Table 2. Analyzed probability distributions.

No.	Distribution	Abbreviation	Parameters
1	Pearsona typu III (Gamma)	P3	3
2	Fischera-Tippeta typ I (Gumbel)	EV1	2
3	Fischera-Tippeta typ II (Frechet)	EV2	3
4	Fischera-Tippeta typ III (Weibull)	EV3	3
5	Logarytmiczno-Normalny	LN	3
6	Uogólniony Pareto	GPA3	3
7	Uogólniony wartości ekstremalnych	GEV	3
8	Uogólniony wykładniczy	GED	3
9	Logarytmiczno-logistyczny	LLOG	3

Estimators of the parameters of individual distributions were determined by the maximum loglikelihood estimation method (MLE), by numerical maximization of the logarithm of the loglikelihood functions, taking into account the range of variability of the scaled parameters of the scale, shape and location - in the case of tree parameters distributions and scale and shape - in the case of the Gumbel (EV1) distribution [1, 2, 16, 17, 18]. In the numerical process of maximizing the logarithm of likelihood function with respect to the parameters of the analyzed distributions, the following assumptions are assumed: the scale parameter must be greater than zero; the shape parameter must be different from zero, and the location parameter as the lower limit assumes the lowest value of the decreasing ordered data serie. Using the logarithm of the likelihood function, the quality of fitting the theoretical distributions to the measured data was described using the *BIC* - Bayesian information criterion of Schwartz and AIC - the Akaike information criterion, givenby the following formulas:

$$BIC = -2\ln L + k\ln N \tag{3}$$

$$AIC = -2\ln L + 2k \tag{4}$$

where: L – analyzed random sample logarithm of the likelihood function, k – number of estimated parameters and N – number of random samples.

Necessary numerical calculations performed using the R- package statistical modules ('FAdistr', 'fExtremes', 'MASS' and 'bbmle').

3 Results

Based on the accepted methodology, the series of maximum daily rainfall totals from the analyzed stations were analyzed (Table 1). The obtained theoretical distributions were compared with the empirical distributions of the measured data using (3) and (4) as shown in Tables 3 and 4, respectively. In both cases, the smallest values were shown, which shows the greatest correlation between the theoretical distributions and the measured data.

Table 3. Values of *AIC* informational criterion for analyzed probability distributions and meteorological stations.

AIC	P3	EV1	LN	EV3	GPA3	GEV	LLOG	GED	EV2
Słubice									
Zielona Góra									
Legnica									
Leszno									
Wrocław									
Jelenia Góra									
Kłodzko									
Opole									
Racibórz									
Cieszyn									
Grabik									
Zgorzelec									

Table 4. Values of *BIC* informational criterion for analyzed probability distributions and meteorological stations.

BIC	P3	EV1	LN	EV3	GPA3	GEV	LLOG	GED	EV2
Słubice	362.44	416.36	374.78	353.94	372.50	379.42	375.17	370.99	379.42
Zielona Góra	344.22	360.12	354.94	344.22	350.60	356.83	356.52	348.84	356.83
Legnica	346.77	369.08	364.39	357.99	358.29	367.35	366.74	358.31	367.35
Leszno	343.39	367.38	354.19	353.12	350.27	357.82	356.06	349.15	357.82
Wrocław	353.25	360.38	355.78	354.58	352.23	358.32	358.14	353.31	358.32
Jelenia Góra	371.30	406.14	389.76	383.96	385.04	393.44	391.23	382.31	393.44
Kłodzko	366.95	406.14	390.09	370.43	395.05	397.20	391.83	377.96	397.20
Opole	313.47	383.88	353.15	325.60	350.28	358.15	353.95	342.89	358.15
Racibórz	299.98	353.61	319.37	310.22	317.07	323.19	320.13	316.39	323.19
Cieszyn	404.19	430.57	425.39	411.61	414.57	429.55	427.75	410.36	429.55
Grabik	340.12	362.29	359.52	349.70	349.22	362.37	366.65	350.02	362.37
Zgorzelec	368.45	391.50	380.52	370.61	373.29	385.72	383.10	368.15	385.72

The AIC and BIC values generally have similar values, and it ranged from 294.25 to 426.76 for AIC's and from 299.98 to 430.57 for BIC's. According to both indicators, the Gamma (P3) distribution most accurately describes the maximum daily precipitation for the analyzed stations. In the vast majority of cases, there is also a relatively small difference in the value of the information criteria between the P3 and GED distributions. In details, in the case of the Slubice station, EV3 distribution became the best described in terms of precipitation data, whereas in case of Wroclaw station, despite the indication of GPA3 distribution as the best, P3 and GED distributions are also possible. For Zielona Gora

DOI: 10.1051/e3sconf/20171700096

station, the AIC and BIC values for P3 and EV3 are identical, indicating the probability of using both probability distributions to describe measurement data and the need for additional ratings indicators. It is noteworthy that the values of the AIC and BIC information criteria for the assumed assumptions give the same results for the GEV and EV2 distributions, moreover, these are high values indicating relatively limited possibilities of their use to describe the measurement data. Figure 3 shows a spatial interpretation of the research results, which were the potential possibility of using the analyzed probability distributions to describe the maximum daily precipitation that occurred in the Upper and Middle Oder basins, based on the data in Tables 3 and 4.

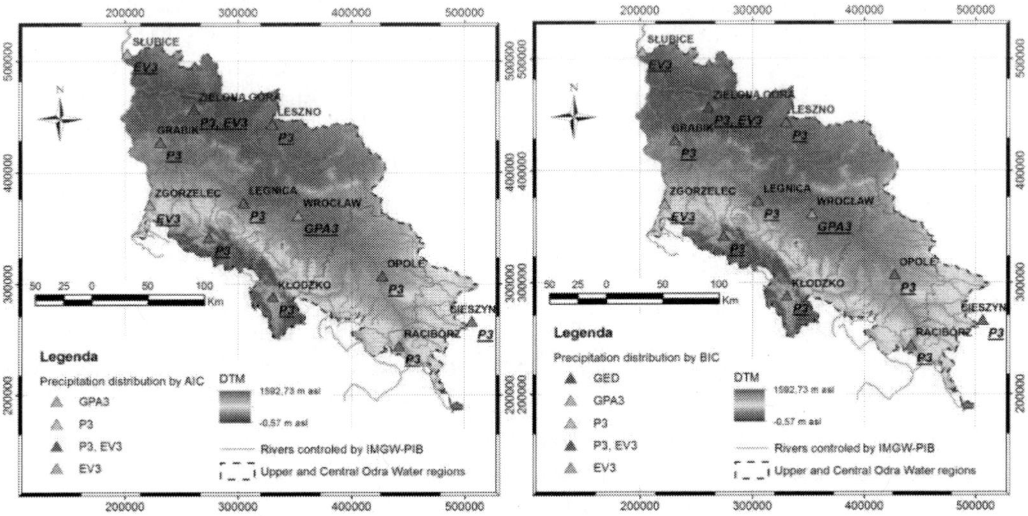

Fig. 3. Best fitted probability distributions for analyzed meteorological stations in the Upper and Middle Odra basin regarding to AIC (from right) and BIC (from left) informational criterion (own elaboration, based on PUWG-92 coordinate system).

4 Conclusions

In order to describe the maximum daily precipitation from 1961 to 2010 (Table 1, Figure 1) at 12 measurement stations in the Upper and Middle Odra basin, a 9 probability distributions (Table 2) were used. For the analyzed area and the rainfall duration of 1440 minutes (measured in rigid intervals, between 6 UTC hours), the theoretical distribution of the most accurate measurement data from the studied stations turned out to be the vast majority of cases (9 out of 12) was Gamma distribution (P3). This allows to formulate a general conclusion that this is the best distribution for describing the daily rainfall precipitation in the Upper and Middle Oder area for the examined time series. Similar results were obtained by Kaźmierczak and Wdowikowski [21] for data from Legnica, Wdowikowski with the team [23] for data in the Lusatian Neisse basin (western part of the Upper and Middle River Basin) and earlier by Krężałek with the team [22] for data from Wroclaw, Legnica, Leszno, Zielona Gora and Slubice. All authors analyzed daily rainfall data from the same period of 1961–2010 [22]. Depending on the accepted information criterion AIC or BIC (Figures 3 and 4), in several cases, it was also possible to use the following:

- three-parameter generalized Pareto distribution (GPA3) for Wroclaw (meets both criteria);
- Weibull (EV3) three-parameter distribution for Slubice and Zielona Góra (meets both criteria), as well as for Zgorzelec (according to AIC),

- three-parameters generalized exponential distribution (GED) for Zgorzelec (according to BIC).

The investigated and presented spatial variation of the suitability of analyzed probability distributions for daily rainfall modeling indicates that it is possible to apply both distribution of the entire river basin district - Gamma distribution (P3), as well as description of measurement data from individual locations with different probability distributions (for example, Gamma (P3) and Weibull (EV3) for Zielona Gora).

However, for practical purposes, ie to achieve the highest quality rainfall modeling results, a similar analysis should be performed for significantly shorter periods of time. Researches from Kotowski and Kaźmierczak [1, 16] done for data from Wroclaw or Kaźmierczak and Wdowikowski [21] for data from Legnica, where the time intervals were analyzed from 5 minutes to 3 or 6 days, clearly indicate that the suitability of individual probability distributions for varied rainfall durations may be different. The authors, however, state that the assessment of the suitability of individual probability distributions within different time intervals for the construction of probabilistic rainfall models is essential for modern way of thinking about maximum rainfall modeling, especially in probabilistic approach. Further and more detailed researches in this matter should be conducted.

References

1. A. Kotowski, *The principles of safe dimensioning of sewer systems* (Seidel-Przywecki, Warsaw, 2015)
2. A. Kotowski, B. Kaźmierczak, A. Dancewicz, *The modeling of precipitations for the dimensioning of sewer systems* (Polish Academy of Sciences, Warsaw, 2010)
3. H. Lorenc, *Climate atlas of Poland* (The Publishing House of the Institute of Meteorology and Water Management, Warsaw, 2005)
4. E. Bogdanowicz, J. Stachý, *Maximum rainfall in Poland. Design characteristics* (The Publishing House of the Institute of Meteorology and Water Management, Warsaw, 1998)
5. A. Żyromski, *Heavy and torrential rain in 1954-1981 in Poland* (Seidel-Przywecki, Warsaw, 2008)
6. I. Pińskwar, *Projection of precipitation extremes in Poland* (The Publishing House of the Water Management Commete in NAoS, Warsaw, 2010).
7. W. Szalińska, I. Otop, T. Tokarczyk: Meteorology Hydrology and Water Management 2, 1 (2014)
8. M. Wdowikowski, *Indicative assessment of precipitation structure in Widawa basin* (The Publishing House of the Institute of Environmental Protection of Technical University of Wroclaw, Wroclaw, 2014)
9. I. Otop, Opera Corcontica, 41 (2004).
10. P. Licznar, J. Łomotowski, Environmental Protection, **27**, 1 (2005)
11. R. Twardosz, *Daily atmospheric precipitation in synoptic and probabilistic terms on the example of Cracow (1886–2002)* (The Publishing House of the Institute of Geography and Spatial Management of the Jagiellonian University, Kraków, 2005)
12. M. Zawilski, A. Brzezińska, Urban Water Journal, 11, 7 (2014)
13. M. Mrowiec, Gaz, Woda i Technika Sanitarna, 7 (2014)

14. R. Suligowski, *Maximum rainfall depth of specified duration and probability of exceedance in Kielce* (The Publishing House of the Hydrological Commission of the Polish Geographical Society, Kielce, 2014)
15. P. Licznar, P. Szeląg, Environmental Protection **36**, 3 (2014)
16. B. Kaźmierczak, A. Kotowski, J Hydrol **525** (2015)
17. S. Coles, *An introduction to statistical modeling of extreme values* (Springer Series in Statistics, Springer-Verlag, London, 2001)
18. W. L. Shinyie, N. Ismail, A. A. Jemain, Water Resour Manag **28**, 11 (2014)
19. P. Migoń, *Exceptional natural events in Lower Silesia and their effects*. (The Publishing House of the Institute of Geography and Regional Development of the University of Wrocław, Wrocław, 2010)
20. J. Kubicz, P.B. Dąbek, Acta Sci. Pol. Formatio Circumiectus **15**, 4 (2016)
21. B. Kaźmierczak, M. Wdowikowski, Periodica Polytechnica Civil Engineering **60**, 2 (2016)
22. K. Krężałek, T. Szymczak, B. Bąk, Woda-Środowisko-Obszary Wiejskie **13**, 4 (2013)
23. M. Wdowikowski, B. Kaźmierczak, O. Ledvinka, Meteorology Hydrology and Water Management, **4**, 1 (2016)

E3S Web of Conferences 17, 00097 (2017) DOI: 10.1051/e3sconf/20171700097
EKO-DOK 2017

The determination of mercury content in the biomass untended for industrial power plant

Magdalena Wiktor[1], *Iwona* Jelonek[1], *Marta* Kasprzyk[1,*], and *Anna* Szpak[1]

[1]University of Silesia, Faculty Earth of Sciences, 41-200 Sosnowiec, Będzińska 60 str., Poland

Abstract. Biomass is one of the oldest and most widely used renewable energy sources. The biomass is the whole organic matter of vegetable or animal origin which is biodegradable. Biomass includes leftovers from agricultural production, forestry residues, and industrial and municipal waste. The use of biomass in the power industry has become a standard and takes place in Poland and other European countries. This paper discusses the correlation of mercury content in different biomass types used in the power industry and in products of biomass combustion. Different biomass types, which are currently burned in a commercial power plant in Poland, were discussed. A photographic documentation of different biomass types, such as straw briquettes, wood briquettes, pellets from energy crops (sunflower husk and wood husk), wood pellets, wood chips, and agro-biomass (seeds) was carried out. The presented paper discusses the results obtained for 15 biomass samples. Five selected biomass samples were burned in controlled conditions in the laboratory at the University of Silesia. The ash resulting from the combustion of five biomass samples was tested for mercury content. A total of twenty biomass samples and its combustion products were tested. Based on the obtained results, it was found that any supply of biomass, regardless of its type, is characterized by variable mercury content in dry matter. In the case of e.g. wood chips, the spread of results reaches 235.1 μm/kg (in dry matter). Meanwhile, the highest mercury content, 472.4 μm/kg (in dry matter) was recorded in the biomass of straw, wood pellets, and pellets from energy crops (sunflower husk). In the case of combustion products of five selected biomass types, a three or four fold increase in the mercury content has been observed.

1 Introduction

Solid biomass, fired or co-fired with fossil fuels, is used in the Polish power sector for many years. According to the European Union and the guidelines of the Central Statistical Office of Poland, biomass is a biodegradable fraction of products, waste and residues (including substances of plant and animal origin), from agriculture, forestry and related industries, as well as a biodegradable fraction of municipal and industrial waste [1, 2]. Solid biomass, included among renewable energy sources (RES), is an alternative to non-renewable fossil fuels, e.g. coal. The demand for biomass from the EU and Polish power

* Corresponding author: marta.kasprzyk@us.edu.pl

industry is constantly increasing. Overall, the EU Member States produce electricity from the following renewable energy sources (Fig. 1).

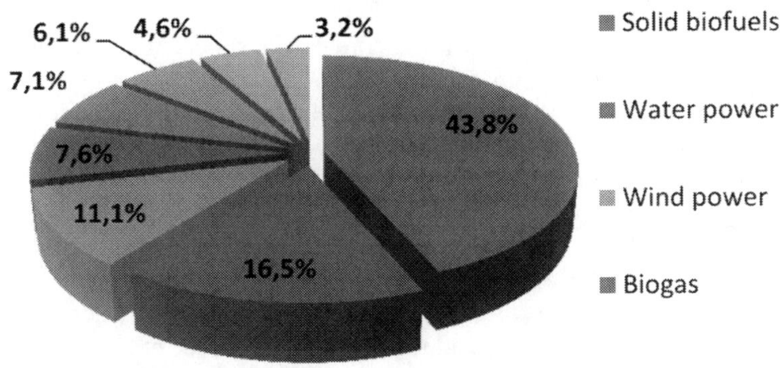

Fig. 1. Energy production from renewable sources in the EU broken down by media in 2014 [2].

However, the electricity generation in Poland is still based on fossil fuels, including bituminous coal and lignite, which occupy the first place with 56% share. Crude oil with the second-largest share is followed by natural gas (13% share). Biomass and waste have the lowest (6%) share in the energy production in Poland (Fig. 2).

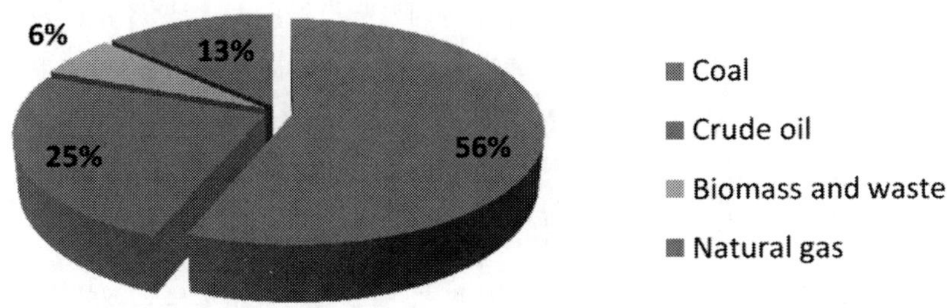

Fig. 2. The demand for primary energy by source [3].

The Polish government forecasts that the energy consumption in Poland in the years 2010-2020 will continue to increase at an average rate of 1.5% per year. The use of renewable energy between 2010 and 2020 should reach 12% [4].

While biomass is regarded as an alternative fuel for industrial power, its mercury content is the reason for increased Hg emissions into the environment resulting from biomass combustion. The level of mercury emissions depends on its content in the combusted fuel and the type of gas treatment system [5].

2 Samples and analytical procedures

The analyzed biomass samples were collected at one of commercial power plants. The samples were collected (according to the PN-90 / G04502 standard) and their qualitative characteristics were assessed. The research material was collected by the power plant employees and stored in sealed glass jars in an amount of 1 to 2 kg (each sample). A total of 15 samples of all types of biomass, burned in the power plant in the period from April to

 DOI: 10.1051/e3sconf/20171700097

November 2015, were collected. The samples were transported to the laboratory, where a detailed description, along with photographic documentation presented in Table 1, was made.

In order to prepare samples for further analysis, the examined biomass samples were grinded and then pulverized in agate mortar. Each sample was divided into two equal portions, one of which was used for the pH and conductivity measurement, and the other of which was used for the mercury content measurement.

The electrical conductivity and pH measurements were carried out on approximately 40g of each sample. The samples were placed in beakers, filled with distilled water to a volume of 100 ml. This solution was allowed to stand for 24 h.

The measurement of total mercury content in the analyzed samples of solid biomass was performed using a DMA-80 (Milestone). Calibration of the analyzer was based on patterns provided by the manufacturer (NIST SRM). The analyzes were performed at the Laboratory of Water Analysis, Department of Earth Sciences, the University of Silesia. Five biomass samples were fired in a muffle furnace, and the resulting ash was further analyzed for total mercury using the same measurement technique as in the case of solid biomass samples.

3 Results and discussion

Macroscopic description of the samples and photographic documentation are summarized in Table 1. The mercury content, analytical moisture, pH, and conductivity of biomass (Table 1) and ash resulting from biomass combustion are shown in Table 2.

Table 1. The mercury content, analytical moisture, pH, and conductivity of the biomass.

No sample	Macroscopic description	Raw samples	Grinded and averaged samples	Mad %	PH	Conductivity	Hg µ/kg
B1	Biomass – straw briquettes	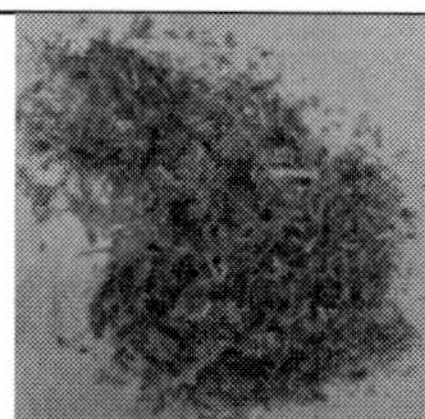	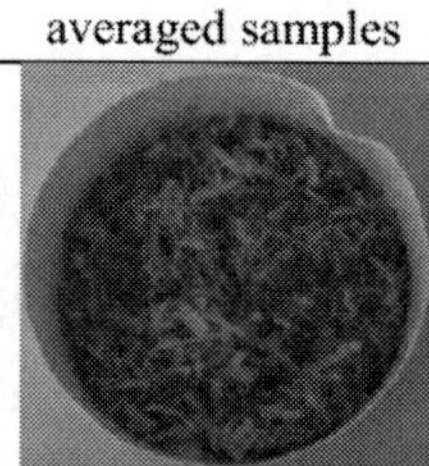	5.38	6.31	52.3	130.4
B2	Biomass – wood chips			6.43	5.91	67.4	295.0
B3	Biomass – (Agro-seeds)			5.30	5.10	119.5	135.6

 DOI: 10.1051/e3sconf/20171700097

B4	Biomass – wood chips	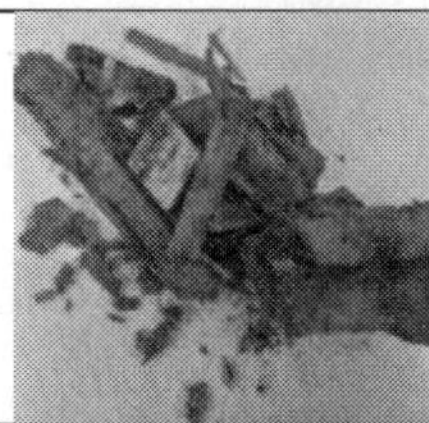	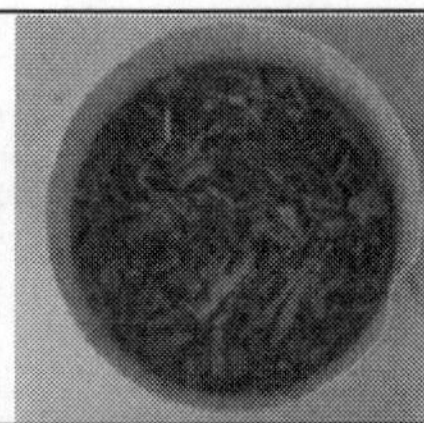	6.66	5.90	62.4	188.4
B5	Biomass – wood briquettes	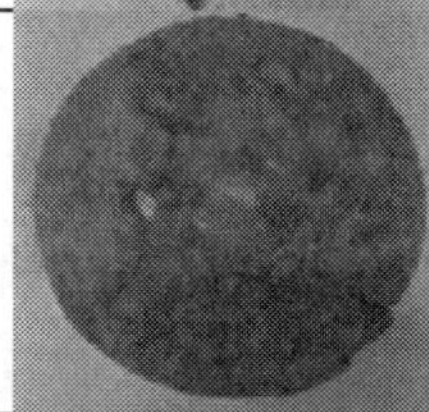	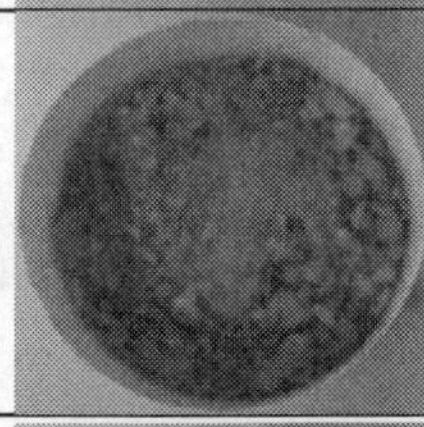	5,13	7.41	13.7	235.2
B6	Biomass – from energy crops (sunflower husk)	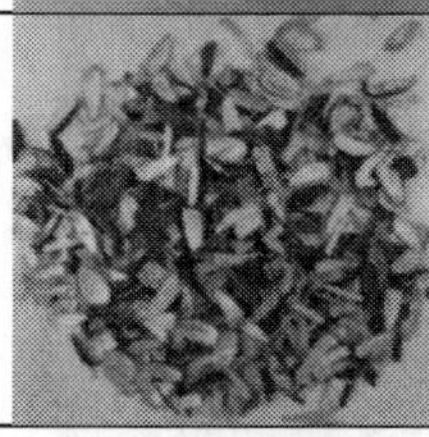		4.10	5.88	74.2	280.2
B7	Biomass Mix – from energy crops (sunflower husk), sawdust, mixed pellets (sunflower husk, wood husk)		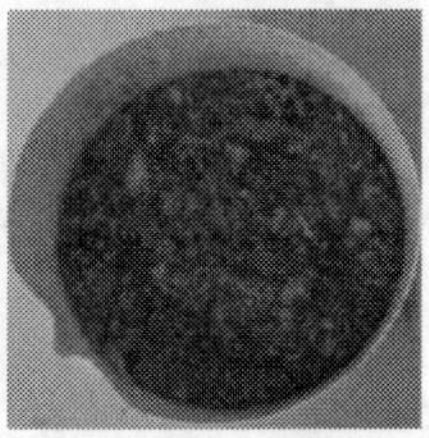	8.94	6.37	45.3	367.2
B8	Biomass Mix – straw, wood pellets, pellets from energy crops (sunflower)			7.84	5.03	119.4	472.4
B9	Biomass – wood chips			7.00	6.39	42.2	143.9
B10	Biomass – pellets from energy crops (sunflower)			7.84	6.28	49.4	43.1

DOI: 10.1051/e3sconf/20171700097

B11	Biomass – wood chips			7.28	6.29	21.9	98.7
B12	Biomass Mix – from energy crops (sunflower husk), wood pellets	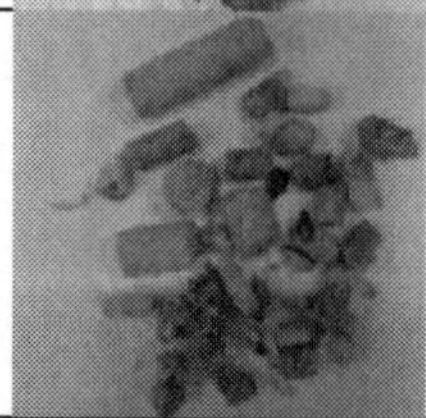		8.22	8.84	104.0	36.5
B13	Biomass – wood chips			6.21	7.79	60.8	59.9
B14	Biomass Mix – straw, wood pellets, pellets from energy crops (sunflower)	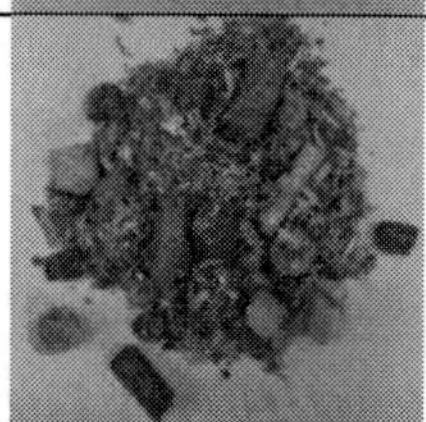		8.64	4.94	133.8	41.6
B15	Biomass – wood chips	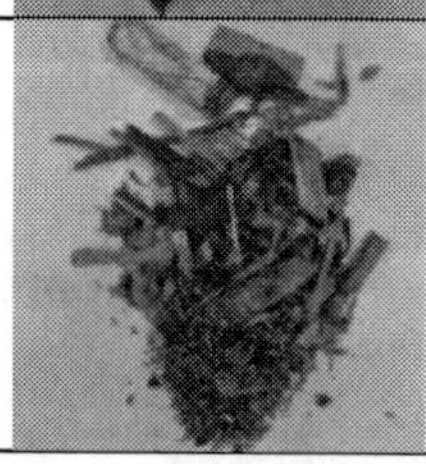	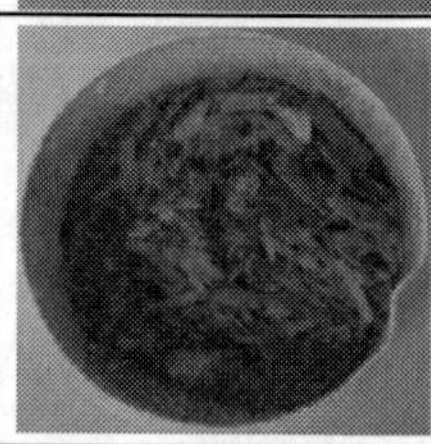	6.02	9.10	88.2	69.8

Mad – Analytical moisture

Table 2. The mercury content in ashes from biomass combustion.

No sample	Macroscopic description	Hg µg/kg
B1-S	Biomass - straw briquettes	297.3
B2-S	Biomass - wood chips	177.5
B3-S	Biomass - Agro-seeds	346.9
B4-S	Biomass - wood chips	298.3
B5-S	Biomass - wood briquettes	339.1

A summary of mercury content in individual biomass types indicates that the mercury content in raw samples of straw briquettes amounts to 130.4 μg/kg and increases up to 297.3 μg/kg after the combustion process. In the case of wood chips, mercury content is in the range from 59.9 μg/kg to 295.0 μg/kg; after the combustion process this value ranges from 177.5 μg/kg to 298.3 μg/kg. For Agro-seeds, mercury content amounts to135.6 μg/kg and 346.9 μg/kg after combustion. The mercury content in wood briquettes is 235.2 μg/kg, while this value increases up to 339.1 μg/kg after the combustion process. In the case of energy crops (sunflower husk) and pellets from energy crops (e.g. sunflower), the mercury content is 280.2 μg/kg and 43.1 μg/kg (the lowest content recorded), respectively. The so-called Biomass Mix from energy crops is characterized by highly variable mercury content in the range from 36.5 μg/kg up to 472.4 μg/kg.

4 Conclusion

Based on the research results presented in this paper, the following conclusions can be drawn:

1. Based on the measured pH and conductivity of biomass, presented in Table 1, it can be concluded that the samples are characterized by low pH (acidic, slightly acidic reaction). In contrast, B15, B13, B12, and B5 samples have alkaline reaction.
2. The conductivity is characterized by relatively low values; this applies particularly to the B11 sample, with electrical conductivity at 21.9 mS.
3. It should be noted that reactions between the products of decomposition of organic matter and mineral substances can be observed in aqueous solutions of biomass. It is suggested that the acidity of the solution is due to free CO2, mineral and organic acids.
4. The differences in pH and conductivity of the samples are resulting from inhomogeneities in the starting material, storage conditions and uneven course of leaching processes.
5. Analytical moisture as received is in the range from 8.94% (sample No B7 – Biomass Mix) to 4.10% (sample No 6 – Biomass from energy crops-sunflower husk).
6. The study has shown significant differences when it comes to the mercury content in individual biomass types, ranging from 36.5 μg/kg to 472.4 μg/kg.
7. The highest concentrations of mercury in the amount of 472.4 μg/kg and the highest conductivity of 119.4 mS were recorded for the Biomass Mix (straw, wood pellets, wood pellets from energy crops - sunflower). It is suggested that such a large accumulation of mercury may be the result of the diverse composition of biomass from various sources.
8. A one-fold, two-fold, and three-fold increase in the mercury content after the combustion of the tested samples were observed in the B1-S, B4-S, B3-S (Agro seeds, up to 346.9 μg/kg, the highest concentration recorded) and B5 (wood briquettes, up to 339.1 μg/kg) samples, respectively.
9. The only exception is the B2-S sample (wood chips), where the mercury content after the combustion process has decreased by approximately 40%.

References

1. Directive 2001/77/EC of the European Parliament and of the Council of 27 September 2001 on the promotion of electricity produced from renewable energy sources in the internal electricity market (2001), publication available at http://orka.sejm.gov.pl/Drektywy.nsf/all/32001L0077/$File/32001L0077.pdf

2. G. Berent-Kowalska, J. Kacprowska, I. Moskal, A. Jurgaś, Central Statistical Office of Poland, Production Department, the Ministry of Energy, Energy Market Agency S.A., (2016), publication available at www.stat.gov.pl

3. Energy Mix 2050. Analysis of Scenarios for Poland. Ministry of Economy, (2011), publication available at http://poznajatom.pl/download/6f/94/17dd23f7acb9bdd550d838a98af9f753b871.pdf.

4. Poland Energy Report, Enerdata, (2012), publication available at http://www.iea.org/publications/freepublications/publication/Poland2011_web.pdf

5. Costs and environmental effectiveness of options for reducing mercury emissions to air from small-scale combustion installations, European Commission DG Environmental (2005), publication available at http://ec.europa.eu/environment/chemicals/mercury/pdf/sci_final_report.pdf

Biofilm biodiversity presented by fluorescent in situ hybridisation

Mirela Wolf[1,*]

[1]Wroclaw University of Science and Technology, Faculty of Environmental Engineering, Department of Sanitary Biology and Ecotechnology, Wybrzeze Wyspiańskiego 27, 50-370 Wroclaw, Poland

Abstract. Numerous microorganisms may be present in the water distribution system. This is associated with the imperfection of purification processes, or secondary water pollution. Not only it results in the deterioration of water quality parameters, but it also increases threat of epidemiological problems. The water that is biologically unstable creates ideal conditions for colonization of the microorganisms to the inner surface of pipelines which may form biofilm. The key issue, enabling prevention and control of the impact of the development of biofilms, is to assess the biodiversity of microbiocenosis. In order to obtain comprehensive characteristics of microorganisms communities on a particular substrate, it is necessary to combine several techniques. Further analysis using molecular biology methods are usually after traditional methods of assessing the microbiological quality of water. Standard methods do not reflect the actual species composition, because they are targeted at the bacteria that can be isolated and cultured in the laboratory. Conventional methods are capable of detecting less than 10% of the organisms in the sample. In order to study the biodiversity of organisms inhabiting a biofilm (apart from the conventional methods) analyses of the diversity of nucleic acids should be used. The first method could be the polymerase chain reaction (PCR) and denaturing gradient gel electrophoresis (DGGE). Another way may be fluorescence in situ hybridization, which allows to detect determined DNA sequence using specially labeled oligonucleotide probes. Visualization of the material is performed using a fluorescence microscope. The main purpose of this article is to present rapid and precise identification groups of microorganisms in their natural habitat in biofilm using fluorescent in situ hybridization method (FISH) . FISH method can be successfully used to visualize these microorganisms, which show difficulties in culturing, as well as to provide knowledge on the phylogenetic structure in different habitats. FISH technique allows the quantitative description of environmental samples.

1 Introduction

Microorganisms in the water supply systems primarily dwell (95% of the total biomass) in the biofilm form. In water distribution systems, the space between the water and the pipe

* Corresponding author: mirela.wolf@pwr.edu.pl

surface is the layer where organic matter is accumulated and microorganisms are settled. Then they multiplied to form biofilms. In this way, the environment for the organisms of different food preferences and oxygen is produced [1].

Mature biofilm is a complex structure comparable to the ecosystem. Microorganisms are suspended in the extracellular polymers (EPS), and the exchange of matter and genes occurs through an extensive system of Casals [2]. First disadvantage is the possibility of detachement and transfer of portions of a microbial biofilm to other parts of the network. Further problem is the possibility of long-term survival or proliferation of pathogens (e.g. *Legionella sp.*) in the structures of the biofilm, among others, because of the limited penetration of disinfectant. The complexity of a mature structure provides protection against disinfectants therefore a potential source of pathogens. A commonly used chlorine disinfection greatly reduce pathogens in the water, but despite of this process their presence in water is still evident [3]. An important aspect is to stimulate destruction processes of metals by microorganisms colonizing the biofim, in particular by their metabolites. Products of metabolism are retained in the EPS layer, which can cause local changes in the physico-chemical properties of materials inhabited by bacteria [4].

Relatively there is little information about the microorganisms biodiversity forming the biofilm in water distribution systems due to the difficult access to the samples. Observations based on laboratory models of distribution indicates that as the maturation of a biofilm, cell density is stabilizing, and the species diversity increases [5]. Premature biofilm surrounded by the EPS provides an excellent basis for the creation of multi-species biocenosis.The species composition of biological membranes is distinguished by a certain subclass of *Proteobacteria*, *Actinobacteria*, *Firmicutes*, *Verrucomicrobia*, *Nitrospirae*, *Bacteriodetes* whose existence depends on the material which made water supply network, the age of the biofilm and methods of disinfection. According to this it can be concluded that the biodiversity of bacteria in the biofilm can be resulted of biotic and abiotic factors in the drinking water distribution system [6].

Traditional methods of assessing the microbiological quality of water are usually just the first step of identifying microorganisms, which directs further analyses using molecular biology methods.The introduction of molecular biology techniques in order to assess the species composition provides information about the biodiversity of organisms forming growths in water supply network. The limited access to the sampling area is a great difficulty. Each distribution system is diverse in terms of bacterial flora. Each water supply system is characterized by its own species composition of microorganisms, so it is not possible to draw general conclusions about the biodiversity of the biofilm [7, 8]. The use of molecular techniques based on the analysis of 16S rRNA/rDNA will provide information about microorganisms, where a standard water analysis do not. One of those methods may be fluorescence in situ hybridization (FISH), which is now widely used to identify bacteria in their natural habitat [9].

The results of research concerning the formation of biofilm and their qualitative composition can be used in the selection of technical materials for the construction of various types of installations in contact with water and the design processes.

2 FISH method

FISH technique is based on hybridizing fluorescently labelled probe to ribosomal rRNA in permeabilized whole microbial cells. Probes are short DNA sequences (16-20 nucleotides) labelled with a fluorescent dye. These sequences are complementary to the 16S rRNA bacteria sequence in bacterial cells. Due to this method microorganisms may be located and identified [10].

FISH analysis consists of the following steps:
1. fixation of biological material,
2. sample dehydration,
3. hybridization of the sample with a probe and washing,
4. microscopic observations,
5. bacteria identification and quantitation using advanced software [11].

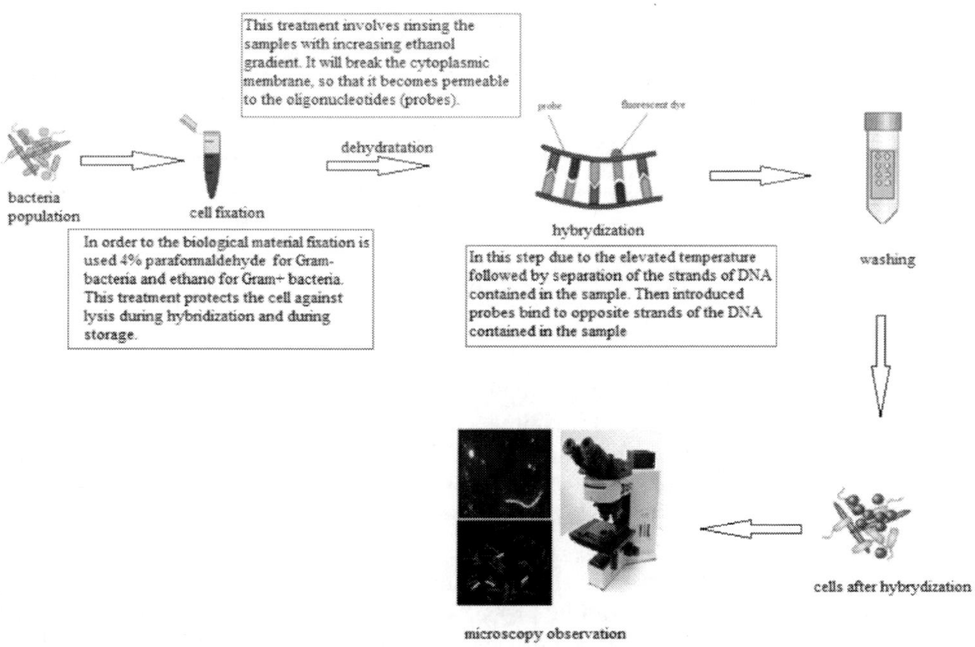

Fig. 1. Main steps of fluorescent in situ hybridization method [11, 12].

The main advantage of the FISH method is direct microorganisms visualization using the microscopy. It has higher sensitivity than the quantitative polymerase chain reaction (PCR) and is not prone to false results e.g. due to contamination with small amounts of DNA. Additionally, with PCR methods one can measure the number of copies of specific genes (not the number of cells), which may differ significantly from the number of particular organisms [12].

2.1 Probes

Microorganisms identification can be used by two probes: one characteristic of the group and second the specific. Simultaneous signal from two probes allows an assignment of the species and can eliminate misinterpretation. Examples of different probes are shown in Table 1 [13].

Depending on the size and use of the fluorescent probe is divided into:

- ✓ whole chromosome painting (WCP)- covering all the selected chromosome, or specific arm; probe covers an area from several to millions base pairs,
- ✓ centromeric enumeration probes (CEP)- specific to individual chromosomes, enable to identify the centromere; they reach a size of tens to thousands base pairs,
- ✓ locus specific identifier (LSI)- complementary to a specific region of a chromosome centromere probes, their size are tens of thousands of base pairs and are visible on both chromosomes as well as in interphase [15].

Table 1. Examples of probe for fluorescent in situ hybridization [16].

Short name	Specificity	Sequence from 5' to 3'
ARC344	Archaea	TCG CGC CTG CTG CIC CCC GT
PLA46	Planctomycetales	GAC TTG CAT GCC TAA TCC
E11	Eubacteria	AGC CAT GCA GCA CCT GTC TC
DELTA495b	Deltaproteobacteria	AGT TAG CCG GCG CTT CCT
UNIVER	all Bacteria	CCG TMT TAC CGC GGC TGC TGG CA

2.2 Methodology

A sample of the biofilm formed on the PCV material was fixed with 12% paraformaldehyde in buffered saline to give 4% as a final concentration of paraformaldehyde. Then it was incubated for 12 hours. The sample was concentrated on a filter with a 0.2 μm pore diameter. The filter was washed with 5ml of sterile water and dried. Then to drain the sample, a number of alcohol dilution (50%, 80%, 96%) was conducted [11, 14].

A filter was placed on a glass slide and covered with 50μl of hybridization buffer (NaCl, Tris-HCl, EDTA, SDS, formamide, probe) amount of each ingredient was depended on the type of the probe (Table 2). Incubation was conducted at 46 °C for 3 hours. Then it was incubated for 30 minutes with a washing solution at 48 °C (NaCl changing concentration, Tris-HCl, EDTA, SDS). In order to visualize all living microorganisms in a sample was stained aqueous DAPI (4',6-diamidyno-2-fenyloindol) solution. The observation was performed by fluorescence microscopy [11, 14].

Probes sequences used to experiment are shown in Table 2.

Table 2. Sequences of probes used to identified different type of bacteria in fluorescent in situ hybridization [17].

Name	Sequence from 5' to 3'	Type	Formamide, %	NaCl in washing buffer, mM
BET42	CCTTCCCA CTTCGTT	*Betaproteo -bactea* class	35	80
HGC69	ATATTACC ACCGCCG	*Actinoba-cteridae class*	35	35

2.3 Results

FISH method determines the total number of viable microorganisms in the test sample as well as a specific type or family of bacteria.

On the tested material bacteria with *Actinobacteridae* class- exhibited red color (8lighter areas), while *Betaproteobacteria* class showed- green color (10 lighter areas). All bacteria was shown using DAPI solution (25 blue lighter areas).

Fig.2. Example of visualisation bacteria in fluorescent microscopy after test measurement on fragment water distribution network made of polyvinyl chloride (lighter places- presence of bacteria), All alive cells-blue colour, *Actinobacteridae* class- red colour, *Betaproteobacteria* class- green colour.

3 Conclusion

Numerous microorganisms may be present in the water distribution system. This is associated with the imperfection of purification processes, or secondary water pollution. Not only it results in the deterioration of water quality parameters, but also increases threat of epidemiological problems.

FISH method can be used to visualize the microorganisms that cause problems in culture and also allows to know the structure of the phylogenetic groups in different habitats. The results of modern molecular methods often require more confirmation tests based on classical microbiology [18].

References

1. M. Lethola, K. Miettinen, M. Keinanen, T. Kekki, O. Laine, A. Hirvonen. Microbiology, chemistry and biofilm development in a pilot drinking water distribution system with copper and plastic pipes. Water Research, **38**, strony 3769–3779 (2004)
2. M. Simoes, L. Simoes, M. Vieira M., A reviev of current and emergent biofilm control strtegies. Food Science and Technology. 4, **43**, 573–583 (2003)
3. M. Batte, B. Appenzeller, D. Grandjean, S. Fass, V. Gauthier, F. Jorand, L. Mathieu, M. Boualam, S. Saby, J. Block, Biofilms in drinking water distribution systems. Reviews in Environmental Science and Biotechnology, **2**, 147–168 (2003)
4. I. Neria-Gonzales, T. Wang, F. Ramirez, J. Romero, C. Hernandez-Rodrigues, Characterization of bacteria community associated to biofilms of corroded oil pipelines from the southeast of Mexico. Anaerobe. 12, **3**, 122–133 (2006)
5. A.Grabińska-Łoniewska, E. Siński, *Mikroorganizmy chorobotwórcze i potencjalnie chorobotwórcze w ekosystemach wodnych i sieciach wodociągowych.* Warszawa : Seidel Przywecki (2010)

6. L. Mathieu, C. Bouteleux, S. Fass, E. Angel, J. Block, Reversible shift in the alfa-, beta-, gamma- proteobacteria populations of drinking water biofilms during discontinous chlorination. Water Research, **56**, 176–184 (2009)
7. D. Berry, C. Xi, L. Raskin, Microbial ecologyof drinking water distributions systems. Current Opinion in Biotechnology, **17**, 297–302 (2006)
8. S. Nevel, K. Roy, N. Boon, Bacterial invasion potential in water is determined by nutrient availability and the indigenous community. FEMS Microbiology Ecology. 2013, Tom 85, strony 593–603.
9. B. Braun, I. Richert, U. Szewczyk, Detecton of ion- depositing Pedomicrobium in native biofilms from the Odertal NAtional Park by a new, species FISH probe. Journal of Microbiological Methods, **79**, strony 37–43 (2009)
10. E.Valpi, J. Bridger, FISH glossary: an overview of the fluorescence in situ hybridization technique, BioTechniques, **45**, 385–409 (2008)
11. P. Nielsen, D. Holger, H. Lemmer, I. Arslan-Alaton, T. Olmez-Hanci, *FISH Handbook for Biological Wastewater Treatment. Identification and quantification of microorganisms in activated sludge and biofilms by FISH.* London, IWA Publishing, 2009. 9781843392316 (2009)
12. B. Brehm-Stecher, E. Johnson, Single-Cell Microbiology: Tools, Technologies and Applications, Microbiology and Molecular Biology Reviews, **65**, 538–559 (2006)
13. M. Jetten, L. Niftrik, M. Storus, B. Kartal, J. Keljtjens, Biochemistry and molecular biology of anammox bacteria. Critical Reviews in Biochemistry and Molecular Biology, **44**, 65–84 (2009)
14. L. Sanz, T. Kochling, Molecuar biology techniques used in wastewater treatment: An overview, Process Biochemistry, **42**, 119–133 (2007)
15. M. Fazli, T. Bjarnsholt, N. Høiby, M. Givskov, T. Tolker-Nielsen, *PNA*-Based Fluorescence In Situ Hybridization for Identification of Bacteria in Clinical Samples, Methods in Molecular Biology, **1211**, 261–271 (2014)
16. T. Liehr, *Classification of FISH probes, Part of the series Springer Protocols Handbooks*, 43–47 (2016)
17. S. Boase, J. Jervis-Bardy, E. Cleland, H.Pant, L.Tan, Bacterial-induced epithelial damage promotes fungal biofilm formation in a sheep model of sinusitis, Allergy and Rhinology, 3, 341–348 (2013)
18. B.Bochner, L. Giovannetti, C. Viti, Important discoveries from analysing bacterial phenotypes, Molecular Microbiology, 70, 274–280 (2008)

Numerical modeling of a downwind-developing mesoscale convective system over the Masurian Lake District

Damian K. Wójcik[1,*]

[1]Institute of Meteorology and Water Management – National Research Institute, ul. Podleśna 61, 01-673 Warsaw, Poland

Abstract. Meteorological data concerning the severe convective system from the 21 August 2007 are analyzed in this study. Compiled information allows to understand the reason for the storm development and to identify its fundamental convective mode. Next, the EULAG model is utilized to perform an idealized test that shows a downwind–developing storm growth in an environment comparable to the one that was observed on the 21 August 2007 in the Masurian Lake District. Finally, the COSMO numerical weather prediction model is applied to reconstruct the storm development. The experiment is carried out for various computational grids having the horizontal grid length between 7.0 and 0.55 km. It turns out that the COSMO model is capable in simulating storms of that type. Since the model is used for operational weather forecasting in Poland the evaluation of this skill contributes to the increase of public safety.

1 Introduction

Deep moist convection brings several hazards to the safety of people and property. Dangerous convective events are commonly named as severe convective storms or systems (SCS). One major mechanism of devastation is linked with precipitation and precipitation-driven downdrafts [1]. In particular, downdraft air spreading over the surface may lead to the formation of severe wind gusts. Therefore, major efforts are undertaken to improve numerical weather prediction (NWP) models capability to forecast development and evolution of severe convection.

The 21 August 2007 SCS in the Masurian Lake District (Masurian LD) distinguished itself by severe surface wind gusts up to 35 m/s that caused extensive damage and loss of life. At that time, severe weather warning informing about the approaching SCS was issued about 50 minutes before the SCS approached Mikołajki town. This short time period was insufficient to provide warning to local community about the heavy thunderstorm.

The NWP models employed in 2007 in the Institute of Meteorology and Water Management – National Research Institute had around 14 km horizontal grid length. That grid length does not allow to simulate deep convective storms. Thus, the then models could

* Corresponding author: damian.wojcik@imgw.pl

not forecast the storm in advance. However, the last decade brought unprecedented improvements to the NWP model dynamics, physics, and computational performance. In particular, the decrease of the horizontal grid length by nearly an order of magnitude resulted in "convection-permitting" non-hydrostatic NWP models. Nowadays, the NWP models are commonly run with horizontal grid length close to 2 km. Such a setup enables resolving basic circulations that are characteristic for deep most convection [2].

The ultimate goal of this study is to perform an assessment of the Consortium for Small-Scale Modeling (COSMO; www.cosmo-model.org) NWP model capabilities in mid-latitude deep moist convection forecasting over Poland. Such contemporary studies that are focused on Poland are not known to the author. Because during warm season severe convective systems occur regularly over Poland, the assessment may contribute to the increase of public safety. Moreover, climate-change-related increase in frequency of severe weather events over Europe is considered to be likely [3], so their skillful forecasting should be a priority. For this purpose, a reconstruction of the severe storm from the 21 August 2007 with COSMO NWP model is undertaken.

Section 2 of this report describes models and simulation strategy used in the experimental part of this study. Section 3 briefly presents meteorological observations related to the storm development. In the last section collected meteorological data are utilized to design an idealized test that models quasi-linear convective system propagation. This approach facilitates better understanding of the 21 August 2007 storm and provides guidelines for semi-realistic numerical reconstruction of the storm.

2 Tools

Two types of numerical experiments: idealized and semi-realistic are utilized in this study. The idealized experiments (Section 4.1) are carried out using the EULAG model. This model solves anelastic version of the Euler equations [4] using an accurate advection and elliptic pressure solvers i.e. MPDATA advection scheme [5] and preconditioned conjugate residuals solver [6], respectively. The framework contains one-moment moist precipitating thermodynamics [7] and turbulence scheme which is based on the turbulent kinetic energy (TKE) prognostic closure of the 1½ order [8].

The basic tool for the semi-realistic experiments is the COSMO model [9, 10] in version 5.01. Its dynamical core solves fully compressible Euler equations using the second-order Runge-Kutta time integrator [11] and the Arakawa C-grid spatial discretization. The model is capable to perform convection-permitting simulations in steep terrain for up to 1.0 km horizontal grid length simulations. Over the rather flat Masurian LD the model handles well simulations with 0.55 km grid length.

The framework of COSMO is equipped with large set of physical parameterizations that contain: radiation, turbulence, cloud microphysics, land-surface model, and surface flux scheme. The operational turbulence scheme [12] is based on the 2nd order turbulence closure with the prognostic equation for TKE. An additional fully three dimensional Smagorinsky-Lilly turbulence scheme is also available in the model [13]. The essential one-moment microphysics parameterization is based on the Kessler warm-rain scheme. Its operational extensions allow to treat basic forms if ice condensate: cloud ice, snow and graupel.

Semi-realistic experiment in Section 4.2 is carried out using a few inclusive computational domains that are depicted in Figure 1. The most coarse domain grid length equals to 7.0 km. The grid lengths of finer domains are equal to 2.2, 1.1 and 0.55 km, respectively. Technical data related to each of the simulations are provided in Table 1.

3 Weather observations

During the early hours of 21 Augusts 2007 weather in Poland was influenced by two synoptic pressure systems: a polar maritime low with the center over Germany and a polar continental high with the center over northern Russia. The cold front, associated with the low, was entering Poland from the south-western direction gradually replacing warm air with colder and drier air. The high and the low were driving upper-air circulation over Poland, creating a favorable condition for the inflow of a sub-tropical air mass from the south and south-eastern directions to the central Poland.

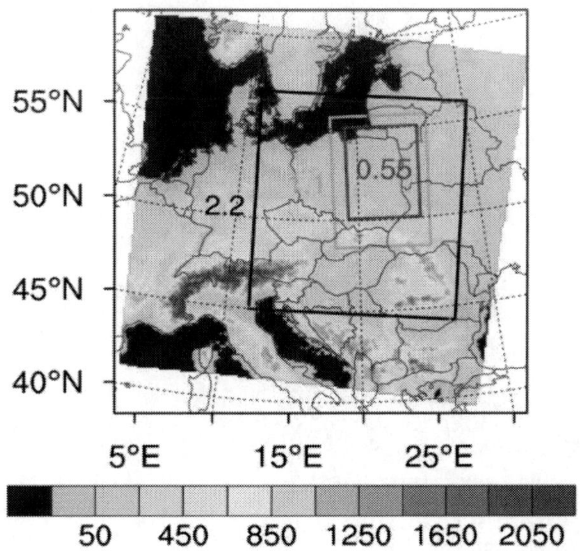

Fig. 1. Computational domains for semi-realistic simulations. The orographic map depicts 7.0 km domain. The 2.2, 1.1 and 0.55 km domains are marked using black, violet and red frames, respectively.

Table 1. Configuration of semi-realistic simulations.

	7.0 km	2.2 km	1.1 km	0.55 km
Domain width x height	310x310	540x600	500x720	750x1000
Simulation start time	0:00 UTC 20.08.2007	18:00 UTC 20.08.2007	03:00 UTC 21.08.2007	07:00 UTC 21.08.2007
Boundary update frequency	3 hours	1 hour	15 minutes	15 minutes
Turbulence scheme	2nd order closure with prognostic TKE			Smag.-Lilly

The Masurian LD was free of deep convective activity between 0:00 and 6:00 UTC. Because of that and owing to the weak south-eastern wind close to the surface, the atmospheric boundary layer (ABL) within the Masurian LD remained quite warm and moist during the night. Shallow fogs were observed in Mikołajki (3:00 UTC) and in Olsztyn (6:00 UTC). At 6:00 UTC, the measured 2-m dew point temperature in Mikołajki was equal to 19.3°C and the 2-m temperature was equal to 21.4°C (relative humidity of 88%). Later, during the day, the ABL air was further heated and moistened till the moment of deep convection development. In other words, the ABL over Masurian LD supported extreme weather formation.

The 12:00 UTC sounding from Legionowo provided basic information about the upper-air conditions in the warm sector of the low. This sounding, depicted in the form of a Skew-T diagram in Fig. 2, shows ABL spanning between the surface, i.e. 96 m above sea

 DOI: 10.1051/e3sconf/20171700099

level (ASL), and around 800 m ASL which is capped by around 4°C inversion layer with about 18% drop in relative humidity. Above the inversion, an elevated mixed layer extends from 1250 to approximately 2620 m ASL. This elevated mixed layer, in combination with warm and humid ABL air over Masurian LD, could foster the build-up of convection available potential energy (CAPE) rich environment over Masurian LD [14].

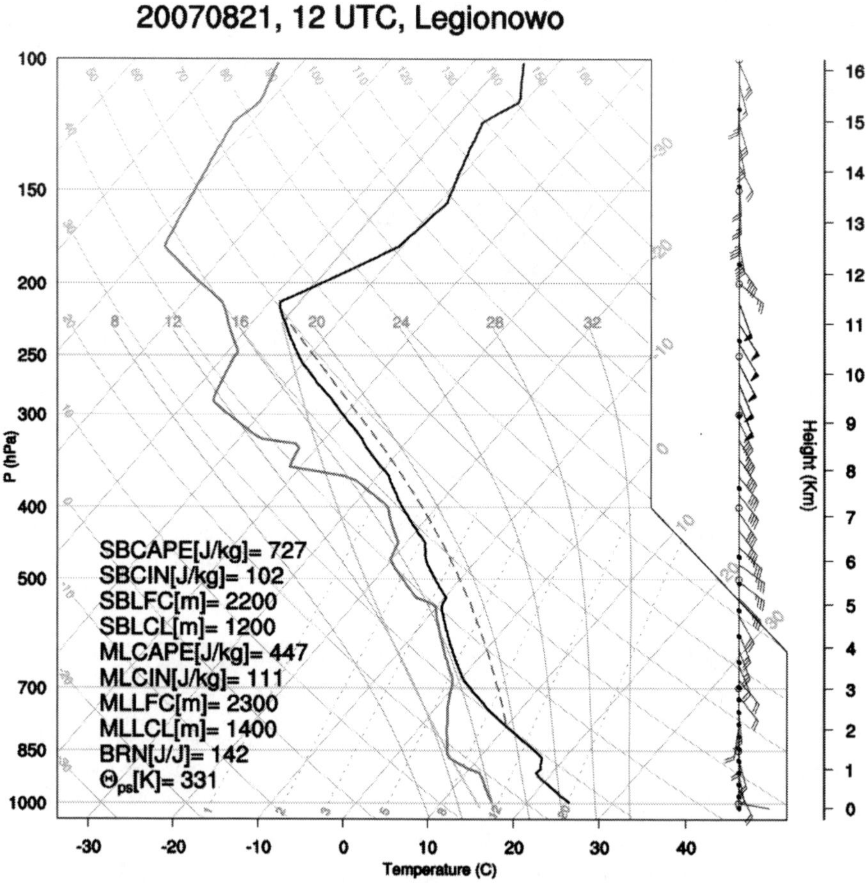

Fig. 2. Skew-T diagram acquired from radiosonde measurement in Legionowo (21 August 2007 12:00 UTC). Blue solid line denotes the vertical profile of dew point temperature. Black solid line denotes the vertical profile of temperature. Red dashed line denotes possible CAPE release. In the bottom left part of the diagram values of several standard convective indices are shown.

 DOI: 10.1051/e3sconf/20171700099

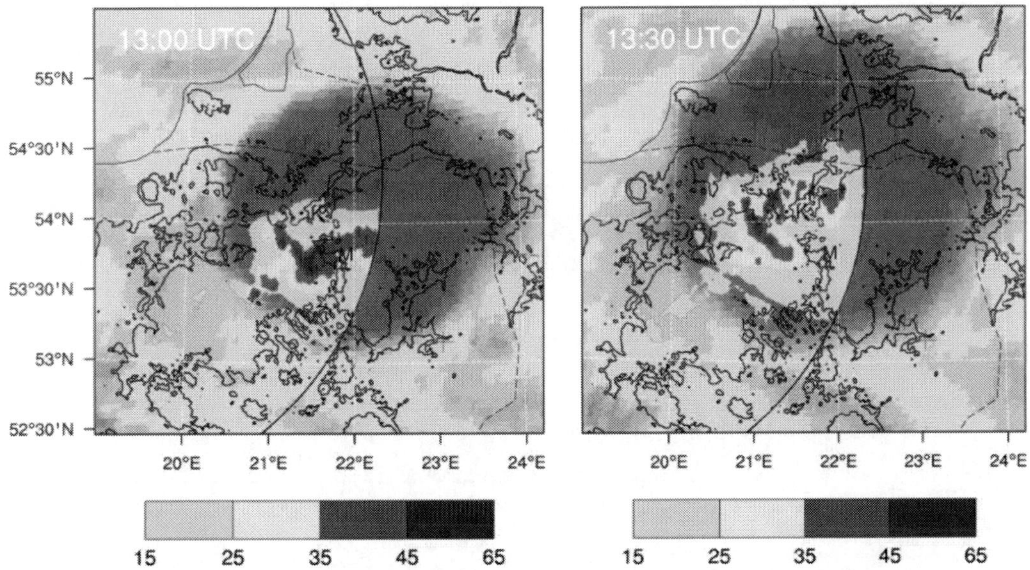

Fig. 3. Radar reflectivity at 6 km height ASL measured by the weather radar in Gdańsk (colors; 21 August 2007) overlaid over satellite image in 7.3 µm wavelength (grayscale). Gray lines mark orography isolines at 40 m and 116 m (the isoline of Śniardwy Lake) height ASL. Black circle marks the radar coverage boundary. The red crosses mark the locations of towering cumulus growth around 11:15 UTC. The capital letters "K" and "M" and the respective black markers denote location of Kętrzyn and Mikołajki towns.

Convection initiation of the severe convective system took place around 11:15 UTC in the northern part of Podlasie Region (see red crosses in Figure 3). The initiation was related to the approach of a moderate convective system from the south. It is worth to notice that according to the satellite images three deep cumulonimbus clouds grew at the windward side of local mesoregion uplands.

The severe convective system represented the class of downwind-developing mesoscale convective systems (MCS) [15] and, to some extent, resembled the squall line convective system [16]. It hosted one bow echo [17] that moved across Masurian LD between 12:45 and 13:50 UTC (Fig. 3). The estimated propagation speed of the bow echo did not exceed 27 m/s. The peak surface wind gusts measured at local WMO-compliant synoptic station in Mikołajki during the storm achieved 35 m/s, however, the average wind speed measured at this station was equal to 16 m/s.

4 Numerical results

4.1 Idealized test

The initial and boundary conditions for this test are based on the sounding from Legionowo (Fig. 2) which is modified within ABL to adjust it to the surface temperature (27.6°C) and dew point temperature (20.2°C) observations from 12:00 UTC from the Mikołajki synoptic station. The adjustment within ABL significantly increases surface-based CAPE value from around 700 J/kg to around 3400 J/kg. The boundary condition for the wind is based on the sounding from Legionowo.

 DOI: 10.1051/e3sconf/20171700099

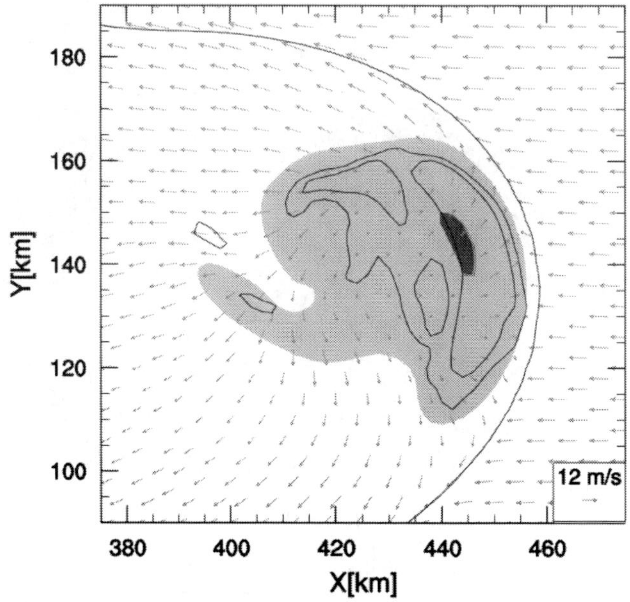

Fig.4. Quasi-linear storm development in 2 km grid length simulation without the Coriolis force for the wind profile as measured in Legionowo (21 August 2007 12:00 UTC). Intermediate storm-relative wind field 160 m above ground level (AGL) at T = 8 hours is shown. Blue line depicts meso-cold front position 160 m AGL. Light and dark blue shading denote rain mixing ratio exceeding 0.1 and 2 g/kg at 160 m AGL. Black contours depict isolines of the vertical velocity field 4108 m AGL: the solid lines mark 1, 2, 5 and 10 m/s values and the dashed lines mark negative values.

Convection initiation utilizes five warm thermals that are released concurrently within the ABL [18]. The initial storms cause the development of a few small cold pools that subsequently merge into a larger one. The presence of the large cold pool subsequently facilitates quasi-linear storm development (Fig. 4). The detailed structure of convective flows depends on the horizontal grid length and the mixing length parameter in the turbulence scheme. In some of the simulations, fast storm propagation is observed and the maximum surface wind speed reaches 30 m/s. Rear-inflow-jet [17] is observed as well.

4.2 Semi-realistic weather simulations

Semi-realistic weather simulations are configured as detailed in Section 2. Initial and boundary data for the most coarse 7.0 km grid length simulations originate from the global model GME [19] with 40.0 km horizontal grid length that was ran by Deutscher Wetterdienst at that time. Simulations utilize nudging (surface and upper-air observations) which is applied until 6:00 UTC of 21 August 2007. It turns out that severe deep convection is not simulated when the basic setup is applied.

In order to improve weather forecasts, the experience gained from Section 4.1 is utilized. In the default COSMO NWP simulations: (i) the elevated mixed layer is not present, and (ii) the surface temperature and dew point temperature over Masurian LD do not reach the observed values. Therefore, these biases are reduced by a nudging-based inclusion of additional meteorological information to the model. This additional nudging includes: (i) nudging of soil temperature (turned on till 2:00 UTC of 21 August 2007), (ii) nudging-based generation of a well-mixed elevated layer (turned on till 6:00 UTC), and

(iii) inhibition of precipitation for areas where available radar reflectivity data do not confirm it (turned on till 12.00 UTC).

The new setup allows to observe the squall line development in 1.1 and 0.55 km horizontal grid length simulations (Figure 5). The squall line has correct timing but is shifted around 30 km to the east. In 2.2 km and 7.0 km horizontal grid length runs the squall line is not simulated.

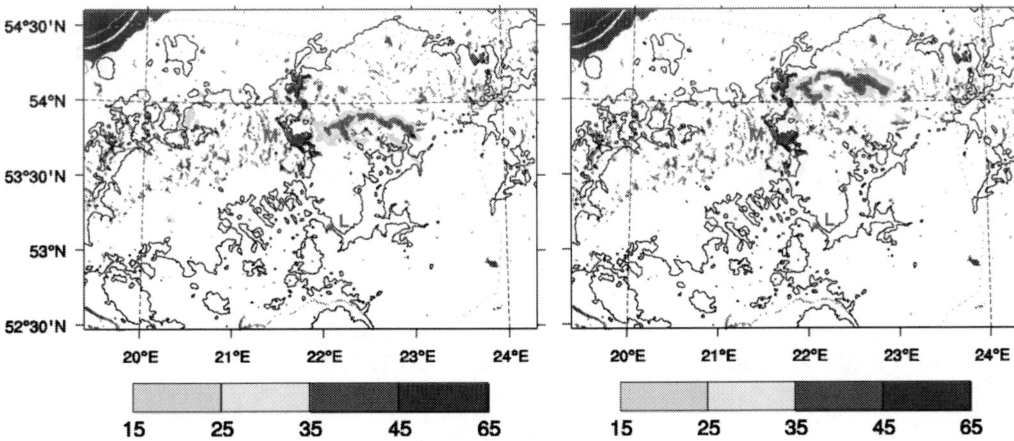

Fig.5. Model-simulated radar reflectivity values at the 850 hPa level height at 13:00 UTC (left) and 13:30 UTC (right). The simulation was performed with COSMO model with horizontal grid length equal to 0.55 km. Black line marks orography isoline at 116 m ASL. Letters "M" and "L" denote an approximate location of Mikołajki and Łomża towns. Lakes and the Baltic Sea are colored in blue.

5 Conclusions

Presented results suggest that COSMO model with 1.1 km and smaller horizontal grid lengths is capable in simulating downwind-developing MCS. Obtained results also confirm that a successful forecast of deep convection requires an accurate information about the initial state of the atmosphere (especially at low levels) and underlying Earth surface (e.g., soil moisture and temperature). Contemporary simulations in the convection-permitting regime provide improved weather forecasts than those available in 2007.

The results of simulations show that the severe convective storm from the 21 August 2007 could not be forecasted using the models available at that time. Nowadays, the probability of a successful numerical forecast of a comparable storm is higher. The quality of such forecasts noticeably depends on the quality of simulation's initial state.

References

1. P. Markowski, Y. Richardson, *Mesoscale Meteorology in Midlatitudes* (JohnWiley & Sons, 2010)
2. G.H. Bryan, J.C. Wyngaard, J. M. Fritsch, Mon Weather Rev **131** (2003)
3. Core Writing Team, R.K. Pachauri, L.A. Meyer, *Synthesis Report. Contribution of Working Groups I, II and III to the Fifth Assessment Report of the Intergovernmental Panel on Climate Change* (IPCC, 2014)
4. F.B. Lipps, R.S. Hemler, J Atmos Sci **39** (1982)
5. P.K. Smolarkiewicz, L.G. Margolin, Mon Weather Rev **121** (1993)

6. P.K. Smolarkiewicz, L.G. Margolin, Mon Weather Rev **124** (1994)
7. W.W. Grabowski, J Atmos Sci **55** (1998)
8. U. Schumann, Theor. Comput. Fluid Dyn. **2** (1991)
9. M. Baldauf, Mon Weather Rev **138** (2010)
10. M. Baldauf, A. Seifert, J. Förstner, D. Majewski, M. Raschendorfer, T. Reinhardt, Mon Weather Rev **139** (2011)
11. L.J. Wicker, W.C. Skamarock, Mon Weather Rev **130** (2002)
12. M. Raschendorfer, Newsl Deutscher Wetterdienst **1** (2001)
13. W. Langhans, J. Schmidli, B. Szintai, Newsl Deutscher Wetterdienst **12** (2013)
14. D.J. Stensrud, J Atmos Sci **50** (1993)
15. S.F. Corfidi, Wea Forecasting, Weather Forecast **18** (2003)
16. R. Rotunno, J.B. Klemp, M.L. Weisman, J Atmos Sci **45** (1988)
17. M.L. Weisman, J Atmos Sci **50** (1993)
18. W. Skamarock, M. Weisman, J. Klemp, J Atmos Sci **51** (1994)
19. D.M. Majewski, D. Liermann, P. Prohl, B. Ritter, M. Buchhold, T. Hanish, G. Paul, W. Wergen, J. Baumgardner, Mon Weather Rev **130** (2002)

The application of biomass ashes for improvement of sewage sludge dewatering

Marta Wójcik[1,*], *Feliks* Stachowicz[1], and *Adam* Masłoń[2]

[1]Rzeszow University of Technology, Department of Materials Forming and Processing, 35–959 Rzeszow Al. Powstańców Warszawy 8, Poland
[2]Rzeszow University of Technology, Department of Chemistry and Environmental Engineering, 35-959 Rzeszow Al. Powstańców Warszawy 6, Poland

Abstract. In compliance with the requirements of the EU climate and energy package (3 x 20 %), the biomass consumption in energetic industry is systematically growing. As a consequence, a large amount of ashes is produced which requires the management in line with law and ecology. Stringent requirements concerning the chemical composition of ashes prevent the biomass ash from being used in building construction. Specific physicochemical properties of ashes allow for their application in sewage sludge treatment. This paper presents the results of conditioning using ashes from willow combustion in sewage sludge dewatering. The findings could provide a new approach on sewage sludge and biomass ashes utilization.

1 Introduction

One of the main sources of atmosphere pollution is energetic industry. As a result of conventional fuel combustion, for example: coal, there are emitted gaseous air pollutants and ashes. The main chemical compounds which are produced during coal combustion are: carbon dioxide, carbon monoxide, SO_x and NO_x. According to the requirements of EU climate and energy package (3 x 20%), renewable raw materials are used as a source of energy more and more often [1]. Biomass has a special importance amongst renewable energy sources. Currently, biomass energy comprises 8÷15 % of the world energy supplies and it is expected to rise to 33–55% by 2050 [2]. In Poland, it is particularly used forest or agricultural biomass [3].

Biomass is a biodegradable fraction contained in municipal and agricultural waste. The biomass consumption is systematically growing in power stations and in electrical power and heating plants [3]. In Poland the biomass consumption was 9 317 763 GJ in coal-burning power plants in 2005 [4]. For comparison, the use of biomass grew to 42 917 011 GJ in 2013. In 2013 the biomass consumption was 26 275 970 GJ in biomass-burning power plants [4]. It is also observed the increasing use of biomass and coal co-combustion. The main advantage of biomass combustion is "zero emission" of carbon dioxide. During the biomass burning, only so much carbon dioxide is emitted as a plant takes from the

* Corresponding author: m.wojcik@prz.edu.pl

atmosphere by photosynthesis when the plants grow [5]. Additionally, the biomass combustion in incinerators causes a hazardous phenomenon concerning the forming of residue and the creating of sinters on heating surface of boilers. Moreover, the biomass contains corrosion components, such as potassium and sodium. Thus, biomass co-combustion is practised more often than the burning of 100% biomass [6].

The main sources of biomass for energetic industry in Poland are energetic crops, for example: straw and *Salix viminalis*. Due to the possibility of long term cultivation cycle, energetic plants are commonly used as a renewable energy source. The largest share of whole energetic plants has a willow tree. Rajczyk et al. [7] state that there are about 780 willow plant plantations with a surface area of 6480 ha, which contains about 95% of perennial plants cultivation.

Biomass is burnt in fluidized-bed incinerators in domestic power plants. The major problem associated with biomass combustion is the forming of combustion products, particularly ashes. Ash is the powdery residue produced as a result of burning of fuel that is captured from exhaust fumes means of electrostatic precipitators (fly ashes) or is discharged from fluidized bed (bottom ashes). According to Jarema-Suchorowska [8], approximately 4200 Mg of biomass ashes were produced in Poland in 2012. Contrary to coal ashes, biomass combustion products are mainly stored. It is estimated that only 29.1% of biomass ashes were recycled in 2013 [8]. Recently, biomass ashes have been used in agriculture as a valuable fertilizer or in road construction [7]. The chemical composition (the content of chlorine and alkalis) and high losses of ignition lead to the fact that most of biomass ashes fail to fulfill standard requirements as a component of cement and concrete production. According to PN-EN 197-1:2012 and PN-EN 450-1:2012, ashes with the chlorine content below 0.1% and with high loss of ignition (>5%) could not be used in production of building materials [9, 10]. Furthermore, the addition of ashes from biomass combustion negatively affects pozzolatic activity index. The application of biomass ashes in mining is also restricted due to exceding the limits of pH, chloride and suplhate leaching, which are determined in PN-G-11011 [4].

The requirements in economic sectors prevent biomass ashes from recycling. For this reason, there have developed new ashes utilization methods. Due to specific sorptive properties, ashes from biomass combustion could be managed in sewage sludge treatment. The addition of biomass ashes influences morphologicfal characteristics of sewage sludge. Sewage sluge has a negative charge and creates a stable system with low-sediment and low-dewatering capacity. After the conditioning with application of biomass ashes, sewage sludge is neutralized and the particles are fused together to from larger flocs [11]. According to Masłoń [12], powder materials below 300 μm diameter are incorporated in floc structure and become a weight of sludge that could improve the sludge settleability and dewatering. This paper presents the results of sewage sludge conditioning with ashes from willow combustion. The findings could provide a new approach for sewage sludge and biomass ashes utilization.

2 Materials and methods

Sewage sludge used in laboratory tests was obtained from the thickening tank from Municipal Wastewater Treatment Plant in Świlcza in the Podkarpacie region. In this study, sewage sludge was taken during summer/autumn and was characterized by grey-brown colour and earthy smell. The detailed physical and chemical properties of raw sewage sludge are presented in the Table 1.

Table 1. The physical and chemical properties of raw sewage sludge.

Parameter	Value
pH	6.65
water content [%]	96.83
dry mass [%]	3.17
concentration of dry mass [g d.m./dm^3]	31.70
CST (capillary suction time) [s]	149.21

In laboratory research, ash from willow combustion in domestic incinerator was used. The willow tree, from which ash was obtained, derived from a small agricultural business, in Wielkopolska region. The granulometric analysis was performed by means of the laser diffraction method. To determine particle size distribution of ash, Malvern Mastersizer 2000 particle size analyser was used. The willow ash was characterized by particles in the range of 0.399–900 μm (Fig.1). Due to the fact that ash was not sieved before laboratory tests, there were some coarse particles in material. The detailed grain size properties of ash from willow trees combustion are presented in Table 2. The results of diffraction analysis XRD show that ash comprises the following oxides: SiO_2, MgO, CaO, Al_2O_3, Fe_2O_3, NaO_2, Na_2O, KO_2. SEM images of ash from willow combustion are shown in Fig. 2.

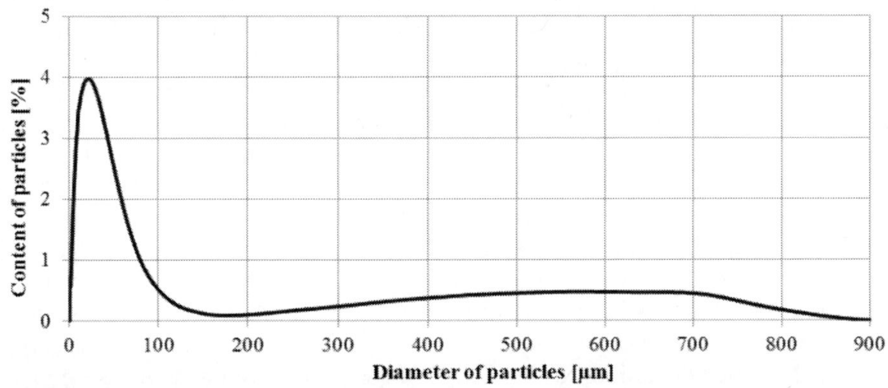

Fig. 1. The granulometric analysis of willow ash.

Fig. 2. SEM images of willow ash.

Table 2. The grain size properties of ash from willow combustion.

Parameter	Value
d_{10} [µm]	2.636
d_{50} [µm]	16.727
d_{90} [µm]	64.314
SMD (Surface Mean Diameter) [µm]	5.896
VMD (Volume Mean Diameter) [µm]	45.092
C_u (Uniformity coefficient) [-]	7.580
C_c (Coefficient of curvature) [-]	1.149

Laboratory tests of sewage sludge conditioning and dewatering with application of willow ash were carried out in three series. The sewage sludge conditioning with ashes was carried out as follows: five beakers with a volume of 1dm^3 were filled with 500 cm^3 of raw sewage sludge. The appropriate dosages of ash (5; 7.5; 15 and 30 g/dm^3) were added to four beakers. The dosages could be expressed as the weight ratio of ash to sewage sludge dry mass: 170 g/kg d.m.; 250 g/kg d.m.; 500 g/kg d.m. and 1000 g/kg d.m., respectively. Firstly, mixtures were rapidly stirred with a speed of 200 rpm for 1 minute and after that, they were stirred with a speed of 50 rpm for 15 minutes. In sewage sludge after conditioning, pH and capillary suction time (CST) were determined. The effectiveness of sewage sludge conditioning with willow ash has been determined by comparing the results for raw sludge.

Then, for prepared samples of sewage sludge, the dewatering capacity was investigated. Mechanical dewatering process was done in a Büchner funnel and in a centrifuge. The sewage sludge dewatering was carried out as follows: 50 cm^3 of sewage sludge was poured into a Büchner funnel. The process was done under 0.01 and 0.02 MPa vacuum pressure for about 15 minutes. Centrifugation of sewage sludge was conducted for two rotation speeds: 2000 and 2500 rpm for 5 minutes. After mechanical dewatering, the moisture content and CST were determined according to PN-EN 15934:2013-02. CST was measured using CST meter (ProLabTech) in accordance with PN-EN 147011:2007. In the conditioned sewage sludge, the pH was analyzed with pH-meter HACH HQ40d according to PN-EN 15933:2013-02.

3 Results and discussion

Firstly, ash from willow combustion was used as a conditioner. Research has shown that the sewage sludge conditioning with biomass ash influenced the improvement of sewage sludge dewatering. The evaluation of conditioning effectiveness (Fig. 3) was done by means of CST measurement. CST of raw sewage sludge was 149.21 s and CST after conditioning with the ash in dosage of 30 g/dm^3 was 66.77 s. The application of ash from willow combustion has resulted in the decline of CST in the whole range of agent used. However, low dosages did not influence the decrease of CST significantly. The best result was achieved for a dose of 30 g/dm^3 and for this reason, this amount of ash was considered as an optimal dosage. For the dosage of 30 g/dm^3, CST was 55.25 % lower than for non-conditioned sewage sludge. In comparison with commonly used polyelectrolytes, ash is a less effective reagent. Research conducted by Piotrowska-Cyplik and Czarnecki [13] has shown that the application of cationic polyelectrolyte F-410 in optimal dosage reduced the CST value of about 80 % (from 70 s to 14 s).

Biomass ashes could be an effective reagent in sewage sludge hygienisation. The influence of ash from willow combustion on the growth of pH was shown in Fig.4 - pH for raw sewage sludge was 6.67 and for sewage sludge after the addition of ash in dosage of 30 g/dm^3 was 9.92. Wang and Viraraghavan [14] reported approximate results of the growth of pH for coal fly in fly ash to sewage sludge ratio 1:1 (the growth of pH from 7.25 to 10.32). The positive impact of ash on the pH is connected with the content of alkaline metals in ash, for example: Ca, Mg and Na, what is acknowledged by the literature review [13, 15]. pH of sewage sludge is one of the major factors that have an influence on the pathogens reduction. According to Marcinkowski [16], most of pathogens are killed at pH > 9 and for this reason, ash from willow combustion could be an effective reagent in sewage sludge hygienisation. It is worth highlighting that biomass ash is less efficient than burnt lime. Own research shown that the dosage of CaO in the amount of 1 kg/1 kg dry mass increased pH of sewage sludge from 6.60 to 13.48.

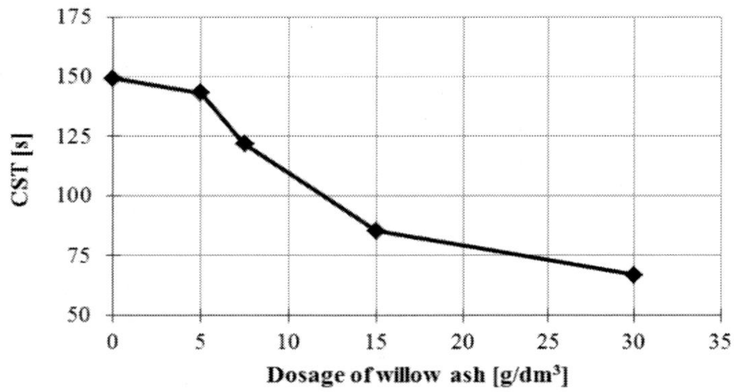

Fig. 3. The influence of sewage sludge conditioning with willow ash on CST.

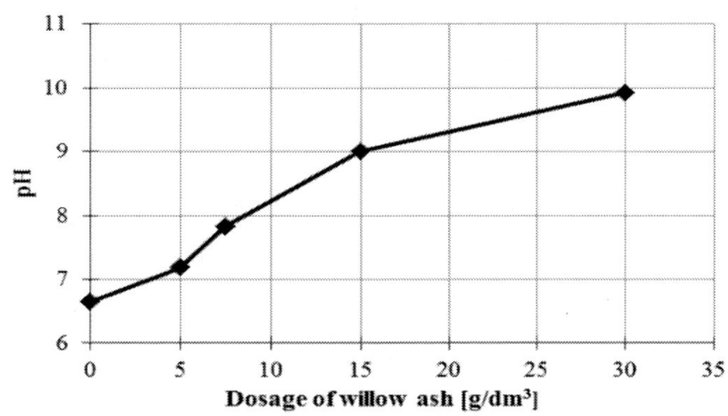

Fig. 4. The influence of willow ash on pH of sewage sludge.

In order to determine the effectiveness of sewage sludge conditioning with biomass ash, conditioned sludge was dewatered with use of Büchner filter and the centrifuge. The effectiveness of dewatering has been determined by comparing the results for raw sewage sludge. The measure of effectiveness was moisture content and CST after dewatering.

Firstly, sewage sludge was investigated using of centrifuge. The moisture content of sewage sludge for two rotation speeds: 2000 and 2500 rpm is shown in Fig. 5. The content of water in sewage sludge decreased with an increase of dosage of ash. With the 30 g/dm^3 dosage of ash, the moisture content decreased from 92.92 % (raw sewage sludge) to 85.77 % (conditioned sewage sludge) for the rotation speed of 2000 rpm. Better results were obtained for higher rotation speed. The application of 30 g/dm^3 dosage of ash from willow combustion reduced the moisture content from 91.80 % (raw sludge) to 84.31% (conditioned sludge). Low dosages did not reduce the moisture content in a significant way.

CST increased with the rise of dosage of ash. It was observed that CST depends on the sewage sludge moisture content. The higher value of CST could mean that sewage sludge has a more consistent structure. The elimination of excessive water, it is resistant to further dewatering. Higher values of CST were obtained for the rotation speed of 2500 rpm what is connected with the better effectiveness of centrifugation (Fig. 6).

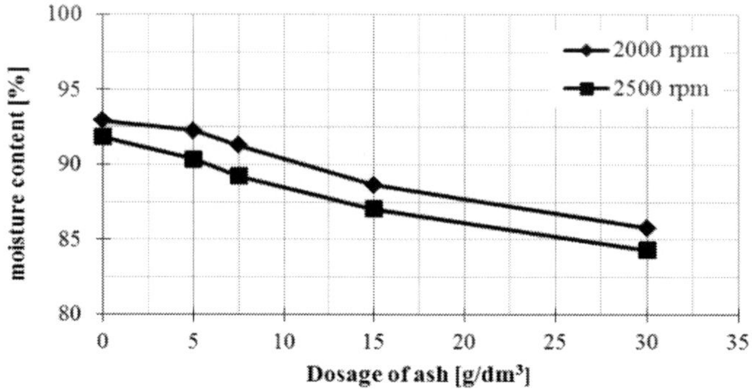

Fig. 5. The influence of willow ash on moisture content after centrifugation.

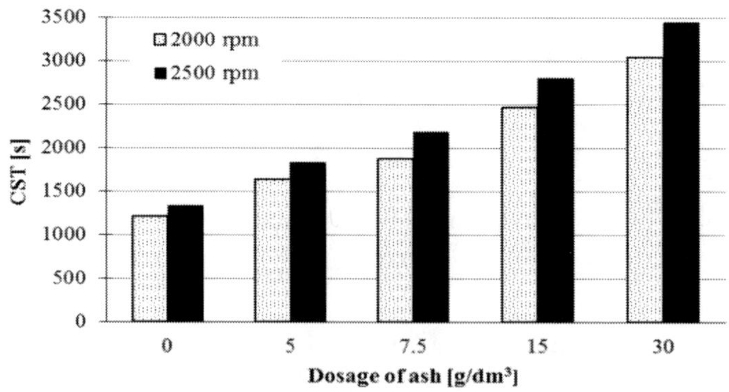

Fig. 6. The influence of willow ash on CST after centrifugation.

The best results of sewage sludge dewatering were obtained for vacuum filtration with Büchner funnel. The filter cake moisture for different dosages of ash after dewatering is presented in Fig. 7. The analysis of the results shows that raw sewage sludge indicated a good susceptibility to dewatering. The filter cake moisture for raw sewage sludge was respectively 89.72 % for 0.01 MPa vacuum pressure and 88.47% for 0.02 MPa vacuum pressure. The sewage sludge content of moisture decreased as the dosage of ash and the vacuum pressure increased. With the 30 g/dm^3 dosage of willow ash, the filter cake moisture content decreased to 78.61 % (0.01 MPa) and to 75.80 % (0.02 MPa). It is worth highlighting that for the vacuum filtration process, the application of ash is effective in a whole range of doses. Similar results of dewatering were obtained by Bień [17] for average cationic polyelectrolyte. Increasing both willow ash dosage and the vacuum pressure causes an increase of CST (Fig.8). In comparison with the results for centrifugation, CST was approximately 48 % higher. The higher value of CST is a result of the better effectiveness of sewage sludge dewatering in vacuum filtration.

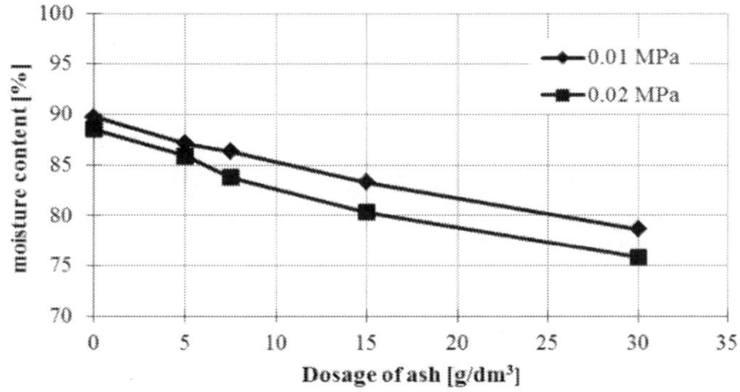

Fig. 7. The influence of willow ash on moisture content after vacuum filtration process.

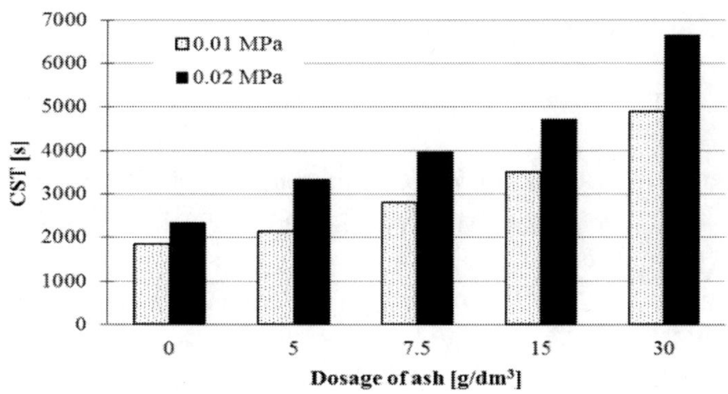

Fig. 8. The influence of willow ash on CST after vacuum filtration process.

4 Summary

The increase in the share of renewable energy in final energy production is associated with the biomass combustion. Biomass fuel is very attractive from the ecological point of view in terms of "zero-emission" of carbon dioxide. During the biomass combustion, there are produced combustion products, such as bottom and fly ashes. Stringent requirements prevent biomass ashes from being used as a component in concrete and cement production. For this reason, biomass ashes are mainly stored.

This study proved that biomass ash could improve sewage sludge dewatering. To evaluate the efficiency of sewage sludge dewatering, the following parameters were used: moisture content and CSK. The obtained results of laboratory tests allow us to draw the following conclusions:

1. The application of willow ash had a positive impact on sewage sludge dewatering what is confirmed by the measurement of CST. This parameter decreases with the increase of dosage of ash.
2. The analysis of results has shown that the addition of willow ash could improve the capacity of sewage sludge dewatering. The application of higher doses of ashes yielded better results which are confirmed by CST and moisture content after dewatering. The best results of sewage sludge dewatering were obtained for vacuum filtration and for

dose of ash of 30 g/dm^3. The addition of low dosages of ash from willow could slightly improve the sewage sludge dewatering.

3. The application of willow ash might raise pH of conditioned sewage sludge. For this reason, it is possible to reduce the amount of pathogens and achieve the partial hygienisation of sewage sludge.

The confirmation of the effectiveness of biomass ashes in sewage sludge conditioning might reduce the consumption of polyelectrolytes and contribute to the reduction of operating costs of treatment plants. After dewatering, sewage sludge containing ash may be considered as a valuable fertilizer in perennial plantations, creating a closed circuit. Due to the application of sewage sludge in energetic plant plantations, heavy metals in sewage sludge do not affect human health [18].

References

1. Law on Renewable Sources of Energy Adopted on 20 February 2015. Journal of Laws, Item 478 (2015)
2. S.V. Vassilev, D. Baxter, L.K. Andersen, C.G. Vassileva, Fuel **105**, 41 (2013)
3. D. Fijałkowska, L. Styszko, Annual Set the Environment Protection **13**, 875 (2011)
4. A. Uliasz-Bocheńczuk, A. Pawluk, J. Skierka, Mineral Resources Management **31**, 145 (2015)
5. T. Czech, A.T. Sobczyk, A. Jaworek, A. Krupa, Conference Materials POL-EMIS, June 2012
6. W. Mokrosz, MOKROSZ Sp. z o.o., Politechnika Śląska 2014
7. K. Rajczyk, E. Giergiczny, M. Szota, Scientific Works of Institute of Ceramics and Building Materials **6**, 73 (2013)
8. [online] [access: 10.12.2016] https.://www.kierunekenergetyka.pl/magazyn, popioły-z-biomasy.html
9. PN-EN 197-1:2012, Cement – Part 1: Composition, specifications and conformity criteria for common cements
10. PN-EN 450-1:2012, Fly Ash For Concrete – Part 1: Definition, Specifications And Conformity Criteria
11. P. Zhang, L. Haifeng, The Chemical Engineering Journal **156**, 620 (2010)
12. A. Masłoń, Instal **4**, 51 (2015).
13. A. Piotrowska-Cyplik, Z. Czarnecki, Journal of Research and Applications in Agricultural Engineering **50**, 23 (2005)
14. S. Wang, T. Viraraghavan, Waste Management **17**, 447 (1997)
15. A. Kowalczyk-Juśko, Proceedings of EC Opole **3**, 159 (2009)
16. T. Marcinkowski, Environmental Pollution Control **25**, 49 (2003)
17. J. Bień, Engineering and Protection of Environment **14**, 329 (2011)
18. F. Stachowicz, T. Trzepieciński, M. Wójcik, A. Masłoń, W. Niemiec, A. Piech, E3S Web Conf. **10**, 1 (2016)

E3S Web of Conferences **17**, 00101 (2017) DOI: 10.1051/e3sconf/20171700101
EKO-DOK 2017

Encouraging "K" strategy nitrifiers over "r" strategists in bioaugmentation reactor

Bartosz Zięba[1,*], and *Kamil* Janiak[1,2]

[1]Faculty of Environmental Engineering, Wroclaw University of Technology, Wybrzeże Stanisława Wyspiańskiego 27, 50-370 Wrocław, Poland
[2]Wroclaw Municipal Water and Sewage Company, Na Grobli 14/16 50-421 Wrocław, Poland

Abstract. Bacteria responsible for ammonia oxidization (AOB) can be differed into two groups which have different bio kinetic parameters. Those groups are slow growing bacteria with lower half saturation coefficient for NH4 (k -strategist) and fast growing bacteria with higher half saturation coefficient for NH4 (r- strategist). By controlling operational conditions of process it is possible to create dominance of one of those colonies to enhance the nitrification rate. Therefore model of sequencing bath reactor was developed to study the correlation of bioaugmentation of the side stream breeder reactor and predominant AOB culture. The course of the study was to find the minimal bioaugmentation influent of AOB rich recycled sludge to create dominance of k-strategist in the side stream reactor with maintaining maximal possible increase of overall AOB load. In addition operating conditions like sludge retention time, NH4 concentration and loading regimes was investigated to determine the optimum for the process.

1 Introduction

Warm and ammonium rich side streams create opportunity to culturenitrifying bacteria with relatively low sludge retention time (SRT) what maximizes their yield (kg bacteria/kg N). Grownbacteria may be next directed to main stream reactor, where they enhance nitrification. It is especially beneficial for wastewater treatments plants (WWTPs), which suffer from insufficient nitrification efficiency in winter period. Success of bioaugmentation depends not only on mass of cultivatedbacteria in bioaugmentation reactor but also on population structure of cultivated nitrifiers, as nitrifiers can be divided into two groups of different strategies [1]. R-strategist are nitrifiers with high maximum growth rate but also requires higher concentration of substrates (bacteria with high $\mu_{A,max}$ and K_{NH}). K-strategists have lower maximum growth rate however they grow reasonable fast in conditions of low concentration of substrates (bacteria with lower $\mu_{A,max}$ and K_{NH}). High temperature and high ammonium concentration in side stream leads to domination of r-strategists while conditions in main stream reactor promote growth of K-strategists. In case of bioaugmentation, r-strategists from bioaugmentation reactor will have hard time in activated sludge reactor as their growth will be slowed down and effect of

* Corresponding author: bartoszieba@gmail.com

DOI: 10.1051/e3sconf/20171700101

bioaugmentationwould be significantly limited. To avoid such situation, bioaugmentation reactor is fed with recirculated activated sludge from main stream reactor. With sufficientstream of activated sludge, despite unfavorable conditions in bioaugmentation reactor, practically only K-strategists from activated sludge are grown. Scheme of bioaugmentation process is show in Fig.1.

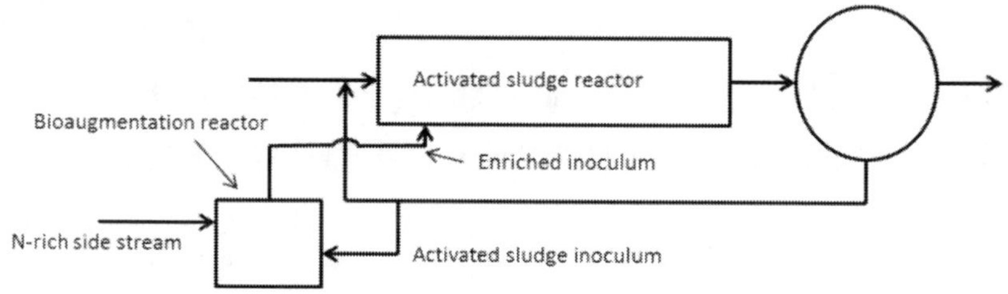

Fig. 1. Scheme of bioaugmentation process.

1.1 The problem and its importance

Main issue of this article is associated with feeding of bioaugmentation reactor with recirculated activated sludge : "How significant recirculated activated sludge stream is needed to create dominance ofK-strategists in bioaugmentation reactor?". Too small amount of recirculated stream of activated sludge will not provide dominance of K-strategists. On the other hand too large amount of that stream will result in drop of yield (kg bacteria/kg N).

Scale of influence of recirculated stream of activated sludge is significant. in Fig. 2. nitrifiers culture balance for a WWTP of 700 000 pe is shown. Assuming that maximal yield is 100% with dosing of 1% of recycled sludge stream. Increasing dose to 3% results in halving the efficiency and above 7>5% balance is negative which means that more nitrifiers is pumped to the bioaugmentation reactor than leaves it – from technological point of view solution ceases to be valid.

It means that dose of seeding stream of activated sludge should be as small as possible but also large enough to assure the dominance of K-strategists. That issue has yet to be solved and only general guidelines where established and are included in B.A.B.E [2] technology patent. According to these guidelines dose of seeding stream should be between 1 to 25% of recycled stream of activated sludge, which means wide spectrum of recommended values.

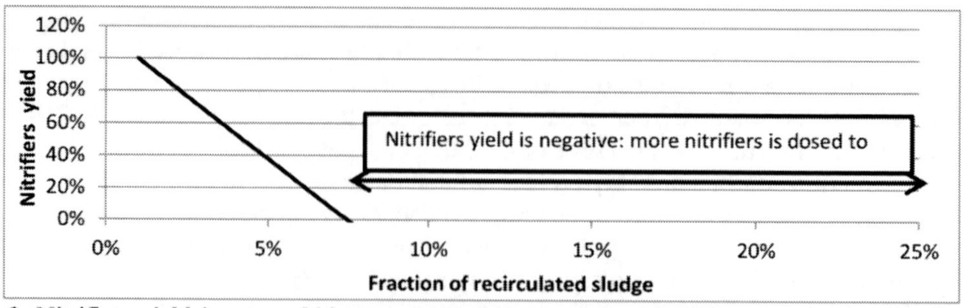

Fig. 2. Nitrifiers yield in case of bioaugmentation reactor fed with different recirculated activated sludge.

1.2 Aim and structure of paper

Aim of this article is to define minimal dose of bioaugmentation stream needed to ensure dominance of activated sludge nitrifiers (K-strategists). In this study simulations of bioaugmentation reactor were carried out. ASM1 model was used, elaborated to simulate two populations of nitrifiers in which one group are K-strategists and second one are r-strategists.

Structure of the article:

- Materials and methods – in this section mathematical model which was used is presented as well as simulated reactor, composition of side stream wastewater and recirculated sludge dosed to bioaugmentation reactor.
- Results – in this section the results of simulations are presented. Wide spectrum of process parameters and 4 configurations of kinetic parameters were studied.
- Discussion – results of simulations are discussed.
- Related work – in this section all the literature reports linked to the subject of this article are presented briefly.
- Assumptions and limitations – presented in this article speculations have limitations and are based on certain assumptions which are presented in this section.

2 Materials and methods

2.1 Model

Computer simulations which were performed to determined dose of recirculated activated sludge directed to bioaugmentation reactor were based on ASM1 model [3]. In this article only transformations including autotrophic bacteria were simulated with complete omission of heterotrophic bacteria. Model was expanded to include second fraction of nitrifying bacteria. By this extension it was possible to investigate interactions between K-strategists and r-strategists and to estimate what dose of recirculated activated sludge is required to create dominance of K-strategists in bioaugmentation reactor. Simulations were conducted until steady state was developed.

2.2 Nitrifiers kinetic coefficients

In the paper it was assumed that differences between K-strategists and r-strategists would be modeled by different $\mu_{A,max}$ and K_{NH} values with rest of coefficients kept at typical values. Table 1 shows assumed values of kinetic coefficients. Two different values of constant K_{NH} for r-strategists and two values of maximal growth rate $\mu_{A,max}$ for K-strategists were used. Values of these kinetic coefficients fit into range stated in literature [4]. More information on those can be found in "related work" section.

Table 1. Kinetic parameters of r-strategists and K-strategists.

Symbol	Unit	Name	r-strategists	K-strategists
$\mu_{A,max}$	d^{-1}	Maximum specific growth rate	1.44	0.96 and 1.2
K_{NH}	g NH_4-N/m^3	Ammonia half-saturation coefficient	5 and 10	1
b_A	d^{-1}	Decay coefficient	0.15	0.15
K_O	gO_2/m^3	Oxygen half-saturation coefficient	2	2
Y_A	g ChZT/g N	Yield of autotrophic biomass	0.15	0.15
θ	-	Temperature sensitivity	1.072	1.072

2.3 Reactor

Studies were performed based on SBR reactor, which volume stand at 1000 m^3. Reactor works in 3 phased regime: filling (dosing side stream wastewater and recirculatedactivated sludge), reaction phase (aeration) and sedimentation and decantation phase (offtake of waste sludge and treated wastewater). Regime of reactor is presented in Fig. 3.

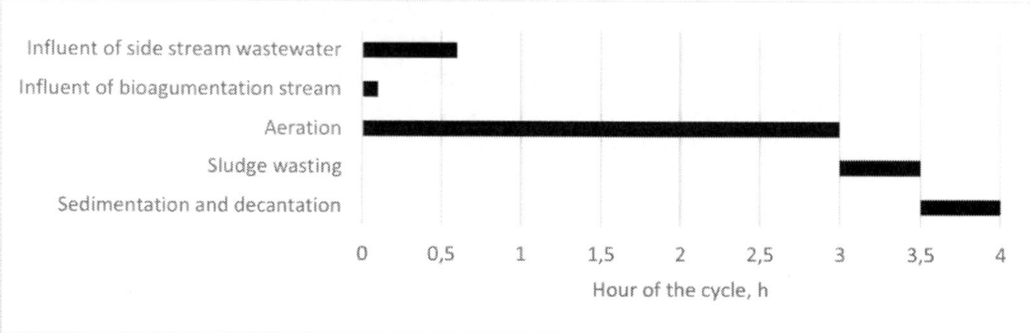

Fig. 3. Reactor phases.

2.4 Composition of bioaugmentation stream

Fraction of particular groups of nitrifiers in recirculated activated sludge was evaluated based on simulations of main stream reactor. In simulations data from Wroclaw's WWTP were used as well as kinetic coefficients from table 1. Based on simulations of this system population structure of bacteria present in main stream reactor was established. Performed studies has shown 95% dominance of K-strategists in recirculated activated sludge. Average concentration in recycled activated sludge stream was assumed at 300 g ChZT/m^3 based on Wroclaw's WWTP simulation studies performed by Piotr Balbierz [5].

2.5 Reactor parameters

Simulations were performed for different values of recirculated activated sludge with simultaneous regulation of hydraulic parameters of the reactor (SRT, time of filling, etc.). Parameters of reactor are shown in Table 2.

Table 2. Reactor parameters.

Parameter	**Value**	**Parameter**	**Value**
Ammonium concentration in side stream [g N/m^3]	1000	Sludge retention time [d]	2,5;4;6;8
Side stream flow [m^3/d]	400	Time of filling [min]	36;120;180
Recirculated activated sludge flow [m^3/d]	0,5;50;100 (0,1%;12,5%;25%)	Temperature [°C]	20;23;26

2.6 Number of simulations

Study consisted of 108 configurations of reactor parameters with additional 4 combinations of kinetic coefficients. Total number of simulation was 432.

DOI: 10.1051/e3sconf/20171700101

3 Results

Simulations show that required dose of recirculated sludge has no constant value and depends on proportion of K-strategists growth rate to r-strategists growth rate with growth rate defined according to equation 1 [6].

$$\mu_A = \mu_{A,max} \frac{S_{NH}}{K_{NH}+S_{NH}} \frac{S_O}{K_{O,A}+S_O} \tag{1}$$

Low average ammonium concentration in reactor promotes growth of K-strategists and if their growth rate is over a dozen percent higher that growth rate of r-strategist then dose of recirculated sludge is of no importance. If average ammonium concentration in reactor is high then situation is reverse and large dose, equivalent to big fraction of recirculated sludge is needed. This is shown on Fig. 4. Y axis represents fraction of K-strategists within all nitrifiers. Closer the fraction to 100% the better composition of sludge. X axis represents proportion of K-strategists growth rate to r-strategists growth. If proportion is higher than 1.0 then K-strategists growth rate is higher. Three data series depicting fractions of K-strategists with different recirculated sludge doses are shown. Two distinct regions can be found in Figure 4. When K-strategists growth rate is higher than r-strategists growth rate by more than 20%, then full domination of K-strategists will occur no matter the sludge dose. With proportion of growth rates reaching unity, smallest dose is not sufficient for K-strategists to dominate sludge and their fraction is getting lower (circle, dotted line).For large doses of recirculated sludge, K-strategists dominate sludge no matter the proportions of growth rates.

In case when K-strategists growth rate is only circa 80% of r-strategists growth rate smallest dose results in no K-strategists in sludge. With 50 m^3/d dose sludge is composed of 80% K-strategists while with dose of 100 m^3/d their fraction increase to 95%.Average ammonium concentration in reactor is therefore decisive factor determining which nitrifiers group will dominate sludge in case of low recirculated activated sludge dose. Fig. 5. presents relation between average ammonium concentration and fraction of K-strategists in sludge in case of two different recirculated activated sludge doses and K_{NH} of r-strategists equal to 5 g N/m^3.Average ammonium concentration lower than K_{NH} value leads to complete dominance of K-strategists while higher value leads to complete dominance of r-strategists if recirculated sludge dose is small. With high dose dominance of K-strategists can be sustained even in case of higher average ammonium concentrations.

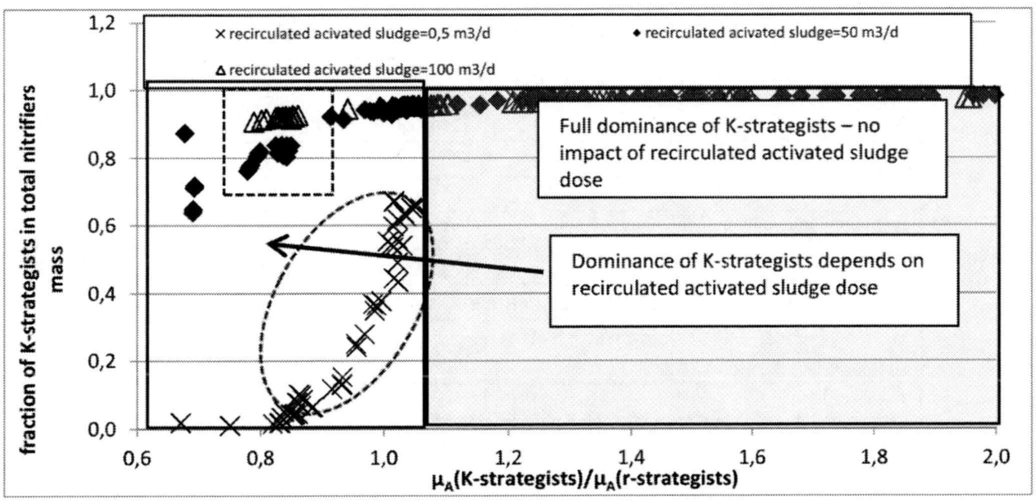

Fig. 4. Fraction of K-strategists in bioaugmentation reactor as a function of growth rate proportion.

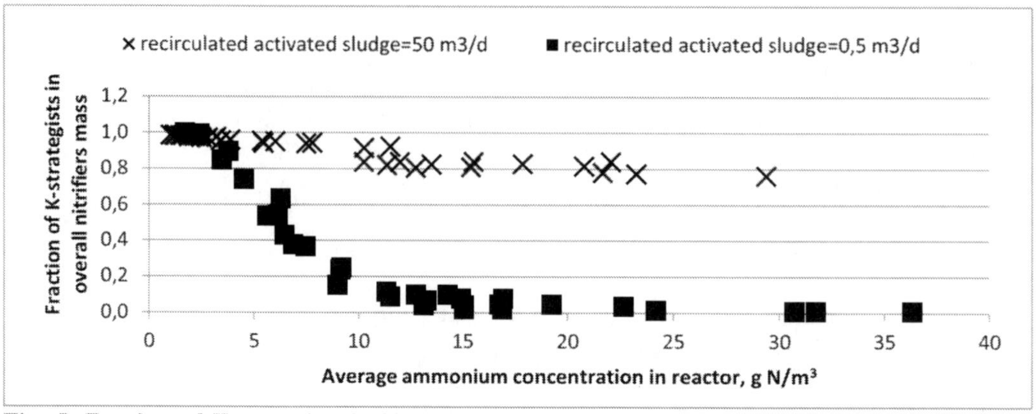

Fig. 5. Fraction of K-strategists in bioaugmentation reactor as a function of average ammonium concentration.

4 Discussion

Results of competition between r-strategists and K-strategists depends on proportion of growth rates and recirculated sludge doses. Growth rates depends on average ammonium concentration among others. As typical values of K_{NH} of r-strategists are in range of 10 g N/m^3, dominance of K-strategists is certain with low recirculated activated sludge flows if average ammonium concentrations is no more than several g N/m^3. Such low value is hard to obtain in reality, as concentration of ammonium in influent is several hundred g N/m^3.

It might be therefore necessary to operate bioaugmentation reactors with higher recirculated activated sludge flows even if that will lead to loss in nitrifiers yield. With low recirculated activated sludge flow overall yield might be higher but yield of K-strategists not.

5 Related work

Careful work has been done to identify kinetic parameters of nitrifiers both treating domestic wastewater as well as side streams. Examples of ammonium affinity constants are presented in Table 4.

Table 3. Ammonium affinity constants (g N/m^3) reported in literature.

Nitrifiers treating domestic wastewater	
Value	Reference
0.6-3.6	[3]
1.0	[6]
0.4	[7]
0.4	[8]
Nitrifiers treating high ammonium concentration streams	
0.75 as N-NH_3 (N-NH_4 approx. 15÷100 depending on pH and T)	[9]
6-14	[10]
0.8	[8]
66	[11]

In case of growth rate of nitrifiers treating domestic wastewater (K-strategists) excellent research paper is available [12]. In this paper results of nitrifier's maximum growth rates measurements from 7 wastewater treatment plants are shown. Bioaugmentation of activated sludge with nitrifiers has been investigated in many articles [13–15].

Bioaugmentation of activated sludge has several practical conceptions: B.A.B.E [16, 17], Maureen [18], InNitri[19], BAR [20]which were implemented in full scale.
ASM1 model has been addressed in hundreds of papers. ASM1 model is well described in paper of Henze et al. [3] and in IWA Scientific and Technical Report No. 9 [6].

6 Assumptions and limitations

1. Analysis was based on single staged nitrification model. Although this assumption is incompatible with reality, equations describing total growth of nitrification bacteria for both first and second phase as well as course of these transformations shows same relations and conclusions based on their analysis are identical.Possible differences would be a result of different coefficients in this model. Additionally the intension of authors was to present relations in full nitrification process.
2. Impact of concentrations of free ammonia and free nitrous acid on nitrifiers growth rate therefore also on SRT is not included. Authors assumed that impact caused by these would be negligible due to low average concentration of ammonia nitrogen in reactor and practical absence of nitrite.
3. Equations of ASM1 model which were used does not include heterotrophic transformations. In authors opinion this simplification does not change results of performed analysis. Possible divergence in results are negligible since organic carbon concentration in side stream wastewater is small in relation to amount of nitrogen and contribution of heterotrophs in nitrogen transformation is also negligible.
4. In performed simulations constant pH of 7.0 and lack of nitrification limitations due to inorganic carbon accessibility was assumed. Therefore dosing of alkalinity to maintain given pH was assumed.
5. In this study sedimentation process was not simulated and it was assumed it was effective in 100%. Additionally it was assumed that dose of recirculated sludge was always sufficient to grant full flocculation of activated sludge in bioaugmentation reactor.
6. Values of $\mu_{A,max}$ of r-strategists were established arbitrary as authors are not familiar with any convincing papers about growth rates of r-strategy nitrifiers.

The authors gratefully acknowledge co-funding from the National Centre for Research and Development (grant no PBS2/B9/25/2014) and Municipal Water and Sewage Company Wroclaw.

References

1. M.A. Dytczak, K.L. Londry, J.A. Oleszkiewicz, Water Environ. Res., **80**, 388–396 (2008)
2. A. Zilverentant,. Patent PCT/NL99/00462,WO0005177 (1999)
3. M. Henze, C.P.Leslie Grady, W. Gujer, G.V.R. Marais, T. Matsuo, Water Res, **21**, 505–515 (1987)
4. L.W. Belser, E.L. Schmidt, FEMS Microbiol Lett, **7**, 213–216 (1980).
5. P.Balbierz, Ph.d.thesis, Wroclaw (2016)

DOI: 10.1051/e3sconf/20171700101

6. M. Henze, W. Gujer, , T. Mino, M. Loosdrecht, IWA Scientific and Technical Report No. 9, IWA (2000)
7. D.Kaelin, R. Manser,L. Rieger, J. Eugster, K.Rottermann, H. Siegrist,Water Res.**43**, 1680–92 (2009)
8. A. Mannucci, G. Munz, G. Mori, J. Makinia, C. Lubello, J.A.Oleszkiewicz, Water SciTechnol,**71**, 15–21 (2015)
9. S. W. H. Van Hulle,E. I. P.Volcke,J. Lo´ pezTeruel, B. Donckels, M. C. M.van Loosdrecht, P.A.Vanrolleghem,J. Chem. Technol. Biotechnol. **82**, 471–480 (2007)
10. R. Manser, K. Muche, W. Gujer, H. Siegrist (2005)
11. N.L.H.Trung, V.T.K.Nghi1, P.T.Nhat, N.P.Dan, Journal of Science and Technology,**52** (2014)
12. R.M. Jones, C.M. Bye, P.L. Dold, Water SciTechnol, **52**, 10–11 (2005)
13. E. Plaza, J. Trela, B. Hultman, Water SciTechnol, **43**, 155–64 (2001)
14. S. Salem, DHJG Berends, HF van der Roest, RJ van der Kuij, MCM van Loosdrecht, Wat Sci Tech, **50,** 87–96 (2004)
15. G. Munz, N. Szoke, J.A. Oleszkiewicz,BioresourceTechnol**125**, 88–96 (2012)
16. B. Hommel, E. van der Zandt, D. Berends, V. Claessen, WEFTEC 2006, No. 6. (2006)
17. S. Salem, PhD thesis, Delft University of Technology, Nederland (2005)
18. D. Katehis, S. Murthy, B. Wett, E. Locke, W. Bailey,WEFTEC 2006 (2006)
19. P. Kos, M.A. Head, J.A. Oleszkiewicz, A. Warakomski, WEFTEC 2000 Los Angeles CA (2000)
20. L. Novák, D. Havrliková, Water SciTechnol, **50**, 139–146 (2004)

The use of fibre-reinforced polymers (FRP) in bridges as a favourable solution for the environment

Artur Żyjewski[1,*], *Jacek* Chróścielewski[1], and *Łukasz* Pyrzowski[1]

[1]Dep. of Mechanics of Materials and Structures, Faculty of Civil and Environmental Engineering, Gdansk University of Technology, G. Narutowicza 11/12, 80-233 Gdańsk, Poland

Abstract. The purpose of this article is to show the modern engineering, in which sustainability and taking care of ecology play a significant role. The authors are focused on FRP composite materials and their applications in civil engineering. Case studies showing renovation and design of new bridges with the use of FRP are presented and discussed to clarify benefits, which this solution provides. Main advantages of FRP materials in comparison with traditional ones, like concrete or steel are showed. The environmental impact of composites is described with respect to all life cycle of a product. Furthermore, some forms of waste management are covered. Last part of the paper refers to scientific description of the pedestrian bridge made of FRP, which was realized under the Fobridge research grant. The group of researchers headed by professor Chróścielewski from Gdansk University of Technology has developed a design solution of the pedestrian bridge manufactured in one production cycle. Moreover, the footbridge construction contains a significant share of a recyclable material commonly called PET. The article contains main characteristics of the structure and production process based on the resin infusion.

1 Introduction

Achievements in the discipline of materials science and engineering allow us to take a new direction, in which the traditional building materials, such as steel or reinforced concrete can be replaced by fibre-reinforced polymers (FRP). About 26% of FRP composites world production is applied in the construction industry, [1]. It means that a large part of market share distribution is dedicated to mentioned area, see Fig. 1. The use of FRP materials in civil engineering have been developed mainly in bridge structures, [2]. The fibre-reinforced polymers first applications were realized in rehabilitation and strengthening of ageing structures. Afterward, the FRP began to be used as a construction material, thereby as a good substitution for steel or concrete. One of the most typical solution of strengthening may be performed by bonding adhesive sheets or plates made of FRP into the soffit of deteriorated concrete or metallic elements to improve flexural

[*] Corresponding author: artzyjew@student.pg.gda.pl

resistance, [2,3]. The example of rehabilitation may be a replacement of degraded concrete decks on FRP ones in ageing steel-concrete bridges. Additional advantage of this solution is obtaining less weight of a structure. The first pedestrian bridge made of FRP was built in 1975 in Tel Aviv, Israel, whereas the first road bridge was created in 1995 in Stonehouse (Surrey), UK (Bonds Mill Lifting) [4].

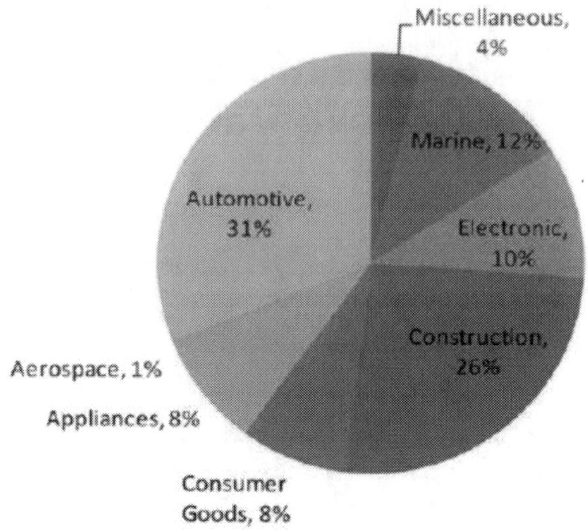

Fig. 1. Market share distribution of fibre-reinforced polymers [1].

Material properties like low volume weight and relatively high strength and stiffness at the same time, are the main aspects for which the FRP composites are increasingly being used in structural applications. Nowadays, not only mechanical parameters of materials have important role in constructions but sustainability of solutions too. To effectively evaluate negative impacts on environment, it is necessary to analyse the entire life cycle of a structure. This analysis includes three main phases: the production phase, the use phase and the end of life phase. Different activities in each phases generate potential sources of environmental impact. It can be, for instance, material extraction, transportation of materials/waste/employees during the construction, on-site activities, traffic disruption and detours during maintenance and construction activities [5]. Summing up, aspects like time or construction cost, energy consumed with production process and carbon emission pose main environmental threat.

2 Sustainability in terms of bridge constructions

Decreasing an impact on the environment caused by bridge constructions is one of the propositions for modern engineering. Current priorities have been established in the European PANTURA Project. The PANTURA is a research program co-financed by the European Commission. The main aim of the project is to elaborate new solutions for more efficient maintenance works, to improve off-site productions and to make bridges more ecological. In general, the efficiency should provide reducing time of construction process, use of materials and minimizing environment pollution. All activities are strongly related with sustainable development [6], which has been defined by World Commission on Environment and Development in 1987.

"..Sustainable development is a process of change in which the exploitation of resources, the direction of investments, the orientation of technological development, and

institutional change are all in harmony and enhance both current and future potential to meet human needs and aspirations."

Therefore, the ideal material has to guarantee recovery of resources and energy. Moreover, this ideal material should not pollute the environment during its life cycle [6]. In the construction industry a large amount of energy consumption and resource utilization cause contamination problems and these factors should be limited at first. The proposed solution is to minimize time of each phases in a construction process by utilization of lighter and stronger materials, see [1,2,5]. A lot of research studies have revealed different advantages with respect to replacement of steel or concrete by composite materials made of fiber-reinforced polymers. Because of very good mechanical properties, FRP composites may have potential benefits for sustainable solutions. The application of bridge structural elements made of fiber-reinforced polymer can provide reduction about 80% of total weight comparing with concrete structures and about 30% taking into account steel ones [1]. Light weight enables fast transportation, allows to reduce traffic disruption, and the installation process become quick and simple. In addition, the carbon emission is also reduced due to traffic reduction. Besides, FRP components require less maintenance work because of high strength and corrosion resistance. All these mentioned aspects provide solutions, which aim is to reach low life cycle cost of the structure [2]. The only disadvantage for widespread application is usually higher initial cost as compared with conventional materials. But this aspect still changes, due to the fact that prices of FRP component materials tend to decrease.

Furthermore, during the production process the energy consumption and carbon emission is relevant higher in case of FRP materials. In spite of these objections FRP composites seems to be environmentally friendly due to reduction of mentioned disadvantages in other life cycle phases, which are: construction, maintenance, and demolition. It is worth to note, that cumulative energy demand for a FRP production process is strictly depend on manufacturing methods and fibre types [1,5]. Table 1 shows various values energy demand in different production techniques. Fig. 2 depicts the comparison of different materials in connection with energy consumption during fabrication process.

Table 1. Cumulative energy demand (CED) [1].

Material	CED (MJ/kg)
Sheet molding compound (SMC)	3.5-3.8
Resin transfer molding (RTM)	12.8
Pultrusion	3.1
Autoclave	21.9
Injection molding	21.1-29.9

3 Examples of FRP application in bridge structures

The paragraph presents some case studies. The aim of this overview is to evaluate environmental impact of FRP materials. Two examples are described.

3.1 Bridge renovation

First example presents a case of bridge renovation. It involves substitution of a deteriorated bridge deck by a new structure made of FRP [5]. The study concerns the small bridge over the *Rokan River* in Sweden. The structure was built in 1948 as a simple supported system of 12 m span length and 6.5 m deck width. It contains two steel girders at 3.6 m spacing. Before renovation the deck was made of a concrete slab without any form of connection with the beams. The expertise from 2002 showed, that the ageing bridge needed a refurbishment. The steel girders were in good conditions. The maintenance work involved replacement of the existing deck, installation of a new barriers and creation of a new surface. Alternative solution was proposed in 2012 during the FRP bridges conference organized in London. The considerations were related to replacing the old bridge deck with a new element made of FRP composites. The analysis included a comparison of all construction activities performed in year 2002 with the proposed alternative solution. The following factors were included during assessment: duration and frequency of work performed, and traffic disturbance caused by these activities.

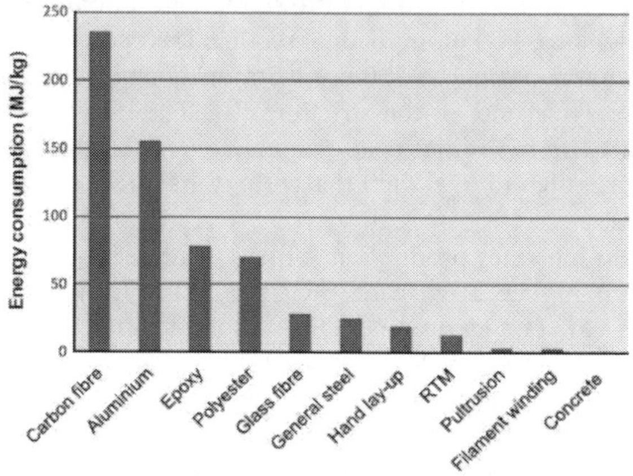

Fig. 2. Energy consumption for production of different materials and manufacturing processes [5].

The carried out investigation shows, that time needed to manufacture a new deck made of FRP is significantly shorter, around two times, in comparison with the production of a concrete slab. This means, that the traffic disturbance would be lower because of diminished temporary closure of the road. The alternative rout in this case is 16 km longer.

In case of the FRP solution, a maintenance works would only concern a replacement of a finished road surface, which could be made from a polymer concrete, typical in such applications. Time needed to accomplish this work is assumed as 24 hours. Such operation should be repeated each 20 years. In case of the traditional solution, it is required to remove and replace insulation and bituminous pavement. Time needed to perform these maintenance works was estimated at 24 hours for replacement of the asphalt and 2 weeks for insulation replacement. In both cases, the superstructure made of FRP composite and the traditional bridge deck, it is necessary to close only one lane to realize all mentioned tasks.

To sum up, the traditional solution causes more disrupted periods during entire life-cycle. The consequence of this, is a diversion of the traffic and finally higher carbon emission. Additionally, total carbon emission generated during entire life-cycle of both bridge structures were compared. Three factors were included in the assessment: production, transportation and traffic detours. The overall results also prove that higher carbon emission is produced in the case of bridge made of traditional materials (90 132 kgCO2). Total carbon emission for the FRP structure is 72 059 kgCO2. It is worth to note, that the carbon emission related to traffic detour will depend on their intensity. The analysed bridge is located in a rural area along a local road.

3.2 Bridge replacement

Second example is based on the replacement project of nineteenth-century highway bridge in North London. It focuses on an environmental impact of fibre reinforced polymers applications in bridges, especially on carbon emission during their entire life-cycle [2]. Two structural solutions are considered. Both are a simply supported single span of 12 m length and 12 m deck width. The difference between two choices is a material used for a construction. First one assumes the use of FRP composites and second one a prestressed concrete. The optional cross sections in analysed bridge are schematically presented in a fig. 3.

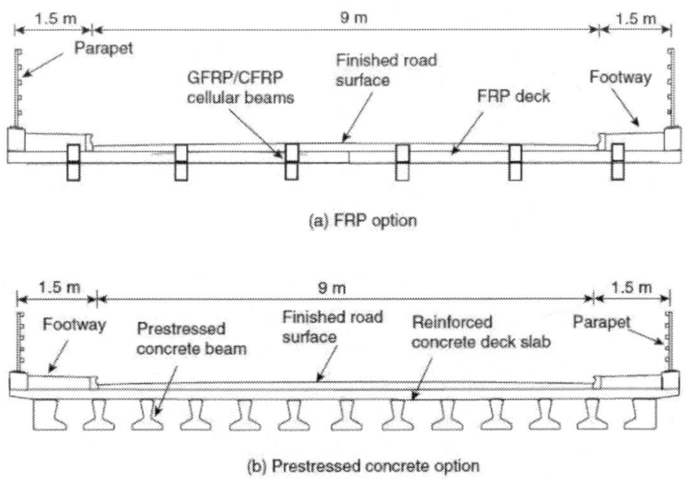

Fig. 3. Optional cross sections in analysed bridge [2].

Both solutions are designed for full highway traffic on both lanes. The carbon emission from three different sources is taken into account during the analysis. The following factors are considered: embodied carbon of any new materials/products, different transportation types (transportation of workers, materials or products, and waste disposal) and traffic diversions. All of these source are considered into three life cycle phases: demolition of the existing bridge, construction and maintenance of a new superstructure. The embodied energy for FRP and steel reinforced concrete is estimated using tables from the Inventory of Carbon and Energy (ICE ver. 1.6a), [7]. In case of glass-fibre reinforced polymer, the value of embodied energy is very limited, because it is a relatively new material in civil engineering application. The quoted value from the ICE table is 100 MJ/kg, but for this investigation it is changed after consulting with one of the major GRP producer in Europe, whose GRP embodied energy is assumed as 33 MJ/kg. An average value is taken as the most objective and representative value for the study. For the reinforced concrete the embodied energy is calculated with a following formula [7]:

$$EE = 2{,}12(EERC+0{,}26n/25) \tag{1}$$

Where, the value of 0,26 means the use of 1 MJ/kg for each 25 kg of the steel per 1 m^3 of concrete. EERC is related with a compressive strength class of concrete. For instance, the embodied energy for C30/37 is 1,08 MJ/kg. As can be seen, production of the 1 kg of

GRP requires much more of the energy use, what means more carbon emission into the environment. Below is the definition of the embodied energy, which is published in [7].

"...The embodied energy (carbon) of a building material can be taken as the total primary energy consumed over its life cycle. This would normally include extraction, manufacturing and transportation. Ideally the boundaries would be set from the extraction of raw materials until the end of the products lifetime."

The transportation of the materials depended on their availability. Traditional materials like steel or concrete can be purchased in local supplies, while companies manufacturing FRP structures are still limited. A transportation distance from supplies/landfills to a construction site for reinforced concrete is assumed to be 10 miles. In case of FRP composites, a water transport from Denmark is considered.

A daily traffic volume, distance and period of disruption are relevant factors in terms of the carbon emission produced during traffic diversion. Time needed to erect a bridge depend on specific construction method, for a FRP system is assumed to be 14 days and for a prestressed concrete structure approximately 85 days.

This case study demonstrates that carbon emission for both cases is similar and that the biggest carbon emission comes from traffic diversions. Considering the last factor, the FRP has an environmental advantage because of shorter time of construction period. Moreover, future improvements for composite materials should decrease theirs embodied emissions.

4 Waste management during the last phase of life-cycle

The FRP composites belong mainly to a group of thermosetting (thermoset) polymers. This means that during the process of chemical reactions called polycondensation, polymerization, or curing, the resin becomes rigid. When the thermosetting polymer are cured once, cannot be remelted or reformed. Relatively good mechanical properties of FRP, makes them an interesting alternative for traditional materials in civil applications. On the other hand the same properties make the recycling process more difficult and provide limited ways of the waste management.

Because of the economic aspects, the most popular form to manage of FRP waste is to disposal it on the landfills. However, landfills areas and raw materials are limited. To protect the environment, it is necessary to use methods of used materials recycle. In general, two categories of the recycling process can be distinguished. First group includes all these methods, in which the mechanical techniques are implemented. Second group is related with the incineration methods for recovery of the embodied energy from waste materials [8]. However, if the FRP composites include glass fibres, the incineration will not be so effective, because fibres are incombustible and consume approximately 1.7 MJ per kilogram of glass-fibre content [1]. Despite of this disadvantage, the incineration is still the most suggested method of recycling. The incineration may be realized, for example, in cement kilns. It has been proved that during the incineration process of unsorted plastics is possible to recovery more than 8 000 thermies per ton, [9] (1 thermie is approximately 4.1855 × 10^6 joules). There are three incineration methods of FRP: pyrolysis, fluidized-bed processing and chemical treatment. The specified methods allow to retrieve individual carbon fibres in CFRPs or glass fibres in GFRPs. The purpose of mechanical processes is to use technique of crushing, milling and/or shredding FRP material into small pieces, from 100 mm to 50 μm. These methods can reduce the mechanical properties of FRPs up to 50%. Therefore, the final product may be used as a fillers for producing panels, mortar and asphaltic concrete. The mechanical methods are mainly used in case of GFRPs composites. Due to the techniques of comminution, entire fibres are impossible to reclaim.

5 The FOBRIDGE project – the case study

The FOBRIDGE is a scientific project co-financed by The National Centre for Research and Development [10]. The aim of this research was to create a novel solution for pedestrian and bicycle bridge which allows easy transportation, short production time and assembling without heavy equipment. In addition, it should be environmentally friendly and easy to recycle. This study was realized during three years (2012-2015). The structure was designed under the research grant realized by consortium of Gdansk University of Technology (leader), Military University of Technology in Warsaw and the private company ROMA Co Ltd. The group of researchers created the single-span, 14,5 m long, shell footbridge made of sandwich panels. These sandwich panels consist of GFRP laminates (outer skins) and a foam (core). The material used for a core is the polyethylene terephthalate, commonly called PET. This makes recycling of a structure much easier in the end of life phase. The proposed structural solution allows the use of already recycled material obtained from plastic packages in the production phase. The infusion process, used for manufacturing, allows to produce the whole structure in one part, without any additional joints or fasteners. Additionally, it minimizes environmental impact, due to the fact that a resin flow is applied in a vacuum. The standardized design and infusion technology also lead to cost reduction, [11]. The final product is a shell-type sandwich structure with U-shape cross section, fig. 4.

Fig. 4. Side view and cross section of the footbridge.

The main advantages of the considered footbridge are durability, dynamic resistance and incombustibility. Moreover, is easy to install because of its light weight (mass of 3,2 t) and does not require any laborious and heavy maintenance work. Also, during the service life any replacement activities are not required. For more information see also [12]. From the ecological point of view the FOBRIDGE solution includes all benefits, which are discussed in previous paragraphs. Also, during the end of life phase, the foam may be crushed and glass fibres can be reclaimed. In general, around 60% of materials which come from the entire structure could be reused.

6 Conclusions

The paper describes some possible applications of FRP composites in bridges from the environmental impact point of view. It is a response for contamination problems caused by a large amount of energy consumption and resource utilization. On the other hand, the FRP solution matches to a modern engineering, which aim is to reduce time of construction process and use of materials, and to minimize environmental pollution. The presented cases prove a huge potential for sustainability of the FRP use in civil engineering. Very good mechanical properties of FRP composites, like low volume weight and relatively high strength and stiffness at the same time, allow to achieve potential benefits for sustainable solution. Light weight enables fast transportation, allows to reduce traffic disruption, and the installation process become quick and simple. Moreover, future improvements for composite materials should decrease theirs embodied emissions.

The carbon emission is another very significant aspect in term of contamination problems. Traffic diversions, transportation types or embodied carbon of any new material are main factors, that should be considered to evaluate carbon emission. However, to make a good assessment of the environmental impact, it is necessary to consider all life cycle of analysed structures. In general, three main phases can by distinguished: construction, maintenance of a new superstructure and demolition. It is worth to note, that the application of FRP structural components in bridge structures do not exceed even 40 years. According to this, the end of life phase never has been reached and investigated.

To protect the environment, it is necessary to use recycling methods. Good waste manage provides very efficient recovery of used materials. FRP structures allow to reuse a huge part of their components by mechanical or incineration methods.

References

1. J.R. Duflou, Y. Deng, K. Van Acker, W. Dewulf, *Do fiber-reinforced polymer composites provide environmentally benign alternatives? A life-cycle-assessment-based study*, (MRS Bulletin, **27** (2012)
2. C. Zhang, *Life cycle assessment (LCA) of fibre reinforced polymer (FRP) composites in civil applications,* Eco-efficient construction and building materials, (2014)
3. R. Sonnenschein, K. Gajdosova, I. Holly, *FRP Composites and Their Using in the Construction of Bridges,* Procedia Engineering, **161**, 447-482 (2016)
4. *http://www.cotswoldcanals.net/photo_page.php?idx=490*
5. V. Mara, R. Haghani, P. Harryson, *Bridge decks of fibre reinforced polymer (FRP): A sustainable solution*, Construction and Building Materials, **50**, 190-199 (2014)
6. L.S. Lee, R. Jain, *The role of FRP composites in a sustainable world,* Springer (published online : 2009)
7. G.P. Hammond, C.I. Jones, *Embodied energy and carbon in construction materials,* (2008)
8. S.J. Pickering, *Recycling technology for thermoset composite materials – current status,* Composites Part A : applied science and manufacturing, **37**, 1206-1215 (2006)
9. R. Liang, G. Hota, *Fiber-reinforced polymer (FRP) composites in environmental engineering application,* Developments in FRP composites for civil engineering, **16** (2013)
10. M. Miśkiewicz, R. Okraszewska, and Ł. Pyrzowski, *Composite footbridge – synergy effect in cooperation between universities and industry,* ICERI2014: 7th International Conference of Education, Research and Innovation, ICERI Proceedings, Seville, Spain, 2897-2903 (2014)
11. M. Miśkiewicz, Ł. Pyrzowski, J. Chróścielewski, K. Wilde, *Structural Health Monitoring of Composite Shell Footbridge for Its Design Validation,* Baltic Geodetic Congress, Gdansk, Poland (2016)
12. J. Chróścielewski, M. Miśkiewicz, Ł. Pyrzowski, B. Sobczyk, K. Wilde, *A novel sandwich footbridge – practical application of laminated composites in bridge design and in situ measurements of static response,* (to be published)

AUTHOR INDEX

AUTHOR INDEX